Biotechnology of Extremophiles Advances and Challenges

极端微生物生物技术

前景与挑战

（巴西）P. H. 兰佩洛托（Pabulo H. Rampelotto） 主编
何海伦 李善妮 主译
陈秀兰 主审

化学工业出版社
·北京·

内容简介

《极端微生物生物技术——前景与挑战》系统阐述极端微生物的研究进展及其在生物技术中的应用，主要涉及极端微生物的多样性、生长与代谢、适应极端环境的分子机制及其生物技术应用，还介绍了组学方法在鉴定极端微生物新型基因和功能蛋白质发掘中的应用、相关结构功能关系的分子动力学研究及其在工业应用中的潜力。

《极端微生物生物技术——前景与挑战》可供从事微生物技术研究的人员查阅，也可供微生物专业师生参考。

图书在版编目（CIP）数据

极端微生物生物技术：前景与挑战 /（巴西）P. H. 兰佩洛托（Pabulo H. Rampelotto）主编；何海伦，李善妮主译. —北京：化学工业出版社，2020. 9

书名原文：Biotechnology of Extremophiles：Advances and Challenges

ISBN 978-7-122-37213-0

Ⅰ. ①极… Ⅱ. ①P… ②何… ③李… Ⅲ. ①微生物-生物工程 Ⅳ. ①TQ92

中国版本图书馆 CIP 数据核字（2020）第 105908 号

责任编辑：傅四周
责任校对：宋　玮　　　　装帧设计：韩　飞

出版发行：化学工业出版社（北京市东城区青年湖南街 13 号　邮政编码 100011）
印　　装：中煤（北京）印务有限公司
787mm×1092mm　1/16　印张 32¼　彩插 4　字数 824 千字　2021 年 1 月北京第 1 版第 1 次印刷

购书咨询：010-64518888　　售后服务：010-64518899
网　　址：http：//www. cip. com. cn
凡购买本书，如有缺损质量问题，本社销售中心负责调换。

定　　价：199. 00 元

关于主编

Pabulo H. Rampelotto 是《生物学和生物技术大挑战系列》(*Grand Challenges in Biology and Biotechnology*)的主编(Springer),也是生命科学和生物技术领域多家科学杂志的主编、副主编、客座编辑以及编委会成员。最近,他的大部分工作集中在专业领域的几个科学期刊以及图书和特刊的编撰整理上。在他编撰的图书和特刊中,一些领域最杰出的团队负责人发表了他们的成果,新的观点和发现,其中包括诺贝尔奖得主和几位 ISI 高引用的科学家。Pabulo 的研究兴趣涵盖了不同的科学领域,从基础到应用研究,包括生物技术、宏基因组学、新一代测序、微生物的生化与分子生物学、极端微生物学和太空生物学。

当不工作时, Pabulo 喜欢在树林、山间和海边散步,一边散步一边思考,总是在不停地思考。

主编寄语

这是由我担任主编的本系列的第一本书,谨以此献给生命中第一个激励我的人——我最亲爱的母亲 Veronete。同时感谢那些教给我科学的真正意义以及科学之美的人,特别是 Stephen J. Gould、Ernst Mayr 和 Carl E. Sagan。

撰稿人名单

Garabed Antranikian Institute of Technical Microbiology, Hamburg University of Technology (TUHH), Hamburg, Germany

Satoshi Akanuma Department of Applied Life Sciences, Tokyo University of Pharmacy and Life Sciences, Hachioji, Tokyo, Japan

Faculty of Human Sciences, Waseda University, Shinjuku, Japan

Vanesa Bautista Department of Agrochemistry and Biochemistry, University of Alicante, Alicante, Spain

Sören Bellenberg Aquatische Biotechnologie, Biofilm Centre, Universität Duisburg-Essen, Essen, Germany

Peter L. Bergquist Department of Molecular Medicine and Pathology, Medical School, University of Auckland, Auckland, New Zealand

Department of Chemistry and Biomolecular Sciences, Macquarie University, Sydney, NSW, Australia

Biomolecular Frontiers Research Centre, Macquarie University, Sydney, NSW, Australia

Aneta Białkowska Institute of Technical Biochemistry, Lodz University of Technology, Lodz, Poland

Jenny M. Blamey Fundación Científica y Cultural Biociencia, José Domingo Cañas, Santiago, Chile

Paul Blum School of Biological Science, University of Nebraska, Lincoln, NE, USA

Freddy Boehmwald Fundación Científica y Cultural Biociencia, José Domingo Cañas, Santiago, Chile

Ivanka Bojadjieva Institute of Microbiology, Bulgarian Academy of Sciences, Sofia, Bulgaria

María José Bonete Department of Agrochemistry and Biochemistry, University of Alicante, Alicante, Spain

Emanuele Bosi Department of Biology, University of Florence, Florence, Italy

Mónica Camacho Department of Agrochemistry and Biochemistry, University of Alicante, Alicante, Spain

Andrew Care ARC Centre of Excellence for Nanoscale BioPhotonics (CNBP), Macquarie University, Sydney, NSW, Australia

Madhab Kumar Chattopadhyay CSIR-Centre for Cellular and Molecular Biology, Hyderabad, India

Chi-Won Choi Division of Bioconvergence Analysis, Daejeon, Republic of Korea

Hubert Cieslinski Faculty of Chemistry, Department of Molecular Biotechnology and Microbiology, Gdańsk University of Technology, Gdańsk, Poland

Priya DasSarma Department of Microbiology and Immunology, University of Maryland School of Medicine, Baltimore, MD, USA

Institute of Marine and Environmental Technology, University System of Maryland, Baltimore, MD, USA

Shiladitya DasSarma Department of Microbiology and Immunology, University of Maryland School of Medicine, Baltimore, MD, USA

Institute of Marine and Environmental Technology, University System of Maryland, Baltimore, MD, USA

Angelina Lo Giudice Institute for Coastal Marine Environment, National Research Council (IAMC-CNR), Messina, Italy

Department of Biological and Environmental Sciences, University of Messina, Messina, Italy

Immacolata Del Giudice Institute of Protein Biochemistry, CNR, Naples, Italy

Paola Di Donato Institute of Biomolecular Chemistry ICB, CNR, Pozzuoli, NA, Italy

Department of Sciences and Technologies, Centro Direzionale, University of Naples 'Parthenope', Naples, Italy

Skander Elleuche Institute of Technical Microbiology, Hamburg University of Technology (TUHH), Hamburg, Germany

Julia Esclapez Department of Agrochemistry and Biochemistry, University of Alicante, Alicante, Spain

Renato Fani Department of Biology, University of Florence, Florence, Italy

Ilaria Finore Institute of Biomolecular Chemistry ICB, CNR, Pozzuoli, NA, Italy

Tomasz Florczak Institute of Technical Biochemistry, Lodz University of Technology, Lodz, Poland

Anna P. Florentino Laboratory of Microbiology, Wageningen University, Wageningen, The Netherlands

Sub-Department of Environmental Technology, Wageningen University, Wageningen, The Netherlands

Patricio Flores Fundación Científica y Cultural Biociencia, José Domingo Cañas, Santiago, Chile

Marco Fondi Department of Biology, University of Florence, Florence, Italy

María Teresa García Faculty of Pharmacy, Department of Microbiology and Parasitology, University of Sevilla, Sevilla, Spain

Dennis W. Grogan Department of Biological Sciences, University of Cincinnati, Cincinnati, OH, USA

Ewa Gromek Institute of Technical Biochemistry, Lodz University of Technology, Lodz, Poland

Gaetano Invernizzi Structural Biology and NMR Laboratory, Department of Biology, University of Copenhagen, Copenhagen, Denmark

Štefan Janecek Laboratory of Protein Evolution, Institute of Molecular Biology, Slovak Academy of Sciences, Bratislava, Slovakia

Faculty of Natural Sciences, Department of Biology, University of SS Cyril and Methodius, Trnava, Slovakia

Margarita Kambourova Institute of Microbiology, Bulgarian Academy of Sciences, Sofia, Bulgaria

Ram Karan Department of Microbiology and Immunology, University of Maryland School of Medicine, Baltimore, MD, USA

Institute of Marine and Environmental Technology, University System of Maryland, Baltimore, MD, USA

Jong-Myoung Kim Department of Microbiology and Immunology, University of Maryland School of Medicine, Baltimore, MD, USA

Institute of Marine and Environmental Technology, University System of Maryland, Baltimore, MD, USA

Department of Marine Bio-Materials and Aquaculture, PuKyong National University, Pusan, Republic of Korea

Seung Il Kim Division of Bioconvergence Analysis, Daejeon, Republic of Korea

Ewelina Krajewska Faculty of Chemistry, Department of Molecular Biotechnology and Microbiology, Gdańsk University of Technology, Gdańsk, Poland

Joanna Krysiak Institute of Technical Biochemistry, Lodz University of Technology, Lodz, Poland

Józef Kur Faculty of Chemistry, Department of Molecular Biotechnology and Microbiology, Gdańsk University of Technology, Gdańsk, Poland

Licia Lama Institute of Biomolecular Chemistry ICB, CNR, Pozzuoli, NA, Italy

Sang-Yeop Lee Division of Bioconvergence Analysis, Daejeon, Republic of Korea

Kesen Ma Department of Biology, University of Waterloo, Waterloo, ON, Canada

Gashaw Mamo Department of Biotechnology, Lund University, Lund, Sweden Indienz AB, Billeberga, Sweden

Giuseppe Manco Institute of Protein Biochemistry, CNR, Naples, Italy

M. Carmen Márquez Faculty of Pharmacy, Department of Microbiology and Parasitology, University of Sevilla, Sevilla, Spain

Rosa María Martínez-Espinosa Department of Agrochemistry and Biochemistry, University of Alicante, Alicante, Spain

Bo Mattiasson Department of Biotechnology, Lund University, Lund, Sweden Indienz AB, Billeberga, Sweden

Samuel McCarthy School of Biological Science, University of Nebraska, Lincoln, NE, USA

Encarnación Mellado Faculty of Pharmacy, Department of Microbiology and Parasitology, University of Sevilla, Sevilla, Spain

María de Lourdes Moreno Faculty of Pharmacy, Department of Microbiology and Parasitology, University of Sevilla, Sevilla, Spain

Patricio Muñoz Fundación Científica y Cultural Biociencia, José Domingo Cañas, Santiago, Chile

Thomas R. Neu Department of River Ecology, Helmholtz Centre for Environmental Research-UFZ, Magdeburg, Germany

Barbara Nicolaus Institute of Biomolecular Chemistry ICB, CNR, Pozzuoli, NA, Italy

Elena Papaleo Structural Biology and NMR Laboratory, Department of Biology, University of Copenhagen, Copenhagen, Denmark

Edmond Changkyun Park Division of Bioconvergence Analysis, Daejeon, Republic of Korea

Deepak Parashar Department of Microbiology, University of Delhi, New Delhi, India

Benjamin Pavlik Department of Chemical and Biomolecular Engineering, University of Nebraska, Lincoln, NE, USA

Anna Pawlak-Szukalska Faculty of Chemistry, Department of Molecular Biotechnology and Microbiology, Gdańsk University of Technology, Gdańsk, Poland

Wolf T. Pecher Department of Microbiology and Immunology, University of Maryland School of Medicine, Baltimore, MD, USA

Institute of Marine and Environmental Technology, University System of Maryland, Baltimore, MD, USA

College of Arts and Sciences, University of Baltimore, Baltimore, MD, USA

Laia Pedro-Roig Department of Agrochemistry and Biochemistry, University of Alicante, Alicante, Spain

Carmen Pire Department of Agrochemistry and Biochemistry, University of Alicante, Alicante, Spain

Francisco Pérez-Pomares Department of Agrochemistry and Biochemistry, University of Alicante, Alicante, Spain

Annarita Poli Institute of Biomolecular Chemistry ICB, CNR, Pozzuoli, NA, Italy

Elena Porzio Institute of Protein Biochemistry, CNR, Naples, Italy

Nadja Radchenkova Institute of Microbiology, Bulgarian Academy of Sciences, Sofia, Bulgaria

Deepak Rudrappa School of Biological Science, University of Nebraska, Lincoln, NE, USA

Irene Sánchez-Andrea Laboratory of Microbiology, Wageningen University, Wageningen, The Netherlands

Wolfgang Sand Aquatische Biotechnologie, Biofilm Centre, Universität Duisburg-Essen, Essen, Germany

Gundlapally Sathyanarayana Reddy CSIR-Centre for Cellular and Molecular Biology, Hyderabad, India

Tulasi Satyanarayana Department of Microbiology, University of Delhi, New Delhi, India

Carola Schröder Institute of Technical Microbiology, Hamburg University of Technology (TUHH), Hamburg, Germany

Archana Sharma Department of Biophysics, University of Delhi, New Delhi, India

Sisinthy Shivaji CSIR-Centre for Cellular and Molecular Biology, Hyderabad, India

Jhaveri Microbiology Centre, L V Prasad Eye Institute, Hyderabad, India

Raghuveer Singh School of Biological Science, University of Nebraska, Lincoln, NE, USA

Alfons J. M. Stams Laboratory of Microbiology, Wageningen University, Wageningen, The Netherlands

CEB-Centre of Biological Engineering, University of Minho, Braga, Portugal

Photini Sinnis Department of Molecular Microbiology and Immunology, Bloomberg School of Public Health, Johns Hopkins University, Baltimore, MD, USA

Anwar Sunna Department of Chemistry and Biomolecular Sciences, Macquarie University, Sydney, NSW, Australia

Biomolecular Frontiers Research Centre, Macquarie University, Sydney, NSW, Australia

Katarzyna Szulczewska Institute of Technical Biochemistry, Lodz University of Technology, Lodz, Poland

Matteo Tiberti Department of Biotechnology and Biosciences, University of Milano-Bicocca, Milan, Italy

Iva Tomova Institute of Microbiology, Bulgarian Academy of Sciences, Sofia, Bulgaria

Annabella Tramice Institute of Biomolecular Chemistry ICB, CNR, Pozzuoli, NA, Italy

Marianna Turkiewicz Institute of Technical Biochemistry, Lodz University of Technology, Lodz, Poland

Ching Tse Department of Biology, University of Waterloo, Waterloo, ON, Canada

Anna Vegara Department of Agrochemistry and Biochemistry, University of Alicante, Alicante, Spain

Mario Vera Institute for Biological and Medical Engineering, Schools of Engineering, Biological Sciences and Medicine, Pontificia Universidad Católica de Chile, Santiago, Chile

Department of Hydraulic and Environmental Engineering, School of Engineering, Pontificia Universidad Católica de Chile, Santiago, Chile

Marta Wanarska Faculty of Chemistry, Department of Molecular Biotechnology and Microbiology, Gdańsk University of Technology, Gdańsk, Poland

Jan Weijma Sub-Department of Environmental Technology, Wageningen University, Wageningen, The Netherlands

Monika Wicka Faculty of Chemistry, Department of Molecular Biotechnology and Microbiology, Gdańsk University of Technology, Gdańsk, Poland

Akihiko Yamagishi Department of Applied Life Sciences, Tokyo University of Pharmacy and Life Sciences, Hachioji, Tokyo, Japan

Sung Ho Yun Division of Bioconvergence Analysis, Daejeon, Republic of Korea

Ruiyong Zhang Aquatische Biotechnologie, Biofilm Centre, Universität Duisburg-Essen, Essen, Germany

翻译人员名单

主译： 何海伦、李善妮

主审： 陈秀兰（山东大学生命科学学院）

译者：（按姓氏笔画排序）

王　猛（中南大学生命科学学院）
刘　丹（中南大学生命科学学院）
刘聪伶（中南大学生命科学学院）
李善妮（中南大学生命科学学院）
肖　潇（中南大学生命科学学院）
吴日帮（中南大学生命科学学院）
何海伦（中南大学生命科学学院）
张　姜（中南大学生命科学学院）
陈　功（中南大学生命科学学院）
武翠玲（长治医学院）
宦　冉（中南大学生命科学学院）
黄嘉丰（中南大学生命科学学院）
雷　鸣（中南大学生命科学学院）
廖斌强（中南大学生命科学学院）

第一章　王猛、刘聪伶　译
第二章　黄嘉丰、李善妮　译
第三章　何海伦、肖潇　译
第四章　何海伦、张姜　译
第五章　何海伦、肖潇　译
第六章　何海伦、张姜　译
第七章　何海伦、陈功　译
第八章　廖斌强、李善妮　译
第九章　武翠玲、刘聪伶　译
第十章　何海伦、刘丹　译
第十一章　吴日帮、王猛　译
第十二章　何海伦、吴日帮　译
第十三章　何海伦、武翠玲　译
第十四章　吴日帮　译
第十五章　何海伦、张姜　译
第十六章　何海伦、雷鸣　译
第十七章　黄嘉丰　译
第十八章　何海伦、廖斌强　译
第十九章　雷鸣、王猛　译
第二十章　何海伦、刘丹　译
第二十一章　宦冉、刘聪伶　译
第二十二章　黄嘉丰、李善妮　译
第二十三章　吴日帮、李善妮　译
第二十四章　黄嘉丰、宦冉　译
第二十五章　刘丹　译

译者的话

极端微生物 (extremophile)是对栖居于地球上所有“非宜居”空间中的微生物的统称，包括嗜热、嗜冷、嗜酸、嗜碱、嗜盐、嗜压等对极端环境耐受的微生物。研究极端环境下极端微生物的生存适应，对于了解“生命本质”和“物种演化”的基础科学问题具有重要的意义。由于特殊的生存环境，极端微生物进化出了独特的环境适应机制，如产生高温酶、嗜冷酶、耐盐蛋白等，这为开发具有特殊用途的新型微生物资源提供了新的研究思路。极端微生物独特的环境适应能力，使其在现代工业化生产中具有巨大的应用潜力。

《极端微生物生物技术——前景与挑战》一书翻译自 Pabulo H. Rampelotto 教授主持撰写的 *Biotechnology of Extremophiles: Advances and Challenges*。本书的翻译工作得到了 Rampelotto 教授的大力支持，在此特别感谢。本书围绕极端微生物的研究进展及其在生物技术中的应用展开，主要涉及极端微生物多样性，生长与代谢，适应极端环境的分子机制及其生物技术应用；还介绍了组学方法在鉴定极端微生物中新型基因和功能蛋白质发掘中的应用，相关结构功能关系的分子动力学研究，及其在工业应用中的潜力。

希望读者通过本书，了解极端微生物的最新发展动向，进一步促进我国极端微生物的研究和发展，加速我国在各类极端环境微生物资源的挖掘和应用方面的开发利用，加速极端微生物的工业化利用进程。

本书适用于所有从事生命科学的科学工作者。由于时间仓促以及译者知识范围和学术水平有限，译文中难免存在疏漏，恳请同行和广大读者不吝指正，译者将不胜感激。

何海伦

中南大学生命科学学院

2020 年 5 月

原版序

“生物体不可能在地球上人类所认定的极端环境中存活。”这一观点在过去很长一段时间被科学界广为接受。直到 20 世纪，人们发现微生物能够在深海、温泉、冰川、沙漠以及火山口等地球恶劣环境中仍保持着旺盛的生命力。一些隶属于细菌、古菌的微生物专门生存于这种极端苛刻的环境中，因此这类极端微生物必须具有充分适应极端环境的代谢过程、生物功能、基因组以及转录组，以克服生命过程中所面临的挑战。时至今日，在自然界或者人造的极端环境中，一些具有独特性状的新物种被先后发现。此外，这类微生物的酶具有独特性并适应极端环境，被称作极端酶，可被应用于不同工业领域的生产过程。

《极端微生物生物技术——前景与挑战》一书编撰的目的是：对当代不同领域中极端微生物的研究以及它们在工业上潜在应用的前景和面临的挑战进行详细的综述。许多著名的科学家在他们各自专业领域为该书贡献了出色的章节。这些章节阐明了关于极端微生物研究的最新的标志性进展，以及在学术和工业上的最新研究成果。

该书的第一部分重点介绍了不同类型的极端环境微生物、环境条件以及在生物技术上的应用潜力这三个基本方面。第一章对极端微生物的生长和代谢进行全面的综述，并对糖类（碳水化合物）和氨基酸代谢进行了重点介绍。第二章和第三章通过介绍相关学者对南极洲、北极圈和喜马拉雅山冰川等寒冷环境的探索，以阐明嗜冷微生物的生物多样性及其在生物技术上应用的潜力。第四章说明嗜热微生物是生产生物活性多糖的一个重要来源。第五章至第七章集中介绍了嗜酸微生物的生活方式，比如嗜酸硫酸盐还原菌、成膜的嗜酸金属氧化菌。而第八章介绍的嗜碱微生物产的酶则被认为有多种工业用途。最后，第九章讲述的嗜盐菌的氮代谢也是一个令人关注的主题。

因为酶类在多种行业上有着巨大的应用潜力（特别是在工业生物技术领域），这些极端微生物就像提供这些酶的“百宝箱”，所以该书的第二部分主要介绍蛋白质组学、生化表征等新的筛选方法，并介绍极端酶主要的类型。为了了解极端酶在新陈代谢、生化功能以及代谢中的作用，人们展开了广泛的研究，并由此发现这些酶能够很好地适应极端温度、压力、高浓度的有毒金属离子和有机溶剂。第十二章至第十四章通过该领域专家的介绍，来阐明脂解酶和蛋白酶在不同工业领域中的应用。第十五章阐述了适冷性 β-半乳糖苷酶在化妆品、医药和食品工业中的巨大应用潜力。第十六章和第十七章作者揭示了农药降解内酯酶和 α-淀粉酶的研究进展和在工业上的应用前景。第十八章和第十九章报道了来自嗜热微生物的酶类，如 DNA 复制相关蛋白质和用于植物废弃物降解的木质纤维素水解酶。

在第三部分中，第二十章介绍了利用生物信息学来改造蛋白质序列的技术，可构建出新的热稳定蛋白质。第二十一章则提出了系统生物学框架，即利用高通量组学技术来研究细胞功能如何响应温度的变化。另外，第二十二章认为应利用最先进的方法对实验微生物进化进行审慎评估。第二十三章描述了固相结合多肽用于分离、纯化、回收嗜热酶的固定化策略。第二十四章说明分子动力学模拟是研究嗜冷微生物极端酶的结构与功能之间关系的一种有效工具。第二十五章则对在嗜盐异源宿主中建立疟疾疫苗的表达系统进行了讨论。

总而言之，该书揭示了极端微生物及其酶类各个方面的特征，以独特的视角全面涵盖了极端微生物生物技术这一研究领域中新颖和令人激动的方向。我向撰稿者和编者表示祝贺。

Garabed Antranikian

德国汉堡

原版前言

在过去数十年中，极端微生物研究突破性的发现挑战了我们对生物化学和分子生物学的认知。在应用方面，极端微生物及其酶类已经孵育出数十亿美元的生物技术产业，比如生物医学、药学、工业、环境和农业等领域。 Taq DNA 聚合酶（分离自黄石国家公园地热泉的栖热水生菌）是极端微生物及其生物分子在生物技术上应用的最著名的例子。确实，极端微生物及其生物活性物质的应用已经开启了生物技术的一个新时代。但是，尽管有这么多的最新进展，我们仍处于探索极端微生物在生物技术潜力上的初始阶段。

出于这个原因，我选择这一主题作为此系列的第一本书。另外，我还为先前在极端微生物和生物技术专刊上取得的成功所激励。更重要的是，极端微生物在商业上的成功应用在科学文献上还未被良好地记载，从而导致工业研究和发展上的激烈竞争。

本书旨在为从事极端微生物及其生物技术应用开发的科研人员、生物技术专家提供一本重要的参考书。通过这本书，我们全面而翔实地论述了这一主题不同方面的最新进展和挑战。本书语言通俗易懂，也适用于任何对这一新兴研究领域感兴趣的读者。

Pabulo H. Rampelotto
巴西南里奥格兰德州阿雷格里港

目录

第一章

极端微生物的生长与代谢

Ching Tse[1]，Kesen Ma[1][2]

1.1 引言

极端生物是适宜在极端环境下生长的一类生物（Rothschild and Mancinelli 2001）。一些极端微生物可以在一种或一种以上的极端环境中生长。这些环境包括极端温度、辐射、干燥、压力等物理条件以及极端盐浓度、pH 值、氧含量、氧化还原电位等化学条件（Rothschild and Mancinelli 2001）。极端生物在三个生物域中均存在。本章将重点介绍其中的微生物，微生物是一类仅在显微镜下可见的微小生物的集合体（Rybicki 1990），包括细菌、原生动物、藻类和真菌（Keller and Friedli 1992）。而那些可以耐受或抵抗诸如低水活度、高辐射和重金属离子等极端条件，但此类条件并不是它们最适生长所需的微生物，被认为极端耐受菌或极端抗性菌。

极端微生物可分为以下几类。①嗜酸菌（acidophiles），生长最适 pH 值≤3（Johnson et al. 2006）。②嗜碱菌（alkaliphiles），最适 pH 值在 9 以上（通常在 10 到 12 之间），在接近中性 pH 值 6.5 的环境中不生长或者生长缓慢（Zhilina et al. 2004）。③嗜冷菌（psychrophiles），最适生长温度≤15℃，生长最高温度＜20℃（Morita 1975）。④嗜热菌（thermophiles），最适生长温度为 45～80℃（Kristjansson 1991）。⑤极端嗜热菌（hyperthermophiles），最适生长温度≥80℃，甚至可在 90℃以上生长（Blumer-Schuette et al. 2008）。⑥嗜盐菌（halophiles），适宜在高浓度 NaCl 条件下生长（＞0.3mol/L）（DasSarma and DasSarma 2012）。⑦嗜压菌（piezophiles），适宜在高压条件下生长，在 50MPa 的气压下，该菌的生长速度比在正常大气压下高出 30%以上（Abe 2007）。⑧嗜旱菌（xerophiles），最适生长条件为水活度（a_w）低于 0.85 的环境（Pitt and Hocking 2009）。⑨嗜高渗菌（osmophiles），适于生长在高浓度有机溶质，特别是高糖的条件下（Baross and Lenovich 1992）。⑩寡营养菌（oligotroph），分离自每升培养基中有机碳含量为 1～15mg 的微生物，并可在相同培养基中再次培养（Poindexter 1981）。

极端耐受菌或极端抗性菌分为以下类型。①重金属耐受菌（metallotolerants），能耐受生长环境中高浓度重金属（＞1mmol/L）（Carapito et al. 2006）。②耐辐射菌（radioresista-

[1] Department of Biology，University of Waterloo，200 University Avenue West，Waterloo，Canada，ON，N2L 3G1.

[2] e-mail：kma@uwaterloo.ca.

nts)，能够在不形成孢子的条件下保护自己的胞浆蛋白不被氧化，可耐受高剂量的电离辐射，并保证DNA双链不断裂（一般辐射吸收剂量大于1kGy，菌落数会减少90%），也可以抵抗长期干燥（Sghaier et al. 2008）。

极端微生物广泛存在于不同的环境条件中，并有独特的代谢能力和（或）物理结构以保证其生存繁殖。极端耐受菌或极端抗性菌也具有一些独特的代谢能力和（或）物理结构来确保其生存。对生物体而言，碳源和氮源是两种必需的营养成分，它们可能会以不同的形式与浓度存在于极端微生物的生活环境中。探索这些微生物如何利用各种底物，并通过怎样的代谢途径使自身生存和繁衍是十分有趣的。极端微生物中存在异养和自养两种生长模式。在本章中，将重点对异养生物及其糖类（俗称碳水化合物）和肽类代谢进行阐述，具体将对该过程中的关键酶、糖类和肽类中心代谢途径中的修饰和改变进行详细讨论；此外将介绍碳代谢和肽代谢之间的相关性。

1.2 生长、底物和转运蛋白

极端微生物可在多种不同类型的底物上生长，包括单糖、双糖、多糖、纤维素、半纤维素、多肽和氨基酸（Kazak et al. 2010）。利用纯培养的方法可测试不同的底物中哪一种营养物质可以满足其生长（表1.1），不同底物分子通过相应的转运体被转运到细胞，从而被细胞利用（Pflüger and Müller 2004；Albers and Driessen 2007；Siebold et al. 2001；Albers et al. 2004）。

ATP结合盒转运蛋白（ABC转运蛋白）属于一级转运蛋白，是古菌（又称古细菌）和细菌主要的转运蛋白（Holland and Blight 1999）。在这两个生物域中，转运蛋白ATP酶亚基中的walker序列、Q环和H区具有相似的序列（Higgins 1995），该体系包括两个跨膜结构域和两个ATP结合结构域及膜外结合蛋白（Albers et al. 2004）。

表1.1 极端条件下异养微生物的生长条件

环境	极端微生物	生长条件			传代时间④/h	参考文献
		底物①	pH②	温度③/℃		
pH	嗜酸微生物（最适pH≤3.0）	细菌：葡萄糖、蔗糖、木糖、半乳糖、果糖、甘油酸、甘油、乙醇、苹果酸、淀粉、酵母抽提物、脯氨酸、谷氨酸、亮氨酸、酪氨酸、胰蛋白胨	1.3～3.0	32～52	2.0～3.0	[1]
		古菌：葡萄糖、半乳糖、淀粉、酵母提取物、蛋白胨、酪氨酸、胰蛋白胨、牛肉膏	1.0～2.2	53～96	4.0～5.0	[2]
	嗜碱微生物（最适pH≥9）	细菌：葡萄糖、阿拉伯糖、半乳糖、乳糖、麦芽糖、木糖、纤维二糖、D-海藻糖、木聚糖、淀粉、糖原、甘露糖、果糖、琼脂、蔗糖、甲酸钠、海藻糖、果胶、甲硫氨酸、酵母抽提物、谷氨酸、蛋白胨	9.0～13.0	45～60	3.0～4.0	[3]

续表

环境	极端微生物	生长条件			传代时间④/h	参考文献
		底物①	pH②	温度③/℃		
温度	嗜冷微生物（最适温度≤15℃，最高温度<20℃）	细菌：葡萄糖、果糖、蔗糖、麦芽糖、木糖、半乳糖、纤维二糖、乳糖、核糖、甘油、甘露醇、*N*-乙酰葡萄糖胺、组氨酸、脯氨酸、丙氨酸、色氨酸	5.0～10.0	5.0～15.0	1.0～6.0	[4]
		古菌：葡萄糖、木糖、酵母抽提物、色氨酸	6.0～9.0	1.0～15.0	5.0～20.0	[5]
	嗜热微生物（最适温度在45～80℃）	细菌：葡萄糖、木糖、阿拉伯糖、半乳糖、纤维二糖、果糖、乳糖、甘露糖、淀粉、甲硫氨酸、亮氨酸、苯丙氨酸、酵母膏、胰蛋白胨	4.0～9.8	45.0～78.0	1.7～5.0	[6]
		古菌：葡萄糖、蔗糖、乳糖、半乳糖、淀粉、麦芽糖、淀粉、酵母抽提物、胰蛋白酶水解物、酪蛋白、蛋白胨	2.7～8.0	50.0～75.0	1.65～4.0	[7]
	超嗜热微生物（最适温度≥80℃或最高温度≥90℃）	细菌：葡萄糖、核糖、木糖、半乳糖、蔗糖、麦芽糖、淀粉、糖原、酵母抽提物	6.5～7.5	80.0～85.0	1.5～2.0	[8]
		古菌：淀粉、糊精、麦芽糖、蔗糖、乳糖、葡聚糖、葡萄糖、木糖、半乳糖、阿拉伯糖、糖原、棉子糖、支链淀粉、纤维二糖、蜜二糖、乳酸、丙酮酸、酵母抽提物、胰蛋白胨、蛋白胨	5.5～7.0	80.0～106.0	0.5～3.2	[9]
盐浓度	嗜盐微生物（氯化钠>0.3mol/L）	细菌：葡萄糖、木糖、阿拉伯糖、棉子糖、蔗糖、半乳糖、乳糖、鼠李糖、海藻糖、酵母抽提物、明胶、淀粉、生物素、谷氨酸、脯氨酸、精氨酸、酪氨酸、酪蛋白氨基酸	6.5～8.0	31.0～50.0	3.0～18.0	[10]
		古菌：葡萄糖、木糖、果糖、阿拉伯糖、乳糖、木酮糖、蔗糖、淀粉、精氨酸、天冬氨酸、谷氨酸、甘氨酸、谷氨酰胺、色氨酸、酵母抽提物	6.5～9.5	30.0～45.0	1.5～3.0	[11]
水活度	嗜高渗生物（高浓度有机溶质、高浓度糖）	真核生物：葡萄糖、蔗糖、果糖	3.5～5.5	28.0～35.0	1.80～12.0	[12]
		细菌：葡萄糖、果糖、蔗糖、谷氨酸	5.0～7.0	25.0～30.0	2.0～3.0	[13]
	旱生微生物（a_w≤0.85）	真核生物：葡萄糖、果糖、葡萄糖、D-半乳糖醛酸酯、山梨醇	5.5～6.8	20.0～41.0	2.0～4.0	[14]

续表

环境	极端微生物	生长条件			传代时间④/h	参考文献
		底物①	pH②	温度③/℃		
高压	耐压菌（最适压力≥50MPa）	细菌：葡萄糖、果糖、麦芽糖、纤维二糖、甘露糖、水杨苷	6.5～8.0	10.0～98.0	0.3～6.2	[15]
低营养浓度	寡营养（1L培养基中含有1～15mg有机碳）	细菌：葡萄糖、木糖、蔗糖、半乳糖、甘露糖、阿拉伯糖、甲硫氨酸、甘氨酸、丝氨酸、谷氨酰胺、半乳糖、甘露糖、甘露醇、乳糖、L-脯氨酸、核糖	6.0～8.5	20.0～37.0	20.0～24.0	[16]
重金属	耐重金属微生物（耐受重金属浓度＞1mmol/L）	细菌：葡萄糖、麦芽糖、蔗糖、乙酸盐、柠檬酸、苹果酸、乙酸苯酯、己二酸、葡萄糖、丙氨酸、蛋白胨	6.0～8.5	25.0～55.0	6.0	[17]
辐射	抗辐射菌（抗辐射剂量≥1kGy）	细菌：葡萄糖、麦芽糖、果糖、阿拉伯糖、甘油、17种氨基酸（半胱氨酸、赖氨酸、丝氨酸、组氨酸、亮氨酸、赖氨酸、甲硫氨酸、苯丙氨酸、苏氨酸、色氨酸、天冬氨酸、缬氨酸、精氨酸、丙氨酸、谷氨酸、谷氨酰胺、丝氨酸）	6.4～9.5	25.0～55.0	2.0～10.0	[18]
		古菌：常见的20种氨基酸、酵母抽提物、胰胨、蛋白胨、丙酮酸盐、淀粉、麦芽糊精、麦芽糖、海藻糖、葡萄糖、乳糖	6.0～7.5	75.0～88.0	4.0～5.0	[19]

注：[1]（Hallberg and Lindström 1994；Hallberg et al. 2010；Harrison Jr 1984）；[2]（Dopson et al. 2004）；[3]（Takai et al. 2001；Blotevogel et al. 1985；Pikuta et al. 2000；Horikoshi 1996；Zhilina et al. 1997）；[4]（Margesin et al. 2003；Yumoto et al. 2003；Novitsky and Morita 1976；Ulitzur 1974）；[5]（Saunders et al. 2003；Preston et al. 1996；Sheridan et al. 2003）；[6]（Rainey et al. 1994；Degryse et al. 1978；Wiegel 1992；Ben-Bassat et al. 1981；Lovitt et al. 1984；Saiki et al. 1985；Huber and Stetter 1991）；[7]（Amend and Shock 2001；Egorova and Antranikian 2005；Huber and Stetter 1991；González et al. 1995）；[8]（Huber et al. 1986）；[9]（Blöchl et al. 1997；González et al. 1998；Huber et al. 2000；Erauso et al. 1993；Sako et al. 1996；Hafenbradl et al. 1996）；[10]（Franzmann et al. 1988；Antón et al. 2002；Gonzalez et al. 1978；Rodriguez-Valera et al. 1983；Oren et al. 1990；Bouchotroch et al. 2001）；[11]（Mormile et al. 2003；Xu et al. 1999；Burns et al. 2007；Robinson et al. 2005）；[12]（Restaino et al. 1983）；[13]（Jojima et al. 2004；Osman et al. 1987）；[14]（Gock et al. 2003；Marco et al. 2009；Marı et al. 2003；Palacios-Cabrera et al. 2005）；[15]（Nogi et al. 2002，2004；Simonato et al. 2006；Alain et al. 2002；Fang et al. 2010；Takai et al. 2009；Kato 1999）；[16]（Merchant et al. 2007；Reddy et al. 2007；Poindexter 1981；Fegatella et al. 1998；Chrzanowski et al. 1996）；[17]（Muller et al. 2006；Margaryan et al. 2010；Brim et al. 1999；Suresh et al. 2004）；[18]（Yoshinaka et al. 1973；Ferreira et al. 1997；Suresh et al. 2004；Rainey et al. 2007；De Groot et al. 2005；Yoo et al. 2010）；[19]（Jolivet et al. 2003，2004）。

①实验室确定的主要生长基质；②相应极端微生物的大多数个体的最适生长 pH 值；③相应极端微生物的大多数个体的最适生长温度；④相应极端微生物大多数个体的传代时间。

在细菌中，ABC 转运蛋白分为两类：糖类吸收转运蛋白和二肽/寡肽转运蛋白（Schneider 2001）。糖类转运的同时也受到磷酸烯醇式丙酮酸（PEP）的调控，目前仍未在古菌中发现糖磷酸转移酶系统（PTS）（Siebold et al. 2001）。PTS 所需的能量来自 PEP，并且通过几种蛋白质把 PEP 的磷酰基团转移到糖上（Postma et al. 1993）。PTS 由两种胞质蛋白（酶Ⅰ和组氨酸蛋白）及特异性糖转运复合体酶Ⅱ组成（Postma et al. 1993）。简而言之，酶Ⅰ将磷酰基团从 PEP 转移到组氨酸蛋白，然后再将磷酰基团转移到不同的转运复合体上（Postma et al. 1993）。例如，在葡萄糖的运输中，酶Ⅱ将葡萄糖磷酸化，运输穿过质膜形成葡萄糖-6-磷酸，随后进入糖酵解（EMP）途径（Bettenbrock et al. 2007）。

在古菌中，除了 ABC 转运蛋白，次级转运蛋白在溶质转运过程中也起到重要作用（Paulsen et al. 2000；Horlacher et al. 1998）。次级转运蛋白通常依赖钠离子或质子跨膜的电化学梯度，而 ABC 转运蛋白则没有这种限制，从而可以在细胞内积累更高浓度的底物（Albers et al. 2004）。古菌和细菌的 ABC 转运蛋白的主要区别是底物结合蛋白（Albers et al. 2004）。在细菌中，膜外蛋白捕获底物，随后底物通过跨膜结构域进而释放（Quiocho and Ledvina 1996）。通过对超嗜热古菌结合蛋白的研究发现：古菌的结合蛋白经过糖基化修饰，可以抗蛋白酶水解（Erra-Pujada et al. 2001；Hettmann et al. 1998；Xavier et al. 1996；Evdokimov et al. 2001；Albers et al. 1999）。在这两类生物中，结合蛋白可与多种长度和组成均不同的底物结合（Elferink et al. 2001；Koning et al. 2002）。古菌 ABC 转运蛋白对单糖和多数双糖的吸收和转运与细菌是类似的（Albers et al. 2004）。同时，极端嗜热菌的一些双糖和低聚糖（纤维二糖、β-葡萄糖苷和低聚纤维）的转运蛋白与细菌二肽/寡肽转运蛋白具有序列同源性（Elferink et al. 2001；Koning 2001）。此外还发现，一些属于二肽/寡肽转运蛋白的操纵子在糖降解酶附近，表明它们可能参与二糖/寡糖的催化和吸收（Nelson et al. 1999）。

对诸如淀粉、糖原、纤维素、半纤维素、木聚糖等多糖的利用，胞外糖基水解酶将其水解成单糖或者二糖，随后通过 ABC 转运蛋白或次级转运蛋白进入细胞（Elferink et al. 2001；Saurin et al. 1999）。除了个别物种，在六碳糖中，葡萄糖作为极端微生物的基础底物（表 1.1）。五碳糖中，木糖是最常用的，但许多极端微生物缺乏相应的转运蛋白，因此只能利用一部分这类糖。

单一氨基酸作为唯一底物不能支持多种极端微生物的生长（表 1.1）。缺乏特定氨基酸生物合成途径的极端微生物往往不能够在缺乏这种氨基酸的条件下生存，例如，激烈火球菌 *Pyrococcus furiosus* DSM3638 的生长需要异亮氨酸和缬氨酸（Hoaki et al. 1994）；超嗜热古菌 *Hyperthermus butylicus*（Zillig et al. 1990）和 *Staphylothermus marinus*（Hao and Ma 2003）偏好利用蛋白质作为它们生长的碳源和氮源（表 1.1）（Schönheit and Schäfer 1995）。总的来说，偏好生长于多糖、寡糖和单糖等底物上的微生物可以利用蛋白质类化合物，但反之则不行，这可能是由于缺乏糖转运蛋白，如在古菌中缺乏糖磷酸转移酶系统（Verhees et al. 2003）。

除了嗜酸微生物和嗜碱微生物，大部分极端微生物适宜生长在中性 pH 范围内，且最佳生长温度范围较广，可能的一个原因是它们的栖息地多样化，具有不同的气候特征。嗜高压微生物适宜生长温度范围最广（10～98℃），以应对深海和火山区的高压（Nogi et al. 2002, 2004；Simonato et al. 2006；Alain et al. 2002；Fang et al. 2010；Takai et al. 2009；Kato 1999）。除了寡营养微生物外，在最佳生长条件下，所有微生物的生长周期都具有相对性，表明微生物通过进化使得其适合在这种极端条件下生长，并且养分是不可或缺的因素（Merchant et al. 2007；Reddy et al. 2007；Poindexter 1981；Fegatella et al. 1998；Chrzanowski et al. 1996）。

不同极端微生物对不同底物的利用，依赖于它们特定的酶和主要代谢途径，这将在下一节中详细阐述，每类极端微生物主要代谢途径对环境压力的适应性也将进行介绍。

1.3 糖类的利用，中心代谢途径和适应极端环境

许多底物都可以被微生物利用。在被运送进入细胞之前，多糖被微生物分泌的水解酶水解（Warren 1996），以下介绍微生物主要的糖类中心代谢途径。

1.3.1 嗜酸微生物和嗜碱微生物

生长在极端 pH 中的微生物称为嗜酸和嗜碱微生物，pH 会影响酸性或碱性氨基酸和酶的电离状态，理解这些物种应对极端条件的机制是十分有意义的（Jaenicke 1981）。研究发现，细胞膜和细胞内环境稳态机制确保极端微生物细胞内 pH 恒定（中性），进而提供了一个稳定的环境，使细胞内的酶功能保持正常（Jolivet et al. 2004）。

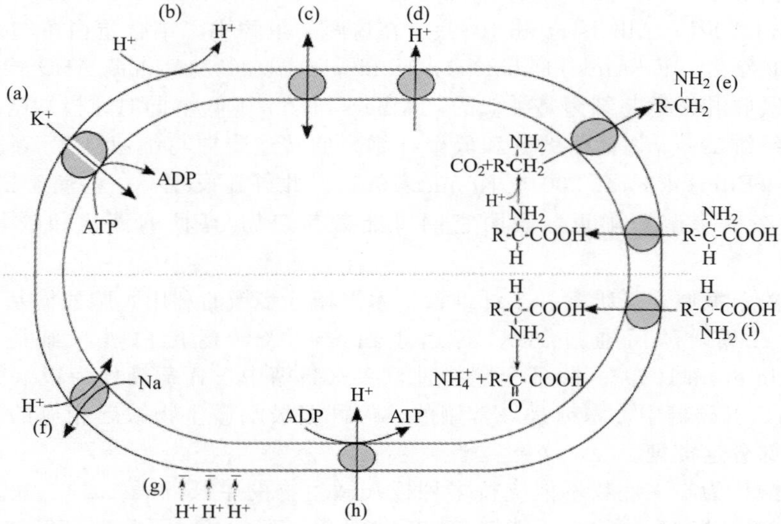

图 1.1 嗜酸和嗜碱微生物的 pH 平衡机制。代表性的嗜酸微生物：（a）钾转运 ATP 酶转运钾产生一个反向膜电位；（b）高度不渗透的细胞膜，防止质子的涌入；（c）丰富的次级转运蛋白减少溶质和营养物质运输到细胞内的能量需求；（d）活性质子由不同的转运蛋白输出；（e）通过消耗质子增加碱性氨基酸脱羧率和提高 pH 代表性的嗜碱微生物；（f）质子利用钠-氢反向转运体流入；（g）质子附着，堆积在酸化的细胞膜上并被运输到细胞内；（h）ATP 合酶增多，偶联质子进入细胞和促进 ATP 合成；（i）增加转运蛋白和氨基酸脱氨酶来产生相应的 α-酮酸

嗜酸微生物中，存在多个机制用于维持酸碱平衡（图 1.1）（Baker-Austin and Dopson 2007），其中一种可能的机制是通过钾离子的内流来产生一种反向膜电位，即膜内正电位，从而建立一种阻止质子内流的化学渗透屏障（Baker-Austin and Dopson 2007）。在嗜酸微生物的基因组中，转运钾离子的 ATP 酶占主导地位（Fütteret et al. 2004；She et al. 2001）。其次它们具有质子难以渗透的细胞膜结构（Konings et al. 2002），嗜酸古菌中，细胞膜是由四醚脂质连接，而不是酯键（Shimada et al. 2002；Batrakov et al. 2002；van de Vossenberg et al. 1998）。此外，多余的质子被及时释放，包括 H^+-ATP 酶质子外排系统、反向转运子和协同转运子（Michels and Bakker 1985；Tyson et al. 2004），还有一个细胞质缓冲体系（Castanie-Cornet et al. 1999），主要是通过碱性氨基酸，如赖氨酸、组氨酸和精氨酸的脱羧反应消耗质子（Castanie-Cornet et al. 1999；Álvarez-Ordóñez et al. 2010）。以赖氨酸为例，

一旦进入细胞，由赖氨酸脱羧酶催化脱羧后转化为尸胺，随后与新的赖氨酸分子交换，被运输到胞外（Álvarez-Ordóñez et al. 2010），胞质 pH 值因质子的消耗而提高（Álvarez-Ordóñez et al. 2010）。除了这些机制，基因组序列分析表明，有机酸降解，大量蛋白质和 DNA 修复相关基因的表达及较小基因组都是对低 pH 的适应（Baker-Austin and Dopson 2007；Ciaramella et al. 2005）。

嗜碱微生物中，维持 pH 稳态的主要有四种机制（图 1.1）（Padan et al. 2005）。许多嗜碱性微生物代谢中，一个重要途径是增加单价阳离子/质子逆向转运蛋白的表达和活性，大部分由 Na^+/H^+ 逆向转运蛋白进行 pH 值的调节（Kitada et al. 2000；Swartz et al. 2005）。另外通过氨基酸脱氨和糖酵解反应以增加酸性代谢产物的生成，也可降低细胞内 pH 值（Blankenhorn et al. 1999；Richard and Foster 2004），ATP 合酶活性增加的同时也可增加 H^+ 数量（Rozen and Belkin 2001；Krulwich et al. 1998）。除此之外，增加细胞表面酸性，使阳离子（H^+ 和 Na^+）结合量增加（Wang et al. 2004）。

糖酵解（EMP）、恩特纳-杜多罗夫（ED）和戊糖磷酸（PP）途径存在于嗜酸微生物和嗜碱微生物中，但是古菌的 EMP 和 ED 途径是经过修饰的。三羧酸循环（TCA）存在于大多数微生物中，包括嗜酸微生物和嗜碱微生物，但有一些微生物由于基因的缺失而不能形成完整的 TCA 循环。

嗜酸微生物中主要途径是 EMP 途径，但是 *Acidiphilium cryptum* Lhet2 没有检测到 EMP 途径关键酶——6-磷酸果糖激酶（Shuttleworth et al. 1985）。研究发现 EMP、ED 和 PP 途径依赖于培养条件（Barrie Johnson and Hallberg 2008）。通过酶活性的测定，混合营养条件下（生长介质中含有亚铁、硫代硫酸盐和有机化合物），三种途径均用于糖类的代谢（Barrie Johnson and Hallberg 2008）；异养条件下，EMP 和 ED 这两个途径起主要作用（Barrie Johnson and Hallberg 2008）；自养条件下，只有 EMP 途径发挥功能，并且检测到较高的二磷酸果糖酶活性，这表明其参与糖异生作用（Barrie Johnson and Hallberg 2008）。嗜酸微生物中存在木聚糖酶，能够利用木聚糖，该酶在工业中有着非常广阔的应用前景（Kumar et al. 2013）。一些细菌，如 *Acidithiobacillus ferrooxidans*、*Acidithiobacillus thiooxidans* 常用于冶炼金属（Mishra et al. 2008；Wang et al. 2009）。*Acidithiobacillus caldus* 中由于缺乏编码琥珀酸脱氢酶和 α-酮戊二酸脱氢酶复合体的基因，导致 TCA 循环不完整（You et al. 2011），不完整的 TCA 循环被认为是一种生物合成途径，而不是产能途径（Wood et al. 2004）。

嗜酸古菌糖代谢也存在 EMP、ED 和 PP 这三种途径（Kengen et al. 1996；Moracci et al. 2000；Budgen and Danson 1986），但古菌有一个修饰的 EMP 途径和两个修饰的 ED 途径，这与嗜酸古菌嗜热或超嗜热的特性相吻合。*Thermoproteus tenax* 为一种超嗜热嗜酸古菌，能够利用三种途径进行糖代谢，并且 EMP 途径的酶是经过特殊修饰的（Kengen et al. 1996）。首先，已糖激酶被修饰为没有调控功能的酶（Kengen et al. 1996）；另外通过双向非变构调节的焦磷酸依赖酶替代 ATP 依赖的磷酸酶和二磷酸果糖酶（Kengen et al. 1996）；同时还含有两个不同的 3-磷酸甘油醛脱氢酶（GAPDH），这两个酶的区别在于对磷酸的依赖性，催化反应的可逆性和变构性质（Kengen et al. 1996；Brunner et al. 2001；Hensel et al. 1987）。NAD^+（氧化型辅酶Ⅰ）依赖型 GAPDH 被用于将 3-磷酸甘油醛（GA3P）转化为 3-磷酸甘油酸，而 $NADP^+$（氧化型辅酶Ⅱ）依赖型则参与合成代谢产物，催化 1,3-二磷酸甘油酸转化为 GA3P（Brunner et al. 2001）。NAD^+ 依赖型 GAPDH 具有变构调节能力（Brunner et al. 1998，2001），它可被抑制，但也可被一系列的代谢物激活，可被 AMP、ADP、葡萄糖-1-磷酸和果糖-6-磷酸激活，但受到 NADP（H）（还原型辅酶Ⅱ）

和 NADH（还原型辅酶Ⅰ）的抑制，这在醛类脱氢酶中很少见（Brunner et al. 2001）。这种复杂的调控表明该酶可能在控制 EMP 途径的代谢中发挥着核心作用，它弥补了可逆的非变构焦磷酸依赖型磷酸果糖激酶调节能力的不足（Brunner et al. 2001）。

ED 途径中存在半磷酸化和非磷酸化两种修饰途径（图 1.2），*T. tenax* 和嗜热古菌 *Thermoplasma acidophilum* 利用半磷酸化的 ED 途径，磷酸化只发生在 2-酮-3-脱氧葡萄糖酸醛缩酶（KDGA）生成甘油酸和甘油醛的水平上，KDGA 具有非磷酸化的中间体（Kengen et al. 1996；Budgen and Danson 1986；van der Oost et al. 2005）。此外，非磷酸化的 ED 途径存在于超嗜热嗜酸古菌 *Sulfolobales* 中，第一个丙酮酸分子的形成中未发生磷酸化，2-酮-3-脱氧葡萄糖糖酸直接裂解成丙酮酸和甘油醛（Budgen and Danson 1986；Kengen et al. 1996；Moracci et al. 2000）。

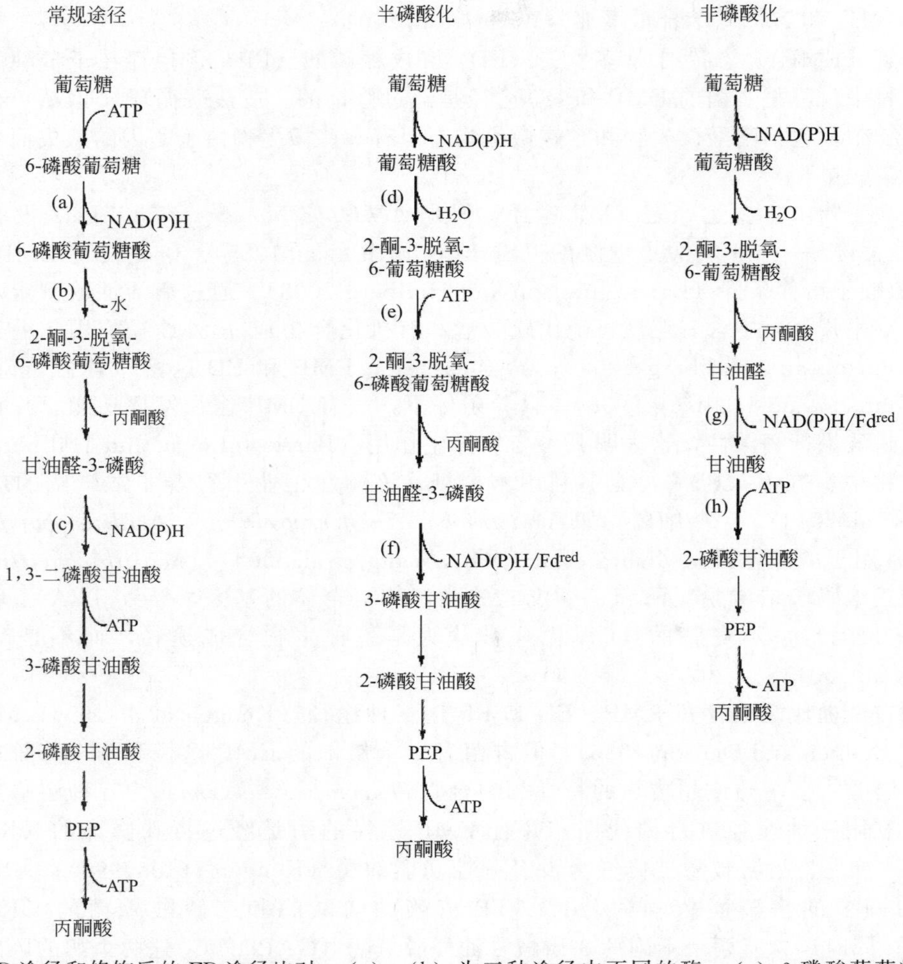

图 1.2 ED 途径和修饰后的 ED 途径比对。(a)～(h) 为三种途径中不同的酶：(a) 6-磷酸葡萄糖脱氢酶；(b) 6-磷酸葡萄糖酸脱水酶；(c) 甘油醛-3-磷酸脱氢酶；(d) 葡萄糖酸脱水酶；(e) 2-酮-3-脱氧葡糖酸激酶；(f) 甘油醛-3-磷酸脱氢酶/甘油醛-3-磷酸氧化还原酶；(g) 醛脱氢酶/醛氧化还原酶；(h) 甘油酸激酶。PEP—磷酸烯醇式丙酮酸

所有的嗜碱细菌 EMP 途径的关键酶都具有高活性，包括已糖激酶、6-磷酸葡萄糖脱氢酶、磷酸果糖激酶、果糖二磷酸醛缩酶和 3-磷酸甘油醛脱氢酶，这表明嗜碱细菌利用 EMP

作为主要代谢途径（Garnova and Krasil'nikova 2003；Zhilina et al. 2004；Paavilainen et al. 1999）。广泛的研究表明嗜碱性细菌 *Amphibacillus*（兼性芽孢杆菌属）、*Halonatronum*、*Alkaliflexus* 等具有降解糖类的功能，同时也具有参与其他中心代谢途径的高活性酶。*Amphibacillus* 具有 ED 途径的关键酶：6-磷酸葡萄糖酸脱水酶和 2-酮-3-脱氧-6-磷酸葡萄糖酸醛缩酶（Garnova and Krasil'nikova 2003）。*Halonatronum* 中发现了参与 PP 途径的酶，*A. fermentum* 中检测到了参与三种葡萄糖代谢途径关键酶的活性（Garnova and Krasil'nikova 2003）。嗜碱嗜盐微生物 *Alkaliflexus imshenetskii* 通过 EMP 和 ED 途径代谢纤维二糖，且没有 PP 途径的相关酶类（Zhilina et al. 2004）。环状芽孢杆菌可通过三种途径代谢葡萄糖，但起主导作用的是 EMP 途径（90%～93%），ED 和 PP 途径占比少于 10%（Paavilainen et al. 1999）。甲烷氧化嗜碱细菌，例如 *Methylobacter alcaliphilus*，在 TCA 循环中没有发现 α-酮戊二酸脱氢酶的存在，从而导致 TCA 循环不完整，该循环多存在于Ⅰ型甲烷氧化菌中（Khmelenina et al. 1997）；其他嗜酸、嗜碱异养微生物基因组分析表明，其具有完整的 TCA 循环系统，甲烷氧化菌和化学自养生物通常具有不完整的 TCA 循环，表明 TCA 途径受到营养模式的影响（Hanson and Hanson 1996）。

极端温度中生长的极端微生物不仅具有独特的代谢途径，而且在古菌中有许多独特的酶。生长在高温下的一些嗜高温极端微生物与嗜酸微生物相似，具有非磷酸化的 ED 途径，表明这种改变是为了适应高温环境，而不是适应极端 pH 条件。为了应对来自环境的压力，极端微生物演化出了多种适应策略。

1.3.2　嗜冷微生物

低温下，微生物面临的两个主要的挑战是低热能和高黏度（D'Amico et al. 2006），该条件下修饰的目标是蛋白质，因为它参与了许多对细胞存活很重要的过程（D'Amico et al. 2006）。

嗜冷微生物中，冷适应/冷休克蛋白和分子伴侣十分丰富，可用于帮助蛋白质正确折叠，同时也含有抗冻蛋白、海藻糖和胞外多糖这类具有抗冻效果的物质（Phadtare 2004；Jia and Davies 2002）。许多参与重要过程的酶，如参与转录和翻译的酶，已经进化为在低温下活性最高（Lim et al. 2000），通过增加它们结构的灵活性和减少在催化过程中必须破坏的由焓驱动的相互作用的数目来实现（Violot et al. 2005；Berger et al. 1996；Russell 2000）。嗜冷微生物的细胞膜中，含有较高的不饱和脂肪酸、多不饱和脂肪酸和甲基支链脂肪酸，脂肪酸具有较短的酰基链长度，以增加膜的流动性从而适应低温（Chintalapati et al. 2004；Russell 1997）。这些机制在低温下维持代谢酶的功能，以利用营养物质，并使嗜冷微生物在低温下生长最佳。

嗜冷细菌和古菌大部分碳源是通过 EMP 途径利用，糖酵解和糖异生途径是完整的（Xu et al. 2003；Bakermans et al. 2003；Médigue et al. 2005；Cavicchioli et al. 2000）。微生物在寒冷环境中所面临的一个挑战是生理反应中酶的活性有所下降，包括中心代谢途径中有关 ATP 产生的酶和中心代谢途径中的辅酶（Amato 2013）。利用蛋白质组学、转录组学和酶检测等方法发现，嗜冷微生物通过上调或增加中心代谢途径中关键酶的活性来克服这个问题（Amato 2013）。嗜冷菌 *Lactobacillus piscium* 中，通过上调甘油醛-3-磷酸脱氢酶活性来克服糖酵解中温度的限制（Garnier et al. 2010），另外果糖二磷酸醛缩酶和磷酸甘油酸激酶也被上调（Garnier et al. 2010）。*Propionibacterium fredenreichii* 中，TCA 循环中所涉及的酶，包括顺乌头酸酶、延胡索酸酶、琥珀酸脱氢酶和柠檬酸合酶均有所上调（Dalmasso et al. 2012）。同样，*Pseudoalteromonas haloplankis* 也通过上调 TCA 循环中的酶活性来克服冷应激（Piette et al. 2010）。增加 TCA 循环中的底物，使得 TCA 中的酶活性提升，以确保催化过程，同时 ATP 的产量可增加。嗜冷古菌 *Methanococcoides burtonii* 具有一个简化的氧化型 TCA 循环，除了存在延胡索酸酶异源二聚体外，草酰乙酸至 α-酮戊二酸方向上的基因几乎完全丢失（Goodchild et al. 2004）。氧化方向上存在柠檬酸合酶和顺乌头酸酶，柠檬酸合酶、顺乌头酸酶以及异柠檬酸/异丙基苹果酸脱氢酶可能是催化异柠檬酸到 α-酮戊二酸剩余步骤的候选酶类（Goodchild et al. 2004）。此外还含有丙酮酸合酶和丙酮酸羧化酶催化乙酰 CoA 合成草酰乙酸（Goodchildet al. 2004），*M. burtonii* 也是一种甲氧基营养菌（Goodchild et al. 2004）。

1.3.3 嗜热微生物

嗜热微生物和超嗜热微生物需要应对高温引起的蛋白质变性（Jaenicke and Böhm 1998）。目前存在两种机制稳定蛋白质和防止蛋白质高温变性，这些机制的选择与极端微生物的起源进化密切相关（Berezovsky and Shakhnovich 2005）。第一种机制是通过对蛋白质结构进行修饰，与中温同系物相比其结构更为紧凑，这是大量分子内相互作用造成的（Berezovsky and Shakhnovich 2005），该机制在生活于高温极端环境的嗜热微生物和超嗜热微生物中发现（England and Shakhnovich 2003；England et al. 2003）。第二种机制是基于序列的，它们的结构与其中温同系物之间没有明显的差异（Dominy et al. 2004；Macedo-Ribeiro et al. 1996），序列的改变能提高热稳定性，增强相互作用，因此在结构上并没有表现出与中温同系物的显著差异（Li et al. 2005）。这一机制是在极端微生物，主要是嗜热微生物中发现的。这种嗜热微生物原是一种中温微生物，后来迁移至炎热环境中，该微生物为了适应环境从而进化出此种策略（Berezovsky and Shakhnovich 2005）。这些微生物还具有分子伴侣、热休克蛋白，能够帮助那些不能自发形成热稳定构象的蛋白质进行折叠（Hendrick and Hartl 1993；Feder and Hofmann 1999）。这些分子伴侣也可以防止非折叠蛋白质在高温下聚集，并且指导错误折叠和变性的蛋白质进入细胞内的蛋白质降解系统（Sterner and Liebl 2001）。常见分子伴侣为 Hsp60 的复合物，也被称为热小体（Sterner and Liebl 2001）。

高温下嗜热细菌主要通过 EMP 途径来利用糖类，而且具有该途径中所有的酶类（Selig et al. 1997）。ED 途径中的关键酶磷酸葡萄糖脱水酶和 2-酮-3-脱氧-6 磷酸葡萄糖酸醛缩酶在嗜热细菌 *Geobacillus thermoglucosidasius*（Tang et al. 2009）和 *Caldicellulosiruptor saccharolyticus*（De Vrije et al. 2007）中并没有发现。研究还发现，在 *G. thermoglucosidasius* 中的 PP 途径利用率在不同氧浓度下会有所差别（Tang et al. 2009），在葡萄糖发酵条件下检测到 PP 途径氧化性部分活性较低，同时发现 6-磷酸葡聚糖酸内酯酶基因的缺失（Tang et al. 2009）。有氧条件下，约三分之二的葡萄糖是通过糖酵解被利用，其余是通过 PP 途径。而在微氧条件下，TCA 循环和 PP 途径会减少到原水平的一半（Tang et al. 2009）。微氧条件下，生长速度下降至 $0.2h^{-1}$，一般来说支持生物质合成的 NADPH 主要来自 PP 途径（Christensen et al. 2002）。嗜热微生物如 *Thermotoga maritima*（海栖热袍菌）的葡萄糖代谢也主要通过常规的 EMP 途径来进行（Schröder et al. 1994）。

嗜热和超嗜热古菌中使用的是修饰后的 ED 和 EMP 途径，而且氧化型 PP 途径被非氧化型的 PP 途径所取代（Schönheit and Schäfer 1995；Van der Oost and Siebers 2007）。下文将用例子来描述每一种被修饰后的酶。

嗜热古菌中糖类代谢的主要途径也是 EMP 途径，但该途径经过一定的修饰（图 1.3）。*Thermococcus zilligii* 和 *Thermococcus stetteri* 具有 ADP 依赖型的磷酸果糖激酶和独特的甘油醛-3-磷酸：铁氧还蛋白氧化还原酶（Ronimus et al. 1999；Xavier et al. 2000；Kengen et al. 1996）。在 *T. zilligii* [^{13}C] 葡萄糖标记实验的基础上提出了一种新的可能涉及戊糖糖酵解的途径（Xavier et al. 2000），该途径如下：

步骤 1　葡萄糖＋ADP ⟶ 葡萄糖-6-磷酸＋AMP（已糖激酶）

步骤 2　葡萄糖-6-磷酸＋$NADP^+$ ⟶ 6-磷酸葡萄糖酸＋NADPH＋H^+（葡萄糖-6-磷酸脱氢酶）

步骤 3　6-磷酸葡萄糖酸 ⟶ 甲酸＋木酮糖-5-磷酸（新型裂解酶）

木酮糖-5-磷酸 ⟶ 甘油醛-3-磷酸＋乙酰-磷酸（戊糖磷酸转酮醇酶）

步骤 4　甘油醛-3-磷酸 ⟶ 丙酮酸 ⟶ 乙酸＋CO_2，乙酰-磷酸 ⟶ 乙酸（戊糖磷酸转酮醇酶途径）

简而言之，葡萄糖-6-磷酸在裂解酶催化下裂解成木酮糖-5-磷酸和甲酸（步骤 3），磷酸戊糖是通过戊糖磷酸转酮醇酶通路进行代谢的（Xavier et al. 2000）。

1.3.4　超嗜热微生物

超嗜热古菌比如 *Desulfurococcales*、*Thermococcales*、*Pyrococcus*、*Archaeoglobus fulgidus* 所利用的是修饰过具有独特酶类的 EMP 途径（图 1.3）（Schönheit and Schäfer 1995；Huber and Stetter 2006；Bertoldo and Antranikian 2006；Kengen et al. 1994；Labes and Schönheit 2001；Siebers and Schönheit 2005）。超嗜热细菌 *Thermotoga maritima* 和一些超嗜热嗜酸微生物也使用 EMP 途径和修饰过的 ED 途径（Siebers and Schönheit 2005）。EMP 途径由一些独特的酶修饰，如 ADP 依赖型激酶，包括 ADP 依赖型葡萄糖激酶和 ADP 依赖型磷酸果糖激酶，而中温微生物是 ATP 依赖型（Sakuraba et al. 2002；Koga et al. 2000；Dörr et al. 2003）。从能量角度看，ADP 不低于 ATP，ADP 水解自由能变化与 ATP 大致相等（Hongo et al. 2006）。ADP 在高温下比 ATP 更稳定，特别是在一些二价金属离子存在的情况下（Hongo et al. 2006；Tetas and Lowenstein1963）。另一个修饰是甘油醛-3-磷酸铁氧还蛋白酶取代了传统糖酵解酶、3-磷酸甘油醛脱氢酶和磷酸甘油酸激酶，它是一种含钨的铁硫蛋白，发现于 *Pyrococcus furiosus*（Mukund and Adams 1991，1995；van der Oost et al. 1998）、*Thermococcales*（Kletzin et al. 1995；Mukund and Adams 1993）和 *Desulfurococcus amylolyticus*（Selig et al. 1997）中。它能够在反应中催化 3 -磷酸甘油酸的产生和利用 GA3P 氧化反应将铁氧还蛋白还原，而在传统的反应中这是两步反应（de Vos et al. 1998）。这个反应中，从磷酸甘油酸激酶中生成 ATP 的步骤被省略，从而将底物水平的 ATP 产量降低为零（Verhees et al. 2004）。这样修饰的原因可能是吡啶核苷酸较铁氧还蛋白不耐高温，同时为了避免生成不耐热的中间产物 1,3-二磷酸甘油酸（Brunner et al. 2001）。为了弥补损失的 ATP 量，*Thermococcus kodakarensis* 中将磷酸烯醇式丙酮酸转换为丙酮酸的过程由 PEP 合酶代替（Sakuraba and Ohshima 2002；Sakuraba et al. 2004）。丙酮酸激酶利用 ADP 并产生一分子 ATP，但 PEP 合酶使用的是 AMP 和磷酸，它可以让每分子葡萄糖产生额外的 2 个 ATP（Sakuraba and Ohshima 2002；Sakuraba et al. 2004；Imanaka et al. 2006）。这也表明 *T. kodakarensis* 中 PEP 合酶基因的缺失可降低其在糖类培养基中的存活率（Imanaka et al. 2006）。

好氧超嗜热嗜酸微生物使用的是非磷酸化的 ED 途径（Siebers and Schöheit 2005），这一途径中葡萄糖被氧化为葡萄糖酸，并形成 2-酮-3-脱氧-葡萄糖酸，同时特定的激酶会将产生的甘油磷酸化（图 1.3c）（Siebers and Schönheit 2005）。

超嗜热古菌中许多酶通过测序发现与之前研究的有所不同，基于基因组序列分析难以对这些酶作出预测。*T. kodakaraensis* KOD1 中，从基因组序列分析中预测了有功能性的 2-脱

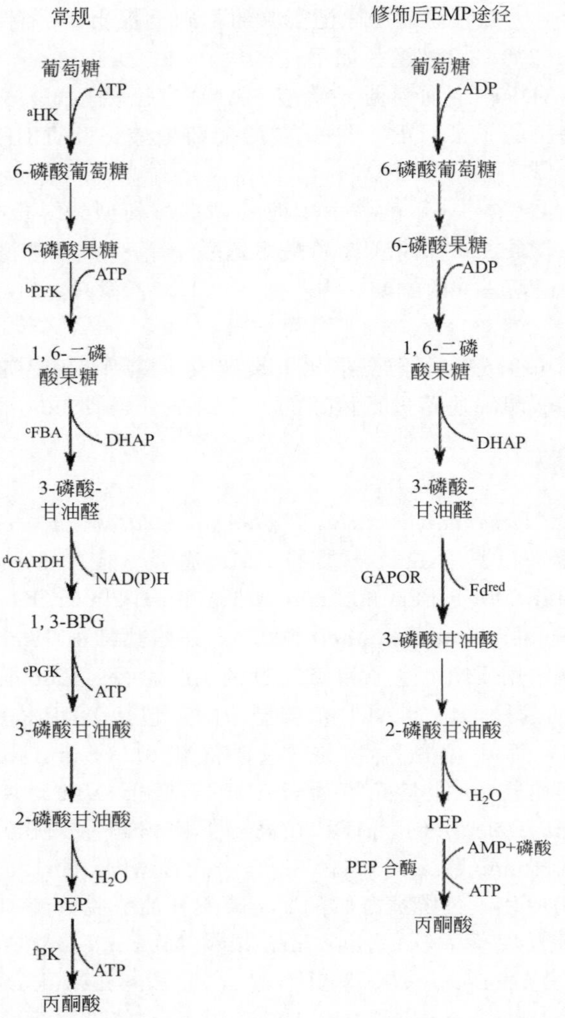

图 1.3 极端微生物中 EMP 途径与修饰后 EMP 途径的比较。极端微生物中传统 EMP 途径的酶经过修饰或替换，并且在超嗜热古菌中存在一个被修饰过的 EMP 途径。被修饰的酶被标记出来。a. HK，已糖激酶；b. PFK，超嗜热/嗜热嗜酸性微生物中修饰过的磷酸果糖激酶；c. FBA，果糖二磷酸酶醛缩酶；d. GAPDH，甘油醛-3-磷酸脱氢酶，在超嗜热嗜酸性微生物中被修饰，在嗜冷微生物中有所上调；e. PGK，磷酸甘油酸激酶在嗜冷微生物中上调；f. PK，在超嗜热微生物被修饰的 EMP 途径中，丙酮酸激酶被 AMP 依赖的 PEP 合酶所替代

GAPOR，在超嗜热微生物修饰过的 EMP 途径中，3-磷酸甘油醛氧化还原酶取代 GAPDH 和 PGK；DHAP，二羟丙酮磷酸；1,3-BPG，1,3-二磷酸甘油酸；PEP，磷酸烯醇式丙酮酸；Fd^{red} 还原性铁氧还蛋白

氧核糖-5-磷酸醛缩酶，还发现了一种新型磷酸戊糖变位酶（PPM），其序列与之前发现的 PPM 并没有显著的同源性（Rashid et al. 2004）。这为戊糖和碳中心代谢途径之间存在的代谢联系提供证据，许多细菌和嗜热微生物中都有这种联系（Rashid et al. 2004），更多在嗜热菌中心代谢途径中起作用的新型酶有望被发现。

1.3.5 嗜盐微生物

高浓度盐导致高渗透压，嗜盐微生物通过排出胞质内的盐分和胞内产生渗透压的物质来

抵御这种压力（Roberts 2000，2004）；所有的嗜盐微生物都含有诸如 Na^+/H^+ 逆向转运蛋白的转运机制以排出细胞内多余的 Na^+（Oren et al. 1990；Obis et al. 1999）。为了让胞质内具有高渗透压，微生物有着不同的策略。第一种策略只有一些古菌和细菌使用，微生物在细胞质中建立起至少比周围的 NaCl 浓度一样高的 KCl 浓度来应对渗透压力（Pflüger and Müller 2004）；另一种普遍使用的策略是积累甘油、甜菜碱、嘧啶、蔗糖和海藻糖这些常见的有机溶质（Imhoff and Rodriguez-Valera 1984；Brown 1976），虽然该策略耗能更高，但其胞内的酶系统不能适应高浓度氯化钾，故采用此策略；此外嗜盐微生物中的酶也经过修饰，可以在高浓度盐中发挥作用。

嗜盐微生物的中心代谢途径中的酶类也被修饰，以应对高盐环境，例如嗜盐细菌 *Salinibacter ruber* 中，葡萄糖通过组成型的、受盐抑制的己糖激酶和组成型的、盐依赖型 NADP 葡萄糖-6-磷酸脱氢酶进行代谢（Oren and Mana 2003）。葡萄糖代谢传统途径是 ED 途径，而不依赖 EMP 途径，因为未检测到葡萄糖脱氢酶和果糖-1，6-二磷酸醛缩酶的活性（Oren and Mana 2003）。

嗜盐古菌 *Halococcus saccharolytucus*、*Haloferax mediteranes* 和 *Haloarcula valismurtis* 中的葡萄糖和半乳糖是通过半磷酸化的 ED 途径进行分解代谢（Johnsenet al. 2001；Falb et al. 2008）；此外它们使用修饰过的 EMP 途径代谢果糖（Johnsen et al. 2001；Falb et al. 2008），它们缺乏编码关键酶 6-磷酸果糖激酶的基因，而经典 EMP 途径的六糖部分，从葡萄糖到果糖-6-磷酸的步骤可能会丢失（Johnsenet al. 2001；Falb et al. 2008）。

1.3.6 嗜高渗和耐旱微生物

生长在高浓度溶质培养基中微生物的渗透压调节机制使得细胞存活并使代谢酶类保持适当的功能以产生能量（Pflüger and Müller 2004）。嗜高渗酵母常见于环境中，酵母中的中心代谢途径是通过 EMP、ED 和 PP 途径来产生丙酮酸进入 TCA 循环（Rodrigues et al. 2006）。甘油在渗透压的调节方面起着至关重要的作用（Nevoigt and Stahl 1997），细胞质内甘油-3-磷酸脱氢酶活性的提升使得甘油产生率显著提高，以应对胞外下降的水活度（Nevoigt and Stahl 1997；Påhlman et al. 2001；Albertyn et al. 1994）；同时甘油跨膜运输蛋白 Fps1p 也关闭，以贮藏细胞内产生的甘油，从而保持内部与外部环境的渗透压相等（Tamás et al. 1999；Luyten et al. 1995）。海藻糖是另一种调节渗透压的兼容性溶质（De Smet et al. 2000；Lippert et al. 1993），常规途径是由 UDP（二磷酸尿苷）-葡萄糖和 6-磷酸葡萄糖合成海藻糖（De Smet et al. 2000；Vogel et al. 1998；Virgilio et al. 1993）。这两种化合物形成海藻糖-6-磷酸，并且游离的海藻糖通过去磷酸化形成（Wolf et al. 2003），途径中的关键酶分别是海藻糖-6-磷酸合酶和海藻糖-6-磷酸磷酸酶（De Smet et al. 2000；Vogel et al. 1998；Virgilioet al. 1993）。面包酵母、酿酒酵母中产生甘油是为应对高浓度的盐，D-阿拉伯糖醇为应对高浓度的葡萄糖（Moran and Witter 1979），甘油和 D-阿拉伯糖醇均可在细胞内积累（Moran and Witter 1979）。

葡萄糖代谢是通过 EMP 途径（约 80%）和 PP（约 20%）途径进行（Moran and Witter 1979），随着葡萄糖浓度的增加（0.6g/mL），PP 途径的参与度随之增加（31%），而 EMP 途径降低（69%）（Moran and Witter 1979）。激活的 PP 途径通过提供足够的 NADPH 来参与高糖耐受，这是细胞生存和合成结构脂类所必需的（Moran and Witter 1979；Tokuoka 1993）。然而当介质为蔗糖时则没有观察到类似现象（Moran and Witter1979）。这是因为 D-阿拉伯糖醇是通过 PP 途径代谢葡萄糖所产生，而 D-阿拉伯糖醇并不参与蔗糖耐受机制（Moran and Witter 1979）。*Zymomonas mobilis* 是一种中温细菌，通过 ED 途径代谢糖并且以不同的糖作为底物时会产生不同的副产物（Sprenger 1996；Rogers et al. 1979；De Graaf et al. 1999）。高浓度糖对ED途径中的酶没有抑制作用，为了应对环境压力，糖的运输是采用一种高速的、运载体介导的促进扩散系统运输的独特方式，其胞内浓度可迅速达到一个接近外部浓度的水平，预防高浓度糖造成的损伤（DiMarco and Romano 1985）。

在高浓度的溶质导致低水活度的条件下，嗜高渗微生物 *Penicillium janczewskii*、*Eurotium chevalieri*、*Xeromyces bisporus* 和 *Wallemia sebi* 产生甘油，将其作为一种重要的细胞渗透调节溶质，并在细胞中积累（Hocking 1986；Huang et al. 2009）。

耐旱真菌通过 EMP 途径代谢糖类（Hocking and Norton 1983），一个极端的例子是 *Xeromyces bisporus* 失去了所有编码催化次生代谢物的酶基因（Leong et al. 2014）。

1.3.7 寡营养微生物

尽管营养物质是生长能量和碳元素的根本来源，但有一类特殊的极端寡营养微生物，可以在低碳环境茁壮成长（1L 培养基中含有 1～15mg 有机碳）（Poindexter 1981），其中心代谢途径成为了人们研究的焦点。它们的代谢途径发生了重排，并且有一些基因由于精简选择而丢失。极端寡营养微生物 *Candidatus pelagibacter* 不具有完整的 EMP 途径，缺乏编码丙酮酸激酶和磷酸果糖激酶的基因（Smith et al. 2013）。然而，目前存在一个假定的操纵子参与编码的 ED 途径修饰基因，以及一个完整的葡萄糖生成途径（Smith et al. 2013）。与之类似，*Caulobacter crescentus* 缺乏磷酸果糖激酶的同源基因（Hottes et al. 2004）。基因组学分析表明该菌在葡萄糖上生长诱导 ED 途径基因的表达，而木糖则无法通过 PP 途径进行利用，这是由于缺乏了关键的木糖异构酶和木酮糖激酶，且在基因组序列中没有找到同源的基因（Hottes et al. 2004）。这种微生物分泌纤维素酶、内-*β*-1,4-葡聚糖酶、*β*-葡萄糖苷酶、木聚糖酶和多糖脱乙酰酶，因此具有降解植物生物高分子的潜力（Hottes et al. 2004）。

1.3.8 抗压菌

所有的极端微生物中，抗压菌碳源利用范围最广（Simonato et al. 2006）。同低温与高温条件一样，高压改变了细胞膜的流动性（Simonato et al. 2006）。抗压菌通过提高不饱和脂肪酸，尤其是脂双层中单不饱和脂肪酸的比例来克服这个问题（Valentine and Valentine 2004；Bartlett and Bidle 1999），这也有助于维持离子的通透性达到具备生物能量的目的（Vossenberg et al. 1995）并调整膜曲率以承受高压下更多的弹性应力（Attard et al. 2000）。抗压菌面临的第二个挑战是转运效率降低（Simonato et al. 2006），通常通过上调转运蛋白来弥补这一点。转运体 Tat2 目前研究较为深入，它是一种高亲和力的色氨酸通透酶，有助于色氨酸的吸收（Abe and Horikoshi 2000；Abe 2003）。在寒冷和炎热的环境中都能找到抗压菌，它们能够产生大量的热休克蛋白和冷休克蛋白。除此之外也产生 *β*-羟基丁酸这种胞内溶质及其寡聚物，以抵抗流体静压力（Martin et al. 2002）。许多抗压菌生活在深海环境中，那里有很多复杂的聚合物，抗压菌有多种酶来降解这些聚合物，具有工业应用潜力（Simonato et al. 2006）。

大部分抗压菌使用 EMP 途径作为糖类代谢的主要代谢途径（表 1.2），然而抗压菌 *Marinitoga piezophila* KA3 只有一条完整的 EMP 途径（Lucas et al. 2012），PP 途径由于缺乏葡萄糖-6-磷酸脱氢酶和 6-磷酸葡萄糖酸脱氢酶而有所缺陷（Lucas et al. 2012）。同时该菌也缺乏 ED 途径必要的酶，如 6-磷酸葡萄糖酸脱水酶和 2-脱氢-3-脱氧-磷酸葡萄糖酸醛缩酶（Lucas et al. 2012）。

表 1.2　极端条件下异养微生物的代谢途径和关键酶活性

环境	微生物类型	代谢途径	酶活性	参考文献
		中心代谢①	肽/氨基酸②	
pH	嗜酸性微生物 (最适 pH≤3.0)	细菌 EMP,ED 和 PP 途径(根据生长环境使用不同的代谢途径) TCA 循环(部分细菌不完整)	大多数参与途径中氨基酸的合成基因已被鉴定	[1]
		古菌 EMP 途径(经过修饰并且只存在于部分古菌)和 ED 途径(主要)	特征蛋白酶和肽酶以及蛋白质降解的完整途径	
	嗜碱性微生物 (最适 pH≥9)	细菌 EMP 途径(主要),ED 和 PP 途径 TCA 循环(部分细菌不完整)	除了丝氨酸降解途径外,发现氨基酸代谢的特征蛋白酶和基因	[2]
温度	嗜冷微生物 (最适温度≤15℃和最高温度<20℃)	细菌 EMP 途径(主要),ED 和 PP 途径 TCA 循环	表征了丝氨酸蛋白酶、肽酶和金属蛋白酶,参与氨基酸降解的基因已被鉴定	[3]
		古菌 EMP 途径(主要),ED 和 PP 途径 TCA 循环(部分不完整)	表征了金属蛋白酶,参与 20 种氨基酸降解途径的基因已被鉴定	
	嗜热微生物 (最适温度 45~80℃)	细菌 EMP 途径(主要),ED(仅存在于部分细菌中)和 PP 途径(缺乏氧化部分) TCA 循环	表征了蛋白酶、肽酶,检测到氨基酸降解途径的典型酶类的基因	[4]
		古菌 EMP 途径(修饰,主要),ED(只在部分古细菌中存在)和 PP 途径 TCA 循环	表征了蛋白酶,参与氨基酸分解代谢途径的基因已被鉴定	
	超嗜热微生物 (最适温度≥80℃或最高温度≥90℃)	细菌 EMP 途径(修饰,主要),ED(只在部分细菌中存在)和 PP 途径 TCA 循环	表征了蛋白酶	[5]
		古菌 EMP 途径(修饰,主要),ED(只在部分古细菌中存在)和 PP 途径 TCA 循环	表征了丝氨酸蛋白酶和吡咯烷酮羧肽酶,所有氨基酸降解的基因已被鉴定	
盐浓度	嗜盐微生物 (>0.3mol/L 氯化钠)	细菌 EMP 途径(糖酵解途径,部分利用),ED(主要)和 PP 途径 TCA 循环	表征了蛋白酶,鉴定了编码转氨酶的基因,表征了丝氨酸肽酶,发现了参与氨基酸降解途径,包括间苯二甲酸途径的基因,但有部分细菌不能合成 8 种以上氨基酸	[6]
		古菌 EMP 途径(修饰),ED(修饰)和 PP 途径 TCA 循环		

续表

环境	微生物类型	代谢途径 中心代谢①	酶活性 肽/氨基酸②	参考文献
水活性	嗜高渗微生物（高浓度有机溶质和糖）	真核生物 EMP途径（主要）和PP途径 TCA循环	表征了碱性蛋白酶	[7]
		细菌 EMP途径，ED途径（主要），PP途径（仅在一些） TCA循环（一些微生物不完整）	鉴定了参与包括氨基酸脱氢酶和氧化酶在内的氨基酸完全降解途径的基因	
	耐旱微生物（a_w＜0.85）	EMP途径（主要），ED和PP途径		[8]
压力	耐高压微生物	细菌 EMP途径 不完整的TCA循环	在基因组中检测到多达46种蛋白酶和肽酶，预测这些物种可以降解多种底物	[9]
低营养浓度	寡营养微生物（1L培养基中含有1～15mg有机碳）	细菌 ED途径（修饰） TCA循环	有丰富的氨基酸的分泌型肽酶将氨基酸用作碳源，并且氨基酸降解途径的基因高表达	[10]
重金属	耐金属蛋白酶（1mmol/L重金属）	细菌 在低碳环境中发现，细菌为自养型	—	[11]
辐射	抗辐射微生物（D≥1kGy）	细菌 EMP途径（不使用，但发现其中关键酶），ED和PP途径 TCA循环	不能利用氨作为氮源，完全依赖外源氨基酸作为氮源，赖氨酸、丝氨酸和半胱氨酸的生物合成途径不完整，但能够在这三种氨基酸中生长，氨基酸降解途径完整	[12]
		古菌 EMP途径（主要）和PP途径 TCA循环	异亮氨酸、脯氨酸、精氨酸、亮氨酸、苯丙氨酸和缬氨酸生物合成途径的基因缺失	

注：[1]（Dopson et al. 2004；Lobos et al. 1986；Kinoshita et al. 1995；Barrie Johnson and Hallberg 2008；Valdés et al. 2008；Karavaĭko et al. 2000；Zhang and Lovitt 2005；Barry et al. 2006；Angelov and Liebl 2006；Ruepp et al. 2000；Kurosawa et al. 1998；Fuchs et al. 1995；Voges et al. 1999）；[2]（Zhilina et al. 1997，2001，2004；Garnova and Krasil'nikova 2003；Detkova and Kevbrin 2009；Paavilainen et al. 1999；Khmelenina et al. 1997；Habib et al. 2012；Rao et al. 1998）；[3]（Herbert and Bell 1977；Zhu et al. 2003；Allen et al. 2009；Villeret et al. 1997；Aghajari et al. 2003；Irwin et al. 2001；Riley et al. 2008；Khan and Sylte 2009）；[4]（Feng et al. 2009；Patel et al. 2006；van Niel et al. 2003；De Vrije et al. 2007；Verhees et al. 2003；Kengen et al. 1996；Budgen and Danson 1986；Guangrong et al. 2006；Gatti et al. 2004；Klingeberg et al. 1995）；[5]（Schröder et al. 1994；Huber et al. 1986；Klingeberg et al. 1991；Morikawa et al. 1994；Eggen et al. 1990；Kaushik et al. 2002）；[6]（Oren and Mana 2003；Gonzalez et al. 2008；Rawal et al. 1988；Baliga et al. 2004；Ng et al. 2000；Anithajothi et al. 2014；Sanchez-Porro et al. 2003；Kamekura et al. 1992；Fine et al. 2006）；[7]（Moran and Witter 1979；Loos et al. 1994；Ogawa et al. 1990；de Montigny et al. 2000；Seo et al. 2005）；[8]（Hocking and Norton 1983；Leong et al. 2014）；[9]（Lucas et al. 2012；Tamegai et al. 2008；Aono et al. 2010）；[10]（Carini et al. 2012；Hottes et al. 2004；Arellano et al. 2010）；[11]（Carapito et al. 2006；Bertin et al. 2011；Bryan et al. 2009）；[12]（Yuan et al. 2012；Sghaier et al. 2008；Zivanovic et al. 2009；Zhang et al. 2000；White et al. 1999）。

① 中心代谢的主要代谢途径，主要是通过酶活性的测定和具有代表性的极端微生物基因组序列分析得到。

② 数据主要通过酶测定和基因组序列的基因鉴定获得，这些基因参与了相应的极端微生物中的氨基酸代谢。

1.3.9 抗辐射菌

在对大多数生物体有害的高强度辐射条件下，由于辐射导致的 DNA 损伤的遗传恢复很大程度取决于能量代谢和蛋白质（Venkateswaran et al. 2000）。一种抗辐射细菌 *Deinococcus* 尽管被发现存在 EMP 途径的关键酶，但该细菌仍通过 ED 和 PP 途径代谢葡萄糖，因为 EMP 途径可能影响对紫外线辐射的抵抗力（White et al. 1999；Zhang et al. 2000；Yuan et al. 2012）。EMP 途径可能会耗尽用于 DNA 修复的代谢产物，并诱导氧化应激和导致抗辐射性减弱（White et al. 1999；Zhang et al. 2000；Yuan et al. 2012）。据报道没有 EMP 途径的 *Deinococcus* 可以在含有果糖、葡萄糖和麦芽糖的培养基上生长（White et al. 1999；Zhang et al. 2000；Yuan et al. 2012）。然而古菌 *Thermococcus gammatolerans* 基因组分析显示，它有一个修饰过的 EMP 途径和非氧化型的 PP 途径（Zivanovic et al. 2009）。除基因组分析以外，还需要进一步的研究来测试它们主要通过哪种途径来利用糖类。

极端细菌的中心代谢途径通常非常保守，只有少量的修饰，但在古菌中存在半磷酸 ED 途径、非磷酸化 ED 途径和 EMP 途径，而且修饰过的 ED 途径在来自不同极端环境中的极端微生物之间彼此共享。除非在特定情况下，极端微生物中的 TCA 循环都是完整的，该循环在提供额外 ATP 分子和分子构建中起着重要的作用（White et al. 2007）。每个途径都对细胞生存起着十分重要的作用，如产生生物合成的前体物质、辅因子和能量。众所周知，ED 途径优于 EMP 途径，ED 途径较 EMP 途径更为快速，并且只存在于原核生物当中（Stettner and Segrè 2013）。尽管每分子葡萄糖通过 EMP 途径比 ED 途径多产生 1 分子 ATP，但是 ED 途径的存在具有另一个优势（Stettner and Segrè 2013）。ED 途径释放更多的自由能，并且反应高度放能，这意味着更利于正向反应（Flamholz et al. 2013；Bar-Even et al. 2012）。相对而言，EMP 途径更接近于平衡反应（Flamholz et al. 2013；Bar-Even et al. 2012）。为了弥补这一缺陷，EMP 途径需要耗费更多的蛋白质来增强酶活性，其蛋白质的合成是 ED 途径的 3.5 倍（Flamholz et al. 2013；Bar-Even et al. 2012）。在厌氧原核生物中，底物水平磷酸化是其节能的主要手段，大部分依赖于 EMP 途径（Stettner and Segrè 2013）。一些极端微生物，如嗜热嗜酸古细菌、嗜盐细菌、嗜高渗菌和寡营养细菌，仅利用 ED 途径，它们不具有 EMP 途径的关键酶，而且人们认为 EMP 途径的进化并非它们在极端环境中存活的关键。目前仍存在一个问题就是古菌中修饰过的 ED 途径，或者大部分微生物中的典型 ED 途径是否为代谢途径的起源，进一步研究可以让人们更好地了解极端微生物共同的中心代谢途径和演化谱系。

1.4 肽与氨基酸代谢

蛋白质、肽和氨基酸被极端微生物在内的许多生物所利用。一般来说，蛋白质由蛋白酶水解成肽，肽进一步水解成氨基酸（图 1.4）（White et al. 2012）。氨基酸被运输到细胞内后，它们可以作为蛋白质的合成原料，和（或）进行脱氨，主要通过氧化脱氨或氨基转移酶催化，并生成相应的 α-酮酸如丙酮酸、草酰乙酸和 α-酮戊二酸等（White et al. 2012）。

因为环境中有不同的蛋白质来源，决定使用哪种类型的蛋白质取决于氮的可利用性（Leigh and Dodsworth 2007）。不同氮源转化为氨需要不同的能量和电子供体，低浓度的氮源利用途径需要更多的能量，通常在氮充足的条件下会被抑制，但细胞发出缺氮的信号时则会被诱导。此外缺氮会通过谷氨酸和 α-酮戊二酸的相互转化以诱导氨的同化（图 1.5）。大部分细菌和古菌使用代谢中间物质 α-酮戊二酸来感应氮的缺乏，使用谷氨酰胺来给出氮充足的信号（Doucette et al. 2011；Magasanik and Kaiser 2002；Kessler et al. 2001；Ehlers et al. 2005；Leigh and Dodsworth 2007）。大多数极端微生物有一个完整的TCA循环，且α-酮戊二酸是碳氮代谢中一种重要的中间产物（Ninfa and Jiang 2005；Fendt et al. 2013）。但在一个含有不完整TCA循环的生物体中，如一些极端微生物，它们通常会具有合成α-酮戊二酸的部分途径，作为氮同化的前体物质（Leigh and Dodsworth 2007）。

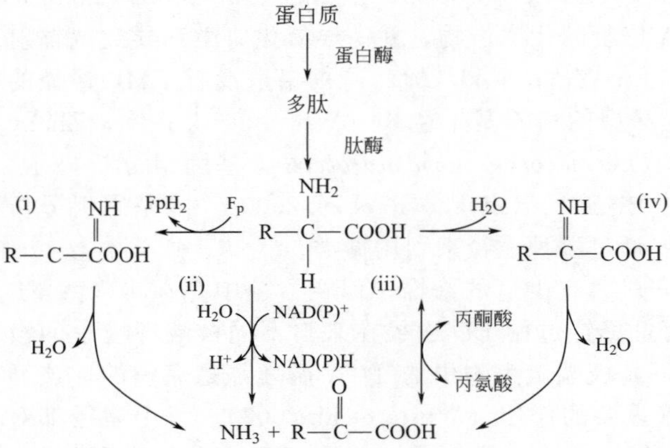

图1.4 极端微生物中蛋白质、多肽和氨基酸的分解代谢。利用酶将氨基酸转化为α-酮酸：（i）氨基酸氧化酶；（ii）氨基酸脱氢酶；（iii）转氨酶；（iv）不同途径中的丝氨酸/苏氨酸脱水酶。F_p：黄素蛋白

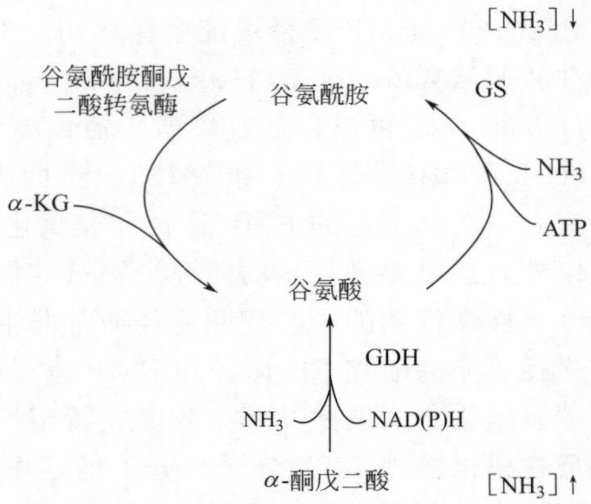

图1.5 用GDH和GOGAT系统合成谷氨酸。高氨浓度下，谷氨酸由谷氨酸脱氢酶合成，在低浓度氨下，谷氨酸通过谷氨酰胺酮戊二酸转氨酶（GOGAT）催化利用谷氨酰胺和α-酮戊二酸合成谷氨酸。α-KG：α-酮戊二酸；GS：谷氨酰胺合酶；GDH：谷氨酸脱氢酶

许多蛋白酶和肽酶在极端微生物中得到良好的表征（表1.2），通过对其代表物种的基因组学研究，可预测不同极端微生物中氨基酸的合成和分解代谢途径。经过脱氨的氨基酸碳骨架将进入中心代谢途径，但主要是作为TCA循环的中间物质，这些中间体也是氨基酸合成的前体物质（White et al. 2007）。一般来说，极端细菌和古菌中，氨基酸合成和分解代谢途径并没有显著的差异。下文对这些途径中具有酶功能的缺失基因进行介绍。

在嗜酸性细菌中，除了参与脯氨酸的生物合成的鸟氨酸环化脱氨酶、芳香族氨基酸转氨酶和参与酪氨酸生物合成的酪氨酸脱氢酶外，所有参与氨基酸合成途径的基因都已被发现（Valdés et al. 2008）。

嗜冷古菌 *Methanococcoidei burtonii* 的谷氨酸主要通过谷氨酰胺合酶和谷氨酰胺氧谷氨酸转氨酶系统进行合成（Allen et al. 2009）。此外没有发现丝氨酸途径的酶，包括羟基丙酮酸还原酶和丝氨酸乙醛酸转氨酶（Allen et al. 2009）。生长在高温环境中的嗜热微生物具有特殊的酶类和代谢途径。已在嗜热细菌 *Thermoanaerobacter* sp. strain X514 中证实，异亮氨酸通过柠檬酸途径合成，以丙酮酸和乙酰辅酶A为前体，以柠檬酸合酶为关键酶，通过柠檬酸途径调节（Feng et al. 2009）。

嗜盐微生物中一些古菌合成氨基酸的数量不超过8种，因此氨基酸是生长所必需的（Ng et al. 2000）。氨基酸同样在嗜高渗微生物的渗透压调节中发挥作用，这类氨基酸能够通过累积或合成获得，包括脯氨酸、甘氨酸和谷氨酸（Galinski 1995）。

寡营养微生物 *Caulobacter crescentus* 可分泌肽酶利用氨基酸作为碳源（Hottes et

al. 2004)，利用不同的途径将不同的氨基酸降解为丙酮酸，并且随后用于能量转化和生长(Hottes et al. 2004)。

抗辐射菌 *Deinococcus* 不能利用氨作为氮源，完全依赖外源性氨基酸作为氮源。此外其赖氨酸、丝氨酸和半胱氨酸的生物合成途径并不完整，但它可以在含这三种氨基酸的培养基中生长(White et al. 1999；Yuan et al. 2012；Zhang et al. 2000)。

没有这三个核心代谢途径的极端微生物可能没有这些氨基酸的生物合成前体，降低了在没有氨基酸的环境中极端微生物的生存能力，这也可能是基因组分析中氨基酸生物合成途径缺失的原因，因为它们不是必需的。

1.5 中心代谢与氨基酸代谢之间的联系

中心代谢途径之间的联系是通过共同的中间物质磷酸甘油醛实现的，该物质能被氧化成丙酮酸(White et al. 2007)。每个途径具有独特的代谢目的，EMP 途径通过一分子葡萄糖产生两分子 ATP，而 ED 途径只产生一分子 ATP，葡萄糖氧化为葡萄糖酸可从生存环境中除去葡萄糖，提供生存竞争优势，而 PP 途径产生戊糖和 NADPH (White et al. 2007)。这些没有一个或多个中心代谢途径的极端微生物往往有独特的酶来补偿缺失途径。例如，没有 PP 途径氧化功能的极端微生物会使用丙酮酸代替：由铁氧还蛋白氧化还原酶和铁氧还蛋白-$NADP^+$ 还原酶来产生额外的所需的 NADPH (Blamey and Adams 1993；Hoek and Rydstrom 1988)。

除了中心代谢途径的联系以外，一些微生物会利用糖类或肽作为碳源和氮源，所以氨基酸和 α-酮酸的可逆性转化变得必不可少(图 1.6)。中心代谢途径和 TCA 循环的几个中间体可作为氨基酸生物合成的前体，而氨基酸的碳骨架也作为 TCA 循环和中心代谢途径的中间体。TCA 循环的中间体草酰乙酸能够用于合成六种氨基酸(Meister 2012)，α-酮戊二酸可用于合成谷氨酰胺和谷氨酸(Peng et al. 1993)，这是用于合成其他氨基酸时最重要的两种氨基酸。琥珀酸 CoA 能够提供一个琥珀酰基团形成中间体物质供赖氨酸和甲硫氨酸的合成(D'Mello and D'Mello 2003)。中心代谢途径的三种糖酵解中间产物，丙酮酸、磷酸甘油醛和磷酸烯醇式丙酮酸是九种氨基酸的前体物质(White et al. 2007)。PP 途径生成的赤藓糖-4-磷酸用于合成芳香族氨基酸(White et al. 2007)。谷氨酸是大部分氨基酸的氨基供体，而且对于无机氮成为细胞物质十分重要。根据氨的浓度，谷氨酸通过两种不同的途径合成(Tempest et al. 1973)。在高浓度氨的环境中(＞1mmol/L)，谷氨酸脱氢酶用于催化 α-酮戊二酸的还原胺化(Tempest et al. 1973)。大多数自然环境下，氨的浓度较低，此时谷氨酸合酶和谷氨酰胺合酶两种酶会参与其中，并且需要 ATP (Tempest et al. 1973)。氨会首先由谷氨酰胺合酶催化，加入到谷氨酰胺中，然后谷氨酰胺的氨会被谷氨酸合酶转移到 α-酮戊二酸上，产生谷氨酸。在氨基酸分解代谢中，所有 20 种氨基酸能被降解为 7 种中间体，包括丙酮酸、乙酰 CoA、乙酰乙酰 CoA、α-酮戊二酸、琥珀酸 CoA、延胡索酸和草酰乙酸，并且进入 TCA 循环(图 1.6)(Massey et al. 1976；Pederson et al. 1999；Barker 1981；Cunin et al. 1986)。氨基酸通过氧化或者转氨酶反应被脱氨，形成相应的 α-酮酸(Massey et al. 1976；Pederson et al. 1999；Barker 1981；Cunin et al. 1986)。第一种方式可以通过氧化酶进行，如 D-氨基酸氧化酶和 L-氨基酸氧化酶(D'aniello et al. 1993)。第二种为特异性更高的酶，NAD (P)$^+$ 依赖的氨基酸脱氢酶(Miflin and Lea 1977)。第三种是通过特异性的脱氨酶使丝氨酸和苏氨酸形成丙酮酸，天冬氨酸形成草酰乙酸和组氨酸(Alföldi et al. 1968)。

对那些只能利用糖类生长的极端微生物而言，包括 TCA 循环在内的中心代谢途径的中间物质会被用于氨基酸的合成（图 1.6)。然而这些微生物中有部分可能缺乏用于合成所有氨基酸的酶或途径，因此它们生长介质中需要额外的肽和/或氨基酸来满足它们的生长需求。还有一些极端微生物无法利用任何糖类生长，而只能利用肽或氨基酸，所以它们必须具有与 TCA 循环和中心代谢途径相关的糖异生途径（图 1.6)。

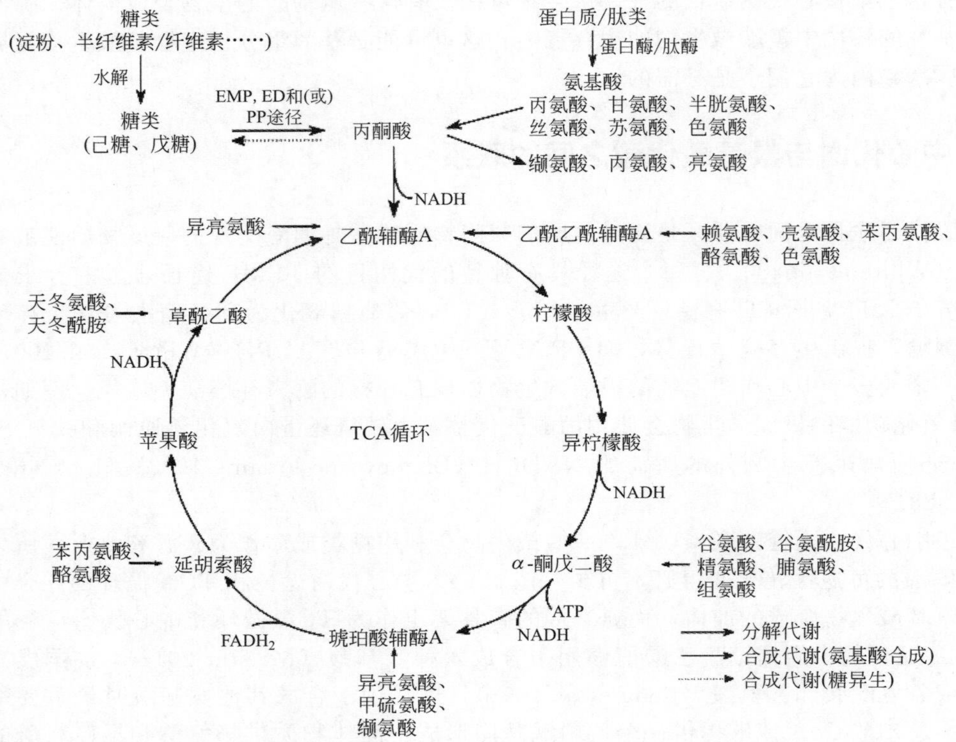

图 1.6 极端微生物中不同底物利用的代谢途径之间的联系。氨基酸进入 TCA 循环的点被标记出来

1.6 结论与前景

糖类和/或肽是所有微生物生长的基础物质，极端微生物也不例外。虽然极端微生物适应极端环境，但是它们仍然利用 EMP、ED、PP 途径中的一个或多个以及 TCA 循环，或经过修饰的这些途径，产生 ATP 和可循环辅因子 NADP（H)/NAD（H)。当使用相同底物时，它们更有竞争力。无论是利用糖类的途径还是利用肽的途径，都具有共同的中间物质，丙酮酸（α-酮酸)，进入随后的反应，进而产生 ATP、结构骨架分子和辅因子。底物特异性的差异主要取决于特定的转运蛋白，因为它们利用糖类和肽/氨基酸的代谢能力非常相似。进一步了解极端微生物的极端耐受和极端抵抗的适应机制是十分有趣的，也是充满挑战的。微生物在各种极端条件下生长的分子基础将是探索极端条件生物催化原理的重要依据。代谢途径中酶的独特性质吸引了许多应用领域的关注，表征这些来自极端微生物的酶是未来开发的第一步。

许多来自不同环境的极端微生物酶类都已经成功应用于工业当中，如来自超嗜热微生物的 DNA 聚合酶目前已广泛应用，嗜热微生物的木聚糖酶也已用于纸浆漂白，嗜酸微生物的纤维素酶和嗜冷微生物的淀粉酶用作洗涤剂（Van Den Burg 2003；Karan et al. 2012），以及一些蛋白酶用于烘焙、酿造和奶制品生产（Bressan et al. 2008）。近年来通过基因组学、蛋白质组学和代谢组学的研究，新酶的发现速度加快，从而促进了新型工业流程的开发。各种基因被用于应对极端环境压力，例如与盐压力相关的脯氨酸合酶和磷酸甘油醛脱氢酶基因，被转入农作物当中。作为工业应用，新发现的酶通常需要进行改造，蛋白质工程的进展将使得极端酶能够被更大规模地使用。

无利益冲突声明

Ching Tse 和 Kesen Ma 声明他们没有利益冲突。

参考文献

Abe F（2003）The Role of Tryptophan Permease Tat2 in Cell Growth of Yeast under High-Pressure Condition. In：Winter R（ed）Advances in High Pressure Bioscience and Biotechnology II，Springer，Berlin，pp 271-274

Abe F（2007）Exploration of the effects of high hydrostatic pressure on microbial growth，physiology and survival：perspectives from piezophysiology. Biosci Biotechnol Biochem 71（10）：2347-2357

Abe F，Horikoshi K（2000）Tryptophan permease gene TAT2 confers high-pressure growth in *Saccharomyces cerevisiae*. Mol Cell Biol 20（21）：8093-8102

Aghajari N，Van Petegem F，Villeret V，Chessa JP，Gerday C，Haser R，Van Beeumen J（2003）Crystal structures of a psychrophilic metalloprotease reveal new insights into catalysis by cold-adapted proteases. Proteins Struct Funct Bioinf 50（4）：636-647

Alain K，Marteinsson VT，Miroshnichenko ML，Bonch-Osmolovskaya EA，Prieur D，Birrien JL（2002）*Marinitoga piezophila* sp. nov.，a rod-shaped，thermo-piezophilic bacterium isolated under high hydrostatic pressure from a deep-sea hydrothermal vent. Int J Syst Evol Microbiol 52（4）：1331-1339

Albers SV，Driessen AJ（2007）Membranes and Transport Proteins of Thermophilic Micro-organisms. In：Robb F，Antranikian G（eds）Thermophiles：Biology and Technology at High Temperatures，CRC Press，New York，pp 39-54

Albers SV，Elferink MG，Charlebois RL，Sensen CW，Driessen AJ，Konings WN（1999）Glucose transport in the extremely thermoacidophilic *Sulfolobus solfataricus* involves a high-affinity membrane-integrated binding protein. J Bacteriol 181（14）：4285-4291

Albers SV，Koning SM，Konings WN，Driessen AJ（2004）Insights into ABC transport in archaea. J Bioenerg Biomembr 36（1）：5-15

Albertyn J，Hohmann S，Thevelein JM，Prior BA（1994）GPD1，which encodes glycerol-3-phosphate dehydrogenase，is essential for growth under osmotic stress in *Saccharomyces cerevisiae*，and its expression is regulated by the high-osmolarity glycerol response pathway. Mol Cell Biol 14（6）：4135-4144

Alföldi L，Raskó I，Kerekes E（1968）L-Serine deaminase of *Escherichia coli*. J Bacteriol 96（5）：1512-1518

Allen MA，Lauro FM，Williams TJ，Burg D，Siddiqui KS，De Francisci D，Chong KW，Pilak O，Chew HH，De Maere MZ（2009）The genome sequence of the psychrophilic archaeon，*Methanococcoides burtonii*：the role of genome evolution in cold adaptation. ISME J 3（9）：1012-1035

Álvarez-Ordóñez A，Fernández A，Bernardo A，López M（2010）Arginine and lysine decarboxylases and the acid tolerance response of *Salmonella Typhimurium*. Int J Food Microbiol 136（3）：278-282

Amato P（2013）Energy Metabolism in Low-temperature and Frozen Conditions in Cold-adapted Microorganism. In：Yumoto I（ed）Cold-adapted microorganisms，Caister Academic Press，UK，pp 71-96

AmendJP，Shock EL（2001）Energetics of overall metabolic reactions of thermophilic and hyper-thermophilic Archaea and Bacteria. FEMS Microbiol Rev 25（2）：175-243

Angelov A，Liebl W（2006）Insights into extreme thermoacidophily based on genome analysis of *Picrophilus torridus* and other thermoacidophilic archaea. J Biotechnol 126（1）：3-10

Anithajothi R，Nagarani N，Umagowsalya G，Duraikannu K，Ramakritinan C（2014）Screening，isolation and characterization of protease producing moderately halophilic microorganism

Halomonas meridiana associated with coral mucus. Toxicol Environ Chem 96（2）：296-306

Antón J，Oren A，Benlloch S，Rodríguez-Valera F，Amann R，Rosselló-Mora R（2002）*Salinibacter ruber* gen. nov.，sp. nov.，a novel，extremely halophilic member of the Bacteria from saltern crystallizer ponds. Int J Syst Evol Microbiol 52（2）：485-491

Aono E，Baba T，Ara T，Nishi T，Nakamichi T，Inamoto E，Toyonaga H，Hasegawa M，Takai Y，Okumura Y（2010）Complete genome sequence and comparative analysis of *Shewanella violacea*，a psychrophilic and piezophilic bacterium from deep sea floor sediments. Mol BioSyst 6（7）：1216-1226

Arellano BH，Ortiz JD，Manzano J，Chen JC（2010）Identification of a dehydrogenase required for lactose metabolism in *Caulobacter crescentus*. Appl Environ Microbiol 76（9）：3004-3014

Attard GS，Templer RH，Smith WS，Hunt AN，Jackowski S（2000）Modulation of CTP：phosphocholine cytidylyltransferase by membrane curvature elastic stress. Proc Natl Acad Sci 97 (16): 9032-9036

Baker-Austin C, Dopson M (2007) Life in acid: pH homeostasis in acidophiles. Trends Microbiol 15 (4): 165-171

Bakermans C, Tsapin AI, Souza-Egipsy V, Gilichinsky DA, Nealson KH (2003) Reproduction and metabolism at—10 C of bacteria isolated from Siberian permafrost. Environ Microbiol 5 (4): 321-326

Baliga NS, Bonneau R, Facciotti MT, Pan M, Glusman G, Deutsch EW, Shannon P, Chiu Y, Weng RS, Gan RR (2004) Genome sequence of *Haloarcula marismortui*: a halophilic archaeon from the Dead Sea. Genome Res 14 (11): 2221-2234

Bar-EvenA, Flamholz A, Noor E, Milo R (2012) Rethinking glycolysis: on the biochemical logic of metabolic pathways. Nat Chem Biol 8 (6): 509-517

Barker H (1981) Amino acid degradation by anaerobic bacteria. Annu Rev Biochem 50 (1): 23-40

Baross JA, Lenovich L (1992) Halophilic and osmophilic microorganisms. Compendium of methods for the microbiological examination of foods. American Public Health Association, Washington, DC, pp 199-212

Barrie Johnson D, Hallberg KB (2008) Carbon, iron and sulfur metabolism in acidophilic microorganisms. Adv Microb Physiol 54: 201-255

Barry RC, Young MJ, Stedman KM, Dratz EA (2006) Proteomic mapping of the hyperthermophilic and acidophilic archaeon *Sulfolobus solfataricus* P2. Electrophoresis 27 (14): 2970-2983

Bartlett D, Bidle K (1999) Membrane-based adaptations of deep-sea piezophiles. In: Seckbach J (ed) Enigmatic Microorganisms and Life in Extreme Environments, Springer, Netherlands, pp 501-512

Batrakov SG, Pivovarova TA, Esipov SE, Sheichenko VI, Karavaiko GI (2002) β-D-Glucopyranosyl caldarchaetidylglycerol is the main lipid of the acidophilic, mesophilic, ferrous iron-oxidising archaeon *Ferroplasma acidiphilum*. BBA-Mol Cell Biol Lipids 1581 (1): 29-35

Ben-Bassat A, Lamed R, Zeikus J (1981) Ethanol production by thermophilic bacteria: meta-bolic control of end product formation in *Thermoanaerobium brockii*. J Bacteriol 146 (1): 192-199

Berezovsky IN, Shakhnovich EI (2005) Physics and evolution of thermophilic adaptation. Proc Natl Acad Sci 102 (36): 12742-12747

Berger F, Morellet N, Menu F, Potier P (1996) Cold shock and cold acclimation proteins in the psychotrophic bacterium *Arthrobacter globiformis* SI55. J Bacteriol 178 (11): 2999-3007

Bertin PN, Heinrich-Salmeron A, Pelletier E, Goulhen-Chollet F, Arsène-Ploetze F, Gallien S, Lauga B, Casiot C, Calteau A, Vallenet D (2011) Metabolic diversity among main microorganisms inside an arsenic-rich ecosystem revealed by meta-and proteo-genomics. ISME J 5 (11): 1735-1747

Bertoldo C, Antranikian G (2006) The order *Thermococcales*. In: Dworkin M, Falkow S, Rosenberg E, Schleifer K, Stackebrandt E (eds) The Prokaryotes, Springer, New York, pp 69-81

Bettenbrock K, Sauter T, Jahreis K, Kremling A, Lengeler JW, Gilles E-D (2007) Correlation between growth rates, EIIACrr phosphorylation, and intracellular cyclic AMP levels in *Escherichia coli* K-12. J Bacteriol 189 (19): 6891-6900

Blamey JM, Adams MW (1993) Purification and characterization of pyruvate ferredoxin oxido-reductase from the hyperthermophilic archaeon *Pyrococcus furiosus*. Biochim Biophys Acta 1161 (1): 19-27

BlankenhornD, Phillips J, Slonczewski JL (1999) Acid-and base-induced proteins during aerobic and anaerobic growth of *Escherichia coli* revealed by two-dimensional gel electrophoresis. J Bacteriol 181 (7): 2209-2216

Blöchl E, Rachel R, Burggraf S, Hafenbradl D, Jannasch HW, Stetter KO (1997) *Pyrolobus fumarii*, gen. and sp. nov., represents a novel group of archaea, extending the upper temperature limit for life to 113℃. Extremophiles 1 (1): 14-21

Blotevogel KH, Fischer U, Mocha M, Jannsen S (1985) *Methanobacterium thermoalcaliphilum* spec. nov., a new moderately alkaliphilic and thermophilic autotrophic methanogen. Arch Microbiol 142 (3): 211-217

Blumer-Schuette SE, Kataeva I, Westpheling J, Adams MW, Kelly RM (2008) Extremely thermophilic microorganisms for biomass conversion: status and prospects. Curr Opin Biotechnol 19 (3): 210-217

Bouchotroch S, Quesada E, del Moral A, Llamas I, Béjar V (2001) *Halomonas maura* sp. nov., a novel moderately halophilic, exopolysaccharide-producing bacterium. Int J Syst Evol Microbiol 51 (5): 1625-1632

Bressan RA, Bohnert HJ, Hasegawa PM (2008) Genetic engineering for salinity stress tolerance. Adv Plant Biochem Mol Biol 1: 347-384

Brim H, Heuer H, Krögerrecklenfort E, Mergeay M, Smalla K (1999) Characterization of the bacterial community of a zinc-polluted soil. Can J Microbiol 45 (4): 326-338

Brown A (1976) Microbial water stress. Bacteriol Rev 40 (4): 803

Brunner NA, Brinkmann H, Siebers B, Hensel R (1998) NAD+-dependent glyceraldehyde-3-phosphate dehydrogenase from *Thermoproteus tenax* the first identified archaeal membrane of the aldehyde dehydrogenase superfamily is a glycolytic enzyme with unusual regulatory properties. J Biol Chem 273 (11): 6149-6156

Brunner NA, Siebers B, Hensel R (2001) Role of two different glyceraldehyde-3-phosphate dehydrogenases in controlling the reversible Embden-Meyerhof-Parnas pathway in *Thermoproteus tenax*: regulation on protein and transcript level. Extremophiles 5 (2): 101-109

Bryan CG, Marchal M, Battaglia-Brunet F, Kugler V, Lemaitre-Guillier C, Lièvremont D, Bertin PN, Arsène-Ploetze F (2009) Carbon and arsenic metabolism in *Thiomonas* strains: differences revealed diverse adaptation processes. BMC Microbiol 9 (1): 127

Budgen N, Danson MJ (1986) Metabolism of glucose via a modified Entner-Doudoroff pathway in the thermoacidophilic archaebacterium *Thermoplasma acidophilum*. FEBS Lett 196 (2): 207-210

Burns DG, Janssen PH, Itoh T, Kamekura M, Li Z, Jensen G, Rodríguez-Valera F, Bolhuis H, Dyall-Smith ML (2007) *Haloquadratum walsbyi* gen. nov., sp. nov., the square haloarchaeon of Walsby, isolated from saltern crystallizers in Australia and Spain. Int J Syst Evol Microbiol 57 (2): 387-392

Carapito C, Muller D, Turlin E, Koechler S, Danchin A, Van Dorsselaer A, Leize-Wagner E, Bertin PN, Lett M-C (2006) Identification of genes and proteins involved in the pleiotropic response to arsenic stress in *Caenibacter arsenoxydans*, a metalloresistant beta-proteobacterium with an unsequenced genome. Biochimie 88 (6): 595-606

Carini P, Steindler L, Beszteri S, Giovannoni SJ (2012) Nutrient requirements for growth of the extreme oligotroph *Candidatus Pelagibacter ubique* HTCC1062 on a defined medium. ISME J 7 (3): 592-602

Castanie-Cornet MP, Penfound TA, Smith D, Elliott JF, Foster JW (1999) Control of acid resistance in *Escherichia coli*. J Bacteriol 181 (11): 3525-3535

Cavicchioli R, Thomas T, Curmi PM (2000) Cold stress response in Archaea. Extremophiles 4 (6): 321-331

Chintalapati S, Kiran M, Shivaji S (2004) Role of membrane lipid fatty acids in cold adaptation. Cell Mol Biol 50 (5): 631-642

Christensen HR, Frokiaer H, Pestka JJ (2002) Lactobacilli differentially modulate expression of cytokines and maturation surface markers in murine dendritic cells. J Immunol 168 (1): 171-178

Chrzanowski T, Kyle M, Elser J, Sterner R (1996) Element ratios and growth dynamics of bacteria in an oligotrophic Canadian shield lake. Aquat Microb Ecol 11 (2): 119-125

Ciaramella M, Napoli A, Rossi M (2005) Another extreme genome: how to live at pH 0. Trends Microbiol 13 (2): 49-51

Cunin R, Glansdorff N, Pierard A, Stalon V (1986) Biosynthesis and metabolism of arginine in bacteria. Microbiol Rev 50 (3): 314

D'Amico S, Collins T, Marx JC, Feller G, Gerday C (2006) Psychrophilic microorganisms: challenges for life. EMBO Rep 7 (4): 385-389

D'aniello A, D'onofrio G, Pischetola M, D'aniello G, Vetere A, Petrucelli L, Fisher G (1993) Biological role of D-amino acid oxidase and D-aspartate oxidase. Effects of D-amino acids. J Biol Chem 268 (36): 26941-26949

D'Mello J, D'Mello J (2003) An outline of pathways in amino acid metabolism. Amino acids in animal nutrition. CABI Publishing, Wallingford, UK, pp 71-86

Dalmasso M, Aubert J, Briard-Bion V, Chuat V, Deutsch SM, Even S, Falentin H, Jan G, Jardin J, Maillard M-B (2012) A Temporal-omic study of *Propionibacterium freudenreichii* CIRMBIA1T adaptation strategies in conditions mimicking cheese ripening in the cold. PLoS One 7 (1), e29083

DasSarma S, DasSarma P (2012) Halophiles. Wiley Online Library

De Graaf AA, Striegel K, Wittig RM, Laufer B, Schmitz G, Wiechert W, Sprenger GA, Sahm H (1999) Metabolic state of *Zymomonas mobilis* in glucose, fructose, and xylose-fed continuous cultures as analysed by 13C-and 31P-NMR spectroscopy. Arch Microbiol 171 (6): 371-385

De Groot A, Chapon V, Servant P, Christen R, Fischer-Le Saux M, Sommer S, Heulin T (2005) *Deinococcus deserti* sp. nov., a gamma-radiation-tolerant bacterium isolated from the Sahara Desert. Int J Syst Evol Microbiol 55 (6): 2441-2446

de Montigny J, Straub ML, Potier S, Tekaia F, Dujon B, Wincker P, Artiguenave F, Souciet JL (2000) Genomic exploration of the hemiascomycetous yeasts: *Zygosaccharomyces rouxii*. FEBS Lett 487 (1): 52-55

De Smet KA, Weston A, Brown IN, Young DB, Robertson BD (2000) Three pathways for trehalose biosynthesis in mycobacteria. Microbiology 146 (1): 199-208

de Vos WM, Kengen SW, Voorhorst WG, van der Oost J (1998) Sugar utilization and its control in hyperthermophiles. Extremophiles 2 (3): 201-205

De Vrije T, Mars A, Budde M, Lai M, Dijkema C, De Waard P, Claassen P (2007) Glycolytic pathway and hydrogen yield studies of the extreme thermophile *Caldicellulosiruptor saccharolyticus*. App Microbiol Biotechnol 74 (6): 1358-1367

Degryse E, Glansdorff N, Piérard A (1978) A comparative analysis of extreme thermophilic bacteria belonging to the genus *Thermus*. Arch Microbiol 117 (2): 189-196

Detkova E, Kevbrin V (2009) Cellobiose catabolism in the haloalkaliphilic hydrolytic bacterium *Alkaliflexus imshenetskii*. Microbiology 78 (3): 267-272

DiMarco AA, Romano AH (1985) D-Glucose transport system of *Zymomonas mobilis*. Appl Environ Microbiol 49 (1): 151-157

Dominy BN, Minoux H, Brooks CL (2004) An electrostatic basis for the stability of thermophilic proteins. Protein Struct Funct Bioinf 57 (1): 128-141

Dopson M, Baker-Austin C, Hind A, Bowman JP, Bond PL (2004) Characterization of *Ferroplasma* isolates and *Ferroplasma acidarmanus* sp. nov., extreme acidophiles from acid mine drainage and industrial bioleaching environments. Appl Environ Microbiol 70 (4): 2079-2088

Dörr C, Zaparty M, Tjaden B, Brinkmann H, Siebers B (2003) The Hexokinase of the Hyperthermophile *Thermoproteus tenax* ATP-Dependent Hexokinases and ADP-Dependent glucokinases, two alternatives for glucose phosphorylation in archaea. J Biol Chem 278 (21): 18744-18753

Doucette CD, Schwab DJ, Wingreen NS, Rabinowitz JD (2011) α-Ketoglutarate coordinates carbon and nitrogen utilization via enzyme I inhibition. Nature Chem Biol 7 (12): 894-901

Eggen R, Geerling A, Watts J, de Vos WM (1990) Characterization of pyrolysin, a hyperthermoactive serine protease from the archaebacterium *Pyrococcus furiosus*. FEMS Microbiol Lett 71 (1-2): 17-20

Egorova K, Antranikian G (2005) Industrial relevance of thermophilic Archaea. Current Opin Microbiol 8 (6): 649-655

Ehlers C, Weidenbach K, Veit K, Forchhammer K, Schmitz RA (2005) Unique mechanistic features of post-translational regulation of glutamine synthetase activity in *Methanosarcina mazei* strain Göl in response to nitrogen availability. Mol Microbiol 55 (6): 1841-1854

Elferink MG, Albers SV, Konings WN, Driessen AJ (2001) Sugar transport in *Sulfolobus solfataricus* is mediated by two families of binding protein-dependent ABC transporters. Mol Microbiol 39 (6): 1494-1503

England JL, Shakhnovich EI (2003) Structural determinant of protein designability. Phys Rev Lett 90 (21): 218101

England JL, Shakhnovich BE, Shakhnovich EI (2003) Natural selection of more designable folds: a mechanism for thermophilic adaptation. Proc Natl Acad Sci 100 (15): 8727-8731

Erauso G, Reysenbach A-L, Godfroy A, Meunier J-R, Crump B, Partensky F, Baross JA, Marteinsson V, Barbier G, Pace NR (1993) *Pyrococcus abyssi* sp. nov., a new hyperthermophilic archaeon isolated from a deep-sea hydrothermal vent. Arch Microbiol 160 (5): 338-349

Erra-Pujada M, Chang-Pi-Hin F, Debeire P, Duchiron F, O'Donohue MJ (2001) Purification and properties of the catalytic domain of the thermostable pullulanase type II from *Thermococcus hydrothermalis*. Biotech Lett 23 (16): 1273-1277

Evdokimov AG, Anderson DE, Routzahn KM, Waugh DS (2001) Structural basis for oligosaccharide recognition by *Pyrococcus furiosus* maltodextrin-binding protein. J Mol Biol 305 (4): 891-904

Falb M, Müller K, Königsmaier L, Oberwinkler T, Horn P, von Gronau S, Gonzalez O, Pfeiffer F, Bornberg-Bauer E, Oesterhelt D (2008) Metabolism of halophilic archaea. Extremophiles 12 (2): 177-196

Fang J, Zhang L, Bazylinski DA (2010) Deep-sea piezosphere and piezophiles: geomicrobiology and biogeochemistry. Trends Microbiol 18 (9): 413-422

Feder ME, Hofmann GE (1999) Heat-shock proteins, molecular chaperones, and the stress response: evolutionary and ecological physiology. Ann Rev Physiol 61 (1): 243-282

Fegatella F, Lim J, Kjelleberg S, Cavicchioli R (1998) Implications of rRNA operon copy number and ribosome content in the marine oligotrophic Ultramicrobacterium Sphingomonassp. Strain RB2256. Appl Environ Microbiol 64 (11): 4433-4438

Fendt SM, Bell EL, Keibler MA, Olenchock BA, Mayers JR, Wasylenko TM, Vokes NI, Guarente L, Vander Heiden MG, Stephanopoulos G (2013) Reductive glutamine metabolism is a function of the α-ketoglutarate to citrate ratio in cells. Nat Commun 4: 2236

Feng X, Mouttaki H, Lin L, Huang R, Wu B, Hemme CL, He Z, Zhang B, Hicks LM, Xu J (2009)
Characterization of the central metabolic pathways in *Thermoanaerobacter* sp. strain X514 via isotopomer-assisted metabolite analysis. Appl Environ Microbiol 75 (15): 5001-5008

Ferreira AC, Nobre MF, Rainey FA, Silva MT, Wait R, Burghardt J, Chung AP, Da Costa MS (1997) *Deinococcus geothermalis* sp. nov. and *Deinococcus murrayi* sp. nov., two extremely radiation-resistant and slightly thermophilic species from hot springs. Int J Syst Bacteriol 47 (4): 939-947

Fine A, Irihimovitch V, Dahan I, Konrad Z, Eichler J (2006) Cloning, expression, and purification of functional Sec11a and Sec11b, type I signal peptidases of the archaeon *Haloferax volcanii*. J Bacteriol 188 (5): 1911-1919

Flamholz A, Noor E, Bar-Even A, Liebermeister W, Milo R (2013) Glycolytic strategy as a tradeoff between energy yield and protein cost. Proc Natl Acad Sci 110 (24): 10039-10044

Franzmann P, Stackebrandt E, Sanderson K, Volkman J, Cameron D, Stevenson P, McMeekin T, Burton H (1988) *Halobacterium lacusprofundi* sp. nov., a halophilic bacterium isolated from Deep Lake, Antarctica. Syst Appl Microbiol 11 (1): 20-27

Fuchs T, Huber H, Teiner K, Burggraf S, Stetter KO (1995) *Metallosphaera prunae*, sp. nov., a novel metal-mobilizing, *Thermoacidophilic Archaeum*, isolated from a uranium mine in Germany. Syst Appl Microbiol 18 (4): 560-566

Fütterer O, Angelov A, Liesegang H, Gottschalk G, Schleper C, Schepers B, Dock C, Antranikian G, Liebl W (2004) Genome sequence of *Picrophilus torridus* and its implications for life around pH 0. Proc Natl Acad Sci 101 (24): 9091-9096

Galinski E (1995) Osmoadaptation in bacteria. Adv MicrobPhysiol 37: 273

Garnier M, Matamoros S, Chevret D, Pilet MF, Leroi F, Tresse O (2010) Adaptation to cold and proteomic responses of the psychrotrophic biopreservative *Lactococcus piscium strain* CNCM I-4031. Appl Environ Microbiol 76 (24): 8011-8018

Garnova E, Krasil'nikova E (2003) Carbohydrate metabolism of the saccharolytic alkaliphilic anaerobes *Halonatronum saccharophilum*, *Amphibacillus fermentum*, and *Amphibacillus tropicus*. Microbiology 72 (5): 558-563

Gatti M, Fornasari M, Lazzi C, Mucchetti G, Neviani E (2004) Peptidase activity in various species of dairy thermophilic lactobacilli. J Appl Microbiol 96 (2): 223-229

Gock MA, Hocking AD, Pitt JI, Poulos PG (2003) Influence of temperature, water activity and pH on growth of some xerophilic fungi. Int J Food Microbiol 81 (1): 11-19

Gonzalez C, Gutierrez C, Ramirez C (1978) *Halobacterium vallismortis* sp. nov. An amylolytic and carbohydrate-metabolizing, extremely halophilic bacterium. Can J Microbiol 24 (6): 710-715

González JM, Kato C, Horikoshi K (1995) *Thermococcus peptonophilus* sp. nov., a fast-growing, extremely thermophilic archaebacterium isolated from deep-sea hydrothermal vents. Arch Microbiol 164 (3): 159-164

González JM, Masuchi Y, Robb FT, Ammerman JW, Maeder DL, Yanagibayashi M, Tamaoka J, Kato C (1998) *Pyrococcus horikoshii* sp. nov., a hyperthermophilic archaeon isolated from a hydrothermal vent at the Okinawa Trough. Extremophiles 2 (2): 123-130

Gonzalez O, Gronau S, Falb M, Pfeiffer F, Mendoza E, Zimmer R, Oesterhelt D (2008) Reconstruction, modeling & analysis of *Halobacterium salinarum* R-1 metabolism. Mol BioSyst 4 (2): 148-159

Goodchild A, SaundersNF, Ertan H, Raftery M, Guilhaus M, Curmi PM, Cavicchioli R (2004) A proteomic determination of cold adaptation in the Antarctic archaeon, *Methanococcoides burtonii*. Mol Microbiol 53 (1): 309-321

Guangrong H, Tiejing Y, Po H, Jiaxing J (2006) Purification and characterization of a protease from thermophilic *Bacillus strain* HS08. Afr J Biotechnol 5 (24): 2433-2438

Habib S, Fakhruddin A, Begum S, Ahmed M (2012) Isolation and screening of thermo stable extracellular alkaline protease producing bacteria from *Tannery Effluents*. J Sci Res 4 (2): 515

Hafenbradl D, Keller M, Dirmeier R, Rachel R, Roßnagel P, Burggraf S, Huber H, Stetter KO (1996) *Ferroglobus placidus* gen. nov., sp. nov., a novel hyperthermophilic archaeum that oxidizes Fe^{2+} at neutral pH under anoxic conditions. Arch Microbiol 166 (5): 308-314

Hallberg KB, Lindström EB (1994) Characterization of *Thiobacillus caldus* sp. nov., a moderately thermophilic acidophile. Microbiology 140 (12): 3451-3456

Hallberg KB, González-Toril E, Johnson DB (2010) *Acidithiobacillus ferrivorans*, sp. nov.; facultatively anaerobic, psychrotolerant iron-, and sulfur-oxidizing acidophiles isolated from metal mine-impacted environments. Extremophiles 14 (1): 9-19

Hanson RS, Hanson TE (1996) Methanotrophic bacteria. Microbiol Rev 60 (2): 439-471

Hao X, Ma K (2003) Minimal sulfur requirement for growth and sulfur-dependent metabolism of the hyperthermophilic archaeon *Staphylothermus marinus*. Archaea 1 (3): 191-197

Harrison AP Jr (1984) The acidophilic *thiobacilli* and other acidophilic bacteria that share their habitat. Annu Rev Microbiol 38 (1): 265-292

Hendrick JP, Hartl F (1993) Molecular chaperone functions of heat-shock proteins. Annu Rev Biochem 62 (1): 349-384

Hensel R, Laumann S, Lang J, Heumann H, Lottspeich F (1987) Characterization of two Dglyceraldehyde-3-phosphate dehydrogenases from the extremely thermophilic archaebacterium *Thermoproteus tenax*. European J Biochem 170 (1-2): 325-333

Herbert R, Bell C (1977) Growth characteristics of an obligately psychrophilic *Vibrio* sp. Arch Microbiol 113 (3): 215-220

Hettmann T, Schmidt CL, Anemüller S, Zähringer U, Moll H, Petersen A, Schäfer G (1998) Cytochrome b 558/566 from the Archaeon *Sulfolobus acidocaldarius* a novel highly glycosyl-ated, membrane-bound b-type hemoprotein. J Biol Chem 273 (20): 12032-12040

Higgins CF (1995) The ABC of channelregulation. Cell 82 (5): 693-696

Hoaki T, Nishijima M, Kato M, Adachi K, Mizobuchi S, Hanzawa N, Maruyama T (1994) Growth requirements of hyperthermophilic sulfur-dependent heterotrophic archaea isolated from a shallow submarine geothermal system with reference to their essential amino acids. Appl Environ Microbiol 60 (8): 2898-2904

Hocking AD (1986) Effects of water activity and culture age on the glycerol accumulation patterns of five fungi. J Gen Microbiol 132 (2): 269-275

Hocking AD, Norton RS (1983) Natural-abundance 13C nuclear magnetic resonance studies on the internal solutes of xerophilic fungi. J Gen Microbiol 129 (9): 2915-2925

Hoek JB, Rydstrom J (1988) Physiological roles of nicotinamide nucleotide transhydrogenase. Biochem J 254: 1-10

Holland IB, Blight MA (1999) ABC-ATPases, adaptable energy generators fuelling transmembrane movement of a variety of molecules in organisms from bacteria to humans. J Mol Biol 293 (2): 381-399

Hongo K, Hirai H, Uemura C, Ono S, Tsunemi J, Higurashi T, Mizobata T, Kawata Y (2006) A novel ATP/ADP hydrolysis activity of hyperthermostable group II chaperonin in the presence of cobalt or manganese ion. FEBS Lett 580 (1): 34-40

Horikoshi K (1996) Alkaliphiles—from an industrial point of view. FEMS Microbiol Rev 18 (2-3): 259-270

Horlacher R, Xavier KB, Santos H, DiRuggiero J, Kossmann M, Boos W (1998) Archaeal binding protein-dependent ABC transporter: molecular and biochemical analysis of the trehalose/maltose transport system of the hyperthermophilic archaeon *Thermococcus litoralis*. J Bacteriol 180 (3): 680-689

Hottes AK, Meewan M, Yang D, Arana N, Romero P, McAdams HH, Stephens C (2004) Transcriptional profiling of *Caulobacter crescentus* during growth on complex and minimal media. J Bacteriol 186 (5): 1448-1461

Huang Y, Chapman B, Wilson M, Hocking AD (2009) Effect of agar concentration on the matric potential of glycerol agar media and the germination and growth of xerophilic and nonxerophilic fungi. Int J Food Microbiol 133 (1): 179-185

Huber G, Stetter KO (1991) *Sulfolobus metallicus*, sp. nov., a novel strictly chemolithoautotrophic thermophilic archaeal species of metal-mobilizers. Syst App Microbiol 14 (4): 372-378

Huber H, Stetter KO (2006) *Desulfurococcales*. In: Dworkin M, Falkow S, Rosenberg E, Schleifer K, Stackebrandt E (eds) The Prokaryotes, Springer, New York, pp 52-68

Huber R, Langworthy TA, König H, Thomm M, Woese CR, Sleytr UB, Stetter KO (1986) *Thermotoga mariti-*

ma sp. nov. represents a new genus of unique extremely thermophilic eubacteria growing up to 90℃. Arch Microbiol 144 (4): 324-333

Huber H, Burggraf S, Mayer T, Wyschkony I, Rachel R, Stetter KO (2000) *Ignicoccus* gen. nov., a novel genus of hyperthermophilic, chemolithoautotrophic Archaea, represented by two new species, *Ignicoccus islandicus* sp nov and Ignicoccus pacificus sp nov. and *Ignicoccus pacificus* sp. nov. Int J Syst Evol Microbiol 50 (6): 2093-2100

Imanaka H, Yamatsu A, Fukui T, Atomi H, Imanaka T (2006) Phosphoenolpyruvate synthase plays an essential role for glycolysis in the modified Embden-Meyerhof pathway in *Thermococcus kodakarensis*. Mol Microbiol 61 (4): 898-909

Imhoff JF, Rodriguez-Valera F (1984) Betaine is the main compatible solute of halophilic eubacteria. J Bacteriol 160 (1): 478-479

Irwin JA, Alfredsson GA, Lanzetti AJ, Gudmundsson HM, Engel PC (2001) Purification and char-acterisation of a serine peptidase from the marine psychrophile strain PA-43. FEMS Microbiol Lett 201 (2): 285-290

Jaenicke R (1981) Enzymes under extremes of physical conditions. Annu Rev Biophys Bioeng 10 (1): 1-67

Jaenicke R, Böhm G (1998) The stability of proteins in extreme environments. Curr Opin Struct Biol 8 (6): 738-748

Jia Z, Davies PL (2002) Antifreeze proteins: an unusual receptor-ligand interaction. Trend Biochem Sci 27 (2): 101-106

Johnsen U, Selig M, Xavier KB, Santos H, Schönheit P (2001) Different glycolytic pathways for glucose and fructose in the halophilic archaeon *Halococcus saccharolyticus*. Arch Microbiol 175 (1): 52-61

Johnson DB, Stallwood B, Kimura S, Hallberg KB (2006) Isolation and characterization of *Acidicaldus organivorus*, gen. nov., sp. nov.: a novel sulfur-oxidizing, ferric iron-reducing thermo-acidophilic heterotrophic *Proteobacterium*. Arch Microbiol 185 (3): 212-221

Jojima Y, Mihara Y, Suzuki S, Yokozeki K, Yamanaka S, Fudou R (2004) *Saccharibacter floricola* gen. nov., sp. nov., a novel osmophilic acetic acid bacterium isolated from pollen. Int J Syst Evol Microbiol 54 (6): 2263-2267

Jolivet E, L'Haridon S, Corre E, Forterre P, Prieur D (2003) *Thermococcus gammatolerans* sp. nov., a hyperthermophilic archaeon from a deep-sea hydrothermal vent that resists ionizing radiation. Int J Syst Evol Microbiol 53 (3): 847-851

Jolivet E, Corre E, L'Haridon S, Forterre P, Prieur D (2004) *Thermococcus marinus* sp. nov. and *Thermococcus radiotolerans* sp. nov., two hyperthermophilic archaea from deep-sea hydrothermal vents that resist ionizing radiation. Extremophiles 8 (3): 219-227

Kamekura M, Seno Y, Holmes M, Dyall-Smith M (1992) Molecular cloning and sequencing of the gene for a halophilic alkaline serine protease (halolysin) from an unidentified halophilic archaea strain (172P1) and expression of the gene in *Haloferax volcanii*. J Bacteriol 174 (3): 736-742

Karan R, Capes MD, DasSarma S (2012) Function and biotechnology of extremophilic enzymes in low water activity. Aquat Biosyst 8 (4): 1-15

Karavaĭko G, Krasil'nikova E, Tsaplina I, Bogdanova T, Zakharchuk L (2000) Growth and carbohydrate metabolism of*sulfobacilli*. Microbiology 70 (3): 293-299

Kato C (1999) Barophiles (Piezophiles). In: Horikoshi K, Tsuji K (eds) Extremophiles in deep-sea environments. Springer, Japan, pp 91-111

Kaushik JK, Ogasahara K, Yutani K (2002) The unusually slow relaxation kinetics of the foldingunfolding of pyrrolidone carboxyl peptidase from a hyperthermophile, *Pyrococcus furiosus*. J Mol Biol 316 (4): 991-1003

Kazak H, Oner E, Dekker RF (2010) Extremophiles as sources of exopolysaccharides. In: Ito R, Matsuo Y (eds) Handbook of Carbohydrate Polymers Development and Properties, Nova Science Publishers, New York, pp 605-619

Keller H, Friedli T (1992) Visibility limit of naked-eye sunspots. Q J R Astron Soc 33: 83-89

KengenS, De Bok F, Van Loo N, Dijkema C, Stams A, De Vos W (1994) Evidence for the operation of a novel Embden-Meyerhof pathway that involves ADP-dependent kinases during sugar fermentation by *Pyrococcus furiosus*. J Biol Chem 269 (26): 17537-17541

Kengen S, Stams AJ, Vos WM (1996) Sugar metabolism of hyperthermophiles. FEMS Microbiol Rev 18 (2-3): 119-137

Kessler PS, Daniel C, Leigh JA (2001) Ammonia switch-off of nitrogen fixation in the methanogenic archaeon *Methanococcus maripaludis*: mechanistic features and requirement for the novel GlnB homologues, NifI1 and NifI2. J Bacteriol 183 (3): 882-889

Khan MTH, Sylte I (2009) Determinants for psychrophilic and thermophilic features of metallo-peptidases of the M4 family. Silico Biol 9 (3): 105-124

Khmelenina VN, Kalyuzhnaya MG, Starostina NG, Suzina NE, Trotsenko YA (1997) Isolation and characterization of halotolerant alkaliphilic methanotrophic bacteria from Tuva soda lakes. Curr Microbiol 35 (5): 257-261

Kinoshita K, Takano M, Koseki T, Ito K, Iwano K (1995) Cloning of the xynNB gene encoding xylanase B from *Aspergillus niger* and its expression in *Aspergillus kawachii*. J Ferment Bioeng 79 (5): 422-428

Kitada M, Kosono S, Kudo T (2000) The Na^+/H^+ antiporter of alkaliphilic *Bacillus* sp. Extremophiles 4 (5): 253-258

Kletzin A, Mukund S, Kelley-Crouse TL, Chan MK, Rees DC, Adams M (1995) Molecular characterization of the genes encoding the tungsten-containing aldehyde ferredoxin oxidoreductase from *Pyrococcus furiosus* and formaldehyde ferredoxin oxidoreductase from *Thermococcus litoralis*. J Bacteriol 177 (16): 4817-4819

Klingeberg M, Hashwa F, Antranikian G (1991) Properties of extremely thermostable proteases from anaerobic hyperthermophilic bacteria. App Microbiol Biotechnol 34 (6): 715-719

Klingeberg M, Galunsky B, Sjoholm C, Kasche V, Antranikian G (1995) Purification and properties of a highly thermostable, sodium dodecyl sulfate-resistant and stereospecific proteinase from the extremely thermophilic archaeon *Thermococcus stetteri*. Appl Environ Microbiol 61 (8): 3098-3104

Koga S, Yoshioka I, Sakuraba H, Takahashi M, Sakasegawa S, Shimizu S, Ohshima T (2000) Biochemical characterization, cloning, and sequencing of ADP-dependent (AMP-forming) glucokinase from two hyperthermophilic archaea, *Pyrococcus furiosus* and *Thermococcus litoralis*. J Biochem 128 (6): 1079-1085

Koning HP (2001) Uptake of pentamidine in *Trypanosoma brucei* is mediated by three distinct transporters: implications for cross-resistance with arsenicals. Mol Pharmacol 59: 586-592

Koning SM, Albers SV, Konings WN, Driessen AJ (2002) Sugar transport in (hyper) thermophilic archaea. Res Microbiol 153 (2): 61-67

Konings WN, Albers SV, Koning S, Driessen AJ (2002) The cell membrane plays a crucial role in survival of bacteria and archaea in extreme environments. Springer 81 (1-4): 61-72

Kristjansson JK (1991) Thermophilic bacteria. CRC Press, Boca Raton, FL

Krulwich TA, Ito M, Hicks DB, Gilmour R, Guffanti AA (1998) pH homeostasis and ATP synthesis: studies of two processes that necessitate inward proton translocation in extremely alkali-philic Bacillus species. Extremophiles 2 (3): 217-222

Kumar V, Verma D, Satyanarayana T (2013) Extremophilic bacterial xylanases: production, characteristics and applications. Curr Biotechnol 2 (4): 380-399

Kurosawa N, Itoh YH, Iwai T, Sugai A, Uda I, Kimura N, Horiuchi T, Itoh T (1998) *Sulfurisphaera ohwakuensis* gen. nov., sp. nov., a novel extremely thermophilic acidophile of the order *Sulfolobales*. Int J Syst Bacteriol 48 (2): 451-456

Labes A, Schönheit P (2001) Sugar utilization in the hyperthermophilic, sulfate-reducing archaeon *Archaeoglobus fulgidus* strain 7324: starch degradation to acetate and CO_2 via a modified Embden-Meyerhof pathway and acetyl-CoA synthetase (ADP-forming). Arch Microbiol 176 (5): 329-338

Leigh JA, Dodsworth JA (2007) Nitrogen regulation in bacteria and archaea. Annu Rev Microbiol 61: 349-377

Leong SL, Lantz H, Pettersson OV, Frisvad JC, Thrane U, Heipieper HJ, Dijksterhuis J, Grabherr M, Pettersson M, Tellgren-Roth C (2014) Genome and physiology of the ascomycete filamentous fungus *Xeromyces bisporus*, the most xerophilic organism isolated to date. Environ Microbiol 17 (2): 496-513

LiW, Zhou X, Lu P (2005) Structural features of thermozymes. Biotechnol Adv 23 (4): 271-281

Lim J, Thomas T, Cavicchioli R (2000) Low temperature regulated DEAD-box RNA helicase from the Antarctic archaeon, *Methanococcoides burtonii*. J Mol Biol 297 (3): 553-567

Lippert K, Galinski EA, Trüper HG (1993) Biosynthesis and function of trehalose in *Ectothiorhodospira halochloris*. Antonie Van Leeuwenhoek 63 (1): 85-91

Lobos JH, Chisolm TE, Bopp LH, Holmes DS (1986) *Acidiphilium organovorum* sp. nov., an acidophilic heterotroph isolated from a *Thiobacillus ferrooxidans* culture. Int J Syst Bacteriol 36 (2): 139-144

Loos H, Krämer R, Sahm H, Sprenger GA (1994) Sorbitol promotes growth of *Zymomonas mobilis* in environments with high concentrations of sugar: evidence for a physiological function of glucose-fructose oxidoreductase in osmoprotection. J Bacteriol 176 (24): 7688-7693

Lovitt R, Longin R, Zeikus J (1984) Ethanol production by thermophilic bacteria: physiological comparison of solvent effects on parent and alcohol-tolerant strains of *Clostridium thermohy drosulfuricum*. Appl Environ Microbiol 48 (1): 171-177

Lucas S, Han J, Lapidus A, Cheng J-F, Goodwin LA, Pitluck S, Peters L, Mikhailova N, Teshima H, Detter JC (2012) Complete genome sequence of the thermophilic, piezophilic, heterotrophic bacterium *Marinitoga piezophila* KA3. Jo Bacteriol 194 (21): 5974-5975

Luyten K, Albertyn J, Skibbe WF, Prior BA, Ramos J, Thevelein JM, Hohmann S (1995) Fps1, a yeast member of the MIP family of channel proteins, is a facilitator for glycerol uptake and efflux and is inactive under osmotic stress. EMBO J 14: 1360-1371

Macedo-Ribeiro S, Darimont B, Sterner R, Huber R (1996) Small structural changes account for the high thermostability of 1 [4Fe-4S] ferredoxin from the hyperthermophilic bacterium *Thermotoga maritima*. Structure 4 (11): 1291-1301

Magasanik B, Kaiser CA (2002) Nitrogen regulation in *Saccharomyces cerevisiae*. Gene 290 (1): 1-18

Marco MG, Rodríguez LV, Ramos EL, Renovato J, Cruz-Hernández MA, Rodríguez R, Contreras J, Aguilar CN (2009) A novel tannase from the xerophilic fungus *Aspergillus niger* GH1. J Microbiol Biotechnol 1: 1-10

Margaryan A, Panosyan H, Popov Y (2010) Isolation and characterization of new metallotolernat *Bacilli* Strains. Biotech Biotechnol Equip 24 (Suppl 1): 450-454

Margesin R, Spröer C, Schumann P, Schinner F (2003) *Pedobacter cryoconitis* sp. nov., a facultative psychrophile from alpine glacier cryoconite. Int J Syst Evol Microbiol 53 (5): 1291-1296

Marı S, Abellana M, Rubinat M, Sanchis V, Ramos A (2003) Efficacy of sorbates on the control of the growth of *Eurotium* species in bakery products with near neutral pH. Int J Food Microbiol 87 (3): 251-258

Martin D, Bartlett DH, Roberts MF (2002) Solute accumulation in the deep-sea bacterium *Photobacteriumprofundum*. Extremophiles 6 (6): 507-514

Massey LK, Sokatch JR, Conrad RS (1976) Branched-chain amino acid catabolism in bacteria. Bacteriol Rev 40 (1): 42

Médigue C, Krin E, Pascal G, Barbe V, Bernsel A, Bertin PN, Cheung F, Cruveiller S, D'Amico S, Duilio A (2005) Coping with cold: the genome of the versatile marine Antarctica bacterium *Pseudoalteromonas haloplanktis* TAC125. Genome Res 15 (10): 1325-1335

Meister A (2012) Biochemistry of the amino acids. Elsevier, New York.

Merchant MM, Welsh AK, McLean RJ (2007) *Rheinheimera texasensis* sp. nov., a halointolerant freshwater oligotroph. Int J Syst Evol Microbiol 57 (10): 2376-2380

Michels M, Bakker EP (1985) Generation of a large, protonophore-sensitive proton motive force and pH difference in the acidophilic bacteria *Thermoplasma acidophilum* and *Bacillus acidocaldarius*. J Bacteriol 161 (1): 231-237

Miflin B, Lea P (1977) Amino acid metabolism. AnnuRev Plant Physiol 28 (1): 299-329

Mishra D, Kim DJ, Ralph D, Ahn JG, Rhee YH (2008) Bioleaching of metals from spent lithium ion secondary batteries using *Acidithiobacillus ferrooxidans*. Waste Manage 28 (2): 333-338

Moracci M, Ponzano BC, Trincone A, Fusco S, De Rosa M, van der Oost J, Sensen CW, Charlebois RL (2000) Identification and molecular characterization of the first α-xylosidase from an archaeon. J Biol Chem 275 (29): 22082-22089

Moran JW, Witter LD (1979) Effect of sugars on D-arabitol production and glucose metabolism in *Saccharomyces rouxii*. J Bacteriol 138 (3): 823-831

Morikawa M, Izawa Y, Rashid N, Hoaki T, Imanaka T (1994) Purification and characterization of a thermostable thiol protease from a newly isolated hyperthermophilic *Pyrococcus* sp. Appl Environ Microbiol 60 (12): 4559-4566

MoritaRY (1975) Psychrophilic bacteria. Bacteriol Rev 39 (2): 144

Mormile MR, Biesen MA, Gutierrez MC, Ventosa A, Pavlovich JB, Onstott TC, Fredrickson JK (2003) Isolation of *Halobacterium salinarum* retrieved directly from halite brine inclusions. Environ Microbiol 5 (11): 1094-1102

Mukund S, Adams M (1991) The novel tungsten-iron-sulfur protein of the hyperthermophilic archaebacterium, *Pyrococcus furiosus*, is an aldehyde ferredoxin oxidoreductase. Evidence for its participation in a unique glycolytic pathway. J Biol Chem 266 (22): 14208-14216

Mukund S, Adams M (1993) Characterization of a novel tungsten-containing formaldehyde ferredoxin oxidoreductase from the hyperthermophilic archaeon, *Thermococcus litoralis*. A role for tungsten in peptide catabolism. J Biol Chem 268 (18): 13592-13600

Mukund S, Adams MW (1995) Glyceraldehyde-3-phosphate ferredoxin oxidoreductase, a novel tungsten-containing enzyme with a potential glycolytic role in the hyperthermophilic archaeon *Pyrococcus furiosus*. J Biol Chem 270 (15): 8389-8392

Muller D, Simeonova DD, Riegel P, Mangenot S, Koechler S, Lièvremont D, Bertin PN, Lett MC (2006) *Herminiimonas arsenicoxydans* sp. nov., a metalloresistant bacterium. Int J Syst Evol Microbiol 56 (8): 1765-1769

Nelson KE, Clayton RA, Gill SR, Gwinn ML, Dodson RJ, Haft DH, Hickey EK, Peterson JD, Nelson WC, Ketchum KA (1999) Evidence for lateral gene transfer between Archaea and bacteria from genome sequence of *Thermotoga maritima*. Nature 399 (6734): 323-329

Nevoigt E, Stahl U (1997) Osmoregulation and glycerol metabolism in the yeast *Saccharomyces cerevisiae*. FEMS Microbiol Rev 21 (3): 231-241

Ng WV, Kennedy SP, Mahairas GG, Berquist B, Pan M, Shukla HD, Lasky SR, Baliga NS, Thorsson V, Sbrogna J (2000) Genome sequence of Halobacterium species NRC-1. Proc Natl Acad Sci 97 (22): 12176-12181

NinfaAJ, Jiang P (2005) PII signal transduction proteins: sensors of α-ketoglutarate that regulate nitrogen metabolism. Curr Opin Microbiol 8 (2): 168-173

Nogi Y, Kato C, Horikoshi K (2002) *Psychromonas kaikoae* sp. nov., a novel from the deepest piezophilic bacterium cold-seep sediments in the Japan Trench. Int J Syst Evol Microbiol 52 (5): 1527-1532

Nogi Y, Hosoya S, Kato C, Horikoshi K (2004) *Colwellia piezophila* sp. nov., a novel piezophilic species from deep-sea sediments of the Japan Trench. Int J Syst Evol Microbiol 54 (5): 1627-1631

Novitsky JA, Morita RY (1976) Morphological characterization of small cells resulting from nutrient starvation of a psychrophilic marine *vibrio*. Appl Environ Microbiol 32 (4): 617-622

Obis D, Guillot A, Gripon JC, Renault P, Bolotin A, Mistou MY (1999) Genetic and biochemical characterization of a high-affinity betaine uptake system (BusA) in *Lactococcus lactis* reveals a new functional organization within bacterial ABC transporters. J Bacteriol 181 (20): 6238-6246

Ogawa Y, Tatsumi H, Murakami S, Ishida Y, Murakami K, Masaki A, Kawabe H, Arimura H, Nakano E, Motai H (1990) Secretion of *Aspergillus oryzae* alkaline protease in an osmophilic yeast, *Zygosaccharomyces rouxii*. Agric Biol Chem 54 (10): 2521-2529

Oren A, Mana L (2003) Sugar metabolism in the extremely halophilic bacterium *Salinibacter ruber*. FEMS Microbiol Lett 223 (1): 83-87

Oren A, Ginzburg M, Ginzburg B, Hochstein L, Volcani B (1990) *Haloarcula marismortui* (*Volcani*) sp. nov., nom. rev., an extremely halophilic bacterium from the Dead Sea. Int J Syst Bacteriol 40 (2): 209-210

Osman Y, Conway T, Bonetti S, Ingram L (1987) Glycolytic flux in *Zymomonas mobilis*: enzyme and metabolite levels during batch fermentation. J Bacteriol 169 (8): 3726-3736

Paavilainen S, Oinonen S, Korpela T (1999) Catabolic pathways of glucose in *Bacillus circulans var. alkalophilus*. Extremophiles 3 (4): 269-276

Padan E, Bibi E, Ito M, Krulwich TA (2005) Alkaline pH homeostasis in bacteria: new insights. BBA Biomem 1717 (2): 67-88

Påhlman A-K, Granath K, Ansell R, Hohmann S, Adler L (2001) The yeast glycerol 3-phosphatases Gpp1p and

Gpp2p are required for glycerol biosynthesis and differentially involved in the cellular responses to osmotic, anaerobic, and oxidative stress. J Biol Chem 276 (5): 3555-3563

Palacios-Cabrera H, Taniwaki MH, Hashimoto JM, Menezes HCd (2005) Growth of *Aspergillus ochraceus*, *A. carbonarius* and *A. niger* on culture media at different water activities and temperatures. Braz J Microbiol 36 (1): 24-28

Patel MA, Ou MS, Harbrucker R, Aldrich HC, Buszko ML, Ingram LO, Shanmugam K (2006) Isolation and characterization of acid-tolerant, thermophilic bacteria for effective fermentation of biomass-derived sugars to lactic acid. Appl Environ Microbiol 72 (5): 3228-3235

Paulsen IT, Nguyen L, Sliwinski MK, Rabus R, Saier MH (2000) Microbial genome analyses: comparative transport capabilities in eighteen prokaryotes. J Mol Biol 301 (1): 75-100

PedersonJA, Steele JL, Christensen JE, Dudley EG (1999) Peptidases and amino acid catabolism in lactic acid bacteria. In: Konings WN, Kuipers OP, Huis In't Veld JHJ (eds) Lactic acid bacteria: genetics, metabolism and applications. Springer, Netherlands, pp 217-246

Peng L, Hertz L, Huang R, Sonnewald U, Petersen S, Westergaard N, Larsson O, Schousboe A (1993) Utilization of glutamine and of TCA cycle constituents as precursors for transmitter glutamate and GABA. Dev Neurosci 15 (3-5): 367-377

Pflüger K, Müller V (2004) Transport of compatible solutes in extremophiles. J Bioenerg Biomem 36 (1): 17-24

Phadtare S (2004) Recent developments in bacterial cold-shock response. Curr Issues Mol Biol 6 (2): 125-136

Piette F, D'Amico S, Struvay C, Mazzucchelli G, Renaut J, Tutino ML, Danchin A, Leprince P, Feller G (2010) Proteomics of life at low temperatures: trigger factor is the primary chaperone in the Antarctic bacterium *Pseudoalteromonas haloplanktis* TAC125. Mol Microbiol 76 (1): 120-132

Pikuta E, Lysenko A, Chuvilskaya N, Mendrock U, Hippe H, Suzina N, Nikitin D, Osipov G, Laurinavichius K (2000) *Anoxybacillus pushchinensis* gen. nov., sp. nov., a novel anaerobic, alkaliphilic, moderately thermophilic bacterium from manure, and description of *Anoxybacillus flavitherms comb.* nov. Int J Syst Evol Microbiol 50 (6): 2109-2117

Pitt JI, Hocking AD (2009) Xerophiles. In: Fungi and food spoilage. Springer, US, pp 339-355 Poindexter JS (1981) Oligotrophy. In: Alexander M (ed) Advances in microbial ecology. Springer, US, pp 63-89

Postma P, Lengeler J, Jacobson G (1993) Phosphoenolpyruvate: carbohydrate phosphotransferase systems of bacteria. Microbiol Rev 57 (3): 543

Preston CM, Wu KY, Molinski TF, DeLong EF (1996) A psychrophilic crenarchaeon inhabits a marine sponge: *Cenarchaeum symbiosum* gen. nov., sp. nov. Proc Natl Acad Sci 93 (13): 6241-6246

Quiocho FA, Ledvina PS (1996) Atomic structure and specificity of bacterial periplasmic receptors for active transport and chemotaxis: variation of common themes. Mol Microbiol 20 (1): 17-25

Rainey F, Donnison A, Janssen P, Saul D, Rodrigo A, Bergquist PL, Daniel RM, Stackebrandt E, Morgan HW (1994) Description of *Caldicellulosiruptor saccharolyticus* gen. nov., sp. nov: an obligately anaerobic, extremely thermophilic, cellulolytic bacterium. FEMS Microbiol Lett 120 (3): 263-266

Rainey FA, Ferreira M, Nobre MF, Ray K, Bagaley D, Earl AM, Battista JR, Gómez-Silva B, McKay CP, da Costa MS (2007) *Deinococcus peraridilitoris* sp. nov., isolated from a coastal desert. Int J Syst Evol Microbiol 57 (7): 1408-1412

Rao MB, Tanksale AM, Ghatge MS, Deshpande VV (1998) Molecular and biotechnological aspects of microbial proteases. Microbiol Mol Biol Rev 62 (3): 597-635

Rashid N, Imanaka H, Fukui T, Atomi H, Imanaka T (2004) Presence of a novel phosphopentomutase and a 2-deoxyribose 5-phosphate aldolase reveals a metabolic link between pentoses and central carbon metabolism in the hyperthermophilic archaeon *Thermococcus kodakaraensis*. J Bacteriol 186 (13): 4185-4191

Rawal N, Kelkar S, Altekar W (1988) Alternative routes of carbohydrate metabolism in halophilic archaebacteria. Indian J Biochem Biophys 25 (6): 674-686

Reddy GS, Potrafka RM, Garcia-Pichel F (2007) *Modestobacter versicolor* sp. nov., an actinobacterium from biological soil crusts that produces melanins under oligotrophy, with emended descriptions of the genus *Modestobacter* and *Modestobacter multiseptatus*. Int J Syst Evol Biol 57 (9): 2014-2020

Restaino L, Bills S, Tscherneff K, Lenovich LM (1983) Growth characteristics of *Saccharomyces rouxii* isolated from chocolate syrup. Appl Environ Microbiol 45 (5): 1614-1621

Richard H, Foster JW (2004) *Escherichia coli* glutamate-and arginine-dependent acid resistance systems increase internal pH and reverse transmembrane potential. J Bacteriol 186 (18): 6032-6041

Riley M, Staley JT, Danchin A, Wang TZ, Brettin TS, Hauser LJ, Land ML, Thompson LS (2008) Genomics of an extreme psychrophile, *Psychromonas ingrahamii*. BMC Genomics 9 (1): 210

Roberts MF (2000) Osmoadaptation and osmoregulation in archaea. Front Biosci 5 (362): 796-812

Roberts MF (2004) Osmoadaptation and osmoregulation in archaea: update 2004. Front Biosci 9: 1999-2019

Robinson JL, Pyzyna B, Atrasz RG, Henderson CA, Morrill KL, Burd AM, DeSoucy E, Fogleman RE, Naylor JB, Steele SM (2005) Growth kinetics of extremely halophilic Archaea (family Halobacteriaceae) as revealed by Arrhenius plots. J Bacteriol 187 (3): 923-929

Rodrigues F, Ludovico P, Leão C (2006) Sugar metabolism in yeasts: an overview of aerobic and anaerobic glucose catabolism. In: Peter G, Rosa C (eds) Biodiversity and ecophysiology of yeasts. Springer, Berlin, pp 101-121

Rodriguez-Valera F, Juez G, Kushner D (1983) *Halobacterium mediterranei* spec, nov., a new carbohydrate-utilizing extreme halophile. Syst App Microbiol 4 (3): 369-381

Rogers P, Lee KJ, Tribe D (1979) Kinetics of alcohol production by *Zymomonas mobilis* at high sugar concentrations. Biotechnol Lett 1 (4): 165-170

Ronimus RS, Koning J, Morgan H (1999) Purification and characterization of an ADP-dependent phosphofructokinase from *Thermococcus zilligii*. Extremophiles 3 (2): 121-129

Rothschild LJ, Mancinelli RL (2001) Life in extreme environments. Nature 409 (6823): 1092-1101

Rozen Y, Belkin S (2001) Survival of enteric bacteria in seawater. FEMS Microbiol Rev 25 (5): 513-529

Ruepp A, Graml W, Santos-Martinez M-L, Koretke KK, Volker C, Mewes HW, Frishman D, Stocker S, Lupas AN, Baumeister W (2000) The genome sequence of the thermoacidophilic scavenger *Thermoplasma acidophilum*. Nature 407 (6803): 508-513

Russell NJ (1997) Psychrophilic bacteria-molecular adaptations of membrane lipids. Comp Biochem Physiol A Physiol 118 (3): 489-493

Russell NJ (2000) Toward a molecular understanding of cold activity of enzymes from psychrophiles. Extremophiles 4 (2): 83-90

Rybicki E (1990) The classification of organisms at the edge of life, or problems with virus systematics. S Afr J Sci 86: 182-186

Saiki T, Kobayashi T, Kawagoe K, Beppu T (1985) *Dictyoglomus thermophilum* gen. nov., sp. nov., a chemoorganotrophic, anaerobic, thermophilic bacterium. Int J Syst Bacteriol 35 (3): 253-259

Sako Y, Nomura N, Uchida A, Ishida Y, Morii H, Koga Y, Hoaki T, Maruyama T (1996) *Aeropyrum pernix* gen. nov., sp. nov., a novel aerobic hyperthermophilic archaeon growing at temperatures up to 100℃. Int J Syst Bacteriol 46 (4): 1070-1077

Sakuraba H, Ohshima T (2002) Novel energy metabolism in anaerobic hyperthermophilic archaea: a modified Embden-Meyerhof pathway. J Biosci Bioeng 93 (5): 441-448

Sakuraba H, Yoshioka I, Koga S, Takahashi M, Kitahama Y, Satomura T, Kawakami R, Ohshima T (2002) ADP-dependent glucokinase/phosphofructokinase, a novel bifunctional enzyme from the hyperthermophilic archaeon *Methanococcus jannaschii*. J Biol Chem 277 (15): 12495-12498

Sakuraba H, Goda S, Ohshima T (2004) Unique sugar metabolism and novel enzymes of hyper thermophilic archaea. Chem Rec 3 (5): 281-287

Sanchez-Porro C, Mellado E, Bertoldo C, Antranikian G, Ventosa A (2003) Screening and charac terization of the protease CP1 produced by the moderately halophilic bacterium *Pseudoalteromonas* sp. strain CP76. Extremophiles 7 (3): 221-228

Saunders NF, Thomas T, Curmi PM, Mattick JS, Kuczek E, Slade R, Davis J, Franzmann PD, Boone D, Rusterholtz K (2003) Mechanisms of thermal adaptation revealed from the genomes of the Antarctic Archaea *Methanogenium frigidum* and *Methanococcoides burtonii*. Genome Res 13 (7): 1580-1588

Saurin W, Hofnung M, Dassa E (1999) Getting in or out: early segregation between importers and exporters in the evolution of ATP-binding cassette (ABC) transporters. J Mol Evol 48 (1): 22-41

Schneider E (2001) ABC transporters catalyzing carbohydrate uptake. Res Microbiol 152 (3): 303-310

Schönheit P, Schäfer T (1995) Metabolism of hyperthermophiles. World J Microbiol Biotechnol 11 (1): 26-57

Schröder C, Selig M, Schönheit P (1994) Glucose fermentation to acetate, CO_2 and H_2 in the anaerobic hyperthermophilic eubacterium *Thermotoga maritima*: involvement of the Embden-Meyerhof pathway. Arch Microbiol 161 (6): 460-470

Selig M, Xavier KB, Santos H, Schonheit P (1997) Comparative analysis of Embden-Meyerhof and Entner-Doudoroff glycolytic pathways in hyperthermophilic archaea and the bacterium *Thermotoga*. Arch Microbiol 167 (4): 217-232

Seo JS, Chong H, Park HS, Yoon KO, Jung C, Kim JJ, Hong JH, Kim H, Kim JH, Kil JI (2005) The genome sequence of the ethanologenic bacterium *Zymomonas mobilis* ZM4. Nat Biotechnol 23 (1): 63-68

Sghaier H, Ghedira K, Benkahla A, Barkallah I (2008) Basal DNA repair machinery is subject to positive selection in ionizing-radiation-resistant bacteria. BMC Genomics 9 (1): 297

She Q, Singh RK, Confalonieri F, Zivanovic Y, Allard G, Awayez MJ, Christina C-Y, Clausen IG, Curtis BA, De Moors A (2001) The complete genome of the crenarchaeon *Sulfolobus solfataricus* P2. Proc Natl Acad Sci 98 (14): 7835-7840

Sheridan PP, Loveland-Curtze J, Miteva VI, Brenchley JE (2003) *Rhodoglobus vestalii* gen. nov., sp. nov., a novel psychrophilic organism isolated from an Antarctic Dry Valley lake. Int J Syst Evol Microbiol 53 (4): 985-994

Shimada H, Nemoto N, Shida Y, Oshima T, Yamagishi A (2002) Complete polar lipid composition of *Thermoplasma acidophilum* HO-62 determined by high-performance liquid chromatography with evaporative light-scattering detection. J Bacteriol 184 (2): 556-563

Shuttleworth KL, Unz RF, Wichlacz PL (1985) Glucose catabolism in strains of acidophilic, heterotrophic bacteria. Appl Environ Microbiol 50 (3): 573-579

Siebers B, Schönheit P (2005) Unusual pathways and enzymes of central carbohydrate metabolism in Archaea. Curr Opin Microbiol 8 (6): 695-705

Siebold C, Flükiger K, Beutler R, Erni B (2001) Carbohydrate transporters of the bacterial phosphoenolpyruvate: sugar phosphotransferase system (PTS). FEBS Lett 504 (3): 104-111

Simonato F, Campanaro S, Lauro FM, Vezzi A, D'Angelo M, Vitulo N, Valle G, Bartlett DH (2006) Piezophilic adaptation: a genomic point of view. J Biotechnol 126 (1): 11-25

Smith DP, Thrash JC, Nicora CD, Lipton MS, Burnum-Johnson KE, Carini P, Smith RD, Giovannoni SJ (2013) Proteomic and transcriptomic analyses of *Candidatus Pelagibacter ubique* describe the first pii-independent response to nitrogen limitation in a free-living Alphaproteobacterium. Mol Biol 6：e00133-00112

Sprenger GA（1996）Carbohydrate metabolism in *Zymomonas mobilis*：a catabolic highway with some scenic routes. FEMS Microbiol Lett 145（3）：301-307

Sterner R，Liebl W（2001）Thermophilic adaptation of proteins. Crit Rev Biochem Mol Biol 36（1）：39-106

Stettner AI，Segrè D（2013）The cost of efficiency in energy metabolism. Proc Natl Acad Sci 110（24）：9629-9630

Suresh K，Reddy G，Sengupta S，Shivaji S（2004）*Deinococcus indicus* sp. nov.，an arsenic-resistant bacterium from an aquifer in West Bengal，India. Int J Syst Evol Microbiol 54（2）：457-461

Swartz TH，Ikewada S，Ishikawa O，Ito M，Krulwich TA（2005）The Mrp system：a giant among monovalent cation/proton antiporters? Extremophiles 9（5）：345-354

Takai K，Moser DP，Onstott TC，Spoelstra N，Pfiffner SM，Dohnalkova A，Fredrickson JK（2001）*Alkaliphilus transvaalensis* gen. nov.，sp. nov.，an extremely alkaliphilic bacterium isolated from a deep South African gold mine. Int J Syst Evol Microbiol 51（4）：1245-1256

Takai K，Miyazaki M，Hirayama H，Nakagawa S，Querellou J，Godfroy A（2009）Isolation and physiological characterization of two novel，piezophilic，thermophilic chemolithoautotrophs from a deep-sea hydrothermal vent chimney. Environ Microbiol 11（8）：1983-1997

Tamás MJ，Luyten K，Sutherland FCW，Hernandez A，Albertyn J，Valadi H，Li H，Prior BA，Kilian SG，Ramos J（1999）Fps1p controls the accumulation and release of the compatible solute glycerol in yeast osmoregulation. Mol Microbiol 31（4）：1087-1104

Tamegai H，Nakasone K，Siddiqui KS（2008）Protein Adaptation to High-Pressure Environments. In：Siddiqui K，Thomas T（eds）Protein Adaptation in Extremophiles，Nova Publishers，New York，pp 167-191

Tang YJ，Sapra R，Joyner D，Hazen TC，Myers S，Reichmuth D，Blanch H，Keasling JD（2009）Analysis of metabolic pathways and fluxes in a newly discovered thermophilic and ethanol-tolerant *Geobacillus* strain. Biotechnol Bioeng 102（5）：1377-1386

Tempest D，Meers J，Brown C（1973）Glutamate synthetase（GOGAT）；a key enzyme in the assimilation of ammonia by prokaryotic organisms. In：Prusiner S，Stadtman ER（eds）The enzymes of glutamine metabolism，Academic Press，UK，pp 167-182

Tetas M，Lowenstein JM（1963）the effect of bivalent metal ions on the hydrolysis of adenosine Di-and triphosphate. Biochemistry 2（2）：350-357

Tokuoka K（1993）Sugar and salt tolerant yeasts. J App Bacteriol 74（2）：101-110

Tyson GW，Chapman J，Hugenholtz P，Allen EE，Ram RJ，Richardson PM，Solovyev VV，Rubin EM，Rokhsar DS，Banfield JF（2004）Community structure and metabolism through reconstruction of microbial genomes from the environment. Nature 428（6978）：37-43

Ulitzur S（1974）*Vibrio parahaemolyticus* and *Vibrio alginolyticus*：short generation-time marine bacteria. Microb Ecol 1（1）：127-135

Valdés J，Pedroso I，Quatrini R，Dodson RJ，Tettelin H，Blake R，Eisen JA，Holmes DS（2008）*Acidithiobacillus ferrooxidans* metabolism：from genome sequence to industrial applications. BMC Genomics 9（1）：597

Valentine RC，Valentine DL（2004）Omega-3 fatty acids in cellular membranes：a unified concept. Prog Lipid Res 43（5）：383-402

van de Vossenberg JL，Driessen AJ，Zillig W，Konings WN（1998）Bioenergetics and cytoplasmic membrane stability of the extremely acidophilic，thermophilic archaeon *Picrophilus oshimae*. Extremophiles 2（2）：67-74

van den Burg B（2003）Extremophiles as a source for novel enzymes. Curr Opin Microbiol 6（3）：213-218

van der Oost J，Siebers B（2007）The glycolytic pathways of Archaea：evolution by tinkering. Archaea：evolution，physiology and molecular biology 22：247-260

van der Oost J，Schut G，Kengen SM，Hagen WR，Thomm M，de Vos WM（1998）The ferredoxin-dependent conversion of glyceraldehyde-3-phosphate in the hyperthermophilic archaeon *Pyrococcus furiosus* represents a novel site of glycolytic regulation. J Biol Chem 273（43）：28149-28154

van der Oost J，Ahmed H，Ettema T，Tjaden B，Geerling A，Siebers B（2005）The semiphosphorylative Entner-Doudoroff pathway in hyperthermophilic archaea：a re-evaluation. Biochem J 390：529-540

van Niel EW，Claassen PA，Stams AJ（2003）Substrate and product inhibition of hydrogen production by the extreme thermophile，*Caldicellulosiruptor saccharolyticus*. Biotechnol Bioeng 81（3）：255-262

Venkateswaran A，McFarlan SC，Ghosal D，Minton KW，Vasilenko A，Makarova K，Wackett LP，Daly MJ（2000）Physiologic determinants of radiation resistance in *Deinococcus radiodurans*. Appl Environ Microbiol 66（6）：2620-2626

Verhees C，Kengen S，Tuininga J，Schut G，Adams M，de Vos W，Van Der Oost J（2003）The unique features of glycolytic pathways in archaea. Biochem J 375：231-246

Verhees C，Kengen S，Tuininga J，Schut G，Adams M，de Vos W，Van der Oost J（2004）The unique features of glycolytic pathways in archaea. Biochem J 377：819-822

Villeret V，Beeumen JV，Chessa JP，Gerday C（1997）Preliminary crystal structure determination of the alkaline protease from the Antarctic psychrophile *Pseudomonas aeruginosa*. Protein Sci 6（11）：2462-2464

Violot S，Aghajari N，Czjzek M，Feller G，Sonan GK，Gouet P，Gerday C，Haser R，Receveur-Bréchot V（2005）Structure of a full length psychrophilic cellulase from *Pseudoalteromonas haloplanktis* revealed by X-ray diffraction and small angle X-ray scattering. J Mol Biol 348（5）：1211-1224

Virgilio C，Burckert N，Bell W，Jeno P，Boller T，Wiemken A（1993）Disruption of TPS2，the gene encoding the

100 kDa subunit of the trehalose-6-phosphate synthase/phosphatase complex in *Saccharomyces cerevisiae*, causes accumulation of trehalose-6-phosphate and loss of trehalose-6-phosphate phosphatase activity. Eur J Biochem 212 (2): 315-323

Vogel G, Aeschbacher RA, Müller J, Boller T, Wiemken A (1998) Trehalose-6-phosphate phosphatases from *Arabidopsis thaliana*: identification by functional complementation of the yeast TPS2 mutant. Plant J 13 (5): 673-683

Voges D, Zwickl P, Baumeister W (1999) The 26S proteasome: a molecular machine designed for controlled proteolysis. Ann Rev Biochem 68 (1): 1015-1068

Vossenberg JL, Ubbink-Kok T, Elferink MG, Driessen AJ, Konings WN (1995) Ion permeability of the cytoplasmic membrane limits the maximum growth temperature of bacteria and archaea. Mol Microbiol 18 (5): 925-932

Wang Z, Hicks DB, Guffanti AA, Baldwin K, Krulwich TA (2004) Replacement of amino acid sequence features of a-and c-subunits of ATP synthases of alkaliphilic *Bacillus* with the *Bacillus* consensus sequence results in defective oxidative phosphorylation and non-fermentative growth at pH 10. 5. J Biol Chem 279 (25): 26546-26554

Wang J, Bai J, Xu J, Liang B (2009) Bioleaching of metals from printed wire boards by *Acidithiobacillus ferrooxidans* and *Acidithiobacillus thiooxidans* and their mixture. J Hazard Mater 172 (2): 1100-1105

Warren R (1996) Microbial hydrolysis of polysaccharides. Ann Rev Microbiol 50 (1): 183-212

White O, Eisen JA, Heidelberg JF, Hickey EK, Peterson JD, Dodson RJ, Haft DH, Gwinn ML, Nelson WC, Richardson DL (1999) Genome sequence of the radioresistant bacterium *Deinococcusradiodurans* R1. Science 286 (5444): 1571-1577

White D, Drummond JT, Fuqua C (2007) The physiology and biochemistry of prokaryotes. Oxford University Press, New York

White D, Drummond JT, Fuqua C (2012) The physiology and biochemistry of prokaryotes, 4th edn. Oxford University Press, New York

Wiegel J (1992) The obligately anaerobic thermophilic bacteria. In: Kristjansson J (ed) Thermophilic Bacteria, CRC Press, Boca Raton, pp 105-184

Wolf A, Krämer R, Morbach S (2003) Three pathways for trehalose metabolism in *Corynebacterium glutamicum* ATCC13032 and their significance in response to osmotic stress. Mol Microbiol 49 (4): 1119-1134

Wood AP, Aurikko JP, Kelly DP (2004) A challenge for 21st century molecular biology and biochemistry: what are the causes of obligate autotrophy and methanotrophy? FEMS Microbiol Rev 28: 335-352

Xavier KB, Martins LO, Peist R, Kossmann M, Boos W, Santos H (1996) High-affinity maltose/trehalose transport system in the hyperthermophilic archaeon *Thermococcus litoralis*. J Bacteriol 178 (16): 4773-4777

Xavier KB, da Costa MS, Santos H (2000) Demonstration of a novel glycolytic pathway in the hyperthermophilic archaeon *Thermococcus zilligii* by 13C-labeling experiments and nuclear magnetic resonance analysis. J Bacteriol 182 (16): 4632-4636

Xu Y, Zhou P, Tian X (1999) Characterization of two novel haloalkaliphilic archaea *Natronorubrum bangense* gen. nov., sp. nov. and *Natronorubrum tibetense* gen. nov., sp. nov. Int J Syst Bacteriol 49 (1): 261-266

Xu Y, Feller G, Gerday C, Glansdorff N (2003) Metabolic enzymes from psychrophilic bacteria: challenge of adaptation to low temperatures in ornithine carbamoyltransferase from *Moritella abyssi*. J Bacteriol 185 (7): 2161-2168

Yoo SH, Weon HY, Kim SJ, Kim YS, Kim BY, Kwon SW (2010) *Deinococcus aerolatus* sp. nov. and *Deinococcus aerophilus* sp. nov., isolated from air samples. Int J Syst Evol Microbiol 60 (5): 1191-1195

Yoshinaka T, Yano K, Yamaguchi H (1973) Isolation of highly radioresistant bacterium, *Arthrobacter radiotolerans* nov sp. Agric Biol Chem 37 (10): 2269-2275

You XY, Guo X, Zheng HJ, Zhang MJ, Liu LJ, Zhu YQ, Zhu B, Wang SY, Zhao GP, Poetsch A (2011) Unraveling the *Acidithiobacillus caldus* complete genome and its central metabolisms for carbon assimilation. J Genet Genomics 38 (6): 243-252

Yuan M, Chen M, Zhang W, Lu W, Wang J, Yang M, Zhao P, Tang R, Li X, Hao Y (2012) Genome sequence and transcriptome analysis of the radioresistant bacterium *Deinococcus gobiensis*: insights into the extreme environmental adaptations. PLoS One 7 (3), e34458

Yumoto I, Hirota K, Sogabe Y, Nodasaka Y, Yokota Y, Hoshino T (2003) *Psychrobacter okhotsken-sis* sp. nov., a lipase-producing facultative psychrophile isolated from the coast of the Okhotsk Sea. Int J Syst Evol Microbiol 53 (6): 1985-1989

Zhang DS, Lovitt R (2005) Studies on growth and metabolism of *Oenococcus oeni* on sugars and sugar mixtures. J Appl Microbiol 99 (3): 565-572

Zhang YM, Wong TY, Chen LY, Lin CS, Liu JK (2000) Induction of a futile Embden-Meyerhof-Parnas pathway in *Deinococcus radiodurans* by Mn: possible role of the pentose phosphate pathway in cell survival. Appl Environ Microbiol 66 (1): 105-112

Zhilina T, Zavarzin G, Rainey F, Pikuta E, Osipov G, Kostrikina N (1997) *Desulfonatronovibrio hydrogenovorans* gen. nov., sp. nov., an alkaliphilic, sulfate-reducing bacterium. Int J Syst Bacteriol 47 (1): 144-149

Zhilina T, Garnova E, Tourova T, Kostrikina N, Zavarzin G (2001) *Halonatronum saccharophilum* gen. nov. sp. nov.: a new haloalkaliphilic bacterium of the order *Haloanaerobiales* from Lake Magadi. Microbiology 70 (1): 64-72

Zhilina TN, Appel R, Probian C, Brossa EL, Harder J, Widdel F, Zavarzin GA (2004) *Alkaliflexus imshenetskii* gen. nov. sp. nov., a new alkaliphilic gliding carbohydrate-fermenting bacterium with propionate formation from a soda lake. Arch Microbiol 182 (2-3): 244-253

Zhu F, Wang S, Zhou P (2003) *Flavobacterium xinjiangense* sp. nov. and *Flavobacterium omnivorum* sp. nov., novel psychrophiles from the China No. 1 glacier. Int J Syst Evol Microbiol 53（3）：853-857

Zillig W，Holz I，Janekovic D，Klenk H，Imsel E，Trent J，Wunderl S，Forjaz V，Coutinho R，Ferreira T（1990）*Hyperthermus butylicus*，a hyperthermophilic sulfur-reducing archaebacterium that ferments peptides. J Bacteriol 172（7）：3959-3965

Zivanovic Y，Armengaud J，Lagorce A，Leplat C，Guérin P，Dutertre M，Anthouard V，Forterre P，Wincker P，Confalonieri F（2009）Genome analysis and genome-wide proteomics of *Thermococcus gammatolerans*，the most radioresistant organism known amongst the archaea. Genome Biol 10（6）

第二章

嗜冷细菌的多样性、冷适应性和生物技术应用

Gundlapally Sathyanarayana Reddy[1], Madhab Kumar Chattopadhyay,[1] Sisinthy Shivaji[2]

2.1 嗜冷细菌的多样性

极地环境包括北极、南极、喜马拉雅山脉、高山冰川、深海、积雪、永久冻土、海冰、湖泊、冻土（特别是半冻土）、寒冷沙漠和洞穴等。这些终年在寒冷的栖息地上生存的微生物被称为嗜冷菌（生长温度从零下到 30℃）（Morita 1975；Helmke and Weyland 2004；Laucks et al. 2005）。嗜冷菌在低温下生存和增殖的能力引起了人们广泛的研究兴趣，然而其确切的机制现在还不是很清楚（见本章 2.2 节）。本章第一部分将介绍南极、北极以及喜马拉雅山脉的原核生物多样性（Chattopadhyay et al. 2014）。

2.1.1 南极洲原核生物多样性

南极洲是地球上最极端的生存环境之一，是已知最寒冷、最干燥多风和冰层最厚（平均厚度约 1.9km）的生存环境（Vincent 1988；Claridge and Campbell 1977；Campbell and Claridge 2000；Smith et al. 1992）。尽管南极洲气候条件极其恶劣，仍有大量不同的生命形式在这个大陆上生存和繁衍（如螨、蜱、海豹、企鹅、苔藓、地衣、细菌、酵母菌、藻类等）。极端嗜冷菌是南极大陆最主要的微生物群落，在维持南极生态系统稳定，特别是在营养循环方面发挥着重要作用（Stokes and Reymond 1966；Herbert and Bell 1977；Tanner and Herbert 1981；Delille and Lagarde 1974；Tanner 1985；Voytek and Ward 1995；Chessa et al. 2000；Kelly et al. 1978；Cavanagh et al. 1996；Denner et al. 2001；Cavicchioli and Thomas 2000）。Ekelof（1908a，b）首次提出细菌、酵母和真菌存在于南极洲的土壤与空气中，Pirie（1904，1912）和 Tsiklinsky（1908）的研究也证实了这一点。50 年后，这些独特的微生物在属和种的水平被鉴定了出来（Boyd 1962；Boyd and Boyd 1962a，b；Friedman 1980；Margini and Castrelos 1963；Marshall and Ohye 1966；Meyer et al. 1962；Pfiser

[1] CSIR-Centre for Cellular and Molecular Biology，Hyderabad，India.

[2] CSIR-Centre for Cellular and Molecular Biology，Hyderabad，India；Jhaveri Microbiology Centre，L V Prasad Eye Institute，Hyderabad，India. e-mail：shivas@ccmb. res. in；shivas@lvpei. org.

and Burkholder 1965；Tsyganov 1970）。随着非培养微生物 16S rRNA 鉴定技术的出现，更多的新型微生物在南极不同生存环境中被发现。NCBI 数据库已经收录了近 6000 条南极细菌的 16S rRNA 基因序列。其中，只有 500 个序列所对应的南极细菌是可培养的（Shivaji and Reddy 2009；Chattopadhyay et al. 2014）。

2.1.1.1 非培养方法研究南极细菌多样性

为了揭示南极地区不同生境中细菌种类的多样性，16S rRNA 基因检测的方法被用于不同环境的细菌多样性分析，包括土壤、蓝藻垫、水、冰水流、沉积物、砂石缝隙、冰河、地热喷口、鸟粪土壤、企鹅、海绵和南极鱼类肠道菌群等（Shivaji and Reddy 2009；Chattopadhyay et al. 2014；Bottos et al. 2014）。南极土壤细菌多样性与土壤类型密切相关（Bottos et al. 2014；Shivaji et al. 2004）。例如，变形菌门（Proteobacteria，47%）是南极半岛土壤中的优势菌，而放线菌门（Actinobacteria，42%）和拟杆菌门（Bacteroidetes，31%）则是南极西部土壤及南极山脉土壤的优势物种。在南极山脉土壤样品中，异常球菌-栖热菌门（Deinococcus-Thermus）细菌属于第三丰富的细菌群落（18%）。Shivaji 等（2011a）观察到南极土壤细菌分层与细菌氧需求量相关，同时还发现存在少量嗜热丝菌门（Caldiserica）细菌。此外，南极土壤中大约 20% 的细菌属于未知物种（Shivaji et al. 2004；Aislabie et al. 2006，2009，2013；Yergeau et al. 2007a，b；Niederberger et al. 2008；Lee et al. 2012；Tiao et al. 2012；Bajerski and Wagner 2013）。这些研究结果表明南极土壤细菌群落随着地理环境、气候、土壤理化性质与当地生物的不同而存在明显差别。

南极淡水、海水及冰川沉积物细菌多样性研究表明，α-变形菌纲（Alphaproteobacteria）、β-变形菌纲（Betaproteobacteria）、γ-变形菌纲（Gammaproteobacteria）、拟杆菌门（Bacteroidetes）、放线菌门（Actinobacteria）、酸杆菌门（Acidobacteria）、芽单胞菌门（Gemmatimonadetes）、厚壁菌门（Firmicutes）和蓝菌门（Cyanobacteria）在土壤和沉积物样品中都存在（Bratina island，Sjöling and Cowan 2003；Ardley island，Li et al. 2006a，b，Vestfold Hills，Bowman et al. 2000；Ross ice shelf，Carr et al. 2013；Wright Glacier，Stibal et al. 2012）。螺旋体科（Spirochaetaceae）、原绿球藻（Prochlorococcus）、δ-变形菌纲（Deltaproteobacteria）[尤其是独特的菌属如脱硫叠球菌属（*Desulfosarcina*）、互营菌属（*Syntrophus*）、地杆菌属（*Geobacter*）、居泥杆菌属（*Pelobacter*）、脱硫单胞菌属（*Desulphuromonas*）]、衣原体目 [副衣原体科（Parachlamydiaceae）]、浮霉菌门（Planctomycete）和 Spirochaetales 也有出现（Bowman et al. 2000；Sjöling and Cowan 2003；Li et al. 2006a，b）。Carr 等（2013）发现 β-变形菌纲中主要（>45%）为硫杆菌属和微热杆菌属（*Teptidiphilus*），以及绿弯菌门（Chloroflexi）出现的比例异常高。除了 γ-变形菌纲在水平面以下 120～125cm 处含量特别高外，上述所有细菌都是均匀分布的。Bowman 等（2000）在缺氧的海洋沉积物中检测到超过 200 种不同的细菌，其中近 31%的物种属于低 GC 含量的革兰氏阳性细菌的新分支。

南极洲只有 0.4%的区域没有或季节性被冰雪覆盖，这限制了南极洲水体微生物的多样性（Wilkins et al. 2013）。水环境包括淡水和高盐水体，永久性冰雪覆盖区和终年无冰区，混合以及分层的湖泊。南极湖泊的水包含了从 150 万年前的泰勒冰川血色瀑布（Blood Falls）冰下流出的水（Mikucki et al. 2009）到有 300 年历史的温德米尔湖（Lake Miers）流出的水（Green et al. 1988），因此这些湖泊的微生物可能是新近产生的也可能起源于远古（Gibson 2006；Cavicchioli 2007）。水环境中微生物种类多样性大体上少于土壤和沉积物样品（Laybourn-Parry，1997），主要的细菌种类为变形菌门（占 47.6%，包括 α-变形菌纲，β-变形菌纲和 δ-变形菌纲），优势种属是红杆菌属（*Rhodobacter*）和鞘氨醇单胞菌属

(*Sphingomonas*) (Huang et al. 2013)。此外，拟杆菌门 (15.1%)、放线菌门 (14.8%)、绿弯菌纲 (10.2%)、酸杆菌门 (6.7%)、厚壁菌门 (3.6%)、芽单胞菌门 (1.3%) 和疣微菌门 (Verrucomicrobia) (0.2%) 为含量较少的种属类型，同时，异常球菌-栖热菌门、硝化螺旋菌纲 (Nitrospira)、浮霉菌门和梭杆菌门 (Fusobacteria) 也有出现。其中，最为普遍的种属是鞘氨醇单胞菌属、柄细菌属 (*Caulobacter*)、短波单胞菌属 (*Brevundimonas*)、詹森菌属 (*Janthinobacterium*)、*Duganella*、极地单胞菌属 (*Polaromonas*)、贪食菌属 (*Variovorax*)、红育菌属 (*Rhodoferax*)、黄杆菌属 (*Flavobacterium*)、土地杆菌属 (*Pedobacter*)、普雷沃菌属 (*Prevotella*)、薄层菌属 (*Hymenobacter*) 以及噬纤维菌属 (*Arcicella*) (Wilkins et al. 2013)。然而，在麦克默多冰架、维多利亚岛和阿蒙森海冰穴中观察到细菌多样性随冰层深度变化而变化，表层优势细菌主要为拟杆菌门、β-变形菌纲和放线菌门；深层区域优势细菌为γ-变形菌纲以及一些未分类的细菌 (10%～20%) (Archer et al. 2014；Kim et al. 2014)。极地杆菌属 (*Polaribacter*) (20%～64%) 和海洋螺菌科 (Oceanospirillaceae) (7%～34%) 在表层水中含量丰富，而远洋杆菌属 (*Pelagibacter*) (7%～42%) 的丰度随着深度的增加而增加。(Kim et al. 2014)。

烃类污染的海水中微生物群落分析发现嗜冷杆菌属 (*Psychrobacter*)、弓形菌属 (*Arcobacter*)、*Formosa*、极地杆菌属 (*Polaribacter*)、棒形杆菌 (*Ulvibacter*) 和黏着杆菌属 (*Tenacibaculum*)、亚硫酸杆菌属 (*Sulfitobacter*) 细菌在污染海水中丰度较无污染海水减少 (Prabagaran et al. 2007)。南极洲海洋和淡水系统之间多样性差异明显。淡水生态系统主要由黄杆菌属 (*Flavobacterium*)、假单胞菌属 (*Pseudomonas*) 和极地单胞菌属 (*Polaromonas*) 构成 (>56%) (Michaud et al. 2012)，而亚硫酸杆菌属 (*Sulfitobacter*)、海旋菌属 (*Thalassospira*)、玫瑰菌属 (*Roseobacter*)、哥利蒂杆菌属 (*Gelidibacter*)、极地杆菌属 (*Polaribacter*)、冷弯菌属 (*Psychroflexus*) 和假交替单胞菌 (*Pseudoalteromonas*) 普遍存在于海水中 (Prabagaran et al. 2007；Guibert et al. 2012；Lo Giudice et al. 2012)。

海冰包含了大量卤水孔洞及通道 (Junge et al. 2004；Søgaard et al. 2010)，其中除了古菌 (archaea) 和一些真核细胞，还包含了丰富的光合细菌、化能细菌和异养菌 (Brown and Bowman 2001；Brinkmeyer et al. 2003；Maas et al. 2012)。γ-变形菌纲、α-变形菌纲和拟杆菌门是主要的细菌组成，放线菌纲 (Actinobacteria) 则基本没有。γ-变形菌纲中 *Colwellia* 和 *Glaciecola* 属最为丰富，海杆菌属 (*Marinobacter*) 亚种则较为少见。α-变形菌纲中玫瑰菌属细菌占主导地位，拟杆菌门中则包括极地杆菌属细菌 (Bowman et al. 1997；Gosink et al. 1998；Staley and Gosink 1999；Brown and Bowman 2001；Junge et al. 2002；Brinkmeyer et al. 2003；Kuhn et al. 2014；Lanoil et al. 2009)。其他海冰中包含的细菌种属包括希瓦氏菌属 (*Shewanella*)、动性球菌属 (*Planococcus*)、交替单胞菌属 (*Alteromonas*)、假交替单胞菌属、嗜冷杆菌属、盐单胞菌属 (*Halomonas*)、假单胞菌属、生丝单胞菌属 (*Hyphomonas*)、鞘氨醇单胞菌属、节细菌属 (*Arthrobacter*) 和盐芽孢杆菌属 (*Halobacillus*) (Bowman et al. 1997)。种群随时间变化分析发现，2 年时间马阔里类芽孢杆菌属 (*Paenisporosarcina*) 细菌降至$\frac{1}{5}$，而芽孢杆菌种群数量增加 4 倍。此外，在两年内不动细菌属 (*Acinetobacter*) 和柯恩氏菌 (*Cohnella*) (厚壁菌门) 替代了类芽孢杆菌属 (*Paenibacillus*) 和 *Jeotgalibacillus* (Doyle et al. 2013)。图 2.1 总结了南极洲细菌多样性的种群分布图。

2.1.1.2 利用功能基因研究南极洲细菌多样性

相同生理功能的细菌功能基因可用于细菌群落多样性研究，如硫酸盐还原菌 (Karr et

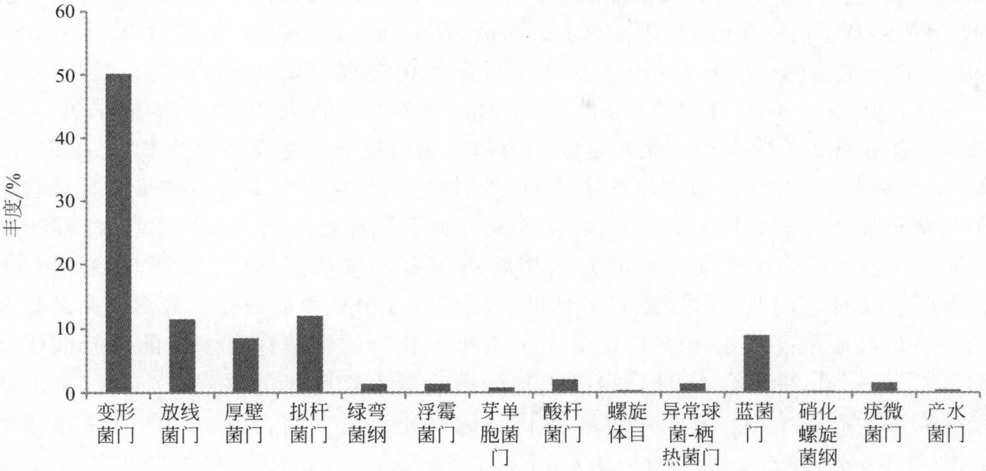

图 2.1 基于非培养方法研究南极洲不同样品细菌多样性及相对丰度。数据由 Shivaji 和 Reddy 统计（2009）

al. 2005）、光合细菌（Karr et al. 2003；Stibal et al. 2012；Kong et al. 2014）、石油烃降解菌（Muangchinda et al. 2014；Marcos et al. 2009）、几丁质酶产生菌、酮基合成菌（Xiao et al. 2005）以及氨氧化细菌（Magalhães et al. 2014）的分类。这些研究通常试图利用引物集来确定所有样品中所有细菌的特定基因。但是，在没有一个共有序列或已知的保守序列的前提下，这种方法可能有失偏颇。这种研究的细菌多样性将由引物的覆盖范围决定。

南极西福尔丘陵超咸水湖含有高浓度的二甲基硫（Yau et al. 2013）。因此，可以利用 *dsrA*（异化亚硫酸盐还原酶）基因对南极弗里克塞尔湖硫酸盐还原菌的种群多样性进行分析（Karr et al. 2005）。研究人员还发现，在冰下湖泊沉积物中超过 70%的硫转化是由铁氧化属细菌（*Sideroxydans*）和产硫酸杆菌属（*Thiobacillus*）完成的（Purcell et al. 2014）。Watanabe 等人（2013）根据 *aprA*（腺苷酰硫酸还原酶 α 亚基）基因在南极淡水湖中鉴定了荚硫菌属（*Thiocapsa*）、硫胞菌属（*Sulfuricella*）、脱硫杆菌属（*Desulfobacterium*）、脱硫豆菌属（*Desulfofaba*）和脱硫肠状菌属（*Desulfotomaculum*）。

南极洲的科考行动导致了以石油为主的碳氢化合物污染问题（Aislabie et al. 2004）。因此，在南极地区分离出可降解多环芳香烃（PAH）的细菌被认为与石油污染相关。PAH 降解基因在革兰氏阳性和阴性菌中均有发现，如变形菌门、放线菌门、疣微菌门、拟杆菌门、厚壁菌门、绿弯菌纲、芽单胞菌门、蓝菌门、Chlorobium 和酸杆菌门。鞘氨醇单胞菌属是自然环境中一种主要的 PAH 降解细菌，在南极部分地区检测到该菌的存在。因此，这些原生的细菌可能用于降解 PAH，从而促进南极洲的生态修复（Muangchinda et al. 2014）。利用 PAH 降解基因对南极细菌鉴定发现，地杆菌属（*Terrabacter*）、分枝杆菌属（*Mycobacterium*）、*Diaphorobacter*、鞘氨醇单胞菌属、伯克氏菌属（*Burkholderia*）、红球菌属（*Rhodococcus*）、诺卡氏菌属（*Nocardia*）和芽孢杆菌属是南极沉积物样品中的主要菌群（Muangchinda et al. 2014；Marcos et al. 2009），而在这些沉积物样品中，假单胞菌起着重要的作用（Ma et al. 2006）。南极半岛金钟湾和乔治王岛样品基于烷烃单加氧酶基因（*alkB*、*alkM*）的细菌多样性分析发现 *Silicibacter pomeroyi*，*Gordonia* sp.，*Prauserella rugosa*，*Nocardioides* sp.，*Rhodococcus* sp.，*Nocardia* sp.，*Pseudomonas putida*，*Acidisphaera* sp. 和 *Alcanivorax borkumensis* 存在 *alkB* 同源基因，*alkM* 基因则存在于 *Acinetobacter* sp. 中（Kuhn et al. 2009）。南极企鹅排泄物中含有大量的几丁质，这些排泄物由土壤微生物进行降解。基于几丁质酶基因（*chi67*，*chi69*，*chiA*，*chiB*，*chiF*）多样性分析显示，革兰氏阴性菌詹森菌属（*Janthinobacterium*）、寡养单胞菌属（*Stenotrophomonas*）、噬细胞菌属（*Cytophaga*）和革兰氏阳性菌链霉菌属（*Streptomyces*）、诺卡土壤菌属（*Norcardiopsis*）在南极洲阿德利岛1600年历史的沉积物中含量丰富（Xiao et al. 2005）。众所周知，微生物，尤其是放线菌纲细菌，是聚酮化合物、生物碱、多肽等天然化合物的资源库，这对于南极洲原核生物而言同样适用。由于细菌群落在聚酮化合物的合成过程中起着重要的作用，因此，科学家尝试探讨基于酮基合成酶（KS）基因的细菌多样性分析。基于KS基因的系统发育研究发现多种细菌群，包括变形菌门、厚壁菌门、浮霉菌门、蓝菌门、酸杆菌门和一些不可培养的共生细菌及5个独立的分支。分离出来的细菌大多数在AA级最高匹配度低于80%，这显示了南极沉积物的细菌存在多样的和新的酮基合成酶基因（Zhao et al. 2008）。此外，尽管大部分湖泊覆盖着厚厚的冰层，光合细菌群落也存在于南极湖泊生态系统中。基于*pufM*基因（编码光合色素结合蛋白）的细菌多样性研究发现，33种*Rubrivivax*、*Acidiphilum*、*Rhodoferax*、*Roseateles*相关的独特表型细菌存在于南极沉积物中（Karr et al. 2003；Stibal et al. 2012）。然而对邦尼湖中*psbA*基因多样性分析发现了表型独特的拟球藻属（*Nannochloropsis*）、棕鞭藻属（*Ochromonas*）、等鞭金藻属（*Isochrysis*）细菌（Kong et al. 2014）。

2.1.1.3 可培养方法研究南极细菌多样性

南极地区冰层的细菌丰度为$0.2\times10^2\sim0.6\times10^{12}$个/g（Carpenter et al. 2000；Priscu et al. 1999；Karl et al. 1999；Delille and Gleizon 2003），水的细菌丰度为$0.2\times10^2\sim10^7$个/mL（Takii et al. 1986；Lo Giudice et al. 2012），沉积物细菌丰度为$8\times10^6\sim2.4\times10^7$个/g（Lanoil et al. 2009；Stibal et al. 2012），土壤细菌丰度为$10^5\sim10^7$个/g（Aislabie et al. 2009）。目前，已经深入研究了南极土壤（Miwa 1975；Yi and Chun 2006；Ruckert 1985；Shivaji et al. 1988，1989a，b；Wery et al. 2003；Bozal et al. 2007）、水体（Lo Giudice et al. 2012；Michaud et al. 2012；Söller et al. 2000；Labrenz et al. 2000；Cristóbal et al. 2011）、蓝藻垫（Reddy et al. 2000，2002a，b，2003a，b，c，d，2004；Van Trappen et al. 2002；Spring et al. 2003；Peeters et al. 2011）、冰（Shivaji et al. 2004，2013c；Antibus et al. 2012；Bowman et al. 1998）以及沉积物（Shivaji et al. 2011a；Yu et al. 2011）可培养细菌的多样性。Shivaji（2009）和Chattopadhyay（2014）等对相关研究进行了综述。在这些栖息地中，水体、土壤以及蓝藻垫的细菌多样性比冰和沉积物的更为丰富（图2.1）。在所有的栖息地中变形菌门、拟杆菌门、放线菌门以及厚壁菌门均广泛存在，而异常球菌-栖热菌门（Deinococcus-thermus）和螺旋体门只在土壤中发现（Hirsch et al. 2004；Antibus et al. 2012；Franzmann and Dobson 1992；Chattopadhyay et al. 2014）。在变形菌纲（Proteobacteria）中，相比于α-变形菌纲、β-变形菌纲和δ-变形菌纲，γ-变形菌纲含量最为丰富，而Epsilonproteobacteria的细菌没有发现。α-、β-、γ-变形菌纲，放线菌门以及Flavobacteria在安第斯（Andean）冰川有出现，Andean冰川因人类难以到达，其人为干扰程度较南北极冰小（Ball et al. 2014）。

对南极可培养细菌相对含量分析发现，土壤中优势细菌及其所占比例分别为厚壁菌门（35%）、放线菌门（25%）和变形菌门（18%）；水体中为变形菌门（42%）、拟杆菌门（31%）和放线菌门（23%）；藻丛中变形菌门（40%）、厚壁菌门（28%）和拟杆菌门（24%）；南极冰中变形菌门（48%）、拟杆菌门（21%）和厚壁菌门（19%）；沉积物中变形菌门（50%）和放线菌门（29%）（图2.2）。以上数据表明变形菌门在南极栖息地中无处不在。从南极分离出来的可培养细菌包括50个新属和170个新种。只有一个新科水平细菌在南极海水中被发现，该科为γ-变形菌科 *Granulosicoccaceae*，且只包含一个种 *Granulosicoccus antarcticus*（Lee et al. 2007）。此外，从墨尔本山和瑞特曼山地热土壤中分离的嗜热细菌吸引着更多科学家的注意，如脂环酸芽孢杆菌（*Alicyclobacillus pohliae*）、解硫胺素芽孢杆菌（*Aneurinibacillus terranovensis*）、厌氧芽孢杆菌（*Anoxybacillus amylolyticus*）、嗜热芽孢杆菌（*Bacillus thermantarcticus*）和短芽孢杆菌（*Brevibacillus levickii*）（Lama et al. 1996；Allan et al. 2005；Poli et al. 2006；Imperio et al. 2008）。尽管如此，也有许多门类细菌在其他寒冷环境中被发现。

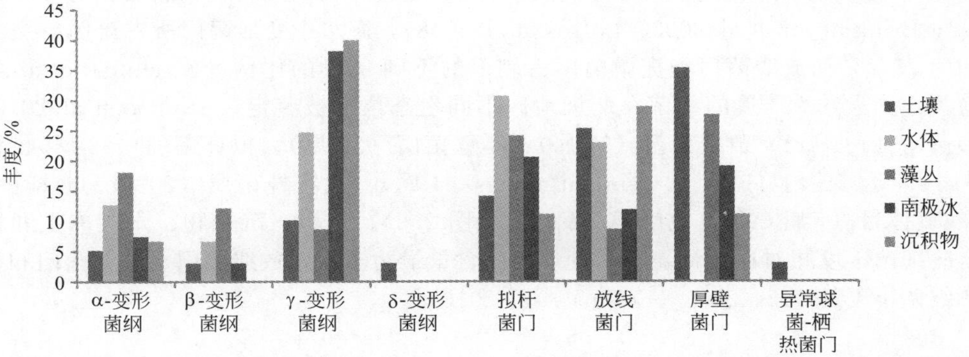

图2.2 南极可培养细菌在土壤、水体、藻丛、南极冰和沉积物样品中的细菌多样性及相对丰度。占比小于3.0%的菌群未列出，本图数据来源于本节参考文献

2.1.2 北极原核生物多样性

北极是指北纬66.5°到北极点之间的地理区域，包括了北冰洋以及加拿大、芬兰、格陵兰岛、冰岛、挪威、俄罗斯、瑞典和美国（阿拉斯加）等地部分陆地区域。北极地区终年严寒，太阳辐射强度弱，常年冻土，缺乏丰富的植被。与南极洲相比，北极微生物的研究起步较晚，北极微生物最早由Kriss（1945）、Pady等（1948）、Polunin和Kelly（1952）、Kelly和Layne（1957）、Boyd和Boyd（1962a，b）等人报道。Zhou等（1997）首次对北极冻土细菌进行非培养方法的研究。到目前为止，NCBI数据库收录了约3429个北极地区分离得到的细菌16S rRNA基因。其中，只有79个序列是新型物种。

2.1.2.1 使用非培养方法研究北极细菌多样性

通过培养、非培养手段研究北极不同生态环境中细菌的丰度及多样性（Chattopadhyay et al. 2014），包括土壤（van Dorst et al. 2014；Chu et al. 2010；Edwards et al. 2011；Zhou et al. 1997；Neufeld and Mohn 2005；Campbell et al. 2010；Schütte et al. 2010），冰（Junge et al. 2002；Brinkmeyer et al. 2003；Yu et al. 2006；Bottos et al. 2008；Collins et al. 2010），永久冻土（Wilhelm et al. 2011；Steven et al. 2007；Hansen et al. 2007）和沉积物（Zhang et al. 2014；Forschner et al. 2009；Li et al. 2006a，b，2009；Lysnes et al. 2004；Perreault et al. 2007；Bienhold et al. 2012）。这些研究表明，北极土壤微生物种群与南极洲的非常相似，其中酸杆菌门、α-变形菌纲、放线菌门、β-变形菌纲和拟杆菌门占种群多样性的83%以上（Chu et al. 2010；Edwards et al. 2011；Zhou et al. 1997；Neufeld and Mohn 2005；Campbell et al. 2010；Schütte et al. 2010）。此外，γ-变形菌纲、疣微菌门、芽单胞菌门和δ-变形菌纲也有报道。土壤和永久冻土的细菌多样性与预期相同，均比较接近。例如，永久冻

土中，变形菌门、放线菌门和酸杆菌门（Steven et al. 2007；Wilhelm et al. 2011）为优势菌群，而拟杆菌门、Gemmatimonas、疣微菌门和浮霉菌门出现较少。研究还表明高纬度极地地表的细菌群落对北极永久冻土上间歇的水流有不一致的响应（Steven et al. 2013）。同时研究人员观察到低 pH 条件下酸杆菌门减少以支持变形菌门的生长，而高 pH 条件则有利于拟杆菌门的生长（Ganzert et al. 2014）。在沉积物中变形菌门的种群含量可占细菌多样性的 50.0%～85.0%（Li et al. 2009；Bienhold et al. 2012；Lysnes et al. 2004；Steven et al. 2008a）。Zhang 等（2014）通过培养或非培养的方法鉴定北极海洋沉积物样品潜在的新物种。

海冰中包含有变形菌门、拟杆菌门、浮霉菌门和绿弯菌门的细菌谱系（Junge et al. 2002；Brinkmeyer et al. 2003；Bottos et al. 2008），海水中变形菌门所占比例 50%，拟杆菌门约为 25%，而放线菌门和疣微菌门占剩下的不到 25%的比例（Kirchman et al. 2010）。到目前为止，大约 284 属的细菌在北极地区不同生态位置被鉴定（Schütte et al. 2010），包括变形菌门（131 属）、放线菌门（57 属）、厚壁菌门（22 属）、拟杆菌门（22 属）、绿弯菌纲（5 属）、浮霉菌门（5 属）、Gemmatimonas（4 属）、丝状杆菌属（2 属）、酸杆菌门（2 属）、螺旋体目（1 属）和其他门类（34 属）（图 2.3）。土壤、沉积物、永久冻土和极地冰中，这些细菌丰度相对比统计显示，变形菌门含量最为丰富，放线菌门、厚壁菌门和拟杆菌门占总细菌丰度的 3%～25%（Collins et al. 2010）。

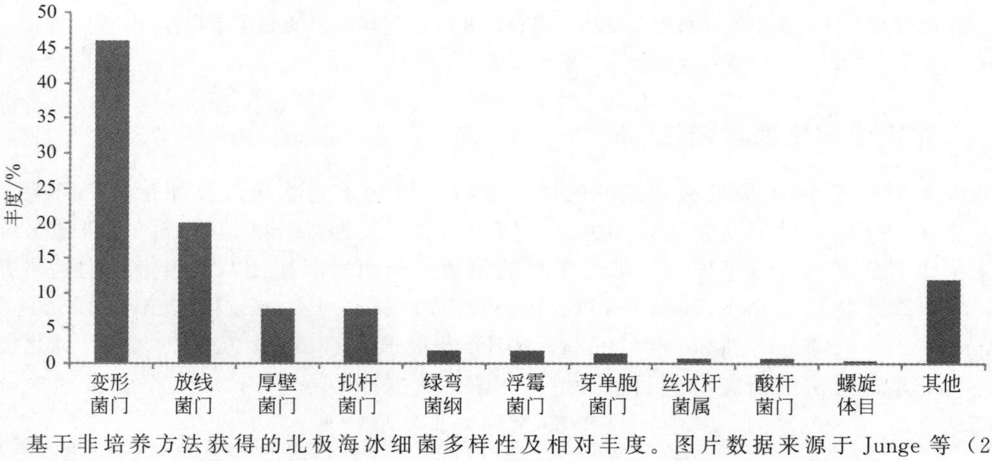

图 2.3 基于非培养方法获得的北极海冰细菌多样性及相对丰度。图片数据来源于 Junge 等（2002），Brinkmeyer 等（2003）和 Bottos 等（2008）

2.1.2.2 利用功能基因研究南极洲细菌多样性

科学家们研究了多个北极地区细菌生理类群多样性，如固氮菌群（Deslippe and Egger 2006）、氨氧化菌群和蓝藻固氮菌群（Díez et al. 2012）的细菌多样性。科学家利用 *nifH* 基因研究固氮菌群相关的细菌群落多样性。约 50%～65%比例的固氮菌群在纲水平来源于 α-变形菌纲、γ-变形菌纲、δ-变形菌纲和厚壁菌门细菌，其中 α-变形菌纲主要含红假单胞菌属（*Rhodopseudomonas*）、固氮螺菌属（*Azospirillum*）、固氮根瘤菌属（*Azorhizobium*）、根瘤菌属（*Rhizobium*）、短根瘤菌属（*Bradyrhizobium*）细菌，β-变形菌纲主要包括草螺菌属（*Herbaspirillum*）和稻属（*Oryza*）细菌，γ-变形菌纲包括 *Pseudomonas*、*Azoarcus*，厚壁菌门（Firmicutes）包括类芽孢杆菌（*Paenibacillus*）。同样利用 *nifH* 基因研究蓝细菌群相关的群落多样性发现，在卤水及海水环境中分布的优势蓝细菌都属于蓝球藻目（Chroococcales)、念珠藻目（Nostocales)、颤藻科（Oscillatoriales）和席藻属（*Phormidium*）（Díez et al. 2012）。进一步利用 *nitAB* 基因研究发现，亚硝化螺菌属（*Nitrosospira*）和亚硝化单胞菌（*Nitrosomonas*）是北极海水中的氨氧化菌群（Bano and Hollibaugh 2000）。此外，气候变化导致的永久冻土融化可导致有机碳通过微生物活动向二氧化碳和甲烷转化。对北极地区制造温室气体的产烷菌群的探索十分令人关注。有研究发现北极地区产烷生物包含 *Methylobacter*（Ⅰ型甲烷氧化菌）、甲基弯曲菌属（*Methylosinus*）和 *Methylocystis*（Ⅱ型甲烷氧化菌）（Wartiainen et al. 2003；Liebner et al. 2009），且种群结构随环境 pH 变化而变化（Martineau et al. 2014）。北极淡水、海水、冰架和沿岸水域均可接收到太阳辐射，因此存在着丰富的光合细菌。Boeuf 等（2013）利用 *pufM* 基因探索北冰洋西边好氧不产氧光合（AAP）菌群发现 α-变形菌纲和 β-变形菌纲各占 49%和 48%。

2.1.2.3 利用培养方法研究北极细菌多样性

北极主要栖息地的细菌丰度与南极地区相近，沉积物细菌丰度为 1.2×10^7～2.3×10^7 个/g（Forschner et al. 2009），永久冻土为 5.2×10^4～1.7×10^9 个/g（Hansen et al. 2007），冰为 0.9×10^3～2.4×10^6 个/mL（Møller et al. 2011；Brinkmeyer et al. 2003）以及水为 5×10^5 个/mL（Møller et al. 2011）。北极多种栖息地的可培养细菌多样性被研究，包括极地冰（Zhang et al. 2008；Auman et al. 2006，2010；Gosink et al. 1998；Brinkmeyer et al. 2003；Groudieva et al. 2004；Bottos et al. 2008；Steven et al. 2008b），沉积物（Prasad et al. 2014；Srinivas et al. 2009；Reddy et al. 2009b；Kim et al. 2009，2012；Knittel et al. 2005；Knoblauch et al. 1999；Vandieken et al. 2006；Shivaji et al. 2012a；Begum et al. 2013），永久冻土（Hansen et al. 2007；Steven et al. 2007，2008a），藻丛（Dong et al. 2012），冻土（Jiang et al. 2012；Jiang et al. 2013；Männistö et al. 2012），水体（Prasad et al. 2014；Yu et al. 2011；Van Trappen et al. 2004；Zhang et al. 2006；Kim et al. 2008；Jang et al. 2011；Al Khudary et al. 2008）以及冰尘穴（Singh et al. 2014）。135 个属近 675 种细菌被发现且永久冻土细菌多样性比藻丛和水中更加丰富（图 2.4）。细菌相对丰度比显示 β-变形菌纲、拟杆菌门和放线菌门为冰和沉积物的主要菌群，而拟杆菌门、厚壁菌门和放线菌门则主要存在于永久冻土和土壤中（图 2.2）。共 675 株细菌从北极样品中被成功分离，其中有 12 个新属（Ren et al. 2015；Storesund and Øvreås 2013；Qu et al. 2014；Prasad et al. 2013；Jiang et al. 2013；Steinsbu et al. 2011；Tamura et al. 2010；Steven et al. 2008b；Lee et al. 2007；Knoblauch et al. 1999；Gosink et al. 1998）以及 70 个新种。这其中，*Parablastomonas*（Ren et al. 2015），*Psychroglaciecola*（Qu et al. 2014），*Arcticibacter*（Prasad et al. 2013），*Huanghella*（Jiang et al. 2013），*Dasania*（Lee et al. 2007）和 *Augustibacter* 属（Tamura et al. 2010）到目前为止只在北极地区发现。北极沉积物中的 δ-变形菌纲中（Knoblauch et al. 1999），发现了 *Desulfofrigus*、*Desulfofaba* 和 *Desulfotalea* 3 个新属共 9 个新种。此外在北极，疣微菌属（*Verrucomicrobium*）中发现 1 个新种（*Luteolibacter luojiensis*）（Jiang et al. 2012）；酸杆菌门中发现 4 个新种（*Granulicella arctica*，*Granulicella mallensis*，*Granulicella tundricola*，*Granulicella sapmiensis*）（Männistö et al. 2012）。除了在热液喷口发现的新的嗜热细菌 *Rhabdothermus arcticus*（37～65℃），北极发现的绝大多数细菌都属于嗜冷细菌，（Steinsbu et al. 2011）。南极洲和北极之间的比较表明，细菌多样性在纲水平上十分相似，在属水平上则相似性较低。

2.1.3 喜马拉雅山脉原核生物多样性

地球上 70%的淡水以冰川形式存在，喜马拉雅山脉是极地地区以外最大的冰川，约

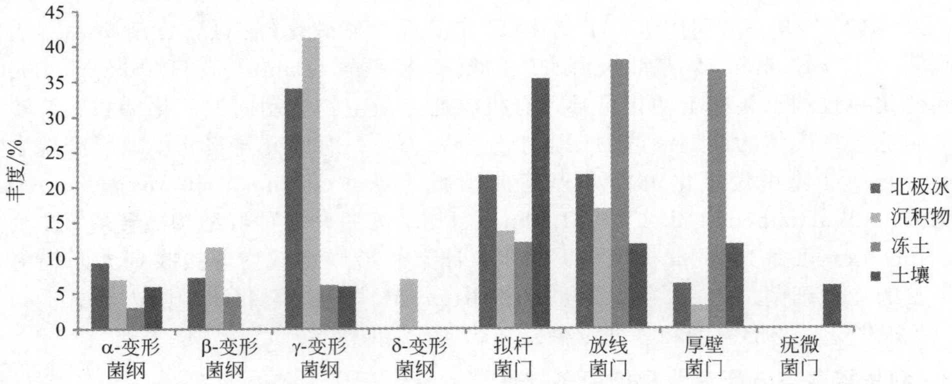

图 2.4 北极冰、沉积物、冻土及土壤样品可培养细菌多样性及相对丰度。去除了丰度低于 3.0%的细菌群落。数据汇集自本节所有文献

$33000km^2$。喜马拉雅山脉被誉为"亚洲水塔"，年平均供水量巨大（Dyurgerov and Meier 1997）。此外，喜马拉雅山地区不同于极地气候的一大特点在于该地区气候随季节及海拔的变化而变化。尽管对有关寒冷地区的研究明显增加，然而喜马拉雅山的细菌多样性仍然在很大程度上未被开发。

2.1.3.1 使用非培养方法研究喜马拉雅山脉细菌多样性

基于培养及非培养手段研究了喜马拉雅三个主要冰川，分别是平德里（Pindari）（Shivaji et al. 2011b）、卡夫尼（Kafni）（Srinivas et al. 2011）和罗普昆德（Roopkund）（Pradhan et al. 2010）的土壤细菌多样性。喜马拉雅山脉西部地区土壤、雪和水体的细菌多样性最近也相继被报道（Gangwar et al. 2009；Liu et al. 2006，2009，2011）。水体细菌丰度较低，约 1.1×10^4～25×10^4 个/mL（Liu et al. 2011）；沉积物样品中细菌丰度比较高，为 0.9×10^7～30.7×10^8 个/g（Shivaji et al. 2011a，b；Pradhan et al. 2010；Srinivas et al. 2011）。水体香农（Shannon）多样性指数为 2.7～3.4，而沉积物为 0.9～4.89（Srinivas et al. 2011）。三个冰川的土壤、水和沉积物样品中具有的共同细菌类别为酸杆菌门、放线菌门、拟杆菌门、绿弯菌纲、Chlamydiae、厚壁菌门、硝化螺旋菌纲、变形菌门和疣微菌门。然而，冰川沉积物的细菌多样性略有不同，Kafni 冰川主要包含螺旋体科和 Tenericutes，而 Pindari 和 Roopkund 冰川主要为芽单胞菌门和浮霉菌门（Srinivas et al. 2011；Shivaji et al. 2011a，b）。Gangwar 等（2009）总结发现土壤中含量最高的为变形菌门细菌（45%～70%），其次是放线菌门（10%～86%）和拟杆菌门（6.4%～28%）（Liu et al. 2006；Gangwar et al. 2009；Srinivas et al. 2011）。三大栖息地种群组成的不同可归因于土壤自然环境的差异。到目前为止，在喜马拉雅山脉不同地区共鉴定出 135 个不同属细菌，群落组成如图 2.5 所示，包括放线菌门（28.0%）、β-变形菌纲（21.5%）、α-变形菌纲（9.6%）、γ-变形菌纲（8.9%）、δ-变形菌纲（11.1%）、拟杆菌门（11.1%）、厚壁菌门（3.7%）、酸杆菌门（1.4%）、疣微菌门（3.0%）和 Epsilonproteobacteria（1.5%）。

2.1.3.2 利用功能基因研究喜马拉雅山脉的细菌多样性

喜马拉雅地区细菌功能基因多样性研究较之南北极而言严重不足。喜马拉雅地区原核生物功能种群的研究只集中在固氮或温室气体产生相关的基因。根据 *nifH* 基因对喜马拉雅地区根际土壤研究发现，固氮种群细胞主要包括 *Rhizobium*（53%）、*Dechloromonas*（7%）、*Bradyrhizobium*（7%）和 29%的不可培养细菌（Suyal et al. 2014）。利用 *mcrA* 基因和 16S

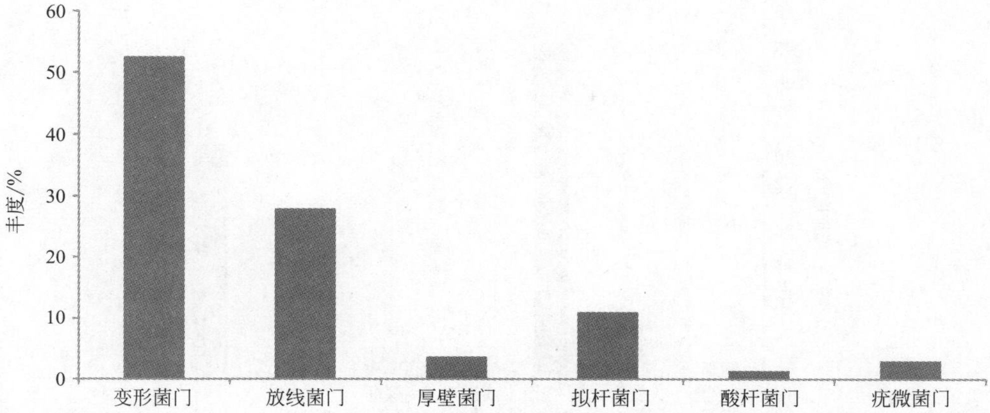

图 2.5　基于非培养方法研究喜马拉雅山脉的卡夫尼、平德里和罗普昆德冰川细菌多样性及相对丰度。数据汇集自本节所有文献

rRNA 基因研究发现，产甲烷微生物群落主要包括 *Methanosarcina*、*Methanocella* 和 *Methanobacterium* 属细菌（Aschenbach et al. 2013）。利用 *pmoA* 基因发现超过 95％的甲烷氧化细菌属于 *Methylohalobius*、*Methylothermus*、*Methylococcus*、*Methylocaldum* 和 *Methylosoma* 属（Zheng et al. 2012）。

2.1.3.3　基于培养方法研究喜马拉雅山脉的细菌多样性

约 466 株细菌从罗普昆德（Roopkund）（Shivaji et al. 2008；Reddy et al. 2009a；Pradhan et al. 2010），平德里（Pindari）（Shivaji et al. 2011a，b；Reddy et al. 2008，2013a，b；Kishore et al. 2009；Pindi et al. 2009），卡夫尼（Kafni）（Kishore et al. 2009；Pindi et al. 2009；Srinivas et al. 2011），哈姆塔（Hamta）冰川（Shivaji et al. 2005；Chaturvedi et al. 2005；Chaturvedi and Shivaji 2007）以及喜马拉雅山西部冰川（Gangwar et al. 2009；Sahay et al. 2013；Yadav et al. 2015）中分离得到，主要包括变形菌门、拟杆菌门、厚壁菌门和放线菌门等的细菌。NCBI 数据库统计数据显示，到目前为止只有 21 个新种在喜马拉雅山脉被发现，一个新属 *Glaciihabitans* 在喜马拉雅地区被发现（Li et al. 2013）。微生物生物勘探研究表明，从 Pindari（Shivaji et al. 2011a，b）和 Kafni（Srinivas et al. 2011）分离的大部分样品中发现淀粉酶、脲酶和脂肪酶在 4℃和 20℃均有活性。Gangwar 等（2009）也表明喜马拉雅高纬度地区的可培养细菌存在脂肪酶活性。另外，喜马拉雅地区可培养微生物的研究发现新的硫还原 *Desulfovibrio psychrotolerans*（Jyothsna et al. 2008）以及光合细菌 *Rhodobacter megalophilus*（Arunasri et al. 2008）（图 2.6）。

2.1.4　极端寒冷环境细菌的一些特性

有趣的是，从南极洲、北极和喜马拉雅山脉分离得到的细菌具有许多相同的与生存密切相关的功能特性：大部分微生物最高生长温度不超过 30℃，也能在 4℃下生长，只有少数细菌除外，例如 *Alicyclobacillus pohliae*（42～60℃）、*Aneurinibacillus terranovensis*（15～55℃）、*Anoxybacillus amylolyticus*（45～65℃）、*Bacillus thermantarcticus*（30～60℃）、*Brevibacillus levickii*（20～55℃）和 *Rhabdothermus arcticus*（37～75℃）（Lama et al. 1996；Allan et al. 2005；Poli et al. 2006；Imperio et al. 2008；Steinsbu et al. 2011）。早期研究表明 16S rRNA 基因（G+C）含量与原核生物生长的环境温度相关。其中，（G+C）主要集中在分子的主干区域，为 rRNA 结构提供稳定性和完整性，而其 rRNA 单链富含嘌呤

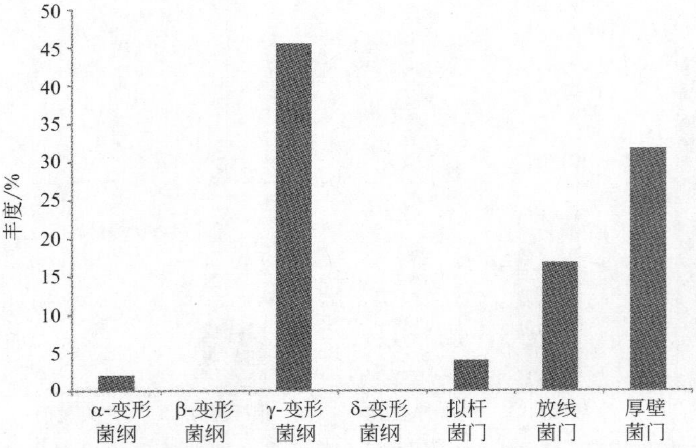

图 2.6 喜马拉雅山脉平德里、卡夫尼、哈姆塔和罗普昆德冰川土壤样品可培养细菌多样性及相对丰度。数据汇集自本节文献

(A，G)。在 20℃以下的生长条件中，8 种冷杆菌（*Cryobacterium*）中有 5 种增长了 17 个螺旋长度（结果未发表）。芽孢杆菌的中温到耐冷的转变研究中，16S rDNA 操纵子产生一系列耐冷信号，包括 C 和 G 转变为 A 和 T、高尿嘧啶含量以及 G：U 错配（Prüss et al. 1999；Lauro et al. 2007）。芽孢杆菌嗜冷种与进化距离相近的中温种属比对发现，中温芽孢杆菌孢子囊较嗜冷种发生明显肿胀（Reddy et al. 2008；Abd El-Rahman et al. 2002；Larkin and Stokes 1967；Priest et al. 1988）。一株分离自喜马拉雅山脉的嗜冷菌 *Bacillus cecembensis* 在其膜中含有 $C_{16:1}$ω7c 醇，然而其进化距离相近的中温种属中则不含（Reddy et al. 2008）。事实上，嗜冷细菌比中温细菌含有更丰富的不饱和脂肪酸。这些嗜冷菌也都拥有低温活性高、高温易变性的酶类（Ray et al. 1992；Reddy et al. 1994；Chattopadhyay et al. 1995），以及其他低温条件生存所必需的基因（Singh et al. 2009；Singh and Shivaji 2010；Sundareswaran et al. 2010）。一些其他冷适应的机制也将在后续章节中提到。

2.2 耐冷细菌的环境适应及重要的生物技术应用

从极端寒冷的栖息地获得细菌形成了具有挑战性的研究领域。在过去的几十年里大量不同类型细菌在南极、北极和喜马拉雅山脉发现，同时，这些细菌耐冷机制在生理生化及基因层面上也有一定的研究（Chattopadhyay et al. 2014）。大部分耐冷细菌为寡营养型和广温性的。本节主要论述到目前为止耐冷细菌或其酶类的主要耐冷机制及其生物技术方面的应用。

2.2.1 耐冷机制：概述

现有的文献证据提供了一些细菌耐冷的机制。北极嗜冷杆菌属细菌通过胞体缩小和分裂的生存策略适应极度低温环境（Ewert and Deming 2014）。另外，与鱼类、昆虫、植物和真菌类似，一些嗜冷细菌通过合成抗冻蛋白进而抵御南极零下低温环境。有研究表明，从南极样品无细胞提取物中发现了高水平的热滞后现象和重结晶抑制活性，这些结果为细菌抗冻蛋白的研究提供了新的思路（Kawahara et al. 2007）。适冷细菌产生一些耐冷或热不稳定蛋白质（Feller and Gerday 2003；Singh et al. 2014），进而维持细胞在低温环境下的代谢活性。细菌对外界低温环境的感知能力也是十分重要的，这个功能由细菌膜上分子传感器完成（Ray et al. 1994）。随温度变化的 DNA 拓扑结构的变化也可能是细菌响应外界环境变化的手段之一（Shivaji and Prakash 2010）。一个双组分的信号转导机制已被证明参与了在环境中的各种应力条件的传感和响应。压力一般由膜感受器组氨酸激酶（HK）探知，在压力环境下，HK 磷酸化从而传递信号给胞质中的反应调节分子，最终上调适应压力相关基因。近来，两种 HK 编码基因，*yyc G* 和 *lis K*，它们在食物致病细菌 *Listeria monocytogenes* 低温下生长的作用被证明（Pöntinen et al. 2015）。保持细菌细胞膜在低温环境下的流动性，是细菌面临极端低温环境的另一挑战。优先合成使膜流动性增强的脂肪酸（不饱和、短链和反式异构脂肪酸）加上一些类胡萝卜素有助于低温环境下细胞膜流动性的恒黏度适应（Chintalapati et al. 2004；Chattopadhyay and Jagannadham 2001）。大多数适冷细菌不会产生内生孢子，因此，细菌可能在低温条件下仍然保持代谢活性。很早之前认为，细菌在极低温环境下只能保持分解代谢活性。但最近一项研究显示，零摄氏度以下细菌仍然具有分解代谢和合成代谢的活性，该研究证据证明了耐冷细菌灵活的代谢机制（Sengupta and Chattopadhyay 2013）。在南极菌群无细胞分离样品中发现转录和翻译在低温条件下仍然进行，然而大肠杆菌无细胞系统在低温条件下的相同代谢过程均停止（Ray et al. 1998）。低温下处理广温嗜冷菌 *Psychrobacter cryohalolentis* K5（由西伯利亚冻土分离得到）后，胞内 ATP 和 ADP 浓度明显提高，表明这可能是有机体在低温条件下代谢反应速率降低的一种代偿机制（Amato and Christner 2009）。在南极细菌中发现大肠杆菌主要的冷休克蛋白基因 *csp A* 的同源基因（Ray et al. 1998）。因此，冷休克蛋白同样在低温细胞代谢过程中起作用。南极细菌低温培养发现，培养基中一些小分子化合物含量明显增加，如脯氨酸和甜菜碱（数据未发表）。因此，像一些两栖动物和昆虫一样，细菌可能通过增加胞内抗冻物质来适应恶劣低温环境。低温环境 DNA 复制时，复制叉频繁受抑制和复制叉翻转被认为细菌无法在低温环境生长的原因之一。南极湖泊分离的嗜冷细菌 *Pseudomonas syringae* Lz4W 可以合成 Rec BCD 蛋白来消除复制叉翻转以及促进 DNA 合成的重启（Sinha et al. 2013）。其他一些蛋白质（如天冬氨酸氨基转移酶和核糖核酸修饰酶）在细菌中的耐冷适应作用已经通过南极细菌冷敏感突变体加以证明（Singh et al. 2009；Sundareswaran et al. 2010）。RNA 降解是调节基因表达的关键步骤，南极细菌 RNA 降解小体的成分与大肠杆菌明显不同（Purusharth et al. 2007）。嗜冷产烷细菌 *Methanolobus psychrophilus* R15 分别在 18℃和 4℃条件下培养，比较不同培养温度的细菌基因组和转录组分析发现，外泌体介导的 RNA 降解在低温条件下加快（Chen et al. 2012）。在自然环境中，压力条件（极端温度和 pH、高盐、氧化压力、紫外辐射等）很少单独存在，通常是多种组合同时存在。这就导致了适冷细菌不得不适应多种压力条件。某些蛋白质在细菌适应低温和其他环境压力因素中都起作用，说明不同的应激反应可能是相互联系的（Chattopadhyay 2008）。南极细菌在低温情况下观察到胞内自由基增加（可造成氧化压力）（Chattopadhyay et al. 2011）。适冷菌使用 Entner Duodoroff 途径（与细胞内 NADPH 水平提高相关）可以帮助它们在低温条件下应对氧化应激的压力（Moreno and Rojo 2014）。聚羟基脂肪酸酯的代谢、储碳聚合物，也似乎能在低温环境中减轻细胞氧化应激压力（Ayub et al. 2009）。最近对 66 个来自不同耐冷和嗜冷环境的细菌（如 *Allivibrio*、*Bacillus*、*Carnobacterium*、*Desulfotalea*、*Exiguobacterium*、*Flavobacterium*、*Pseudomonas*、*Psychrobacter*、*Runella*、*Spingopyxis*）的质粒进行 meta 分析发现抗低温及紫外辐射的基因同时也都可能具有抗重金属、类金属功能和抗生素活性（Dziewit and Bartosik 2014）。对该领域的深入研究也将有利于揭示细菌细胞耐受机制。

DNA 测序技术为基因的调控及功能鉴定提供新的方法，在了解人类健康与疾病、农业以及其他许多生物参与的环境过程等中发挥着重要的作用。因此，任何低成本、高效快速的全基因组测序技术将大大丰富我们对基因及其功能的认识。二代测序技术（NGS）就是一种可高效得到不同样品多个物种全基因组序列的技术，这有助于我们获得极端环境适应的基因组信息，了解物种独特功能。最近有研究通过基因组测序技术研究了 *Pseudomonas extremaustralis* 的环境适应性及极端环境耐受性（Raiger Iustman et al. 2015）。渗透调节、冷适应、胞外多糖产生和复杂化合物的降解所需的基因在这株分离自南极的细菌均有发现，通过水平转移获得的基因也有发现。基于比较基因组分析发现，编码冷休克蛋白（Csps）和冷驯化蛋白（Caps）的基因只在嗜冷细菌 *Pseudomonas extremaustralis* 中发现。相比之下，中温细菌 *P. Syringae* pv. *syringae* B278a 和 *P. aeruginosa* PAO1 含有最低拷贝的 *Csps* 基因并且不含有 *Cap* 基因。这些结果表明，上述蛋白质在 *P. extremaustralis* 冷适应过程中发挥重要的作用。一株分离自湖泊，具有耐受低温、紫外线辐射和氧化应激能力的细菌 *Hymenobacter* 的全基因组草图被发表（Koo et al. 2014）。有研究也报告了不同的低温环境中一些嗜冷菌的基因组序列，如南极洲（Sreenivas et al. 2014；Reddy et al. 2013a，b，2014；Kumar et al. 2013a，b），北极（Shivaji et al. 2013a，b；Kumar et al. 2013b），喜马拉雅冰川（Reddy et al. 2014）和 大气平流层（Shivaji et al. 2012b）。

针对嗜冷细菌 *Sphingobacterium antarcticum*、*Oceanisphaera arctica* 以及 *Exiguobacterium indicum* 的 *Csps* 基因的比较基因组分析发现，*CspA* 在上述细菌中均有出现，*Csp C*、*Csp D*、*Csp E* 和 *Csp G* 至少在一种细菌中有存在，而 *Csp B*、*Csp I* 则不存在（结果未发表）。这种适冷相关的现象需要进一步研究。嗜冷菌比较基因组分析除了能提供对特定基因在冷适应的作用信息外，也为研究细菌面对其他生存压力的适应性反应（即高盐、高和低 pH 值）提供实验数据支持。

2.2.2 潜在的生物技术应用

耐冷生物产生的低温活性、热不稳定性酶类在生物技术方面具有巨大的应用潜力。一种分离自南极细菌的热不稳定磷酸酶在中性和碱性环境下有活性（中国香港商标局专利保护）。该酶在体外顺序实验中非常有用，包括限制性酶切、去磷酸化、酶失活、酶连接或末端标记等，因此可进一步简化实验步骤以及减少对核酸的实验需求量。适冷细菌降解烃化物的能力可用于污水处理中（Margesin and Schinner 2001）。研究发现一些分离自北极冰川的冰尘穴中的种属在低温环境中具有更高的纤维素酶活性（Singh et al. 2014）。耐冷微生物分离得到的嗜冷蛋白酶非常适合应用于食品加工和皮革制造业，因为它们在自来水的自然温度下仍保存着较高的酶活，因此，可以忽略加工过程的热处理步骤。低温酶也被用于衣服清洗剂中。通过在大肠杆菌中克隆表达北极细菌 *Rhodococcus* sp. AW25M09 编码的低温脂肪酶在高 pH、有机溶剂（乙腈、二乙醚）和高盐浓度（1mol/L NaCl）条件下仍保持活性。因此，该酶可用于工业大规模生产（De Santi et al. 2014）。许多 *Pseudomonas* 适冷菌种都可产生脂肪酶，在中温或低温环境中有较高活性。这些细菌可应用于脂肪酶介导的生物柴油生产过程（Moreno and Rojo 2014）。鉴于目前病原微生物对现有抗生素的广泛耐药性，人们越来越关注于抗生素的替代来源。极地微生物就是用于此目的的潜在候选菌之一，尽管还没有实质性的证据证明极地细菌在抗生素开发方面的作用（Liu et al. 2013；Maida et al. 2014；Fondi et al. 2014）。

寒冷气候下人类活动导致大量垃圾的积累，对环境造成严重威胁。在这样极端恶劣的气候下需要使用石油产品用于取暖。石油产品的消耗将引发环境的污染问题。此外，飞机以及储油地的石油泄漏将造成严重的环境污染。尽管采取了一系列的预防措施，但由于原生态环境中人为干预的增加，石油意外泄漏也时有发生。通过空运污染土壤对污染地区进行清理是非常昂贵的，而利用当地微生物的原位生物修复方法目前是石油污染处理最优先的选择。原位生物修复包括通过微生物或酶类转化有机污染物为低毒产品并进入生物地球化学循环中。

高温生物降解有许多优势（如增强溶解性和生物利用率），但不适用于低温环境下的细菌。然而，有报道发现，南北极土壤分离到具有降解碳氢化合物能力的细菌。这些细菌不止存在于石油污染地区，在未被污染的地区也有发现。一株南极细菌假单胞菌（*Pseudomonas*）被发现在含有硫酸铵、硝酸盐、亚硝酸盐和3.5%柴油，pH 7.0，温度10～20℃条件下生长最优（Shukor et al. 2009）。在一般情况下，该菌能迅速地适应污染物以及随着污染的发生菌体数目能迅速增加。自然界中石油相关的烃化合物种类繁多，如异、环、直链烷烃，单芳香有机化合物和多环芳烃等。所以，针对不同类型烃化物，有不同生化机制的微生物是污染物生物降解所必需的（Timmis et al. 2010）。在一项研究中，暴露于石油污染一段时间的微生物，其对污染物的降解速度较正常菌群更快，表明微生物环境适应的重要性。因此，提前暴露于污染物环境似乎是快速降解必不可少的条件。过去的十年实时定量PCR技术的成功应用，揭示了不同石油污染地区，一些负责烃化物的生物降解的基因（如单加氧酶、双加氧酶）的重要作用（Powell et al. 2006）。然而，尽管它们有降解多种烃化物的能力，但寒冷环境发现的所有微生物菌群无法降解所有石油污染物。此外，一些污染物（如高取代度高分子化合物）是结构稳定且微生物难以降解的。例如，沥青是一种高黏性的液体或半固体石油产品。沥青路面（矿料与沥青结合）在智利和加拿大发生溢出污染后每年都要进行检测。2010年8月深水地平线（Deepwater Horizon）油矿爆炸之后，尽管很大一部分污染物由微生物代谢降解，以及人为干预清除，仍然有大量的烃化物残留（Kimes et al. 2014），在低温情况下降解速率十分缓慢。通过喷洒含有石油降解微生物的表面活性剂可以显著改善石油降解速率。表面活性剂有利于从土壤中溶解石油，它的隔热效果可保持土壤适宜的温度（Jeong et al. 2015）。除了石油污染物，其他类型的污染（重金属污染）由于人为活动也在南极土壤中发生（Claridge et al. 1995；Lohan et al. 2001；Chaparro et al. 2007）。一株分离自南极南设得兰群岛的假单胞菌能将钼酸钠或Mo^{6+}转化为钼蓝，该转化的最适温度范围为15～20℃。因此，该菌株适用于在寒冷和温带地区的生物修复（Ahmad et al. 2013）。多氯联苯（PCBs）广泛用于工业和商业，由于在研究站的航空运输、河流输入（在北极地区）以及含PCBs物品的处置不当（如电器等），PCBs污染在极地地区也有发生。具有降解PCBs的微生物在南北极也时有发现（Master and Mohn 1998；De Domenico et al. 2004）。

除了低温，土壤中营养素的缺乏是微生物土壤修复的另一个限制因素。因此，有时需要通过添加营养成分来激活土壤细菌的生物活性（Ruberto et al. 2009）。有时结合物理清除（直接撤去污染物等）进行生物修复是现实可行的。提高中温大肠杆菌耐冷性，可通过克隆来源于南极微生物的两个分子伴侣蛋白基因（Ferrer et al. 2003）以及一个耐寒植物中编码抗冻蛋白的基因（Deng et al. 2014）来实现。使用具有强降解污染物能力的基因工程微生物似乎是一个有吸引力的选择，但在南极洲引入外来微生物是被禁止的。

致谢　作者感谢印度科学与发展委员会和工业研究会（CSIR）、生物技术部和印度政府科学技术部为各种项目提供资金。

无利益冲突声明

作者声明他们没有利益冲突。

参考文献

Abd El-Rahman HA，Fritze D，Spröer C，Claus D（2002）Two novel psychrotolerant species，*Bacillus psychrotolerans* sp. nov. and *Bacillus psychrodurans* sp. nov.，which contain ornithine in their cell walls. Int J Syst Evol Microbiol 52：2127-2133

Ahmad SA, Shukor MY, Shamaan NA, Mac Cormack WP, Syed MA (2013) Molybdate reduction to molybdenum blue by an Antarctic bacterium. Biomed Res Int 2013: 871-941

Aislabie J, Balks M, Foght J, Waterhouse E (2004) Hydrocarbon spills on Antarctic soils: Effects and management. Environ Sci Technol 38: 1265-1274

Aislabie J, Saul DJ, Foght JM (2006) Bioremediation of hydrocarbon-contaminated polar soils. Extremophiles 10: 171-179

Aislabie J, Jordan S, Ayton J, Klassen JL, Barker GM, Turner S (2009) Bacterial diversity associated with ornithogenic soil of the Ross Sea region, Antarctica. Can J Microbiol 55: 21-36

Aislabie JM, Lau A, Dsouza M, Shepherd C, Rhodes P, Turner SJ (2013) Bacterial composition of soils of the Lake Wellman area, Darwin Mountains, Antarctica. Extremophiles 17: 775-786

Al Khudary R, Stösser NI, Qoura F, Antranikian G (2008) *Pseudoalteromonas arctica* sp. nov., an aerobic, psychrotolerant, marine bacterium isolated from Spitzbergen. Int J Syst Evol Microbiol 58: 2018-2024

Allan RN, Lebbe L, Heyrman J, De Vos P, Buchanan CJ, Logan NA (2005) *Brevibacillus levickii* sp. nov. and *Aneurinibacillus terranovensis* sp. nov., two novel thermoacidophiles isolated from geothermal soils of northern Victoria Land, Antarctica. Int J Syst Evol Microbiol 55: 1039-1050

Amato P, Christner BC (2009) Energy metabolism response to low-temperature and frozen conditions in *Psychrobacter cryohalolentis*. Appl Environ Microbiol 75: 711-718

Antibus DE, Leff LG, Hall BL, Baeseman JL, Blackwood CB (2012) Cultivable bacteria from ancient algal mats from the McMurdo Dry alleys, Antarctica. Extremophiles 16: 105-114

Archer SD, McDonald IR, Herbold CW, Cary SC (2014) Characterisation of bacterioplankton communities in the meltwater ponds of Bratina Island, Victoria Land, Antarctica. FEMS Microbiol Ecol 89: 451-464

Arunasri K, Venkata Ramana V, Spröer C, Sasikala C, Ramana CV (2008) *Rhodobacter megalophilus* sp. nov., a phototroph from the Indian Himalayas possessing a wide temperature range for growth. Int J Syst Evol Microbiol 58: 1792-1796

Aschenbach K, Conrad R, Reháková K, Doležal J, Janatková K, Angel R (2013) Methanogens at the top of the world: occurrence and potential activity of methanogens in newly deglaciated soils in high-altitude cold deserts in the Western Himalayas. Front Microbiol 4: 359

Auman AJ, Breezee JL, Gosink JJ, Kämpfer P, Staley JT (2006) *Psychromonas ingrahamii* sp. nov., a novel gas vacuolate, psychrophilic bacterium isolated from Arctic polar sea ice. Int J Syst Evol Microbiol 56: 1001-1007

Auman AJ, Breezee JL, Gosink JJ, Schumann P, Barnes CR, Kämpfer P, Staley JT (2010) *Psychromonas boydii* sp. nov., a gas-vacuolate, psychrophilic bacterium isolated from an Arctic sea-ice core. Int J Syst Evol Microbiol 60: 84-92

Ayub ND, Tribelli PM, Lopez NI (2009) Polyhydroxyalkanoates are essential for maintenance of redox state in the Antarctic bacterium *Pseudomonas* sp. 14-3 during low temperature adaptation. Extremophiles 13: 59-66

Bajerski F, Wagner D (2013) Bacterial succession in Antarctic soils of two glacier forefields on Larsemann Hills, East Antarctica. FEMS Microbiol Ecol 85: 128-142

Ball MM, GómezW, Magallanes X, Rosales R, Melfo A, Yarzábal LA (2014) Bacteria recovered from a high-altitude, tropical glacier in Venezuelan Andes. World J Microbiol Biotechnol 30: 931-941

Bano N, Hollibaugh JT (2000) Diversity and distribution of DNA sequences with affinity to ammonia-oxidizing bacteria of the beta subdivision of the class Proteobacteria in the Arctic Ocean. Appl Environ Microbiol 66: 1960-1969

Begum Z, Srinivas TN, Manasa P, Sailaja B, Sunil B, Prasad S, Shivaji S (2013) *Winogradskyella psychrotolerans* sp. nov., a marine bacterium of the family Flavobacteriaceae isolated from Arctic sediment. Int J Syst Evol Microbiol 63: 1646-1652

Bienhold C, Boetius A, Ramette A (2012) The energy-diversity relationship of complex bacterial communities in Arctic deep-sea sediments. ISME J 6: 724-773

Boeuf D, Cottrell MT, Kirchman DL, Lebaron P, Jeanthon C (2013) Summer community structure of aerobic anoxygenic phototrophic bacteria in the western Arctic Ocean. FEMS Microbiol Ecol 85: 417-432

Bottos EM, Vincent WF, Greer CW, Whyte LG (2008) Prokaryotic diversity of arctic ice shelf microbial mats. Environ Microbiol 10: 950-966

Bottos EM, Woo AC, Zawar-Reza P, Pointing SB, Cary SC (2014) Airborne bacterial populations above desert soils of the McMurdo Dry Valleys, Antarctica. Microb Ecol 67: 120-128

Bowman JP, McCammon SA, Brown MV, Nichols DS, McMeekin TA (1997) Diversity and association of psychrophilic bacteria in Antarctic sea ice. Appl Environ Microbiol 63: 3068-3078

Bowman JP, Gosink JJ, McCammon SA, Lewis TE, Nichols DS, Nichols PD, Skerratt JH, Staley JT, McMeekin TA (1998) *Colwe/lia demingiae* sp. nov., *Colwe/lia hornerae* sp. nov., *Colwellia rossensis* sp. nov. and *Colwellia psychrotropica* sp. nov.: psychrophilic Antarctic species with the ability to synthesize docosahexanoic acid (22: 6 omega 3). Int J Syst Bacterial 48: 1171-1180

Bowman JP, Rea SM, McCammon SA, McMeekin TA (2000) Diversity and community structure within anoxic sediment from marine salinity meromictic lakes and a coastal meromictic marine basin, Vestfold Hills, Eastern Antarctica. Environ Microbiol 2: 227-237

Boyd WL (1962) Comparison of soil bacteria and their metabolic activities in Arctic and Antarctic regions. Polar Rec 11: 319

Boyd WL, Boyd JW (1962a) Presence of *Azotobacter* species in polar regions. J Bacterial 85: 1121-1123

Boyd WL, Boyd JW (1962b) Viability of thermophiles and coliform bacteria in arctic soils and water. Can J Microbiol

8：189-192

Bozal N，Montes MJ，Mercadé E（2007）*Pseudomonas guinea* sp. nov.，a novel psychrotolerant bacterium from an Antarctic environment. Int J Syst Evol Microbiol 57：2609-2612

Brinkmeyer R，Knittel K，Jurgens J，Weyland H，Amann R，Helmke E（2003）Diversity and structure of bacterial communities in Arctic versus Antarctic pack ice. Appl Environ Microbiol 69：6610-6619

Brown MV，Bowman JP（2001）A molecular phylogenetic survey of sea-ice microbial communities（SIMCO）. FEMS Microbiol Ecol 35：267-275

Campbell IB，Claridge GGC（2000）Soil temperature，moisture and salinity patterns in Transantarctic Mountain cold desert ecosystems. In：Davidson W，Howard-Williams C，Broady P（eds）Antarctic ecosystems：models for wider ecological understanding. New Zealand Natural Sciences，Canterbury University，Christchurch，New Zealand，pp 233-240

Campbell BJ，Polson SW，Hanson TE，Mack MC，Schuur EA（2010）The effect of nutrient deposition on bacterial communities in Arctic tundra soil. Environ Microbiol 12：1842-1854

Carpenter EJ，Lin S，Capone DG（2000）Bacterial activity in South Pole snow. Appl Environ Microbiol 66：4514-4517

Carr SA，Vogel SW，Dunbar RB，Brandes J，Spear JR，Levy R，Naish TR，Powell RD，Wakeham SG，Mandernack KW（2013）Bacterial abundance and composition in marine sediments beneath the Ross Ice Shelf，Antarctica. Geobiology 11：377-395

Cavanagh J，Austin JJ，Sanderson K（1996）Novel *Psychrobacter* species from Antarctic ornithogenic soils. Int J Syst Bacterial 46：841-848

Cavicchioli R（2007）Antarctic metagenomics. Microbiol Aust 28：98-103

Cavicchioli R，Thomas T（2000）Extremophiles. In：Lederberg J（ed）Encyclopedia of microbiology，2nd edn. vol 2. Academic Press，San Diego，pp 317-337

Chaparro MAE，Nunez H，Lirio JM，Gogorza CSG，Sinito AM（2007）Magnetic screening and heavy metal pollution studies in soils from Marambio Station，Antarctica. Antarctic Sci 19：379-393

Chattopadhyay MK（2008）Cryotolerance in bacteria：Interlink with adaptation to other stress factors. Trends Microbiol 16：455

Chattopadhyay MK，Jagannadham MV（2001）Maintenance of membrane fluidity in Antarctic bacteria. Polar Biol 24：386-388

Chattopadhyay MK，Uma Devi K，Gopisankar Y，Shivaji S（1995）Thermolabile alkaline phosphatase from *Sphingobacterium antarcticus*，a psychrotrophic bacterium from Antarctica. Polar Biol 15：215-219

Chattopadhyay MK，Raghu G，Sharma YVRK，Biju AR，Rajasekharan MVR，Shivaji S（2011）Increase in oxidative stress at low temperature in an Antarctic bacterium. Curr Microbiol 62：544-546

Chattopadhyay MK，Reddy GSN，Shivaji S（2014）Psychrophilic bacteria：Biodiversity，Molecular basis of cold adaptation and biotechnological implications. Curr Biotech 3：100-116

Chaturvedi P，Shivaji S（2007）*Exiguobacterium indicum* sp. nov.，a psychrophilic bacterium from the Hamta glacier of the Himalayan mountain ranges in India. Int J Syst Evol Microbiol 56：2765-2770

Chaturvedi P，Reddy GSN，Shivaji S（2005）*Dyadobacter hamtensis* sp. nov.，from Hamta Glacier，Himalayas，India. Int J Syst Evol Microbiol 55：2113-2117

Chen Z，Yu H，Li L，Hu S，Dong X（2012）The genome and transcriptome of a newly described psychrophilic archaeon，*Methanolobus psychrophilus* R15，reveal its cold adaptive characteristics. Environ Microbiol Rep 4：633-641

Chessa JP，Petrescu I，Bentahir M，Van Beeumen J，Gerday C（2000）Purification，physico-chemical characterization and sequence of a heat labile alkaline metalloprotease isolated from a psychrophilic *Pseudomonas* species. Biochim Biophys Acta 1479：265-274

Chintalapati S，Kiran MD，Shivaji S（2004）Role of membrane lipid fatty acids in cold adaptation. Cell Mol Biol（Noisy-le-Grand）50：631-642

Chu HY，Fierer N，Lauber CL，Caporaso JG，Knight R，Grogan P（2010）Soil bacterial diversity in the Arctic is not fundamentally different from that found in other biomes. Environ Microbiol 12：2998-3006

Claridge GGC，Campbell IB（1977）The salts in Antarctic soils，their distribution and relationship to soil processes. Soil Sci 123：377-384

Claridge GGC，Campbell IB，Powell HKJ，Amin ZH，Balks MR（1995）Heavy metal contamination in some soils of the McMurdo Sound region，Antarctica. Antarctic Sci 7：9-14

Collins RE，Rocap G，Deming JW（2010）Persistence of bacterial and archaeal communities in sea ice through an Arctic winter. Environ Microbiol 12：1828-1841

Cristóbal HA，López MA，Kothe E，Abate CM（2011）Diversity of protease-producing marine bacteria from sub-antarctic environments. J Basic Microbiol 51：590-600

De Domenico M，Lo Giudice A，Michaud L，Saitta M，Bruni V（2004）Diesel oil and PCB-degrading psychrotrophic bacteria isolated from Antarctic seawaters（Terra Nova Bay，Ross Sea）. Polar Res 23：141-146

De Santi C，Tedesco P，Ambrosino L，Altermark B，Willassen NP，de Pascale D（2014）A new alkaliphilic cold-active esterase from the psychrophilic marine bacterium *Rhodococcus* sp.：functional and structural studies and biotechnological potential. Appl Biochem Biotechnol 172：3054-3068

Delille D，Gleizon F（2003）Distribution of enteric bacteria in Antarctic seawater surrounding the Port-aux-Français permanent station（Kerguelen Island）. Mar Pollut Bull 46：1179-1183

Delille D，Lagarde E（1974）Contribution al'etude ecologique des Milieux subantarctiques. V Rev lnst Pasteur Lyon 7：149-165

Deng LQ, Yu HQ, Liu YP, Jiao PP, Zhou SF, Zhang SZ, Li WC, Fu FL (2014) Heterologous expression of anti-freeze protein gene AnAFP from *Ammopiptanthusnanus* enhances cold tolerance in *Escherichia coli* and tobacco. Gene 539: 132-140

Denner EBM, Mark B, Busse HJ, Turkiewicz M, Lubitz W (2001) *Psychrobacter proteolyticus* sp. nov., a psychrotrophic, halotolerant bacterium isolated from the Antarctic krill *Euphausia superba* Dana, excreting a cold-adapted metalloprotease. Syst Appl Microbiol 24: 44-53

Deslippe JR, Egger KN (2006) Molecular diversity of nifH genes from bacteria associated with high arctic dwarf shrubs. Microb Ecol 51: 516-525

Díez B, Bergman B, Pedrós-Alió C, Antó M, Snoeijs P (2012) High cyanobacterial nifH gene diversity in Arctic seawater and sea ice brine. Environ Microbiol Rep 4: 360-366

Dong S, Yang J, Zhang XY, Shi M, Song XY, Chen XL, Zhang YZ (2012) Cultivable alginate lyaseexcreting bacteria associated with the Arctic brown alga Laminaria. Mar Drugs 10: 2481-2491

Doyle SM, Montross SN, Skidmore ML, Christner BC (2013) Characterizing microbial diversity and the potential for metabolic function at -15℃ in the Basal ice of Taylor Glacier, antarctica. Biology (Basel) 2: 1034-1053

Dyurgerov MB, Meier MF (1997) Mass balance of mountain and subpolar glaciers: a new global assessment for 1961-1990. Arctic Alpine Res 29: 379-391

Dziewit L, Bartosik D (2014) Plasmids of psychrophilic and psychrotolerant bacteria and their role in adaptation to cold environments. Front Microbiol 5: 596

Edwards A, Anesio AM, Rassner SM, Sattler B, Hubbard B, Perkins WT, Young M, Griffith GW (2011) Possible interactions between bacterial diversity, microbial activity and supraglacial hydrology of cryoconite holes in Svalbard. ISME J 5: 150-160

Ekelof E (1908a) Bakteriologische studien wahrend der Schwedishen Sudpolar expedition 1901-1903. In: Wissenschaftiche Ergebnisse der Schwedischen sudpolar Expedition 1901-1903. 0. Nordenskjold (ed) Lithogr. lnst. Generalstabs, Stockholm, p 210

Ekelof E (1908b) Studien uber den Bakteriengehalt der luft und des Erdbodeus der antarktishen Gegenden, ausgefiihrt wahrend der schwedischen sudpolar expedition 1901-1903. Zhyg lnfekt 56: 344-370

Friedman E (1980) Endolithic microbial life in hot and cold deserts. Orig Life 10: 223-235

Ewert M, Deming JW (2014) Bacterial responses to fluctuations andextremes in temperature and brine salinity at the surface of Arctic winter sea ice. FEMS Microbiol Ecol 89: 476-489

Feller G, Gerday C (2003) Psychrophilic enzymes: hot topics in cold adaptation. Nat Rev Microbiol 1: 200-208

Ferrer M, Chernikova TN, Yakimov MM, Golyshin PN, Timmis KN (2003) Chaperonins govern growth of *Escherichia coli* at low temperatures. Nat Biotechnol 21: 1266-1267

Fondi M, Orlandini V, Perrin E, Maida I, Bosi E, Papaleo MC, Michaud L, Lo Giudice A, de Pascale D, Tutino ML, Liò P, Fani R (2014) Draft genomes of three Antarctic *Psychrobacter* strains producing antimicrobial compounds against *Burkholderiacepacia* complex, opportunistic human pathogens. Mar Genomics 13: 37-38

Forschner SR, Sheffer R, Rowley DC, Smith DC (2009) Microbial diversity in Cenozoic sediments recovered from the Lomonosov Ridge in the Central Arctic basin. Environ Microbiol 11: 630-639

Franzmann PD, Dobson SJ (1992) Cell wall-less, free-living spirochetes in Antarctica. FEMS Microbiol Lett 76: 289-292

GangwarP, Alam SI, Bansod S, Singh L (2009) Bacterial diversity of soil samples from the western Himalayas, India. Can J Microbiol 55: 564-577

Ganzert L, Bajerski F, Wagner D (2014) Bacterial community composition and diversity of five different permafrost-affected soils of Northeast Greenland. FEMS Microbiol Ecol 89: 426-441

Gibson JAE (2006) Limnology of Epiglacial Lakes of the Framnes Mountains, Antarctica: insights into a widespread but poorly Studied Lake Type. Scientific Committee on Antarctic Research (SCAR), vol XXIX, Hobart, 11-15 July, p 422

Gosink JJ, Woese CR, Staley JT (1998) *Polaribacter* gen. nov., with three new species, *P. irgensii* sp. nov., *P. franzmannii* sp. nov., and *P. filamentus* sp. nov., gas vacuolate polar marine bacteria of the Cytophaga-Flavobacterium-Bacteroides group and reclassification of '*Flectobacillus glomeratus*'as *Polaribacter glomeratus* comb. nov. Int J Syst Bacteriol 48: 223-235

Green WJ, Angle MP, Chave KE (1988) The geochemistry of Antarctic streams and their role in the evolution of four lakes of the McMurdo Dry Valleys. Geochim Cosmochim Acta 52: 1265-1274

Groudieva T, Kambourova M, Yusef H, Royter M, Grote R, Trinks H, Antranikian G (2004) Diversity and cold-active hydrolytic enzymes of culturable bacteria associated with Arctic sea ice, Spitzbergen. Extremophiles 8: 475-488

Guibert LM, Loviso CL, Marcos MS, Commendatore MG, Dionisi HM, Lozada M (2012) Alkane biodegradation genes from chronically polluted subantarctic coastal sediments and their shifts in response to oil exposure. Microb Ecol 64: 605-616

Hansen AA, Herbert RA, Mikkelsen K et al (2007) Viability, diversity and composition of the bacterial community in a high Arctic permafrost soil from Spitsbergen, Northern Norway. Environ Microbiol 9: 2870-2884

Helmke E, Weyland H (2004) Psychrophilic versus psychrotolerant bacteria-occurrence and significance in polar and temperate marine habitats. Cell Mol Biol 50: 553-561

Herbert RA, Bell CR (1977) Growth characteristics of an obligately psychrophilic Vibrio sp. Arch Microbiol 113: 215-220

Hirsch P, Mevs U, Kroppenstedt RM, Schumann P, Stackebrandt E (2004) Cryptoendolithic actinomycetes from antarctic sandstone rock samples: *Micromonospora endolithica* sp. nov. and two isolates related to *Micromonospora coerulea* Jensen 1932. Syst Appl Microbiol 27: 166-174

Huang J, Hoover RB, Swain A, Murdock C, Bej AK (2013) Comparison of the microbial diversity and abundance between the freshwater land-locked lakes of Schirmacher Oasis and the perennially ice-covered Lake Untersee in East Antarctica. In: Proceedings of International Society for Optical Engineering, Instruments, Methods and Mission for Astrobiology (SPIE) 7819, 78190W

Imperio T, Viti C, Marri L (2008) *Alicyclobacillus pohliae* sp. nov., a thermophilic, endosporeforming bacterium isolated from geothermal soil of the north-west slope of Mount Melbourne (Antarctica). Int J Syst Evol Microbiol 58: 221-225

Jang GI, Hwang CY, Choi HG, Kang SH, Cho BC (2011) Description of *Spongiibacter borealis* sp. nov., isolated from Arctic seawater, and reclassification of *Melitea salexigens* Urios et al. 2008 as a later heterotypic synonym of *Spongiibacter marinus* Graeber et al. 2008 with emended descriptions of the genus *Spongiibacter* and *Spongiibacter marinus*. Int J Syst Evol Microbiol 61: 2895-2900

Jeong SW, Jeong J, Kim J (2015) Simple surface foam application enhances bioremediation of oil-contaminated soil in cold conditions. J Hazard Mater 286C: 164-170

Jiang F, Li W, Xiao M, Dai J, Kan W, Chen L, Li W, Fang C, Peng F (2012) *Luteolibacter luojiensis* sp. nov., isolated from Arctic tundra soil, and emended description of the genus *Luteolibacter*. Int J Syst Evol Microbiol 62: 2259-2263

Jiang F, Xiao M, Chen L, Kan W, Peng F, Dai J, Chang X, Li W, Fang C (2013) *Huanghella arctica* gen. nov., sp. nov., a bacterium of the family *Cytophagaceae* isolated from Arctic tundra soil. Int J Syst Evol Microbiol 63: 696-702

Junge K, Imhoff F, Staley T, Deming JW (2002) Phylogenetic diversity of numerically important Arctic sea-ice bacteria cultured at subzero temperature. Microb Ecol 43: 315-328

Junge K, Eicken H, Deming JW (2004) Bacterial activity at −2 to −20℃ in Arctic wintertime sea ice. Appl Environ Microbiol 70: 550-557

Jyothsna TSS, Sasikala C, Ramana CV (2008) *Desulfovibrio psychrotolerans* sp. nov., a psychrotolerant and moderately alkaliphilic sulfate-reducing deltaproteobacterium from the Himalayas. Int J Syst Evol Microbiol 58: 821-825

Karl DM, Bird DF, Björkman K, Houlihan T, Shackelford R, Tupas L (1999) Microorganisms in the accreted ice of Lake Vostok, Antarctica. Science 286: 2144-2147

Karr EA, Sattley WM, Jung DO, Madigan MT, Achenbach LA (2003) Remarkable diversity of phototrophic purple bacteria in a permanently frozen Antarctic lake. Appl Environ Microbiol 69: 4910-4914

Karr EA, Sattley WM, Rice MR, Jung DO, Madigan MT, Achenbach LA (2005) Diversity and distribution of sulfate-reducing bacteria in permanently frozen Lake Fryxell, McMurdo Dry Valleys, Antarctica. Appl Environ Microbiol 71: 6353-6359

Kawahara H, Iwanaka Y, Higa S, Muryoi N, Sato M, Honda M, Omura H, Obata H (2007) A novel, intracellular antifreeze protein in an antarctic bacterium, *Flavobacterium xanthum*. Cryo Lett 28: 39-49

Kelly CD, Layne S (1957) Bacteria found in the air over Canada and the American Arctic. Can J Microbiol 3: 447-455

Kelly MD, Lukaschewsky S, Anderson CG (1978) Bacterial flora*of* Antarctic krill and some of their enzymatic properties. J Food Sci 43: 1196-1197

Kim HJ, Park S, Lee JM, Park S, Jung W, Kang JS, Joo HM, Seo KW, Kang SH (2008) *Moritella dasanensis* sp. nov., a psychrophilic bacterium isolated from the Arctic ocean. Int J Syst Evol Microbiol 58: 817-820

Kim SJ, Park SJ, Yoon DN, Park BJ, Choi BR, Lee DH, Roh Y, Rhee SK (2009) *Marinobacterium maritimum* sp. nov., a marine bacterium isolated from Arctic sediment. Int J Syst Evol Microbiol 59: 3030-3034

Kim SJ, Park SJ, Oh YS, Lee SA, Shin KS, Roh DH, Rhee SK (2012) *Shewanella arctica* sp. nov., an iron-reducing bacterium isolated from Arctic marine sediment. Int J Syst Evol Microbiol 62: 1128-1133

Kim JG, Park SJ, Quan ZX, Jung MY, Cha IT, Kim SJ, Kim KH, Yang EJ, Kim YN, Lee SH, Rhee SK (2014) Unveiling abundance and distribution of planktonic Bacteria and Archaea in a polynya in Amundsen Sea, Antarctica. Environ Microbiol 16: 566-578

Kimes NE, Callaghan AV, Suflita JM, Morris PJ (2014) Microbial transformation of the Deepwater Horizon oil spill-past, present, and future perspectives. Front Microbiol 5: 603

Kirchman DL, Cottrell MT, Lovejoy C (2010) The structure of bacterial communities in the western Arctic Ocean as revealed by pyrosequencing of 16S rRNA genes. Environ Microbiol 12: 1132-1143

Kishore KH, Begum Z, Pathan AAK, Shivaji S (2009) *Paenibacillus glacialis* sp. nov. isolated from Kafni glacier of Himalayas, India. Int J Syst Evol Microbiol 60: 1909-1913

Knittel K, Kuever J, Meyerdierks A, Meinke R, Amann R, Brinkhoff T (2005) *Thiomicrospira arctica* sp. nov. and *Thiomicrospira psychrophila* sp. nov., psychrophilic, obligately chemo-lithoautotrophic, sulfur-oxidizing bacteria isolated from marine Arctic sediments. Int J Syst Evol Microbiol 55: 781-786

Knoblauch C, Sahm K, Jørgensen BB (1999) Psychrophilic sulfate-reducing bacteria isolated from permanently cold arctic marine sediments: description of *Desulfofrigus oceanense* gen. nov., sp. nov., *Desulfofrigus fragile* sp. nov., *Desulfofaba gelida* gen. nov., sp. nov., *Desulfotalea psychrophila* gen. nov., sp. nov. and *Desulfotalea arctica* sp. nov. Int J Syst Bacteriol 49: 1631-1643

Kong W, Li W, Romancova I, Prášil O, Morgan-Kiss RM (2014) An integrated study of photochemical function and expression of a key photochemical gene (psbA) in photosynthetic communities of Lake Bonney (McMurdo Dry Valleys, Antarctica). FEMS Microbiol Ecol 89: 293-302

Koo H, Ptacek T, Crowley M, Swain AK, Osborne JD, Bej AK, Andersen DT (2014) Draft genome sequence of *Hymenobacter* sp. Strain IS2118, Isolated from a freshwater lake in Schirmacher Oasis, Antarctica, reveals diverse genes for adaptation to cold ecosystems. Genome Announc 2: e00739-14

Kriss AE (1945) Microorganisms of the eastern part of the Arctic Ocean. Mikrobiologiia 14: 268-276

Kuhn E, Bellicanta GS, Pellizari VH (2009) New alk genes detected in Antarctic marine sediments. Environ Microbiol 11: 669-673

Kuhn E, Ichimura AS, Peng V, Fritsen CH, Trubl G, Doran PT, Murray AE (2014) Brine assemblages of ultrasmall microbial cells within the ice cover of Lake Vida, Antarctica. Appl Environ Microbiol 80: 3687-3698

Kumar PA, Sreenivas A, Singh A, Shivaji S (2013a) Draft genome sequence of *Winogradskyella psychrotolerans* RS-3 T, isolated from the marine transect of Kongsfjorden, Ny-Ålesund, Svalbard, Arctic. Genome Announc 1: e00630-13

Kumar PA, Singh A, Sreenivas A, Begum Z, Reddy GSN, Shivaji S (2013b) Draft genome sequence of *Leifsonia rubra* CMS 76R T, isolated from a cyanobacterial mat sample from a pond in Wright Valley, McMurdo, Antarctica. Genome Announc 1: e00633-13

Labrenz M, Tindall BJ, Lawson PA, Collins MD, Schumann P, Hirsch P (2000) *Staleya guttiformis* gen. nov., sp. nov. and *Sulfitobacter brevis* sp. nov., alpha-3-Proteobacteria from hypersaline, heliothermal and meromictic Antarctic Ekho Lake. Int J Syst Evol Microbial 50: 303-313

Lama L, Nicolaus B, Calandrelli V, Esposito E, Gambacorta A (1996) Xylanase produced by *Bacillusthermoantarcticus*, a new thermophilic bacillus. Enzyme Eng XIII: 284-286

Lanoil B, Skidmore M, Priscu JC, Han S, Foo W, Vogel SW, Tulaczyk S, Engelhardt H (2009) Bacteria beneath the West Antarctic ice sheet. Environ Microbiol 11: 609-615

Larkin JM, Stokes JL (1967) Taxonomy of psychrophilic strains of Bacillus. J Bacteriol 94: 889-895

Laucks ML, Sengupta A, Junge K, Davis EJ, Swan-son BD (2005) Comparison of psychroactive Arctic marine bacteria and common mesophilic bacteria using surface-enhanced Raman spectroscopy. Appl Spectrosc 10: 1222-1228

Lauro FM, Chastain RA, Blankenship LE, Yayanos AA, Bartlett DH (2007) The unique 16S rRNA genes of piezophiles reflect both phylogeny and adaptation. Appl Environ Microbiol 73: 838-845

Laybourn-Parry J (1997) The microbial loop in Antarctic Lakes. In: Howards-Williams C, Lyons WB, Hawes I (eds) Ecosystem processes in Antarctic ice-free landscapes. Balkema, Rotterdam/Brookfield, pp 231-240

Lee K, Lee HK, Choi TH, Kim KM, Cho JC (2007) *Granulosicoccaceae* fam. nov., to include *Granulosicoccus antarcticus* gen. nov., sp. nov., a non-phototrophic, obligately aerobic chemoheterotroph in the order Chromatiales, isolated from Antarctic seawater. J Microbiol Biotechnol 17: 1483-1490

Lee CK, Barbier BA, Bottos EM, McDonald IR, Cary SC (2012) The inter-valley soil comparative survey: the ecology of dry valley edaphic microbial communities. ISME J 6: 1046-1057

Li S, Xiao X, Yin X, Wang F (2006a) Bacterial community along a historic lake sediment core of Ardley Island, west Antarctica. Extremophiles 10: 461-467

Li HR, Yu Y, Zeng YX, Chen B, Ren DM (2006b) Phylogenetic analysis of bacterial diversity in Pacific Arctic sediments. Wei Sheng Wu Xue Bao 46: 177-183

Li H, Yu Y, Luo W, Zeng Y, Chen B (2009) Bacterial diversity in surface sediments from the Pacific Arctic Ocean. Extremophiles 13: 233-246

Li AH, Liu HC, Xin YH, Kim SG, Zhou YG (2013) *Glaciihabitans tibetensis* gen. nov., sp. nov., a psychrotolerant bacterium of the family *Microbacteriaceae*, isolated from glacier ice water. Int J Syst Evol Microbiol 64: 579-587

Liebner S, Rublack K, StuehrmannT, Wagner D (2009) Diversity of aerobic methanotrophic bacteria in a permafrost active layer soil of the Lena Delta, Siberia. Microbiol Ecol 57: 25-35

Liu Y, Yao T, Jiao N, Kang S, Zeng Y, Huang S (2006) Microbial community structure in moraine lakes and glacial meltwaters, Mount Everest. FEMS Microbiol Lett 265: 98-105

Liu YQ, Yao TD, Jiao NZ, Kang SC, Xu BQ, Zeng YH, Huang SJ, Liu XB (2009) Bacterial diversity in the snow over Tibetan Plateau Glaciers. Extremophiles 13: 411-423

LiuY, Yao T, Jiao N, Tian L, Hu A, Yu W, Li S (2011) Microbial diversity in the snow, a moraine lake and a stream in Himalayan glacier. Extremophiles 15: 411-421

Liu JT, Lu XL, Liu XY, Gao Y, Hu B, Jiao BH, Zheng H (2013) Bioactive natural products from the antarctic and arctic organisms. Mini Rev Med Chem 13: 617-626

Lo Giudice A, Caruso C, Mangano S, Bruni V, De Domenico M, Michaud L (2012) Marine bacterioplankton diversity and community composition in an antarctic coastal environment. Microb Ecol 63: 210-223

Lohan MC, Statham PJ, Peck L (2001) Trace metals in the Antarctic soft-shelled clam *Laternulaelliptica*: implications for metal pollution from Antarctic research stations. Polar Biol 24: 808-817

Lysnes K, Thorseth IH, Steinsbu BO, Øvreås L, Torsvik T, Pedersen RB (2004) Microbial community diversity in seafloor basalt from the Arctic spreading ridges. FEMS Microbiol Ecol 50: 213-230

Ma Y, Wang L, Shao Z (2006) Pseudomonas, the dominant polycyclic aromatic hydrocarbon-degrading bacteria isolated from Antarctic soils and the role of large plasmids in horizontal gene transfer. Environ Microbiol 8: 455-465

Maas EW, Simpson AM, Martin A, Thompson S, Koh EY, Davy SK, Ryan KG, O'Toole RF (2012) Phylogenetic analyses of bacteria in sea ice at Cape Hallett, Antarctica. N Z J Mar Freshw Res 46: 3-12

Magalhães CM, Machado A, Frank-Fahle B, Lee CK, Cary SC (2014) The ecological dichotomy of ammonia-oxidizing archaea and bacteria in the hyper-arid soils of the Antarctic Dry Valleys. Front Microbiol 5: 515

Maida I, Fondi M, Papaleo MC, Perrin E, Orlandini V, Emiliani G, de Pascale D, Parrilli E, Tutino ML, Michaud L, Lo Giudice A, Romoli R, Bartolucci G, Fani R (2014) Phenotypic and genomic characterization of the Antarctic bacterium *Gillisia* sp. CAL575, a producer of antimicrobial compounds. Extremophiles 18: 35-49

Männistö MK, Rawat S, Starovoytov V, Häggblom MM (2012) *Granulicella arctica* sp. nov., *Granulicella mallensis* sp. nov., *Granulicella tundricola* sp. nov. and *Granulicella sapmiensis* sp. nov., novel acidobacteria from tundra-soil. Int J Syst Evol Microbiol 62: 2097-2106

Marcos MS, Lozada M, Dionisi HM (2009) Aromatic hydrocarbon degradation genes from chronically polluted sub-antarctic marine sediments. Lett Appl Microbiol 49: 602-608

Margesin R, Schinner F (2001) Biodegradation and bioremediation of hydrocarbons in extreme environments. Appl Microbiol Biotechnol 56: 650-663

Margini RA, Castrelos OD (1963) Examenes bacteriologicos de aire, rieve ysuelo de carbo primaruera y Estacion cientifica Ellsworth. Inst Antarcti Argent Pub 76: 1-15

Marshall BJ, Ohye DF (1966) *Bacillus macquariensis* sp. nov, a psychrotrophic bacterium from sub-antarctic soil. J Gen Microbiol 44: 41-46

Martineau C, Pan Y, Bodrossy L, Yergeau E, Whyte LG, Greer CW (2014) Atmospheric methane oxidizers are present and active in Canadian high Arctic soils. FEMS Microbiol Ecol 89: 257-269

Master ER, Mohn WW (1998) Psychrotolerant bacteria Isolated from Arctic Soil that degrade polychlorinated biphenyls at low temperatures. Appl Environ Microbiol 64: 4823-4829

Meyer GH, Morrow MB, Wyss O, Berg TE, Littlepage JL (1962) Antarctica: the microbiology of an unfrozen saline pond. Science 138: 1103-1104

Michaud L, Caruso C, Mangano S, Interdonato F, Bruni V, Lo Giudice A (2012) Predominance of *Flavobacterium*, *Pseudomonas*, and *Polaromonas* within the prokaryotic community of freshwater shallow lakes in the northern Victoria Land, East Antarctica. FEMS Microbiol Ecol 82: 391-404

Mikucki JA, Pearson A, Johnston DT, Turchyn AV, Farquhar J, Schrag DP, Anbar AD, Priscu JC, Lee PA (2009) A contemporary microbially maintained subglacial ferrous "ocean". Science 324: 397-400

Miwa T (1975) Clostridia in soil of the Antarctica. Jpn J Med Sci Biol 28: 201-213

Møller AK, Barkay T, Abu Al-Soud W, Sørensen SJ, Skov H, Kroer N (2011) Diversity and characterization of mercury-resistant bacteria in snow, freshwater and sea-ice brine from the High Arctic. FEMS Microbiol Ecol 75: 390-401

Moreno R, Rojo F (2014) Features of pseudomonads growing at low temperatures: another facet of their versatility. Environ Microbiol Rep 6: 417-426

MoritaRY (1975) Psychrophilic bacteria. Bacteriol Rev 39: 144-167

Muangchinda C, Chavanich S, Viyakarn V, Watanabe K, Imura S, Vangnai AS, Pinyakong O (2014) Abundance and diversity of functional genes involved in the degradation of aromatic hydrocarbons in Antarctic soils and sediments around Syowa Station. Environ Sci Pollut Res 22: 4725-4735

Neufeld JD, Mohn WW (2005) Unexpectedly high bacterial diversity in Arctic tundra relative to boreal forest soils, revealed by serial analysis of ribosomal sequence tags. Appl Environ Microbiol 71: 5710-5718

Niederberger TD, McDonald IR, Hacker AL, Soo RM, Barrett JE, Wall DH, Cary SC (2008) Microbial community composition in soils of Northern Victoria Land, Antarctica. Environ Microbiol 10: 1713-1724

Pady SM, Kelly CD, Polunin N (1948) Arctic aerobiology; preliminary report on fungi and bacteria isolated from the air in 1947. Nature 162: 379-381

Peeters K, Hodgson DA, Convey P, Willems A (2011) Culturable diversity of heterotrophic bacteria in Forlidas Pond (Pensacola Mountains) and Lundström Lake (Shackleton Range), Antarctica. Microb Ecol 62: 399-413

Perreault NN, Andersen DT, Pollard WH, Greer CW, Whyte LG (2007) Characterization of the prokaryotic diversity in cold saline perennial springs of the Canadian high Arctic. Appl Environ Microbiol 73: 1532-1543

Pfiser RM, Burkholder PR (1965) Numerical taxonomy of some bacteria isolated from Antarctic and tropical sea waters. J Bacterial 90: 863-872

Pindi PK, Hara Kishore K, Reddy GSN, Shivaji S (2009) Description of *Leifsonia kafniensis* sp. nov. and *Leifsonia antarctica* sp. nov. Int J Syst Evol Microbiol 59: 1348-1352

Pirie JHH (1904) First Antarcticvoyage of the Scotia. J Bacterial Scot Geogr Mag 20: 129-132

Pirie JHH (1912) Notes on Antarctic bacteriology. Rep Sci Res SY Scotia 3: 157-168

Poli A, Esposito E, Lama L, Orlando P, Nicolaus G, de Appolonia F, Gambacorta A, Nicolaus B (2006) *Anoxybacillus amylolyticus* sp. nov., a thermophilic amylase producing bacterium isolated from Mount Rittmann (Antarctica). Syst Appl Microbiol 29: 300-307

Polunin N, Kelly CD (1952) Arctic aerobiology; fungi and bacteria, etc., caught in the air during flights over the geographical North Pole. Nature 170: 314-316

Pöntinen A, Markkula A, Lindström M, Korkeala H (2015) Two-component system histidine kinases involved in growth of *Listeria monocytogenes* EGD-e at low temperatures. Appl Environ Microbiol 81: 3994-4004

Powell SM, Ferguson SH, Bowman JP, Snape I (2006) Using real-time PCR to assess changes in the hydrocarbon-degrading microbial community in Antarctic soil during bioremediation. Microb Ecol 52: 523-532

Prabagaran SR, Manorama R, Delille D, Shivaji S (2007) Predominance of *Roseobacter*, *Sulfitobacter*, *Glaciecola* and *Psychrobacter* in seawater collected off Ushuaia, Argentina, Sub-Antarctica. FEMS Microbiol Ecol 59: 342-355

Pradhan S, Srinivas TNR, Pindi PK, Hara Kishore K, Begum Z, Singh PK, Singh AK, Pratibha MS, Yasala AK, Reddy GSN, Shivaji S (2010) Bacterial biodiversity from Roopkund Glacier, Himalayan mountain ranges, India. Extremophiles 14: 377-395

Prasad S, Manasa BP, Buddhi S, Pratibha MS, Begum Z, Bandi S, Tirunagari P, Shivaji S (2013) *Arcticibacter svalbardensis* gen. nov., sp. nov., of the family *Sphingobacteriaceae* in the phylum *Bacteroidetes*, isolated from Arctic soil. Int J Syst Evol Microbiol 63: 1627-1632

Prasad S, Manasa P, Buddhi S, Tirunagari P, Begum Z, Rajan S, Shivaji S (2014) Diversity and bioprospective potential (cold-active enzymes) of cultivable marine bacteria from the subarctic glacial Fjord, Kongsfjorden. Curr Microbiol 68: 233-238

Priest FG, Goodfellow M, Todd C (1988) A numerical classification of the genus Bacillus. J Gen Microbiol 134: 1847-1882

Priscu JC, Adams EE, Lyons WB, Voytek MA, Mogk DW, Brown RL, McKay CP, Takacs CD, Welch KA, Wolf CF, Kirshtein JD, Avci R (1999) Geomicrobiology of subglacial ice above Lake Vostok, Antarctica. Science 286: 2141-2144

Prüss BM, Francis KP, von Stetten F, Scherer S (1999) Correlation of 16S ribosomal DNA signature sequences with temperature-dependent growth rates of mesophilic and psychrotolerant strains of the *Bacillus cereus* group. J Bacteriol 181: 2624-2630

Purcell AM, Mikucki JA, Achberger AM, Alekhina IA, Barbante C, Christner BC, Ghosh D, Michaud AB, Mitchell AC, Priscu JC, Scherer R, Skidmore ML, Vick-Majors TJ, The Wissard Science Team (2014) Microbial sulfur transformations in sediments from Subglacial Lake Whillans. Front Microbiol 5: 594

Purusharth RI, Madhuri B, Ray MK (2007) Exoribonuclease R in Pseudomonas syringae is essential for growth at low temperature and plays a novel role in the 3′-end processing of 16 and 5 S ribosomal RNA. J Biol Chem 282: 16267-16277

Qu Z, Jiang F, Chang X, Qiu X, Ren L, Fang C, Peng F (2014) *Psychroglaciecola arctica* gen. nov., sp. nov., isolated from Arctic glacial foreland soil. Int J Syst Evol Microbiol 64: 1817-1824

Raiger Iustman LJ, Tribelli PM, Ibarra JG, Catone MV, Solar Venero EC, López NI (2015) Genome sequence analysis of *Pseudomonas extremaustralis* provides new insights into environmental adaptability and extreme conditions resistance. Extremophiles 19: 207-220

Ray MK, Kumar GS, Shivaji S (1994) Phosphorylation of membrane proteins in response to temperature in an Antarctic *Pseudomonas syringae*. Microbiology 140: 3217-3723

Ray MK, Devi KU, Kumar GS, Shivaji S (1992) Extracellular protease from the antarctic yeast Candida humicola. Appl Environ Microbiol 58: 1918-1923

Ray MK, Kumar GS, Janiyani K, Kannan K, Jagtap P, Basu MK, Shivaji S (1998) Adaptation to low temperature and regulation of gene expression in Antarctic psychrotrophic bacteria. J Biosci 23: 423-435

Reddy GSN, Rajagopalan G, Shivaji S (1994) Thermolabile ribonuclease from Antarctic psychrotropic bacteria: detection of the enzyme in various bacteria and purification from *Pseudomonas fluorescens*. FEMS Microbiol Lett 122: 211-216

Reddy GSN, Agarwal RK, MatsumotoGI SS (2000) *Arthrobacter flavus* sp. nov., a psychrotropic bacterium isolated from a pond in Mc Murdo dry valley, Antarctica. Int J Syst Evol Microbiol 50: 1553-1561

Reddy GSN, Prakash JSS, Matsumoto GI, Stackebrandt E, Shivaji S (2002a) *Arthrobacter roseus* sp. nov., a psychrotropic bacterium isolated from an Antarctic cyanobacterial mat sample. Int J Syst Evol Microbiol 52: 1017-1021

Reddy GSN, Prakash JSS, Vairamani M, Prabhakar S, Matsumoto GI, Shivaji S (2002b) *Planococcus antarcticus* and *Planococcus psychrophilus* spp. nov. isolated from cyanobacterial mat samples collected from ponds in Antarctica. Extremophiles 6: 253-261

Reddy GSN, Matsumoto GI, Shivaji S (2003a) *Sporosarcina macmurdoensis* sp. nov. from a cyanobacterial mat samples from a pond in the McMurdo dry valley, Antarctica. Int J Syst Evol Microbiol 53: 1363-1367

Reddy GSN, Prakash JSS, Srinivas R, Matsumoto GI, Shivaji S (2003b) *Leifsonia rubra* sp. nov. and *Leifsonia aurea* sp. nov. psychrophiles from a Pond in Antarctica. Int J Syst Evol Microbiol 53: 977-984

Reddy GSN, Raghavan PMU, Sarita NB, Prakash JSS, Nagesh N, Delille D, Shivaji S (2003c) *Halomonas glacies* sp. nov. isolated from fast ice of Adelie Land, Antarctica. Extremophiles 7: 55-61

Reddy GSN, Prakash JSS, Prabahar V, Matsumoto GI, Stackebrandt E, Shivaji S (2003d) *Kocuria polaris* sp. nov., an orange pigmented psychrotrophic bacterium isolated from an Antarctic cyanobacterial mat sample. Int J Syst Evol Microbiol 53: 183-187

Reddy GSN, Matsumoto GI, Shuman P, Stackebrandt E, Shivaji S (2004) Psychrophilic *Pseudomonas* from Antarctica: *Pseudomonasantarctica* sp. nov., *Pseudomonas meridianae* sp. nov. and *Pseudomonas proteolytica* sp. nov. Int J Syst Evol Microbiol 54: 713-719

Reddy GSN, Uttam A, Shivaji S (2008) *Bacillus cecembensis* sp. nov., a bacterium isolated from the Pindari glacier of the Himalayan mountain ranges, India. Int J Syst Evol Microbiol 58: 2330-2335

Reddy GSN, Pradhan S, Manorama R, Shivaji S (2009a) *Cryobacterium roopkundensis* sp. nov., a psychrophilic bacterium from Roopkund Glacier of the Himalayan mountain ranges. Int J Syst Evol Microbiol 60: 866-870

Reddy PVV, Shiva Nageswara Rao SS, Pratibha MS, Sailaja B, Kavya B, Manorama RR, Singh SM, Radha Srinivas TN, Shivaji S (2009b) Bacterial diversity and bioprospecting for cold-active enzymes from culturable bacteria associated with sediment from a melt water stream of Midtre Lovenbreen glacier, an Arctic glacier. Res Microbiol 160: 538-546

Reddy GSN, Poorna Manasa B, Singh SK, Shivaji S (2013a) *Paenisporosarcina indica* sp. nov., a psychrophilic bacterium from Pindari Glacier of the Himalayan mountain ranges and reclassification of *Sporosarcina antarctica* Yu et al.,

2008 as *Paenisporosarcina antarctica* comb. nov. and emended description of the genus *Paenisporosarcina*. Int J Syst Evol Microbiol IJS/2012/047514

Reddy GSN, Ara S, Singh A, Kumar Pinnaka A, Shivaji S (2013b) Draft genome sequence of *Psychrobacter aquaticus* CMS 56^T, isolated from a cyanobacterial mat sample collected from water bodies in the McMurdo Dry Valley region of Antarctica. Genome Announc 1: e00918-13

Reddy GSN, Sreenivas A, Shivaji S (2014) Draft genome sequence *Cryobacterium roopkundensis* strain RuGl7, isolated from a soil sample in the vicinity of Roopkund Lake, Himalayas, India. Genome Announc 2: e01206-e01214

Ren L, Chang X, Jiang F, Kan W, Qu Z, Qiu X, Fang C, Peng F (2015) *Parablastomonas arctica* gen. nov., sp. nov., isolated from high Arctic glacial till. Int J Syst Evol Microbiol 65: 260-266

Ruberto L, Dias R, Lo Balbo A, Vazquez SC, Hernandez EA, Mac Cormack WP (2009) Influence of nutrients addition and bioaugmentation on the hydrocarbon biodegradation of a chronically contaminated Antarctic soil. J Appl Microbiol 106: 1101-1110

Ruckert G (1985) *Myxobacteria* from Antarctic soils. Biol Fert Soil 1: 215-216

Sahay H, Babu BK, Singh S, Kaushik R, Saxena AK, Arora DK (2013) Cold-active hydrolases producing bacteria from two different sub-glacial Himalayan lakes. J Basic Microbiol 53: 703-714

Schütte UM, Abdo Z, Foster J, Ravel J, Bunge J, Solheim B, Forney LJ (2010) Bacterial diversity in a glacier foreland of the high Arctic. Mol Ecol 1: 54-66

Sengupta D, Chattopadhyay MK (2013) Metabolism in bacteria at low temperature: a recent report. J Biosci 38: 409-12

Shivaji S, Prakash JS (2010) How do bacteria sense and respond to low temperature? Arch Microbiol 192: 85-95

Shivaji S, Reddy GSN (2009) Chapter 2. Bacterial biodiversity of Antarctica: conventional polyphasic and rRNA approaches. In: Bej AK, Aislabie J, Atlas RM (eds) Polar microbiology: the ecology, biodiversity and bioremediation potential of microorganisms in extremely cold environments. Taylor & Francis, Boca Raton

Shivaji S, Rao NS, Saisree L, Sheth V, Reddy GSN, Bhargava PM (1988) Isolation and identification of *Micrococcus* roseus and *Planococcus* sp. from Schirmacher Oasis, Antarctica. J Biosci 113: 409-414

Shivaji S, Rao NS, Saisree L, Sheth V, Reddy GSN, Bhargava PM (1989a) Isolation and identifica-tion of *Pseudomonas* sp. from Schirmacher Oasis, Antarctica. Appl Environ Microbiol 55: 767-770

Shivaji S, Rao NS, Saisree L, Reddy GSN, Seshu Kumar G, Bhargava PM (1989b) Isolates of *Arthrobacter* from the soils of Schirmacher Oasis, Antarctica. Polar Biol 10: 225-229

Shivaji S, Reddy GSN, Prasad RA, Kutty R, Ravenschlag K (2004) Bacterial diversity of a soil sample from Schirmacher Oasis, Antarctica. Cell Mol Biol 50: 525-536

Shivaji S, Chaturvedi P, Reddy GSN, Suresh K (2005) *Pedobacter himalayensis* sp. nov. from Hamta glacier located in the Himalayan mountain ranges of India. Int J Syst Evol Microbiol 55: 1083-1088

Shivaji S, Bhadra B, Rao RS, Pradhan S (2008) *Rhodotorula himalayensis* sp. nov., a novel psychrophilic yeast isolated from Roopkund lake of the Himalayan mountain ranges, India. Extremophiles 12: 375-381

Shivaji S, Kumari K, Kishore KH, Pindi PK, Rao PS, Radha Srinivas TN, Asthana R, Ravindra R (2011a) Vertical distribution of bacteria in a lake sediment from Antarctica by culture-independent and culture-dependent approaches. Res Microbiol 162: 191-203

Shivaji S, Pratibha MS, Sailaja B, Hara Kishore K, Singh AK, Begum Z, Anarasi U, Prabagaran SR, Reddy GSN, Srinivas TN (2011b) Bacterial diversity of soil in the vicinity of Pindari gla-cier, Himalayan mountain ranges, India, using culturable bacteria and soil 16S rRNA gene clones. Extremophiles 15: 1-22

Shivaji S, Reddy PV, Rao SS, Begum Z, Manasa P, Srinivas TN (2012a) *Cyclobacterium qasimii* sp. nov., a psychrotolerant bacterium isolated from Arctic marine sediment. Int J Syst Evol Microbiol 62: 2133-2139

Shivaji S, Ara S, Singh SK, Bandi S, Singh A, Pinnaka AK (2012b) Draft genome sequence of *Bacillus isronensis* strain B3W22 isolated from the upper atmosphere. J Bacteriol 194: 6624-6625

Shivaji S, Ara S, Singh A, Kumar Pinnaka A (2013a) Draft genome sequence of *Cyclobacterium qasimii* M12-11B T, isolated from an Arctic marine sediment. Genome Announc 1: e00642-13

Shivaji S, Ara S, Prasad S, Manasa BP, Begum Z, Singh A, Kumar Pinnaka A (2013b) Draft genome sequence of *Arcticibacter svalbardensis* MN12-7 T, a member of the family Sphingobacteriaceae isolated from an Arctic soil sample. Genome Announc 1: e00484-13

Shivaji S, Begum Z, Shiva Nageswara Rao SS, Vishnu Vardhan Reddy PV, Manasa P, Sailaja B, Prathiba MS, Thamban M, Krishnan KP, Singh SM, Srinivas TN (2013c) Antarctic ice core samples: culturable bacterial diversity. Res Microbiol 164: 70-82

Shukor MY, Hassan NA, Jusoh AZ, Perumal N, Shamaan NA, MacCormack WP, Syed MA (2009) Isolation and characterization of a *Pseudomonas* diesel-degrading strain from Antarctica. J Environ Biol 30: 1-6

Singh AK, Shivaji S (2010) A cold-active and a heat-labile t-RNA modification GTPase from a psychrophilic bacterium *Pseudomonas syringae* (Lz4W). Res Microbiol 161: 46-50

Singh AK, Pindi PK, Dube S, Sundareswaran VR, Shivaji S (2009) In the psychrophilic *Pseudomonas syringae*, trmE is important for low temperature growth. Appl Environ Microbiol 75: 4419-4426

Singh P, Singh SM, Dhakephalkar P (2014) Diversity, cold active enzymes and adaptation strategies of bacteria inhabiting glacier cryoconite holes of High Arctic. Extremophiles 18: 229-242

Sinha AK, Pavankumar TL, Kamisetty S, Mittal P, Ray MK (2013) Replication arrest is a major threat to growth at low temperature in Antarctic *Pseudomonas syringae* Lz4W. Mol Microbiol 89: 792-810

Sjöling S, Cowan DA (2003) High 16S rDNA bacterial diversity in glacial meltwater lake sediment, Bratina Island,

Antarctica. Extremophiles 7：275-282

Smith RC，Prezelin，BB，Baker KS，Bidigare RR，Boucher NP，Coley T，Karentz D，MacIntyre S，Matlick HA，Menzies D，Ondrusek M，Wan Z，Waters KJ（1992）Ozone depletion：ultraviolet radiation and phytoplankton biology in Antarctic waters. Science 255：952-959

Søgaard DH，Kristensen M，Rysgaard S，Glud RN，Hansen PJ，Hilligsøe KM（2010）Autotrophic and heterotrophic activity in Arctic first-year sea ice：Seasonal study from Malene Bight，SW Greenland. Mar Ecol Prog Ser 419：31-45

Söller R，Hirsch P，Blohm D，Labrenz M（2000）Differentiation of newly described antarctic bacterial isolates related to Roseobacter species based on 16S-23S rDNA internal transcribed spacer sequences. Int J Syst Evol Microbiol 50：909-915

Spring S，Merkhoffer B，Weiss N，Kroppenstedt RM，Hippe H，Stackebrandt E（2003）Characterization of novel psychrophilic clostridia from an Antarctic microbial mat：description of *Clostridium frigoris* sp. nov.，*Clostridium lacusfryxellense* sp. nov.，*Clostridium bowmanii* sp. nov.，and *Clostridium psychrophilum* sp. nov.，and reclassification of *Clostridium laramiense* as *Clostridium estertheticum* subsp. *laramiense* subsp. nov. Int J Syst Evol Microbiol 53：1019-1029

Sreenivas A，Reddy GSN，Shivaji S（2014）Draft genome sequence of a psychrophilic bacterium *Sphingomonas antarcticum* 4BY，isolated from the soils of Schirmacher Oasis，Antarctica. Genome Announc 2：e00696-14

Srinivas TN，Nageswara Rao SS，Vishnu Vardhan Reddy P，Pratibha MS，Sailaja B，Kavya B，Hara Kishore K，Begum Z，Singh SM，Shivaji S（2009）Bacterial diversity and bioprospecting for cold-active lipases，amylases and proteases，from culturable bacteria of kongsfjorden and Ny-alesund，Svalbard，Arctic. Curr Microbiol 59：537-547

Srinivas TN，Singh SM，Pradhan S，Pratibha MS，Kishore KH，Singh AK，Begum Z，Prabagaran SR，Reddy GSN，Shivaji S（2011）Comparison of bacterial diversity in proglacial soil from Kafni Glacier，Himalayan Mountain ranges，India，with the bacterial diversity of other glaciers in the world. Extremophiles 15：673-690

Staley JT，Gosink JJ（1999）Poles apart：biodiversity and biogeography of sea ice bacteria. Annu Rev Microbiol 53：189-215

Steinsbu BO，Tindall BJ，Torsvik VL，Thorseth IH，Daae FL，Pedersen RB（2011）*Rhabdothermus arcticus* gen. nov.，sp. nov.，a member of the family Thermaceae isolated from a hydrothermal vent chimney in the Soria Moria vent field on the Arctic Mid-Ocean Ridge. Int J Syst Evol Microbiol 61：2197-2204

Steven B，Briggs G，McKay CP，Pollard WH，Greer CW，Whyte LG（2007）Characterization of the microbial diversity in a permafrost sample from the Canadian high Arctic using culturedependent and culture-independent methods. FEMS Microbiol Ecol 59：513-523

Steven B，Pollard WH，Greer CW，Whyte LG（2008a）Microbial diversity and activity through a permafrost/ground ice core profile from the Canadian high Arctic. Environ Microbiol 10：3388-3403

Steven B，Chen MQ，Greer CW，Whyte LG，Niederberger TD（2008b）*Tumebacillus permanentifrigoris* gen. nov.，sp. nov.，an aerobic，spore-forming bacterium isolated from Canadian high Arctic permafrost. Int J Syst Evol Microbiol 58：1497-1501

Steven B，Lionard M，Kuske CR，Vincent WF（2013）High bacterial diversity of biological soil crusts in water tracks over permafrost in the high arctic polar desert. PLoS One 8，e71489

Stibal M，Hasan F，Wadham JL，Sharp MJ，Anesio AM（2012）Prokaryotic diversity in sediments beneath two polar glaciers with contrasting organic carbon substrates. Extremophiles 16：255-265

Stokes JL，Reymond ML（1966）Quantitative ecology of psychrophilic microorganisms. Appl Microbiol 14：74-78

Storesund JE，Øvreås L（2013）Diversity of Planctomycetes in iron-hydroxide deposits from the Arctic Mid Ocean Ridge（AMOR）and description of *Bythopirellula goksoyri* gen. nov.，sp. nov.，a novel Planctomycete from deep sea iron-hydroxide deposits. Antonie Van Leeuwenhoek 104：569-584

Sundareswaran VR，Singh AK，Dube S，Shivaji S（2010）Aspartate aminotransferase is involved in cold adaptation in psychrophilic *Pseudomonas syringae*. Arch Microbiol 192：663-672

Suyal DC，Yadav A，Shouche Y，Goel R（2014）Differential proteomics in response to low temperature diazotrophy of Himalayan psychrophilic nitrogen fixing Pseudomonas migulae S10724 strain. Curr Microbiol 68：543-550

Takii S，Kondal T，Hiraishi A，Matsumoto GI，Kawano T，Torii T（1986）Vertical distribution in and isolation of bacteria from Lake Vanda：an Antarctic lake. Hydrobiologia 135：15-21

Tamura T，Ishida Y，Otoguro M，Yamamura H，Hayakawa M，Suzuki K（2010）*Angustibacter luteus* gen. nov.，sp. nov.，isolated from subarctic forest soil. Int J Syst Evol Microbiol 60：2441-2445

Tanner AC（1985）The role of bacteria in the cycling of nutrients within the maritime Antarctic environment. In：Siegfried WR，Condy PR，Laws RM（eds）Antarctic nutrient cycles and food webs. Springer，Berlin，pp 123-127

Tanner AC，Herbert RA（1981）Nutrient regeneration in Antarctic marine sediments. Kiel Meeresforsch Sanderh 5：390-395

Tiao G，Lee CK，McDonald IR，Cowan DA，Cary SC（2012）Rapid microbial response to the presence of an ancient relic in the Antarctic Dry Valleys. Nat Commun 3：660

Timmis KN，McGenity TJ，Meer JR，deLorenzo V（eds）（2010）Handbook of hydrocarbon and lipid microbiology. Springer，Berlin. doi：10.1007/978-3-540-77587-4

Tsiklinsky M（1908）La flore microbieene dans les regions due pole Sud. In：Expedition Antarctique Francais 1903-1905. 3：1-33

Tsyganov VA（1970）Detection and morphological cultural characteristics of *Actinomycetes* from the antarctic. Mikrobiologia 39：821-826 van Dorst J，Bissett A，Palmer AS，Brown M，Snape I，Stark JS，Raymond B，McKinlay J，Ji M，Winsley T，Ferrari BC（2014）Community fingerprinting in a sequencing world. FEMS Microbiol Ecol 89：316-330

Van Trappen S，Mergaert J，Van Eygen S，Dawyndt P，Cnockaert MC，Swings J（2002）Diversity of 746 heterotrophic bacteria isolated from microbial mats from ten Antarctic lakes. Syst Appl Microbiol 25：603-610

Van Trappen S，Tan TL，Yang J，Mergaert J，Swings J（2004）*Glaciecola polaris* sp. nov.，a novel budding and prosthecate bacterium from the Arctic Ocean，and emended description of the genus Glaciecola. Int J Syst Evol Microbiol 54：1765-1771

Vandieken V，Mussmann M，Niemann H，Jørgensen BB（2006）*Desulfuromonas svalbardensis* sp. nov. and *Desulfuromusa ferrireducens* sp. nov.，psychrophilic，Fe（III）-reducing bacteria isolated from Arctic sediments，Svalbard. Int J Syst Evol Microbiol 56：1133-1139

Vincent CF（1988）Microbial ecosystems of Antarctica. Cambridge University Press，Cambridge，p 303

Voytek MA，Ward BB（1995）Detection of ammonium-oxidizing bacteria of the beta-subclass of the class Proteobacteria in aquatic samples with the PCR. Appl Environ Microbiol 61：1444-1450

Wartiainen I，Hestnes AG，Svenning MM（2003）Methanotrophic diversity in high Arctic wetlands on the islands of Svalbard（Norway）-denaturing gradient gel electrophoresis analysis of soil DNA and enrichment cultures. Can J Microbiol 49：602-612

Watanabe T，Kojima H，Takano Y，Fukui M（2013）Diversity of sulfur-cycle prokaryotes in fresh-water lake sediments investigated using aprA as the functional marker gene. Syst Appl Microbiol 36：436-443

Wery N，Gerike U，Sharman A，Chaudhuri JB，Hough DW，Danson MJ（2003）Use of a packed-column bioreactor for isolation of diverse protease-producing bacteria from antarctic soil. Appl Environ Microbiol 69：1457-1464

Wilhelm RC，Niederberger TD，Greer C，Whyte LG（2011）Microbial diversity of active layer and permafrost in an acidic wetland from the Canadian High Arctic. Can J Microbiol 57：303-315

Wilkins D，Yau S，Williams TJ，Allen MA，Brown MV，DeMaere MZ，Lauro FM，Cavicchioli R（2013）Key microbial drivers in Antarctic aquatic environments. FEMS Microbiol Rev 37：303-335

Xiao X，Yin X，Lin J，Sun L，You Z，Wang P，Wang F（2005）Chitinase genes in lake sediments of Ardley Island，Antarctica. Appl Environ Microbiol 71：7904-7909

Yadav AN，Sachan SG，Verma P，Tyagi SP，Kaushik R，Saxena AK（2015）Culturable diversity and functional annotation of psychrotrophic bacteria from cold desert of Leh Ladakh（India）. World J Microbiol Biotechnol 31：95-108

Yau S，Lauro FM，Williams TJ，Demaere MZ，Brown MV，Rich J，Gibson JA，Cavicchioli R（2013）Metagenomic insights into strategies of carbon conservation and unusual sulfur biogeochemistry in a hypersaline Antarctic lake. ISME J 7：1944-1961

Yergeau E，Newsham KK，Pearce DA，Kowalchuk GA（2007a）Patterns of bacterial diversity across a range of Antarctic terrestrial habitats. Environ Microbiol 9：2670-2682

Yergeau E，Bokhorst S，Huiskes AH，Boschker HT，Aerts R，Kowalchuk GA（2007b）Size and structure of bacterial，fungal and nematode communities along an Antarctic environmental gradient. FEMS Microbiol Ecol 59：436-451

Yi H，Chun J（2006）*Flavobacterium weaverense* sp. nov. and *Flavobacterium segetis* sp. nov.，novel psychrophiles isolated from the Antarctic. Int J Syst Evol Microbiol 56：1239-1244

Yu Y，Li HR，Chen B，Zeng YX，He JF（2006）Phylogenetic diversity and cold-adaptive hydrolytic enzymes of culturable psychrophilic bacteria associated with sea ice from high latitude ocean，Arctic. Wei Sheng Wu Xue Bao 46：184-190

Yu Y，Li HR，Zeng YX，Chen B（2011）Bacterial diversity and bioprospecting for cold-active hydrolytic enzymes from culturable bacteria associated with sediment from Nella Fjord，Eastern Antarctica. Mar Drugs 9：184-195

Zhang DC，Yu Y，Chen B，Wang HX，Liu HC，Dong XZ，Zhou PJ（2006）*Glaciecola psychrophila* sp. nov.，a novel psychrophilic bacterium isolated from the Arctic. Int J Syst Evol Microbiol 56：2867-2869

Zhang DC，Li HR，Xin YH，Liu HC，Chen B，Chi ZM，Zhou PJ，Yu Y（2008）*Marinomonas arctica* sp. nov.，a psychrotolerant bacterium isolated from the Arctic. Int J Syst Evol Microbiol 58：1715-1718

Zhang G，Cao T，Ying J，Yang Y，Ma L（2014）Diversity and novelty of actinobacteria in Arctic marine sediments. A Van Leeuw 105：743-754

Zhao J，Yang N，Zeng R（2008）Phylogenetic analysis of type I polyketide synthase and nonribosomal peptide synthetase genes in Antarctic sediment. Extremophiles 12：97-105

Zheng Y，Yang W，Sun X，Wang SP，Rui YC，Luo CY，Guo LD（2012）Methanotrophic community structure and activity under warming and grazing of alpine meadow on the Tibetan Plateau Appl Microbiol Biotechnol. doi：10. 1007/s00253-011-3535-5

Zhou J，Davey ME，Figueras JB，Rivkina E，Gilichinsky D，Tiedje JM（1997）Phylogenetic diversity of a bacterial community determined from Siberian tundra soil DNA. Microbiology 143：3913-3919

第三章

极地适冷细菌和真菌的抗菌潜力

Angelina Lo Giudice[1]，Renato Fani[2]

3.1 引言

地球的生物圈，一年大多数时间（约 85%）都处在低于 5℃的寒冷环境中。在地球寒冷的栖息地中，极地占 14%。而在如此寒冷的条件下常常会产生一系列环境压力，包括干燥、营养匮乏、盐度过高、有害太阳辐射和低生化活性等（Pearce 2012）。虽然如此恶劣的条件并不能满足大部分生命形式的生存条件，但还是存在数量丰富的有机体在这样寒冷的栖息地中成功繁衍，尤其是微生物。其无论是在生物多样性上，还是在生物量上都远胜于其他有机体（Feller and Gerday 2003；Margesin 2007；Pearce 2012）。根据它们的主要生存温度的不同，适冷微生物通常又被分为嗜冷微生物和耐冷微生物（Morita 1975）。嗜冷微生物的最佳生长温度低于 15℃，甚至不能在超过 20℃的环境下生长。而耐冷微生物的生长温度范围较广，超过 20℃时会达到最快生长速度。因此，热敏感的嗜冷微生物必须永远生活在寒冷的环境中，而耐冷微生物绝大多数存在于季节交替和昼夜温度波动的环境中（Margesin 2007）。在全球范围内，耐冷微生物所占比例大于嗜冷菌（Pearce 2012）。

因为适冷微生物长时间遭受巨大的环境压力，所以它们进化出了各种各样的生理结构和生理特性，以确保能够在特殊的环境下存活（Pearce 2012）。如低温下冷活性酶的三维结构灵活性会增强，产生冷适应蛋白（Caps）和冷休克蛋白（Csps），细胞膜中不饱和脂肪酸和类胡萝卜素的含量也会升高，用来维持膜的流动性和渗透性，此外还能合成一些抗冻物质（Margesin et al. 2007；Russell 2008）。除了细胞修饰以外，这些适冷微生物的冷活性拮抗性质还可以影响其他竞争微生物的生存，从而有助于微生物适应永久的低温（Lo Giudiceet al. 2007a；Manganoet al. 2009；Prasadet al. 2011；Bellet al. 2013）。与此同时，这些特性还预示了适冷微生物可用于新型抗菌化合物的工业化开发。因此，各种复杂和深奥的与适冷微生物生态学息息相关的生存策略研究为生物技术发展提供了宝贵的理论基础（Cavicchioli et

[1] Institute for Coastal Marine Environment，National Research Council（IAMC-CNR），Spianata San Raineri 86，I-98124 Messina，Italy；Department of Biological and Environmental Sciences，University of Messina，Viale F. Stagno d'Alcontrès 31，I-98166 Messina，Italy. e-mail：angelina. logiudice@iamc. cnr. it.

[2] Department of Biology，University of Florence，Via Madonna del Piano 6，50019 Sesto Fiorentino，Firenze，Italy.

al. 2002)。

3.2 极地适冷微生物的抗生素潜能

公共健康面临的严峻问题主要包括全球对抗生素的耐药性增强、多重耐药性病原体的发展和新型传染病的出现。此外，在最近几十年里，很少有新型抗生素被发现。为了解决这一急迫的问题，开发新型有效抗生素已成为天然药物化学重要的探索方向。在历史上，大多数具有生物活性的微生物产物都是从放线菌和丝状真菌中获得的，并且主要来自陆地环境(Biondi et al. 2008)。目前，探索不常见、尚未被充分开发的、有医学价值的资源，筛选以前较少被利用的微生物，是发现新型天然药物的有效方法。有趣的是，在极端环境中生存的微生物会在特殊的条件下产生活性生物分子，可见极端环境下的微生物是新型代谢产物的宝贵来源，包括具有独特结构和特殊生物活性的抗菌化合物（Hemala et al. 2014)。由于这些原因，生产抗菌化合物（主要是抗细菌和抗真菌）的适冷微生物（异养和自养的细菌和真菌）已从南北极中各种水生与陆生环境中被分离出来。一些针对人类病原体的研究已经开展。在有些研究中，微生物抑制剂也已经被提取（部分或完全），并研究了其特性。来自极地地区的微生物的多重抗微生物潜力将在以下部分讨论。

3.2.1 陆地环境

土壤一直是开发程度最高的，发现有价值天然生物产品最多的生态环境，研究主要针对放线菌和真菌。这些放线菌和真菌广泛存在于土壤中，并能产生多种的次生代谢产物和化合物，这些次生代谢产物和化合物表现出不同的生物特性。本节概述了来自极地土壤的具有抗菌可能性的细菌和真菌。

3.2.1.1 放线菌

对来自极地土壤的生物活性筛选已经主要集中在放线菌上（Moncheva et al. 2002；Nedialkova and Naidenova 2005；Gesheva 2010；Lee et al. 2012a；Pan et al. 2013)。这类细菌代表了几十年来最有成效的抗生素来源。新发现的抗生素大部分（80%）来源于链霉菌属(放线菌目)，它们具有的生物合成能力在微生物领域中没有竞争对手（到目前为止）(Bérdy 2005；Bull and Stach 2007；Manivasagan et al. 2014)。

然而，现在从常见链霉菌属中发现新的商业化次生代谢产物变得越来越困难，因此迫切需要分离和测试这个属中的新成员（Lyutskanova et al. 2009)。如来自未完全开发的栖息地的链霉菌。正如下文讨论结果所强调的，极地土壤无疑是新型生物活性化合物的丰富来源。活性放线菌的分离株（包括链霉菌属）如表 3.1 所示。

Pan 等（2013）选择与耐冷链霉菌 *Streptomyces beijiangensis*（西格尼岛，南奥克尼群岛）密切相关的 46 株南极放线菌菌株进行检测，发现它们对变形杆菌和金黄色葡萄球菌有抗菌活性，然而对大肠杆菌没有表现出活性（在表 3.1 中 46 株被分离出的白色链霉菌中，*Streptomyces* sp. PSY097 作为代表菌被报道)。发现所有的菌株都含有非核糖体肽合成酶基因（*NRPS*)。据 Gesheva（2010）报道，在能合成胞外活性产物的放线菌中，来自 *Streptomyces* sp. 5 的生物质的甲醇提取物表现出对革兰氏阳性菌和植物病原真菌的抗菌活性。另一种嗜冷链霉菌（菌株 8）能够抑制革兰氏阳性菌、酵母和植物病原真菌的生长（Gesheva 2009)。Gesheva 和 Negoita（2012）进一步报道了 *Streptomyces* spp. 的抗细菌或/和抗真菌活性（针对枯草芽孢杆菌、热带假丝酵母和枝孢菌)，其中 *Streptomyces* spp. 10 和 21 来自南极洲的哈斯韦尔岛土壤，但没有报道它们各自的抗菌谱。

表 3.1　来自极地土壤中的活性放线菌

地域	分离物	来源地	目标菌株①	生物活性分子的主要特征	参考文献
南极	*Arthrobacter* sp. HPG8	南极洲东部	**7,14,15,21,25**	蛋白质性质	O'Brien et al.(2004)
	Arthrobacter sp. HPH17	南极洲东部	**7,14,25**	蛋白质性质	O'Brien et al.(2004)
	Arthrobacter sp. 1	南极洲东部	**5,9,11,23,25**		Gesheva(2010)
	Arthrobacter sp. 9	南极洲东部	**5,23,25**		Gesheva(2010)
	Brevibacterium spp. BV2, BV34 and BV35	巴里恩托斯岛(Barrientos Island)	**8**		Lee et al.(2012a)
	Brevibacterium sp. BV37	巴里恩托斯岛	**17,25**		Lee et al.(2012a)
	Demetria spp. DT40 and DT41	巴里恩托斯岛	**25**		Lee et al.(2012a)
	Gordonia spp. G3 and G48	巴里恩托斯岛	**8,25**		Lee et al.(2012a)
	Janibacter sp. JB26	巴里恩托斯岛	**17**		Lee et al.(2012a)
	Kocuria sp. KC21	巴里恩托斯岛	**17,25**		Lee et al.(2012a)
	Lapillicoccus sp. LC31	巴里恩托斯岛	**8**		Lee et al.(2012a)
	Micromonospora sp. 18	南极洲东部	**5,23,25**		Gesheva(2010)
	Micromonospora sp. MM6	巴里恩托斯岛	**17**		Lee et al.(2012a)
	Micromonospora sp. MM32	巴里恩托斯岛	**20**		Lee et al.(2012a)
	Nocardioides sp. ND52	巴里恩托斯岛	**8**		Lee et al.(2012a)
	Nocardioides sp. A-1	南极洲东部	**5,16,24,27**	糖脂和/或脂肽性质	Gesheva and Vasileva-Tonkova(2012)
	Rhodococcus sp. 2	南极洲东部	**23**		Gesheva(2010)
	Rhodococcus sp. RC56	巴里恩托斯岛	**25**		Lee et al.(2012a)
	Streptomyces sp. PSY097	西格尼岛(Signy Island)	**19,25**		Pan et al.(2013)
	Streptomyces sp. NTK 97	特拉诺瓦湾(Terra Nova Bay)	**5,25**	frigocyclinone	Bruntner et al.(2005)
	Streptomyces flavovirens 6[7]	利文斯顿岛(Livingston Island)			Ivanova et al.(2002)
	Streptomyces sp. 5	南极洲东部	**5,9,11,23,25**	放线菌素	Gesheva(2010)
	Streptomyces sp. 8	南极洲东部	**2,5,6,9,10,11,13,18,22,23,25**	非多烯大环内酯类抗生素,阿扎霉素(azalomycin)B和尼日利亚菌素(nigericin)	Gesheva(2009)

续表

地域	分离物	来源地	目标菌株①	生物活性分子的主要特征	参考文献
北极	*Streptomyces* spp. SB9, SB72 and SB81	斯瓦尔巴群岛 (Svalbard Islands)	**1,3,4,5,9,12,18,22,23,25,26**		Lyutskanova et al. (2009)
	Streptomyces spp. SB33 and SB47	斯瓦尔巴群岛	**1,3,4,5,9,18,22,23,25,26**		Lyutskanova et al. (2009)

① 目标菌株：**1.** *Acinetobacter johnsonii*；**2.** *Aspergillus niger*；**3.** *Bacillus megatherium*；**4.** *B. mycoides*；**5.** *B. subtilis*；**6.** *Botrytis cinerea*；**7.** *Brochothrix thermosphacta*；**8.** *Candida albicans*；**9.** *C. tropicalis*；**10.** *C. utilis*；**11.** *Cladosporium cladosporioides*；**12.** *Escherichia coli*；**13.** *Fusarium oxysporum*；**14.** *Listeria innocua*；**15.** *L. monocytogenes*；**16.** *Micrococcus* sp.；**17.** MRSA；**18.** *Penicillium chrysogenum*；**19.** *Proteus vulgaris*；**20.** *Pseudomonas aeruginosa*；**21.** *P. fragi*；**22.** *Saccharomyces cerevisiae*；**23.** *Sarcina lutea*；**24.** *Staphylococcus* sp.；**25.** *S. aureus*；**26.** *Trichosporon cutaneum*；**27.** *Xanthomonas oryzae*。

在斯匹次卑尔根群岛（北冰洋）采集的多年冻土中的91种耐冷链霉菌菌株中，有五分之一对革兰氏阳性和革兰氏阴性细菌、酵母以及真菌表现出很强的抗菌活性（Lyutskanova et al. 2009）。其无细胞上清液的薄层色谱（TLC）结果显示出这是由三种主要化合物组成的抗菌复合物，从 *Streptomyces* sp. SB9 的培养物中获得的上清液中抗菌物质浓度最高，但是没有报道介绍该抗菌复合物的生物化学特性。

除链霉菌属以外，活性放线菌也已从极地土壤中被分离了出来。来自 *Micromonospora* sp. 18 甲醇提取物显示出良好的抗菌活性（Gesheva 2010）。Lee 等（2012）从来自南极巴里恩托斯岛不同位置的放线菌菌株中筛选出能产生具有抗菌和抗真菌活性的次生代谢物的菌株。使用四种筛选模型，用于鉴定七种针对白念珠菌、七种针对金黄色葡萄球菌、四种针对耐甲氧西林抗性金黄色葡萄球菌（MRSA）和一种针对铜绿假单胞菌的有活性的分离物。最具生物活性的属是具有四种生物活性分离物的短杆菌属。通过16S rRNA基因的序列同源性比较，结合系统发育分析，分离的五个物种（分别属于 *Demetria*，*Nocardioides*，*Lapillicoccus* 和 *Rhodococcus*）中可能被鉴定到新的属或种，这意味着可能发现具有抗菌活性的新型菌株。

Gesheva 和 Vasileva-Tonkova（2012）测试了类诺卡氏菌 *Nocardioides* sp. A-1 在有不同碳源补充的矿物盐培养基的无细胞上清液培养物的抗菌活性。后者有利于生产对革兰氏阳性和革兰氏阴性细菌，特别是金黄色葡萄球菌和米氏黄单胞菌，具有抗菌活性的广谱化合物。通过 TLC 的初步分析显示，具有抗菌活性的化合物主要是糖脂和/或脂肽，这取决于培养该菌所使用的碳源。

3.2.1.2 其他细菌

表3.2中报道了极地土壤中除放线菌外产生抗生素的异养菌，最初的研究由 O'Brien 等（2004）进行，通过筛选来自南极洲东部土壤的4496个细菌分离株，用于生产可能用于冷冻食品保藏的冷活性抗菌物质。有四种抑菌株（占总分离株的0.29%）隶属于节杆菌属、平球菌属和假单胞菌属。

这些菌株合成的抑制剂对蛋白酶的敏感性揭示了它们的本质即蛋白质。来自地杆菌 *Pedobacter* sp. BG5、假单胞菌 *Pseudomonas* spp. CG21 和 MTC3 的抗菌剂也是如此。它们从分离自南极土壤的2465个细菌分离株当中被筛选出来，对一种或多种食源性病原体的生长具有抑制作用（Wong et al. 2011）。而来自假单胞菌的另外三种有活性的抑制剂（即

WEK1，WEA1 和 MA2）对过氧化氢酶、脂肪酶、α-淀粉酶和蛋白酶不敏感。

表 3.2 来自南极土壤的活性蓝细菌和异养细菌

门或类	分离物	来源地	目标菌株①	备注	参考文献
Alphaproteobacteria	*Bradyrhizobium* sp. BR45	巴里恩托斯岛	**5**		Lee et al.（2012b）
	Bradyrhizobium sp. BR42	巴里恩托斯岛	**5**		Lee et al.（2012b）
	Bradyrhizobium sp. BR62	巴里恩托斯岛	**5**		Lee et al.（2012b）
	Bradyrhizobium sp. BR65	巴里恩托斯岛	**5**		Lee et al.（2012b）
	Bradyrhizobium sp. BR82	巴里恩托斯岛	**5**		Lee et al.（2012b）
	Bradyrhizobium sp. BR88	巴里恩托斯岛	**5**		Lee et al.（2012b）
	Bradyrhizobium sp. BR96	巴里恩托斯岛	**5**		Lee et al.（2012b）
	Bradyrhizobium sp. BR100	巴里恩托斯岛	**5**		Lee et al.（2012b）
	Bradyrhizobium sp. BR105	巴里恩托斯岛	**5**		Lee et al.（2012b）
	Methylobacterium sp. MB63	巴里恩托斯岛	**18**		Lee et al.（2012b）
	Methylobacterium sp. MB104	巴里恩托斯岛	**5**		Lee et al.（2012b）
	Methylobacterium sp. MB20	巴里恩托斯岛	**5**		Lee et al.（2012b）
	Paracoccus sp. PC101	巴里恩托斯岛	**5**		Lee et al.（2012b）
	Sphingomonas sp. SM14	巴里恩托斯岛	**5**		Lee et al.（2012b）
Betaproteobacteria	*Janthinobacterium* sp. SMN33.6	菲尔德斯半岛（Fildes Peninsula）	**1,10,11,14,18**		Asencio et al.（2009）
Gammaproteobacteria	*Pseudomonas* sp. CrCD21	南极洲东部	**4,12,19**	蛋白质	O'Brien et al.（2004）
	Pseudomonas sp. CG21	乔治王岛（King George Island）	**9,10**	蛋白质	Wong et al.（2011）
	Pseudomonas sp. MTC3	乔治王岛	**9,10,11,16**	蛋白质	Wong et al.（2011）
	Pseudomonas sp. WEK1	乔治王岛	**20**		Wong et al.（2011）
	Pseudomonas sp. WEA1	乔治王岛	**20**		Wong et al.（2011）
	Pseudomonas sp. MA2	乔治王岛	**20**		Wong et al.（2011）
Bacteroidetes	*Pedobacter* sp. BG5	乔治王岛	**2,9,10,11,17**	蛋白质	Wong et al.（2011）
Firmicutes	*Planococcus* sp. CHF8	南极洲东部	**4,12,13**	蛋白质	O'Brien et al.（2004）
	Enterococcus sp. APR 210	施尔马赫绿洲（Schirmacher Oasis）	**5,6,11,12,13**	类细菌素	Shekh et al.（2011）
Cyanobacteria	*Fischerella* sp.	哈斯韦尔岛	**3,7,8**		Gesheva and Negoita（2012）

① 目标菌株：**1.** *Acinetobacter baumannii*；**2.** *Bacillus cereus*；**3.** *B. subtilis*；**4.** *Brochothrix thermosphacta*；**5.** *Candida albicans*；**6.** *C. krusei*；**7.** *C. tropicalis*；**8.** *Cladosporium cladosporioides*；**9.** *Enterobacter cloacae*；**10.** *Escherichia coli*；**11.** *Klebsiella pneumoniae*；**12.** *Listeria innocua*；**13.** *L. monocytogenes*；**14.** *Pseudomonas aeruginosa*；**15.** *P. putida*；**16.** *Salmonella enterica*；**17.** *Salmonella* spp.；**18.** *Serratia marcescens*；**19.** *Staphylococcus aureus*；**20.** *Vibrio parahaemolyticus*。

Shekh 等人（2011）选择了来自南极企鹅群的肠球菌 *Enterococcus* sp. APR 210，因为它具有抑制多重耐药性真菌病原性白念珠菌 *C. albicans* NCIM 3471 和克鲁氏假丝酵母

(*Candida krusei*) 生长的能力，并具有抗细菌活性。他们研究了酶、温度和 pH 对肠球菌 APR 210 无菌上清液的抗真菌活性的影响。在 100℃和 121℃下肠球菌 APR 210 的生物活性完全丧失，而在反复冻融和在−20℃以及−80℃长期储存后仍能保持生物活性。在 pH 值为 2、4 和 10 时，活性完全丧失。然而，在 pH 值 6、6.9 和 8 处没有观察到活性损失。抑制念珠菌致病原理是由于其对蛋白酶 K（其引起抗菌活性的完全丧失）和链霉蛋白酶 E（其导致生物活性部分丧失）敏感。用胰蛋白酶处理对生物活性没有影响。结果表明，肠球菌 APR 210 产生的抗菌化合物 APR 210 属于Ⅱ类细菌素或类细菌素抑制剂的蛋白质。Lee 等人（2012）测试了来自南极巴里恩托斯岛（Barrientos Island）土壤的 57 株变形菌的抗细菌和抗真菌能力，筛选后，共鉴定了 14 种产生生物活性代谢物的菌株。*Bradyrhizobium*、*Paracoccus* 和 *Sphingomonas* 属的菌株对白念珠菌有活性，来自甲基杆菌属（*Methylobacterium*）的三株分离菌显示出对金黄色葡萄球菌和白念珠菌有生物活性，而对铜绿假单胞菌或 MRSA（耐甲氧西林金黄色葡萄球菌）没有生物活性。

来自南极土壤的紫色杆菌属 *Janthinobacterium* sp. SMN 33.6 的乙醇提取物被证实对医源性多抗性分离株有抗菌活性，例如黏质沙雷氏菌［MIC（最低抑菌浓度）为 0.5～2μg/mL］、铜绿假单胞菌（MIC＝1μg/mL）、肺炎克雷伯菌（MIC＝16μg/mL）、大肠杆菌（MIC 为 0.5～1μg/mL）和鲍氏不动杆菌（MIC＝1μg/mL）（Asencio et al. 2009）。

在细菌中，来自极地土壤的蓝细菌（cyanobacteria）很少被报道为抗菌剂的生产者。Gesheva 和 Negoita（2012）发现侧生藻 *Fischerella* sp.（哈斯韦尔群岛，南极）可抑制枯草芽孢杆菌（*B. subtilis*）、热带假丝酵母菌（*C. tropicalis*）和子囊菌（*C. cladosporioides*）的生长。

3.2.1.3　真菌

真菌可能代表了一种新的天然生物活性分子的重要来源，在真菌中，青霉属（*Penicillium*）和曲霉属（*Aspergillus*）的菌株通常在土壤中大量存在，它们的抗菌活性经常被报道，包括极地土壤发现的青霉属和曲霉属（表 3.3）。

表 3.3　来离自极地土壤的具有抗菌活性的子囊菌

分离物	来源地	目标菌株[①]	备注	参考文献
Aspergillus nidulans	哈斯韦尔岛，南极洲	未注明		Gesheva and Negoita (2012)
Aspergillus sydowii 9541	埃尔斯沃思山(Ellsworth Mountains)，南极洲	**4,7**		Godinho et al. (2015)
Geomyces sp. 2481	乔治王岛，南极洲	**1,5,8**	地霉素(geomycin)A～C	Li et al. (2008)
Penicillium allii-sativi 9451	埃尔斯沃思山，南极洲	**8**		Godinho et al. (2015)
P. allii-sativi 9458	埃尔斯沃思山，南极洲	**4**		Godinho et al. (2015)
P. allii-sativi 9508	埃尔斯沃思山，南极洲	**8**		Godinho et al. (2015)
P. allii-sativi 9524	埃尔斯沃思山，南极洲	**8**		Godinho et al. (2015)
P, brevicompactum 9446	埃尔斯沃思山，南极洲	**8**		Godinho et al. (2015)
P, brevicompactum 9448	埃尔斯沃思山，南极洲	**4,8**		Godinho et al. (2015)
P. chrysogenum 9466	埃尔斯沃思山，南极洲	**4**		Godinho et al. (2015)

续表

分离物	来源地	目标菌株①	备注	参考文献
P. chrysogenum 9534	埃尔斯沃思山,南极洲	**8**		Godinho et al.(2015)
P. griseofulvum	格陵兰		灰黄霉素富里酸菌丝酰胺(mycelianamide)、异烟棒曲霉素(roquefortine)C和D、裸麦角碱(chanoclavine)I、野麦角碱(elymoclavine)	Frisvad et al.(2004)
P. griseofulvum strain VKM FW-2251	科雷马低地(Kolyma lowland),俄罗斯	未注明	灰黄霉素	Kozlovsky et al.(2012)
P. nalgiovense Laxa	南极洲	**3,5,8**	两性霉素(amphotericin)B	Svahn et al.(2015)
P. rubens 9496	埃尔斯沃思山,南极洲	**4**		Godinho et al.(2015)
P. verrucosum	哈斯韦尔岛(Haswell Island),南极洲	未注明		Gesheva and Negoita (2012)
Phoma herbarum CCFEE 5015	干燥谷(Dry Valleys),南极洲	**2,5,6,7**		Onofri et al.(2000)
Phoma herbarum CFEE 5007	干燥谷,南极洲	**2,5,6,7**		Onofri et al.(2000)
Phoma herbarum CCFEE 5020	干燥谷,南极洲	**2,5,6,7**		Onofri et al.(2000)
Phoma herbarum CCFEE 459	干燥谷,南极洲	**2,5,6,7**		Onofri et al.(2000)

① 目标菌株:**1.** *Aspergillus fumigatus*;**2.** *Bacillus subtilis*;**3.** *Candida albicans*;**4.** *Cladosporium sphaerospermum*;**5.** *Escherichia coli*;**6.** *Pseudomonas putida*;**7.** *Sarcina* sp.;**8.** *Staphylococcus aureus*。

根据从南北极长久冻土得到的青霉亚属(青霉属)菌株的次级代谢产物的特性,来研究它们的抗菌作用,并阐明它们的分类学地位(Kozlovsky et al. 2012)。从科雷马河(Kolyma)洼地(俄罗斯)得到的 *Penicillum griseofulvum* VKM FW-2251 菌株能合成一种与灰黄霉素理化性质相同的代谢物,这种聚酮化合物代谢物是一种抗真菌抗生素,但由于其高毒性,目前很少使用。Gesheva 和 Negoita(2012)报道了来自南极土壤的 *P. verrucosum* 和 *A. nidulans* 菌株可产生抗生素,但没有详细报道它们各自的抗菌谱。从南极土壤和岩石中分离的另外 14 株真菌也被筛选用于生产抗生素物质(Onofri et al. 2000)。其中来自土壤的草茎点霉菌 *Phoma herbarum* CCFEE 5015 和 CCFEE 5007,来自岩石的菌株 CCFEE 5020 和 CCFEE 459 都显示出抑制活性。通过进一步实验表明,在 5~25℃ 的温度范围内,生物活性最高的菌株 CCFEE 5020 的抗生素活性较高,而在 30℃ 会降低。搅拌和通气会对菌株的生长和抗生素生产产生强烈影响,这表明该种菌株需要高浓度的氧。

最近,Godinho 等人(2015)报道了来自寒冷寡营养的南极土壤(Ellsworth Mountains)

的具有抗细菌和抗真菌活性的 17 种的真菌分离物（表 3.3），主要隶属于曲霉属和青霉属。生物活性提取物通常含有脂肪酸官能团和甘油三酯。由于在芳香族和烯烃区域中存在活性质子，因此形成了高度功能化的次生代谢物。

3.2.2 极地湖泊和池塘

对于栖息在极地地区湖泊的微生物的抗菌活性的研究较少，主要是对南极湖泊底栖生物研究。由于极端的气候条件和没有更高级的后生动物，它们已经生存了几千年，几乎未受干扰，属于密集群落的微生物，其合成的抗生素和/或毒素可能给它们带来了生存优势（Biondi et al. 2008）。

3.2.2.1 异养菌

Rojas 等人（2009）从在南极湖底生长的微生物群落中分离了细菌菌株（表 3.4）。其中，122 株显示出对革兰氏阳性金黄色葡萄球菌有抗菌活性，并对屎肠球菌（*E. faecium*）有较低程度的抗菌活性。同时对革兰氏阴性大肠杆菌具有抑菌活性，这些菌株中的少数菌株还对新型隐球菌（*Cryptococcus neoformans*）、烟曲霉（*A. fumigatus*）有活性，并对白念珠菌（*C. albicans*）具有较低的活性。活性菌株隶属于 α-、β-和 γ-变形菌纲，拟杆菌科，(G+C) 含量不同的革兰氏阳性菌的几个谱系。具有相对较强抗菌活性的菌株的提取物，其 LC-MS 组分分离证明了其化学结构的新颖性，并对其进行了进一步研究。其中两株细菌，属于节杆菌属的 R-7513［来自弗里克塞尔湖（Lake Fryxell）］和 R-7941［来自霍尔湖（Lake Hoare)］，会产生对革兰氏阳性细菌有活性的有效抗菌化合物。由于这些化合物与盐屋霉素（siomycin）、环噻唑霉素或硫肽霉素（具有与革兰氏阳性细菌如耐甲氧西林金黄色葡萄球菌相似的抗菌谱）有关，它们可能代表新的环状噻唑肽类抗生素。从紫色杆菌 *Janthinobacterium* sp. R-7687 和假单胞菌 *Pseudomonas* sp. R-12535 中获得的生物活性提取物，初步质谱结果显示，它们的结构与任何已知的化合物都不相同，表明这些南极细菌的抗菌活性可能依赖于新型的化合物的产生。

表 3.4 极地湖泊和池塘中的活性细菌、蓝细菌和真菌

微生物	门或类	分离物	来源地	目标菌株①	备注	参考文献
细菌	Bacteroidetes	*Flavobacterium* sp. Ant342	施尔马赫绿洲	**7**	黄色素(flexirubin)	Mojib et al. (2010)
	Betaproteobacteria	*Janthinobacterium* sp. Ant5-2	施尔马赫绿洲	**7**	紫菌素(violacin)	Mojib et al. (2010)
		Janthinobacterium sp. R-7687	干燥谷	**5,6,10**		Rojas et al. (2009)
	Gammaproteobacteria	*Shewanella* sp. R-8990	维斯特福丘陵(Vestfold Hills)	**6,10**		Rojas et al. (2009)
		Pseudomonas sp. R-12565	拉斯曼丘陵(Larsemann Hills)	**5,6,10**		Rojas et al. (2009)
		Pseudomonas sp. R-12533	拉斯曼丘陵	**5,6,9,10**		Rojas et al. (2009)
		Pseudomonas sp. R-12535	拉斯曼丘陵	**6,9,10**		Rojas et al. (2009)
		Psychrobacter sp. R-12597	拉斯曼丘陵	**6,10**		Rojas et al. (2009)

续表

微生物	门或类	分离物	来源地	目标菌株①	备注	参考文献
细菌	Actinobacteria	*Arthrobacter* spp. R-7513 and R-7941	干燥谷	**5,10**	环状噻唑肽(cxyclic thiazolyl peptide)	Rojas et al.(2009)
	Cyanobacteria	*Leptolyngbyaantarctica* ANT. LG2. 3	未标明	**10**		Biondi et al.(2008)
		Leptolyngbyaantarctica ANT. LG2. 5	未标明	**10**		Biondi et al.(2008)
		Leptolyngbyaantarctica ANT. L18. 2	未标明	**10**		Biondi et al.(2008)
		Nostoc sp. ANT. L34. 1	未标明	**10**		Biondi et al.(2008)
		Nostoc sp. ANT. LPR. 1	未标明	**4**		Biondi et al.(2008)
		Nostoc ANT. L52B. 1	未标明	**4,10**		Biondi et al.(2008)
		Nostoc ANT. L36. 1	未标明	**4,10**		Biondi et al.(2008)
		Nostoc ANT. LG2. 6	未标明	**4,10**		Biondi et al.(2008)
		Phormidiummurray ANT. PE. 1	未标明	**10**		Biondi et al.(2008)
		Phormidiumpriesteyi ANT. LPR. 6	未标明	**10**		Biondi et al.(2008)
		Phormidiumpriesteyi ANT. L61. 2	未标明	**10**		Biondi et al.(2008)
		Phormidiumpriesteyi ANT. L52. 4	未标明	**1,4**		Biondi et al.(2008)
		Phormidiumpriesteyi ANT. L52. 6	未标明	**1,4,10**		Biondi et al.(2008)
		Pseudophormidium ANT. LG2. 1	未标明	**4**		Biondi et al.(2008)
		Pseudophormidium ANT. LPR. 2	未标明	**10**		Biondi et al.(2008)
		Pseudophormidium ANT. LPR. 3	未标明	**4,10**		Biondi et al.(2008)
		Pseudophormidium ANT. LG2. 2	未标明	**4,10**		Biondi et al.(2008)
真菌	Ascomycota	*Aspergillus clavatus* IWW 447	萨拉塔恩湖(Lake Sarah Tarn)	**2,3,5,6,10**		Brunati et al.(2009)
		Aspergillus niger IWW 1026	Pendant 湖	**10**		Brunati et al.(2009)
		Beauveria sp. IWW 1017	Pendant 湖	**3**		Brunati et al.(2009)
		Cladosporium sp. IWW 1019	Pendant 湖	**6**		Brunati et al.(2009)
		Penicillium sp. IWW 1054	Pendant 湖	**2,3,6**		Brunati et al.(2009)
		Penicillium sp. IWW 1059	埃斯湖(Lake Ace)	**2,3,6**		Brunati et al.(2009)
		P. chrysogenum IWW 1053	Highway 湖	**5,6,10**		Brunati et al.(2009)
		P. chrysogenum IWW 1055	Pendant 湖	**1,6,10**		Brunati et al.(2009)
		P. chrysogenum TF 3/3	Tarn Flat	**6,8,10**	β-内酰胺	Montemartini Corte et al.(2000)

续表

微生物	门或类	分离物	来源地	目标菌株①	备注	参考文献
真菌	Ascomycota	*P. chrysogenum* G 3/2	冈瓦纳(Gondwana)	**8,10**		Montemartini Corte et al. (2000)
		P. citrinum S1/4R	斯库亚湖(Skua Lake)	**8,10**		Montemartini Corte et al. (2000)
		P. citrinum S1/4bis	斯库亚湖	**8,10**		Montemartini Corte et al. (2000)
		P. crustosum IWW 1023	瓦茨湖(Lake Watts)	**6**		Brunati et al. (2009)
		P. roseopurpureum S1/3 bis	斯库亚湖	**8,10**		Montemartini Corte et al. (2000)
		P. waksmanii S1/5	斯库亚湖	**6,10**		Montemartini Corte et al. (2000)
		P. waksmanii II/3	Inexepress 岛	**6,8,10**		Montemartini Corte et al. (2000)
		P. waksmanii G3/17	冈瓦纳	**6,10**		Montemartini Corte et al. (2000)
		P. waksmanii G3/18	冈瓦纳	**6,10**		Montemartini Corte et al. (2000)

①目标菌株：**1.** *Aspergillus fumigatus*；**2.** *Candida albicans*；**3.** *C. neoformans*；**4.** *Cryptococcus neoformans*；**5.** *Enterococcus faecium*；**6.** *Escherichia coli*；**7.** *Mycobacterium tubercolosis*；**8.** *Micrococcus luteus*；**9.** *Pseudomonas aeruginosa*；**10.** *Staphylococcus aureus*。

3.2.2.2 蓝细菌

蓝细菌的成员代表着来自底栖垫的微生物群落。Taton 等人（2006）和 Biondi 等人（2008）首次报道了一些南极蓝细菌的大规模培养和药物筛选，用于开发和生产新型药物先导分子（表 3.4）。从拉斯曼丘陵、伯林根群岛（Bølingen Islands）、韦斯特福尔山脉、劳尔群岛（Rauer Islands）和麦克马多干河谷（McMurdo Dry Valley）的 23 个湖泊和池塘收集的 27 个底栖微生物样品的分离菌株。其中 17 种蓝细菌显示出对金黄色葡萄球菌、丝状真菌烟曲霉（*A. fumigatus*）或新型隐球酵母菌（*C. neoformans*）的抗菌活性。对于假席藻属（*Pseudophormidium*）和念珠藻属（*Nostoc*），表现出的抗菌活性最高。7 个分离株显示仅对金黄色葡萄球菌的抗菌活性，3 个分离株显示仅抗真菌活性和 7 株分离株既具有抗细菌活性也具有抗真菌活性。从同一个湖泊中分离的 *P. priestleyi* 菌株 ANT. L52.4 和 ANT. L52.6 显示出有效的抗真菌和/或抗细菌活性，并展开了进一步分析。在相同的洗脱时间下，这两株菌对烟曲霉有抑制活性的部分，其色谱图非常相似。在同一 LC-MS 系统中，*Pseudophormidium* sp. ANT. LPR. 2、*L. antarctica* ANT. LG2. 3 和 *Nostoc* sp. L34. 1 显示，对金黄色葡萄球菌有抑制活性的组分有相似的洗脱时间，表明 3 种菌产生了在化学性质上相似的抗菌化合物，可见生物活性是菌株特异性而不是物种特异性的。现在还没有鉴定出这些抗菌代谢物，因此推测这些南极蓝细菌的抗菌活性是由于产生了新型的化合物。

3.2.2.3 真菌

Brunati 等人（2009）从南极湖泊微生物垫中分离出了 47 种具有抗菌活性的丝状真菌（表 3.4）。它们中的大多数抑制金黄色葡萄球菌（14%）、大肠杆菌（10%）、白念珠菌（11%）和新生隐球菌（8%）的生长，而对肠杆菌和丝状真菌都不太敏感。最具生物活性的真菌是耐冷的丝状真菌，如青霉属、曲霉属、白僵菌属和枝孢霉属。适冷的南极丝状真菌 *Thelebolus* 产生的抗菌活性物质比青霉属、曲霉属和枝孢霉属少。从棒曲霉 *A. clavatus* IWW 447 和黑曲霉 *A. niger* IWW 1026 提取物的 LC-MS 分析结果表明其具有一种新型的抗生素分子。从韦斯特福尔山脉（Vestfold Hills）的盐水湖中分离的产黄青霉菌株 *P. chrysogenum* IWW 1056 中的主要产物是两种生物活性双蒽醌，为细皱青霉素和醌茜素（具有多种医学和杀虫作用）。它们能抑制金黄色葡萄球菌（MIC 分别为 8μg/mL 和大于 128μg/mL）、卡他莫拉菌（*Moraxella catharrhalis*）（MIC 分别为 0.25μg/mL 和 8μg/mL）、大肠杆菌（MIC 分别为 32μg/mL 和大于 128μg/mL）、铜绿假单胞菌（MIC 分别为大于 128μg/mL 和大于 128μg/mL）和白念珠菌（MIC 分别为大于 128μg/mL 和大于 128μg/mL）。

Montemartini Corte 等人（2000）从南极大陆池塘的沉积物中筛选到了具有抗菌活性青霉菌（*Penicillium*），确定了青霉属菌株的液体培养物的 MIC，并鉴定了九种分离物。其中，*P. chrysogenum* TF 3/3 与这种类型菌株的不同之处在于其产生红棕色色素，并且还对革兰氏阴性菌有抗菌活性。这些数据初步表明了 β-内酰胺抗生素的合成。

3.2.3 海洋环境

地球表面超过 70%被水覆盖，主要是海洋，因此，海洋环境代表了世界上一些最未知的极端环境。时至今日，来自海洋环境的药物发现的成功率是 1/3140，比行业平均值（即，5000～10000 个测试化合物中发现 1 个）高大约两倍到三倍（Gerwick and Moore 2012; Giddings and Newman 2015）。因此，海洋环境是寻找具有潜力，可生产新型药物的微生物资源的重要来源（Giddings and Newman 2015）。目前已经从北极和南极海洋环境中分离出大量能够合成具有抗菌活性物质的微生物，如以下部分所述。

3.2.3.1 细菌

主要针对极地海水和沉积物中的放线菌，及其他异养细菌（表 3.5）。Lo Giudice 等人（2007b）从南极海水和沉积物中分离了具有抗菌活性的不同细菌。与杆菌属、巢杆菌属、节杆菌属和红球菌属有系统发育关系的十六种放线菌对大肠杆菌和奇异变形杆菌菌株具有活性，对藤黄微球菌和枯草芽孢杆菌活性较弱。在除了放线菌外的异养细菌中，来自特拉诺瓦湾（Terra Nova Bay）（南极洲）的海水柱（五个假交替单胞菌）的分离株和贝氏肩孔南极鱼 *Trematomus bernacchii* 的肠内容物（一个假单胞菌分离株）的分离株，显示出对大肠杆菌和奇异变形杆菌有很强的抗菌活性（Lo Giudice et al. 2007b）。此外，不同的（即使密切相关）分离物表现出不同的抑制模式，该发现表明了抗菌活性很可能具有菌株特异性，即同一属的菌株成员可能合成多个化合物作用于不同的目标。

Wietz 等人（2012）从 511 个不同的、随机选择的北冰洋来源（例如海冰、表面海水、浮游动物、深海和融水）的菌株中，筛选出了 16 株细菌，通过反复测试证明其对鳗弧菌和金黄色葡萄球菌具有较高的抗菌活性。生产抗生素的菌株分别属于节杆菌属（7 个菌株）、假交替单胞菌属（4 个菌株）、嗜冷菌属（2 个菌株）和弧菌属（3 个菌株）。来自节杆菌属的乙醇提取物 WX11 可抑制鲑气单胞菌（*Aeromonas salmonicida*）、蜡状芽孢杆菌（*B. cereus*）、产单核细胞李斯特菌（*Listeria monocytogenes*）、金黄色葡萄球菌（*S. aureus*）、

表 3.5　海洋极地环境中的活性细菌

门或类	分离物	来源	目标菌株[①]	备注	参考文献
Gammaproteobacteria	*Pseudoalteromonas* sp. F26	南极洲海水	**6,14**		Lo Giudice et al. (2007b)
	Pseudoalteromonas sp. G24	南极洲海水	**6,14**		Lo Giudice et al. (2007b)
	Pseudoalteromonas sp. 59	南极洲海水	**6**		Lo Giudice et al. (2007b)
	Pseudoalteromonas sp. 129	南极洲海水	**6,14**		Lo Giudice et al. (2007b)
	Pseudoalteromonas sp. 131	南极洲海水	**6,14**		Lo Giudice et al. (2007b)
	Pseudomonas sp. 65/3	南极洲鱼	**14**		Lo Giudice et al. (2007b)
	Pseudoalteromonas sp. MB33	北冰洋桡足动物	**19**		Wietz et al. (2012)
	Pseudoalteromonas sp. MB205	北冰洋表层水	**19**		Wietz et al. (2012)
	Pseudoalteromonas sp. MB220	北冰洋海冰	**19**		Wietz et al. (2012)
	Pseudoalteromonas sp. MB240	北冰洋海冰	**19**		Wietz et al. (2012)
	Psychrobacter sp. XX5	北冰洋海冰	**19**		Wietz et al. (2012)
	Psychrobacter sp. ST4	北冰洋海冰	**19**		Wietz et al. (2012)
	Vibrio sp. RR12	北冰洋等足动物	**19**		Wietz et al. (2012)
	Vibrio sp. EF14	北冰洋深海水	**17,19**		Wietz et al. (2012)
	Vibrio sp. RS9	北冰洋海冰	**17,19**		Wietz et al. (2012)
Bacteroidetes	*Salegentibacter* sp. T436	北冰洋海冰	**2,4,5,6,8,9,10,11,12,13,15,18**	硝基化合物	Al-Zereini et al. (2007)
Actinobacteria	*Arthrobacter* sp. B20	南极洲海水	**9,14**		Lo Giudice et al. (2007b)
	Arthrobacter sp. F40	南极洲海水	**4**		Lo Giudice et al. (2007b)
	Arthrobacter sp. G18	南极洲海水	**6,9**		Lo Giudice et al. (2007b)
	Arthrobacter sp. G75	南极洲海水	**6,9,14**		Lo Giudice et al. (2007b)
	Arthrobacter sp. PP12	北冰洋海冰	**17,19**		Wietz et al. (2012)
	Arthrobacter sp. MB182	北冰洋海冰	**17,19**		Wietz et al. (2012)
	Arthrobacter sp. SS14	北冰洋桡足动物	**17,19**		Wietz et al. (2012)
	Arthrobacter sp. TT4	北极融水	**17,19**		Wietz et al. (2012)
	Arthrobacter sp. ZZ3 18	北冰洋海冰	**17,19**		Wietz et al. (2012)
	Arthrobacter sp. LM7	北冰洋表层水	**17,19**		Wietz et al. (2012)
	Arthrobacter sp. WX11	北冰洋深海水	**1,3,7,16,17,19,20,21,22,23,24**	节杆菌素 (arthrobacilins)	Wietz et al. (2012)
	Janibacter sp. B8	南极洲海水	**6,9,14**		Lo Giudice et al. (2007b)
	Janibacter sp. F21	南极洲海水	**14**		Lo Giudice et al. (2007b)
	Janibacter sp. F34	南极洲海水	**6,9,14**		Lo Giudice et al. (2007b)
	Janibacter sp. F39	南极洲海水	**14**		Lo Giudice et al. (2007b)
	Janibacter sp. G4	南极洲海水	**6,14**		Lo Giudice et al. (2007b)
	Janibacter sp. G5	南极洲海水	**6,14**		Lo Giudice et al. (2007b)
	Janibacter sp. I44	南极洲海水	**6,9,14**		Lo Giudice et al. (2007b)

续表

门或类	分离物	来源	目标菌株①	备注	参考文献
Actinobacteria	*Microlunatus* sp. y400	南极洲沉积物	**4**		Yuan et al. (2014)
	Nesterenkonia sp. S1-21	南极洲沉积物	**6**		Lo Giudice et al. (2007b)
	Nesterenkonia sp. S1-40	南极洲沉积物	**6**		Lo Giudice et al. (2007b)
	Nocardiopsis sp. y4	南极洲沉积物	**4**	普那米星，tetarimycin A	Yuan et al. (2014)
	Nocardiopsis sp. y47	南极洲沉积物	**4**	普那米星，tetarimycin A	Yuan et al. (2014)
	Nocardiopsis sp. y17	南极洲沉积物	**4,5**	普那米星，tetarimycin A	Yuan et al. (2014)
	Nocardiopsis sp. y18	南极洲沉积物	**4,5**	普那米星，tetarimycin A	Yuan et al. (2014)
	Nocardiopsis sp. y64	南极洲沉积物	**4,5**	普那米星，tetarimycin A	Yuan et al. (2014)
	Rhodococcus sp. B7	南极洲沉积物	**6,9,14**		Lo Giudice et al. (2007b)
	Rhodococcus sp. G77	南极洲沉积物	**6**		Lo Giudice et al. (2007b)
	Rhodococcus sp. W4-5	南极洲沉积物	**4,6**		Lo Giudice et al. (2007b)
	Streptomyces sp. ART5	北冰洋沉积物	**5**	色苷 C-1027 发色团 V	Moon et al. (2014)
	Streptomyces sp. y146	北冰洋沉积物	**4,5,6**		Yuan et al. (2014)
	Streptomyces sp. y2	北冰洋沉积物	**4,5,17**		Yuan et al. (2014)
	Streptomyces sp. y23	北冰洋沉积物	**4,5**		Yuan et al. (2014)
	Streptomyces sp. y222	北冰洋沉积物	**4,5**		Yuan et al. (2014)
	Streptomyces sp. y481	北冰洋沉积物	**4**		Yuan et al. (2014)

① 目标菌株：**1.** *Aeromonas salmonicida*；**2.** *Bacillus brevis*；**3.** *B. cereus*；**4.** *B. subtilis*；**5.** *Candida albicans*；**6.** *Escherichia coli*；**7.** *Listeria monocytogenes*；**8.** *Magnaporthe grisea*；**9.** *Micrococcus luteus*；**10.** *Mucor miehei*；**11.** *Nematospora coryli*；**12.** *Paecilomyces variotii*；**13.** *Penicillium notatum*；**14.** *Proteus mirabilis*；**15.** *P. vulgaris*；**16.** *Salmonella enterica*；**17.** *Staphylococcus aureus*；**18.** *Ustilago nuda*；**19.** *Vibrio anguillarum*；**20.** *V. parahaemolyticus*；**21.** *V. harveyi*；**22.** *V. vulnificus*；**23.** *Yersinia enterolitica*；**24.** *Y. rucke*。

肠炎沙门氏菌（*Salmonella enterica*）、创伤弧菌（*Vibrio vulnificus*）、副溶血弧菌（*V. parahaemolyticus*）、哈氏弧菌（*V. harveyi*）、鳗弧菌（*V. anguillarum*）、肠耶尔森氏菌（*Yersinia enterolitica*）和克氏杆菌（*Y. ruckeri.*）的生长。在乙醇和乙酸乙酯萃取物中检测到节杆菌素（arthrobacilins）A、B 和 C，但后者缺乏抗菌活性，表明抗菌可能依赖于不同化合物的协同作用。

Yuan 等人（2014）报道了来自北冰洋深海沉积物样品中显示出具有抗细菌和/或抗真菌活性的 11 种放线菌菌株。其中，7 种显示对枯草芽孢杆菌和白念珠菌有活性。然而，链霉菌属 y146 和链霉菌属 y2 分别具有对大肠杆菌和金黄色葡萄球菌的活性。活性菌株属于链霉菌属、诺卡氏菌属和小月菌属（*Microlunatus*）。存在编码聚酮合酶Ⅰ（PKS Ⅰ）、聚酮合酶Ⅱ（PKS Ⅱ）、非核糖体肽合成酶（NRPS）、氨基脱氧异分支酸合成酶（phzE）、dTDP-葡萄糖-4,6-脱水酶（dTGD）、卤化酶（Halo）或细胞色素 P450 羟化酶（CYP）的基因，通过

PCR扩增检测，显示出所有分离菌株的基因组具有至少两个参与次生代谢物的生物合成的基因簇。对五个来自诺卡氏菌属菌株基因组中聚酮合酶Ⅱ（PKS Ⅱ）酮缩酶结构域（KS）的PCR产物进行了测序。来自这些菌株的KS序列显示其与参与帕地霉素生物合成［来自马杜拉放线菌（*Actinomadura hibisca*）的抗真菌抗生素］的fabF编码蛋白的KS结构域具有67%和69%的氨基酸序列一致性，并且与TamM具有68%的相似性，这与具有四环素MRSA活性抗生素芳香聚酮A的生物合成有关（Yuan et al. 2014）。

3.2.3.2 真菌

Montemartini Corte等人（2000）从南极洲罗斯海50m深的木饵中分离出青霉菌。测定了青霉菌株液体培养物的MIC，选择了七种分离物，主要是产黄青霉，其具有抗金黄色葡萄球菌和藤黄微球菌的活性。然而，我们观察到*P. melinii* R55的优良活性可能依赖于展青霉素的合成，已经证实的该类型菌株见表3.6。此外，表3.6中还列出了从南极大型藻类和海绵中分离出来的其他活性真菌（见本章第3.2.4.2和3.2.4.3节）。

表3.6 具有抗菌活性的海洋真菌

微生物	分离物	来源	目标菌株[①]	备注	参考文献
Ascomycota	*Dipodascusaustraliensis* 6031	南极洲海藻	**1,2**		Furbino et al.（2014）
	Cladosporium sp. F09-T13-2	南极洲海绵	**7**		Henríquez et al.（2014）
	Epicoccum sp. F09-T15-1	南极洲海绵	**7,8**		Henríquez et al.（2014）
	Epicoccum sp. F09-T15-4	南极洲海绵	**8,9**		Henríquez et al.（2014）
	Geomyces sp. F09-T1-8	南极洲海绵	**4,7,8,9**		Henríquez et al.（2014）
	Geomyces sp. F09-T3-19	南极洲海绵	**4,7,8,9**		Henríquez et al.（2014）
	Geomyces sp. F09-T3-5	南极洲海绵	**7,8,9**		Henríquez et al.（2014）
	Metschnikowia australis MH47.1.2	南极洲海藻	**1,2**		Furbino et al.（2014）
	Penicillium chrysogenum R9	Sunk	**6,8**		Montemartini Corte et al.（2000）
	Penicillium chrysogenum R28	Sunk	**6,8**		Montemartini Corte et al.（2000）
	Penicillium chrysogenum R31	Sunk	**6,8**		Montemartini Corte et al.（2000）
	Penicillium chrysogenum R34	Sunk	**6,8**		Montemartini Corte et al.（2000）
	Penicillium chrysogenum R36	Sunk	**6,8**		Montemartini Corte et al.（2000）
	Penicillium chrysogenum R38	Sunk	**6,8**		Montemartini Corte et al.（2000）
	P. commune	南极洲海藻	**8,9**		Henríquez et al.（2014）
	P. commune	南极洲海藻	**4,8**		Henríquez et al.（2014）
	P. melinii R55	Sunk	**5,6,8**	Patulin	Montemartini Corte et al.（2000）
	P. polonicum	南极洲海藻	**4,8,9**		Henríquez et al.（2014）

续表

微生物	分离物	来源	目标菌株①	备注	参考文献
Ascomycota	*Penicillium steckii*	南极洲海藻	**1,2**		Furbino et al.(2014)
	Penicillium sp. 6034	南极洲海藻	**3**		Godinho et al.(2013)
	Penicillium sp. 6120	南极洲海藻	**3**		Godinho et al.(2013)
	Pseudogymnoascus	南极洲海藻	**1,2**		Furbino et al.(2014)
	Pseudogymnoascus	南极洲海藻	**1,2,3**		Furbino et al.(2014)
	Pseudeurotium	南极洲海绵	**8**		Henríquez et al.(2014)
	Thelebolus	南极洲海绵	**8**		Henríquez et al.(2014)
	Trichocladium sp. F09-T24-1	南极洲海绵	**4,8,9**		Henríquez et al.(2014)
Basidiomycota	*Guehomyes pullulans* MH33.1	南极洲海藻	**1,2,3**		Furbino et al.(2014)

① 目标菌株：**1.** *Candida albicans*；**2.** *C. krusei*；**3.** *Cladosporium sphaerospermum*；**4.** *Clavibacter michiganensis*；**5.** *Escherichia coli*；**6.** *Micrococcus luteus*；**7.** *Pseudomonas aeruginosa*；**8.** *Staphylococcus aureus*；**9.** *Xanthomonas campestris*。

3.2.4 与其他生物相关的微生物

微生物特殊的遗传和代谢可塑性使它们可以与许多真核生物（微生物）相互作用。实际上，它们通常在植物和/或动物上定植，这些动植物具有使它们能够存活，并合成多种化合物的生理结构特征，能够保护自身，抵抗病原体和污损生物的侵害。由于它们具有不同的药物潜力（抗病毒、抗增殖、抗炎、抗肿瘤和抗分枝杆菌活性），一些有机体（主要是无脊椎动物）已经吸引了许多研究者的注意。然而，最近一些报告表明，从藻类、无脊椎动物和地衣获得的一些代谢物可能是由它们的真菌和细菌共生体合成的。已报道了极地地区地衣、藻类、海绵和与之相对应的细菌和/或真菌具有抗菌活性。

3.2.4.1 和地衣苔藓相关的微生物

Kim 等人（2012，2013，2014a，b）已对与北极和南极地衣相关的细菌菌株的抗菌潜力进行评估。活性细菌主要属于 α-、β-、γ-变形菌纲中的鞘氨醇单胞菌属（*Sphingomonas*）、伯克霍尔德氏菌属（*Burkholderia*）和红杆菌属（*Rhodanobacter*）（表 3.7）。使用不同溶剂（即丙酮、水、氯仿、乙醚、乙醇、甲醇和石油醚）获得的细菌提取物显示出抗革兰氏阳性（金黄色葡萄球菌、枯草芽孢杆菌和藤黄微球菌）和革兰氏阴性（阴沟肠杆菌、铜绿假单胞菌和大肠杆菌）靶病原体。关于溶剂，乙酸乙酯的萃取比高极性（例如乙醇）或极低极性（例如石油醚）的溶剂更有效。所有水相提取物对靶细菌都没有表现出任何抗菌活性，可能是由于活性化合物水溶性较差。

表 3.7 与地衣共生的具有抗菌活性的菌株

门或类	分离物	来源地	目标菌株	参考文献
Alphaproteobacteria	*Sphingomonas* sp. PAM26605	斯瓦尔巴群岛(Svalbard Islands),北极	**1,3,5,6**	Kim et al.(2014a)
	Sphingomonas sp. PAM26625	斯瓦尔巴群岛,北极	**3,5,6**	Kim et al.(2014a)
	Sphingomonas sp. PAMC26556	乔治王岛,南极洲	**5**	Kim et al.(2014b)
	Sphingomonas sp. PAMC26561	乔治王岛,南极洲	**3,5**	Kim et al.(2014b)
	Sphingomonas sp. KOPRI26645	斯瓦尔巴群岛,北极	**1,2,3,4.5**	Kim et al.(2012)

续表

门或类	分离物	来源地	目标菌株	参考文献
Betaproteobacteria	*Burkholderia* sp. PAM26606	斯瓦尔巴群岛,北极	**1,3,4,5**	Kim et al. (2014a)
	Burkholderia sp. PAM26607	斯瓦尔巴群岛,北极	**3,6**	Kim et al. (2014a)
	Burkholderia sp. PAM26608	斯瓦尔巴群岛,北极	**1,3,5,6**	Kim et al. (2014a)
	Burkholderia sp. PAM26507	乔治王岛,南极洲	**4**	Kim et al. (2014b)
	Burkholderia sp. PAM26537	乔治王岛,南极洲	**4**	Kim et al. (2014b)
	Burkholderia sp. PAM26633	乔治王岛,南极洲	**1,3,5,6**	Kim et al. (2014b)
	Burkholderia sp. KOPRI26643	斯瓦尔巴群岛,北极	**1,2,3,4,5,6**	Kim et al. (2012)
	Burkholderia sp. KOPRI26644	斯瓦尔巴群岛,北极	**1,2,3,4,5,6**	Kim et al. (2012)
	Burkholderia sp. KOPRI26646	斯瓦尔巴群岛,北极	**1,2,3,4,5,6**	Kim et al. (2012)
	Burkholderia sp. KOPRI26647	斯瓦尔巴群岛,北极	**1,2,4**	Kim et al. (2012)
Gammaproteobacteria	*Rhodanobacter* sp. PAMC26515	乔治王岛,南极洲	**4**	Kim et al. (2014b)
	Rhodanobacter sp. PAMC26518	乔治王岛,南极洲	**4**	Kim et al. (2014b)
	Rhodanobacter Sp. PAMC26538	乔治王岛,南极洲	**1,2,3,5,6**	Kim et al. (2014b)
	Rhodanobacter sp. PAMC26551	乔治王岛,南极洲	**1,3,5,6**	Kim et al (2014b)
	Rhodanobacter sp. PAMC26552	乔治王岛,南极洲	**4**	Kim et al. (2014b)
	Rhodanobacter sp. PAMC26557	乔治王岛,南极洲	**3,4**	Kim et al. (2014b)
Bacteroidetes	*Hymenobacter* sp. PAMC26554	乔治王岛,南极洲	**1,3,6**	Kim et al. (2014b)
Actinobacteria	*Streptomyces* sp. PAMC26508	乔治王岛,南极洲	**1,2,3,4,5,6**	Kim et al. (2014b)
	Frigoribacterium sp. PAMC26555	乔治王岛,南极洲	**1,4,6**	Kim et al. (2014b)
Firmicutes	*Paenibacillus* sp. PAMC26517	乔治王岛,南极洲	**1,2,6**	Kim et al. (2014b)

① 目标菌株：**1.** *Bacillus subtilis*；**2.** *Enterobacter cloacae*；**3.** *Escherichia coli*；**4.** *Micrococcus luteus*；**5.** *Pseudomonas aeruginosa*；**6.** *Staphylococcus aureus*。

Melo 等（2014）报道了南极内生真菌高山被孢霉菌 *Mortierella alpina* ITA1-CCMA 952 的抗菌活性，该菌株分离自南极苔藓（*Schistidium antarctici*）［海军湾（Admiralty Bay)，乔治王岛］。菌株能抑制大肠杆菌（MIC 为 26.9μg/mL）、铜绿假单胞菌和粪肠球菌（MIC 均为 107μg/mL）的生长，但不具有抗真菌活性。通过 GC-MS 分析表明，抗菌代谢物被鉴定为吡咯并吡嗪生物碱。

目前还没有关于从地衣中分离的真菌或来自苔藓的细菌的抗微生物活性的研究。

3.2.4.2 与海绵相关的微生物

海洋海绵是栖息多种生物体的动物，在一些情况下，其高达 35%的湿重来自微生物。此外，微生物丰度高的海绵可达 10^8～10^{10} 个/g，即使微生物丰度低的海绵其微生物含量也可达到 10^5～10^6 个/g，这表明了从与其他有机体共存的环境中分离的微生物中发现的多种新型次生代谢物具有巨大潜力（Hentschel et al. 2002)。已知与海绵相关的微生物具有很强的生物活性，涵盖了广泛的生物学功能。海绵-微生物共生体是潜在的化学和生态现象，为开发新的药物先导化合物提供可持续的资源。充分认识与海洋海绵相关微生物的抗菌潜力，海绵微生物共栖系统无疑是当今一个重要的研究热点。在此基础上，南极海绵代表了潜在的、丰富的、未开发的新型抗菌化合物的重要来源，Papaleo 等人（2012）的研究旨在表征来自特拉诺瓦湾的三种不同的南极海绵（*Lissodendoryx nobilis*、*Anoxycalyx joubini* 和

Haliclonissa verrucosa）中的可培养细菌群落（共 132 株），并研究这些菌株中的部分菌株是否可以抑制超过 70 种机会性病原体的生长，包括影响囊性纤维化（CF）患者的病原菌。获得的数据显示，大多数与海绵相关的南极细菌能够完全抑制伯克霍尔德氏菌（*Burkholderia cepacia*）复合体（Bcc）的生长，它是最重要的 CF 病原体之一。另一方面，相同的菌株不干扰其他病原细菌，如铜绿假单胞菌或金黄色葡萄球菌的生长，表明了其对 Bcc 细菌的特异性抑制活性。对活性最强的分离株（即假交替单胞菌属 TB41 和 AC163，希瓦氏菌属 TB4，嗜冷杆菌 TB47 和 TB67）进行的进一步研究表明，部分抗菌化合物可能是 mVOCs，mVOCs 是构成微生物生态系统中不同生物体之间相互关系的重要调节因子。这一发现通过在具有代表性的一组南极细菌中进行的固相微萃取气相色谱质谱联用（SPME-GC-MS）实验中得到证实，揭示了一系列微生物挥发性有机物 mVOCs 的产生，其合成很可能是组成型的，因为它不是由目标菌株的存在而诱导产生的（Romoli et al. 2011）。更有趣的是，VOCs 似乎比大多数常用的抗生素更能有效地抑制 Bcc 细菌的生长。应用假交替单胞菌 TB41 的代谢组学方法选择了 30 种化合物，其中一些可能负责抑制分离自南极的 *B. cenocepacia*（Bc）LMG16654（Romoli et al. 2014）的生长。在单独生长或在 Bc-LMG16654 存在下，从菌株 TB41 获得的 mVOCs 合成曲线没有显著的不同，表明这种分子的产生是组成型的，并且不会因为目标菌株的存在而改变，因为后者既不干扰产生，也不诱导不同 mVOCs 的合成。最近，Papaleo 等人（2013）也报告了在需氧条件下嗜冷菌（分离自 *A. joubini*）的挥发性组分（Mangano et al. 2009）。结果表明，南极细菌具有抗菌能力，其活性可能依赖于 mVOCs 的复杂混合物，而该混合物的相对浓度则取决于生长条件（存在/不存在氧和所使用的生长培养基）。结果还显示，只有假交替单胞菌 TB41（Mangano et al. 2009）拥有 *nrps-pks* 基因，序列类似于报道基因组的节杆菌 *Arthrobacter* sp. TB23（分离自南极海绵线虫）（Mangano et al. 2009，Fondi et al. 2012）（参见本章 3.3 节）。

关于真菌，Henríquez 等（2014）首次描述了与南极海洋海绵相关的真菌的生物多样性和代谢潜力，测定了真菌培养物提取物可能的抗菌活性（除了抗肿瘤和抗氧化剂外，没有报道）。其中 52 种提取物，主要来自地丝霉属（*Geomyces*）和未鉴定的相关物质，对一些受试的细菌显示出抗菌活性，详见表 3.6。通常，真菌提取物对耐革兰氏阴性细菌要比对抗革兰氏阳性细菌（特别是对金黄色葡萄球菌）更有活性。5 个分离株（属于地丝霉属和附球菌属）表现出对铜绿假单胞菌的抗菌活性，44 种分离株显示对金黄色葡萄球菌的抑制活性，11 种分离株（属于地丝霉属和青霉属，或未鉴定）具有针对密执安棒状杆菌属的抗菌活性，以及 22 种分离株（属于地丝霉属、青霉属、附球菌属和枝孢属，或未鉴定）表现出对黄单胞菌的抗菌活性。具有相同 ITS 类型的几种真菌分离物显示不同的抗菌活性。地丝霉属是最多产的真菌属，具有抗菌活性的分离株有 18 种（占 54.5%）。其中，地丝霉属 F09-T1-8 和 F09-T3-19 活性最强。Henríquez 等（2014）首次描述了假散囊菌属（*Pseudeurotium*）中的抗菌活性。更令人感兴趣的是，大多数未鉴定的分离株（锤舌菌纲）也是抗菌物质的生产者。

3.2.4.3 与大型藻类相关的真菌

海洋藻类在有机矿物质循环中，特别是在南极浅水区的沿岸和沿海生态系统中发挥关键作用。南极藻类具有高度地方性，在作为主要生产者、海洋食草动物的食物以及构成栖息地结构方面发挥着重要作用。迄今为止，只报道了两个与南极藻类相关的真菌的抗菌活性物质（表 3.6）。

Godinho 等人（2013）分析了与八个南极大型藻类相关的真菌的分布和多样性以及合成生物活性物质的能力。青霉菌菌株 6034 和 6120 分别分离自地方性物种南极红藻 *Palmaria decipiens*［红藻植物门（Rhodophyta）］和礁膜属 *Monostroma hariotii*［绿藻门（Chlorophyta）］，获得的提取物具有较高选择性抗真菌（和/或锥虫）活性。青霉菌 6120 对丝状真菌球孢枝孢菌（*Cladosporium sphaerospermum*）显示出抗菌活性，抑制率达到 96%，MIC 值为 250mg/mL，初步核磁共振分析表明其存在高度官能化的芳香化合物。

最近，Furbino 等人（2014 年）描述了与地方性的南极大型藻类 *Pyropia endiviifolia*（红藻植物门）和礁膜属 *Monostroma hariotii*（绿藻门）相关的真菌群落的特点，这是南极半岛较为丰富的物种，共六个海藻真菌类群能够产生具有生物活性的化合物。所有提取物显示对白念珠菌（*Candida albicans*）和克鲁斯念珠菌（*C. krusei*）有选择性抗真菌活性，而通过假裸囊菌 *Pseudogymnoascus* sp. 2 和普鲁兰久浩酵母 *Guehomyes pullulans* MH33. 1 也能抑制球孢枝孢菌。另外，歧皱青霉 *Penicillum steckii* 6012 的提取物对黄热病病毒的抑制率为 96%。目尚未发现提取物的抗菌或抗锥虫活性。

据我们所知，迄今为止还没有关于分离自北极大型藻类的真菌的抗菌活性的相关研究报道。

3.3 具有抗生素活性的海绵相关南极细菌的遗传与基因组

最近获得了部分海绵相关细菌菌株基因组序列草图，通过对不同属的南极细菌菌株的分子和基因组分析，对其代谢潜力有了进一步的了解（Fondi et al. 2012；Papaleo et al. 2013；Maida et al. 2014；Orlandini et al. 2014；Bosi et al. 2015；Maida et al. 2015）。

在 Papaleo 等人（2012）描述的分离株中，通过基因组学法进一步分析了 *Gillisia* 属（即分离株 CAL575）、嗜冷杆菌属（*Psychrobacter*）（即分离株 TB2、TB15 和 AC24）和节杆菌属（*Arthrobacter*）（即分离株 TB26、CAL618 和 TB23）中的代表菌株，分析了其对 Bcc（伯克霍尔德菌）的抑制活性（Fondi et al. 2014；Maida et al. 2014；Orlandini et al. 2014）。Maida 等人（2014）运用不同的技术，如基因组学、表型表征和 mVOCs 分析，系统研究了从海绵 *H. verrucosa* 中分离的菌株 *Gillisia* sp. CAL575 的特性。Fondi 等（2014）报道了嗜冷杆菌属分离株 TB2、TB15（来自海绵 *L. nobilis*）和 AC24（来自海绵 *H. verrucosa*）的基因组序列草图。特别是嗜冷杆菌 AC24，在任何生长培养基当中，都能有效抑制几乎所有测试的 Bcc 菌株的生长。与 AC24 相比，TB2 和 TB15 抑制能力有所下降，并且在有些情况下，对 Bcc 菌株生长的抑制受相应的生长培养基的影响。此外，节杆菌属 TB26（来自海绵 *L. nobilis*）和 CAL 618（来自海绵 *H. verrucosa*）显示出非常类似的抑制模式。这表明抗菌化合物的生物合成的遗传因素属于核心基因组（Orlandini et al. 2014）。此外，交叉划线培养的结果表明，扩散性的有机分子（与挥发性化合物组合）也可能干扰 Bcc 菌株的生长，假交替单胞菌能够干扰 Bcc 菌株生长也证实了这一点（Maida et al.，2015）。

检索 38 种南极细菌菌株（包括上文已经提及的）的基因组序列草图，以寻找具有抗菌活性的次生代谢物生物合成的基因。根据之前的实验数据（Papaleo et al. 2012），在 *Gillisa* CAL575 菌株中，除了存在 *pks* Ⅲ型基因和萜烯的生物合成簇之外，并没有涉及次生代谢物生物合成的基因（*pks* 或 *nrps*）。此外，在节杆菌 TB23 的基因组中，分别包括 *pks* Ⅲ、*nrps* 基因和萜烯生物合成基因的三个基因簇，负责抗菌化合物的生物合成，可能靶向作用于 Bcc 细菌。这些菌株的基因组的比较分析显示参与次生代谢物生物合成的核心基因组中只有少数基因的存在（Papaleo et al. 2013）。此外，还表明这些化合物的生物合成可以通过仍然未知的代谢途径合成。

结果表明，三种嗜冷杆菌所利用的抑制伯克霍尔德菌属（*Burkholderia*）代表菌株的生长的代谢策略超出了已经表征的生化系统的范围。

3.4 具有抗菌活性的适冷微生物的生物活性代谢产物

来自适冷微生物的新生物活性分子的数量在过去十年中有了显著的增加，其中一些具有抗菌活性（Bratchkova and Ivanova 2011；Liu et al. 2013）。在本章的第二部分中已经介绍了几种生物活性分子，其他信息如下。

Bruntner 等人（2005）从南极灰色链霉菌菌株 NTK 97（来自土壤）中分离了一种新的芳香聚酮类抗生素，称为氟戈环酮（frigocyclinone），该抗生素由四角霉素的一部分通过 C-糖苷键链接到氨基脱氧糖胺上（图 3.1）。弗瑞霉素对革兰氏阳性菌（例如枯草芽孢杆菌，MIC 值为 4.6μg/mL；金黄色葡萄球菌，MIC 值为 15μg/mL）显示出良好的抑菌活性，而

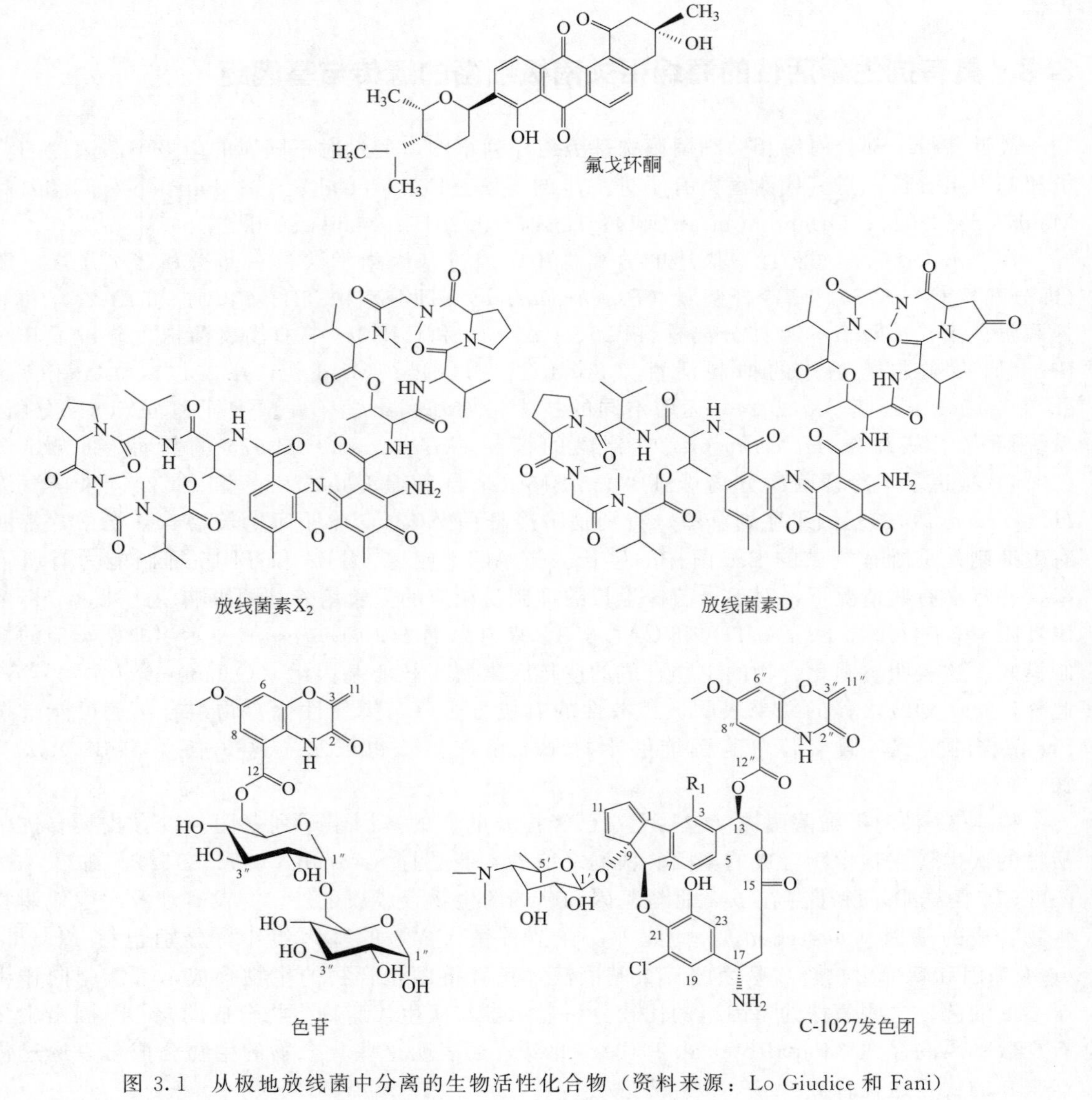

图 3.1 从极地放线菌中分离的生物活性化合物（资料来源：Lo Giudice 和 Fani）

对革兰氏阴性菌（即大肠杆菌、荧光假单胞菌、奇异变形杆菌）、丝状真菌（即灰葡萄孢、臭曲霉、青霉和拟青霉）和酵母菌（即酿酒酵母和白念珠菌）不敏感。

从南极洲利文斯顿地区的土壤样品中分离的黄色链霉菌 *Streptomyces flavovirens* 6^7（Ivanova et al.，2002）（图 3.1），合成了具有抗细菌活性和抗肿瘤活性的分子，属肽类抗生素中的放线菌素。从南极洲利文斯顿地区的水样本中分离得到的菌株链霉菌 *Streptomyces* sp. 1010，在其培养液中发现了新物质 2-氨基-9,13-二甲基十七烷酸、邻苯二甲酸二乙酯、1,3-双（3-苯氧基苯氧基）苯、己二酸二辛酯。邻苯二甲酸二乙酯具有对藤黄微球菌 *M. luteus*（MIC 为 3μg/mL）、枯草芽孢杆菌（MIC 为 12μg/mL）和金黄色葡萄球菌（MIC 为 25μg/mL）的抗菌活性（Ivanova et al. 2001）。

Gesheva（2010）通过与不同的标准抗生素进行比较，表明了 *Streptomyces* sp. 8 可产生三种抗生素：非多烯大环内酯抗生素（non-polyenic macrolide antibiotic）（由两种成分组成）、阿扎毒素 B（azalomycin B）和尼日利亚菌素（nigericin）。

Moon 等人（2014）还从分离自北极海洋表层沉积物（东西伯利亚大陆边缘，深度 354m）链霉菌菌株 ART5 中获得了新的苯并噁嗪次生代谢物（benzoxazine secondary metabolites）、色苷（articoside）和 C-1027 发色团 V（C-1027 chromophore V）（2014）（图 3.1）。色苷是苯并噁嗪二糖，是一种以前没有报道的结构类型，而 C-1027 发色团 V 具有氯原子，氨基糖、环戊二烯［a］茚和 3′-氯-5′-羟基-β-酪氨酸部分。色苷和 C-1027 发色团 V 可抑制白念珠菌的异柠檬酸裂解酶，这是在该菌致病性中起重要作用的一种酶，IC_{50} 分别为 30.4μmol/L 和 37.9μmol/L。

在细菌中，除了放线菌外，Mojib 等人（2010）还测试了两种活性色素的抗分杆杆菌活性，紫色杆菌素，来自紫色杆菌 *Janthinobacterium* sp. Ant5-2（J-PVP）分泌的一种紫色色素和黄色素，来自黄杆菌 *Flavobacterium* sp. Ant342（F-YOP）的一种橙黄色色素。发现两种细菌均分离自南极洲东部的内陆淡水湖施尔马赫奥西斯（Schirmacher Oasis）。结果表明，J-PVP 和 F-YOP 针对耻垢分枝杆菌 *Mycobacterium smegmatis* mc^2155 的 MIC 值分别为 8.6μg/mL 和 3.6μg/mL，无毒性结核分枝杆菌 *M. tuberculosis* mc^26230 的 MIC 值分别为 5μg/mL 和 2.6μg/mL，对有毒性结核分枝杆菌 *M. tuberculosis* H37Rv 的 MIC 值分别为 34.4μg/mL 和 10.8μg/mL。Huang 等人（2012）进一步研究了 J-PVP 对多重耐药（MDR）和耐甲氧西林金黄色葡萄球菌（MRSA）菌株的有效性，J-PVP 和 F-YOP 的结构目前尚未阐明。

从来自北极冰块的耐盐杆菌 *Salegentibacter* sp. strain T436 中分离出 19 种芳香族硝基化合物，作为来自自然资源的化合物，其中 4 种是新型化合物，6 种为从未报道过的化合物（Al-Zereini et al. 2007；Schuhmann et al. 2009）。新的天然产物显示较弱的抗真菌、抗细菌活性和细胞毒性。而 2-硝基-4-（2′-硝基乙烯基）-苯酚是最有效的抗菌和细胞毒性物质，对所有目标都有活性。

发现来自南极海绵 *Isodictya setifera* 的铜绿假单胞菌含有抑制枯草芽孢杆菌、金黄色葡萄球菌和藤黄微球菌生长的代谢物（Jayatilake et al. 1996）。该菌的发酵液中含有一系列二酮哌嗪，包括一种新的天然产物和两种已知的吩嗪生物碱抗生素。

在蓝细菌中，Asthana 等（2009）报道了一种新型的抗细菌分子，该分子含有 4-[(5-羧基-2-羟基)-苄基]-1,10-二羟基-3,4,7,11,11-五甲基-八氢环戊二烯的结构。这种细胞内生物分子在结构上与二萜类化合物 noscomin 相似，具有耐酸性，对结核分枝杆菌（*M. tuberculosis* H37Rv）（MIC 为 2.5μg/mL）、金黄色葡萄球菌（*S. aureus*）（MIC 为 0.5μg/mL）、产气肠杆菌（*Enterobacter aerogenes*）（MIC 为 4.0μg/mL）、伤寒沙门氏菌（*Salmonella typhi*）

（MIC 为 2.0μg/mL）、铜绿假单胞菌（*P. aeruginosa*）（MIC 为 2.0μg/mL）和具有多重耐药性的大肠杆菌（MIC 为 16μg/mL）有抗菌活性。

在真菌中，获自土壤样品［菲尔德斯（Fildes）半岛、乔治王岛］的南极子囊菌 *Geomyces* sp. 2481，在固体基质发酵培养基中生长（Li et al. 2008）。其有机溶剂提取物含有五种新的曲地酸衍生物，分别为曲地酸乙酯（ethyl asterrate）、正丁基曲地酸（*n*-butyl asterrate）和土霉素（geomycins）A～C。土霉素 B 显示出显著的对烟曲霉（*A. fumigatus*）的抗真菌活性，MIC 值为 20μg/mL。土霉素 C 显示对金黄色葡萄球菌（MIC 为 24μg/mL）和大肠杆菌（MIC 为 20μg/mL）的抗菌活性。

从灰绿青霉（*Penicillium griseofulvum*）（格陵兰岛）中分离出许多具有抗菌活性的代谢物：灰黄霉素（griseofulvin）、富里酸（fulvic acid）、菌丝酰胺（mycelianamide）、娄地青霉素（roquefortine）C 和 D、裸麦角碱（chanoclavine）Ⅰ和野麦角碱（elymoclavine）（Frisvad et al. 2004）。

最近，Svahn 等人（2015）从废弃的企鹅巢的土壤样品中分离出的青霉菌 *Penicillium nalgiovense* Laxa。两性霉素 B 是一种全球范围内用于抗真菌感染的抗真菌剂，它是由青霉菌 *P. nalgiovense* Laxa 分泌的唯一代谢物，对白念珠菌、金黄色葡萄球菌和大肠杆菌具有显著的抑制活性。

3.5 结论

该综述强调了来自极地区域的细菌和真菌的多功能抗菌潜力，在这种极端和（经常）未被勘探的栖息地环境中有望发现制药行业感兴趣的新的天然化合物。

有趣的是，属于相同物种（分离自相同或不同栖息地）的微生物通常显示不同的抗菌谱。这表明了：（1）抑制活性特异性更可能是菌株特异性，而不是种属特异性；（2）单一物种很可能合成一系列且作用于多个目标的化合物；（3）抗菌谱可能取决于初始的栖息地。因此，如果要研究次生代谢产物的多样性，在培养物收集过程中保证相同的细菌/真菌物种的不同分离株变得非常重要。为此，我们对来自极地的微生物的抗菌活性的了解来自于可培养菌株的筛选。这种方法是大多数微生物生物深入研究的基础，可以获得微生物中全部的基因组信息，使得在实验室中研究它们的表型成为可能（de Pascale et al. 2012）。然而，来自特定环境的总微生物群落的菌株只有有限部分可以在实验室条件下培养，因此留下了独特的未探索的不可培养微生物的多样性的生物勘探潜力（Vester et al. 2015）。高通量分子生物学技术的最新发展为使用不依赖培养的方法（包括宏基因组学和单细胞基因组测序）进行生物勘探铺平了道路。结合改进的培育方法和宏基因组学方法用于适冷微生物的功能筛选可以富集关于抗菌活性的群落的目标信息。在极地地区的微生物开发的这种新兴趋势下，极地微生物值得在未来的研究计划中被认真考虑。

无利益冲突声明

Angelina Lo Giudice 和 Renato Fani 声明他们没有利益冲突。

参考文献

Al-Zereini W，Schuhmann I，Laatsch H，Helmke E，Anke H（2007）New aromatic nitro compounds from *Salegentibacter* sp. T436，an Arctic sea ice bacterium：taxonomy，fermentation，isolation and biological activities. J Antibiot 60：301-308

Asencio G，Lavina P，Alegría K，Domínguez M，Bello H，González-Rocha G，González-Aravena M（2009）Antibacterial activity of the Antarctic bacterium *Janthinobacterium* sp. SMN 33. 6 against multi-resistant Gram-negative bacteria. Electron J Biotechnol 17：1-5

Asthana RK，Deepali TMK，Srivastava A，Singh AP，Singh SP，Nath G，Srivastava R，Srivastava BS（2009）Isolation and identification of a new antibacterial entity from the Antarctic cyanobacterium *Nostoc* CCC 537. J Phycol 21：81-88

Bell TH，Callender KL，Whyte LG，Greer CW（2013）Microbial competition in polar soils：a review of an understudied but potentially important control on productivity. Biology 2：533-554

Bérdy J（2005）Bioactive microbial metabolites. J Antibiot 58：1-26

Biondi N，Tredici MR，Taton A，Wilmotte A，Hodgson DA，Losi D，Marinelli F（2008）Cyanobacteria from benthic mats of Antarctic lakes as a source of new bioactivities. J Appl Microbiol 105：105-115

Bosi E，Fondi M，Maida I，Perrin E，de Pascale D，Tutino ML，Parrilli E，Lo Giudice A，Filloux A，Fani R（2015）Genome-scale phylogenetic and DNA composition analyses of Antarcti *Pseudoalteromonas* bacteria reveal inconsistencies in current taxonomic affiliation. Hydrobiologia 761：85-95.

Bratchkova A，Ivanova V（2011）Bioactive metabolites produced by microorganisms collected in Antarctica and the Arctic. Biotechnol Biotechnol Equip 25：1-7

Brunati M，Rojas JL，Sponga F，Ciciliato I，Losi D，Göttlich E，de Hoog S，Genilloud O，Marinelli F（2009）Diversity and pharmaceutical screening of fungi from benthic mats of Antarctic lakes. Mar Genomics 2：43-50

Bruntner C，Binder T，Pathom-aree W，Goodfellow M，Bull AT，Potterat O，Puder C，Hörer S，Schmid A，Bolek W，Wagner K，Mihm G，Fiedler H-P（2005）Frigocyclinone，a novel angucyclinone antibiotic by a *Streptomyces griseus* strain from Antarctica. J Antibiot 58：346-349

Bull AT，Stach JEM（2007）Marine actinobacteria：new opportunities for natural product search and discovery. Trends Microbiol 15：491-499

Cavicchioli R，Siddiqui KS，Andrews D，Sowers KR（2002）Low-temperature extremophiles and their applications. Curr Opin Biotechnol 13：253-261

de Pascale D，de Santi C，Fu J，Landfald B（2012）The microbial diversity of Polar environments is a fertile ground for bioprospecting. Mar Genomics 8：15-22

Feller G，Gerday C（2003）Psychrophilic enzymes：hot topics in cold adaptation. Nat Rev Microbiol 1：200-208

Fondi M，Orlandini V，Maida I，Perrin E，Papaleo MC，Emiliani G，de Pascale D，Parrilli E，Tutino ML，Michaud L，Lo Giudice A，Fani R（2012）The draft genome of the VOCs-producing Antarctic bacterium *Arthrobacter* sp. TB23 able to inhibit Cystic Fibrosis pathogens belonging to the *Burkholderia cepacia* complex. J Bacteriol 194：6334-6335

Fondi M，Orlandini V，Perrin E，Maida I，Bosi E，Papaleo MC，Michaud L，Lo Giudice A，de Pascale D，Tutino ML，Liò P，Fani R（2014）Draft genomes of three Antarctic *Psychrobacter* strains known to have antimicrobial activity against *Burkholderia cepacia* complex opportunistic pathogens. Mar Genomics 13：37-38

Frisvad JC，Smedsgaard J，Larsen TO，Samson RA（2004）Mycotoxins，drugs and other extrolites produced by species in *Penicillium* subgenus. Stud Mycol 49：201-241

Furbino LE，Godinho VM，Santiago IF，Pellizari FM，Alves TM，Zani CL，Junior PA，Romanha AJ，Carvalho AG，Gil LH，Rosa CA，Minnis AM，Rosa LH（2014）Diversity patterns，ecology and biological activities of fungal communities associated with the endemic macroalgae across the Antarctic peninsula. Microb Ecol 67：775-787

Gerwick WH，Moore BS（2012）Lessons from the past and charting the future of marine natural products drug discovery and chemical biology. Chem Biol 19：85-98

Gesheva V（2009）Distribution of psychrophilic microorganisms in soils of Terra Nova Bay and Edmonson Point，Victoria Land and their biosynthetic capabilities. Polar Biol 32：1287-1291

Gesheva V（2010）Production of antibiotics and enzymes by soil microorganisms from the Windmill Islands Region，Wilkes Land，East Antarctica. Polar Biol 33：1351-1357

Gesheva V，Negoita T（2012）Psychrotrophic microorganism communities in soils of Haswell Island，Antarctica，and their biosynthetic potential. Polar Biol 35：291-297

Gesheva V，Vasileva-Tonkova E（2012）Production of enzymes and antimicrobial compounds by halophilic Antarctic *Nocardioides* sp. grown on different carbon sources. World J Microbiol Biotechnol 28：2069-2076

Giddings L-A，Newman DJ（2015）Bioactive compounds from marine extremophiles. Springer Briefs Microbiol 1-124

Godinho VM，Furbino LE，Santiago IF，Pellizzari FM，Yokoya NS，Pupo D，Alves TM，Junior PA，Romanha AJ，Zani CL，Cantrell CL，Rosa CA，Rosa LH（2013）Diversity and bioprospecting of fungal communities associated with endemic and cold-adapted macroalgae in Antarctica. ISME J 7：1434-1451

Godinho VM，Gonçalves VN，Santiago IF，Figueredo HM，Vitoreli GA，Schaefer CE，Barbosa EC，Oliveira JG，Alves TM，Zani CL，Junior PA，Murta SM，Romanha AJ，Kroon EG，Cantrell CL，Wedge DE，Duke SO，Ali A，Rosa CA，Rosa LH（2015）Diversity and bioprospection of fungal community present in oligotrophic soil of continental Antarctica. Extremophiles 19：585-596

Hemala L，Zhanga D，Margesin R（2014）Cold-active antibacterial and antifungal activities and antibiotic resistance of bacteria isolated from an alpine hydrocarbon-contaminated industrial site. Res Microbiol 165：447-456

Henríquez M，Vergara K，Norambuena J，Beiza A，Maza F，Ubilla P，Araya I，Chávez R，San-Martín A，Darias J，Darias MJ，Vaca I（2014）Diversity of cultivable fungi associated with Antarctic marine sponges and screening for their antimicrobial，antitumoral and antioxidant potential. World J Microbiol Biotechnol 30：65-76

Hentschel U，Hopke J，Horn M，Friedrich AB，Wagner M，Hacker J，Moore BS（2002）Molecular evidence for a uniform microbial community in sponges from different oceans. Appl Environ Microbiol 68：4431-4440

Huang JP, Mojib N, Rr G, Watkins S, Waites KB, Ravindra R, Andersen DT, Bej AK (2012) Antimicrobial activity of PVP from an Antarctic bacterium, *Janthinobacterium* sp. Ant5-2, on multi-drug and methicillin resistant *Staphylococcus aureus*. Nat Prod Bioprospect 2：104-110

Ivanova V, Oriol M, Montes M-J, García A, Guinea J (2001) Secondary metabolites from a *Streptomyces* strain isolated from Livingston Island. Antarctica Z Naturforsch C 56：1-5

Ivanova V, Yocheva L, Schlegel R, Graefe U, Kolarova M, Aleksieva K, Naidenova M (2002) Antibiotic complex from *Streptomyces flavovirens* 6^7, isolated from Livingston Island, Antarctica. Bulgarian Antarct Res Life Sci 3：35-42

Jayatilake GS, Thornton MP, Leonard AC, Grimwade JE, Baker BJ (1996) Metabolites from an Antarctic sponge-associated bacterium, *Pseudomonas aeruginosa*. J Nat Prod 59：293-296

Kim M-K, Park H, Oh T-J (2012) Antibacterial properties associated with microorganisms isolated from Arctic lichens. Korean J Microbiol Biotechnol 40：380-388

Kim M-K, Park H, Oh T-J (2013) Antibacterial properties of the bacterial associates of the Arctic lichen *Stereocaulon* sp. Afr J Microbiol Res 7：3651-3657

Kim M-K, Park H, Oh T-J (2014a) Antibacterial and antioxidant capacity of polar microorganisms isolated from Arctic lichen *Ochrolechia* sp. Pol J Microbiol 63：317-322

Kim M-K, Park H, Oh T-J (2014b) Antibacterial and antioxidant potential of polar microorganisms isolated from Antarctic lichen *Psoroma* sp. Afr J Microbiol Res 8：3529-3535

Kozlovsky AG, Zhelifonova VP, Antipova TV, Baskunov BP, Kochkina GA, Ozerskaya SM (2012) Secondary metabolite profiles of the *Penicillium* fungi isolated from the Arctic and Antarctic permafrost as elements of polyphase taxonomy. Microbiology 81：308-313

Lee L-H, Cheah Y-K, Sidik SM, Ab Mutalib N-S, Tang Y-L, Lin H-P, Hong K (2012a) Molecular characterization of Antarctic actinobacteria and screening for antimicrobial metabolite production. World J Microbiol Biotechnol 28：2125-2137

Lee L-H, Cheah Y-K, Syakima AMN, Shiran MS, Tang Y-L, Lin H-P, Hong K (2012b) Analysis of Antarctic proteobacteria by PCR fingerprinting and screening for antimicrobial secondary metabolites. Genet Mol Res 11：1627-1641

Li Y, Sun B, Liu S, Jiang L, Liu X, Zhang H, Che Y (2008) Bioactive asterric acid derivatives from the Antarctic ascomycete fungus *Geomyces* sp. J Nat Prod 1：1643-1646

Liu J-T, Lu X-L, Liu X-Y, Yun G, Bo H, Jiao B-H, Zheng H (2013) Bioactive natural products from the Antarctic and Arctic organisms source. Mini Rev Med Chem 13：617-626

Lo Giudice A, Brilli M, Bruni V, De Domenico M, Fani R, Michaud L (2007a) Bacterium bacterium inhibitory interactions among psychrotrophic bacteria isolated from Antarctic sea-waters (Terra Nova Bay, Ross Sea). FEMS Microbiol Ecol 60：383-396

Lo Giudice A, Bruni V, Michaud L (2007b) Characterization of Antarctic psychrotrophic bacteria with antibacterial activities against terrestrial microorganisms. J Basic Microbiol 47：496-505

Lyutskanova D, Ivanova V, Stoilova-Disheva M, Kolarova M, Aleksieva K, Raykovska V, Peltekovavska V, Peltekova V (2009) Isolation, characterization and screening for antimicro-bial activities of psychrotolerant Streptomycetes isolated from polar permafrost soil. Biotechnol Biotechnol Eq 23.

Maida I, Fondi M, Papaleo MC, Perrin E, Orlandini V, Emiliani G, de Pascale D, Parrilli E, Tutino ML, Michaud L, Lo Giudice A, Romoli R, Bartolucci G, Fani R (2014) Phenotypic and genomic characterization of the Antarctic bacterium *Gillisia* sp. CAL575, a producer of antimicrobial compounds. Extremophiles 18：35-49

Maida I, Bosi E, Fondi M, Perrin E, Orlandini V, Papaleo MC, Mengoni A, de Pascale D, Tutino ML, Michaud L, Lo Giudice A, Fani R (2015. Antimicrobial activity of Pseudoalteromonas strains isolated from the Ross Sea (Antarctica) vs Cystic Fibrosis opportunistic pathogens. Hydrobiologia 761：443-457

Mangano S, Michaud L, Caruso C, Brilli M, Bruni V, Fani R, Lo Giudice A (2009) Antagonistic interactions among psychrotrophic cultivable bacteria isolated from Antarctic sponges: a preliminary analysis. Res Microbiol 160：27-37

Manivasagan P, Venkatesan J, Sivakumar K, Kim S-K (2014) Pharmaceutically active secondary metabolites of marine actinobacteria. Microbiol Res 169：262-278

Margesin R (2007) Alpine microorganism: useful tools for low-temperature bioremediation.
J Microbiol 45：281-285

Melo IS, Santos SN, Rosa LH, Parma MM, Silva LJ, Queiroz SC, Pellizari VH (2014) Isolation and biological activities of an endophytic *Mortierella alpina* strain from the Antarctic moss *Schistidium antarctici*. Extremophiles 18：15-23

Mojib N, Philpott R, Huang JP, Niederweis M, Bej AK (2010) Antimycobacterial activity in vitro of pigments isolated from Antarctic bacteria. Antonie Van Leeuwenhoek 98：531-540

Moncheva P, Tishkov S, Dimitrova N, Chipeva V, Antonova-Nikolova S, Bogatzevska N (2002) Characteristics of soil Actinomycetes from Antarctica. J Cult Collect 3：3-14

Montemartini Corte A, Liotta M, Venturi CB, Calegari L (2000) Antibacterial activity of *Penicillium* spp. strains isolated in extreme environments. Polar Biol 23：294-297

Moon K, Ahn C-H, Shin Y, Won TH, Ko K, Lee SK, Oh K-B, Shin J, Nam SI, Oh DC (2014) New benzoxazine secondary metabolites from an Arctic actinomycetes. Mar Drugs 12：2526-2538

MoritaRY (1975) Psychrophilic bacteria. Bacteriol Rev 39：144-167

Nedialkova D, Naidenova M (2005) Screening the antimicrobial activity of Actinomycetes strains isolated from Antarctica. J Cult Collect 4：29-35

O'Brien A, Sharp R, Russell NJ, Roller S (2004) Antarctic bacteria inhibit growth of food-borne microorganisms at low temperatures. FEMS Microbiol Ecol 48：157-167

Onofri S，Fenice M，Cicalini AR，Tosi S，Magrino A，Pagano S，Selbmann L，Zucconi L，Vishniac HS，Ocampo-Friedmann R，Friedmann EI（2000）Ecology and biology of microfungi from Antarctic rocks and soils. Ital J Zool 67：163-167

Orlandini V，Maida I，Fondi M，Perrin E，Papaleo MC，Bosi E，de Pascale D，Tutino ML，Michaud L，Lo Giudice A，Fani R（2014）Genomic analysis of three sponge-associated *Arthrobacter* Antarctic strains，inhibiting the growth of *Burkholderia cepacia* complex bacteria by synthesizing volatile organic compounds. Microbiol Res 169：593-601

Pan SY，Tan GYA，Convey P，Pearce DA，Tan IKP（2013）Diversity and bioactivity of actinomy-cetes from Signy Island terrestrial soils，maritime Antarctic. Adv Polar Sci 24：208-212

Papaleo MC，Fondi M，Maida I，Perrin E，Lo Giudice A，Michaud L，Mangano S，Bartolucci G，Romoli R，Fani R（2012）Sponge-associated microbial Antarctic communities exhibiting antimicrobial activity *against Burkholderia cepacia* complex bacteria. Biotechnol Adv 30：272-293

Papaleo MC，Romoli R，Bartolucci G，Maida I，Perrin E，Fondi M，Orlandini V，Mengoni A，Emiliani G，Tutino ML，Parrilli E，de Pascale D，Michaud L，Lo Giudice A，Fani R（2013）Bioactive volatile organic compounds from Antarctic（sponges）bacteria. New Biotechnol 30：824-838

Pearce DA（2012）Extremophiles in Antarctica：life at low temperatures. In：Stan-Lotter H，Fendrihan S（eds）Adaption of microbial life to environmental extremes. Springer，Vienna，pp 87-118

Prasad S，Manasa P，Buddhi S，Singh SM，Shivaji S（2011）Antagonistic interaction networks among bacteria from a cold soil environment. FEMS Microbiol Ecol 78：376-385

Rojas JL，Martín J，Tormo JR，Vicente F，Brunati M，Ciciliato I，Losi D，Van Trappen S，Mergaert J，Swings J，Marinelli F，Genilloud O（2009）Bacterial diversity from benthic mats of Antarctic lakes as a source of new bioactive metabolites. Mar Genomics 2：33-41

Romoli R，Papaleo MC，de Pascale D，Tutino ML，Michaud L，Lo Giudice A，Fani R，Bartolucci G（2011）Characterization of the volatile profile of Antarctic bacteria by using solid-phase microextraction -gas chromatography mass spectrometry. J Mass Spectrom 46：1051-1059

Romoli R，Papaleo MC，de Pascale D，Tutino ML，Michaud L，Lo Giudice A，Fani R，Bartolucci G（2014）GC-MS volatolomic approach to study the antimicrobial activity of the Antarctic bacterium *Pseudoalteromonas* sp. TB41. Metabolomics 10：42-51

Russell NJ（2008）Membrane components and cold sensing. In：Margesin R，Schinner F，Marx JC，Gerday C（eds）Psychrophiles：from biodiversity to biotechnology. Springer，Berlin，Heidelberg，pp 177-190

Schuhmann I，Yao CB，Al-Zereini W，Anke H，Helmke E，Laatsch H（2009）Nitro derivatives from the Arctic ice bacterium *Salegentibacter* sp. isolate T436. J Antibiot 62：453-460

Shekh RM，Singh P，Singh SM，Roy U（2011）Antifungal activity of Arctic and Antarctic bacteria isolates. Polar Biol 34：139-143

Svahn KS，Chryssanthou E，Olsen B，Bohlin L，Göransson U（2015）*Penicillium nalgiovense* Laxa isolated from Antarctica is a new source of the antifungal metabolite amphotericin B. Fungal Biol Biotechnol 2：1. doi：10. 1186/s40694-014-0011-x

Taton A，Grubisic S，Ertz D，Hodgson DA，Piccardi R，Biondi N，Tredici MR，Mainini M，Losi D，Marinelli F，Wilmotte A（2006）Polyphasic study of antarctic cyanobacterial strains. J Phycol 42：1257-1270

Vester JK，Glaring MA，Stougaard P（2015）Improved cultivation and metagenomics as new tools for bioprospecting in cold environments. Extremophiles 19：17-29

Wietz M，Månsson M，Bowman JS，Blom N，Ng Y，Gram L（2012）Wide distribution of closely related，antibiotic-producing *Arthrobacter* strains throughout the Arctic Ocean. Appl Environ Microbiol 78：2039-2042

Wong CMV，Tam HK，Alias SA，González M，González-Rocha G，Domínguez-Yévenes M（2011）*Pseudomonas* and *Pedobacter* isolates from King George Island inhibited the growth of foodborne pathogens. Pol Polar Res 32：3-14

Yuan M，Yu Y，Li H-R，Dong N，Zhang X-H（2014）Phylogenetic diversity and biological activity of Actinobacteria isolated from the Chukchi Shelf marine sediments in the Arctic Ocean. Mar Drugs 12：1281-1297

第四章
嗜热菌：特殊性质胞外多糖的重要来源

Margarita Kambourova[1,2], Nadja Radchenkova[1],
Iva Tomova[1], Ivanka Bojadjieva[1]

4.1 引言

嗜热菌作为一种极端微生物，能够在45～122℃之间相对较高的温度中生存（Takai et al. 2008）。极端微生物不仅有耐受极端环境的能力，而且能够在恶劣的生存条件下保持功能活性。“极端”是一个相对的术语，是指相对于人类生存而言条件太过于恶劣的环境（Satyanarayana et al. 2005）。20世纪末，以细菌域和古菌域为代表的嗜热菌从地热和热液口以及其他更高的温度环境中分离，生命能承受的温度上限已经发生了数次变化。嗜热菌组群包括生长温度能达到50℃的兼性嗜热菌；生长温度在50～70℃，最佳生长温度在55～65℃之间的专性嗜热菌；生长温度在65～80℃之间的极端嗜热菌；以及最适生长温度高于80℃的超嗜热菌（Wiegel and Canganella 2001）。

嗜热生态环境包括火山和地热区（陆地、地下和海洋热泉）、温泉、太阳加热处理的废弃物、储油库和人造栖息地等（图4.1）。高温环境限制真核生物的生长，因此在高温环境中细菌、古菌以及病毒占主导地位（López-López et al. 2013）。许多观点认为古菌和细菌的祖先可能是（超）嗜热菌（Di Giulio 2003）。全球大量研究致力于揭示在极端条件下存在的意想不到的类群和新物种的多样性，及其在生物技术上的开发潜力。极端微生物拥有新的代谢特征，这使得它们能在恶劣环境下维持生理活性并且产生独特的代谢物（Rozanov et al. 2014）。这些代谢化合物近几年来显著地促进了生物技术的发展。

嗜热微生物及其耐热性酶在生物技术领域是不可或缺的，这是因为它们在高温下有着独特的机能，并能在各种恶劣的工业环境中保持稳定。参与嗜热适应过程的这些化合物中，胞外多糖（EPS）由于其结构和功能的多样性，以及在细胞中的生理学作用引起了研究者们广泛的关注。胞外多糖是微生物分泌到周围环境中的高分子。它们是由相同或不同的糖基以重复单元形式排列组成。同多糖是由同种糖基组成；杂多糖是由不同的糖基组成。

[1] Institute of Microbiology, Bulgarian Academy of Sciences, Acad. G. Bonchev str. 26, 1113 Sofi a, Bulgaria.
[2] e-mail: margikam@microbio. bas. bg.

图 4.1　在保加利亚的巴尼亚萨帕雷瓦间歇泉，温泉的温度高达 103℃

4.2　细胞胞外多糖的生理作用

细胞表面胞外多糖的堆积是极端微生物适应极端环境常见的策略，包括稳定膜结构和参与细胞保护。细胞表面的多糖层通常是细胞大小的几倍。然而，细菌胞外多糖的生理作用可能比现阶段已知的更加多样化和复杂。这些生物聚合物多糖根据其分类学及生态学地位的不同具有不同的功能（Poli et al. 2011；Nwodo et al. 2012）。尽管胞外多糖的生产需要细胞总储备能量的 70%，但是它对微生物生长有显著影响，并且还取决于环境因素。细菌胞外多糖的一些功能如图 4.2 所示。

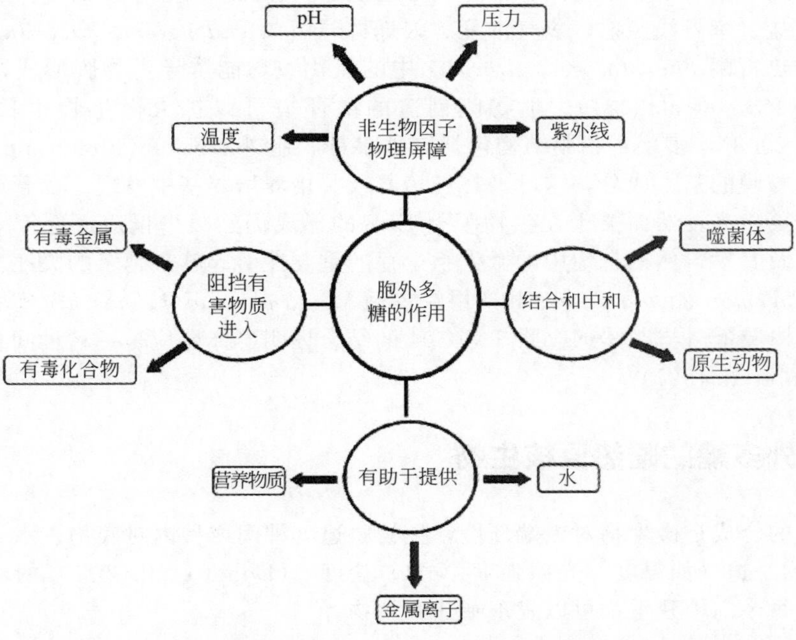

图 4.2　胞外多糖有助于微生物细胞在极端条件下生存的可能的生理学作用

① EPS作为一种普通的物理屏障保护微生物。EPS产生是对选择性环境压力，包括渗透压、温度、pH、大气压和光照强度的一种适应性（Otero and Vincenzini 2003）。

② 包裹的EPS有利于维持细胞的水合微环境，在缺水的环境中可以保持生存所需水分。

③ 在极端环境如寡营养条件下，EPS可以捕获周围环境的营养物质，EPS基质结合并富集水体中可生物降解的化合物和阳离子。胞外多糖中硫酸盐和糖醛酸等阴离子残基的存在，使其可以与阳离子相互作用，如金属离子，并获得必需的矿物质和营养物质。

④ EPS还可增强微生物的表面黏附性，并且可以作为絮凝剂或者乳化剂。许多微生物合成胞外多糖覆盖细胞膜表面的疏水位点，以达到促进增长和黏附到固体表面的目的。

⑤ 细胞表面EPS的存在可能影响有害物质的扩散，如抗生素或者有毒化合物（例如有毒金属离子、二氧化硫和乙醇）进出细胞。由于多糖的阴离子特性，它们可以增强离子如Pb^{2+}和Cu^{2+}的固定，因此具有重要的生态学意义。EPSs通过絮凝和结合溶液中的金属离子来清除重金属（Nicolaus et al. 2010）。嗜热地芽孢杆菌 *Geobacillus tepidamans* 的胞外聚合物具有对阿瓦醇的抗细胞毒性作用（Kambourova et al. 2009）。

⑥ EPS通过结合和中和噬菌体参与细胞保护（Vu et al. 2009）。

⑦ 虽然普遍认为EPS不能作为能量储备物质，且微生物不能降解其自身的EPS（Donot et al. 2012），然而，一些超嗜热菌种［硫化叶菌（*Sulfolobus*）、热球菌属（*Thermococcus*）和热袍菌属（*Thermotoga*）］产生的EPSs可以在其他有机物质来源缺乏的极端环境中间接作为细胞外能源存储物来起作用（Nicolaus et al. 1993；Rinker and Kelly 2000）。

4.2.1 生物膜

在自然环境下，细菌产生胞外聚合物与界相表面结合下细菌细胞的功能联合体，即为生物膜。生物膜的完整性对于细菌生存至关重要（Moons et al. 2009）。生物膜在微生物对抗环境压力、生物膜表面黏附、细胞间相互作用等均起重要作用。环境pH的升高、降低和升高生长温度、高盐、紫外线照射、氧气浓度、闪烁古球菌 *Archaeoglobus fulgidus* 产生的抗生素以及深海嗜热古菌 *Thermococcus litoralis* 中的氯化铵均能诱导生物膜形成，证明生物膜的保护作用（Pysz et al. 2004）。生物膜创造的微环境可以扩大微生物生长的限制范围（Lowell et al. 2008）。多层生物膜结构包括胞外多糖、胞外DNAs（Flemming et al. 2007）。胞外多糖是生物膜的基本成分，这已经在纯培养或共培养培养基中发现。尽管许多嗜热菌已经从温泉中分离出来，然而关于在生物膜上EPS的形成仍然很少报道。近年来，生物膜的形成机制，以及其对自然栖息地中的微生物生存的重要性引起越来越多的关注。通过共培养极端嗜热菌 *Thermotoga maritima* 和产甲烷古菌 *Methanococcus jannaschii* 实验发现，一种含GGDEF结构域蛋白与群体感应肽TM0504的结合证明群体感应是一种种间相互作用方式（Muralidharan et al. 1997）。

4.3 产胞外多糖的嗜热原核生物

胞外多糖的合成是微生物对极端环境，生物胁迫（种间或种内对底物、水或生长因子的竞争），非生物胁迫（如温度、光照强度、pH、盐度）（Donot et al. 2012）的响应结果。在古菌和细菌的每个门中几乎都可以发现嗜热微生物。

4.3.1 细菌

研究发现，嗜热菌包括专性嗜热属的 *Bacillus*、*Geobacillus*、*Brevibacillus* 和 *Aeribacillus*，极端嗜热属的 *Thermus* 以及超嗜热属的 *Thermotoga* 能够产生嗜热胞外多糖。它们从大陆温泉或浅海热液喷口中被分离出来（表 4.1）。极端嗜热细菌 *Thermotoga maritima* 最初是从海底地热区域中分离出来（Huber et al. 1986）。这种微生物是严格厌氧的异养菌，最适生长温度为 80℃。由于微生物很少独立存在，种属间的互利关系可能扩大极端微生物对热环境的适应范围（Kolter and Losick 1998）。产甲烷菌对氢气的堆积是自然环境中异养菌和产甲烷菌共同生长的前提（Muralidharan et al. 1997）。产甲烷古菌 *Methanococcus jannaschii* 和极端嗜热厌氧发酵菌 *Thermotoga maritima* 共培养可进一步增加菌群密度。从 *Thermus aquaticus* YT-1 的生物膜分离得到一种新型胞外多糖（Lin et al. 2011）。

表 4.1 嗜热原核生物产生的胞外多糖

域名	微生物	分离来源	生长温度/℃	碳源	最大产量/(mg/L)	功能特征和潜在应用	文献
细菌	*Thermus aquaticus* YT-1	美国黄石温泉	60	无数据	无数据	免疫调节作用	Lin et al. (2011)
	Thermotoga maritima	意大利弗卡诺地热海洋沉积物	88	麦芽糖	120	无数据	Rinker and Kelly (2000)
	Geobacillus thermoantarcticus	南极洲墨尔本山火山口	65	甘露糖	400	无数据	Manca et al. (1996); Nicolaus et al. (2004)
	Bacillus licheniformis B3-15	意大利弗卡诺浅海温泉	45	葡萄糖	165	抗病毒和免疫调节作用	Maugeri et al. (2002); Arena et al. (2006)
	Bacillus licheniformis T14	意大利帕纳雷亚岛浅热液喷口	50	蔗糖	366	抗病毒和免疫调节作用	Spanò et al. (2013); Gugliandolo et al. (2014)
	Geobacillus sp. 4001	意大利弗雷格兰雷斯浅海热液喷口	65	半乳糖，蔗糖	55	无数据	Nicolaus et al. (2002); Nicolaus et al. (2004)
	Geobacillus sp. 4004	意大利弗雷格兰雷斯浅海热液喷口	65	海藻糖，蔗糖	65	无数据	Nicolaus et al. (2002); Nicolaus et al. (2004)
	Geobacillus thermodenitrificans B3-72	意大利弗卡诺岛浅热液喷口	65	葡萄糖，蔗糖	70	EPS2:抗病毒和免疫调节作用	Nicolaus et al. (2000); Arena et al. (2009)
	Geobacillus tepidamans V264	韦林格勒温泉	60	麦芽糖	111.4	高热稳定性，抗细胞毒活性	Kambourova et al. (2009)
	Aeribacillus pallidus 418	保加利亚热泉盆地温泉	55	麦芽糖	170	高热稳定性，乳化性能	Radchenkova et al. (2013); Radchenkova et al. (2014)
	Brevibacillus thermoruber 423	保加利亚布拉格耶夫格勒地区温泉	55	麦芽糖	897	生物医学应用	Yasar Yildiz et al. (2014)

续表

域名	微生物	分离来源	生长温度/℃	碳源	最大产量/(mg/L)	功能特征和潜在应用	文献
古菌	*Thermococcus litoralis*	意大利那不勒斯浅海温泉	88	麦芽糖	180	通常由真核生物产生甘露聚糖化合物	Rinker and Kelly (1996); Rinker and Kelly(2000)
	Sulfolobus solfataricus MT4	意大利阿尼亚诺热酸性泉	88	葡萄糖	8.4	无数据	Nicolaus et al. (1993)
	Sulfolobus solfataricus MT3	意大利阿尼亚诺热酸性泉	75	葡萄糖	7	无数据	Nicolaus et al. (1993)
	Sulfolobus tokodaii	日本别府温泉	76	无数据	无数据	无数据	Koerdt et al. (2010)
	Sulfolobus acidocaldarius	美国黄石温泉	76	无数据	无数据	无数据	Koerdt et al. (2010)

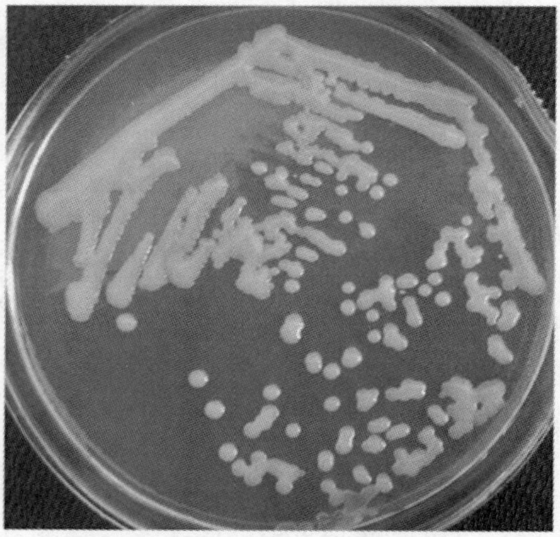

图 4.3 专性嗜热菌节芽孢杆菌 *Aeribacillus pallidus* 418 在 LB 琼脂培养平板上培养 24h 产生的黏液

尽管对嗜热杆菌来源的嗜热酶进行了大量的探索，但是关于它们 EPS 的研究仍然处于探索阶段。*Bacillus licheniformis* B3-15 和 T14（Maugeri et al. 2002；Spanò et al. 2013），以及从浅海热液喷口和地热温泉分离的地芽孢杆菌属 *Geobacillus* 菌株（Manca et al. 1996；Nicolaus et al. 2000，2002，2003，2004；Kambourova et al. 2009）均具备有分泌胞外多糖的能力。苍白空气芽孢杆菌 *Aeribacillus pallidus* 418 可产生两种不同的胞外多糖（图 4.3）（Radchenkova et al. 2013）。

4.3.2 古菌

极端古菌的发现对生物催化具有重要意义，这是因为它们分泌的酶可以改善工业中的多个领域（Węgrzyn and Żukrowski，2014）。来自古细菌的另一类对生物技术工业具有潜在影响的生物聚合物就是 EPS。

在许多不同的环境中，古菌通常被认为生物膜群落的主要组成成分（Krüger et al. 2008；Zhang et al. 2008），很少有研究报道古菌生物膜的形成。首次报道关于硫磺矿硫化叶菌 *Sulfolobus solfataricus* 产生 EPS 的研究距今已经超过 20 年（Nicolaus et al. 1993），之后进一步研究发现 *Sulfolobus acidocaldarius* 和 *Sulfolobus tokodaii* 能够形成生物膜（Koerdt et al. 2010）。由于在 *Archaeoglobus profundus*、*Archaeoglobus fulgidus*（Lapaglia and Hartzell 1997）、*Thermococcus litoralis*（Rinker and Kelly 1996）、*Methanococcus jannaschii*（Lapaglia and Hartzell 1997）以及 *Methanothermobacter thermoautotrophicus*（Thoma et al. 2008）中均观察到类似的生物膜，可见，生物膜的形成可能是古菌常见的应激反应（Hartzell et al. 1999）。

Sulfolobus（*Crenarchaeota*）在世界各地均有分布，在温度、pH 和地理条件急剧变化

的酸性泥泞的温泉中广泛存在，该细菌需要快速适应不断变化的外界条件（Koerdt et al. 2010）。

EPS的产生不仅局限于泉古菌门（*Crenarchaeota*）细菌，一些广古菌门（*Euryarchaeota*）细菌也具有产生EPS的能力。广古菌门强烈炽热球菌 *Pyrococcus furiosus*（最适生长温度为100℃，常用作超嗜热菌的模式微生物）通过鞭毛在生物膜状结构中相互连接（Näther et al. 2006）。这种微生物能够黏附到生物表面。目前已经证明它能与另一种超嗜热古菌 *Methanopyrus kandleri* 共培养，并形成生物膜，*Methanopyrus kandleri* 在类似的生态环境中可以存活（接近沸水的生存温度和厌氧条件）。*Pyrococcus furiosus* 和 *Methanopyrus kandleri* 在实验室条件下能够在不到24h内形成古菌双物种生物膜（Schopf et al. 2008）。从浅海温泉中分离的异养硫依赖性极端嗜热菌 *Thermococcus litoralis*（广古菌门）在多种条件下也能在亲水表面上形成生物膜。Baker-Austin 等（2010）在极端嗜盐菌 *Ferroplasma acidarmanu*（Euryarchaeotes）中观察到两种不同的生物膜形态。厌氧海洋超嗜热菌 *Archaeoglobus fulgidus* 生物膜的产生与其在寡营养、高温和潜在有毒化合物等极端环境中广泛定植的能力密切相关（Lapaglia and Hartzell 1997）。

4.4 耐热性胞外多糖的化学和结构组成

与植物多糖相反，细菌胞外多糖，尤其是嗜热微生物胞外多糖的结构特性的研究很少。不同微生物合成的胞外多糖组成不同，因此其化学和物理性质也不同。比较多个数据库里的哺乳动物和细菌的碳水化合物结构数据发现，细菌中单糖的多样性较哺乳动物的高出10倍，糖苷键类型则高出9倍（Herget et al. 2008）。大部分的胞外多糖是由3种或4种不同单糖形成的10个或更少数目的重复单元，而构成的杂多糖（Poli et al. 2011）。最常见的单糖是通过强的1,4-β、1,3-β 或 2,6-β 键和较为灵活的1,2-α-或1,6-α-键的方式进行连接。多种连接方式可以同时发生在一个多糖分子上。重复糖单元主要由葡萄糖、半乳糖、甘露糖、糖醛酸、*N*-乙酰葡糖胺、*N*-乙酰半乳糖胺和鼠李糖以不同的比例组成。多糖的组成和结构决定其主要构象。二级结构包括聚合螺旋，酰基取代基影响从无规卷曲到有序螺旋聚集体的转变（Sutherland 1994）。EPS中糖胺聚糖的刚性结构保证了分子和细胞结构的完整性。Lin 等（2011）观察 *Thermus aquaticus* YT-1产生的EPS发现，无规卷曲倾向于通过 *N*-乙酰半乳糖胺的乙酰基形成螺旋结构。它的存在有助于形成规则和稳定的结构。通常杂多糖含有非糖成分，如乙酸盐、丙酮酸盐、琥珀酸盐、磷酸盐、硫酸盐、甲基酯、蛋白质、核酸和脂质（Nicolaus et al. 2010），这些组分可能在多糖分子中发挥重要的作用，例如EPS的过硫酸化可以改变其生物活性（Courtois et al. 2014）。一些EPS是中性大分子，还有些EPS，由于含有糖醛酸或缩酮基团，形成了多聚阴离子长链（Nicolaus et al. 2010）。

来自 *Thermococcus litoralis* 和 *Geobacillus thermoantarcticus* 的多糖是由甘露糖组成的同多糖（Rinker and Kelly，1996；Manca et al. 1996）。*Geobacillus tepidamans* 合成的EPS大部分为纯的葡聚糖（Kambourova et al. 2009）。

由嗜热微生物合成的大多数杂多糖的主要单糖组分是葡萄糖或甘露糖，也有果糖或半乳糖（表4.2）。从浅海热液口中分离的三种 *Geobacillus* 产生了不同的EPS（Nicolaus et al. 2003）。两株 *Geobacillus* 产生的EPS主要为葡萄糖，半乳糖和甘露糖以不同比例组合，而第三株菌的EPS主要包含葡糖胺、阿拉伯糖、半乳糖和甘露糖。从嗜热菌 *Brevibacillus thermoruber* 423分离的EPS由五种不同的糖组成（葡萄糖为主要单体单元）（Yasar Yildiz et al. 2014）。*Aeribacillus pallidus* 418产生两种含多种单糖组分的EPS（EPS1有6种，EPS2

表 4.2 嗜热微生物合成的胞外多糖特性

域名	微生物	生长温度/℃	糖类和蛋白质含量	分子质量/Da	EPS 组成	文献
细菌	*Thermus aquaticus* YT-1	60	无数据	500000	呋喃半乳糖/半乳糖/*N*-乙酰半乳糖(1∶1∶2)	Lin et al.(2011)
	Thermotoga maritima(与 *Methanococcus jannaschii* 共培养)	80	5%糖类	未检出	葡萄糖/核糖/甘露糖(1∶0.06∶0.03)	Johnson et al.(2005)
	Geobacillus thermoantarcticus	65	96%的糖类,0.2%蛋白质	EPS1:300000 EPS2:300000	EPS1:甘露糖/葡萄糖(1.0∶0.7);tEPS2:甘露糖/葡萄糖(1∶tr)	Manca et al.(1996);Nicolaus et al.(2004)
	Bacillus licheniformis B3-15	45	66%的糖类,5%蛋白质	600000	甘露糖/葡萄糖(1∶3)	Maugeri et al.(2002)
	Bacillus licheniformis T14	50	99%的糖类,1%的蛋白质	1000000	果糖/岩藻糖/葡萄糖/半乳糖/甘露糖(1.0∶0.75∶0.28∶tr∶tr)	Spanò et al.(2013);
	Geobacillus sp. 4001	65	81%的糖类,7%蛋白质	380000	甘露糖/葡萄糖/半乳糖/甘露糖胺(1∶0.1∶tr∶tr)	Nicolaus et al.(2002);Nicolaus et al.(2004)
	Geobacillus sp. 4004	65	65%的糖类,2.6%的蛋白质	>1000000	半乳糖/甘露糖/葡萄糖/阿拉伯糖(1∶0.8∶0.4∶0.2)	Nicolaus et al.(2002);Nicolaus et al.(2004)
	Geobacillus thermodenitrificans B3-72	65	80%的糖类,3%蛋白质	EPS2:400000	EPS1:甘露糖/葡萄糖(0.3∶1) EPS2:甘露糖/葡萄糖(1∶0.2)	Nicolaus et al.(2000)
	Geobacillus tepidamans V264	60	100%糖类	>1000000	葡萄糖/半乳糖/岩藻糖/果糖(1∶0.07∶0.04∶0.02)	Kambourova et al.(2009)
	Aeribacillus pallidus 418	55	80.5%糖类,19 %蛋白质	EPS1:700000 EPS2:1000000	EPS1:甘露糖/葡萄糖/半乳糖/葡糖胺/半乳糖/核糖(1∶0.16∶0.1∶0.09∶0.07∶0.06∶0.04) EPS2:甘露糖/半乳糖/葡萄糖/半乳糖胺/葡糖胺/核糖/阿拉伯糖(1∶0.5∶0.46∶0.35∶0.24∶0.16∶0.14)	Radchenkova et al.(2013)
	Brevibacillus thermoruber 423	55	92%糖类,6%蛋白质	无数据	葡萄糖/半乳糖/半乳糖胺/甘露糖/甘露糖胺(1∶0.3∶0.25∶0.16∶0.04)	Yasar Yildiz et al.(2014)

续表

域名	微生物	生长温度/℃	糖类和蛋白质含量	分子质量/Da	EPS组成	文献
古菌	*Thermococcus litoralis*	88	无数据	41000	甘露糖	Rinker and Kelly(1996)
	Sulfolobus solfataricus MT4	88	75%糖类，0.1%蛋白质	无数据	葡萄糖/甘露糖/葡萄糖/半乳糖（1.2：1.0：0.18：0.13）	Nicolaus et al. (1993)
	Sulfolobus solfataricus MT3	75	75%糖类，0.1%蛋白质	无数据	葡萄糖/甘露糖/葡萄糖/半乳糖（1.2：1.0：0.77：0.73）	Nicolaus et al. (1993)
	Sulfolobus tokodaii	76	无数据	无数据	甘露糖/葡萄糖/半乳糖/*N*-乙酰氨基葡萄糖	Koerdt et al. (2010)
	Sulfolobus acidocaldarius	76	无数据	无数据	甘露糖/葡萄糖/半乳糖/*N*-乙酰氨基葡萄糖	Koerdt et al. (2010)

注：tr表示微量。

有7种)，其中甘露糖为主要成分（Radchenkova et al. 2013)。最近，*Aeribacillus pallidus* YM-1被报道能产生一种由脂质（47.6%)、糖类（41.1%）和蛋白质（11.3%）组成的新型生物乳化剂（Zheng et al. 2012)。糖类部分由葡萄糖（36.6%)、阿卓糖（30.9%)、甘露糖（24.4%）和半乳糖（8.1%）组成。

有研究发现由 *Bacillus thermantarticus* 产生的EPS中存在丙酮酸和硫酸盐（Nicolaus et al. 2004)。Lin等（2011）解析了由 *Thermus aquaticus* 分泌的EPS TA-1的主要结构。该聚合物由半乳糖呋喃糖、吡喃半乳糖和*N*-乙酰半乳糖胺的四糖重复单元组成，并且缺乏酸性糖。*Thermotoga maritima* 的5%干重是胞外多糖，且多糖由91.2%葡萄糖、5.2%核糖和2.7%甘露糖组成（Johnson et al. 2005)。*Sulfolobus* 产生含有甘露糖、葡萄糖、半乳糖和*N*-乙酰葡糖胺的胞外多糖（Koerdt et al. 2010)。

嗜热菌产生的EPS的另一个特征是它们的高分子质量（从几十万道尔顿开始)。嗜热杆菌产生的EPS分子质量为380kDa、400kDa、600kDa和1000kDa（Nicolaus et al. 2003)。*Aeribacillus pallidus* 418产生EPS的分子质量约为700kDa和1000kDa（Radchenkova et al. 2013)。从另一种嗜热菌 *Geobacillus thermoantarcticus* 产生EPS的分子质量约为300kDa（Manca et al. 1996)。*Geobacillus tepidamans* V264产生多糖的分子质量高于1000kDa（Kambourova et al. 2009)。根据Kumar等人（2007）的报道发现嗜温细菌合成的多糖分子质量多为10～30kDa，这些值显著低于嗜热细菌合成的多糖的分子质量。嗜温菌也能产生高分子质量的EPS，例如由 *Alteromonas macleodii* 产生的EPS（300kDa和1500kDa）(Raguénès et al. 1996，2003)；由假单胞菌属（Conti et al. 1994）产生的藻酸盐（34～500kDa）和胶凝糖（250kDa和490kDa）(Milas et al. 1990)。根据Patel等人（2010）的研究，乳杆菌所产的EPS分子质量为10～1000kDa。而Kralj等（2004）发现由乳杆菌分泌的EPS分子质量为1000～5000kDa。

嗜热细菌产生的EPS是热稳定性的。目前报道热稳定性最强的多糖来自 *Geobacillus tepidamans* V264（280℃)、*Geobacillus thermodenitrificans* strain B3-72（240℃）(Arena et al. 2006）和 *Bacillus licheniformis*（240℃）(Spanò et al. 2013）(Kambourova et al. 2009)。关于嗜温细菌胞外多糖热稳定性的研究却很少。在一株中度嗜盐细菌中发现，其

能产生高度热稳定性的胞外多糖（熔点为 207℃）（Cojoc et al. 2009）。

4.5 嗜热菌合成胞外多糖的特定条件

细菌 EPS 产量的增加传统上是通过菌株的筛选和/或优化培养条件实现，然而每种选定的细菌可能具有难以克服的生理极限（Freitas et al. 2011a）。细菌 EPS 生物合成的调控是一个复杂的过程，涉及大量的酶和调控蛋白（Jaiswal et al. 2014）。尽管 EPS 具有结构多样性，目前已知的细菌具有四种聚合机制，即细胞外生物合成、合酶依赖性生物合成、ABC-转运蛋白依赖性和最常见的 wzx/wzy 依赖性途径。在嗜温微生物中涉及糖核苷酸合成、重复单元合成和重复单元聚合的 EPS 生物合成机制得到了充分研究（De Vuyst et al. 2001；Freitas et al. 2011b）。最近一个嗜热菌 *Brevibacillus thermoruber* 423 多糖的合成过程中糖类的摄取假定机制被提出，该机制与嗜温菌 EPS 合成过程中糖摄取的机制类似（Yildiz et al. 2015）。通过基因组注释发现与 EPS 生物合成相关的关键基因，同时显示存在 NDP-糖的生物合成相关基因。基因组信息揭示了 *Brevibacillus thermoruber* 423 的 EPS 生物合成中存在 ABC-转运蛋白依赖性途径（Bth. peg. 2228，Bth. peg. 4273，Bth. peg. 3612，Bth. peg. 3618，Bth. peg. 4275）。此外，在极端嗜热细菌 *Thermotoga maritima* 中检测到一条胞外多糖合成途径（Johnson et al. 2005）。在与 *Methanococcus jannaschii* 共培养条件下，利用转录组学技术分析 *Thermotoga maritima* 的产糖过程发现一条编码多肽的基因明显上调。这条多肽分子包含一段与嗜温菌多肽信号分子相同的序列。*Streptococcus thermophilus* 完整的 15kb St Sfi6 eps 基因簇的鉴定表明，产物与已知的糖基转移酶具有高度的相似性，提示它们在合成重复单体中的潜在作用（Delcour et al. 2000）。编码 EPS 生物合成所需的蛋白质或酶的基因位于嗜热菌基因 DNA 上（Yildiz et al. 2015）而对于多数乳酸菌则位于质粒 DNA 上（Laws et al. 2001）。相反的是，嗜热 LAB *Streptococcus thermophilus*（Harutoshi 2013）缺乏编码黏液所需的质粒，而 *Lactobacillus fermentum* TDS030603（Dan et al. 2009）的基因组 DNA 中存在 EPS 基因簇。

尽管微生物 EPS 的组成和含量是由遗传决定的，但它们也取决于几个因素，例如菌株类型、碳源、氮源、矿物盐、微量元素、介质组分比例、发酵条件（温度、pH、搅拌和通气量）（Nicolaus et al. 2010）。微生物合成的 EPS 主要取决于培养基中可利用的碳源和氮源，能够生产 EPS 的微生物利用糖类作为其碳源和能量来源，铵盐和氨基酸作为它们的氮源（Gandhi et al. 1997；Czaczyk and Wojciechowska 2003）。这些微生物通常使用较低成本的糖类如葡萄糖、麦芽糖或蔗糖作为碳源，尽管在其他糖类存在的条件下它们可以产生更多的胞外多糖。一些嗜热产糖菌在含有葡萄糖的培养基中多糖的产生较麦芽糖培养基低（Rinker and Kelly 2000；Kambourova et al. 2009；Radchenkova et al. 2013，Yasar Yildiz et al. 2014）。有研究者指出，丰富的碳源和较少的氮含量能使细菌多糖产量增加（Radchenkova et al. 2013；Yasar Yildiz et al. 2014）。添加额外氮源有利于菌体量的增加，但会减少 EPS 的产量。EPS 的产生是否与细菌生长相关还不能确定。EPS 产生与 *Aeribacillus pallidus* 418（Radchenkova et al. 2013），*Brevibacillus thermoruber* 423（Yasar Yildiz et al. 2014），*Geobacillus tepidamans*（Kambourova et al. 2009）的生长相关。不少嗜温产糖菌在整个生长稳定期都能合成 EPS（Conti et al. 1994；Raguénès et al. 1997）。

氧气是好氧生物生产 EPS 的关键底物，但由于氧气在溶液中溶解度低，因此需要保证氧气持续供应。好氧微生物多糖产物可以通过优化通气量和增加搅拌速率来实现。通过反应器的体积控制气泡大小及其分散程度是细菌产多糖性能的关键。气泡尺寸越小，溶液与气体接触的表面积就越大，这样可以有效提高氧气传递速率。氧气在介质中的溶解度低，尤其是在高温工艺中，氧气（空气）传递成为需氧反应的限速步骤。高温导致氧气的溶解度急剧降低，这是在生物反应器中大规模生产 EPS 最重要的参数。近几年许多学者把在生物反应器中氧气的传递过程作为研究对象（Garcia-Ochoa and Gomez，2009）。然而，关于优化嗜热过程中 EPS 生产的搅拌和通气条件的研究信息仍然非常缺乏。通气量和搅拌对于 *Aeribacillus pallidus* 418 达到好氧工艺中最大的生产力是至关重要的（Radchenkova et al. 2014）。根据一些学者的报道（Rau et al. 1992；Radchenkova et al. 2014），限制溶解氧含量是增加多聚物产生的理想方法。众所周知，EPS 的重要生理作用是帮助细胞适应恶劣生境，对于极端微生物，它可以作为在极端环境下生存的增强剂（Nicolaus et al. 2010）。短暂的氧限制可以促进微生物合成 EPS 以应对这种恶劣条件。在搅拌式生物反应器中，大量的变量如搅拌速度、搅拌器类型、搅拌器数量和气体流速等影响反应液的混合和物质传递（Garcia-Ochoa and Gomez，2009）。大量的叶轮能确保反应底物的均匀分布，以及保证反应液中高热量和物质的传递比率（Chhabra，2003）。径向流涡轮机是非常受欢迎的一类实验室规模搅拌设备，它能够改变介质的剪切力。搅拌和通气量影响的研究显示这两个参数均能影响特定的 EPS 产量（Radchenkova et al. 2014），而搅拌比通气量更有效。

极端嗜热菌生产 EPS 通常不需要厌氧环境，即使在 EPS 作为生物膜累积时也是如此。通过在 $D=0.25h^{-1}$ 的厌氧恒化器的尼龙网上连续培养 *Thermotoga maritima* 可形成足够多的生物膜（Pysz et al. 2004）。在没有搅拌和气体喷射的间歇和连续培养中，使用超嗜热古菌 *Thermococcus litoralis* 和超嗜热菌 *Thermotoga maritima* 来比较 EPS 合成过程中的特定生理特征（Rinker and Kelly 2000），亦或者用于优化 *Thermococcus hydrothermalis* 的培养条件（Postec et al. 2005）。在其他情况下，低速搅拌对于微生物发酵培养是最好的，如 *Thermotoga maritima* 的最佳搅拌速率为 100r/min（Johnson et al. 2005），两种硫叶菌属（*Sulfolobus*）的嗜热古菌也要求低速搅拌（Nicolaus et al. 1993）。

生物反应器中嗜热菌的培养具有一些特征。尽管氧气的传输系数增加，但氧气在较高温度下的溶解度较低，这就决定了需要通气和搅拌（Shih and Pan 2011；Kennes and Veiga 2013）。另一个缺点是高热反应器中压缩机和热交换器所需的投资成本高（Van Groenestijn et al. 2002）。高温导致较高的蒸发热，且这些热量需要被回收利用。Sharp 和 Raven（1997）推荐使用添加镍、钼、铬或特氟龙涂层防止高温、盐离子和硫化物引起的腐蚀。厌氧膜生物反应器中，陶瓷膜的热稳定性高和使用寿命较长，可以在高温条件下使用（Abeynayaka and Visvanathan 2011）。

嗜热菌 EPS 的产量低于已报道的嗜温菌，它们 EPS 的产量通常在 50～200μg/mL 范围内变化（Kambourova et al. 2009；Manca et al. 1996；Nicolaus et al. 2003）。已报道的兼性嗜热菌 *Bacillus licheniformis* 在复杂培养基中 50℃ 培养 48h 有更高的 EPS 产量（366μg/mL）（Spanò et al. 2013）。比起其他的嗜热微生物，嗜热短杆菌 *Brevibacillus thermoruber* 423 的 EPS 产量最高（863μg/mL）（Yasar Yildiz et al. 2014）。

4.6 嗜热菌胞外多糖的潜在生物技术应用

EPS 的生物技术应用十分广泛，从传统领域如食品、制药和化妆品行业到新型生物医学领域都有涉及。新型微生物多糖较传统的多糖具有明显的市场优势，如廉价的底物、低成本下游工艺的开发以及具有更好的功能性质或新的特性。根据 Belsito 等人（2012）的报道，目前在化妆品制备领域已经使用了 19 种微生物多糖，而这些微生物多糖均不是由嗜热微生物产生的。

大约有 12 种已报道的嗜热菌 EPS 多糖组成成分和理化特性是已知的。利用嗜热微生物合成 EPS 具有以下优点：

① 在高温及寡营养条件下，微生物生长速率高，发酵周期短（通常持续几个小时）(Kambourova et al. 2009；Radchenkova et al. 2013；Yasar Yildiz et al. 2014)。

② 在高温下培养传质效果好（Turner et al. 2007；Kumar et al. 2011）。

③ 培养液的黏度在高温下较低，能量消耗较低（Haki and Rakshit 2003）。

④ 高温条件下受污染的风险低（Turner et al. 2007；Kikani et al. 2010；Xiao et al. 2015）。

⑤ 嗜热菌产生的非致病性产品适用于食品和化妆品工业（Nicolaus et al. 2010）。

⑥ 在食品工业加工过程中，嗜热菌和嗜热古菌产生的 EPS 可以保存在高温下，以保持其乳化和流变性能（Sajna et al. 2013）。

⑦ 由于分子具有热稳定的刚性结构，因此在极端 pH、温度和盐度也能保持活性。

⑧ 它们构成化妆品行业所需的稳定的油/水乳液（Radchenkova et al. 2014）。

热泉中的嗜热菌生物量低，且 EPS 产量也低（Krebs et al. 2014）。此外，极端环境中不可培养微生物的比例非常高（Lorenz et al. 2002）。这些缺点在一定程度上可以通过探索更好的 EPS 生产菌株和优化栽培条件来克服。遗传工程在提高胞外多糖产量方面，具有广阔的发展前景。细菌的遗传操作比高等生物更容易（Morris and Harding，2014）。通过遗传操作可以改变 EPS 的化学组成和结构，从而进一步对它们进行具体的探索。

另一个缺点是一些情况下使用的底物类型导致产品回收成本高。在嗜热菌合成胞外多糖过程中底物利用效率比中温细菌更低，因为其生物量相对较低。利用农业废弃物或乳品废弃物可以降低 EPS 的成本。大型发酵罐的固定成本明显高于简单植物多糖的提取成本，但在细菌大规模生产过程中，利用生物技术的优势可产生大量 EPS，从而降低了产品的相对成本。

嗜热微生物合成 EPS 量较少，生产成本较高，传统的多聚物在性质和纯度上达不到市场要求，为市场开发适销的 EPS 引起人们极大的兴趣（Kumar et al. 2007）。由嗜热细菌产生的新型生物多聚物具有不同的化学结构和理化性质，在医疗和药物应用领域均有应用价值，特别是在药物运输、组织工程方面，也可作为免疫刺激、免疫调节、抗肿瘤、抗病毒、抗炎和抗氧化制剂（Arena et al. 2009；Sam et al. 2011；Lee and Mooney 2012；Freitas et al. 2014）。

过去几年，已经有学者报道了具有抗肿瘤活性的微生物胞外多糖（Khalikova et al. 2006；Nwodo et al. 2012）。从热液喷口分离的海洋细菌产生的新型 EPS（Courtois et al. 2014），在其自然或过硫酸化状态下可以调节补体系统，这表明它可以有效治疗由失调引起的免疫系统疾病和补体系统的过度活化。由 *Bacillus licheniformis* T14 和 *Geobacillus thermodenitrificans* B3-72 产生的胞外多糖具有剂量依赖性免疫调节和抗病毒作用，并且观察到用这种 EPS 治疗后的免疫系统能够部分恢复（Arena et al. 2006，2009）。新型多糖 EPS1-T14 能够作为免疫调节剂通过触发 Th1 型细胞因子的产生，进而抑制 2 型单纯疱疹病毒（HSV-2）(Gugliandolo et al. 2014)。从帕纳雷阿岛分离的 *Bacillus licheniformis* T14 合成了一种新型的 EPS 岩藻糖，具有特殊的化学和流变学特性（Spanò et al. 2013）。

Thermus aquaticus YT-1 产生的新型胞外多糖 TA-1 能够刺激巨噬细胞产生细胞因子，增强免疫反应（Lin et al. 2011）。胞外多糖 TA-1 中的 D-半乳糖呋喃糖残基可能与 TA-1 在巨噬细胞内的免疫调节活性有关，该免疫活性是细菌感染宿主细胞的第一道防线。来自

Aeribacillus pallidus 418 的 EPS 具有良好的乳化性能，在化妆品工业中具有潜在的应用前景（Radchenkova et al. 2014）。EPS 与其他生物多聚物混合可以显著地改变多糖的性质。*Aeribacillus pallidus* 418 产生的 EPS 与黄原胶混合可作为稳定的乳液，特别是在化妆品工业中有较大的应用价值（Radchenkova et al. 2014）。

由于许多食品生产过程都是在高温条件下进行的，因此嗜热菌合成的热稳定型胞外多糖在食品工业有着重要的应用（Sajna et al. 2013）。这些 EPS 高温耐受性表明它们可以在较高温度下作为不同食品制剂和美容霜乳液制剂，且黏度较低，易混合。即使在室温下，它也能作为长期保存产品的介质。热稳定型的 EPS 溶液能够在石油钻井高温流体中保持高黏度。它们作为絮凝剂在城市废水热处理中具有巨大的应用潜力。

对嗜热微生物胞外多糖的研究揭示了其性质，如高分子量、良好的乳化性能和增效作用、抗细胞毒性化合物的生物活性、抗病毒和免疫调节活性等。胞外多糖高分子量表明其具有良好的黏度，及其在恶劣的工业条件下分子的稳定性，这些特性表明嗜热微生物合成的 EPS 在生物技术和生物医学多个领域中具有巨大的应用潜力。本章综述了关于嗜热微生物合成 EPS 的研究进展，新型生物多聚物在工业上的应用价值及生物活性在未来潜在应用趋势。

无利益冲突声明

Margarita Kambourova，Nadja Radchenkova，Iva Tomova 和 Ivanka Bojadjieva 声明没有利益冲突。

参考文献

Abeynayaka A，Visvanathan C（2011）Performance comparison of mesophilic and thermophilic aerobic sidestream membrane bioreactors treating high strength wastewater. Bioresour Technol 102：5345-5352

Arena A，Maugeri TL，Pavone B，Iannello D，Gugliandolo C，Bisignano G（2006）Antiviral and immunoregulatory effect of a novel exopolysaccharide from a marine thermotolerant *Bacillus licheniformis*. Int Immunopharmacol 6：8-13

Arena A，Gugliandolo C，Stassi G，Pavone B，Iannello D，Bisignano G，Maugeri TL（2009）An exopolysaccharide produced by *Geobacillus thermodenitrificans* strain B3-72：Antiviral activity on immunocompetent cells. Immunol Lett 123：132-137

Baker-Austin C，Potrykus J，Wexler M，Bond PL，Dopson M（2010）Biofilm development in the extremely acidophilic archaeon '*Ferroplasma acidarmanus*' Fer1. Extremophiles 14：485-491

Belsito MD，Hill RA，Klaassen CD，Liebler D，Marks Jr JG，Ronald C（2012）Safety Assessment of Microbial Polysaccharide Gums as Used in Cosmetics. http：//www. cir-safety. org/sites/default/files/microb092012rep. pdf

Chhabra RP（2003）Fluid mechanics and heat transfer with non-Newtonian liquids in mechanically agitated vessels. Adv Heat Transfer 37：77-176

Cojoc R，Merciu S，Oancea P，Pincu E，Dumitru L，Enache M（2009）Highly thermostable exo-polysaccharide produced by the moderately halophilic bacterium isolated from a man-made young salt lake in Romania. Pol J Microbiol 58：289-294

Conti E，Flaibani A，O'Regan M，Sutherland IW（1994）Alginate from *Pseudomonas fluorescence* and *P. putida*：production and properties. Microbiology 140：1125-1132

Courtois A，Berthou C，Guézennec J，Boisset C，Bordron A（2014）Exopolysaccharides isolated from hydrothermal vent bacteria can modulate the complement system. PLos One 9（4），e94965 Czaczyk K，Wojciechowska K（2003）Formation of bacterial biofilms—the essence of the matter and mechanisms of interactions. Biotechnologia 3：180-192

Dan T，Fukuda K，Sugai-Bannai M，Takakuwa N，Motoshima H，Urashima T（2009）Characterization and expression analysis of the exopolysaccharide gene cluster in *Lactobacillus fermentum* TDS030603. Biosci Biotechnol Biochem 73：2656-2664

De Vuyst L，De Vin F，Vaningelgem F，Degeest B（2001）Recent developments in the biosynthesis and applications of heteropolysaccharides from lactic acid bacteria. Int Dairy J 11：687-707

Delcour J，Ferain T，Hols P（2000）Advances in the genetics of thermophilic lactic acid bacteria. Curr Opin Biotech 11：497-504

Di Giulio M（2003）The universal ancestor was a thermophile or a hyperthermophile：tests and further evidence. J Theor Biol 221：425-436

Donot F，Fontana A，Baccou JC，Schorr-Galindo S（2012）Microbial exopolysaccharides：main examples of synthesis, excretion, genetics and extraction. Carbohydr Polym 87: 951-962

Flemming HC, Neu TR, Wozniak DJ (2007) The EPS matrix: the "house of biofilm cells". J Bacteriol 189: 7945-7947

Freitas F, Alves VD, Reis MA (2011a) Advances in bacterial exopolysaccharides: from production to biotechnological applications. Trends Biotechnol 29: 388-398

Freitas F, Alves VD, Torres CAV, Cruz M, Sousa I, Melo MJ, Ramos AM, Reis MAM (2011b) Fucose-containing exopolysaccharide produced by the newly isolated *Enterobacter* strain A47 DSM23139. Carbohydr Polym 1: 159-165

Freitas F, Alves VD, Reis M, Crespo J, Coelhoso I (2014) Microbial polysaccharide-based membranes: Current and future applications. J Appl Polym Sci 131: doi: 10.1002/app.40047

Gandhi HP, Ray RM, Patel RM (1997) Exopolymer production by *Bacillus* species. Carbohydr Polym 34: 323-327

Garcia-Ochoa F, Gomez E (2009) Bioreactor scale-up and oxygen transfer rate in microbial processes: an overview. Biotech Adv 27: 153-176

Gugliandolo C, Spanò A, Lentini V, Arena A, Maugeri TL (2014) Antiviral and immunomodulatory effects of a novel bacterial exopolysaccharide of shallow marine vent origin. J Appl Microbiol 116: 1028-1034

Haki GD, Rakshit SK (2003) Developments in industrially important thermostable enzymes: a review. Bioresour Technol 89: 17-34

Hartzell PL, Millstein J, Lapaglia C (1999) Biofilm formation in hyperthermophilic archaea. Methods Enzymol 310: 335-349

Harutoshi T (2013) Exopolysaccharides of lactic acid bacteria for food and colon health applica-tions. In: Kongo M (ed) Exopolysaccharides of lactic acid bacteria for food and colon health applications. INTECH Open Access Publisher, pp 515-538

Herget S, Toukach P, Ranzinger R, Hull W, Knirel Y, Von Der Lieth C-W (2008) Statistical analysis of the bacterial carbohydrate structure data base (BCSDB): characteristics and diversity of bacterial carbohydrates in comparison with mammalian glycans. BMC Struct Biol 8: 35

Huber R, Langworthy TA, König H, Thomm M, Woese CR, Sleytr UB, Stetter KO (1986) *Thermotoga maritima* sp. nov. represents a new genus of unique extremely thermophilic eubacteria growing up to 90℃. Arch Microbiol 144: 324-333

Jaiswal P, Sharma R, Sanodiya BS, Bisen PS (2014) Microbial exopolysaccharides: natural modulators of dairy products. J Appl Pharm Sci 4: 105-109

Johnson MR, Montero CI, Conners SB, Shockley KR, Bridger SL, Kelly RM (2005) Population density-dependent regulation of exopolysaccharide formation in the hyperthermophilic bacterium *Thermotoga maritima*. Mol Microbiol 55: 664-674

Kambourova M, Mandeva R, Dimova D, Poli A, Nicolaus B, Tommonaro G (2009) Production and characterization of a microbial glucan, synthesized by *Geobacillus tepidamans* V264 isolated from Bulgarian hot spring. Carbohydr Polym 77: 338-343

Kennes C, Veiga MC (2013) In: Kennes C, Veiga MC (eds) Bioreactors for waste gas treatment, vol. 4: Springer Science & Business Media, Dordrecht, pp 47-98

Khalikova TA, Korolenko TA, Zhanaeva SY, Kaledin VI, Kogan G (2006) Enhancing effect of new biological response modifier sulfoethylated (1→3)-beta-D-glucan on antitumor activity of cyclophosphamide in the treatment of experimental murine leukoses. Exp Oncol 28: 308-313

Kikani BA, Shukla RJ, Singh SP (2010) Biocatalytic potential of thermophilic bacteria and actinomycetes. In: Méndez-Vilas A (ed) Current research, technology and education. Topics in applied microbiology and microbial biotechnology, FORMATEX C/Zurbarán 1, 2° -Oficina 1 06002 Badajoz Spain vol 2, pp 1000-1007

Koerdt A, Gödeke J, Berger J, Thormann KM, Albers S-V (2010) Crenarchaeal biofilm formation under extreme conditions. PLos One 5 (11), e14104

Kolter R, Losick R (1998) One for all and all for one. Science 280: 226-227

Kralj S, van Geel-Schutten GH, Dondorff MMG, Kirsanovs S, van der Maarel MJEC, Dijkhuizen L (2004) Glucan synthesis in the genus *Lactobacillus*: isolation and characterization of glucansucrase genes, enzymes and glucan products from six different strains. Microbiology 150: 3681-3690

Krebs JE, Vaishampayan P, Probst AJ, Tom LM, Marteinsson VG, Andersen GL, Venkateswaran K (2014) Microbial community structures of novel icelandic hot spring systems revealed by PhyloChip G3 Analysis. Astrobiology 14: 229-240

Krüger M, Blumenberg M, Kasten S, Wieland A, Känel L, Klock JH, Michaelis W, Seifert R (2008) A novel, multi-layered methanotrophic microbial mat system growing on the sediment of the Black Sea. Environ Microbiol 10: 1934-1947

Kumar AS, Mody K, Jha B (2007) Bacterial exopolysaccharides: A perception. J Basic Microbiol 47: 103-117

Kumar L, Awasthi G, Singh B (2011) Extremophiles: a novel source of industrially important enzymes. Biotechnology 10: 121-135

Lapaglia C, Hartzell PL (1997) Stress-induced production of biofilm in the hyperthermophile *Archaeoglobus fulgidus*. Appl Environ Microbiol 63: 3158-3163

Laws A, Gu Y, Marshall V (2001) Biosynthesis, characterisation, and design of bacterial exopolysaccharides from lactic acid bacteria. Biotechnol Adv 19: 597-625

Lee KY, Mooney DJ (2012) Alginate: Properties and biomedical applications. Prog Polym Sci 37: 106-126

Lin M-H，Yan Y-L，Chen Y-P，Hua K-F，Lu C-P，Sheu F，Lin G-H，Tsay S-S，Liang S-M，Wu S-H（2011）A novel exopolysaccharide from the biofilm of *Thermus aquaticus* YT-1 induces the immune response through toll-like receptor 2. J Biol Chem 286：17736-17745

López-López O，Cerdán ME，González-Siso MI（2013）Hot spring metagenomics. Life 3：308-320 Lorenz P，Liebeton K，Niehaus F，Eck J（2002）Screening for novel enzymes for biocatalytic processes：accessing the metagenome as a resource of novel functional sequence space. Curr Opin Biotechnol 13：572-577

Lowell RP，Seewald JS，Metaxas A，Perfit MR（2008）In：Lowell RP，Seewald JS，Metaxas A，Perfit，MR（eds）Modeling hydrothermal processes at ocean spreading centers：magma to microbe-an overview. Magma to Microbe. American Geophysical Union，Washington，DC. doi：10. 1029/178GM02

Manca MC，Lama L，Improta R，Esposito E，Gambacorta A，Nicolaus B（1996）Chemical composition of two exopolysaccharides from *Bacillus thermoantarcticus*. Appl Environ Microbiol 62：3265-3269

Maugeri TL，Gugliandolo C，Caccamo D，Panico A，Lama L，Gambacorta A，Nicolaus B（2002）A halophilic thermotolerant *Bacillus* isolated from a marine hot spring able to produce a new exopolysaccharide. Biotechnol Lett 24：515-519

Milas M，Shi X，Rinaudo M（1990）On the physicochemical properties of gellan gum. Biopolymers 30：451-464

Moons P，Michiels CW，Aertsen A（2009）Bacterial interactions in biofilms. Crit Rev Microbiol 35：157-168

Morris GA，Harding SE（2014）Production of polysaccharides. In：Panesar PS，Marwaha SS（eds）Biotechnology in agriculture and food processing：opportunities and challenges. Taylor & Francis Group CRC Press，Boca Raton，Fl. pp 355-386

Muralidharan V，Rinker KD，Hirsh IS，Bouwer EJ，Kelly RM（1997）Hydrogen transfer between methanogens and fermentative heterotrophs in hyperthermophilic cocultures. Biotechnol Bioeng 56：268-278

Näther DJ，Rachel R，Wanner G，Wirth R（2006）Flagella of *Pyrococcus furiosus*：multifunctional organelles，made for swimming，adhesion to various surfaces，and cell-cell contacts. J Bacteriol 188：6915-6923

Nicolaus B，Manca MC，Romano I，Lama L（1993）Production of an exopolysaccharide from two thermophilic archaea belonging to the genus *Sulfolobus*. FEMS Microbiol Lett 109：203-206

Nicolaus B，Panico A，Manca MC，Lama L，Gambacorta A，Maugeri T，Gugliandolo C，Caccamo D（2000）A thermophilic *Bacillus* isolated from an Eolian shallow hydrothermal vent，able to produce exopolysaccharides. Syst Appl Microbiol 23：426-432

Nicolaus B，Lama L，Panico A，Gambacorta A（2002）Production and characterization of exopolysaccharides excreted by thermophilic bacteria from shallow，marine hydrothermal vents of Flegrean areas（Italy）. Syst Appl Microbiol 25：319-325

Nicolaus B，Moriello V，Maugeri T，Gugliandolo C，Gambacorta A（2003）Bacilli from shallow mediterranean marine vents producers of exopolysaccharides. Recent Res Devel Microbiol 7：197-208

Nicolaus B，Schiano Moriello V，Lama L，Poli A，Gambacorta A（2004）Polysaccharides from extremophilic microorganisms. Orig Life Evol Biosph 34：159-169

Nicolaus B，Kambourova M，Oner ET（2010）Exopolysaccharides from extremophiles：from fundamentals to biotechnology. Environ Technol 31：1145-1158

Nwodo UU，Green E，Okoh AL（2012）Bacterial exopolysaccharides：functionality and prospects. Int J Mol Sci 13：14002-14015

Otero A，Vincenzini M（2003）Extracellular polysaccharide synthesis by *Nostoc* strains as affected by N source and light intensity. J Biotechnol 102：143-152

Patel AK，Michaud P，Singhania RR，Soccol CR，Pandey A（2010）Polysaccharides from probiotics：new developments as food additives. Food Technol Biotechnol 48：451-463

Poli A，Di Donato P，Abbamondi GR，Nicolaus B（2011）Synthesis，production，and biotechnological applications of exopolysaccharides and polyhydroxyalkanoates by *Archaea*. Archaea，Article ID 693253

Postec A，Pignet P，Cueff-Gauchard V，Schmitt A，Querellou J，Godfroy A（2005）Optimisation of growth conditions for continuous culture of the hyperthermophilic archaeon *Thermococcus hydrothermalis* and development of sulphur-free defined and minimal media. Res Microbiol 156：82-87

Pysz MA，Conners SB，Montero CI，Shockley KR，Johnson MR，Ward DE，Kelly RM（2004）Transcriptional analysis of biofilm formation processes in the anaerobic，hyperthermophilic bacterium *Thermotoga maritima*. Appl Environ Microbiol 70：6098-6112

Radchenkova N，Vassilev S，Panchev I，Anzelmo G，Tomova I，Nicolaus B，Kuncheva M，Petrov K，Kambourova M（2013）Production and properties of two novel exopolysaccharides synthesized by a thermophilic bacterium *Aeribacillus pallidus* 418. Appl Biochem Biotechnol 171：31-43

Radchenkova N，Vassilev S，Martinov M，Kuncheva M，Panchev I，Vlaev S，Kambourova M（2014）Optimization of the aeration and agitation speed of *Aeribacillus palidus* 418 exopolysaccharide production and the emulsifying properties of the product. Process Biochem 49：576-582

Raguénès G，Pignet P，Gauthier G，Peres A，Christen R，Rougeaux H，Guezennec J（1996）Description of a new polymer-secreting bacterium from a deep-sea hydrothermal vent，*Alteromonas macleodii* subsp. fijiensis，and preliminary characterization of the polymer. Appl Environ Microbiol 62：67-73

Raguénès G，Christen R，Guezennec J，Pignet P，Barbier G（1997）*Vibrio diabolicus* sp. nov.，a new polysaccharide-secreting organism isolated from a deep-sea hydrothermal vent polychaete annelid，*Alvinella pompejana*. Int J Syst Bacteriol 47：989-995

Raguénès G，Cambon-Bonavita MA，Lohier JF，Boisset C，Guezennec J（2003）A novel，highly viscous polysaccharide excreted by an *Alteromonas* isolated from a deep-sea hydrothermal vent shrimp. Curr Microbiol 46：0448-0452

Rau U，Gure E，Olszewski E，Wagner F（1992）Enhanced glucan formation of filamentous fungi by effective mixing，oxygen limitation and fed-batch processing. J Ind Microbiol 9：19-26

Rinker KD，Kelly RM（1996）Growth physiology of the hyperthermophilic archaeon *Thermococcus litoralis*：development of a sulfur-free defined medium，characterization of an exopolysaccharide，and evidence of biofilm formation. Appl Environ Microbiol 62：4478-4485

Rinker KD，Kelly RM（2000）Effect of carbon and nitrogen sources on growth dynamics and exopolysaccharide production for the hyperthermophilic archaeon *Thermococcus litoralis* and bacterium *Thermotoga maritima*. Biotechnol Bioeng 69：537-547

Rozanov AS，Bryanskaya AV，Malup TK，Meshcheryakova IA，Lazareva EV，Taran OP，Ivanisenko TV，Ivanisenko VA，Zhmodik SM，Kolchanov NA，Peltek SE（2014）Molecular analysis of the benthos microbial community in Zavarzin thermal spring（Uzon Caldera，Kamchatka，Russia）. BMC Genomics 15（Suppl 12）：S12

SajnaKV，Sukumaran RK，Gottumukkala LD，Jayamurthy H，Dhar KS，Pandey A（2013）Studies on structural and physical characteristics of a novel exopolysaccharide from *Pseudozyma* sp. NII 08165. Int J Biol Macromol 59：84-89

Sam S，Kucukasik F，Yenigun O，Nicolaus B，Toksoy Öner E，Yukselen MA（2011）Flocculating performances of exopolysaccharides produced by a halophilic bacterial strain cultivated on agro-industrial waste. Biores Technol 102：1788-1794

Satyanarayana T，Raghukumar C，Shivaji S（2005）Extremophilic microbes：diversity and perspectives. Curr Sci 89：78-90

Schopf S，Wanner G，Rachel R，Wirth R（2008）An archaeal bi-species biofilm formed by *Pyrococcus furiosus* and *Methanopyrus kandleri*. Arch Microbiol 190：371-377

Sharp RJ，Raven NDH（1997）In：Rhodes PM，Stanbury PF（eds）Isolation and growth of hyperthermophiles. Applied microbial physiology. IRL Press，New York，pp 23-52

Shih TW，Pan TM（2011）Stress responses of thermophilic Geobacillus sp. NTU 03 caused by heat and heat-induced stress. Microbiol Res 166：346-359

Spanò A，Concetta Gugliandolo C，Lentini V，Maugeri TL，Anzelmo G，Poli A，Nicolaus B（2013）A novel EPS-producing strain of *Bacillus licheniformis* isolated from a shallow vent off Panarea Island（Italy）. Curr Microbiol 67：21-29

Sutherland IW（1994）Structure-function relationships in microbial exopolysaccharides. Biotech Adv 12：393-448

Takai K，Nakamura K，Toki T，Tsunogai U，Miyazaki M，Miyazaki J，Hirayama H，Nakagawa S，Nunoura T，Horikoshi K（2008）Cell proliferation at 122℃ and isotopically heavy CH_4 production by a hyperthermophilic methanogen under high-pressure cultivation. Proc Natl Acad Sci U S A 105：10949-10954

Thoma C，Frank M，Rachel R，Schmid S，Näther D，Wanner G，Wirth R（2008）The Mth60 fimbriae of *Methanothermobacter thermoautotrophicus* are functional adhesins. Environ Microbiol 10：2785-2795

Turner P，Mamo G，Karlsson EN（2007）Potential and utilization of thermophiles and thermostable enzymes in biorefining. Microb Cell Fact 6：1-23

Van Groenestijn JW，Hazewinkel JHO，Nienoord M，Bussmann PJT（2002）Energy aspects of biological hydrogen production in high rate bioreactors operated in the thermophilic temperature range. Int J Hydrogen Energy 27：1141-1147

Vu B，Chen M，Crawford RJ，Ivanova EP（2009）Bacterial extracellular polysaccharides involved in biofilm formation. Molecules 14：2535-2554

Węgrzyn A，Żukrowski K（2014）Biotechnological applications of archaeal extremozymes. Chemik 68：717-722

Wiegel J，Canganella F（2001）Extreme thermophiles. In：Encyclopedia of life sciences，article Wiley，Chichester. http：//www. els. net/WileyCDA/ElsArticle/refId-a0000392. html

Xiao Z，Zhang Y，Xi L，Huo F，Zhao JY，Li J（2015）Thermophilic production of polyhydroxyalkanoates by a novel *Aneurinibacillus* strain isolated from Gudao oilfield. China J Basic Microbiol doi：. doi：10. 1002/jobm. 201400843

Yasar Yildiz S，Anzelmo G，Ozer T，Radchenkova N，Genc S，Di Donato P，Nicolaus B，Toksoy Oner E，Kambourova M（2014）Brevibacillus themoruber：a promising microbial cell factory for exopolysaccharide production. J Appl Microbiol 116（2）：314-324. ISSN：1364-5072.

Yildiz SY，Radchenkova N，Arga KY，Kambourova M，Toksoy Oner E（2015）Genomic analysis of *Brevibacillus thermoruber* 423 reveals its biotechnological and industrial potential. Appl Microbiol Biotechnol 99：2277-2289

Zhang CL，Ye Q，Huang Z，Li W，Chen J，Song Z，Zhao W，Bagwell C，Inskeep WP，Ross C，Gao L，Wiegel J，Romanek CS，Shock EL，Hedlund BP（2008）Global occurrence of archaeal amoA genes in terrestrial hot springs. Appl Environ Microbiol 74：6417-6426

Zheng C，Li Z，Su J，Zhang R，Liu C，Zhao M（2012）Characterization and emulsifying property of a novel bioemulsifier by *Aeribacillus pallidus* YM-1. J Appl Microbiol 113：44-51

第五章
嗜酸硫还原微生物的生态生理学和应用

Anna P. Florentino[1], Jan Weijma[2], Alfons J. M. Stams[3], Irene Sánchez-Andrea[4]

5.1 自然界中的硫化合物

硫是地壳中的重要元素，在岩石圈中所占质量分数约 0.05%（Steudel and Eckert 2003）。硫集中分布在各种陆壳岩石中，如金属硫化物矿床［黄铁矿（FeS_2）、黄铜矿（$CuFeS_2$）、磁黄铁矿（FeS）］或硫酸盐矿物［石膏（$CaSO_4 \cdot 2H_2O$）、重晶石（$BaSO_4$）］。硫有九种不同的价态，自然界中最丰富的是－2 价的硫（硫化物和还原的有机硫）、0 价的单质硫和＋6 价的硫酸盐（Steudel 2000；Tang et al. 2009）。

硫化物生成硫酸盐的氧化反应涉及八个电子的得失，几步反应分步完成，其中单质硫、硫代硫酸盐、亚硫酸盐和多硫化物（Hedderich et al. 1999）作为中间产物出现。这些中间产物在溶液中的重要性和稳定性取决于 pH、温度以及溶液中它们的种类（Knickerbocker et al. 2000）。

环境中，不同形态的硫的转化主要依赖于微生物（Steudel 2000）。微生物参与的有机和无机硫化合物的转化极大地影响了生物圈的物理、化学及生物性质。

硫循环能通过两种观点来分析（Canfield and Farquhar 2012）。从地质学观点来看，海洋地壳的生成与硫从地幔向地表和海洋的转移有关（Canfield 2004），通过火山喷发释放 SO_2

[1] Laboratory of Microbiology, Wageningen University, Stippeneng 4, 6708 WE Wageningen, The Netherlands; Sub-Department of Environmental Technology, Wageningen University, Bornse Weilanden 9, 6708 WG Wageningen, The Netherlands.

[2] Sub-Department of Environmental Technology, Wageningen University, Bornse Weilanden 9, 6708 WG Wageningen, The Netherlands.

[3] Laboratory of Microbiology, Wageningen University, Stippeneng 4, 6708 WE Wageningen, The Netherlands; CEB-Centre of Biological Engineering, University of Minho, Campus de Gualtar, 4710-057 Braga, Portugal.

[4] Laboratory of Microbiology, Wageningen University, Stippeneng 4, 6708 WE Wageningen, The Netherlands. e-mail: irene. sanchezandrea@wur. nl.

和 H_2S，在热液循环期间释放 H_2S，以及岩浆硫化物矿的侵蚀（Canfield and Farquhar 2012）。从生物学的观点看，当生成的硫化物用于合成反应时，硫酸盐和/或还原硫被同化，当用于能量转化或生长时，硫酸盐和/或还原硫被异化（Tang et al. 2009；Canfield and Farquhar 2012）。

几种不产氧的光合细菌在光合作用中可以将硫化物作为电子供体，生成单质硫、硫酸盐（Ghosh and Dam 2009），而有些情况下生成硫代硫酸盐（Pfennig 1975）（式 5.1～式 5.3）。化能自养原核生物通过与 O_2、硝酸盐、锰或铁的还原偶联来使硫化物氧化（Hedderich et al. 1999；Ohmura et al. 2002）。

$$H_2S+\frac{1}{2}O_2 \longrightarrow S^0+H_2O \tag{5.1}$$

$$H_2S+2O_2 \longrightarrow H_2SO_4 \tag{5.2}$$

$$4H_2S+5O_2 \longrightarrow 2S_2O_3+4H_2O \tag{5.3}$$

硫化物氧化产物的单质硫（S^0）、硫代硫酸盐（$S_2O_3^{2-}$）和亚硫酸盐（SO_3^{2-}）能被微生物氧化、还原或歧化成硫酸盐和硫化物。单质硫的歧化在环境中具有重要意义（Steudel 2000；Tang et al. 2009；Canfield and Farquhar 2012）。本节中描述的生物反应总结在图 5.1 中。化学反应在下一部分中描述。

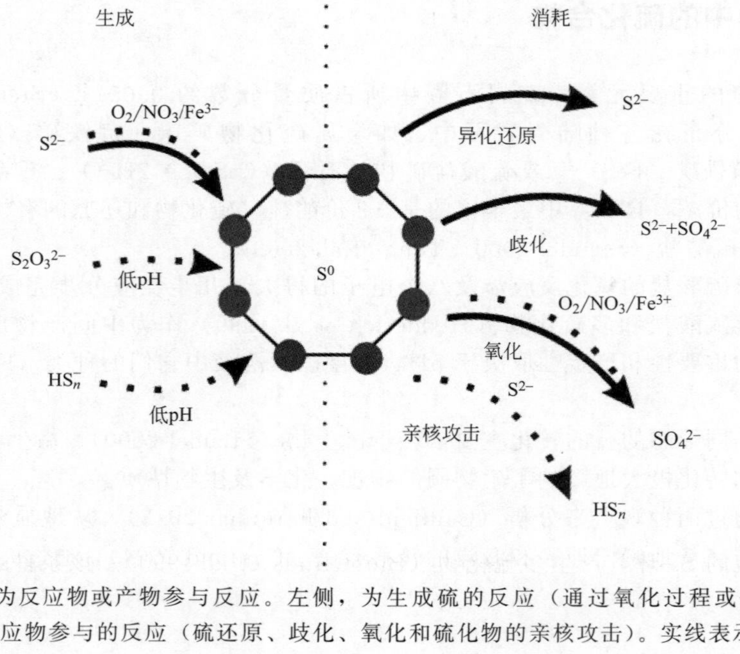

图 5.1 单质硫作为反应物或产物参与反应。左侧，为生成硫的反应（通过氧化过程或介质的酸化）。右侧，显示硫作为反应物参与的反应（硫还原、歧化、氧化和硫化物的亲核攻击）。实线表示生物反应，虚线表示化学反应

5.2 单质硫的化学性质

单质硫（S_8^0）是固体结构形式最多的分子，具有常规大气压和高压两种同素异形体。虽然存在超过 180 种不同的同素异形体和多晶型（知识框 5.1），但是在标准温度和压力条件（273.15K 和 10^5 Pa）下，单质硫的唯一稳定形式是斜方晶 α-S_8^0（Steudel and Eckert 2003）。

知识框 5.1

同素异形体：材料在不同温度和压力条件下具有不同的结构形式，并且具有在相反条件下时重新恢复这些结构的能力。因此，同素异形体是一个可逆的多态性。

多晶型：固体材料以多种晶体结构形式存在的现象。如果温度和压力发生变化，并且不伴随固体的熔化或蒸发，则会导致固体原子内部结构的改变。

氧代化合物：氧原子与碳或另一种元素（═O）结合的化合物。

硫几乎不溶于水，α-S_8^0 在 20℃的溶解度仅为 5μg/L（Boulegue 1978）。一般来说，硫同素异形体的分子量越大则在有机溶剂中的溶解度越低。二硫化碳、甲苯和二氯甲烷是最好的硫溶剂，而在室温下环烷烃只能溶解较小的环分子（Steudel and Eckert 2003）。在较高温度（65～140℃）下，单质硫也可溶于压缩气体如氮气、甲烷、二氧化碳和硫化氢，这对于气体工业很重要，因为许多天然气储层也含有 H_2S 和单质硫。例如，在 10～30MPa 的压力范围内，单质硫在硫化氢中的溶解度可从 65℃下 38.6mg/L（Roof 1971）增加至 90℃下 65.7mg/L（Gu et al. 1993），100℃下 68.1mg/L，110℃下 91.2mg/L（Roof 1971），140℃下 110.8mg/L（Brunner and Woll 1980）。

通常单质硫交联的常规形式称为硫华，主要由 S_8^0-环状和链状大分子构成的一些聚合硫组成（Steudel and Eckert 2003）（图 5.2）。从 S_8^0（环）到 S（链）的反应热为每个硫原子 115.14kJ/mol，比聚合硫中的 S—S 键之间的键强 2.3kJ/mol（Franz et al. 2007）。因此，聚合硫可能更容易通过硫还原或硫氧化微生物获得。

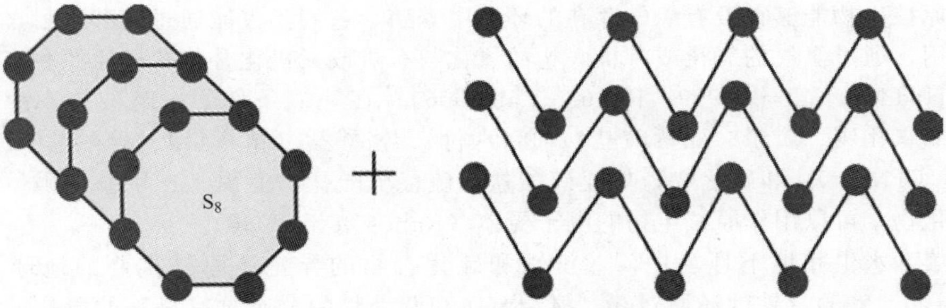

图 5.2 金属硫化合物的环状和链状大分子

当硫化物（S^{2-}）存在于与单质硫相同的环境中时，通常在高 pH 值下，HS^- 阴离子的亲核裂解单质硫的 S_8^0-环，产生多硫化物（Blumentals et al. 1990）。该条件下，由于多硫化物在高温和中高 pH 值下具有较高的溶解性，因此认为多硫化物比单质硫更适于作电子受体（Schauder and Müller1993）。四硫化物（S_4^{2-}）和五硫化物（S_5^{2-}）（Rabus et al. 2006），它们可以在中性环境下相互快速转化，用于嗜中性硫还原微生物的生长（Schauder and Müller1993）。

在硫化物溶液中多硫化物（S_n^{2-}）的平衡浓度取决于 pH、温度和硫化物浓度。由于在低 pH 下 S_n^{2-} 不稳定，当 pH 降低时，多硫化物的 pH 平衡浓度急剧下降，反应向单质硫和硫化物方向进行，反应见式 5.4（Schauder and Müller 1993）。

$$2S_5{}^{2-}+4H^+ \rightleftharpoons \alpha S_8+2H_2S \tag{5.4}$$

平衡浓度也随温度的增加而增加。因此，0.1mmol/L 多硫化物可以在 30℃，pH 6.7 时溶解，而在 pH 5.5 时，相同的量要在 90℃下溶解。由于解离常数，在硫化物溶液中，pH

8.0 和 37℃下 S_8^0 转化为多硫化物的最大量大致相当于硫化物浓度（Klimmek et al. 1991）。然而，当 pH 值低于 H_2S 的 pK_a1（在 25℃下为 7.0）时，多硫化物在浓度较低的溶液中形成（Hedderich et al. 1999）。

此外，硫代硫酸盐在酸性 pH 条件下不稳定，易分解成硫氧化物、硫化物和胶体硫/溶解硫，可作为纳米晶体（式 5.5）（Wang et al. 1998），其溶液为乳状悬浮液。在自然环境中，改变单质硫的表面性质增加其亲水性时，有机聚合物与胶体硫颗粒会发生黏附（Breher 2004）。由于硫颗粒随着硫化物一起生成，它们可以反应生成多硫化物离子的水溶液，这会影响环境中硫的流动性，有利于其生物氧化性和多硫化物及硫化物的形成动力学（Breher 2004）。此外，胶体硫在热力学上不稳定并最终会形成小的可沉降晶体（Kleinjan et al. 2005）。

$$3S_2O_3^{2-}+4H^+ \rightleftharpoons 2H_2S+SO_3^{2-}+3SO_2 \tag{5.5}$$

另一种形式的单质硫，比斜方晶形式更加亲水，被称为生物硫（Steudel and Eckert 2003）。生物硫是硫化物被生物氧化时形成的，并且可以作为硫微球体在细胞间或细胞外储存，（Kleinjan et al. 2005）。研究表明生物硫可吸附有机聚合物如蛋白质或末端有机基团，因而表现更为亲水。此外，生物硫结构可能在不同硫细菌之间有所不同（Steudel et al. 2003）。有机基团还能使光养细菌产生的长硫链更稳定。然而，化能营养细菌主要形成由 8 个硫原子组成的环（Kleinjan et al. 2005）。

5.3 硫还原微生物

许多原核生物能够在没有氧气存在的环境中繁殖，它们不仅能通过发酵途径，还可以通过呼吸作用，通过氢气的氧化或有机底物的氧化与有机或无机化合物的还原偶联来为无氧生长供能（Hedderich et al. 1999；Rabus et al. 2006）。在缺氧条件下，硝酸盐、锰（四价）、三价铁、二氧化碳、质子、亚硒酸盐、铀（六价）、铬酸盐（铬六价）、砷酸盐、三甲基胺-*N*-氧化物（TMAO）和硫化物，如亚硫酸盐、硫代硫酸盐、亚砜、二甲基亚砜（DMSO）、有机二硫化物，可以用作原核生物的电子受体（Rabus et al. 2006）。

在土壤、水生和地下环境中，三价铁和硫化合物的异化还原是重要的地球生物反应（Lovley et al. 1995）。铁的还原对铁、痕量金属和营养物的分布具有显著的影响。此外，它参与有机物的降解，并且可以对有机物环境污染和金属环境污染进行生物修复（Lovley et al. 2004）。以硫化合物作为能量来源时，一些微生物可以进行三价铁的还原。

硫化氢作为硫化物还原的主要终产物而为人们所关注。硫化物因其对环境化学的显著影响而为人所知，此外，在很多微生物中，硫化物可以作为电子供体（Rabus et al. 2006）。研究发现，由于硫酸盐的丰度及热稳定性，在厌氧呼吸中常作为电子受体。

在深海热液口、温泉和其他极端环境中，单质硫的还原是非常重要的，在这些地方经常分离到很多微生物，它们的多样性与硫酸盐还原菌多样性相当（Stetter 1996）。

5.3.1 硫还原菌的生态生理学

目前已知的硫还原菌分布于细菌的（图 5.3 a,b）9 个门，69 个属和在古菌的 2 个门，37 个属中（图 5.4）。它们使用单质硫作为电子受体氧化有机化合物或 H_2。

虽然微生物硫还原已经在早期研究中广为发现，例如 Beijerink（1895）和 Pelsh（1936）发现分离自泥质区的弧菌以硫的还原作为能量的唯一来源，硫和 H_2 分别作为电子的受体和供体。通过硫还原生长的第一种纯培养菌是氧化乙酸脱硫单胞菌（*Desulfuromonas acetoxi-*

dans），它是一种专性厌氧乙酸盐降解嗜温菌，不能利用硫酸盐（SO_4^{2-}）（Pfennig 和 Biebl 1976）。

随后，许多硫还原菌被分离出来，这些硫还原菌还有还原其他化合物如硫代硫酸盐，三价铁、硝酸盐和氧的能力，并且发现缺氧更有利于还原过程（Rabus et al. 2006）。以硫酸盐（Biebl and Pfennig 1977）、三价铁（Caccavo et al. 1994）和四价锰（Myers and Nealson 1988）作为电子受体的微生物也具有硫还原的能力。只有少数种类的硫酸盐还原菌能够通过还原硫生长，并且有时单质硫会抑制其生长（Bak and Pfennig 1987；Burggraf et al. 1990）。

硫还原原核生物能够在较宽的温度（−2～110℃）和 pH（1～10.5）范围生长。大多数硫还原菌生长在中性环境下。据报道，从硫质喷气口田中分离出的一些极端嗜热古菌，在 pH 低至 1 时仍能生长，例如酸菌属（*Acidianus ambivalens*）、布式酸菌（*Acidianus brierleyi*）、*Styogiolobus azoricus*、火山热原体菌（*Thermoplasma volcanium*）和嗜酸热原体菌（*Thermoplasma acidophilum*）（Segerer et al. 1986，1988，1991）。迄今为止报道的硫还原菌酸氧化硫杆菌（*Acidithiobacillus ferrooxidans*）生长的最低 pH 为 1.3（Ohmura et al. 2002），细菌中也发现了几种嗜酸性和耐酸性物种，例如脱硫芽孢弯曲菌（*Desulfosporosinus acididurans*）（pH 3.8）、脱硫杆状菌（*Desulfurobacterium thermolithotrophum*）、*Marinitoga hydrogenitolerans* 和热厌氧弧菌（*Thermanaerovibrio velox*）（pH 4.5）（L'Haridon et al. 1998；Zavarzina et al. 2000；Postec et al. 2005；Sánchez-Andrea et

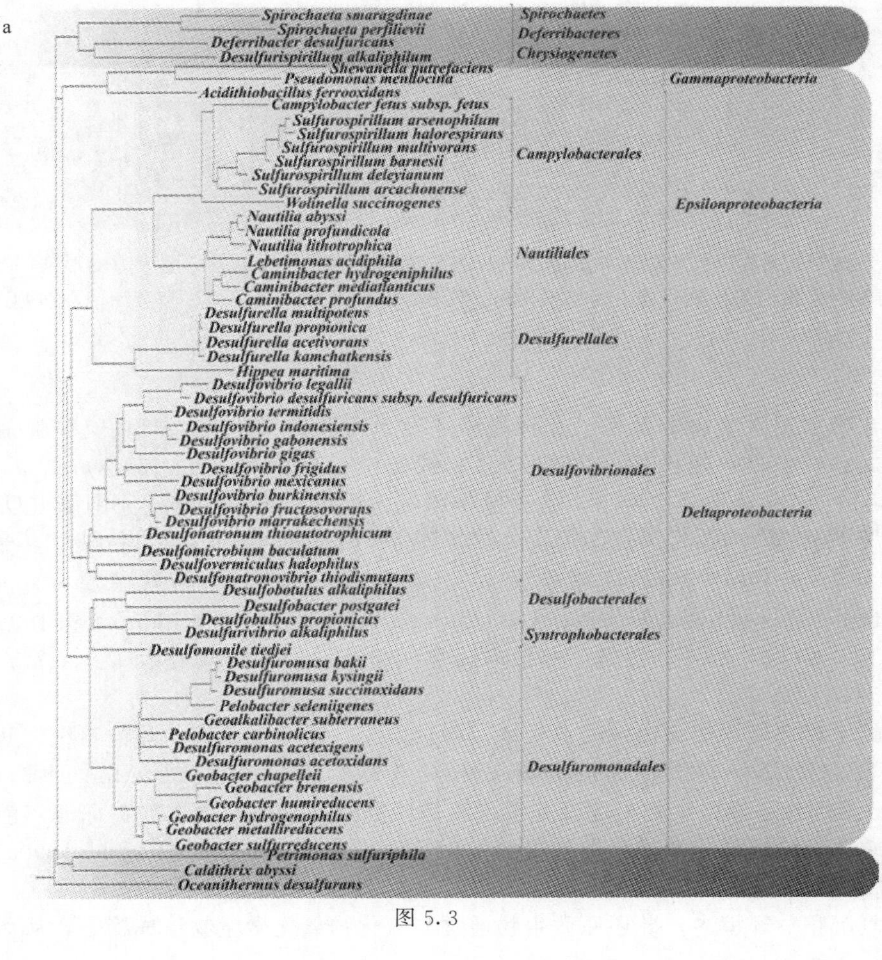

图 5.3

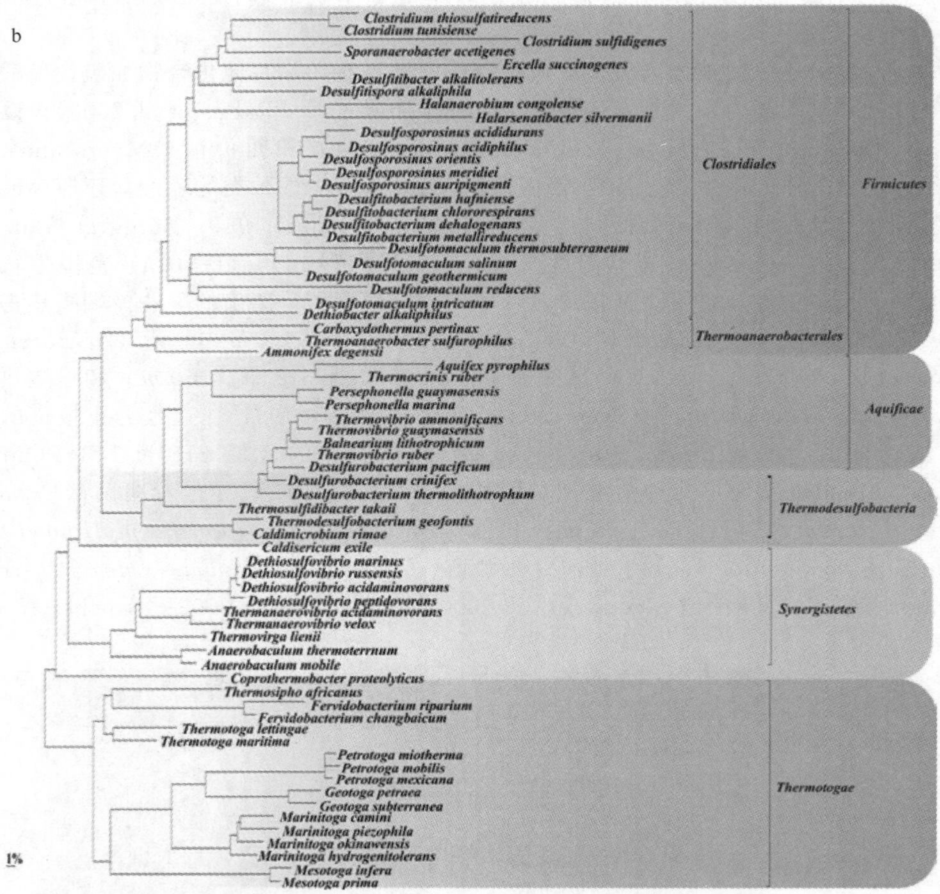

图 5.3 在全物种系统发育树中硫还原菌的16S rRNA基因序列的种属关系。图 a 为变形菌门、螺旋体门、脱铁杆菌门和产金菌门的序列；图 b 为厚壁菌门、产水菌门、热脱硫杆菌门、互养菌门和热袍菌门的序列。大约有 1%序列替代率

al. 2015）。

虽然在嗜中温微生物中发现了几种能够还原单质硫的微生物，如脱硫单胞菌属（*Desulfuromonas*）、贝日阿托氏菌（*Beggiatoa*）或硫磺单胞菌属（*Sulfurospirillum*）（Pfennig and Biebl 1976），但是高温下，硫还原作用更为广泛。比如脱硫还原菌属（*Desulfurella*）和热厌氧杆菌（*Thermoanaerobacter*）（Bonch-Osmolovskaya et al. 1990b；Bonch-Osmolovskaya et al. 1997）等轻度嗜热菌（T_{OPT} 为 40～60℃）；制氨菌属（*Ammonifex*）（Huber et al. 1996）和除硫杆菌属（*Desulfurobacterium*）（L'Haridon et al. 1998）等中度嗜热菌（T_{OPT} 为 60～80℃）以及一些极端嗜热的硫还原菌，如产液菌属（*Aquifex*）（Huber et al.，1992）。

极端的生存环境，如火山区的温水池、酸性温泉、浅海和深海的热液系统、盐湖和缺氧泥沙沉积物海港都存在能在高温和低 pH 下生长的硫还原菌（Stetter 1996；Rabus et al. 2006）。由于硫还原微生物丰度高并且代谢活性独特，硫还原微生物在深海热液口、温泉和其他极端环境的地球化学循环中起重要作用（Bonch-Osmolovskaya et al. 1990a；Alain et al.，2009；Birrien et al. 2011）。

在缺氧沉积物环境中，硫还原微生物通常与硫化物氧化微生物形成缔合，从而为它们提

供单质硫。硫还原微生物将单质硫还原成硫化物，硫化物作为硫氧化微生物的电子供体（Pfennig 1975）。在热液口，能发现一些硫还原菌在火山喷发口或热液羽流中自由存活或与动物共生，例如管虫和虾，它们像其他生物一样在火山口扮演相同的角色，即还原和氧化硫化合物（Alain et al. 2009）。

上述的硫还原细菌广泛存在于生命系统发育进化树中，属于变形菌门（Proteobacteria）（δ-，ε-和 γ-变形菌门）、热脱硫杆菌门（Thermodesulfobacteria）、螺旋体门（Spirochaetes）、脱铁杆菌门（Deferribacteres）、产金菌门（Chrysiogenetes）、厚壁菌门（Firmicutes）、产水菌门（Aquificiae）、互养菌门（Synergistetes）和热袍菌门（Thermotogae）（图 5. 3a，b）。在梭菌和嗜热厌氧菌中，硫还原菌具有相当广泛的代谢特征（Hernandez-Eugenio et al. 2002；Sallam and Steinbüchel 2009）。在古细菌中，硫还原菌分布在个古菌门（Euryarchaeota）（Fiala and Stetter 1986；Burggraf et al. 1990）和泉古菌门（*Crenarchaeota*）中（图 5. 4）（Itoh et al. 1998；Prokofeva et al. 2000；Itoh et al. 2003）。

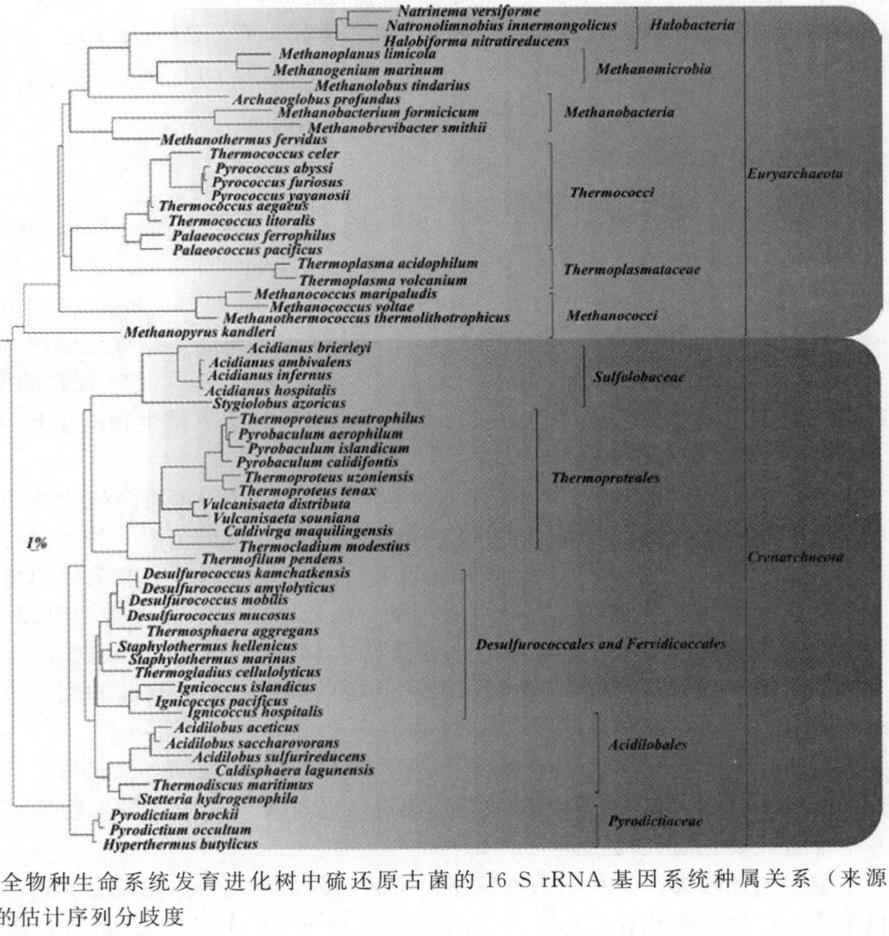

图 5. 4　在全物种生命系统发育进化树中硫还原古菌的 16 S rRNA 基因系统种属关系（来源：Yarza 等 2008）1%的估计序列分歧度

除了产琥珀酸沃廉菌（*Wolinella succinogenes*）和古菌火球菌（*Pyrococcus furiosus*）等少数微生物，硫还原微生物硫代谢机制研究很少。此外，除了硫还原的生物化学和生物能量学研究，很少有人注意到有关硫还原菌的电子供体的转换。大多数集中于硫还原微生物代谢途径和能量守恒相关的以氢或甲酸作为电子供体的无机营养生长上。仅在少数细菌中研究了利用酯酸盐的异养生长（Schröder et al. 1988；Klimmek et al. 1991；Kreis-Kleinschmidt et al. 1995)。例如，通过柠檬酸循环，在脱硫脲和脱硫单胞菌中研究了硫作为电子受体的乙酸盐的氧化。电子传递通过铁氧化还原蛋白（铁氧还蛋白）进行，铁氧化还原蛋白在 α-酮戊二酸脱氢酶催化的反应中通过 NADP 接受来自 α-酮戊二酸的电子，而甲基萘醌介导电子流向硫还原酶（Schmitz et al. 1990；Rosenberg et al. 2013）。不同细菌中酯酸盐活化和琥珀酸盐形成的机制有所不同。乙酸氧化脱硫单胞菌（*D. acetoxidans*）中，有可能是仅有一种酶（琥珀酰 CoA：乙酰 CoA 转移酶）参与乙酸和琥珀酰 CoA 形成乙酰 CoA 和琥珀酸的反应。在 *D. acetivorans* 中酯酸盐通过乙酰磷酸酯形成乙酰 CoA，这个过程这涉及乙酸激酶和磷酸乙酰转移酶，而琥珀酰 CoA 则通过琥珀酰 CoA 合成酶形成琥珀酸（Schmitz et al. 1990）。

醇如甲醇和乙醇；有机酸如丙酸、丁酸和乳酸；糖如葡萄糖、果糖、纤维二糖、纤维素、乳糖、阿拉伯糖、鼠李糖、麦芽糖；淀粉和糖蜜为硫还原菌的有机底物（Bonch-Osmolovskaya et al. 1990b；Finster et al. 1997；Dirmeier et al. 1998；Boyd et al. 2007）。

硫还原菌利用碳源底物可发生完全氧化也可发生不完全氧化。第一种情况下，产物只有 CO_2 [脱硫单胞菌属（*Desulfuromonas*）和硫还原菌属（*Desulfurella*）]（Pfennig and Biebl 1976；Rainey and Hollen 2005）；第二种情况，醋酸盐和 CO_2 则作为最终产物（*Wolinella* 和 *Shewanella*）（Macy et al. 1986）。

5.3.2 硫的新陈代谢

α-S_8^0 的低溶解度限制了硫还原菌的快速生长（Bonch-Osmolovskaya et al. 1990b；Schauder and Müller1993；Miroshnichenko et al. 1998；Prokofeva et al. 2000）。提高硫溶解度有两种可行的方法（Cammack et al. 1984；Zöphel et al. 1991；Schauder and Müller 1993）。一种是将硫转化为更亲水和/或可溶的形式，例如多硫化物，这样能够促进硫还原菌更快生长（Blumentals et al. 1990；Schauder and Müller 1993）。有可能在较高的温度和 pH 条件下，硫的溶解度增加和多硫化物的形成有利于嗜热和极端嗜热微生物的生长（Belkin et al. 1985）。

然而，由于多硫化物在低 pH 下不稳定，我们需要通过硫还原菌将多硫化物与蛋白质合成结合蛋白，例如多硫化物硫转移酶，能够在多硫化物浓度较低的情况下快速进行多硫化物呼吸（Klimmek 2005），因此多硫化物仍能作为底物。或者，可能是嗜酸菌使用由多硫化物分解形成的纳米结晶作为电子受体。迄今为止，在多硫化物或纳米晶体是否可以作为嗜酸/耐酸微生物的电子受体的问题上，仍然没有一致意见（Boyd and Druschel 2013）。除了多硫化物，通过单质硫与小部分羰基化合物（框 5.1）如醛、羧酸、酮、酰胺和酯缔合形成的亲水性硫（Steudel et al. 1989），也可以作为微生物的电子受体。

值得注意的是，当多硫化物无法溶解时，这些微生物能够利用单质硫生长（Thamdrup et al. 1993；Finster et al. 1998）。作为一种替代机制，通过物理吸附，微生物能够直接吸附单质硫，并把硫直接转化成硫化物。

尽管仍不清楚硫还原细菌是利用哪种硫还原机制，但是极端嗜热化能自养古菌可能通过物理吸附将单质硫还原为硫化物（Pihl et al. 1989；Stetter et al. 1993）。此外，由于多硫化物在低 pH 下不稳定，并且快速解离成硫和硫化物，由此可以推测，在极端嗜酸性菌（*A. ambivalens*）中，单质硫可能是硫还原微生物的真正底物（Laska et al. 2003）。

硫还原酶已经从一些硫还原菌中被纯化了出来，并进行了表征（Schröder et al. 1988；Childers and Noll 1994；Ng et al. 2000；Laska et al. 2003）。但是目前通过多硫化物进行硫还原只在产琥珀酸沃林氏菌（*W. succinogenes*）（Klimmek et al. 1991）、火球菌（*P. furiosus*）

（Blumentals et al. 1990）和一些梭菌属（*Clostridium*）中得到证实（Takahashi et al. 2010）。

5.3.3　参与硫还原的酶

一般来说，硫还原酶的命名还没有一个标准。当酶被首次分离时会按照某一特征命名，随着更多性质被发现，其名称也会改变。如分离自 *P. furiosus* 的硫氢化酶（sulfhydrogenase，Shy）。然而，这些酶还受除硫以外的代谢物调节，所以硫氢化酶这个命名变得较为混乱；因此将其命名为来自极端嗜热菌的氢化酶（Hyh）（Vignais et al. 2001）。然而，数据库中仍然是将硫氢化酶作为酶的主要名称，因此在该书中也延用此名称。

基因组数据库中，硫还原菌中的酶命名为硫还原酶，而没有对其所属类群进行相关说明。这样可能会导致两个酶名称一样，但事实上是不同的酶。

目前为止，涉及将单质硫和多硫化物还原为硫化氢的三种酶在文献中均有描述：比如从产琥珀酸沃林氏菌（*Wolinella succinogenes*）分离的多硫化物还原酶（Hedderich et al. 1999）、从 *P. furiosus* 分离得到的硫化物脱氢酶和硫氢化酶（Ma and Adams 1994）。

5.3.3.1　多聚硫化物还原酶

膜结合酶是钼蝶呤结合蛋白，通过操纵子 *psrABC* 预测可能由三个亚基组成（Krafft et al. 1995）。钼蝶呤辅因子位于催化亚基 PsrA 上，其具有［4Fe-4S］铁硫中心。每摩尔纯化的酶含有 20mol 的游离铁和硫。因为亚基 PsrB 含有 4 个［4Fe-4S］铁硫中心，所以上述量与酶的整体含量一致（Hedderich et al. 1999）。

多硫化物还原酶（PsrC）的疏水性亚基将酶锚定在膜中。膜锚的定位点与酶催化亚基之间的电子转移极有可能通过亚基 PsrB 介导。PsrB 可能与膜的外周质的其他亚基结合（Dietrich and Klimmek 2002）。纯化的酶含有甲基萘醌作为辅因子。由于其亲脂性，甲基萘醌可能与酶的亚基 PsrC 结合。

多硫化物催化亚基 PsrA 的可能还原机制为：多硫化物在最后一个硫原子处裂解，硫原子被释放并进一步与氧化的钼蝶呤辅因子结合。PsrA 中的钼辅因子很可能是由两个钼蝶呤鸟嘌呤核苷酸分子协调的。因此，在吸收质子（可能通过硫化脱氢酶）和两个电子之后，HS^- 被释放并且钼被还原（图 5.5）（Klimmek et al.，1991）。

在 JGI 基因组数据库中的基因亚基的序列登录号为 PsrA：NP906381；PsrB：NP906382；PsrC：NP906383。

5.3.3.2　硫化物脱氢酶

硫化物脱氢酶（sulfide dehydrogenase），也称为黄素蛋白 c 硫化脱氢酶，是一种双功能细胞质酶，利用 NADPH 作为电子供体，催化多硫化物还原为硫化物（Ma and Adams 1994）。但它也能起到铁氧还蛋白的功能（$NADP^+$ 氧化还原酶）（Ma and Adams 1994）。$NADP^+$ 的还原被认为 H_2 还原处理的必要一步。

该蛋白质含有两个黄素和三个不同的［Fe-S］中心，包含了一种天冬氨酸和三个半胱氨酸［Asp（Cys）$_3$］的修饰结构，其结合了被组氨酸配位的蛋白质簇，即一个具有高还原潜力的调控［3Fe-4S］簇和一个具有不常见还原特性的［4Fe-4S］簇（Hagen et al. 2000）。两个簇的高还原性电势尚不明确，但黄素的氧化还原势能与硫化脱氢酶和铁氧还蛋白（$NADP^+$ 氧化还原酶）保持一致。

由于硫化物脱氢酶的亚基中的铁-硫簇的性质还没有完全被解读，作用机制尚不清楚。

在 JGI 基因组数据库中的基因亚基的序列登录号为 SudHA：AAL81451/AAL82034；

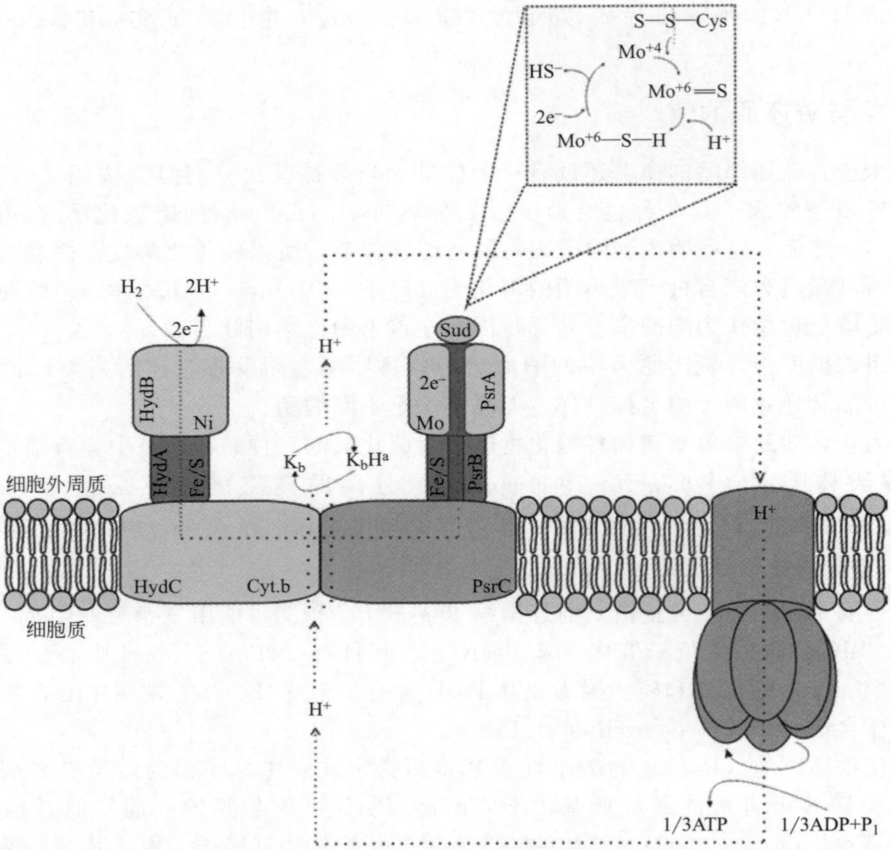

图 5.5　假丝酵母中单质硫还原（通过多硫化物）和厌氧电子传递链的假设。为了在酶之间发生电子转移，假定酶的碰撞是必要的，甲基萘醌似乎结合到多硫化物还原酶的亚基 C 上。质子也被假定通过甲基萘醌转移到外周质。氢化酶的亚基标记为 HydA、HydB 和 HydC，多硫化物还原酶的亚基标记为 PsrA、PsrB 和 PsrC。K 代表醌，Sud 代表硫/多硫化物转移酶。模型引自 Hedderich 等人（1999）和 Rosenberg 等人（2013）

SudHB：AAL81452/AAL82035。

5.3.3.3　硫氢化酶

从 *P. furiosus* 中纯化出两种不同的细胞质氢代谢酶，并显示硫还原酶活性。两者分别被称为硫氢化酶（sulfhydrogenase）Ⅰ和Ⅱ，也称为 NAD（P）H：硫氧化还原酶或辅酶 A（CoA)-依赖性 NADP（H）硫氧化还原酶（Bryant and Adams 1989；Ma et al. 1993，2000）。

硫氢化酶Ⅰ（Bryant and Adams 1989）和Ⅱ（Ma et al. 2000）都可以使用 H_2 作为电子供体将 S_8^0 和多硫化物还原为 H_2S。这两种蛋白质都有四个亚基，镍、铁硫中心和黄素腺嘌呤二核苷酸，但其亚基的催化活性和排列不同。硫氢化酶Ⅰ是异四聚体（αβγδ），并且硫氢化酶Ⅱ被认为由两个异四聚体 $(\alpha\beta\gamma\delta)_2$ 构成的二聚体（Bryant and Adams 1989）。在这两种情况下，β 和 γ 亚基都具有硫还原酶的作用，而 α 和 δ 具有氢化酶的作用。

酶之间存在三个主要差异。(1) Ma 等人（2000）发现硫氢化酶Ⅱ对于氢的生成、摄取和硫还原的活性较低。(2) 硫氢化酶Ⅱ对单质硫和多硫化物具有更高的亲和力，阐明了在当硫浓度很低时该酶的生理相关性。(3) 硫化氢酶Ⅱ对 NAD（H）亲和力较大，硫化氢酶Ⅰ

对 NADP（H）的亲和力较大，可能使用具有相同效率的两种核苷酸。

JGI基因组数据库中的两种复合物的基因亚基的序列登录号为 shyA：AAL81018/AAL81456；shyB：AAL81015/AAL81453；shyC：AAL81016/AAL81454；shyD：AAL81017/AAL81455。

从嗜酸古菌 *A. ambivalens* 中（Laska et al. 2003）纯化出可能与单质硫还原相关的一种新酶，这种酶用 H_2 或 $NADPH_2$ 作为电子供体还原单质硫。该菌株硫还原酶与来自 *W. succinogene* 的一个膜结合蛋白相关，其亚基具有相似的结构和性质。至少三种蛋白质构成核心酶的主要结构：催化亚基，可能包含一个钼蝶呤（SreA）、铁硫蛋白（SreB）和膜锚定蛋白（SreC）。研究显示膜锚定蛋白在系统发育上与 *W. succinogenes* 中的类似蛋白质无关。因为酶是在不存在硫化物的情况下分离的，所以它最有可能还原单质硫而不是多硫化物。因为硫还原酶不能在没有氢化酶的情况下纯化，因此很难对其进行更深入研究（Laska et al. 2003）。酶的完整特性对于酶的表征十分重要，只有这样才能揭示它是否是硫还原微生物中的真正的新酶，也将有助于阐明酶的作用机制。

类似的酶存在于古菌和细菌的多种微生物中，例如脱铁脱硫弧菌（*Deferribacter desulfuricans*）、脱硫杆菌（*Desulfitobacterium dehalogenans*）、暗杆菌（*Pelobacter carbinolicus*）、脱硫弧菌（*Desulfovibrio frigidus*）、硫还原酸叶菌（*Acidilobus sulfurireducens*）、硫还原菌（*Desulfurella acetivorans*）、热厌氧弧菌（*Thermanaerovibrio acidaminovorans*）、热脱硫杆菌（*Thermodesulfobacterium geofontis*）、暖球形菌（*Caldisphaera lagunensis*）、火山鬃菌属（*Vulcanisaeta distribute*）、冰岛热棒菌（*Pyrobaculum islandicum*）、海洋甲烷球菌（*Methanococcus maripaludis*）和盐碱湖菌属（*Natronolimnobius innermongolicus*）。

通过数据库的检索表明，上述酶存在于许多微生物的基因组中，其中很多至今尚未作为硫还原菌被报道。这些潜在的硫还原原核生物广泛分布于系统发育进化树中，也包括一些尚无硫还原菌报道的门，例如绿弯菌门（Chloroflexi）、放线菌门（Actinobacteria）、硝化螺菌门（Nitrospira）、绿菌门（Chlorobi）或理研菌科（Rikenellaceae）。古菌中，硫还原菌仅分布于泉古菌门（Crenarchaeota）和广古菌门（Euryarchaeota）。尽管已经对其中的一些微生物进行了检测，却没有发现具有硫还原活性，不确定是否是因为培养条件对于菌的生长和/或硫还原不适宜。在某些情况下，例如脱硫弧菌（*Desulfonatronovibrio thiodismutans*）、脱硫菌（*Desulfonatronum thioautotrophicum*）和嗜碱性脱硫杆菌（*Desulfobotulus alkaliphilus*）的休眠细胞中单质硫减少，但硫不支持其生长。一个合理的解释是硫化物和单质硫反应生成多硫化物抑制部分微生物的生长（Sorokin et al. 2011）。

5.3.4　通过多硫化物还原硫

Zöphel 等人（1991）研究硫化螺旋菌（*Sulfurospirillum deleyianum*）（以前称为 *Spirillum* 5175）发现，当向培养基中加入硫醇类物质，例如谷胱甘肽和硫化物，可以促进细胞提取物膜组分单质硫的还原，这可能是由于亲核攻击使二硫键裂解增强。他们还提出硫化物和硫形成的多硫化物链，是硫酸盐脱硫弧菌（*Desulfovibrio desulfuricans*）的细胞色素 c_3 还原硫的中间体（Cammack et al. 1984）。硫化物是由多硫化物还原形成的（S^{2-}），通过亲核攻击切割 S_8^0-环，产生新的多硫化物分子，之后通过细胞色素 c_3 快速还原为 S^{2-}（Cammack et al. 1984）。

Macy 等人（1986）发现了 *W. succinogenes* 可利用甲酸盐和单质硫生长，生成 H_2S 和 CO_2。后来，Klimmek 等（1991）也发现了 *W. succinogenes* 的生长依赖甲酸盐和多硫化物。

Ringel 等人（1996）推测在 *W. succinogenes* 中，多硫化物作为硫呼吸作用的中间体，在厌氧条件下向培养基中加入 Fe^{2+} 可以沉淀细菌产生的所有硫化物，生成 FeS。在这种情况下，多硫化物的形成被阻止，并观察到利用甲酸盐和单质硫的 *W. succinogenes* 生长，由此得出结论，单质硫是 *W. succinogenes* 硫还原中的最终电子受体。三年后，Hedderich 等人（1999）从加入 Fe^{2+} 的培养物中分离到可溶性含硫组分和细胞周质的硫化脱氢酶，即所谓的 Sud 蛋白。当他们用 CN^- 和硫代硫酸盐处理分离的蛋白质时，没有观察到反应现象；但当将多硫化物加入到培养基中时，形成硫氰酸盐（式 5.6）。

$$S_n^{2-} + CN^- \longrightarrow SCN^- + S_{n-1}^{2-} \tag{5.6}$$

研究发现 Sud 蛋白协助多硫化物中的硫从溶液中转移到多硫化物还原酶（Psr）的催化位点（Klimmek et al. 1991）。多硫化物/硫还原中，存在于 Psr 中的甲基萘醌被认为氢化酶的电子受体（Rosenberg et al. 2013）。氢或甲酸盐组成的多硫化物还原电子传递链由多硫化物还原酶和氢化酶或甲酸脱氢酶构成，推测氢化酶和多硫化物还原酶随机分布在 *W. succinogenes* 膜中（Jankielewicz et al. 1995）。

后来的研究表明，*W. succinogenes* 中的 8-甲基-甲基萘醌对于硫还原是必不可少的（Jankielewicz et al. 1995；Hedderich et al. 1999）。大多数甲基萘醌溶解在膜的脂双层中，并且通过扩散在电子转移中起作用，这是它参与硫/多硫化物还原机制的第一个假设。然而，溶解在膜中的甲基萘醌的氧化还原电位比多硫化物中的高很多，这使得电子不可能直接从脱氢酶流向多硫化物还原酶（Hedderich et al. 1999）。亦或者，甲基萘醌可能与多硫化物还原酶结合，并且是氢化酶细胞色素 b 亚基的主要电子受体（Hedde-rich et al. 1999）。因此，从氢化酶到多硫化物还原酶的电子转移可能需要膜内两种酶的碰撞或聚集（图 5.5）。甲基萘醌在膜内，推测其还原与氢化酶从细胞质摄取质子有关，并且其氧化是通过多硫化物还原酶催化，同时偶联质子释放到细胞周质中（Dietrich and Klimmek 2002）。

从 *W. succinogenes* 基因组文库中对一些基因进行亚克隆，如编码甲酸脱氢酶的 *frh* 基因（Bokranz et al. 1991）、编码多硫化物还原酶的 *psr* 基因（Krafft et al. 1995））和编码硫化物脱氢酶的 *sud* 基因（Kreis-Kleinschmidt et al. 1995）。

Blumentals 等人（1990）研究古菌 *P. factiosus* 中硫还原的机理，发现硫化物和多硫化物在培养物中的形成，单质硫与微生物物理隔离，这表明古菌和单质硫之间的接触对于代谢不是必需的，并且可溶性多硫化物可用作硫还原的底物。*P. furiosus* 中的硫还原是否与节约能量相关尚不清楚。硫可以仅作为电子阱，使有机化合物发酵更为有效（Rosenberg et al. 2013）。

P. furiosus 可以将质子作为末端电子受体，直接将 H_2 的产生与 ATP 的合成偶联。多蛋白膜将氢化酶复合物和铁氧还蛋白结合，作为低电位的电子供体，将电子转移与质子还原和质子易位偶联（Sapra et al. 2003）。

5.3.5 通过物理附着将硫还原固定化

由于单质硫在水中的溶解度低，一些微生物在外膜的表面对其进行还原，具体机制研究较少。由于一些原核生物也能够还原位于外膜处的不溶性矿物氧化物（Lovley 1991；Lovley et al. 2004；Hartshorne et al. 2009），不同电子转移策略被先后提出，这也与硫还原菌密切相关。

例如，在希瓦氏菌（*Shewanella*）和地杆菌（*Geobacter*）等催化铁还原的种属中，发现了一些硫还原菌，据报道外部不溶性氧化铁的还原通过四种不同的机理：（1）细胞色素 c 将呼吸链延伸至细胞表面（Richardson 2000；Lovley et al. 2004；Richter et al. 2012）；（2）细胞外还原介质，如腐殖酸、醌、吩嗪和半胱氨酸，可以在电子传递链的末端提供电子和不溶性受体之间传递电子（Lovley et al. 1998；Scott et al. 1998；Newman and Kolter 2000；Hernandez and Newman 2001）；（3）在不存在细胞色素 c 的情况下，微生物可以产生修饰的菌毛，即所谓的纳米线，用作细胞和氧化物表面之间的电连接（Reguera et al. 2005）；（4）一些菌株可以利用纳米晶体结晶、导体或半导体矿物质（例如氧化铁）来构建导电网络（Kato et al.，2010）。

据报道，一些微生物可将单质硫直接还原为硫化物，例如嗜酸菌（*A. obivalens*）、嗜酸氧化亚铁硫杆菌（*A. ferrooxidans*）、热网菌（*Pyrodictium abyssi*）和 *Pyrodictium brockii*，从中进行了系列研究，并在此对其进行概述。

在布氏热网菌的膜中检测到氢化酶、醌和细胞色素 c（Pihl and Maier 1991；Pihl et al. 1992）。氢化酶是 Ni-Fe 型，具有两个亚基（Pihl and Maier 1991）。Pihl 等人（1992）对氢化酶进行了核磁共振分析，证明其与所有的醌比较均有所不同。当醌暴露于 UV 而失活时，电子传递活性丧失。加入醌则重新激活该过程，意味着电子转移顺序是：氢化酶→醌→细胞色素 c。由此可见，细胞色素 c 应该是尚未鉴定的硫还原酶的电子供体。

Dirmeier 等人（1998）从 *P. abyssi* TAG11 菌株的膜组分中分离了硫氧化还原酶复合物，其电子传递链不同于布氏热网菌（*P. brockii*）。生物体的完整呼吸链由一系列多酶复合物组成，其中电子传递的组分，氢化酶和硫还原酶有序排列。还原酶由至少九个亚基组成，具有两个 b 型细胞色素和一个 c 型细胞色素。在热网菌属 *P. abyssi* 菌株的膜复合物中没有检测到醌。如从 *P. brockii* 提取出的硫氧化还原酶中有镍的存在，表明其氢化酶是 Ni-Fe 型（Rosenberg et al. 2013）。

在假丝酵母中，当醌作为电子载体的氢化酶存在时，硫还原酶则用氢作为电子供体还原单质硫（Laska et al. 2003）。氢化酶具有与从苏云金芽孢杆菌纯化的氢化酶类似的亚基，一个同源的含 Ni 的催化亚基（HynL/HydB），一个同源的含 Fe-S 的电子转移亚基（HynS/HydB）和一个非同源膜锚定蛋白（IspI/HydC）（Laska et al. 2003）。因此，该微生物中的电子传递链最可能由通过醌连接的两种酶组成（图 5.6）。

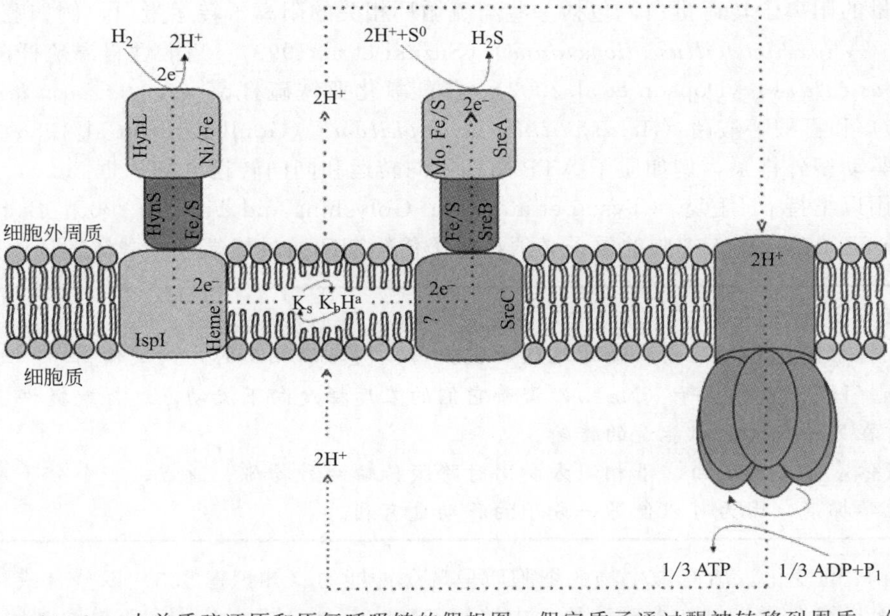

图 5.6　*A. ambivalens* 中单质硫还原和厌氧呼吸链的假想图。假定质子通过醌被转移到周质。仅表示主要亚基结构。氢化酶的亚基标记为 HynL、HynS 和 IspI，硫还原酶的亚基标记为 SreA、SreB 和 SreC。K 代表醌。模型改编自 Laska 等人（2003）

5.4 适应酸性条件的机制

许多硫还原微生物生长最适 pH 是中性。尽管如此，已经鉴定了能够在酸性环境中生长的几个物种（Stetter 1996；Hedderich et al. 1999；Yoneda et al. 2012）。相较于嗜中性微生物，嗜酸性或耐酸性微生物能够耐受较大的跨细胞质膜的 pH 梯度。这些微生物通常面临超出细胞膜的质子动力，其可以通过驱动能量依赖性过程来达到 pH 稳态（Baker-Austin and Dopson 2007）。在酸性条件下，为了保持生理 pH，微生物采取几种策略。Baker-Austin 和 Dopson（2007）提出了微生物在低 pH 条件下生长的途径和机制，这一部分将在本节中总结：利用特异性转运蛋白和酶进行质子输出，具有特殊渗透性能，增加缓冲能力和增强表面正电荷。

一般来说，嗜酸菌和耐酸菌细胞膜流动性低且高度不可渗，以限制质子流入细胞质（Benjamin and Datta 1995；Dilworth and Glenn 1999；Konings et al. 2002）。一些嗜酸性古菌的膜由四醚脂类组成，这使得它们对质子高度不渗透。此外，通常在细菌和真核细胞膜中发现，醚键相较于酯键对酸水解更不敏感（Macalady and Banfield 2003；Golyshina and Timmis 2005）。此外，来自膜的脂质，糖脂含量更高，一个或多个糖单元暴露于细胞的外表面（De Rosa et al. 1983；Chong 2010）。虽然仍然缺乏直接证据，但有人提出，古菌细胞表面上的糖的大量修饰可以对抗质子的流入（Shimada et al. 2008；Wang et al. 2012）。

嗜酸菌中减少膜通道的大小和渗透性是维持 pH 稳态的另一种机制。膜孔隙减小，根据电荷和大小对进入膜孔蛋白的离子进行选择（Amaro et al. 1991）。

嗜酸菌减少质子流入的另一种机制是维持细胞内和细胞外环境之间的电势差，而没有电流流过膜，在嗜中性菌中产生内部正 $\Delta\Psi$ 对抗内部负 $\Delta\Psi$，即所谓的唐南平衡。这种唐南电位可能是由于更多的钾离子流入产生的。这种机制的重要性表现为：在几种嗜酸菌基因组中鉴定到大量的阳离子转运蛋白，包括一些与硫循环相关的阳离子转运蛋白，例如嗜酸性氧化硫硫杆菌（*Acidithiobacillus thiooxidans*）（Suzuki et al. 1999）、嗜酸性喜温硫杆菌（*Acidithiobacillus caldus*）（Dopson et al. 2002）、嗜酸氧化亚铁硫杆菌（*A. ferrooxidans*）（Cox et al. 1979）和喜酸嗜酸菌（*Acidiphilium acidophilum*）（Goulbourne et al. 1986）。

质子驱动型外排泵，例如质子 ATP 酶、逆向转运和同向转运（知识框 5.2），也被一些嗜酸菌使用以维持 pH 稳态（Tyson et al. 2004；Golyshina and Timmis 2005；Baker-Austin and Dopson 2007）。进入细胞的质子必须在电子传输期间通过终端电子受体的还原来平衡。

> **知识框 5.2**
>
> **反转录器：** 积极运输物质通过膜的膜蛋白，同时以相反的方向运输离子。通常是氢离子（H^+）或钠离子（Na^+），沿着它们的浓度梯度向下流动，这样就提供了另一种物质在另一个方向上能量的传输。
>
> **载体：** 整合膜蛋白，沿相同方向同时跨膜传输两种物质。通常，一个分子可以上移电化学梯度，因为可以使另一分子的移动更有利。

根据 pH 的变化，所有微生物的细胞质呈现缓冲能力（知识框 5.3）以螯合或释放质子。蛋白质和无机聚合物（例如多磷酸盐）中的氨基酸或其他有机小分子和可电离基团具有这种缓冲能力（Slonczewski et al. 1982；Zychlinsky and Matin 1983；Krulwich et al. 1985；Leone et al. 2007）。Zychlinsky 和 Matin（1983）比较了喜酸嗜酸菌和大肠杆菌的缓冲能力，结果显示嗜酸菌的缓冲能力较高，每个 pH 单位分别为 97mmol H^+ 和 85mmol H^+。Krulwich 等人（1985）也发现，嗜热杆菌在中性条件下具有比其他杆菌（每 pH 单位约 400～550mmol H^+）有更高的缓冲能力（每 pH 单位约 600mmol H^+）。然而，两项研究中获得的结果表明，嗜酸菌的缓冲能力不一定高于嗜中性菌的缓冲能力。这表明缓冲能力只有与其他机制一起时才有助于 pH 稳态。

知识框 5.3

缓冲能力：溶液通过吸收或解吸 H^+ 和 OH^- 离子来抵抗 pH 变化的能力。它由将溶液的 pH 值改变 1 所需的酸或碱的物质的量除以 pH 变化和缓冲液的体积来表示。

低 pH 环境可以损害细胞中的生物分子，这需要修复机制。这可以解释为什么嗜酸菌的基因组中存在大量的 DNA 和蛋白质修复基因（Crossman et al. 2004）。在低 pH 下，嗜酸菌中参与蛋白质重折叠的分子伴侣高度表达，表明它们可以在酸性条件下微生物的存活中起作用。

对生长活跃期间细胞内 pH 为 5.6 的专性嗜酸菌的研究显示，几种酶在 pH 1.7～4.0 下起作用，当细胞在极低的 pH 值下生长时，这些酶需要发挥作用以使新陈代谢开始。已经在许多嗜酸菌的蛋白质组中检测到较高含量的铁蛋白，这有助于酶在低 pH 下维持 pH 稳定性（Ferrer et al. 2007）。从这些嗜酸菌的纯化蛋白质中除去铁，会使二级结构破坏，酶失活。铁被认为在维持蛋白质的三维结构中起到“铁铆钉”的作用，在酸性条件下稳定蛋白质（Ferrer et al. 2007）。

大多数有机酸，例如乙酸和乳酸促进质子在低 pH 下跨膜转移。在这种条件下，酸以质子化形式扩散到细胞中，因此质子在细胞质中解离，pH 升高（Baker-Austin and Dopson 2007）。因此，一些嗜酸菌可通过降解有机酸起到解毒作用。

5.5 生物技术的应用

5.5.1 工业废水和酸性矿井排水

硫化矿物的生物氧化和酸性富金属矿物废水的形成之间存在关系（Hoffert 1947；Johnson 1995，2003）。简而言之，由于它们暴露于氧化环境（O_2 或 Fe^{3+}）中，金属硫化物如黄铁矿的氧化是酸性矿井排水（AMD）形成的根本原因（Johnson and Hallberg 2005）。在大多数情况下，三价铁是矿石氧化的主要氧化剂（式 5.7 和式 5.8），其生物再生（式 5.9）维持矿物的开放式氧化（Schippers and Sand 1999；Johnson and Hallberg 2005；Vera et al. 2013）和酸性环境形成，在此过程中金属通常会被溶解。

$$FeS_2 + 6Fe^{3+} + 3H_2O \longrightarrow S_2O_3^{2-} + 7Fe^{2+} + 6H^+ \tag{5.7}$$

$$S_2O_3^{2-} + 8Fe^{3+} + 5H_2O \longrightarrow 2SO_4^{2-} + 8Fe^{2+} + 10H^+ \tag{5.8}$$

$$4Fe^{2+} + O_2 + 4H^+ \longrightarrow 4Fe^{3+} + 2H_2O \tag{5.9}$$

铜、锌、镉、砷、锰、铝、铅、镍、银、汞、铬和铁是在酸性矿井排水和工业废水中含量较为丰富的金属，因为它们可以在较宽浓度范围 10^{-6}～10^2g/L 内存在（Huisman et al. 2006）。例如，受天然酸性岩石排水影响的廷托河（Tinto River）中，铁的浓度高达 20.2g/L，铜的浓度高达 0.7g/L，锌的浓度高达 0.56g/L（Lopez-Archilla et al. 2001）；而在纺织工业的废水中，铁浓度可达到 0.11g/L，铜和锌的浓度达到了 0.01g/L（Joshi and

Santani 2012)。

5.5.2 最先进的金属去除和回收方法

5.5.2.1 化学/物理方法

许多化学/物理方法通过添加化学品到污染的废水中去除重金属，例如吸附、离子交换、复合物形成和沉淀，这是用于处理酸性矿井排水和其他金属污染物的最广泛应用的化学/物理方法（Johnson and Hallberg 2005）。

为了在提高 pH 的过程中沉淀金属，可将一些中和剂加入到介质中，例如碳酸钙、氧化钙、氢氧化钙或氢氧化钠（Weijma et al. 2002）。尽管可以有效处理，但这些方法相对昂贵，并产生大量残留金属沉渣，金属回收再利用率低（Gallegos-Garcia et al. 2009；Tekerlekopoulou et al. 2010）。

5.5.2.2 微生物学方法

微生物过程，如甲烷生成、反硝化、铁和锰的还原、产生碱性化合物，这可能导致金属形成氢氧化物沉淀（Johnson and Hallberg 2005）。即可以从流出物中除去氢氧化物，所有金属一起沉淀，所产生的废弃物需要处理，这导致该方法的额外成本增加。金属还可以通过生物电化学系统回收，其中有机底物在阳极处被生物氧化，从而产生电子，用于在阴极处还原金属离子如 Cu^{2+}（Heijne et al. 2010）。过去的许多研究使用过金属生物吸附的概念，即金属离子吸附到生物物质如细菌细胞和植物的表面。但该方法没有广泛应用，可能是由于金属负载能力较低，并且几乎不能进行回收金属残余物。

基于硫化物沉淀金属的生物反应器系统与物理方法一样有效，其操作成本低，并产生较少的残余沉渣（Johnson and Hallberg 2005）。硫化作用是基于厌氧条件下微生物对简单有机化合物或氢的氧化，从含硫物质如硫酸盐、亚硫酸盐、硫代硫酸盐、有机亚砜、单质硫、多硫化物和有机二硫化物的还原中产生硫化物。多功能性的产硫微生物允许电子供体和硫源的多种组合，而且该方法的操作条件比较宽泛（温度、盐度、pH）。

5.5.3 金属去除和回收

在用于金属去除和回收的方法中，生物产生的硫化物与溶解的重金属结合，例如作为不溶性金属硫化物沉淀的 Cu^{2+}、Zn^{2+} 和 Ni^{2+}（Hulshof et al. 2006；Neculita et al. 2007）。大多数金属硫化物在中性至碱性 pH 下的溶解度极低，远低于相应的金属氢氧化物。因此，可以得到更好的出水效果，并且可以回收更多的金属。此外，反应速率更高，能产生酸稳定的金属硫化物，例如 Co、Ni 和 Cu，具有良好的沉降性质和高的再利用潜力（Tsukamoto et al. 2004；Gallegos-Garcia et al. 2009；Lewis 2010；Sánchez-Andrea et al. 2014）。冶炼厂用于金属生产的原料通常是含有其硫化物形式的金属精矿，例如含 ZnS 的闪锌矿。这有利于使用生物沉淀的金属硫化物作为冶炼厂的原料。ZnS 已在荷兰的尼尔斯塔（Nyrstar）的锌精炼厂中应用（Weijma et al. 2002）。

硫酸盐还原是处理采矿和冶金废水最常用的生物工艺。然而，只有少数的中度嗜酸性硫酸盐还原菌种：热脱硫化菌（*Thermodesulfobium narugense*），可以在 pH 4 下生长（Mori et al. 2003）；脱硫芽孢弯曲菌（*Desulfosporosinus acidiphilus*），可以在 pH 3 下生长（Jameson et al. 2010）和脱硫芽孢菌（*Desulfosporosinus acididurans*），可以在 pH 3.8 下生长（Sánchez-Andrea et al. 2015）。对于酸性矿井和酸性岩石排水的处理来说，硫还原微生物的使用也备受关注（Hoffert 1947；Johnson 1995，2003）。

为了处理金属污染的硫，例如酸性矿井/岩石排水，已经提出了两种设计的硫化生物反应器。一个是基于独立运行的生物和化学室（Tabak et al. 2003）。在生物室中，硫化氢通过气体循环产生并转移到化学回路中，该回路接收原始进水（图 5.7a）。因此，金属硫化物的产生和金属的沉淀是通过载气（氮）从生物溶液中提取硫化氢来分离的，然后硫化氢气体溶解在金属污染的废水中。在该装置中，生物硫化物和金属污染流之间没有接触。这是该设计的主要优点，因为其防止了高酸度和金属浓度而引起的可能的生物质毒性反应（Johnson and Hallberg 2005）。缺点是载气再循环需要高能量输入。这种技术已经用于金属如 Cu 和 Zn 研究（Foucher et al. 2001；Al-Tarazi et al. 2005；Gramp et al. 2009）。由于单独的硫化物产生和金属硫化物沉淀，两个工艺流程可以在最佳条件下进行控制。例如，通过控制沉淀器中的 pH 和 pS（$-\log[S^{2-}]$），可以实现各个重金属的选择性沉淀（Veeken and Rulkens 2003；König et al. 2006；Sampaio et al. 2009）。这些产生相对较纯净的金属硫化物沉淀，在冶金工业中作为精矿原料的补充具有更高的价值（Grootscholten et al. 2008）。

其他设计的系统只有一个反应体系，其中生物硫化物产生和金属沉淀同时发生（图 5.7b）。

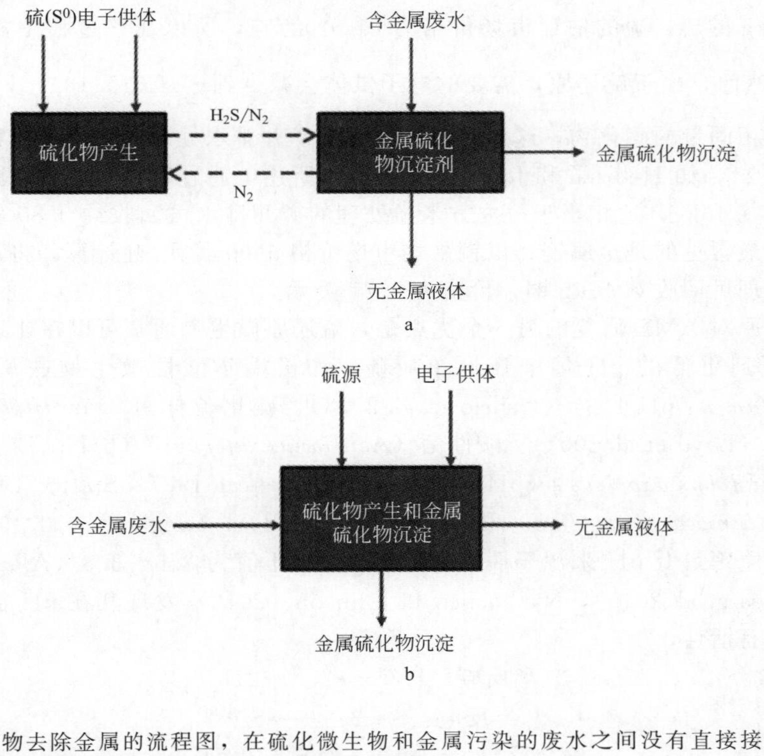

图 5.7 两步法生物去除金属的流程图，在硫化微生物和金属污染的废水之间没有直接接触（a）。一步法生物金属去除，硫化微生物和金属污染的废水之间直接接触（b）

在这种构造中，由于硫化培养物与来自流入物溶解的金属接触，因此金属毒性是设计和操作方面要考虑的问题。通过保持相对于金属过量的硫化物，产生可适应金属负载和生物活性波动的“硫化物缓冲液”。这种配置的优点是硫化物产生和金属硫化物沉淀发生在单个单元中，从而消除了对载气的能量密集型再循环的需要。这种工艺流程方案已经被研究过（Labrenz et al. 2000；Steed et al. 2000；Kaksonen et al. 2003；Johnson and Hallberg 2005；Sierra-Alvarez et al. 2006；Gallegos-Garcia et al. 2009；Sánchez-Andrea et al. 2012）。Weijma 等人（2002）和 Möbiuset 等人（2015）描述了生物硫化物产生的全面操作过程。

5.5.4 硫酸盐还原过程与硫还原过程成本的比较分析

采矿或金属工业的废水通常含有少量的有机物。为了将硫酸盐完全还原为硫化物，需要添加电子供体（Liamleam and Annachhatre 2007）。基于反应的化学计量，单质硫作为电子受体比硫酸盐更适合，因为在该方法中，每摩尔硫化物仅需要两个电子（式 5.10），而硫酸盐需要八个（式 5.11）。所产生的硫化物决定了待回收的金属的量（式 5.12），因此用相同量的金属沉淀，与硫酸盐还原相比，该方法仅需要$\frac{1}{4}$的电子供体用于硫还原。

Hedrich 和 Johnson（2014）对模块化反应器的成本进行了分析，以便氧化铁还原硫酸盐，来沉淀低 pH 环境中的金属。用甘油作为电子供体处理 $1m^3$ 矿井水所需的 $42m^3$ 硫磺反应器，能够产生 3.96mol 的硫化物。由于甘油与硫酸盐的物质的量之比为 4∶7（式 5.12），在反应器中将需要 2.26mol（208.52g）甘油。如文中所述，假设甘油的市场价格为 2400 美元/吨，那么此过程中该试剂的成本为 0.5 美元。

如果代替硫酸盐，使用硫作为电子受体，在 $42m^3$ 反应器中达到相同量的硫化物，则需要加入 0.126kg 的硫。硫的估计市场价格为 61 美元/吨，所以在该过程中需要额外的费用 0.008 美元。然而，由于硫还原，需要的电子供体会减少到$\frac{1}{4}$（式 5.13），只需要 52.13g 的甘油就可获得相同量的硫化物，这意味着每立方米矿井排水处理的成本会降低 0.37 美元。

根据式 5.11，如 Hedrich 和 Johnson（2014）所述，通过硫或硫酸盐还原产生的硫化物的量可以回收 0.46kg，这相当于每立方米经处理的矿井排水可回收约 0.80 美元。考虑到酸性矿井排水中最常见的是金属铜，以铜平均市场价格 5900 美元/吨计算，将意味着每立方米矿井排水经处理可回收 0.46kg 铜，价格为 2.71 美元。

单质硫还原对 AMD 修复的另一个优点是，硫还原微生物通常可以在比迄今为止所描述的硫酸盐还原剂更低的 pH 值下还原单质硫。如极端嗜酸性微生物铁氧化亚铁硫杆菌（*A. ferrooxidans*）（pH 1.8）（Osorio et al. 2013）、硫化硫杆菌（*Acidilobus sulfurireducens*）（pH 2）（Boyd et al. 2007）、酸性菌（*Acidianus infernus*）（pH 1.5）（Stetter 1996）、憎叶菌（*Stygiolobus azoricus*）（pH 1）（Svetlichnyi et al. 1987；Stetter 1996）、嗜酸热原体（*Thermoplasma acidophilum*）和 *volcanicum*（pH 1）（Segerer et al. 1988）。目前分离到的硫酸盐还原菌最低 pH 来源于脱硫芽孢菌属，pH 值为 3.6～3.8（Alazard et al. 2010；Sánchez-Andrea et al. 2015），Ñancucheo 和 Johnson（2012）发现其在 pH 低至 2.5 的生物反应器中仍具有活性。

$$\text{半反应：} H_2 \longrightarrow 2e^- + 2H^+$$
$$\text{半反应：} S + 2e^- \longrightarrow S^{2-}$$
$$\text{全反应：} S + H_2 \longrightarrow S^{2-} + 2H^+ \qquad (5.10)$$
$$\text{半反应：} 4H_2 \longrightarrow 8e^- + 8H^+$$
$$\text{半反应：} SO_4^{2-} + 8e^- + 8H^+ \longrightarrow S^{2-} + 4H_2O$$
$$\text{全反应：} SO_4^{2-} + 4H_2 \longrightarrow S^{2-} + 4H_2O \qquad (5.11)$$
$$H_2S + Me^{2+} \longrightarrow MeS_{(s)} + 2H^+ \qquad (5.12)$$

Me^{2+} 代表金属离子，例如 Zn^{2+}、Cu^{2+}、Pb^{2+} 和 Ni^{2+}。

$$4C_3H_8O_3 + 7SO_4^{2-} \longrightarrow 7H_2S + 12CO_2 + 16H_2O \qquad (5.13)$$
$$C_3H_8O_3 + 7S + 3H_2O \longrightarrow 7H_2S + 3CO_2 \qquad (5.14)$$

此外，许多硫酸盐还原微生物如脱硫肠状菌属、脱硫叶菌属、古生球菌（Castro et al. 2002）、脱硫弧菌属、热脱硫杆菌属（Widdel and Pfennig 1981；Widdel 1988；Widdel and Pfennig 1991）、脱硫孢菌属（Sánchez-Andrea et al. 2015）不能进行完全的硫还原。这意味着它们有助于乙酸在培养基中的积累。对于大多数硫还原微生物，特别是属于δ-变形菌纲类的硫还原微生物，不能将有机底物氧化为 CO_2，例如脱硫单胞菌属、地杆菌属、变形杆菌属和脱硫菌属，后者通常在酸性环境中发现（Bonch-Osmolovskaya et al. 1990b；Miroshnichenko et al. 1998）。

在冶金过程中，硫的还原用来处理金属料液更具有前景，这些料液不含硫酸盐，通常是酸性的，有时是高温的。然而，由于一些原因，如硫酸盐在AMD水中自然存在，硫酸盐还原可能仍然是原位系统如可渗透反应性屏障的最简单的选择。

5.6 结论与前景

参与硫循环的微生物在工业和环境的角度来看是非常重要的，特别是进行硫化反应的微生物。硫还原的原核生物广泛分布在海洋和陆地环境中，并且能够在较宽的温度和 pH 范围内生长。能够在酸性环境中生长的硫还原微生物，对于选择性金属沉淀和生物除污过程来说是具有发展前景的。

虽然对几种嗜酸硫还原菌进行了研究，但是它们的生理学和对极端条件的具体适应机制仍然知之甚少。对这些微生物研究将有助于阐明硫还原菌利用底物类型，硫还原菌的生理学和生态学特性，以及它们在工程生态系统中的行为，也将有利于选择性沉淀和从采矿及金属冶炼中回收重金属反应器的发展。

致谢

A. P. Florentino 的博士研究项目得到了巴西政府科学和技术发展组织即巴西国家科学技术委员会（Consueho Nacional de Desenvolvimento Científico e Tecnológico）的支持。I. Sánchez-Andrea 和 A. J. M. Stams 的研究由 ERC 拨款项目 323009 和荷兰教育、文化和科学部的 Gravitation 拨款项目 024.002.002 资助。

无利益冲突声明

Anna P. Florentino，Jan Weijma，Alfons J. M. Stams 和 IreneSánchez-Andrea 声明他们没有利益冲突。

参考文献

Alain K，Callac N，Guégan M，Lesongeur F，Crassous P，Cambon-Bonavita MA，Querellou J，Prieur D（2009）*Nautilia abyssi* sp. nov.，a thermophilic，chemolithoautotrophic，sulfurreducing bacterium isolated from an East Pacific Rise hydrothermal vent. Int J Syst Evol Microbiol 59（6）：1310-1315

Alazard D，Joseph M，Battaglia-Brunet F，Cayol JL，Ollivier B（2010）*Desulfosporosinus acidiphilus* sp. nov.：a moderately acidophilic sulfate-reducing bacterium isolated from acid mining drainage sediments：New taxa：Firmicutes（Class Clostridia，Order Clostridiales，Family Peptococcaceae）. Extremophiles 14（3）：305-312

Al-Tarazi M，Heesink ABM，Versteeg GF，Azzam MOJ，Azzam K（2005）Precipitation of CuS and ZnS in a bubble column reactor. AIChE J 51（1）：235-246

Amaro AM，Chamorro D，Seeger M，Arredondo R，Peirano I，Jerez CA（1991）Effect of external pH perturbations on in vivo protein synthesis by the acidophilic bacterium*Thiobacillus ferro-oxidans*. J Bacteriol 173（2）：910-915

Bak F，Pfennig N（1987）Chemolithotrophic growth of *Desulfovibrio sulfodismutans* sp. nov. by disproportionation of inorganic sulfur compounds. Arch Microbiol 147（2）：184-189

Baker-Austin C，Dopson M（2007）Life in acid：pH homeostasis in acidophiles. Trends Microbiol 15（4）：165-171

Beijerink WM（1895）Über Spirillum desulfuricans als Ursache von Sulfatreduction. Centralb Bakteriol II 1：49-59

Belkin S，Wirsen CO，Jannasch HW（1985）Biological and abiological sulfur reduction at high temperatures. Appl Environ Microbiol 49：1057-1061

Benjamin MM，Datta AR（1995）Acid tolerance of enterohemorrhagic*Escherichia coli*. Appl Environ Microbiol 61（4）：1669-1672

Biebl H，Pfennig N（1977）Growth of sulfate-reducing bacteria with sulfur as electron acceptor. Arch Microbiol 112（1）：115-117

Birrien JL，Zeng X，Jebbar M，Cambon-Bonavita MA，Quérellou J，Oger P，Bienvenu N，Xiao X，Prieur D（2011）*Pyrococcus yayanosii* sp. nov.，an obligate piezophilic hyperthermophilic archaeon isolated from a deep-sea hydrothermal vent. Int J Syst Evol Microbiol 61（12）：2827-2881

Blumentals II，Itoh M，Olson GJ，Kelly RM（1990）Role of polysulfides in reduction of elemental sulfur by the hyperthermophilic archaebacterium*Pyrococcus furiosus*. Appl Environ Microbiol 56（5）：1255-1262

Bokranz M，Gutmann M，Körtner C，Kojro E，Fahrenholz F，Lauterbach F，Kröger A（1991）Cloning and nucleotide sequence of the structural genes encoding the formate dehydrogenase of*Wolinella succinogenes*. Arch Microbiol 156（2）：119-128

Bonch-Osmolovskaya EA，Miroshnichenko ML，Kostrikina NA，Chernych NA，Zavarzin GA（1990a）*Thermoproteus uzoniensis* sp. nov.，a new extremely thermophilic archaebacterium from Kamchatka continental hot springs. J Bacteriol 154（6）：556-559

Bonch-Osmolovskaya EA，Sokolova TG，Kostrikina NA，Zavarzin GA（1990b）*Desulfurella ace-tivorans* gen. nov. and sp. nov.-a new thermophilic sulfur-reducing eubacterium. Arch Microbiol 153：151-155

Bonch-Osmolovskaya EA，Miroshnichenko ML，Chernykh NA，Kostrikina NA，Pikuta EV，Rainey FA（1997）Reduction of elemental sulfur by moderately thermophilic organotrophic bacteria and the description of thermoanaerobacter*Sulfurophilus* sp. nov. Microbiol 66（5）：581-587

Boulegue J（1978）Solubility of elemental sulfur in water at 298 K. Phosphorus Sulfur Silicon Relat Elem 5（1）：127-128

Boyd ES，Druschel GK（2013）Involvement of intermediate sulfur species in biological reduction of elemental sulfur under acidic，hydrothermal conditions. Appl Environ Microbiol 79（6）：2061-2068

Boyd ES，Jackson RA，Encarnacion G，Zahn JA，Beard T，Leavitt WD，Pi Y，Zhang CL，Pearson A，Geesey GG（2007）Isolation，characterization，and ecology of sulfur-respiring crenarchaea inhabiting acid-sulfate-chloride-containing geothermal springs in Yellowstone National Park. Appl Environ Microbiol 73（20）：6669-6677

Breher F（2004）Elemental sulfur and sulfur-rich compounds，Angewandte Chemie. International ed.；WILEY-VCH Verlag，43（28）：3629a-3630

Brunner E，Woll W（1980）Solubility of sulfur in hydrogen sulfide and sour gases C1-SPE. Soc Petrol Eng J 8 doi：10. 2118/8778-pa

Bryant FO，Adams MW（1989）Characterization of hydrogenase from the hyperthermophilic archaebacterium，*Pyrococcus furiosus*. J Biol Chem 264（9）：5070-5079

Burggraf S，Jannasch HW，Nicolaus B，Stetter KO（1990）*Archaeoglobus profundus* sp. nov.，represents a new species within the sulfate-reducing archaebacteria. Syst Appl Microbiol 13（1）：24-28

Caccavo F，Lonergan DJ，Lovley DR，Davis M，Stolz JF，McInerney MJ（1994）*Geobacter sulfurreducens* sp. nov.，a hydrogen-and acetate-oxidizing dissimilatory metal-reducing microorganism. Appl Environ Microbiol 60（10）：3752-3759

Cammack R，Fauque G，Moura JJG，LeGall J（1984）ESR studies of cytochrome c3 from *Desulfovibrio desulfuricans* strain Norway 4：Midpoint potentials of the four haems，and interactions with ferredoxin and colloidal sulfur. Biochim Biophys Acta 784（1）：68-74

Canfield DE（2004）The evolution of the Earth surface sulfur reservoir. Am J Sci 304（10）：839-861

Canfield DE，Farquhar J（2012）The global sulfur cycle. Fundamentals of geobiology，John Wiley & Sons，Ltd，Chichester，UK，pp 49-64

Castro H，Reddy KR，Ogram A（2002）Composition and function of sulfate-reducing prokaryotes in eutrophic and pristine areas of the Florida everglades. Appl Environ Microbiol 68（12）：6129-6137

Childers SE，Noll KM（1994）Characterization and regulation of sulfur reductase activity in*Thermotoga neapolitana*. Appl Environ Microbiol 60（7）：2622-2626

Chong PLG（2010）Archaebacterial bipolar tetraether lipids：physico-chemical and membrane properties. Chem Phys Lipids 163（3）：253-265

Cox JC，Nicholls DG，Ingledew WJ（1979）Transmembrane electrical potential and transmem-brane pH gradient in the acidophile *Thiobacillus ferrooxidans*. Biochem J 178（1）：195-200 Crossman L，Holden M，Pain A，Parkhill J（2004）Genomes beyond compare. Nat Rev Microbiol 2（8）：616-617

De Rosa M，Gambacorta A，Nicolaus B（1983）A New type of cell membrane，in thermophilic archaebacteria，based on bipolar ether lipids. J Membr Sci 16：287-294

Dietrich W，Klimmek O（2002）The function of methyl-menaquinone-6 and polysulfide reductase membrane anchor（PsrC）in polysulfide respiration of*Wolinella succinogenes*. Eur J Biochem 269（4）：1086-1095

Dilworth MJ，Glenn AR（1999）Problems of adverse pH and bacterial strategies to combat it. Novartis Found Symp 221：4-14，discussion 14-18

Dirmeier R，Keller M，Frey G，Huber H，Stetter KO（1998）Purification and properties of an extremely thermostable membrane-bound sulfur-reducing complex from the hyperthermophilic*Pyrodictium abyssi*. Eur J Biochem 252（3）：

486-491

Dopson M, Lindstrom EB, Hallberg KB (2002) ATP generation during reduced inorganic sulfur compound oxidation by*Acidithiobacillus caldus* is exclusively due to electron transport phos-phorylation. Extremophiles 6 (2): 123-129

Ferrer M, Golyshina OV, Beloqui A, Golyshin PN, Timmis KN (2007) The cellular machinery of*Ferroplasma acidiphilum* is iron-protein-dominated. Nature 445 (7123): 91-94

Fiala G, Stetter K (1986) *Pyrococcus furiosus* sp. nov. represents a novel genus of marine hetero-trophic archaebacteria growing optimally at 100℃. Arch Microbiol 145 (1): 56-61

Finster K, Coates JD, Liesack W, Pfennig N (1997) *Desulfuromonas thiophila* sp. nov., a new obligately sulfur-reducing bacterium from anoxic freshwater sediment. Int J Syst Bacteriol 47 (3): 754-758

Finster K, Leiesack W, Thamdrup BO (1998) Elemental sulfur and thiosulfate disproportionation by *Desulfocapsa sulfoexigens* sp. nov., a new anaerobic bacterium isolated from marine surface sediment. Appl Environ Microbiol 64 (1): 119-125

Foucher S, Battaglia-Brunet F, Ignatiadis I, Morin D (2001) Treatment by sulfate-reducing bacteria of Chessy acid-mine drainage and metals recovery. Chem Eng Sci 56 (4): 1639-1645

Franz B, Lichtenberg H, Hormes J, Modrow H, Dahl C, Prange A (2007) Utilization of solid 'elemental' sulfur by the phototrophic purple sulfur bacterium *Allochromatium vinosum*: a sul-fur K-edge X-ray absorption spectroscopy study. Microbiology 153 (4): 1268-1274

Gallegos-Garcia M, Celis LB, Rangel-Mendez R, Razo-Flores E (2009) Precipitation and recovery of metal sulfides from metal containing acidic wastewater in a sulfidogenic down-flow fluidized bed reactor. Biotechnol Bioeng 102 (1): 91-99

Ghosh W, Dam B (2009) Biochemistry and molecular biology of lithotrophic sulfur oxidation by taxonomically and ecologically diverse bacteria and archaea. FEMS Microbiol Rev 33 (6): 999-1043

Golyshina OV, Timmis KN (2005) *Ferroplasma* and relatives, recently discovered cell wall-lacking archaea making a living in extremely acid, heavy metal-rich environments. Environ Microbiol 7 (9): 1277-1288

Goulbourne E, Matin M, Zychlinsky E, Matin A (1986) Mechanism of delta pH maintenance in active and inactive cells of an obligately acidophilic bacterium. J Bacteriol 166 (1): 59-65

Gramp JP, Wang H, Bigham JM, Jones FS, Tuovinen OH (2009) Biogenic synthesis and reduction of Fe (III)-hydroxysulfates. Geomicrobiol J 26 (4): 275-280

Grootscholten T, Keesman K, Lens P (2008) Modelling and on-line estimation of zinc sulfide precipitation in a continuously stirred tank reactor. Sep Purif Technol 63 (3): 654-660

Gu MX, Li Q, Zhou SY, Chen WD, Guo TM (1993) Experimental and modeling studies on the phase behavior of high H_2S-content natural gas mixtures. Fluid Phase Equilib 82: 173-182

Hagen WR, Silva PJ, Amorim MA, Hagedoorn PL, Wassink H, Haaker H, Robb FT (2000) Novel structure and redox chemistry of the prosthetic groups of the iron-sulfur flavoprotein sulfide dehydrogenase from*Pyrococcus furiosus*; evidence for a [2Fe-2S] cluster with Asp $(Cys)_3$ ligands. J Biol Inorg Chem 5 (4): 527-534

Hartshorne RS, Reardon CL, Ross D, Nuester J, Clarke TA, Gates AJ, Mills PC, Fredrickson JK, Zachara JM, Shi L, Beliaev AS, Marshall MJ, Tien M, Brantley S, Butt JN, Richardson DJ (2009) Characterization of an electron conduit between bacteria and the extracellular environment. Proc Natl Acad Sci USA 106 (52): 22169-22174

Hedderich R, Klimmek O, Kröger A, Dirmeier R, Keller M, Stetter KO (1999) Anaerobic respira-tion with elemental sulfur and with disulfides. FEMS Microbiol Rev 22: 29

Hedrich S, Johnson DB (2014) Remediation and selective recovery of metals from acidic mine waters using novel modular bioreactors. Environ Sci Technol 48 (20): 12206-12212

Heijne AT, Liu F, Weijden RVD, Weijma J, Buisman CJN, Hamelers HVM (2010) Copper recov-ery combined with electricity production in a microbial fuel cell. Environ Sci Technol 44 (11): 4376-4381

Hernandez ME, Newman DK (2001) Extracellular electron transfer. Cell Mol Life Sci 58 (11): 1562-1571

Hernandez-Eugenio G, Fardeau ML, Cayol JL, Patel BKC, Thomas P, Macarie H, Garcia JL, Ollivier B (2002) *Sporanaerobacter acetigenes* gen. nov., sp. nov., a novel acetogenic, faculta-tively sulfur-reducing bacterium. Int J Syst Evol Microbiol 52 (4): 1217-1223

Hoffert JR (1947) Acid mine drainage. Ind Eng Chem 39 (5): 642-646

Huber R, Wilharm T, Huber D, Trincone A, Burggraf S, König H, Reinhard R, Rockinger I, Fricke H, Stetter KO (1992) *Aquifex pyrophilus* gen. nov. sp. nov., represents a novel group of marine hyperthermophilic hydrogen-oxidizing bacteria. Syst Appl Microbiol 15 (3): 340-351

Huber R, Rossnagel P, Woese CR, Rachel R, Langworthy TA, Stetter KO (1996) Formation of ammonium from nitrate during chemolithoautotrophic growth of the extremely thermophilic bacterium*Ammonifex degensii* gen. nov. sp. nov. Syst Appl Microbiol 19 (1): 40-49

Huisman JL, Schouten G, Schultz CE (2006) Biologically produced sulfide for purification of process streams, effluent treatment and recovery of metals in the metal and mining industry. Hydrometallurgy 83 (1-4): 106-113

Hulshof AH, Blowes DW, Gould WD (2006) Evaluation of in situ layers for treatment of acid mine drainage: a field comparison. Water Res 40 (9): 1816-1826

Itoh T, Suzuki K, Nakase T (1998) *Thermocladium modestius* gen. nov., sp. nov., a new genus of rod-shaped, extremely thermophilic crenarchaeote. Int J Syst Bacteriol 48 (3): 879-887

Itoh T, Suzuki K, Sanchez PC, Nakase T (2003) *Caldisphaera lagunensis* gen. nov., sp. nov., a novel thermoacidophilic crenarchaeote isolated from a hot spring at Mt Maquiling, Philippines.

Int J Syst Evol Microbiol 53 (4): 1149-1154

Jameson E, Rowe OF, Hallberg KB, Johnson DB (2010) Sulfidogenesis and selective precipitation of metals at low pH mediated by *Acidithiobacillus* spp. and acidophilic sulfate-reducing bacteria. Hydrometallurgy 104 (3-4): 488-493

Jankielewicz A, Klimmek O, Kröger A (1995) The electron transfer from hydrogenase and formate dehydrogenase to polysulfide reductase in the membrane of *Wolinella succinogenes*. Biochim Biophys Acta 1231 (2): 157-162

Johnson DB (1995) Acidophilic microbial communities: candidates for bioremediation of acidic mine effluents. Int Biodeter Biodegr 35 (1-3): 41-58

Johnson DB (2003) Chemical and microbiological characteristics of mineral spoils and drainage waters at abandoned coal and metal mines. Water Air Soil Pollut 3 (1): 47-66

Johnson DB, Hallberg KB (2005) Acid mine drainage remediation options: a review. Sci Total Environ 338 (1-2): 3-14

Joshi VJ, Santani DD (2012) Physicochemical characterization and heavy meta concentration in effluent of textile industry. Univers J Environ Res Technol 2 (2): 93-96

Kaksonen AH, Riekkola-Vanhanen ML, Puhakka JA (2003) Optimization of metal sulfide precipi-tation in fluidized-bed treatment of acidic wastewater. Water Res 37 (2): 255-266

Kato S, Nakamura R, Kai F, Watanabe K, Hashimoto K (2010) Respiratory interactions of soil bacteria with (semi) conductive iron-oxide minerals. Environ Microbiol 12 (12): 3114-3123

Kleinjan WE, de Keizer A, Janssen AJH (2005) Equilibrium of the reaction between dissolved sodium sulfide and biologically produced sulfur. Colloids Surf B 43 (3-4): 228-237

Klimmek O (2005) The biological cycle of sulfur. Metal ions in biological systems, CRC Press. 43: 105-130

Klimmek O, Kröger A, Steudel R, Holdt G (1991) Growth of *Wolinella succinogenes* with polysulfide as terminal acceptor of phosphorylative electron transport. Arch Microbiol 155 (2): 177-182

Knickerbocker C, Nordstrom DK, Southam G (2000) The role of "blebbing" in overcoming the hydrophobic barrier during biooxidation of elemental sulfur by Thiobacillus thiooxidans. Chem Geol 169 (3-4): 425-433

König J, Keesman KJ, Veeken AHM, Lens PNL (2006) Dynamic modelling and process control of ZnS precipitation. Separ Sci Technol 41 (6): 1025-1042

Konings WN, Albers SV, Koning S, Driessen AJM (2002) The cell membrane plays a crucial role in survival of bacteria and archaea in extreme environments. A Van Leeuw 81: 61-72

Krafft T, Gross R, Kröger A (1995) The function of*Wolinella succinogenes* psr genes in electron transport with polysulfide as the terminal electron acceptor. Eur J Biochem 230 (2): 601-606 Kreis-Kleinschmidt V, Fahrenholz F, Kojro E, Kröger A (1995) Periplasmic sulfide dehydrogenase (sud) from *Wolinella succinogenes*: Isolation, nucleotide sequence of the sud gene and its expression in *Escherichia coli*. Eur J Biochem 227 (1-2): 137-142

Krulwich TA, Agus R, Schneier M, Guffanti AA (1985) Buffering capacity of bacilli that grow at different pH ranges. J Bacteriol 162 (2): 768-772

Labrenz M, Druschel GK, Thomsen-Ebert T, Gilbert B, Welch SA, Kemner KM, Logan GA, Summons RE, De Stasio G, Bond PL, Lai B, Kelly SD, Banfield JF (2000) Formation of sphalerite (ZnS) deposits in natural biofilms of sulfate-reducing bacteria. Science 290 (5497): 1744-1747

Laska S, Lottspeich F, Kletzin A (2003) Membrane-bound hydrogenase and sulfur reductase of the hyperthermophilic and acidophilic archaeon*Acidianus ambivalens*. Microbiology Pt 149 (9): 2357-2371

Leone L, Ferri D, Manfredi C, Persson P, Shchukarev A, Sjöberg S, Loring J (2007) Modeling the acid-base properties of bacterial surfaces: a combined spectroscopic and potentiometric study of the gram-positive bacterium*Bacillus subtilis*. Environ Sci Technol 41 (18): 6465-6471

Lewis AE (2010) Review of metal sulfide precipitation. Hydrometallurgy 104 (2): 222-234

L'Haridon S, Cilia V, Messner P, Raguénès G, Gambacorta A, Sleytr UB, Prieur D, Jeanthon C (1998) *Desulfurobacterium thermolithotrophum* gen. nov., sp. nov., a novel autotrophic, sulfur-reducing bacterium isolated from a deep-sea hydrothermal vent. Int J Syst Bacteriol 48 (3): 701-711

Liamleam W, Annachhatre AP (2007) Electron donors for biological sulfate reduction. Biotechnol Adv 25 (5): 452-463

Lopez-Archilla AI, Marin I, Amils R (2001) Microbial community composition and ecology of an acidic aquatic environment: The Tinto River, Spain. Microbiol Ecol 41 (1): 20-35

Lovley DR (1991) Dissimilatory Fe (III) and Mn (IV) reduction. Microbiol Rev 55 (2): 259-287

Lovley DR, Phillips EJP, Lonergan DJ, Widman PK (1995) Fe (III) and S0 reduction by *Pelobacter carbinolicus*. Appl Environ Microbiol 61 (6): 2132-2138

Lovley DR, Fraga JL, Blunt-Harris EL, Hayes LA, Phillips EJP, Coates JD (1998) Humic substances as a mediator for microbially catalyzed metal reduction. Acta Hydrochim Hydrobiol 26 (3): 152-157

Lovley DR, Holmes DE, Nevin KP (2004) Dissimilatory Fe (III) and Mn (IV) reduction. Adv Microb Physiol 49: 219-286

Ma K, Adams MW (1994) Sulfide dehydrogenase from the hyperthermophilic archaeon *Pyrococcus furiosus*: a new multifunctional enzyme involved in the reduction of elemental sulfur. J Bacteriol 176 (21): 6509-6517

Ma K, Schicho RN, Kelly RM, Adams MW (1993) Hydrogenase of the hyperthermophile *Pyrococcus furiosus*is an elemental sulfur reductase or sulfhydrogenase: evidence for a sulfur-reducing hydrogenase ancestor. Proc Natl Acad Sci USA 90 (11): 5341-5344

Ma K, Weiss R, Adams MW (2000) Characterization of hydrogenase II from the hyperthermophilic archaeon *Pyrococcus furiosus* and assessment of its role in sulfur reduction. J Bacteriol 182 (7): 1864-1871

Macalady J, Banfield JF (2003) Molecular geomicrobiology: genes and geochemical cycling. Earth Planet Sci Lett 209 (1-2): 1-17

Macy JM, Schröder IE, Thauer RK, Kröger A (1986) Growth the *Wolinella succinogenes* on H_2S plus fumarate and on formate plus sulfur as energy sources. Arch Microbiol 144 (2): 147-150 Miroshnichenko ML, Rainey FA, Hippe H, Chernyh NA, Kostrikina NA, Bonch-Osmolovskaya EA (1998) *Desulfurella karnchatkensis* sp. nov. and *Desulfurella propionica* sp. nov. , new sulfur respiring thermophilic bacteria from Kamchatka thermal environments. Int J Syst Bacteriol 48: 475-479

Möbius CH, Demel I, Schmid A, Temper U, Carozzi A, Buisman C, Weijma J, Rijnaarts HHM, Hendrickx TLG, Brockmann M (2015) Anwendung für organisch und anorganisch belastete Abwässer anderer Industriebereiche. In: Rosenwinkel K-H, Kroiss H, Dichtl N, Seyfried C-F, Weiland P (eds) Anaerobtechnik. Springer, Berlin, pp 541-601

Mori K, Kim H, Kakegawa T, Hanada S (2003) A novel lineage of sulfate-reducing microor-ganisms: *Thermodesulfobiaceae* fam. nov. , *Thermodesulfobium narugense*, gen. nov. , sp. nov. , a new thermophilic isolate from a hot spring. Extremophiles 7 (4): 283-290

Myers CR, Nealson KH (1988) Bacterial manganese reduction and growth with manganese oxide as the sole electron acceptor. Science 240 (4857): 1319-1321

Năncucheo I, Johnson DB (2012) Selective removal of transition metals from acidic mine waters by novel consortia of acidophilic sulfidogenic bacteria. Microbial Biotech 5 (1): 34-44

Neculita CM, Zagury GJ, Bussière B (2007) Passive treatment of acid mine drainage in bioreactors using sulfate-reducing bacteria. J Environ Qual 36 (1): 1-16

Newman DK, Kolter R (2000) A role for excreted quinones in extracellular electron transfer. Nature 405 (6782): 94-97

Ng KY, Sawada R, Inoue S, Kamimura K, Sugio T (2000) Purification and some properties of sulfur reductase from the iron-oxidizing bacterium*Thiobacillus ferrooxidans* NASF-1. J Biosci Bioeng 90 (2): 199-203

Ohmura N, Sasaki K, Matsumoto N, Saiki H (2002) Anaerobic respiration using Fe^{3+}, S^0, and H_2 in the chemolithoautotrophic bacterium *Acidithiobacillus ferrooxidans*. J Bacteriol 184 (8): 2081-2087

Osorio H, Mangold S, Denis Y, Ñancucheo I, Esparza M, Johnson DB, Bonnefoy V, Dopson M, Holmes DS (2013) Anaerobic sulfur metabolism coupled to dissimilatory iron reduction in the extremophile *Acidithiobacillus ferrooxidans*. Appl Environ Microbiol 79 (7): 2172-2181

Pelsh AO (1936) Hydrobiology of Karabugaz bay of the Caspian sea. Trudy Solyan Laboratory Akademie Nauk 5 (49): 31

Pfennig N (1975) The phototrophic bacteria and their role in the sulfur cycle. Plant Soil 43 (1-3): 1-16

Pfennig N, Biebl H (1976) *Desulfuromonas acetoxidans* gen. nov. and sp. nov. , a new anaerobic, sulfur-reducing, acetate-oxidizing bacterium. Arch Microbiol 110 (1): 3-12

Pihl TD, Maier RJ (1991) Purification and characterization of the hydrogen uptake hydrogenase from the hyperthermophilic archaebacterium *Pyrodictium brockii*. J Bacteriol 173 (6): 1839-1844

Pihl TD, Schicho RN, Kelly RM, Maier RJ (1989) Characterization of hydrogen-uptake activity in the hyperthermophile *Pyrodictium brockii*. Proc Natl Acad Sci USA 86 (1): 138-141

Pihl TD, Black LK, Schulman BA, Maier RJ (1992) Hydrogen-oxidizing electron transport com-ponents in the hyperthermophilic archaebacterium *Pyrodictium brockii*. J Bacteriol 174 (1): 137-143

Postec A, Breton CL, Fardeau ML, Lesongeur F, Pignet P, Querellou J, Ollivier B, Godfroy A (2005) *Marinitoga hydrogenitolerans* sp. nov. , a novel member of the order Thermotogales isolated from a black smoker chimney on the Mid-Atlantic Ridge. Int J Syst Evol Microbiol 55 (3): 1217-1221

Prokofeva MI, Miroshnichenko ML, Kostrikina NA, Chernyh NA, Kuznetsov BB, Tourova TP, Bonch-Osmolovskaya EA (2000) *Acidilobus aceticus* gen. nov. , sp. nov. , a novel anaerobic thermoacidophilic archaeon from continental hot vents in Kamchatka. Int J Syst Evol Microbiol 50 (6): 2001-2008

Rabus R, Hansen TA, Widdel F (2006) Dissimilatory sulfate-and sulfur-reducing prokaryotes. In: Dworkin M, Falkow S, Rosenberg E, Schleifer K-H, Stackebrandt E (eds) The prokaryotes, vol 2. Springer, New York, pp 659-768

Rainey FA, Hollen B (2005) *Desulfurella* Bonch-Osmolovskaya, Sokolova, Kostrikina, and Zavarzin 1993, 624 VP emend. Miroshnichenko, Rainey, Hippe, Chernyh, Kostrikina and Bonch-Osmolovskaya 1998, 478 (Effective publication: Bonch-Osmolovskaya, Sokolova, Kostrikina, and Zavarzin 1990, 155). Bergey's Manual® of Systematic Bacteriology, Springer 923-924

Reguera G, McCarthy KD, Mehta T, Nicoll JS, Tuominen MT, Lovley DR (2005) Extracellular electron transfer via microbial nanowires. Nature 435 (7045): 1098-1101

Richardson DJ (2000) Bacterial respiration: a flexible process for a changing environment. Microbiol 146 (3): 551-571

Richter K, Schicklberger M, Gescher J (2012) Dissimilatory reduction of extracellular electron acceptors in anaerobic respiration. Appl Environ Microbiol 78 (4): 913-921

Ringel M, Gross R, Krafft T, Kröger A, Schauder R (1996) Growth of *Wolinella succinogenes* with elemental sulfur in the absence of polysulfide. Arch Microbiol 165 (1): 62-64

Roof JG (1971) Solubility of sulfur in hydrogen sulfide and in carbon disulfide at elevated temperature and pressure C1-SPE. Soc Petrol Eng J 11 (3): 5

Rosenberg E, DeLong E, Lory S, Stackebrandt E, Thompson F, Rabus R, Hansen TA, Widdel F (2013) Dissimilatory sulfate-and sulfur-reducing prokaryotes. In: The prokaryotes. Springer, Berlin, pp 309-404

Sallam A, Steinbüchel A (2009) *Clostridium sulfidigenes* sp. nov., a mesophilic, proteolytic, thio-sulfate-and sulfur-reducing bacterium isolated from pond sediment. Int J Syst Evol Microbiol 59 (7): 1661-1665

Sampaio RMM, Timmers RA, Xu Y, Keesman KJ, Lens PNL (2009) Selective precipitation of Cu from Zn in a pS controlled continuously stirred tank reactor. J Hazard Mater 165 (1): 256-265

Sánchez-Andrea I, Triana D, Sanz JL (2012) Bioremediation of acid mine drainage coupled with domestic wastewater treatment. Water Sci Technol 66 (11): 2425-2431

Sánchez-Andrea I, Sanz JL, Bijmans MFM, Stams AJM (2014) Sulfate reduction at low pH to remediate acid mine drainage. J Hazard Mater 269: 98-109

Sánchez-Andrea I, Stams AJM, Hedrich S, Nancucheo I, Johnson DB (2015) *Desulfosporosinus acididurans* sp. nov.: an acidophilic sulfate-reducing bacterium isolated from acidic sediments. Extremophiles 19 (1): 39-47

Sapra R, Bagramyan K, Adams MWW (2003) A simple energy-conserving system: proton reduc-tion coupled to proton translocation. Proc Natl Acad Sci USA 100 (13): 7545-7550

Schauder R, Müller E (1993) Polysulfide as a possible substrate for sulfur-reducing bacteria. Arch Microbiol 160 (5): 377-382

Schippers A, Sand W (1999) Bacterial leaching of metal sulfides proceeds by two indirect mecha-nisms via thiosulfate or via polysulfides and sulfur. Appl Environ Microbiol 65 (1): 319-321 Schmitz RA, Bonch-Osmolovskaya EA, Thauer RK (1990) Different mechanisms of acetate acti-vation in *Desulfurella acetivorans* and *Desulfuromonas acetoxidans*. Arch Microbiol 154 (3): 274-279

Schröder I, Kröger A, Macy JM (1988) Isolation of the sulfur reductase and reconstitution of the sulfur respiration of *Wolinella succinogenes*. Arch Microbiol 149 (6): 572-579

Scott DT, McKnight DM, Blunt-Harris EL, Kolesar SE, Lovley DR (1998) Quinone moieties act as electron acceptors in the reduction of humic substances by humics-reducing microorgan-isms. Environ Sci Technol 32 (19): 2984-2989

Segerer A, Neuner A, Kristjansson JK, Stetter KO (1986) *Acidianus infernus* gen. nov., sp. nov., and *Acidianus brierleyi* comb. nov.: Facultatively aerobic, extremely acidophilic thermophilic sulfur-metabolizing archaebacteria. Int J Syst Bacteriol 36 (4): 559-564

Segerer A, Langworthy TA, Stetter KO (1988) *Thermoplasma acidophilum* and *Thermoplasma volcanium* sp. nov. from Solfatara Fields. Syst Appl Microbiol 10 (2): 161-171

Segerer AH, Trincone A, Gahrtz M, Stetter KO (1991) *Stygiolobus azoricus* gen. nov., sp. nov. represents a novel genus of anaerobic, extremely thermoacidophilic archaebacteria of the order *Sulfolobales*. Int J Syst Bacteriol 41 (4): 495-501

Shimada H, Nemoto N, Shida Y, Oshima T, Yamagishi A (2008) Effects of pH and temperature on the composition of polar lipids in *Thermoplasma acidophilum* HO-62. J Bacteriol 190 (15): 5404-5411

Sierra-Alvarez R, Karri S, Freeman S, Field JA (2006) Biological treatment of heavy metals in acid mine drainage using sulfate reducing bioreactors. Water Sci Technol 54 (2): 179-185

Slonczewski JL, Macnab RM, Alger JR, Castle AM (1982) Effects of pH and repellent tactic stimuli on protein methylation levels in *Escherichia coli*. J Bacteriol 152 (1): 384-399

Sorokin DY, Tourova TP, Kolganova TV, Detkova EN, Galinski EA, Muyzer G (2011) Culturable diversity of lithotrophic haloalkaliphilic sulfate-reducing bacteria in soda lakes and the descrip-tion of *Desulfonatronum thioautotrophicum* sp. nov., *Desulfonatronum thiosulfatophilum* sp. nov., *Desulfonatronovibrio thiodismutans* sp. nov., and *Desulfonatronovibrio magnus* sp. nov. Extremophiles 15 (3): 391-401

Steed VS, Suidan MT, Gupta M, Miyahara T, Acheson CM, Sayles GD (2000) Development of a sulfate-reducing biological process to remove heavy metals from acid mine drainage. Water Environ Res 72 (5): 530-535

Stetter KO (1996) Hyperthermophilic prokaryotes. FEMS Microbiol Rev 18: 149-158

Stetter KO, Huber R, Blochl E, Kurr M, Eden RD, Fielder M, Cash H, Vance I (1993) Hyperthermophilic archaea are thriving in deep North Sea and Alaskan oil reservoirs. Nature 365 (6448): 743-745

Steudel R (2000) The sulfur cycle. In: Lens PNL, Hulshoff Pol L (eds) Environmental technologies to treat sulfur pollution. IWA Publishing, London, pp 1-31

Steudel R, Eckert B (2003) Solid sulfur allotropes sulfur allotropes. Elemental sulfur and sulfur-rich compounds I. Springer, Berlin Heidelberg. 230: 1-80

Steudel R, Göbel T, Holdt G (1989) The molecular nature of the hydrophilic sulfur prepared from aqueous sulfide and sulfite (selmi sulfur sol). Z Naturforsch B 44 (5): 526-530

Steudel R, Kleinjan WE, de Keizer A, Janssen AJH (2003) Biologically produced sulfur. Elemental sulfur and sulfur-rich compounds I. Springer, Berlin Heidelberg, 230: 167-188

Suzuki I, Lee D, Mackay B, Harahuc L, Oh JK (1999) Effect of various ions, pH, and osmotic pressure on oxidation of elemental sulfur by *Thiobacillus thiooxidans*. Appl Environ Microbiol 65 (11): 5163-5168

Svetlichnyi VA, Slesarev AI, Svetlichnaya TP, Zavarzin GA (1987) *Caldococcus litoralis*, gen. nov. sp. nov. -a new marine, extremely thermophilic, sulfur-reducing archaebacterium. Mikrobiologiya 56: 831-838

Tabak H, Scharp R, Burckle J, Kawahara F, Govind R (2003) Advances in biotreatment of acid mine drainage and biorecovery of metals: 1. Metal precipitation for recovery and recycle. Biodegradation 14 (6): 423-436

Takahashi Y, Suto K, Inoue C (2010) Polysulfide reduction by *Clostridium* relatives isolated from sulfate-reducing enrichment cultures. J Biosci Bioeng 109 (4): 372-380

Tang K, Baskaran V, Nemati M (2009) Bacteria of the sulfur cycle: An overview of microbiology, biokinetics and their role in petroleum and mining industries. Biochem Eng J 44 (1): 73-94

Tekerlekopoulou AG, Tsiamis G, Dermou E, Siozios S, Bourtzis K, Vayenas DV (2010) The effect of carbon source on microbial community structure and Cr (VI) reduction rate. Biotechnol Bioeng 107 (3): 478-487

Thamdrup B, Finster K, Hansen JW, Bak F (1993) Bacterial disproportionation of elemental sulfur coupled to chemical reduction of iron or manganese. Appl Environ Microbiol 59 (1): 101-108

Tsukamoto TK, Killion HA, Miller GC (2004) Column experiments for microbiological treatment of acid mine drainage: low-temperature, low-pH and matrix investigations. Water Res 38 (6): 1405-1418

Tyson GW, Chapman J, Hugenholtz P, Allen EE, Ram RJ, Richardson PM, Solovyev VV, Rubin EM, Rokhsar DS, Banfield JF (2004) Community structure and metabolism through reconstruction of microbial genomes from the environment. Nature 428 (6978): 37-43

Veeken AHM, Rulkens WH (2003) Innovative developments in the selective removal and reuse of heavy metals from wastewaters. Water Sci Technol 47 (10): 9-16

Vera M, Schippers A, Sand W (2013) Progress in bioleaching: fundamentals and mechanisms of bacterial metal sulfide oxidation-part A. Appl Microbiol Biotechnol 97 (17): 7529-7541

Vignais PM, Billoud B, Meyer J (2001) Classification and phylogeny of hydrogenases I. FEMS Microbiol Rev 25 (4): 455-501

Wang F, Tessier A, Buffle J (1998) Voltammetric determination of elemental sulfur in pore waters. Limnol Oceanogr 43 (6): 1353-1361

Wang X, Lv B, Cai G, Fu L, Wu Y, Wang X, Ren B, Ma H (2012) A proton shelter inspired by the sugar coating of acidophilic archaea. Sci Rep 2 (892): 1-5

Weijma J, Copini CFM, Buisman CJN, Schultz CE (2002) Biological recovery of metals, sulfur and water in the mining and metallurgical industry. In: Lens PNL, Hulshoff Pol LW, Wilderer P, Asano T (eds) Water recycling and recovery in Industry. IWA Publishing, London, pp 605-622

Widdel F (1988) Microbiology and ecology of sulfate-and sulfur-reducing bacteria. Willey-Interscience Publication, New York Widdel F, Pfennig N (1981) Studies on dissimilatory sulfate-reducing bacteria that decompose fatty acids. I. Isolation of new sulfate-reducing bacteria enriched with acetate from saline envi-ronments. Description of *Desulfobacter postgatei* gen. nov., sp. nov. Arch Microbiol 129 (5): 395-400

Widdel F, Pfennig N (1991). The genus *Desulfuromonas* and other Gram-negative sulfur-reducing eubacteria. In: Balows A, Trüper HG, Dworkin M, Harder W, Schleifer KH (eds) The prokary-otes. Springer, Berlin Yarza P, Richter M, Peplies J, Euzeby J, Amann R, Schleifer KH, Ludwig W, Glockner FO, Rossello-Mora R (2008) The all-species living tree project: a 16S rRNA-based phylogenetic tree of all sequenced type strains. Syst Appl Microbiol 31 (4): 241-250

Yoneda Y, Yoshida T, Kawaichi S, Daifuku T, Takabe K, Sako Y (2012) *Carboxydothermus pertinax* sp. nov., a thermophilic, hydrogenogenic, Fe (III)-reducing, sulfur-reducing carboxydotro-phic bacterium from an acidic hot spring. Int J Syst Evol Microbiol 62 (7): 1692-1697

Zavarzina DG, Zhilina TN, Tourova TP, Kuznetsov BB, Kostrikina NA, Bonch-Osmolovskaya EA (2000) *Thermanaerovibrio velox* sp. nov., a new anaerobic, thermophilic, organotrophic bacte-rium that reduces elemental sulfur, and emended description of the genus *Thermanaerovibrio*. Int J Syst Evol Microbiol 50 (3): 1287-1295

Zöphel A, Kennedy MC, Beinert H, Kroneck PMH (1991) Investigations on microbial sulfur res-piration. Eur J Biochem 195 (3): 849-856

Zychlinsky E, Matin A (1983) Cytoplasmic pH homeostasis in an acidophilic bacterium, *Thiobacillus acidophilus*. J Bacteriol 156 (3): 1352-1355

第六章

嗜酸性金属/硫氧化微生物的生物膜生活方式

Ruiyong Zhang[1,2], S ren Bellenberg[1], Thomas R. Neu[3], Wolfgang Sand[1], Mario Vera[4]

6.1 引言

形成生物膜是地球上的微生物最普遍和广泛存在的生活方式。不同类群的微生物聚集在细胞胞外聚合物（EPS）的基质中。EPS主要由多糖、蛋白质、核酸和亲脂性化合物组成。生物膜可以看作是空气-水界面的表面相关膜或“浮动垫”。生物膜能帮助细胞抵抗恶劣的环境压力，如干燥、营养匮乏、辐射和/或氧化应激（Flemming and Wingender 2010）。

不同种类的嗜酸性金属/硫氧化微生物（AMOM）可以加快诸如黄铁矿（FeS_2）、黄铜矿（$CuFeS_2$）等金属硫化物（MS）的生物浸出。这些微生物可以通过氧化亚铁离子和/或还原无机硫化合物（$RISC_S$）影响金属硫化物的溶解程度，从而得到三价铁离子、质子和金属硫化物攻击剂。但是这样就会产生酸性矿井/岩石排水（AMD/ARD），重金属和硫酸造成的水或土壤污染便成了一个严重的问题。嗜酸硫杆菌属（*Acidithiobacillus*）、钩端螺菌属（*Leptospirillum*）、酸性杆菌属（*Acidiferrobacter*）、嗜酸菌属（*Acidiphilium*）、铁原体属（*Ferroplasma*）、酸微菌属（*Acidimicrobium*）、铁微菌属（*Ferrimicrobium*）、硫化杆菌属（*Sulfobacillus*）、金属球菌属（*Metallosphaera*）、硫化叶菌（*Sulfolobus*）和酸菌属（*Acidianus*）这几种嗜酸性微生物类群常见于酸性矿井排水，以及嗜中温和嗜高温条件下的生物矿化操作中。在“接触-浸出”模式中，微生物繁殖并附着在金属硫化物表面，导致硫化物矿物溶解的电化学过程在细菌细胞和金属硫化物的交界面发生。这个空间充满着胞外聚

[1] Aquatische Biotechnologie，Biofi lm Centre，Universität Duisburg-Essen，Universitätsstraße 5，45141 Essen，Germany.

[2] e-mail：ruiyong. zhang@uni-due. de.

[3] Department of River Ecology，Helmholtz Centre for Environmental Research - UFZ，Brückstrasse 3a，39114 Magdeburg，Germany.

[4] Institute for Biological and Medical Engineering，Schools of Engineering，Biological Sciences and Medicine，Pontifi cia Universidad Católica de Chile，Vicuña Mackenna 4860，7820436 Macul，Santiago，Chile；Department of Hydraulic and Environmental Engineering，School of Engineering，Pontificia Universidad Católica de Chile，Vicuña Mackenna 4860，7820436 Macul，Santiago，Chile. e-mail：mariovera@uc. cl.

合物，嗜酸氧化亚铁硫杆菌（*Acidithiobacillus ferrooxidans*）的 EPS 主要由多糖、脂质和糖醛酸组成（Gehrke et al. 1998）。

目前对嗜酸性金属/硫氧化微生物介导的金属硫化物的生物膜形成过程并不完全了解。在实验室条件下，大多数嗜酸性金属/硫氧化微生物能够在金属硫化物上形成单层生物膜，金属硫化物也作为它们的能量底物。由于受到酸性 pH、高浓度的重金属和有限的电子供体和受体的限制，环境中的嗜酸性生物膜物种丰度通常较低。一般来说，这些生物膜主要由铁氧化的化能自养细菌，如钩端螺菌属Ⅱ类和由异养或混合营养型细菌、古菌和真核生物的混合群体构成（Denef et al. 2010）。

嗜酸性金属/硫氧化细菌在不同的水平上调控生物膜形成和 EPS 生物合成，包括能量底物、无机磷酸（Pi）的限制和细胞-细胞间“群体感应”（QS）的交流机制。本章概述了嗜酸性金属/硫氧化微生物的生物膜生存方式，包括表面科学、显微技术、细胞间交流、种间和分子间相互作用以及高通量研究。这一领域的未来发展前景包括了阐明 EPS 生物合成途径和 EPS 聚合物化学性质的系统研究。多种嗜酸菌生物膜内的细胞间交流和微生物相互作用被认为嗜酸性金属/硫氧化微生物代谢活性的关键决定因素。对嗜酸性金属/硫氧化微生物生物膜方面的全面解析将有利于其在生物技术领域的应用，例如生物矿化（强化生物膜的形成和浸出速率），并将用于缓解酸性矿井排水的生成（抑制生物膜形成或生物膜细胞的失活在其中可能发挥重要的作用）。

6.2　嗜酸性金属/硫氧化微生物生物膜

溶解金属硫化物的微生物主要包括嗜酸性细菌和古菌（即在 pH 值低于 3 时生长的微生物）。这些微生物能够氧化还原无机硫化合物和/或亚铁离子。浸出细菌分布于变形菌门［酸硫杆菌属（*Acidithiobacillus*）、嗜酸菌属（*Acidiphilium*）、酸性杆菌属（*Acidiferrobacter*）、*Ferrovum*］，硝化螺旋菌门［钩端螺旋菌属（*Leptospirillum*）］、厚壁菌门［脂环酸芽孢杆菌（*Alicyclobacillus*）、硫化杆菌（*Sulfobacillus*）］和放线菌门［铁微菌属（*Ferrimicrobium*）、酸微菌属（*Acidimicrobium*）、高铁丝菌属（*Ferrithrix*）］。最近几个综述概述了采矿生物群落中微生物多样性（Schippers 2007；Hedrich et al. 2011；Dopson and Johnson 2012；Johnson 2014）。在所有细菌群中，已描述了嗜温和中度嗜热微生物（Clark and Norris 1996；Norris and Johnson 1998；Norris et al. 2000）。大多数浸出古菌属于嗜热硫化叶菌目，它可以氧化硫和/或亚铁，例如硫化叶菌（*Sulfolobus*）、酸菌属（*Acidianus*）、金属球菌属（*Metallosphaera*）和硫球形菌属（*Sulfurisphaera*）（Norris et al. 2000；Wheaton et al. 2015）。另外，在热原体嗜中温亚铁氧化物种中，已知的有嗜酸铁原体菌（*Ferroplasma acidiphilum*）（Golyshina et al. 2000）、酸阿尔曼铁原体（*Ferroplasma acidarmanus*）（Edwards et al. 2000b；Dopson et al. 2004）和嗜热铁原体（*Ferroplasma thermophilum*）（Zhou et al. 2008）。

6.2.1　金属硫化物溶解机制

基于金属硫化物的酸溶解有两种不同的生物浸出途径。硫代硫酸盐途径，参与酸不溶性金属硫化物的溶解，例如黄铁矿、辉钼矿（MoS_2）和钨矿（WS_2）。多硫化物途径，与酸溶性金属硫化物的溶解相关，如闪锌矿（ZnS）、方铅矿（PbS）或黄铜矿（Schippers et al. 1996；Schippers and Sand 1999；Sand et al. 2001）。这两种途径是间接作用机制，因为在酸性溶液中，酸不溶性金属硫化物仅能被铁离子进行氧化，而铁离子是通过微生物氧化亚铁离子所提供的（Schippers et al. 1996；Rodriguez et al. 2003；Gleisner et al. 2006）。酸溶性金属硫化物也可通过质子侵蚀而溶解。目前研究人员已提出了两种浸出方式："接触"和"非接触"浸出（Sand et al. 2001；Rawlings 2002）。非接触浸出是通过浮游细胞进行的，主要是对亚铁离子进行氧化。所得的铁离子在矿物表面被还原，并且部分硫化物被氧化，这样，亚铁离子再次进入循环。从严格意义上讲，这与"间接机制"是相对应的（Sand et al. 1995）。相比之下，接触浸出方式考虑了金属硫化物表面形成生物膜的细胞。在这种情况下，导致金属硫化物溶解的电化学过程发生在细菌细胞和金属硫化物表面交界处的 EPS 中。

人们已经检测到黄铁矿中阳极和阴极反应。首先，电子从黄铁矿的表面转移到水中的氧化剂上，通常是 O_2 或铁离子（阴极反应）。其次，电荷从阳极反应位点转移以替换从阴极位点损失的电子。随后，在阳极位点上，水分子中的氧原子与硫原子相互作用，产生硫氧化物。该步骤将一个电子释放到固体中，并将一个或两个氢离子释放到溶液中（Rimstidt and Vaughan 2003）。细胞很可能会被趋化吸引到这些阳极区域上。阳极和阴极硫铁比例的不平衡可能会导致晶体晶格出现缺陷，在结晶过程中包含其他金属原子和/或结晶过程中温度的变化都会导致铁与硫的比例不平衡（导致非晶态到高度结晶的结构）。最近，Crundwell（2013）提出了一个新的假说，矿物溶解的新假说是基于电解质的化学机制，这个化学机制更新了矿物溶解动力学的内容。它提出：（1）不将表面分离成阳极位点和阴极位点；（2）在矿物的整体上没有电子流动；（3）第一步溶解反应不是通过酸；（4）矿物溶液界面上的电荷交换是限速步骤；（5）细菌浸出不改变反应机理。这意味着生物效应仅局限在那些不影响限速步骤的反应。

6.2.2 微生物 EPS 吸附金属硫化物

由于在细菌和金属硫化物表面之间形成了独特的微环境（图 6.1），微生物对矿石的附着和随后的生物膜形成增加了浸出活性。氧化亚铁硫杆菌中的铁离子存在于 EPS 中，每一个铁离子都可能与两个糖醛酸残基络合。络合的铁离子的第一功能是介导细胞附着，其中细胞通过静电附着到带负电的黄铁矿上。络合的铁离子的第二个功能是氧化溶解金属硫化物，类似于游离铁离子在"非接触"浸出中的作用。在微生物中，从金属硫化物中获取的电子通过氧化还原链降低分子氧，这样便形成了跨周质空间，并连接外膜和内膜的超级复合物（Castelle et al. 2008）。

一般来说，大多数嗜酸性金属/硫氧化微生物可以在金属硫化物上形成生物膜。假设不限制附着的表面，在 24h 内将会有高达 80%～90%的接种微生物附着在金属硫化物上（Dispirito et al. 1983；Bagdigian and Myerson 1986；Gehrke et al. 1998；Harneit et al. 2006）。要注意的是，初始附着率强烈依赖于物种的性质以及它们的预培养条件。一般来说，钩端螺旋体菌属比铁氧化性嗜酸杆菌属有着更高的黄铁矿附着率。在后者中，亚铁或黄铁矿培养的细胞中 EPS 进行了相当大的化学修饰，而显示出比硫-培养的细胞对黄铁矿更高的附着率（Gehrke et al. 1998）。尽管附着表面积不受限制，然而一些细胞总是保持在悬浮状态。例如，当氧化亚铁硫杆菌 *At. ferrooxidans* ATCC 23270^T 或酸性硫杆菌 *Acidithiobacillus ferrivorans* SS3 与黄铁矿孵育 24h，只有约 50%的细胞附着于矿石上（Bellenberg et al. 2015），其原因仍然未知。

在氧化亚铁硫杆菌 *At. ferrooxidans* strain R1 中证明了黄铁矿培养的细胞与亚铁培养细胞具有类似的 EPS 组成。这些 EPS 包括葡萄糖、鼠李糖、岩藻糖、木糖、甘露糖、C_{12}～C_{20} 饱和脂肪酸、葡萄糖醛酸和铁离子（Gehrke et al. 1998；Gehrke et al. 2001）。然而，黄铁矿培养的细胞具有超过十倍量的 EPS。正如已经提及的，初始附着主要由静电相互作用驱

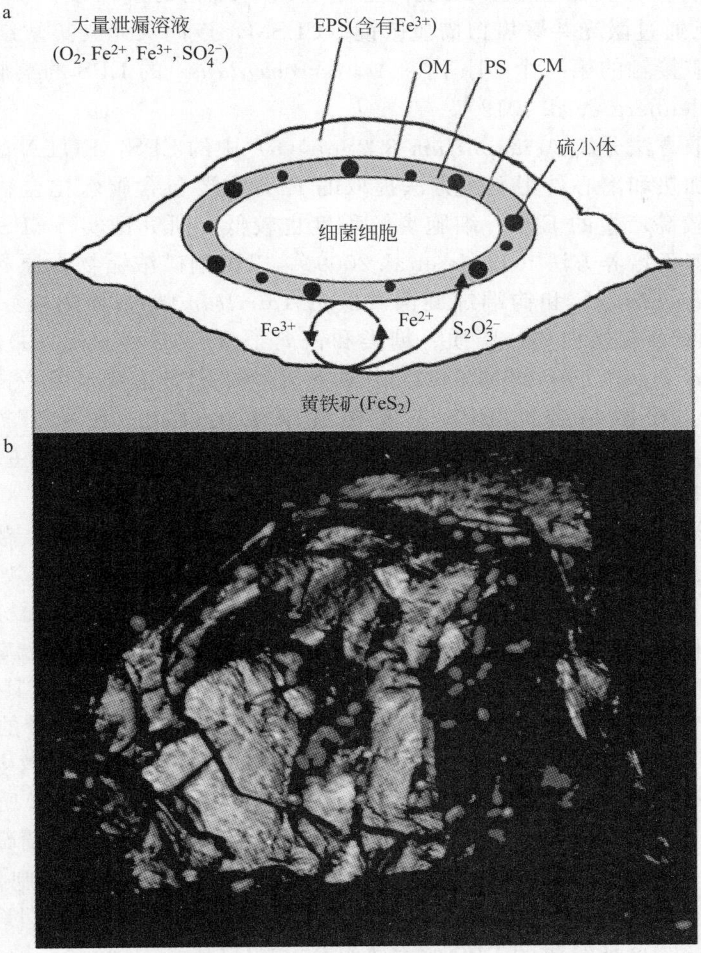

图 6.1 细菌细胞催化的接触浸出模型（来自 Vera 等，2013b，见彩图）。(a) 表示嵌入在附着于黄铁矿上的胞外聚合物（EPS）层中的生物膜单元。图中显示在金属硫化物（MS）溶解过程中存在的化合物，如铁(二价)/(三价)、硫代硫酸盐。CM 表示细胞质膜；PS 表示周质空间；OM 表示外膜。(b) CLSM 图像显示的是孵育 1 周后氧化亚铁硫杆菌细胞定植的黄铁矿颗粒（50～100μm）的 3D 投影。采用 Syto 9（绿色）对细胞中的核酸和伴刀豆球蛋白 A（Con A）进行双重染色。绿色为 Syto 9，红色为 Con A-四甲基罗丹明异硫氰酸酯（TRITC），灰色为反射。图中显示了三个通道的合并图像。细菌定植模式与表面缺陷密切相关

动（最可能是 2mol 带负电的葡萄糖醛酸残基和 1mol 带正电的铁离子作用，导致净正电荷），并在硫酸溶液中（pH＝2）带负电的黄铁矿表面发生（Solari et al. 1992；Blake et al. 1994）。疏水相互作用也促进了对金属硫化物表面的附着（Gehrke et al. 1998；Sampson et al. 2000b），这尤其适用于疏水性强的表面，例如单质硫（S^0）。疏水相互作用以及共价键似乎介导次级（紧密）表面附着。硫培养的细胞不能很好地附着到黄铁矿上，因为它们的EPS组成与用黄铁矿培养的细胞不同。它们的 EPS 含有相当少的单糖，并缺少糖醛酸，完全不存在络合的铁离子或其他带正电荷的离子。相比之下，来自硫培养细胞的 EPS 比从黄铁矿培养的细胞提取的 EPS 含有更多的脂肪酸。因此，疏水性相互作用似乎完全与氧化亚铁硫杆菌的细胞附着于单质硫相关（Gehrke et al. 1998）。FTIR 分析表明 *At. ferrooxidans* 可溶性亚铁培养细胞的表面电荷与固体底物培养的细胞明显不同。与用亚铁离子培养细胞相

比，当在不溶性底物（例如黄铁矿或硫）上培养时，细胞的蛋白质产量更高（Sharma et al. 2003）。我们已通过激光共聚焦扫描显微镜（CLSM）连同荧光凝集素结合分析（FLBA）揭示了在与黄铁矿接触的第一个 24h 内，*At. ferrooxidans*^{T} 的 EPS 的荚膜多糖（CPS）是如何形成的（Bellenberg et al. 2012）。

比较五种提取方法分离 *Acidiphilum* 3. 2Sup（5）中的 EPS：EDTA 法、NaOH 法、离子交换树脂法、加热和离心。用上述方法提取的 EPS 主要包含碳水化合物和蛋白质。采用 EDTA 法能获得较高产量的 EPS，细胞裂解程度也较低。研究证实，EPS 的产量和化学组成很大程度上依赖于培养方法（Tapia et al. 2009）。以黄铜矿浓缩物为原料，嗜酸硫化杆菌（*Acidithiobacillus caldus*）和钩端螺旋菌（*Leptospirillum ferriphilum*）的混合培养物所形成的 EPS 中主要含有蛋白质、脂质、糖类和铁离子（Zeng et al. 2010）。最近研究表明，泉古菌 *Sulfolobus metallicus* DSM 6482^T 中紧密结合的 EPS 主要是由蛋白质和碳水化合物组成。相比之下，松散结合的 EPS 主要含有碳水化合物。在硫培养金属硫化叶菌（*S. metallicus*^{T}）的生物膜上的 EPS 中检测到胞外 DNA 和蛋白质，主要成分是蛋白质（Zhang et al. 2015b）。

在搅拌反应器中进行 *L. ferriphilum* BRGM1 占主导的发酵培养，产生 EPS，以研究影响最优生物浸出过程的相关因素。在氮气限制条件下，细菌附着和浸出效率都随着 EPS 产量的降低而降低。CO_2 限制导致胞外多糖产量的显著减少（d'Hugues et al. 2008）。另一个研究是关于在若干连续操作的生物浸出系统中，对几种金属硫化物（例如黄铁矿、闪锌矿和黄铜矿）上生长的中温、中等嗜热和嗜热浸出菌株进行混合培养，分析其 EPS 的组成。结果发现提取的 EPS，70%主要是糖类，其中还有少量蛋白质，以及痕量的腐殖酸和糖醛酸（Govender and Gericke 2011）。来自加利福尼亚州铁矿山的两种嗜酸性微生物生物膜的研究表明，它们的 EPS 是由碳水化合物、金属、蛋白质、少量 DNA 和脂质组成的（Jiao et al. 2010）。另外，还报道了对生长在有机化合物上的嗜酸菌的研究。在葡萄糖中生长的两种硫化叶菌株分泌一种含有单糖，如葡萄糖、甘露糖、葡糖胺和半乳糖的胞外多糖（Nicolaus et al. 1993）。蛋白质和碳水化合物也是生长在胰蛋白胨、N-Z-胺或葡萄糖培养基中的 *Sulfolobus solfataricus* P2 所分泌的 EPS 的主要组分（Koerdt et al. 2012）。

嗜酸性金属/硫氧化微生物（AMOM）在金属硫化物上的吸附位点以及如何检测/感知这个特定位点仍然为大家所关注。文献中的一些证据（Andrews 1988；Ohmura et al. 1993；Shrihari et al. 1995；Dziurla et al. 1998；Sanhueza et al. 1999；Edwards et al. 2001）和笔者的成果表明微生物并非随机附着到金属硫化物上（Gehrke et al. 1998；Sand et al. 1998；Gehrke et al. 2001；Noël et al.，2010；Zhang et al. 2014，2015a）。通过原子力显微镜（AFM）以及激光共聚焦显微镜（CLSM）已经可以观察到氧化亚铁硫杆菌 *At. ferrooxidans* 和其他 AMOM 优先（>90%）附着的位点表面具有可见缺陷，例如孔隙、划痕等（也可见本章第 6. 4 节）。细胞似乎更容易附着到具有低结晶度的区域，从而导致细胞沿结晶轴定向延伸，并沿着该方向进行氧化（Sanhueza et al. 1999）。细胞趋向于对孔隙和划痕的黏附，可能是由于增大了接触面积和保护其免受弱剪切力。相比之下，细胞附着到具有低结晶度和结晶轴的区域通常与附着表面的形貌变化并不相关。因此，一些诱导剂的存在可以解释微生物附着到矿物表面上特定位点的问题，这很可能是由表面上的电荷不平衡引起的，例如通过氧化过程造成的电荷不平衡。氧化亚铁硫杆菌（*At. ferrooxidans*）和铁氧化螺菌（*Leptospirillum ferrooxidans*）的几个菌株具有化学感应系统（Acuña et al. 1992；Meyer et al. 2002）。在金属硫化物浸出过程中可能强制产生趋化作用和亚铁离子、铁离子、硫代硫酸盐和其他化合物的吸附梯度（Edwards et al. 2000a）。发生在局部阳极的溶出，能给溶液中带来亚铁离子和硫代硫酸盐。最近利用配备有用于力映射的 Kelvin 探针的原子力显微镜的研究表明，附着到黄铁矿表面的土壤杆菌 *L. ferrooxidans* 细胞比周围表面带更多的负电荷（约 100～200mV）（即从黄铁矿中提取电子）（Vera et al. 2013b）。在类似的情况下，Little 和他的同事们证实硫酸盐还原细菌附着在钢表面上的阳极附近（纳米范围）。亚铁离子释放后，后者便带上了负电。作为细菌附着的结果，阳极和阴极变为永久性（证明），并且钢开始溶解（Little et al. 2000）。该观察结果也可能与金属硫化物的生物浸出有关。总之，细胞似乎被它们的趋化系统瞬时吸附（带电）到溶解位点，并且导致金属硫化物表面上的阳极和阴极变成永久性的。在生物膜形成的情况下，溶解过程发生在胞外多糖层内（图 6.1），被认为填充细胞外膜和金属硫化物表面之间的空隙体积的反应。Tributsch 及他的同事证明了这个距离宽 10～100nm（Rodriguez-Leiva and Tributsch 1988）。通过体内 AFM 估测氧化亚铁硫杆菌（*At. ferrooxidans*）的 EPS 厚度为（28.7±13.5）nm（Taylor and Lower 2008）。体内硫-或黄铁矿-培养细胞的 EPS 厚度值仍有待阐明。据推测，这些值将高于上述值，因为当细菌与这些固体底物一起生长时 EPS 水平升高（Sand et al. 1998）。

新开发的涂有嗜酸性细菌的原子力显微镜探针能够测量嗜酸性金属/硫氧化微生物和矿物表面之间的相互作用力（Diao et al. 2014a，b）。发现与同一底物相互作用，在黄铜矿上生长的氧化亚铁硫杆菌显示出较强的相互作用力。在亚铁离子或硫中生长的细胞，它们的相互作用力较低。此外，研究发现黄铜矿浸出率和细胞附着力与细胞黏附力成正相关（Zhu et al. 2015）。在氧化亚铁硫杆菌、硫氧化硫杆菌和铁氧化硫杆菌的三个物种中，后者对黄铜矿表现出最高的黏附力。EPS 缺陷细胞黏附力和初始细胞附着到金属硫化物表面的能力显著降低（Zhu et al. 2012）。

还有几个问题仍然存在。如是否有些菌株会受益于已存在原代定植菌以便黏附这些以前存在的生物膜？已知发生在某些物种之间的拮抗或协同相互作用的生物化学/分子机制是什么？最近我们已经在含有两个物种的生物膜中发现，存在具有铁-氧化剂活性的生物膜可能会影响其他物种随后的细胞附着。*At. thiooxidans* 细胞在含有 *At. ferrooxidans* 或者 *L. ferrooxidans* 预生长的生物膜黄铁矿上有超过 40%的附着率。有趣的是，该细菌在预生长了 *L. ferrooxidans* 的黄铁矿上的附着，快于预生长了 *At. ferrooxidans* 的黄铁矿。*L. ferrooxidans* 能比 *At. ferrooxidans* 更有效地浸出黄铁矿，*At. thiooxidans* 更快的附着可能与对 RISCs 的趋化反应有关，就像黄铁矿浸出后会释放硫代硫酸盐（Bellenberg et al. 2014）。完整地分析 *At. thiooxidans* 基因组序列，揭示了鞭毛形成和趋化性的完整基因（Valdes et al. 2011）。相比之下，*At. ferrooxidans* 的细胞附着到在黄铁矿晶粒上预生长的 *L. ferroxidans* 强烈依赖于其预培养。硫代硫酸盐培养的细胞受到 *L. ferroxidans* 存在的影响，而亚铁培养的细胞不受影响（Bellenberg et al. 2014）。总而言之，铁氧化微生物的存在已被描述为天然酸性矿井排水生物膜主要定植者（Wilmes et al. 2009），这可能是使硫氧化微生物有效地附着金属硫化物的相关因素。

6.3　铁氧化硫杆菌对生物膜生存方式的生理学适应

6.3.1　氧化应激反应

嗜酸性金属/硫氧化微生物在矿物/液体界面处形成高水平的细胞外活性氧（ROS）。氧（在酸性水溶液中）与矿物晶格结合的铁反应形成过氧化氢（H_2O_2）（Nooshabadi and Rao 2014），它是由黄铁矿中的金属硫化物生成，尤其是矿石在生物湿法冶金加工之前被压碎时形成（Jones et al. 2011）。当存在黄铁矿时，H_2O_2 与亚铁发生芬顿反应，并形成高活性羟自由基。此外，在酸性 pH 值限制扩散速率的情况下，活性氧如 O_2^- 和 H_2O_2 容易穿透细胞膜，当这些化合物的浓度升高时，可能导致 DNA 和酶的损伤（Imlay 2013）。最近研究显示了生长在黄铁矿中的 *At. ferrooxidans* 或 *At. ferrivorans* 细菌与亚铁培养细菌相反，它们能够在缺铁 24h 后和孵育在 1mmol/L H_2O_2 的情况下氧化亚铁离子或黄铁矿。这表明了这些细胞已经适应了在金属硫化物表面上增多的活性氧。结果还证明了在亚铁培养细胞转移到含有黄铁矿的培养基后，只有黄铁矿存在的情况下才会显著抑制浮游亚群（PP）和矿物附着的生物膜细胞亚群（BP）转移到亚铁培养基后的铁氧化活性，且对 BP 的抑制作用更强（Bellenberg et al. 2015）。原因可能是多方面的：（1）BP 细胞向生物膜生活方式的代谢转变；（2）抑制性活性氧的释放，包括造成代谢抑制或活性代谢细胞数量的减少（Jones et al. 2011）。在这种情况下，嗜酸性金属/硫氧化微生物中生物膜形成和 EPS 产生的另一功能，可能与抵抗活性氧造成更强的损伤相关。生物膜细胞 EPS 水平的升高可能起到阻断矿物上的反应位点的作用，并作为这些产生活性氧位点的活性氧清除剂（Bellenberg et al. 2015）。

6.3.2 无机磷酸盐和多磷酸盐代谢

每个微生物生物膜都是通过 EPS 或其他附着因子与微生物的表面发生物理相互作用的结果。EPS 生物合成由环境信号介导，例如营养物可利用性和胁迫影响（Janczarek 2011）。无机磷酸盐（P_i）是所有活细胞的必需营养素。由于铁离子存在时 P_i 会沉淀，酸性浸出细菌需要应对 P_i 缺失问题（Tuovinen 1990）。对于存活在 P_i 缺乏的大肠杆菌和许多其他细菌而言都会引起磷调节子（Pho regulon），这是由 PhoB/PhoR 双组分调节系统调控的遗传系统，并且在 P_i 稳态中起关键作用（Wanner 1996a，b）。文献已经描述了 *At. ferrooxidans* 中 P_i 的缺乏反应和磷调节子的存在（Seeger and Jerez 1993；Vera et al. 2003；Alvarez and Jerez 2004）。它除了在细菌 P_i 稳态中起作用，该系统也是连接稳定期反应、毒力、群体感应，特别是生物膜形成的复合调节网络的一部分，在几种细菌物种中被诱导作为针对 P_i 缺乏的应答（Lamarche et al. 2008）。例如在苜蓿中华根瘤菌（*Sinorizo bium melliloti*）中，当 P_i 受限时，DNA 结合转录因子 PhoB 正向调节参与根瘤菌附着和根瘤结核感染的胞外多糖合成（Rüberg et al.，1999）。同时文献还阐述了根癌农杆菌（*Agrobacterium tumefaciens*）在 P_i 缺乏的生长条件下，生物膜形成会增强（Danhorn et al. 2004）。已经观察到 *At. ferrooxidans* 细胞在缺乏 P_i 的亚铁培养基中继代培养的铁氧化细胞能更有效地附着于金属硫化物和 S^0 表面（Amaro et al. 1993）。已经通过荧光显微镜（EFM）和 FLBA 技术研究发现，与生长在 P_i 充足的环境中的细胞相比，在 P_i 缺乏的情况下生长的 *At. ferrooxidans* 铁氧化细胞 CPS 含量增加（Bellenberg et al. 2012）。

嗜酸性金属/硫氧化微生物积累了高水平的无机聚磷酸盐（polyP），无机聚磷酸盐是来自三大类生命体的细胞中长达 1000 个残基的链，是磷酸酐连接的磷酸盐残基聚合物。polyP 可以作为 P_i 储备库、应力响应调节器以及嗜酸性金属/硫氧化微生物中重要的铜抵抗因子（Alvarez and Jerez 2004；Remonsellez et al. 2006；Kanao et al. 2007；Orell et al. 2010）。然而，它最重要的功能可能是参与细菌对环境变化和压力条件的生理调节。缺乏 polyP 的突变株在稳定期更加敏感，并且显示对热、氧化剂、渗透攻击、抗生素和 UV 辐射的抗性的降低（Rao and Kornberg 1996；Kim et al. 2002）。最近的研究指出，polyP 还具有在保护免受氧化应激上的作用。该作用可以通过直接和间接机制实现，例如：（1）蛋白伴侣，因为不同于蛋白质，polyP 不与活性氧反应；（2）参与 Mn^{2+}/polyP 复合物的形成，它可以在 Mn^{2+}/polyP 复合物水解后将 O^{2-} 解毒，或通过非催化机制参与形成 Mn^{3+}-多聚 P（Mn^{3+}-polyP）；（3）为了防止芬顿反应，其中最可能的机制是促进铜离子的螯合和排出（Grillo-Puertas et al. 2014)；(4) 在细菌进入稳定期的分子机制中发挥重要作用，其中细菌可以高度抗压；(5) 直接参与一般压力应答网络的调节（Gray and Jakob 2015）。生物膜形成过程中浸出菌的 polyP 含量变化的研究，以及构建 polyP 缺陷菌株，将有助于更好地理解 polyP 多功能聚合物在嗜酸性金属/硫氧化微生物的生物膜生活方式中的作用。

6.3.3 影响铁氧化硫杆菌生物膜组成的其他因素

对于影响黄铁矿中生长的铁氧化硫杆菌形成生物膜的其他因素目前也已得到确定。通常，在非最佳生长温度，或在离子强度增加的情况下培养，会导致菌体在黄铁矿上的定植减少。与 *At. ferrooxidans*^{T} 相比，*At. ferrivorans* SS3 在黄铁矿上生长生物膜的形成有所增强。铁离子的存在增加了黄铁矿定植率，特别是当使用黄铁矿培养菌体细胞时。而在上述二者存在的情况下加入 20mmol/L 铜离子，会导致黄铁矿培养细胞的生物膜形成减少。这项观察与黄铁矿培养物暴露在铜中细胞所产生的 EPS 组成不同有关。另外还鉴定了一些附着的效应物。添加 1mmol/L 葡萄糖醛酸，特别是与 1mmol/L 铁离子结合，对所有测试菌株的附着具有较强的促进作用。当 1mmol/L 葡萄糖醛酸加入到 *At. ferrooxidans* 中，对黄铁矿定植起到了促进作用，这与黄铁矿溶解增强约 25%和浮游细胞数量的显著升高相关（Bellenberg et al. 2015)。

目前仍然有几个问题存在。如是否有其他的生物和非生物因素影响生物膜形成？这些因素对铁氧化硫杆菌的影响是否不同？PP 的存在是否影响金属硫化物溶解的开始？黄铁矿培养基中 BP 和 PP 之间的铁氧化活性是否有差异？附着细胞初始数量的增加与黄铁矿溶解的提高有关吗？为了回答这些问题，笔者比较了从 *At. ferrooxidans*^{T} 中的铁氧化活性到黄铁矿培养基中 BP 和 PP 的贡献值。结果表明，在前 4～5 天内，只有 BP 对黄铁矿的溶解有影响（Bellenberg et al. 2015)。

这些浸出细菌根据能量来源（硫酸亚铁培养的浮游细胞，几乎不产生 EPS）调整 EPS 的组成和量的分子机制，以及对固着生活方式的生理调整仍有待详细阐明。最近的结果显示在其他细菌系统中一些相似性的生物膜形成方式也可能发生在嗜酸性金属/硫氧化细菌中，例如，参与 P_i/polyP 代谢、EPS 生物合成途径、协调压力反应、改变代谢速率和细胞信号网络。我们对这一领域的最新认识在本章第 6.5 节进行了汇总。

6.4 嗜酸性金属/硫氧化微生物的生物膜可视化

微生物生物膜生活方式通常包括附着、生物膜发育、成熟和分散（Stoodley et al. 2002；Hall-Stoodley et al. 2004)。EPS 组分能够形成三维生物膜结构。几个先进显微技术已被应用于揭示生物膜形成的过程和动力学研究。CLSM 结合荧光探针提供了详细的三维结构和组成信息（Lawrence et al. 2003；Neu and Lawrence 2014a)。EPS 代表微生物生物膜的关键部分，是生物膜功能的关键要素（Neu and Lawrence 2009)。由于 EPS 的复杂性，开发了通过 FLBA 原位分析 EPS 糖复合体的方法（Staudt et al. 2003；Peltola et al. 2008；Zippel and Neu 2011；Bennke et al. 2013；Castro et al. 2014)。FLBA 允许同时可视化和表征 EPS 糖复合体。如果与蛋白质、核酸或脂质等特异性的其他物质结合，可以在视觉上表征额外的 EPS 组分（Neu and Lawrence 2014a,b)。透射电子显微镜（TEM）以及原子力显微镜（AFM），为生物膜结构的观测提供了最高的分辨率（Dufrêne 2003；Lawrence et al. 2003)。此外，扫描电子显微镜（SEM）已广泛用于生物膜可视化和监测领域（Baldensperger et al. 1974；Bryant et al. 1984；Mikkelsen et al. 2007)。环境扫描电子显微镜（ESEM）可以尽可能减少生物膜脱水，并保留天然生物膜形态和支撑面（Priester et al. 2007）。这些技术以及磁共振成像（MRI）、扫描透射 X 射线显微镜（STXM）、拉曼显微镜（RM）和表面增强拉曼散射仪（SERS）可以原位、非破坏性地分析结构、种类和 EPS 组成以及微生物生物膜变化的动态过程（Lawrence et al. 2003；Manz et al. 2003；Ivleva et al. 2009；Wagner et al. 2009）。拉曼显微镜为具有纳米级空间分辨率的生物样品提供单一生物指纹，并且能够制成光学和化学图像。由于水是生物膜基质的主要成分，拉曼显微镜是原位研究的理想选择。相比之下，扫描透射 X 射线显微镜允许通过生化和元素成像使生物膜结构可视化，并且这些技术不涉及使用额外的探针。然而，虽然扫描透射 X 射线显微镜显示独特的数据集，但它需要同步辐射光源时间，并且在样品装载方面有局限性。

如上所述，嗜酸性金属/硫氧化微生物具有极其特殊的生态地位，通常表现为高含量的金属离子、低 pH 和高温度梯度。虽然生物膜的产生和架构已经研究了几十年，但如果与医疗和工业领域相关的微生物生物膜相比，关于嗜酸性生物膜的详细了解仍然相当有限。当然，近年来，对于由嗜酸微生物产生的生物膜的可视化和表征已经取得了一些进展。

6.4.1 实验室条件下培养的嗜酸性金属/硫氧化微生物的生物膜组成

6.4.1.1 硫元素附着的生物膜组成

S^0 在水中具有非常低的溶解度（$<5\mu g/L$）。因此，直接接触是微生物氧化的前提条件（Vogler and Umbreit 1941）。Waksman 首次使用光学显微镜观察到 S^0 颗粒存在于被细菌包围的 *At. Thiooxidans* 的生长培养基中（Waksman 1932）。可视化显示由 S^0 培养的嗜酸菌生物膜形成的细胞组成是单层的，并且与细胞相关的侵蚀位点均匀分布在 S^0 表面上。一般来说，嗜酸细菌和古菌显示优先附着到 S^0 表面上存在的缺陷位点。另外还观察了 *At. thiooxidans*（Schaeffer et al. 1963），硫化叶菌（*Sulfolobus* sp.）（Weiss 1973），脱氮硫杆菌（*Thiobacillus denitrificans*）(Baldensperger et al. 1974）和 *At. ferrooxidans*（Espejo and Romero 1987）。此外，发现菌毛和细胞壁组分如糖被（glycocalyx）参与连接硫化叶菌、硫杆菌（*Thiobacillus albertis*）和 S^0（Weiss 1973；Bryant et al. 1983；Bryant et al. 1984；Laishley et al. 1986）。通过钌红染色显示了 EPS 中糖胺聚糖的存在（Bryant et al. 1983），并且使膜泡可视化，提出的假说认为这些是为了帮助细胞克服它们在 S^0 上生长所必需的疏水屏障（Knickerbocker et al. 2000；Crescenzi et al. 2006）。最近，我们通过 FLBA 研究了在 S^0 上培养的嗜酸热古菌 *Acidianus* sp. DSM 29099 的生物膜形成，发现生物膜细胞作为细胞簇和小菌落的单独组群呈现不均匀分布。凝集素显示可用于研究在古老的 S^0 表面产生的 EPS 糖缀合物。此外，在这些生物膜中检测到包含单糖（例如岩藻糖、葡萄糖、半乳糖、甘露糖、*N*-乙酰葡糖胺和 *N*-乙酰半乳糖胺）的各种糖缀合物（糖复合物）（Zhang et al. 2015a）。图 6.2 通过 CLSM 的可视化显示了 *Acidianus* sp. DSM 29099 在 S^0 上的生物膜基质的一个实例。

除了标记的凝集素，其他荧光染料可以同时应用于显现和表征额外的 EPS 化合物。例如，Syto 和 Sypro 系列分别通过它们的核酸和细胞蛋白质被用于检测细胞。FM 染料（FM1-43 和 FM4-64）和尼罗红是特异性染色膜和亲脂性化合物。这些染色剂也可用于生物膜中的细胞外化合物染色（Lawrence et al. 2007；Neu and Lawrence 2014a）。DDAO（7-羟基-9H-1,3-二氯-9,9-二甲基吖啶-2-酮）用于核酸染色通常不穿透细胞膜。因此，它被选择作为用于细胞外 DNA 染色（eDNA）的首选荧光染料（Koerdt et al. 2010）。几种类型的着色剂组合使我们获得关于不同能量基质上的嗜酸菌生物膜的详细信息。例如，通过结合

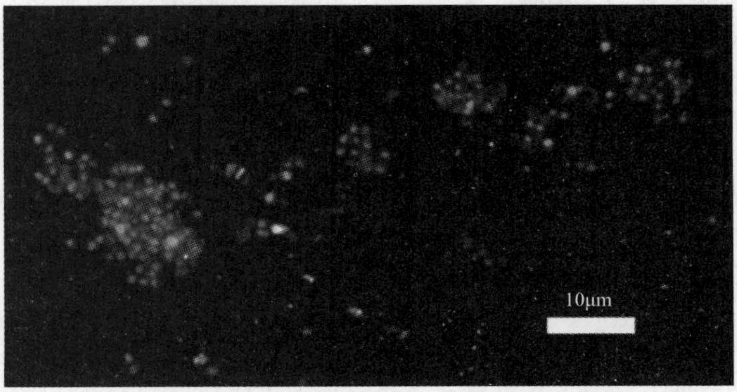

图 6.2 附着硫元素的嗜酸热古菌 DSM 29099 生物膜的最大强度投影（见彩图）。采用 Con A 和 SybrGreen 对细胞和生物膜基质进行双重染色。颜色分配：绿色为 SybrGreen，红色为 Con A-四甲基罗丹明异硫氰酸酯（TRITC），灰色为反射。微生物形式的生物膜细胞嵌入含有甘露糖和葡萄糖的 EPS 基质中

Sypro、DDAO 和几种凝集素，*S. metallicus*^{T} 的生物膜被包埋在含蛋白质和 eDNA 的 EPS 基质中（Zhang et al. 2015b）。图 6.3 显示了生长在 S^0 上的 *S. metallicus*^{T} 的生物膜中存在 eDNA。

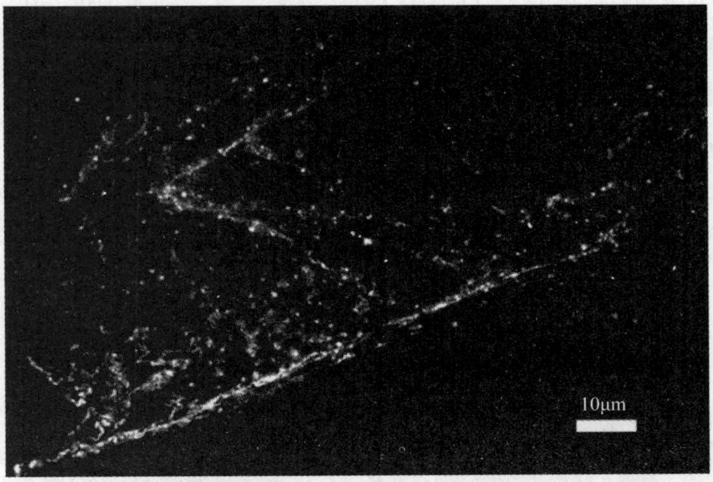

图 6.3 附着单质硫的 *S. metallicus*^{T} 生物膜的 CLSM 图像。采用 SybrGreen 对细胞和生物膜基质进行染色。弥散性和线性的 DNA 信号在细胞周围或与细胞连接处是可见的

6.4.1.2 矿物硫化物上生物膜的形成

一般来说，细胞形成单层生物膜，并表现出选择性附着，优先附着在具有晶体缺陷、断裂和有孔的位点。这些例子包括可视化观察生长在黄铁矿或低品级铜矿黄铜矿上的 *Caldariella*（嗜热古菌）细胞（Murr and Berry，1976），在黄铁矿或黄铜矿上生长的 *At. ferrooxidans*（Wakao et al. 1984；Gehrke et al. 1998；Sanhueza et al. 1999；Sampson et al. 2000a；Tributsch and RojasChapana 2000；Lei et al. 2009；Noël et al. 2010），在黄铁矿上生长的 *A. caldus*（Edwards et al. 2000a）和在加利福尼亚的铁山的黄铁矿表面富集培养物（Edwards et al. 1998，1999b）。相比之下，吸附到黄铁矿的硫化热硫氧还杆菌（*Sulfobacillus thermosulfidooxidans*）的原子力显微镜原位分析表明，表面上的细胞分布在小

集群中，而不是形成连续的生物膜。没有找到证据表明微生物优选附着到某个位点或有定位偏好（Becker et al. 2011）。此外，金属球丝菌和 *S. metallicus* 附着显示没有优先取向。然而，黄铁矿氧化和凹坑蚀刻受表面对称性的影响（Etzel et al. 2008）。有趣的是，两种不同的生物膜形态认为属于中度嗜热古菌 *F. acidarmanus* Fer1 的。在黄铁矿表面上发现了多层生物膜，并且在生物反应器的烧结玻璃喷雾器上发现长达 5mm 的丝（Baker-Austin et al. 2010）。*M. sedula* 通过 EFM 沿着金属矿石摆动，这表明细菌附属物参与细胞附着到矿石上的过程（Huber et al. 1989）。另外经常观察到 EPS（黏液或荚膜）的产生。这种现象也已经在 *Sb. thermosulfidooxidans*（Golovacheva 1978）、*At. ferrooxidans*，*At. thiooxidans* 和/或 *L. ferrooxidans*（Rojas-Chapana et al. 1996；Telegdi et al. 1998；Tributsch and Rojas-Chapana 2000；Bellenberg et al. 2012；González et al. 2012）菌中报道过。EPS 的产生可以由几个因素诱导，例如直接与固体底物或其溶解产物接触（Bellenberg et al. 2012；Zhang et al. 2015a）。使用 SEM 和 TEM 首次记录了在黄铁矿上的嗜热菌 *Acidianus brierleyi*、*M. sedula* 和 *S. metallicus* 的无菌培养物与黄铁矿相互作用。在黄铁矿表面上形成几个沉积结构，包括亚微米沉淀物和盘状结构（Mikkelsen et al. 2007）。

AFM 和 EFM 的组合使得人们可以高分辨率地确认生物的结构或化学本质（Mangold et al. 2008）。它们可以记录矿物的表面性质及其由于微生物活性而引起的改性。此外，使用的探针（例如荧光标记的凝集素）可以提供生物膜细胞的额外的生物化学信息。还发现了在黄铁矿上生长的 *F. acidiphilum*（Zhang et al. 2014），在黄铜矿上生长的 *Metallosphaera hakonensis*（Africa et al. 2013），在黄铁矿上生长的 *At. ferrooxidans*，和混合生长的 *At. thiooxidans* 和 *At. ferrooxidans*（Harneit et al. 2006；Florian et al. 2010；Noël et al. 2010；Florian et al. 2011；Gonzalez et al. 2013）等细菌可形成单层生物膜，并产生 EPS。生物膜形成伴随着产生含有甘露糖或葡萄糖的 EPS。通过与其他技术结合，例如开尔文探针，可以测量表面上的相对电位差和电荷分布（Vera et al. 2013b）。我们应用 FLBA 来分析 *F. acidiphilum*、*S. metallicus* 和 *Acidianus* sp. DSM 29099 的纯培养在黄铁矿上生长所形成的生物膜。在这些生物膜中检测到含有单糖，例如岩藻糖、葡萄糖、半乳糖、甘露糖、唾液酸、*N*-乙酰葡糖胺和 *N*-乙酰半乳糖胺的糖复合体。检测到 EPS 有两种主要结合模式，紧密结合或松散结合（Zhang et al. 2015a）。微生物细胞优先附着于黄铁矿表面的特定位点（图 6.4）。

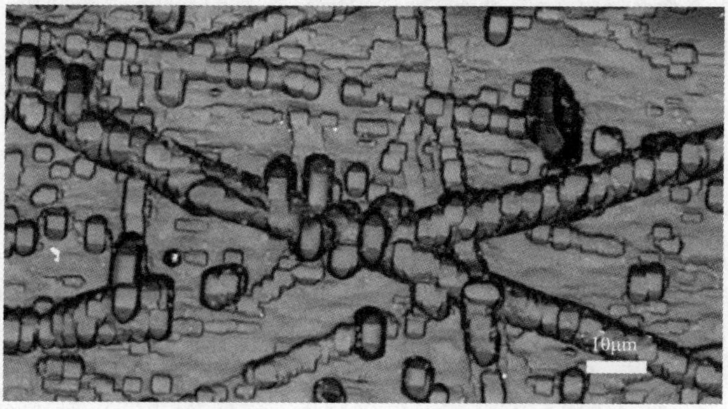

图 6.4　附着于黄铁矿上的 *Acidianus* sp. DSM 29099 生物膜的 CLSM 图像（见彩图）。采用凝集素 AAL 对细胞进行染色。颜色分配：绿色为 AAL-异硫氰酸荧光素（FITC）；黄铁矿表面以反射模式（灰色）显示。细胞优先附着到黄铁矿表面有缺损的位点

在嗜酸性金属/硫氧化古菌中控制生物膜形成的分子机制还很少被探索（Orell etal. 2013）。对其他表面的泉古菌生物膜的形态、结构和化学成分进行了初步研究。首次对三种密切相关的磺胺苯磺酸钠生物膜进行分析。检测到 *S. solfataricus* 和 *S. tokodaii* 的生物膜具有“地毯状”结构，以及 *S. acidocaldarius* 高密度的“塔状”结构（Koerdt et al. 2010）。细胞附体，如毛或鞭毛，被证明参与了 *S. solfataricusus* 最初附着在各种表面，包括玻璃、云母、黄铁矿和碳涂层金网格（Zolghadr et al. 2010）。利用古菌中荧光蛋白的表达，我们发现三种嗜酸葡萄球菌 IV 型黏液样细胞附属物参与其中，但在细胞定植和玻璃表面生物膜形成方面具有不同的功能（Henche et al. 2012）。通过荧光显微镜和 CSLM 的比较分析，发现 *S. solfataricus* 菌中甘露糖苷酶在古菌生物膜的形成和 EPS 组成的调节中起到重要作用（Koerdt et al. 2012）。

6.4.2 自然环境中嗜酸微生物的生物膜组成

与低 pH 和高金属浓度相关的微生物生物膜常见于自然环境中，如在西班牙红河或铁山（里士满，加利福尼亚）的酸性矿井位点。与实验室生物膜相反，环境中的生物膜通常形成不同形态延伸的宏观结构。在这些生物膜中已经检测到细菌、古菌和真核微生物，如在铁山酸性矿井位点的泥里（Edwards et al. 1999a，2000b），已经发现了主要由钩端螺旋体主导的大约 1cm 直径的肉眼可见的菌膜带（Bond et al. 2000b）。几项研究都集中在菌膜带和黏液中的微生物群落（Tyson et al. 2004；Wilmes et al. 2009；Jones et al. 2012；Yelton et al. 2013；Aliaga Goltsman et al. 2015）。通过 CLSM 分析酸性（pH≤2）液体中的底栖群落。观察到主要的微生物金藻 *Gloeochrysis* 在石头上形成肉眼可见的棕色的生物膜，黏稠的基质中包含无机颗粒和真菌菌丝，所有这些都由 EPS 聚集在一起构成（Baffico et al. 2004）。西班牙红河研究证实了 *At. ferrooxidans*、*L. ferrooxidans* 和 *Acidiphilum* spp. 作为优势物种可分泌可见长丝（García-Moyano et al. 2007）。从科尼希施泰因（Konigstein，萨克森，德国）的一个地下矿观察到生物膜和黏液用四种酸稳定荧光染料染色。证明这些染料适合在环境 pH 条件下用于微生物的嗜酸性生物膜染色（Brockmann et al. 2010）。

6.4.3 嗜酸生物膜的分离

生物膜脱离以及细胞从嗜酸生物膜中扩散仍然是一个未被探索的领域，这些过程还没被很好理解。目前已经在金属硫化物上观察到微生物的印迹，这表明嗜酸性金属/硫氧化微生物会有生物膜分离的行为。这些有机物主要是由微生物在与矿物表面短暂接触时遗留下来的，或者是作为生物膜生活周期一部分的细胞程序性分离所造成的。微生物标记相关表面的这种能力首先报道为“足迹”，如通过剪切力分离假单胞菌细胞后留在玻璃表面上的聚合物物质（Marshall et al. 1971）。微生物足迹主要由在细胞脱离或机械去除后留在表面上的 EPS 组成（Neu and Marshall 1990，1991）。在铁上生长的 *T. intermedius* 中间体的足迹（Telegdi et al. 1998）。*At. ferrooxidans*（Rojas-Chapana et al. 1996；Mangold et al. 2008）、*St. thermosulfidooxidans*（Becker et al. 2011）和中温嗜酸化能无机自养生物的混合培养物中占优势的 *L. ferriphilum* 足迹已经被描述（Ghorbani et al. 2012）。我们检测到 *S. metallicus* 和 *Acidianus* sp. DSM 29099 在细胞脱离后表面残留甘露糖或葡萄糖（Zhang and Vera，未发表的数据）。我们还观察到定植、生物膜形成和铁氧化酸硫杆菌和钩端螺旋体的细胞分离的不同动力学过程。钩端螺旋体形成的生物膜比铁氧化酸硫杆菌形成的生物膜能稳定更长的时间。这些脱离过程的形成可能是由于活性氧的存在、离子强度的增加和 pH 的降低引起的氧化应激反应。已有研究表明 EPS 化合物如脂多糖（LPS）在其成熟期间可能从生物膜分离（Jiao et al. 2010）。

已经发现几种细胞胞外酶存在于生物膜的 EPS 中。这些胞外酶可能参与生物聚合物的降解，导致细胞脱离和生物膜分散（Flemming and Wingender 2010）。但目前已鉴定的酶中并没有发现能参与嗜酸性生物膜脱离的酶。转录组学研究表明，有几个基因可能与嗜酸性生物膜的维持和脱离/分散有关（Moreno-Paz et al. 2010）。

6.4.4 胞外组分的化学图谱

最近，基于 STXM 成像和 μ-XRF 图谱的同步加速器辐射已被应用于 *At. ferrooxidans* 的胞外巯基（SH）的原位比较分析。发现生长在 S^0 上的 *At. ferrooxidans* 的巯基含量大约是在亚铁上生长细胞的巯基含量的四倍。这些数据表明细胞胞外巯基在氧化之前对硫活化起重要作用（Xia et al. 2013）。STXM 已被用于对在黄铁矿生长上的 *At. ferrooxidans* 形成的生物膜和 EPS 进行可视化。这些生物膜中多糖和蛋白质的分布可通过光学显微技术直接观察。富含多糖的生物分子位于黄铁矿-细胞的交界处，富含脂质和蛋白质的区域位于细胞中心区域（Mitsunobu et al. 2015）。

6.5 嗜酸性金属/硫氧化微生物生物膜生活方式的分子研究

6.5.1 *At. ferrooxidans*^{T} 生物膜组成的蛋白质组学

如前所述，不同能量底物生长的细胞，EPS 的含量和组成不同，这适应于活性氧水平的增强并且作为转变成生物膜生活方式所必需的生理适应性，浮游和固着细胞在其基因表达方面不同，它们的蛋白质组模式有所不同。*At. ferrooxidans*^{T} 具备革兰氏阴性细菌同源的 CPS 生物合成和输出系统（Barreto et al. 2005）。基因 *Afe* _ 1339 和 *Afe* _ 2975 编码与大肠杆菌 Wzy 和 KpsD 同源的两种外膜多糖输出蛋白（Whitfield 2006）。将在亚铁生长的细胞转移到黄铁矿培养物后，*Afe* _ 2975 在 PP 和 BP 中的表达水平增强（Bellenberg et al. 2011）。对 *At. ferrivorans* SS3 完整基因组序列中假定 CPS 合成基因的生物信息分析，揭示了 Wz CPS 生物合成和输出系统的同源基因的存在，而 Kps 系统的同源基因缺失。大部分的假定基因包括 *wza*（多糖生物合成输出结构域，*Acife* _ 1181），*wzc*（酪氨酸蛋白激酶，*Acife* _ 1180），*wzb*（酪氨酸-蛋白-磷酸酶，*Acife* _ 1179）和 *wzx*（多糖生物合成蛋白，*Acife* _ 1172）可能被聚集到这种细菌的操纵子中。也存在 *wzy* 基因（多糖生物合成蛋白，*Acife* _ 0130）和两个拷贝的 *wbaP* 基因（十一烷酰基-半乳糖-磷酸转移酶，*Acife* _ 1166 和 *Acife* _ 1194），但不在操纵子中聚集。根据这一发现可以推测，如果这些基因簇主导 EPS 多糖聚合物的生物合成，这两个物种之间将是不同的。而且 Wz 系统在 *At. ferrivorans* 中的功能仍有待考证。

在 *At. ferrooxidans*^{T} 中进行生物膜形成的高通量蛋白质组学研究，以便映射其在早期生物膜形成过程期间变化。它通过使用鸟枪定量蛋白质组学法比较了黄铁矿生物膜形成 24h 后 PP 和 BP 的蛋白质组学。总共有 1319 种蛋白质（预测 *At. ferrooxidans*^{T} 蛋白质组的 42%），发现在两个细胞亚群中至少占检测到的蛋白质总量的 16%蛋白质呈现升高或降低的水平。发现在 BP 中有关增强了 EPS 合成和含有电子载体的分子的生物合成，还原的无机硫化合物的转运以及渗透压力应答的功能。

在 BP 中发现三种转录因子处于增加（MerR，IclR）或降低（AbrB）水平，表明它们参与 *At. ferrooxidans* 生物膜形成过程的调控。此外，膜和外膜运输功能，包括 CPS 生物合成蛋白质的增加，外排泵、脂蛋白、ABC 转运蛋白和几种在 BP 中与应激反应相关的蛋白质水平都呈增加水平。人们普遍认为，细菌在生物膜形成过程中表现出应激反应（Otto and Silhavy 2002）。

如第 3 章所述，BP 必须应对渗透压的变化以及在黄铁矿存在下产生的活性氧水平的增加。*At. ferrooxidans*^{T} 的基因组具有几种基因编码与抗氧化应激反应相关的蛋白质。这些蛋白包括一种 Mn-依赖性超氧化物歧化酶（SodA），以及烷基氢过氧化物酶家族（AhpC 和 AhpD）的成员。此外，编码了几种用于活性氧的非酶中和的组分。这些包括硫氧还蛋白，用于谷胱甘肽（GSH）的生物合成和输出的基因，以及维持其他分子的硫醇/二硫键平衡的二硫化物还原酶（Valdes et al. 2008）。在这种情况下对与渗透和氧化应激反应相关的蛋白质的变化进行了检测。除了渗透压传感蛋白 EnvZ，发现编码铁和 2-氧戊二酸（2-OG）依赖性双加氧酶（AFE_3138）的蛋白质在生物膜细胞中呈上升水平。这个蛋白质家族的成员作为对高渗透压的反应可能参与了四氢嘧啶的生物合成（Reuter et al. 2010）。谷胱甘肽是主要的生物抗氧化剂，这有助于维持原核生物的氧化还原平衡（Smirnova and Oktyabrsky 2005）。GSH 对于重建蛋白质和脂质的氧化还原平衡也特别重要。此外，GSH 在 *At. ferrooxidans* 中具有双重作用。因为结果已经表明，在胞质硫二加氧酶（SDO）催化反应中，低分子量的含硫醇化合物如 GSH 与元素硫氧化成亚硫酸盐有关（Rohwerder and Sand 2003）。参与 GSH 代谢的几种蛋白质在 BP 中进行诱导。这些蛋白质包括 AFE_1390（CydC）、GSH 转运蛋白、AFE_0366、AFE_2773、GSH 还原酶（GR）和 GSH S-转移酶的 YghU 家族的蛋白质。这表明周质 GSH 水平的增加，这是对氧化应激或相关硫和 RISCs 氧化的 GSH 驱动应答的一部分（Vera et al. 2013a）。在这种情况下，显示 GR 编码基因在 *At. ferrooxidans* 暴露于铜后被诱导，表明它参与回收 GSH，来应对氧化应激（Xia et al. 2011）。

某些分子可以调控生物膜形成。多胺是广泛存在于细菌中的有机聚阳离子。多胺的共同功能似乎不存在，但最近的证据表明它们促进或抑制不同细菌胞外或胞内生物膜的形成（Karatan and Michael 2013）。不对称多胺亚精胺广泛存在于细菌中。GSH 主要在大肠杆菌需氧生长时存在，谷胱甘肽-亚精胺（GspSH）在大肠杆菌厌氧生长和稳定期时占优势（Smith et al. 1995）。发现可能与 GspSH（AFE_2773）的合成相关的 GSH S-转移酶 YghU 家族的蛋白质在 *At. ferrooxidans*^{T} 中有所增强。这预示着 YghU 属于 GSH S-转移酶的新家族（Stourman et al. 2011）。亚精胺在不同能量源，如硫酸亚铁、硫和黄铜矿上生长的 *At. thiooxidans* 细胞中普遍存在（Martínez et al. 2013）。

生物膜细胞具有“改变代谢速率”的特征（Seneviratne et al. 2012）。在 *At. ferrooxidans* BP 中涉及细胞组分再循环的蛋白质含量增加，而参与呼吸功能的蛋白质含量降低。这些包括几种核苷酸酶和 ATP 酶，它可以增强用于大分子从头生物合成的结构单元库。在 BP 中还发生了辅酶和辅酶生物合成的增强，包括含有铁-硫簇、血红素、吡咯并喹啉醌（PQQ）和与泛醌代谢相关功能的蛋白质（Vera et al. 2013a）。所检测的总蛋白质中约五分之一是假定蛋白质（231 种蛋白质，其中 BP 和 PP 中 12% 的蛋白质含量有所改变）。因此，*At. ferrooxidans* 中仍然未知的代谢途径与生物膜形成可能是相关的。在这些分子中，可以考虑一些具有潜在的抗生素或抑制剂作用的“生物膜特异性分子”。后者已在一些细菌物种中被证明可能参与对抗竞争者和捕食者的生存斗争（Beloin et al. 2004；Rendueles and Ghigo 2012）。随着生物膜蛋白组中假想蛋白质包括 AMD 生物蛋白质的增多，需要进一步研究异源表达以及随后的生化筛选，以确定其功能（Denef et al. 2010）。

众所周知，生物膜发育是一系列有序的事件：浮游微生物黏附到表面；随后这些黏附的生物体产生一组离散的微生物菌落，其中 EPS 生产与三维群落结构开始一起出现。在成熟阶段，一些生物膜细胞离开去寻找新的表面（Flemming and Wingender 2010）。如前所述，在嗜酸性生物膜中，我们有微观证据证实有细胞脱离过程的存在（见本章第 6.4 节）。然而，据我们所知，没有关于可能涉及细胞脱离过程的物理因素、生理反应和分子机制的详细认知。

6.5.2 酸性矿井排水生物膜的分子多样性研究

Jillian Banfield 及其同事发表了 AMD 生物膜对生态主要贡献和分子机制研究。这些研究包括对天然酸性矿井排水群落及其生物膜的宏基因组、宏蛋白质组学和蛋白质组学分析（Ram et al. 2005；Denef et al. 2009；Wilmes et al. 2009）。从加利福尼亚北部的铁山里士满矿中获得宏基因组测序可以重建近似完整的钩端螺旋菌群Ⅱ（*Leptospirillum* Group Ⅱ，其是主要微生物）的基因组，随后表征为嗜铁钩端螺旋菌（*Leptospirillum rubarum*）（Goltsman et al. 2009）和铁原体型Ⅱ（Tyson et al. 2004）。随后的研究对这些群落中存在的一些其他微生物进行基因组重建，如 *Leptospirillum* group Ⅲ，后来被定性为 *L. ferrodiazotrophum*（Goltsman et al. 2009）和 *Leptospirillum* group Ⅳ UBA BS（Goltsman et al. 2013）以及一些不可培养的新型古菌（Yelton et al. 2013）。

酸性矿井排水生物膜在溪流和池塘的气液界面处生长，它覆盖在含有黄铁矿的沉积物上。生物膜生长（百微米厚）产生三价铁，随后催化黄铁矿氧化，偶联无机和生物氧化过程。这个系统的主要特点是物种多样性较低和生物膜发展的可再生阶段随着时间推移不断延续（Wilmes et al. 2009；Denef et al. 2010）。在酸性矿井排水生物膜的发育初期，*Leptospirillum* group Ⅱ占主导地位，而 *Leptospirillum* group Ⅲ和一些古菌存在比例较小。当生物膜进入成熟阶段，*Leptospirillum* group Ⅲ和古菌的比例增加，并且一些真核生物出现。有趣的是，虽然 *Leptospirillum* groups Ⅱ和Ⅲ存在于所有的生物膜发展过程中，但它们在空间排布上有很大的不同。*Leptospirillum* groups Ⅱ组形成紧密的细胞团，而大部分 *Leptospirillum* groups Ⅲ形成微集落或单细胞（Wilmes et al. 2009）。

6.5.3 酸性矿井排水生物膜的蛋白质组学

群落样本中自然生物膜群体的基因组类型是由蛋白质组学数据推断的。样品来自里士满矿（Richmond）区的两个位置（UBA 和 5-way GC），每个位置都有一个独特的 *Leptospirillum* group Ⅱ型，在 16S rRNA 序列水平上有 0.3%的差异。两种基因组在基因水平上均具有 80%的同源性，氨基酸平均同源性为 95.2%。蛋白质组学推断的基因组分型（PIGT）提供了两个密切相关的 *Leptospirillum* group Ⅱ群体之间重组的证据（Lo et al. 2007）。随后，利用 PIGT 对里士满矿 4 年多来的 27 个生物膜样本进行分析，对优势 *Leptospirillum group* Ⅱ群体进行基因型分析。鉴定出了 6 种不同的表型，它们是来自两种亲本基因型的重组体，证实同源重组是这些生物膜中用于精细环境适应的主要策略（Denef et al. 2009）。

另外，对从环境样品接种的实验室培养的生物膜与酸性矿井排水生物膜进行了蛋白质组分析比较（Belnap et al. 2010，2011）。有趣的是，实验室培养的生物膜也主要由 *Leptospirillum* group Ⅱ产生，*Leptospirillum* group Ⅲ在实验室培养的生物膜丰度较低。生物膜样品显示与功能类别相关的蛋白质丰度更高，这些功能类别如能量产生和转化，细胞运动，细胞壁、膜和包膜的发生，细胞内分泌以及囊泡运输功能。相比之下，转录蛋白以及可能参与防御机制的蛋白质在实验室生长的生物膜中更丰富（Belnap et al. 2010）。进一步的研究确定参与氨基酸和核苷酸代谢的 *Leptospirillum* group Ⅱ蛋白质以及参与细胞膜和包膜发生的蛋白质在高 pH 下过表达。此外，一些低丰度的细菌和古菌出现了 pH 特异性划分，因为 *Lep-*

tospirillum group Ⅲ在较高 pH 值下生长的生物膜更丰富，而古菌在较低 pH 值下更丰富（Belnap et al. 2011）。

通过高通量蛋白质组学和代谢组学研究已经表明，来自 *Leptospirillum* group Ⅱ和 *Leptospirillum* group Ⅲ的鉴定的蛋白质和代谢物显示出生物相关模式，这表明重构它们的代谢和/或调节网络可以降低它们的竞争力，并且允许它们占据不同的生态位置（Wilmes et al. 2010）。在评估酸性矿井排水蛋白质组的重复性后，一组可靠的分类蛋白质被鉴定，可用于预测生物膜群落的生长阶段。在生物膜生长的早期阶段，因为这个阶段蛋白质的生物合成机能、细胞分裂以及 1 碳或 2 碳化合物的代谢等功能更为旺盛，*Leptospirillum* group Ⅱ细胞群通过代谢重组来应对一些非生物胁迫。应激反应在早期生物膜中是丰富的，因为参与到金属流中的蛋白质增加。在该生长阶段增加的蛋白质中，发现了细胞色素 572，具有高 O_2 亲和力的细胞色素氧化酶以及细胞色素 c-553。细胞分裂功能在早期生物膜中也更丰富，并且这些功能由于 DNA 突变的积累而进一步减少。与成熟的生物膜相比，这与在早期生物膜中检测到的核糖体蛋白的丰度增加 26 倍是一致的。而参与更复杂的碳水化合物代谢、氨基酸生物合成、氨基酸和核苷酸糖代谢、脂多糖生物合成、淀粉和海藻糖代谢的几种酶在成熟生物膜中更丰富。来自晚期生物膜的蛋白质组也显示出与 Pi 和钼转运有关的蛋白质水平增加，表明必需的营养物质例如氧、磷、氮和钼在这些生物膜中达到了极限（Mueller et al. 2011）。

最近研究了酸性矿井微生物生物膜群落对温度梯度的响应。温度升高，如从 40℃升高到 46℃，能抑制两种钩端螺旋菌基因型的碳固定，而可能遭受病毒压力，而这三种可能受到病毒威胁；这将通过释放病毒裂解物增加碳转换（Mosier et al. 2015）。几种上调的酶可能参与酸性矿井排水生物膜中 EPS 组成的变化，为产生碳水化合物，例如葡萄糖、半乳糖、鼠李糖、庚糖和甘露糖（Jiao et al. 2010）。最近，在显性钩端螺旋菌Ⅱ型生物膜群体中检测到约 29%的蛋白质具有翻译后修饰（PTM）。其中，43%携带多种 PTM。在早期和成熟生物膜之间，以及在两个生态分化的 *Leptospirillum* group Ⅱ细菌的直系同源之间，PTMs 有很大差异（Li et al. 2014）。通过代谢组学分析，发现由钩端螺菌属支配的天然嗜酸性微生物生物膜含有较高丰度的异常溶血磷脂酰乙醇胺（PE）脂质。这些脂质的特殊的极性头部基团结构可能与其对铁和钙离子的亲和力有关（Fischer et al. 2012）。在 pH＝0.9 的溶液中生长的生物膜中发现了约 3500 个代谢特征。通过稳定同位素标记，在这些生物膜群落中发现了几个非典型的代谢产物，这可能表示有新的化合物产生。在已知的代谢产物中，牛磺酸、四氢嘧啶和羟基四氢嘧啶能提供保护，而免受渗透压胁迫。牛磺酸由细菌或古菌合成尚未明确（但可能被 AMD 菌群中一些细菌同化）。据推测，里士满矿区酸性矿井排水主要的群落嗜酸性真菌 *Acydomyces richmondensis* 可能负责牛磺酸的生物合成（Mosier et al. 2013）。钩端螺旋菌Ⅱ的基因组具有四氢嘧啶或羟基四氢嘧啶生物合成的基因。在成熟的生物膜中，四氢嘧啶和羟基四氢嘧啶的产生减少，这表明在早期生物膜发育过程中，四氢嘧啶和羟基四氢嘧啶更丰富，有助于细菌暴露于高离子强度的酸性矿井排水中。在后期，生物膜结构可以提供一些保护，以防止高金属离子和质子浓度（Mosier et al. 2013）。

6.6 嗜酸性金属/硫氧化细菌的细胞间通讯

微生物生物膜发育是一个复杂的过程，它通过一系列不同的机制在不同的水平上进行调控。微生物不仅作为单细胞存在，并且它们能够以类似于多细胞生物体的协调方式去协调它们的活性。一些细菌利用复杂的细胞间通讯系统，这依赖于小信号分子的分泌和感知，从而控制多目标基因的表达。革兰氏阴性菌生物膜的形成和 EPS 的产生受群体感应（QS）的细胞-细胞通讯机制的调控（Marketon et al. 2003）。QS 细胞-细胞信号是由一种名为自诱导因子（AIs）的可扩散分子介导的，它有助于以细胞密度依赖的方式来调控细胞过程（Labbate et al. 2007）。

6.6.1 *At. ferrooxidans*^{T} 的群体感应研究

At. ferrooxidans^T 具有一个 AI-1-型群体感应系统，涉及生产几种酰基高丝氨酸内酯（AHL）类型的 AI（Farah et al. 2005；Rivas et al. 2005）。该系统由 AHL 合酶（AfeI），结合 AHL 的转录调控子（AfeR），作为二元复合物［AfeR-AHL］的靶标的 *afe* 盒和不同的 AHL 信号分子组成。在 *At. ferrooxidans*^{T} 中合成了至少 9 个具有不同的 C-3 取代基（氧代或羟基）并且酰基链中的碳原子范围在 8 和 16 之间的 AHL。铁氧化菌 Pi 的匮乏增加了 *afe*I 和 AHL 的转录（Farah et al. 2005；Valenzuela et al. 2007）。此外，与亚铁生长细胞相比，在 S^0 生长的细胞中 *afeI*-转录物水平也增加。总而言之，这些信息表明 *At. ferrooxidans*^{T}的群体感应系统（*afeI*/*afeR*）可参与 EPS 生产的调控，并因此参与生物膜形成的调控，这已在其他的革兰氏阴性菌中被证实（Marketon et al. 2003；Parsek and Greenberg 2005；Labbate et al. 2007；Decho et al. 2010）。如果是这样，合成的 AHL 和 AHL 类似物（Choudhary and Schmidt-Dannert 2010；Stevens et al. 2010；Galloway et al. 2012）可用于 *At. ferrooxidans*^{T} 增强或抑制群体感应-调控的一些性状，例如 EPS 产生和生物膜形成。已经证实，通过外部添加合成长链 AHL 可刺激在黄铁矿或 S^0 表面上生物膜形成 EPS（Gonzalez et al. 2013）。在长链 AHL 存在下，在聚碳酸酯滤器上生长的生物膜中观察到 CPS 水平的增加。与这些发现一致，生物信息学研究预测至少 75 个基因具有 *afe* 盒，暗示它们的表达受 *At. ferrooxidans*^{T} 中的 AfeR-AHL 复合体调控。其中，基因编码糖基转移酶、金属-β 内酰胺酶可能与 RNA 代谢的蛋白质和主动转运的相关蛋白质有关，暗示它们直接关系到 EPS 的生物合成和输出（Banderas and Guiliani 2013）。

在生物浸出过程中，细胞附着到金属硫化物是一种选择性过程，其中微生物优先附着在黄铁矿表面上的某些位点。因此，在这些界面处可能发生细胞密度的增加。这种类型的细胞捕获会导致附着细胞中的群体感应自诱导，但不导致浮游细胞中的群体感应自诱导。据推测这种激活可能有助于建立 EPS/CPS 基质和生物膜表型，并可以在固体底物培养物中观察到。相比之下，亚铁离子的培养物作为可溶性底物 AHL 检测物，或者不能指示，或者其指示水平低于自动诱导的阈值。一个相似的调控机制已经被假设存在于霍乱弧菌生物膜形成的起始期间（Waters et al. 2008）。

6.6.2 其他嗜酸性浸出细菌中的细胞间通讯

细菌在自然环境中的相互作用是一个复杂的研究领域。目前，人们对这个领域知之甚少。然而，微生物相互作用，如竞争性和合作性在实验室菌株中得到证实，并假设对实验室培养的单个物种的繁殖或在自然生态位中的微生物群落组成有强烈影响（Johnson 1998）。这些相互作用还起到了除了针对非生物胁迫（例如适宜温度范围、营养物、电子受体、重金属浓度或增强的渗透压）的特异性适应或应变特异性适应之外的作用。如前所述，与中性环境相比，嗜酸性金属/硫氧化微生物的自然栖息地的微生物多样性更低，这是由于它们需要适应极端 pH、重金属浓度和电子受体等（Bond et al. 2000a）。然而，由于它们涉及的数量、微生物的种间和种间相互作用是复杂的，因此有大量的环境变量要考虑进来。这些变量包括温度、水合作用、pH、多种无机离子、有机化合物（分泌的代谢物、EPS、蛋白质、具有抗生素活性的次生代谢物或源自细胞裂解或同种异体生物质的其他化合物）。这与嗜酸性金属/硫氧化微生物生态地位的特殊性尤其相关，因为在混合物种培养中，这些常见的营养物、电子受体和碳源被认为具有竞争力的。嗜酸性异养菌、真菌、藻类和噬菌体也可能影响细菌群落组成的生态过程，如协同作用、共生性和自然捕食。

为了了解在生物浸出生境中共存的不同物种之间的细胞通讯，通过质谱法和根癌土壤杆菌 *Agrobacterium tumefaciens* NTL4 生物报告分析了 AHL 的存在（Cha et al. 1998）。分析了 *At. ferrivorans* SS3、*At. thiooxidans* DSM 14887、*L. ferrooxidans* DSM 2391 和两种酸性杆菌菌株的培养液。类似于 *At. ferrooxidans*^{T}，酸性杆菌菌株产生大量多样性的 AHLs：C10-AHL、C12-AHL、3-羟基-C12-AHL、C14-AHL、3-羟基-C14-AHL 和 C16-AHL。在 *At. thiooxidans* 中检测到两种未被取代的 AHLs（C10 和 C12-AHL），而在铁氧化弧菌 *L. ferrooxidans* DSM 2391 和 *At. ferrivorans* SS3 的培养物中没有检测到 AHLs（Bellenberg et al. 2014）。

测定的可合成 AHLs 的混合物，包括 *At. ferrooxidans*^{T}、*At. ferrivorans* SS3、嗜酸杆菌 SPⅢ/3（*Acidiferrobacter* sp. SPⅢ/3）和氧化亚铁硫杆菌 DSM 2391（*L. ferrooxidans* DSM 2391）的纯培养和二元混合培养物，被添加到黄铁矿浸出实验中。在 AHL 混合物添加时，与黄铁矿的连接黏附降低，与矿物溶解度降低有关。因此，基于 AHL 的种间相互作用可能发生在自然环境中，并且也在 *L. ferrooxidans* DSM 2391 和 *Acidiferrobacter* sp. SPⅢ/3 的二元混合培养物中观察到相互依赖的抑制。

在浸出细菌的全基因组中寻找 QS 系统的结果显示，在 *At. ferrivorans* SS3（Acife_1471，56%同一性）和氧化亚铁硫杆菌 C2-3（*L. ferrooxidans* C2-3）（LFE_1606，29%同一性）存在基因中编码 LuxR 样蛋白。有趣的是，在 *L. ferrooxidans* C2-3 或 *At. ferrivorans* SS3 的培养物中没有检测到 AHLs。这些菌株具有 LuxR 样受体的事实可以部分地解释在添加某些 AHL 混合物后会观察到的一些抑制效应。有趣的是，*At. thiooxidans* 菌株在 C8 至 C10 的范围内产生不同的 AHL，并且也可以感应它们，因为添加 C8、氧代-C8 和 C10-AHL 导致 S^0 上的生物膜形成能力增强。然而，在 *At. thiooxidans* DSM 14887^T 的基因组中没有编码与 *luxI/R* 蛋白的同源基因（Valdes et al. 2008）。目前还不知道 *At. thiooxidans* 在没有任何 LuxR 样受体情况下如何感知 AHL_S，并在它们添加后存在差异响应。这暗示着可能存在参与 AHL 生物合成和感应的新途径。这些途径可能广泛存在于一些酸性硫杆菌的物种中。总之，这些结果清楚表明，必须彻底了解细胞-细胞通讯机制及其与生物膜形成表型的关联，才能进一步影响生物技术过程并制订对抗 MS 不必要的自然浸出的对策。

6.6.3 酸性硫杆菌的 c-di-GMP 通路

在许多细菌物种中，向生物膜生活方式的转变由第二信使 c-di-GMP 控制。它由两个 GTP 分子通过二鸟苷酸环化酶（DGC）和磷酸二酯酶（PDE）合成。有几种蛋白质被称为 c-di-GMP 效应因子，其中具有 PilZ 结构域的蛋白质最为典型（Amikam and Galperin 2006；Hengge 2009）。这些调控表型如鞭毛、抽搐运动和 EPS 生物合成。在 *At. ferrooxidans*^{T} 中，c-di-GMP 在 S^0 或黄铁矿上的生物膜细胞中增加，强烈表明它参与 *At. ferrooxidans*^{T} 生物膜的形成（Ruiz et al. 2011）。最近研究表明，c-di-GMP 通路在 *At. caldus* ATCC 51756^T 中也起作用。在其基因组序列中发现了编码 DGC 和 PDE 效应蛋白的基因，并且通过质谱法证实了 c-di-GMP 的存在。此外，几种具有 DGC 结构域的酶的基因，在肠道沙门氏菌血清型鼠伤寒沙门氏菌突变体中的异源遗传互补所证实。最近构建的由基因 *ACAty*_

C1319 编码的 DGC 的缺失突变体，可能是 *At. caldus* 中的主要 DGC 酶。该突变菌株的 c-di-GMP 水平是野生型菌株的 $\frac{1}{14}$。有趣的是，突变株运动性的增强和附着 S^0 能力的降低，表明 c-di-GMP 途径参与 *At. caldus* 的群集运动和细胞附着于 S^0 表面的调控（Castro et al. 2015）。

6.7 未来前景

目前关于嗜酸性生物膜的化学组成和微生物动力学的认识仍然有限。来自这些生物膜的大分子和代谢物在 EPS 内的存在和功能作用仍有待研究。高通量 FLBA 凝集素研究可以更深入、更详细地了解的生物膜形成和界面相互作用。酸稳定荧光染料，如开发从金属/硫氧化嗜酸菌中分离凝集素将会有较好的发展。新型的物理和化学显微技术的组合，例如，拉曼显微技术和纳米技术（STED 和闪视显微镜）可进行详细的研究。与分子技术联合，它可以在不同程度上大大延伸对嗜酸性生物膜功能、化学结构和动力学的了解（Neu and Lawrence 2015）。未来的分子研究应包括聚合物化学、代谢组学以及蛋白质的分子克隆和蛋白质化学，这些均是“未知”的。这些可能在控制嗜酸性生物膜表观动力学以及与其他微生物群体的相互作用中具有重要作用。群体感应、细胞信号转导和群体感应类似物的化学稳定和增强活动的发展的详细研究，也可以有助于提出新的方法来影响金属硫化物的生物浸出。对生物膜生活方式和细胞-细胞相互作用的详细理解，将有助于制订在生物浸出操作的策略，包括提高金属浸出率和减轻酸性矿井排水污染，减少或失活生物膜可以减轻环境问题。

致谢

我们要感谢 Ute Kuhlicke（马格德堡 Helmholtz 环境研究中心，UFDE，河流生态学）在 CLSM 和图像处理方面的技术支持。张瑞勇感谢中国国家留学基金委（CSC）的财政支持（第 2010637124 号）。

无利益冲突声明

Ruiyong Zhang，Sören Bellenberg，Thomas R. Neu，Wolfgang Sand 和 Mario Vera 声明没有利益冲突。

参考文献

Acuña J，Rojas J，Amaro AM，Toledo H，Jerez CA（1992）Chemotaxis of *Leptospirillum ferrooxidans* and other acidophilic chemolithotrophs：comparison with the Escherichia coli chemosensorysystem. FEMS Microbiol Lett 75：37-42

Africa C-J，van Hille RP，Sand W，Harrison ST（2013）Investigation and in situ visualisation of interfacial interactions of thermophilic microorganisms with metal-sulphides in a simulated heap environment. Miner Eng 48：100-107

Aliaga Goltsman DS，Comolli LR，Thomas BC，Banfield JF（2015）Community transcriptomics reveals unexpected high microbial diversity in acidophilic biofilm communities. ISME J 9：1014-1023

Alvarez S，Jerez CA（2004）Copper ions stimulate polyphosphate degradation and phosphate efflux in *Acidithiobacillus ferrooxidans*. Appl Environ Microbiol 70：5177-5182

Amaro AM，Seeger M，Arredondo R，Moreno M，Jerez CA（1993）The growth conditions affect *Thiobacillus ferrooxidans* attachment to solids. In：Torma AE，Apel ML，Brierley CL（eds）Biohydrometallurgical technologies. The Minerals，Metals & Materials Society，Warrendale，PA pp 577-585

Amikam D，Galperin MY（2006）PilZ domain is part of the bacterial c-di-GMP binding protein. Bioinformatics 22：3-6

Andrews GF（1988）The selective adsorption of Thiobacilli to dislocation sites on pyrite surfaces. Biotechnol Bioeng 31：378-381

Baffico GD，Diaz MM，Wenzel MT，Koschorreck M，Schimmele M，Neu TR，Pedrozo F（2004）Community structure and photosynthetic activity of epilithon from a highly acidic（pH≤2）mountain stream in Patagonia，Argentina. Extremophiles 8：463-473

Bagdigian RM，Myerson AS（1986）The adsorption of *Thiobacillus ferrooxidans* on coal surfaces. Biotechnol Bioeng 28：467-479

Baker-Austin C，Potrykus J，Wexler M，Bond P，Dopson M（2010）Biofilm development in the extremely acidophilic archaeon '*Ferroplasma acidarmanus*' Fer1. Extremophiles 14：485-491 Baldensperger J，Guarraia L，Humphreys W（1974）Scanning electron microscopy of thiobacilli grown on colloidal sulfur. Arch Microbiol 99：323-329

Banderas A，Guiliani N（2013）Bioinformatic prediction of gene functions regulated by quorum sensing in the bioleaching bacterium *Acidithiobacillus ferrooxidans*. Int J Mol Sci 14：16901-16916

Barreto M，Jedlicki E，Holmes DS（2005）Identification of a gene cluster for the formation of extracellular polysaccharide precursors in the chemolithoautotroph *Acidithiobacillus ferrooxi-dans*. Appl Environ Microbiol 71：2902-2909

Becker T，Gorham N，Shiers D，Watling H（2011）In situ imaging of *Sulfobacillus thermosulfido oxidans* on pyrite under conditions of variable pH using tapping mode atomic force micros-copy. Process Biochem 46：966-976

Bellenberg S，Vera M，Sand W（2011）Transcriptomic studies of capsular polysaccharide export systems involved in biofilm formation by *Acidithiobacillus ferrooxidans*. In：Qiu GZ，Jiang T，Qin WQ，Liu XD，Yang Y，Wang HD（eds）Biohydrometallurgy：Biotech Key to Unlock Mineral Resources Value：Proceedings of the 19th International Biohydrometallurgy Symposium. Central South University Press，Changsha，China，pp 460-464

Bellenberg S，Leon-Morales C-F，Sand W，Vera M（2012）Visualization of capsular polysaccharide induction in *Acidithiobacillus ferrooxidans*. Hydrometallurgy 129-130：82-89

Bellenberg S，Diaz M，Noel N，Sand W，Poetsch A，Guiliani N，Vera M（2014）Biofilm formation，communication and interactions of leaching bacteria during colonization of pyrite and sulfur surfaces. Res Microbiol 165：773-781

Bellenberg S，Barthen R，Boretska M，Zhang R，Sand W，Vera M（2015）Manipulation of pyrite colonization and leaching by iron-oxidizing *Acidithiobacillus* species. Appl Microbiol Biotechnol 99：1435-1449

Belnap CP，Pan C，VerBerkmoes NC，Power ME，Samatova NF，Carver RL，Hettich RL，Banfield JF（2010）Cultivation and quantitative proteomic analyses of acidophilic microbial communi-ties. ISME J 4：520-530

Belnap CP，Pan C，Denef VJ，Samatova NF，Hettich RL，Banfield JF（2011）Quantitative proteomic analyses of the response of acidophilic microbial communities to different pH conditions. ISME J 5：1152-1161

Beloin C，Valle J，Latour-Lambert P，Faure P，Kzreminski M，Balestrino D，Haagensen JA，Molin S，Prensier G，Arbeille B，Ghigo JM（2004）Global impact of mature biofilm lifestyle on Escherichia coli K-12 gene expression. Mol Microbiol 51：659-674

Bennke CM，Neu TR，Fuchs BM，Amann R（2013）Mapping glycoconjugate-mediated interactions of marine Bacteroidetes with diatoms. Syst Appl Microbiol 36：417-425

Blake RC，Howard GT，McGinness S（1994）Enhanced yields of iron-oxidizing bacteria by in situ electrochemical reduction of soluble iron in the growth medium. Appl Environ Microbiol 60：2704-2710

Bond PL，Druschel GK，Banfield JF（2000a）Comparison of acid mine drainage microbial communities in physically and geochemically distinct ecosystems. Appl Environ Microbiol 66：4962-4971

Bond PL，Smriga SP，Banfield JF（2000b）Phylogeny of microorganisms populating a thick，subaerial，predominantly lithotrophic biofilm at an extreme acid mine drainage site. Appl Environ Microbiol 66：3842-3849

Brockmann S，Arnold T，Schweder B，Bernhard G（2010）Visualizing acidophilic microorganisms in biofilm communities using acid stable fluorescence dyes. J Fluoresc 20：943-951

Bryant R，McGroarty K，Costerton J，Laishley E（1983）Isolation and characterization of a new acidophilic Thiobacillus species（*T. albertis*）. Can J Microbiol 29：1159-1170

Bryant R，Costerton J，Laishley E（1984）The role of Thiobacillus albertis glycocalyx in the adhesion of cells to elemental sulfur. Can J Microbiol 30：81-90

Castelle C，Guiral M，Malarte G，Ledgham F，Leroy G，Brugna M，Giudici-Orticoni M-T（2008）A new iron-oxidizing/O2-reducing supercomplex spanning both inner and outer membranes，isolated from the extreme acidophile *Acidithiobacillus ferrooxidans*. J Biol Chem 283：25803-25811

Castro L，Zhang R，Muñoz JA，González F，Blázquez ML，Sand W，Ballester A（2014）Characterization of exopolymeric substances（EPS）produced by *Aeromonas hydrophila* under reducing conditions. Biofouling 30：501-511

Castro M，Deane SM，Ruiz L，Rawlings DE，Guiliani N（2015）Diguanylate cyclase null mutant reveals that C-Di-GMP pathway regulates the motility and adherence of the extremophile bac-terium *Acidithiobacillus caldus*. PLoS One 10，e0116399

Cha C，Gao P，Chen Y-C，Shaw PD，Farrand SK（1998）Production of Acyl-homoserine lactone Quorum-sensing signals by Gram-negative plant-associated bacteria. Mol Plant-Microbe Interact 11：1119-1129

Choudhary S，Schmidt-Dannert C（2010）Applications of quorum sensing in biotechnology. Appl Microbiol Biotechnol 86：1267-1279

Clark DA，Norris PR（1996）*Acidimicrobium ferrooxidans* gen. nov.，sp. nov.：mixed-culture ferrous iron oxidation with *Sulfobacillus* species. Microbiology 142：785-790

Crescenzi F，Crisari A，D'Angel E，Nardella A（2006）Control of acidity development on solid sulfur due to bacterial action. Environ Sci Technol 40：6782-6786

Crundwell F（2013）The dissolution and leaching of minerals：mechanisms，myths and misunderstandings. Hydrometallurgy 139：132-148

d'Hugues P，Joulian C，Spolaore P，Michel C，Garrido F，Morin D（2008）Continuous bioleaching of a pyrite concentrate in stirred reactors：Population dynamics and exopolysaccharide production vs. bioleaching performance. Hydrometallurgy 94: 34-41

Danhorn T, Hentzer M, Givskov M, Parsek MR, Fuqua C (2004) Phosphorus limitation enhances biofilm formation of the plant pathogen *Agrobacterium tumefaciens* through the PhoR-PhoB regulatory system. J Bacteriol 186: 4492-4501

Decho AW, Frey RL, Ferry JL (2010) Chemical challenges to bacterial AHL signaling in the environment. Chem Rev 111: 86-99

Denef VJ, VerBerkmoes NC, Shah MB, Abraham P, Lefsrud M, Hettich RL, Banfield JF (2009) Proteomics-inferred genome typing (PIGT) demonstrates inter-population recombination as a strategy for environmental adaptation. Environ Microbiol 11: 313-325

Denef VJ, Mueller RS, Banfield JF (2010) AMD biofilms: using model communities to study microbial evolution and ecological complexity in nature. ISME J 4: 599-610

Diao M, Taran E, Mahler S, Nguyen AV (2014a) A concise review of nanoscopic aspects of bioleaching bacteria-mineral interactions. Adv Colloid Interface Sci 212: 45-63

Diao M, Taran E, Mahler SM, Nguyen AV (2014b) Comparison and evaluation of immobilization methods for preparing bacterial probes using acidophilic bioleaching bacteria *Acidithiobacillus thiooxidans* for AFM studies. J Microbiol Methods 102: 12-14

Dispirito AA, Dugan PR, Tuovinen OH (1983) Sorption of *Thiobacillus ferrooxidans* to particulate material. Biotechnol Bioeng 25: 1163-1168

Dopson M, Johnson DB (2012) Biodiversity, metabolism and applications of acidophilic sulfur metabolizing microorganisms. Environ Microbiol 14: 2620-2631

Dopson M, Baker-Austin C, Hind A, Bowman JP, Bond PL (2004) Characterization of Ferroplasma isolates and *Ferroplasma acidarmanus* sp. nov., extreme acidophiles from acid mine drainage and industrial bioleaching environments. Appl Environ Microbiol 70: 2079-2088

Dufrêne YF (2003) Recent progress in the application of atomic force microscopy imaging and force spectroscopy to microbiology. Curr Opin Microbiol 6: 317-323

Dziurla MA, Achouak W, Lam BT, Heulin T, Berthelin J (1998) Enzyme-linked immunofiltration assay to estimate attachment of thiobacilli to pyrite. Appl Environ Microbiol 64: 2937-2942 Edwards KJ, Schrenk MO, Hamers R, Banfield JF (1998) Microbial oxidation of pyrite: experiments using microorganisms from an extreme acidic environment. Am Mineral 83: 1444-1453

Edwards KJ, Gihring TM, Banfield JF (1999a) Seasonal variations in microbial populations and environmental conditions in an extreme acid mine drainage environment. Appl Environ Microbiol 65: 3627-3632

Edwards KJ, Goebel BM, Rodgers TM, Schrenk MO, Gihring TM, Cardona MM, Mcguire MM, Hamers RJ, Pace NR, Banfield JF (1999b) Geomicrobiology of pyrite (FeS_2) dissolution: case study at Iron Mountain, California. Geomicrobiol J 16: 155-179

Edwards KJ, Bond PL, Banfield JF (2000a) Characteristics of attachment and growth of *Thiobacillus caldus* on sulphide minerals: a chemotactic response to sulphur minerals? Environ Microbiol 2: 324-332

Edwards KJ, Bond PL, Gihring TM, Banfield JF (2000b) An archaeal iron-oxidizing extreme acidophile important in acid mine drainage. Science 287: 1796-1799

Edwards KJ, Hu B, Hamers RJ, Banfield JF (2001) A new look at microbial leaching patterns on sulfide minerals. FEMS Microbiol Ecol 34: 197-206

Espejo RT, Romero P (1987) Growth of Thiobacillus *ferrooxidans* on elemental sulfur. Appl Environ Microbiol 53: 1907-1912

Etzel K, Klingl A, Huber H, Rachel R, Schmalz G, Thomm M, Depmeier W (2008) Etching of {111} and {210} synthetic pyrite surfaces by two archaeal strains, *Metallosphaera sedula* and *Sulfolobus metallicus*. Hydrometallurgy 94: 116-120

Farah C, Vera M, Morin D, Haras D, Jerez CA, Guiliani N (2005) Evidence for a functional quorum-sensing type AI-1 system in the extremophilic bacterium *Acidithiobacillus ferrooxidans*. Appl Environ Microbiol 71: 7033-7040

Fischer CR, Wilmes P, Bowen BP, Northen TR, Banfield JF (2012) Deuterium-exchange metabo lomics identifies N-methyl lyso phosphatidylethanolamines as abundant lipids in acidophilic mixed microbial communities. Metabolomics 8: 566-578

Flemming H-C, Wingender J (2010) The biofilm matrix. Nat Rev Microbiol 8: 623-633

Florian B, Noël N, Sand W (2010) Visualization of initial attachment of bioleaching bacteria using combined atomic force and epifluorescence microscopy. Miner Eng 23: 532-535

Florian B, Noël N, Thyssen C, Felschau I, Sand W (2011) Some quantitative data on bacterial attachment to pyrite. Miner Eng 24: 1132-1138

Galloway WR, Hodgkinson JT, Bowden S, Welch M, Spring DR (2012) Applications of small molecule activators and inhibitors of quorum sensing in Gram-negative bacteria. Trends Microbiol 20: 449-458

García-Moyano A, González-Toril E, Aguilera A, Amils R (2007) Prokaryotic community composition and ecology of floating macroscopic filaments from an extreme acidic environment, Rio Tinto (SW, Spain). Syst Appl Microbiol 30: 601-614

Gehrke T, Telegdi J, Thierry D, Sand W (1998) Importance of extracellular polymeric substances from *Thiobacillus ferrooxidans* for bioleaching. Appl Environ Microbiol 64: 2743-2747

Gehrke T, Hallmann R, Kinzler K, Sand W (2001) The EPS of *Acidithiobacillus ferrooxidans*-a model for structure-function relationships of attached bacteria and their physiology. Water Sci Technol 43: 159-167

Ghorbani Y, Petersen J, Harrison ST, Tupikina OV, Becker M, Mainza AN, Franzidis J-P (2012) An experimental study of the long-term bioleaching of large sphalerite ore particles in a circulating fluid fixed-bed reactor. Hydrometallurgy 129: 161-171

Gleisner M, Herbert RB, Kockum PCF (2006) Pyrite oxidation by *Acidithiobacillus ferrooxidans* at various concentrations of dissolved oxygen. Chem Geol 225: 16-29

Golovacheva R (1978) Attachment of *Sulfobacillus thermosulfidooxidans* cells to the surface of sulfide minerals. Mikrobiologiia 48: 528-533

Goltsman DSA, Denef VJ, Singer SW, VerBerkmoes NC, Lefsrud M, Mueller RS, Dick GJ, Sun CL, Wheeler KE, Zemla A (2009) Community genomic and proteomic analyses of chemoau totrophic iron-oxidizing "*Leptospirillum rubarum*" (Group II) and "*Leptospirillum ferrodiazotrophum*" (Group III) bacteria in acid mine drainage biofilms. Appl Environ Microbiol 75: 4599-4615

Goltsman DS, Dasari M, Thomas BC, Shah MB, VerBerkmoes NC, Hettich RL, Banfield JF (2013) New group in the *Leptospirillum* clade: cultivation-independent community genomics, proteomics, and transcriptomics of the new species "*Leptospirillum* group IV UBA BS". Appl Environ Microbiol 79: 5384-5393

Golyshina OV, Pivovarova TA, Karavaiko GI, Kondrateva TF, Moore ER, Abraham WR, Lunsdorf H, Timmis KN, Yakimov MM, Golyshin PN (2000) *Ferroplasma acidiphilum* gen. nov., sp. nov., an acidophilic, autotrophic, ferrous-iron-oxidizing, cell-wall-lacking, mesophilic member of the *Ferroplasmaceae* fam. nov., comprising a distinct lineage of the Archaea. Int J Syst Evol Microbiol 50: 997-1006

González DM, Lara RH, Alvarado KN, Valdez-Pérez D, Navarro-Contreras HR, Cruz R, García-Meza JV (2012) Evolution of biofilms during the colonization process of pyrite by *Acidithiobacillus thiooxidans*. Appl Microbiol Biotechnol 93: 763-775

Gonzalez A, Bellenberg S, Mamani S, Ruiz L, Echeverria A, Soulere L, Doutheau A, Demergasso C, Sand W, Queneau Y, Vera M, Guiliani N (2013) AHL signaling molecules with a large acyl chain enhance biofilm formation on sulfur and metal sulfides by the bioleaching bacterium *Acidithiobacillus ferrooxidans*. Appl Microbiol Biotechnol 97: 3729-3737

Govender Y, Gericke M (2011) Extracellular polymeric substances (EPS) from bioleaching systems and its application in bioflotation. Miner Eng 24: 1122-1127

Gray MJ, Jakob U (2015) Oxidative stress protection by polyphosphate—new roles for an old player. Curr Opin Microbiol 24: 1-6

Grillo-Puertas M, Schurig-Briccio LA, Rodríguez-Montelongo L, Rintoul MR, Rapisarda VA (2014) Copper tolerance mediated by polyphosphate degradation and low-affinity inorganic phosphate transport system in *Escherichia coli*. BMC Microbiol 14: 72

Hall-Stoodley L, Costerton JW, Stoodley P (2004) Bacterial biofilms: from the natural environment to infectious diseases. Nat Rev Microbiol 2: 95-108

Harneit K, Göksel A, Kock D, Klock J-H, Gehrke T, Sand W (2006) Adhesion to metal sulfide surfaces by cells of *Acidithiobacillus ferrooxidans*, *Acidithiobacillus thiooxidans* and *Leptospirillum ferrooxidans*. Hydrometallurgy 83: 245-254

Hedrich S, Schlomann M, Johnson DB (2011) The iron-oxidizing proteobacteria. Microbiology 157: 1551-1564

Henche AL, Koerdt A, Ghosh A, Albers SV (2012) Influence of cell surface structures on crenar chaeal biofilm formation using a thermostable green fluorescent protein. Environ Microbiol 14: 779-793

Hengge R (2009) Principles of c-di-GMP signalling in bacteria. Nat Rev Microbiol 7: 263-273 Huber G, Spinnler C, Gambacorta A, Stetter KO (1989) *Metallosphaera sedula* gen, and sp. nov. represents a new genus of aerobic, metal-mobilizing, thermoacidophilic archaebacteria. Syst Appl Microbiol 12: 38-47

Imlay JA (2013) The molecular mechanisms and physiological consequences of oxidative stress: lessons from a model bacterium. Nat Rev Microbiol 11: 443-454

Ivleva N, Wagner M, Horn H, Niessner R, Haisch C (2009) Towards a nondestructive chemical characterization of biofilm matrix by Raman microscopy. Anal Bioanal Chem 393: 197-206

Janczarek M (2011) Environmental signals and regulatory pathways that influence exopolysac charide production in rhizobia. Int J Mol Sci 12: 7898-7933

Jiao Y, Cody GD, Harding AK, Wilmes P, Schrenk M, Wheeler KE, Banfield JF, Thelen MP (2010) Characterization of extracellular polymeric substances from acidophilic microbial bio-films. Appl Environ Microbiol 76: 2916-2922

Johnson DB (1998) Biodiversity and ecology of acidophilic microorganisms. FEMS Microbiol Ecol 27: 307-317

Johnson DB (2014) Biomining—biotechnologies for extracting and recovering metals from ores and waste materials. Curr Opin Biotechnol 30: 24-31

Jones GC, Corin KC, van Hille RP, Harrison STL (2011) The generation of toxic reactive oxygen species (ROS) from mechanically activated sulphide concentrates and its effect on thermo-philic bioleaching. Miner Eng 24: 1198-1208

Jones DS, Albrecht HL, Dawson KS, Schaperdoth I, Freeman KH, Pi Y, Pearson A, Macalady JL (2012) Community genomic analysis of an extremely acidophilic sulfur-oxidizing biofilm. ISME J 6: 158-170

Kanao T, Kamimura K, Sugio T (2007) Identification of a gene encoding a tetrathionate hydrolase in *Acidithiobacillus ferrooxidans*. J Biotechnol 132: 16-22

Karatan E, Michael AJ (2013) A wider role for polyamines in biofilm formation. Biotechnol Lett 35: 1715-1717

Kim KS, Rao NN, Fraley CD, Kornberg A (2002) Inorganic polyphosphate is essential for long term survival and virulence factors in *Shigella* and *Salmonella* spp. Proc Natl Acad Sci U S A 99: 7675-7680

Knickerbocker C, Nordstrom D, Southam G (2000) The role of "blebbing" in overcoming the hydrophobic barrier during biooxidation of elemental sulfur by *Thiobacillus thiooxidans*. Chem Geol 169：425-433

Koerdt A，Gödeke J，Berger J，Thormann KM，Albers S-V（2010）Crenarchaeal biofilm formation under extreme conditions. PLoS One 5，e14104

Koerdt A，Jachlewski S，Ghosh A，Wingender J，Siebers B，Albers S-V（2012）Complementation of *Sulfolobus solfataricus* PBL2025 with an α-mannosidase：effects on surface attachment and biofilm formation. Extremophiles 16：115-125

Labbate M，Zhu H，Thung L，Bandara R，Larsen MR，Willcox MD，Givskov M，Rice SA，Kjelleberg S（2007）Quorum-sensing regulation of adhesion in *Serratia marcescens* MG1 is surface dependent. J Bacteriol 189：2702-2711

Laishley E，Bryant R，Kobryn B，Hyne J（1986）Microcrystalline structure and surface area of elemental sulphur as factors influencing its oxidation by *Thiobacillus albertis*. Can J Microbiol 32：237-242

Lamarche MG，Wanner BL，Crepin S，Harel J（2008）The phosphate regulon and bacterial virulence：a regulatory network connecting phosphate homeostasis and pathogenesis. FEMS Microbiol Rev 32：461-473

Lawrence J，Swerhone G，Leppard G，Araki T，Zhang X，West M，Hitchcock A（2003）Scanning transmission X-ray，laser scanning，and transmission electron microscopy mapping of the exopolymeric matrix of microbial biofilms. Appl Environ Microbiol 69：5543-5554

Lawrence JR，Korber DR，Neu TR（2007）Analytical imaging and microscopy techniques. In：Hurst CJ，Crawford RL，Garland JL，Lipson DA，Mills AL，Stetzenbach LD（eds）Manual of environmental microbiology. ASM Press，Washington，DC，pp 40-68

Lei J，Huaiyang Z，Xiaotong P，Zhonghao D（2009）The use of microscopy techniques to analyze microbial biofilm of the bio-oxidized chalcopyrite surface. Miner Eng 22：37-42

Li Z，Wang Y，Yao Q，Justice NB，Ahn TH，Xu D，Hettich RL，Banfield JF，Pan C（2014）Diverse and divergent protein post-translational modifications in two growth stages of a natural micro-bial community. Nat Commun 5：4405

Little B，Ray B，Pope R，Franklin M，White DC（2000）Spatial and temporal relationships between localised corrosion and bacterial activity on iron-containing substrata. In：Sequeira CAC（ed）Microbial corrosion. European Federation of Corrosion Publications；Institute of Materials，London，pp 21-35

Lo I，Denef VJ，VerBerkmoes NC，Shah MB，Goltsman D，DiBartolo G，Tyson GW，Allen EE，Ram RJ，Detter JC（2007）Strain-resolved community proteomics reveals recombining genomes of acidophilic bacteria. Nature 446：537-541

Mangold S，Harneit K，Rohwerder T，Claus G，Sand W（2008）Novel combination of atomic force microscopy and epifluorescence microscopy for visualization of leaching bacteria on pyrite. Appl Environ Microbiol 74：410-415

Manz B，Volke F，Goll D，Horn H（2003）Measuring local flow velocities and biofilm structure in biofilm systems with Magnetic Resonance Imaging（MRI）. Biotechnol Bioeng 84：424-432 Marketon MM，Glenn SA，Eberhard A，González JE（2003）Quorum sensing controls exopolysaccharide production in Sinorhizobium meliloti. J Bacteriol 185：325-331

Marshall K，Stout R，Mitchell R（1971）Mechanism of the initial events in the sorption of marine bacteria to surfaces. J Gen Microbiol 68：337-348

Martínez P，Gálvez S，Ohtsuka N，Budinich M，Cortés MP，Serpell C，Nakahigashi K，Hirayama A，Tomita M，Soga T（2013）Metabolomic study of Chilean biomining bacteria Acidithiobacillus ferrooxidans strain Wenelen and *Acidithiobacillus thiooxidans* strain Licanantay. Metabolomics 9：247-257

Meyer G，Schneider-Merck T，Böhme S，Sand W（2002）A simple method for investigations on the chemotaxis of *A. ferrooxidans* and *D. vulgaris*. Acta Biotechnol 22：391-399

Mikkelsen D，Kappler U，Webb R，Rasch R，McEwan A，Sly L（2007）Visualisation of pyrite leaching by selected thermophilic archaea：nature of microorganism-ore interactions during bioleaching. Hydrometallurgy 88：143-153

Mitsunobu S，Zhu M，Takeichi Y，Ohigashi T，Suga H，Makita H，Sakata M，Ono K，Mase K，Takahashi Y（2015）Nanoscale identification of extracellular organic substances at the microbe-mineral interface by scanning transmission X-ray microscopy. Chem Lett 44：91-93

Moreno-Paz M，Gomez M，Arcas A，Parro V（2010）. Environmental transcriptome analysis reveals physiological differences between biofilm and planktonic modes of life of the iron oxidizing bacteria *Leptospirillum* spp. in their natural microbial community. BMC Genomics 11：404.

Mosier AC，Justice NB，Bowen BP，Baran R，Thomas BC，Northen TR，Banfield JF（2013）Metabolites associated with adaptation of microorganisms to an acidophilic，metal-rich environment identified by stable-isotope-enabled metabolomics. MBio 4：e00484-12

Mosier AC，Li Z，Thomas BC，Hettich RL，Pan C，Banfield JF（2015）Elevated temperature alters proteomic responses of individual organisms within a biofilm community. ISME J 9：180-194 Mueller RS，Dill BD，Pan C，Belnap CP，Thomas BC，Verberkmoes NC，Hettich RL，Banfield JF（2011）Proteome changes in the initial bacterial colonist during ecological succession in an acid mine drainage biofilm community. Environ Microbiol 13：2279-2292

Murr L，Berry V（1976）Direct observations of selective attachment of bacteria on low-grade sulfide ores and other mineral surfaces. Hydrometallurgy 2：11-24

Neu T，Lawrence J（2009）Extracellular polymeric substances in microbial biofilms. In：Moran AP，Holst O，Brennan PJ，von Itzstein M（eds）Microbial glycobiology：structures，relevance and applications. Elsevier，San Diego，pp 735-758

Neu TR，Lawrence JR（2014a）Advanced techniques for in situ analysis of the biofilm matrix（structure，composition，dynamics）by means of laser scanning microscopy. In：Donelli G（ed）Microbial biofilms：methods and protocols，methods in molecular biology. Springer，New York，pp 43-64

Neu TR, Lawrence JR (2014b) Investigation of microbial biofilm structure by laser scanning microscopy. Adv Biochem Eng Biotechnol 146: 1-51

Neu TR, Lawrence JR (2015) Innovative techniques, sensors, and approaches for imaging biofilms at different scales. Trends Microbiol 23: 233-242

Neu TR, Marshall KC (1990) Bacterial polymers: physicochemical aspects of their interactions at interfaces. J Biomater Appl 5: 107-133

Neu TR, Marshall KC (1991) Microbial "footprints" —a new approach to adhesive polymers. Biofouling 3: 101-112

Nicolaus B, Manca MC, Romano I, Lama L (1993) Production of an exopolysaccharide from two thermophilic archaea belonging to the genus *Sulfolobus*. FEMS Microbiol Lett 109: 203-206

Noël N, Florian B, Sand W (2010) AFM & EFM study on attachment of acidophilic leaching organisms. Hydrometallurgy 104: 370-375

Nooshabadi AJ, Rao KH (2014) Formation of hydrogen peroxide by sulphide minerals. Hydrometallurgy 141: 82-88

Norris PR, Johnson DB (1998) Acidophilic microorganisms. In: Horikoshi K, Grant WD (eds) Extremophiles: microbial life in extreme environments. Wiley-Liss, New York, pp 133-153

Norris PR, Burton NP, Foulis NA (2000) Acidophiles in bioreactor mineral processing. Extremophiles 4: 71-76

Ohmura N, Kitamura K, Saiki H (1993) Selective adhesion of *Thiobacillus ferrooxidans* to pyrite. Appl Environ Microbiol 59: 4044-4050

Orell A, Navarro CA, Arancibia R, Mobarec JC, Jerez CA (2010) Life in blue: copper resistance mechanisms of bacteria and archaea used in industrial biomining of minerals. Biotechnol Adv 28: 839-848

Orell A, Fröls S, Albers S-V (2013) Archaeal biofilms: the great unexplored. Annu Rev Microbiol 67: 337-354

Otto K, Silhavy TJ (2002) Surface sensing and adhesion of Escherichia coli controlled by the Cpx-signaling pathway. Proc Natl Acad Sci U S A 99: 2287-2292

Parsek MR, Greenberg E (2005) Sociomicrobiology: the connections between quorum sensing and biofilms. Trends Microbiol 13: 27-33

Peltola M, Neu TR, Raulio M, Kolari M, Salkinoja-Salonen MS (2008) Architecture of Deinococcus geothermalis biofilms on glass and steel: a lectin study. Environ Microbiol 10: 1752-1759

Priester JH, Horst AM, Van De Werfhorst LC, Saleta JL, Mertes LA, Holden PA (2007) Enhanced visualization of microbial biofilms by staining and environmental scanning electron micros-copy. J Microbiol Methods 68: 577-587

Ram RJ, VerBerkmoes NC, Thelen MP, Tyson GW, Baker BJ, Blake RC, Shah M, Hettich RL, Banfield JF (2005) Community proteomics of a natural microbial biofilm. Science 308: 1915-1920

Rao NN, Kornberg A (1996) Inorganic polyphosphate supports resistance and survival of stationary-phase *Escherichia coli*. J Bacteriol 178: 1394-1400

Rawlings DE (2002) Heavy metal mining using microbes. Annu Rev Microbiol 56: 65-91 Remonsellez F, Orell A, Jerez CA (2006) Copper tolerance of the thermoacidophilic archaeon *Sulfolobus metallicus*: possible role of polyphosphate metabolism. Microbiology 152: 59-66 Rendueles O, Ghigo J-M (2012) Multi-species biofilms: how to avoid unfriendly neighbors. FEMS Microbiol Rev 36: 972-989

Reuter K, Pittelkow M, Bursy J, Heine A, Craan T, Bremer E (2010) Synthesis of 5-hydroxyectoine from ectoine: crystal structure of the non-heme iron (II) and 2-oxoglutarate-dependent dioxy-genase EctD. PLoS One 5, e10647

Rimstidt JD, Vaughan DJ (2003) Pyrite oxidation: A state-of-the-art assessment of the reaction mechanism. Geochim Cosmochim Acta 67: 873-880

Rivas M, Seeger M, Holmes DS, Jedlicki E (2005) A Lux-like quorum sensing system in the extreme acidophile *Acidithiobacillus ferrooxidans*. Biol Res 38: 283-297

Rodriguez Y, Ballester A, Blazquez ML, Gonzalez F, Munoz JA (2003) New information on the pyrite bioleaching mechanism at low and high temperature. Hydrometallurgy 71: 37-46

Rodriguez-Leiva M, Tributsch H (1988) Morphology of bacterial leaching patterns by *Thiobacillus ferrooxidans* on synthetic pyrite. Arch Microbiol 149: 401-405

Rohwerder T, Sand W (2003) The sulfane sulfur of persulfides is the actual substrate of the sulfur oxidizing enzymes from *Acidithiobacillus* and *Acidiphilium* spp. Microbiology 149: 1699-1710

Rojas-Chapana JA, Giersig M, Tributsch H (1996) The path of sulfur during the bio-oxidation of pyrite by *Thiobacillus ferrooxidans*. Fuel 75: 923-930

Rüberg S, Pühler A, Becker A (1999) Biosynthesis of the exopolysaccharide galactoglucan in *Sinorhizobium meliloti* is subject to a complex control by the phosphate-dependent regulator PhoB and the proteins ExpG and MucR. Microbiology 145: 603-611

Ruiz LM, Castro M, Barriga A, Jerez CA, Guiliani N (2011) The extremophile *Acidithiobacillus ferrooxidans* possesses a c-di-GMP signalling pathway that could play a significant role during bioleaching of minerals. Lett Appl Microbiol 54: 133-139

Sampson M, Phillips C, Ball A (2000a) Investigation of the attachment of *Thiobacillus ferrooxidans* to mineral sulfides using scanning electron microscopy analysis. Miner Eng 13: 643-656 Sampson MI, Phillips CV, Blake RCI (2000b) Influence of the attachment of acidophilic bacteria during the oxidation of mineral sulfides. Miner Eng 13: 373-389

Sand W, Gerke T, Hallmann R, Schippers A (1995) Sulfur chemistry, biofilm, and the (in) direct attack mechanism—a critical evaluation of bacterial leaching. Appl Microbiol Biotechnol 43: 961-966

Sand W, Gehrke T, Hallmann R, Schippers A (1998) Towards a novel bioleaching mechanism. Miner Process Extract Metall Rev 19: 97-106

Sand W, Gehrke T, Jozsa PG, Schippers A (2001) (Bio) chemistry of bacterial leaching—direct vs. indirect bioleaching. Hydrometallurgy 59: 159-175

Sanhueza A, Ferrer I, Vargas T, Amils R, Sánchez C (1999) Attachment of *Thiobacillus ferrooxidans* on synthetic pyrite of varying structural and electronic properties. Hydrometallurgy 51: 115-129

Schaeffer W, Holbert P, Umbreit W (1963) Attachment of Thiobacillus thiooxidans to sulfur crystals. J Bacteriol 85: 137-140

Schippers A (2007) Microorganisms involved in bioleaching and nucleic acid-based molecular methods for their identification and quantification. In: Donati ER, Sand W (eds) Microbial processing of metal sulfides. Springer, Dordrecht, pp 3-33

Schippers A, Sand W (1999) Bacterial leaching of metal sulfides proceeds by two indirect mechanisms via thiosulfate or via polysulfides and sulfur. Appl Environ Microbiol 65: 319-321

Schippers A, Jozsa P, Sand W (1996) Sulfur chemistry in bacterial leaching of pyrite. Appl Environ Microbiol 62: 3424-3431

Seeger M, Jerez CA (1993) Phosphate-starvation induced changes in *Thiobacillus ferrooxidans*. FEMS Microbiol Lett 108: 35-41

Seneviratne CJ, Wang Y, Jin L, Wong SS, Herath TD, Samaranayake LP (2012) Unraveling the resistance of microbial biofilms: has proteomics been helpful? Proteomics 12: 651-665

Sharma P, Das A, Rao KH, Forssberg K (2003) Surface characterization of *Acidithiobacillus ferrooxidans* cells grown under different conditions. Hydrometallurgy 71: 285-292

Shrihari RK, Modak JM, Kumar R, Gandhi KS (1995) Dissolution of particles of pyrite mineral by direct attachment of *Thiobacillus ferrooxidans*. Hydrometallurgy 38: 175-187

Smirnova GV, Oktyabrsky ON (2005) Glutathione in bacteria. Biochem (Mosc) 70: 1199-1211 Smith K, Borges A, Ariyanayagam MR, Fairlamb AH (1995) Glutathionylspermidine metabolism in Escherichia coli. Biochem J 312: 465-469

Solari JA, Huerta G, Escobar B, Vargas T, Badilla-Ohlbaum R, Rubio J (1992) Interfacial phenom ena affecting the adhesion of *Thiobacillus ferrooxidans* to sulphide mineral surfaces. Colloid Surf 69: 159-166

Staudt C, Horn H, Hempel D, Neu T (2003) Screening of lectins for staining lectin-specific glyco-conjugates in the EPS of biofilms. In: Lens P, O'Flaherty V, Moran AP, Stoodley P, Mahony T (eds) Biofilms in medicine, industry and environmental technology. IWA Publishing, London, pp 308-327

Stevens AM, Queneau Y, Soulere L, Sv B, Doutheau A (2010) Mechanisms and synthetic modula-tors of AHL-dependent gene regulation. Chem Rev 111: 4-27

Stoodley P, Sauer K, Davies D, Costerton JW (2002) Biofilms as complex differentiated communi-ties. Annu Rev Microbiol 56: 187-209

Stourman NV, Branch MC, Schaab MR, Harp JM, Ladner JE, Armstrong RN (2011) Structure and function of YghU, a nu-class glutathione transferase related to YfcG from *Escherichia coli*. Biochemistry 50: 1274-1281

Tapia J, Munoz J, Gonzalez F, Blazquez M, Malki M, Ballester A (2009) Extraction of extracel lular polymeric substances from the acidophilic bacterium *Acidiphilium* 3.2 Sup (5). Water Sci Technol 59: 1959-1967

Taylor ES, Lower SK (2008) Thickness and surface density of extracellular polymers on *Acidithiobacillus ferrooxidans*. Appl Environ Microbiol 74: 309-311

Telegdi J, Keresztes Z, Pálinkás G, Kálmán E, Sand W (1998) Microbially influenced corrosion visualized by atomic force microscopy. Appl Phys A Mater Sci Process 66: S639-S642

Tributsch H, Rojas-Chapana JA (2000) Metal sulfide semiconductor electrochemical mechanisms induced by bacterial activity. Electrochim Acta 45: 4705-4716

Tuovinen O (1990) Biological fundamentals of mineral leaching processes. In: Ehrich HL, Brierley CL (eds) Microbial mineral recovery. McGraw-Hill, New York, pp 55-77

Tyson GW, Chapman J, Hugenholtz P, Allen EE, Ram RJ, Richardson PM, Solovyev VV, Rubin EM, Rokhsar DS, Banfield JF (2004) Community structure and metabolism through reconstruction of microbial genomes from the environment. Nature 428: 37-43

Valdes J, Pedroso I, Quatrini R, Holmes DS (2008) Comparative genome analysis of *Acidithiobacillus ferrooxidans*, *A. thiooxidans* and *A. caldus*: insights into their metabolism and ecophysiology. Hydrometallurgy 94: 180-184

Valdes J, Ossandon F, Quatrini R, Dopson M, Holmes DS (2011) Draft genome sequence of the extremely acidophilic biomining bacterium *Acidithiobacillus thiooxidans* ATCC 19377 provides insights into the evolution of the Acidithiobacillus genus. J Bacteriol 193: 7003-7004

Valenzuela S, Banderas A, Jerez CA, Guiliani N (2007) Cell-cell communication in bacteria. A promising new approach to improve bioleaching efficiency? In: Donati ER, Sand W (eds) Microbial processing of metal sulfides. Springer, Dordrecht, pp 253-264

Vera M, Guiliani N, Jerez CA (2003) Proteomic and genomic analysis of the phosphate starvation response of *Acidithiobacillus ferrooxidans*. Hydrometallurgy 71: 125-132

Vera M, Krok B, Bellenberg S, Sand W, Poetsch A (2013a) Shotgun proteomics study of early biofilm formation process of *Acidithiobacillus ferrooxidans* ATCC 23270 on pyrite. Proteomics 13: 1133-1144

Vera M, Schippers A, Sand W (2013b) Progress in bioleaching: fundamentals and mechanisms of bacterial metal sulfide oxidation-part A. Appl Microbiol Biotechnol 97: 7529-7541

Vogler K, Umbreit W (1941) The necessity for direct contact in sulfur oxidation by *Thiobacillus thiooxidans*. Soil Sci 51: 331-338

Wagner M, Ivleva NP, Haisch C, Niessner R, Horn H (2009) Combined use of confocal laser scan-ning microscopy (CLSM) and Raman microscopy (RM): Investigations on EPS-matrix. Water Res 43: 63-76

Wakao N, Mishina M, Sakurai Y, Shiota H (1984) Bacterial pyrite oxidation III. Adsorption of *Thiobacillus ferrooxidans* cells on solid surfaces and its effect on iron release from pyrite. J Gen Appl Microbiol 30: 63-77

Waksman SA (1932) Principles of soil microbiology. Tindall & Cox, Bailliere, London

Wanner BL (1996a) Phosphorus assimilation and control of phosphate regulon. In: Neidhardt FC, Ingraham JL, Low KB, Magasanik B, Schaechter M, Umbarger HE (eds) *Escherichia coli* and *Salmonella typhimurium*: cellular and molecular biology. ASM Press, Washington, DC, pp 1357-1381

Wanner BL (1996b) Signal transduction in the control of phosphate-regulated genes of Escherichia coli. Kidney Int 49: 964-967

Waters CM, Lu W, Rabinowitz JD, Bassler BL (2008) Quorum sensing controls biofilm formation in Vibrio *cholerae* through modulation of cyclic di-GMP levels and repression of vpsT. J Bacteriol 190: 2527-2536

Weiss R (1973) Attachment of bacteria to sulphur in extreme environments. J Gen Microbiol 77: 501-507

Wheaton G, Counts J, Mukherjee A, Kruh J, Kelly R (2015) The confluence of heavy metal biooxi-dation and heavy metal resistance: Implications for bioleaching by extreme thermoacidophiles. Minerals 5: 397-451

Whitfield C (2006) Biosynthesis and assembly of capsular polysaccharides in *Escherichia coli*. Annu Rev Biochem 75: 39-68

Wilmes P, Remis JP, Hwang M, Auer M, Thelen MP, Banfield JF (2009) Natural acidophilic biofilm communities reflect distinct organismal and functional organization. ISME J 3: 266-270

Wilmes P, Bowen BP, Thomas BC, Mueller RS, Denef VJ, Verberkmoes NC, Hettich RL, Northen TR, Banfield JF (2010) Metabolome-proteome differentiation coupled to microbial divergence. MBio 1: e00246-10

Xia JL, Wu S, Zhang RY, Zhang CG, He H, Jiang HC, Nie ZY, Qiu GZ (2011) Effects of copper exposure on expression of glutathione-related genes in *Acidithiobacillus ferrooxidans*. Curr Microbiol 62: 1460-1466

Xia JL, Liu HC, Nie ZY, Peng AA, Zhen XJ, Yang Y, Zhang XL (2013) Synchrotron radiation based STXM analysis and micro-XRF mapping of differential expression of extracellular thiol groups by Acidithiobacillus ferrooxidans grown on Fe^{2+} and S^0. J Microbiol Methods 94: 257-261

Yelton AP, Comolli LR, Justice NB, Castelle C, Denef VJ, Thomas BC, Banfield JF (2013) Comparative genomics in acid mine drainage biofilm communities reveals metabolic and structural differentiation of co-occurring archaea. BMC Genomics 14: 485

Zeng W, Qiu G, Zhou H, Liu X, Chen M, Chao W, Zhang C, Peng J (2010) Characterization of extracellular polymeric substances extracted during the bioleaching of chalcopyrite concen-trate. Hydrometallurgy 100: 177-180

Zhang R, Bellenberg S, Castro L, Neu TR, Sand W, Vera M (2014) Colonization and biofilm for-mation of the extremely acidophilic archaeon *Ferroplasma acidiphilum*. Hydrometallurgy 150: 245-252

Zhang R, Neu T, Bellenberg S, Kuhlicke U, Sand W, Vera M (2015a) Use of lectins to in situ visualize glycoconjugates of extracellular polymeric substances in acidophilic archaeal biofilms. Microb Biotechnol 8: 448-461

Zhang R, Neu T, Zhang Y, Bellenberg S, Kuhlicke U, Li Q, Sand W, Vera M (2015b) Visualization and analysis of EPS glycoconjugates of the thermoacidophilic archaeon *Sulfolobus metallicus*. Appl Microbiol Biotechnol 99: 7343-7356

Zhou H, Zhang R, Hu P, Zeng W, Xie Y, Wu C, Qiu G (2008) Isolation and characterization of *Ferroplasma thermophilum* sp. nov., a novel extremely acidophilic, moderately thermophilic archaeon and its role in bioleaching of chalcopyrite. J Appl Microbiol 105: 591-601

Zhu J, Li Q, Jiao W, Jiang H, Sand W, Xia J, Liu X, Qin W, Qiu G, Hu Y (2012) Adhesion forces between cells of *Acidithiobacillus ferrooxidans*, *Acidithiobacillus thiooxidans* or *Leptospirillum ferrooxidans* and chalcopyrite. Colloids Surf B 94: 95-100

Zhu J, Wang Q, Zhou S, Li Q, Gan M, Jiang H, Qin W, Liu X, Hu Y, Qiu G (2015) Insights into the relation between adhesion force and chalcopyrite-bioleaching by *Acidithiobacillus ferrooxidans*. Colloids Surf B 126: 351-357

Zippel B, Neu T (2011) Characterization of glycoconjugates of extracellular polymeric substances in tufa-associated biofilms by using fluorescence lectin-binding analysis. Appl Environ Microbiol 77: 505-516

Zolghadr B, Klingl A, Koerdt A, Driessen AJ, Rachel R, Albers S-V (2010) Appendage-mediated surface adherence of *Sulfolobus solfataricus*. J Bacteriol 192: 104-110

第七章

嗜酸细菌：生物学研究和应用

Archana Sharma[1], Deepak Parashar[2], Tulasi Satyanarayana[2][3]

7.1 引言

微生物是地球上生物多样性的主要来源，许多微生物生长在地球上的极端环境中，比如地热区、极地、酸性或碱性水域、深海寒冷高压等环境。极端环境的确切定义目前仍存在争议，它是一种只有有限细菌能够生存并生长的环境。极端环境在多个层面上与人们认为的“正常”不同，适中条件的 pH 值在中性附近，温度在20℃至40℃之间，气压约为 0.1 MPa（一个大气压），以及含有适量的营养物质和盐分（Satyanarayana et al. 2005；Thiel 2011）。极端环境中生长着适应性良好的生物，被称作极端生物。其可以在含有重金属、酸、石油和天然气等对人类有毒的条件下存活。微生物能够在恶劣环境中生长，其对极端环境的适应机制引起了科学家的兴趣，另外极端微生物以及它们的特殊的产物在生物技术领域中可能大有用处。

地球上存在不同酸度的天然环境和人工环境，pH 值低于 3.0 的大部分区域是由于当地大量的硫或黄铁矿被空气中的氧气氧化所造成的。硫和黄铁矿都是通过放热反应进行非生物氧化。前者被氧化成硫酸，后者的亚铁被氧化成三价铁。这两个非生物过程，其速率可被嗜酸细菌提升 10^6 倍。大多数酸性黄铁矿区域是通过采矿产生的，通常在煤矿和硫矿周围形成。这些地区的硫化物浓度非常高，pH 值低至 1.0，有机物含量非常低，并且由于含有高浓度的重金属而具有强烈的毒性。在所有的酸性环境中，酸主要来源于硫酸。自燃反应使垃圾堆自身加热，并提供了嗜热微生物所必需的高温环境。在冶矿废水和尾矿等阳光能照射到的地方，某些光合藻类可以在此生长（Satyanarayana et al. 2005）。在酸性环境条件下，极端嗜酸和嗜酸的明确界限仍有待商榷。Johnson（1998，2008）给极端嗜酸生物一个确切的定义，即它们的最适生长 pH 值为 3 或者更低，包含了许多古菌、细菌、真核生物三大生物域中的自养和异养生物。这些生物存在于诸如硫酸池、硫质温泉等天然酸性环境以及由于煤矿、金属矿的开采所产生的人工酸性环境。大部分已知的嗜热嗜酸菌为古菌，通常存在于最

[1] Department of Biophysics，University of Delhi，New Delhi，India.

[2] Department of Microbiology，University of Delhi，South Campus，Benito Juarez Road，New Delhi 110021，India.

[3] e-mail：tsnarayana@gmail. com.

极端的酸性环境中。

嗜酸细菌的应用范围如图 7.1 所示。嗜酸细菌在低品质矿石的生物冶炼中起着重要的作用，并且其分泌的酶在食品、饲料生产中已有广泛应用。除此之外，嗜酸细菌还可用于生物修复和生物发电。AMD 微生物的宏基因组和宏转录组分析为微生物在寡营养和极端酸性环境中的反应和适应机制的研究提供了线索（Chen et al. 2014）。有关嗜酸性细菌的生物学研究和应用的近期进展将在本章进行详细阐述。

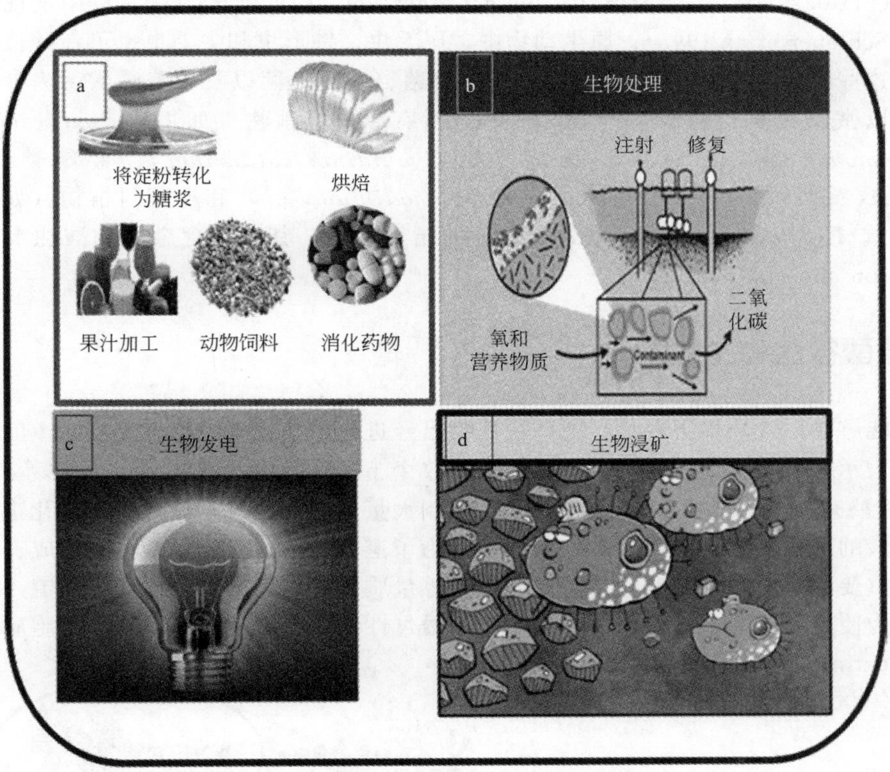

图 7.1 嗜酸细菌及其酶的潜在应用

7.2 嗜酸微生物的起源

地球上有许多天然酸性环境，这些环境中生长着许多在极低 pH 值下具有独特生存和适应机制的嗜酸细菌。火山口和地热区等富含硫的区域已被发现富含嗜酸性微生物，比如 *Sulfolobus solfataricus*（黄石国家公园，美国）（Jeffries et al. 2001），*Sulfolobus hakonensis*（箱根地热区，日本）（Takayanagi et al. 1996），*Acidianus sulfidivorans*（利希尔岛，新几内亚）（Plumb et al. 2007），*Vulcanisaeta thermophila*（马荣火山，菲律宾共和国）（Yim et al. 2015）。*Picrophilus torridus* 和 *Picrophilus oshimae* 是从日本北部火山口中所分离出的中度嗜热异养古菌，其最适生长 pH 值（pH 0.7）是所有已知的非矿物氧化嗜酸性细菌中最低的（Schleper et al. 1996）。在嗜酸性古菌中，*Picrophilus*、*Sulfolobus*、*Metallosphaera* 和 *Sulfurococcus* 是专性好氧菌，*Thermoplasma* 和 *Acidianus* 是兼性厌氧菌，*Stygiolobus azoricus* 是专性厌氧菌（Johnson 1998）。

有关嗜中温、高温的嗜酸细菌研究报道较多；而嗜冷和耐寒的嗜酸细菌的相关研究非常少，尽管许多中高纬度地区存在着像地下矿井排水这样的极端酸性、低温的环境。分离出的许多铁氧化菌和异养菌是耐冷的，但是其中并没有真正的嗜冷细菌（Berthelot et al. 1994）。

酸性矿井排水（AMD）伴随着金属、煤的开采而产生。目前关于西班牙力拓河，挪威希灵达尔矿山、皇家矿山，威尔士帕雷斯矿场和加利福尼亚铁矿山的 AMD 研究较为深入。螺菌属、硫杆菌属、铁原体属、杆菌属的细菌是 AMD 中的主要微生物；许多种类的酵母菌比如红酵母属、假丝酵母属、隐球菌属、*Trichosporon* 是 AMD 中的优势物种；*Acontium velatum*（pH 0.2～0.7）和 *Scytalidium acidophilum*（pH 0）是从 AMD 中发现的耐铜嗜酸真菌（Schleper et al. 1995）；原生动物中，尾毛虫、钟形虫和尖毛虫也可在酸性矿井排水中生存。这些嗜酸/耐酸原生动物以铁化能自养菌、硫氧化菌以及异养嗜酸菌为食。极端酸性水中也以微藻类充当初级生产者，其中包括有绿藻门植物（如 *Dunaliella acidophila*、*Chlamydomonas* spp.）、红藻门植物（如 *Cyanidium caldarium*）、金藻类植物（如 *Ochromonas* spp.）、双鞭毛虫门植物（如 *Gymnodinium* sp.）和裸藻门植物（如 *Euglena mutabilis*）。*E. mutabilis* 可作为 AMD 污染的指示物种，其在富含金属的酸性水域中占优势（Johnson and Hallberg 2003）。

7.3 嗜酸微生物生物学

为了能够在酸性环境下存活，嗜酸微生物已经进化出独特的维持细胞内 pH 值为中性的策略，这表示着在穿过细胞膜的过程中存在数个 pH 梯度单元。在理论上该策略可利用 F_0F_1 ATP 酶来产生大量 ATP，但是质子无节制大量涌入细胞质，导致快速酸化进而损害蛋白质和核酸的正常功能，最终导致细胞死亡。为了避免 DNA 转录、蛋白质合成、酶的活性等细胞内重要的生命过程的紊乱，需要有一个等量质子流出机制以平衡质子的流入。嗜酸微生物被认为使用了一些组合机制，比如质子不透过性细胞膜、反向膜电位和细胞质的缓冲等（Baker-Austin and Dopson 2007）（图 7.2）。

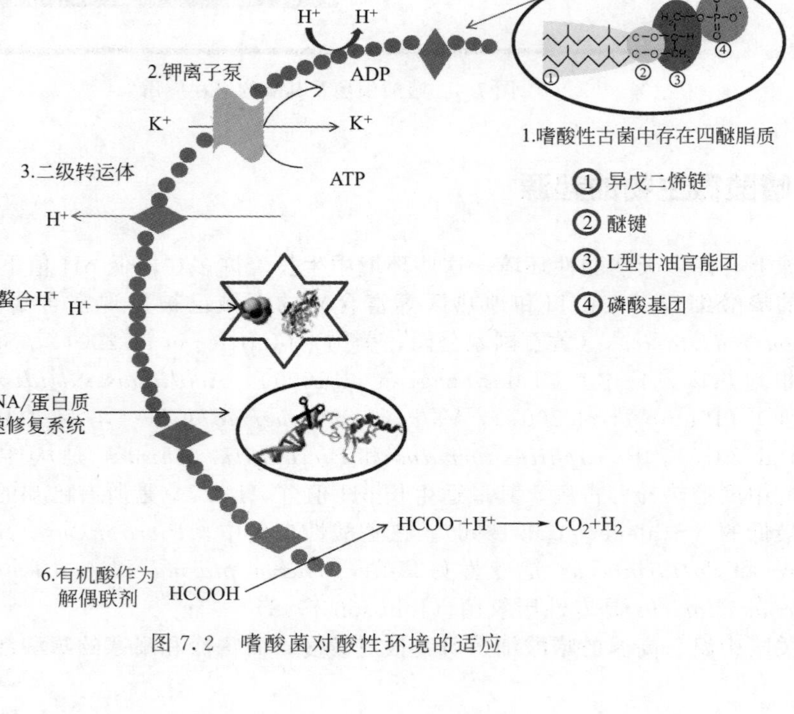

图 7.2 嗜酸菌对酸性环境的适应

细胞膜作为生长在酸性环境中嗜酸细菌抵御低 pH 的主要防线，嗜酸性细菌细胞膜的脂肪酸和脂质组成与嗜中性细菌有很大的不同。嗜酸菌中古菌所占的比例相对较高，可能是由于古菌的细胞膜对质子渗透性低。古菌细胞膜特征是存在四醚脂质（Batrakov et al. 2002；Pivovarova et al. 2002；Shimada et al. 2002；Macalady et al. 2004），这些醚键与细菌和真核生物中的脂键相比使得膜脂质更不容易被酸解（Golyshina and Timmis 2005）。对从 *P. oshimae* 膜脂质中提取的脂质体的研究表明，质子不透过性可能也是由于刚性单层的存在阻止了膜和庞大的类异戊二烯核的分裂（Van de Vossenberg et al. 1998a，b）。有些古菌比如 *Ferroplasma* 和 *Thermoplasma* 无细胞壁（Golyshina et al. 2000），*Thermoplasma* spp. 的细胞膜没有固醇、糖蛋白和脂聚糖。对古菌的细胞膜结构的详细分析显示，膜脂质对保持膜内恒定 pH、膜质子梯度和 ATP 合成酶的正常功能至关重要。

嗜酸性细菌利用反向膜电位（正 $\Delta\Psi$）作为 pH 动态平衡的机制，这与嗜中性细菌中的负 $\Delta\Psi$ 相反。这个正 $\Delta\Psi$ 是通过唐南电势产生的正离子产生，通常是钾离子，抑制质子进入细胞内。嗜酸细菌比如 *P. torridus*（Futterer et al. 2004）、*F. acidarmanus*、*S. solfataricus*（Jeffries et al. 2001）和 *Leptospirillum*（Tyson et al. 2004）的基因组分析表明，大量的不成比例的阳离子转运体的存在可能与唐南电势的产生相关（Donnan et al. 2004）。

嗜酸性细菌的细胞膜具有高度不透过性，如果突然有质子涌入，细胞质的缓冲能力可以隔离这些质子，从而阻止继而产生的损害。所有的嗜酸性细菌都含有细胞质缓冲分子，包含有基本氨基酸比如赖氨酸、组氨酸、精氨酸，以及其他能够捕获质子的氨基酸。极端微生物 *A. acidophilum* 与嗜中性微生物 *E. coli* 的细胞质缓冲能力相比，后者具有更好的缓冲能力（Zychlinsky and Matin 1983）。这表明了嗜酸性细菌细胞质对 pH 的缓冲能力并不比嗜中性微生物高。其他的缓冲分子还包括磷酸盐和钾离子（Spijkerman et al. 2007）。

对大部分诸如 *Ferroplasma*、*Leptospirillum*、*Acidithiobacillus ferrooxidans*、*A. thioxidans*、*A. caldus* 等嗜酸性细菌的基因组分析证实了质子流出系统的存在，其包含有 H^+-ATP 酶、逆向转运体和协同转运体（Tyson et al. 2004）。关于嗜酸性细菌中次级转运体的报道已经非常多。激活的次级转运体是利用跨膜质子或钠离子的电化学梯度来驱动运输的膜蛋白。在 *P. torridus* 和 *T. acidophilum* 中，这些次级转运体大量存在，表明这些嗜酸性细菌适应低 pH（Futterer et al. 2004）。

异养嗜酸细菌普遍能够降解有机酸，比如乙酸和乳酸（Alexander et al. 1987；Ciaramella et al. 2005），这些酸对嗜酸性细菌来说是有害的，因为它们在酸性 pH 下可以作为呼吸链的解偶联剂。这些质子化酸/共轭碱分离的质子能够轻易穿过细胞膜（Baker-Austin and Dopson 2007）。极端微生物的基因组中含有编码这些有机酸降解通路的酶的基因；但是这些基因与低 pH 的联系尚不明确。有趣的是，所有能够生长在极端酸性下的嗜酸性细菌都是异养的，并且能十分高效地降解有机酸。

分子伴侣是与其他蛋白质正确折叠有关的蛋白质（Crossman et al. 2004）。有趣的是，在嗜酸性细菌中，人们发现了热休克蛋白/分子伴侣的高表达。这些分子伴侣能使细胞快速高效地修复受损蛋白质（Laksanalamai and Robb 2004）。

7.4 酸稳定蛋白质的分子适应

嗜酸性酶在酸性环境条件下具有正确的折叠结构和稳定性，并且在 pH 低至 1.0 时也有催化活性。该适应性对蛋白质在低 pH 值条件下发挥功能是必要的，因为酸能阻碍氨基酸的残基带电荷，可能会使蛋白质的天然结构不稳定。酸稳定蛋白的确切适应机制尚不清楚，但是这些酶表面的酸性氨基酸（在中性 pH 带负电）似乎能使酶在低 pH 时具有活性。*S. solfataricus* 的 β-内切葡聚糖酶表面有许多谷氨酸和天冬氨酸残基，从而在中性 pH 下产生高的负表面电荷，这是在低 pH 下酸稳定酶的一个重要适应性调节机制。大量的酸性残基也与 β-内切葡聚糖酶的低等电点（pI）相吻合，但是 *S. solfataricus* 的中性 β-葡聚糖酶和酸稳定 β-葡聚糖酶具有相似的等电点，尽管前者在中性至弱酸性条件下表现出最佳活性，这意味着大量酸性表面残基不是决定 β-内切葡聚糖酶酸稳定性的唯一因素（Huang et al. 2005）。

A. acidocaldarius 的酸稳定 α-淀粉酶表面的正负电荷密度都较低。这避免了在酸性 pH 条件下带电基团的静电排斥作用，并且有可能是酸稳定的一个适应性调整（Schwermann et al. 1994）。

对 *F. acidiphilum* 的数个蛋白质的分析表明，其在低于细胞质 pH 值的条件下依然有活性。这可能是由于这些酶胞内区室化以及胞质内存在 pH 梯度；另一个可能的原因是这些酶形成了多酶复合物，能够使最适 pH 值提升至接近细胞质的 pH。*F. acidiphilum* 的蛋白质组分析发现有高比例铁蛋白存在，有利于酶的 pH 稳定性。这个铁有"铁铆钉"的功能，稳定了蛋白质的三维结构（Golyshina and Timmis 2005）。

7.5 嗜酸细菌的元基因组和元转录组分析

目前已报道的嗜酸性细菌的基因组草图或全图总共有 56 个，包括 30 个细菌，26 个古菌。多重基因组序列的获得为预测浸矿微生物群落之间基因与代谢的相互作用和分析基因组结构和进化的发展方向提供了可能。Chen 等（2014）利用比较基因组学和转录组学方法研究了地球上不同地点的 AMD 的微生物群落，*Acidithiobacillus*、*Leptospirillum* 和 *Acidiphilium* 是微生物群落中的主要物种并表现出高转录活性。

AMD 的微生物群落比较分析表明，微生物通过调节参与低 pH 适应性，碳、氮、磷同化，能量代谢，抗环境胁迫等多种功能的基因的表达来适应不同的环境条件。嗜酸细菌 *Acidthiobacillus ferrivorans* 和 *Leptospirillum ferrodiazotrophum* 的比较分析表明，为了在不同条件下生存，它们在营养吸收和能量代谢方面具有多种不同的策略。Ram 等（2005）通过对 AMD 的微生物群落的分析，发现了与蛋白质折叠和氧化应激反应相关的优势蛋白质，这说明生物分子损伤是微生物在极端环境下生存面临的主要挑战。

7.6 嗜酸细菌和古菌进行微生物湿法冶金

大部分极端酸性环境是人类活动所导致的，比如金属和煤矿的开采。微生物通过对单质硫的氧化以及对硫化物（RSCs）和亚铁的还原而产酸。

单质硫主要蕴含在由二氧化硫和硫化氢凝结形成的地热区域。

$$SO_2 + 2H_2S \longrightarrow 2H_2O + 3S^0$$

单质硫被自养和异养微生物氧化为硫酸，如果硫酸不能进一步为碱矿石所中和，就会导致酸的产生。

$$S^0 + H_2O + \frac{3}{2}O_2 \longrightarrow H_2SO_4$$

二硫化铁矿物黄铁矿是地球上最常见、储存量最为丰富的含硫矿物，并常与其他金属硫化矿共生。三价铁具有强氧化性，可将矿石中的硫氧化成硫代硫酸盐，自身被还原成亚铁。

$$FeS_2 + 6Fe^{3+} + 3H_2O \longrightarrow 7Fe^{2+} + S_2O_3{}^{2-} + 6H^+$$

亚铁可被多种嗜酸性细菌和古菌氧化。

$$4Fe^{2+} + O_2 + 4H^+ \longrightarrow 4Fe^{3+} + 2H_2O$$

硫代硫酸盐在酸性溶液下不能和三价铁共存，会被进一步氧化成其他还原态的无机硫化物（RISCs），比如三聚硫氰酸根（$S_3O_6^{2-}$）、连四硫酸根（$S_4O_6^{2-}$）和单质硫（S^0），这些RISCs可作为硫氧化细菌和古菌的底物。

大部分嗜酸铁硫氧化菌是自养微生物，*A. ferrooxidans* 和 *A. thiooxidans* 这两个化能自养菌是目前研究较为透彻的具有铁/硫氧化能力的嗜酸性微生物，其利用无机电子供体固定 CO_2。除了 AMD 区域，它们通常还可以从河流、运河和硫酸酸化土壤中分离出来。

能够进行铁和/或 RSCs 异化氧化的原核生物的营养类型既有混合营养型又有专性异养型。目前从极端酸性环境中分离出许多异养微生物，专性嗜酸异养菌包含有古菌、细菌、真菌、酵母和原生生物。一些嗜中温原核生物如 *Ferromicrobium acidophilum* 和 *L. ferrooxidans* 是铁氧化细菌，并在铁的异化氧化中起直接作用（Pronk and Johnson 1992；Johnson 1998）。极端嗜热嗜酸铁氧化细菌包括 *Acidianus brierleyi*、*A. infernus*、*A. ambivalens*、*Metallosphaera sedula* 和 *Sulfurococcus yellowstonii*。硫氧化菌中的 *Sulfolobus shibatae*（兼养型）、*S. Solfataricus*（兼养型）、*S. hakonensis*（兼养型）、*S. metallicus*（自养型）、*Metallosphaera prunae*（兼养型）和 *Sulfurococcus mirabilis*（兼养型）为嗜热嗜酸性菌（Johnson 1998）。

长期以来，微生物对从矿石和废料中提取和回收金属有着重要影响，但是它们在其中扮演的角色尚未被认知。18～19 世纪，力拓矿场（南西班牙）和帕雷斯矿场（安格尔西岛，北威尔士）建造的“沉降池”，用来从已冶炼的矿石中回收铜。20 世纪中期，发现了细菌介导含金属硫化矿物溶解的现象，由此产生了生物冶矿的概念，并实现了采矿过程中微生物技术的应用（Rawlings and Johnson 2007；Johnson 2008）。

在矿石和精矿的冶炼方面，生物法较传统的火法冶金有如下几个优势：最主要的优势为能够处理低品质矿石、能耗低；有害废物的产生量非常少因而具有环境效益；并可从冶金废料中回收金属，从而减少尾矿的产量（Johnson 2008）。

硫化矿物可以通过生物浸提或生物氧化处理，使目的金属溶解（如黄铜矿和靛铜矿中的铜）；生物浸提和氧化还被用来溶解与细金粒伴生的黄铁矿和砷黄铁矿（毒砂），并通过氰化法提取贵金属（Johnson 2008）。许多生物浸矿的商业化方法为大家所熟知，包括原位浸出、堆积浸出和桶式浸取。原位浸出即在压力作用下把溶液和空气泵进矿体，或者通过爆破使矿体可渗透，金属富集液通过矿体下面的钻井回收（Bampton et al. 1983；Brierley and Rrierley 2000）。堆积浸出中需要使用未粉碎的废石，这些废料中含有极少量的铜（0.1%～0.5%），利用传统方法难以进行回收。堆积浸出的矿石需要预处理，比如减小体积以提升矿物和浸出剂的相互作用、建造防渗地基防止浸出剂损失和水体的污染（Rawlings 1997）。浸出剂是湿法冶金中用来从矿石中选择性提取目的金属的液体介质，它能够使目的金属快速浸出，金属在浸取后以浓缩态回收。原位浸出和堆积浸出中，浸出剂可用于浸堆的顶端和表面，富含金属的浸出液可重复利用。浸堆的顶端可喷洒稀硫酸以降低浸堆的 pH 值，并促进嗜酸性细菌的生长；在浸堆底部收集酸液，并输送到回收站。金属通过多种方法从这些酸液中提取出来，比如黏结、溶剂萃取和电解冶金。桶式浸取用于在罐槽或生物反应器中溶解已粉碎的氧化矿（Siddiqui et al. 2009）（图 7.3）。

生物采矿已被应用到铜、金、铀、钴，以及一些其他的金属如镍和锌的提取；一些重要金属，如铜、金和铀的生物采矿方法将在下文详细讨论。

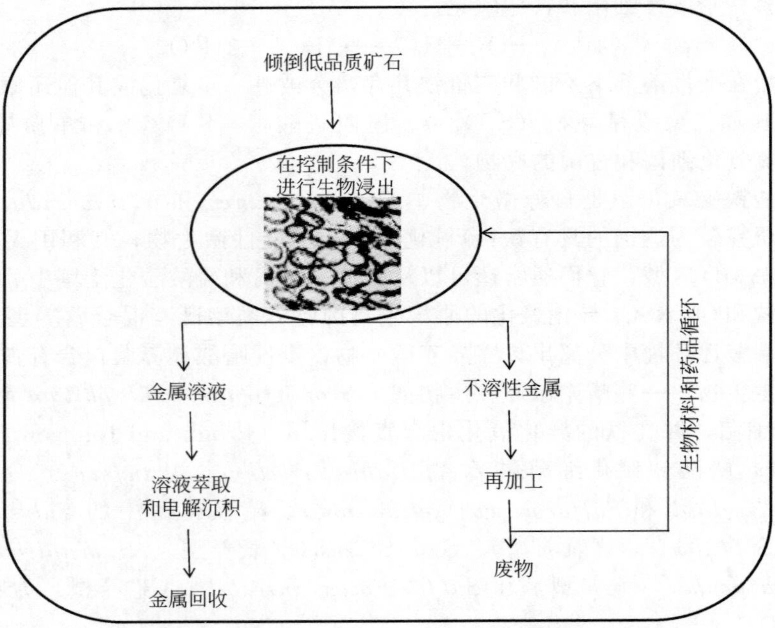

图 7.3　从低品位矿石中浸出金属的步骤

7.6.1　铜

铜矿石，比如黄铜矿（Cu_2S）或靛铜矿（CuS）在堆成堆前需要经过粉碎、硫酸酸化、在转筒中凝聚以使细料与粗颗粒结合等预处理（Schnell 1997）。当含铁溶液流过矿堆时，生长在矿石表面和溶液里的嗜酸微生物使之转化为三价铁离子，并在硫酸铜生成过程中起重要作用，可溶性铜和铁收集后泵入回收装置，最终回收铜。

7.6.2　金

在难冶炼的矿石中，金颗粒被不溶的硫化物覆盖，通过氰化物溶液溶解可从矿石中回收金。在生物氧化过程中，细菌氧化覆盖在金颗粒矿石或精矿上的部分硫（Dew et al. 1997）。首先，细菌通过将硫和金属氧化至更高的氧化态以催化毒砂（FeAsS）的分解，同时利用 H_2 和 Fe^{3+} 还原双氧，从而得到可溶性产物。

$$FeAsS_{(S)} \longrightarrow Fe^{2+}_{(aq)} + As^{3+}_{(aq)} + S^{6+}_{(aq)}$$

该过程发生在细菌细胞膜上，穿过细胞的电子被用到生化过程中来产生能量，并将氧分子还原成水。在第二阶段，细菌将亚铁（Fe^{2+}）氧化成铁离子（Fe^{3+}），并进一步将金属氧化到活性更高氧化态。随着电子的获得，Fe^{3+} 在连续循环中被还原成 Fe^{2+}，最终金从矿石中分离获得，生物氧化后金的回收率更高（Siddiqui et al. 2009）。

7.6.3　铀

铀的回收与铜类似，铀是通过微生物产生的三价铁和硫酸将不溶的铀氧化物转化为可溶性硫酸盐从而实现回收（Siddiqui et al. 2009）。

$$UO_2 + Fe_2(SO_4)_3 \longrightarrow UO_2SO_4 + 2FeSO_4$$

$$UO_3 + H_2SO_4 \longrightarrow UO_2SO_4 + H_2O$$

金属与还原态硫共存时不溶，比如NiS、ZnS和含钴黄铁矿，但是当其被氧化成硫酸盐时则可溶，并且硫酸盐可通过生物冶矿回收。铅可从含有醋酸铅的溶液中获得，溶液可被回收用于浸取铅硫矿或含铅硫化物颗粒（Geisler and Pudington 1996）。生物采矿有助于低品质矿石的回收利用，其应用主要取决于所要回收金属的重要程度（Rawlings 2002）。目前生物采矿技术面临的一个主要挑战是矿体和生物采矿技术间的最佳匹配，并找到具有经济价值的最佳浓度和规模。

7.7 酸稳定性酶的应用

许多酶从嗜酸细菌和古菌中获得，从极端嗜热嗜酸菌中获取的酶在低pH和高温下仍具有活性。在许多工业中具有重要应用潜力，比如淀粉、果汁、饲料和烘培行业。目前从嗜酸细菌中发现了淀粉水解酶、木聚糖酶、蛋白水解酶、纤维素酶、酸性磷酸酶和麦芽糖结合蛋白（Sharma et al. 2012），下面将对这些酶及其应用领域进行讨论。

7.7.1 淀粉工业

淀粉是各种糖浆的主要来源，是制药和制糖工业的基础。淀粉酶是淀粉糖化、烘焙、造纸和纺织工业中应用广泛的几种重要酶之一。淀粉糖化和烘焙通过使用酸性α-淀粉酶而受益良多。目前淀粉工业中使用的α-淀粉酶最适酶活条件为95℃，pH6.8，并且Ca^{2+}可提升其稳定性。因此利用该酶进行工业生产时不能在天然淀粉的pH条件（3.2～4.5）下进行（Shivaramakrishnan et al. 2006；Sharma et al. 2012）。为了适应液化所需酶的最佳pH值，淀粉浆的pH要从其天然pH 3.2～4.5调整到5.8～6.2，并需要补充Ca^{2+}来提高酶的活性和/或稳定性；下一个糖化步骤需要将pH值再次调整到4.2～4.5。这两个步骤（调整pH和除盐）最好可以省略，因为它们既消耗时间，又提高了生产成本。因此，需要从极端微生物中提取符合特殊工业应用需求的天然极端酶（Sharma et al. 2012）；*A. acidocaldarius* 的α-淀粉酶是第一个最适温度为75℃，pH为3.0的热稳定和酸稳定蛋白质（Matzke et al. 1997；Bertoldo et al. 2004）。关于嗜热酸稳定α-淀粉酶的报道非常少（Bai et al. 2012；Sharma and Satyanarayana 2010；Liu and Xu 2008）；*A. acidocaldarius* 和 *Bacillus acidicola* 的酸稳定α-淀粉酶在淀粉工业中应用广泛（Sharma and Satyanarayana 2010；Bai et al. 2012）。淀粉工业中使用的另一类淀粉酶是葡萄糖淀粉酶，这些酶发现于 *P. torridus*、*P. oshimae* 和 *T. acidophilum* 等嗜热嗜酸古菌。古菌的葡糖淀粉酶在pH 2.0，温度90℃时具有最佳活性；真菌、酵母和细菌产生的葡糖淀粉酶在70℃，pH 3.5～6时活性最高。

7.7.2 烘焙工业

尽管烘焙工业也用到淀粉酶，但是酶的特性与淀粉工业中要求不同。烘培工业中的麦芽糖淀粉酶需要具有中等强度的热稳定性。麦芽糖具有天然保鲜剂的作用，而中等强度的热稳定酶可在烘焙结束后失活，阻止反应继续进行，防止产品变质。嗜酸细菌 *B. acidicola* 的酸性α-淀粉酶最适pH值为4.5，因此可用于烘焙工业。

近年来，木聚糖酶的应用越来越广泛。烘焙工业中木聚糖酶的关键作用是使面粉中的半纤维素分解和调整水分，从而使面团更柔软易于揉捏。面团中加入木聚糖酶可以帮助吸收水分，阻止发酵，增加面包体积。鉴于面团是酸性的，所以烘焙工业需要使用在酸性条件下稳定的木聚糖酶。Shal 等（2006）发现嗜酸真菌 *Aspergillus foetidus* 中的酸稳定木聚糖酶（最适 pH 5.3）可作为全麦面包的面包改进剂。

7.7.3 果汁工业

果汁和蔬菜汁的生产过程中需要用到提取、澄清和稳定的方法。过去，刚开始生产柑橘类果汁时，过滤问题导致浑浊，果汁产量很低。现在随着果胶酶、木聚糖酶、α-淀粉酶、纤维素酶和其他酶的使用，果汁黏度和浊度降低，产量有所提高，香气、香精油、维生素和矿物质的含量有所提升。

在果汁生产过程中的所有酶中，最重要的是果胶酶。它可降解果胶，果胶是存在于胞间层和幼植物初级细胞壁的结构性多糖。酸性果胶酶被用于商业化苹果汁、梨汁和葡萄汁的生产中。果胶酶具有高水平的多聚半乳糖醛酸酶活性，可用于稳定灌装橙汁、李子汁、番茄汁、果泥、蜜饯和单细胞产品。单细胞产品是通过将组织转变为全细胞的悬浊液而产生，这种类型的产品可用来作为果肉饮料、婴儿食品、奶制品的基本原料，比如布丁和酸奶，该过程被称作浸渍，所用的酶称作“软化酶”，通常是由纤维素酶、半纤维素酶和果胶酶混合而成。果汁生产和酿酒中所用到的酸性果胶酶主要来自真菌，特别是 *A. niger*；其他酸稳定果胶酶的来源见表 7.1。

表 7.1 工业用到的酸稳定酶的来源

酶	菌株	pH	参考文献
α-淀粉酶	*Alicyclobacillus acidocaldarius*	3.0	Matzke et al. (1997)
	Bacillus acidicola	4.0	Sharma and Satyanarayana(2010)
	Bacillus sp. YX1	5.0	Liu and Xu(2008)
葡萄糖淀粉酶	*Thermoplasma acidophilum*	2.0	Serour and Antranikian(2002)
	Picrophilus torridus	2.0	Serour and Antranikian(2002)
	P. oshimae	2.0	Serour and Antranikian(2002)
蛋白酶	*Xanthomonas* sp.	2.7	Oda et al. (1987a)
	Pseudomonas sp.	3.0	Oda et al. (1987b)
	Sulfolobus acidocaldarius	2.0	Murao et al. (1988)
	Thermoplasma volcanium	3.0	Fusek et al. (1990)
内切葡聚糖酶	*A. acidocaldarius*	4.0	Eckert and Schneider(2003)
果胶酶	*Aspergillus niger* CH4	4.5～6.0	Acuna-Arguelles et al. (1995)
	Penicillium frequentans	4.5～4.7	Borin et al. (1996)
	Sclerotium rolfsii	3.5	Channe and Shewal(1995)
	Rhizoctonia solani	4.8	Marcus et al. (1986)
	Mucor pusillus	5.0	Al-Obaidi et al. (1987)
木聚糖酶	*A. foetidus*	5.3	Shah et al. (2006)
	A. awamori	5.0	Do et al. (2012)
植酸酶	*A. niger*	5.0	Soni et al. (2010)

7.7.4 动物饲料

动物饲料中添加酶，比如木聚糖酶、淀粉酶、纤维素酶、果胶酶、植物酶和蛋白酶等，可以减少排泄物中的磷、氮、铜、锌等不必要的残留，在减少环境污染方面起着重要作用。在这些酶当中，酸性木聚糖酶和肌醇六磷酸酶在动物饲料中大量使用。在动物饲料中添加木聚糖酶可水解其中的阿拉伯木聚糖，阿拉伯木聚糖存在于谷类的细胞壁中，并且在家禽中具有抗营养作用。肌醇六磷酸酶是另一类在动物饲料中添加的酶，微生物的肌醇六磷酸酶主要添加到动物（猪和家禽）饲料和人类食品中，以提高矿物质的生物利用度和改善食品加工。木聚糖酶和肌醇六磷酸酶主要来源于真菌和酵母，目前关于嗜酸性细菌和古菌中的木聚糖酶和肌醇六磷酸酶的报道非常少。一些关于酸稳定木聚糖酶（包括 *S. solfataricus*）的报道表明其在 pH 3.5 和 95℃时对羟甲纤维素的降解活性最高，*Acidobacterium capsulatum* 的木聚糖酶在酸性范围内具有最佳活性和稳定性（Inagaki et al. 1998）。据我们所知，目前尚未有嗜酸细菌和古菌的肌醇六磷酸酶的报道，鉴于细菌酸性肌醇六磷酸酶比真菌肌醇六磷酸酶具有更高的底物特异性和更好的催化效率，因此有必要对可用于动物饲料的细菌酸性肌醇六磷酸酶进行探索（Rodriguez et al. 1999；Kim et al. 2003）。

7.7.5 制药工业

天冬氨酸蛋白酶，也被称为羧基蛋白酶，在酸性溶液中，与能够消化蛋白质和多肽的蛋白水解酶相对应。其来自于多种生物，包括哺乳动物、真菌、植物和逆转录病毒，最近在古菌和细菌中也有发现。酸性蛋白酶在食品、饮料和制药工业中有着重要的应用。由于其在活性位点存在两个天冬氨酸残基（Asp32、Asp35，根据胃蛋白酶编号），其在低 pH 值条件下酶的最适 pH 发生改变（Davies 1990）。*S. acidocaldarius* 的嗜热蛋白酶（thermopsin）是一种酸性蛋白酶，其活性中心缺乏天冬酰基残基，最适酶活条件为 pH 2.0 和 90℃。

胶原酶是一种与胶原蛋白水解有关的蛋白酶，胶原可作为药品和化妆品的非过敏性防腐剂（Gaffney et al. 1996；Honda 1998）。*Bacillus* NATP-1 和 *Alicyclobacillus sendaiensis* NATP-1 的胶原酶最适 pH 均在酸性范围，且这两株菌均为嗜酸细菌（Nakayama et al. 2000）。

7.8 商品化酸稳定酶

酶的历史起源于 1811 年，Kirchhoff 发现了第一个淀粉水解酶（Gupta et al. 2003）。酶的商业化历史起源于 1830 年，法国市场上首次出现了用于生产糊精、烘焙食品、啤酒和果酒的淀粉酶。1874 年丹麦的 Christian Hansen 创立了第一家生产奶酪凝乳酶的公司（Christian Hansen's Laboratory）（Chandel et al. 2007）。1894 年，用作助消化的真菌淀粉酶上市（Pandey et al. 2000），但是直到微生物蛋白酶作为洗涤剂添加剂时，酶才成为了家庭日用品。第一个细菌蛋白酶于 1959 年上市，随着诺维信（Novozyme）公司对酶的生产和研发，酶的价值日益提升（Leisola et al. 2002）。

调查显示，2013 年，全球的酶的市场价值为 44 亿美元，并且有望在 2020 年达到 76.5 亿美元。目前，在所有商业酶中，糖酶占据了主要市场，因为它们在食品、饮料行业中应用广泛。其次是蛋白酶，2013 年占据了全球酶市场的 27%。酶的重要性日益凸显，它们不仅降低了产品成本，并且有利于环境保护。

商业上大多数用于工业生产的酶在中性 pH 值和适中温度条件下活性最高，极端条件比如低 pH 会使它们快速失活，因此有必要寻找酸稳定性酶。淀粉工业中有着对酸稳定性酶（比如淀粉酶、葡萄糖异构酶）的需求，因为天然淀粉 pH 为 3.0～4.5。酸稳定性酶具有巨大的发展潜力，可以彻底改变现有的工业生产过程，使许多新的应用成为可能。现今，努力降低纤维质生物燃料生产成本，使其替代汽油燃料已成为当今一个最重要的趋势，酶的主要生产厂商比如 Novozyme 和 Danisco 已在这一领域投入了巨资。随着其他生物科技巨头如美国维莱尼姆（Verenium）公司、爱尔兰奎斯特（Quest）国际有限公司、丹麦健赞（Genzyme）公司、DSM 等在酶学领域投资的增加，酸稳定性酶市场有望迎来改观。

为了开发具有经济效益、环境友好的酸稳定性酶，科学知识和生产技术之间的广泛协作势在必行。表 7.2 为主要的市售的酸稳定性酶。如今，随着技术进步，市场竞争日趋激烈，酸稳定性酶的利润率也随之大幅下降，高利润的酶的生产已经成为新的挑战。尽管前景美好，但是由于制度的不健全，全球的酶市场仍面临着风险与挑战，发达国家尤为显著。尽管如此，酸稳定性酶的应用在未来有望增长。

表 7.2　商业用途的酸稳定性酶的详细信息

来源	酶	商业名称	最适 pH	应用	公司
Trichoderma longibrachiatum	β-葡聚糖酶	BrewZyme LP	4.2	烘焙、淀粉麸质分离、酒精发酵和动物饲料	Danisco
NA①	木聚糖酶	CeluStar XL	3.5～7.5		Danisco
NA	纤维素酶	HTec3	4.8～5.2	纤维素生物燃料中生物质降解的应用	Novozymes
NA	淀粉酶	SAN™Extra	3.0～4.5	淀粉水解	Novozymes
NA	果胶酶	Pectinex® Ultra Mash	3.0～4.2	磨碎苹果、梨和葡萄用来生产果汁	Novozymes
Aspergillus niger	葡萄糖淀粉酶	Boli GA-150	5	乙醇、酿造、谷氨酸盐和抗生素发酵等工业	Boli bioproducts
Bacillus deramificans	普鲁兰酶	OPTIMAX L-1000.	5	高果葡玉米糖浆（HFCS）以及啤酒和酒精饮料的生产	Genencor
A. niger	淀粉酶	Hazyme	4.5～6	果汁澄清	DSM
	蛋白酶	Enzyclean	2.5～8.0	茶叶制造业中工厂机械和地板的清理	Nivshakti Bioenergy Pvt Ltd.
NA	纤维素酶	ROCKSOFT™ ACL CONC	4.5	应用在牛仔服饰的浮石打磨法中	Dyadic International, Inc.

① 详细信息不可查。

7.9 利用嗜酸微生物发电

微生物燃料电池（MFC）可利用废水中存在的生物可降解化合物，通过微生物催化反应，以生物电的形式产生能量（Habermann and Pommer 1991；Logan et al. 2006；Wen et al. 2009；Raghavulu et al. 2009）。该装置在产生电的同时还能治理废水，因而具有非常广阔的发展前景（图 7.4）。微生物燃料电池中，细菌被放置于阳极上，与阴极的电子受体相分离，细菌只有通过向阳极转运电子才能完成呼吸作用。细菌在阳极氧化废水中的有机或无机物质，从而产生二氧化碳、质子和电子。

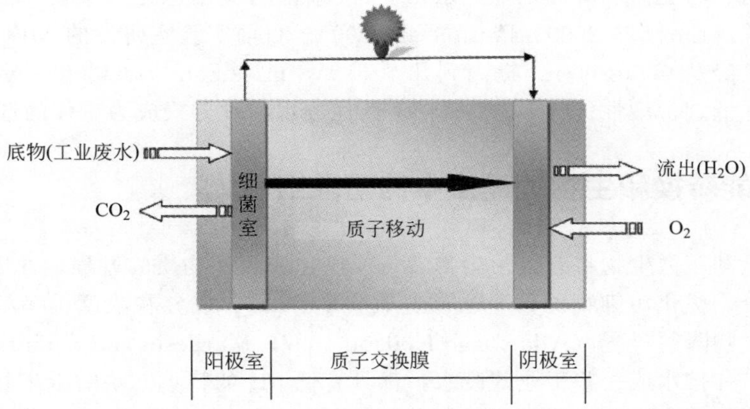

图 7.4　微生物燃料电池的原理图

所产生的质子和电子分别通过质子交换膜和外部电路向阴极移动（Oh et al. 2010；Rabey and Verstraete 2005），阴极发生氧化反应（通常是氧气）。反应方程如下所示（Jadhav and Ghangrekar 2009）。然而，该系统因内阻过大而产生的电能有限，故目前尚未实现商业化（Cheng and Logan 2011；Feng et al. 2008），可在微生物代谢反应方面作出改进，以提升系统所产生的能量。

$$\text{阳极：} C_{12}H_{22}O_{11} + 13H_2O \longrightarrow 12CO_2 + 48H^+ + 48e^-$$

$$\text{阴极：} 48H^+ + 48e^- + 12O_2 \longrightarrow 24H_2O$$

$$\text{总反应：} C_{12}H_{22}O_{11} + 12O_2 \longrightarrow 12CO_2 + 11H_2O$$

$$\Delta G = -5792.2\text{kJ/mol}$$

最近，利用微生物燃料电池发电受到越来越多的关注，嗜酸细菌可在 MFC 中起到重要作用，因为这些微生物在酸性条件下比嗜中性微生物催化效率更高（Jadhav and Ghangrekar 2009；Raghavulu et al. 2009）。电解质的 pH 值直接关系到 MFC 的效率，阳极和阴极溶液的 pH 相差越大，MFC 内部阻力的改变，发电效率越高。内部阻力是阳极电解质阻力、阴极电解质阻力和质子交换膜（PEM）产生阻力的总和。MFC 的内部阻力随着阳极和阴极溶液的 pH 差的增大而降低，因为高 pH 差使通过 PEM 的质子流加速。因此，当阳极和阴极的 pH 差越大，MFC 的产能效率越高（Jadhav and Ghangrekar 2009）。Borole 等（2008）在低于 pH 4 的环境中成功利用嗜酸细菌 *Acidiphilium cryptum* 作为生物催化剂，实现在酸性条件下发电，同样 Jadhav 和 Ghangrekar（2009）也发现了高 pH 差与最大发电功率相关。

为了促进细菌生长，大部分 MFCs 在中性 pH 条件下进行，但该过程面临一个问题，即在此 pH 值下质子浓度较低，导致系统内部阻力高。鉴于此嗜酸性细菌与低 pH 的电解液结合使用，MFC 的效率可显著提升（Ieropoulos et al. 2005；Daniel et al. 2009）。此外，当 MFC 使用酸性废水时，嗜酸性系统可以克服细菌效率的限制（Biffinger et al. 2008；Erable et al. 2009）。力拓河（维尔瓦，西班牙）是一个天然酸性生态系统，平均 pH 值为 2.3±0.6，水中大量的三价铁离子作为缓冲剂维持河的酸性 pH。微生物生态学研究已经证实水中 80%的原核生物由三种细菌 *A. ferrooxidans*、*Acidiphilium* spp. 和 *Leptospirillum* spp. 组成，三者均为铁循环中主要的成员（García-Muñoz et al. 2011；González-Toril et al. 2003；Amaral-Zettler et al. 2002）。研究表明在不考虑使用阴极电解质本身的情况下，酸性 pH 相对于中性和碱性 pH 能产生更多的能量。Sulonen 等（2015）发现在 pH 1.2～2.5 时，可使用四硫酸根作为 *Acidithiobacillus* 菌和 *Ferroplasma* 菌的电子供体。当废水为酸性时（比如淀粉、巧克力和啤酒工业废水），嗜酸细菌的应用也尤为重要，因为可以省去添加缓冲剂的步骤（Lu et al. 2009；Patil et al. 2009）。目前实验室研发的 MFC 功率输出在不到十年时间内已经从 0.001W/m^2 提高到几个 6.9W/m^2（Fan et al. 2008；Oh et al. 2010），随着材料成本降低，价廉性优的 MFC 相较于其他废水治理方式成为更佳的选择。

7.10 生物能转换和生物处理法中的嗜酸细菌

生物修复是利用微生物将有机污染物转化或矿化成无害物质的过程，并成为地球生物化学循环的一部分。优化污染物的营养物质、氧含量、pH、成分和浓度等多种因素，可加快有毒化合物的生物降解过程（Allard and Neilson 1997；Margesin and Schinner 2001）。许多环境，如酸性矿井排水或一些工业的废水，都具有低 pH 的特点。嗜酸微生物适于在极端环境下生长，烃降解或者重金属累积嗜酸细菌（比如 *Acidiphilium rubrum*）具有从被污染的极端生长环境中去除污染物的能力（Johnson 1995；Stapleton et al. 1998；Roling et al. 2006）。

酸性矿井排水（AMD）是世界上水污染最严重的形式之一，对环境造成了严重的危害。AMD 是全球采矿业面临的主要环境问题，AMD 不仅与地表水和地下水的污染有关，还是重金属散布到环境中的主要原因。不仅如此，金属矿采废水中含有的氰化物和重金属等有毒物质，对人类健康和生态平衡造成了严重的威胁（Sheoran and Sheoran 2006；Hallberg 2010）。除此之外，AMD 的高酸性使其他金属矿物进一步溶解，致使其中的金属含量比中性环境更高，其中铝、铜、铅、锌、镉、镍和砷等元素浓度通常较高（Sullivan and Yelton 1988；Johnson 1995）。

AMD 可使被污染水体 pH 下降，毒性增加，不适于水生生物生长。不仅如此，AMD 中氧化和沉淀的金属也导致中和作用下降，并由此使被污染水体的 pH 降低。pH 值的下降导致有毒金属的溶解度提高，对水生生物产生毒害作用。除此之外，金属沉淀，尤其是铁和铝的沉淀，导致被污染水体底部的金属聚集，破坏水生生物食物链，抑制水底生物的生殖，进而扰乱了水生生物圈（Hallberg 2010）。

AMD 治理的传统方法是通过加入外源碱以提高 pH 值至铁氧化细菌所需 pH 之上，从而降低酸生成的速率，但该方法用于治理酸性矿井排水成本过高（Sheoran and Sheoran 2006）。然而，可利用与重金属元素相互作用的嗜酸性微生物从污染土壤中去除有毒金属。细菌具有能够耐受高浓度金属分子离子的特性（比如主动外排或通过金属分子伴侣捕获）；此外，基因重复、基因组岛的存在，以及无机多磷酸盐驱动的金属抗性机制，使嗜酸菌适合进行生物修复（Dopson et al. 2003；Franke and Rensing 2007；Navarro et al. 2009；Krulwich et al. 2011）。

许多情况下，酸性矿井排水和含有多环芳烃的工业泄漏物污染环境，这些化合物通常认为对生物和环境都是有害的（Sutherland 1992；Pothuluri and Cerniglia 1994；Stapleton et al. 1998）。因此，人们对有助于污染修复的极端嗜酸菌产生了极大的兴趣。

据报道嗜酸菌可降解多种碳氢化合物，包括脂肪族、芳香族、卤化物和硝化物，适合在这种环境条件下生长的烃类降解嗜酸菌在污染环境的生物修复方面发挥了重要作用。不同类型碳氢化合物的生物降解（转化或矿化）在多种极端栖息地中均有发现，石油碳氢化合物中的多种成分可利用嗜酸细菌进行降解（Stapleton et al. 1998；Christen et al. 2012）。

许多重金属耐受嗜酸性细菌能够代谢多种脂肪族碳氢化合物，这些细菌已在酸性条件下被分离。此外，大多对嗜酸性细菌有毒的脂肪族有机酸（Alexander et al. 1987）被用来作为产能和生长的底物。一个与 *Acidocella* 菌相近的潜在的嗜酸性细菌已被分离出来，如果连续少量地加入乙酸（1mmol/L，体积分数大约为 0.006%），可耐受高浓度的乙酸（Gemmen and Knowles 2000）。此外，Roling 等（2006）从天然的石油表面渗出物中提取出了 DNA，并且做了可培养分离的实验，表明了嗜酸性细菌占主导地位，尤其是 α-Proteobacteria 居多（主要是 *Acidiphilium* 和 *Acidocella*）。古菌的存在未被证实，但是有真菌的存在，样本 pH 范围是 3.0 到 5.0。

Stapleton 等（1998）研究发现，从 pH 为 2.0 的煤存放地的土壤样品中分离出来的嗜酸性细菌具有降解芳香族化合物的能力。在如此低的 pH 下，也有超过 40%的碳氢化合物、萘和甲苯被降解为二氧化碳和水。DNA 杂交分析表明，从这些样品的整个菌群样本中分离出来的核酸不与属于嗜中性细菌的基因（*nahA*、*nahG*、*nahH*、*todC1C2* 和 *tomA*）杂交。这些数据结果表明了芳香族化合物的降解能发生在具有极端低 pH 值的环境中。同样 Hamamura 等（2005）发现细菌可在直链烷烃中生长（比如链长度 C_{15} 到 C_{30} 的正烷烃，包括支链烷烃），在低 pH 值下（pH 2.8～3.8）主要为姥鲛烷和植烷，这是硫酸盐地热活动的特征。细菌群落通过 16S rRNA 基因克隆数据库来分类，显示了其与变形菌门的异养嗜酸性细菌 *Acidisphaera*、*Acidiphilium* 菌和化能无机自养菌 *Acidithiobacillus* spp 序列之间存在相关性。富含碳氢化合物的沙土混合物的 16S rRNA 基因片段的变性梯度凝胶电泳（DGGE）显示出与异养嗜酸性细菌相关序列为主的 DGGE 图谱。除此以外，有人培养了一个烷烃降解菌株，证实了在一个酸性碳氢化合物渗透土壤中存在具有烷烃降解能力的天然菌群。最近，Christen 等（2012）报道了 *S. solfataricus*，一个嗜热嗜酸性古菌，它是一个环境适应良好的菌株，可在 80℃和 pH 3.2 下降解酚。酚是来自于多种工业比如精炼、焦化、煤矿生产和石油化学生产的废水中的一种有机污染物。在一些工业废水中，酚的浓度可以达到 6.8g/L。嗜酸性细菌可在高温，低 pH 下生长，其降解酚的潜力对于被酚污染的热酸性地区的生物修复来说具有重要意义。

鉴于嗜酸细菌具有还原铁、硫的能力，其在酸性废水的治理中扮演着非常重要的角色，烃类降解嗜酸菌在被石油污染的酸性废水处理过程中起着重要的作用。

7.11 食品中的嗜酸微生物

酸性食物通常具有防腐特性，但是仍有一些微生物可在酸性条件下生长，并且是重要的腐败微生物，例如 *Alicyclobacillus* 是果汁工业中的主要污染菌。1984 年，德国一起严重的苹果汁污染事件就是由 *Alicyclobacillus* 引起的，随后又有许多报道 *Alicyclobacillus* 导致果汁、混合果汁、浓缩果汁、碳酸果汁和货架期冰红茶的腐败。*Alicyclobacillus* 的孢子具有高度耐酸耐热的特性（Walker and Phillips 2008），它们能够在巴氏灭菌（92℃，10s）中存活，并引起食物腐败。鉴于此 *A. acidoterrestris* 被作为酸性食品巴氏灭菌中的指示生物（Sliva and Gibbs 2001）。

7.12 结论与前景

嗜酸细菌是一组微生物的集合体，鉴于嗜酸微生物应用的广泛性，其多样性目前仍亟待探索。尽管有一些尝试去解释嗜酸微生物在酸性环境下进化出的生存生长策略，但这方面仍然需要进一步探索。嗜酸微生物的酸稳定性酶/蛋白在工业生产中大有用处，并且嗜酸细菌已被广泛应用于低品质矿石的生物浸取，微生物燃料中嗜酸细菌的利用也使该产业的前景被广为看好。

无利益冲突声明

Archana Sharma，Deepak Parashar，和 Tulasi Satyanarayana 声明不存在利益冲突。

参考文献

Acuna-Arguelles ME，Gutierrez-Rajas M，Viniegra-Gonzalez G，Favela-Toress E（1995）Production and properties of three pectinolytic activities produced by *A. niger* in submerged and solid state fermentation. Appl Microbiol Biotechnol 43：808-814

Alexander B，Leach S，Ingledew WJ（1987）The relationship between chemiosmotic parameters and sensitivity to anions and organic acids in the acidophile *Thiobacillus ferrooxidans*. J Gen Microbiol 133：1171-1179

Allard AS，Neilson AH（1997）Bioremediation of organic waste sites：a critical review of micro-biological aspects. Int Biodeter Biodegr 39：253-285

Al-Obaidi ZS，Aziz GM，Al-Bakir AY（1987）Screening of fungal strains for polygalacturonase production. J Agric Water Resour Res 6：125-182

Amaral-Zettler LA，Gómez F，Zettler E，Keenan BG，Amils R，Sogin ML（2002）Eukaryotic diver-sity in Spain's river of fire：this ancient and hostile ecosystem hosts a surprising variety of microbial organisms. Nature 417：37

Bai Y，Huang H，Meng K，Shi P，Yang P，Luo H，Luo C，Feng Y，Zhang W，Yao B（2012）Identification of an acidic α-amylase from *Alicyclobacillus* sp. A4 and assessment of its appli-cation in the starch industry. Food Chem 131：1473-1478

Baker-Austin C，Dopson M（2007）Life in acid：pH homeostasis in acidophiles. Trends Microbiol 15（4）：165-171

Bampton KF，Bologiannis F，Canterford JH，Smith AN（1983）Development of experimental insitu leaching at the Mutooroo copper mine，South Australia. In：Australasian IMM Annual Conference，Broken Hill，NSW，pp 371-379

Batrakov SG，Pivovarova TA，Esipov SE，Sheichenko VI，Karavaiko GI（2002）Beta-D-glycopyranosyl caldarchaetidylglycerol is the main lipid of the acidophilic，mesophilic，ferrous iron-oxidizing archaeon *Ferroplasm acidiphilum*. Biochim Biophys Acta 1581：29-35

Berthelot D，Leduc LG，Ferroni GD（1994）The absence of psychrophilic *Thiobacillus ferrooxidans* and acidophilic heterotrophic bacteria in cold，tailings effluents from a uranium mine. Can J Microbiol 40：60-63

Bertoldo C，Dock C，Antranikian G（2004）Thermoacidophilic microorganisms and their novel biocatalysts. Eng Life Sci 4：521-531

Biffinger JC，Pietron J，Bretschger O，Nadeau LJ，Johnson GR，Williams CC，Nealson KH，Ringeisen BR（2008）The influence of acidity on microbial fuel cells containing *Shewanella oneidensis*. Biosens Bioelectron 24：900-905

Borin MDF，Said S，Fonseca MJV（1996）Purification and biochemical characterization of an extracellular endopolygalacturonase from *Penicillium frequentans*. J Agric Food Chem 44：1616-1620

Borole AP，Neill HO，Tsouris C，Cesar S（2008）A microbial fuel cell operating at low pH using the acidophile *Acidiphilium cryptum*. Biotechnol Lett 30：1367-1372

Brierley CL，Brierley JA（2000）Bioheap processes：operational requirements and techniques. In：Proceedings of Randol Copper Hydromet Roundtable，Tucson，Ariz. pp 95-103

Chandel AK，Rudravaram R，Rao LV，Ravindra P，Narasu ML（2007）Industrial enzymes in bio-industrial sector development：An Indian perspective. J Commer Biotechnol 13：283-291

Channe PS，Shewal JG（1995）Pectinase production by *Sclerotium rolfsii*：Effect of culture condi-tions. Folia Microbiol 40：111-117

Chen L-X，Hu M，Huang L-N，Hua Z-S，Kuang J-L，Li S-J，Shu W-S（2014）Comparative metage-nomic and metatranscriptomic analyses of microbial communities in acid mine drainage. Int Soc Microb Ecol 9：1-14

Cheng SA，Logan BE（2011）Increasing power generation for scaling up single-chamber air cath-ode microbial fuel cells. Biores Technol 102：4468-4473

Christen P，Vega A，Casalot L，Simon G，Auria R（2012）Kinetics of aerobic phenol biodegradation by the acidophilic and hyperthermophilic archaeon *Sulfolobus solfataricus* 98/2. Biochem Eng J 62：56-61

Ciaramella M，Napoli A，Rossi M（2005）Another extreme genome：how to live at pH 0. Trends Microbiol 13：49-51

Crossman L，Holden M，Pain A，Parkhill J（2004）Genomes beyond compare. Nat Rev Microbiol 2：616-617

Daniel DK，Mankidy BD，Ambarish K，Manogari R（2009）Construction and operation of a micro-bial fuel cell for electricity generation from wastewater. Int J Hydrogen Energy 34：7555-7560

Davies DR（1990）The structure and function of the aspartic proteinases. Annu Rev Biophys Chem 19：189-215

Dew DW，Lawson EN，Broadhurst JL（1997）The Biox® process for biooxidation of gold-bearing ores or concentrates. In：Rawlings DE（ed）Biomining：theory microbes and Industrial pro-cesses. Springer，Berlin，pp 45-80

Do TT，Quyen DT，Dam TH（2012）Purification and characterization of an acid-stable and organic solvent-tolerant xylanase from *Aspergillus awamori* VTCC-F312. Sci Asia 38：157-165

Dopson M，Baker-Austin C，Koppineedi PR，Bond PL（2003）Growth in sulfidic mineral environ-ments：metal resistance mechanisms in acidophilic micro-organisms. Microbiology 149：1959-1970

Dopson M，Baker-Austin C，Hind A，Bowman JP，Bond PL（2004）Characterization of *Ferroplasma* isolates and *Ferroplasma acidarmanus* sp. nov.，extreme acidophiles from acid mine drainage and industrial bioleaching environments. Appl Environ Microbiol 70：2079-2088

Eckert K，Schneider E（2003）A thermoacidophilic endoglucanase（CelB）from *Alicyclo bacillusacidocaldarius* displays high sequence similarity to arabinofuranosidases belonging to family 51 of glycoside hydrolases. Eur J Biochem 270：3593-3602

Erable B，Etcheverry L，Bergel A（2009）Increased power from a two chamber microbial fuel cell with a low pH air-cathode compartment. Electrochem Commun 11：619-622

Fan Y，Sharbrough E，Liu H（2008）Quantification of the internal resistance distribution of micro-bial fuel cells. Environ Sci Technol 42：8101-8107

Feng Y，Wang X，Logan BE，Lee H（2008）Brewery wastewater treatment using air cathode micro-bial fuel cells. Appl Microbiol Biotechnol 78：873-880

Franke S，Rensing C（2007）Acidophiles. Mechanisms to tolerate metal and acid toxicity. In：Gerday C，Glansdorff N（eds）Physiology and biochemistry of extremophiles. ASM Press，Washington，DC，pp 271-278

Fusek M，Lin XL，Tang J（1990）Enzymic properties of thermopsin. J Biol Chem 265：1496-1501

Futterer O，Angelov A，Liesegang H，Gottschalk G，Schleper C，Schepers B，Dock C，Antranikian G，Liebl W（2004）Genome sequence of *Picrophilus torridus* and its implications for life around pH 0. Proc Natl Acad Sci USA 101：9091-9096

Gaffney PJ，Edgell TA，Dawson PA，Ford AW，Stocker E（1996）A pig collagen peptide fraction. A unique material for maintaining biological activity during lyophilization and during storage in the liquid state. J Pharm Pharmacol 48：896-898

García-Muñoz J，Amils R，Fernández VM，Lacey ALD，Malki M（2011）Electricity generation by microorganisms in the sediment-water interface of an extreme acidic microorganism. Int Microbiol 14：73-81

Geisler RA，Pudington IE（1996）Treatment of lead sulfide bearing minerals. US Patent 5：523-066

Gemmell RT，Knowles CJ（2000）Utilisation of aliphatic compounds by acidophilic heterotrophic bacteria：the potential for bioremediation of acidic wastewaters contaminated with toxic organic compounds and heavy metals. FEMS Microbiol Lett 192：185-190

Golyshina OV，Timmis KN（2005）*Ferroplasma* and relatives，recently discovered cell wall lacking archaea making a living in extremely acid，heavy metal-rich environments. Environ Microbiol 7：1277-1288

Golyshina OV，Pivovarova TA，Karavaiko GI，Moore ER，Abracham WR，Lunsdorf H，Timmis KN，Yakimov MM，Golyshin PN（2000）*Ferroplasma acidiphilum gen. Nov.，sp. Nov.*，an aci dophilic autotrophic，ferrous-iron oxidizing，cell wall lacking，mesophilic member of the Ferroplasmaceae fam. Nov.，comprising distinct lineage of Archaea. Int J Syst Evol Microbiol 50：997-1006

González-Toril E，Llobet-Brossa E，Casamayor EO，Amann R，Amils R（2003）Microbial ecology of an extreme acidic environment，the Tinto River. Appl Environ Microbiol 69：4853-4865

Gupta R，Gigras P，Mohapatra H，Goswami VK，Chauhan B（2003）Microbial α-amylase：a bio technological perspective. Process Biochem 38：1599-1616

Habermann W，Pommer EH（1991）Biological fuel cells with sulphide storage capacity. Appl Microbiol Biotechnol 35：128-133

Hallberg KB（2010）New perspectives in acid mine drainage microbiology. Hydrometallurgy 104：448-453

Hamamura N，Olson SH，Ward DM，Inskeep WP（2005）Diversity and functional analysis of bacterial communities associated with natural hydrocarbon seeps in acidic soils at Rainbow Springs，Yellowstone National Park. Appl Environ Microbiol 71：5943-5950

Honda S（1998）Dietary use of collagen and collagen peptides for cosmetics. Food Style 21：54-60

Huang Y，Krauss G，Cottaz S，Driguez H，Lipps G（2005）A highly acid-stable and thermostable endo-b-glucanase from the thermoacidophilic archaeon *Sulfolobus solfataricus*. Biochem J 385：581-588

Ieropoulos IA，Greenman J，Melhuish C，Hart J（2005）Comparative study of three types of micro-bial fuel cell. Enzyme Microb Tech 37：238-245

Inagaki K，Nakahira K，Mukai K，Tamura T，Tanaka H（1998）Gene cloning and characterization of an acidic xylanase from *Acidobacterium capsulatum*. Biosci Biotechnol Biochem 62：1061-1067

Jadhav GS，Ghangrekar MM（2009）Performance of microbial fuel cell subjected to variation in pH，temperature，external load and substrate concentration. Biores Technol 100：717-723

Jeffries AC，Kozera CJ，Medina N，Peng X，Thi-Ngoc HP，Redder P，Schenk ME，Theriault C，Tolstrup N，Charlebois RL，Doolittle WF，Duguet M，Gaasterland T，Garrett RA，Ragan MA，Sensen CW，Vander Oost J（2001）The complete genome of the Crenarchaeon *Sulfolobus sol fataricus* P2. Proc Natl Acad Sci USA 98：7835-7840

Johnson DB（1995）Acidophilic microbial communities：candidates for bioremediation of acidic mine effluents. Int Biodeter Biodegrad 35：41-58

Johnson D（1998）Biodiversity and ecology of acidophilic microorganism. FEMS Microbiol Ecol 27：307

Johnson D（2008）Biodiversity and interactions of acidophiles：Key to understanding and optimizing microbial processing of ores and concentrates. Trans of Nonferr Metal Soc China 18：1367-1373

Johnson D，Hallberg K（2003）The microbiology of acidic mine waters. Res Microbiol 154：466-473

Kim HW, Kim YO, Lee JH, Kim KK, Kim YJ (2003) Isolation and characterization of a phytase with improved properties from *Citrobacter braakii*. Biotechnol Lett 25: 1231-1234

Krulwich TA, Sachs G, Padan E (2011) Molecular aspects of bacterial pH sensing and homeosta-sis. Nat Rev Microbiol 9: 330-343

Laksanalamai P, Robb FT (2004) Small heat shock proteins from extremophiles: a review. Extremophiles 8: 1-11

Leisola M, Jokela J, Pastinen O, Turunen O, Schoemaker H (2002) Industrial use of enzymes. In: Encyclopedia of life support systems (EOLSS), EOLSS Publishers, Oxford

Liu XD, Xu Y (2008) A novel raw starch digesting a-amylase from a newly isolated *Bacillus* sp.

YX-1: purification and characterization. BioresTechnol 99: 4315-4320

Logan BE, Hamelers B, Rozendal R, Schrorder U, Keller J, Freguia S, Aelterman P, Verstraete W, Rabaey K (2006) Microbial fuel cells: methodology and technology. Environ Sci Technol 40: 5181

Lu N, Zhou SG, Zhuang L, Zhnag JT, Ni JR (2009) Electricity generation from starch processing wastewater using microbial fuel cell technology. Biochem Eng J 43: 246-251

Macalady JL, Vestling MM, Baumler D, Boekelheide N, Kaspar CW, Banfield JF (2004) Tetraether-linked membrane monolayers in *Ferroplasma* spp: a key to survival in acid. Extremophiles 8: 411-419

Marcus L, Barash I, Sneh B, Koltin Y, Finker A (1986) Purification and characterization of pecto-lytic enzymes produced by virulent and hypovirulent isolates of *Rhizoctonia solani* Kuhn. Physiol Mol Plant Pathol 29: 325-336

Margesin R, Schinner F (2001) Biodegradation and bioremediation of hydrocarbons in extreme environments. Appl Microbiol Biotechnol 56: 650-663

Matzke J, Schwermann B, Baker EP (1997) Acidstable and acidophilic proteins: the example of the alpha amylase from *Alicyclobacillus acidocaldarius*. Comp Biochem Physiol A Mol Integr Physiol 118: 411-419

Murao S, Okhuni K, Naganao M (1988) A novel thermostable S-PI (pepstatin Ac)-insensitive acid proteinase from thermophilic *Bacillus* novo sp. strain Mn-32. Agric Biol Chem 52: 1029-1031

Nakayama T, Tsuruoka N, Akai M, Nishino T (2000) Thermostable collagenolytic activity of a novel thermophilic isolate, *Bacillus* sp. Strain NTAP-1. J Biosci Bioeng 89: 612-614

Navarro CA, Orellana LH, Mauriaca C, Jerez CA (2009) Transcriptional and functional studies of *Acidithiobacillus ferrooxidans* genes related to survival in the presence of copper. Appl Environ Microbiol 75: 6102-6109

Oda K, Nakazima T, Terashita T, Suziki KA, Murao S (1987a) Purification and properties of an S-PI (Pepstatin Ac) insensitive carboxyl proteinase from a *Xanthomonas* sp. bacterium. Agric Biol Chem 51: 3073-3080

Oda K, Sugitani M, Fukuhara K, Murao S (1987b) Purification and properties of a pepstatin-insensitive carboxyl proteinase from a gram negative bacterium. Biochim Biophys Acta 923: 463-469

Oh JR, Kim GC, Premier TH, Lee C, Kim WT (2010) Sloan Sustainable wastewater treatment: how might microbial fuel cells contribute. Biotechnol Adv 28: 871-881

Pandey A, Nigam P, Soccol CR, Soccol VT, Singh D, Mohan R (2000) Advances in microbial amylases. Biotechnol Appl Biochem 31: 135-152

Patil SA, Surakasi VP, Koul S, Ijmulwar S, Vivek A, Shouche YS, Kapadnis BP (2009) Electricity generation using chocolate industry wastewater and its treatment in activated sludge based microbial fuel cell and analysis of developed microbial community in the anode chamber. Biores Technol 100: 5132-5139

Pivovarova TA, Kondrateva TF, Batrakov SG, Esipov SE, Sheichenko VI, Bykova SA (2002) Phenotypic features of *Ferroplasma acidiphilum* strains Yt and Y-2. Microbiol (Moscow) 71: 809-818

Plumb JJ, Haddad CM, Gibson JA, Franzmann PD (2007) *Acidianus sulfidivorans* sp. nov., an extremely acidophilic, thermophilic archaeon isolated from a solfatara on Lihir Island, Papua New Guinea, and emendation of the genus description. Int J Syst Evol Microbiol 57: 1418-1423

Pothuluri JV, Cerniglia CE (1994) Microbial metabolism of polycyclic aromatic hydrocarbons. In: Chaundry GR (ed) Biological degradation and bioremediation of toxic chemicals. Dioscorides Press, Portland, OR, pp 92-124

Pronk JT, Johnson DB (1992) Oxidation and reduction of iron by acidophilic bacteria. Geomicrobiol J 10 (3-4): 153-171

Rabaey K, Verstraete W (2005) Microbial fuel cells: novel biotechnology for energy generation.

Trends Biotechnol 23: 291-298

Raghavulu SV, Mohan SV, Goud RK, Sarma PN (2009) Effect of anodic pH microenvironment on microbial fuel cell (MFC) performance in concurrence with aerated and ferricyanide catho-lytes. Electrochem Commun 11: 371-375

Ram RJ, VerBerkmoes NC, Thelen MP, Tyson GW, Baker BJ, Blake RC II, Shah M, Hettich RL, Banfield JF (2005) Community proteomics of a natural microbial biofilm. Science 308: 1915-1999

Rawlings DE (ed) (1997) Biomining: theory, microbes and industrial processes. Springer, Berlin Rawlings DE (2002) Heavy metal mining using microbes. Annu Rev Microbiol 56: 65-91

Rawlings DE, Johnson DB (2007) The microbiology of biomining: development and optimization of mineral-oxidizing microbial consortia. Microbiology 153: 315-324

Rodriguez E, Han Y, Lei XG (1999) Cloning, sequencing and expression of an *Escherichia coli* acid phosphatase, phytase gene (*appA2*) isolated from pig colon. Biochem Biophys Res Commun 257: 117-123

Roling WFM, Ortega-Lucach S, Larter SR, Head IM (2006) Acidophilic microbial communities associated with a natural, biodegraded hydrocarbon seepage. J Appl Microbiol 101: 290-299

Satyanarayana T, Raghukumar C, Shivaji S (2005) Extremophilic microbes: diversity and perspec-tives. Curr Sci 89: 78-90

Schleper C，Püehler G，Kuhlmorgen B，Ziling W（1995）Life at extremely low pH. Nature 375：741-742

Schleper C，Puehler G，Klenk HP，Zillig W（1996）*Picrophilus oshimae Picrophilus torridus fam. nov.*，*gen. nov.*，*sp. nov.*，two species of hyperacidophilic，thermophilic，heterotrophic aerobic archaea. Int J Syst Bacteriol 46：814-816

Schnell HA（1997）Bioleaching of copper. In：Rawlings DE（ed）Biomining：theory microbes and industrial processes. Springer-Verlag and Landes Bioscience，Berlin，Germany，pp 21-43

Schwermann B，Pfau K，Liliensiek B，Schleyer M，Fischer T，Bakker EP（1994）Purification，properties and structural aspects of the thermoacidophilic α-amylase from *Alicyclo-bacillusacidocaldarius* ATCC 27009. Insight into acidostability of proteins. Eur J Biochem 226：981-991

Serour E，Antranikian G（2002）Novel thermoactive glucamylases from the thermoacidophilic archaea *Thermoplasma acidophilum*，*Picrophilus torridus* and *Picrophilus oshimae*. Antonie van Leewenhoek 81：73-83

Shah AR，Shah RK，Madamwar D（2006）Improvement of the quality of whole wheat bread by supplementation of xylanase from *Aspergillus foetidus*. Biores Technol 97：2047-2053

Sharma A，Satyanarayana T（2010）High maltose-forming，Ca^{2+}-independent and acid stable α-amylase from a novel acidophilic bacterium *Bacillus acidicola* TSAS1. Biotechnol Lett 32：1503-1507

Sharma A，Satyanarayana T（2012）Production of acid-stable and high-maltose-forming α-amylase of *Bacillus acidicola* by solid-state fermentation and immobilized cells and its applicability in baking. Appl Biochem Biotechnol 168：1025-1034

Sharma A，Kawarabayasi Y，Satyanarayana T（2012）Acidophilic bacteria and archaea：acid stable biocatalysts and their potential applications. Extremophiles 16：1-19

Sheoran AS，Sheoran V（2006）Heavy metal removal mechanism of acid mine drainage in wet-lands：A critical review. Miner Eng 19：105-116

Shimada H，Nemoto N，Shida Y，Oshima T，Yamagishi A（2002）Complete polar lipid composition of *Thermoplasma acidophilum* HO-62 determined by high performance liquid chromatogra-phy with evaporative light scattering detection. J Bacteriol 184：556-563

Shivaramakrishnan S，Gangadharan D，Nampoothiri KM，Soccol CR，Pandey A（2006）α-Amylases from microbial sources-an overview on recent developments. Food Technol Biotechnol 44：173-184

Siddiqui MH，Kumar A，Kesari KK，Arif JM（2009）Biomining-A useful approach toward metal extraction. American-Eurasian J Agron 2：84-88

Silva FVM，Gibbs P（2001）*Alicyclobacillus acidoterrestris* spores in fruit products and design of pasteurization processes. Trends Food Sci Technol 12：68-74

Soni SK，Magdum A，Khire JM（2010）Purification and characterization of two distinct acidic phytases with broad pH stability from *Aspergillus niger* NCIM 563. World J Microbiol Biotechnol 26：2009-2018

Spijkerman E，Bissinger V，Meister A，Gaedke U（2007）Low potassium and inorganic carbon concentrations influence a possible phosphorus limitation in *Chlamydomonas acidophila* (Chlorophyceae). Eur J Phycol 42：327-339

Stapleton RD，Savage DC，Sayler GS，Stacey G（1998）Biodegradation of aromatic hydrocarbons in an extremely acidic environment. Appl Environ Microbiol 64：4180-4184

Sullivan PJ，Yelton JL（1988）An evaluation of trace element release associated with acid mine drainage. Environ Geol Water Sci 12：181-186

Sulonen ML，Kokko ME，Lakaniemi AM，Puhakka JA（2015）Electricity generation from tetrathi-onate in microbial fuel cells by acidophiles. J Hazard Mater 284：182-189

Sutherland JB（1992）Detoxification of polycyclic aromatic hydrocarbons by fungi. J Ind Microbiol 9：53-62

Takayanagi S，Kawasaki H，Sugimori K，Yamada T，Sugai A，Ito T，Yamasato K，Shioda M（1996）*Sulfolobus hakonensis* sp. nov.，a novel species of acidothermophilic archaeon. Int J Syst Bacteriol 46：377-382

Thiel V（2011）Extremeenvironments. Encyclopedia of Geobiology 362-366

Tyson GW，Chapman J，Hugenholtz P，Allen EE，Ram JR，Richardson MP，Solovyev VV，Rubin ME，Rokhsar SD，Banfield FJ（2004）Community structure and metabolism through recon struction of microbial genomes from the environment. Nature 428：37-43

Vande Vossenberg JL，Driessen AJ，Konings WN（1998a）The essence of being extremophilic：the role of the unique archaeal membrane lipids. Extremophiles 2：163-170

Van de Vossenberg JL，Driessen AJ，Zillig W，Konings WN（1998b）Bioenergetics and cytoplasmic membrane stability of the extremely acidophilic，thermophilic archaeon *Picrophilus oshimae*. Extremophiles 2：67-74

Walker M，Phillips CA（2008）*Alicyclobacillus acidoterrestris*：an increasing threat to the fruit juice industry. Int J Food Sci Technol 43：250-260

Wen Q，Wu Y，Cao D，Zhao L，Sun Q（2009）Electricity generation and modeling of microbial fuel cell from continuous beer brewery wastewater. Biores Technol 100：4171-4175

Yim KJ，Cha IT，Rhee JK，Song HS，Hyun DW，Lee HW，Kim D，Kim KN，Nam YD，Seo MJ，Bae JW，Roh SW（2015）*Vulcanisaeta thermophila* sp. nov.，a hyperthermophilic and acidophilic crenarchaeon isolated from solfataric soil. Int J Syst Evol Microbiol 65：201-205

Zychlinsky E，Matin A（1983）Effect of starvation on cytoplasmic pH，proton motive force，and viability of an acidophilic bacterium *Thiobacillus acidophilus*. J Bacteriol 153：371-374

第八章

嗜碱微生物在生物技术中的应用

Gashaw Mamo[1]，Bo Mattiasson[2]

8.1 引言

嗜碱微生物通常是指生长在碱性条件下的微生物。这类微生物中，一部分以高 pH 作为最佳生长条件，而另一部分则可以在高 pH 环境下生长，但并非最佳生长条件。普遍认为最佳生长条件为 pH 值 9 或以上的微生物被定义为嗜碱微生物，进一步地可将其再分为专性嗜碱微生物（不能在中性 pH 下生长）和兼性嗜碱微生物（其能够在中性条件下生长）。最适生长条件为 pH 9 以下的但又能在高 pH 环境下存活的微生物被称为耐碱微生物。营养物质的类型、浓度以及培养温度等几个因素都可以影响微生物生长的最佳 pH（Horikoshi 1999）。因此，微生物生长的最佳 pH 是根据生长条件的变化而变化的。至少在某一种特定的培养条件（例如培养基和温度）中具有在 pH 9 以上最适生长能力，似乎是嗜碱性微生物的唯一区分特征。

8.1.1 嗜碱微生物的生存环境及多样性分析

目前已从不同种类的天然和人造碱性环境中分离出嗜碱性微生物，这些环境对于其他类型的有机体而言是一种极端环境。碱湖和碱性沙漠是天然存在的最稳定的碱性环境，其 pH 值通常超过 10。这些生境是嗜碱微生物的主要来源，并且有几种新物种从中被分离和描述（Jones et al. 1999；Horikoshi 1999；Delgado et al. 2006；Canganella and Wiegel 2011；Grant and Sorokin 2011；Sorokin et al. 2014；Glaring et al. 2015）。具有高 pH 的碱性微环境也可以由人类活动造成，例如水泥制造、纸和纸浆生产、采矿活动、靛蓝染料制备、食品和纺织品加工、电镀加工等（Grant and Tindall 1986）。这些工业活动的废液流域很可能富集嗜碱微生物。此外，在白蚁的内脏（Taksawan et al. 2005）、粪便（Horikoshi 1999）甚至中性土壤中也发现有嗜碱微生物的存在（Horikoshi 1991）。

---

[1] Department of Biotechnology，Lund University，P. O. Box 124，Lund，Sweden；Indienz AB，Billeberga，Sweden. e-mail：gashaw. mamo@biotek. lu. se.

[2] Department of Biotechnology，Lund University，P. O. Box 124，Lund，Sweden；Indienz AB，Billeberga，Sweden. bo. mattiasson@biotek. lu. se.

已知在碱性环境中栖息着不同的生物物种，它们分别属于细菌、古菌和真核生物（Hayashi et al. 1996；Horikoshi 1999；Duckworth et al. 1996；Canganella and Wiegel 2011）。迄今为止鉴定的绝大多数嗜碱微生物属于真细菌，其中很大一部分是从碱湖样品中分离得到的。这些极端环境微生物不仅在分类学上存在不同，而且它们在生理学上也具有多样性，并且属于不同的类群，例如好氧菌、兼性厌氧菌、专性厌氧菌、嗜盐菌、嗜热菌、嗜冷菌、嗜温菌等（Grant and Sorokin 2011；Sorokin et al. 2015）。

8.1.2 高 pH 值环境的适应

微生物进化出适应的策略使得它们能在不同的环境中繁衍生息。嗜碱微生物在进化的过程中产生了在极高的 pH 环境中生存的适应机制。有些嗜碱微生物甚至可以在 pH 13 以上正常生长（Roadcap et al. 2006）。目前已有大量研究致力于解开微生物适应极高 pH 环境的可能机制。这些研究大多集中在 Na^+/H^+ 反向转运蛋白降低细胞质 pH 的机制上（Kitada et al. 1994；Krulwieh et al. 1997），细胞壁保护细胞免受高 pH 的影响（Aono et al. 1999）以及膜内外生物能转运的影响（Sturr et al. 1994；Krulwieh et al. 1998）。

转录组和蛋白质组分析结果补充了嗜碱微生物群体高 pH 耐受的生理和遗传信息。研究发现微生物在碱性环境下的正常生长与代谢改变密切相关，如促进酸的产生、促进捕获和保留质子的转运蛋白和酶的表达（如 ATP 合酶和单价阳离子/质子逆向转运蛋白）以及有助于保留细胞质质子的细胞表面成分的改变等（Krulwich et al. 2001a，b；Padan et al. 2005）。这些研究表明，嗜碱微生物综合使用不同的策略使其在高 pH 环境中得以繁衍。嗜碱微生物能产生在碱性条件下保持活性和稳定的酶的能力是另一个重要的策略，这一点在涉及嗜碱微生物的适应性机制时，通常不太明显。

8.1.3 嗜碱微生物在生物技术开发领域的应用前景

这些极端微生物的特殊适应性机制，我们可以对其进行改造从而为己所用（图 8.1）。嗜碱微生物的适应机制，在耐碱胞外酶在生物技术的应用中非常有用，引起了研究者们广泛的关注（Fujinami and Fujisawa 2010）。嗜碱微生物能够产生高 pH 条件下仍具有高活性的胞外酶（图 8.2）。许多工业应用中需要能在较高的 pH 下保持活性和稳定性的酶。另一方面，嗜碱微生物胞内酶的 pH 耐受性通常与来自非嗜碱微生物类似，这主要是因为嗜碱适应机制保持细胞质中 pH 接近中性。人们发现来源于极端微生物的胞内酶并不具有与细胞生长环境条件完全相关的特性，无论在嗜碱微生物还是嗜酸微生物中都是如此（Baker-Austin and Dopson 2007），并且在一定程度上对于来自嗜热微生物胞内酶的热稳定性也是如此（Turner et al. 2007）。

与其他大多数极端微生物不同，嗜碱微生物被认为优秀的胞外酶生产者。嗜碱微生物生长速率快，甚至很多时候要比中性微生物更快（Niehaus et al. 1999 ；Horikoshi 2006），并且菌体密度相对较高。这些都是微生物在生物技术应用中所期望的特性。

除了酶以外，嗜碱微生物还具有显著的有机酸生产能力，而有机酸在工业生产过程中被大量使用，因而在生物技术中也具有潜在的应用。已知一些嗜碱微生物也产生铁载体和类胡萝卜素，在众多应用领域中具有巨大的潜力。嗜碱微生物作为生物活性化合物的来源的研究较少，然而，这些研究均表明嗜碱微生物作为新型生物活性化合物来源的应用潜力。

在本章中，我们主要讨论了胞外耐碱性酶的一些研究现状和应用，同时评估了嗜碱微生物在有机酸、类胡萝卜素、生物活性化合物和铁载体生产中的潜力。然而，这些并不是嗜碱菌在生物技术中的全部应用，它们只是作为实例被选择出来进行介绍。

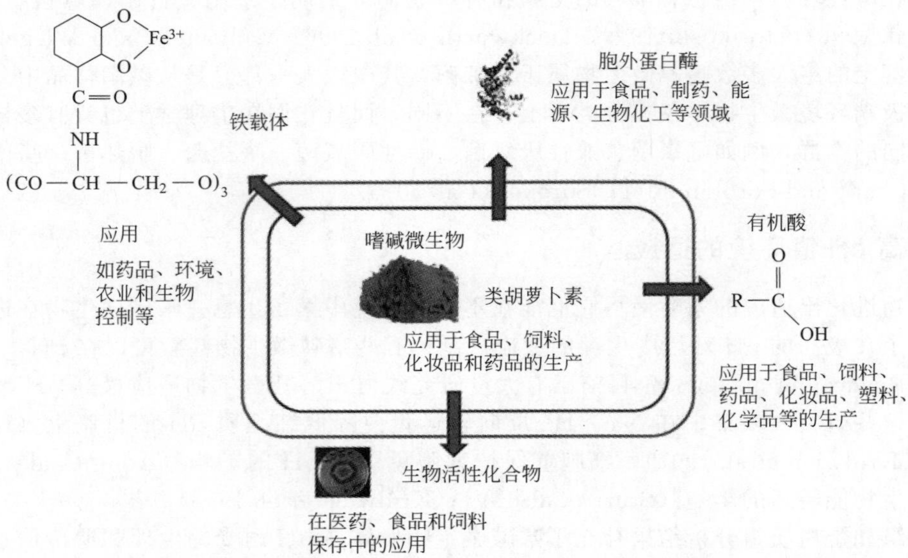

图 8.1　一些嗜碱微生物的适应机制及其生物技术重要性

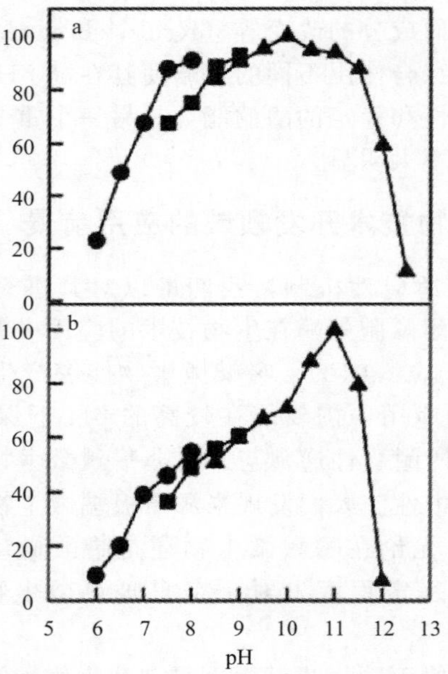

图 8.2　pH 对由两种嗜碱性菌株产生的蛋白酶 AL20（a）和 AL89（b）酶活性的影响。活性在 50℃下测定。缓冲液为磷酸盐缓冲液，pH 6～8（实心圆）；Tris-HCl 缓冲液，pH 7.5～9（方块）；以及甘氨酸-NaOH 缓冲液，pH 8.5～12（实心三角形）。各种缓冲液溶度均为 50mmol/L（授权转载自 Gessesse 等人 2003）

8.2　嗜碱微生物的耐碱性酶及其工业化应用

酶的应用是工业生物技术最重要的组成部分之一。随着绿色化学的出现和推广，大量的酶被用于工业生产的各个领域。这些应用中需要一部分能在碱性条件下具有活性的酶（图

8.3)。而嗜碱微生物的酶，特别是其胞外酶能够很好地满足该领域工业酶的需求。嗜碱微生物已被认为耐碱性酶的可靠来源，目前已经分离出了大量来自嗜碱微生物的耐碱性酶，对其进行了表征、评估，并且其中一些被应用于不同的行业。大多数商业上重要的耐碱性酶是生物聚合物降解酶。本章我们将着重总结这些酶在一些特定领域中的应用。

8.2.1　碱性蛋白酶的应用

蛋白酶，一类可应用于各种领域的降解蛋白质的酶，包括食品加工、医药、化妆品、合成化学、废物处理等。2013 年全球工业酶销售量估计约为 37 亿美元，蛋白酶占了这一市场的近一半。碱性蛋白酶还被用于洗涤和皮革鞣制工业中，并且它们是商业蛋白酶的重要组成之一。

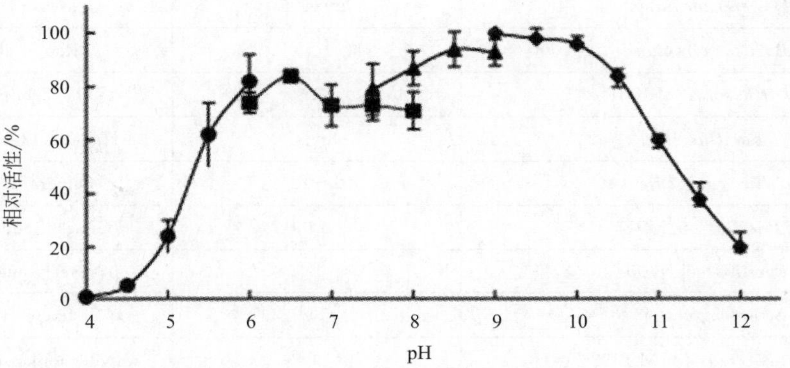

图 8.3　pH 对嗜碱芽孢杆菌 S7（*Bacillus halodurans*）木聚糖酶活性的影响。将酶在 70℃下与溶解在浓度均为 50mmol/L 的醋酸钠缓冲液（实心圆）、磷酸钠缓冲液（方块）、Tris-HCl 缓冲液（三角形）以及甘氨酸-NaOH 缓冲液（菱形）中的含量为 10g/L 桦木木聚糖一起温育。将约为 0.5U/mL 的木聚糖酶在 pH 9 条件下的活性作为 100%（授权转载自 Mamo 等 2006）

8.2.1.1　清洁剂

含酶洗涤制剂已经广泛使用了五十多年。这些生物催化剂广泛存在于普通洗涤剂、去污剂、自动洗碗清洁剂、洗衣预清洁剂以及工业和医疗清洁剂的配方中。如今在发达国家，市场上超过一半的洗涤剂中至少含有一种酶，仅洗涤剂工业就约占全球酶销售总量的 30%。在洗涤剂中添加酶之所以成功，主要是由于加入酶后洗涤性能更佳，以及其在节能减排中的突出贡献。根据从 Novozymes 网页检索的信息，衣物洗涤是普通家庭最耗能的活动之一。仅仅在欧洲，将洗涤温度从 40～60℃降低到 30℃，就可以减少 1200 万吨二氧化碳的排放，这相当于 300 万辆汽车每年的排放量。在洗衣洗涤剂中使用酶的一个好处就是降低了洗涤温度却不降低清洗效率。这表明在该应用中酶的使用具有明显的经济和环境效益。

不同种类的洗涤剂用于各种清洁过程。在清洁剂中，碱性洗涤剂（其 pH 范围为 9～12.5）是餐具和衣物洗涤中最常用的洗涤剂。碱性增强了污垢去除能力从而获得更好的洗涤性能。碱性洗涤剂含有有助于增强洗涤剂碱性的钠衍生物，例如三聚磷酸盐、碳酸盐、磷酸盐、碳酸氢盐和硫酸盐等。虽然已知用蛋白酶处理可以容易且有效地除去蛋白质污垢，但并非所有的蛋白酶对去污剂的配制都是重要的，使用与洗涤剂成分及其高 pH 相容的蛋白酶才是关键。用于碱性洗涤剂配方的理想蛋白酶应具有以下特点：(a) 在碱性条件下保持稳定；(b) 螯合剂和漂白剂耐受；(c) 在 30～60℃具有相当高的活性和稳定性；(d) 在室温下储存数月而不失活。然而，到目前为止的绝大多数蛋白酶不具有所有这些性质，因此需要通过筛选新的蛋白酶来寻找更适合的碱性蛋白酶，或改造现有的蛋白酶来调整其用于洗涤剂的性质。目前已经研究了大量耐碱性蛋白酶，并且在表8.1中给出了由嗜碱菌产生的耐碱性蛋白酶的一些实例。

表 8.1 一些耐碱性蛋白酶

有机体	蛋白酶活性最佳pH值	参考文献
Paenibacillus tezpurensis AS-S24-II	9.5	Rai et al. (2010)
Nocardiopsis prasina HA-4	7	Ningthoujam et al. (2009)
Bacillus sp. GUS1	6～12	Seifzadeh et al. (2008)
Stenotrophomonas maltophilia-MTCC7528	10	Kuddus and Ramteke(2009)
Nesterenkornia sp.	7～11.5	Gessesse et al. (2003)
Bacillus circulans		Rao et al. (2009)
Bacillus sp. KSM-K 16	12.3	Kobayashi et al. (1995)
Bacillus sp.	8～9	Haile and Gessesse(2012)
Bacillus patagoniensis	9～12	Olivera et al. (2006)
Bacillus sp. L21	11	Genckal and Tari(2006)
Bacillus sp. JB-99	11	Johnvesly and Naik(2001)
Streptomyces sp.	11	Tatineni et al. (2008)
Bacillus cereus MCM B-326	9	Nilegaonkar et al. (2007)
Bacillus halodurans JB 99	11	Shrinivas and Naik(2011)

大多数市售的洗涤剂蛋白酶来自芽孢杆菌属细菌。已知主要洗涤剂蛋白酶供应商如Novozymes、GistBrocades、Nagase Chemtex、Showa Denko K.K.、Kao和Genencor开发了第一代和第二代用于洗涤剂的芽孢杆菌蛋白酶。Novozymes公司为市场提供了Alcalase、Esparase和Savinase三种蛋白酶。碱性蛋白酶Alcalase来自地衣芽孢杆菌，Esperase和Savinase分别来自地衣芽孢杆菌和解淀粉芽孢杆菌的嗜碱性菌株。另一个知名的蛋白酶品牌Maxatase的酶来自地衣芽孢杆菌，其专利由GistBrocades公司拥有。来自地衣芽孢杆菌的Alcalase和Maxatase可以在pH 7～10.5和10～65℃范围内有效使用。Savinase和Esperase可以分别在pH 11和12碱性更强的条件下使用。由于高pH稳定性（图8.4）和对洗涤介质中存在的螯合剂的抗性，来自*Nesterencornia*菌的碱性蛋白酶显示出作为洗涤剂添加剂的有趣特性（Bakhtiar et al. 2003）。近年来，来自Novozymes（Kannase®、Durazyme®）和Genencor（Properase®、PurafectOxP®）的第三代洗涤剂蛋白酶（改造的芽孢杆菌蛋白酶）也已经在市场上使用。

8.2.1.2 脱毛

皮革鞣制涉及几个步骤中，脱毛是皮革加工的基本预鞣步骤之一。脱毛是指去除毛发与表皮的非胶原蛋白。常规的脱毛方法使用石灰和硫化物，这也是皮革鞣制工业的主要污染源之一。硫化物的使用可能引起硫化氢（H_2S）的形成，其具有令人讨厌的气味和毒性，如果不进行处理则会对环境造成严重的威胁。对环境的日益关注以及关于环境污染的严格立法已经迫使皮革制造行业寻找更环保的替代工艺。其中，使用耐碱性蛋白酶进行皮革脱毛则具有极强的吸引力，这些蛋白酶被称为角蛋白酶（Anbu et al. 2005；Brandelli 2008；Cai et

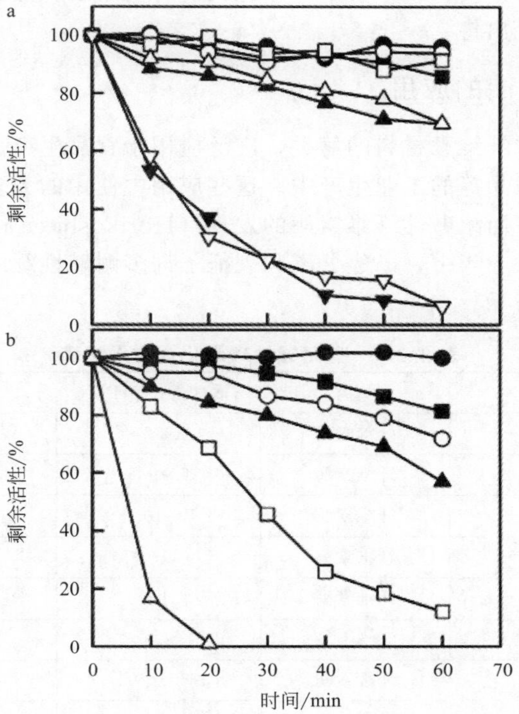

图 8.4 在不存在（空心部分）或存在（实心部分）5mmol/L Ca^{2+} 的情况下测定蛋白酶 AL-20(a) 和蛋白酶 AL-89(b) 的热稳定性。将酶在 55℃（空心圆圈、实心圆圈）、60℃（空心方块、实心方块）、65℃（空心三角形、实心三角形）以及 70℃（空心倒三角、实心倒三角）下在 pH 10 浓度为 50mmol/L 的甘氨酸-NaOH 缓冲液中孵育。在各个时间间隔取出样品，用标准测定程序测量残留活性（授权转载自 Gessesse 等 2003）

al. 2008)。许多酶的脱毛效率已经得到研究（Shrinivas and Naik 2011；Senthilvelan et al. 2012；Haile and Gessesse 2012)。酶法脱毛通常在 pH 8 至 10 的范围内进行，并且目前已有一些市售的脱毛蛋白酶。

使用酶/偏硅酸钠代替石灰/硫化物系统证明酶促脱毛的技术和经济优势。酶的使用使加工过程化学需氧量（COD）和总固体负载分别降低了 55%和 25%（Bhavan et al. 2008)。在脱毛步骤中使用酶将减少一半的石灰消耗，并且还缩短了处理时间（Soerensen et al. 2011；de Souza and Gutterres 2012)。这有利于减少脱毛步骤中脱灰剂（例如氯化铵或硫酸盐）的使用量，并进一步有助于降低工艺成本和环境负担。

8.2.1.3 软化

通过除去非胶原类蛋白质例如弹性蛋白、白蛋白、球蛋白和黏蛋白来软化皮革的过程称为软化。该过程使胶原纤维断裂并增强鞣制化学品的吸收，从而赋予成品革所需要的性能。在皮革加工步骤中，软化是使用酶的第一步。研究表明，酶的使用显著降低了鞣剂的量（Kamini et al. 1999)，从环保的观点看也是有吸引力的。

从牛和猪胰腺中提取的蛋白水解酶已用于软化。提取物含有糜蛋白酶原、胰蛋白酶原、羧肽酶原，随后分别被激活成糜蛋白酶、胰蛋白酶和羧肽酶。多年来，细菌和真菌酶已经取代了胰腺制剂，主要是由于细菌和真菌酶易于生产制备以及其理想的性能优点。用耐碱蛋白酶软化的皮革显示出良好的理化性质（Thanikaivelan et al. 2004；Choudhary et al. 2004)。

微生物来源的耐碱蛋白酶不损害胶原蛋白，因此人们对使用耐碱蛋白酶更感兴趣，并且目前已有许多这类酶在市场上出售。

8.2.2 耐碱性纤维素酶的应用

纤维素酶是一类分解纤维素材料的酶类，广泛应用于食品生产、饲料加工、生物能源载体开发、生化试剂、纺织品等的工业生产中。这些应用中使用的大多数酶通常是从真菌获得的酸性或中性纤维素酶。而耐碱性纤维素酶的发现（Horikoshi et al. 1984）开辟了纤维素酶应用的新窗口。在过去三十年中，已经分离和表征了许多耐碱性纤维素酶，绝大多数来自芽孢杆菌菌株，如表 8.2 所示。

表 8.2 一些耐碱性碳水化合物水解酶

有机体	酶类	活性最佳 pH 值	参考文献
Bacillus sp. KSM-635	纤维素酶	9.5	Ito et al. (1989)
Cephalosporium sp. Rym-202	纤维素酶	7～9.5	Kang and Rhe(1995)
Bacillus sp.	纤维素酶	8～10.9	Sashihara et al. (1984)
Bacillus licheniformis C108	纤维素酶	10	Aygan et al. (2011)
Bacillus sp. HSH-810	纤维素酶	10	Kim et al. (2005)
B. circulans	纤维素酶	8.5	Hakamada et al. (2002)
Bacillus sp. KSM-S237	纤维素酶	8.6～9	Hakamada et al. (1997)
B. halodurans S7	木聚糖酶	9～10	Mamo et al. (2006)
Bacillus sp. AR-009	木聚糖酶	9～10	Gessesse(1998)
Micrococcus sp. AR-135	木聚糖酶	7.5～9	Gessesse and Mamo(1998)
B. pumilus 13a	木聚糖酶	9	Duarte et al. (2000)
Bacillus sp.	木聚糖酶	9～10	Balakrishnan et al. (1992)
Bacillus sp.	木聚糖酶	9～10	Nakamura et al. (1994)
Bacillus sp. NT-33	果胶酶	10.5	Cao et al. (1992)
Bacillus sp. MG-cp-2	果胶酶	10	Kapoor et al. (2001)
Amycolata sp.	果胶酶	10.25	Bruhlmann et al. (1994)
B. licheniformis	果胶酶	11	Singh et al. (1999)
B. halodurans LBK 34	淀粉酶	10.5～11.5	Hashim et al. (2005)
Bacillus sp. KSM-K38	淀粉酶	8～9.5	Hagihara et al. (2001)
Bacillus sp. TS-23	淀粉酶	9	Lin et al. (1998)
B. agaradhaerens LS-3C	环糊精糖基转移酶	9	Martins and Hatti-Kaul(2002)
B. pseudalcaliphilus 20RF	环糊精糖基转移酶	5～10	Atanasova et al. (2011)

8.2.2.1 清洁剂

耐碱性纤维素酶的发现使得含有纤维素酶的洗涤剂得以出现。与其他洗涤剂酶不同，纤维素酶不降解油类或蛋白质类污渍，而是改性衣服纤维素纤维的表面结构以改善整体洗涤性能。用于碱性洗涤制剂的纤维素酶必须在碱性条件下是有活性的并且稳定的，并且改性无定形纤维素结构（微绒毛），而不是织物的主要结晶纤维结构。目前已知的碱性纤维素酶不降解或仅轻微降解结晶纤维素（Horikoshi 2011）。这些酶作用于无定形纤维素并增强去除污垢（包括在自然界中为油性的污垢）的能力（Ito et al. 1989；Murata et al. 1991）。从织物表面除去非结晶纤维素并清除附着在其上的污垢，因而具有很好的清洁效果。除了去除污垢，在洗涤剂中使用耐碱性纤维素酶还具有增亮色泽的效果。由天然或混合纤维素纤维制成的服装，当使用和洗涤多次时会变得“蓬松”和褪色，这是由于结晶纤维素分散成微纤维从纤维素表面分离出来。当光照在该表面上时，一部分光将被反射回来，从而使观察者留下服装颜色暗淡的印象。由洗涤剂纤维素酶引起的微纤维的降解可以消除表面模糊性，使原有的衣服呈现出光滑和有光泽的外观，并恢复其原始的色彩亮度。此外，微纤维的降解还改善了织物的亲水性和吸湿性（Bhat 2000），这对洗涤性能有积极贡献。在洗涤剂中使用耐碱性纤维素酶对纤维素材质的衣服提供了另一个理想的效果——更好的柔软性。

来源于嗜碱芽孢杆菌 KSM-635 的耐碱性纤维素酶 EglK 是第一种用于洗涤制剂的纤维素酶（Ito et al. 1989）。从那时起，大量的耐碱性纤维素酶被报道并评估了其潜在的洗涤剂应用（Ito et al. 1998）。包括花王公司和 Novozymes 公司在内的几家公司一直在开发和销售用于洗涤剂应用的耐碱性纤维素酶（Olsen and Falholt 1998）。

8.2.2.2　生物石磨

纤维素酶在纺织工业中还有另一个有趣的应用。这些酶被用于纤维素纺织物的最终步骤，改善纺织品的手感和外观（Hebeish and Ibrahim 2007；Karmakar and Ray 2011）。在处理棉质织物中使用纤维素酶具有技术和经济优点。例如，牛仔裤通常要在具有浮石（1～2kg/件）的洗衣机中洗涤。这种石洗会损伤机器和影响服装质量。为了缓解这些问题，在已知的生物石磨方法中使用纤维素酶代替浮石。在该方法中使用纤维素酶降解棉织物纱线表面上的微纤维，这有助于在洗涤循环期间去除染料。事实证明，这个过程在解决洗衣机耗损问题上十分有效。此外，生物石磨通过最小化磨损来改善织物质量，同时它还具有经济、便捷以及环境友好等优点（Sukumaran et al. 2005；Singh et al. 2007；Araújo et al. 2008；Andreaus et al. 2014）。

虽然中性和酸性纤维素酶也可用于生物石磨过程中，但由于回染的需要，因而耐碱性纤维素酶更适用于生物石磨（Cavaco-Paulo et al. 1998）。然而，尽管很具有潜力，但是迄今还没有关于碱性纤维素酶在生物石磨工业化使用的报道。

8.2.2.3　墨渍清除

世界上大量的纸张被回收利用，与玻璃和塑料相比，纸张的回收数量更多。根据环境保护局（EPA）的报告，每回收/t 纸可节省约 17 棵树，26500L 水和 1750L 石油。这使得更多的纸张回收再利用，并不断努力改善回收过程的技术。

纤维素酶已经被用于提高纸张回收的效率。超过 70％的纸张是未涂画的或者是用墨粉打印的（Thakur 2006），去除这种墨粉是再循环过程的重要步骤。然而，常规的碱性脱墨过程难以去除墨粉，且该过程消耗大量化学物质和水（Shrinath et al. 1991；Thomas 1994）。纸张再循环要在碱性条件下进行，碱性条件使纤维素纤维膨胀并增加其柔软性。这种 pH 引起的变化促进了纤维表面油墨的清除（Wielen et al. 1999）。除了纤维的膨胀，碱也可以直接与染料相互作用，这可能削弱其与纤维的相互作用，促进其清除（Shrinath et al. 1991）。因此，如果考虑将酶用于脱墨过程，则酶在碱性处理过程中应当是有活性的且稳定的。

人们已经研究了耐碱性纤维素酶在墨粉去除中的用途，结果显示，这些酶能极大地改善从废纸中除去墨粉并促进纸张再循环的过程（Gubitz et al. 1998；Bajpai and Bajpai 1998；Lee et al. 2007；Vyas and Lachke 2003；Ibarra et al. 2012）。因此，不论是从经济、技术还是环境的角度来看，酶的使用都是令人满意的。虽然我们认为纤维素酶对微纤维的水解有助于在浮选/洗涤步骤期间的染料分离（Ibarra et al. 2012），但是还不知道这些酶去除染料的确切机制。

8.2.3 耐碱性木聚糖酶及其在牛皮纸浆去木质素作用中的应用

1973年首次报道了来自嗜碱性微生物的木聚糖降解酶（Horikoshi and Atsukawa 1973）。该木聚糖酶来自嗜碱芽孢杆菌C-59-2，在pH 6～8范围内具有最佳活性。随后，纯化和研究了来自各种嗜碱微生物的多种木聚糖酶。这些木聚糖酶大多数酶促反应的最适pH为中性，但在碱性范围内也具有相对良好的活性。但是，也有报道显示少数木聚糖酶以碱性，甚至更高的pH为酶活性最适pH值。迄今为止报道的大多数碱性木聚糖酶在pH 9～10下具有最佳酶活性，包括来自芽孢杆菌AR-009（*Bacillus* sp. AR-009）的XylB（Gessesse 1998），芽孢杆菌TAR-1（*Bacillus* sp. TAR-1）的木聚糖酶（Nakamura et al. 1994），嗜碱芽孢杆菌C-125（*Bacillushalo-durans* C-125）（Honda et al. 1985），嗜碱芽孢杆菌S7（*Bacillus halodurans* S7）（Mamo et al. 2006）和芽孢杆菌NCL-86-6-10（*Bacillus* sp. NCL-86-6-10）（Balakrishnan et al. 1992）。其他高嗜碱性木聚糖酶包括来自芽孢杆菌属菌株41M-1（*Bacillus* sp. strain 41M-1）的木聚糖酶J（Nakamura et al. 1993）和来自短小芽孢杆菌13a（*Bacillus pumilus* 13a）的木聚糖酶（Duarte et al. 2000），两者在pH 9下具有最佳酶活性。

芽孢杆菌是几种耐碱性酶的可靠来源，该属的成员也是迄今最有潜力的木聚糖酶来源。这些酶在高pH下能够稳定工作。到目前为止，极少数在pH值高达10，温度在70℃或高于70℃时具有最佳酶活性的木聚糖酶得到表征（Honda et al. 1985；Gessesse 1998；Chang et al. 2004；Mamo et al. 2006）。此外，这些木聚糖酶在高达65℃的温度下相当稳定，在底物存在的情况下则具有更好的热稳定性。

耐碱性木聚糖酶的最重要应用体现在纸浆的去木质素过程中。现代纸浆和造纸工业使用不同的方法从木质纤维素生物质中回收纤维素纤维，其中木质纤维素生物质大部分来自木材。制浆过程中的主要方法为化学制浆，包括硫酸盐法、硫化法和半化学法。硫酸盐法制浆占总化学制浆的90%以上。硫酸盐法需要在约170℃下用硫氢化钠和氢氧化钠溶液碱化处理2h，使木质素降解和溶解。由这种碱性蒸煮得到的浆料主要由于残余的木质素和木质素衍生物共价键合到半纤维素部分而呈棕色，除去这种木质素通常需要使用氯来漂白。虽然氯的使用能有效地漂白纸浆，但是会导致形成不必要的高毒性和容易导致氯化反应的有机副产物（Nagarathnamma and Bajpai 1999；Pokhrel and Viraraghavan 2004）。然而这些方法对环境的污染较为严重，因此有必要寻找替代方法避免或最小化使用氯。目前，不同的环保漂白化学品如H_2O_2、O_2和O_3代替了有毒的氯漂白化学品（Abad et al. 2001）。

漂白前对牛皮纸浆进行酶处理，水解存在于纸浆中的木聚糖。木聚糖水解以促进木质素去除，因此木聚糖酶处理将大幅减少氯的使用（Vicuna et al. 1997；Ximenes et al. 1999；Haarhoff et al. 1999）。木聚糖酶的使用也有利于提高漂白效率和降低H_2O_2、O_2和O_3的消耗（Srinivasan and Rele 1999；Abad et al. 2001）。人们已经进行了几项研究以了解木聚糖酶增强纸浆漂白的机制。这些研究的结果围绕着三个主要机制：（1）提高木质素提取率；（2）改变碳水化合物和木质素结合；（3）裂解重新沉积的木聚糖（Kenealy and Jeffries 2003）。

到目前为止，使用木聚糖酶用于牛皮纸浆的预漂白成为纸浆和造纸工业中酶应用的最大成功案例之一。由于牛皮纸浆温度较高，且是碱性的，所以在加入酶之前需要进行冷却和中和。然而，这是费时且昂贵的步骤。如果使用热稳定性碱性木聚糖酶，则能显著降低冷却和调整pH的成本。这无疑是人们积极寻找碱性热稳定木聚糖酶的主要动力。

8.2.4 耐碱性果胶酶

果胶是由 α-(1-4) 连接的D-半乳糖醛酸残基的骨架与一些鼠李糖残基一起构建的酸性大分子多糖，并且侧链富含阿拉伯糖、半乳糖和木糖。果胶质是更复杂的聚合碳水化合物之一。许多改变植物生物量的过程涉及一定程度的果胶去除，这些过程可以通过化学或/和酶处理来完成。然而，使用果胶酶相对于化学方法处理具有明显的优势。酸性和中性活性果胶酶已用于加工食品、果汁、葡萄酒等工业（Alkorta et al. 1998）。而耐碱性果胶酶的发现则为其新的应用铺平了道路。如今，这些酶是工业酶中最重要的组成部分之一，并且用于纺织品加工、植物韧皮脱胶、废物管理和造纸等不同的应用领域。

8.2.4.1 生物煮练棉花

纺织工业严重依赖于棉纤维。将棉花加工成可用于纺织品生产的纱线需要几个步骤。其中一个重要的步骤就是去除不溶于水的纤维保护结构成分果胶。从棉花中除去果胶被称为煮练，常规方法是使用氢氧化钠溶液和高温溶解果胶。碱性条件增溶植物材料的果胶、半纤维素和木质素部分最终留下纤维素纤维。除了促进棉的非纤维素组分的去除之外，高 pH 能够防止降解纤维素纤维的微生物（特别是真菌）的生长。然而，该方法成本较高并且对环境污染严重，需要使用大量水漂洗以及需要大量能量来加热悬浮液。生物法煮练作为常用的替代方法，在低温下选择性使用可水解棉花果胶酶制剂，可显著降低成本，且更加环保（Bruhlmann et al. 1994）。此外，耐碱性果胶酶的使用减少了氢氧化钠的消耗，并获得比常规方法质量更好的纤维（Hoondal et al. 2002）。

8.2.4.2 韧皮纤维的脱胶

去除覆盖在植物纤维上纤维素胶状物质的部分称为脱胶，这是生产用于不同功能的纤维所必需的步骤（Said et al. 1991）。传统上，韧皮纤维在含有润湿剂和还原剂的质量浓度为 0.02～0.2kg/L 氢氧化钠溶液中脱胶（Cao et al. 1992）。浸泡 24h 后，将悬浮液煮沸 4h 以上，漂洗、中和并彻底洗涤。这个过程不仅昂贵，而且对环境也有负面影响。

由于成本较低且环境友好等特点，与从棉花中回收纤维的生物脱胶一样，其他植物来源的纤维也可以使用果胶酶进行加工。迄今为止，已经尝试了用果胶酶来使亚麻、苎麻、黄麻和大麻的纤维脱胶（Cao et al. 1992；Bruhlmann et al. 1994；Henriksson et al. 1997；Kapoor et al. 2001）。碱性果胶酶对于植物纤维的脱胶是理想的，因为高 pH 防止了真菌的污染，因此可以采用开放式发酵系统，显著降低预期的加工成本。此外，它还能除去木质素，因此可以使后续步骤中所需的漂白剂的量最小化。

8.2.5 耐碱性淀粉降解酶

淀粉降解酶是食品、纺织品、乙醇燃料等应用加工中最重要的酶类之一。淀粉酶有多种类型，并且由不同微生物产生。尽管它们在一些应用中十分重要，但对其中耐碱性淀粉酶相对研究较少。下面介绍耐碱性淀粉修饰酶的两个重要应用实例。

8.2.5.1 耐碱性α-淀粉酶

在淀粉和纺织工业中，耐碱性 α-淀粉酶可用于碱性条件下水解淀粉。此外，它还可以用作自动洗碗机和洗衣店的洗涤剂成分。虽然不是来自嗜碱性微生物并且在碱性条件下不具有活性，但 Novozymes 开发了相应的淀粉酶产品，如 Termamyl® 及其突变体 Duramyl® 已用于配制清洁剂以去除基于食物的淀粉污渍（Olsen and Falholt 1998）。人们对于开发在碱性洗涤剂中具有最佳活性和稳定性的淀粉酶的兴趣，促进了许多耐碱性淀粉酶的研究发现

(Hashim et al. 2004；Burhan et al. 2003；Hashim et al. 2005；Saxena et al. 2007)。

已知淀粉酶是一种活性中心需要钙离子的金属酶类，而洗涤剂中含有螯合剂，容易使钙离子从淀粉酶剥离并使其失活。因此，开发用于洗涤剂应用的淀粉酶目标之一就是寻找在洗涤剂螯合剂存在的情况下仍能起作用的碱性淀粉酶。在这方面，已经报道了来自芽孢杆菌KSMK38的一种独特的淀粉酶AmyK38，其在高碱性条件下不仅表现出对螯合剂的耐受能力，而且对氧化剂和表面活性剂的影响有着显著的耐受性（Hagihara et al. 2001）。更详细研究表明，该淀粉酶不含有钙离子，其活性与钙无关（Nonaka et al. 2003）。

8.2.5.2 环糊精糖基转移酶

环糊精广泛用于食品、化妆品、药品等领域，可用于稳定挥发性化合物的制备，可作为抗氧化剂和防紫外线辐射的分子盾牌，也可作为除臭剂和非水溶性物质的增溶剂（Szente and Szejtli 2004；Kurkov and Loftsson 2013；Crini 2014）。环糊精由环糊精糖基转移酶（CGT酶，一种具有多功能活性的酶）作用于淀粉产生。CGT酶催化通常包含6、7或8个葡萄糖单元的环状寡糖的形成，对应的分别被称为α-、β-和γ-环糊精。这些分子形成过程中的闭环反应由酶催化，它通过线性寡糖的非还原端与酶结合作为分子内糖基转移反应的受体。来自不同微生物的CGT酶通常产生不同水平的α-、β-和γ-环糊精。

虽然目前已经深入研究了几种CGT酶，但来自嗜碱芽孢杆菌的CGT酶是目前发现的最有希望用于生产环糊精的酶（Matzuzawa et al. 1975；Martins et al. 2001；Niehaus et al. 1999）。大多数耐碱性CGT酶将淀粉转化为β-环糊精作为主要产物，其中含有少量其他类型的环糊精（Horikoshi 1999；Atanasova et al. 2011；Martins and Hatti-Kaul 2002）。来自嗜碱微生物的CGT酶的催化性质十分具有吸引力，因为它简化了主要产物的后续下游加工步骤。因此目前用于工业生产环糊精的主要商业酶均来自嗜碱芽孢杆菌菌株。

8.3 其他嗜碱微生物活性产物

除了作为工业和环境应用酶的来源之外，嗜碱微生物还可以提供多种生物活性产物。嗜碱微生物易于生长、生长速度快、菌体密度高，从而可以使发酵过程中的污染风险降至最低，因此它们对于多种生产工艺都是具有极强吸引力的。此外，高度生理多样性可以允许研究者选择理想的产品以及合适的微生物来开发生产过程。然而，在开发嗜碱微生物活性产物应用方面，迄今为止只取得了有限的成果。下面讨论一些选定的产品实例和使用嗜碱微生物制备这些产品的潜力。

8.3.1 嗜碱微生物的有机酸生产

有机酸是生产食品、药品等多种产品的非常有用的原料。有机酸可以通过化学或生物（微生物）途径生产，然而微生物生产作为化学生产重要的替代选择近年来不断受到科研工作者的重视。这些生物活性产物占全球发酵市场的很大一部分。如丝状真菌、乳酸菌和一些遗传修饰的大肠杆菌和酵母已被用于生产各种不同的有机酸。然而，人们很少关注去挖掘其他微生物菌群的生产潜力。例如，在嗜碱性生物的发酵液中检测到有机酸并不少见，但是对嗜碱微生物产生有机酸的研究却十分罕见。以下提供了关于嗜碱微生物与有机酸生产相关的研究摘要，以突出其在该领域的显著潜力。

当在由糖类组成的培养基中生长时，嗜碱微生物可显著地改变培养基的pH，而这种pH的变化主要是由于酸的产生导致。嗜碱微生物生产有机酸可能与其高pH环境的适应性有关。从生物技术的角度来看，嗜碱微生物生产有机酸似乎非常有吸引力。糖分解代谢的研究表明，一些嗜碱菌从不同的碳水化合物中产生乙酸、甲酸、琥珀酸、丙酸、丁酸和乳酸等有机酸（Paavilainen et al. 1994；Garnova and Krasilnikova 2003；Zhilina et al. 2004；Kulshreshtha et al. 2012；Yokaryo and Tokiwa 2014）。这种发酵性能可以允许使用廉价的混合碳水化合物为底物如植物生物质来水解生产有机酸。此外，大多数碱性微生物是耐盐的，实际上在不存在钠或其他一价离子的情况下是不能生长，因此可以使用碱如 NaOH 来维持生产介质的 pH 而不是 $CaCO_3$。因为使用 $CaCO_3$ 会使得在发酵过程中难以混合。使用嗜碱性微生物生产有机酸的另一个优点是由于介质的高 pH 和盐含量，降低了被其他菌污染的风险（Calabia et al. 2011；Jiang et al. 2013）。因此，可以推测，嗜碱微生物可能是有机酸生产十分有前景的候选工程菌株（Paavilainen et al. 1994）。

能够产生有机酸的嗜盐和嗜碱性乳酸菌已经从乳酪、腐烂的海藻、鱼和发酵的咸虾酱中被分离得到（Ishikawa et al. 2007，2009，2013）。从海洋环境中分离的嗜碱性微生物在 pH 9 下产生约 66g/L 的 L-乳酸（Calabia et al. 2011）。已经有报道称，嗜碱性微小杆菌菌株可产生 125g/L 100%光学纯度的 L-乳酸（Jiang et al. 2013）。也已报道嗜碱芽孢杆菌菌株在含碳水化合物的培养基中产生不同的有机酸，并且还观察到酸的形成随着 pH 和缓冲液浓度的增加而提高（Paavilainen et al. 1994）。在迄今报道的嗜碱性菌中，芽孢杆菌 WL-S20 能够获得高水平的酸积累（Meng et al. 2012）。而使用氢氧化钠作为中和剂进行检测，该菌能够产生 225g/L 的 L-乳酸，产率为 99.3%。更有趣的是，可以使用花生粕作为芽孢杆菌的经济氮源，并且在发酵液中没有检测到 D-乳酸的存在。在这项工作中取得的结果与已报道的鼠李糖乳杆菌（Berry et al. 1999）、干酪乳杆菌（Ding and Tan 2006）和乳酸乳杆菌（Bai et al. 2003）的结果相当。而且这些乳杆菌菌株都没有获得像嗜碱芽孢杆菌如此惊人的光学纯度。

虽然已知许多嗜碱微生物可产生有机酸，但除了乳酸生产以外，对其他有机酸如甲酸、琥珀酸或乙酸的生产研究非常有限。目前可获得的研究报告简单地描述了在嗜碱微生物的发酵液中存在不同的有机酸。在未优化的情况下，可产生高达 5g/L 乙酸（Paavilainen et al. 1994）和 2g/L 甲酸的产量（Kulshreshtha et al. 2012）。如进一步研究不同有机酸的筛选和生产优化则预计将会有更高的生产率。

8.3.2　类胡萝卜素

类胡萝卜素是由植物和大量微生物产生的色素分子。这些化合物是四萜类化合物，并且包含 40 个碳原子。迄今为止，已知获得了超过 700 种类胡萝卜素结构，大多数是通过氧化、羟基化和其他反应的碳链改性而产生的。类胡萝卜素有黄色、橙色或红色，已知其可保护微生物免受氧化损伤。这些性质使类胡萝卜素可作为食品着色剂、维生素 A 的来源、饲料添加剂和用于化妆品和药用的营养品等。预计未来几年，类胡萝卜素的全球年销售额将远高于 2010 年 12 亿美元的销售额（BCC Research 2011）。研究还表明类胡萝卜素具有显著的抗癌作用（Palozza et al. 2009；Miyashita 2009）。

由于市场（从现有的和新的应用）的扩大，亟需寻找传统类胡萝卜素生产方式（合成反应和从植物中提取）的替代方法。由于低成本、产量高和更安全，使用微生物的生产类胡萝卜素正在成为新的趋势（Lee and Schmidt-Dannert 2002；Mata-Gómez et al. 2014）。

已知的有几种嗜碱性细菌可产生类胡萝卜素，有研究者提出它可能对嗜碱性生物的生存具有重要意义（Aono and Horikoshi 1991），然而，这还需要进一步研究来确认。在嗜碱微生物中，具有生产类胡萝卜素能力的细菌似乎广泛分布于不同属。Aono 和 Horikoshi 分离得到一株产类胡萝卜素嗜碱微生物芽孢杆菌菌株（Aono and Horikoshi 1991）。据报道，一种树状微杆菌（*Microbacterium arborescens*）在 pH 10.5 下生长时会产生番茄红素样类胡萝卜素（Godinho and Bhosle 2008）。硫氧化性的 *Roseinatronobacter thiooxidans* 在含有有机氮的碱性介质中可产生类胡萝卜素（Sorokin et al. 2000）。合成类胡萝卜素的副球菌 *Paracoccus bogoriensis* 每克湿细胞可分泌 0.4 mg 虾青素（Osanjo et al. 2009），虾青素以其抗癌作用而闻名（Palozza et al. 2009；Miyashita 2009），并且具有显著的抗氧化和免疫调节活性（Jyonouchi et al. 1995）。据报道，*Heliorestis* 的不同嗜碱性菌株可以产生新型类胡萝卜素葡萄糖苷酯（Takaichi et al. 2003），这也需要进一步研究以评估其应用潜力。具有生产这些有价值产品能力的物种广泛存在于嗜碱性微生物中，进一步的研究可能可以从这类生物中发现新型类胡萝卜素。此外，由于嗜碱性微生物耐盐、耐高 pH，因此以嗜碱性微生物用作生产菌株时，预期污染可以显著降低。

8.3.3 抗菌物质

当今世界，病原菌耐药性的问题十分严重，已严重威胁人类社会。与需求相反，新抗生素的发现率一直在下降，部分可能是由于抗生素生产筛选研究只集中在某些微生物群落上。对于很少探索的一些微生物类群，特别是一些可产生新型抗菌剂的极端微生物十分值得去研究。事实上，对极端微生物的筛选也可产生用于治疗非传染性疾病，包括癌症和心血管问题的其他生物活性化合物。

目前已知链霉菌的嗜碱性菌株可产生不同的抗微生物药物。例如，链霉菌 AK（*Streptomyces* sp. AK）可产生对抗真菌、原生动物和人类不同癌细胞系的活性药物（Dietera et al. 2003）。*S. ananensis* 和 *S. aburaviensis* 的嗜碱性菌株可产生有效针对革兰氏阳性菌的未知化合物（Vasavada et al. 2006；Thumar et al. 2010）。此外，据报道桑那西沙门菌 *S. tanashiensis* 可分泌抗真菌和抗细菌的代谢物（Singh et al. 2009）。同时也在其他嗜碱菌属中观察到抗微生物物质的产生。达松维尔拟诺卡氏菌（*Nocardiopsis dassonvillei*）可合成和积累细胞内吩嗪抗生素（Tsujibo et al. 1988），淡紫色拟青霉菌（*Paecilomyces lilacinus*）菌株可产生两种抗真菌肽（Sato et al. 1980）。诺卡氏菌属（*Nocardiopsis*）的一些未知成员可产生独特的高度功能化的多环代谢物萘螺环酮（naphthospironone），其表现出抗细菌活性以及细胞毒性（Ding et al. 2010）。一种研究最多的嗜碱芽孢杆菌，耐盐芽孢杆菌可分泌一种被称为 haloduracin 的新型双组分细菌素（Lawton et al. 2007；Danesh et al. 2011）。

对于通过嗜碱性细菌生产抗微生物药物的筛选研究仍十分有限，并且迄今为止还没有从这类微生物开发出相应的商用抗生素。然而，少数研究表明，如果进行设计良好的广泛筛选，嗜碱微生物可能是潜在的新型抗生素的重要来源。这有助于满足对新型抗微生物药物的日益增长的需求。

8.3.4 铁载体

微生物需要铁元素用于它们的细胞代谢过程，包括电子传递链和作为介导一系列生物化学反应的酶的辅因子。微生物已经进化出让它们自身获得这种重要资源的机制。在细菌中发现的策略之一是产生称为铁载体的低分子量螯合化合物。当菌株经常生长在铁十分有限的条件下时，就会产生大量铁载体。目前，已知有大约 500 种类型的铁载体。

微生物铁载体在农业、医药、环境等方面具有巨大的应用潜力。在农业中，它有助于提高土壤肥力、减少重金属的过度积累、作为抑制许多植物病原真菌生长的生物控制剂等。在医学中，这些引人注目的化合物可用于治疗与人体中铁过量有关的各种疾病，例如血色素沉着、含铁血黄素沉积和铁中毒等需要从人体中去除过量的铁，可以用基于铁载体的药物治疗这种疾病（Pietrangelo 2002）。在自然界中存在许多铁载体-抗生素偶联物，例如沙利霉素（Vértesy et al. 1995）和白霉素（Benz et al. 1982）。如果我们成功地模仿这种天然功能制备出特异性铁载体-抗生素偶联物，可能有机会克服目前世界上面临的一些耐药性问题。此外，铁载体可用于磁共振成像（MRI）（Doble et al. 2003）、癌症治疗（Chua et al. 2003；Miethke and Marahiel 2007）和疟疾治疗（Gysin et al. 1991）。在环境应用领域中，铁载体可以用于去除或固定重金属污染物，如在饮用水和土壤中。类似地，它可用于从污染土壤和水中去除放射性污染物（Ruggiero et al. 2000；Von Gunten and Benes 1995）。

无机铁和其他金属离子在碱性环境中极度难溶。在碱性湖泊中，除了高 pH 之外，由光合蓝细菌释放的丰富的氧可促进铁的沉淀，因此铁离子的浓度低于细菌最佳生长所需的浓度。为了确保获得铁和其他金属离子，许多嗜碱微生物都可以产生铁载体（Gascoyne et al. 1991；McMillan et al. 2010）。因此，嗜碱微生物可能是一种具有吸引力的铁载体的来源。通过进一步研究，可以从嗜碱微生物中鉴定出多种新型铁载体，这些铁载体可以适用于多种应用。

8.4 嗜碱微生物在环境保护中的应用

微生物是生物圈中重要组成部分，几乎可以调节地球上的大多数生物地球化学系统。微生物已被直接或间接用于环境保护。在许多行业中，微生物酶或工程菌已经取代了有毒催化剂和工艺的使用，并对环境产生积极影响。微生物已被用于清洁和恢复受污染的土壤和水。它们还被用于替代化肥、不易降解的化学和有毒的杀虫剂等。微生物已经被用于从石油生产的可再生资源中生产化学品。这些有机体已经实现了可行和具有成本效益的加工工艺来保护我们的环境，如果没有它们的参与是不可能达到这种效果的。下面总结了在特定的环境应用中使用嗜碱微生物的例子。

8.4.1 中和碱性污水

许多人类活动导致碱性废物的产生，其在释放到环境之前必须被中和。例如工业清洁、电镀、制浆、采矿、纺织加工等通常都会产生碱性废液。这种废弃物应该被中和后才能排放，并且通常是利用无机酸进行的。人们已经知道嗜碱微生物能够产生有机酸并降低其介质的 pH。目前已知嗜碱微生物在生产有机酸时会降低其培养基的 pH，即使它具有强烈缓冲性。一个有趣的例子是使用微小杆菌（*Exiguobacterium* sp.）可将工业废水的 pH 从 12 降低至 7.5（Kulshreshtha et al. 2012）。如果碱性菌被用于中和碱性废物，最重要的事情是确保废物中含有糖类，如果没有，需要补充一些廉价的糖类，如植物生物质水解产物。

8.4.2 从纺织工艺介质中去除过氧化氢

纺织物的漂白通常使用过氧化氢，在碱性条件（pH＞9）下进行（Spiro and Griffith 1997）。未反应的过氧化氢需要在漂白过程之后除去，否则会干扰纺织品材料的后续加工。一种方法是洗涤纺织物并除去过氧化氢，这种方法耗时并且需要大量的水。此外，废水是碱性的，并且含有过氧化物（Weck 1991），这意味着排放前需要进一步处理，这增加了额外的成本。另一种选择是降解过氧化氢，例如通过加入亚硫酸氢钠（Fruhwirth et al. 2002；Paar et al. 2001），但会导致盐的形成，增加了溶液的离子强度，如果在后续步骤中使用水，则该盐可能对纤维加工产生负面影响。因此，人们期望使用过氧化氢酶将过氧化氢分解成水和氧。已知的绝大多数过氧化氢酶在酸性至中性条件下是具有活性的，但是使用过氧化氢漂白是在高 pH 下进行，因此使用耐碱性过氧化氢酶才是合乎需要的，因为它避免了 pH 再调节和相关的盐形成。目前已经报道了来自嗜碱微生物和耐碱性嗜热微生物的少量耐碱性过氧化氢酶（Gudelj et al. 2001；Kagawa et al. 1999；Michaud-Soret et al. 1998；Thompson et al. 2003）。由于蛋白质和染料分子之间的相互作用，因此不推荐使用游离过氧化氢酶（Tzanov et al. 2001）。为此，固定化酶是重要的且已证明是有用的（Costa et al. 2002）。另一种方法是使用完整微生物，它可以保护细胞内过氧化氢酶不与染料分子相互作用，但仍然暴露于扩散到细胞内的过氧化氢当中（Oluoch et al. 2006）。这项研究的结果（Oluoch et al. 2006）显示了嗜碱性微生物全细胞在去除过氧化氢和其他污染物中的潜力。

8.5 结论

嗜碱微生物广泛分布在天然和人造碱性环境中。它们高度多样化且相对容易培养，这使得它们在生物技术应用中很有吸引力。虽然已经有许多使用嗜碱微生物或其产物的应用，但是在生物技术应用中的巨大潜力仍应该被意识到。即使在本章中描述的一些具有巨大应用潜力的例子，仍然还需要进一步研究才能达到商业水平。总的来说，针对这类微生物的研究相对较少，这是迄今为止它们在生物技术中尚未广泛使用的主要原因之一。然而，从现有研究来看，新技术的出现和可用性以及对基于微生物的产品和方法的需求日益增长，无疑将会扩展这些代表性微生物在生物技术中的应用。

致谢

这项工作的一部分由瑞典研究委员会通过常规研究基金和研究关联基金资助。

无利益冲突声明

Gashaw Mamo 和 Bo Mattiasson 声明没有利益冲突。

参考文献

Abad S，Santos V，Parajó JC（2001）Totally chlorine-free bleaching of Acetosolv pulps：a clean approach to dissolving pulp manufacture. J Chem Technol Biotechnol 76：1117-1123

Alkorta I，Garbisu C，Llama M，Serra J（1998）Industrial application of pectic enzymes. Process Biochem 33：21-28

Anbu P，Gopinath SCB，Hilda A，Priya TL，Annadurai G（2005）Purification of keratinase from poultry farm isolate-*Scopulariopsis brevicaulis* and statistical optimization of enzyme activity. Enzyme Microb Technol 36：639-647

Andreaus J，Olekszyszen DN，Silveira MHL（2014）Processing of cellulosic textile materials with cellulases in cellulose and other naturally occurring polymers. In：Fontana JD，Tiboni M，Grzybowski A（eds）Cellulose and other naturally occurring polymers. Research Signpost，Kerala，India，pp 11-19 ISBN：978-81-308-0543-6

Aono R，Horikoshi K（1991）Carotenes produced by alkaliphilic yellow pigmented strains of *Bacillus*. Agric Biol Chem 55：2643-2645

Aono R，Ito M，Maehida T（1999）Contribution of the cell wall component teichuronopeptide to pH homeostasis and alkaliphily in the alkaliphile *Bacillus lentus* C-125. J Bacteriol 181：6600-6606

Araújo R，Casal M，Cavaco-Paulo A（2008）Application of enzymes for textile fibres processing. Biocatal Biotransform 26：332-349

Atanasova N，Kitayska T，Bojadjieva I，Yankov D，Tonkova A（2011）A novel cyclodextrin glu-canotransferase from alkaliphilic *Bacillus pseudalcaliphilus* 20RF：purification and properties. Process Biochem 46：116-122

Aygan A，Karcioglu L，Arikan B（2011）Alkaline thermostable and halophilic endoglucanase from *Bacillus licheniformis* C108. Afr J Biotechnol 10：789-796

Bai DM，Wei Q，Yan ZH，Zhao XM，Li XG，Xu SM（2003）Fed-batch fermentation of *Lactobacillus lactis* for hyper-production of L-lactic acid. Biotechnol Lett 25：1833-1835

BajpaiP，Bajpai PK（1998）Deinking with enzymes：a review. Tappi J 81：111-117

Baker-Austin C，Dopson M（2007）Life in acid：pH homeostasis in acidophiles. Trends Microbiol 15：165-171

Bakhtiar S，Andersson MM，Gessesse A，Mattiasson B，Hatti-Kaul R（2003）Stability characteris-tics of a calcium-independent alkaline protease from *Nesterenkornia* sp. Enzyme Microb Technol 32：525-531

Balakrishnan H，Dutta-Choudhary N，Srinivasan MC，Rele MV (1992) Cellulase-free xylanase production from an alkalophilic *Bacillus* sp. NCL-87-6-10. World J Microbiol Biotechnol 8：627-631

BCC Research：The global market for carotenoids (2011) http：//www. bccresearch. com/market-research/food-and-beverage/carotenoids-global-market-fod025d. html.

Benz G，Schroder T，Kurz J，Wunsche C，Karl W，Steffens G，Pfitzner J，Schmidt D (1982) Konstitution der desferriform der albomycine d1，d2 and e. Angew Chem 94：552-553 and Suppl 1322-1335.

Berry AR，Franco CMM，Zhang W，Middelberg APJ (1999) Growth and lactic acid production in batch culture of *Lactobacillus rhamnosus* in a defined medium. Biotechnol Lett 21：163-167

Bhat MK (2000) Cellulases and related enzymes in biotechnology. Biotechnol Adv 18：355-383

Bhavan S，Rao JR，Nair BU (2008) A potential new commercial method for processing leather to reduce environmental impact. Environ Sci Pollut Res Int 15：293-295

Brandelli A (2008) Bacterial keratinases：useful enzymes for bioprocessing agroindustrial wastes and beyond. Food Bioprocess Technol 1：105-116

Bruhlmann F，Kim KS，Zimmerman W，Fletcher A (1994) Pectinolytic enzymes from *Actinomycetes* for degumming of ramie bast fibers. Appl Environ Microbiol 60：2107-2112

Burhan A，Nisa U，Gökhan C，Ömer C，Ashabil A，Osman G (2003) Enzymatic properties of a novel thermostable，thermophilic，alkaline and chelator resistant amylase from an alkaliphilic *Bacillus* sp. isolate ANT-6. Process Biochem 38：1397-1403

Cai C-G，Chen J-S，Ql J-J，Yin Y，Zheng X-d (2008) Purification and characterization of keratinase from a new *Bacillus subtilis* strain. J Zhejiang Univ Sci B 9：713-720

Calabia BP，Tokiwa Y，Aiba S (2011) Fermentative production of L-(＋)-lactic acid by an alkaliphi-lic marine microorganism. Biotechnol Lett 33：1429-1433

Canganella F，Wiegel J (2011) Extremophiles：from abyssal to terrestrial ecosystems and possibly beyond. Naturwissenschaften 98：253-279

Cao J，Zheng L，Chen S (1992) Screening of pectinase producer from alkalophilic bacteria and study on its potential application in degumming of ramie. Enzyme Microbiol Technol 14：1013-1016

Cavaco-Paulo A，Morgado J，Almeida L，Kilburn D (1998) Indigo backstaining during cellulase washing. Text Res J 68：398-401

Chang P，Tsai WS，Tsai CL，Tseng MJ (2004) Cloning and characterization of two thermostable xylanases from an alkaliphilic *Bacillus* firmus. Biochem Biophys Res Commun 319：1017-1025

Choudhary RB，Jana AK，Jha MK (2004) Enzyme technology applications in leather processing. Indian J Chem Technol 11：659-671

Chua AC，Ingram HA，Raymond KN，Baker E (2003) Multidentate pyridinones inhibit the metab-olism of nontransferrin-bound iron by hepatocytes and hepatoma cells. Eur J Biochem 270：1689-1698

Costa SA，Tzanov T，Carneiro F，Gübitz GM，Cavaco-Paulo A (2002) Recycling of textile bleaching effluents for dyeing using immobilized catalase. Biotechnol Lett 24：173-176

Crini G (2014) A history of cyclodextrins. Chem Rev 114：10940-10975

Danesh A，Mamo G，Mattiasson B (2011) Production of haloduracin by *Bacillus halodurans* using solid-state fermentation. Biotechnol Lett 33：1339-1344

de Souza FR，Gutterres M (2012) Application of enzymes in leather processing：a comparison between chemical and coenzymatic processes. Braz J Chem Eng 29：473-481

Delgado O，Quillaguamán J，Bakhtiar S，Mattiasson B，Gessesse A，Hatti-Kaul R (2006) *Nesterenkonia aethiopica* sp. nov.，an alkaliphilic，moderate halophile isolated from an Ethiopian soda lake. Int J Syst Evol Microbiol 56：1229-1232

Dietera A，Hamm A，Fiedler HP，Goodfellow M，Muller WE，Brun R，Bringmann G (2003) Pyrocoll，an antibiotic，antiparasitic and antitumor compound produced by a novel alkaliphilic *Streptomyces* strain. J Antibiot 56：639-646

Ding S，Tan T (2006) L-Lactic acid production by *Lactobacillus casei* fermentation using different fed-batch feeding strategies. Process Biochem 41：1451-1454

Ding ZG，Li MG，Zhao JY，Ren J，Huang R，Xie MJ，Cui XL，Zhu HJ，Wen ML (2010) Naphthospironone a：an unprecedented and highly functionalized polycyclic metabolite from an alkaline mine waste extremophile. Chemistry 16：3902-3905

Doble DMJ，Melchior M，OSullivan B，Siering C，Xu J，Pierre VC，Raymond KN (2003) Toward optimized high-relaxivity MRI agents：the effect of ligand basicity on the thermodynamic sta-bility of hexadentate hydroxypyridonate/catecholate gadolinium (III) complexes. Inorg Chem 42：4930-4937

Duarte MC，Pellegrino AC，Portugal EP，Ponezi AN，Franco TT (2000) Characterization of alka-line xylanases from *Bacillus pumilus*. Braz J Microbiol 31：90-94

Duckworth AW，Grant WD，Jones BE，van Steenbergen R (1996) Phylogenetic diversity of soda lake alkaliphiles. FEMS Microbiol Lett 19：181-191

Fruhwirth GO，Paar A，Gudelj M，Cavaco-Paulo A，Robra K-H，Gübitz GM (2002) An immobi-lized catalase peroxidase from the alkalothermophile *Bacillus* SF for the treatment of textile-bleaching effluents. Appl Microbiol Biotechnol 60：313-319

Fujinami S，Fujisawa M (2010) Industrial applications of alkaliphiles and their enzymes-past，present and future. Environ Technol 31：845-856

Garnova ES，Krasilnikova EN (2003) Carbohydrate metabolism of the saccharolytic alkaliphilic anaerobes *Halonatronum saccharophilum*，*Amphibacillus fermentum*，and *Amphibacillus trop-icus*. Mikrobiologiia 72：558-563

Gascoyne DJ，Connor JA，Bull AT (1991) Capacity of siderophore-producing alkalophilic bacteria to accumulate iron， gallium and aluminium. Appl Microbiol Biotechnol 36：136-141

Genckal H，Tari C（2006）Alkaline protease production from alkalophilic *Bacillus* sp. isolated from natural habitats. Enzyme Microb Technol 39：703-710

Gessesse A，Mamo G（1998）Purification and characterization of an alkaline xylanase from alkali-philic Micrococcus sp AR-135. J Ind Microbiol Biotechnol 20：210-214

Gessesse A，Hatti-Kaul R，Gashe BA，Mattiasson B（2003）Novel alkaline proteases from alkali-philic bacteria grown on chicken feather. Enzyme Microb Technol 32：519-524

Glaring MA，Vester JK，Lylloffl JE，Al-Soud WA，Sørensen SJ，Stougaard P（2015）Microbial diversity in a permanently cold and alkaline environment in Greenland. PLoS One. doi：10. 1371/ journal. pone. 0124863

Godinho A，Bhosle S（2008）Carotenes produced by alkaliphilic orange-pigmented strain of *Microbacterium arborescens*—AGSB isolated from coastal sand dunes. Indian J Mar Sci 37：207-312

Grant WD，Sorokin DY（2011）Distribution and diversity of soda lake alkaliphiles In：Horikoshi K（ed）Extremophiles handbook. Springer，Tokyo，pp 28-54

Grant WD，Tindall BJ（1986）The alkaline saline environment. In：Herbert RA，Codd GA（eds）Microbes in extreme environments. Academic Press，London，pp 25-54

Gubitz GM，Mansfield SD，Bo¨hm D，Saddler JN（1998）Effect of endoglucanases and hemicel-lulases in magnetic and flotation deinking of xerographic and laser-printed papers. J Biotechnol 65：209-215

Gudelj M，Fruiwirth GO，Paar A，Lottspeich F，Robra K-H，Cavaco-Paulo A，Gübitz GM（2001）A catalase-peroxidase from a newly isolated thermoalkaliphilic *Bacillus* sp. with potential for the treatment of textile bleaching effluents. Extremophiles 5：423-429

Gysin J，Crenn Y，Pereira da Silva L，Breton C（1991）Siderophores as anti parasitic agents. US Patent 5：192-807

Haarhoff J，Moes CJ，Cerff C，Wyk WJV，Gerischer G，Janse BJH（1999）Characterization and biobleaching effect of hemicellulases produced by thermophilic fungi. Biotechnol Lett 21：415-420

Hagihara H，Igarashi K，Hayashi Y，Endo K，Ikawa-Kitayama K，Ozaki K，Kawai S，Ito S（2001）Novel alpha-amylase that is highly resistant to chelating reagents and chemical oxidants from the alkaliphilic *Bacillus* isolate KSM-K38. Appl Environ Microbiol 67：1744-1750

Haile G，Gessesse A（2012）Properties of alkaline protease C45 Produced by alkaliphilic *Bacillus* sp. isolated from Chitu，Ethiopian SodaLake. J Biotechnol Biomater 2：136

Hakamada Y，Koike K，Yoshimatsu T，Mori H，Kobayashi T，Ito S（1997）Thermostable alkaline cellulase from an alkaliphilic isolate，*Bacillus* sp. KSM-S237. Extremophiles 1：151-156

Hakamada Y，Endo K，Takizawa S，Kobayashi T，Shirai T，Yamane T，Ito S（2002）Enzymatic properties，crystallization and deduced aminoacid sequence of an alkaline endoglucanase from *Bacillus circulans*. Biochim Biophys Acta 1570：174-180

Hashim SO，Delgado O，Hatti-Kaul R，Mulaa FJ，Mattiasson B（2004）Starch hydrolysing *Bacillus halodurans* isolates from a Kenyan soda lake. Biotechnol Lett 26：823-828

Hashim SO，Delgado OD，Martínez MA，Kaul RH，Mulaa FJ，Mattiasson B（2005）Alkaline active maltohexaose-forming α-amylase from *Bacillus halodurans* LBK 34. Enzyme Microb Technol 36：139-146

Hayashi K，Youichi N，Ohara N，Uichimura T，Suzuki H，Komagata K，Kozaki M（1996）Low-temperature-active cellulase produced by *Acremonium alcalophilum* JCM 7366. J Ferment Bioeng 81：185-190

Hebeish A，Ibrahim NA（2007）The impact of frontier sciences on textile industry. Colourage 54：41-55

Henriksson G，Akin DE，Hanlin RT，Rodriguez C，Archibald DD，Rigsby LL，Eriksson KEL（1997）Identification and retting efficiencies of fungi isolated from dew-retted flax in the United States and Europe. Appl Environ Microbiol 63：3950-3956

Honda H，Kudo T，Ikura Y，Horikoshi K（1985）Two types of xylanases of alkalophilic *Bacillus* sp. No. C-125. Can J Microbiol 31：538-542

Hoondal GS，Tiwari RP，Tewari R，Dahiya N，Beg QK（2002）Microbial alkaline pectinases and their industrial applications：a review. Appl Microbiol Biotechnol 59：409-418

Horikoshi K（1991）Microorganisms in alkaline environments. Kodansha，Tokyo，VCH，Weinheim，New York

Horikoshi K（1999）Alkaliphilies：some applications of their products for biotechnology. Microbiol. Molecul Biol Rev 63：735-750

Horikoshi K（2006）Alkaliphiles：genetic properties and applications of enzymes. Springer，Berlin Horikoshi K（2011）Enzymes isolated from alkaliphiles. In：Extremophiles handbook. Springer，Japan，pp 164-177

Horikoshi K，Atsukawa Y（1973）Xylanase produced by alkalophilic *Bacillus* no C-59-2. Agric Biol Chem 37：2097-2103

Horikoshi K，Nakao M，Kurono Y，Saschihara N（1984）Cellulases of an alkalophilic *Bacillus* strain isolated from soil. Can J Microbiol 30：774-779

Ibarra D，Monte MC，Blanco A，Martinez AT，Martinez MJ（2012）Enzymatic deinking of secondary fibers：cellulases/hemicellulases versus laccase-mediator system. J Ind Microbiol Biotechnol 39：1-9

Ishikawa M，Kodama K，Yasuda H，Okamoto-Kainuma A，Koizumi K，Yamasato K（2007）Presence of halophilic and alkaliphilic lactic acid bacteria in various cheese. Lett Appl Microbiol 44：308-313

Ishikawa M，Tanasupawat S，Nakajima K，Kanamori H，Ishizaki S，Kodama K，Okamoto-Kainuma A，Koizumi Y，Yamamoto Y，Yamasato K（2009）*Alkalibacterium thalassium* sp. nov.，*Alkalibacterium pelagium* sp. nov.，*Alkalibacterium putridalgicola* sp. nov. and *Alkalibacterium kapii* sp. nov.，slightly halophilic and alkaliphilic marine lactic acid bacteria isolated from marine organisms and salted foods collected in Japan and Thailand. Int J Syst Evol Microbiol 59：1215-1226

Ishikawa M，Yamasato K，Kodama K，Yasuda H，Matsuyama M，Okamoto-Kainuma A，Koizumi Y（2013）*Alkalibacterium gilvum* sp. nov.，slightly halophilic and alkaliphilic lactic acid bacte-rium isolated from soft and semi-hard cheeses. Int J Syst Evol Microbiol 63：1471-1478

Ito S，Shikata S，Ozaki K，Kawai S，Okamoto K，Inoue S，Takei A，Ohta Y，Satoh T（1989）Alkaline cellulase for laundry detergents：production by *Bacillus* sp. KSM-635 and enzymatic properties. Agric Biol Chem 53：1275-1281

Ito S，Kobayashi T，Ara K，Ozaki K，Kawai S，Hatada Y（1998）Alkaline detergent enzymes from alkaliphiles：enzymatic properties，genetics，and structures. Extremophiles 2：185-190

Jiang X，Xue Y，Wang A，Wang L，Zhang G，Zeng Q，Yu B，Ma Y（2013）Efficient production of polymer-grade L-lactate by an alkaliphilic *Exiguobacterium* sp. strain under nonsterile open fermentation conditions. Bioresour Technol 143：665-668

Johnvesly B，Naik GR（2001）Studies on production of thermostable alkaline protease from ther-mophilic and alkaliphilic *Bacillus* sp. JB-99 in a chemically defined medium. Process Biochem 37：139-144

Jones BE，Grant WD，Duckworth AW，Owenson GG（1999）Microbial Diversity of soda lakes. Extremophiles 2：191-200

Jyonouchi H，Sun S，Gross M（1995）Effect of carotenoids on in vitro immunoglobulin production by human peripheral blood mononuclear cells：Astaxanthin，a carotenoid without vitamin A activity，enhances in vitro immunoglobulin production in response to a T-dependent stimulant and antigen. Nutr Cancer 23：171-183

Kagawa M，Murakoshi N，Nishikawa Y，Matsumoto G，Kurata Y，Mizobata T，Kawata Y，Nagai J（1999）Purification and cloning of a thermostable manganese catalase from thermophilic bacterium. Arch Biochem Biophys 362：346-355

Kamini NR，Hemachander C，Mala JGS，Puvanakrishnan R（1999）Microbial enzyme technology as an alternative to conventional chemicals in leather industry. Curr Sci 77：80-86

Kang MK，Rhe YH（1995）Carboxymethyl cellulases active and stable at alkaline pH from alkalo-philic *Cephalosporium* sp. Rym-202. Biotechnol Lett 17：507-512

Kapoor M，Beg QK，Bhushan B，Singh K，Dadhich KS，Hoondal GS（2001）Application of an alkaline and thermostable polygalacturonase from *Bacillus* sp. MG-cp-2 in degumming of ramie（*Boehmeria nivea*）and sunn hemp（*Crotalaria juncea*）bast fibers. Process Biochem 36：803-807

Karmakar M，Ray RR（2011）Current trends in research and application of microbial cellulases. Res J Microbiol 6：41-53

Kenealy WR，TW Jeffries（2003）Enzyme processes for pulp and paper：a review of recent devel-opments. In：Goodell B，Nicholas DD，Schultz TP（eds）Wood deterioration and preservation：advances in our changing world. Oxford University Press，Oxford，pp 210-239

Kim JY，Hur SH，Hong JH（2005）Purification and characterization of an alkaline cellulase from a newly isolated alkalophilic *Bacillus* sp. HSH-810. Biotechnol Lett 27：313-316

Kitada M，Hashimoto M，Kudo T，Horikoshi K（1994）Properties of two different Na^+/H^+ antiport systems in alkaliphilic *Bacillus* sp. strain C-125. J Bacteriol 176：6464-6469

Kobayashi T，Hakamada Y，Adachi S，Hitomi J，Yoshimatsu T，Koike K，Kawai S，Ito S（1995）Purification and properties of an alkaline protease from alkalophilic *Bacillus* sp. KSM-K16. Appl Microbiol Biotechnol 43：473-481

Krulwich TA，Ito M，Gilmour R，Hicks DB，Guffanti AA（1998）Energetics of alkaliphilic *Bacillus* species：physiology and molecules. Adv Microb Physiol 40：401-438

Krulwich TA，Ito M，Guffani AA（2001a）The Na-dependence of alkaliphily in *Bacillus*. Biochim Biophys Acta 1501：158-168

Krulwich TA，Ito M，Guffanti AA（2001b）The Na^+-dependency of alkaliphiliy in *Bacillus*. Biochem Biophys Acta 1505：158-168

Krulwieh TA，Ito M，Gilmour R，Guffanti AA（1997）Mechanisms of cytoplasmic pH regulation in alkaliphilic strains of *Bacillus*. Extremophiles 1：163-169

Kuddus M，Ramteke PW（2009）Cold-active extracellular alkaline protease from an alkaliphilic *Stenotrophomonas maltophilia*：production of enzyme and its industrial applications. Can J Microbiol 55：1294-1301

Kulshreshtha NM，Kumar A，Bisht G，Pasha S and Kumar R（2012）Usefulness of organic acid produced by *Exiguobacterium* sp. 12/1 on neutralization of alkaline wastewater. Sci World J 2012：345101 doi：10. 1100/2012/345101

Kurkov SV，Loftsson T（2013）Cyclodextrins. Int J Pharma 453：167-180

Lawton EM，Cotter PD，Hill C，Ross RP（2007）Identification of a novel two-peptide lantibiotic，haloduracin，produced by the alkaliphile *Bacillus halodurans* C-125. FEMS Microbiol Lett 267：64-71

Lee PC，Schmidt-Dannert C（2002）Metabolic engineering towards biotechnological production of carotenoids in microorganisms. Appl Microbiol Biotechnol 60：1-11

Lee CK，Darah I，Ibrahim CO（2007）Enzymatic deinking of laser printed office waste papers：some governing parameters on deinking efficiency. Bioresour Technol 98：1684-1689

Lin LL，Chyau CC，Hsu WH（1998）Production and properties of a raw-starch-degrading amylase from the thermophilic and alkaliphilic *Bacillus* sp. TS-23. Biotechnol Appl Biochem 28：61-68

Mamo G，Hatti-Kaul R，Mattiasson B（2006）A thermostable alkaline active endo-β-1-4-xylanase from *Bacillus halodurans* S7：Purification and characterization. Enzyme Microb Technol 39：1492-1498

Martins RF，Hatti-Kaul R（2002）A new cyclodextrin glycosyltransferase from an alkaliphilic *Bacillus agaradhaerens* isolate：purification and characterization. Enzyme Microb Technol 30：116-124

Martins RF，Davids W，Abu Al-Soud W，Levander F，Rådström P，Hatti-Kaul R（2001）Starch-hydrolyzing bacteria from Ethiopian soda lakes. Extremophiles 5：135-144

Mata-Gómez LC，Montañez JC，Méndez-Zavala A，Aguilar CN（2014）Biotechnological produc-tion of carotenoids by yeasts：an overview. Microb Cell Fact 13：12

Matzuzawa M，Kawano M，Nakamura N and Horikoshi K（1975）An improved method for the preparation of Schardinger-Dextrin on an industrial scale by cyclodextrin glycosyl transferase of an alkalophilic *Bacillus* sp.（ATCC 21783）. Starch-Stärke 27：410-413

McMillan DGG, Velasquez I, Nunn BL, Goodlett DR, Hunter KA, Lamont I, Sander SG, Cook GM (2010) Acquisition of iron by alkaliphilic *Bacillus* species. Appl Environ Microbiol 76: 6955-6961

Meng Y, Xue Y, Yu B, Gao C, Ma Y (2012) Efficient production of L-lactic acid with high optical purity by alkaliphilic *Bacillus* sp. WL-S20. Bioresour Technol 116: 334-339

Michaud-Soret I, Jacquamet L, Debaecker-Petit N, Le Pape L, Barynin VV, Latour J-M (1998) The existence of two oxidized Mn (III) forms from *Thermus thermophilus* manganese catalase. Inorg Chem 37: 3874-3876

Miethke M, Marahiel MA (2007) Siderophore-based iron acquisition and pathogen control. Microbiol. Mol Biol Rev 71: 413-451

Miyashita K (2009) Function of marine carotenoids. Forum Nutr 61: 136-146

Murata M, Hoshino E, Yokosuka M, Suzuki A (1991) New detergent mechanism with use of novel alkaline cellulose. J Am Oil Chem Soc 68: 553-558

Nagarathnamma R, Bajpai P (1999) Decolorization anddetoxification of extraction-stage effluent from chlorine bleaching of Kraft pulp by *Rhizopus oryzae*. Appl Environ Microbiol 65: 1078-1082

Nakamura S, Wakabayashi K, Nakai R, Aono R, Horikoshi K (1993) Purification and some prop-erties of an alkaline xylanase from alkaliphilic *Bacillus* sp. strain 41M-1. Appl Environ Microbiol 59: 2311-2316

Nakamura S, Nakai R, Wajabatacgu K, Ishiguro Y, Aono R, Horikoshi K (1994) Thermophilic alkaline xylanase from newly isolated alkaliphilic and thermophilic *Bacillus* sp. strain TAR-1. Biosci Biotechnol Biochem 58: 78-81

Niehaus F, Bertoldo C, Kähler M, Antranikian G (1999) Extremophiles as a source of novel enzymes for industrial applications. Appl Mircrobiol Biotechnol 51: 711-729

Nilegaonkar SS, Zambare VP, Kanekar PP, Dharephalkar PK, Sarnaik SS (2007) Production and partial characterization of dehairing protease from *Bacillus cereus* MCM B-326. Bioresour Technol 98: 1238-1245

Ningthoujam DS, Kshetri P, Sanasam S, Nimaichand S (2009) Screening, identification of best producers and optimization of extracellular proteases from moderately halophilic alkalither-motolerant indigenous actinomycetes. World Appl Sci J 7: 907-916

Nonaka T, Fujihashi M, Kita A, Hagihara H, Ozaki K, Ito S, Miki K (2003) Crystal structure of calcium-free alpha-amylase from *Bacillus* sp. strain KSM-K38 (AmyK38) and its sodium ion binding sites. J Biol Chem 278: 24818-24824

Olivera N, Sequeiros C, Sineriz F, Breccia J (2006) Characterization of alkaline proteases from novel alkali-tolerant bacterium *Bacillus patagoniensis*. World J Microbiol Biotechnol 22: 737-743

Olsen HS, Falholt P (1998) The role of enzymes in modern detergency. J Surfactants Deterg 1: 555-566

Oluoch KR, Welander U, Andersson MM, Mulaa FJ, Mattiasson B, Hatti-Kaul R (2006) Hydrogen peroxide degradation by immobilized cells of alkaliphilic *Bacillus halodurans*. Biocat Biotransfor 24: 215-222

Osanjo GO, Muthike EW, Tsuma L, Okoth MW, Bulimo WD, Lünsdorf H, Abraham WR, Dion M, Timmis KN, Golyshin PN, Mulaa FJ (2009) A salt lake extremophile, *Paracoccus bogoriensis* sp. nov., efficiently produces xanthophyll carotenoids. Afr J Microbiol Res 3: 426-433

Paar A, Costa S, Tzanov T, Gudelj M, Robra K-H, Cavaco-Paulo A, Gübitz GM (2001) Thermo-alkali-stable catalases from newly isolated *Bacillus* sp. for treatment and recycling of textile bleaching effluents. J Biotechnol 89: 147-153

Paavilainen S, Helistö P, Korpela T (1994) Conversion of carbohydrates to organic acids by alka-liphilic bacilli. J Ferment Bioeng 78: 217-222

Padan E, Bibi E, Ito M, Krulwich TA (2005) Alkaline pH homeostasis in bacteria: new insights. Biochim Biophys Acta 1717: 67-88

Palozza P, Torelli C, Boninsegna A, Simone R, Catalano A, Mele MC, Picci N (2009) Growth-inhibitory effects of the astaxanthin-rich alga *Haematococcus pluvialis* in human colon cancer cells. Cancer Lett 283: 108-117

Pietrangelo A (2002) Mechanism of iron toxicity. In: Hershko C (ed) Iron chelation theraphy, vol. 509. Kluwer Academic/Plenum Publishers, New York, pp 19-43

Pokhrel D, Viraraghavan T (2004) Treatment of pulp and paper mill wastewater-a review. Sci Total Environ 333: 37-58

Rai SK, Roy JK, Mukherjee AK (2010) Characterisation of a detergent-stable alkaline protease from a novel thermophilic strain *Paenibacillus tezpurensis* sp. nov. AS-S24-II. Appl Microbiol Biotechnol 85: 1437-1450

Rao CS, Sathish T, Ravichandra P, Prakasham RS (2009) Characterization of thermo-and deter-gent stable serine protease from isolated *Bacillus circulans* and evaluation of eco-friendly applications. Process Biochem 44: 262-268

Roadcap GS, Sanford RA, Jin Q, Pardinas JR, Bethke CM (2006) Extremely alkaline (pH>12) ground water hosts diverse microbial community. Groundwater 44: 511-517

Ruggiero CE, Neu MP, Matonic JH, Reilly SD (2000) Interactions of Pu with desferrioxamine siderophores can affect bioavailability and mobility. Actinide Res Q 2000: 16-18

Said S, Fonseca MJV, Siessere V (1991) Pectinase production by *Penicillium frequentans*. World J Microbiol Biotechnol 7: 607-608

Sashihara N, Kudo T, Horikoshi K (1984) Molecular cloning and expression of cellulase genes of alkalophilic *Bacillus* sp. strain N-4 in *Escherichia coli*. J Bacteriol 158: 503-506

Sato M, BeppuT, Arima K (1980) Properties and structure of a novel peptide antibiotic no. 1970.
Agric Biol Chem 44: 3037-3040

Saxena RK, Dutt K, Agarwal L, Nayyar P (2007) A highly thermostable and alkaline amylase from a *Bacillus* sp. PN5. Bioresour Technol 98: 260-265

Seifzadeh S, Sajedi RH, Sariri R (2008) Isolation and characterization of thermophilic alkaline proteases resistant to sodium dodecyl sulfate and ethylene diamine tetraacetic acid from *Bacillus* sp. GUS1. Iranian J Biotechnol 6: 214-221

Senthilvelan T, Kanagaraj J, Mandal AB (2012) Application of enzymes for dehairing of skins: cleaner leather processing. Clean Technol Envir 14：889-897

Shrinath A，Szewczak JT，Bowen IJ（1991）A review of ink removal techniques in current deinking technology. Tappi J 74：85-93

Shrinivas D，Naik GR（2011）Characterization of alkaline thermostable keratinolytic protease from thermoallkalophilic *Bacillus halodurans* JB 99 exhibiting dehairing activity. Int Biodeterior Biodegrad 65：29-35

Singh SA，Plattner H，Diekmann H（1999）Exopolygalacturonate lyase from a thermophilic *Bacillus* sp. Enzyme Microb Technol 25：420-425

Singh A，Kuhad RC，Ward OP（2007）Industrial application of microbial cellulases. In：Kuhad RC，Singh A（eds）Lignocellulose biotechnology：future prospects. I. K. International Publishing House，New Delhi，pp 345-358

Singh LS，Mazumder S，Bora TC（2009）Optimisation of process parameters for growth and bioactive metabolite produced by a salt-tolerant and alkaliphilic actinomycete，*Streptomyces tanashiensis* strain A2D. J Mycol Med 19：225-233

Soerensen NH，Hoff T，Oestergaard PR，Cassland P（2011）Enzyme dehairing of skins and hides. WO 2011161135 A1

Sorokin DY，Turova TP，Kuznetsov BB，Briantseva IA，Gorlenko VM（2000）*Roseinatronobacter thiooxidans* gen. nov.，sp. nov.，a new alkaliphilic aerobic bacteriochlorophyll a-containing bacterium isolated from a soda lake. Mikrobiologia 69：89-97

Sorokin DY，Berben T，Melton ED，Overmars L，Vavourakis CD，Muyzer G（2014）Microbial diversity and biogeochemical cycling in soda lakes. Extremophiles 18：791-809

Sorokin DY，Banciu HL，Muyzer G（2015）Functional microbiology of soda lakes. Curr Opin Microbiol 25：88-96

Spiro MC，Griffith WP（1997）The mechanism of hydrogen peroxide bleaching. Text Chem Color 29：12-13

Srinivasan MC，Rele MV（1999）Microbial xylanases for paper industry. Curr Sci 77：137-142

Sturr MG，Guffanti AA，Krulwieh TA（1994）Growth and bioenergetics of alkaliphilic *Bacillus firmus* OF4 in continuous culture at high pH. J Bacteriol 176：3111-3116

Sukumaran RK，Singhania RR，Pandey A（2005）Microbial cellulases-production，applications and challenges. J Sci Ind Res 64：832-844

Szente L，Szejtli J（2004）Cyclodextrins as food ingredients. Trends Food Sci Tech 15：137-142

Takaichi S，Oh-Oka H，Maoka T，Jung DO，Madigan MT（2003）Novel carotenoid glucoside esters from alkaliphilic heliobacteria. Arch Microbiol 179：95-100

Taksawan T，Yuichi H，Saori K，Moriya O，Savitr T，Napavarn N，Toshiaki K（2005）Comparison of bacterial communities in the alkaline gut segment among various species of higher termites. Extremophiles 9：229-238

Tatineni R，Doddapaneni KK，Potumarthi RC，Vellanki RN，Kandathil MT，Kolli N，Mangamoori LN（2008）Purification and characterization of an alkaline keratinase from *Streptomyces* sp. Bioresour Technol 99：1596-1602

Thakur IS（2006）Industrial biotechnology：problems and remedies. I. K. International，New Delhi，p 36 Thanikaivelan P，Rao JR，Nair BU，Ramasami T（2004）Progress and recent trends in biotechnological methods for leather processing. Trends Biotechnol 22：181-188

Thomas WJ（1994）Comparison of enzyme-enhanced with conventional deinking of xerographic and laser-printed paper. Tappi J 77：173-179

Thompson VS，Schaller KD，Apel WA（2003）Purification and characterization of a novel thermo-alkali stable catalase from *Thermus brokianus*. Biotechnol Prog 19：1292-1299

Thumar JT，Dhulia K，Singh SP（2010）Isolation and partial purification of an antimicrobial agent from halotolerant alkaliphilic *Streptomyces aburaviensis* strain Kut-8. World J Microbiol Biotechnol 26：2081-2087

Tsujibo H，Sato T，Inui M，Yamamoto H，Inamori Y（1988）Intracellular accumulation of phen-azine antibiotics production by an alkalophilic actinomycete. Agric Biol Chem 52：301-306 Turner P，Mamo G，Karlsson EN（2007）Potential and utilization of thermophiles and thermostable enzymes in biorefining. Microb CellFact 6：9-10

Tzanov T，Costa S，Gübitz GM，Cavaco-Paulo A（2001）Dyeing in catalase treated bleaching baths. Color Technol 117：1-5

Vasavada SH，Thumar JT，Singh SP（2006）Secretion of a potent antibiotic by salt-tolerant and alkaliphilic actinomycete *Streptomyces sannanensis* strain RJT-1. Curr Sci 91：1393-1397

Vértesy L，Aretz W，Fehlhaber HW，Kogler H（1995）Salimycin A-D，antibiotoka aus *Streptomyces violaveus*，DSM 8286，mit siderophor-aminoglycosid-struktur. Helv Chim Acta 78：46-60

Vicuna R，Escobar F，Osses M，Jara A（1997）Bleaching of Eucalyptus Kraft pulp with commercial xylanases. Biotechnol Lett 19：575-578

Von Gunten HR，Benes P（1995）Speciation of radionuclides in the environment. Radiochim Acta 69：1-29

Vyas S，Lachke A（2003）Biodeinking of mixed office waste paper by alkaline active cellulases from alkalotolerant *Fusarium* sp. Enzyme Microb Technol 32：236-245

Weck M（1991）Hydrogen peroxide-an environmentally safe textile bleaching agent. Text Prax Int 46：144-147

Wielen LCV，Panek JC，Pfromm PH（1999）Fracture of toner due to paper swelling. Tappi J 82：115-121

Ximenes FA，Sousa MV，Puls J，Silva FG，Filho EXF（1999）Purification and characterization of a low-molecular-weight xylanase produced by *Acrophialophora nainiana*. Curr Microbiol 38：18-21

Yokaryo H，Tokiwa Y（2014）Isolation of alkaliphilic bacteria for production of high optically pure L-（+）-lactic acid. J Gen Appl Microbiol 60：270-275

Zhilina TA，Appel R，Probian C，Brossa EL，Harder J，Widdel F，Zavarzin GA（2004）*Alkaliflexus imshenetskii* gen. nov. sp. nov.，a new alkaliphilic gliding carbohydrate-fermenting bacterium with propionate formation from a soda lake. Arch Microbiol 182：244-253

第九章

嗜盐古菌氮代谢研究进展及其在生物技术领域的应用

Julia Esclapez[1]，Mónica Camacho[1]，Carmen Pire[1]，Vanesa Bautista[1]，Anna Vegara[1]，Laia Pedro-Roig[1]，Francisco Pérez-Pomares[1]，Rosa María Martínez-Espinosa[1]，María José Bonete[1][2]

9.1 引言

在生物圈中，硝酸盐同化是将无机氮转化为有机氮的重要途径。这一途径需要硝酸还原酶和亚硝酸还原酶的催化。首先，NO_3^- 被高亲和转运体运入细胞，在硝酸还原酶（Nas；EC 1. 6. 6. 2）和亚硝酸还原酶（NiR；EC 1. 7. 7. 1）的催化下被还原成 NH_4^+。然后通过谷氨酰胺合成酶/谷氨酸合酶途径（GS-GOGAT；EC 6. 3. 1. 2，EC 1. 4. 7. 1）或通过谷氨酸脱氢酶（GDH；EC 1. 4. 1. 2）催化铵与碳骨架结合。其中 GS/GOGAT 途径尤其重要，因为它可在细胞内氨浓度低的情况下将氨同化为 L-谷氨酸（Glu），并且在该条件下可有效地替代其他谷氨酸生物合成反应（如 GDH 催化的 Glu 的合成）（Bonete et al. 2008）。

在古菌的两个主类群泉古菌（Crenarchaeota）和广古菌（Euryarchaeota）的基因组中已经发现参与编码硝酸盐及氨同化蛋白质的相关基因（Feng et al. 2012）。另一方面，对土壤中的嗜盐古菌群落进行研究发现，在盐碱环境中，嗜盐古菌同化硝酸盐还原作用是非常重要的（Alcántara-Hernández et al. 2009）。然而，对嗜盐古菌中参与 NO_3^-、NO_2^- 和 NH^{4+} 同化过程的酶的性质的研究还很欠缺，直至本章撰写，研究主要集中在地中海富盐菌（*Haloferax mediterranei*）（图 9. 1）。以下内容对已经报道的嗜盐古菌蛋白转运体及同化性硝酸盐还原相关酶进行综述。

---

[1] Department of Agrochemistry and Biochemistry，University of Alicante，Carretera；San Vicente del Raspeid s/n-03690 San Vicente del Raspeid，Alicante，Spain.

[2] e-mail：mjbonete@ua. es.

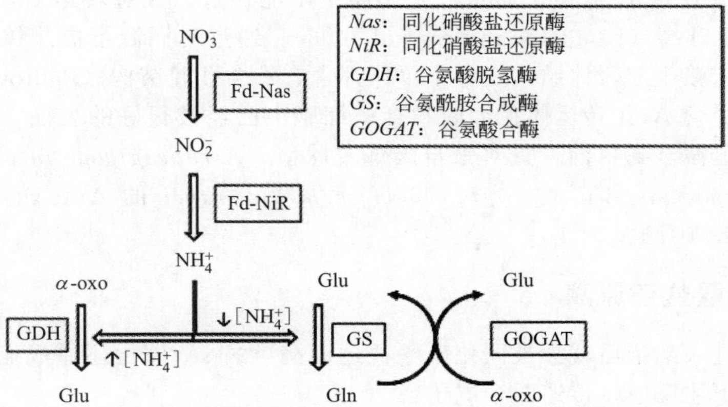

图 9.1 *Haloferax mediterranei* 中硝酸盐及氨同化途径。α-oxo：α-酮戊二酸；Glu：L-谷氨酸；Gln：L-谷氨酰胺

9.2 硝酸盐/亚硝酸盐和氨同化相关酶

9.2.1 NO_3^- 和 NO_2^- 转运体

硝酸盐或亚硝酸盐等含氮化合物首先要被吸收才能被作为氮源利用，一般来讲，在原核生物中，编码 NO_3^- 摄取及还原的相关基因通常聚集成簇（More-Vivián and Flores 2007）。此外，NO_3^- 通常通过主动转运进入细胞，截止到目前，原核生物中，发现两种硝酸盐转运蛋白：ATP 依赖的 ABC 转运体（由整合膜蛋白亚基和胞质 ATP 结合蛋白及细胞周质的底物结合蛋白组成）及主要易化子超家族（MFS 透性酶）的单体 NarK 转运体。后者依赖质子动力进行转运。在古菌、细菌、真核生物中均存在 ABC 转运体（Wanner and Soppa 1999）。细菌 NarK 转运体可以分为两类，NarK1（质子：硝酸盐同向转运体，反硝化作用的起始）和 NarK2（硝酸盐：亚硝酸盐逆向转运体，维持转化速率）（Wood et al. 2002）。这些蛋白质参与 NO_3^- / NO_2^- 的转化，而不是简单地摄取某一单一离子。然而，MFS 家族转运体的转运机制还未被阐明。有研究认为细菌硝酸盐同化需要依赖 ATP 的 ABC 转运体完成，而反硝化作用主要由质子动力驱动的 NarK 转运体完成（Wood et al. 2002）。通过数据库比对嗜盐古菌硝酸盐同化相关基因，发现在极端嗜盐古菌 *Haloferax mediterranei* 中 *nasB* 基因（Q703N4）编码 NO_3^- 转运体是膜蛋白（分子质量约 46.1kDa），含有 12 个 α 螺旋，属于 NarK1 转运体。该 *nasB* 基因与嗜热菌属（*Thermus*）、副球菌属（*Paracoccus*）和假单胞菌属（*Pseudomonas*）基因同源性较高（Lledó et al. 2005）。*Haloferax mediterranei* 来源的 NarK 是目前第一个被报道的古菌 NarK 转运体，并不是像最初认为的那样只存在于细菌及真核生物。研究表明，NarK 转运体不仅参与同化过程中硝酸盐的转运，而且还参与反硝化过程，该转运体负责硝酸盐/亚硝酸盐的转换。然而，还需要对这种转运体进一步研究才能更好地理解嗜盐古菌对硝酸盐和亚硝酸盐的摄取。

9.2.2 氨转运体

胺要作为氮源用于生长，第一步是通过铵转运体（Amt）来对其进行摄取。Amt 转运体通常的特点是：对胺具有高度亲和性和选择性，对碱性阳离子的非通透性，对甲胺的通透性及可在低物质的量浓度下快速饱和。这种转运体为三聚体蛋白，含有 11 个跨膜域。每个单体含有一个疏水的 NH_3 通道，而水分子和离子不能通过。经过转运后，NH_3 被质子化，同时导致该侧 pH 升高（Khademi and Stroud 2006）。通过对同源蛋白晶体结构进行研究发现转运体的功能依赖于三聚体结构，每个单体都含有一个孔腔结构（Pantoja 2012）。从生物化学的角度讲，尽管 Amt 转运体在细菌和真核细胞中已经被很好的表征，但在古菌中还不是很清楚。实际上截至到目前，只对来自闪烁古球菌（*Archaeoglobus fulgidus*）（Andrade et al. 2005；Andrade and Einsle 2007）和 *Hfx. mediterranei* 的 Amt 转运体进行了研究（Pedro-Roig et al. 2013）。

9.2.3 同化硝酸盐还原酶

一旦硝酸盐进入细胞后就会被同化硝酸盐还原酶（Nas）催化还原成亚硝酸盐，Nas 属于胞内酶，催化下述反应，传递两个电子：

NO_3^- ＋NADH（或还原型铁氧还蛋白）＋H^+ ⟶ NO_2^- ＋ NAD^+（或氧化型铁氧还蛋白）＋ H_2O

生理状态下，大多数 Nas 既可以利用 NADH 也可以利用铁氧还蛋白作为电子的供体，当然也有利用黄素氧还蛋白代替铁氧还蛋白的，来自于细菌的 Nas 在有氧的条件下才可以检测到活性，且受硝酸盐的诱导，受铵盐的抑制（Richardson et al. 2001；Luque-Almagro et al. 2011）。

铁氧还蛋白依赖的 Nas（Fd-Nas）通常为单体酶，NADH 依赖的 Nas（NADH-Nas）为异二聚体（Richardson et al. 2001）。Fd-Nas 和 NADH-Nas 的结构和功能与很多原核生物中的间质异化型硝酸还原酶（Nap；EC 1.7.99.4）及膜结合呼吸性硝酸还原酶（Nar；EC 1.7.99.4）是不同的。基因序列分析揭示了 *Hfx. mediterranei* 铁氧还蛋白依赖的 Nas（Q703N5）是钼酶，钼-二-吡喃蝶呤鸟苷二核苷辅因子（Mo-bis-PGD）的合成由 MobA 催化，MobA 催化 GTP 与亚钼蝶呤的结合，从而产生活性辅助因子 PGD。另一方面，光谱学结果显示还存在一个［4Fe-4S］簇。到目前为止，该单体酶是从嗜盐古菌中分离鉴定到的唯一的 Nas（分子质量大约 75kDa），其电子流方向为：由铁氧还蛋白中的［2Fe-2S］簇（低氧化还原电位电子供体）到［4Fe-4S］簇，再到钼-辅因子使 NO^{3-} 还原（Martínez-Espinosa et al. 2001a；Lledó et al. 2005；Bonete et al. 2008）。

与 *Hfx. mediterranei* 中 Nas 蛋白最相似的是假单胞菌属（*Pseudomonas*）、黄单胞菌属（*Xanthomonas*）和聚球藻属（*Synechococcus*）的 Nas 蛋白。将它们与 *Hfx. mediterranei* 中可能编码 Nas 的基因序列进行比较，发现总体相似度很低，保守序列一般位于辅因子结合位点。在不同底物浓度下（硝酸盐），改变还原型甲基紫精（MV）的浓度来分析 Nas 的动力学参数。硝酸盐及 MV 的表观 K_m 值分别为（0.95±0.12）mmol/L 和（0.07±0.12）mmol/L。这种嗜盐的 Nas 对温度和 NaCl 的依赖性很高（最适温度为 80℃，NaCl 的浓度为 3.1～2.2mol/L）。*Hfx. mediterranei* 中 Nas 蛋白可以接受来自甲基紫精和苄基紫精的电子，但是不能接受来自 NADH/NADPH 的电子。一般有氧生长条件下可检测到 Nas，当生长达平稳期时，其活性下降甚至完全消失。当铵根离子（或硝酸盐的替代氮源）存在时，检测不到 Nas 的活性（Martínez-Espinosa et al. 2001a）。

9.2.4 同化亚硝酸盐还原酶

上述反应中生成的亚硝酸盐经铁氧还蛋白依赖的同化亚硝酸还原酶（NiR，Q703N2）催化还原成铵，涉及六个电子的转移：

NO_2^- ＋3NADH（或 3 还原型铁氧还蛋白）＋$5H^+$ ⟶ NH_4^+ ＋$3NAD^+$（或 3 氧化型铁

氧还蛋白）$+2H_2O$

同化亚硝酸盐还原酶按照特异性电子供体可以分为：铁氧还蛋白依赖的 NiR[Fd-NiR，来自原核和真核光合生物的研究] 和 NAD(P)H 依赖的 NiR [NAP(P)H-NiR，来自放线菌及大多数异养细菌]，Fd-NiRs 是细胞质单体蛋白，含有血红素和 [4Fe-4S] 簇作为氧化还原中心。电子由铁氧还蛋白的 [2Fe-2S] 传递到 [4Fe-4S]，然后再传递到血红素催化 NO^{2-} 的还原。NADH-NiRs 二聚体包括非共价结合的 FAD、[4Fe-4S] 簇和血红素辅基（Fernández et al. 1998）。

来自 *Hfx. mediterranei* 的 NiR，是第一个从该类微生物中纯化出来的 NiR，分子质量约为 66kDa 的单体蛋白。与已知的，如来自于蓝藻和细菌的铁氧还蛋白依赖的亚硝酸盐还原酶有很高的序列相似性。它的部分性质已经被表征：K_m 值约为 8.6（亚硝酸盐）；当 NaCl 浓度为 3.3mol/L，温度为 60℃时酶活最高。同目前研究的大多数细菌同化亚硝酸盐还原酶一样，来自 *Hfx. mediterranei* 的 NiR 包括血红素辅基和 Fe-S 中心作为氧化还原中心（Martínez-Espinosa et al. 2001b；Lledó et al. 2005）。

9.2.5　谷氨酰胺合成酶-谷氨酸合酶循环及谷氨酸脱氢酶

对大部分细菌和古菌而言，合成 L-Glu 是氨和氮同化的最主要的方式。*Hfx. mediterranei* 中该过程涉及三个酶的催化：Gln 合成酶（GS），催化铵生成 Gln；Glu 合酶（GOGAT），催化 Gln 生成 Glu；Glu 脱氢酶（GDH），催化 α-酮戊二酸还原氨化生成 Glu 的可逆还原（Ferrer et al. 1996；Díaz et al. 2006；Martínez-Espinosa et al. 2006；Pire et al. 2014）。GS-GOGAT 循环需要 ATP，且对氨的亲和性很高。然而，GDH 不消耗 ATP，但是当细胞处于氮限制条件下其效率较低。这些酶在生物三大系统中都存在。

9.2.5.1　Gln 合成酶

Gln 合成酶（GS）存在于 GS-GOGAT 途径，已经对一些来自极端嗜热及产甲烷古菌、原核生物及哺乳动物的 GS 蛋白进行了生化与分子水平的研究（Adul Rahman et al. 1997；Reitzer 2003）。该蛋白质具有双重功能，一方面可以生成 Gln，另一方面可以与谷氨酸合酶协同同化铵。根据分子质量和亚基数量可以将 GS 分为 3 类：GSⅠ包含 12 个亚基，分子质量范围是 44～60kDa，存在于古菌和细菌（Brown et al. 1994；Robertson and Alberte 1996）；GSⅡ，八聚体酶，亚基分子质量范围是 35～50kDa，分布于少数的土壤细菌和真核生物中（Kumada et al. 1993）；GSⅢ是六聚体蛋白，亚基分子质量为 75kDa，分布于厌氧细菌和蓝细菌中（Reyes and Florencio 1994）。

有关古菌来源的 GS 的报道非常少，仅有的关于 GS 的研究来自于嗜盐古菌 *Hfx. mediterranei* 和盐沼盐杆菌（*Hbt. salinarium*），均为八聚体，属于Ⅱ型 GS（Manitz and Holldorf 1993；Martínez-Espinosa et al. 2006）。然而，一些来自极端嗜热或产甲烷古菌的 GS 是十二聚体，分子质量约为 600kDa。嗜盐古菌来源的 GS 与来自真核生物及土壤细菌的 GS 具有相似的性质。来自 *Hfx. mediterranei* 的研究表明，嗜盐古菌来源的 GS 能够同化亚硝酸还原酶催化生成铵（Martínez-Espinosa et al. 2006，2007）。然而，当培养基中氮源浓度很高时，是由 GDH 催化铵的同化。*Hfx. mediterranei* 来源的 GS 的理化性质研究显示，其最适 pH 大约 8，当 NaCl 或 KCl 浓度分别达到 3.5mol/L 和 2.5mol/L 时，稳定性最高，无法计算 ADP 对应的 K_m 值（转换酶活性），因为不同浓度测得的结果几乎都是一样的 [K_m 值（3.1 ± 0.5）mmol/L]。NH_2OH、Gln、ATP 及 Glu 表观 K_m 分别为（10.5 ± 3.5）mmol/L，（25 ± 1.8）mmol/L、（0.30 ± 0.08）mmol/L 及（4.9 ± 1.5）mmol/L。

这些值与来自盐沼盐杆菌（*Hbt. salinarium*）的 GS 相似，与来自蓝细菌及古菌 GS 的值所在范围一致（Manitz and Holldorf 1993）。金属离子对 GS 的影响的结果显示 Mn^{2+} 可能作为辅因子起到激活转换酶活性的作用。

9.2.5.2 谷氨酸合酶

谷氨酸合酶（GOGAT 或 GltS）是 GS/GOGAT 循环的第二种酶，催化 L-谷氨酰胺的酰胺氮转移给 α-酮戊二酸，形成两分子 L-谷氨酸。不同生物和组织中分布着不同类型的 GOGAT，其亚基组成、辅因子含量和还原剂都有所差异（Vanon and Curti，2005）。GOGAT 主要可以分为三类。(a) 细菌 NADPH 依赖型 GOGAT（NADPH-GOGAT），由两个不同的亚基组成（α 大亚基，β 小亚基，分别是 *gltB* 和 *gltD* 基因的编码产物）。α 亚基由四个结构域组成，包括 N 末端酰胺转移酶结构域、FMN 结合结构域和 C 末端结构域。β 亚基含有 NADPH 结合位点。(b) 来自光合细胞的铁氧还蛋白依赖型 GOGAT（Fd-GOGAT），由单条多肽链组成，与细菌 NADPH-GOGAT 的 α 亚基类似。(c) 来自酵母、真菌和低等动物的 NADH 依赖型 GOGAT（NADH-GOGAT）由单条多肽链组成，N 末端区域与细菌 α 亚基类似，C 末端区域与细菌 β 亚基类似。古菌有许多不同类型的 GOGAT。

在古菌中还存在不属于上述任何类型的 GOGAT。在 *Archaeoglobus fulgidus*、詹氏甲烷球菌（*Methanococcus jannaschii*）和嗜热自养甲烷杆菌（*Methanobacterium thermoautotrophicum*）中，存在截短的 *gltB* 基因，极端嗜热古菌火球菌属（*Pyrococcus* sp. KOD1）中的 GOGAT 与细菌小亚基同源（*gltD* 基因产物），可单独发挥功能，*Pyrococcus* 属其他菌株基因组中也只包含 *gltD* 基因（Nesbo et al. 2001）。来自 *Hfx. mediterranei* 的 GOGAT 与植物和蓝细菌来源的 GOGAT 一样，属于铁氧还蛋白依赖型酶，其与细菌 NADPH-GOGAT 中 α 亚基相似。通过色谱分离的 *Hfx. mediterranei* 中的两种不同的［2Fe-2S］铁氧还蛋白，仅其中一种具有活性。该酶也显示出嗜盐的特性，且非常稳定，当 NaCl 浓度为 3～4mol/L，pH 7.5，温度为 50℃时酶活性最大（Pire et al. 2014）。在其他嗜盐古菌基因组中可以找到类似的基因。

之前唯一的关于 Fd-GOGAT 的研究是来自非光合生物体嗜热氢杆菌（*Hydrogenobacter thermophilus*），一种氢氧化的化能自养型细菌（Kameya et al. 2007）。在已知的 Fd-GOGATs 中，存在被称为 Fd 环的保守序列，该序列不存在于 NADPH-GOGAT 中，可能参与与铁氧还蛋白的相互作用，但是来自 *H. thermophilus* 和 *Hfx. mediterranei* 的 Fd-GOGAT 不具有这样的保守序列（Kameya et al. 2007；Pire et al. 2014），表明在这些菌中该保守序列与铁氧还蛋白的相互作用可能是不必要的。来自 *H. thermophilus* 的酶可以被一些在 TCA 循环代谢中的琥珀酸、草酰乙酸、苹果酸或柠檬酸等有机酸活化。

9.2.5.3 谷氨酸脱氢酶

谷氨酸脱氢酶（GDH）可将 α-酮戊二酸转化为 L-谷氨酸，同样可以催化逆反应的进行，该酶是联系碳氮代谢的重要组成部分。细胞可以通过调节谷氨酸和 α-酮戊二酸的浓度来协调氮和碳代谢。*Hfx. mediterranei* 中铵的同化由 GDH 催化（Ferrer et al. 1996）。在有机体中也已经证明存在另外控制谷氨酸和 α-酮戊二酸浓度的途径，如 GS-GOGAT 途径，我们需要研究在什么条件下哪条途径优先，或者细胞在某一特定条件，特定时间下每条代谢途径的功能。在细菌中已经证明有效氮是决定因素。当铵的浓度达到一定值时，谷氨酸脱氢酶催化 α-酮戊二酸和铵转化为 L-谷氨酸。当铵限制时，主要由 GS/GOGAT 途径生成 L-谷氨酸，GS 对铵的亲和力很高，当铵达不到要求的浓度时，由于其对铵的 K_m 值低，所以优先进行这条途径。然而，也有例外，一些细菌，当铵浓度低时，GDH 也具有很高的活性（Hochman et al. 1988）。

不同来源的 GDHs 具有不同的辅酶偏好性。按照该特征将其分为三组：NAD 依赖的谷氨酸脱氢酶，NADP 依赖的谷氨酸脱氢酶；不依赖 NAD 或 NADP 的 GDHs；双辅酶依赖的 GDHs。据报道，嗜盐古菌 *Hfx. mediterranei* 具有至少两种不同的 GDH、NADP-GDH 和 NAD-GDH，已经对其进行了分离并表征了其性质（Ferrer et al. 1996；Díaz et al. 2006）。来自 *Hfx. mediterranei* 的 NADP 依赖型 GDH 推测其为分子质量 320kDa 的六聚体酶，由六个分子质量约为 55kDa 的单体组成。pH 8.5 时发生氨化反应，该酶对 $NADP^+$ 的 K_m 值为 0.18mmol/L；对 α-酮戊二酸的 K_m 值为 0.34mmol/L；对铵的 K_m 值为 4.2mmol/L。

研究发现多种代谢物对不同来源的谷氨酸脱氢酶均有调节作用，代谢物对活性的影响取决于谷氨酸脱氢酶的种类。例如 GTP、ATP、ADP 和 AMP 是来自哺乳动物的 GDH 的变构调节剂，对来自盐生盐杆菌（*Hbt. halobium*）来源的 NAD-GDH 没有影响（Bonete et al. 1996）。然而，对该嗜盐酶纯化后发现其活性受到 TCA 中间体（如延胡索酸、草酰乙酸盐、琥珀酸和苹果酸）的高度影响，这些代谢物对氨基化和脱氨基反应都具有强烈的抑制作用，表明该酶与 TCA 循环密切相关。

对这种嗜盐酶氧化脱氨反应的初始速率进行研究发现除了上述提及的 TCA 代谢通路中的代谢产物，还有许多代谢物如 $NADP^+$、D-Glu、戊二酸酯等，以及二羧酸化合物如己二酸可以起到调节作用。另一方面，由于 D-Glu 与底物 L-Glu 结构具有相似性，起到竞争性抑制作用，其他氨基酸则作对该酶有激活作用。

9.2.6 铁氧还蛋白作为硝酸盐/亚硝酸盐同化和 GS/GOGAT 循环中的电子供体

铁氧还蛋白是含有由铁和硫原子组成的铁硫簇的小分子蛋白，铁硫簇可以通过接受或释放电子（+2 价或+3 价）来改变氧化还原状态。铁氧还蛋白在生物氧化还原反应中作为电子传递体（Beinert 2000）。*Hfx. mediterranei* 中，硝酸盐和亚硝酸盐同化过程中参与电子传递的铁氧还蛋白是一种小蛋白质（SDS-PAGE 显示分子质量约为 21kDa）（Martínez-Espinosa et al. 2003；Zafrilla et al. 2011），其紫外可见光谱、核磁共振（NMR）、电子顺磁共振（EPR）光谱与植物和细菌来源的 [2Fe-2S] 铁氧还蛋白类似。利用紫外可见光谱分析法测定了 [2Fe-2S] 2+/1+转化的中点电位为−285mV，这种铁氧还蛋白热稳定性非常好，可耐受的最高盐浓度为 4mol/L，耐受的最高温度为 80℃。对来自其他嗜盐古菌如 *Halobacterium halobium*（Geiger et al. 1978）、极端嗜盐古菌 *Haloarcula marismortui*（Frolow et al. 1996）和日本盐盒菌（*Haloarcula japonica*）（Sugimori et al. 2000）的铁氧还蛋白进行研究发现均包含由 2 个 Fe 原子和 2 个 S 原子组成的铁硫簇。

9.3 硝酸盐/亚硝酸盐和氨同化相关酶的调节

9.3.1 硝酸盐和亚硝酸盐还原酶的调节

基于分子生物学的手段，Lledó 等人于 2005 年第一次对嗜盐古菌 *Hfx. mediterranei* 中的硝酸盐同化基因进行了鉴定和转录分析。该研究对 *Hfx. mediterranei* 基因组片段进行了序列测定和分析。生物信息学分析显示其存在四个开放阅读框（ORF）：*nasA*（2144 bp）、*nasB*（1509bp）、*nasC*（633bp）和 *nasD*（1761bp）。这些基因序列与细菌和真核生物来源的 *nas* 高度同源。

① *nasA* 具有 707 个氨基酸残基，被鉴定为同化硝酸盐还原酶。包含 Fe-S 簇和 Mo-MGD（Campbell and Kinghorn 1990）。基于蛋白质序列分析，将嗜盐硝酸盐还原酶分为单体铁氧还蛋白 Nas 组。

② *nasB* 具有 503 个氨基酸残基。它对应于 NarK 家族的硝酸盐转运蛋白，这是以前

在古菌中未鉴定的一组蛋白质。具体来说，与Ⅰ型 NarK 相对应（Moir and Wood 2001），参与了同化途径中的硝酸盐摄取。

③ *nasC* 具有 211 个氨基酸残基的蛋白质，其对应于钼蝶呤鸟嘌呤二核苷酸生物合成蛋白。这表明嗜盐硝酸盐还原酶中存在 Mo 辅因子。

④ *nasD* 具有 587 个氨基酸残基的蛋白质，被鉴定为单体铁氧还蛋白依赖型亚硝酸盐还原酶。它还具有参与 Fe-S 簇形成的残基。

位于 *Hfx. mediterranei* 基因组中的该 6720 bp 的片段代表了硝酸盐同化基因的新型结构。此外，与细菌不同（Lin and Stewart 1998），这些基因被转录为两个独立的信使 RNA，第一个为多顺反子（*nasABC*），第二个为单顺反子（*nasD*）（Lledó et al. 2005）。因为每个操纵子都由不同的启动子控制，所以这种基因排列使操纵子的调节产生了不同问题。两个启动子都具有回文序列，其可能与其他古菌域的微生物一样是转录调控因子结合区域。

通过 RT-PCR 对硝酸盐同化调控进行初步分析，目的是确定氮源对 *nasABC* 和 *nasD* 表达的影响（Lledó et al. 2005；Martínez-Espinosa et al. 2007）。以铵作为氮源时，*Hfx. mediterranei* 中没有检测到 *nasABC* 或 *nasD* 的转录。然而，当 *Hfx. mediterranei* 生长在有硝酸盐存在的培养基中时，检测到这些基因的表达（图 9.2），表明这些基因是硝酸盐同化所必需的。当培养物的光密度（OD）升高到 0.4 以上时，*nasA* 的表达较高，这与之前研究发现的当 OD＝0.9 时 Nas 活性最高的研究结果一致。然而，*Hfx. mediterranei* 生长在亚硝酸盐培养基中较生长在硝酸盐培养基的 *nasD* 表达要高，这表明亚硝酸盐对 *nasD* 有激活作用。

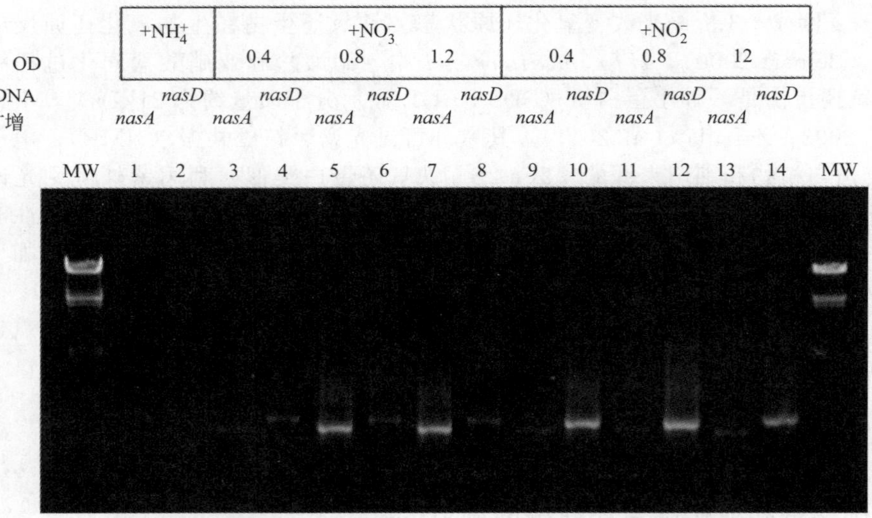

图 9.2 琼脂糖凝胶电泳分析 *nasA* 和 *nasD* 基因表达情况（*nasA* 和 *nasD* 序列来自 RT-PCR）

从不同光密度菌体中提取总 RNA，逆转录生成 cDNA，设计相应引物扩增 *nasA* 和 *nasD*。培养基中添加铵（泳道 1，泳道 2）；培养基添加硝酸盐（泳道 3～泳道 8）；培养基添加亚硝酸盐（泳道 9～泳道 14）；泳道 MW：分子量标准（Lledó 等 2005）

研究表明，两个操纵子 *nasABC* 和 *nasD* 都在转录水平受到调控，铵是抑制硝酸盐还原酶和亚硝酸还原酶基因转录的关键因素。此外，因为硝酸盐和亚硝酸盐可以提高 *nasABC* 和 *nasD* 的表达水平，所以硝酸盐和亚硝酸盐参与了同化途径的特异性调控。

Martínez-Espinosa 等（2009）在不同浓度的硝酸盐（0.5～100mmol/L）和亚硝酸盐（0.5～2mmol/L）存在下，对 *Hfx. mediterranei* 进行生理生化研究。研究表明，在硝酸盐浓

度高于 5mmol/L 的情况下，首先检测到硝酸还原酶活性，其次是亚硝酸还原酶活性。该条件下，在检测到亚硝酸还原酶活性之前，细胞内亚硝酸盐浓度缓慢增加，表明亚硝酸盐积累可以作为该酶活性增加的信号。通过实时定量 PCR 实验（图 9.3a 和 b）分析了所述条件下的 *nasA* 和 *nasD* 的表达，显示它们的 mRNA 浓度在指数期增加，而在稳定期初期降低。在所有条件下，在酶活性达到最大之前 *nasA* 和 *nasD* mRNA 表达量先达到最大。特别地，*nasA* mRNA 的表达水平在 5～25mmol/L 硝酸盐培养物中无显著差异，然而，硝酸盐浓度为 25～100mmol/L 范围内，*nasA* mRNA 的表达水平逐渐增加，当硝酸盐达到 100mmol/L 时，mRNA 表达量达到最大（图 9.3a）。此外，在含有 100mmol/L NO^{3-} 的培养基中 *nasA* mRNA 表达量达到最大后迅速降低。这可能是由于硝酸盐同化产物能够抑制 *nasA* 的表达。另一方面，*nasD* mRNA 水平显示对硝酸盐浓度具有明显的依赖性（图 9.3b），并且总是在 Nas 活性达到最高后检测到 *nasD* 表达达到最大值。因此，细胞内亚硝酸盐的积累可以诱导 *nasD* 表达。

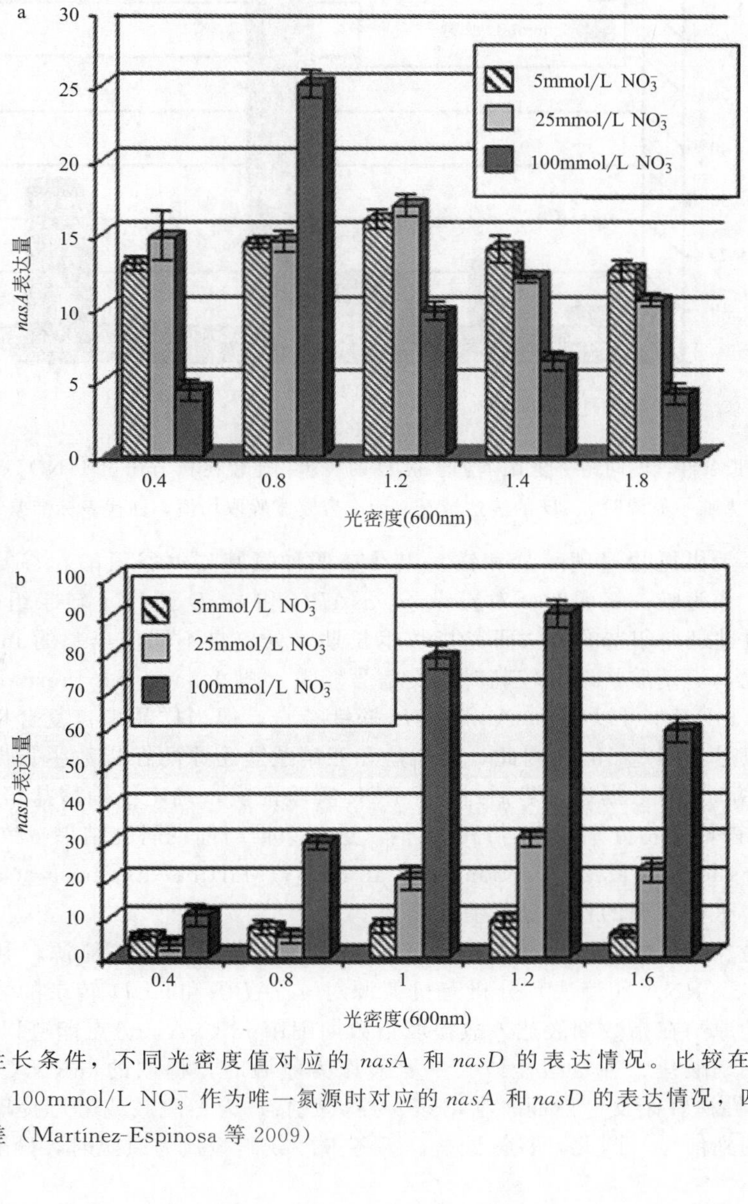

图 9.3 在不同生长条件，不同光密度值对应的 *nasA* 和 *nasD* 的表达情况。比较在 5mmol/L NO_3^-、25mmol/L NO_3^-，100mmol/L NO_3^- 作为唯一氮源时对应的 *nasA* 和 *nasD* 的表达情况，四次重复试验取均值，杠代表标准差（Martínez-Espinosa 等 2009）

菌株 *Hfx. mediterranei* 在以亚硝酸盐作为唯一氮源时，观察到检测的所有亚硝酸盐浓度均可被细胞消耗，并且在光密度相对较低的情况下就可以达到稳定期（Martínez-Espinosa et al. 2009）。令人惊讶的是，在亚硝酸盐作为氮源的培养基中检测到 *nasA* 的表达，其表达与亚硝酸盐浓度无关。该条件下，*nasA* 的表达量低于硝酸盐作为氮源时的表达量，因此硝酸盐对 *nasA* 的表达具有促进作用。如 Chai 和 Stewart（1998）所述，在具有不同于硝酸盐的氮源培养基中，*nasA* 基因的基本表达并不频繁，因此，这一数据表明，在这种嗜盐微生物中可能存在其他类型的调节。在以亚硝酸盐为氮源的培养物中检测到 *nasD* 的表达。图 9.4 显示其表达取决于亚硝酸盐浓度以及光密度。

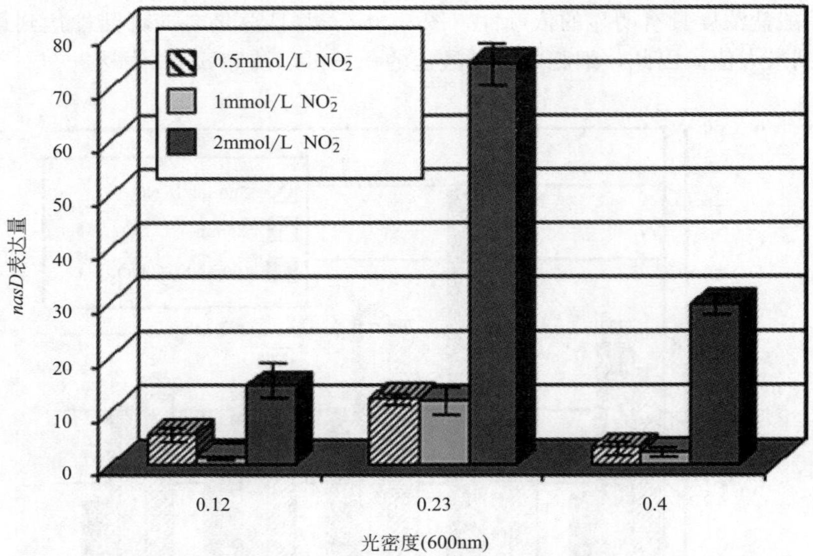

图 9.4　在不同生长条件，不同光密度值对应的 *nasD* 的表达。比较在 0.5mmol/L NO_2^-、1mmol/L NO_2^-、2mmol/L NO_2^- 作为唯一氮源时 *nasD* 的表达情况，四次重复试验取均值，杠代表标准差

最近，在以无机氮盐（硝酸盐和铵）以及在四种氨基酸（谷氨酸、谷氨酰胺、天冬氨酸、天冬酰胺）作为唯一氮源时，对操纵子 *nasABC* 和 *nasD* 进行了转录组分析（Esclapez et al. 2014）。通过斑点印迹的方法研究指数生长期，分离自不同培养基的 mRNA 与特异性探针杂交的情况，当硝酸盐浓度较高时杂交信号较强（图 9.5a）。Northern 分析显示以硝酸盐作为氮源时，mRNA 可以与 *nasA* 和 *nasD* 探针杂交，但当样品来自复合培养基时，未观察到任何杂交信号（图 9.5b）。因此，硝酸盐和亚硝酸盐还原酶在指数生长期，硝酸盐存在条件下可以表达。这些数据进一步通过 RT-PCR 实验证实，确定了硝酸盐培养物中的 *nasA* 和 *nasD* 表达比铵培养物分别高 10 倍和 15 倍。这些数据与前期研究结果一致（McCarty and Bremner 1992；Lledó et al. 2005；Bonete et al. 2008；Martínez-Espinosa et al. 2009），但是在嗜盐微生物中有机氮源的作用还没有被研究。

以谷氨酸盐、谷氨酰胺、天冬氨酸或天冬酰胺作为培养基唯一氮源，利用斑点印迹和 Northern 印迹法（RNA 印迹法）分析有机氮源对 *nasABC* 和 *nasD* 转录的影响，因为硝酸盐和亚硝酸盐还原酶在指数期表达，故提取指数期细菌 mRNA，斑点印迹以及 Northern 印迹分析显示，用硝酸盐、谷氨酸盐或天冬氨酸作为氮源时，提取的 mRNA 样品能与 *nasA* 和 *nasD* 的特异性探针杂交。然而，在以铵、谷氨酰胺、天冬酰胺或酵母提取物作为培养基的样品中未检测到信号。因此，谷氨酸盐、天冬氨酸或硝酸盐为氮源时，硝酸盐还原酶和亚

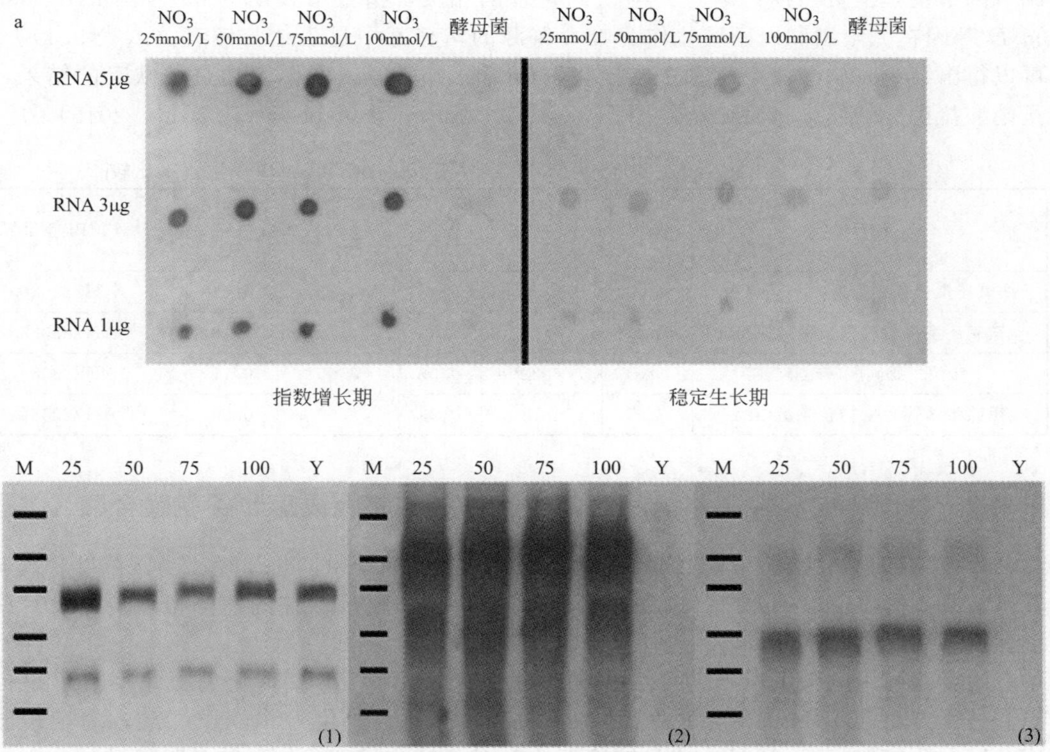

图 9.5　(a) 利用 *nasA* 特异的探针进行斑点印迹，从不同硝酸盐浓度下，提取处于指数增长期及稳定期的菌体中 mRNA 样品；(b) Northern 印迹法检测硝酸盐浓度为 25～100mmol/L 时，位于指数期的菌体中的基因表达水平

(1) 23S 和 16S rRNA 亚甲蓝染色作为对照控制 RNA 的质量和数量。(2) 利用 *nasA* 探针进行 Northern 分析。(3) 利用 *nasD* 探针进行 Northern 分析（Esclapez 等 2014）

硝酸盐还原酶均有表达，以硝酸盐为氮源时信号最强。而铵、谷氨酰胺或天冬酰胺则是硝酸盐同化途径中相关基因转录的阻遏物。在植物方面的研究表明，甘氨酸、谷氨酰胺和天冬酰胺抑制硝酸盐对硝酸还原酶的激活，而只有一个氨基的谷氨酸和天冬氨酸则诱导其表达（Radin 1977）。后来，也有研究表明谷氨酰胺、铵或类似物对土壤中同化硝酸还原酶活性具有抑制作用（McCarty and Bremner 1992）。以氨基酸作为氮源的最新研究表明，内部氨基酸池可以反映细胞内氮水平（Miller et al. 2008）。根据从嗜盐微生物 *Hfx. mediterra* 获得的数据可以推断，抑制硝酸盐和亚硝酸盐还原酶表达的氮源是那些铵参与合成的氨基酸，如谷氨酰胺和天冬酰胺。

在三种分别含有不同氮源的培养基中，通过微阵列技术对 *nasABC* 和 *nasD* 基因表达进行研究：(a) 铵培养基，稳定期（AmSt）和指数增长期（AmEx）的菌体；(b) 硝酸盐培养基，指数生长期（NiEx）的菌体；(c) 菌体处于氮限制（NSta）条件（Esclapez et al. 2015）。结果显示以铵作为氮源与以硝酸盐或氮限制条件下基因表达呈现显著的差异。该结果与以前使用不同方法进行分析得到的结果一致（Lledó et al. 2005；Martínez-Espinosa et al. 2009；Esclapez et al. 2014）。因此，可以得出结论，同化硝酸盐还原酶和亚硝酸盐还原酶表达的真正原因是铵的缺乏。此外，铵的缺乏不仅改变了参与硝酸盐同化途径的基因的表达，还改变了参与氮代谢的蛋白质的转录水平，如转运蛋白、蛋白质生物合成相关酶、转录

调节因子及其他蛋白质的表达。与氮同化基因相关，指数增长期的 *nasA*、*nasB*、*nasC* 和 *nasD* 基因在氮限制和以硝酸盐作为唯一氮源的培养基中过表达与铵有关（表 9.1）。因此，可以得出结论，硝酸盐不是真正诱导物（如氮限制）。*nas* 基因表达是铵缺乏的结果，而不是硝酸盐诱导的结果（Martínez-Espinosa et al. 2007；Esclapez et al. 2014，2015）。

表 9.1 NiEx-AmEx 和 NSta-AmEx 条件下 *nasABC* 和 *nasD* 表达水平比较

注释	基因	NSta-AmEx		NiEx-AmEx		NCBI 登录号
		Log_2FC	SD	Log_2FC	SD	
铁氧还蛋白依赖型亚硝酸还原酶	*nasD*	5.54	0.03	4.12	0.02	AHZ23086.1
铁氧还蛋白依赖型硝酸还原酶	*nasA*	5.46	0.50	4.70	0.50	AHZ23083.1
硝酸/亚硝酸转运体	*nasB*	4.22	0.07	3.60	0.11	AHZ23084.1
钼蝶呤鸟嘌呤生物合成蛋白 A	*nasC*	4.12	0.23	2.50	0.30	AHZ23085.1

9.3.2 氨同化的调控：谷氨酸脱氢酶、谷氨酰胺合成酶和谷氨酸合酶

各种生物为了应对不断变化的生存环境，产生了两种氨同化途径。当铵浓度较高时，是谷氨酸脱氢酶（GDH）及其代表的途径，其首要作用要确保铵的适当摄取。当没有铵或其浓度不够高时，GS/GOGAT 途径是细胞摄取氮的主要方式。

同时，因为氨基可以参与许多生物合成途径，所以 GDH 催化的氨化反应可以提供氮源，另一方面，GDH 的氧化脱氨反应可以将 L-谷氨酸转化为 α-酮戊二酸，为三羧酸循环（TCA）提供碳骨架，通过生成足够浓度的 L-谷氨酸来平衡谷氨酰胺与谷氨酸的比例对于保持 α-酮戊二酸的必需水平是非常关键的，因为 α-酮戊二酸参与 TCA 循环和转氨反应。GDH 催化的这两种不同反应，我们分别称为同化和异化作用，NADH 或 NADPH 通常作为辅因子发挥作用，二者的性质和调节均不同（Bonete et al. 1986，1987；Ferrer et al. 1996；Ingoldsby et al. 2005；Díaz et al. 2006；Tomita et al. 2010）。如前所述，氨基酸对来自 *Hbt salinarum* 的 NAD-GDH 具有激活作用，同时 TCA 中间产物和其他代谢物抑制其活性（Bonete et al. 1996；Pérez-Pomares et al. 1999）。

在生物体代谢中 GDH 两种催化活性的存在可能与前面提到的 GDH 的基本作用相关（Smith et al. 1975；Santero et al. 2012），极端微生物如嗜盐古菌中不同 GDH 的确切作用仍然需要进一步研究。在 *Hbt salinarum* 基因组中发现 GDH 不仅仅只有两种，推测至少有四种 GDH 基因，该发现使研究变得更加有趣（Ingoldsby et al. 2005）。

最近 Tomita 等人（2010）研究发现，嗜热栖热菌（*Thermus thermophilus*）的谷氨酸脱氢酶在酶的调节方面显示出有趣的特征。对这种极端嗜热菌的基因组进行分析发现，有两个 GDH 基因，即 *gdhA* 和 *gdhB* 推测位于一个操纵子，它们分别对应于两种不同谷氨酸脱氢酶。为了了解它们真正的功能，Tomita 等人（2010）纯化和表征了这两个基因表达的产物。GdhA 显示没有 GDH 活性，GdhB 显示具有 GDH 活性，其还原氨化活性比氧化脱氨活性高 1.3 倍。GdhA 与 His-标签融合的 GdhB 的共表达并纯化也显示具有谷氨酸脱氢酶活性，我们将其命名为 GdhA-GdhB。GdhA-GdhB 的还原氨化活性低于单独的 GdhB，但氧化脱氨活性则较高。GdhA-GdhB 氧化脱氨活性是还原氨化活性的 3.1 倍。

正如报道的嗜盐古菌 *Hbt salinarum* 的 NAD-谷氨酸脱氢酶（Bonete et al. 1996），氨基酸对复合物 GdhA-GdhB 具有活化作用，疏水性氨基酸作用更强。亮氨酸对 GdhA-GdhB 的活化作用最强，存在 1mmol/L 亮氨酸比无亮氨酸时的还原氨化活性高 9.74 倍，氧化脱氨活

性高 2.45 倍。亮氨酸对 GdhB 的激活作用就要低很多。这种活化作用的动力学分析表明，亮氨酸提高了谷氨酸脱氢酶的转化数。GDH 的异源寡聚体中，GdhA 是调节亚基，GdhB 是催化亚基，GdhA 通过与 GdhB 形成异源复合物来调节 GdhB 的活性，该异源复合物的形成依赖于疏水性氨基酸的浓度。Tomita 等人证实了异源寡聚体谷氨酸脱氢酶受到变构调控作用（Tomita et al. 2011）。

关于谷氨酸合酶调节的报道很少，不同的生物体调控的方式不同。在植物中有两种形式的 GOGAT 酶：Fd-GOGAT，其通常以高活性存在叶绿体光合组织中；而 NADH-GOGAT 存在于非光合质体中（Bowsher et al. 2007）。这两种酶在氮同化过程中都可以催化谷氨酸的生成，谷氨酸脱氢酶并不是生成谷氨酸的重要方式。GDH 可能主要通过脱氨生成 α-酮戊二酸和氨，分别用于呼吸作用和酰胺的形成（Miftin 2002；Forde and Lea 2007），并且还将与 NADH-GOGAT 共同作用，以控制植物体内 Glu 的稳态（Labboun et al. 2009）。

在大肠杆菌中，当细胞生长不受能量限制时，使用 GS/GOGAT 途径。铵浓度较低时，主要通过 GS-GOGAT 途径合成谷氨酸，这也是调控谷氨酰胺池的重要方式。当铵和磷酸盐浓度较高及细胞受到能量和碳的限制时，通过 GDH 途径合成谷氨酸（Helling 1994）。

在细菌中，有许多转录调控因子调控 GOGAT 的合成。*gltB* 和 *gltD* 基因分别编码 GOGAT 的大亚基和小亚基。在大肠杆菌中，*gltBD* 的表达受到包括亮氨酸反应蛋白 LRP 和 CRP（Reitzer 2003）的调控。集胞藻属（*Synechocystis* sp. PCC6803）主要通过 GS-GOGAT 途径同化氨。但非指数期生长的蓝细菌更倾向于通过 GDH 途径合成谷氨酸（Chávez et al. 1999）。在谷氨酸棒状杆菌（*Corynebacterium glutamicum*）中，基因 *gltB* 和 *gltD* 形成的操纵子的转录受到氮代谢调控蛋白 AmtR 的调控，该生物中，AmtR 对氮调节基因具有转录抑制作用（Beckers et al. 2001），可调节其他氮相关基因如 *amtB* 和 *glnK* 的转录。Schulz 等人（2001）发现，在富氮和氮限制条件下，*Corynebacterium glutamicum* 菌中 *gltBD* 上游编码组氨酸激酶的 *hkm* 基因被破坏，GOGAT 活性降低$\frac{1}{2}$。*hkm* 的转录受到氮缺乏的诱导，表明 Hkm 蛋白可以反映生长培养基的营养状态（Schulz et al. 2001）。

在枯草芽孢杆菌（*B. subtilis*）中没有 GDH，负责氨同化途径的两种酶是谷氨酰胺合成酶和谷氨酸合酶。*gltAB* 操纵子需要特异性的正调节因子 *GltC*，其表达受到 *TnrA* 的抑制，*TnrA* 是氮代谢调控的全局性转录调节因子（Belitsky et al. 2000）。

在转录水平研究嗜盐古菌中，阐明了不同途径中不同酶的作用。图 9.6a 显示了不同培养基（分别以铵、硝酸盐、谷氨酸或谷氨酰胺作为氮源），指数期 *gdh-1*、*glnA* 和 *gltS* 表达谱的比较。结果表明，硝酸盐培养基中 *gltS* 表达比铵培养基高 70 倍。硝酸盐的存在增强了 *glnA* 的表达，但抑制了 *gdh-1* 的表达。这与铵缺乏时依赖 GS/GOGAT 途径的假设是一致的。但是在铵浓度较高情况下，尽管 *gdh-1*、*gltS* 和 *glnA* 的转录水平非常相似，但仍然优先进行 GDH 催化的氨同化途径（图 9.6 b）。在以谷氨酸作为唯一氮源的培养基中，*gdh-1*、*glnA* 和 *gltS* 表达低于铵培养基。然而，谷氨酰胺培养基中，*glnA* 表达被抑制，而 *gltS* 和 *gdh-1* 的表达增强，这表明细胞中高水平的谷氨酸是抑制氨同化的信号，但高水平的谷氨酰胺易于升高铵浓度。实际上，*gltS* 和 *gdh-1* 的表达都高于含有铵的培养基。Pire 等人（2014）的 RT-qPCR 结果表明 GS/GOGAT 途径可能是在铵缺乏条件下氨同化的首选途径。在以铵为氮源的培养基中，*gdhA-1* 和 *gdh1* 转录水平相似，*gdhA-1* 相对于 *gdh1* 的表达变化为 1.1±0.3。两种酶均在氨化反应中起作用（Ferrer et al. 1996；Díaz et al. 2006），但是当氮源为硝酸盐而不是铵时，*gdhA-1* 的基因表达没有变化，*gdh1* 的表达被高度抑制，*gdhA-1* 似乎是组成型表达，以往研究表明，铵培养基中 NADPH-GDH 的活性高于硝酸盐培养基，表明硝酸盐抑制其活性（Ferrer et al. 1996）。

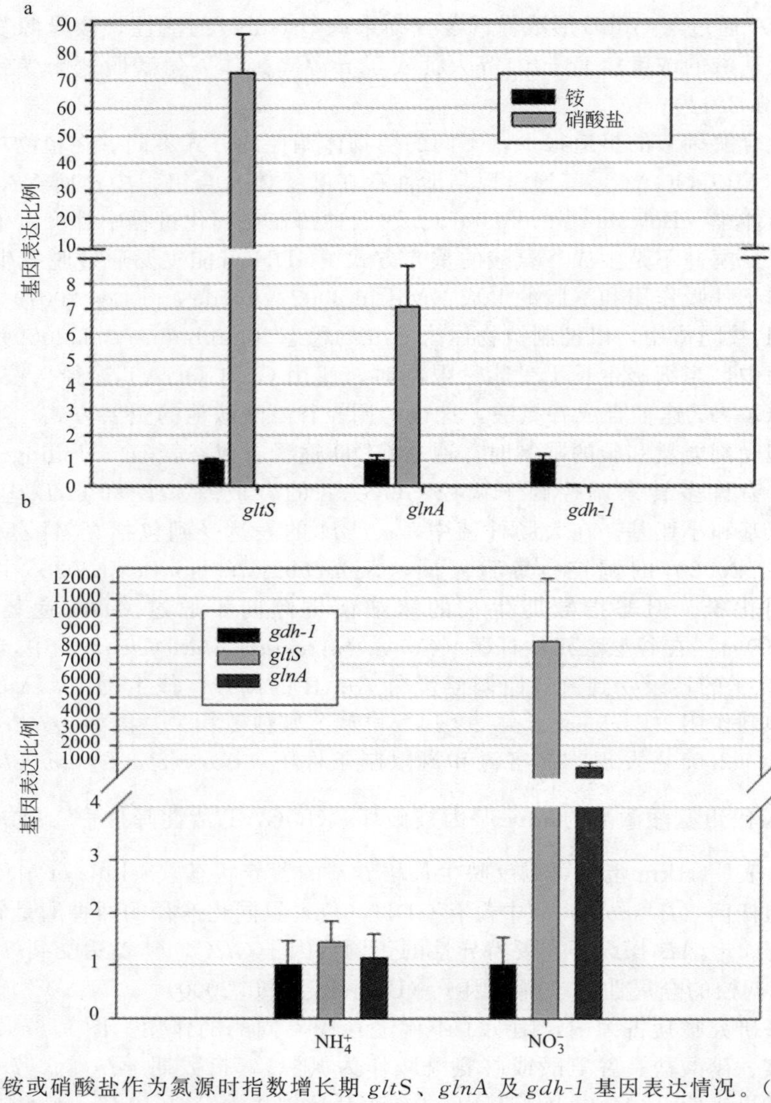

图 9.6　研究以铵或硝酸盐作为氮源时指数增长期 *gltS*、*glnA* 及 *gdh-1* 基因表达情况。(a) 以铵为氮源时的基因表达作为内参；(b) *gdh-1* 为内参基因（Pire 等 2014）

在硝酸盐培养基中，*gdh-1* 的转录被高度抑制，*gltS* 和 *glnA* 的转录被增强。当培养基中有铵存在时，GDH 反应首先摄取铵，因为铵是相较于硝酸盐或其他含氮化合物的优选氮源（Bonete et al. 2008）。然而，Northern 印迹分析以及 RT-qPCR 结果表明即使铵存在时，Fd-GOGAT 仍然有基础水平的转录。

如图 9.6b 所示，在含有铵的培养基中 *gltS* 的转录水平与 *gdh-1* 相似。该条件下，因为 GDH 对 α-酮戊二酸和铵的亲和力较高，故其活性占主导地位。这些结果表明，α-酮戊二酸是必不可少的。已经提出，“体内”GOGAT 催化反应的速率由 GS 产生的谷氨酰胺来调节，而 GS 活性受细胞内谷氨酰胺和 α-酮戊二酸的水平调节，而 GOGAT 又可以通过调节 α-酮戊二酸浓度调节 GS（Vanoni and Curti 1999）。在我们看来，根据 K_m 值，α-酮戊二酸是决定通过 GS/GOGAT 途径进行同化的关键代谢物（Pire et al. 2014）。

所有这些结果表明在 *Hfx. mediterranei* 中，硝酸盐还原酶和亚硝酸盐还原酶催化反应生成的氨主要通过 GS/GOGAT 循环进入碳骨架中。*Hfx. mediterranei* 在不同氮源（铵、硝酸盐和氮限制）中的全基因表达谱已经被分析（Esclapez et al. 2015）。对以铵为氮源和以硝酸盐为氮源或氮缺乏培养条件下的差异转录因子进行了鉴定。已经鉴定了一些转录调控因子，如两种 ArsR 型转录因子及一些未知的转录调控因子。对这些转录调控因子的分析，将有助于鉴定嗜盐古菌适应铵缺乏的氮调节因子。

9.4　嗜盐古菌中的氮调控蛋白

1969 年，当研究大肠杆菌谷氨酰胺合成酶翻译后修饰时发现了氮调控蛋白（PⅡ）信号转导蛋白（Shapiro 1969）。目前已知 PⅡ蛋白参与氮同化的调节过程（Ninfa and Jiang 2005；Leigh and Dodsworth 2007；Forchhammer 2008）。它可以同时结合 α-酮戊二酸和 ATP（Forchhammer 2010）。受 PII 调节的蛋白质有：AmtB 铵转运蛋白，当细胞氮水平足够高时，其能够保持 GlnK 与其结合（Javelle et al. 2004）；谷氨酰胺合成酶腺苷酰转移酶（GlnE），其共价修饰谷氨酰胺合成酶的活性（Jiang et al. 2007）；DraG/DraT，调节固氮菌中的固氮酶（Dixon and Kahn 2004）和蓝细菌中精氨酸生物合成的重要酶 NAGK（Heinrich et al. 2004）等。

有文章已经报道了古菌中的 GlnB-K 蛋白质家族，如马氏甲烷八叠球菌（*Methanosarcina mazei*）（Ehlers et al. 2002，2005）和 *Archaeoglobus fulgidus*（Helfmann et al. 2010；Litz et al. 2011）。对于嗜盐古菌，这些蛋白质的研究始于 2011 年，在 *Hfx. mediterranei* 基因组中存在两个 PⅡ蛋白的同源基因（Pedro-Roig et al. 2011），氨基酸相似性为 84%，归为 GlnKs（$GlnK_1$ 和 $GlnK_2$），均与铵转运蛋白（*amt*）基因连锁。两对基因连续地位于基因组中，由 300 个核苷酸组成的非编码序列隔开。两个 GlnKs（PS00638、PROSITE）C-末端区域高度保守，但 PⅡ蛋白只有尿苷酰化位点是保守的（PS00496，PROSITE）。*Hfx. mediterranei* 的两个 GlnK 蛋白中存在氨基酸序列 WRGEEY。与其他非嗜盐蛋白质比较有两个变化，第一个残基 W 在非嗜盐菌中对应的是 Y，第四个残基 E 在非嗜盐菌中对应的是 A 或 S，第四个残基的替代可能与嗜盐菌的适应性相关。与其他 PⅡ蛋白一样，$GlnK_1$ 的 60 位酪氨酸可以参与尿苷酰化，$GlnK_2$ 的 61 位酪氨酸可以参与尿苷酰化。通过使用戊二醛交联技术进行研究发现，GlnK 蛋白可以形成三聚体，具有四级结构（分子质量 45kDa）（PedroRoig et al. 2013b）。$GlnK_2$ 蛋白晶体结构已被确定，这是第一个有关嗜盐古菌 PⅡ蛋白的结构研究，并确认所有生物体的 PⅡ蛋白质结构高度保守（Palanca et al. 2014）。

9.4.1　不同氮源的 GlnK 蛋白及其在 *Hfx. mediterranei* 氮代谢调控中的作用

不同培养条件下的 *Hfx. mediterranei*：用酵母提取物作为有机氮源，或以铵或硝酸盐作为无机氮源（加入过量葡萄糖作为这些合成培养基的碳源）。在硝酸盐存在下，检测到两种 GlnK，而在铵或酵母存在时则无法检测到 GlnK（Pedro-Roig et al. 2011）。对于 GS，存在同样的情况（Martínez-Espinosa et al. 2006），因此可以推断出 PⅡ和 GS 之间存在一定关系，PⅡ对 GS 有激活作用。*Hfx mediterranei* 生长在 75mmol/L 硝酸盐为氮源的培养基中，在指数增长期，GS 活性和 GlnK 表达较低，而到达稳定期后，GS 和 GlnK 高表达（图 9.7b，泳道 9）。这表明在 *Hfx mediterranei* 中，GlnK 通过 GS/GOGAT 途径在氨同化中发挥激活功能（Pedro-Roig et al. 2011），该研究在体外也已经被证实。在存在和不存在 GlnK 的条件下分别测定 GS 活性，GlnK 存在时 GS 酶活性增加了 50%。此外，SDS-PAGE

分析显示，细胞在硝酸盐培养基中生长时只有对应于 $GlnK_1$ 的一条带，$GlnK_1$ 和 $GlnK_2$ 具有不同的电泳迁移率（Pedro-Roig et al. 2013b）。

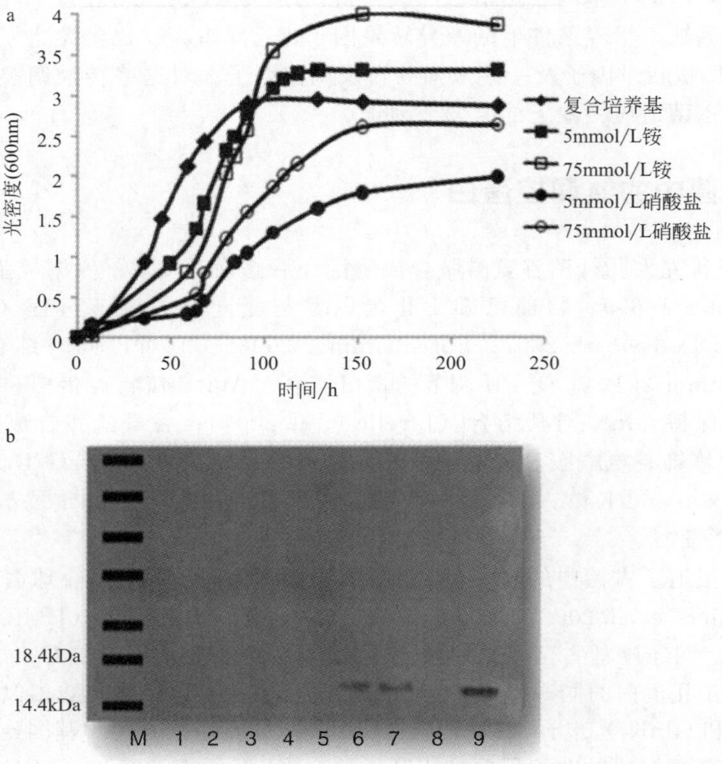

图 9.7 (a) 不同氮源对应的 *Hfx. mediterranei* 的生长曲线；(b) SDS-PAGE 及 Western blot 分析 GlnK 的表达：泳道 M，标准分子质量；泳道 1，复合培养基，指数增长期；泳道 2，复合培养基，稳定期；泳道 3，5mmol/L 铵，指数增长期；泳道 4，5mmol/L 铵，稳定期；泳道 5，75mmol/L 铵，指数增长期；泳道 6，5mmol/L 硝酸盐，指数增长期；泳道 7，5mmol/L 硝酸盐，稳定期；泳道 8，75mmol/L 硝酸盐，指数增长期；泳道 9，75mmol/L 硝酸盐，稳定期（Pedro-Roig 等 2011）

添加 10mmol/L α-酮戊二酸，GS 的生物合成活性增加 12 倍，在 PⅡ蛋白也存在时则可增加 18 倍。α-酮戊二酸是一种代谢物，当氮缺乏时，可以通过 GS 来促进铵同化。当 GlnK 存在时，GS 活性增加，这是由于两种蛋白质之间形成了复合物，已经通过凝胶过滤色谱法、SDS-PAGE 和免疫印迹等方法（数据未显示）对此进行了证明（Pedro-Roig et al. 2013b）。α-酮戊二酸的存在对于形成 GS-GlnK 而言是必需的，可增加 GS 活性，并且当细胞处在低氮时，应通过 GS / GOGAT 途径进行氨同化。来自 *Hfx. mediterranei* 的 GlnK-GS 复合物的化学计量仍不清楚，但已经通过计算获得了一个十二聚体的 GS 与四个三聚体 PⅡ的化学计量比。

9.4.2 *Hfx. mediterranei* 中 PⅡ同源蛋白的翻译后修饰

使用二维电泳（2DE），应用免疫印迹技术及 MALDI-TOF-MS 技术研究嗜盐古菌中的 PⅡ蛋白的翻译后修饰。利用这些研究手段可以确定这些蛋白质是否进行了翻译后修饰（PTM），因为经翻译后修饰的蛋白质在等电点时会发生较小的位移。*Hfx. mediterranei* 中 $GlnK_1$ 的理论 pI 为 5.15，$GlnK_2$ 为 5.11。制备针对 GlnK T 环区的特异性抗体，利用免疫印迹技术，获得了 pI 值和分子量与预期大小一致的五个不同信号（图 9.8 b）（Pedro-Roig et al. 2013a）。斑点 1，2 和 3（图 9.8b）属于 $GlnK_1$，斑点 4 和 5 属于 $GlnK_2$。来自 *Hfx. mediterranei* 的两个 GlnK 一级结构高度同源（84%）会掩盖两种蛋白质之间的区别。根据两种蛋白质所含肽的质量不同，通过质谱可以进行特异性鉴定。

前面研究发现，以硝酸盐作为培养基时，在 *Hfx. mediterranei* 中检测不到 $GlnK_2$ 表达（Pedro-Roig et al. 2013b）；当细胞处于氮缺乏时，通过免疫印迹技术可以检测 $GlnK_2$（Pedro-Roig et al. 2013a）。这具有重要生理学意义，当氮源充足时，仅需要两种 GlnK 其中之一进行代谢调节；当没有氮源时，则两种 GlnK 都必须存在。这表明 GlnKs 是氮缺乏时氮同化的激活剂，这与 GlnKs 体外激活 GS 是一致的（Pedro-Roig et al. 2013b）。

在图 9.8 b 中的斑点 2 和 4 的，相较理论分子质量增加了 306 Da，该质量与 UMP 组（单同位素质量为 306.03）匹配，斑点 3 和 5 中质量不发生变化。因此可以推断在斑点 2 和 4 对应的 $GlnK_1$ 和 $GlnK_2$ 蛋白分别是尿苷酰化的，而在斑点 3 和 5 中，蛋白质没被修饰。推断该 GlnKs 修饰发生在保守的酪氨酸残基（Y60 $GlnK_1$，Y61 $GlnK_2$）。因为如前所述，只有 PⅡ蛋白的尿苷酰化位点的片段是保守的（PS00496，PROSITE）。由此得到重要结论：来自古菌域的 PⅡ蛋白经过了翻译后修饰。但是，在 *Hfx. mediterranei* 基因组中，尿苷酸转移酶/去尿苷酸酶（由 *glnD* 编码）的同源物尚未发现（Pedro-Roig et al. 2013a）。因此，应该有其他类似的可以发挥该功能的嗜盐古菌蛋白质。

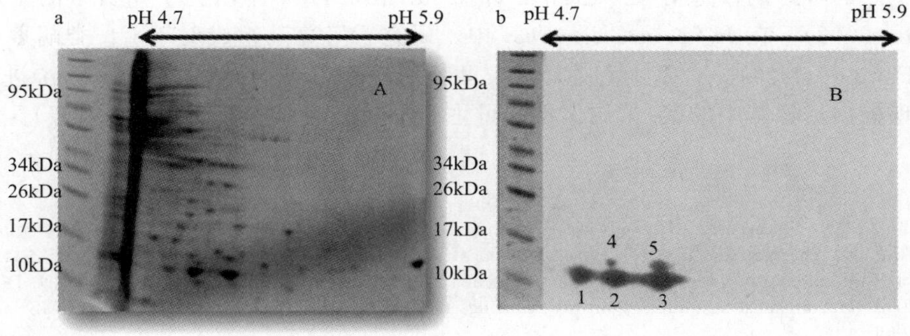

图 9.8 (a) 细胞进行 20h 氮限制培养，提取总蛋白质，取 500g 总蛋白质进行二维电泳，采用的 pH 4.7～5.9 的 IPG 胶条及 Any kD TGX 胶均为伯乐公司生产，考马斯亮蓝染色。(b) 利用 GlnK 特异性抗体进行免疫反应，HRP 标记二抗，鲁米诺作为底物进行化学发光检测（Pedro-Roig 等 2013a）

9.4.3 铵限制条件下 *Hfx. mediterranei* 中 *glnK* 和 *amtB* 基因的共转录

对来自 *Hfx. mediterranei* 的 *amtB-glnK* 基因的转录水平进行研究发现（Pedro-Roig et al. 2013）：当培养基富含氨或酵母提取物时，没有检测到 *amtB-glnK* 转录物，推断 *amtB-glnK* 基因没有发生转录或转录物因氨过量而降解；当培养基富含硝酸盐时，*amtB-glnK* 基因发生转录（数据未显示）。Northern 印迹发现 *amtB-glnK* 基因是成对共转录的，并且培养基中存在硝酸盐时 *amtB1-glnK1* 表达，*amtB2-glnK2* 不表达。此外，*amtB-glnK* 转录水平受培养基中可利用氮的调节，当培养基含铵时，不表达；在培养基中铵缺乏时，表达量最大。

9.5 生物技术应用

人类活动导致了土壤和地下水中盐类，如硝酸盐和亚硝酸盐的浓度增加。例如从事农药、除草剂、爆炸物和染料制造的公司均会带来这类问题。硝酸盐和亚硝酸盐对农业、环境和公共卫生有着重大的影响。即使是低浓度的硝酸盐和亚硝酸盐也会对大多数生物产生影响。*Hfx. mediterranei* 是最先被报道的具有同化硝酸盐和亚硝酸盐能力的嗜盐菌，这取决于高盐环境中存在硝酸还原酶和亚硝酸还原酶，*Hfx. mediterranei* 生理特征显示，这种微生物能够在硝酸盐（高达 2mol/L）或亚硝酸盐（高达 50mmol/L）作为唯一氮源（Martínez-Espinosa et al. 2009）的培养基上进行生长。*Hfx. mediterranei* 能够消耗培养基中的硝酸盐和亚硝酸盐，所以可以在这种对大多数微生物有害的环境中生长。废水中硝酸盐和亚硝酸盐的浓度通常低于 *Hfx. mediterranei* 中检测的浓度。因此，推断这种微生物可用于高盐废水或海水的生物修复。以相同的方式，对 Nas 和 NiR 酶进行固定可以用于生物修复或在传感器中用于硝酸盐和亚硝酸盐的检测。

另一方面，在过去几年，类胡萝卜素由于其在营养学和人类健康方面的潜在作用而被广泛研究。大量结果表明，该色素可以通过猝灭单线态氧或自由基，增加抗体的产生来延缓癌症和心脏疾病（Edge et al. 1997；Carpenter et al. 1997；Palozza et al. 1998；Chew et al. 1999）。在营养学方面，类胡萝卜素被用作食品加工的着色剂（Bauernfeind 1981）。对产类胡萝卜素的微生物，如红斑球菌（*Haematococcus pluvialis*）、三鞭毛虫（*Blaskeslea trispora*）和杜氏盐藻（*Dunaliella salina*）（Olaizola 2000；Mehta et al. 2003；Raja et al. 2007）等也进行了研究。然而，关于极端嗜盐微生物的研究却很少，其独特的性质使其成为色素生产的极好的候选对象。已经在 *Hfx. alexandrinu* 中研究了角黄素的生产（Asker and Ohta 2002），在 *Hfx. mediterranei* 中，研究了培养基组成对于 C_{50} 类胡萝卜素生产的影响（Fang et al. 2010）。基于上述研究，我们研究小组对 *Hfx. mediterranei* 进行了全面分析，发现根据培养基中的碳/氮比例不同可生产不同的胡萝卜素（数据未发表）。

参考文献

Adul Rahman RN，Jongsareejit B，Fujiwara S，Imanaka T（1997）Characterization of recombinant glutamine synthetase from the hyperthermophilic archaeon *Pyrococcus* sp. strain KODl. Appl Environ Microbiol 63：2472-2476

Alcántara-Hernández RJ，Valenzuela-Encinas C，Zavala-Díaz de la Serna FJ，Rodriguez-Revilla J，Dendooven L，Marsch R（2009）Haloarchaeal assimilatory nitrate-reducing communities from a saline alkaline soil. FEMS Microbiol Lett 298（1）：56-66

Andrade SL，EinsleO（2007）The Amt/Mep/Rh family of ammonium transport proteins. Mol Membr Biol 24：357-365

Andrade SL，Dickmanns A，Ficner R，EinsleO（2005）Crystal structure of the archaeal ammonium transporter Amt-1 from *Archaeoglobus fulgidus*. Proc Natl Acad Sci U S A 102：14994-14999

Asker D，Ohta Y（2002）Production of canthaxanthin by *Haloferax alexarulrines* under non-aseptic conditions and a simple，rapid method for its extraction. Appl Microbiol Biotechnol 58：743-750

Bauernfeind JC（1981）Natural food colors. In：Bauernfeind JC，Stewart GF，Schweigert BS，Hawthorn J（eds）Carotenoids as colorants and vitamin A precursors. Technological and nutritional applications. Academic Press，New York，pp 1-45. ISBN：978-0-12-082850-0

Beckers G，Nolden L，Burkovski A（2001）Glutamate synthase of *Corynebacterium glutamicum* is not essential for glutamate synthesis and is regulated by the nitrogen status. Microbiology 147（11）：296 1-2970

Beinert H（2000）Iron-sulfur proteins：ancient structures，still full of surprises. J Biol Inorg Chem 5：2-15

Belitsky BR，Wray LV，Fisher SH，Bohannon DE，Sonenshein AL（2000）Role of TnrA in nitrogen source-dependent repression of bacillus subtilis glutamate synthase gene expression. J Bacteriol 182（21）：5939-5947

BoneteMJ，Camacho ML，Cadenas E（1986）Purification and some properties of NAD^+-dependent glutamate dehydrogenase from *Halobacterium halobium*. Int J Biochem 18：785-789

Bonete MJ，Camacho ML，Cadenas E（1987）A new glutamate dehydrogenase from *Halobacterium halobium* with different coenzyme specificity. Int J Biochem 19：1149-1155

Bonete MJ，Perez-Pomares F，Ferrer J，Camacho ML（1996）NAD-glutamate dehydrogenase from *Halobacterium halobium*：inhibition and activation by TCA intermediates and amino acids. Biochim Biophys Acta 1289（1）：14-24

Bonete MJ，Martínez-Espinosa RM，Pire C，Zafrilla B，Richardson DJ（2008）Nitrogen metabolism in haloarchaea. Saline Syst 4：9. doi：10. 1186/1746-1448-4-9

Bowsher CG，Lacey AE，Hanke GT，Clarkson DT，Saker LR，Stulen I，Emes MJ（2007）The effect of Glc6P uptake and its subsequent oxidation within pea root plastids on nitrite reduction and glutamate synthesis . J Exp Bot 58 (5): 1109-1118

Brown JR, Masuchi Y, RobbFT, Doolittle WF (1994) Evolutionary relationships of bacterial and archaeal glutamine synthetase genes. J Mol Evol 38 (6): 566-576

Campbell WH, Kinghorn JR (1990) Functional domains of assimilatory nitrate reductase and nitrite reductases. Trends Biochem Sci 15: 315-319

Carpenter KLH, van der Veen C, HirdR, Dennis IF, Ding T, Mitchinson MJ (1997) The carotenoids: β-carotene, canthaxanthin and zeaxanthin inhibit macrophage-mediated LDL oxidation. FEBS Lett 401 : 262-266

Chai W, Stewart V (1998) NasR, a novel RNA-binding protein, mediates nitrate-responsive transcription antitermination of the *Klebsiella oxytoca* M5al nasF operon leader in vitro. J Mol Biol 283: 339-351

Chávez S, Lucena JM, Reyes JC, Florencio FJ, Candau P (1999) The presence of glutamate dehydrogenase is a selective advantage for the cyanobacterium *Synechocystis* sp. strain PCC 6803 under nonexponential growth conditions. J Bacteriol 181 (3): 808-813

Chew BP, Park JS, Wong MW, Wong TSA (1999) Comparison of the anticancer activities of dietary β-carotene, canthaxanthin and astaxanthin in mice in vivo. Anticancer Res 19: 1849-1853

Díaz S, Pérez-Pomares F, Pire C, Ferrer J, Bonete MJ (2006) Gene cloning, heterologous overexpression and optimized refolding of the NAD-glutamate dehydrogenase from *Haloferax mediterranei*. Extremophiles 10 (2): 105-115

Dixon R, Kahn D (2004) Genetic regulation of biological nitrogen fixation. Nat Rev Microbiol 2 (8): 621-631

Edge R, McGarvey DJ, Truscott TG (1997) The carotenoids as antioxidants -a review. J Photoe hem Photo biol B Biol 4 1: 189-200

Ehlers C, Grabbe R, Veit K, SchmitzRA (2002) Characterization of GlnKl from *Methanosarcina mazei* Strain Gol: complementation of an Escherichia coli glnK Mutant Strain by GlnK1 . J Bacteriol 184: 1028-1040

Ehlers C, Weidenbach K, Veit K, Forchhammer K, SchmitzRA (2005) Unique mechanistic features of post-translational regulation of glutamine synthetase activity in *Methanosarcina mazei* strain G61 in response to nitrogen availability. Mol Micro biol 55: 1841-1854

Esclapez J, Bravo-Barrales G, Bautista V, Pire C, Camacho M, Bonete MJ (2014) Effects of nitrogen sources on the nitrate assimilation in *Haloferax mediterranei*: growth kinetics and transcriptomic analysis. FEMS Microbiol Lett 350: 168-174

Esclapez J, Pire C, Camacho M, Bautista V, Martínez-Espinosa RM, Zafrilla B, Vegara A, AlcarazLA, Bonete MJ (2015) Transcriptional profiles of *Haloferax mediterranei* based on nitrogen availability. J Biotechnol 193: 100-107

Fang CJ, Ku KL, Lee MH, Su NW (2010) Influence ofnutritive factors on C50 carotenoids production by *Haloferax mediterranei* ATCC 33500 with two-stage cultivation. Bioresour Technol 101 (16): 6487-6493

Feng J, Liu B, Zhang Z, Ren Y, Li Y, Gan F, Huang Y, Chen X, Shen P, Wang L, Tang B, Tang XF (2012) The complete genome sequence of *Natrinema* sp. 17-2, a haloarchaeon capable of growth on synthetic media without amino acid supplements. PLoS One 7 (7), e41621

Fernández E, Galvan A, Quesada A (1998) Nitrogen assimilation and its regulation. In: The molecular biology of chloroplast and mitochondria in *Chlamydomonas*. Kluwer Academic Publishers, The Netherlands, pp 637-659

Ferrer J, Pérez-Pomares F, Bonete MJ (1996) NADP-glutamate dehydrogenase from the halophilic archaeon *Haloferax mediterranei*: enzyme purification, N-terrninal sequence and stability. FEMS Microbiol Lett 141 (1): 59-63

Forchharnmer K (2008) P (II) signal transducers: novel functional and structural insights. Trends Microbiol 16 (2): 65-72

Forchharnmer K (2010) The network of PII signalling protein interactions in unicellular cyanobacteria. Adv Exp Med Biol 675: 71-90. Chap 5. doi: l0. 1007/978-1-4419-1528-3 _ 5

Forde BG, Lea PJ (2007) Glutamate in plants: metabolism, regulation, and signalling. J Exp Bot 58 (9): 2339-2358

Frolow F, Harel M, Sussman JL, Mevarech M, Shoham M (1996) Insights into protein adaptation to a saturated salt environment from the crystal structure of a halophilic 2Fe-2S ferredoxin. Nat Struct Biol 3 (5): 452-458

Geiger B, Mevarech M, Werber MM (1978) Immunochemical characterization of ferredoxin from *Halobacterium* of the Dead Sea. Eur J Biochem 84 (2): 449-455

Heinrich A, Maheswaran M, Ruppert U, Forchharnmer K (2004) The *Synechococcus elongates* PII signal transduction protein controls arginine synthesis by complex formation with N-acetyl-Lglutamate kinase. Mol Microbiol 52: 1303-1314

Helfmann S, LÜ W, Litz C, Andrade SLA (2010) Cooperative binding of MgATP and MgADP in the trimeric PII protein GlnK2 from *Archaeoglobus fulgidus*. J Mol Biol 402: 165-177

Helling RB (1994) Why does Escherichia coli have two primary pathways for synthesis of glutamate? J Bacteriol 176: 4664-4668

Hochman A, Nissany A, Aruizur M (1988) Nitrate reduction and assimilation by a moderately halophilic, halotolerant bacterium Bal. Biochim Biophys Acta 965 (1): 82-89

Ingoldsby LM, Geoghegan KF, Hayden BM, Engel PC (2005) The discovery of four distinct glutamate dehydrogenase genes in a strain of *Halobacterium salinarum*. Gene 349: 237-244

Javelle A, Severi E, Thornton J, Merrick M (2004) Ammonium sensing in *Escherichia coli*. Role of the ammonium transporter AmtB and AmtB-GlnK complex formation. J Biol Chem 279: 8530-8538

Jiang P, Mayo AE, NinfaAJ (2007) Escherichia coli glutaruine synthetase adenylyltransferase (ATase, EC 2. 7. 7. 49): kinetic characterization of regulation by PII, PII-UMP, glutaruine, and cx-ketoglutarate. Biochemistry 46: 4133-4146

Kameya M, Ikeda T, Nakamura M, Arai H, Ishii M, Igarashi Y (2007) A novel ferredoxindependent glutamate synthase from the hydrogen-oxidizing chemoautotrophic bacterium *Hydrogenobacter thermophilus* TK-6. J Bacteriol 189 (7): 2805-2812

Khademi S, Stroud RM (2006) The Amt/MEP/Rh family: structure of AmtB and the mechanism of ammonia gas conduction. Physiology (Bethesda) 21: 419-429

Kumada Y, Benson DR, Hillemann D, Hosted TJ, Rochefort DA, Thompson CJ, Wohlleben W, Tateno Y (1993) Evolution of the glutamine synthetase gene, one of the oldest existing and functioning genes . Proc Natl Acad Sci USA 90: 3009-3013

Labboun S, Terce-Laforgue T, Roscher A, Bedu M, Restivo FM, Velanis CN et al (2009) Resolving the role of plant glutamate dehydrogenase. I. In vivo real time nuclear magnetic resonance spectroscopy experiments. Plant Cell Physiol 50: 1761-1773

Leigh JA, Dodsworth JA (2007) Nitrogen regulation in bacteria and archaea. Annu Rev Microbiol 61: 349-377

Lin JT, Stewart V (1998) Nitrate assimilation by bacteria. Adv Microb Physiol 39: 330-379

Litz C, Helfmann S, Gerhardt S, Andrade SLA (2011) Structure of GlnKl, a signaling protein fromArchaeoglobus fulgidus. Acta Crystallogr 67: 178-181

Lledó B, Marhuenda-Egea FC, Martinez-Espinosa RM, Bonete MJ (2005) Identification and transcriptional analysis of nitrate assimilation genes in the halophilic archaeon *Haloferax mediterranei*. Gene 361: 80-88

Luque-Almagro VM, GatesAJ, Moreno-Vivian C, Ferguson SJ, Richardson DJ, Roldan MD (2011) Bacterial nitrate assimilation: gene distribution and regulation. Biochem Soc Trans 39: 1838-1843

Manitz B, Holldorf AW (1993) Purification and properties of glutamine synthetase from the archaebacterium *Halobacterium salinarium*. Arch Microbiol 159: 90-97

Martínez-Espinosa RM, Marhuenda-Egea FC, Bonete MJ (200la) Assimilatory nitrate reductase from the haloarchaeon *Haloferax mediterranei*: purification and characterisation. FEMS Microbiol Lett 204 (2): 381-385

Martínez-Espinosa RM, Marhuenda-Egea FC, Bonete MJ (2001 b) Purification and characterization of a possible assimilatory nitrite reductase from the halophile archaeon *Haloferax mediterranei*. FEMS Microbiol Lett 196 (2): 113-118

Martínez-Espinosa RM, Marhuenda-Egea FC, Donaire A, Bonete MJ (2003) NMR studies of a ferredoxin from *Haloferax mediterranei* and its physiological role in nitrate assimilatory pathway. Biochim Biophys Acta 1623 (1): 47-51

Martínez-Espinosa RM, Esclapez J, Bautista V, Bonete MJ (2006) An octameric prokaryotic glutamine synthetase from the haloarchaeon *Haloferax mediterranei*. FEMS Microbiol Lett 264 (1): 110-116

Martínez-Espinosa RM, Lled6 B, Marhuenda-Egea FC, Bonete MJ (2007) The effect of ammonium on assimilatory nitrate reduction in the haloarchaeon *Haloferax mediterranei*. Extremophiles 11 (6): 759-767

Martínez-Espinosa RM, Lledo B, Marhuenda-Egea FC, Dfaz S, Bonete MJ (2009) NO^{3-}/NO^{2-} assimilation in halophilic archaea: physiological analysis, *nasA* and *nasD* expressions. Extremophiles 13: 785-792

McCarty GW, Bremner J (1992) Regulation of assimilatory nitrate reductase activity in soil by microbial assimilation of ammonium. Proc Natl Acad Sci USA 89: 453-456

Mehta BJ, Obraztsova IN, Cerda-Olmedo E (2003) Mutants and intersexual heterokaryons of *Blakeslea trispora* for production of beta-carotene and lycopene. Appl Environ Microbiol 69: 4043-4048

Millin BJ (2002) The role of glutamine synthetase and glutamate dehydrogenase in nitrogen assimilation and possibilities for improvement in the nitrogen utilization of crops. J Exp Bot 53 (370): 979-987

Miller A, Fan X, Shen Q, Smith SJ (2008) Amino acids and nitrate as signals for the regulation of nitrogen acquisition. J Exp Bot59: 111-119

Moir W, Wood NJ (2001) Nitrate and nitrite transport in bacteria. Cell Mol Life Sci 58: 215-224

Moreno-Vivián C, Flores E (2007) Nitrate assimilation in Bacteria. In: Bothe H, Ferguson SF, Newton WE (eds) Biology of the nitrogen cycle. Elsevier BV, Amsterdam, pp 263-282

Nesbo CL, L'Haridon S, Stetter KO, Doolittle WF (2001) Phylogenetic analyses of two 'archaeal' genes in thermotoga maritima reveal multiple transfers between archaea and bacteria. Mol Biol Evol 18 (3): 362-375

NinfaAJ, Jiang P (2005) PII signal transduction proteins: sensors of [alpha] -ketoglutarate that regulate nitrogen metabolism. Curr Opin Microbiol 8: 168-173

Olaizola M (2000) Commercial production of astaxanthin from *Haematococcus pluvialis* using 25, 000-liter outdoor photobioreactors. J Appl Phycol 12: 499-506

Palanca C, Pedro-Roig L, Llacer JL, Camacho M, Bonete MJ, Rubio V (2014) The structure of a PII signaling protein from a halophilic archaeon reveals novel traits and high-salt adaptations. FEBS J 281 : 3299-3314

Palozza P, Maggiano N, Calviello G, Lanza P, Piccioni E, Ranelletti FO, Bartoli GM (1 998) Canthaxanthin induces apoptosis in human cancer cell lines . Carcinogenesis 19: 373-376

Pantoja 0 (2012) High affinity ammonium transporters: molecular mechanism of action. Front Plant Sci 3: 34. doi: 10. 3389/fpls. 2012. 00034

Pedro-Roig L, Camacho M, Bonete M J (2011) In vitro proof of direct regulation o f glutaruine synthetase by GlnK proteins in the extreme halophilic archaeon *Haloferax mediterranei*. Biochem Soc Trans 39: 259-262

Pedro-Roig L, Camacho M, Bonete MJ (2013a) *Haloferax mediterranei* GlnK proteins are posttranslationally modified by uridylylation . Proteomics 13: 1371-1374

Pedro-Roig L, Camacho M, Bonete MJ (2013b) Regulation of ammonium assimilation in *Haloferax mediterranei*: Interaction between glutamine synthetase and two GlnK proteins. Biochim Biophys Acta 1834: 16-23

Pedro-Roig L, Lange C, Bonete MJ, Soppa J, Maupin-Furlow J (2013c) Nitrogen regulation of protein-protein interactions and transcript levels of GlnK PII regulator and AmtB ammonium transporter homologs in archaea . Microbiologyopen 2 (5): 826-840

Pérez-Pomares F, Ferrer J, Camacho M, Pire C, Llorca F, Bonete MJ (1999) Amino acid residues implied in the catalytic mechanism of NAD-dependent glutamate dehydrogenase from *Halobacterium salinarum*. Biochim Biophys Acta 1426: 513-525

Pire C, Martinez-Espinosa RM, Perez-Pom ares F, Esclapez J, Bonete MJ (2014) Ferredoxindependent glutamate synthase : involvement in ammonium assimilation in *Haloferax mediterranei*. Extremophiles 18 (1): 147-159

Radin JW (1977) Amino acid interactions in the regulation of nitrate reductase induction in cotton root tips . Plant Physiol 60: 467-469

Raja R, Hemaiswarya S, Rengasamy R (2007) Exploitation of *Dunaliella* for betacarotene production. Appl Microbiol Biotechnol 74: 517-523

Reitzer L (2003) Nitrogen assimilation and global regulation in *Escherichia coli*. Annu Rev Microbiol 57: 155-176

Reyes JC, FlorencioFJ (1994) A mutant lacking the glutamine synthetase gene (*glnA*) is impaired in the regulation of the nitrate assimilation system in the cyanobacterium *Synechocystis* sp. strain PCC 6803. J Bacteriol 176: 7516-7523

Richardson DJ, Berks BC, Russell DA, Spiro S, Taylor CJ (2001) Functional, biochemical and genetic diversity of prokaryotic nitrate reductase s . Cell Mol Life Sci 58 (2): 165-178

Robertson DL, Alberte RS (1996) Isolation and characterization of glutarnine synthetase from the marine diatom *Skeletonema costatun*. Plant Physiol 111 (4): 1169-1175

Santero E, Hervas A B, Canosa I, Govantes F (2012) Glutamate dehydrogenases: enzymology, physiological role and biotechnological relevance; Chapter 12. In: CanutoRA (ed) Dehydrogenases. InTech, under CC BY 3. 0 license . doi: 10. 5772/ 2903. ISBN 9 78-953-307-019-3

SchulzAA, Collett HJ, Reid SJ (2001) Nitrogen and carbon regulation of glutaruine synthetase and glutamate synthase in *Corynebacterium glutamicum* ATCC 13032. FEMS Microbiol Lett 205 (2): 361-367

Shapiro BM (19 69) The glutamine synthetase deadenylylating enzym e system from *Escherichia coli*. Resolution into two components, specific nucleotide stimulation, and cofactor requirements. Biochemistry 8 (2): 659-670

Smith EL, Austen BM, Blumenthal KM, Nye JF (1975) Glutamate dehydrogenase . In: Boyer PD (ed) The enzym es, vol 11, 3rd edn. Academic Press, New York, pp 293-367

Sugimori D, Ichimata T, Ikeda A, Nakamura S (2000) Purification and characterization of a ferredoxin from *Haloarcula japonica* strain TR-1. Biometals 13 (1): 23-28

Tomita T, Miyazaki T, Miyazaki J, Kuzuyama T, Nishiyama M (2010) Hetero-oligomeric glutamate dehydrogenase from *Thermus thermophilus*. Microbiology 156: 3801-3813

Tomita T, Kuzuyama T, Nishiyama M (2011) Structural basis for leucine-induced allosteric activation of glutamate dehydrogenase . J Biol Chem 286: 37406-37413

Vanoni MA, Curti B (1999) Glutamate synthase: a complex iron-sulfur ftavoprotein . Cell Mol Life Sci 55: 617-638

Vanoni MA, Curti B (2005) Structure and function studies on the ironsu lfur ftavoenzyme glutamate synthase : an unexpectedly complex self-regulated enzyme . Arch Biochem Biophys 433 (1): 193-211

Wanner C, Soppa J (1999) Genetic Identification of three ABC transporters as essential elements for nitrate respiration in *Haloferax volcanii*. Genetics 152 (4): 1417-1428

Wood NJ, Alizadeh T, Richardson DJ, Ferguson SJ, Moir JW (2002) Two domains of a dual-function NarK protein are required for nitrate uptake, the first step of denitrification in *Paracoccus pantotrophus*. Mol Microbiol 44 (1): 157-170

Zafrilla B, Martinez-Espinosa RM, Bonete MJ, Butt JN, Richardson DJ, Gates AJ (2011) A haloarchaeal ferredoxin electron donor that plays an essential role in nitrate assimilation. Biochem Soc Trans 39 (6): 1844-1848

第十章
蛋白质组学方法鉴定极端微生物新型蛋白质

Sung Ho Yun[1]，Chi-Won Choi[1]，Sang-Yeop Lee[1]，Edmond Changkyun Park[1]，Seung Il Kim[1,2]

10.1 引言

极端微生物是指生活在极端环境中（如极端高温或低温、强酸、高盐、强辐射、高压及寡营养等）的有机体（Burg et al. 2011）。与自然环境中大多数微生物相比，极端微生物更具特性，鉴定极端微生物的新型基因或蛋白质是目前的一个研究热点。水生栖热菌 *Thermus aquaticus*（Chien et al. 1976）中提取的 Taq DNA 聚合酶和南极湖泊细菌（Gilbert et al. 2004）分泌的抗冻蛋白（AFPs）就是极端微生物独特蛋白质的典型代表。

通过基因组测序技术和功能基因组学研究人们已经成功积累了大量极端微生物的基因组数据，包括基因在 DNA 水平上的信息（Lee et al. 2008；Vezzi et al. 2005）。近年来，蛋白质组学技术已经成为鉴定极端微生物中新型蛋白质的一种新工具。由于蛋白质的表达和翻译后修饰是动态的，并且对细胞功能有深远的影响，因此蛋白质组学成为了功能基因组学中必不可少的一部分。然而，在天然生长条件下培养极端微生物并得到纯培养的难度较大，致使蛋白质组学在极端微生物中的应用受限。随着蛋白质组学方法以及培养技术的快速发展，这些困难逐渐被克服。虽然用于普通细菌蛋白质组学分析的基本技术通常也适用于极端微生物样本，但仍应该研究开发样品制备的具体优化方案（Burg et al. 2011；Ellen et al. 2009；Yun et al. 2011a）。

极端微生物蛋白质组学研究可以提供关于在特定条件下诱导的蛋白质数量的信息（Burg et al. 2011）。蛋白质组学数据可用于阐明极端生物的生理特性，利用生物信息学，还能根据其相应的生物学作用对蛋白质进行分类。极端微生物具有独特的新陈代谢途径和生存机制，是发现新型酶和生物化合物的资源宝库，尤其是它们分泌的酶，如嗜冷酶、碱性纤维素酶、耐热 DNA 聚合酶等，在生物技术行业有着广泛的应用（Antranikian et al. 2005；Podar and Reysenbach 2006）。因此，蛋白质组学技术在筛选鉴定极端微生物的新蛋白质方面起着至关重要的作用。

[1] Division of Bioconvergence Analysis，Daejeon 305-806，Republic of Korea.

[2] e-mail：ksi@kbsi. re. kr.

在本章的第一部分，我们介绍蛋白质组学的基本原理，其中包括蛋白质分离方法、蛋白质鉴定方法、生物信息学的应用和高通量筛选。在第二部分，我们将概述用于鉴定极端微生物新型蛋白质的筛选方法。

10.2　蛋白质组学技术的发展

本节介绍了蛋白质分离鉴定的基本原理和基本概念。此外还描述了蛋白质组学分析的必要工具——质谱法和生物信息学技术。

10.2.1　蛋白质的分离和制备

制备用于蛋白质组学分析的样品的第一步是破碎细胞，获得可溶性样品。可使用反复冻融、超声破碎和高压均质等多种破碎方法。综合研究对象的脆性和坚固性，选择最合适的方法。

根据可溶性样品的溶解度、净电荷（等电点）、大小（分子量）和疏水性等理化性质进行分离和富集。在凝胶方法中，一般是运用各种聚丙烯酰胺凝胶电泳（PAGE）分离蛋白质，最常用于总蛋白质分离的技术是一维 SDS-聚丙烯酰胺凝胶电泳（1D SDS-PAGE）、双向凝胶电泳（2-DE）和等电聚焦电泳（IEF）。1D SDS-聚丙烯酰胺凝胶电泳是最简单最有效的分离制备蛋白质用于蛋白质组学分析的方法。该方法通常需要将蛋白质样品与含有 SDS 和巯基还原剂（β-巯基乙醇或二硫苏糖醇）的上样缓冲液混合。双向凝胶电泳则是最佳的凝胶电泳方法，通常用于复杂样品的整体分析。该方法的第一向是基于蛋白质的等电点不同，用等电聚焦分离；第二向则按分子量的不同，用 SDS-PAGE 分离（图 10.1a）。

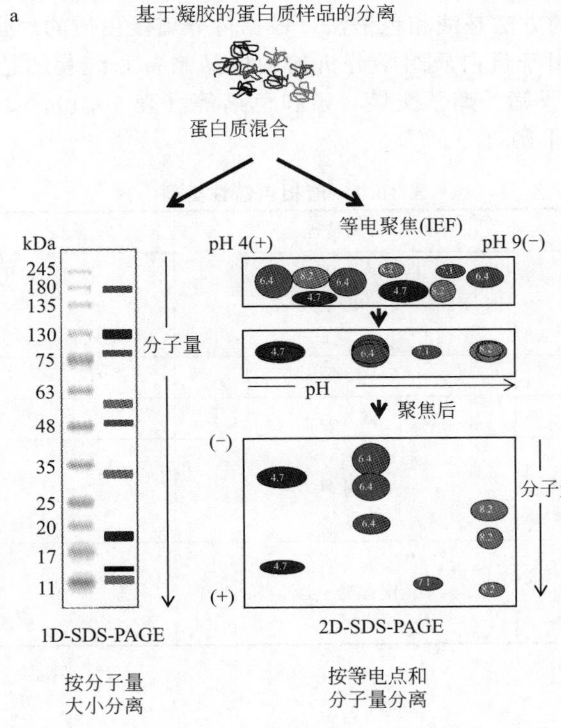

图 10.1

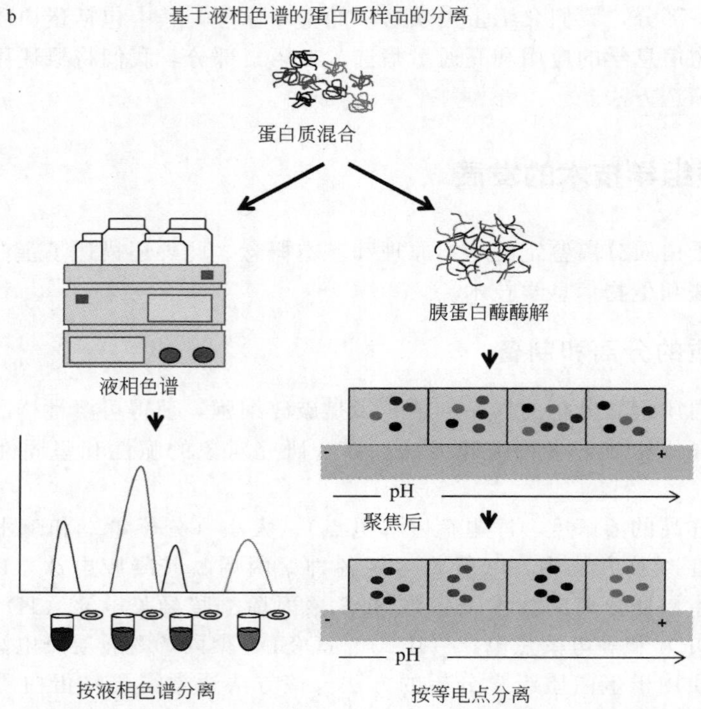

图 10.1　两种蛋白质分离技术。(a) 有凝胶：根据蛋白质混合物的分子量和等电点进行电泳分离；(b) 无凝胶：根据蛋白质混合物（肽混合物）在 LC 柱上的理化性质（如分子量大小、电荷、整体疏水性和等电点等）进行分级分离

另一种分离蛋白质的方法是液相色谱法，根据待分离蛋白质的理化性质，选择不同的液相色谱柱进行分离。常用于蛋白质组学分析前的样品制备和纯化的色谱柱包括：反向色谱 (C-4、C-8 和 C-18)、分子筛、离子交换、亲和色谱等（表 10.1）。高效液相色谱法则常用于分离蛋白质或多肽（图 10.1b）。

表 10.1　液相色谱柱类型

色谱柱类型	分离特性	优点
离子交换	电荷	高分辨率 载量大 速度快
分子筛/排阻色谱	大小	高分辨率
正/反相色谱	疏水性	高分辨率
疏水色谱	疏水性	较高分辨率 载量较大 速度较快
亲和色谱	配体专一性	高分辨率 载量大 速度快

10.2.2　蛋白质鉴定

常用于鉴定未知蛋白质的方法有两种：Edman 降解法（埃德曼降解法）和串联质谱分

析（MS/MS）（Choi et al. 2012）。Edman 降解法是通过标记和切割蛋白质的 N-末端残基来鉴定肽序列。首先用异硫氰酸酯标记蛋白质或肽游离的 N-末端，然后在适当条件下，切割被标记的 N-末端的氨基酸残基，产生氨基酸衍生物和余下的多肽链（少了一个残基）的游离氨基末端。但是 Edman 降解法不能运用于 N-末端被修饰的蛋白质。如果 N-末端氨基酸已被化学修饰，则用于裂解的化学物（异硫氰酸酯）不能与 N-末端氨基酸的氨基相互作用，从而导致 Edman 降解终止，这一过程被称为 N-末端阻断。此外，Edman 降解法只能够精确地测定 30 个以内氨基酸的序列。但 Edman 降解法仍然是用于蛋白质鉴定和阐明蛋白质中二硫键有用的方法（图 10.2a）。

质谱（MS）可以精确测量肽片段的质荷比，因此能够精确地确定蛋白质的分子量以及鉴定蛋白质，是一种灵敏度高、结果准确的方法。进行蛋白质鉴定，则先要用胰蛋白酶水解蛋白质，再进行肽质量指纹图谱（PMF）或 MS 分析。如果该信息不能提供用于蛋白质鉴定的明确数据，则可以使用 MS/MS 对蛋白质的胰蛋白酶水解肽段进行从头测序（图 10.2b）。

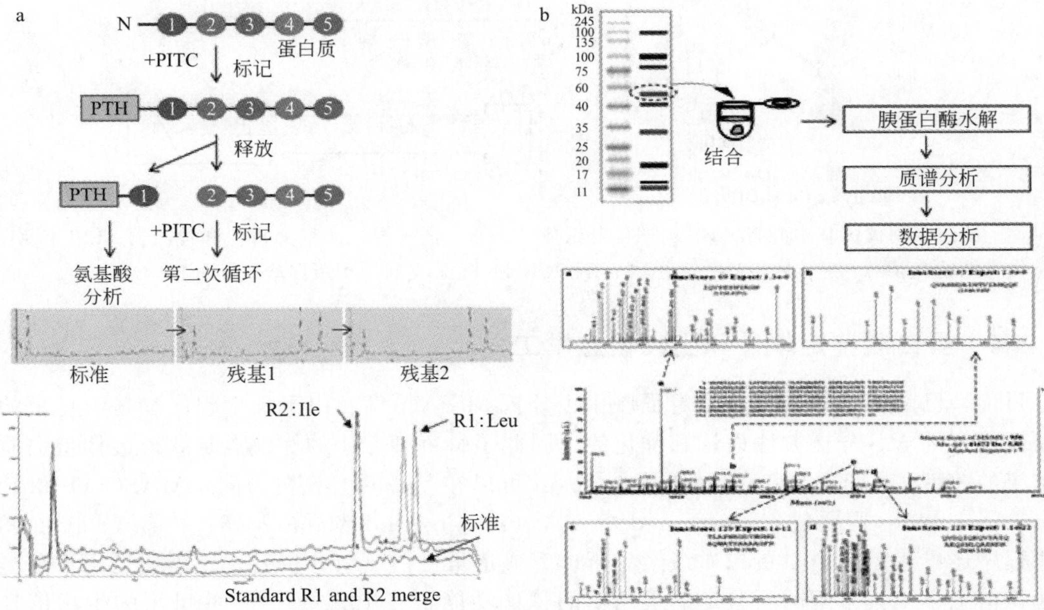

图 10.2　N-端测序示意图。(a) Edman 降解法：蛋白质的 N-端氨基用异硫氰酸苯酯（PITC）标记，标记的 N-端氨基被选择性裂解，最后用质谱法鉴定提取的乙内酰苯硫脲（PTH-氨基酸）。可以再次重复该过程以鉴定下一个氨基酸。(b) 肽质量指纹图谱（PMF）和 MS / MS 分析法：在 PMF 分析中，未知蛋白质被限制酶如胰蛋白酶切割成较小的肽，并通过 MS 法精确测量其分子量，然后将肽质量与蛋白质序列数据库进行比较。在 MS / MS 分析中，使用碰撞诱导解离（CDI）进一步分离肽，然后从 MS / MS 谱中的分子量列表中确定其序列。部分图片内容来自先前文献报道（Choi 等 2012）

10.2.3　质谱仪

2002 年诺贝尔化学奖授予 John Bennett Fenn 的电喷雾电离法（ESI）和 Tanichi Koichi 的基质辅助激光解吸电离法（MALDI）以及这两种方法在生物大分子（尤其是蛋白质）电离中的应用。MS 分析法是将这两种离子化方法与各种质量分析程序结合使用。MALDI 方法通常与飞行时间（TOF）质谱分析仪组合使用。MALDI-TOF MS 已经被用于测量蛋白质和多肽的分子量以及肽质量指纹图谱（PMF）。MALDI-TOF/TOF MS 常用于胰蛋白酶水解肽段的 MS/MS 分析以及氨基酸的从头测序。ESI 方法通常与四极杆和 TOF 组合。ESI-Q TOF MS 也用于 MS/MS 分析。最近，由于其易用性，电喷雾离子阱质谱（LTQ/Obitrap）已成为一种流行的用于高通量蛋白质组学分析的质谱分析仪（图 10.3）。

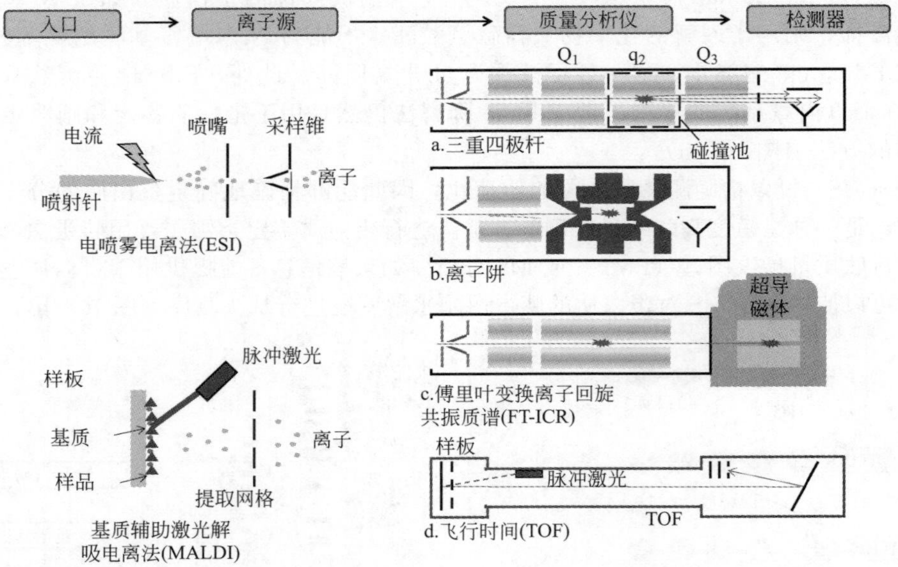

图 10.3 质谱联用仪的基本配置。MS 由四部分组成：入口、离子源、质量分析仪和检测器。ESI 和 MALDI 是代表性的离子源。三重四极杆、离子阱、FT-ICR 和 TOF 主要用于蛋白质组学分析

10.2.4 蛋白质鉴定软件和生物信息学工具

目前，已经有几种蛋白质鉴定程序用于分析 MS/MS 谱。由于每个程序都是为特定的质谱仪或特定的 MS 分析方法设计和优化的，因此了解每种方法的优缺点是获得最佳蛋白质组学结果的关键。APEX、Census 和 MaxQuant 可以在互联网上免费访问，MASCOT 和 Proteome Discoverer 需要付费（Braisted et al. 2008；Cox and Mann 2008；Park et al. 2008）。这些程序还可用于高通量蛋白质组学分析。完成定量蛋白质组分析后，还需要运用不同类型的生物信息学工具来阐明所鉴定的蛋白质的具体功能，表 10.2 列了一些可用的生物信息学工具。

表 10.2 可用的生物信息学工具

程序	网址	功能
TMHMM Server v. 2.0	http://www.cbs.dtu.dk/services/TMHMM	预测蛋白质跨膜拓扑结构
Phobius	http://phobius.sbc.su.se/	预测蛋白质跨膜拓扑结构
SignalP 4.0	http://www.cbs.dtu.dk/services/ SignalP/	预测信号肽切割位点的存在和位置
MASCOT	http://www.matrixscience.com/	蛋白质鉴定
SEQUEST	http://fields.scripps.edu/sequest/	蛋白质鉴定
PSORTdb 3.0	http://www.psort.org/psortb	预测亚细胞定位
CELLO v 2.5	http://cello.life.nctu.edu.tw	预测亚细胞定位
KEGGPATHWAY Database	http://www.genome.jp/kegg/ pathway.html	通路分析

续表

程序	网址	功能
FIGfams	http://www.nmpdr.org/FIG/wiki/ view.cgi/FIG/FigFam	功能分类
COGs	http://www.ncbi.nlm.nih.gov/ COG/	功能分类
The Gene Ontology	http://www.geneontology.org/	功能分类
UM-BBD	https://umbbd.ethz.ch/	生物催化/生物降解数据库

Cello 和 PsortDB 是用于识别蛋白质的细胞定位的常用软件。Cello 基于以 FASTA 格式提供的序列信息预测蛋白质的定位（Yu et al. 2006，2011）。然而，Cello 不能利用可访问的基因组数据库，而这些基因组数据库通常包括细胞定位的信息。另一方面，当需要了解基因组数据库中细菌和古菌的蛋白质定位信息时，PsortDB 是一个不错的选择，因为它可以与 BlastP 结合使用，用来预测新型蛋白质的细胞定位。另一个众所周知的生物信息学工具是 THMHH 服务器（Moller et al. 2001），它可用于识别和预测蛋白质的跨膜位置。它还能预测膜内和膜外以及跨膜螺旋区的蛋白质序列的范围。另一个程序 SignalP 主要用于预测蛋白质序列的信号肽切割位点（Petersen et al. 2011）。TMHMM 服务器和 SignalP 都以 FASTA 格式作为蛋白质序列的输入格式。

上述程序只适用单个蛋白质的分析和表征。全面的蛋白质组学研究需要更高级的生物信息学工具来阐明蛋白质复合体及其生物学功能。COGs 数据库和 GO 数据库应运而生（Ashburner et al. 2000；Tatusov et al. 2003）。应用该程序可以预测目的蛋白质的生物学功能，从而对其进行生物学分类。STRING 数据库主要用于分析蛋白质的相互作用（Franceschini et al. 2013）。特别的，STRING 数据库可以将蛋白质相互作用可视化。KEGG 数据库是确定蛋白质功能的最受欢迎的网络分析数据库（Kanehisa and Goto 2000）。

10.3 极端微生物新型蛋白质的筛选

本小节展示了从极端微生物中筛选出新型蛋白质的几个研究案例。通常研究者是使用 2DE-MALDI-TOF/TOF 和 1D-LC-MS/MS 两种高通量筛选方法进行蛋白质筛选的。

10.3.1 极端微生物蛋白质的高通量筛选

2DE-MALDI-TOF/TOF 和 1D-LC-MS/MS 是常用于高通量筛选的蛋白质组学工具。2DE-MALDI-TOF/TOF 由两种技术组成：2-DE 和 MALDI-TOF/TOF。2-DE 的优点是蛋白质样品在凝胶上的可视化。根据蛋白质的分子量、等电点（pI）和丰度将其可视化。翻译后修饰和蛋白质降解的信息也可以从凝胶中获得。然而，2-DE 不能检测低丰度蛋白质。此外，该方法在分离极酸极碱蛋白质以及疏水性蛋白质（如膜蛋白）方面存在困难。但 2DE-MALDI-TOF/TOF 仍是一个非常有用的细菌和极端微生物蛋白质组学分析的工具。通常是通过肽谱、MS/MS 分析或氨基酸的从头测序来鉴定双向电泳分离出的蛋白质斑点（Chong and Wright 2005；Yun et al. 2014）。与 LC-MS/MS 相比，MALDI-TOF 的蛋白质鉴定方法相对更快速、经济，对样品的相对纯度要求更高。基因组测序预测嗜盐古菌 *Halobacterium salinarum*（菌株 R1，DSM 671）包含 2784 个蛋白质编码基因。在 pH 3.5～5.5 条件下，运用双向电泳分离 *H. salinarum* 的胞浆蛋白（图 10.4）。分析每个银染点，结果发现 661 个蛋白质形成了 1800 个不同的斑点，其中有 94 个蛋白质点是蛋白质翻译后修饰导致的。通过这个实验，约 40% 的 *H. salinarum* 的胞浆蛋白能被确定（Tebbe et al. 2005）。

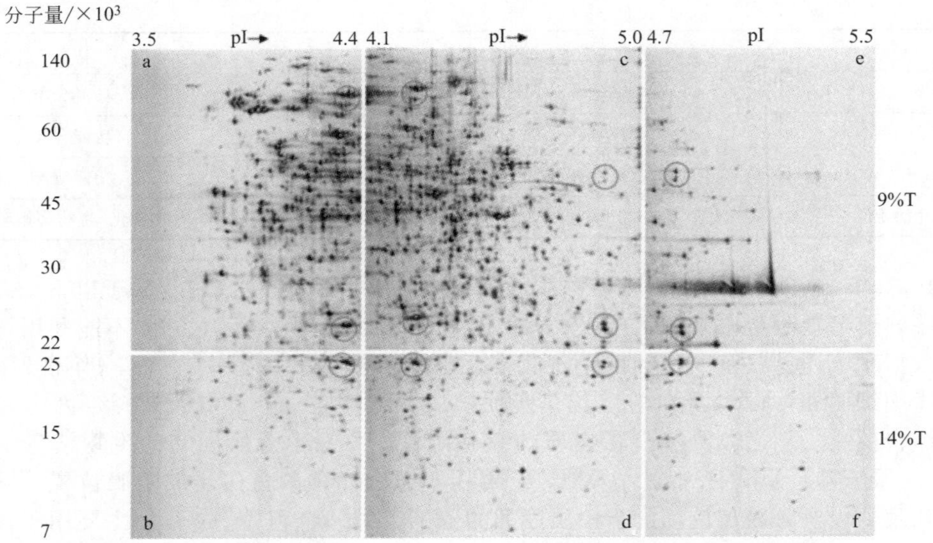

图 10.4 嗜盐古菌的双向电泳示意图。在 pI 3.5～5.5 范围的重叠变焦凝胶上电泳分离 *H. salinarum* 的胞浆蛋白并用硝酸银染色。每个 pI 范围都覆盖有两种不同的丙烯酰胺浓度（9%和 14%T）。标记不同的参考区域（a～f）（摘自 Tebbe 等 2005）

另一种高通量蛋白质组学工具是 1D-LC-MS/MS 技术，它将 SDS-PAGE 和 LC-MS/MS 结合并用于蛋白质分离和鉴定。该方法首先是利用 SDS-PAGE 电泳分离蛋白质样品。通常，根据分子量大小，蛋白质样品可分成 7～20 个部分。然后用蛋白质水解酶如胰蛋白酶处理每个凝胶片段，并在 MS/MS 之前对胰蛋白酶水解肽段混合物进行反相（C-18 或 C-8）色谱纯化。SDS-PAGE 尤其适用于膜蛋白和疏水蛋白质的分离（Yun et al. 2014）。LC-MS/MS 蛋白质鉴定方法高度灵敏且可靠，提供的实验结果也最为可信。然而，与 MALDI-TOF/TOF MS 相比，该方法所用时间相对较长且操作难度较大。LC-MS/MS 已被用于在一氧化碳条件下培养的超嗜热古菌 *Thermococcus onnurineus* NA1 的蛋白质组分析（图 10.5）。从 *T. onnurineus* NA1 1976 个预测蛋白质序列中，通过两次 LC-MS/MS 分析鉴定出了 1395 个蛋白质序列，占了总数的 70.6%（Yun et al. 2011b）。此外还能利用串联 LC [强阳离子交换（SCX）和反相（RP）色谱] 方法代替 1D-LC 技术，来分离鉴定胰蛋白酶水解肽段混合物。这种技术通常被称为多维蛋白质鉴定技术（MudPIT）。蛋白质混合物（在蛋白质水解之前）也可通过液相 IEF、LC 或 SDS-PAGE 分离（Chong and Wright 2005）。

10.3.1.1 极端嗜热古菌的热稳定蛋白质的分析

蛋白质组学方法尚未广泛用于鉴定极端嗜热古菌的热稳定蛋白质。近年来，热及化学摄动方法已用于热稳定蛋白质的筛选（Prosinecki et al. 2006）。这些方法降低了可溶性蛋白质的复杂性，使胞浆超热稳定蛋白质的富集成为可能。最适生长温度高于 85℃ 的 *T. onnurineus* NA1 菌株的热稳定蛋白质就是用类似的方法来筛选的。*T. onnurineus* NA1 菌株的超热稳定蛋白质的筛选鉴定则需要将上述方法联合使用。首先将超热稳定蛋白质在 100℃下浓缩 2 h，然后通过 2-DE 和 1-DE/MS-MS 鉴定（Yun et al. 2011a）（图 10.6）。这种方法从 *T. onnurineus* NA1 菌株中鉴定出了热稳定的 DNA 聚合酶、氨基肽酶和 α-淀粉酶。

10.3.1.2 耐辐射菌的蛋白质组学分析

Deinococcus deserti 和 *Deinococcus radiodurans* 这些耐辐射细菌可以在几个小时内修复

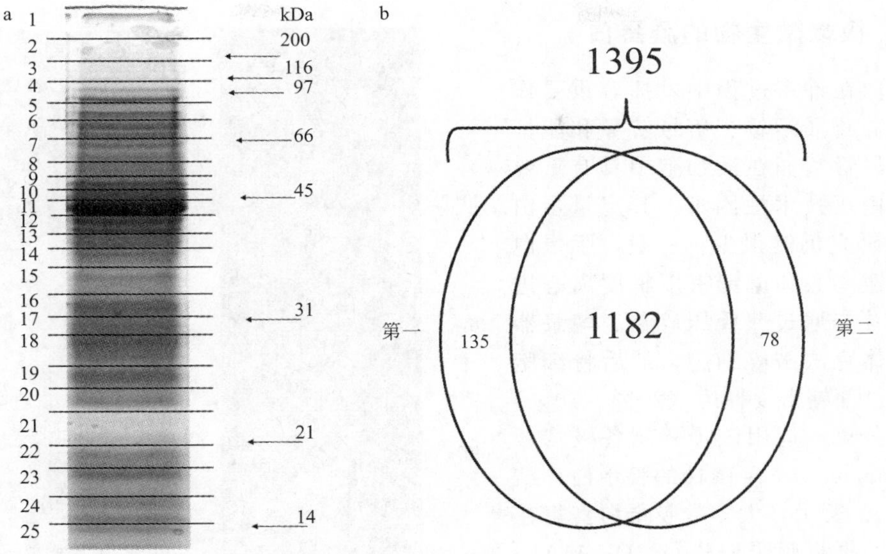

图 10.5 (a) 提取 100μg 在酵母提取物/蛋白胨/硫磺培养基 (YPS) 和 CO 条件下培养的 *Thermococcus onnurineus* NA1 的蛋白质，并在 12%SDS-PAGE 上进行电泳。凝胶用考马斯亮蓝 R-250 染色，根据分子量分成 25 份，再用胰蛋白酶水解每个凝胶片段。(b) 通过 1D-LC-MS / MS 分析确定 NA1 蛋白质的维恩图 (摘自 Yun 等 2011b)

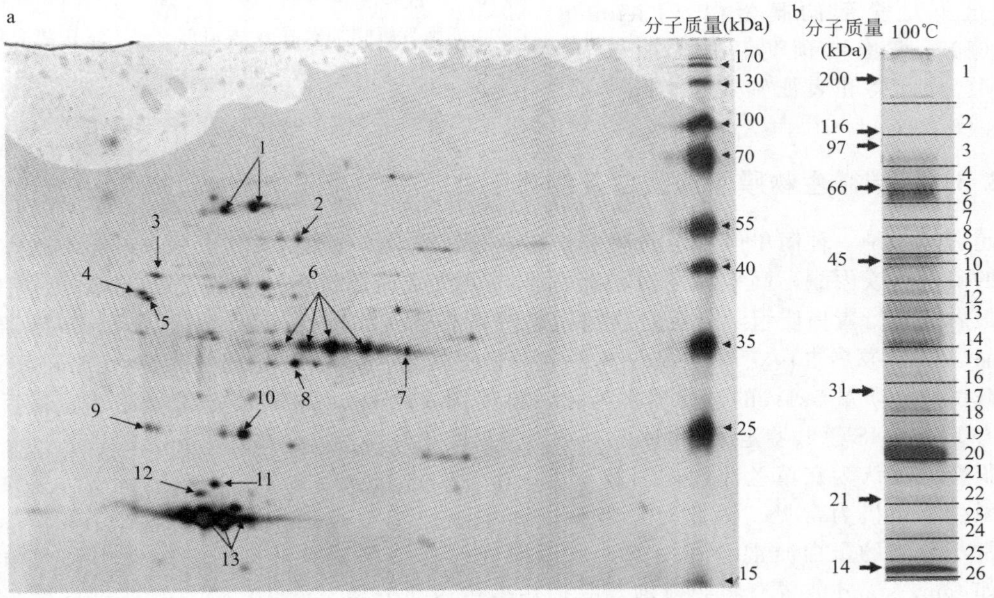

图 10.6 *T. onnurineus* NA1 菌株的超热稳定蛋白质的双向电泳图。蛋白质样品在 100℃ 下热处理 10min，通过离心去除凝固的蛋白质，富集可溶性的超热稳定蛋白质，进行双向电泳 (a) 和 12%SDS-PAGE 电泳 (b) (摘自 Yun 等 2011a)

受损的 DNA。最近，研究人员运用蛋白质组学分析了它们的辐射驻留蛋白。2DE-MALDI-TOF/TOF 结果表明经过剧烈辐射处理后，*Deinococcus deserti* 中与 DNA 损伤应答相关的蛋白质表达上调 (Dedieu et al. 2013)。而 *Deinococcus radiodurans* 在辐射条件下，也能检测到与 DNA 修复、缓解氧化应激和蛋白质翻译/折叠相关的蛋白质 (Basu and Apte 2012)。

10.3.2 极端微生物的膜蛋白

膜蛋白在许多过程中发挥着重要作用，如营养物质运输、信号转导和能量转换等。尽管它们在活细胞中具有重要功能，但由于技术性困难，其定量蛋白质组学的研究仍然很少。一般，膜蛋白是通过超速离心和蔗糖密度梯度离心进行分离。首先通过弗氏压碎器、匀浆器或渗透压休克法破碎细胞，然后将细胞裂解物通过蔗糖密度梯度（20%～50%）进行分级分离。运用该步骤制备极端微生物如 *Thermococcus* 菌株的膜蛋白，结果如图 10.7 所示，大多数膜蛋白富集在重密度层，再将膜蛋白进行 2DE-MALDI-TOF/TOF 或 1D-LC-MS/MS 分析。Klein 等人通过超速离心和蔗糖密度梯度离心纯化嗜盐古菌 *Halobacterium* 的膜蛋白，一共鉴定出了 165 种蛋白质，虽然这不是全部的膜蛋白（Klein et al. 2005）。膜蛋白预测程序，如 TMHMM，是分类和表征鉴定蛋白质所必需的。

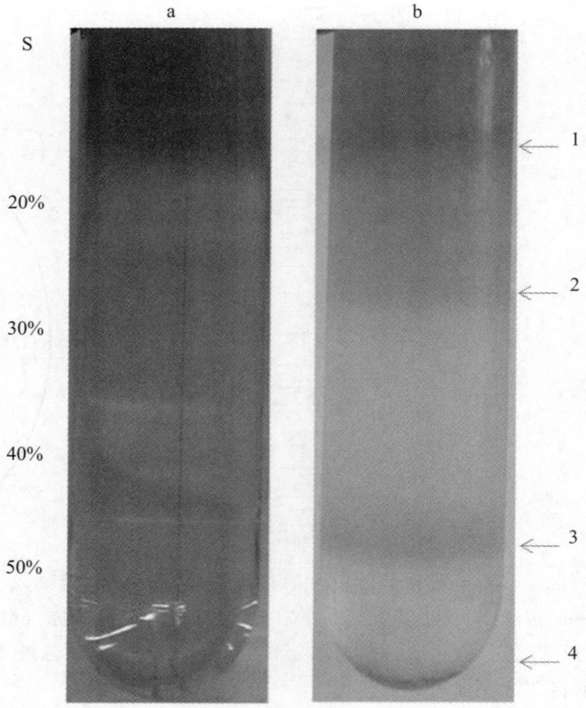

图 10.7 蔗糖密度梯度（20%～50%）离心（见彩图）。细胞裂解物加载在液体顶部（a）。超速离心后，膜蛋白集中在第 3 和第 4 部分（b）

10.3.3 极端微生物膜囊泡蛋白及胞外蛋白

超速离心是一种简单但常用的从培养基中富集分泌蛋白质的方法。通常用沉淀法（80%以上的硫酸铵或丙酮）高效制备分泌蛋白质，再将蛋白质沉淀重悬在适当的缓冲液中并透析，最后在进行蛋白质组学分析之前利用超滤离心管（如 viva spin 或 centricon）浓缩蛋白质样品。极端嗜热古菌 *Pyrococcus furiosus* 分泌的蛋白质也是先超滤（Sartorius AG，德国，截留分子质量 5kDa 的过滤器）纯化，再使用 Vivaspin 20 离心浓缩器浓缩，最后通过 1DE-LC-MS/MS 总共鉴定了 58 种蛋白质，包括淀粉降解酶（Schmid et al. 2013）。

膜囊泡被认为在革兰氏阴性菌和革兰氏阳性菌中发生的新型分泌过程。由于膜囊泡在极端微生物中极为常见，因此对膜囊泡中新型蛋白质的鉴定和功能研究十分重要。膜囊泡通常与分泌的蛋白质混合在一起，所以研究时应该将两者分开。一般使用过滤浓缩装置（如 Quix-stand 装置）将膜囊泡与其他污染蛋白质分离（图 10.8）。再通过鸟枪法蛋白质组学（1D-LC-MS / MS）或 2DE-MALDI-TOF/TOF 分析纯化膜囊泡蛋白。Ellen 等人使用 MALDI-TOF 质谱仪分析研究嗜酸热硫化叶菌 *Sulfolobus acidocaldarius*、硫矿硫化叶菌 *Sulfolobus solfataricus* 和硫化叶菌 *Sulfolobus tokodaii* 中的膜囊泡蛋白（Ellen et al. 2009）。在电子显微镜下，三种硫化叶菌的膜囊泡的直径范围约为 90nm 到 230nm，并且被可能由表层蛋白质形成的蛋白质层包围。LC-MS 分析鉴定了三种硫化叶菌的膜囊泡蛋白。

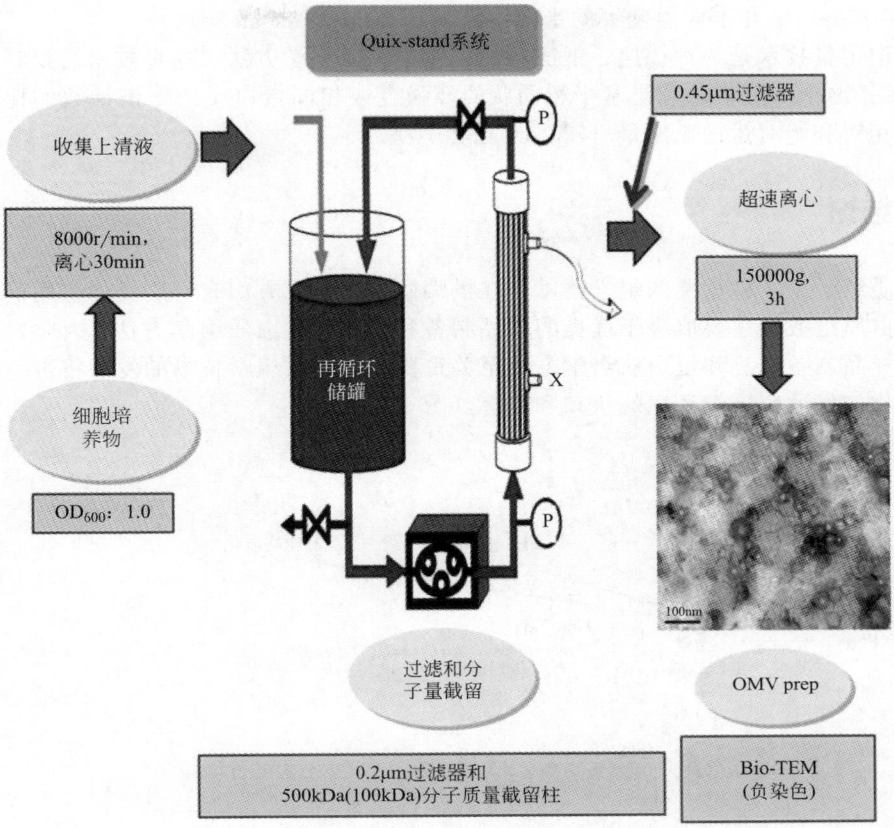

图 10.8　Quix-stand 装置分离膜囊泡步骤。通过过滤器（0.2μm）和分子质量截留柱（500kDa 或 100kDa）浓缩膜囊泡。通过透射电镜技术确定超速离心（150000g，3 h）纯化后的膜囊泡。纯化的膜囊泡在深度冷冻下储存直至进行蛋白质组学分析

10.3.4　定量蛋白质组学分析

近年来，体外标记法被用于定量蛋白质组学分析。一般主要是使用同位素标记进行比较分析。同位素亲和标记（ICAT）技术是指在胰蛋白酶水解之前用同位素^{13}C或^{12}C标记样品蛋白质的半胱氨酸残基。同位素标记相对和绝对定量（iTRAQ）技术是利用多种同位素试剂标记蛋白质多肽 N 末端或赖氨酸侧链基团。然后通过 LC/MS-MS 分析标记的肽混合物。目前，有几项蛋白质组学研究应用了 iTRAQ 技术筛选极端微生物的新型蛋白质。其中一个例子是 *Methanococcoides burtonii*（Goodchild et al. 2005）。研究者运用两个定量蛋白质组学研究方法来评估低温对该菌的影响。iTRAQ 分析结果表明，鉴定的 163 个蛋白质中有 14 个蛋白质在 4℃和 23℃条件下的表达有显著差异。另一个例子是极端嗜热古菌 *Sulfolobus solfataricus*，Pham 等人从在三种不同温度（65℃、70℃和 80℃）下生长的 *Sulfolobus solfataricus* 中定量分析鉴定出 246 种膜蛋白（Pham et al. 2010）。

双向差异凝胶电泳（2D-DIGE）技术可以用来比较不同样品之间的蛋白质量。先用不同的荧光染料标记不同的蛋白质样品，再将样品混合在一起，最后通过 2-DE 技术分离。电泳后的凝胶用每种染料的激发波长逐个扫描，因此我们能够分别看到每个蛋白质点的差异（Unlu et al. 1997）。这种技术可用于观察蛋白质丰度、翻译后修饰、截断和任何可能改变蛋白质大小或等电点的变化。Piette 等应用 2D-DIGE 方法比较了假交替单胞菌 *Pseudoaltero-*

monas haloplanktis 在不同温度下的蛋白质组表达（Piette et al. 2011）。

无标记定量技术是一种快速、低成本的定量蛋白质组学方法。这种技术需要将样品进行单独的 LC-MS/MS 分析，它是基于相同肽的峰强度或相同蛋白质的光谱计数的比较，最终可以获得相对和绝对蛋白质定量（Zhu et al. 2010）。

10.4 结论

蛋白质组学是一种重要的组学技术，在极端微生物研究方面有着广泛的应用前景。获得优质的蛋白质组数据主要取决于优良的样品制备和合适的蛋白质组学方法。因此，进一步开发高效的样品制备方法和蛋白质组学方法至关重要。蛋白质组学技术的发展将进一步加深对极端微生物如何适应极端环境的认知和了解（图 10.9）。

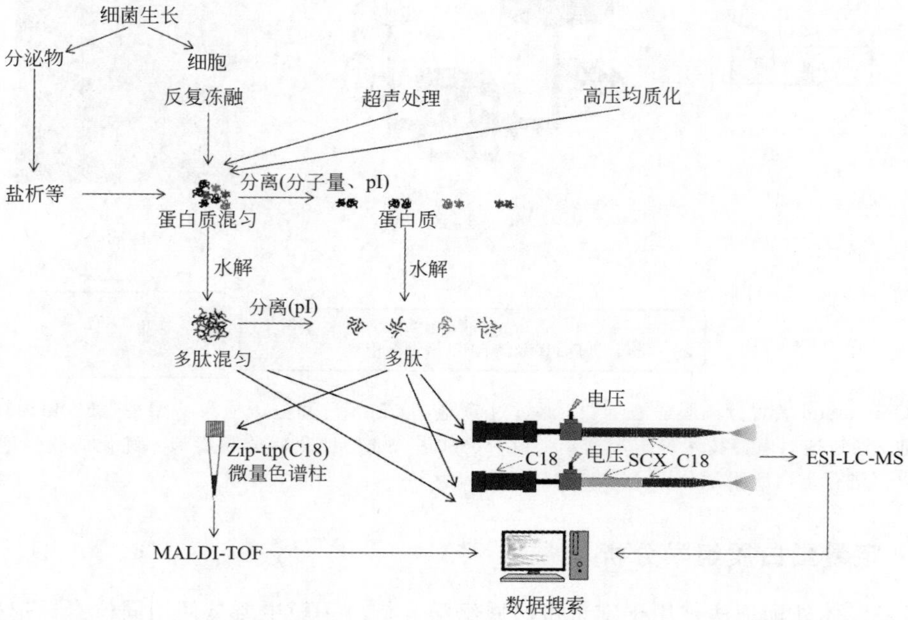

图 10.9 从极端微生物中寻找新蛋白的蛋白质组学方法的概述

参考文献

Antranikian G，Vorgias CE，Bertoldo C（2005）Extreme environments as a resource for microor-ganisms and novel biocatalysts. Adv Biochem Eng Biotechnol 96：219-262

Ashburner M，Ball CA，Blake JA，Botstein D，Butler H，Cherry JM，Davis AP，Dolinski K，Dwight SS，Eppig JT，Harris MA，Hill DP，Issel-Tarver L，Kasarskis A，Lewis S，Matese JC，Richardson JE，Ringwald M，Rubin GM，Sherlock G（2000）Gene ontology：tool for the unification of biol-ogy. The gene ontology consortium. Nat Genet 25：25-29

Basu B，Apte SK（2012）Gamma radiation-induced proteome of Deinococcus radiodurans primar-ily targets DNA repair and oxidative stress alleviation. Mol Cell Proteomics 11（M111）：011734

Braisted JC，Kuntumalla S，Vogel C，Marcotte EM，Rodrigues AR，Wang R，Huang ST，Ferlanti ES，Saeed AI，Fleischmann RD，Peterson SN，Pieper R（2008）The APEX quantitative pro-teomics tool：generating protein quantitation estimates from LC-MS/MSproteomics results. BMC Bioinformatics 9：529

Burg D，Ng C，Ting L，Cavicchioli R（2011）Proteomics of extremophiles. Environ Microbiol 13：1934-1955

Chien A，Edgar DB，Trela JM（1976）Deoxyribonucleic acid polymerase from the extreme thermo-phile Thermus aquaticus. J Bacteriol 127：1550-1557

Choi CW，Lee YG，Kwon SO，Kim HY，Lee JC，Chung YH，Yun CY，Kim SI（2012）Analysis of Streptococcus pneumoniae secreted antigens by immuno-proteomic approach. Diagn Microbiol Infect Dis 72：318-327

Chong PK，Wright PC（2005）Identification and characterization of the Sulfolobus solfataricus P2 proteome. J Proteome Res 4：1789-1798

Cox J，Mann M（2008）MaxQuant enables high peptide identification rates，individualized p. p. b. -range mass accuracies and proteome-wide protein quantification. Nat Biotechnol 26：1367-1372

Dedieu A，Sahinovic E，Guerin P，Blanchard L，Fochesato S，Meunier B，de Groot A，Armengaud J（2013）Major soluble proteome changes in Deinococcus deserti over the earliest stages following gamma-ray irradiation. Proteome Sci 11：3

Ellen AF，Albers SV，Huibers W，Pitcher A，Hobel CF，Schwarz H，Folea M，Schouten S，Boekema EJ，Poolman B，Driessen AJ（2009）Proteomic analysis of secreted membrane vesicles of archaeal Sulfolobus species reveals the presence of endosome sorting complex components. Extremophiles 13：67-79

Franceschini A，Szklarczyk D，Frankild S，Kuhn M，Simonovic M，Roth A，Lin J，Minguez P，Bork P，von Mering C，Jensen LJ（2013）STRING v9. 1：protein-protein interaction networks，with increased coverage and integration. Nucleic Acids Res 41：D808-D815

Gilbert JA，Hill PJ，Dodd CE，Laybourn-Parry J（2004）Demonstration of antifreeze protein activity in Antarctic lake bacteria. Microbiology 150：171-180

Goodchild A，Raftery M，Saunders NF，Guilhaus M，Cavicchioli R（2005）Cold adaptation of the Antarctic archaeon，Methanococcoides burtonii assessed by proteomics using ICAT. J Proteome Res 4：473-480

Kanehisa M，Goto S（2000）KEGG：Kyoto encyclopedia of genes and genomes. Nucleic Acids Res 28：27-30

Klein C，Garcia-Rizo C，Bisle B，Scheffer B，Zischka H，Pfeiffer F，Siedler F，Oesterhelt D（2005）The membrane proteome of Halobacterium salinarum. Proteomics 5：180-197

Lee HS，Kang SG，Bae SS，Lim JK，Cho Y，Kim YJ，Jeon JH，Cha SS，Kwon KK，Kim HT，Park CJ，Lee HW，Kim SI，Chun J，Colwell RR，Kim SJ，Lee JH（2008）The complete genome sequence of Thermococcus onnurineus NA1 reveals a mixed heterotrophic and carboxydotro-phic metabolism. J Bacteriol 190：7491-7499

Moller S，Croning MD，Apweiler R（2001）Evaluation of methods for the prediction of membrane spanning regions. Bioinformatics 17：646-653

Park SK，Venable JD，Xu T，Yates JR 3rd（2008）A quantitative analysis software tool for mass spectrometry-based proteomics. Nat Methods 5：319-322

Petersen TN，Brunak S，von Heijne G，Nielsen H（2011）SignalP 4. 0：discriminating signal peptides from transmembrane regions. Nat Methods 8：785-786

Pham TK，Sierocinski P，van der Oost J，Wright PC（2010）Quantitative proteomic analysis of Sulfolobus solfataricus membrane proteins. J Proteome Res 9：1165-1172

Piette F，D' Amico S，Mazzucchelli G，Danchin A，Leprince P，Feller G（2011）Life in the cold：a proteomic study of cold-repressed proteins in the antarctic bacterium pseudoalteromonas haloplanktis TAC125. Appl Environ Microbiol 77：3881-3883

Podar M，Reysenbach AL（2006）New opportunities revealed by biotechnological explorations of extremophiles. Curr Opin Biotechnol 17：250-255

Prosinecki V，Botelho HM，Francese S，Mastrobuoni G，Moneti G，Urich T，Kletzin A，Gomes CM（2006）A proteomic approach toward the selection of proteins with enhanced intrinsic confor-mational stability. J Proteome Res 5：2720-2726

Schmid G，Mathiesen G，Arntzen MO，Eijsink VG，Thomm M（2013）Experimental and computa-tional analysis of the secretome of the hyperthermophilic archaeon Pyrococcus furiosus. Extremophiles 17：921-930

Tatusov RL，Fedorova ND，Jackson JD，Jacobs AR，Kiryutin B，Koonin EV，Krylov DM，Mazumder R，Mekhedov SL，Nikolskaya AN，Rao BS，Smirnov S，Sverdlov AV，Vasudevan S，Wolf YI，Yin JJ，Natale DA（2003）The COG database：an updated version includes eukaryotes. BMC Bioinformatics 4：41

Tebbe A，Klein C，Bisle B，Siedler F，Scheffer B，Garcia-Rizo C，Wolfertz J，Hickmann V，Pfeiffer F，Oesterhelt D（2005）Analysis of the cytosolic proteome of Halobacterium salinarum and its implication for genome annotation. Proteomics 5：168-179

Unlu M，Morgan ME，Minden JS（1997）Difference gel electrophoresis：a single gel method for detecting changes in protein extracts. Electrophoresis 18：2071-2077

Vezzi A，Campanaro S，D'Angelo M，Simonato F，Vitulo N，Lauro FM，Cestaro A，Malacrida G，Simionati B，Cannata N，Romualdi C，Bartlett DH，Valle G（2005）Life at depth：Photobacterium profundum genome sequence and expression analysis. Science 307：1459-1461

Yu CS，Chen YC，Lu CH，Hwang JK（2006）Prediction of protein subcellular localization. Proteins 64：643-651

Yu NY，Laird MR，Spencer C，Brinkman FS（2011）PSORTdb—an expanded，auto-updated，user-friendly protein subcellular localization database for Bacteria and Archaea. Nucleic Acids Res 39：D241-D244

Yun SH，Choi CW，Kwon SO，Lee YG，Chung YH，Jung HJ，Kim YJ，Lee JH，Choi JS，Kim S，Kim SI（2011a）Enrichment and proteome analysis of a hyperthermostable protein set of archaeon Thermococcus onnurineus NA1. Extremophiles 15：451-461

Yun SH，Kwon SO，Park GW，Kim JY，Kang SG，Lee JH，Chung YH，Kim S，Choi JS，Kim SI（2011b）Proteome analysis of Thermococcus onnurineus NA1 reveals the expression of hydrogen gene cluster under carboxydotrophic growth. J Proteomics 74：1926-1933

Yun SH，Lee YG，Choi CW，Lee SY，Kim SI（2014）Proteomic exploration of extremophiles. Curr Biotechnol 3：87-99

Zhu W，Smith JW，Huang CM（2010）Mass spectrometry-based label-free quantitative proteomics. J Biomed Biotechnol 2010：840518

第十一章
功能筛选发掘新型极端酶

Freddy Boehmwald[1], Patricio Muñoz[1],
Patricio Flores[1], Jenny M.Blamey[1,2]

11.1 引言

在全世界范围内，目前每年在传统制造业中从生物质中生产出来的商品超过 4000 亿美元（USDA 2008），包括无机与有机类化学制剂、医药、肥皂、去垢剂、纸张、木材、燃料、润滑油脂、油漆等。如今，这种新型生物产业正从以石油原料为基础的制造业转向以生物质为基础的制造业，其中生物催化已经被证明是一种将生物质转化为高价值产品的重要工具，生物酶使得原材料和能源得到了有效利用，减少了废物的产生，简化了生产工艺。

用于制造生物质产品的酶可以分为：大宗酶（有时候被称为“工业酶”）和特种酶。到目前为止，生物质产品的制造工艺对于性能卓越、功能精细的特种酶的需求不断增加，从而获得更优质的产品和降低酶消耗（图 11.1）。因此，发掘新酶、开发生产酶的新工艺，以及优化具有特定酶催化步骤的微生物系统，将会推动未来生产各种高效绿色生物产品的技术革新，尤其是在特种化工和精细化工领域。

在化工领域中，酶类的使用表现出更多的优势，如具有更高的选择性，提高可持续性，更高的安全性，减少了合成过程中有毒成分的产生，生产工艺更加环保，并且减少了对环境的污染。随着社会环保意识不断增强，以及新的环境法规的实施，这种趋势在未来若干年里也将不断加强，酶的应用和生物催化技术将成为迈向环保与可持续发展经济的奠基石。

尽管生物质产品如此重要，但是由于缺乏相应的酶类将其整合到已有产品工艺中，它的应用仍然受到制约。在酶的开发方面主要的技术挑战包括：

① 发掘更有效和更强力的酶（WEF 2010）。目前可以获得的酶类由于稳定性与效率不理想而难以整合到完整的工艺流程中，在工业应用中受到底物浓度、剪切力、温度、pH 值、有机溶剂和工业环境中所有的常规条件所限制。由于酶通常是生物催化过程中最为昂贵的试剂，要在生物工艺中有效控制成本就需要更加稳定的酶。

② 发掘新型酶。目前能够获得的工业酶的数量和种类并不能覆盖工业中生物催化的化学转化过程。到目前为止，只有大约 20 种微生物酶能够进行工业化规模的生产和销售（Li

❶ Fundación Cientifi ca y Cultural Biociencia，José Domingo Cañas 2280，Ñuñoa，Santiago，Chile.
❷ e-mail：jblamey@bioscience.cl.

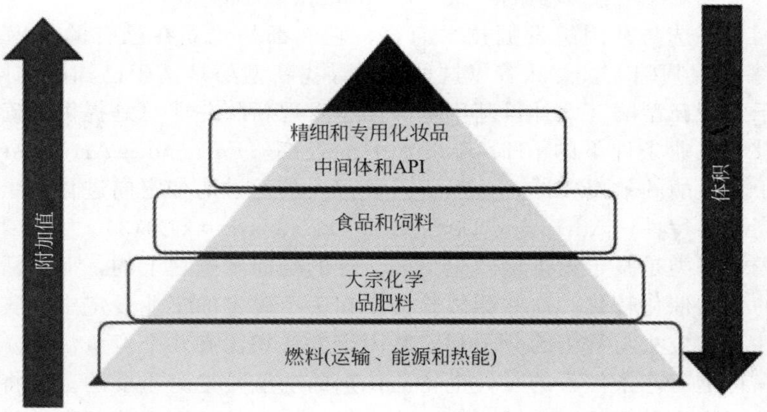

图 11.1　与生物质生物加工相关的产品及领域。酶参与到各种产品的制造过程中。只要生物产品的制造过程需要特种酶和更多的复杂特性，终产品的附加值就会更高，并且酶的需求量也越少。API，即先进的医药中间体

et al. 2012)，而且工业酶的数量也不能满足能够生产出光学纯度的化合物的特殊需求。

目前特种化工与精细化工行业正寻求能够替代传统化学转化工艺的生物催化技术。然而，如果想在经济上可行，就需要提高可获得的工业酶的种类，从而发掘出高稳定性（能够承受工业生产时的极端条件）、高活性、对映选择性，以及能够催化正常范围内底物转化的酶（Andexer et al. 2009）。

不同于大多数普通的工业酶制剂，从极端微生物中分离的生物催化剂（尤其是嗜热酶、超嗜热酶）为工业生物催化提供了更有吸引力的解决办法，一些更稳定可靠的极端酶成为新型工业酶的理想候选对象。

极端微生物能够在极端温度、pH、压力、盐度等环境中自然快速生长，它们能够在极端环境中生存的关键在于发展出具有高度特异性的分子来适应极端环境。此外，极端微生物显然具有一些全新的代谢途径，为现有的和新的应用提供了大量具有新活性的酶。因此，对于在开发工业生物技术问题解决方案中起着重要作用的酶类而言，极端微生物是一个尚未被开发的来源之一。

11.2　筛选极端酶的方法

大自然通过进化发展出大量的生物催化剂资源。然而，能否寻找到适合特定工业应用的酶，取决于我们是否拥有有效评估生物多样性的技术与能力，如目前使用的宏基因组筛选、基因组发掘以及定向开发极端酶（Adrio and Demain 2014；Bachmann 2014；Leis et al. 2013）。

在某一个特定的环境中可培养的微生物数目往往不及 1%，而从极端环境中培养得到的微生物甚至更少，因此利用宏基因组筛选的方式寻找新酶尤为重要。这种不依赖于培养的技术是通过从环境样品中制备庞大的基因组文库以及搜索潜在的能够编码酶类的开放阅读框而实现的。要获得这种结果，需要依赖于两种技术方法：对异源表达蛋白质功能活性的检测或者通过测序确定是否具有同源基因序列（Adrio and Demain 2014；Schmitz et al. 2008）。

另一方面，基因组发掘还可以通过与数据库中的注释序列进行比对，寻找目标酶的开放阅读框，并且进行基因克隆和异源表达。尽管之前提到这种技术存在难点，但通过对注释序

列进行挖掘能够提供更多意想不到的结果（Bachmann 2014）。

对于宏基因组筛选与基因组发掘技术而言，它们都是建立在已有的编码某种特定酶的DNA或RNA序列的基础上，意味着以这种方式寻找新型酶是基于已知酶的同源基因测序，实际上这不利于发现新型酶。此外，基于基因测序发现新酶并非总是提供准确的信息，如底物特异性，以及在工业条件下的pH、温度等酶的效率（Fernández-Arrojo et al. 2010）。通常情况下，由于重组酶的效率较低，这就要求研究人员通过诸如定向进化和蛋白质工程的方法学来研究提高酶的特性（Adrio and Demain 2014；Leis et al. 2013）。

定向开发极端酶类是基于大批量微生物酶活性的功能筛选之上的，而酶活性的功能检测则是基于现有的生物催化转化。这些测试验证了在一个确定的工业设定中真实存在的生物催化反应。功能性筛选方法与常用的分子和宏基因组方法相比有几个竞争优势。第一，对于工业而言，确定在特定工艺条件下的某种生物催化剂的功能是尤为重要的，它将影响着能否获得所需要的目的产物。尽管可以采用分子方法克隆感兴趣的基因来获得大批量蛋白质，但是它并不能控制酶的特性，如特异性、pH和热稳定性等，而这些问题可以通过酶功能检测的方法得到解决。

利用这个方法，通过搜索针对特定工业需求的适当的酶类，可能可以解决一些问题，如由底物或产物产生的抑制作用、稳定性、底物特异性和对映体选择性。尽管听起来这是一个解决所有工业问题的理想办法，但是这并不简单，因为为了找到催化工业领域感兴趣的特定转化的合适酶类，需要对大量微生物/粗提物或者样品进行筛选，并且应该在每次筛选新型酶活性时都要开发新的酶分析方法。不仅如此，为了让这个方法在工业上变得可行，还需要研究酶的小试工艺，以便工业化放大和快速转化为最有效的用于特定工业用途的生物催化剂。

多年来，这种功能酶的生物发现过程，其分析速度、体积和精度明显受到人为能力和局限性的影响。这种技术壁垒增加了特定酶的开发时间。然而，随着重组DNA、DNA测序和其他技术的发展，对于微生物生理学的认知不断加深，新酶活性的发展得以加速和提升（Hallam et al. 2014；Schallmey et al. 2014）。同时，极端微生物分类描述以及宏基因组学等先进分子技术快速发展，通过合理、半合理和随机的定向进化提高或改变酶活，有助于发现新型的微生物酶。例如，有多种抗芽孢杆菌溶素从噬菌体基因组中克隆获得，以及通过宏基因组学方法克隆获得一种气菌溶胞蛋白（Adrio and Demain 2014；Schmitz et al 2008）。此外，同样通过宏基因组学的方法还获得多种具有工业和生物技术应用价值的极端脂肪酶和酯酶（López-López et al. 2014）。

11.3 生物催化剂的高通量功能筛选

正如前面所说，功能筛选具有多个优势，但这些优势都基于大量的样品检测。因此，人为地对如此庞大数目的样品进行检测是不可能也是不实际的，同时人工检测容易出错，还需要更长的检测时间。

在过去十年，高通量筛选已经被用于发掘工业领域感兴趣的不同种类的化合物。这种技术成果也已用于筛选新型的生物催化剂。高通量筛选能够对大型的化学和生物样本库的活性进行自动检测（Glaser and Venus 2014；Inglese et al. 2006）。在实践中，高通量筛选是一种实验设备可以同时分析多个生物学过程，实验使用的有96孔、384孔或1536孔板，所有的成分依照特定的实验计划加入到板中。目前大多数公司使用这种技术的目标是追求更高的密度和更小的体积，从而降低成本。试剂保存在微孔板或二维码标记的小管中。通过液体分配器、处理设备、灵敏探测器、机器人设备、数据加工以及主控软件，高通量筛选能够每天自动执行成百上千个化学或生物学检测实验（Klumpp et al. 2006；Omtaa et al. 2012；Zhu et al. 2010）。

目前大部分高通量筛选技术围绕制药公司寻找新型药物或化合物（Zhu et al. 2010）。但是，这种技术实际上能够应用在不同的研究领域，包括寻找新型生物催化剂（Donadio et al. 2009；Glaser and Venus 2014；Inglese et al. 2006；King et al. 2010）。而高通量技术在极端微生物酶筛选中的应用发展却十分缓慢，需要进一步努力使该技术适当地适应于功能上搜索新酶的需求。

高通量筛选技术的目标是提供有用的信息应用于上百万种样品中并得到高度可重复性的鉴定结果。在筛选大量来源于极端微生物的提取物时，高通量筛选能够正确地筛选具有特定酶活性的微生物，因此要求的严格程度要超过实验室的检测（Acker and Auldn 2014；Brooks et al. 2012）。

通常情况下，利用高通量筛选酶活性有三个主要的测定因素：检测底物的消耗、产物的生成以及配体与酶的直接结合（Acker and Auldn 2014）。

开发一个高通量筛选标准的实验并非一件容易的事，每一个阶段的实验设计都要极其谨慎。一个完整的高通量筛选极端微生物催化剂的常规流程由几个相关的阶段组成。而高通量筛选的核心依赖于试剂的设置、检测技术的开发以及验证。优化的酶/底物比率、辅因子、缓冲体系、检测方法、成分干扰以及结果的验证都是实施过程中的关键要素（Liauda et al. 2014）。

尽管从极端微生物中获得粗提取物文库不需要复杂的制备，但是生物催化剂的鉴定步骤却是缓慢而复杂的过程。由于提取物具有化学上的复杂性，生物样品的稳健性筛选试验具有较大的挑战（Zhu et al. 2010）。商品化的酶类通常会给出设定的初始反应条件以检测特定的酶活性，这可以获悉酶活测定所要求的缓冲体系、pH 值、盐浓度以及其他因素（Acker and Auldn 2014；Liauda et al. 2014），这是创建更强大且更灵敏的筛选方法的开始。然而所用提取物中酶的纯度可能会改变其稳定性，并影响最终的选择。此外，底物浓度与测定的信号强度有直接的关系。更高浓度的底物能够增加试验的转化率，直到酶变得饱和。尽管通常会推荐使用由酶的米氏常数所计算获得的底物浓度来进行试验，但增加底物浓度可能会降低检测出酶的竞争性抑制剂的概率。此外，某些酶需要辅助因子以发挥其功能，对于这些酶而言，即使明确了辅助因子不会在反应中被消耗，但和底物一样，辅因子的可用性同样需要被考虑。缓冲体系和 pH 值的选择也是非常重要的，不仅是因为依赖于酶最适 pH 值的试验信号会受到 pH 的影响，还因为其他组分也会受 pH 影响。例如，磷酸盐缓冲液与需要 Mg^{2+} 和 Ca^{2+} 作为辅助因子的反应是不兼容的，因为这些金属离子的磷酸盐很难溶于水相缓冲液中（Acker and Auldn 2014；Liauda et al. 2014）。化合物干扰是在高通量筛选技术中常用的术语，它指的是一些化合物的特性可能会对检测产生干扰，从而产生假阳性的结果。例如，有色或荧光化合物可以产生广泛的分析干扰，可溶性较差或不溶性底物可以聚集并干扰光学检测系统（Johnston et al. 2008；Moger et al. 2006）。如果在高通量筛选试验中存在的化合物显著干扰了真实生物活性的测定，这将导致很难准确分析所获得的结果（Noah 2010）。

在大型化学库和/或生物库中快速和准确地识别活性化合物一直是开展高通量筛选试验的最终目标。然而，一个特定的高通量筛选试验的成功很大程度上取决于采用这种技术的专业性或适用性，为此定义了 Z 因子（Zhang et al. 1999）。这个比率反映了检测信号与测量数据相关的变量，适用于评价试验的质量。Z 因子是一个无量纲统计值，是每个高通量筛选试验的简单特性。这就提供了一个有利的工具进行比较和质量评估试验、关键参数优化及后续统计验证。Z值介于0～0.5之间时被定义为指示双测试，或者阳性（酶活性）和阴性对照之间存在低信号分离，这些均意味着此试验效果不好。若Z值介于0.5～1之间则意味着高通量方法是一种良好的筛选工具，其中Z的最大值为1。

检测体系的小型化是当前高通量筛选发展的应用策略。检测体积的减少会使成本、材料、时间消耗降低，这种降低可以通过缩小微孔板的孔体积得以实现（Brooks et al. 2012）。我们设计了一个经典的实验，在96孔板中以高通量方法筛选谷氨酸脱氢酶（图11.2a），在板中，第1列与第12列分别为阳性与阴性对照组，其他80个孔为不同嗜热菌的提取物（未发表的结果）。用于特定检测的孔的数目将决定不同提取物和重复试验的数目。然而，如果微孔板的排列增加到384～1536个孔，那将会使用更少的微孔板，但是需要更加准确和灵敏的分液器去操作小体积液体。在图11.2中，Z因子是通过谷氨酸脱氢酶实验手动检测（b）与自动检测（c）计算获得的。起初两种检测方法在高通量筛选标准下都能够被验证，Z因子的值都超过0.5（Zhang et al. 1999）。但是手动检验的离散程度要更高，从而影响了与自动化检测比较时的Z因子的值。手动检测每一块微孔板所花费的时间要高于自动检测所花费的时间，这使得通过手动操作进行成千上万次的检测与重复的效益十分低下。因此通过高通量筛选技术筛选数目庞大的微生物提取物，无论是在减少筛选时间还是试剂消耗上都具有优势。

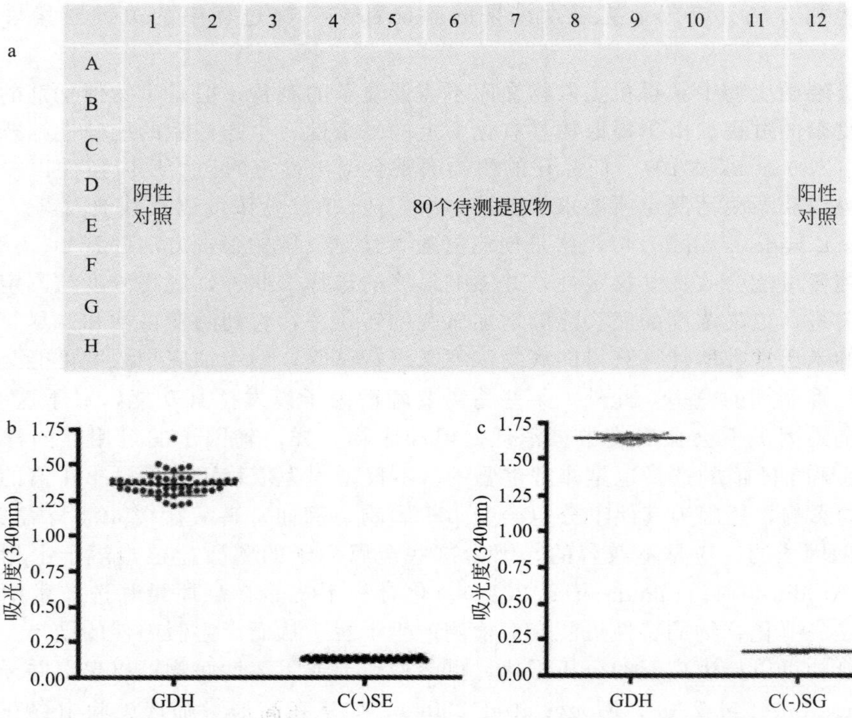

图11.2 高通量筛选谷氨酸脱氢酶的构建与验证。(a) 以96孔板构建的典型的高通量筛选。(b) 验证手动操作的高通量筛选（Z因子＝0.78；Window：10.5）。(c) 验证自动系统操作的高通量筛选（Z因子＝0.94；Window：9.9）

高通量筛选的最终目标是将从特定筛选的输出结果中获得所有信息及时放入合适的生物或化学环境中进行决策与鉴定。这些工作主要由人工完成，但同样具有实现自动化的可能性。其中的关键是拥有一个良好的实验室信息管理系统（LIMS）。因为分析过程会生成庞大的数据，即使前期所有的条件都设定准确，但数据的恢复与管理也经常遇到问题（Omtaa et al. 2012）。

最后，高通量筛选试验结果的准确性需要运用多种已被认可的统计方法进行统计分析。另一方面，使用传统的或不太常用的统计方法，人工操作最大的困难之一就是重复次数低。而高通量筛选试验包含了大量对照数据的重复检测，这意味着样本和对照具有严谨的误差检测。

11.4　生物技术中具有潜力的极端微生物酶

尽管从环境角度来看，以生物催化剂作为化学工艺的替代品往往具有非常吸引人的优势，但在工业应用上需要从成本与效益的角度来评判它们与传统化学催化剂的竞争力（Jaeger 2004）。然而，对于一个特定的生物催化过程而言，天然酶通常不能满足大规模应用的要求，同时，它们的理化性质也需要进行一定的调整。实际上，工业加工经常在极端的条件下进行，包括高压高温、极端 pH、非水系溶液以及氧化性，这些条件会使酶失去活性。一种成功的工业生物催化剂要求它是经济实惠的并且具有合适的高活性、特异性和稳定性，以提升在加工所需要的条件下的催化效果和成本效益（Jemli et al. 2014）。

目前大部分酶都是从中温微生物中获得的，尽管它们具有很多优势，但是由于在极端温度、pH 和离子强度下它们的稳定性有限，应用上受到了限制。与之相反的是，极端微生物是极端酶类的天然来源，这些酶在极端的条件下具有理想的稳定性。因此，极端微生物以及其极端酶的应用正从基础科学被转化为工业可行的技术。每种极端微生物的独特性都能够被用来为特定应用提供酶类。如今在世界范围内已有大量资源投入到开发工业与生物医药用途的极端酶当中（Bhattacharya and Pletschke 2014）。

根据来源于极端微生物与中温微生物酶类催化反应的类型，可以把他们分成六类（表 11.1）（MacDonald and Tipton 2014）。

表 11.1　极端酶类的生化分类

分类	极端酶例子	应用	来源/获得方式	参考文献
氧化还原酶	嗜热和嗜冷性乙醛脱氢酶	手性合成	*Sulfolobus solfataricus*（嗜热硫矿硫化叶菌）/未明确	Egorova and Antranikian (2005)
		化工产品	*Thermococcus litoralis*（栖热球菌）/未明确	Egorova and Antranikian (2005)
			Thermococcus hydrothermalis（嗜热菌）/重组	Antoine et al. (1999)
			Flavobacterium frigimaris KUC-1（黄杆菌）/细菌自身	Kazuoka et al. (2007)
	嗜热与嗜冷超氧化物歧化酶	医疗	*Thermus thermophilus* HB27（嗜热栖热菌）/重组	Liu et al. (2011)
		化妆品工业	*Alicyclobacillus* sp. CC2（脂环酸芽孢杆菌）/细菌自身	Correa-Llantén et al. (2014)
		食品工业	*Exiguobacterium* sp. OS-77（微小杆菌）/细菌自身	Nonaka et al. (2014)

续表

分类	极端酶例子	应用	来源/获得方式	参考文献
氧化还原酶	嗜热与嗜冷超氧化物歧化酶	化工产品	*Pseudoalteromonas haloplanktis*(河豚毒素假交替单胞菌)/细菌自身	Castellano et al. (2006)
	嗜冷过氧化氢酶	化妆品工业	*Vibrio salmonicida*(杀鲑弧菌)/细菌自身	Lorentzen et al. (2006)
			Serratia sp. I1P(沙雷氏菌)/市售,重组	未发表
	嗜盐过氧化氢酶	食品工业	*Halobacterium halobium*(盐生盐杆菌)/细菌自身	Brown-Pterson and Salin (1995)
	嗜热漆酶	纺织工业	*Thermus thermophilus* HB27(嗜热栖热菌)/细菌自身或重组	Miyazaki(2005)
		纸浆与造纸工业 食品工业 生物治理	*Bacillus* sp. FNT(芽孢杆菌)/细菌自身	未发表
转移酶	嗜热与超嗜热DNA聚合酶	分子生物学	*Thermus aquaticus*(栖热水生菌)(Tag聚合酶)/市售,重组	Chien et al. (1976),Peake (1989)
			Pyrococcus furiosus(火球菌)(Pfu聚合酶)/市售,细菌自身或重组	Lundberg et al. (1991)
			Thermotoga maritima(海栖热袍菌)(ULTIMA DNA聚合酶)/市售,已停产	Diaz and Sabino(1998)
			Sulfolobus solfataricus(嗜热硫矿硫化叶菌)/突变库高通量筛选	Kardashliev et al. (2014)
	嗜热转氨酶	手性合成	*Sulfolobus solfataricus*(嗜热硫矿硫化叶菌)/细菌自身	Marino et al. (1988)
水解酶	嗜热和嗜冷碱性脂肪酶	洗涤剂	*Geobacillus* sp. ID17(地衣芽孢杆菌)/细菌自身	Muñoz et al. (2013)
		手性合成	*Thermosyntropha lipolytica*(嗜热解脂互营杆菌)/细菌自身	Salameh and Wiegel (2007)
		化工产品	*Pseudomonas fluorescens* B68(荧光假单胞菌)/重组	Luo et al. (2006)
		生物柴油生产	*Thermus* sp. P1074(栖热菌属)/重组库高通量筛选	Lagarde et al. (2002)
	嗜热性腈水解酶	手性合成	*Pyrococcus abyssi*(火球菌)/重组	Mueller et al. (2006)
		—	*Pyrococcus* sp. M24(火球菌)/细菌自身	未发表

续表

分类	极端酶例子	应用	来源/获得方式	参考文献
水解酶	嗜热木聚糖酶	纸浆与纸张漂白	*Pyrodictium abyssi*（火球菌）/细菌自身	Egorova and Antranikian (2005)
	嗜热与嗜冷蛋白酶	洗涤剂	*Aeropyrum pernix* K1（嗜热泉生古菌）/细菌自身和重组	Bouzas et al. (2006)
		手性合成	*Serratia rubidaea*（悬钩子沙雷菌）/细菌自身	Cavicchioli et al. (2011)
	嗜热淀粉酶	淀粉加工	*Bacillu slicheniformis*（地衣形芽孢杆菌）/细菌自身	Bouzas et al. (2006)
			Pyrococcus woesei（乌兹炽热球菌）/重组	Bouzas et al. (2006)
	嗜热与超嗜热糖苷水解酶	木质纤维素解构	*Thermophilic Clostridia*（乌兹炽热球菌）/重组	Blumer-Schuette et al. (2014)
			Caldicellulosiruptor saccharolyticus（极端嗜热纤维素降解菌）/重组	Blumer-Schuette et al. (2014)
异构酶	嗜盐木糖异构酶	高果糖糖浆	*Halothermothrix orenii*（嗜热盐丝菌）/未生产	Bhattacharya and Pletschke(2014)
	嗜热消旋酶	D-氨基酸合成	*Geobacillus stearothermophilus*（嗜热脂肪地芽孢杆菌）/细菌自身	Soda et al. (1988)
裂合酶	碱性海藻酸裂合酶	制药	*Agarivorans* sp. JAM-A1m（噬琼胶菌）/细菌自身	Kobayashi et al. (2009)
	嗜冷碱性果胶酸裂合酶	废水处理	*Pseudoalteromonas haloplanktis*（河豚毒素假交替单胞菌）/重组	van Truong et al. (2001)
			Mrakia frigida（耐冷海洋嗜杀酵母）/细菌自身	Margesin et al. (2005)
连接酶	嗜热与嗜冷DNA连接酶	分子生物学	*Aquifex pyrophilus*（嗜火液菌）/重组	Lim et al. (2001)
			Pseudoalteromonas haloplanktis（河豚毒素假交替单胞菌）/重组	Georlette et al. (2000)

到目前为止，大部分研究极端微生物酶类的尝试都专注于筛选可培养的微生物，这些微生物具有商业价值的独特酶类。此外，宏基因组学方法使得从可培养与不可培养微生物中获得多种酶类成为可能。然而这两种方法中一次筛选试验也会耗费大量时间和试剂。为了解决这些问题，使用高通量筛选技术快速从极端微生物库中鉴定新型酶是非常必要的。下一部分将会介绍多种具有工业潜力的极端微生物酶以及如何采用高通量技术筛选特定的酶类。

11.4.1　氧化还原酶

氧化还原酶能催化氧化还原反应，把电子从电子供体（还原剂）中转移到电子受体（氧化剂）中。这类型的酶都会借助诸如 NAD/NADH 或 FAD/$FADH_2$ 辅因子。这些酶类，如漆酶、过氧化氢酶、过氧化物酶以及脱氢酶，在多种工艺中都有所应用，因而备受关注，提高它们的稳定性将会提高其生物技术的价值（Guzik et al. 2014）。

在氧化还原反应中，脱氢酶能够从底物中去除或转移氢原子，用于还原醛或酮中的羰基，以及化合物中的碳碳双键。乙醇脱氢酶（ADH）是这类酶的代表，因为该酶能够选择性地合成光学活性的醇，而光学活性的醇是用于构建药用化合物的关键原料。ADH 的生理功能是催化乙醇转化为醛或酮，并把 NAD（P）$^+$ 还原为 NAD（P）H。ADH 广泛分布于微生物当中，包括极端微生物。如来源于嗜热硫矿硫化叶菌的 ADH，以 NAD^+ 为辅因子并且含有锌离子。另一方面，来源于栖热球菌 *Thermococcus litoralis* 的 ADH 不需要金属离子，并使用 $NADP^+$ 氧化伯醇。这个酶具有较高的热稳定性，在 98℃下的半衰期为 15min，85℃下半衰期为 2 h（Egorova and Antranikian 2005）。另外还有其他热稳定的 ADH，如从 *Thermococcus hydrothermalis*（超嗜热古菌）中纯化的依赖于 $NADP^+$ 的 ADH，来源于 *Alicyclobacillus acidocaldarius*（酸热脂环酸杆菌）（60℃下具有最佳活性）的依赖于 NAD^+ 的乙醇-乙醛氧化还原酶，来源于布氏嗜热厌氧杆菌 *Thermoanaerobacter brockii* 的依赖于 NADH 的含 Zn^{2+} 的 ADH 等（Radianingtyas and Wright 2003）。嗜热微生物与极端嗜热微生物的 ADHs 在多种反应过程与生产过程中起着重要的作用，例如生产乙醇、有机溶剂以及乙酸。工业乙醇和酶制剂的生产领域对来源于嗜热微生物的 ADH 十分感兴趣，因为唯有这类微生物适合用于生物质的定向发酵，通过真空蒸馏生产乙醇以及获得热稳定的有活性的商业酶（Andrade et al. 2001）。这些酶也被用于化学组分的合成，尤其是药物分子合成，如合成手性药物中手性分子（Radianingtyas and Wright 2003）。

另一方面，嗜冷微生物的 ADH 也在南极微生物中被发现，其中包括莫拉克斯氏菌属 *Moraxella*（莫拉氏菌）与 *Flavobacterium* 属（黄杆菌属）（Tsigos et al. 1998；Kazuoka et al. 2007）。耐低温细菌 *Flavobacterium frigidimaris* KUC-1（冷海水黄杆菌）能够产生一种低温下具有活性的、热稳定的、依赖于 NAD^+ 的 ADH，这种酶的活性温度范围为 0～85℃，并且其最适温度为 70℃。然而在 20℃以下仍然表现出与其他嗜冷酶相似的高催化活性，在 60℃以及存在 NAD^+ 的情况下半衰期达 143min，是缺乏辅因子时的半衰期（50min）的 3 倍（Kazuoka et al. 2007）。这些特性使得它们在药物的手性合成中具有应用潜力，因为该反应过程需要使用在低温条件下稳定的酶类来减少副产物的形成。

大部分 ADHs 都需要 NAD^+ 或 $NADP^+$ 作为辅因子，辅因子的还原可以通过在 340nm 处吸光度的升高来检测。但这种方法通常不适用于高通量筛选，因为细胞裂解以及 96 孔板会产生检测的背景噪声信号，所以需要使用特殊的不具有紫外吸收特性的微孔板，但是大规模使用这种微孔板往往成本较高。比色法一定程度上解决了这些困难，并且十分适用于高通量筛选。比如硝基蓝四氮唑（NBT）类的四氮唑盐，会被还原为甲臜染料，这种成分具有可见光吸收特性，反应可以很容易在过滤盘上进行可视化检测或者在 96 孔板上读取数值。许多生成有色甲臜的反应都与 ADH 催化生成 NAD（P）H 有关（Johannes et al. 2006）。

目前已经开发出不同的高通量检测 NBT/吩嗪硫酸二甲酯（PMS）的方法，用于鉴定新型 ADH，并且无需对酶进行纯化。这种新的筛选方法采用生物信息学、分子生物学技术和直接体外表达酶的方式快速检测与分析 ADHs 的特性，已有 18 种新型的具有广泛底物特异性的耐高温的 ADHs 被发现和鉴定出来（Ravot et al. 2003）。这种方法可以拓展到任何一种需要使用 NAD（P）$^+$ 作为辅因子的脱氢酶类，并且可以替换为不同种类的极端微生物脱氢酶。

另一种具有生物技术应用潜力的氧化还原酶类是超氧化物歧化酶（SOD），这种酶能够催化超氧阴离子歧化为过氧化氢以及氧分子，同时也是氧化应激的第一道防线。所有 SODs 的活性位点都含有氧化还原类的金属元素，例如 Cu^{2+}、Zn^{2+}、Mn^{2+} 和 Fe^{2+}。SOD 因具有显著的抗氧化功效已被广泛用于医学、美容、食品、农业和化工领域（Liu et al. 2011）。

热稳定性是商业采用 SOD 的主要需求，因为高温变性是酶失活的普遍诱因。如今，人们对来源于嗜热微生物和极端嗜热微生物的 SOD 越来越感兴趣。一种纯化自 *Thermus thermophilus* HB27（嗜热栖热菌）的 Mn-SOD 在 100℃高温处理后依然能够保持 57%的活性，被认为一种高度稳定的酶（Liu et al. 2011）。另一种来源于 *Alicyclobacillus* sp. CC2（脂环酸芽孢杆菌）的热稳定 SOD 在 50℃条件下孵育 6 h 后仍具有初始活性的 80%，并且推测该酶是细菌 CC2 抗氧化防御系统的一部分，使细菌能够在极端环境下存活，避免在南极环境下产生氧化应激反应（Correa-Llantén et al. 2014）。

一种从抗紫外线的细菌嗜放射异常球菌 *Deinococcus radiophilus* 中纯化的 Mn/Fe-SOD 能够在 40℃，pH 为 5～11 的范围内保持稳定，但在酸性条件下则十分不稳定（Yun and Lee 2004）。另一种热稳定 Mn-SOD 从嗜冷菌 *Exiguobacterium* sp. OS-77（微小杆菌）中获得，它的最适反应条件分别为 pH 9.0 和 5℃，而纯化后的 Mn-SOD 热稳定温度高达 45℃，并且在 60℃中处理 21.2min 后仍保留约 50%的活力（Nonaka et al. 2014）。来源于嗜冷菌 *Pseudoalteromonas haloplanktis*（河豚毒素假交替单胞菌）的 PhSOD 则表现出更高的热耐受性，其半数失活温度为 54.2℃（Castellano et al. 2006），该酶的生物化学特性如能够减少自由基对皮肤的损伤，以及减少乳腺癌放射治疗时的纤维化，表明它在医学和美容行业中具有一定的应用潜力（Shafey et al. 2010），通过使用这些热稳定的 SODs 很有可能可以减少皮肤中脂质过氧化。

目前仍然没有筛选微生物细胞 SOD 活性的高通量技术，然而，Trevigen® 公司已经开发出一个系统，能够使用 96 孔板高通量检测哺乳动物组织和细胞裂解液中的 SOD 活性（HT Superoxide Dismutase Assay Kit）。检测时，黄嘌呤氧化酶催化黄嘌呤转化为尿酸，同时产生超氧阴离子自由基和过氧化氢，进一步反应生成具有 450nm 特征吸收的臜盐。SODs 可以降低超氧阴离子的浓度，从而降低臜盐的浓度，臜盐生成量的下降程度可反映实验样品中 SOD 活性，这种方法可有效地用于筛选具有相应酶活力的极端微生物样品。

过氧化氢酶（catalase，CAT）是肝细胞脱毒系统的一部分，用于对抗代谢过程中产生的活性氧簇。该酶能够清除过氧化氢生成氧和水，在分析以及诊断方法中扮演着生物传感器和生物标记物的角色，另外该酶还广泛应用于纺织、造纸、食品和制药工业（Sooch et al. 2014）。

CATs 已在嗜冷弧菌属中发现，一种来源于 *Vibrio salmonicida*（杀鲑弧菌）的 CAT 的最适酶活反应条件为 0～10℃和 pH 8.8，即使在 60℃下该酶依然具有活性。它的热稳定性在 50℃时开始下降，50℃下的半衰期为 34min，并在 70℃时迅速失活，但是在 37℃下孵育 6.5 h 后仍能保持 50%的活性（Lorentzen et al. 2006）。另一种热稳定性更强的 CAT 由一个南极的生物科学基金会从 *Serratia* 属（沙雷氏菌）耐冷菌 I1P 中纯化得到，从 I1P 纯化的 CAT 经过 50℃孵育 8 h 仍保持 60%的初始活力，而利用 *E. coli* 异源表达时，其最佳催化条件为 pH 6～7 和 50℃，重组的 I1P CAT 在 50℃中孵育 6 h 仍具有 50%的活性，并且该酶在 20℃到 70℃范围内都具有活性。这种来源于 I1P 的具有更高热稳定性和较宽温度范围的 CAT 很可能可以在美容业上用作抗衰老乳霜，降低过氧化氢对皮肤的氧化。

耐盐细菌 *Halobacterium halobium*（盐生盐杆菌）可以产生一种 CAT，该酶需要在 NaCl 存在的条件下才具有最佳活性，缓冲液中缺盐将导致酶失活（Brown-Pterson and Salin 1995）。这种类型的酶在食品工业中可与葡萄糖氧化酶混合作为保藏系统。而且，由于 CAT

具有高转化速率，相对简单和明确的反应机制，对于 CAT 的新型应用方式不断涌现出来。

目前仍缺乏从极端微生物中筛选 CAT 的高通量方法。Li 和 Schellhorn（2007）曾经提出一种快速动力学的微试验方法，它是根据 Beers 和 Sizer 分析方法（1952）进行了改进，从而在短时间内可以平行筛选大量样本，因此很适合用于高通量筛选极端微生物中的 CAT。

漆酶是一种蓝色多铜氧化酶，能够将氧的四电子还原与不同有机底物的氧化进行耦合，比如酚、多酚、苯胺，甚至是单电子转移机制的无机化合物。漆酶因为它们广泛的催化反应能力以及底物特异性，所以具有巨大的生物技术应用潜力。其中具有较好前景的包括纺织染料漂白、纸浆漂白、食品加工、水土生物治理、聚合物的合成、生物传感器开发以及生物燃料细胞反应器（Kunamneni et al. 2008）。

虽然一些漆酶已应用于纺织、食品和其他工业，但为了满足未来商业需求，还急需开发新的漆酶。更准确地说，需要一些节能、可生物降解的漆酶以更好地适应高效的、可持续的、环保的工业发展。然而，这类酶最主要的瓶颈问题是在极端的工业条件下稳定性低，储存过程中降解过快。来源于嗜热微生物的漆酶很有可能会突破这些限制。

一种从 *T. thermophilus* HB27（嗜热栖热菌）极端嗜热菌中纯化得到的漆酶具有惊人的嗜热特性，其最佳酶活反应温度为 92℃，80℃下半衰期超过 14 h，成为目前已报道的嗜热性最强的漆酶，甚至超过来源于 *Bacillus subtilis*（枯草芽孢杆菌）的 CotA（Miyazaki 2005）。这些特性使得该酶可用于食品工业中消除溶解氧，以提高植物油风味和品质。此外，该酶的稳定性对于有机催化反应十分重要，因为溶剂的存在会使中温酶失活，但不会影响嗜热酶。但是该酶的最适 pH 是 5.0，这限制了它在造纸和纺织业中的应用，因为这些加工过程需要在碱性 pH 值下进行。为了突破该限制，生物科学基金会从一种名叫 FNT 的嗜热 *Bacillus* 中开发出一种碱性漆酶，其最适温度为 70℃，且在 100℃下仍然具有活性。它的最适 pH 值在 7.0 到 8.0 之间，在 pH 9.5 条件下也依然具有活性，因此更适用于造纸工业。

ABTS 试验是一种适用于漆酶的灵活的检测方式。漆酶催化后的绿色阳离子自由基产物能够通过分光光度法进行检测（Johannes et al. 2006）。该试验用于定量检测定向进化的真菌 *Saccharomyces cerevisiae*（酿酒酵母）中漆酶的功能表达。这种漆酶的蛋白质表达量提高了 8 倍，转化数提升了 22 倍（Bulter et al. 2003）。本检测方法可用于漆酶功能性选择和增强稳定性后的筛选，并有利于未来嗜热性漆酶的开发。

11.4.2 转移酶

转移酶是一类催化特定功能基团从一分子（供体）转移到另一分子（受体）的酶。它们参与不同类型的代谢反应，包括转移酰基链、核苷酸（DNA 与 RNA 聚合酶）、单糖、酮、醛和胺等。

Taq 聚合酶（来源于 *Thermus aquaticus* 的 I 型 DNA 聚合酶）是热稳定 DNA 聚合酶中最著名与最有代表性的酶。该酶来自分离于美国黄石国家公园的嗜热菌 *T. aquaticus*（Chien et al. 1976）。Taq 酶在聚合酶链式反应（PCR）的高温下扩增特定 DNA 片段时非常有用（Peake 1989）。

通常嗜热微生物都具有热稳定的 DNA 聚合酶。极端嗜热微生物比嗜热微生物能提供更多的热稳定酶类，例如来源于 *Pyrococcus furiosus* 的 Pfu 聚合酶比 Taq 聚合酶更加稳定，并具有高保真性（Lundberg et al. 1991）。

来源于 *Thermococcus kodakaraensis* 的重组 KOD1 DNA 聚合酶具有低错误率、高持续合成能力和高延伸率的特性，这使得目的 DNA 序列可以精确扩增 6 kb 的长度（Egorova and Antranikian 2005）。另外一个 DNA 聚合酶的例子是从 *Thermococcus marinus* 中获得的

Tma DNA 聚合酶也适用于 PCR（Bae et al. 2009）。

第一个来源于极端嗜热菌的市售酶是 *Thermotoga maritima*（海栖热袍菌）的 ULTIMA DNA 聚合酶，它拥有 3′～5′外切酶活性，由于它具有校对功能，认为它会使得 PCR 结果更加准确，然而该酶在商业上的应用并不成功，因为测序显示并没有显著的保真性（Diaz and Sabino 1998）。

目前有一种筛选定向进化的 DNA 聚合酶的系统利用蝎形荧光探针作为报告分子，在 96 孔板中进行操作。这种系统目前已在 *Sulfolobus solfataricus*（嗜热硫矿硫化叶菌）的 DNA 聚合酶定向进化中开发和得到验证，提高了其在连续错配下的延伸效率。该探针筛选系统能够对持续合成能力和保真性低的聚合酶进行重新设计，甚至使该酶不具有次级活性，如外切酶或链置换活性，进而满足定向进化产生的多样性的需求（Kardashliev et al. 2014）。这种方法可用于从极端嗜热微生物中筛选对分子生物学实验有用的新型 DNA 聚合酶。

转氨酶（TAs）是一种广泛存在的酶类，它能够催化氨基酸与胺类的合成与降解。氨基能够从氨基供体中转移到 α-酮酸、酮或醛类氨基受体的羰基碳原子上（Höhne and Bornscheuer 2012）。

基于转氨酶的底物特异性可以把酶分成 3 类：α-转氨酶，能转化含羧基的底物的 α 位点，并可合成非天然氨基酸（非蛋白质组成氨基酸和/或 D 构象氨基酸）；ω-转氨酶，转移连接与羧基相隔至少一个碳原子的伯碳上的末端氨基；转氨酶或氨基转移酶，作用于缺乏羧基的底物上，如酮和胺，利用丙酮酸作为通用的氨基受体，可以合成手性伯胺，这些氨基转移酶还能用于合成手性有机分子（Höhne and Bornscheuer 2012）。

α-转氨酶和氨基转移酶显示出高度的对映体选择性，可以用于合成不同的氨基酸和药学领域感兴趣的化合物。然而，用于合成手性化合物的大部分底物具有灭活作用，而且有机溶剂的存在会使目前用于手性合成的转氨酶失活，因此需要寻找新型稳定的转氨酶。

目前已发现含有两个吡哆醛磷酸盐的转氨酶：来自 *S. solfataricus* 的天冬氨酸转氨酶（Marino et al. 1988）以及丝氨酸转氨酶（Littlechild 2011）。天冬氨酸转氨酶可逆地在天冬氨酸与谷氨酸之间转移 α-氨基，是氨基酸代谢中一种重要的酶，但还没有被开发应用于生物技术领域。丝氨酸转氨酶参与丝氨酸合成，在细菌中还未被发现。这种转氨酶可催化 L-丝氨酸和丙酮酸转化成为 3-羟基丙酮酸和丙氨酸。这种酶显示出针对甲硫氨酸、天冬酰胺、谷氨酰胺、苯丙氨酸、组氨酸和色氨酸的催化活性。该酶也能与转酮酶混合用于合成手性化合物（Chen et al. 2007）。该酶是嗜热古菌 α-转氨酶中具有商业应用潜力的例子之一。

Walton 和 Chica（2013）报道称开发出一种利用 96 孔板的高通量筛选 α-酮戊二酸依赖的转氨酶突变库的筛选平台。它们适用于 L-谷氨酸脱氢酶偶联试验，可以用连续的方法筛选特异性针对 L-或 D-氨基酸的转氨酶突变库。该检测可重复、快速且灵敏，具有从嗜热微生物和极端嗜热微生物中发掘新型转氨酶的潜力。

11.4.3 水解酶

水解酶通过添加水分子催化断裂化学键，它们水解剪切复杂的大分子，如蛋白质、淀粉、脂肪和核酸，生成更加简单的分子。

这些酶已经在各种各样的应用领域进行了广泛研究，在这些酶当中，脂肪酶是催化长链酰基甘油酯键水解与合成的酶类。它们是一种多功能的生物催化剂，应用于诸如食品、乳业、制药、去垢剂、纺织、制浆造纸、动物饲料、皮革和美容等领域。目前对于脂肪酶的关注正在不断上升，因为它不仅具有催化酯键水解的特性，它还可以选择性合成单一化合物。许多脂肪酶都可以从商业渠道获得，部分有单一特定或是多个用途（Houde et al. 2004）。然而，尽管已有大量商业脂肪酶，但它们在极端条件下稳定性较差，因此在工业上的应用受到了限制。

嗜热微生物的脂肪酶通常在高温下和有机溶剂中仍保持稳定性和活性。例如 *Thermosyntropha lipolytica*（嗜热解脂互营杆菌）分泌的脂肪酶 LipA 和 LipB，这些酶具有很高的活性温度（最佳温度为 96℃）并且高温下稳定。两种酶都可以添加到高温漂洗时使用的洗涤剂中，也可以在高温下产生脂肪酸、二酰甘油和不同的酯类化合物（Salameh and Wiegel 2007）。

Lip1 是南极细菌 *Geobacillus* sp. ID17（地衣芽孢杆菌）分泌的四种脂肪酶之一，该酶最佳活性条件为 65℃和 pH 9.0，因此可用于洗涤剂。而且，它在 70℃中孵育 8 h 后仍保持 70%活性（Muñoz et al. 2013），该酶的催化特性还可以通过离子溶液进行调整，如在 1-丁基-3-甲基咪唑六氟磷酸盐和 1-丁基-3-甲基咪唑甲基硫酸盐中可以提高其催化活性（Muñoz et al. 2015）。此外，Lip1 可以从薄荷醇异构体中选择性合成酯类化合物，它的对映体选择性能够通过改变反应媒介进行调整。这些特性都显示 Lip1 具有良好的生物技术开发潜力。

为了寻找到用于工业生物转化的新型酶，Lagarde 等（2002）使用新一代热稳定报告底物 CLIPS-O（催化剂的单位氧化鉴定过程）与不同长度的酰基链进行酯化，来筛选嗜热菌属微生物中的酯酶和脂肪酶。CLIPS-O 是一种多功能检测方法，检测带有发光或荧光标记而又不会因为化学修饰失活的酶。*Thermus* sp. P1074（水生栖热菌）具有高水平的胞内酯酶活性作用于 CLIPS-O 的短长链酯。通过高通量筛选菌株基因组文库，发现了三个阳性克隆表达嗜热酯酶，而这些酯酶都是在筛选链长特异性时被发现的，使用 CLIPS-O 底物可能实现高通量筛选嗜热菌和极端嗜热菌的脂肪酶与酯酶。同时，使用 CLIPS-O 酯化不同长度的酰基链或许能确定底物特异性。

特异性腈水解酶变得越来越重要，尤其是在制药工业中手性药物的生物合成。人们从极端嗜热古菌 *Pyrococcus abyssi*（火球菌）获得了一种热稳定的腈水解酶。该酶在 60～90℃之间具有活性，70℃下半衰期为 9h（Mueller et al. 2006），这些特性表明了它在工业中具有巨大潜力。然而，在丙酮（一种常规工业中常用的有机溶剂）存在的情况下该酶会受到抑制。最近，研究人员从 *Pyrococcus* 属的新成员中纯化鉴别出一种热稳定的腈水解酶，这个腈水解酶的最佳温度为 85℃并且具有针对（S）-苯乙醇腈异构体的对映体选择性，这意味着该酶具有合成（S）-苯乙醇酸的潜力，而（S）-苯乙醇酸是一种具有多种药理特性的化合物（未发表）。

许多高通量筛选方法已被开发出来用于筛选腈水解酶。例如 He 等（2011）报道了一种快速简单的高通量筛选方法，一种基于铁氧肟酸盐的分光光度法。Coady 等（2013）开发了一种使用腈水解酶筛选手性 β 羟基酸产物的高通量方法。该策略包括毒性、饥饿和诱导研究与随后的使用感兴趣的底物进行比色筛选活性。这个高通量策略使用 96 孔板，可以快速筛选 256 个新型细菌针对 β 羟基酸的生物催化。此外，*Rhodococcus erythropolis*（红串红球菌）SET1 已被鉴定和发现可以催化 3-羟基丁腈水解，并且在温和条件下具有高度对映体选择性。因此，这将很可能用于筛选高度对映体选择性的腈水解酶以及使用高通量的方式来选择具有高度对映体选择性的嗜热微生物酶和极端嗜热微生物酶。

另一种用于纸浆和纸张漂白的水解酶是耐热木聚糖酶，该酶可降解半纤维素中成分最丰富的木聚糖。在嗜热古菌中，*Pyrodictium abyssi* 和 *Thermococcus zilliggi* AN1 均产木聚糖酶。两种酶都能在超过 100℃条件下保持活性，适合在纸浆和造纸工业中使用（Egorova and Antranikian 2005）。

市售蛋白酶的生产远多于与生物技术相关的其他酶类。它们是最重要的工业酶类并占到全球市场的65%。生物技术的投入进一步推动了能够耐受极端工业条件的耐热蛋白酶的研究。蛋白酶可以分为两类：内肽酶，剪切蛋白质内部的肽键；外肽酶，可切割蛋白质末端氨基酸。许多来源于极端嗜热古菌和细菌的蛋白质水解酶已被鉴定出来，可在高温、去垢剂和其他变性剂（如尿素、盐酸胍、二硫苏糖醇或2-巯基乙醇）存在的情况下具有活性。人们从*T. maritima*、*T. aggregans*、*T. celer*、*T. litoralis*和*Pyrococcus* sp. KODI这些微生物中鉴定出了极端嗜热蛋白酶。*Pyrococcus* sp. KODI中硫醇蛋白酶的最适反应温度是110℃，是最嗜热的蛋白酶。来源于好氧极端嗜热菌*Aeropyrum pernix* K1（嗜热泉生古菌）的蛋白酶已被分离出来，这种微生物生长的最适温度为90℃，并且它的好氧生长模式以及大规模培养的营养需求容易控制。这些特性使得这种极端嗜热微生物成为一种极具吸引力的耐热蛋白酶来源（Bouzas et al. 2006）。

淀粉的工业酶法加工是基于（部分）麦芽糊精、麦芽糖和葡萄糖糖浆的水解。这种工艺要求酶类在超过65～70℃下保持稳定性和活性。主要作用淀粉的酶是α-淀粉酶和支链淀粉酶，它们能把淀粉降解成麦芽寡糖和葡萄糖（Kaper et al. 2003）。

淀粉的酶法工艺包含三步：糊化（淀粉颗粒的溶解，通过用水高温加热淀粉形成黏性悬浮液）；液化（部分水解淀粉，失去黏性）；以及糖化（通过进一步水解产生葡萄糖和麦芽糖）（Bouzas et al. 2006）。

最早发现的耐热α-淀粉酶是从*Bacillus subtilis*（枯草芽孢杆菌）、*Bacillus amyloliquefaciens*（解淀粉芽孢杆菌）和*Bacillus licheniformis*（地衣芽孢杆菌）中分离的。然而，大部分α-淀粉酶需要钙来保持活性，并且添加到反应中的钙会以草酸钙的形式沉淀出来，所以大部分α-淀粉酶的工业应用受到限制（Saha et al. 2014）。降低草酸钙浓度就需要更低的反应pH值，因此寻找耐酸的α-淀粉酶就变得尤为重要。目前人们已经从*Pyrococcus woesei*、*Pyrococcus furiosus*、*Thermococcus profundus*和*Thermococcus hydrothermalis*这些微生物中获得了极端嗜热和嗜酸性α-淀粉酶。这些酶最适反应温度都是100℃，而最适反应pH比从*Bacillus*中发现的α-淀粉酶更低（Bouzas et al. 2006）。

α-葡萄糖苷酶通常用于淀粉降解的最后一步。一种分离自*T. maritime*的极端嗜热α-葡萄糖苷酶，需要NAD^+和Mn^{2+}来保持活性。在*P. furiosus*、*P. woesei*、*Sulfolobus shibatae*、*Sulfolobus solfataricus*、*Thermococcus* strain AN1和*Thermococcus hydrothermalis*中也已经被检测出另一种极端嗜热α-葡萄糖苷酶。从*P. furiosus*中得到的胞内α-葡萄糖苷酶是最耐高温的极端酶，它的最适酶活反应温度为105～115℃，pH范围为5.0～6.0（Bouzas et al. 2006）。

普鲁兰多糖降解酶同样属于α-淀粉酶家族，它普遍存在于各物种中，包括极端嗜热古菌。这些酶已经从*Thermococcusceler*（速生热球菌）、*Desulfurococcus mucosus*（除硫球菌）、*Staphylothermus marinus*（葡萄热菌）、*Thermococcus agregans*、*P. furiosus*、*Thermococcus litoralis*、*T. hydrothermalis*、*Pyrococcus* strain ES4和*T. maritima*中分离得到。它们在超过100℃下都具有良好的稳定性，甚至在缺乏底物或钙离子的情况下也是如此（Bouzas et al. 2006）。

目前对木质纤维素酶的要求是能够在极端条件下使用，如高盐浓度、有机溶剂、极端pH值和温度。这些极端的条件都不是目前使用的酶的最适反应条件，会降低酶的转化率与寿命。嗜热微生物是具有吸引力的候选者，它可以用于把木质纤维素转化为生物燃料，生产稳定有效的碳水化合物降解酶。*Clostridium*（梭菌）和*Caldicellulosiruptor*（热解纤维素果汁杆菌）属的嗜热微生物分别产生纤维素酶与非纤维素酶，它们能够把植物生物质转化为可发酵糖类，这为生物燃料生产中要求减少预处理或无需预处理提供了可能性（Blumer-

Schuette et al. 2014）。

有研究人员构建了来源于 *Halomonas* sp. 的纤维素酶 Cel8H 随机突变文库，同时将 RoboLector 用于选择和鉴别文库中的候选者。这种高通量筛选系统是 Kunze 等（2012）开发的，RoboLector 是由用于微生物微型规模培养/反应的在线显示系统（BioLector）和自动液体处理机器人组成。机器人与 BioLector 之间的作用可以生成大量数据，并且进行系统的高通量实验。RoboLector 可以获得酶生产过程中大量的动力学数据，包括指定宿主的培养、表达步骤以及酶的最终鉴别。这个自动化筛选系统可以支持以最少的人力来搜索可用于木质纤维素处理的新型极端酶的开发和生产过程。

对有合适性能和新颖功能的新型热稳定水解酶可以节约工业加工中的时间、金钱和能源成本。哥本哈根大学的热酶计划（Hot Zyme project）旨在利用代谢组学的方法鉴定这类酶。许多高通量筛选技术都用于鉴定从地热环境中分离的微生物中的新型水解酶。然而，目前仍没有这个计划相关产品的报道。

另外一种存在于极端温度的是嗜冷水解酶。适冷酶具有较高的生物技术价值，它们在低温下具有较高的催化效率（k_{cat} 值）。它们的热稳定性较低，在有机溶剂中活性高，这些特点都使得它们更具经济优势。此外适冷酶一般比中温同源的酶类在低温下产量更高，并且使用适冷酶时的工艺流程更加节能。因此，适冷水解酶已经发现被用于如家用洗涤剂等产品中。

这些酶在低温下降解底物的能力被应用于清洁，包括洗衣皂、食品、乳制品、医疗设备和水处理等行业中，并且在乳制品、水处理、洗涤剂和酿造业中对于这种清洁方式的生命周期评估（life cycle assessment，指分析追踪能源的需求、原材料的使用和废料流的处理）已有所报道。洗涤剂制造商意识到了降低洗涤温度和节能改进之间的关系，洗涤温度从 40℃降到 30℃，可以减少 30%的用电量，这等于每次洗涤降低了 100 g 的二氧化碳排放。蛋白酶、淀粉酶、脂肪酶和纤维素酶，如来源于诺维信（Novazymes）公司的碱性蛋白酶（Alcalase）、淀粉酶（Natalase）和脂肪酶（Lipolase Ultra）已经被用于低温洗涤。这些酶低温清洗的能力降低了部分工业洗涤时的温度，包括自动洗碗机、水处理膜的清洗、酿造和奶制品设备的清洗。来源于嗜冷微生物的酶，如来自 *Serratia rubidaea* 和 *Stenotrophomonas maltophilia*（嗜麦芽窄食单胞菌）的蛋白酶有利于提高基于酶的低温清洗模式的效率（Cavicchioli et al. 2011）。

在环境温度下的物体表面，如建筑、地毯和长凳都无法轻易加热或者浸泡在清洗溶液中，所以常常采用喷雾和擦拭的清洁方式，这为适冷酶提供了用武之地。一种脂肪酶和葡萄糖苷酶已被用于建筑节能工程中，来清洗石头中的霉菌，降低了使用常规清洁剂造成的损伤（Cavicchioli et al. 2011）。

11.4.4 异构酶

异构酶能催化一个分子从一种构象转化成另一种构象，实现分子内的重排，这些酶催化一个底物生成一个产物。异构酶催化反应贯穿着许多代谢途径，如糖酵解。这类酶包括消旋酶、差向异构酶、顺反异构酶、分子内氧化还原酶、分子内裂解酶和分子内转移酶。

丙氨酸消旋酶是一种依赖于吡哆醛 5′-磷酸的酶，催化 L-和 D-丙氨酸消旋，细菌利用丙氨酸消旋酶来提供 D-消旋体合成细胞壁的肽聚糖（Okubo et al. 1999）。

Soda 等（1988）开发了一种通过偶联嗜热脂肪土芽孢杆菌 *Geobacillus stearothermophilus* 的丙氨酸消旋酶生成的产物与不同的酶来合成多种 D 型氨基酸的工艺。然而，这个生产工艺的产量远低于工业中使用的化学合成方法。

果糖在商业上用于食品和饮料行业，这种糖能够在许多食物中找到并且有相对较高的甜度。它能够通过葡萄糖异构来获得，这是一个由葡萄糖/木糖异构酶催化的可逆反应。该过程对于果糖糖浆的生产十分重要。工业上使用淀粉来生产果糖，该过程中依赖淀粉水解成高浓度葡萄糖糖浆，然后加入异构酶来生成果糖。这种生产果糖糖浆的经典工艺使用了 α-淀粉酶来液化淀粉，然后使用糖化酶糖化水解的淀粉。这使得产物可以定向进行异构（Gaily et al. 2013）。

目前已经从中温微生物中分离得到许多木糖异构酶，它们的最适酶活反应 pH 值在 7.5 和 9.0 之间，高温和碱性 pH 值均会导致果糖溶液中形成副产物。所以，工业制备果糖温度通常限制在 60℃，并需要中性或弱酸性 pH 值。

Geobacillus thermodenitrificans TH2 产生了一种非常稳定的木糖异构酶，它在 4℃ 和 50℃ 中分别孵育 96 h 后仍保持活性。而它的最适酶活反应温度为 80℃，最适 pH 值为 7.5，不适合于高果糖玉米糖浆的生产（Konak et al. 2014）。

Halothermothrix orenii 基因组编码了一种高度热稳定的木糖异构酶，通过对它在 pH 6.0 和 7.0 之间的活性与稳定性评估，显示它是一种用于高果糖玉米糖浆生产的候选酶类（Bhattacharya and Pletschke 2014）。然而，目前仍需要更多的研究去确定该生物技术在工艺流程的应用。

目前甚至在中温微生物中，仍没有一种高通量筛选的方法能随意地筛选各种类型的异构酶。

11.4.5　裂合酶

裂合酶可以催化 C—C 键、C—O 键、C—N 键的断裂，但不是通过水解或氧化的方式。这种断裂化学键的过程会导致形成双键或成环。这类酶在一个反应方向中利用两种底物，但只有一种底物参与到另一个反应方向中。

海藻酸裂合酶通过糖苷键的 β 消除反应来降解海藻酸，并在非还原末端产生含有不饱和糖醛酸的寡糖以及不饱和糖醛酸单体。海藻酸寡糖具有特殊的生物活性，它们可刺激细胞因子的产生。因此海藻酸裂合酶已用作功能性寡糖生产的生物催化剂，引起人们的广泛关注。海藻酸裂合酶能够作为囊性纤维病患者的药物使用，因为它们可以提高抗生素杀死黏液中的铜绿假单胞菌的能力（Kim et al. 2011）。人们从深海细菌 *Agarivorans* sp. JAM-A1m 中纯化出一种在高碱性和含盐条件下才有活性的海藻酸裂合酶，该酶在碱性 pH 10.0 下具有高活性，当反应缓冲液中加入 200mmol/L NaCl 后具有最高活性 pH 值变成 9.0。该海藻酸裂合酶能够酶解海藻酸产生含有甘露糖醛酸-古罗糖醛酸和富含古罗糖醛酸的寡糖片段（Kobayashi et al. 2009）。这些片段可以刺激人内皮细胞生长以及巨噬细胞细胞毒性因子的分泌。

果胶酶可应用于食品和饮料行业，以及植物纤维的浸渍过程。果胶是植物细胞壁中含量丰富的一种结构成分，并作为纤维素和半纤维素的支撑基质而发挥作用。果胶由一条甲基酯化的 α-1,4-D-聚半乳糖醛酸主链和高度分支的鼠李半乳糖醛酸组成，后者可以作为 1,4-连接侧链的结合位点，比如阿拉伯聚糖、半乳聚糖和阿拉伯半乳聚糖（Kluskens et al. 2003）。在这些酶中，果胶酸裂合酶是一种果胶解聚酶，它能够通过反式消除机制剪切聚半乳糖醛酸的 α-1,4-半乳糖苷链接，该机制需要 Ca^{2+} 来激活（Kobayashi et al. 2000）。

耐热果胶酸裂合酶已在嗜碱微生物 *Bacillus* sp. P-4-N 中发现。该酶在含有 100mmol/L NaCl 和 pH 5.0～11.5 范围内（50℃，60 h）的条件下十分稳定（Kobayashi et al. 2000）。来源于 *T. maritima* 的耐热果胶酸裂合酶具有高度的热激活性与热稳定性，它的最适反应条件为 90℃和 pH 9.0，95℃下半衰期大约为 2 h（Kluskens et al. 2003）。

果胶酸裂合酶可用于清除废水中的果胶物质，这是一种经济、环保的处理方式。嗜冷酶的使用对于这种降解过程则更为理想，它可进一步降低能量消耗。从嗜冷菌 *Pseudoalteromonas haloplanktis* strain ANT/505（van Truong et al. 2001）和 *Mrakia frigida*（Margesin et al. 2005）中已纯化出果胶酸裂合酶，这些酶的最佳酶活反应条件为温度在 30℃左右且偏碱性环境，它们适用于清除胶质中的果胶成分。

目前还没有使用高通量检测技术对极端微生物裂合酶进行筛选，开发这类技术筛选新型裂合酶和异构酶具有巨大的工业应用潜力。

11.4.6 连接酶

连接酶能通过形成化学键把两个大分子连接在一起，这类酶在自然界中广泛分布，但分析方法限制了它们的应用。

DNA 连接酶是细胞过程所需的酶类，如 DNA 复制、DNA 重组和 DNA 修复。它们催化 DNA 双链中相邻的 5′-磷酸基和 3′-羟基形成磷酸二酯键（Georlette et al. 2000）。

DNA 连接酶在分子生物学的常规实验中广泛使用，例如表达载体上的基因克隆。热稳定 DNA 连接酶可通过高温六聚体引物连接来构建测序引物，通过重复扩增检测法检测三核苷酸重复序列，或者通过寡核苷酸环化进行 DNA 检测（Egorova and Antranikian 2005）。

Aquifex pyrophilus 的 DNA 连接酶基因成功在 *E. coli* 中表达和纯化，该酶在 65℃和 pH 值在 8.0～8.6 范围内具有较高闭合断裂的活性。在 95℃中孵育 60min 后仍保持超过 75%的活性，而 *T. aquaticus* 和 *E. coli* 的 DNA 连接酶在 95℃中半衰期分别为 15min 和 5min。*A. pyrophilus* 的耐热 DNA 连接酶可应用于重复扩增检测，也可用于 DNA 诊断（Lim et al. 2001）。

另一方面，据报道一种来自 *P. haloplanktis* 的适冷 DNA 连接酶在低温下显示出比 *E. coli* DNA 连接酶更高的催化效率。*P. haloplanktis* DNA 连接酶用于酶适应低温的研究，这为生物技术提供了新的工具。而市售 DNA 连接酶在低于 15℃的温度下活性较低，且需要延长孵育时间，这些条件使得残留的核酸酶可能干扰连接反应。嗜冷连接酶的另一个优势是灭活温度相对较低，避免了 DNA 变性（Georlette et al. 2000）。

11.5 结论

尽管极端酶类有着与生俱来的优势，但实际上，可以获得的极端生物催化工具非常有限。因此，只有在一些科学性的挑战中有所突破，极端酶类的潜力才能得到充分体现。虽然分子方法让人们可以快速克隆，获得大量感兴趣的基因，但不可能只使用基因组学的方法来获得蛋白质。辨别特定的酶特性，例如热稳定性、特定活性，这些只能通过功能性实验进行研究。此外，还要强调整合功能性方法和大规模生物技术的相关性，进一步加强在极端工业条件下执行生物催化作用的能力。

无利益冲突声明

Freddy Boehmwald，Patricio Muñoz，Patricio Flores 和 Jenny M. Blamey 声明他们没有利益冲突。

参考文献

Acker M, Auldn D (2014) Considerations for the design and reporting of enzyme assays in high-throughput screening applications. Perspect Sci 1: 56-73

Adrio J, Demain A (2014) Microbial enzymes: tools for biotechnological processes. Biomolecules 4: 117-139

Andexer J, Langermann J, Kragl U, Pohl M (2009) How to overcome limitations in biotechnological processes examples from hydroxynitrile lyase applications. Trends Biotechnol 27: 599-607

Andrade C, Aguiar W, Antranikian G (2001) Physiological aspects involved in production of xyl-anolytic enzymes by deep-sea hyperthermophilic archaeon *Pyrodictium abyssi*. Appl BiochemBiotechnol 91-93: 655-669

Antoine E, Rolland J, Raffin J, Dietrich J (1999) Cloning and over-expression in *Escherichia coli* of the gene encoding NADPH group III alcohol dehydrogenase from *Thermococcus hydrothermalis*. Characterization and comparison of the native and the recombinant enzymes. Eur J Biochem 264: 880-889

Bachmann B (2014) Microbial genome mining for accelerated natural products discovery: is a renaissance in the making? J Ind Microbiol Biotechnol 41: 175-184

Bae H, Kim K, Lee J, Song J, Kil E, Kim J, Kwon S (2009) Characterization of DNA polymerase from the hyperthermophilic archaeon *Thermococcus marinus* and its application to PCR. Extremophiles 13: 657-667

Beers R, Sizer I (1952) A spectrophotometric method for measuring the breakdown of hydrogen peroxide by catalase. J Biol Chem 195: 133-140

Bhattacharya A, Pletschke B (2014) Review of the enzymatic machinery of *Halothermothrix orenii* with special reference to industrial applications. Enzyme Microb Technol 55: 159-169

Blumer-Schuette S, Brown S, Sander K, Bayer E, Kataeva I, Zurawski J, Conway J, Adams M, Kelly R (2014) Thermophilic lignocellulose deconstruction. FEMS Microbiol Rev 38: 393-448 Bouzas T, Barros-Velázquez J, González T (2006) Industrial applications of hyperthermophilic enzymes: areview. Protein Pept Lett 13: 645-651

Brooks H, Geeganage S, Kahl S, Montrose C, Sittampalam S, Smith M, Weidner J (2012) Basics of enzymatic assays for HTS. In: Sittampalam G, Coussens N, Nelson H et al (eds) Assay guidance manual. Eli Lilly & Company and The National Center for Advancing Translational Sciences, Indianapolis, Bethesda, pp 1-13

Brown-Pterson N, Salin M (1995) Purification and characterization of a mesohalic catalase from the halophilic bacterium *Halobacterium halobium*. J Bacteriol 177: 378-384

Bulter T, Alcalde M, Sieber V, Meinhold P, Schlachtbauer C, Arnold F (2003) Functional expression of a fungal laccase in *Saccharomyces cerevisiae* by directed evolution. Appl Environ Microbiol 69: 987-995

Castellano I, Di Maro A, Ruocco M, Chambery A, Parente A, Di Martino M, Parlato G, Masullo M, De Vendittis E (2006) Psychrophilic superoxide dismutase from *Pseudoalteromonas haloplanktis*: biochemical characterization and identification of a highly reactive cysteine residue. Biochimie 88: 1377-1389

Cavicchioli R, Charlton T, Ertan H, Mohd Omar S, Siddiqui K, Williams T (2011) Biotechnological used of enzymes from psychrophiles. Microb Biotechnol 4: 449-460

Chen B, Sayar A, Kaulmann U, Dalby P, Ward J, Woodley J (2007) Reaction modelling and simulation to assess the integrated use of transketolase and ω-transaminase for the synthesis of an aminotriol. Biocatal Biotransform 24: 449-457

Chien A, Edgar D, Trela J (1976) Deoxyribonucleic acid polymerase from the extreme thermohile *Thermus aquaticus*. J Bacteriol 127: 1550-1557

Coady T, Coffey L, O' Reilly C, Owens E, Lennon C (2013) A high throughput screening strategy for the assessment of nitrile-hydrolyzing activity towards the production of enantiopure β-hydroxy acids. J Mol Catal B Enzym 97: 150-155

Correa-Llantén D, Amenábar M, Muñoz P, Monsalves M, Castro M, Blamey J (2014) *Alicyclobacillus* sp. strain CC2, a thermo-acidophilic bacterium isolated from Deception Island (Antarctica) containing a thermostable superoxide dismutase enzyme. Adv Pol Sci 25: 92-96

Diaz R, Sabino E (1998) Accuracy of replication in the polymerase chain reaction. Comparison between *Thermotoga maritima* DNA polymerase and *Thermus aquaticus* DNA polymerase. Braz J Med Biol Res 31: 1239-1242

Donadio S, Monciardini P, Sosio M (2009) Approaches to discovering novel antibacterial and antifungal agents. Methods Enzymol 458: 3-28

Egorova K, Antranikian G (2005) Industrial relevance of thermophilic Archaea. Curr Opin Microbiol 8: 649-655

Fernández-Arrojo L, Guazzaroni M, López-Cortés N, Beloqui A, Ferrer M (2010) Metagenomic era for biocatalyst identification. Curr Opin Biotechnol 21: 725-733

Gaily M, Sulieman A, Abasaeed A (2013) Kinetics of a three-step isomerization of glucose to fructose using immobilized enzyme. Int J Chem Eng Appl 4: 31-34

Georlette D, Jónsson Z, van Petegem F, Chessa J, van Beeumen J, Hübscher U, Gerday C (2000) A DNA ligase from the psychrophile *Pseudoalteromonas haloplanktis* gives insights into the adaptation of proteins to low temperatures. Eur J Biochem 267: 3502-3512

Glaser R, Venus J (2014) Screening of *Bacillus coagulans* strains in lignin supplemented minimal medium with high throughput turbidity measurements. Biotechnol Rep 4: 60-65

Guzik U, Hupert-Kocurek K, Wojcieszyńska D (2014) Immobilization as a strategy for improving enzyme properties-application to oxidoreductases. Molecules 19: 8995-9018

Hallam S, Nelson H, Greger V, Perreault-Micale C, Davie J, Faulkner N, Neitzel D, Casey K, Umbarger M, Chennagiri N, Kramer A, Porreca G, Kennedy C (2014) Validation for clinical use of and initial clinical experience with a novel approach to population-based carrier screen-ing using high-throughput, next-generation DNA sequencing. J Mol Diagn 2: 180-189

He Y, Ma C, Xu J, Zhou L (2011) A high-throughput screening strategy for nitrile-hydrolyzing enzymes based on ferric hydroxamate spectrophotometry. Appl Microbiol Biotechnol 89: 817-823

Höhne M, Bornscheuer U (2012) Application of transaminases. In: Drauz K, Gröger H, May O (eds) Enzyme catalysis in organic synthesis, 3rd edn. Wiley-VCH Verlag GmbH & Co, Weinheim, pp 779-820

Hotzyme Project. TheUniversity of Copenhagen. http: //hotzyme. com/project/

Houde A, Kademi A, Leblanc D (2004) Lipases and their industrial applications. Appl Biochem Biotechnol 118: 155-170

Inglese J, Auld D, Jadhav A, Johnson R, Simeonov A, Yasgar A, Zheng W, Austin C (2006) Quantitative high-throughput screening: a titration-based approach that efficiently identifies biological activities in large chemical libraries. Proc Natl Acad Sci USA 31: 11473-11478

Jaeger K (2004) Protein technologies and commercial enzymes: white is the hype-biocatalyst on the move. Curr Opin Biotechnol 15: 269-271

Jemli S, Ayadi-Zouari D, Hlima H, Bejar S (2014) Biocatalyst: application and engineering for industrial purposes. Crit Rev Biotechnol 6: 1-13

Johannes T, Woodyer R, Zhao H (2006) High-throughput screening methods developed for oxido-reductases. In: Reymond J (ed) Enzyme assays: high-throughput screening, genetic selection and fingerprinting. John Wiley & Sons, New York, pp 77-93

Johnston P, Soares K, Shinde S, Foster C, Shun T, Takyi H, Wipf P, Lazo J (2008) Development of a 384-well colorimetric assay to quantify hydrogen peroxide generated by the redox cycling of compounds in the presence of reducing agents. Assay Drug Dev Technol 4: 505-518

Kaper T, van der Marel M, Euverink G, Dijkhuizen L (2003) Exploring and exploiting starchmodifying amylomaltases from thermophiles. Biochem Soc Trans 32: 279-282

Kardashliev T, Ruff A, Zhao J, Schwaneberg U (2014) A high-throughput screening method to reengineer DNA polymerases for random mutagenesis. Mol Biotechnol 56: 274-283

KazuokaT, Oikawa T, Maraoka I, Kuroda S, Soda K (2007) A cold-active and thermostable alco-hol dehydrogenase of a psychrotorelant from Antarctic seawater, *Flavobacterium frigidimaris* KUC-1. Extremophiles 11: 257-267

Kim H, Lee C, Lee E (2011) Alginate lyase: structure, property, and application. Biotechnol Bioprocess Eng 16: 843-851

King O, Li X, Sakurai M, Kawamura A, Rose N, Quinn A, Rai G, Mott B, Beswick P, Klose R, Oppermann U, Jadhav A, Heightman T, Maloney D, Schofield C, Simeonov A (2010) Quantitative high-throughput screening identifies 8-hydroxyquinolines as cell-active histone demethylase inhibitors. PLoS One 5: 1-12

Klumpp M, Boettcher A, Becker D, Meder G, Blank J, Leder L, Forstner M, Ottl J, Mayr L (2006) Readout technologies for highly miniaturized kinase assays applicable to high-throughput screening in a 1536-well format. J Biomol Screen 11: 617-633

Kluskens L, van Alebeek G, Voragen A, de Vos W, van der Oost J (2003) Molecular and biochemical characterization of the thermoactive family 1 pectate lyase from the hyperthermophilic bacterium *Thermotoga maritima*. Biochem J 370: 651-659

Kobayashi T, Hatada Y, Suzumatsu A, Saeki K, Hakamada Y, Ito S (2000) Highly alkaline pectate lyase Pel-4A from alkaliphilic *Bacillus* sp. strain P-4-N: its catalytic properties and deduced amino acid sequence. Extremophiles 4: 377-383

Kobayashi T, Uchimura K, Miyazaki M, Nogi Y, Horikoshi K (2009) A new high-alkaline alginate lyase from a deep-sea bacterium *Agarivorans* sp. Extremophiles 13: 121-129

Konak L, Kolcuoglu Y, Ozbek E, Colak A, Ergenoglu B (2014) Purification and characterization of an extremely stable glucose isomerase from *Geobacillus thermodenitrificans* TH2. Appl Biochem Microbiol 50: 25-29

Kunamneni A, Plou F, Ballesteros A, Alcalde M (2008) Laccases and their applications: a patent review. Recent Pat Biotechnol 2: 10-24

Kunze M, Schmidt T, Fischer R, Commandeur U, Büchs (2012) Application of high-throughput screening methods for the development of improved enzymes for cellulose degradation. In: Abstracts of the ACHEMA conference 2012, ACHEMA, Frankfurt am Main, 18-22 June 2012

Lagarde D, Nguyen H, Ravot G, Wahler D, Reymond J, Hills G, Veit T, Lefevre F (2002) Highthroughput screening of thermostable esterases for industrial bioconversions. Org Process Res Dev 6: 441-445

Leis B, Angelov A, Liebl W (2013) Screening and expression of genes from metagenomes. Adv Appl Microbiol 83: 1-68

Li Y, Schellhorn H (2007) Rapid kinetic microassay for catalase activity. J Biomol Tech 18: 185-187

Li S, Yang X, Yang S, Zhu M, Wang X (2012) Technology prospecting on enzymes: applications, marketing and engineering. Comput Struct Biotechnol J 2: 1-11

Liauda N, Navarroa D, Vidald N, Sigoillota JC, Raouche S (2014) High throughput automated colorimetric method for the screening of L-lactic acid producing microorganisms. MethodsX 1: 254-257

Lim J, Choi J, Han S, Kim S, Hwang H, Jin D, Ahn B, Han Y (2001) Molecular cloning and characterization of thermostable DNA ligase from *Aquifex pyrophilus*, a hyperthermophilic bacte-rium. Extremophiles 5: 161-168

Littlechild J (2011) Thermophilic archaeal enzymes and applications in biocatalysis. Biochem Soc Trans 39: 155-158

Liu J, Yin M, Zhu H, Lu J, Cui Z (2011) Purification and characterization of a hyperthermostable Mn-superoxide dismutase from *Thermus thermophiles* HB27. Extremophiles 15: 221-226

López-López O, Cerdán M, González M (2014) New extremophilic lipases and esterases from metagenomics. Curr Protein Pept Sci 15: 445-455

Lorentzen M, Moe E, Jouve M, Willanssen N (2006) Cold adapted features of *Vibrio salmonicida* catalase: characterization and comparison to the mesophilic counterpart from *Proteus mirabi-lis*. Extremophiles 10: 427-440

Lundberg K, Shoemaker D, Adams M, Short J, Sorge J, Mathur E (1991) High-fidelity amplifica-tion using a thermostable DNA polymerase isolated from *Pyrococcus furiosus*. Gene 198：1-6

Luo Y，Zheng Y，Jiang Z，Ma Y，Wei D（2006）A novel psychrophilic lipase from *Pseudomonas fluorescens* with unique property in chiral resolution and biodiesel production via transesterification. Appl Microbiol Biotechnol 73：349-355

MacDonald A，Tipton K（2014）Fifty-five years of enzyme classification：advances and difficul-ties. FEBS J 281：583-592

Margesin R，Fauster V，Fonteyne P（2005）Characterization of cold-active pectate lyases from psychrophilic *Mrakia frigida*. Lett Appl Microbiol 40：453-459

Marino G，Nitti G，Arcone M，Sannia G，Gambacorta A，De Rosa M（1988）Purification and characterization of aspartate aminotransferase from thermoacidophilic archaebacterium *Sulfolobus solfataricus*. J Biol Chem 263：12305-12309

Miyazaki K（2005）A hyperthermophilic laccase from *Thermus thermophiles* HB27. Extremophiles 9：415-425

Moger J，Gribbon P，Sewing A，Winlove P（2006）The application of fluorescence lifetime readouts in high-throughput screening. J Biomol Screen 7：765-772

Mueller P，Egorova K，Vorgias C，Boutou E，Trauthwein H，Verseck S，Antranikian G（2006）Cloning，overexpression，and characterization of a thermoactive nitrilase from the hyperther-mophilic archaeon *Pyrococcus abyssi*. Protein Expr Purif 47：672-681

Muñoz P，Correa-Llantén D，Blamey J（2013）Production，purification and partial characterization of four lipases from a thermophile isolated from Deception Island. Lipids 48：527-533

Muñoz P，Correa-Llantén D，Blamey J（2015）Ionic liquids increase the catalytic efficiency of a lipase（Lipl）from an Antarctic thermophilic bacterium. Lipids 50：49-55

Noah J（2010）New developments and emerging trends in high-throughput screening methods for lead compound identification. Int J High Throughput Screen 1：141-149

Nonaka K，Yoon K，Ogo S（2014）Biochemical characterization of psychrophilic Mn-superoxide dismutase from newly isolated *Exiguobacterium* sp. OS-77. Extremophiles 18：363-373

Okubo Y，Yokoigawa K，Esaki N，Soda K，Kawai H（1999）Characterization of psychrophilic alanine racemase from *Bacillus psychrosaccharolyticus*. Biochem Biophys Res Commun 256：333-340

Omtaa W，Egana D，Spruitb M，Brinkkemperb S（2012）Information architecture in high through-put screening. Procedia Technol 5：696-705

Peake I（1989）The polymerase chain reaction. J Clin Pathol 42：673-676

Radianingtyas H，Wright P（2003）Alcohol dehydrogenases from thermophilic and hyperthermophilic archaea and bacteria. FEMS Microbiol Rev 27：593-616

Ravot G，Wahler D，Favre-Bulle O，Cilia V，Lefevre F（2003）High throughput discovery of alcohol dehydrogenases for industrial biocatalysis. Adv Synth Catal 345：691-694

Saha K，Maity S，Roy S，Pahan K，Pathak R，Majumdar S，Gupta S（2014）Optimization of amy-lase production from *Bacillus amyloliquefaciens*（MTCC 1270）using solid state fermentation. Int J Microbiol 2014：7640-7646

Salameh M，Wiegel J（2007）Lipases from extremophiles and potential for industrial applications. Adv Appl Microbiol 61：253-283

Schallmey M，Frunzke J，Eggeling L，Marienhagen J（2014）Looking for the pick of the bunch：high-throughput screening of producing microorganisms with biosensors. Curr Opin Biotechnol 26：148-154

Schmitz J，Daniel A，Collin M，Schuch R，Fischetti V（2008）RapidDNA library construction for functional genomic and metagenomic screening. Appl Environ Microbiol 74：1649-1652

Shafey H，Bahashwan S，Alghaithy A，Ghanem S（2010）Microbial superoxide dismutase enzyme as therapeutic agent and future gene therapy. Cur Res Technol Edu Top Appl Microbiol Microbial Biotech 435-440

Soda K，Tanaka H，Tanizawa K（1988）Thermostable alanine racemase and its application to D-amino acid synthesis. Ann N Y Acad Sci 542：375-382

Sooch B，Kauldhar B，Puri M（2014）Recent insights into microbial catalases：isolation，production and purification. Biotechnol Adv 32：1429-1447

Tsigos I，Velonia K，Smonou I，Bouriotis V（1998）Purification and characterization of an alcohol dehydrogenase from the Antarctic psychrophile *Moraxella* sp. TAE123. Eur J Biochem 254：356-362

USDA（2008）U. S. Biobased Products：Market Potential and Projection Through 2025. United States Department of Agriculture（OCE-2008-1）www. usda. gov/oce/reports/energy/index. htm

van Truong L，Tuyen H，Helmke E，Binh L，Schweder T（2001）Cloning of two pectate lyase genes from the marine Antarctic bacterium *Pseudoalteromonas haloplanktis* strain ANT/505 and characterization of the enzymes. Extremophiles 5：35-44

Walton C，Chica R（2013）A high-throughput assay for screening L-or D-amino acid specific aminotransferase mutant libraries. Anal Biochem 441：190-198

WEF（2010）The Future of Industrial Biorefineries. World Economic Forum. REF：210610. http：//www3. weforum. org/docs/IP/2012/CH/WEF _ CollaborativeInnovation _ FutureofIndustrialBiorefineries. pdf

Yun Y，Lee Y（2004）Purification and some properties of superoxide dismutase from *Deinococcus radiophilus*，the UV-resistant bacterium. Extremophiles 8：237-242

Zhang J，Chung T，Oldenburg K（1999）A simple statistical parameter for use in evaluation and validation of high throughput screening assays. J Biomol Screen 4：67-73

Zhu Y，Zhang Z，Zhang M，Mais D，Wang M（2010）High throughput screening for bioactive com-ponents from traditional Chinese medicine. Comb Chem High T Screen 13：837-848

第十二章 来源于嗜冷和（极端）嗜热原核生物的极端脂解酶及其在工业应用中的潜力

Skander Elleuche[1,2]，Carola Schröder[1]，
Garabed Antranikian[1]

12.1 引言

非水溶性的脂类和脂肪组成了地球上绝大部分天然生物质能。脂解酶（脂肪酶和酯酶）是一种自催化三酰甘油水解酶，从三酰甘油中催化释放游离脂肪酸和甘油、单酰甘油以及二酰甘油（图 12.1a）。

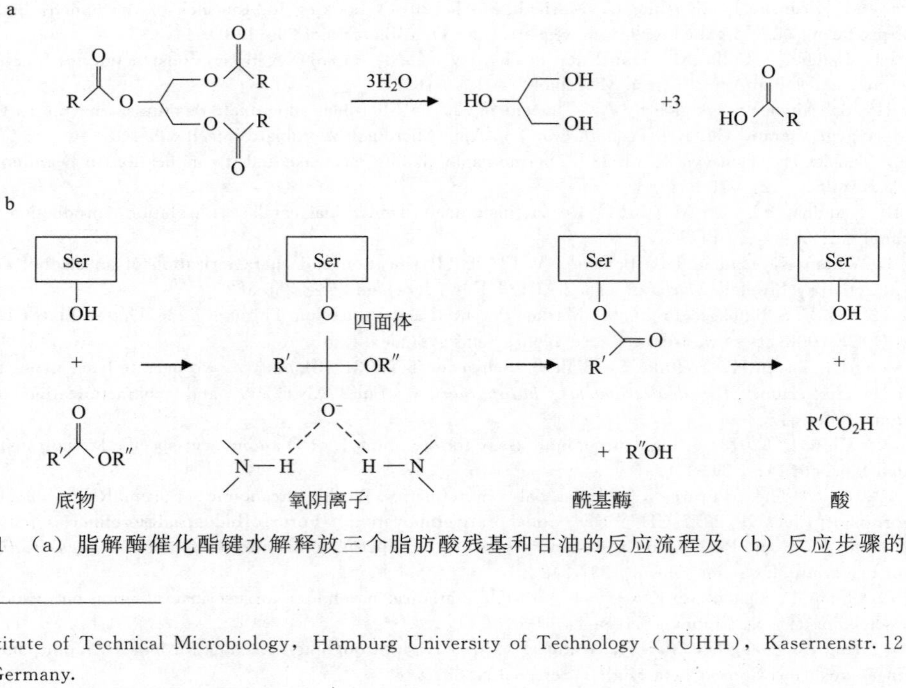

图 12.1 （a）脂解酶催化酯键水解释放三个脂肪酸残基和甘油的反应流程及（b）反应步骤的说明

[1] Institute of Technical Microbiology，Hamburg University of Technology（TUHH），Kasernenstr. 12，D-21073 Hamburg，Germany.

[2] e-mail：skander. elleuche@tuhh. de.

脂肪酶（EC 3.1.1.3）更易催化非水溶性的长链酰酯（多于 10 个碳原子），而酯酶（EC 3.1.1.1）则催化水溶性短链脂肪酸（少于 10 个碳原子）的水解（图 12.2）（Bornscheuer 2002）。在文献中，这类酶大部分都根据其偏好的底物而命名为“酯酶”或“脂肪酶”。作者更倾向于根据偏好酶解中等长度链（10 个碳原子）为界限区分酯酶或脂肪酶。目前没有统一标准认为底物链多长就归类为脂肪酶或者酯酶（Fuciños et al. 2011，2014；Chow et al. 2012）。因此，基于底物链长度的差异，脂肪酶和酯酶甚至可以通过结构改造相互转化。

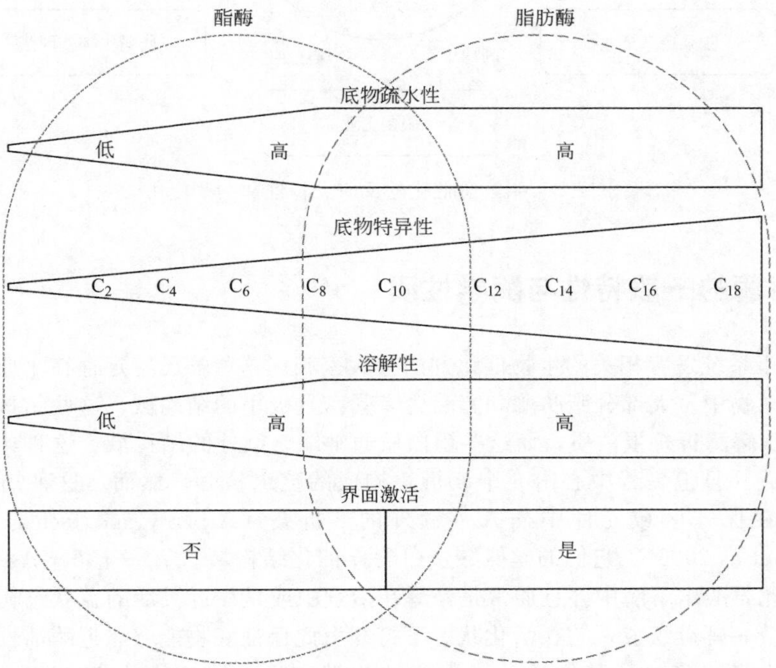

图 12.2　酯酶、脂肪酶在特性上的重叠，无法被明确分配（修改自 Bornscheuer 2012）

脂肪酶是天然疏水的，在它的活性位点区域存在着巨大的疏水表面。脂肪酶的另一个特点是它的活性界面，使得酶对于存在于水-胶束界面中的底物具有相对于溶解在液体中的底物具有更高的活性。此外，脂肪酶不仅可以水解还可以修饰酯键，包括反酯化、酯化以及氨解、醇解（酯交换）和酸解（Wicka et al. 2013）。脂解酶通常具有对映体、化学和区域选择性，这使得它们优于有机化学方法，并且在多领域的应用中有着巨大的潜能。原核与真核生物内的脂肪酶的生理作用主要是动员脂肪。这些酶具有底物特异性和较广的溶剂耐受性（有机溶剂、离子溶液、非常规溶液如水-有机两相系统或微乳液有机凝胶），它们与洗涤、食品、化工、制药、农业和调味品行业息息相关。脂肪酶在利用脂类生产生物燃料中也起着主导作用（Joseph et al. 2008；Hwang et al. 2014）。如今，脂解酶被认为生物技术应用中最重要的生物催化剂之一（图 12.3）（Wicka et al. 2013），特别是来源于极端微生物的酶可以提供多种用途，包括手性化合物的有机合成，食品行业和生物医药行业的添加剂。

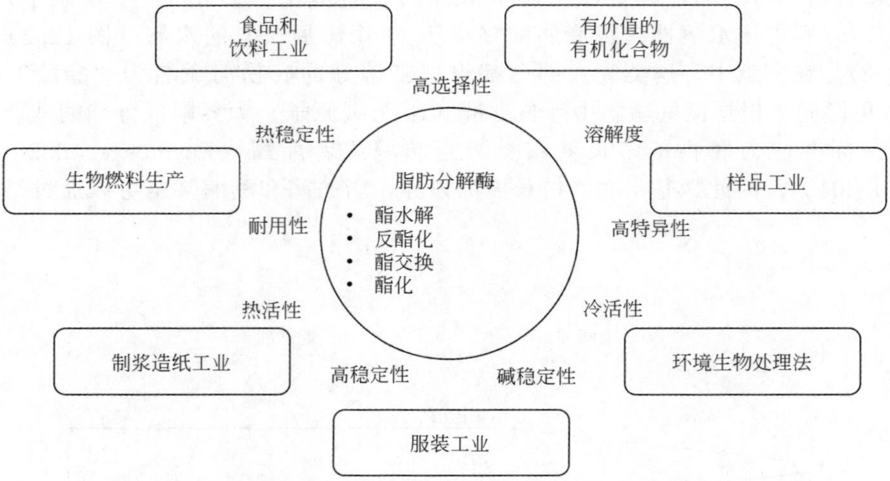

图 12.3 脂解酶催化的反应、特性和一些应用

12.2 脂解酶的一般特性与酶学性质

脂解酶属于系统发育相关的生物催化剂的一大家族。这些酶已鉴定存在于原核和真核微生物、植物以及动物中。大部分脂肪酶和酯酶含有紧密且极小的结构域，这些结构域可以被清晰地分配给 α/β 水解酶折叠蛋白质，而这些蛋白质通常缺乏额外的结构域。这种典型的折叠包含两层两亲 α-螺旋，且包裹着中心由 8 个 β-折叠构成的疏水核心。然而，这种描述最近有所改变，其中包括在保守区域元件中插入了额外的 β-折叠（Arpigny and Jaeger 1999；Shaw et al. 2002；Siew et al. 2005）。它们的丝氨酸蛋白酶样催化结构域含有一个 Ser-Asp-His 组合，广泛存在于所有丝氨酸水解酶中并且通常是暴露在溶剂中或是靠近灵活的盖状结构。这个盖状结构是由 1 或 2 个 α-螺旋组成，它在活化状态下打开由此保证底物能够靠近酶活性位点。这个盖状结构被描述为脂肪酶的独特性质，因为该结构域的运动出现在水-脂界面中（活性界面）（Bornscheuer 2002）。催化区域中的丝氨酸扮演亲核基团的角色，组氨酸作为碱性基团，天冬氨酸或谷氨酸作为酸性残基（Bornscheuer 2002）。此外，丝氨酸嵌入高度保守的五肽 Gly-X-Ser-X-Gly 当中，这属于典型的 α/β 水解酶超家族（Bornscheuer 2002）。

1999 年提出了一种脂解酶分类方式，该方法是基于脂解酶氨基酸序列对比以及一些生物学特性进行的（Arpigny and Jaeger 1999）。根据这个方法可将脂解酶分为 8 个家族（Ⅰ-Ⅷ）。家族Ⅰ包括真正的脂肪酶，催化性丝氨酸嵌入到高度保守基序 Gly-His-Ser-X-Gly 中。该家族又分为 6 个亚族（I.1-I.6），I.3 亚族包括来源于 *Pseudomonas* sp. 和其他菌属的酶，而来源于 *Geobacillus* sp. 的脂解酶则属于 I.5 亚族（图 12.4）。家族Ⅱ（GDSL 家族）中，Gly-Asp-Ser-(Leu) 代替了五肽序列。家族Ⅲ则包含来源于 *Streptomyces* 和 *Moraxella* 属的胞外脂肪酶。家族Ⅳ含有典型的具有丝氨酸的基序 Gly-Asp-Ser-Ala-Gly-(Gly)，该家族也被称为 HSL（hormone-sensitive lipase，激素敏感脂肪酶）家族，因为它与哺乳动物的 HSL 高度相似。来源于嗜冷微生物（0～30℃）、中温微生物（30～45℃）、嗜热微生物（50～80℃）和极端嗜热古菌（80～110℃）的脂解酶都归属于 HSL（图 12.4）。家族Ⅴ中的五肽被修饰为 Gly-X-Ser-(Met)-Gly-Gly，家族Ⅵ为 Gly-Phe-Ser-Gln-Gly，家族Ⅶ为 Gly-Phe-Ser-Gln-Gly-Gly，家族Ⅷ为 Ser-X-X-Lys（Arpigny and Jaeger 1999）。此外，还描述了一些不同家族的其他保守氨基酸排列。从 1999 年起该分类得以延伸，通过宏基因组测序检测编码新型脂解酶，发现了与已知的酶相似度较低的新酶（Kim et al. 2008；Nacke et al. 2011）。新提出的家族很少有具体家族名称，于是提出了将已建立的家族分类延续下去，家族Ⅸ（PhaZ7）、Ⅹ（EstD）、Ⅺ（LipG）、Ⅻ（LipEH166）、XⅢ（Est30）和XⅣ（EstA3）。这些家族的保守五肽序列分别为 Ala-His-Ser-Met-Gly、Gly-His-Ser-Leu-Gly、Gly-His-Ser-Leu-Gly-Gly、Gly-His-Ser-Leu-Gly、Gly-Leu-Ser-Leu-Gly-Gly 和 Cys-His-Ser-Met-Gly（Rao et al. 2011）。同时，家族Ⅹ被鉴定为 LipR-簇，且包含了带有 Gly-Tyr-Ser-Gly-Gly 五肽的酶（Bassegoda et al. 2012b）。另一个新家族（XⅤ）也存在含有丝氨酸的基序 Ser-His-Ser-Gln-Gly（Bayer et al. 2010）。家族XⅥ（Bacterial _ Est97）进一步扩展了分类（Lenfant et al. 2013）。此外，在来源于不同的 *Thermus* 种（生长于 50～70℃）的蛋白质序列中发现了可能代表另一个家族的五肽 Gly-Cys-Ser-Ala-Gly（Fuciños et al. 2011）。这些发表的内容反映了来源于中温微生物和极端微生物的脂解酶的数目繁多且具有多样性。

```
家族Ⅰ
Pfl   GLSGKDVVVSGHSLGGLAVNSMAD 218   YENDPVFRALDGSSF 266   VNLPTWVSHLPT 316  | 家族Ⅰ.3
Psp   GLSGKDVLVSGHSLGGLAVNSLAD 218   YENDPVFRALDGSSF 266   VNLPTWVSHLPT 316  |
Pfr   QVGAQRVNLIGHSQGALTARYVAA 94    RENDGMVGRFSSHLG 214   IRSDYPLDHLDT 236
Gza   LKRGGRIHIIAHSQGGQTARMLVS 152   LENDGIVNTVSMNGP 356   DMGTYNVDHLEI 389  | 家族Ⅰ.5
Gth   LKRGGRIHIIAHSQGGQTARMLVS 152   LENDGIVNTISMNGP 356   DMGTYNVDHLEI 389  |
                 ** *                 ***                          **

家族Ⅳ
Mse   LVYYHGGG-FVFGS 84   AVAGDSAGGNLSAVV 159   LVITAEYDPLRDQGE 251   QGMIHGFLSFY 280
Pca   VVYYHGGG-FVLGS 91   AVAGDSAGGNLAAVT 166   LVITAEYDPLRDEGE 261   NGVIHGFVNFY 290
Psp   LVFFHGGG-FVMGN 89   ALAGDSAGGNLALAV 164   TLITAEFDPLRDEGE 257   EGMIHGFISMA 286
Sac   LVFYHGGG-FVFGD 89   VVAGDSAGGNLAAVV 163   LVITAEYDPLRDEGE 255   DGMIHGFMTMP 284
Uba   VVYIHGGGPFVYGQ 70   QKTGDSAGGNLAAVV 175   VIITAELDPLRDQGE 248   NGMIHGADVIF 285
       *  **** ** *          ********            **** ***** *           ***
```

图 12.4　Ⅰ家族中Ⅰ.3 和Ⅰ.5 亚家族，以及Ⅳ家族脂解酶保守序列的部分氨基酸序列比对。Ⅰ家族的酶来源于 *Pseudomonas fluorescens*（*Pfl*，AY694785）、*Pseudomonas* sp. YY31（*Psp*，AB642679）、*Pseudomonas fragi*（*Pfr*，AJ250176）、*Geobacillus zalihae* T1（*Gza*，AY260764）和 *Geobacillus thermoleovorans* YN（*Gth*，DQ298518）。Ⅳ家族的酶来源于 *Metallosphaera sedula* DSM5348（*Mse*，YP001191160）、*Pyrobaculum calidifontis*（*Pca*，AB078331）、*Pseudomonas* sp. B11-1（*Psp*，AF034088），*Sulfolobus acidophilus* DSM10332（*Sac*，AEW03609）以及一种不可培养的细菌（Uba，EF563989）。（•）标记的为参与催化的三个氨基酸残基，(-) 标记的为氧阴离子孔形成的位点

α/β 水解酶的催化机制包括 5 个反应步骤（图 12.1b），首先结合酯类底物并由催化性丝氨酸介导亲核攻击形成具有四面体结构的中间体。在酯键被剪切前，含氧阴离子孔由两到三个氢键维持稳定。然后部分醇从酶中释放，最后酰基化的酶被水解（Joseph et al. 2008）。脂解酶在工业应用的特性中，最重要的就是其热稳定性以及最佳活性的温度范围。此外，不同的脂解酶具有多种特性，最适 pH、对阻断剂和金属离子耐受性、动力学性质、脂肪酸特异性以及位点特异性。在对适冷和适热脂解酶的研究中发现，结构修饰与酶在特定条件下的柔性有关联。在此背景下，人们猜想嗜冷菌的酶中赖氨酸残基被精氨酸残基替代，脯氨酸的数量、环状结构、二硫键和盐桥、芳香环-芳香环及疏水作用都会降低以适应寒冷环境与热不稳定性。此外，人们还推测了甘氨酸残基数目对于多肽区域的局部移动有重要的作用（Joseph et al. 2008），与整个酶的结构相比，适冷酶灵活的催化域对于热失活更加敏感，而来源于嗜热微生物和大部分中温微生物的酶通常在蛋白质变性温度附近显示出高活性（Feller and Gerday 2003；Wi et al. 2014）。

12.3 嗜冷微生物的脂解酶

嗜冷微生物生存在寒冷的环境中，如海冰、深海和雪域。与来源于中温微生物和嗜热微生物相比，这些微生物的酶功能在寒冷温度中更加高效（Elleuche et al. 2014）。来源于半子囊酵母 *Candida antarctica*，具有高度对映选择性的 CalB 是研究最多的嗜冷微生物脂肪酶。CalB 只以全酶显示，因为它来源于真核细胞并且是一个重要的脂肪酶模型（Joseph et al. 2008）。

12.3.1 适冷酯酶与脂肪酶的多样性

尽管地球生物圈中大部分环境都是寒冷的并且栖息着嗜冷微生物，但只有一些适冷酯酶和脂肪酶从生长在永冻海冰、冻土、冰川、深海或山区的细菌中分离出来（Al Khudary et al. 2010；Wu et al. 2013；Wi et al. 2014；Parra et al. 2015；Tchigvintsev et al. 2015）。其他适冷细菌的脂解酶都是从食物样品分离，如 *Serratia marcescens* 脂肪酶（Abdou 2003）。

大部分适冷脂解酶的最佳活性在 30℃左右，并且在超过 45℃时会迅速失活。但也有一些例外，例如来源于阿拉斯加嗜冷细菌 *Pseudomonas* sp. B11-1 的一种脂肪酶，它的最适温度在 45℃（Choo et al. 1998）。此外，这种酶通常能够在水的凝固点附近仍保持活性。从被石油污染的土壤的宏基因组文库中得到的一种适冷酯酶在－5℃下仍具有活性（Elend et al. 2007）。大部分适冷脂解酶都从天然宿主中纯化和鉴别出来，而有少量报道它们可以通过异源表达编码脂解酶的基因获得（表 12.1）。为了提供大量的适冷脂解酶，人们已经可以通过大肠杆菌和其他中温异源宿主来产生重组酶（Feller et al. 1996；Trincone 2011）。现代 DNA 重组技术提供了提高同工酶产量的工具，但这些同工酶只能以低产率来进行生产纯化。克隆适冷微生物编码脂肪酶的基因在大肠杆菌中表达已经在 20 世纪 80 年代末和 90 年代初被报道出来。*P. fragi* 和 *Moraxella* TA144 的脂肪酶已经可以被克隆和测序（Aoyama et al. 1988；Feller et al. 1991）。与野生型的酶比较表明，这种适冷脂肪酶在重组蛋白质中仍保存着主要的特性（Feller et al. 1991）。目前已显示寒冷的生态环境是多种适冷酶的丰富来源。为了从寒冷环境中鉴定新型的脂解酶，建立了基于活性的筛选方法，起初需要用含脂的物料诱导基因表达，如橄榄油或豆油，三油酸甘油酯、三丁酸甘油酯或三辛酸甘油酯（Henne et al. 2000；Joseph et al. 2008；Wi et al. 2014）。最近的研究中，有 23 个独特的活性克隆体从 274000 个基因文库的克隆中筛选分离出来，该基因库是采集自巴伦支海（俄罗斯）以及地中海（意大利）中温度在 3～15℃之间的采样点的海水样品制备而成。其中 5 种高活性羧酸酯酶的催化特性被检测出来，它们显示出广泛的生化多功能性，包括最佳温度为 15℃或 30℃，底物偏好分别为对硝基苯酚-戊酸酯（C_5），对硝基苯酚-丙酸酯（C_3）或 α-萘基丙酸酯（C_3），以及对硝基苯酚-辛酸酯（C_8）（Tchigvintsev et al. 2015）。最近有不同的策略应用于鉴定南极海水中编码脂肪酶的基因，人们设计了一致-简并杂交寡核苷酸引物用于扩增基因组 DNA 中的新型基因，它可以在 15～25℃之间鉴定底物偏好为对硝基苯酚-己酸酯（C_{10}）的典型适冷脂肪酶（Parra et al. 2015）。

表 12.1 来源于嗜冷细菌的重组脂解酶

分类	来源	最适温度/℃	最适pH	分子质量/kDa	偏好底物	参考文献
酯酶	不可培养细菌(Lipo1)	10	7.5	35.6	*p*NP-丁酸(C_4)	(Roh and Villatte 2008)
	Bacillus pumilus ArcL5(BpL5)	20	9.0	19.2	*p*NP-辛酸(C_8)	(Wi et al. 2014)

续表

分类	来源	最适温度/℃	最适pH	分子质量/kDa	偏好底物	参考文献
酯酶	不可培养细菌(CHA2)	20	11	34.7	*p*NP-丙酸酯(C_3)	(Hu et al. 2012)
	Pseudoalteromonas sp. 643A(EstA)	20	7.5	23.0	*p*NP-丁酸酯(C_4)	(Dlugolecka et al. 2009)
	Pseudoalteromonas arctica (EstO)	25	7.5	44.1	*p*NP-丁酸酯(C_4)	(Al Khudary et al. 2010)
	Psychrobacter pacificensis (Est10)	25	7.5	24.6	*p*NP-丁酸酯(C_4)	(Wu et al. 2013)
	Pseudomonas fragi (rPFL)	29	8.0	32.0	三丁酸(C_4)	(Alquati et al. 2002)
	Pseudomonas fluorescens (LipA)	30	9.5	42.0	(S)-酮洛芬乙基酯(C_3)	(Choi et al. 2003)
	Rhodococcus sp. (RhLip)	30	11.0	38.0	*p*NP-丁酸酯(C_4)	(De Santi et al. 2014)
	不可培养细菌(LipA)	35	8.0	32.2	*p*NP-丁酸酯(C_4)	(Couto et al. 2010)
	Pseudomonas sp. B11-1 (LipP)	45	8.0	33.7	*p*NP-丁酸酯(C_4)	(Choo et al. 1998)
脂肪酶	*Pseudomonas fluorescens* (LipB68)	20	8.0	50.2	*p*NP-醛酸(C_{10})	(Luo et al. 2006)
	Photobacterium lipolyticum M37(M37 脂肪酶)	25	8.0	38.0	*p*NP-醛酸(C_{10})	(Ryu et al. 2006)
	不可培养细菌(rEML1)	25	8.0	33.6	月桂酸甘油酯(C_{12})	(Jeon et al. 2009)
	Colwellia psychrerythraea 34H(CpsLip)	25	7.0	35.0	*p*NP-月桂酸酯(C_{12})	(Do et al. 2013)
	Pseudomonas sp. YY31 (LipYY31)	25～30	8.0	49.5	*p*NP-醛酸(C_{10})	(Yamashiro et al. 2013)
	Pseudomonas sp. TK-3(LipTK-3)	25～30	8.0	50.1	*p*NP-醛酸(C_{10})	(Tanaka et al. 2012)
	Psychrobacter sp. 7195 (LipA1)	30	9.0	35.2	*p*NP-肉豆蔻酸(C_{14})	(Zhang et al. 2007)
	Pseudomonas sp. 7323 (LipA)	30	9.0	64.4	*p*NP-醛酸(C_{10})	(Zhang and Zeng 2008)
	不可培养细菌 (LipCE)	30	7.0	53.2	*p*NP-醛酸(C_{10})	(Elend et al. 2007)
	Pseudomonas sp. KB700A(KB-Lip)	35	8.0～8.5	49.9	*p*NP-醛酸(C_{10})	(Rashid et al. 2001)

目前最适酶活温度最低的一种酯酶（Lipo1）是来源于一种不可培养的微生物，最适温度在10℃（Roh and Villatte 2008）。此外，Lipo1在10℃中比40℃中的稳定性更高，因此它在有机化学应用中具有极大的应用潜力。在墨西哥湾沉积物中分离得到的深海嗜冷菌*Psychrobacter pacificensis*，对其基因组文库进行筛选，显示它存在适冷和耐盐的酯酶。Est10的最佳活性在25℃，在0℃下可保持55%的活性。此外，在高浓度盐（2～5mol/L NaCl）条件下它的催化活

性和稳定性甚至有所升高。作为一种典型的酯酶，该酶偏好以对硝基苯酚-丁酸酯为底物（Wu et al. 2013）。除了来源于真正的适冷细菌的脂解酶以外，有很多中温微生物也存在低温活性酶，包括来源于 *Bacillus* 属的几种脂肪酶，如 *Bacillus* sp. HH-01（Kamijo et al. 2011）。还有一个例子是来源于 *Streptomyces coelicolor* A3（2）的适冷酯酶，基因组挖掘显示这株细菌有至少 50 个基因可能编码脂解酶。EstC 偏好以对硝基苯酚-戊酸酯为底物，它的最适活性温度为 35℃。同时，该水解酶在 10℃中孵育时仍保持 25%的活力（Brault et al. 2012）。

12.3.2 通过遗传工程改良嗜冷菌的酶学性质

冷活性酶可通过分子生物学技术（包括定向进化和合理的蛋白质设计）来达到提高热鲁棒性（robustness，又称稳健性）或效率的目标（Bassegoda et al. 2012a；Joshi and Satyanarayana 2015）。适冷酶天然不耐热，非常适合于序列以及结构修饰。对 CalB 进行定点突变造成单氨基酸突变（T103G）可以增加其在高温下的稳定性，但同时也会使得酶活下降 50%（Patkar et al. 1998）。来源于北极嗜冷菌 *Bacillus pumilus* ArcL5 的脂解酶 BpL5 最适活性在 20℃，并在 5℃下保持约 85%的活性。基于结构模型预测定点突变丝氨酸可显著提升酶对对硝基苯酚-辛酸酯和三辛酸甘油酯的催化活性，同时不会影响它的最适 pH 和温度（Wi et al. 2014）。另外一个方法是利用定向进化增强 *Pseudomonas aeruginosa* LST-03 的适冷脂肪酶 Lip9 在有机溶剂中的稳定性，该微生物自身对有机溶剂有耐受性，并且可分泌高度稳定的酶，该酶已在大肠杆菌中重组表达。在第一个方法中，Lip9 是从包涵体中纯化所得，并在同种细菌的天然脂肪酶特异性折叠酶的帮助下进行重新折叠（Ogino et al. 2007）。使用二甲基亚砜（除了底物 tri-*n*-butyrin）评价酶抵抗有机溶剂的稳定性的增强程度，鉴定出来的候选酶将进行测序以及使用不同的有机溶剂进行测试。在环己烷和正癸烷存在的情况下，通过这种方法鉴定出突变酶 LST-03-R65 的稳定性提高了 9 到 11 倍。该酶在蛋白质表面上累积氨基酸突变，推测这些氨基酸的替换可能有助于阻止有机溶剂进入蛋白质内部（Kawata and Ogino 2009）。研究者在提升酶的热稳定性方面开展了大量的工作，因为热变性是工业加工条件中酶失活的主要原因。一致认为突变能够改变酶在有机溶剂中的稳定性，蛋白质表面的氨基酸残基同样也显示与热稳定性有密切联系。一种来源于 *Bacillus subtilis* 的脂肪酶替换 9 个氨基酸后，除了最适温度改变了 20℃外，溶解温度也上升了 15℃（Ahmad et al. 2008）。嗜冷菌 *Pseudomonas fragi* 产生了一种适冷脂肪酶 PFL，它的最适温度为 29℃（Alquati et al. 2002）。PFL 甚至在 10℃下仍保持高度稳定性与活性，但在温和的温度中短时间孵育，如在 42℃中孵育 19min，会完全失活。易错 PCR 技术开发出了一种含有四个氨基酸替换的酶的变异体，这使得它在 42℃中的半衰期提高了 5 倍，最适温度变化了 10℃（Gatti-Lafranconi et al. 2008）。另一种预测稳定突变的方法是对比嗜冷微生物与同源中温微生物或嗜热微生物的脂解酶（Bassegoda et al. 2012a）。例如来源于北极嗜冷菌 *P. arctica* 的多结构域羧酸酯酶 EstO，它通过缺失非催化结构域 OsmC 可提升 α/β 水解酶功能活性折叠区域的热稳定性。对一级和二级结构的详细调查结合表征实验显示 OsmC 结构域很可能对酶的柔性与低温适应性十分重要（Al Khudary et al. 2010；Elleuche et al. 2011）。这种基因修饰后的候选酶在多个工业流程中对生物技术应用具有吸引力（图 12.3）。

12.3.3 适冷脂解酶的应用

罕见的特异性以及低温下的高活性使得适冷脂解酶在多个工业领域具有开发和应用的潜力。来源于嗜冷微生物的脂肪酶和酯酶主要用于不稳定手性化合物的有机合成，也同样用作食品工业中（奶酪制造和面包业）的添加剂、环境生物转化（生物治理）中的去污剂（节能，冷洗）、生物医药（制药）或分子生物学方法（Joseph et al. 2008；Joshi and Satyanarayana 2015）。此外，脂肪酶在嗜冷宿主菌中还可以防止包涵体的形成（Feller et al. 1996）。目前，低特异活性和再循环能力差以及无法获得纯品，因此从经济可行性来看，冷活性脂肪酶在工业应用中受到了一定程度的阻碍。然而，也有几个适冷脂解酶用于工业加工的过程，或者已申请了各种应用的专利以及正在调查其工业生物技术的潜力的例子。这种应用不仅使用了更低的温度从而节省能源，还减少使用对环境有害的化学品和洗涤剂，如与传统用酸和氧化剂相比，材料的酶促退浆更加环保。另外，非水系溶剂可以在脂解酶催化生产高价值化合物时使用，在油脂修饰和糖基聚合物或生产表面活性剂、乳化剂、结构脂类、蜡酯或芳香化合物中都已使用到（Joseph et al. 2008）。工业中使用的低温活性原型酶 CalB 来源于半子囊酵母 *C. antarctica*，该酶已被申请专利应用于食品、化学和制药。有趣的是，当该酶固定在固相载体时，其在非水系溶液中的热稳定性很高（Koops et al. 1999）。

原则上，脂肪酶作为洗衣房和家用洗碗机中的去污剂成分是最为成功的。脂解酶适用于去污剂，它在碱性条件下（pH 10～11）以及适当的温度下仍保持活性和稳定（Wicka et al. 2013）。Novozyme 在 1998 年开发了第一个遗传工程改造的工业酶——Lipolase®，其来源于真菌 *Thermomyces lanuginose*。该酶适用于洗衣业帮助去除衣物上的污渍，如油脂、黄油、口红、酱油和色拉油。此外，该酶经过工程改造后，Novozyme 开发出第二代和第三代突变体，同样在市场上存在于不同的洗涤剂配方史。另外一个著名的例子是一种洗涤酶 Lipomax®，它是一种 *Pseudomonas alcaligenes* 脂肪酶的工程突变体，并在 1995 年由 Gist-Brocades 开发（Joshi and Satyanarayana 2015）。脂解酶也可作为漂白液隐形眼镜清洗、液体皮革清洗剂的添加成分（Joseph et al. 2008）。

嗜冷脂解酶同样可用于食品工业中需要低温进行的反应当中，为此可以避免原料与风味组分的污染和变化。尤其是来源于霉菌的脂肪酶也已用于改进食品制造中的传统化学工艺。原核生物来源的脂解酶也被认为具有增加食品香气与风味的潜力，如来源于嗜冷菌 *Pseudomonas* strain P38 的脂肪酶可在低温有机溶剂中催化风味酯类的合成（Tan et al. 1996）。此外，对于从动植物脂类中富集多种不饱和脂肪酸以及改变或替代甘油骨架上的脂肪酸的位置，脂解酶同样发挥着十分重要的作用（Wicka et al. 2013）。

低温活性脂肪酶也在通过酯交换反应制备生物柴油上显示出巨大潜力。一种来源于嗜冷菌 *Pseudomonas fluorescens* B68 的酶 LipB68 在低温下具有高活性，并且在 20℃豆油中孵育 120 h 后生产出 92%的生物柴油。在低温下进行该工艺可节省大量的能源消耗（Luo et al. 2006）。

由于制药工业底物混杂以及要在多种有机溶剂中催化具有区域选择性的反应，脂解酶对于制药工业同样具有吸引力。与有机化学方法相比，酶催化工艺更加环保，并且在生产最佳活性的手性药物方面具有高度选择性（图 12.5），而非外消旋手性构建的化学物质是精细化工行业中有价值的化合物。最普遍的抗炎

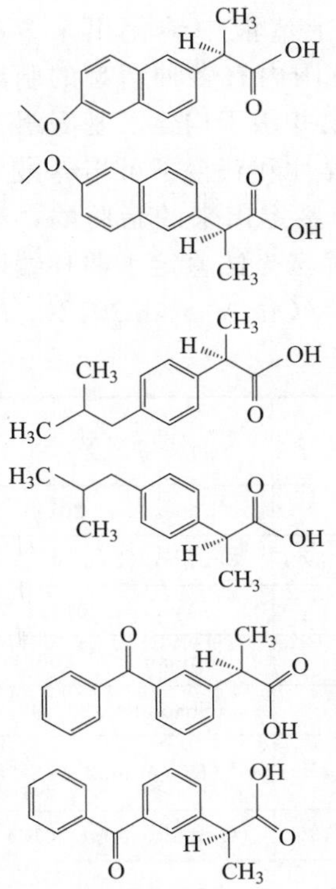

R-甲氧萘丙酸

S-甲氧萘丙酸

R-异丁苯丙酸

S-异丁苯丙酸

R-苯酮苯丙酸

S-苯酮苯丙酸

图 12.5　通过脂解酶催化的酯化反应生产的手性药物的分子结构

药物之一是酮洛芬［(R，S)-2-(3-苯甲酰基苯基）丙酸］，其中S-酮洛芬具有药理活性（de O. Carvalho et al. 2006）。其他用于对抗由脂解酶加工的人类疾病的非甾体抗炎药物的实例是萘普生和布洛芬（Hess et al. 2008）。此外，来源于*S. marcescensis*的脂肪酶可被用于生产钙通道阻断剂药物——地尔硫卓（Wicka et al. 2013）。尽管低温活性脂解酶蛋白质工程已经能够稳定提升酶的特性，但它的工业应用与高温活性酶相比并没有得到快速发展（Joseph et al. 2008）。

12.4 来源于（极端）嗜热微生物的脂解酶

嗜热微生物生长在50℃到80℃之间，主要成员有细菌和古菌，只有一些属于中温嗜热的真核生物（生存在60℃以下）被鉴定出来。此外古菌占据着极端微生物群体的绝大部分，它们甚至能适应超过110℃的高温条件。嗜热微生物栖息的典型环境包括热泉、热液喷口和火山湖。尽管热稳定酶在中温微生物和嗜热微生物中都有被发现，但嗜热微生物所产生的酶相对更加强力，因此特别适用于工业（Hasan et al. 2005）。

12.4.1 热活性脂肪酶和酯酶的多样性

最近对来源于嗜热微生物的酶的需求与关注不断上升，使得大量热稳定脂肪酶被鉴别出来。表12.2重点描述了来源于嗜热细菌的重组脂解酶。这些酶在55℃到80℃之间有最佳的催化效果，分子质量位于19kDa到55kDa之间。在已鉴别的重组酯酶中最适温度最高的是来源于嗜热细菌*Thermus scotoductus*的酶，其最适温度高达80℃（du Plessis et al. 2010）。一个最适温度为78℃的重组脂肪酶分离自*Fervidobacterium changbaicum*（Cai et al. 2011）。两种脂肪酶直接从*Thermosyntropha lipolytica* DSM1103的培养上清液中纯化出来，在96℃中的活性最高（Salameh and Wiegel 2007）。在宏基因组方法范围内有多种新型的脂解酶被发现。存在于环境样品中的嗜热微生物能够在蒸馏的温度下富集并用于构建宏基因组文库。宏基因组能够直接从合适的高温环境中分离出来，用于挖掘多样性和获得不可培养微生物。目前仅有的与已知蛋白质相似度很低的酶是从土壤宏基因组中鉴定出来的脂肪酶，有32%～45%的可能性为α/β水解酶（Choi et al. 2013）。另一个功能筛选方法鉴定了两种编码脂肪酶的基因，这两种脂肪酶的最适活性的温度分别为70℃和75℃（Chow et al. 2012）。

表12.2 来源于嗜热细菌的重组脂解酶

分类	来源	最适温度/℃	最适pH值	分子质量/kDa	偏好底物	参考文献
酯酶	*Acidicaldus* USBA-GBX-499	55	9.0	34	*p*NP-己酸酯(C_6)	(Lopez et al. 2014)
	Thermus thermophiles HB27	58.2	6.3	37.7	*p*NP-己酸脂(C_{10})	(Fuciños et al. 2014)
	Anoxybacillus sp. PDF1	60	8.0	26	*p*NP-丁酸酯(C_4)	(Ay et al. 2011)
	Geobacillus thermoleovorans YN	60～65	9.5	29	*p*NP-乙酯(C_2)	(Soliman et al. 2007)
	Metagenome(活性污泥)	70	8.5	53	*p*NP-乙酯(C_2)	(Shao et al. 2013)
	Thermoanaerobacter tengcongensis MB4	70	9.5	50	*p*NP-乙酸酯(C_6)	(Rao et al. 2011)
	Thermosyntropha lipolytica	70	8.0	—	*p*NP-丁酸酯(C_4)	(Gumerov et al. 2012)
	Thermotoga maritime	70	5.0～5.5	35.2	*p*NP-丁酸酯(C_4)	(Tao et al. 2013)
	Thermus scotoductus SA-01	80	7.0	28.6	*p*NP-丁酸酯(C_4)	(du Plessis et al. 2010)

续表

分类	来源	最适温度/℃	最适pH值	分子质量/kDa	偏好底物	参考文献
脂肪酶	*Geobacillus* sp. EPT9	55	8.5	44.8	*p*NP-棕榈酸酯(C_{16})	(Zhu et al. 2015)
	Bacillusthermoleovorans ID-1	60	8.0～9.0	43	*p*NP-醛酸(C_{10})	(Lee et al. 2001)
	Bacillus thermoleovorans ID-1	60～65	9.0	19	*p*NP-辛酸酯(C_8)	(Lee et al. 2001)
	Geobacillus sp. strain T1	70	9.0	43	月桂酸甘油酯(C_{12})	(Leow et al. 2007)
	Geobacillus thermoleovorans YN	70	9.5	43	*p*NP-醛(C_{10})	(Soliman et al. 2007)
	富集培养(土壤和水样)	70	8.0	31.7	*p*NP-辛酸酯(C_8)	(Chow et al. 2012)
	富集培养(土壤和水样)	75	8.0	36	*p*NP-醛酸(C_{10})	(Chow et al. 2012)
	Thermoanaerobacter thermohydrosulfuricus SOL1	75	8.0	34.2	*p*NP-醛酸(C_{10})	(Royter et al. 2009)
	Caldanaerobacter subterraneus subsp. *tengcongensis*	75	7.0	32.1	*p*NP-醛酸(C_{10})	(Royter et al. 2009)
	Fervidobacterium changbaicum	78	7.8	33	*p*NP-醛酸(C_{10})	(Cai et al. 2011)

大部分已鉴别的来源于嗜热细菌的羧酸酯水解酶都被认定为酯酶。脂解酶根据底物范围通常可归类为酯酶或脂肪酶。对于人工合成底物上所连接的脂肪酸链的长度而言，有些脂解酶同时显示出酯酶和脂肪酶的活性（表 12.2）。来自 *Thermosynthropha lipolytica* 的 TSLip1 在 70℃下催化对硝基苯酚-丁酸酯的活性最高，对 C_{10} 底物活性最弱（<10%），当底物换为对硝基苯酚-棕榈酸酯（C_{16}）底物时活性又上升（Gumerov et al. 2012）。此外，多种植物油，如豆油、橄榄油、玉米油和葵花油都能被水解。据报道 *Geobacillus* sp. T1 的重组脂肪酶在 70℃下同样具有广泛的作用底物。多种含有 2～18 个碳的脂肪酸链的天然油类和三酰甘油底物都能被水解（Leow et al. 2007）。酶的稳定性、最高活性、底物特异性都能通过点突变被改变。如一种脂肪酶突变体（N355K）对对硝基苯酚-棕榈酸酯（C_{16}）具有最高的活性，而野生型的酶则对月桂酸甘油酯（C_{12}）活性最高，对 C_{16} 酯活性只有约 15%（Sharma et al. 2014）。一种重组的古菌羧酸酯酶对对硝基苯酚-辛酸酯（C_8）活性最高而对对硝基苯酚-丁酸酯（C_4）则没有活性，该酶在 C 末端带有 His 标签时也显示出类似的结果，而当克隆基因带有 N 端 His 标签编码区域时，该酶具有了更广泛的作用底物，包括对硝基苯酚-丁酸酯（C_4）和对硝基苯酚-辛酸酯（C_8）。由于底物结合口袋位于 N 端，因此连接标签会导致某些改变（Killens-Cade et al. 2014）。与天然的酶相比重组蛋白质可能会导致该酶特性变化（Leow et al. 2007）。

一般而言，大部分细菌脂解酶都在中性或碱性的 pH 中活性最高，尽管它们具有广谱的 pH 值范围（Gupta et al. 2004）。据报道许多热稳定脂解酶都只有在碱性 pH 值中才具有活性和保持稳定。从非极端生境的宏基因组中分离的酯酶和脂肪酶通常也偏向于碱性的条件（Choi et al. 2013）。比如，来源于嗜热嗜酸性细菌 *Acidicaldus* sp. 的酯酶在 pH 9.0 中活性最高，而它在 pH 7.0 中只有不到 20%的活性。文献报道只有两种热稳定酶在弱酸性条件下活性最佳，包括 *Thermus thermophilus* 的酯酶（最适 pH 值为 6.3）以及 *Thermotoga maritima* 的酯酶（最适 pH 值为 5.0～5.5）（Tao et al. 2013；Fuciños et al. 2014）。后者是首次从嗜热细菌中分离的耐酸酯酶。

12.4.2 极端嗜热古菌作为脂解酶的来源

嗜热有机体生长在50～80℃中，而极端嗜热菌则生长在80～113℃中。古菌就是其中的代表（Egorova and Antranikian 2005）。据称，来源于古菌的热活性酶的最适温度通常超过来源于细菌的热活性酶。此外，它们与细菌的蛋白质相比本身具有很高的热稳定性与化学稳定性（Levisson et al. 2009）。一些已生产的或鉴别的重组古菌脂解酶如表12.3所示。来源于 *Pyrococcus furiosus* 的重组酯酶的最适温度最高可达100℃，它在100℃下的半衰期为34 h（Ikeda and Clark 1998）。*Sulfolobus solfataricus* P1的酶在80℃中活性最高，在该温度下孵育5d后仍保持41%的活性（Park et al. 2006）。来源于 *Metallosphaera sedula* 的酯酶在95℃下对 C_8 底物的活性最高，而在37℃下表现出广泛的催化活性，而且在90℃下孵育6h后仍保持超过70%的活性（Killens-Cade et al. 2014）。*Sulfolobus acidophilum* 的酯酶最适温度为70℃，相对而言比较低，在90℃下残余的活性超过20%（Zhang et al. 2014）。*Pyrobaculum* 以及 *Aeropyrum pernix* 的酶在80℃到90℃之间活性最高（Hotta et al. 2002；Gao et al. 2003；Shao et al. 2014）。所有的酯酶在中性或弱碱性pH范围内有活性，除了有一个酯酶在pH 10～11之间有活性（表12.3），该酶甚至在pH 12.0中仍具有活性，它是碱性最高的水解酶（Rusnak et al. 2005）。目前仍没找到极端嗜热微生物来源的最适pH低于7.0的重组酯酶。据报道一种来源于极端嗜热嗜酸古菌 *Picrophilus torridus*（生长在60℃和pH 1.0～2.0）的酯酶显示在pH 6.5和70℃中具有最高活性（Hess et al. 2008）。然而，有一个例外被报道出来，来源于 *Ferroplasma acidophilum* 的酯酶最适条件为pH 2.0和50℃，它在pH 5.0下不具有活性，适合生长在温和的温度和pH 1.7的环境下（Golyshina et al. 2006）。

在目前的文献报道中，人们还没有在极端嗜热古菌中找到一种脂肪酶格外偏好于长链的脂肪酸底物（>C_8）。在表12.3中列出的一些脂肪酶在某些程度上具有水解长链底物的能力。来源于 *Pyrococcus furiosus* 的酶对对硝基苯酚-棕榈酸酯（C_{16}）具有15%的残余活性（Alqueres et al. 2011）。*Aeropyrum pernix* 的酶水解 C_{18} 酯的活性为19%而对对硝基苯酚-辛酸酯（C_8）的活性为100%（Gao et al. 2003）。来源于 *Archaeglobus fulgidus* 的脂解酶AFL含有一个小的盖子结构域和一个额外的C末端脂结合结构域（Chen et al. 2009）。AFL先前被鉴别和认定为一种酯酶，是由于它对对硝基苯酚-乙酸酯比棕榈酸酯具有更高的活性（Rusnak et al. 2005）。报道显示AFL的奇特的C末端结构域是结合长链底物所必需的，为此该酶由于这个结构发现随后被归类为真正的脂肪酶（Chen et al. 2009）。通过脂解酶的结构解析，酶的分类比人工底物偏好性评价更加可靠。然而，最近从中温古菌 *Haloarcula* sp. G41中鉴定出了一种对长链底物对硝基苯酚-肉豆蔻酸酯（C_{14}）具有偏好性的脂肪酶，该酶在70℃，pH 8.0的条件下具有最佳活性（Li and Yu 2014）。这意味着古菌作为新型独特脂肪酶的来源有着巨大的潜力。

表12.3 来源于超嗜热古菌的重组脂解酶

来源	最适温度/℃	最适pH	分子质量/kDa	偏好底物	参考文献
Archaeoglobus fulgidus DSM 4304	70	10.0～11.0	52.8	丁酸甲酯(C_4)/对硝基苯酚-乙酯(C_2)	(Rusnak et al. 2005)

续表

来源	最适温度/℃	最适pH	分子质量/kDa	偏好底物	参考文献
Sulfolobus acidophilus DSM10332	70	8.0	34.1	对硝基苯酚-丁酸酯(C_4)	(Zhang et al. 2014)
Pyrobaculum sp. 1860	80	9.0	23	对硝基苯酚-乙酸酯(C_2)	(Shao et al. 2014)
Pyrococcus furiosus	80	7.0	26.5	庚酸 4-甲基伞形醇酯(C_7)	(Alqueres et al. 2011)
Archaeoglobus fulgidus DSM 4304	80	7.0～8.0	27.5	对硝基苯酚-丁酸酯(C_4)	(Kim et al. 2008)
Thermococcus kodakarensis KOD1	85	8.0	29	对硝基苯酚-丁酸酯(C_4)	(Cui et al. 2012)
Sulfolobus solfataricus P1	80	8.0	34	对硝基苯酚-辛酸酯(C_8)	(Park et al. 2006)
Sulfolobus solfataricus	90	6.5～7.0	32	对硝基苯酚-戊酸酯(C_5)	(Morana et al. 2002)
Aeropyrum pernix K1	90	8.0	63	对硝基苯酚-辛酸酯(C_8)	(Gao et al. 2003)
Pyrobaculum calidifontis VA1	90	7.0	34	对硝基苯酚-己酸酯(C_6)	(Hotta et al. 2002)
Metallosphaera sedula DSM 5348	95	7.0	33.2	对硝基苯酚-辛酸酯(C_8)	(Killens-Cade et al. 2014)
Pyrococcus furiosus	100	7.6	—	乙酸 4-甲基伞形醇酯(C_2)	(Ikeda and Clark 1998)

12.4.3　酶工程

蛋白质的热变性会由于更高的刚性或紧凑的结构等特点得以降低。然而，更高的刚性伴随着更差的柔性，被认为会降低酶的活性，但这种假设最近被驳回了。人们鉴定出一种脂肪酶突变体，该酶比野生型的活性区域刚性更强，但同时活性与稳定性都得到了提升（Kamal et al. 2012）。不同的因素似乎对热活性酶的稳定性有不同的作用，如二硫键的数量、疏水作用、氢键和金属结合（Vieille and Zeikus 2001）。对一个热稳定脂肪酶的单个氨基酸进行替换（E315G）后显示，如果残基靠近活性位点，将会导致热稳定性下降。与此相反的是N355K 突变体能够使环的柔性变强，N355K 突变体通过形成额外的氢键增强了热稳定性(Sharma et al. 2014)。*Geobacillus* sp. EPT9 产生的脂肪酶含有一个锌结合域和比中温 *Bacillus* 菌株更高比例的脯氨酸与精氨酸（Zhu et al. 2015）。脯氨酸是最坚固的氨基酸，而精氨酸则参与多种非共价相互作用。破坏 *Geobacillus stearothermophilus* 脂肪酶 L1 的锌结合位点可导致酶最佳活性温度由 60℃变成 45～50℃，并且特异活性与热稳定性都会降低(Choi et al. 2005)。因此，锌结合域可能参与活性位点结构的构建。目前并不能确定在高温下活性和稳定性的一个普遍规律，然而，变换一个残基可能会导致蛋白质内相互作用的一个小小改变，构象的误差可能导致稳定性和/或酶活性的显著下降。

目前已构建了几种用于增强热稳定性的酶工程改造方法，使得突变体在较高的温度下具有最高的活性。尽管也存在一些相反的例子，如与一种从土壤宏基因组中鉴定的野生型脂肪酶相比，一个氨基酸的替换导致最适温度从 50℃下降到 40℃（Sharma et al. 2014）。因此，来源于嗜热微生物和极端嗜热微生物的强力酶通过工程改造后可适用于在极端条件和温和温度中进行的工艺。例如洗涤工艺热衷于在较低温度下达到节能的功效，同时在碱性的苛刻条件下要求脂肪酶足够稳定（Jaeger and Reetz 1998）。脂解酶开发和改进的最新成果之一是掺入了非标准氨基酸。用大量的非标准氨基酸替代 *Thermoanaerobacter thermohydrosulfuri-*

cus（TLL）来源的热活性脂肪酶中的氨基酸可形成一种对热不敏感的“冷洗”酶（Hoesl et al. 2011）。

12.4.4 热稳定脂肪酶和酯酶的应用

大部分工业加工过程都在高于45℃温度中进行（Sharma et al. 2002），更高的操作温度可以导致更高的酶促反应速率、底物溶解度以及更低的黏度和污染风险。此外，来自嗜热微生物和极端嗜热微生物的酶在这种苛刻的变性条件下更能保持稳定（Mozhaev 1993）。细菌和古菌酶类的高水平重组生产已经达到足够工业应用的生物催化剂所需的产量。使用脂解酶进行转化反应比化学、蒸汽或压力反应的条件更加温和。酯酶和脂肪酶的特异性可以减少或排除不希望的副产品以及不必要的分离步骤从而减低了成本。据显示T1脂肪酶在无溶剂体系中，反应温度和孵育温度是影响薄荷醇丁酸酯产量的重要原因（Wahab et al. 2014）。用重组酶进行高产率合成，不需要随后从产物中分出易燃和有毒溶剂。不过如果必须使用溶剂时，脂解酶就需要具有高耐受性。来自 *T. thermohydrosulfuricus* SOL1 和 *Caldanaerobacter subterraneus* subsp. *tengcongensis* 的脂解酶能够抵抗几种浓度高达99%的溶剂和去垢剂（Royter et al. 2009）。另外，据报道来自 *T. thermohydrosulfuricus* SOL1 的酯酶对于仲醇酯具有较高的偏好性，以及对药物相关底物的R-对映体具有较高的选择性（Royter et al. 2009）。该酶在高温下具有活性且能够抵抗有机溶剂，因此它可作为生物转化领域非水相介质应用的候选酶。来自 *Thermotoga maritima* 的酯酶 Tm1160 在水解消旋酮洛芬乙酯时显示出具有对映体选择性（Tao et al. 2013）。同样，来自 *Acidicaldus* sp. 的酯酶可以在苛刻条件下合成相关的S-对映体萘普生和布洛芬酯（图12.5）（Lopez et al. 2014）。

一个工业内酯化工艺的例子是把棕榈油转化为代可可脂，这是一种具有高熔点并且使得食品、糖果和化妆品更加诱人的添加剂物质。廉价油能够通过酶的转化升级成富有营养的结构脂肪。高值产品，如人乳脂替代品和类可可脂，在市场上有售（Hasan et al. 2005）。选择性水解三酰甘油也可用于开发风味或形成风味前体（Jaeger and Reetz 1998）。这些成分很难从本源物质中分离，因此工业提取工艺通常不划算。萜烯酯，如乙酸松油酯、香茅醇或香叶醇酯，含有短链脂肪酸，能够通过脂解酶生产用在饮料、食品和制药工业中。通过绿色途径合成昂贵成分产品比化学合成更受欢迎。此外，通过阐明脂解酶对多种底物的广泛作用，人们还发现了新的气味分子（Dhake et al. 2013）。

热稳定脂肪酶也能够用在以藻类为基础的生物燃料生产领域。通过剪切储存的三酰甘油能够提高游离脂肪酸数目。在高温中运用热稳定脂解酶把脂类转化为燃料可以节省时间和能源（Killens-Cade et al. 2014）。此外，在制浆造纸工业中脂解酶可用于在高温中清除木材中的疏水成分，如三酰甘油和蜡质（Jaeger and Reetz 1998）。

12.5 结论

由于极端脂解酶的多功能酶学性质，它们作为理想的候选酶可用于各种工业当中，包括手性分子和药物的有机合成、食品、洗涤业以及环境修复。一系列冷活性和热活性酯酶与脂肪酶已经通过现代分子生物学和微生物学方法从极端古菌和细菌当中被鉴定出来，如今这些酶已被应用于特定的工艺和工程方法当中。然而，在开发某些工业应用所需的特质生物催化剂时，仍然还有一些障碍需要解决。建立新型的和改良的技术来鉴定和工程改造编码生物催化剂的基因来大批量生产有活性的重组蛋白质已经取得了进步。最近在组学时代中的一些例子是开发和分配合适的表达宿主、定向进化、基因转移、有效的克隆方法、含非标准氨基酸的酶同系物、融合酶以及定点突变（Sharma et al. 2002；Gatti-Lafranconi et al. 2008；Al Khudary et al. 2010；Merkel et al. 2010；Hoesl et al. 2011；Bassegoda et al. 2012a；Marquardt et al. 2014；Elleuche 2015；Joshi and Satyanarayana 2015）。

参考文献

Abdou AM（2003）Purification and partial characterization of psychrotrophic *Serratia marcescens* lipase. J Dairy Sci 86：127-132

Ahmad S，Kamal MZ，Sankaranarayanan R，Rao NM（2008）Thermostable *Bacillus subtilis* lipases：in vitro evolution and structural insight. J Mol Biol 381：324-340

Al Khudary R，Venkatachalam R，Katzer M，Elleuche S，Antranikian G（2010）A cold-adapted esterase of a novel marine isolate，*Pseudoalteromonas arctica*：gene cloning，enzyme purification and characterization. Extremophiles 14：273-285

Alquati C，De Gioia L，Santarossa G，Alberghina L，Fantucci P，Lotti M（2002）The cold-active lipase of *Pseudomonas fragi*. Heterologous expression，biochemical characterization and molecular modeling. Eur J Biochem 269：3321-3328

Alqueres SM，Branco RV，Freire DM，Alves TL，Martins OB，Almeida RV（2011）Characterization of the recombinant thermostable lipase（Pf2001）from *Pyrococcus furiosus*：effects of thiore-doxin fusion tag and triton X-100. Enzyme Res 2011：316939

Aoyama S，Yoshida N，Inouye S（1988）Cloning，sequencing and expression of the lipase gene from *Pseudomonas fragi* IFO-12049 in E. coli. FEBS Lett 242：36-40

Arpigny JL，Jaeger KE（1999）Bacterial lipolytic enzymes：classification and properties. Biochem J 343（Pt 1）：177-183

Ay F，Karaoglu H，Inan K，Canakci S，Belduz AO（2011）Cloning，purification and characterization of a thermostable carboxylesterase from *Anoxybacillus* sp. PDF1. Protein Expr Purif 80：74-79

Bassegoda A，Cesarini S，Diaz P（2012a）Lipase improvement：goals and strategies. Comput Struct Biotechnol J 2，e201209005

Bassegoda A，Pastor FI，Diaz P（2012b）*Rhodococcus* sp. strain CR-53 LipR，the first member of a new bacterial lipase family（family X）displaying an unusual Y-type oxyanion hole，similar to the *Candida antarctica* lipase clan. Appl Environ Microbiol 78：1724-1732

Bayer S，Kunert A，Ballschmiter M，Greiner-Stoeffele T（2010）Indication for a new lipolytic enzyme family：isolation and characterization of two esterases from a metagenomic library. J Mol Microbiol Biotechnol 18：181-187

Bornscheuer UT（2002）Microbial carboxyl esterases：classification，properties and application in biocatalysis. FEMS Microbiol Rev 26：73-81

Brault G，Shareck F，Hurtubise Y，Lepine F，Doucet N（2012）Isolation and characterization of EstC，a new cold-active esterase from *Streptomyces coelicolor* A3（2）. PLoS ONE 7，e32041

Cai J，Xie Y，Song BK，Wang Y，Zhang Z，Feng Y（2011）*Fervidobacterium changbaicum* Lip1：identification，cloning，and characterization of the thermophilic lipase as a new member of bacterial lipase family V. Appl Microbiol Biotechnol 89：1463-1473

Chen CK，Lee GC，Ko TP，Guo RT，Huang LM，Liu HJ，Ho YF，Shaw JF，Wang AH（2009）Structure of the alkalohyperthermophilic *Archaeoglobus fulgidus* lipase contains a unique C-terminal domain essential for long-chain substrate binding. J Mol Biol 390：672-685

Choi GS，Kim JY，Kim JH，Ryu YW，Kim GJ（2003）Construction and characterization of a recom-binant esterase with high activity and enantioselectivity to（S）-ketoprofen ethyl ester. Protein Expr Purif 29：85-93

Choi WC，Kim MH，Ro HS，Ryu SR，Oh TK，Lee JK（2005）Zinc in lipase L1 from *Geobacillus stearothermophilus* L1 and structural implications on thermal stability. FEBS Lett 579：3461-3466

Choi JE，Kwon MA，Na HY，Hahm DH，Song JK（2013）Isolation and characterization of a metagenome-derived thermoalkaliphilic esterase with high stability over a broad pH range. Extremophiles 17：1013-1021

Choo DW，Kurihara T，Suzuki T，Soda K，Esaki N（1998）A cold-adapted lipase of an Alaskan psychrotroph，*Pseudomonas* sp. strain B11-1：gene cloning and enzyme purification and char-acterization. Appl Environ Microbiol 64：486-491

Chow J，Kovacic F，Dall Antonia Y，Krauss U，Fersini F，Schmeisser C，Lauinger B，Bongen P，Pietruszka J，Schmidt M，Menyes I，Bornscheuer UT，Eckstein M，Thum O，Liese A，Mueller-Dieckmann J，Jaeger KE，Streit WR（2012）The metagenome-derived enzymes LipS and LipT increase the diversity of known lipases. PLoS ONE 7，e47665

Couto GH，Glogauer A，Faoro H，Chubatsu LS，Souza EM，Pedrosa FO（2010）Isolation of a novel lipase from a metagenomic library derived from mangrove sediment from the south Brazilian coast. Genet Mol Res 9：514-523

Cui Z，Wang Y，Pham BP，Ping F，Pan H，Cheong GW，Zhang S，Jia B（2012）High level expression and characterization of a thermostable lysophospholipase from *Thermococcus kodakarensis* KOD1. Extremophiles 16：619-625

de O. Carvalho P，Contesini FJ，Bizaco R，Calafatti SA，Macedo GA（2006）Optimization of enan-tioselective resolution of racemic ibuprofen by native lipase from *Aspergillus niger*. J Ind Microbiol Biotechnol 33：713-718

De Santi C，Tedesco P，Ambrosino L，Altermark B，Willassen NP，de Pascale D（2014）A new alkaliphilic cold-active esterase from the psychrophilic marine bacterium *Rhodococcus* sp.：functional and structural studies and biotechnological potential. Appl Biochem Biotechnol 172：3054-3068

Dhake KP, Thakare DD, Bhanage BM (2013) Lipase: a potential biocatalyst for the synthesis of valuable flavour and fragrance ester compounds. Falvour Fragr J 28: 71-83

Dlugolecka A, Cieslinski H, Bruzdziak P, Gottfried K, Turkiewicz M, Kur J (2009) Purification and biochemical characteristic of a cold-active recombinant esterase from *Pseudoalteromonas* sp. 643A under denaturing conditions. Pol J Microbiol 58: 211-218

Do H, Lee JH, Kwon MH, Song HE, An JY, Eom SH, Lee SG, Kim HJ (2013) Purification, char-acterization and preliminary X-ray diffraction analysis of a cold-active lipase (CpsLip) from the psychrophilic bacterium *Colwellia psychrerythraea* 34H. Acta Crystallogr Sect F Struct Biol Cryst Commun 69: 920-924

du Plessis EM, Berger E, Stark T, Louw ME, Visser D (2010) Characterization of a novel thermo-stable esterase from *Thermus scotoductus* SA-01: evidence of a new family of lipolytic ester-ases. Curr Microbiol 60: 248-253

Egorova K, Antranikian G (2005) Industrial relevance of thermophilic Archaea. Curr Opin Microbiol 8: 649-655

Elend C, Schmeisser C, Hoebenreich H, Steele HL, Streit WR (2007) Isolation and characteriza-tion of a metagenome-derived and cold-active lipase with high stereospecificity for (R) -ibuprofen esters. J Biotechnol 130: 370-377

Elleuche S (2015) Bringing functions together with fusion enzymes-from nature' s inventions to biotechnological applications. Appl Microbiol Biotechnol 99: 1545-1556

Elleuche S, Piascheck H, Antranikian G (2011) Fusion of the OsmC domain from esterase EstO confers thermolability to the cold-active xylanase Xyn8 from *Pseudoalteromonas arctica*. Extremophiles 15: 311-317

Elleuche S, Schröder C, Sahm K, Antranikian G (2014) Extremozymes—biocatalysts with unique properties from extremophilic microorganisms. Curr Opin Biotechnol 29: 116-123

Feller G, Gerday C (2003) Psychrophilic enzymes: hot topics in cold adaptation. Nat Rev Microbiol 1: 200-208

Feller G, Thiry M, Arpigny JL, Gerday C (1991) Cloning and expression in *Escherichia coli* of three lipase-encoding genes from the psychrotrophic antarctic strain *Moraxella* TA144. Gene 102: 111-115

Feller G, Narinx E, Arpigny JL, Aittaleb M, Baise E, Genicot S, Gerday C (1996) Enzymes from psychrophilic organisms. FEMS Microbiol Rev 18: 189-202

Fuciños P, Pastrana L, Sanroman A, Longo MA, Hermoso JA, Rua ML (2011) An esterase from *Thermus thermophilus* HB27 with hyper-thermoalkalophilic properties: purification, charac-terisation and structural modelling. J Mol Catal B Enzym 70: 127-137

Fuciños P, Atanes E, Lopez-lopez O, Solaroli M, Cerdan ME, Gonzalez-Siso MI, Pastrana L, Rua ML (2014) Cloning, expression, purification and characterization of an oligomeric His-tagged thermophilic esterase from *Thermus thermophilus* HB27. Process Biochem 49: 927-935

Gao R, Feng Y, Ishikawa K, Ishida H, Ando S, Kosugi Y, Cao S (2003) Cloning, purification and properties of a hyperthermophilic esterase from archaeon *Aeropyrum pernix* K1. J Mol Catal B Enzym 24-25: 1-8

Gatti-Lafranconi P, Caldarazzo SM, Villa A, Alberghina L, Lotti M (2008) Unscrambling thermal stability and temperature adaptation in evolved variants of a cold-active lipase. FEBS Lett 582: 2313-2318

Golyshina OV, Golyshin PN, Timmis KN, Ferrer M (2006) The 'pH optimum anomaly' of intra-cellular enzymes of *Ferroplasma acidiphilum*. Environ Microbiol 8: 416-425

Gumerov VM, Mardanov AV, Kolosov PM, Ravin NV (2012) Isolation and functional character-ization of lipase from the thermophilic alkali-tolerant bacterium *Thermosyntropha lipolytica*. Prikl Biokhim Mikrobiol 48: 376-382

Gupta R, Gupta N, Rathi P (2004) Bacterial lipases: an overview of production, purification and biochemical properties. Appl Microbiol Biotechnol 64: 763-781

Hasan F, Shah AA, Hameed A (2005) Industrial applications of microbial lipases. Enzyme Microb Technol 39: 235-251

Henne A, Schmitz RA, Bomeke M, Gottschalk G, Daniel R (2000) Screening of environmental DNA libraries for the presence of genes conferring lipolytic activity on *Escherichia coli*. Appl Environ Microbiol 66: 3113-3116

Hess M, Katzer M, Antranikian G (2008) Extremely thermostable esterases from the thermoacido-philic euryarchaeon *Picrophilus torridus*. Extremophiles 12: 351-364

Hoesl MG, Acevedo-Rocha CG, Nehring S, Royter M, Wolschner C, Wiltschi B, Budisa N, Antranikian G (2011) Lipase congeners designed by genetic code engineering. Chemcatchem 3: 213-221

Hotta Y, Ezaki S, Atomi H, Imanaka T (2002) Extremely stable and versatile carboxylesterase from a hyperthermophilic archaeon. Appl Environ Microbiol 68: 3925-3931

Hu XP, Heath C, Taylor MP, Tuffin M, Cowan D (2012) A novel, extremely alkaliphilic and cold-active esterase from Antarctic desert soil. Extremophiles 16: 79-86

Hwang HT, Qi F, Yuan C, Zhao X, Ramkrishna D, Liu D, Varma A (2014) Lipase-catalyzed pro-cess for biodiesel production: protein engineering and lipase production. Biotechnol Bioeng 111: 639-653

Ikeda M, Clark DS (1998) Molecular cloning of extremely thermostable esterase gene from hyper-thermophilic archaeon *Pyrococcus furiosus* in *Escherichia coli*. Biotechnol Bioeng 57: 624-629

Jaeger KE, Reetz MT (1998) Microbial lipases form versatile tools for biotechnology. Trends Biotechnol 16: 396-403

Jeon JH, Kim JT, Kim YJ, Kim HK, Lee HS, Kang SG, Kim SJ, Lee JH (2009) Cloning and char-acterization of a new cold-active lipase from a deep-sea sediment metagenome. Appl Microbiol Biotechnol 81: 865-874

Joseph B, Ramteke PW, Thomas G (2008) Cold active microbial lipases: some hot issues and recent developments. Biotechnol Adv 26: 457-470

Joshi S, Satyanarayana T (2015) In vitro engineering of microbial enzymes with multifarious applications: prospects and perspectives. Bioresour Technol 176: 273-283

Kamal MZ, Mohammad TA, Krishnamoorthy G, Rao NM (2012) Role of active site rigidity in activity: MD simulation and fluorescence study on a lipase mutant. PLoS One 7, e35188

KamijoT, Saito A, Ema S, Yoh I, Hayashi H, Nagata R, Nagata Y, Ando A (2011) Molecular and enzymatic characterization of a subfamily I. 4 lipase from an edible oil-degrader *Bacillus* sp. HH-01. Antonie Van Leeuwenhoek 99: 179-187

Kawata T, Ogino H (2009) Enhancement of the organic solvent-stability of the LST-03 lipase by directed evolution. Biotechnol Prog 25: 1605-1611

Killens-Cade R, Turner R, MacInnes C, Grunden A (2014) Characterization of a thermostable, recombinant carboxylesterase from the hyperthermophilic archaeon *Metallosphaera sedula* DSM5348. Adv Enzym Res 2: 1-13

Kim SB, Lee W, Ryu YW (2008) Cloning and characterization of thermostable esterase from *Archaeoglobus fulgidus*. J Microbiol 46: 100-107

Koops BC, Papadimou E, Verheij HM, Slotboom AJ, Egmond MR (1999) Activity and stability of chemically modified *Candida antarctica* lipase B adsorbed on solid supports. Appl Microbiol Biotechnol 52: 791-796

Lee D-W, Kim H-W, Lee K-W, Kim B-C, Choe E-A, Lee H-S, Kim D-S, Pyun Y-R (2001) Purification and characterization of two distinct thermostable lipases from the gram-positive thermophilic bacterium *Bacillus thermoleovorans* ID-1. Enzyme Microb Technol 29: 363-371

Lenfant N, Hotelier T, Velluet E, Bourne Y, Marchot P, Chatonnet A (2013) ESTHER, the database of the alpha/beta-hydrolase fold superfamily of proteins: tools to explore diversity of functions. Nucleic Acids Res 41: D423-D429

Leow TC, Rahman RN, Basri M, Salleh AB (2007) A thermoalkaliphilic lipase of *Geobacillus* sp. T1. Extremophiles 11: 527-535

Levisson M, van der Oost J, Kengen SW (2009) Carboxylic ester hydrolases from hyperthermo-philes. Extremophiles 13: 567-581

Li X, Yu HY (2014) Characterization of an organic solvent-tolerant lipase from *Haloarcula* sp. G41 and its application for biodiesel production. Folia Microbiol (Praha) 59: 455-463

Lopez G, Chow J, Bongen P, Lauinger B, Pietruszka J, Streit WR, Baena S (2014) A novel ther-moalkalostable esterase from *Acidicaldus* sp. strain USBA-GBX-499 with enantioselectivity isolated from an acidic hot springs of Colombian Andes. Appl Microbiol Biotechnol 98: 8603-8616

Luo Y, Zheng Y, Jiang Z, Ma Y, Wei D (2006) A novel psychrophilic lipase from *Pseudomonas fluorescens* with unique property in chiral resolution and biodiesel production via transesterifi-cation. Appl Microbiol Biotechnol 73: 349-355

Marquardt T, von der Heyde A, Elleuche S (2014) Design and establishment of a vector system that enables production of multifusion proteins and easy purification by a two-step affinity chromatography approach. J Microbiol Methods 105: 47-50

Merkel L, Schauer M, Antranikian G, Budisa N (2010) Parallel incorporation of different fluori-nated amino acids: on the way to "teflon" proteins. Chembiochem 11: 1505-1507

Morana A, Di Prizito N, Aurilia V, Rossi M, Cannio R (2002) A carboxylesterase from the hyper-thermophilic archaeon Sulfolobus solfataricus: cloning of the gene, characterization of the pro-tein. Gene 283: 107-115

Mozhaev VV (1993) Mechanism-based strategies for protein thermostabilization. Trends Biotechnol 11: 88-95

Nacke H, Will C, Herzog S, Nowka B, Engelhaupt M, Daniel R (2011) Identification of novel lipolytic genes and gene families by screening of metagenomic libraries derived from soil samples of the German Biodiversity Exploratories. FEMS Microbiol Ecol 78: 188-201

Ogino H, Katou Y, Akagi R, Mimitsuka T, Hiroshima S, Gemba Y, Doukyu N, Yasuda M, Ishimi K, Ishikawa H (2007) Cloning and expression of gene, and activation of an organic solvent-stable lipase from *Pseudomonas aeruginosa* LST-03. Extremophiles 11: 809-817

Park YJ, Choi SY, Lee HB (2006) A carboxylesterase from the thermoacidophilic archaeon *Sulfolobus solfataricus* P1; purification, characterization, and expression. Biochim Biophys Acta 1760: 820-828

Parra LP, Espina G, Devia J, Salazar O, Andrews B, Asenjo JA (2015) Identification of lipase encoding genes from Antarctic seawater bacteria using degenerate primers: expression of a cold-active lipase with high specific activity. Enzyme Microb Technol 68: 56-61

Patkar S, Vind J, Kelstrup E, Christensen MW, Svendsen A, Borch K, Kirk O (1998) Effect of mutations in *Candida antarctica* B lipase. Chem Phys Lipids 93: 95-101

Rao L, Xue Y, Zhou C, Tao J, Li G, Lu JR, Ma Y (2011) A thermostable esterase from *Thermoanaerobacter tengcongensis* opening up a new family of bacterial lipolytic enzymes. Biochim Biophys Acta 1814: 1695-1702

Rashid N, Shimada Y, Ezaki S, Atomi H, Imanaka T (2001) Low-temperature lipase from psychro-trophic *Pseudomonas* sp. strain KB700A. Appl Environ Microbiol 67: 4064-4069

Roh C, Villatte F (2008) Isolation of a low-temperature adapted lipolytic enzyme from uncultivated microorganism. J Appl Microbiol 105: 116-123

Royter M, Schmidt M, Elend C, Hobenreich H, Schafer T, Bornscheuer UT, Antranikian G (2009) Thermostable lipases from the extreme thermophilic anaerobic bacteria *Thermoanaerobacter thermohydrosulfuricus* SOL1 and *Caldanaerobacter subterraneus* subsp. *tengcongensis*. Extremophiles 13: 769-783

Rusnak M, Nieveler J, Schmid RD, Petri R (2005) The putative lipase, AF1763, from *Archaeoglobus fulgidusis* is a carboxylesterase with a very high pH optimum. Biotechnol Lett 27: 743-748

Ryu HS, Kim HK, Choi WC, Kim MH, Park SY, Han NS, Oh TK, Lee JK (2006) New cold-adapted lipase from *Photobacterium lipolyticum* sp. nov. that is closely related to filamentous fungal lipases. Appl Microbiol Biotechnol 70: 321-

326

Salameh MA，Wiegel J（2007）Purification and characterization of two highly thermophilic alka-line lipases from *Thermosyntropha lipolytica*. Appl Environ Microbiol 73：7725-7731

Shao H，Xu L，Yan Y（2013）Isolation and characterization of a thermostable esterase from a metagenomic library. J Ind Microbiol Biotechnol 40：1211-1222

Shao H，Xu L，Yan Y（2014）Biochemical characterization of a carboxylesterase from the archaeon *Pyrobaculum* sp. 1860 and a rational explanation of its substrate specificity and thermostability. Int J Mol Sci 15：16885-16910

Sharma R，Sona SK，Vohra RM，Gupta LK，Gupta JK（2002）Purification and characterisation of a thermostable alkaline lipase from a new thermophilic *Bacillus* sp. RSJ-1. Process Biochem 37：1075-1084

Sharma PK，Kumar R，Garg P，Kaur J（2014）Insights into controlling role of substitution mutation，E315G on thermostability of a lipase cloned from metagenome of hot spring soil. 3 Biotech 4：189-196

Shaw E，McCue LA，Lawrence CE，Dordick JS（2002）Identification of a novel class in the alpha/beta hydrolase fold superfamily：the N-myc differentiation-related proteins. Proteins 47：163-168

Siew N，Saini HK，Fischer D（2005）A putative novel alpha/beta hydrolase ORFan family in *Bacillus*. FEBS Lett 579：3175-3182

Soliman NA，Knoll M，Abdel-Fattah YR，Schmid RD，Lange S（2007）Molecular cloning and characterization of thermostable esterase and lipase from *Geobacillus thermoleovorans* YN isolated from desert soil in Egypt. Process Biochem 42：1090-1100

Tan S，Owusu Apenten RK，Knapp J（1996）Low temperature organic phase biocatalysis using cold-adapted lipase from psychrotrophic *Pseudomonas* P38. Food Chem 57：415-418

Tanaka D，Yoneda S，Yamashiro Y，Sakatoku A，Kayashima T，Yamakawa K，Nakamura S（2012）Characterization of a new cold-adapted lipase from *Pseudomonas* sp. TK-3. Appl Biochem Biotechnol 168：327-338

Tao W，Shengxue F，Duobin M，Xuan Y，Congcong D，Xihua W（2013）Characterization of a new thermophilic and acid tolerant esterase from *Thermotoga maritima* capable of hydrolytic reso-lution of racemic ketoprofen ethyl ester. J Mol Catal B Enzym 85-86：23-30

Tchigvintsev A，Tran H，Popovic A，Kovacic F，Brown G，Flick R，Hajighasemi M，Egorova O，Somody JC，Tchigvintsev D，Khusnutdinova A，Chernikova TN，Golyshina OV，Yakimov MM，Savchenko A，Golyshin PN，Jaeger KE，Yakunin AF（2015）The environment shapes microbial enzymes：five cold-active and salt-resistant carboxylesterases from marine metagenomes. Appl Microbiol Biotechnol 99：2165-2178

Trincone A（2011）Marine biocatalysts：enzymatic features and applications. Mar Drugs 9：478-499

Vieille C，Zeikus GJ（2001）Hyperthermophilic enzymes：sources，uses，and molecular mecha-nisms for thermostability. Microbiol Mol Biol Rev 65：1-43

Wahab RA，Basri M，Rahman RNZRA，Salleh AB，Rahman MBA，Chaibakhsh N，Leow TC（2014）Enzymatic production of a solvent-free menthyl butyrate via response surface method-ology catalyzed by a novel thermostable lipase from *Geobacillus zalihae*. Biotechnol Biotechnol Equip 28：1065-1072

Wi AR，Jeon SJ，Kim S，Park HJ，Kim D，Han SJ，Yim JH，Kim HW（2014）Characterization and a point mutational approach of a psychrophilic lipase from an arctic bacterium，*Bacillus pumilus*. Biotechnol Lett 36：1295-1302

Wicka M，Krajewska E，Pawlak A（2013）Cold-adapted bacterial lipolytic enzymes and their appli-cations. PhD Interdisp J 2：107-112

Wu G，Wu G，Zhan T，Shao Z，Liu Z（2013）Characterization of a cold-adapted and salt-tolerant esterase from a psychrotrophic bacterium *Psychrobacter pacificensis*. Extremophiles 17：809-819

Yamashiro Y，Sakatoku A，Tanaka D，Nakamura S（2013）A cold-adapted and organic solvent-tolerant lipase from a psychrotrophic bacterium *Pseudomonas* sp. strain YY31：identification，cloning，and characterization. Appl Biochem Biotechnol 171：989-1000

Zhang JW，Zeng RY（2008）Molecular cloning and expression of a cold-adapted lipase gene from an Antarctic deep sea psychrotrophic bacterium *Pseudomonas* sp. 7323. Mar Biotechnol 10：612-621

Zhang J，Lin S，Zeng R（2007）Cloning，expression，and characterization of a cold-adapted lipase gene from an antarctic deep-sea psychrotrophic bacterium，*Psychrobacter* sp 7195. J Microbiol Biotechnol 17：604-610

Zhang XY，Fan X，Qiu YJ，Li CY，Xing S，Zheng YT，Xu JH（2014）Newly identified thermostable esterase from *Sulfobacillus acidophilus*：properties and performance in phthalate ester degrada-tion. Appl Environ Microbiol 80：6870-6878

Zhu Y，Li H，Ni H，Xiao A，Li L，Cai H（2015）Molecular cloning and characterization of a ther-mostable lipase from deep-sea thermophile *Geobacillus* sp. EPT9. World J Microbiol Biotechnol 31：295-306

第十三章
脂解酶生产菌——嗜盐细菌和古菌

María de Lourdes Moreno[1]，M-Carmen Márquez[1]，
María Teresa García[1]，Encarnación Mellado[1][2]

13.1 引言

极端微生物是指生活在极端环境中，诸如高盐、极端 pH 值、极端温度和压力等环境中的微生物（Pikuta et al. 2007）。高盐环境，例如在晒盐田中，栖息着众多的嗜盐菌（Grant et al. 1998；Ventosa. 2006；Ventosa et al. 2014）。

嗜盐微生物具有较高的系统发育多样性，包括如下代表类别：原核生物，包括古菌和细菌；真核生物，包括真菌和酵母；原生生物，包括原生动物和藻类。对其进行了有关生态学、生理学和遗传学等方面的研究（Ventosa et al. 1998；Uratani et al. 2014）。本章将介绍高盐环境中的两大类代表微生物：嗜盐古菌和中度嗜盐细菌。嗜盐古菌为极端好氧嗜盐古菌，生活在 NaCl 浓度至少为 10％的环境中，最适盐浓度为 20％～25％。它们属于盐杆菌科（Halobacteriaceae），70％以上分属于不同的属。中度嗜盐细菌生长的盐浓度范围较宽，为 5％～15％（大多数为 10％），隶属于很多不同的属，这些属有些只有嗜盐菌，有些既含有嗜盐菌也含非嗜盐菌（Ventosa 2006）。

嗜盐菌通过以下策略适应高盐生境："盐内"策略，通过增加胞质中无机阳离子的物质的量浓度来适应高盐环境；"盐外"策略，合成不干扰细胞代谢的低分子量有机相容性溶质来提高对高渗环境的适应性（Empadinhas and Da Costa 2008；Goh et al. 2011；Ceylan et al. 2012）。

因为嗜盐菌生长在极端环境条件下，所以它们分泌的酶类具有一些优良的性质，通常被称为极端酶（Antranikian et al. 2005）。来源于嗜盐菌的酶不仅可以应用于盐浓度较高的工业化生产中，同时还对有机溶剂和高温具有较好的耐受性（Oren 2010）。在水解酶中，脂肪酶（E. C. 3. 1. 1. 3）和羧酸酯酶（E. C. 3. 1. 1. 1）是最具有代表性的类别，在自然界从细菌到高等真核生物均有分布。这些酶可催化不同的反应，如酯化、酯交换和氨解（Jaeger et

---

[1] Faculty of Pharmacy，Department of Microbiology and Parasitology，University of Sevilla，Profesor García González，2，41012 Sevilla，Spain.

[2] e-mail：emellado@us. es.

al. 1999)。脂解酶包括脂肪酶和酯酶。脂肪酶水解长链脂酰基甘油（$\geqslant C_{10}$），酯酶水解短链脂肪酸酯（$\leqslant C_{10}$）（Jaeger et al. 1999）。两种酶之间的差异基于两个方面，即能够识别的底物以及酶作用的位点。脂肪酶可在油-水界面处催化三酰甘油的水解，而酯酶作用于水溶性底物。脂肪酶大都具有界面激活效应（油、水界面）（Brockerhoff and Jensen 1974；Jaeger et al. 1999）。脂肪酶和酯酶包含在α/β水解酶超家族中（Ollis et al. 1992），该家族的酶通常含有由Ser、His和Asp残基组成的催化三联体，Ser残基通常包含在活性位点基序G-x-S-x-G内（Arpigny and Jaeger 1999）。该基序通常位于亲核试剂（nucleophile elbow）（Ollis et al. 1992）。相比之下，大多数酯酶存在基序S-x-x-K，该序列作为活性中心在C类β-内酰胺酶（Knox et al，1996）及青霉素结合蛋白（PBPs）中是保守的（Joris et al. 1988）。

Arpigny和Jaeger（1999）对脂解酶进行分类的方法普遍为人们所接受，根据其保守的氨基酸残基和生物学性质将微生物脂肪酶分成8个家族。之后，使用宏基因组方法对其他几种酶进行表征，又提出六个新的家族（Handrick et al. 2001；Ewis et al. 2004；Lee et al. 2006，2010；Kim et al. 2009；Bayer et al. 2010；Rao et al. 2011；Bassegoda et al. 2012；Fu et al. 2013）。

1935年，首次发现来自于草酸青霉菌（*Penicillium oxalicum*）和黄曲霉（*Aspergillus flavus*）的脂肪酶（Kirsh 1935），自此，脂肪酶被逐渐应用于工业，成为继蛋白酶和糖酶之后的第三大类商业酶（Hasan 2006）。由于脂解酶的多功能性，脂解酶被用于不同的工业过程，它能催化三酰甘油的水解及合成（Jaeger et al. 1999）。此外，脂肪酶具有区域选择性，这一性质对其工业化应用具有重要意义，因为要对复杂底物分子实现选择性极为困难。此外，大多数工业条件较为苛刻，因此，分离那些在极端pH、高温、低温、高盐和有机化合物作为溶剂存在时仍具有较高活性的脂肪酶非常重要（Mellado et al. 2005）。从这个意义上讲，来自嗜盐菌的脂解酶可以用于工业化过程，也是未来生物催化剂发展的趋势。

有关极端微生物来源的脂肪酶和酯酶（如嗜盐菌）的综述主要集中于其在生物技术领域的应用（Mellado et al. 2005；Ventosa et al. 2005；Salameh and Wiegel 2007；Litchfield 2011；Schreck and Grunden 2014；Dalmaso et al. 2015）。在本章中，将对高盐环境来源的产脂解酶嗜盐微生物进行综述，将重点介绍这些脂解酶生产菌——古菌和细菌的多样性，包括酶学性质的表征及它们在生物技术领域的应用。

13.2 脂解酶的重要来源——高盐环境

高盐环境可以为工业生产提供不同类型的酶，但是相较于其他极端环境并没有引起人们足够的重视。在过去几十年中，人们利用不同的策略从高盐环境中分离到一些新的脂肪酶。大多数策略是基于不同的培养方法。最近，基于功能基因组的方法也得以实施（López-López et al. 2014）。常规方法是使用含有适当乳化底物的高盐固体培养基进行脂肪酶筛选，这也是迄今为止最通用的筛选方法。该方法通过观察菌落周围的降解圈的形成，来检测脂肪酶活性。吐温和三丁酸甘油酯是检测脂肪酶最常用的两种底物。以吐温为底物，菌落周围的降解圈是由生成的脂肪酸盐形成的；而三丁酸甘油酯的降解也是通过菌落周围形成的清晰的降解圈来检测的。由于酯酶不能水解长链底物，故可以通过延长底物的链长来筛选脂肪酶，例如含维多利亚蓝的橄榄油乳化底物（细菌周围会出现蓝色），含罗丹明B的橄榄油乳化底物（UV 350nm，菌落周围出现橙色荧光）（Hasan et al. 2009；Babavalian et al. 2014）。

在全球范围内开展了针对高盐环境中微生物的多样性以及胞外脂肪酶的科学研究，这些研究主要集中在晒盐田和盐湖。

2003年，Sánchez-Porro 等人研究了位于西班牙南部各盐场的水样和沉积物样品中嗜盐菌分泌的酶的多样性（Sánchez-Porro et al. 2003）。分离到892株可分泌不同水解酶的中度嗜盐菌。不同来源的（水样和沉积物）菌株分泌的水解酶活性非常相似。作者还对保藏的21株中度嗜盐菌进行研究发现大多数没有水解活性。从这些水样和沉积物中分离到菌株中有217株菌可以分泌脂肪酶，有23株菌系统发育关系较远，其中大多数菌株为革兰氏阴性杆菌（15株），分别属于色盐杆菌属（*Chromohalobacter*）、盐单胞菌属（*Halomonas*）和盐弧菌属（*Salinibrivio*）。这三个属包括大量的γ变形菌门中度嗜盐菌，绝大多数分离自高盐环境（Amoozegar et al. 2008b；de la Haba et al. 2012）。另一方面，革兰氏阳性菌主要有芽孢杆菌（*Bacillus*）、需盐芽孢杆菌属（*Salibacillus*）、盐水球菌属（*Salinicoccus*）和海球菌属（*Marinococcus*）。其中菌株SM19可能为海球菌属新成员（Martín et al. 2003），而且进一步研究了其胞内嗜盐酶LipBL（Pérez et al. 2011，2012），此外，菌株SM19基因组框架图已经被绘制，发现不仅含有脂肪酶基因，而且还包括淀粉酶、蛋白酶以及DNA酶基因（Papke et al. 2013）。另一方面，Moreno等人（2009）对位于西班牙西南地区的两个盐池也进行了研究，分离到150株能够分泌不同类型的水解酶的细菌，可分泌蛋白酶、淀粉酶、脂肪酶及DNA酶。在该研究中，通过16S rRNA序列分析发现，其中43株菌属于极端嗜盐菌，仅3株属于*Salicola*、盐水杆菌属（*Salinibacter*）和假单胞菌属（*Pseudomonas*），其余的属于嗜盐古菌，包括盐盒菌属（*Haloarcula*）、盐深红菌属（*Halorubrum*）和嗜盐杆菌（*Halobacterium*）属成员。43株菌中，4株具有脂肪酶活性，且这4株菌具有复合水解酶活性（3株具有淀粉酶和脂肪酶活性，1株具有蛋白酶和脂肪酶活性）。从系统发育角度讲，其中6株分离菌与*Salicola*属、嗜盐假单胞菌（*Pseudomonas halophila*）、特腊帕尼盐红菌（*Halorubrum trapanicum*）、盐沼盐杆菌（*Halobacterium salinarum*）进化关系接近。因为菌株*Salicola* sp. IC10分泌的蛋白酶和脂肪酶活性在生物技术领域具有潜在应用价值，所以对其进行了进一步的研究（Moreno et al. 2009，2013）。这些研究表明，西班牙南部的晒盐田蕴藏着很多具有重要经济价值的酶。

Birbir等从土耳其盐矿中分离到12种极端嗜盐微生物（2004），可以水解吐温80。多相分类学研究显示它们属于盐杆菌科（Halobacteriaceae），是高盐环境中的主要类群，包括*Haloarcula*、*Halobacterium*、钠线菌属（*Natrinema*）和*Halorubrum*四个属（Birbir et al. 2004）。对土耳其的三个盐湖进行研究，目的是从嗜盐古菌中分离脂解酶。从最初分离到的118株菌中筛选出5株菌具有高酯酶活性（Ozcan et al. 2009）。进一步经分类学鉴定显示5株菌分别属于盐长寿菌属（*Halovivax*）、*Natrinema*和盐陆生菌属（*Haloterrigena*），而其中菌株E49可能为新种（Ozcan et al. 2012）。这项研究表明，嗜盐古菌通常具有脂肪酶和酯酶活性。

阿尔及利亚拥有许多天然的高盐环境，通常位于干旱和半干旱地区。2004年Hacěne等人研究了位于阿尔及利亚中心的埃尔果累阿（El Golea）盐湖的微生物多样性。一共分离到471株菌，分属于古菌域和细菌域的31个属（Hacěne et al. 2004）。Bhatnagar等（2005）选择了其中35株嗜盐细菌进行脂解酶活性的筛选。研究了不同的温度、pH值、底物和盐浓度对极端嗜盐古菌——嗜盐碱球菌属（*Natronococcus*）菌株TC6分泌的脂肪酶活性的影响（Bhatnagar et al. 2005）。随后，TC6分泌的酶被表征，这是古菌域中第一个被鉴定的脂肪酶（Boutaiba et al. 2006）。另一方面，Kharroub等人（2014年）从阿尔及利亚三个盐湖（Ezzemoul、Bethiaa和Melghir）进行了产淀粉酶、明胶酶和脂肪酶的极端嗜盐菌的筛选。这些湖泊盐度为0.3～0.36kg/L，夏季温度可达50℃，因此，是嗜盐微生物生长的最佳环境，其中多样的微生物群落在工业上具有很大的应用价值。研究者从中分离出44株极端嗜盐古菌，发现其中大多数具有淀粉酶活性（22株），然而仅在4株菌中检测到了胞外脂肪酶活性，系统发育分析显示与 *Halorubrum*、盐微菌属（*Halomicrobium*）和富盐菌属（*Haloferax*）亲缘关系相近（Kharroub et al. 2014）。

Sana 等人（2007）从孟加拉湾的洛锡安岛的沉积物中分离出 400 株菌。从中筛选出 5 株具有高活性胞外酯酶的耐盐细菌，研究了其对盐、pH、温度和有机溶剂的耐受性。其中芽孢杆菌属（*Bacillus* sp.）酯酶活性较高。Karpushova 等人在 2005 年首次报道了从海洋芽孢杆菌分离到两种酯酶。芽孢杆菌属的成员具有代谢多样性的特征，分布极其广泛，在不同的环境（包括高盐环境）都能分离得到，其分泌的酶类有些已经被应用到工业领域（Karpushova et al. 2005；Ventosa et al. 2008；Liu 2013）。

在伊朗也进行了产水解酶嗜盐微生物的多样性的相关研究。2008 年，Amoozegar 等人从各种高盐环境中分离出 55 株生产脂肪酶的中度嗜盐菌（Amoozegar et al. 2008a），并对 *Salinivibrio* sp. 菌株 SA-2 进行进一步研究（Amoozegar et al. 2008b）。另一方面，Rohban 等人（2009）分离并表征了来自伊朗盐湖 Howz Soltan Lake 的 231 株中度嗜盐菌，并检测了胞外脂肪酶活性。在这项研究中，观察到海洋芽孢杆菌属（*Oceanobacillus*）、*Halomonas* 和薄壁芽孢杆菌属（*Gracilibacillus*）成员的脂解酶活性较高。两年后，从 Maharlu 盐湖中分离出 33 株具有脂肪酶活性的嗜盐细菌，并选择革兰氏阳性杆菌属和葡萄球菌属（*Staphylococcus*）的 13 株中度嗜盐菌株进行活性研究（Ghamesi et al. 2011）。最近，Babavalian 等人（2014）从伊朗三个不同的盐湖中分离出 581 株菌，为耐盐菌、中度和极度嗜盐细菌，能够产生 9 种不同水解酶（淀粉酶、蛋白酶、脂肪酶、DNA 酶、菊粉酶、木聚糖酶、羧甲基纤维素酶、果胶酶和支链淀粉酶）。研究发现这些高盐环境中的大多数脂肪酶生产菌属于厚壁菌门（Firmicutes），包括鱼芽孢杆菌属（*Piscibacillus*）、*Bacillus*、*Oceanobacillus*、*Gracilibacillus*、盐芽孢杆菌属（*Halobacillus*）和海洋枯草芽孢杆菌属（*Thalasobacillus*）等，其次是 γ 变形菌门海源菌属（*Idiomarina*）、*Halomonas* 和 *Salicola* 属。然而，对三个湖泊的菌群进行比较，发现这些属的分布存在差异。

Kumar 等人（2012）对印度西海岸和南海岸以及位于拉贾斯坦邦地区的 Sambhar 盐湖中产酶嗜盐细菌的多样性进行了研究。筛选到了 108 株具有不同水解酶活性的中度嗜盐细菌（淀粉酶、脂肪酶和蛋白酶）。通过表型和 16S rRNA 序列分析对 8 种脂肪酶生产菌进行鉴定，发现它们与海杆菌（*Marinobacter*）、*Halomonas*、色盐杆菌属（*Chromohalobacter*）和地微杆菌属（*Geomicrobium*）亲缘关系较近。如前所述，从其他高盐环境中分离到的解脂海杆菌（*Marinobacter lipolyticus*），*Halomonas* sp. 和 *Chromohalobacter* sp. 细菌也可分泌脂肪酶（Martín et al. 2003；Sánchez-Porro et al. 2003）。

2012 年，Ardakani 等在波斯湾进行细菌胞外水解酶的筛选。分离出 5 株中度嗜盐的假交替单胞菌（*Pseudoalteromonas*），并通过在琼脂培养基上中分别添加牛奶、吐温 80 和淀粉的方法进行蛋白酶、脂肪酶和淀粉酶的筛选。其中 2 株菌显示具有 3 种水解酶活性，可应用于生物技术领域（Ardakini et al. 2012）。

总之，所有这些研究证明高盐环境有利于嗜盐细菌和古菌生长，多样化的微生物群落有利于发现新的脂解酶，其新的酶学特征在工业上具有潜在的应用价值。虽然在所研究的水解酶中，产脂肪酶微生物不如产蛋白酶和淀粉酶微生物丰富，但筛选方法的进一步优化将有助于对其进行更好的开发。

13.3 嗜盐酶和耐盐酶特性

如前所述，嗜盐菌的生长依赖 NaCl，从嗜盐菌中分泌的酶被称为嗜盐酶，其通常需要依赖较高的 NaCl 浓度才能维持其催化活性。

然而，有一些蛋白质虽然不是从高盐环境中分离出来的，但是仍然可以耐受很高的盐浓度。这些蛋白质通常具有耐盐性，但它们不具有盐依赖性，它们在较宽范围的 NaCl 浓度下具有活性（Graziano and Merlino 2014）。

嗜盐菌通过两种机制来维持渗透平衡：在细胞质中积累无机盐离子和有机相容性溶质的方式（Goh et al. 2011；Ceylan et al. 2012）。

嗜盐古菌通过在细胞质中积累 Na^+ 和 K^+ 来对抗外部的高渗环境。嗜盐菌主要从结构和功能两个水平来适应高盐环境（Madern et al. 2000；Joo and Kim 2005）。不同的纯化手段和晶体学研究有助于了解嗜盐蛋白质渗透调节功能和独特的理化性质（Detkova and Boltyanskaya 2007）。晶体学研究发现嗜盐酶的表面有两个典型的特点：一方面，酶表面酸性氨基酸过量，易形成水化膜，有助于稳定性的维持（Sinha and Khare 2014；Graziano and Merlino 2014）；另一方面，表面疏水残基数量较少（Siglioccolo et al. 2011）。酶蛋白中的酸性氨基酸残基通过离子键与碱性氨基酸残基相连（Lanyi 1974；Eisenberg et al. 1992；Danson and Hough 1997；Madern et al. 2000；Detkova and Boltyanskaya 2007；Bonete and Martínez-Espinosa 2011；Sinha and Khare 2013）。Oren（2013）通过对嗜盐菌的蛋白质组进行研究，阐明其对高盐环境的适应机制。

一般来讲，嗜盐古菌的蛋白质往往需要高盐浓度来维持其活性，如需要高达 2～4mol/L 的 Na^+ 或 K^+ 来维持其活性。这些蛋白质不仅需要高盐浓度来维持其活性，而且高盐浓度也有利于维持其稳定性。

因此，大多数这类蛋白质在 NaCl 浓度低于 1.2mol/L 时发生变性，此外，这些蛋白质所含谷氨酸的比例较高，该氨基酸具有较高的亲水性。嗜盐古菌的几种酶诸如来自嗜盐富饶菌属（*Haloferax volcanii*）的二氢叶酸还原酶和异柠檬酸脱氢酶（Pieper et al. 1998），来自死海盐盒菌（*Haloarcula marismortui*）的苹果酸脱氢酶和 2Fe-2S 铁氧还蛋白（Zaccai et al. 1989；Madern et al. 1995；Mevarech et al. 2000）及来自盐生沼杆菌（*Halobacterium salinarum*）的谷氨酸脱氢酶（Britton et al. 1998）等都具有以上特征。对来自 *Haloarcula marismortui* 的酯酶进行研究，通过三维建模发现其表面酸性氨基酸残基增多而碱性氨基酸残基减少（Müller-Santos et al. 2009）。此外，对一些代表性的嗜盐古菌进行了基因组分析，也证实了上述特征（Ng et al. 2000；Lee et al. 2014）。

尽管 *Salinibacter ruber*（Antón et al，2002）是一种极端嗜盐细菌，但是这种细菌与嗜盐古菌应对渗透胁迫的策略相同，即通过在其细胞质中积累 K^+ 和 Na^+ 来对抗高渗环境。然而，对来自 *Salinibacter ruber* 的酶的研究发现这些酶不遵循一般模式。对它们在不同盐浓度下的性质进行研究发现，脂肪酸合成酶和葡萄糖-6-P-脱氢酶表现出类似于嗜盐酶的行为，其活性依赖高盐浓度（Oren and Mana 2002；oren et al. 2003）。然而，另外一些酶无论在有无盐存在的情况下都是有活性的，例如异柠檬酸脱氢酶（Oren and Mana 2002）和甘油激酶（Sher et al. 2004）。

大多数的中度嗜盐菌和某些极端嗜盐菌通过积累相容性溶质来应对高盐环境中的渗透胁迫。因此，这些细菌细胞质中离子浓度不高。有研究者发现该类细菌蛋白质中的酸性残基略有增加，但不如嗜盐古菌中的那样高（Gandbhir et al. 1995；Coronado et al. 2000；Moreno et al. 2010）。

相容性溶质包括有机化合物如甘油、糖及其衍生物、季胺盐（甘氨酸甜菜碱）及四氢嘧啶。此外，生物在不同的胁迫条件下（温度或盐度）积累的相容性溶质是不同的（例如甘露糖甘油酸酯或磷酸二肌醇酯）（Esteves et al. 2014），即使抗辐射胁迫也是如此（Webb and

DiRuggiero 2012)。嗜盐盐芽孢杆菌（*Halobacillus halophilus*）在不同生长时期和不同盐浓度下合成不同的相容性溶质以应对渗透胁迫（Saum et al. 2013)。

总而言之，来自极端嗜盐菌的酶与来自中度嗜盐菌的酶对盐的适应机理不同。极端嗜盐菌分泌的酶必须耐受内部（细胞质）和细胞外培养基中的高盐浓度，但中度嗜盐菌的酶的细胞质中盐浓度较低，而细胞外培养基中盐浓度增加，它们必须适应这种盐度的变化。

已经确定了 100 种脂肪酶的三维结构，但是对嗜盐菌脂肪酶结构研究的很少（Jiang et al. 2012)。本文中提到的一些有较好应用前景的酶的晶体结构的解析将为研究该类蛋白质盐适应性的分子结构基础提供线索。

13.4 嗜盐和耐盐脂解酶在生物技术领域的应用

13.4.1 细菌来源的脂解酶

如前所述，关于嗜盐菌脂肪酶的报道有很多（Boutaiba et al. 2006；Amoozegar et al. 2008a；Niño de Guzman et al. 2008；Ozcan et al. 2009；Rohban et al. 2009；Ghamesi et al. 2011；Moreno et al. 2012)，然而对其分离纯化和性质表征的研究很少。表 13.1 列出了不同来源的酶及其相关性质。

不同来源的脂肪酶在非水相催化反应中的区域选择性、热稳定性、底物特异性、最适 pH 和动力学各不相同（Gupta et al. 2004)。对映体选择性、催化底物广泛及非水相中的高催化是新型脂肪酶的必备特征（Doukyu and Ogino 2010)。

酶在有无盐存在的情况下均可发挥其作用，是其在生物技术领域应用的要求之一。该类酶在工业生产中无论高盐或低盐情况均可使用，而一些脂解酶当处于低盐浓度时，酶活迅速下降，且不可逆转。如表 13.1 所示，某些嗜盐菌脂肪酶对盐具有极高的耐受性。*Salinivibrio* 属的细菌胞外脂肪酶 SA-2 虽然表现出高盐耐受性（0～3mol/L NaCl)，但无盐条件下酶活性最高（Amoozegar et al. 2008a)。据报道 *Halobacillus* sp. LY5 分泌的胞外酯酶，在高盐和低盐浓度下，甚至无盐条件下都具有酶活性（Li et al. 2012)。

Idiomarina sp. W33 的脂肪酶在较宽的 NaCl 浓度范围下（2.5%～25%）仍具有较高的活性和稳定性，在无 NaCl 存在时可保留 70%的酶活（Li et al. 2014)。深海耐盐杆菌 *Pelagibacterium halotolerans* $B2^T$ 的酯酶 PE10 显示出对 NaCl 的强耐受性，在 0～4mol/L NaCl 中均具有活性，在 3mol/L NaCl 下酶活性最高。对 PE10 的三维建模显示其对高盐环境的耐受性与该酶表面负静电势相关（Jiang et al. 2012)。在无 NaCl 时，*Marinobacter lipolyticus* SM19 的耐盐脂肪酶 LipBL 具有最大活性，但该酶在高达 4mol/L 的 NaCl 浓度下仍保留 20%的酶活（Pérez et al. 2011)。

一些工业生产过程对温度要求较高，该过程中应用到的脂肪酶应该具有较好的热稳定性(Janssen et al. 1994)。蛋白质中的热稳定性归因于离子相互间作用的增加、疏水表面积减少、结构更致密、稳定 α-螺旋和降低构型应变等因素（Sinha and Khare 2013)。已经证明盐的存在可以调节嗜盐蛋白质在高温下的稳定性（Boutaiba et al. 2006)。据报道，大多数嗜盐脂肪酶在 45～55℃的温度范围内活性最大（表 13.1)，某些脂肪酶热稳定性非常好。来自 *Chromohalobacter* sp. LY7-8 的胞外脂肪酶在较宽范围的温度范围（30～90℃）具有高活性，最适温度为 60℃（Li and Yu 2012)。*Idiomarina* sp. W33 来源的脂肪酶在 20～90℃范围内热稳定性较好，酶的最适温度为 60℃，90℃下处理 24h，酶仍然保持大于 80%活性（Li et al. 2014)。*Marinobacter lipolyticus* SM19 的脂肪酶因为其优于其他嗜盐和非嗜盐脂肪酶，所以可用于许多工业过程。它的最适温度为 80℃，在较宽的温度范围内具有活性（5～

90℃）（Pérez et al. 2011）。盐渍微菌（*Salimicrobium* sp. LY19）的胞外酯酶具有极好的热稳定性。该酶在低于 70℃ 的温度下稳定性较高，并且在 80℃ 下孵育 2h，仍然具有活性（Xin and Hui-Ying 2013）。

表 13.1 嗜盐/耐盐脂解酶的特征及潜在应用

菌株	细菌门	酶	最适 NaCl 浓度/(mol/L)	最适温度/℃	最适 pH	底物特异性（天然底物）	酶学性质	参考文献
Salinivibrio sp. strain SA-2	γ-Proteobacteria	胞外脂肪酶	0	50	7.5	短链对硝基苯酚(*p*NP)	热稳定较好（60℃处理 4h 或 80℃处理 30min，仍保留 90%的酶活）；高盐耐受性（3mol/L NaCl、KCl、Na_2SO_4和 $NaNO_3$）	Amoozegar et al. (2008a)
Thalassobacillus sp. strain DF-E4	Firmicutes	胞外羧酸酯酶	0.5	40	8.5	短链对硝基苯酚丁酸酯(*p*-NPB)	碱稳定性（pH 6～9.5）	Lv et al. (2011)
Halobacillus sp. strain LY5	Firmicutes	胞外酯酶	1.7	50	10	短链对硝基苯酚丁酸酯(*p*-NPB)	热稳定性（30～90℃）；碱稳定性（pH 6～12）；高盐耐受性（0～20% NaCl）；对表面活性剂具有很好的耐受性（SDS 和 Triton X-100）	Li et al. (2012)
Pelagibacterium halotolerans B2T	α-Proteobacteria	酯酶	3	45	7.5	短链对硝基苯酚乙酸酯(*p*-NPA)	高盐耐受性（0～4mol/L NaCl）	Jiang et al. (2012)
Chromohalobacter sp. strain LY7-8	γ-Proteobacteria	胞外脂肪酶	2.1	60	9.0	长链对硝基苯酚豆蔻酸酯（*p*-NPM）	热稳定性（30～90℃）；碱稳定性（pH 6～12）；耐盐性（0～20% NaCl）；存在表面活性剂时酶活较高（SDS 和 Triton X-100）	Li and Yu (2012)
Marinobacter lipolyticus strain SM19	γ-Proteobacteria	胞内脂肪酶	0	80	8.0	短链/中链对硝基苯酚己酸酯（*p*-NPH）	在有机溶剂中具有较高的活性和稳定性；碱稳定性（pH 6～10）；热稳定性（5～90℃）；耐盐性（0～3mol/L NaCl）；对油脂的水解效率很高（富集 PUFAs）	Pérez et al. (2011)
Salimicrobium sp. strain LY19	Bacteoidetes/Chlorobi	胞外酯酶	0.8	50	7.0	短链对硝基苯酚丁酸酯(*p*-NPB)	当有机溶剂存在时具有高活性和稳定性（在异辛烷中酶活性增加）；耐盐性（2.5%～25% NaCl）；热稳定性（20～70℃）；碱稳定性 pH 7～11	Xin and Hui-Ying (2013)

续表

菌株	细菌门	酶	最适 NaCl 浓度/(mol/L)	最适温度/℃	最适 pH	底物特异性(天然底物)	酶学性质	参考文献
Staphylococcus sp. strain CJ3	Firmicutes	胞外脂肪酶	0	45	8.0	短链对硝基苯酚酯	耐盐性(0～5mol/L NaCl);在有机溶剂中稳定性较好(长链醇)	Daoud et al. (2013)
Idiomarina sp. strain W33	γ-Proteobacteria	胞外脂肪酶	1.7	60	7.0～9.0	长链对硝基苯酚豆蔻酸酯(*p*-NPM)	很好的热稳定性(20～90℃);碱稳定性(pH 7～11);耐盐性(2.5%～25%);耐有机溶剂,可用于非水相催化合成生物柴油	Li et al. (2014)
Halobacillus sp. AP-MSU8	Firmicutes	胞外脂肪酶	2.5	40	9.0	未测定	未测定	Esakkiraj et al. (2014)

用于有机合成的脂解酶需要较高的活性和稳定性（Doukyu and Ogino 2010)。众所周知，有机溶剂是温和的离液剂，可以破坏蛋白质亚基之间的氢键，并通过影响活性中心的临界水浓度来降低催化效率。然而，嗜盐酶通常在有机溶剂中较为稳定（Zaccai 2004；Gupta and Khare 2009；Kumar and Khare 2012)。

此外，在极端酶水解动力学中，水的作用最基本的，因为水解和合成反应之间的热力学平衡在有机介质中趋向于酯合成，或者控制酶的水合度使反应趋向于水解反应（Ebel and Zaccai，2004)。据报道高盐环境是产生水解酶的极端微生物的极好来源，该类酶在有机溶剂中非常稳定（Ventosa et al. 2005)。

大多数脂肪酶通过覆盖在酶活性位点的“盖子”上的疏水性氨基酸残基与有机溶剂分子的相互作用而形成开放构象来发挥酶的催化作用。有机溶剂极性参数（log P 值）低于 2 的溶剂相较于大于 2 的溶剂更容易从酶分子中去除水，使酶变性（Laane et al. 1987)。与一般趋势相反，*Marinobacter lipolyticus* sp. SM19 分泌的脂肪酶 LipBL 在各种工业相关的亲水性和疏水性有机溶剂存在时是有活性的，在丙二醇或甲苯等存在下，酶活几乎增加了两倍（Pérez et al. 2012)。此外，*Idiomarina* sp. W33 脂肪酶在甘油、DMSO、甲苯、环己烷或正己烷的存在下，可保持大于 80%的活性，并且在亲水性有机溶剂（甲醇和乙腈）存在下，可保留大于 60%的活性（Li et al. 2014)。然而，*Salimicrobium* sp. LY19 源酯酶在极性溶剂中不稳定，在非极性疏水溶剂（即异辛烷）存在下活性增加（Xin and Hui-Ying 2013)。*Halobacillus* sp. AP-MSU8 脂肪酶在浓度 10%的有机溶剂如苯和丙酮中有活性（Esakkiraj et al. 2014)。深海耐盐杆菌 *Pelagibacterium halotolerans* PE10 来源的脂酶在一些洗涤剂和有机溶剂的存在下显示出稳定性。该酶在含有 1%Triton X-100、15%DMSO 或 15%甲醇的条件下中保持 50%以上活性（Jiang et al. 2012)。

13.4.2 古菌来源的脂解酶

生存在极端环境中的嗜盐古菌，特别是在太阳能盐蒸发池，其高含盐量、强日照和寡营养的特点导致其水活度降低（Litchfield 2011)。由于嗜盐古菌分泌的酶类具有很多特殊的酶学特性，故可被用于诸多催化反应（Sellek and Chaudhuri 1999)。早期嗜盐古菌的应用主要集中在生物塑料、废水处理、农业和医疗等不同领域（Margesin and Schinner 2001)。此外，对微生物源脂肪酶和酯酶的综述较多，但是嗜盐古菌（Jaeger and Eggert 2002）通常没有被提及。

对分离自阿尔及利亚撒哈拉沙漠的样品进行脂肪酶活性初筛，其中嗜盐碱古菌盐碱球菌属（*Natronococcus* sp.）具有脂解酶活性（Bhatnagar et al. 2005）。它是在古菌域中发现的第一个真正意义上的脂肪酶，对其酶学性质并不是很了解。*Natronococcus* sp. TC6 的耐热脂肪酶显示在 4mol/L NaCl 下具有最高活性，在无 NaCl 存在情况下无活性，该脂肪酶能够在高盐浓度下水解橄榄油（Boutaiba et al. 2006）。

对嗜盐古菌死海盐盒菌（*Haloarcula marismortui*）的基因组进行了序列测定和分析，确定了可能编码脂肪酶和酯酶的基因（Baliga et al. 2004）。后来，Camacho 等人（2009）证实了该推测，同时对酶学性质进行了研究。其酯酶活性在 45℃时最高，脂肪酶活性在 50℃最高，然而，有关于其他嗜盐古菌酶的研究显示在 50～60℃之间酶活性最大（Studdert et al. 2001；Johnsen and Schonheit 2004；Hutcheon et al. 2005）。*Haloarcula marismortui* 的粗酶液在不同 NaCl 浓度（0.5～5mol/L）下具有酯酶和脂肪酶活性，意味着在极端高盐条件可能通过构象的改变以维持其活性（Camacho et al. 2009）。

酯酶是胞内酶，主要参与水解各种类型的酯键。Camacho 等人（2010）研究了 *Halobacterium* sp. NRC-1 来源的酯酶。与其他嗜盐酶相比，*Halobacterium* sp. NRC-1 胞内酯酶活性在 80℃最高。

盐盒菌属（*Haloarcula* sp. G41）来源的脂肪酶具有耐有机溶剂、耐盐、耐热和碱稳定性的特征。该脂肪酶在 30～80℃范围内，pH 6～11 和 10%～25%的 NaCl 浓度下具有高活性和稳定性，在 70℃，pH 8 和 15%NaCl 条件下酶活性最高。该脂肪酶在各种有机溶剂存在下仍具有较好的活性和稳定性。有趣的是，甘油和正己烷增加了脂肪酶的活性，菌株 G41 脂肪酶在亲水性溶剂存在下其活性仍可保持 50%，这与在亲水性溶剂中较易失活的趋势相反。此外，在叔丁醇体系催化生物柴油的合成反应中，对游离酶和固定化酶进行了改性。菌株 G41 固定化脂肪酶的催化效率较高（Li and Yu 2014）。

13.5　未来前景

如本章前面所述，嗜盐菌源脂解酶在生物技术领域的应用潜力很大，且具有很好的生态效益，脂解酶在水性和非水性介质中的区域选择性和多功能性使其成为不同反应中性质优良的生物催化剂，如生物聚合物的合成、手性药物的生物合成、食品工业中的应用等（Hasan et al. 2006）。此外，脂解酶还可以用于食品加工及农业废弃物处理（Jordan and Mullen 2007）。

到目前为止，用于上述工业应用的大多数商业化酶已经从嗜中温微生物中分离出来（Jaeger and Eggert 2002）。但是由于这些酶对有机溶剂的耐受性低，在高温下不起作用，反应过程中易失活，所以在工业生产中往往不能很好地发挥作用。以上因素不仅增加了生产成本，而且降低了工业用酶的有效性。绝大多数酶在某些特定条件下容易变性失活，不稳定。而文中所述的嗜盐脂解酶在苛刻的条件下仍具有较高活性，这为工业化应用提供了可能性。

虽然嗜盐菌产生的新型脂肪酶具有特殊的酶学性质，在生物技术领域具有潜在的应用价值，但是这些酶还未商品化，它们在生物技术领域的应用尚未广泛开发。事实上，目前仅有几种来自嗜盐细菌的脂肪酶被商业化，如海杆菌源脂解酶（*M. lipolyticus*）LipBL（Pérez et al. 2011）。

已经发现对嗜盐菌进行商业化的大规模培养非常困难。虽然，从商业化角度讲，大规模培养可以有效降低酶的分离成本，但是在大多数情况下，并没有有效的规模化培养方法。酶的表达量往往较低，一般低于细胞总蛋白质的5%（Pire et al. 2001）。虽然用不同的实验方法优化了嗜中温脂肪酶生产菌的培养条件（Turki 2013），但对嗜盐微生物产脂解酶进行优化的系统的研究较少（Chauhan et al. 2013）。基因工程则利用原核生物（大肠杆菌）和真核生物酿酒酵母（*Sacharomyces cerevisiae*）和毕赤酵母（*Pichia Pastoris*）表达脂肪酶基因以提高脂肪酶产量（Valero 2012）。定点诱变的方法通常被用来提高酶的稳定性及改良酶的催化特性。总之，产酶优化及通过新型异源宿主对嗜盐酶进行大量表达面临巨大的挑战。

近年来，环境基因组和全基因组测序为嗜盐酶生产菌在生物技术领域的应用提供了新的机遇。在不可培养嗜盐细菌和古菌中，编码脂解酶的基因是非必需基因，比较可培养和不可培养微生物中脂肪酶基因，分析它们的系统进化关系是非常重要的。

最后，我们将重点介绍嗜盐菌脂肪酶的应用。

13.5.1 生物柴油工业

近年来，由于石油价格上涨，生物柴油的生产逐渐增加，通常工业上通过碱催化方法制备生物柴油（脂肪酸烷基酯），其特征是反应速率高。然而，油中存在的水和游离脂肪酸往往产生较多的皂化物，使得下游工业进程变得困难（Uthoff et al. 2009）。利用脂肪酶进行柴油生产克服了许多利用化学法生产生物柴油过程中遇到的问题（Ribeiro et al. 2011；Hama Kondo. 2013）。然而，酶法生产仍处于初级阶段，必须进一步寻找足够稳定的，在酯交换反应过程中不至于失活的新型脂肪酶，选择更为经济的油性底物，设计特殊的生物反应器则有利于最终产品的回收。

13.5.2 制药业

关于将嗜盐脂肪酶用于制药工业的化合物生产中，研究侧重于开发新的环境友好型反应，脂肪酶的高选择性有利于合成具有生物特性的复杂分子的中间体，脂肪酶作为生物催化剂具有区域选择性和化学选择性，可选择性酰化碳水化合物。这些反应的产物是合成低聚糖和具有医疗用途的糖复合物的中间体，例如抗癌剂（Morris et al. 2011）、抗病毒药（Howe et al. 2013）或疫苗（Peri 2013）。

因此，利用烯糖组装的方法，通过不同的二糖结构单元制备糖复合体。由于这些碳水化合物中类似的官能团数量很多，合适的保护和去保护序列在实现这一目标方面起决定性的作用。新型脂肪酶以不同的速度进行乙酰化和去乙酰化，而且可以作用于不同位点的糖基，使每一位置的糖基均形成一系列区域异构体，可以在药物制备中快速合成所需要的低聚糖。

13.5.3 营养行业

天然多酚作为有效的抗氧化剂与人类疾病预防的关系是众所周知的（Díaz-Rubio et al. 2014）。一些研究强调多酚亲脂特性的重要性，它具有显著的抗氧化能力和改进的理化性质。因此，对天然多酚进行改性，使其成为新型脂溶性抗氧化剂，对于食品工业是非常有意义的。预期这些改良的类似物具有比天然多酚更好的生物利用率，可用于制备有益健康的“保健食物”（Lucas et al. 2010；Gupta et al. 2013）。天然多酚具有多功能性，导致化学合成法不仅复杂，而且还需要克服许多技术难题。通常需要长时间进行保护化和去保护化以在特定位置获得单酰化。而脂肪酶具有区域选择性和化学选择性，所以可以利用酶催化的方法替代化学方法。寻找能够对天然多酚进行化学选择性和区域选择性酰化的脂肪酶将具有重要意义。

致谢

本研究得到君塔·安达卢西亚（Junta de Andalucía）的资助（P08-RMN-3515 和 P11-CVI-7427）

参考文献

Amoozegar MA，Salehghamari E，Khajeh K，Kabiri M，Naddaf S（2008a）Production of an extracellular thermohalophilic lipase from a moderately halophilic bacterium，*Salinivibrio* sp. strain SA-2. J Basic Microbiol 48：160-167

Amoozegar MA，Schumann P，Hajighasemi M，Fatemi AZ，Karbalaei-Heidari HR（2008b）*Salinivibrio proteolyticus* sp. nov. a moderately halophilic and proteolytic species from a hypersaline lake in Iran. Int J Syst Evol Microbiol 58：1159-1163

Antón J，Oren A，Benlloch S，Rodríguez-Valera F，Amann R，Rosselló-Mora R（2002）*Salinibacterruber* gen. nov.，sp. nov.，a novel，extremely halophilic member of the Bacteria from saltern crystallizer ponds. Int J Syst Evol Microbiol 52：485-491

Antranikian G，Vorgias CE，Bertoldo C（2005）Extreme environments as a resource for microorganisms and novel biocatalysts. Adv Biochem Eng Biotechnol 96：219-262

Ardakini MR，Roayaie M，Poshtkouhian A，Amozeegar MA，Zolgharnein H（2012）Isolation of moderately halophilic*Pseudoalteromonas* producing extracellular hydrolytic enzymes from Persian Gulf. Indian J Microbiol 52：94-98

Arpigny LL，Jaeger K-E（1999）Bacterial lipolytic enzymes：classification and properties. Biochem J 343：177-183

Babavalian H，Amoozegar MA，Zahraei S，Rohban R，Shakeri F，Moghaddam MM（2014）Comparison of bacterial biodiversity and enzyme production in three hypersaline lakes；Urmia，Howz-Soltan and Aran-Bidgol. Indian J Microbiol 54：444-449

Baliga NS，Bonneau R，Facciotti MT，Pan M，Glusman G，Deutsch EW，Shannon P，Chiu Y，Weng RS，Gan RR，Hung P，Date SV，Marcotte E，Hood L，Ng WV（2004）Genome sequence of *Haloarcula marismortui*：a halophilic archaeon from the Dead Sea. Genome Res 14：2221-2234

Bassegoda A，Pastor FI，Diaz P（2012）*Rhodococcus* sp. strain CR-53 LipR，the first member of a new bacterial lipase family（family X）displaying an unusual Y-type oxyanion hole，similar to the *Candida antarctica* lipase clan. Appl Environ Microbiol 78：1724-1832

Bayer S，Kunert A，Ballschmiter M，Greiner-Stoeffele T（2010）Indication for a new lipolytic enzyme family：isolation and characterization of two esterases from a metagenomic library. J Mol Microbiol Biotechnol 18：181-187

Bhatnagar T，Boutaiba S，Hacène H，Cayol JL，Fardeau ML，Ollivier B，Baratti JC（2005）Lipolytic activity from *Halobacteria*：screening and hydrolase production. FEMS Microbiol Lett 248：133-140

Birbir M，Ogan A，Calli B，Mertoglu B（2004）Enzyme characteristics of extremely halophilic archaeal community in Tuzkoy salt mine，Turkey. World J Microbiol Biotechnol 20：613-621

Bonete MJ，Martínez-Espinosa RM（2011）Enzymes from halophilic archaea：open questions. In：Ventosa A，Oren A，Ma Y（eds）Halophiles hypersaline environments. Springer，Berlin/Heidelberg，pp 359-371

Boutaiba S，Bhatnagar T，Hacène H，Mitchell DA，Baratti JC（2006）Preliminary characterization of a lipolytic activity from an extremely halophilic archaeon，*Natronococcus* sp. J Mol Catal B：Enzym 41：21 -26

Britton KL，Stillman TJ，Yip KS，Forterre P，Engel PC，Rice DW（1998）Insights into the molecular basis of salt tolerance from the study of glutamate dehydrogenase from *Halobacterium salinarum*. J Biol Chem 273：9023-9030

Brockerhoff J，Jensen RG（1974）Lipolytic enzymes. In：Brockerhoff J，Jensen RG（eds）Lipases. Academic，New York，pp 25-175

Camacho RM，Mateos JC，González-Reynoso O，Prado LA，Córdova J（2009）Production and characterization of esterase and lipase from *Haloarcula marismortui*. J Ind Microbiol Biotechnol 36：901-909

Camacho RM，Mateos-Díaz JC，Diaz-Monta ño DM，González-Reynoso O，Córdova J（2010）Carboxyl ester hydrolases production and growth of a halophilic archaeon，*Halobacterium* sp. NRC-1. Extremophiles 14：99-106

Ceylan S，Yilan G，Akbulut BS，Poli A，Kazan D（2012）Interplay of adaptive capabilities of *Halomonas* sp. AAD12 under salt stress. J Biosci Bioeng 114：45-52

Chauhan M，Chauhan RS，Garlapati VK（2013）Modelling and optimization studies on a novel lipase production by *Staphylococcus arlettae* through submerged fermentation. Enzyme Res 2013：353954，8p

Coronado MJ，Vargas C，Mellado E，Tegos G，Drainas C，Nieto JJ，Ventosa A（2000）The α-amylase gene *amy*H of the moderate halophile *Halomonas meridiana*：cloning and molecular characterization. Microbiology 146：861-868

Dalmaso GZL，Ferreira D，Vermelho AB（2015）Marine extremophiles：a source of hydrolases for biotechnological applications. Mar Drugs 13：1925-1965

Danson MJ，Hough DW（1997）The structural basis of protein halophilicity. Comp Biochem Physiol A Physiol 117：307-312

Daoud L，Kamoun J，Ali MB，Jallouli R，Bradai R，Mechichi T，Gargouri Y，Ali YB，Aloulou A（2013）Purification and biochemical characterization of a halotolerant *Staphylococcus* sp. extracellular lipase. Int J Biol Macromol 57：232-237

De la Haba RR，Márquez MC，Papke RT，Ventosa A（2012）Multilocus sequence analysis of the family *Halomona-*

daceae. Int J Syst Evol Microbiol 62：520-538

Detkova EN，Boltyanskaya YV（2007）Osmoadaptation of haloalkaliphilic bacteria：role of osmoregulators and their possible practical application. Microbiology 76：511-522

Ribeiro BD，de Castro AM，Coelho MA，Freire DM（2011）Production and use of lipases in bioenergy：a review from the feedstocks to biodiesel production. Enzyme Res 2011：615803，16 p

Díaz-Rubio ME，Pérez-Jiménez J，Martínez-Bartolomé MA，álvarez I，Saura-Calixto F（2014）Regular consumption of an antioxidant-rich juice improves oxidative status and causes metabolome changes in healthy adults. Plant Foods Hum Nutr 70：9-14

Doukyu N，Ogino H（2010）Organic solvent-tolerant enzymes. Biochem Eng J 48：270-282

Ebel C，Zaccai G（2004）Crowding in extremophiles：linkage between solvation and weak protein-protein interactions，stability and dynamics，provides insight into molecular adaptation. J Mol Recognit 17：382-389

Eisenberg H，Mevarech M，Zaccai G（1992）Biochemical，structural，and molecular genetic aspects of halophilism. Adv Protein Chem 43：1-6

Empadinhas N，da Costa MS（2008）Osmoadaptation mechanisms in prokaryotes：distribution of compatible solutes. Int Microbiol 11：151-161

Esakkiraj P，Prabakaran G，Maruthiah T，Immanuel G，Palavesam A（2014）Purification and characterization of halophilic alkaline lipase from *Halobacillus* sp. Proc Natl Acad Sci India Sect B Biol Sci. doi：10. 1007/s40011-014-0437-1

Esteves AM，Chandrayan SK，McTernan PM，Borges N，Adams MWW，Santos H（2014）Mannosylglycerate and Di-myo-Inositol phosphate have interchangeable roles during adaptation of *Pyrococcus furiosus* to heat stress. Appl Environ Microbiol 80：4226-4233

Ewis HE，Abdelal AT，Lu CD（2004）Molecular cloning and characterization of two thermostable carboxyl esterases from *Geobacillus stearothermophilus*. Gene 329：187-195

Fu J，Leiros HK，de Pascale D，Johnson KA，Blencke HM，Landfald B（2013）Functional and structural studies of a novel cold-adapted esterase from an Arctic intertidal metagenomic library. Appl Microbiol Biotechnol 97：3965-3978

Gandbhir M，Rasched I，Marliere P，Mutzel R（1995）Convergent evolution of amino acid usage in archaebacterial and eubacterial lineages adapted to high salt. Res Microbiol 146：113-120

Ghamesi Y，Rasoul-Amini S，Kazemi A，Zarrinic G，Morowvat MH，Kargar M（2011）Isolation and characterization of some moderately halophilic bacteria with lipase activity. Mikrobiologiia 80：477-481

Goh F，Jeon YJ，Barrow K，Neilan BA，Burns BP（2011）Osmoadaptive strategies of the archaeon *Halococcus hamelinensis* isolated from a hypersaline stromatolite environment. Astrobiology 6：529-536

Grant WD，Gemmel RT，GcGenity TJ（1998）Halophiles. In：Horikoshi K，Grant WD（eds）Extremophiles：microbial life in extreme environments. Wiley-Liss Inc.，New York，pp 93-132

Graziano G，Merlino A（2014）Molecular bases of protein halotolerance. Biochim Biophys Acta 4：850-858

Gupta A，Khare SK（2009）Enzymes from solvent tolerant microbes：useful biocatalysts for nonaqueous enzymology. Crit Rev Biotechnol 29：44-54

Gupta R，Gupta N，Rathi P（2004）Bacterial lipases：an overview of production，purifi cation and biochemical properties. Appl Microbiol Biotechnol 64：763-781

Gupta A，Kagliwal LD，Singhal RS（2013）Biotransformation of polyphenols for improved bioavailability and processing stability. Adv Food Nutr Res 69：183-217

Hacěne H，Rafaa F，Chebbhounia N，Boutaibaa S，Bhatnagarb T，Baratti JC，Ollivier B（2004）Biodiversity of prokaryotic microflora in El Golea Salt Lake，Algerian Sahara. J Arid Environ 58：273-284

Hama S，Kondo A（2013）Enzymatic biodiesel production：an overview of potential feedstocks and process development. Bioresour Technol 135：386-395

Handrick R，Reinhardt S，Focarete ML，Scandola M，Adamus G，Kowalczuk M，Jendrossek D（2001）A new type of thermoalkalophilic hydrolase of *Paucimonas lemoignei* with high specificity for amorphous polyesters of short chain-length hydroxyalkanoic acids. J Biol Chem 276：36215-36224

Hasan F，Shah AA，Hameed A（2006）Industrial applications of microbial lipases. Enzyme Microbiol Technol 39：235-251

Hasan F，Shah AA，Hameed A（2009）Methods for detection and characterization of lipases：a comprehensive review. Biotechnol Adv 27：782-798

Howe JD，Smith N，Lee MJ，Ardes-Guisot N，Vauzeilles B，Désiré J，Baron A，Blériot Y，Sollogoub M，Alonzi DS，Butters TD（2013）Novel imino sugar α-glucosidase inhibitors as antiviral compounds. Bioorg Med Chem 21：4831-4838

Hutcheon GW，Vasisht N，Bolhuis A（2005）Characterisation of a highly stable alpha-amylase from the halophilic archaeon *Haloarcula hispanica*. Extremophiles 9：487-495

Jaeger KE，Eggert T（2002）Lipases for biotechnology. Curr Opin Biotechnol 13：390-397

Jaeger KE，Dijkstra BW，Reetz MT（1999）Bacterial biocatalysts：molecular biology，three dimensional structures，and biotechnological applications of lipases. Annu Rev Microbiol 53：315-351

Janssen PH，Monk CR，Morgan HW（1994）A thermophilic，lipolytic *Bacillus* sp.，and continuous assay of its p-nitro-phenyl-palmitate esterase activity. FEMS Microbiol Lett 120：195-200

Jiang X，Huo Y，Cheng H，Zhang X，Zhu X，Wu M（2012）Cloning，expression and characterization of a halotolerant esterase from a marine bacterium *Pelagibacterium halotolerans* B2^{T}. Extremophiles 16：427-435

Johnsen U，Schonheit P（2004）Novel xylose dehydrogenase in the halophilic archaeon *Haloarcula marismortui*. J

Bacteriol 186：6198-6207

Joo WA，Kim CW（2005）Proteomics of halophilic archaea. J Chromatogr B Analyt Technol Biomed Life Sci 815：237-250

Jordan SN，Mullen GJ（2007）Enzymatic hydrolysis of organic waste materials in a solid-liquid system. Waste Manag 27：1820-1828

Joris B，Ghuysen JM，Dive G，Renard A，Dideberg O，Charlier P，Frère JM，Kelly JA，Boyington JC，Moews PC，Knox JR（1988）The active-site-serine penicillin-recognizing enzymes as members of the *Streptomyces* R61 DD-peptidase family. Biochem J 250：313-324

Karpushova A，Brümmer F，Barth S，Lange S，Schmid RD（2005）Cloning，recombinant expression and biochemical characterisation of novel esterases from *Bacillus* sp. associated with the marine sponge *Aplysina aerophoba*. Appl Microbiol Biotechnol 67：59-69

Kharroub K，Gomri MA，Aguilera M，Monteoliva-Sánchez M（2014）Diversity of hydrolytic enzymes in haloarchaea isolated from Algerian sabkhas. Afr J Microbiol Res 8：3992-4001

Kim EY，Oh KH，Lee MH，Kang CH，Oh TK，Yoon JH（2009）Novel cold-adapted alkaline lipase from an intertidal flat metagenome and proposal for a new family of bacterial lipases. Appl Environ Microbiol 75：257-260

Kirsh D（1935）Lipase production by *Penicillium oxalicum* and *Aspergillus flavus*. Bot Gaz 97：321-333

Knox JR，Moews PC，Frere JM（1996）Molecular evolution of bacterial β-lactam resistance. Chem Biol 3：937-947

Kumar S，Khare SK（2012）Purifi cation and characterization of maltooligosaccharide-forming alpha-amylase from moderately halophilic *Marinobacter* sp. EMB8. Bioresour Technol 116：247-251

Kumar S，Karan R，Kapoor S，Singh SP，Khare SK（2012）Screening and isolation of halophilic bacteria producing industrially important enzymes. Braz J Microbiol 43：1595-1603

Laane C，Boeren S，Vos K，Veeger C（1987）Rules for optimization of biocatalysis in organic solvents. Biotechnol Bioeng 30：81-87

Lanyi JK（1974）Salt-dependent properties of protein from extremely halophilic bacteria. Bacteriol Rev 38：272-290

Lee MH，Lee CH，Oh TK，Song JK，Yoon JH（2006）Isolation and characterization of a novel lipase from a metagenomic library of tidal flat sediments：evidence for a new family of bacterial lipases. Appl Environ Microbiol 72：7406-7409

Lee MH，Hong KS，Malhotra S，Park JH，Hwang EC，Choi HK，Kim YS，Tao W，Lee SW（2010）A new esterase EstD2 isolated from plant rhizosphere soil metagenome. Appl Microbiol Biotechnol 88：1125-1134

Lee HW，Yim KJ，Song HS，Nam YD，Choi HJ，Seo MJ，Kim KN，Kim D，Roh SW，Rhee JK（2014）Draft genome sequence of *Halorubrum halophilum* $B8^T$，an extremely halophilic archaeon isolated from salt-fermented seafood. Mar Genomics 18：117-118

Li X，Yu H-Y（2012）Characterization of a novel extracellular lipase from a halophilic isolate，*Chromohalobacter* sp. LY7-8. Afr J Microbiol Res 14：3516-3522

Li X，Yu H-Y（2014）Characterization of an organic solvent-tolerant lipase from *Haloarcula* sp. G41 and its application for biodiesel production. Folia Microbiol 59：455-463

Li X，Yu H-U，Lin Y-F（2012）Purification and characterization of an extracellular esterase from a moderately halophilic bacterium，*Halobacillus* sp. strain LY5. Afr J Biotechnol 23：6327-6334

Li X，Qian P，Wu SG，Yu HY（2014）Characterization of an organic solvent-tolerant lipase from *Idiomarina* sp. W33 and its application for biodiesel production using Jatropha oil. Extremophiles 18：171-178

Litchfield CD（2011）Potential for industrial products from the halophilic *Archaea*. J Ind Microbiol Biotechnol 38：1635-1647

Liu L，Liu Y，Shin HD，Chen RR，Wang NS，Li J，Du G，Chen J（2013）Developing *Bacillus* spp. as a cell factory for production of microbial enzymes and industrially important biochemicals in the context of systems and synthetic biology. Appl Microbiol Biotechnol 97：6113-6127

López-López O，Cerdán ME，González Siso MI（2014）New extremophilic lipases and esterases from metagenomics. Curr Protein Pept Sci 15：445-455

Lucas R，Comelles F，Alcántara D，Maldonado OS，Curcuroze M，Parra JL，Morales JC（2010）Surface-active properties of lipophilic antioxidants tyrosol and hydroxytyrosol fatty acid esters：a potential explanation for the nonlinear hypothesis of the antioxidant activity in oil-inwater emulsions. J Agric Food Chem 14：8021-8026

Lv X-Y，Guo L-Z，Song L，Fu Q，Zhao K，Li A-X，Luo X-L，Lu W-D（2011）Purification and characterization of a novel extracellular carboxylesterase from the moderately halophilic bacterium *Thalassobacillus* sp. strain DF-E4. Ann Microbiol 61：281-290

Madern D，Pfister C，Zaccai G（1995）Mutation at behaviour of malate deshydrogenase from *Haloarcula marismortui* in physiological salts. Eur J Biochem 230：1088-1095

Madern D，Ebel C，Zaccai G（2000）Halophilic adaptation of enzymes. Extremophiles 4：91-98

Margesin R，Schinner F（2001）Potential of halotolerant and halophilic microorganisms for biotechnology. Extremophiles 5：73-83

Martín S，Márquez MC，Sánchez-Porro C，Mellado E，Arahal DR，Ventosa A（2003）*Marinobacter lipolyticus* sp. nov.，a novel moderate halophile with lipolytic activity. Int J Syst Evol Microbiol 53：1383-1387

Mellado E，Martín S，Sánchez-Porro C，Ventosa A（2005）Lipolytic enzymes from extremophilic microorganisms. In：Mellado E，Barredo JL（eds）Microorganisms for industrial enzymes and biocontrol. Research Signpost，Kerala/India，pp 25-43

Mevarech M，Frolow F，Gloss LM（2000）Halophilic enzymes：proteins with a grain of salt. Biophys Chem 86：155-

164

Moreno ML，Garcia MT，Ventosa A，Mellado E（2009）Characterization of *Salicola* sp. IC10，a lipase-and protease producing extreme halophile. FEMS Microbiol Ecol 68：59-71

Moreno ML，García MT，Ventosa A，Iglesias-Guerra F，Mellado E（2010）The extremely halophilic bacterium *Salicola marasensis* IC10 accumulates the compatible solute betaine. Syst Appl Microbiol 33：308-310

Moreno ML，Piubeli F，Bonfá MRL，García MT，Durrant LR，Mellado E（2012）Analysis and characterization of cultivable extremophilic hydrolytic bacterial community in heavy-metalcontaminated soils from the Atacama Desert and their biotechnological potentials. J Appl Microbiol 113：550-559

Moreno ML，Pérez D，Garcia MT，Mellado E（2013）Halophilic bacteria as a source of novel hydrolytic enzymes. Life 3：38-51

Morris JC，Chiche J，Grellier C，Lopez M，Bornaghi LF，Maresca A，Supuran CT，Pouyssegur J，Poulsen S-A（2011）Targeting hypoxic tumor cell viability with carbohydrate-based carbonic anhydrase IX and XII inhibitors. J Med Chem 54：6905-6918

Müller-Santos M，de Souza EM，Pedrosa Fde O，Mitchell DA，Longhi S，Carrière F，Canaan S，Krieger N（2009）First evidence for the salt-dependent folding and activity of an esterase from the halophilic archaea *Haloarcula marismortui*. Biochim Biophys Acta 1791：719-729

Ng WV，Kennedy SP，Mahairas GG，Berquist B，Pan M，Shukla HD，Lasky SR，Baliga NS，Thorsson V，Sbrogna J，Swartzell S，Weir D，May J，Dahl TA，Welti R，Goo YA，Leithauser B，Séller K，Cruz R，Danson MJ，Hough DW，Maddocks DG，Jablonski PE，Krebs MP，Angevine CM，Dale H，Isenbarger TA，Peck RF，Pohlschroder M，Spudich JL，Jung KW，Alam M，Freitas T，Hou S，Daniels CJ，Dennis PP，Omer AD，Ebhardt H，Lowe TM，Liang P，Riley M，Hood L，DasSarma S（2000）Genome sequence of *Halobacterium* species NRC-1. Proc Natl Acad Sci USA 97：12176-12181

Ni ño de Guzman M，Virginia A，Antezana H，Svoboda M（2008）Lipolytic enzyme production by halophilic/halotolerant microorganisms isolated from Laguna Verde，Bolivia. Rev Boliv Quim 25：14-23

Ollis DL，Cheah E，Cygler M，Dijkstra B，Frolow F，Franken SM，Harel M，Remington SJ，Silman I，Schrag J，Sussman JL，Verschueren KHG，Goldman A（1992）The alpha/beta hydrolase fold. Protein Eng 5：197-211

Oren A（2010）Industrial and environmental applications of halophilic microorganisms. Environ Technol 31：825-834

Oren A（2013）Life at high salt concentrations，intracellular KCl concentrations，and acidic proteomes. Front Microbiol 4：315

Oren A，Mana L（2002）Amino acid composition of bulk protein and salt relationships of selected enzymes of *Salinibacter ruber*，an extremely halophilic bacterium. Extremophiles 6：217-223

Oren A，Rodríguez-Valera F，Antón J，Benlloch S，Rosselló-Mora R，Amann R，Coleman J，Russell NJ（2003）Red，extremely halophilic，but not archaeal：the physiology and ecology of *Salinibacter ruber*，a bacterium isolated from saltern crystallizer ponds. In：Ventosa A（ed）Halophilic microorganisms. Springer，Berlin/Heidelberg，pp 63-76

Ozcan B，Ozyilmaz G，Cokmus C，Caliskan M（2009）Characterization of extracellular esterase and lipase activities from five halophilic archaeal strains. J Ind Microbiol Biotechnol 36：105-110

Ozcan B，Ozyilmaz G，Cihan A，Cokmus C，Caliskan M（2012）Phylogenetic analysis and characterization of lipolytic activity of halophilic archaeal isolates. Mikrobiologiia 81：205-213

Papke RT，de la Haba R，Infante-Domínguez C，Pérez D，Sánchez-Porro C，Lapierre P，Ventosa A（2013）Draft genome sequence of the moderately halophilic bacterium *Marinobacter lipolyticus* strain SM19. Genome Announc 1：e00379-13

Pérez D，Martin S，Fernández-Lorente G，Filice M，Guisán JM，Ventosa A，Garcia MT，Mellado E（2011）A novel halophilic lipase，LipBL，showing high efficiency in the production of eicosapentaenoic acid（EPA）. PLoS One 6，e23325

Pérez D，Kovacic F，Wilhelm S，Jaeger KE，Garcia MT，Ventosa A，Mellado E（2012）Identification of amino acids involved in hydrolytic activity of lipase LipBL from *Marinobacter lipolyticus*. Microbiology 158：2192-2203

Peri F（2013）Clustered carbohydrates in synthetic vaccines. Chem Soc Rev 42：4543-4556

Pieper U，Kapadia G，Mevarech M，Herzberg O（1998）Structural features of halophilicity derived from the crystal structure of dihydrofolate reductase from the Dead Sea halophilic archaeon，*Haloferax volcanii*. Structure 6：75-88

Pikuta EV，Hoover RB，Tang J（2007）Microbial extremophiles at the limits of life. Crit Rev Microbiol 33：183-209

Pire C，Esclapez J，Ferrer J，Bonete MJ（2001）Heterologous overexpression of glucose dehydrogenase from the halophilic archaeon *Haloferax mediterranei*，an enzyme of the medium chain dehydrogenase/reductase family. FEMS Microbiol Lett 200：221-227

Rao L，Xue Y，Zhou C，Tao J，Li G，Lu JR，Ma Y（2011）A thermostable esterase from *Thermoanaerobacter tengcongensis* opening up a new family of bacterial lipolytic enzymes. Biochim Biophys Acta 1814：1695-1702

Rohban R，Amoozegar MA，Ventosa A（2009）Screening and isolation of halophilic bacteria producing extracellular hydrolyses from Howz Soltan Lake，Iran. J Ind Microbiol Biotechnol 36：333-340

Salameh M，Wiegel L（2007）Lipases from extremophiles and potential for industrial applications. Adv Appl Microbiol 61：253-283

Sana BGD，Saha M，Mukherjee J（2007）Purification and characterization of an extremely dimethylsulfoxide tolerant esterase from a salt-tolerant *Bacillus* species isolated from the marine environment of the Sundarbans. Process Biochem 42：1571-1578

Sánchez-Porro C，Martín S，Mellado E，Ventosa A（2003）Diversity of moderately halophilic bacteria producing extracellular hydrolytic enzymes. J Appl Microbiol 94：295-300

Saum SH，Pfeiffer F，Palm P，Rampp M，Schuster SC，Müller V，Oesterhelt D（2013）Chloride and organic osmolytes：a hybrid strategy to cope with elevated salinities by the moderately halophilic，chloride-dependent bacterium *Halobacillus halophilus*. Environ Microbiol 15：1619-1633

Schreck SD，Grunden AM（2014）Biotechnological applications of halophilic lipases and thioesterases. Appl Microbiol Biotechnol 98：1011-1021

Sellek GA，Chaudhuri JB（1999）Biocatalysis in organic media using enzymes from extramophiles. Enzyme Microbiol Technol 25：471-482

Sher J，Elevi R，Mana L，Oren A（2004）Glycerol metabolism in the extremely halophilic bacterium *Salinibacter ruber*. FEMS Microbiol Lett 232：211-215

Siglioccolo A，Paiardini A，Piscitelli M，Pascarella S（2011）Structural adaptation of extreme halophilic proteins through decrease of conserved hydrophobic contact surface. BMC Struct Biol 11：50

Sinha R，Khare SK（2013）Thermostable proteases. In：Satyanarayana T，Littlechild J，Kawarabayasi Y（eds）Thermophilic microbes in environmental and industrial biotechnology. Springer，Dordrecht，pp 859-880

Sinha R，Khare SK（2014）Protective role of salt in catalysis and maintaining structure of halophilic proteins against denaturation. Front Microbiol 5：165

Studdert CA，Herrera-Seitz MK，Plasencia-Gil MI，Sanchez JJ，De Castro RE（2001）Purification and biochemical characterization of the haloalkaliphilic archaeon *Natronococcus occultus* extracellular serine protease. J Basic Microbiol 41：375-383

Turki S（2013）Towards the development of systems for high-yield production of microbial lipases. Biotechnol Lett 35：1551-1560

Uratani JM，Kumaraswamy R，Rodriguez J（2014）A systematic strain selection approach for halotolerant and halophilic bioprocess development：a review. Extremophiles 18：629-639

Uthoff S，Bröker D，Steinbüchel A（2009）Current state and perspectives of producing biodiesel like compounds by biotechnology. Microb Biotechnol 2：551-565

Valero F（2012）Heterologous expression systems for lipases：a review. In：Sandoval G（ed）Lipases and phospholipases，Methods in molecular biology. Humana Press，New York，pp 161-178

Ventosa A（2006）Unusual micro-organisms from unusual habitats：hypersaline environments. In：Logan NA，Lappin-Scott HM，Oyston PCF（eds）Prokaryotic diversity：mechanisms and signifycance. Cambridge University Press，Cambridge，pp 223-253

Ventosa A，Nieto JJ，Oren A（1998）Biology of moderately halophilic aerobic bacteria. Microbiol Mol Biol Rev 62：504-544

Ventosa A，Sánchez-Porro C，Martin S，Mellado E（2005）Halophilic achaea and bacteria as a source of extracellular hydrolytic enzymes. In：Gunde-Cimerman N，Oren A，Plemenitas A（eds）Adaptation of life at high salt concentrations in *Archaea*，*Bacteria* and *Eukarya*. Springer，Berlin/Heidelberg，pp 337-355

Ventosa A，Mellado E，Sánchez-Porro C，Márquez MC（2008）Halophilic and halotolerant microorganisms from soils. In：Dion P，Nautiyal CS（eds）Microbiology of extreme soils. Soil biology. Springer，Berlin/Heidelberg，pp 87-115

Ventosa A，Fernández AB，León MJ，Sánchez-Porro C，Rodriguez-Valera F（2014）The Santa Pola saltern as a model for studying the microbiota of hypersaline environments. Extremophiles 18：811-824

Webb KM，DiRuggiero J（2012）Role of Mn^{2+} and compatible solutes in the radiation resistance of thermophilic bacteria and archaea. Archaea 2012：845756，11 p

Xin L，Hui-Ying Y（2013）Purification and characterization of an extracellular esterase with organic solvent tolerance from a halotolerant isolate，*Salimicrobium* sp. LY19. BMC Biotechnol 13：108

Zaccai G（2004）The effect of water on protein dynamics. Philos Trans R Soc Lond B Biol Sci 359：1269-1275

Zaccai G，Cendrin F，Haik Y，Borochov N，Eisenberg H（1989）Stabilization of halophilic malate dehydrogenase. J Mol Biol 208：491-500

第十四章

极端蛋白酶：特殊功能的开发、潜在来源以及生物技术中的应用

Aneta Bialkowska[1][2]，Ewa Gromek[1]，
Tomasz Florczak[1]，Joanna Krysiak[1]，
Katarzyna Szulczewska[1]，Marianna Turkiewicz[1]

14.1 引言：蛋白酶的一般特性与分类

蛋白酶（蛋白水解酶、肽酶）是过去几十年里第一种被用于工业实际生产的酶类。一开始，在洗涤剂中发现了它们的用途（Rao et al. 1998），直到今天这仍是蛋白酶制剂的最大使用领域（Kumar et al. 2008；Khan 2013）。如今，蛋白酶，尤其是微生物来源的蛋白酶，被广泛地应用于食品、饲料、纺织、皮革以及制药工业。它们被越来越广泛地用于医疗和美容业当中。在生产方面，蛋白酶占据着全球酶制剂市场的核心，如今几十年里正高速发展着（Sarethy et al. 2011）。

从病毒和细菌到人类，蛋白酶在所有生命体中有各种突变体。编码蛋白酶的基因大约占据人类全基因组的 2%（Li et al. 2013）。它们不仅参与膳食蛋白质的消化和细胞内蛋白质的转化，还参与到各种代谢与调控过程，对有机体的正常运作有重要意义。蛋白酶是最有特征的酶之一，而专门针对蛋白酶的研究为现阶段研究蛋白质结构域与功能的关系作出了贡献。

在正式意义上，IUBMB（生物化学与分子生物学国际联盟）的分类委员会把蛋白酶归类为水解酶（EC 3），可作用于肽键（亚纲 4，EC 3.4），因此也被称作肽酶。这类酶的分类到 20 世纪 50 年代下半叶还在不断变化，直到官方发表了第一份酶的列表，水解肽键的新蛋白酶还不断被描述，并且对在催化过程中使用的酶的多样化及分子机制还在不断地研究当中。如今，EC 3.4 类含有 15 个亚类，包括外肽酶（氨基肽酶、羧基肽酶和二肽酶）以及内肽酶（天冬氨酸蛋白酶、半胱氨酸蛋白酶、丝氨酸蛋白酶、苏氨酸蛋白酶、金属蛋白酶和催化机制未知的蛋白酶，EC 3.4.21～25 和 EC 3.4.99）。同时，对于蛋白酶一级与二级结构的研究，拓展了对这些蛋白质之间的进化相关性的认识，在这基础上分为天冬氨酸蛋白酶、

[1] Institute of Technical Biochemistry，Lodz University of Technology，Stefanowskiego 4/10，90-924. Lodz，Poland.

[2] e-mail：aneta. bialkowska@p. lodz. pl.

半胱氨酸蛋白酶、谷氨酸蛋白酶、丝氨酸蛋白酶、苏氨酸蛋白酶、金属蛋白酶、混合或催化机制未知的蛋白酶，包括水解酶、天冬酰胺肽裂解酶。每一组又被进一步分成不同的包含了序列高度同源的酶家族，以及3D结构或分子中催化氨基酸序列很相似的簇。目前，MEROPS数据库把肽酶分成不同家族，并且提供了成千上万种单独肽酶的描述，一共列出了353个家族，106个亚家族和49个簇，显示了这些蛋白酶结构的多样性。此外，尽管实际上它们都水解肽键，但是也参与到了各种各样的代谢过程中。

基于对极端微生物来源蛋白酶的研究，对于蛋白酶的认识已经得到了极大的扩展。从更微观的角度来说，通过对极端和温和环境微生物的宏基因组进行比较分析，那些嗜极端特性的蛋白酶基因可能会被展现出来。接下来将介绍目前在这一领域的研究现状，包括极端蛋白酶在经济上的实际使用以及在生物技术中可能的用途。

14.2 极端蛋白酶的来源、特性和结构上的适应性

14.2.1 嗜热蛋白酶

嗜热微生物适应了在高温中生存，是嗜热蛋白酶最大的来源。它们被分为可生存在60～65℃与37℃的兼性嗜热菌；生长在65～70℃但不能在低于40℃生长的专性嗜热菌；生存在40～70℃，最适生长温度为65℃的极端嗜热菌；以及在80～115℃中生长的超嗜热菌（Kikani et al. 2010）。生产蛋白酶的嗜热微生物能够在各种生境中被发现，例如热带土壤（De Azeredo et al. 2004；Jaouadi et al. 2010a），海洋地热泉（Klingeberg et al. 1995），包括海平面2500m以下，压力超过250atm（1atm＝101325Pa）的热液喷口和间歇泉（Matsuzawa et al. 1988；van den Burg et al. 1991），地热沉积物，地热蒸汽（Jang et al. 2002），硫气孔（Morikawa et al. 1994；Chavez Croocker et al. 1999），火山和发酵堆肥（Hasbay Ifrij and Ogel 2002），地热发电厂沸腾的排水，热水管道系统（Dib et al. 1998），皮革厂废水和牲畜粪肥的高热蒸煮锅（Majeed et al. 2013），以及其他多种环境中。适应高温的蛋白酶是从超嗜热和嗜热古菌、细菌和丝状真菌中分离的。它们在超过60～70℃中保持稳定并具有活性，有些细菌甚至可以在超过100℃的环境中保持稳定（Sako et al. 1997；Morikawa et al. 1994）。与中温微生物相比，它们更能抵抗有机溶剂和去污剂（Jaouadi et al. 2010a；Lagzian and Asoodeh 2012），pH值以及其他变性因子（Synowiecki 2010）的影响。由于底物在中温环境中溶解度低或者酶促反应吸热的特点，它们多在高温中使用。高温能够提升底物的溶解度，同样因为环境黏度低可以加速扩散。高温还能够减少中温和适冷微生物污染的风险，有助于控制工艺的消毒。

提升热活性和热稳定性是由多个结构适应所决定的，甚至氨基酸序列中小的改变，都能够促进对高温的抗性。在平行比较试验中，来源于热容芽孢杆菌（*Bacillus caldolyticus*）突变体的中性蛋白酶中仅仅三个氨基酸位点的差异，使得该酶能在比 *Bacillus stearothermophilus*（嗜热脂肪芽孢杆菌）Nrp蛋白酶高8.2℃的温度下孵育30min后仍保持50%的活性（van den Burg et al. 1991）。在更高温度下的稳定性取决于分子折叠的水平。温度上升导致结构松散，分子的柔性增加，不可逆的构象变化最终导致原来属性的损失。嗜热蛋白酶有着比中温和适冷蛋白酶更加紧密的排列和坚固的结构，这有助于维持它们在高温中的催化特性。分子的刚性是由额外的离子键、二硫键、氢键和疏水作用产生。

此外，结合了金属离子的嗜热酶，其空间、表面上亲水区域减少，具有较短的环和较少的残基，且更倾向于脱氨与被氧化（Li et al. 2005）。就金属而言，尤其是Ca^{2+}、Mg^{2+}、Mn^{2+}、Zn^{2+}或Sr^{2+}，能够特异性或非特异性地与酶分子结合，离子的结合对于在更高的温

度中稳定结构和防止错误折叠是必需的（De Azeredo et al. 2004；Li et al. 2007）。

嗜热蛋白酶分子中的静电作用不仅多种多样，而且更加稳定。与中温和嗜冷的枯草杆菌蛋白酶相比，一种来自 *Thermoactinomyces vulgaris*（普通高温放线菌）（Kleine 1982）的耐热丝氨酸蛋白酶，在 85～90℃中具有最高活性，以及在 70℃中稳定性最高，该酶具有非常复杂和稳定的盐桥网络，通常是由精氨酸残基形成（Tiberti and Papaleo 2011）。如果通过适当的突变把 *Thermus aquaticus* YT-1（水生栖热菌）的枯草溶菌素Ⅰ中的一个盐桥去掉，蛋白酶的热稳定性明显降低，80℃中半衰期从 120min 降低到 40min（Matsuzawa et al. 1988）。

二硫键对于高温中枯草溶菌素Ⅰ的稳定性和催化活性同样重要。通过适当的突变弱化蛋白酶会导致酶热稳定性急剧下降。此外，已有研究证明，蛋白酶表面的环状结构中的脯氨酸有利于其稳定性（Sakaguchi et al. 2008）。所有这些修饰保证了蛋白酶在高温中的高活性，但在低温中的活性明显降低，这有利于加工完成后失活嗜热蛋白酶，无需使用昂贵的抑制剂。

到目前为止，已经分离和鉴定多种嗜热蛋白酶（表 14.1），这些酶在 pH 值 2.0～12.0 的范围内具有活性，最适温度为 45～115℃。目前已经证明由古菌产生的蛋白酶是嗜热性最强的蛋白酶，其中包括丝氨酸蛋白酶、半胱氨酸蛋白酶、金属蛋白酶和天冬氨酸蛋白酶。超嗜热古菌 *Pyrococcus* sp.（火球菌属）产生一种丝氨酸蛋白酶，其最适酶活性温度超过 100℃。分离自 *Pyrococcus furiosus* 的 S66 蛋白酶在 105℃和 pH 7.0 中活性最强，在 98℃中测量的半衰期为 33 h（Blumentals et al. 1990）。Pyrolysin 有比 S66 更高的最适温度，达 115℃，但热稳定性较低（95℃中半衰期只有 9h）（Eggen et al. 1990；Voorhorst et al. 1996）。*Aeropyrum pernix* K1（嗜热泉生古菌）产生的丝氨酸蛋白酶同样有着高活性和热稳定性。其中一个丝氨酸蛋白酶最适酶活温度为 90℃，最适 pH 值为 8.0～9.0，是目前发现的所有蛋白酶中热稳定性最强的丝氨酸蛋白酶，在 90℃孵育 4h 依旧保持着活性，120℃中半衰期为 30min。若 Ca^{2+} 浓度为 1mmol/L 时其热稳定性更进一步提升，120℃中保持 4h，其活性没有损失（Catara et al. 2003）。*Sulfolobus solfataricus*（硫化叶菌）产生的 SsMTP 蛋白酶，最适温度和 pH 值分别为 70℃和 2.0，有着较高的热稳定性（80℃中半衰期为 20h）。

表 14.1 已表征的嗜热蛋白酶的例子

来源	蛋白酶(PDB码)	生物体	最适温度/℃	最适pH	热稳定性	参考文献
古菌	pernisine，枯草杆菌素样的丝氨酸蛋白酶	*Aeropyrum pernix* K1	90	8.0	在 90℃中稳定，100℃中半衰期为 60min，含有 1mmol/L $CaCl_2$ 时在 120℃中保持稳定 4h	Catara et al.(2003)
	pernilase，丝氨酸蛋白酶（1VE6，1VE7，2QR5，2QZP，3O4G，3O4H，3O4I，3O4J，4RE5，4RE6）	*Aeropyrum pernix* K1	90	9.0	在 90℃中保持稳定 4h；在 100℃和 110℃中半衰期分别为 85min 和 12min	Chavez Croocker et al.(1999)
	pyrolysin，丝氨酸蛋白酶	*Pyrococcus furiosus*	115	6.5～10.5	在 80℃中稳定；半衰期：95℃中为 9h，100℃中为 4h，105℃中为 20min，110℃中为 3min	Eggen et al.(1990)，Voorhorst et al.(1996)

续表

来源	蛋白酶(PDB码)	生物体	最适温度/℃	最适pH	热稳定性	参考文献
古菌	S66丝氨酸蛋白酶	*Pyrococcus furiosus*	105	7.0	在98℃中半衰期为33h	Blumentals et al.(1990)
	aeropyrolysin，金属蛋白酶	*Aeropyrum pernix* K1	在1mmol/L $CaCl_2$中为110℃	6.0～8.0	在100℃，1mmol/L $CaCl_2$中稳定；120℃中半衰期为2.5h；在125℃，1mmol/L $CaCl_2$中为1.2h	Sako et al.(1997)
	半胱氨酸蛋白酶	*Pyrococcus* sp. KOD1	110	7.0	90℃中保持稳定2h；100℃中半衰期为60min	Morikawa et al.(1994)
	thermopsin，酸性蛋白酶	*Sulfolobus acidocaldarius*	90	2.0	在80℃和pH 4.5条件下保持稳定48h	Lin and Tang (1990)，Fusek et al.(1990)
	SsMTP，thermopsin样蛋白酶，酸性蛋白酶	*Sulfolobus solfataricus* P2	70 (40～90)	2.0	在50～80℃之间稳定，80℃中半衰期为20d	Cannio et al.(2010)
细菌	枯草杆菌素样蛋白酶	*Bacillus* sp. MLA64	95	9.5	90℃中保持稳定1h，110℃中半衰期为25min，120℃中半衰期为10min	Lagzian and Asoodeh(2012)
	thermicin，枯草杆菌素样蛋白酶	*Thermoanaerobacter yonseiensis* KB-1	92.5	9.0	80℃中半衰期为30h	Jang et al.(2002)
	caldolysin(枯草溶菌素Ⅰ)，丝氨酸蛋白酶	*Thermus aquaticus* T-351	75	8.0	在80℃中稳定；在90℃，10mmol/L $CaCl_2$中半衰期为1h	Cowan and Daniel(1982)
	Rt41A蛋白酶，丝氨酸蛋白酶	*Thermus* sp. Rt41A	含有5mmol/L $CaCl_2$时为90℃	8.0	在70℃中保持稳定24h；含有5mmol/L $CaCl_2$时在80℃中半衰期为13.5h	Peek et al.(1992)
	proteolysin，丝氨酸蛋白酶	*Coprothermobacter proteolyticus*（在大肠杆菌中表达）	85	9.5	70℃中保持稳定22h，80℃中半衰期为40min	Toplak et al.(2013)
	丝氨酸蛋白酶	*Aquifex pyrophilus*	85～95	7.0～9.0	85℃中半衰期为90h，105℃中半衰期为6h	Choi et al.(1999)
	碱性丝氨酸蛋白酶	*Geobacillus stearothermophilus* B-1172(在大肠杆菌中表达)	60	8.0	90℃中稳定1h，100℃中半衰期为1h	Iqbal et al.(2015)

续表

来源	蛋白酶(PDB码)	生物体	最适温度/℃	最适pH	热稳定性	参考文献
细菌	中性蛋白酶	*Bacillus* sp. HUTBS62	80	6.8	70℃中半衰期为6h，90℃中半衰期为3h	Aqel et al.(2012)
放线菌	TfpA,丝氨酸蛋白酶	*Thermomonospora fusca* YX(在毕赤酵母中表达)	80	8.5	80℃中稳定,85℃孵育15min后保持50%的活性	Kim and Lei(2005)
真菌	丝氨酸蛋白酶	*Chaetomium thermophilum*	60	8.0	在60℃中稳定,70℃中半衰期为60min,90℃中为10min	Li and Li(2009)
	碱性角蛋白水解酶	*Chrysosporium keratinophilum*	90	9.0	90℃中半衰期为30min	Dozie et al.(1994)

嗜热菌蛋白酶包含有多个种类。目前为止，已经分离和报道了40多种在50～95℃之间具有活性的蛋白酶。据报道，来自*Bacillus* sp. MLA64的枯草杆菌素样蛋白酶最适活性温度为95℃（Lagzian and Asoodeh 2012）。与其他细菌蛋白酶相比，枯草杆菌素样蛋白酶已被证明具有极端热稳定性，其在100℃和120℃中半衰期分别为150min和10min。在吐温-80、SDS和Triton X-100存在的情况下仍保持活性，这与*Coprothermobacter proteolyticus*（溶蛋白栖热粪杆菌）的碱性丝氨酸蛋白酶类似，但后者还耐受过氧化氢，并且SDS还能增强其活性（Majeed et al. 2013）。另外一个来自*C. proteolyticus*的碱性蛋白酶在85℃和pH 9.5的条件下活性最高，市售的枯草杆菌蛋白酶A更能耐受有机溶剂（Toplak et al. 2013）。另一种来自*B. laterosporus* AK1的丝氨酸蛋白酶是在去污剂存在的情况下仍保持稳定的酶，最适温度和pH达75℃和9.0，在市售去污剂中孵育1h后仍保持38%～75%的活性，在80℃和pH 7.0～11.0范围内也能保持稳定，这意味着该酶可能可以用作新一代去污剂的添加剂（Arulmani et al. 2007）。一些细菌的嗜热蛋白酶显示出对变性物质如尿素和盐酸胍等较高的耐受能力。来自*T. aquaticus* YT-1的枯草溶菌素Ⅰ，80℃下最适pH为10.0，在7mol/L尿素，6mol/L盐酸胍和1% SDS存在的情况下仍保持活性（Matsuzawa et al. 1988），与来自*T. aquaticus* T-351的caldolysin相似（Cowan and Daniel 1982）。

由嗜热真菌和放线菌生产的蛋白酶数量相对而言较少。就真菌酶而言，它们包括来自*Chaetomium thermophilum*（嗜热毛壳菌）的PRO33和PRO66（Li et al. 2007），樟绒枝霉*Malbranchea pulchella* var. *sulfurea*所产的嗜热酶thermomycolase（Ong and Gaucher 1976），和*Thermoascus aurantiacus* var. *levisporus* Ⅰ（嗜热子囊菌）的蛋白酶Ⅰ（Marcy et al. 1984）。在放线菌中，我们可以辨别出来自*Streptomyces* sp. AB1（链霉素属）的KERAB（Jaouadi et al. 2010a），来自*Thermoactinomyces candidus*（纯白嗜热放线菌）的keratinase（Ignatova et al. 1999）和来自*Thermomonospora fusca* YX（褐色热单胞菌）的TfrA（Kim and Lei 2005）。它们的最适催化温度要低于来自古菌和细菌的酶，在45～70℃范围内。第一个特例是来自*Chrysosporium keratinophilum*（嗜角质金孢子菌）的碱性角蛋白酶，在90℃和pH 9.0下活性最高（Dozie et al. 1994）。第二个特例则是来自*T. fusca* YX的一种丝氨酸蛋白酶，TfpA，在80℃和pH 8.5中活性最高，该酶在该反应条件下能保持稳定，85℃中半衰期为15min（Kim and Lei 2005）。

14.2.2 嗜冷蛋白酶

适冷蛋白酶由嗜冷和耐冷微生物产生。这两种微生物都能够在0℃中生长。对真正的嗜冷微生物而言，其最适生长温度不超过15℃，最高生长温度也只有20℃。而对于耐冷微生物而言，其最适生长温度要高，大约在20～35℃之间（Morita1975）。这两种微生物都栖息在永久寒冷的区域和有周期性温度下降的环境中，可以在海水、北极土壤、冰川、高山土壤和永久冻土中找到它们。它们构成了寒冷地区植物和动物中的微生物群落，以及成为能够在冷却和冷冻食物中检测到的污染物。

嗜冷微生物最重要的适应能力之一，是它们所产的酶在动力学和结构上能够适应寒冷的环境，并得以生存。这是酶动力学与结构适应了低温生存后的结果。动力学的适应涉及相对较低的最适活性温度，通常为20～30℃，比位于30～40℃之间的中温微生物要低。这类酶的另一个特点是在0～30℃之间催化常数 k_{cat} 和催化效率 k_{cat}/K_m 要更高，热稳定性更低。与同类型的中温酶和嗜热酶相比，嗜冷微生物合成的酶具有更好的结构柔性，这是由于其分子内作用较弱，特别是在活性位点处，此外还与存在于酶周围的与极性环境相互作用的表面环有关。这些蛋白酶几乎不含增强天然构象刚性的脯氨酸残基，但含更多的甘氨酸残基，尤其会成簇的存在，这对柔性存在产生有利的影响。还可观察到碱性氨基酸残基含量降低，如精氨酸和赖氨酸。在适当的pH环境中，这些氨基酸能够形成离子键增强蛋白酶分子的刚性。该结构的一个重要特点是疏水作用降低，这得益于其分子占据的空间比同源中温蛋白酶少。对于冷活性酶，分子与溶剂之间的亲水作用则会增强，这是由于其表面存在大量极性氨基酸残基。此外，在这种情况下，通常更容易接近活性位点，从而抵消了底物扩散降低的效应（Feller 2003，2013）。

在迄今为止已报道的能生产冷活性蛋白酶的细菌当中，最占优势的细菌类型是假单胞菌（*Pseudomonas*）、假交替单胞菌（*Pseudoalteromonas*）、芽孢杆菌（*Bacillus*）、产气荚膜杆菌（*Clostridium*）、*Colwellia*、沙雷氏菌（*Serratia*）、希瓦氏菌（*Shewanella*）、脱硫杆菌属（*Sulfitobacter*）、盐单胞菌属（*Halomonas*）和海单胞菌（*Marinomonas*）。对于嗜冷酵母和丝状真菌蛋白酶的关注则较少（表14.2）。真核生物中最有趣的酶之一是来源于嗜冷真菌 *Aspergillus ustus*（焦曲霉）的丝氨酸蛋白酶，它分离自中印度河盆地的深海沉积物，甚至在300bar（1bar＝10^5Pa）压力下仍能保持其活性（Damare et al. 2006）。也有其他产蛋白酶的细菌来源于深海沉积物，例如 *Pseudoalteromonas* sp. SM9913（Chen et al. 2003）、*Halobacillus* sp. SCSIO 20089（盐杆菌）（Yang et al. 2013）、*Pseudomonas lundensis*（Yang et al. 2010）和 *Colwellia psychrerythraea*（Huston et al. 2004）。通常，这些菌株也存在于海水中。例如Yuan等（2009）从日本海分离了一株 *Enterococcus faecalis* TN-9（粪肠球菌），该细菌能够产生嗜冷金属蛋白酶。另外一株从黄海发现的 *Pseudomonas aeruginosa* HY1215 能够分泌丝氨酸蛋白酶（Hao and Sun 2015）。Zhu等人从黄海分离的 *Penicillium chrysogenum* FS010（产黄青霉）能够产生一种类似的酶（Zhu et al. 2009）。另一个在冷环境中富含蛋白水解酶菌株的例子是从南极土壤中鉴定出的土生隐球菌（*Candida humicola*），它能分泌目前已在文献中描述的唯一来源于酵母的一种嗜冷胞外天冬氨酸蛋白酶。也存在能产生丝氨酸蛋白酶的细菌，如 *Stenotrophomonas maltophilia*（嗜麦芽窄食单胞菌）（Vazquez et al. 2005）和 *Clostridium* sp.（Alam et al. 2005）。另一个重要的嗜冷蛋白酶来源是高原微生物，例如 *Serratia marcescens* TS1（灵杆菌）（Tariq et al. 2011）、*S. maltophilia* MTCC 7528（Kuddus and Ramteke 2009）和 *Pedobacter cryoconitis*（Margesin et al. 2005）。

表 14.2 来自嗜冷微生物的胞外蛋白酶

来源	蛋白酶（PDB 码）	生物体	最适温度/℃	最适 pH	热稳定性	参考文献
细菌	丝氨酸蛋白酶	*Pseudoalteromonas* sp. SM9913	30	7.0	在 30℃ 中能保持稳定 20min	Chen et al. (2003)
	丝氨酸蛋白酶	*Pseudoalteromonas* sp. NJ276	30	8.0	在 40℃ 中能保持稳定 400min	Wang et al. (2008)
	丝氨酸蛋白酶	*Clostridium* sp.	37	7.0	60℃ 孵育 1h 后保持超过 40% 的活性	Alam et al. (2005)
	丝氨酸蛋白酶	*Clostridium* sp. LP3	45	7.0～8.0	在 50℃ 中能保持稳定 30min	Alam et al. (2006)
	丝氨酸蛋白酶	*Colwellia* sp. NJ341	35	8.0	40℃ 中孵育 50min 后保持 50% 活性	Wang et al. (2005)
	丝氨酸蛋白酶	*Pseudomonas aeruginosa* HY1215	25	10.0	在 35℃ 中能保持稳定 30min	Hao and Sun (2015)
	丝氨酸蛋白酶	*Pseudomonas* sp. strain DY-A	40	10.0	在 30℃ 中能保持稳定 90min	Zeng et al. (2003)
	丝氨酸蛋白酶	*Bacillus amyloliquefaciens* S94	45	对于蛋白质底物而言为 10.0；对于合成底物而言为 8.0	暂无数据	Son and Kim (2003)
	丝氨酸蛋白酶	*Planomicrobium* sp. 547	35	9.0	50℃ 中孵育 2h 后保持 40% 活性	Yang et al. (2011)
	丝氨酸蛋白酶	*Bacillus* TA41	40	9.0	在 50℃ 中暴露 15min 后完全失去活性	Davail et al. (1994)
	丝氨酸蛋白酶	*Bacillus* TA39	暂无数据	9.0～10.0	50℃ 中半衰期 6min	Narinx et al. (1997)
	金属蛋白酶	*Enterococcus faecalis* TN-9	30	7.5～8.0	在 45℃ 中能保持稳定 10min	Yuan et al. (2009)
	金属蛋白酶	*Serratia marcescens* TS1	20	8.5	暂无数据	Tariq et al. (2011)
	金属蛋白酶	*Pseudomonas lundensis*	30	10.4	含有 Ca^{2+} 时在 25～40℃ 中保持稳定 2h	Yang et al. (2010)
	金属蛋白酶	*Stenotrophomonas maltophilia* MTCC 7528	20	10.0	在 20℃ 中能保持稳定 3h	Kuddus and Ramteke (2009)
	金属蛋白酶	*Pedobacter cryoconitis*	40	8.0	在 20～30℃ 中能保持稳定 1h	Margesinet al. (2005)
	金属蛋白酶 (3CIA)	*Colwellia psychrerythraea* strain 34H	19	7.0	在 30℃、40℃、45℃ 和 50℃ 中半衰期分别为 67min、38min、10min 和 5min	Huston et al. (2004)

续表

来源	蛋白酶（PDB码）	生物体	最适温度/℃	最适 pH	热稳定性	参考文献
酵母与丝状真菌	丝氨酸蛋白酶	*Aspergillus ustus*	45	9.0	50℃中孵育10min保持80%活性；60℃中孵育10min保持55%活性	Damare et al.(2006)
	丝氨酸蛋白酶	*Leucosporidium antarcticum*（now *Glaciozyma antarctica*）	25	8.0～8.5	在30℃中保持稳定	Turkiewicz et al.(2003)
	丝氨酸蛋白酶	*Penicillium chrysogenum* FS010	35	9.0	60℃中孵育5min保持10%活性	Zhu et al.(2009)
	天冬氨酸蛋白酶	*Candida humicola*（现在为*Cryptococcus humicola*）	37	1.0～1.2	在37℃中保持稳定2h	Ray et al.(1992)

文献中描述的大部分嗜冷蛋白酶都在30～40℃活性最高（表14.2），然而，还有一些酶有着更低的最适温度，例如从克什米尔地区的塔吉瓦斯冰川的土壤样本中分离的*Stenotrophomonas* sp. IIIM-ST045（寡养单胞菌）所分泌的碱性蛋白酶，在4～37℃范围内有显著降解酪蛋白的能力，并且在15℃和pH 10.0时活性最高（Saba et al. 2012）。而在喜马拉雅山脉西部的甘戈特里冰川中分离的*S. maltophilia* MTCC 7528生产的金属蛋白酶，在20℃和pH 10.0条件下水解偶氮酪蛋白的活性最高，该酶在市售洗涤剂中能保持稳定（孵育3h后能保持65%～80%活力），是一种极佳的在低温下去除含蛋白质污渍的清除剂，同时它能够耐受反复冻融，这可能与该菌的栖息环境有关（Kuddus and Ramteke 2009）。

来自高山冰川的冰雪中的*P. cryoconitis*（Margesin et al. 2005）和来源于黄海的*P. lundensis* HW08（Yang et al. 2010）所产的金属蛋白酶也显示出类似的特征。*P. aeruginosa* HY1215（Hao and Sun 2015）和*Glaciozyma antarctica*（*Leucosporidium antarcticum*）（Turkiewicz et al. 2003）分泌的丝氨酸蛋白酶是另外两种能够在相对低温中存活的酶，它们的最适活性温度为25℃，在0℃中分别保持最高活性的30%和20%。适冷酶的另一个重要特性是其热稳定性与产酶菌最适生存温度下的高催化效率有关。大部分文献中描述的适冷蛋白酶热稳定性都较低，在40～50℃中就会部分或完全失活。关于这点，值得一提的是分别来源于*Clostridium* sp.（Alam et al. 2005）、*Clostridium* sp. LP3（Alam et al. 2006）和*Planomicrobium* sp. 547（动性杆菌）（Yang et al. 2011）的三种丝氨酸蛋白酶，第一个酶在60℃中孵育1h后能维持最高活性的40%，第二个酶在50℃中保温30min后可保持80%的活性，而第三个酶在50℃中孵育2h后可保持40%的活性。

钙离子也在稳定特定蛋白酶结构中发挥着重要的作用，当存在这些离子时通常会引起蛋白酶的热稳定性升高，例如*P. lundensis* HW08的金属蛋白酶在20～35℃时，缺乏钙离子与存在5mmol/L钙离子的酶活性相比会损失20%的活性，而在40℃下则损失50%（Yang et al. 2010）。*Flavobacterium psychrophilum*（嗜冷黄杆菌）的金属蛋白酶Fpp2，在含有5mmol/L $CaCl_2$时其最适温度由24℃变为33～37℃，而且酶的热稳定性会上升，在40℃中暴露5min后仍能够维持95%的活性，但在钙离子缺乏时则完全失活（Secades et al. 2003）。

文献中报道的绝大部分嗜冷蛋白酶显示在中性或碱性环境中有最高活性。目前为止，只有一种最适pH为1.0的胞外天冬氨酸蛋白酶被报道（Ray et al. 1992）。对于去污剂工业而言，似乎对在高pH值中有活性的蛋白酶更感兴趣。其中包括来自*Pseudomonas* sp. strain

DY-A（Zeng et al. 2003）、*Bacillus amyloliquefaciens* S94（解淀粉芽孢杆菌）（Son and Kim 2003）、*Bacillus* TA39（Narinx et al. 1997）的丝氨酸蛋白酶以及来源于 *P. lundensis*（Yang et al. 2010）和 *S. maltophilia* MTCC 7528（Kuddus and Ramteke 2009）的金属蛋白酶。来自 *P. lundensis*（Yang et al. 2010）的金属蛋白酶 Ps5 具有应用于洗涤剂行业的潜力，因为该酶能够在 30℃和 pH 10.4 下维持水解酪蛋白的最高活性。此外，在 H_2O_2 溶液中有着显著的稳定性，在 1% H_2O_2 溶液中 Ps5 比对照样品的稳定性高 24%，而在 10%浓度下其活性仅下降 27%。另外该酶在 1%非离子表面活性剂溶液中（吐温 20 和吐温 80）能够保持稳定，在 5%浓度下其稳定性仅下降一半。从黄海中分离的 *P. aeruginosa* HY1215 所产的丝氨酸蛋白酶也有类似的稳定性（Hao and Sun 2015），在浓度为 0.2%～1.2%的 H_2O_2 溶液中，其活性比对照样品高 45%，但在 2% H_2O_2 溶液中其活性下降了 60%，另外，该酶在非离子表面活性剂，如吐温 40、吐温 80 和 Triton X-100 中具有高稳定性，在低于 1.8%浓度下能够维持其最大活性长达 1h。

14.2.3 嗜碱蛋白酶

嗜碱蛋白酶在碱性环境中具有活性且能够保持稳定。它们由嗜碱（包括嗜热、嗜冷和嗜盐碱性）微生物所产生，这些微生物在 pH 值高于 9.0 的环境生长最佳或生长得很好，但在中性 pH 值中无法生长或生长缓慢（Horikoshi 1999；Sarethy et al. 2011）。这些酶也可以由其他无需在高 pH 值中生长的极端微生物产生。

分泌嗜碱蛋白酶的嗜碱微生物是从各种环境的泥土、碱性苏打湖、盐湖、地表 3.2km 以下超深金矿的矿水围坝、堆肥、浴室瓷砖缝隙以及湖或海岸采集的羽毛样品中分离的（Gessesse et al. 2003；Saeki et al. 2002；Kobayashi et al. 2007；Karan et al. 2011；Dastager et al. 2008；Deng et al. 2010；Mitsuiki et al. 2002；Bakhtiar et al. 2002；Raval et al. 2014）。这些微生物主要为细菌，包括放线菌（图 14.1）和本章 14.2.4 中描述的嗜盐碱性古菌。

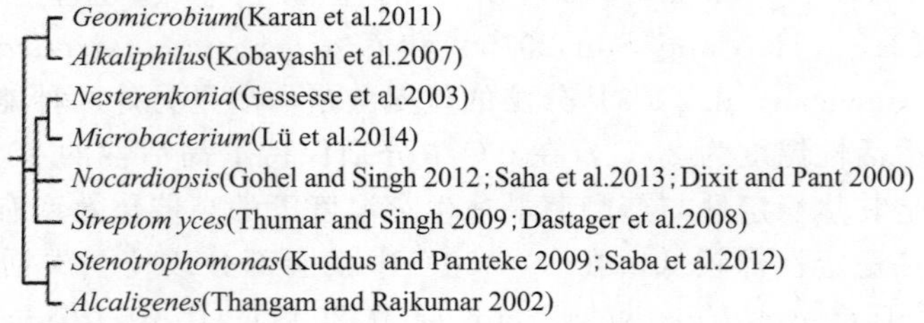

图 14.1 利用数据库 http：//www.ncbi.nlm.nih.gov/Taxonomy/ CommonTree/wwwcmt.cgi 制作的系统进化树，显示了本章 14.2.3 节中描述的具有代表性的嗜碱性微生物之间的关系（Sayers 等 2009；Benson 等 2009）

嗜碱蛋白酶主要为胞外丝氨酸蛋白酶（表 14.3），能够被 PMSF（phenylmethanesulfonyl fluoride）或 3,4-DCl（3,4-dichlorocoumarin）抑制，例如来自 *Nesterenkonia* sp.（涅斯捷连科氏菌）的蛋白酶 AL-20 对 PMSF 不敏感（Gessesse et al. 2003）。Kuddus 和 Ramteke（2009）描述的一种独特的金属蛋白酶被归类为嗜碱蛋白酶，因为该酶在 pH 10.0 中水解偶氮酪蛋白的活性最高，在 pH 8.0～11.0 中可保持大约 84%的活性，并且能够在 pH 8.0～10.0，20℃条件下保持稳定 1h。由于产嗜碱蛋白酶的微生物是从多种生境中分离的，这些酶的最适温度较宽，如来自 *Stenotrophomonas* sp. IIIM-ST045 的蛋白酶最适温度为 15℃（Saba et al. 2012），来自 *Nesterenkonia* sp. AL-20 的蛋白酶 AL-20 最适温度为 70℃（Gessesse et al. 2003）。影响酶活性与稳定性最重要的因素是环境的碱性。通常这些酶最适的 pH

值在 9.0～11.0 之间，而 Kobayashi 等（2007）从严格厌氧和极端嗜碱性微生物嗜碱菌 *Alkaliphilus transvaalensis* 中分离的高碱性蛋白酶 ALTP 在 pH 12.6 和 40℃下水解酪蛋白的能力最强。嗜碱蛋白酶在 pH 5.0～12.0 范围内能够保持稳定，如来源于 *Bacillus clausii* Ⅰ-52 的蛋白酶 BCAP 在该条件下能够保持稳定 72 h（Joo et al. 2003）。与那些最适 pH 较低的酶不同，由于一些细微结构的差异，嗜碱蛋白酶才能够在高 pH 值中保持活性。强嗜碱蛋白酶（最适 pH 10.0～11.0）分子中碱性氨基酸 Arg 和 His 残基的数目较高，而酸性氨基酸 Asp 和 Glu 残基数目较少。高含量的 Arg 可以形成更多的氢键和离子对（主要为 Arg-Asp），能够稳定蛋白质在碱性环境中的结构。这种氨基酸组成使得这些蛋白酶的等电点更高（Fujinami and Fujisawa 2010）。

表 14.3 来自嗜碱微生物的胞外碱性蛋白酶

来源	蛋白酶（PDB码）	生物体	最适温度/℃	最适 pH	pH 稳定性	参考文献
细菌	丝氨酸蛋白酶 AL-89	*Bacillus pseudofirmus* AL-89	60	11.0	pH6.5～10.5 时在 50℃中保持稳定 1h	Gessesse et al.(2003)
	丝氨酸蛋白酶 BCAP	*Bacillus clausii* I-52	60～65	11.0	pH5.0～12.0 中保持稳定 72h	Joo et al.(2003), Joo and Chang(2005)
	丝氨酸蛋白酶 KP-43	*Bacillus* sp. KSM-KP43	60℃(不含 Ca^{2+}) 70℃(含 Ca^{2+})	11.0～12.0	25℃，pH 6.0～12.0 中保持稳定 24h	Saeki et al.(2002)
	丝氨酸蛋白酶 SABP	*Bacillus pumilus* CBS	65	9.0～10.6	40℃，pH 9.0～10.6 中保持稳定 72h	Jaouadi et al.(2008)
	丝氨酸蛋白酶 ALTP	*Alkaliphilus transvaalensis*	70	12.6	50℃，pH 5.0～11.0 中保持稳定 10min	Kobayashi et al.(2007)
	碱性蛋白酶	*Stenotrophomonas* sp. IIIM-ST045	15	10.0	pH 6.8～12.0 中保持稳定 1h	Saba et al.(2012)
	丝氨酸蛋白酶 Ve_2-20-9_1	Strain Ve_2-20-9_1	50	10.0	50℃，pH 9.0～11.0 中保持稳定 24h	Raval et al.(2014)
	碱性蛋白酶	*Stenotrophomonas maltophilia* strain SK	40	9.0	暂无数据	Waghmare et al.(2015)
	丝氨酸蛋白酶 AprB	*Bacillus* sp. B001	60℃(不含 Ca^{2+}) 70℃(含 Ca^{2+})	10.0	30℃，pH 5.0～12.0 中保持稳定 6h	Deng et al.(2010)
	丝氨酸蛋白酶 AP-2	*Bacillus* spp. NCDC 180	55	12.0	30℃，pH 6.0～12.0 中保持稳定 4h（不含 5mmol/L Ca^{2+}）；含有 5mmol/L Ca^{2+} 时，30℃ pH 6.0～9.0 中保持稳定 24h	Kumar et al.(1999)

续表

来源	蛋白酶（PDB码）	生物体	最适温度/℃	最适pH	pH稳定性	参考文献
细菌	丝氨酸蛋白酶	*Bacillus subtilis* VSG-4	50	9.0	37℃，pH 9.0中保持稳定15h	Giri et al.(2011)
	丝氨酸蛋白酶蛋白酶M	*Bacillus* sp. KSM-K16	55℃（不含5mmol/L Ca^{2+}）70℃（含5mmol/L Ca^{2+}）	12.3	pH 6.0～11.0（不含2mmol/L Ca^{2+}）；35℃，pH 5.0～12.0（含2mmol/L Ca^{2+}）保持稳定10min	Kobayashi et al.(1995)
	丝氨酸蛋白酶	*Bacillus* sp. SSR1	40℃（不含5mmol/L Ca^{2+}）45℃（含5mmol/L Ca^{2+}）	10.0	40℃，pH 8.0～11.0中保持稳定1h	Singh et al.(2001)
	丝氨酸蛋白酶rBLAP	*Bacillus lehensis*（在*E. coli*中表达）	50	12.8	pH 10.0～12.8中保持稳定26h	Joshi and Satyanarayana(2013b)
	丝氨酸蛋白酶	Strain AH-6	37	9.5～11.0	37℃，pH 10.5～13.0中保持稳定1h	Dodia et al.(2008)
放线菌	丝氨酸蛋白酶AL-20	*Nesterenkonia* sp. AL-20	70	10.0	50℃，pH 6.0～11.0中保持稳定1h	Gessesse et al.(2003)
	丝氨酸蛋白酶OM-6	*Brachystreptospora xinjiangensis* OM-6	60	10.0～11.0	40℃，pH 6.0～12.0中保持稳定6h	Gohel and Singh(2013)
	丝氨酸蛋白酶	*Nocardiopsis alba* strain OK-5	70	10.0～11.0	40℃，pH 6.0～12.0中保持稳定6h	Gohel and Singh(2012)
	丝氨酸蛋白酶蛋白酶Mit-1	*Streptomyces clavuligerus* strain Mit-1	70	10.0～11.0	pH 8.0～10.0中保持稳定6h	Thumar and Singh(2007)
	丝氨酸蛋白酶NAPase	*Nocardiopsis* sp. TOA-1	70～75	11.0～11.5	30℃，pH 1.5～12.0中保持稳定24h	Mitsuiki et al.(2002)
	丝氨酸蛋白酶蛋白酶2	*Nocardiopsis* sp. NCIM 5124	60	10.0	pH 4.0～10.0中保持稳定1h	Dixit and Pant(2000)

在嗜碱蛋白酶中，有一类被鉴定为氧化稳定性的丝氨酸蛋白酶（OSPs），能够抵抗氧化剂的影响（Saeki et al. 2002）。如来自 *Bacillus pumilus* CBS 的蛋白酶 SAPB，在 40℃下，5% 的 H_2O_2 中孵育 24h 后，其活性与对照组相比上升了 28%（Jaouadi et al. 2008）。来源于 *B. clausii* Ⅰ-52 的一种蛋白酶在 2.5%过硼酸盐中室温孵育 72h 后仍能保持活性，并且在 5%的 H_2O_2 中其活性与对照组相比上升了 16%（Joo et al. 2003）。还有些蛋白酶能够在使用去污剂（SDS、Triton X-100、吐温 20、西曲溴铵）的实验室条件下，以及市售洗衣粉存在的情况下保持活性，如蛋白酶 SABP。在存在 5%吐温 20、吐温 80、Triton X-100 和 1% SDS 的情况下，SABP 在 65℃中孵育 15min 仍保持活性，在 40℃中孵育 24h 也不会丧失其稳定性（Jaouadi et al. 2008）。而存在市售液体去污剂（Dinol、Lav＋、Nadhif）和固体去

污剂（OMO、Dixan 或 Det）的情况下，在 40℃ 中孵育 1h 仍保持稳定（Jaouadi et al. 2009）。一种来源于 *Bacillus subtilis* VSG-4 的丝氨酸蛋白酶，在 37℃，1%离子型去污剂（吐温 80 和 Triton X-100）中孵育 1h 仍没有失去活性，而在 5mmol/L SDS 存在时仅损失 25%的活性。在相同的处理条件下，该酶在 1%固体去污剂（Surf excel、Rin、Tide 和 Ariel）存在下仍能保持 68%～100%的活性（Giri et al. 2011）。

通常，嗜碱蛋白酶能够有效水解酪蛋白、偶氮酪蛋白或血红蛋白（Gupta et al. 2005；Jaouadi et al. 2008；Kobayashi et al. 2007；Deng et al. 2010），但是也有一些能降解角蛋白的酶十分受关注，如来源于 *Nocardiopsis* sp. SD5（诺卡氏土壤菌）的蛋白酶能够在 45～50℃中降解天然羽毛（Saha et al. 2013）。Gessesse 等（2003）分离鉴定的细菌 *Nesterenkonia* sp. AL-20 和 *Bacillus pseudofirmus* AL-89（嗜碱芽孢杆菌）能够利用未经研磨或其他机械预处理的羽毛作为唯一的氮源，生长并分泌胞外蛋白酶。特别是 *Bacillus pseudofirmus* AL-89 在无额外添加养分的培养基中，表现出高产蛋白酶和快速有效降解羽毛的优异特性。具有角质降解活性的嗜碱性蛋白酶能够结合和水解固体形态的角蛋白。能够作用于结合在固体表面上的蛋白质底物是去污剂中使用酶的一个重要特性，例如如何去除衬衫领子上的角蛋白污渍，这是目前去污剂使用时的一个大问题（Gessesse et al. 2003）。

14.2.4 嗜盐蛋白酶

嗜盐蛋白酶是由嗜盐生物生产的蛋白酶，其催化活性通常需要 NaCl 存在。当一个蛋白酶是耐盐蛋白酶时，它们不一定来源于嗜盐微生物，但是它们能够在较宽的 NaCl 浓度范围内具有活性，而对 NaCl 无任何依赖（Graziano and Merlino 2014）。

分泌盐依赖性胞外蛋白酶的微生物有古菌、细菌和真核细胞（图 14.2），这些微生物可以从盐湖、高盐湖、盐锅、盐沼泽、洼地、地下盐卤水矿床的泉水、盐田、盐墨海、碱湖、深海沉积物、海水、发酵鱼酱和其他地方分离获得（Yin et al. 2014；Setati 2010；Sinha and Khare 2013）。以古菌为例，它们通常极端嗜盐，对于某些菌属，它们需要培养基中含 3.4～5.1mol/L NaCl 才生长（Graziano and Merlino 2014）。细菌菌株包括嗜盐菌、嗜盐碱菌以及不需要在生长培养基中增加 NaCl 浓度的细菌。嗜盐菌可分为在 0.5mol/L NaCl 中最适生长的轻度嗜盐菌，在 0.5～2.6mol/L 的 NaCl 中生长最佳的中度嗜盐菌，以及需要在 4.3mol/L 以上 NaCl 浓度中生长的极少数极端嗜盐菌（Setati 2010）。

到目前为止，所有已发现的来源于嗜盐古菌的胞外蛋白酶都是能够被 PMSF 抑制的丝氨酸蛋白酶。细菌所产的蛋白酶，丝氨酸蛋白酶占据了最大一部分，此外也发现了一些金属蛋白酶，如来自 *Pseudoalteromonas* sp. 的蛋白酶（Sánchez-Porro et al. 2003；Xiong et al. 2007）。这两类蛋白酶的活性都能被对氯汞苯甲酸和 β-巯基乙醇或二硫苏糖醇抑制，例如来自 *Pseudoalteromonas* sp.（Sánchez-Porro et al. 2003）、*Geomicrobium* sp. EMB2（Karan and Khare 2010）、*Natrialba magadii*（马加蒂嗜盐碱杆菌）（Giménez et al. 2000）、*Haloferax lucentensis* VKMM 007（卢森坦富盐菌）（Manikandan et al. 2009）的蛋白酶，这暗示着二硫键和/或半胱氨酸残基在维持蛋白酶生物活性构象中有重要作用。在适当的 NaCl 浓度中，嗜盐蛋白酶的最适温度根据不同酶而不同，如来自太平洋阿拉斯加阿留申边缘海域沉积物中的 *Pseudoalteromonas* sp. 所产的金属蛋白酶最适温度为 35℃（Xiong et al. 2007），*Chromohalobacter* sp. TVSP101（色盐杆菌）的丝氨酸蛋白酶最适温度为 75℃（Vidyasagar et al. 2009）。尽管分离自碱性环境（pH 7.0～11.0），但其最适 pH 值范围仍是相当广，而蛋白酶在 pH 5.0 到 12.0 之间能够保持稳定时，其最适 pH 值范围就更宽了。

图 14.2　利用数据库 http://www.ncbi.nlm.nih.gov/Taxonomy/CommonTree/wwwcmt.cgi 制作的系统进化树，显示了本章 14.2.4 节中介绍的代表性嗜盐微生物的进化关系（Sayer 等 2009；Benson 等 2009）。* 标记的属于古菌

影响这些酶的活性和稳定性的最重要因素是反应环境的盐度。毫无疑问，这与嗜盐微生物的栖息环境有联系，为了在高渗透压环境中生存，这些微生物需要维持等渗，因此，它们需要保持胞内高浓度的 NaCl（Manikandan et al. 2009）。与其他酶相比，嗜盐蛋白酶改变了其氨基酸序列。与非嗜盐酶相比，通常它们甚至有 20%以上的酸性氨基酸残基（天冬氨酸和谷氨酸）以簇的形式位于蛋白酶表面。同时，它们含更少的赖氨酸残基和更多小的疏水残基，如丙氨酸、甘氨酸，以及极性残基，如丝氨酸和苏氨酸，缺乏大型非极性残基，如亮氨酸、异亮氨酸、甲硫氨酸和苯丙氨酸。由于负电荷的增加，嗜盐蛋白酶可能结合更多氢离子，降低了分子表面的疏水性以及在高浓度 NaCl 中聚集的趋势。这是嗜盐微生物适应因高浓度盐而缺水的环境的基本机制（Graziano and Merlino 2014；Gomes and Steiner 2004）。分离自 *Halobacterium* sp. TuA4（盐杆菌）（Schmitt et al. 1990）的蛋白酶和未鉴定的嗜盐碱菌株 A2 的蛋白酶（Yu 1991）需要不超过 0.3mol/L 的 NaCl（De Castro et al. 2005），而大部分酶在更高的 NaCl 浓度（从 1.0mol/L 到 5.1mol/L 不等）下才显示其最大活性。当合成底物被水解时，盐浓度甚至更高。例如分离自 *Natrialba asiatica*（亚洲嗜盐碱杆菌）的蛋白酶 F11（Kamekura and Seno 1990），在 5.1mol/L 的 NaCl 浓度中降解 N-succ-AAPF-pNa 最快；而降解偶氮酪蛋白时 NaCl 浓度为 1.7～2.4mol/L。推测可能是在更高的 NaCl 浓度中，偶氮酪蛋白更加亲水，从而失去了合适的构象，使得水解更加困难。而在相同条件下，寡肽并没有改变它们的构象，所以能够被蛋白酶有效水解。研究人员（Nordberg and von Hofsten 1969；Studdert et al. 2001；Shi et al. 2006）比较了在存在 NaCl 和 KCl 时蛋白酶的催化效率，发现存在 NaCl 时活性更高，因为这些酶的生产者一般生活在富含 NaCl 的环境中。从蛋白酶 Nep 的 NaCl 浓度 1.0mol/L（Giménez et al. 2000）到来自 *Haloferax mediterranei* strain 1583（地中海富盐菌）（Stepanow et al. 1992）的蛋白酶的 4.5mol/L 的 NaCl，可见，相对较高的盐度是维持蛋白酶稳定所必需的。研究显示，从环境中去除 NaCl 会导致来自 *Halobacterium halobium*（盐生盐杆菌）（Izotova et al. 1983）的蛋白酶以及来自 *Hfx. mediterranei* strain 1583（Stepanow et al. 1992）的蛋白酶不可逆的失活，这是由于低离子强度环境加速了这些蛋白酶的变性。值得注意的是蛋白酶 SptA（Shi et al. 2006）

在缺乏 NaCl 时仍保持了 20%的活性，一旦 NaCl 浓度增加到 2.5mol/L，将获得 60%的活性。据作者所言，这意味着在缺乏 NaCl 时酶经历了可逆的变性。另一方面，来自 *Halogeometricum borinquense* TSS101（伯林盐几何菌）（Vidyasagar et al. 2006）的蛋白酶在缺乏 NaCl 和分别提供 10%蔗糖和 20%甜菜碱的环境中能维持其总活力。这表明了为了维持其活性，该酶需要一个适当的渗透压或者需要降低环境中水活度（a_w），而并非需要 Na^+ 或 Cl^-（Vidyasagar et al. 2006；Litchfield 2011）。

目前为止发现的嗜盐和耐盐细菌蛋白酶都比来源于古菌的蛋白酶的耐盐性更好。来自 *Bacillus* sp. EMB9（Sinaha and Khare 2013），*Pseudoalteromonas* sp. CP76（Sánchez-Porro et al. 2003），*Geomicrobium* sp. EMB2（地杆菌）（Karan and Khare 2010）或 *B. subtilis* FP-133（Setyorini et al. 2006）的酶的最适盐度条件从 0mol/L 到 1.0mol/L NaCl 不等。相反，来自 *Virgibacillus* sp. SK33（枝芽孢杆菌）的丝氨酸蛋白酶显示在浓度为 1.7mol/L 到 4.3mol/L NaCl 时活性最高（Sinsuwan et al. 2010）。来自 *Chromohalobacter* sp. TVSP101 丝氨酸蛋白酶的最适盐度为 4.5mol/L NaCl（Vidyasagar et al. 2009）。Purohit 和 Singh（2011）发现一个具有有趣特性的酶，该酶分离自 *Oceanobacillus iheyensis* O. M. A_{18}（伊平屋海洋杆菌），它能根据环境盐度改变自身最适温度。在 0.25～0.5mol/L NaCl 中，其最适温度为 60℃，当 NaCl 浓度上升到 2.0～3.0mol/L 时，其最适温度上升了 20℃。尽管这些酶在缺乏盐的环境中也能水解底物，但它们仍需求少量的 NaCl（如 90mmol/L）来维持稳定。然而，目前发现的大部分蛋白酶都能够在含 2.0～4.5mol/L NaCl 的溶液中保持稳定（表 14.4）。例如 *Chromohalobacter* sp. TVSP101 的蛋白酶，该菌是少数极端嗜盐细菌之一，其酶活性在盐度低于 1.0mol/L 会完全不可逆丧失。但是，用甘氨酸、蔗糖（10%）或甘油（20%）替代 NaCl 时，该酶能够保持其全部活性（Vidyasagar et al. 2009）。另一个来源于中度嗜盐碱细菌 *Geomicrobium* sp. EMB2 的蛋白酶，仅需要 0.17mol/L NaCl 来维持其活性，当去除环境中的 NaCl 时，其活性会丧失一半以上。然而，当 NaCl 或其他渗透物（甘露糖、蔗糖、甘油或甜菜碱）重新添加进去直到终浓度达 5%时，该酶能够重新获得大部分活力（Karan and Khare 2010）。

表 14.4 来自嗜盐微生物的胞外嗜盐蛋白酶和耐盐蛋白酶

来源	蛋白酶	生物体	最适温度/℃	最适 pH	最适盐浓度以及在 NaCl/KCl 中的稳定性	参考文献
古菌	丝氨酸蛋白酶 Nep（halolysin）	*Natrialba magadii*	60	8.0～10.0	1.0～1.5mol/L NaCl；4℃下，在 1.0～3.0mol/L NaCl 或 3.0mol/L KCl 中保持稳定 7d	Giménez et al.（2000），Ruiz and De Castro（2007）
	丝氨酸蛋白酶 EP（胰凝乳蛋白样）	*Natronococcus occultus* NCBM 2192	60	7.0～9.0	1.0mol/L NaCl 或 KCl；4℃下，在 3.0mol/L NaCl 或 KCl 中至少保持稳定 7d	Studdert et al.（2001）
	丝氨酸内肽酶	*Halobacterium halobium* S9	40	8.7	4.0mol/L N-琥珀酰 Ala-Ala-Pro-Phe-对硝基苯胺	Capiralla et al.（2002）
	丝氨酸蛋白酶 SptA（halolysin）	*Natrinema* sp. J7	50	8.0	2.5mol/L NaCl 和 2.0mol/L KCl，在 NaCl 中活性更高；4℃下在 2.5mol/L NaCl 中至少保持稳定 1 个月；37℃下为 2h	Shi et al.（2006）

续表

来源	蛋白酶	生物体	最适温度/℃	最适pH	最适盐浓度以及在NaCl/KCl中的稳定性	参考文献
古菌	丝氨酸蛋白酶	*Halogeometricum borinquense* TSS101	60	10.0	3.4～4.3mol/L；在60℃，3.4mol/L NaCl中稳定2h，4℃下30d	Vidyasagar et al.(2006)
	丝氨酸蛋白酶(trypsin-like)	*Haloferax lucentensis* VKMM 007	60	8.0	4.3～5.13mol/L	Manikandan et al.(2009)
细菌	金属蛋白酶	*Pseudoalter omonas* sp. CP76	55	8.5	0～1.0mol/L；在2.0mol/L NaCl，4℃下稳定24h	Sánchez-Porro et al.(2003)
	金属蛋白酶	*Pseudoalteromonas* spp.	35	7.5	2.0mol/L	Xiong et al.(2007)
	丝氨酸蛋白酶	*Bacillus* sp. EMB9	55	9.0	0.17mol/L NaCl/0.13mol/L KCl；在4℃下，0.8mol/L NaCl和0.67mol/L KCl下保持稳定48h	Sinha and Khare(2013)
	丝氨酸蛋白酶	*Chromohal obacter* sp. TVSP101	75	8.0	4.5mol/L NaCl；在4.5mol/L NaCl，60～70℃下保持稳定2h；在4.5mol/L NaCl室温下保持稳定20天，在−4℃下保持稳定45d	Vidyasagar et al.(2009)
	丝氨酸蛋白酶	*Filobacillus* sp. RF2-5	60	10.0～11.0	0～0.86mol/L(酪蛋白作为底物)，2.6～4.3mol/L(*N*-琥珀酰Ala-Ala-Phe-对硝基苯胺作为底物)；在4.3mol/L NaCl，30℃下保持稳定24h	Hiraga et al.(2005)
	丝氨酸蛋白酶	*Halobacillus* sp. SR5-3	50	10.0	4.3mol/L(*N*-琥珀酰Ala-Ala-Pro-Phe-甲基香豆素为底物)；37℃下，在3.4～6.0mol/L NaCl中稳定24h	Namwong et al.(2006)
	丝氨酸蛋白酶	*Virgibacillus* sp. SK33	55	7.5	1.7～3.4mol/L；在0～4.3mol/L NaCl，55℃下保持稳定60min	Sinsuwan et al.(2010)
	丝氨酸蛋白酶	*Geomicrobium* sp. EMB2	50	10.0	0.17mol/L	Karan and Khare(2010)
	丝氨酸蛋白酶	*Oceanobacillus iheyensis* O.M.A18	50	11.0	60℃下为0.25～0.5mol/L，80℃下为2.0～3.0mol/L	Purohit and Singh(2011)

14.3 极端蛋白酶的应用

如今，全球酶制剂市场值估计达大约51亿美元，年增长率为7%，是世界范围内发展最好的市场之一（Sarethy et al. 2011）。蛋白酶作为主要的酶制剂，占据了市场的关键位置，约为所有市售酶类的60%。它们主要的来源是动物（如牛胃）、植物（如菠萝、无花果和木瓜）以及微生物（如 *Bacillus* sp.，*Pseudomonas* spp.）。在近几十年里，蛋白酶，尤其是微生物蛋白酶已经占据了市场的重要位置并被用于工业当中。对植物蛋白酶而言，其生产很大程度上受制于气候条件，而动物蛋白酶则会激起伦理上的争议。然而，不管其来源是什么，蛋白酶都可以以工业规模被使用，主要是因为它们能够水解肽键。许多蛋白酶也可以有效地催化微水环境中的合成反应，该特性主要用于制药和营养。

在最开始，蛋白酶市场主要是中温蛋白酶（通常大部分为耐碱），但随着极端微生物研究的深入发展，证明了用极端酶代替目前使用的酶可以提高生物催化技术的效率并降低成本。取决于其来源，极端蛋白酶适合在低或高温以及酸性、碱性或极端盐环境中生存繁殖。它们要么能表现出理想的抵抗热变性能力，要么极度不耐热。因此，它们能够在加工条件中被选择性失活。蛋白酶中取得最大商业成功的例子包括枯草杆菌蛋白酶（subtilisin）（用作工业目的，主要有嗜碱和耐碱菌株生产）以及其他碱性蛋白酶，占据了蛋白酶市场约50%的份额（表14.5）。蛋白酶有着广泛的用途，这得益于它们在本章14.2.3中详细分析的特点。表14.5中给出了用于去污、丝绸精炼、食品与饲料工业、皮革脱毛、化妆品以及制药的市售蛋白酶清单。大部分嗜碱性蛋白酶是从枯草杆菌中分离的，这些酶主要作为洗衣粉和自动餐具去污剂的添加成分，应用于洗涤剂行业，它们的功能是降解蛋白质污渍，包括血渍、牛奶、蛋液、草渍和酱汁。

表14.5 市售细菌碱性蛋白酶、来源、应用以及它们的工业供应商

供应商	产品商标	来源细菌	应用
Novo Nordisk，丹麦	Alcalase	*B. licheniformis*	洗涤剂，蚕丝脱胶
	Savinase	*Bacillus* sp.	洗涤剂，纺织业
	Esperase	*B. lentus*	洗涤剂，食品工业，蚕丝脱胶
	Everlase[①]	*B. clausii*	洗涤剂
	Kannase[①]	*B. clausii*	洗涤剂
	Durazym	*Bacillus* sp.	洗涤剂
	Novozyme 243	*B. licheniformis*	假牙清洁剂
	Biofeed pro	*B. licheniformis*	饲料业
	Nue	*Bacillus* sp.	皮革业
Gist-Brocades，荷兰	Subtilisin	*B. alcalophilus*	洗涤剂
	Maxacal	*Bacillus* sp.	洗涤剂
	Maxatase	*Bacillus* sp.	洗涤剂
Genencor International，美国	Purafect	*B. lentus*	洗涤剂
	PurafectOxP[①]	*B. lentus*	洗涤剂
	Properase[①]	*B. alkalophilus* PB92	洗涤剂
	Primatan	Bacterial sources	皮革业
	FNA[①]	*B. amyloliquefaciens*	洗涤剂
	FN4[①]	*B. lentus*	洗涤剂
Solvay Enzymes，德国	Opticlean	*B. alcalophilus*	洗涤剂
	Optimase	*B. licheniformis*	洗涤剂
	Maxapem[①]	*Bacillus* sp.	洗涤剂

续表

供应商	产品商标	来源细菌	应用
Amano Pharmaceuticals，日本	PROTIN SD-NY10	*B. amyloliquefaciens*	食品工业
	PROTIN SD AY10F	*B. licheniformis*	食品工业
	Proleather	*Bacillus* sp.	食品工业
	Collagenase	*Clostridium* sp.	技术性使用
Nagase Biochemicals，日本	Bioprase SP-10	*B. subtilis*	食品工业
	Bioprase concentrate	*B. subtilis*	美容，制药
	Bioprase	*B. subtilis*	化工
Godo Shusei，日本	Godo-Bap	*B. licheniformis*	洗涤剂，食品
Enzyme Development，美国	Enzeco alkaline protease—L FG	*B. licheniformis*	食品工业
AB Enzymes，德国	Corolase 7089	*B. subtilis*	食品工业
	Biotouch® ROC 250L	*Bacillus* sp.	食品工业
无锡赛德生物工程有限公司，中国	Wuxi	*Bacillus* sp.	洗涤剂
Advance Biochemicals，印度	Protosol	*Bacillus* sp.	洗涤剂
Henkel，德国	BLAPS① or BLAP X①	*B. lentus*	洗涤剂

资料来源：修订自 Gupta 等 2002。
①蛋白质工程。

一旦人们意识到去污酶类的重要性，就会有多种多样有着特定用途的去污蛋白酶被发现。其中之一是 Alkazym（Novodan，Copenhagen，Denmark），这是一种在膜清洗工艺中起着重要作用的酶类。Tergazyme（Alconox，New York，USA）、Ultrasil（Henkel，Dusseldorf，Germany）和 P3-paradigm（Henkel-Ecolab，Dusseldorf，Germany）是用于这一目的的其他酶。另外一个例子是一种以酶为基础的清洁剂 Pronod 153L，主要用于清除外科器械中的血渍蛋白质。蛋白酶在印度用于光学仪器的清洗（Gupta et al. 2002）。

14.3.1 洗涤剂工业

第一个含有细菌来源的丝氨酸蛋白酶的洗涤剂是在 1956 年被称为 BIO-40 的洗涤剂。七年之后，一家丹麦公司 Novo Industry A/S 从耐碱菌株 *Bacillus licheniformis* 中发现碱性蛋白酶，其商品名称为 Biotex（Maurer 2004），这是一种丝氨酸内切蛋白酶，对阴离子或非离子表面活性剂有抵抗能力，并且在 60℃和 pH 8.3 中显示出最高活性。

如上所述，碱性丝氨酸蛋白酶，又被称为枯草杆菌素（subtilisin），由芽孢杆菌属菌株产生，几乎占市场上生产的所有蛋白水解酶的一半。在 2002 年，它们的纯蛋白酶在欧盟的生产量达 900 吨（Maurer 2004）。其耐高碱性、在碱性环境中的稳定性、广泛的特异性以及由细胞分泌的特性，使得生产制备更加简易，保证了该酶类在商业上成功应用。从 1984 年起，研究者为了改良枯草杆菌素和其他蛋白酶进行了各种尝试。然而，成百上千的突变体当中，只有少数进入工业实践，第一个是在 1990 年，这些仅仅是改良了特性的枯草杆菌素突变体。在 2004 年，通过定点突变或定向进化的方式改良了 8 个这样的枯草杆菌素，并在市场上获得使用。改良的特性包括提升其热稳定性、在有机溶剂中的活性、改变其底物特异性和最适 pH（表 14.6）。在第一个工程酶枯草杆菌素中，Met222（一个位于催化性 Ser221 旁的容易被氧化的氨基酸）被替换成不可被氧化的氨基酸，如丙氨酸、丝氨酸或亮氨酸，以获得氧化耐受能力上升的蛋白质。

表 14.6 枯草杆菌素工程的工业用途

突变类型	蛋白酶及其来源微生物	应用策略	结果	参考文献
合理突变	subtilisin E，*B. subtilis*	第 61 和 98 个半胱氨酸之间增加二硫键	热稳定性显著提升	Takagi et al.（1990）
	subtilisin BPN'，*B. subtilis*	第 22 和 87 个半胱氨酸之间增加二硫键	提高稳定性，尤其在缺乏钙离子的情况下	Pantoliano et al.（1987）
	subtilisin，*B. amyloliquefaciens*	活性中心旁的第 222 个甲硫氨酸替代为不可被氧化的氨基酸：丙氨酸、丝氨酸或亮氨酸	提高氧化耐受性	Estell et al.（1985）
	subtilisin，*B. lentus*	第 222 个甲硫氨酸突变为甲基甲烷硫代磺酸酯修饰的半胱酰氨基团（Me-MTS）	增强稳定性和氧化条件下的催化活性	Grøn et al.（1990）
	aprA-subtilisin，*B. amyloliquefaciens*	突变位点：109 位天冬酰胺突变为丝氨酸，218 位天冬酰胺突变为丝氨酸，76 位天冬酰胺突变为天冬氨酸	提高稳定性，脆性转变温度提高 7℃	Narhi et al.（1991）
	subtilisin，*Bacillus* sp. AK. 1	通过 131 位甘氨酸突变为天冬氨酸以及 172 位脯氨酸突变为天冬氨酸将负电性的天冬氨酸引入到钙离子亲和力较低的结合位点	通过增强枯草杆菌蛋白酶与钙离子之间的静电作用提高稳定性	Pantoliano et al.（1988）
	subtilisin，*B. gibsonii*（alkaline protease BgAP）	21 位异亮氨酸突变为缬氨酸，39 位丝氨酸突变为谷氨酸，74 位天冬酰胺突变为天冬氨酸，87 位天冬氨酸突变为谷氨酸，122 位甲硫氨酸突变为亮氨酸，253 位天冬酰胺突变为天冬氨酸	提高 15℃ 下的活性（k_{cat} 值由 23.2/s 增加到 35.3/s）；60℃ 下的半衰期从 2min 提升到 224min	Martinez et al.（2013）
	SABP，*B. pumilus*	突变：（1）单突变——99 位天冬酰胺突变为酪氨酸；（2）双突变——31 位亮氨酸突变为异亮氨酸，99 位天冬酰胺突变为酪氨酸；（3）三突变——31 位亮氨酸突变为异亮氨酸，33 位苏氨酸突变为丝氨酸，99 位天冬酰胺突变为酪氨酸	提高 50℃ 和 60℃ 下的半衰期；最适温度由 65℃ 变成 75℃；对 *N*-琥珀酰-Ala-Ala-Pro-Phe-对硝基苯胺的 k_{cat}/K_m 值提升 31 倍	Jaouadi et al.（2010b）
	subtilisin BPN'，*B. subtilis*	129 位脯氨酸突变为甘氨酸以及 156 位谷氨酸突变为精氨酸	双突变后能够剪切多肽中磷酸化的酪氨酸（选择性提升 2500 倍）	Knight et al.（2007）
	subtilisin，*Bacillus* YaB	将影响枯草杆菌素底物特异性的 S1 口袋中 124 位甘氨酸突变为丙氨酸或缬氨酸以及 151 位甘氨酸突变为丙氨酸	提高降解弹性蛋白质的活性	Mei et al.（1998）

续表

突变类型	蛋白酶及其来源微生物	应用策略	结果	参考文献
合理突变	subtilisin BPN', *B. subtilis*	64 位具有催化活性的组氨酸替换为丙氨酸	提高对含组氨酸的底物的特异性	Carter et al.(1989)
	subtilisin BPN', *B. subtilis*	144 位酪氨酸突变为丙氨酸,107 位异亮氨酸突变为甘氨酸,126 位亮氨酸突变为甘氨酸	提高对含有强疏水性 P_4 侧链的底物的特异性	Rheinnecker et al.(1993, 1994)
随机突变	subtilisin E, *B. subtilis*	易错 PCR,筛选	在二甲基甲酰胺中活性提升 170 倍	Chen and Arnold(1993)
	subtilisin BPN', *B. subtilis*	去环,盒式诱变,筛选	半衰期提升 1000 倍	Strausberg et al.(1995)
	subtilisin, *B. lentus*	易错 PCR,在空心纤维中富集	酶分泌量提升 50%	Naki et al.(1998)
	subtilisin E, *B. subtilis*	随机引物法,筛选	65℃ 中半衰期提高 8 倍	Shao et al.(1998)
	subtilisin E, *B. subtilis*	易错 PCR,DNA 改组,筛选	提高在不同温度下的活性,以及最大温度提高 17℃	Zhao and Arnold(1999)
	subtilisin, *Bacillus* sp. TA41	饱和诱变以及体外重组(DNA 改组)	提升在低温下的活性和高温中的稳定性	Miyazaki et al.(2000)

目前用于洗涤剂的来源于枯草芽孢杆菌的枯草杆菌素是许多引起广义索赔专利的主要问题。因此经典的筛选方法正不断被用于从中温和极端微生物中分离新型的碱性蛋白酶。为了这个目标，也对基因组与宏基因组文库进行了搜索。例如，Rai 等（2010）发现一种潜在的枯草芽孢杆菌蛋白酶的替代品，可以在洗衣业中广泛使用，该酶是一种在洗涤剂中稳定的丝氨酸蛋白酶，分离自嗜热细菌 *Paenibacillus tezpurensis* sp. Nov（提斯浦尔类芽孢杆菌）。该酶是 Ca^{2+} 非依赖性的，并且在 pH 9.5 和 45～50℃条件下具有最大活性。该酶在室温中也表现出显著的稳定性以及对表面活性剂和绝大部分测试的市售洗涤剂的兼容性。来自嗜盐菌的蛋白酶，例如 *Geomicrobium* sp. EMB2（Karan and Khare 2010）或 *Virgibacillus* sp. SK33（Sinsuwan et al. 2010），在洗涤剂工业中也有着巨大的应用潜力。应强调的是，极端蛋白酶通常是在中温菌株选择过程中获得的。例如，来自中温丝状真菌 *Conidiobolus coronatus* 的嗜碱性蛋白酶能够兼容印度使用的市售洗涤剂，在 50℃，Ca^{2+}（25mmol/L）和甘氨酸（1mol/L）中孵育 56min 能保持 43%的活性（Phadatare et al. 1993）。同样，据 Khan（2013）报道，来自 *Aspergillus clavatus* ES1（棒曲霉）的碱性丝氨酸蛋白酶能够用于洗涤剂工业，因为该酶对盐、溶剂、洗涤剂和漂白剂耐受。

在宏基因组研究中，Neveu 等（2011）分别从戈壁滩和死亡谷沙漠表层沙样的宏基因组文库分离到两株丝氨酸蛋白酶 DV1 和 M30，这两种酶在碱性环境中（DV1 最适 pH 8.0，M30 最适 pH 为 11.0）都具有活性，并且在存在非离子去污剂和 SDS 情况下保持稳定，这使得它们可用于日用化工。通常，对传统环境的宏基因组筛选会发现能够在极端条件中存活的蛋白酶，例如极端碱性环境是一个决定酶在洗涤剂中应用的决定性因素之一。这些酶的例子包括从比利时落叶林的 A 层中分离的碱性蛋白酶 KP-43（Biver et al. 2013），或从山羊表皮宏基因组中分离的丝氨酸蛋白酶（AS-protease）（Pushpam et al. 2011）。

Toplak 等（2013）提出了另一种方法寻找主要用于洗涤剂的枯草杆菌素样蛋白酶。他们通过基因组挖掘寻找新的枯草杆菌素突变体以及关于编码这些蛋白水解酶的基因。利用这种方法，作者选择了嗜热细菌和古菌基因组中编码潜在同源枯草杆菌素 E 的基因，寻找至少与枯草杆菌素有 30％相同的基因序列，找到了 24 个可能与枯草杆菌素 E 同源的基因，并通过 PCR 从选定的菌株中分离出其中的 16 个基因，使用的引物是在枯草杆菌素编码基因基础上设计的。然后，他们把基因克隆到 pBAD 载体中并在 *E. coli* 中表达，通过含有选择性生长成分（脱脂牛奶、抗生素、诱导物）的琼脂培养基中进行筛选，获得含有来自 *Thermus aquaticus*、*Pseudomonas mendocina*（门多萨假单胞菌）、*Geobacillus thermodenitrificans*（嗜热脱氮土芽孢杆菌）、*Deinococcus geothermalis*（地热奇异球菌）和 *Coprothermobacter proteolyticus*（栖热粪杆菌）的 6 种不同蛋白酶基因的功能克隆体，含有来自 *C. proteolyticus* 的蛋白酶基因的克隆体在 80℃中具有最高生长活性。该蛋白酶命名为蛋白溶解素（proteolysin），在 *E. coli* 中表达效率达 20mg/L。该酶显示出热稳定性，并且在 85℃和 pH 8.0 条件下活性最高。在 70℃中孵育 20h 后能保持 35％的活性。蛋白溶解素的另一个特性是对 DMSO、DMF、乙醇和盐酸胍有极强的耐受性，能被 10％非离子去污剂吐温 20 激活（活性上升了 80％）。

洗涤剂工业同样会使用在有机溶剂中稳定并且能够在低温中存活的嗜冷嗜碱性蛋白酶。尤其是在各种工业的清洗用途中尤为重要，包括洗涤、洗碗、食品、乳制品和酿造、医疗用具和水处理等（Cavicchioli et al. 2011）。为此需要同时利用这些蛋白酶的不耐热性与其在低温中的高催化效率。它们的不耐热性可以导致反应过程完成后通过温度的细微增加而让蛋白酶选择性失活。通过这种方法能够降低最终产物被酶降解的风险。

通常认为含有适冷蛋白酶的洗涤剂比不含酶的洗涤剂效果更好，并且它们能够更有效清除衣服上的血液、牛奶、草渍或汗渍等蛋白质污渍（Kuddus and Ramteke 2011）。Novozymes 设计了一种在低温中清洗衣服的适冷酶，命名为 Polarzyme。假设洗涤温度从 40℃降低到 30℃，那么每家每户的能源都能得到节约，可见该产品的重要。通过这种方式，能够降低 30％的能耗并且每洗 100g 衣物都可减少 CO_2 的排放（Cavicchioli et al. 2011）。除了经济和环境方面的优势外，在洗涤剂中使用适冷蛋白酶可能会改善纺织材料的质量，尤其是那些含有合成纤维不能耐受超过 50～60℃的织物（Joshi and Satyanarayana 2013a）。因此，越来越多的注意力集中在这类蛋白酶的生产当中。例如，Baghel 等（2005）从 *B. subtilis* 发现了一种适冷蛋白酶，在 SDS 溶液中保持稳定并且在吐温 80 和车轮洗涤剂中活性更高。最近，申请了来自 *Flavobacterium balustinum*（大比目鱼黄杆菌）的丝氨酸蛋白酶（CP70）的专利，其最适温度比 Savinase 这种传统上用于洗涤业的蛋白酶的最适温度低 20℃。该酶在 30℃中 1h 能保持稳定，且最适 pH 值范围为是 6.5～10.0，而且，表面活性剂和漂白剂也不影响其活性（Hasan and Tamiya 2001）。另一种碱性适冷蛋白酶分离自 *Stenotrophomonas maltophilia*，在 20℃和 pH 10.0 中活性最高，该酶其他特性包括在市售的洗涤剂中有良好的稳定性和兼容性，使得它能在低温中高效地清除各种蛋白质污渍（Kuddus and Ramteke 2009）。根据 Kuddus 和 Ramteke（2011）的研究，来自 *S. maltophilia* 的适冷蛋白酶能够完全清除血渍和草渍，并分别将其反射率提高 26％和 23％。据作者所言，蛋白酶能够用在为衣服精细清洁而设计的温和洗涤剂中，在纺织业中则可用于清除生丝中的丝胶，因为丝胶的蛋白质含量会增加生丝的刚度和暗淡度。Doddapaneni 等（2007）也看到了低温洗涤剂的应用潜力。他们从 *Serratia rubidaea*（悬钩子沙雷菌）中发现了两种蛋白酶，能够在 30～40℃和 pH 8.0～10.0 中保持活性。Pawar 等（2009）从 *Bacillus* sp. 158 中分离出一种耐热蛋白酶（在 30～40℃和 pH6.0～7.0 下稳定），该酶能够用于隐形眼镜的清理，提高镜片的透光率。嗜盐酶也显示出在洗衣业中的应用潜力，除了它们能在极端盐度环境中的存活能力外，它们还能在洗涤剂和有机溶剂中维持稳定。例如来自嗜盐菌 *Bacillus* sp. EMB9（Sinha and Khare 2013），*Geomicrobium* sp. EMB2（Karan and Khare 2010）或 *Virgibacillus* sp. EMB13（Sinha and Khare 2012）的蛋白酶。研究者测试了后者对多种洗涤剂的兼容性（将商业洗衣粉中的酶热变性后）以及与这些洗涤剂结合清除织物上的血渍的能力，结果非常鼓舞人心。

14.3.2 食品工业

微生物蛋白酶广泛应用于食品工业，尤其是极端微生物蛋白酶，常用于提升工艺效率以及终产物的质量。它们主要优势是有更高的特异性，更低的副反应风险，降低了中温微生物污染的风险，对于嗜冷和嗜热酶，通过选择性热失活的可能性更大。毫无疑问，在食品加工方面，嗜碱性蛋白酶是家用化学品的佼佼者。嗜碱性蛋白酶的初级用途是水解天然蛋白质底物。40 多年来，碱性蛋白酶的使用，使得高营养价值的蛋白质水解物被发现。市售的蛋白质水解物能够从如酪蛋白（Miprodan；MD Foods，Germany）、乳清（Lacprodan；MD Foods，Germany）、大豆（Proup；Novo Nordisk，Denmark）和肉类（Flavourzyme，Novo Nordisk，Denmark）这些底物中分离（Gupta et al. 2002）。它们在血压调节中有重要作用，也被用于婴幼儿食品配方、特定食疗产品以及果汁和软饮料中的强化剂。碱性蛋白酶也可以用于肉类加工，例如市售蛋白酶 SEB Tender 70 被广泛用于分解胶原蛋白嫩化肉类，使得肉类更加可口（Singhal et al. 2012）。来自 *B. licheniformis* 的嗜碱性蛋白酶常被应用于食品工业中，该酶被用于加工豆粕来获得可溶无苦味的水解产物，这些产物可用作蛋白质强化软饮料及食疗食品的成分。碱性蛋白酶也可用于从鱼和肉类行业的副产品以及甲壳质生产过程中的甲壳类贝壳废料中回收蛋白质（Synowiecki 2010）。冷活性蛋白酶也能够用于相似的目的，但这仅针对高度不耐热的底物或者产物。这些酶在低温下有效且具有高度特异性，因此减少了副产物的数量，促进不稳定蛋白酶在整个过程中或者在完成后的选择性失活，而无需耗费大量能源。它们用于啤酒处理、面包、烘焙和加速奶酪成熟。与 alkalozymes 相似，适冷蛋白酶，如适冷胶原蛋白酶 MCP-01，能够用于嫩化肉类或改善冻肉风味，该酶在 4℃中能使剪切肉类的力度降低 23%，并且将肉类相对肌纤维片段的离子指数提升 91.7%。MCP-01 还有助于保持肉类的新鲜颜色和水分，与同样用于嫩化的木瓜蛋白酶和菠萝蛋白酶相比，该酶显示出独特的嫩化机制，在 4℃中对胶原蛋白的降解有很强的选择性（Zhao et al. 2012）。另外，嗜冷蛋白酶能促进鱼卵脱膜，鱼/肉胶质的蒸发以及脂肪熔炼。适冷蛋白酶可能也能够用于可消化产品的生产，将这些产品包裹或混合在宠物干粮中可以改善其味道（Kuddus and Ramteke 2012）。

嗜盐和耐盐细菌也能产生蛋白酶，而且更适用于食品工业的盐条件或无盐系统中。另外一些例子还包括富含蛋白质的食品的盐发酵过程，例如鱼和肉类产品加工，以及酱油的生产（Setati 2010）。来自 *Halobacillus* sp. SR5-3（Namwong et al. 2006）和 *Halobacterium* 的酶就是用于鱼酱生产的蛋白酶的例子（Akolkar et al. 2010）。

14.3.3 皮革与纺织工业

分离自极端微生物的蛋白酶的嗜碱特性已经在皮革工业中使用了很长一段时间。这些酶可以代替浸泡、脱毛和软化（如 NaCl）中使用的有害物质来处理皮革。如今，碱性蛋白酶可与熟石灰和 NaCl 一同参与皮肤中非胶原成分的降解和去除球状蛋白，如白蛋白和球蛋白。在脱毛和软化时更多地使用酶不仅可以防止污染，还能节省时间和提升皮革质量。目前胰蛋白酶已被用于处理皮革，但仍有一些嗜碱性的替代品。例如 Varela 等（1997）用来自 *B. subtilis* HQDB32 的碱性蛋白酶给羊皮脱毛。George 等（1995）用类似的来自 *B. amyloliquefaciens* 的蛋白酶给兽皮去毛，而 Hameed 等（1999）则将其用于软化和皮革加工。在过去几年里，有几种新型的极端蛋白酶被发现具有脱毛能力，同时不会降解胶原和角质，因此不会破坏隐藏的胶原结构和阻碍角蛋白回收。这些酶包括来自嗜碱细菌 *B. pumilus* CBS（Jaouadi et al. 2009）和中温细菌 *P. aeruginosa* PD 100（Najafi et al. 2005）的蛋白酶。后者尽管是分离自中温菌株，但该酶具有多种特性，如耐碱性、热稳定性以及耐有机溶剂和去污剂，是一种独特的蛋白酶，可以被认为一个多极端性酶。此外，该酶有广泛的底物特异性，不仅能够用于脱毛，还能用于动物食品工业、澄清饮料、皮肤软化、生产氨基酸和多肽以及分子生物学中 DNA 纯化。另一种耐去污剂的丝氨酸碱性蛋白酶来源于 *B. pumilus* CBS，该酶能够以对胶原损伤最少的方式清除皮肤毛发，可以有效降解羽毛、鸡毛、羊毛和牛毛（Jaouadi et al. 2009）。

嗜碱性适冷蛋白酶还能够用于纺织业改良生产方式和纺织成品。它们能有效去除生丝中丝胶，丝胶是一种在蚕丝纤维表面的蛋白质物质。传统上，脱胶是在含肥皂的碱性溶液中完成的，这是一个强烈的过程，因为不仅是丝胶，连纤维也会受到损害。相反，蛋白酶可以清除丝胶而不破坏蚕丝纤维。因此，丝线会比传统方法处理得到的更加强韧。通过适冷蛋白酶处理后，羊毛和丝绸表面的纤维能够提供新的独特的产品（Najafi et al. 2005）。例如 Puri（2001）发现的 *Bacillus* sp. RGR-14 的碱性蛋白酶具有脱胶效果，纤维的扫描电镜结果显示，一些簇团已经分解，而被处理的纤维仍保持光滑和致密的结构。

14.3.4 家禽业

使用具有降解角蛋白活性的蛋白酶来分解处理各种各样的废料（如肉类行业的副产品或来自家禽的羽毛），每年大约会产生 10000t 这样的废料（Suzuki et al. 2006）。如今，这些粗料是通过非酶的碱性水解和蒸汽压力蒸煮的方式进行处理的，此过程中特定的氨基酸会被破坏并产生无营养价值的赖丙氨酸和羊毛硫氨酸。酶法降解家禽业的废料可形成能够用于动物的肥料和膳食蛋白质补充剂的水解物。此外，角蛋白酶能够用作皮革和美容业中的脱毛剂，以及用作去污剂和可食用膜的成分。目前已发现的大部分角蛋白酶尽管有着较高的热稳定性，但都是由中温微生物生产的。例如，在 37℃中培养的中温菌株 *B. licheniformis* K-19 产生一种在 30～90℃中有活性的角蛋白酶。该酶显示在 60℃和 pH 7.5～8.0 中有着最高活性（Xu et al. 2009）。相反，分离自嗜热细菌 *Fervidobacterium islandicum* AW-1（海岛闪烁杆菌）的蛋白酶在 100℃和 pH 9.0 中显示水解角蛋白的活性，而在最适温度中，其半衰期达 90min（Nam et al. 2002）。另一种角蛋白酶由 Paul 等（2014）从嗜盐菌 *Bacillus tequilensis* hsTKB2 中发现，是一种嗜碱（pH 9.0～10.5）、热稳定（50～80℃）和嗜盐（0～30% NaCl）的酶。

14.3.5 医疗和制药工业

如今几乎没有信息报道极端微生物的蛋白酶在医疗行业的应用。在 20 世纪 90 年代，有一些关于利用具有降解弹性蛋白质能力的 *B. subtilis* 316 M 制备弹性蛋白酶的讨论，这些酶可用于处理烧伤、化脓的伤口、痈、疖和深部脓肿（Kudrya and Simonenko 1994）。另一个想法是用来自 *Bacillus* sp. strain CK 11-4 的具有溶纤维活性的碱性蛋白酶作为血栓溶解剂（Kim et al. 1996）。然而，如果按照 Feller（2013）认为的冷水鱼和甲壳纲动物可能可以归类为极端生物的观点，那就不能忘记来源于大西洋鳕鱼（*Gadus morhua*）的胰蛋白酶，其商品名称为 ColdZyme®。胰蛋白酶可用于口腔清洁和防止牙菌斑的形成，牙菌斑是一种与龋齿相关的细菌的生物膜，主要是突变链球菌（*Streptococcus mutans*）和嗜酸乳杆菌（*Lactobacillus acidophilus*）。另一个用于防止蛀牙和牙周病的产品是 Krillase®，它含有分离自南极鳞虾 *Euphausia superba* Dana 的胰蛋白酶、胰凝乳蛋白酶样蛋白酶和羧肽酶 A 和 B。该产品也能用作处理难以愈合的伤口和溃疡的伤口敷料。它不仅可以清除坏死组织，还能加速健康组织肉芽的形成（Fornbacke and Clarsund 2013）。

在最近几年，已经出现了运用肽酶，尤其是有角蛋白水解活性的嗜热蛋白酶，降解感染的朊病毒蛋白质亚型（PrPSc），这些蛋白质往往聚集形成不规则结构的淀粉样蛋白。淀粉样蛋白的沉积可能会引起人体患上疯牛病和克罗伊茨费尔特-雅各布病。许多研究证明阮病毒聚集对于传统蛋白酶有极强的抵抗能力，甚至在 121℃ 灭菌条件中也不会降解（Langeveld et al. 2003）。首例不使用去污剂和非碱性环境降解 PrPSc 的报道是与分离自链霉菌属（*Streptomyces*）、热厌氧杆菌属（*Thermoanaerobacter*）和巴斯德毕赤酵母（*Thermococcus*）（Hui et al. 2004）的角蛋白酶应用有关。后两种极端微生物水解蛋白质的潜力被用于水解热预处理的朊病毒聚集体，这两种菌株也被提到用于动物粪便消毒（Suzuki et al. 2006）。

14.3.6 蛋白酶在非常规介质中的应用

蛋白酶水解反应的动力学和热力学研究已经显示，很可能能够将其反应平衡转向合成反应从而形成肽键（Glass 1981；Bongers and Heimer 1994；Elmore 2002）。反应的动力学控制包括维持缩合产物的浓度低于平衡浓度，只有这样才能在水存在下进行蛋白酶催化的肽合成（Morihara and Oka 1981），但考虑到底物的电离常数（热力学控制），大部分酶都需要一个较低的极性环境。通常而言，目前有三种这样的环境在使用：水-有机溶剂系统（水和水溶性有机溶剂系统），两相系统（水和非水溶性有机溶剂系统）（Doukyu and Ogino 2010）和有固定水活度（a_w）的微水相溶剂系统（Halling 1994）。然而，这种有利于合成反应的条件可能会对酶的活性产生负面影响。因此为了在水受限系统中使用特定的蛋白酶催化，该酶应该在这种环境中显示出高活性和稳定性（Deetz and Rozzell 1988）。这些参数的分析对于正确评估酶在非常规介质中进行的合成反应中的可行性是必不可少的。

14.3.6.1 极端蛋白酶在低极性介质中的特性

酶在非极性环境中失去活性有很多原因，如蛋白酶构象的改变、构象柔性的降低、失去关键的水或界面失活（Doukyu and Ogino 2010）。这些变化可能可通过基因工程、化学酶工程和酶的固定化及混合优化得以改善（Kumar and Bhalla 2005）。另一个方法是使用能够在非极性环境中自然起催化作用的极端蛋白酶，主要是嗜冷和嗜盐蛋白酶，这得益于它们的结构适应性（参考本章 14.2.2 和本章 14.2.4）。

嗜盐微生物的胞外蛋白酶是在低极性介质中极其有价值的合成工具，由于它们在高浓度盐中的具有吸湿性，因此能够适用低水活度环境。来自 γ-Proteobacterium（γ-变形菌）的丝氨酸蛋白酶就是这样的酶，在 35% NaCl 中孵育 18h 后仍能保持约 80% 的初始活力。当在水-有机溶剂系统（乙二醇、乙醇、正丁醇、丙酮、二甲基亚砜、二甲苯、氯乙烯）中测试时，无论是单相还是两相，该酶都比没有溶剂时具有更高的活性（甚至提升两倍）和稳定性（高达 10d）（Sana et al. 2006）。

与来自 γ-Proteobacterium 的蛋白酶不一样，其他一些酶，比如来自古菌 *Natrialba magadii* 的嗜盐碱蛋白酶就需要更高的盐浓度才能在有机介质中起作用。在含不同 logP 值（辛醇/水分配系数；Laane et al. 1987）范围的介质中测试该酶，介质 logP 值范围为 −1.76（甘油）到 0.5（异丙醇）且含有 0.5mol/L 或 1.5mol/L NaCl，发现在所有含高浓度盐的溶剂中它都显示出更高的活性（Ruiz and De Castro 2007）。对来自 *Halobacterium halobium*（Kim and Dordick 1997）的极端嗜盐蛋白酶进行稳定性研究，结果显示该酶在含 0.2mol/L NaCl 的 40% DMSO 中，比在 0.2mol/L NaCl 中稳定 40 倍。其稳定性会随着 40% DMSO 中 NaCl 的浓度上升而上升，比单独在 2mol/L NaCl 中要高 150 倍。然而，当 DMSO 被 40% 1,4-二氧杂环己烷替代时，尽管 NaCl 浓度维持不变，蛋白酶的稳定性也会显著下降。含有 0.2mol/L NaCl 的四氢呋喃也能提升蛋白酶的稳定性，但是仅仅在浓度高达 40%的范围里有效，当浓度达到 80%，其稳定性就会比在无有机溶剂时还低。*Hbt. Halobium* 蛋白酶在多种有机溶剂中稳定性的显著差异可以归根于它们天然的盐溶盐析特性（Kim and Dordick 1997）。Sellek 和 Chaudhuri（1999）也研究了 *Hbt. Halobium* 蛋白酶的特性，他们发现在 32%二甲基甲酰胺（DMF）中该酶有最高的合成活性，其中的蛋白酶水解效率，以催化效率（k_{cat}/K_m）表示，要低$\frac{2}{3}$。与之对比，该常数在 subtilisin Carlsberg 中没有改变。这两种蛋白酶之间的差异表明嗜盐酶尤其适应在有机介质中进行合成。

尽管应用受到限制，嗜冷蛋白酶依旧在低极性介质催化多肽合成领域有着吸引力。首先，温度的降低可导致平衡从水解向合成转变，因为酰基-酶中间产物的形成减慢。其次，降低温度能耗更低。最后，嗜冷酶在有机介质中更稳定。此外，由于嗜冷蛋白酶的柔性更好，可能可以用于疏水性更强的溶剂，降低蛋白酶构象的流动性以及其催化效率（Sellek and Chaudhuri 1999）。

除了嗜盐和嗜冷微生物以外，其他许多嗜热微生物也产生能在有机溶剂中稳定的蛋白酶。来自 *Thermus* sp. Rt4A2 的蛋白酶，在 4℃以及 90%的乙腈中仅损失 25%的活力，然而，当乙腈被相同浓度的丁醇替代后，其活性下降 59%（Freeman et al. 1993）。来自 *P. aeruginosa* PST-01 和 *Thermus* sp. Rt4A2 的蛋白酶有着不同的稳定性，证明了在为这些酶选择合适的溶剂进行合成反应时并没有一个通用的规则可用。在酶的结构中，细微而不易察觉的差异可能会导致它们的稳定性和合适的溶剂不同。来自 *Coprothermobacter proteolyticus* 的蛋白溶解素（proteolysin）是可能能被用于多肽合成的热稳定蛋白酶之一（Toplak et al. 2013）。该酶显示对 DMSO、DMF 和乙醇等溶剂有较好的耐受性，尽管这些溶剂不会使酶稳定。来自 *B. thermoproteolyticus* 的固定化酶嗜热菌素（thermolysin）是催化多肽合成最常用的酶之一，对其稳定性的研究明确了尽管嗜热蛋白酶在有机溶剂中不稳定，但它们在含有乙腈（嗜热菌素在其中孵育 5h 后保持 72%的活性）、叔丁醇（93%）或叔戊醇（98%）之类的物质中仍能保持活性。

值得注意的不仅仅是蛋白酶可能在有机溶剂中更稳定，对有机溶剂耐受的菌株也能被归类为另外一种类型的极端微生物。它们不仅包含嗜盐菌株，如古菌 *Halobacterium* ssp.（Akolkar et al. 2008），还包含非嗜盐菌株，如 *P. aeruginosa* PST-01。PST-01 蛋白酶在 25%有机溶剂中，如乙二醇、1,4-丁二醇、1,5-戊二醇、乙醇、*n*-己醇、甲醇、丁醇、DMSO 和其他溶剂，比在水相环境中的稳定性更高（5 到 10 倍）。其稳定性远远高于一般用于生物转化的 subtilisin Carlsberg（Ogino et al. 1999a，b).

14.3.6.2 用蛋白酶进行合成

只有少数来自极端微生物的蛋白酶已被用于多肽合成工艺，这些例子在表 14.7 中列出。需要注意的是，在低水含量的疏水环境中，蛋白酶主要合成二肽和三肽。因此，需要一些有着不同底物特异性的蛋白酶来合成更长的多肽。例如，Kimura 等（1990b，表 14.7）需要三步来合成 Z-L-Tyr-Gly-Gly-L-Phe-L-Leu-OEt，第一步它们使用木瓜蛋白酶和 α-糜蛋白酶获得 Z-Gly-Gly-OBut 和 Z-L-Tyr-Gly-Gly-OBut，后者再被转化为 Z-L-Phe-L-Leu-OEt，然后用嗜热菌素转化为目的多肽。Kullman（1982，表 14.7）使用的是另一种方法：利用木瓜蛋白酶、α-糜蛋白酶、芳香硫酸酯酶（EC 3.1.6.1；为 *O*-硫酸酪氨酸脱硫）、嗜热菌素和氨肽酶 M 进行多肽的化学缩合，构建合成胆囊收缩素。

表 14.7 利用极端蛋白酶催化多肽合成反应的例子

蛋白酶	合成多肽	参考文献
Thermus sp. RT41A protease	Bz-Ala-Tyr-NH_2	Wilson et al. (1994)
Thermolysin	Asp-Phe(天冬苯丙二肽酯)	Kühn et al. (2002)
	α-L-Asp-Phe-OM	Rao et al. (1998)
	Z-L-Tyr-Gly-Gly-L-Phe-L-Leu-OEt①	Kimura et al. (1990a,b,c)
	Asp-Tyr-Met-Gly-Trp-Met-Asp-Phe①(缩胆囊肽)	Kullmann(1982)
	Asp-Tyr-Thr-Gly-Trp-Met-Asp-Phe①(雨蛙肽)	Takai et al. (1981)
PST-01 蛋白酶	Cbz-Lys-Leu-NH_2	Ogino et al. (1999a,b)
	Cbz-Ala-Leu-NH_2	
	Cbz-Ala-Phe-NH_2	
Natrialba magadii protease	Ac-Phe-Gly-Phe-NH_2	Ruiz et al. (2010)

在动力学控制多肽合成中，来自嗜碱枯草芽孢杆菌的在碱性 pH 中有着最高活性的枯草杆菌素（subtilisin）扮演着重要角色。这些酶合成肽键的速率要比水解的速率快（Stepanov 1996）。其中枯草杆菌素 Carlsberg，在有机溶剂中最常用，据 Klein 等（2000）的报道该酶有着巨大的潜力。在 pH 9.5 含有体积分数 50% DMF 和强碱哌啶类化合物（在胺类底物中加入中和酸）的介质中，枯草杆菌素作为亲核试剂催化酰基供体（Z-Val-Trp-OMe）与酰胺类化合物（Gly-NH_2 产量超过 70%），以及二肽 Gly-Xaa（Xaa＝Gly，Ala，Phe，Gln，Ser，Val，Lys，Trp）之间形成肽键。该过程没有观察到水解反应。

枯草杆菌素在水解反应中具有广泛的底物特异性。在合成反应中，其特异性可以通过改变介质来变换。例如，Klein 等（2000）研究枯草杆菌素 Carlsberg 发现添加乙腈（乙腈：DMF：水＝8：1：1）可提高单核氨基酸亲核剂的合成产量（除了 Gly-NH_2）以及可以在 Z-Val-Trp-OMe、Lys-NH_2 和 Ala-NH_2 的缩合反应中获得最高产量（至少 70%）。对于亲核二肽 Xaa-Gly，添加乙腈可使 Xaa＝Ser，Lys，Thr，Asn，Gln，Met，His，Val，Glu，Asp 和 Arg 合成产量提升三倍。然而，对于 Gly-Xaa，其产量的变化明显更低，只有 Xaa＝Asp 和 Pro 能观察到产量有明显提升。

一些极端蛋白酶能够合成非蛋白质类成分，如通过酯交换合成酯类。来自 *B. pseudofirmus* AL-89 的碱性蛋白酶（最适 pH 10.0）就是一个好例子。该酶在最适溶剂系统（DMF-DMSO）中可催化合成蔗糖月桂酸酯（Pedersen et al. 2003）。Subtilisin Carlsberg 是另外一个用蛋白酶合成酯类的例子。该酶可用于含有 D-果糖部分的二糖和寡糖的酰基化（Riva et al. 1998）。Riva 等（1998）在含有活化酯类-三氟乙醇丁酮酸酯的无水 DMF 中获得了乳果糖、异构麦芽糖、水苏糖和蔗糖的 1-*O*-丁酰基衍生物，产量达 50%。这些通过选择性酯化反应生产的糖酯是极有价值的表面活性剂。

14.4 结论和挑战

蛋白酶，尤其是微生物来源的蛋白酶，在使用酶催化制造多种高附加值产品的工业流程中起着主导作用。这些酶独特的多样性为此提供了基础，毫无疑问的，在未来也会如此，这是因为它们在各种工业分支和医学中有着越来越多的用途。有些用途目前甚至已经很难去定义，因为已知的由微生物合成的蛋白酶数量太少。然而，对于来自极端微生物的蛋白酶的研究使得在过去十年里蛋白酶商业制备的数量不断上升，并且根据目前的趋势预测，在未来仍会上升。尽管如此，但目前为止只有少部分极端生境的微生物被认知，这意味着大部分由微生物合成的催化蛋白质，包括最有价值的、独特的酶，在工业中还无法获得。分析极端环境的宏基因组和宏转录组已经能够分离编码独特蛋白酶的基因，该方法也成为了大量合成这种酶的理想方法，尤其是那些来自目前在实验室条件下不可培养的微生物的酶，这些研究应该加强。然而，采用传统方法对合成新型蛋白酶的新型极端微生物进行勘测也应该继续。经过改良的快速 DNA 测序技术使数据库中大量微生物全基因组快速增加，这些基因组挖掘可能可以寻找到蛋白酶新的突变体，同样对于极端蛋白酶分子适应的研究也是非常有前途的。他们的研究结果可以促进对已知蛋白酶的合理工程设计，并且从更深入的角度来看，可能会被用于设计这些酶的新型突变体来针对特定的应用和工业条件。同样地专门为极端微生物来源的酶类建立新型表达系统也是一种迫切需求，因为在中温宿主中编码这些酶类的基因表达通常不够高效，或者不能表达，制约了许多在工业和医药上受到青睐特性的相关酶类的广泛应用。

无利益冲突声明

Aneta Biatkowska、Ewa Gromek、Tomasz Florczak、Joanna Krysiak、Katarzyna Szulczewska 和 Marianna Turkiewicz 声明他们没有利益冲突。

参考文献

Akolkar AV，Desai AJ（2010）Catalytic and thermodynamic characterization of protease from *Halobacterium* sp. SP1（1）. Res Microbiol 161（5）：355-362

Akolkar AV，Deshpande GM，Raval KN，Durai D，Nerurkar AS，Desai AJ（2008）Organic solvent tolerance of *Halobacterium* sp. SP1（1）and its extracellular protease. J Basic Microbiol 48（5）：421-425

Akolkar AV，Durai D，Desai AJ（2010）*Halobacterium* sp. SP1（1）as a starter culture for accelerating fish sauce fermentation. J Appl Microbiol 109（1）：44-53

Alam SI，Dube S，Reddy GSN，Bhattacharya BK，Shivaji S，Singh L（2005）Purification and characterization of extracellular protease produced by *Clostridium* sp. from Schirmacher oasis，Antarctica. Enzyme Microb Technol 36（5-6）：824-831

Alam S，Dube S，Agarwal M，Singh L（2006）Purification and characterization of an extracellular protease produced by psychrotolerant *Clostridium* sp. LP3 from lake sediment of Leh，India. Can J Microbiol 52（12）：1238-1246

Amoozegar MA，Fatemi ZA，Karbalaei-Heidari HR，Razavi MR（2007）Production of an extracellular alkaline metalloprotease from a newly isolated，moderately halophile，*Salinivibrio* sp. strain AF-2004. Microbiol Res 162（4）：369-377

Aqel H，Al-Quadan F，Yousef TK（2012）A novel neutral protease from thermophilic *Bacillus* strain HUTBS62. J Biosci Biotechnol 1（2）：117-123

Arulmani M，Aparanjini K，Vasanthi K，Arumugam P，Arivuchelvi M，Kalaichelvan PT（2007）Purification and partial characterization of serine protease from thermostable alkalophilic *Bacillus laterosporus*-AK1. World J Microbiol Biotechnol 23（4）：475-481

Baghel VS，Tripathi RD，Ramteke RW，Gopal K，Dwivedi S，Jain RK，Rai UN，Singh SN（2005）Psychrotrophic proteolytic bacteria from cold environments of Gangotri glacier，Western Himalaya，India. Enzyme Microb Technol 36（5-6）：654-659

Bakhtiar S，Andersson MM，Gessesse A，Mattiasson B，Hatti-Kaul R（2002）Stability characteristics of a calcium-independent alkaline protease from *Nesterenkonia* sp. Enzyme Microb Technol 32（5）：525-531

Benson DA，Karsch-Mizrachi I，Lipman DJ，Ostell J，Sayers EW（2009）GenBank. Nucleic Acids Res 37：D26-D31

Biver S，Portetelle D，Vandenbol M（2013）Characterization of a new oxidant-stable serine protease isolated by functional metagenomics. SpringerPlus 2：410

Blumentals II，Robinson AS，Kelly RM（1990）Characterization of sodium dodecyl sulfateresistant proteolytic activity in the hyperthermophilic archaebacterium *Pyrococcus furiosus*. Appl Environ Microbiol 56（7）：1992-1998

Bongers J，Heimer EP（1994）Recent applications of enzymatic peptide synthesis. Peptides 15（1）：183-193

Cannio R，Catara G，Fiume I，Balestrieri M，Rossi M，Palmieri G（2010）Identification of a cellbound extracellular protease overproduced by *Sulfolobus solfataricus* in peptide-rich media. Protein Pept Lett 17（1）：78-85

Capiralla H，Hiroi T，Hirokawa T，Maeda S（2002）Purification and characterization of a hydrophobic amino acid-specific endopeptidase from *Halobacterium halobium* S9 with potential application in debittering of protein hydrolysates. Process Biochem 38 (4): 571-579

CarterP, Nilsson B, Burnier JP, Burdick D, Wells JA (1989) Engineering subtilisin BPN' for site-specific proteolysis. Proteins: Struct Funct Bioinf 6 (3): 240-248

Catara G, Ruggiero G, La Cara F, Digilio FA, Capasso A, Rossi M (2003) A novel extracellular subtilisin-like protease from the hyperthermophile *Aeropyrum pernix* K1: biochemical properties, cloning, and expression. Extremophiles 7 (5): 391-399

Cavicchioli R, Charlton T, Ertan H, Mohd Omar S, Siddiqui KS, Williams TJ (2011) Biotechnological uses of enzymes from psychrophiles. Microb Biotechnol 4 (4): 449-460

Chavez Croocker P, Sako Y, Uchida A (1999) Purification and characterization of an intracellular heat-stable proteinase (pernilase) from the marine hyperthermophilic archaeon *Aeropyrum pernix* K1. Extremophiles 3 (1): 3-9

Chen K, Arnold FH (1993) Tuning the activity of an enzyme for unusual environments: sequential random mutagenesis of subtilisin E for catalysis in dimethylformamide. Proc Natl Acad Sci U S A 90 (12): 5618-5622

Chen XL, Zhang YZ, Gao PJ, Luan XW (2003) Two different proteases produced by a deep-sea psychrotrophic bacterial strain, *Pseudoaltermonas* sp. SM9913. Mar Biol 143 (5): 989-993

Choi IG, Bang WG, Kim SH, Yu YG (1999) Extremely thermostable serine-type protease from *Aquifex pyrophilus*. Molecular cloning, expression, and characterization. J Biol Chem 274 (2): 881-888

Cowan DA, Daniel RM (1982) Purification and some properties of an extracellular protease (caldolysin) from an extreme thermophile. Biochim Biophys Acta 705 (3): 293-305

Damare S, Raghukumar C, Muraleedharan U, Raghukumar S (2006) Deep-sea fungi as a source of alkaline and cold-tolerant proteases. Enzyme Microb Technol 39 (2): 172-181

Dastager SG, Dayanand A, Li WJ, Kim CJ, Lee JC, Park DJ, Tian XP, Raziuddin QS (2008) Proteolytic activity from an alkali-thermotolerant *Streptomyces gulbargensis* sp. nov. Curr Microbiol 57 (6): 638-642

Davail S, Feller G, Narinx E, Gerday C (1994) Cold adaptation of proteins. Purification, characterization, and sequence of the heat-labile subtilisin from the Antarctic psychrophile *Bacillus* TA41. J Biol Chem 269 (26): 17448-17453

De Azeredo L, Freire D, Soares R, Leite S, Coelho R (2004) Production and partial characterization of thermophilic proteases from *Streptomyces* sp. isolated from Brazilian cerrado soil. Enzyme Microb Technol 34 (3-4): 354-358

De Castro RE, Maupin-Furlow JA, Giménez MI, Herrera Seitz MK, Sánchez JJ (2005) Haloarchaeal proteases and proteolytic systems. FEMS Microbiol Rev 30 (1): 17-35

Deetz JS, Rozzell JD (1988) Enzyme-catalysed reactions in non-aqueous media. Trends Biotechnol 6 (1): 15-19

Deng A, Wu J, Zhang Y, Zhang G, Wen T (2010) Purification and characterization of a surfactantstable high-alkaline protease from *Bacillus* sp. B001. Bioresour Technol 101 (18): 7100-7106

Dib R, Chobert JM, Dalgalarrondo M, Barbier G, Haertlé T (1998) Purification, molecular properties and specificity of a thermoactive and thermostable proteinase from *Pyrococcus abyssi*, strain st 549, hyperthermophilic archaea from deep-sea hydrothermal ecosystem. FEBS Lett 431 (2): 279-284

Dixit VS, Pant A (2000) Comparative characterization of two serine endopeptidases from *Nocardiopsis* sp. NCIM 5124. Biochim Biophys Acta 1523 (2-3): 261-268

Doddapaneni KK, Tatineni R, Vellanki RV, Gandu B, Panyala NR, Chakali B, Mangamoori LN (2007) Purification and characterization of two novel extra cellular proteases from *Serratia rubidaea*. Process Biochem 42 (8): 1229-1236

Dodia MS, Rawal CM, Bhimani HG, Joshi RH, Khare SK, Singh SP (2008) Purification and stability characteristics of an alkaline serine protease from a newly isolated Haloalkaliphilic bacterium sp. AH-6. J Ind Microbiol Biotechnol 35 (2): 121-131

Doukyu N, Ogino H (2010) Organic solvent-tolerant enzymes. Biochem Eng J 48 (13): 270-282. doi: 10.1016/j.bej.2009.09.009

Dozie IN, Okeke CN, Unaeze NC (1994) A thermostable, alkaline-active, keratinolytic proteinase from *Chrysosporium keratinophilum*. World J Microbiol Biotechnol 10 (5): 563-567

Eggen R, Geerling A, Watts J, de Vos WM (1990) Characterization of pyrolysin, a hyperthermoactive serine protease from the archaebacterium *Pyrococcus furiosus*. FEMS Microbiol Lett 71 (1-2): 17-20

Elmore DT (2002) Peptide synthesis. In: Barret GC, Davies JS (eds) Amino acids, peptides and proteins: Volume 33. RSC Publishing, UK, pp. 83-134

Estell DA, Graycar TP, Wells JA (1985) Engineering an enzyme by site-directed mutagenesis to be resistant to chemical oxidation. J Biol Chem 260 (11): 6518-6521

Feller G (2003) Molecular adaptations to cold in psychrophilic enzymes. Cell Mol Life Sci 60 (4): 648-662

Feller G (2013) Psychrophilic enzymes: from folding to function and biotechnology. Scientifica. Article ID 512840

Fornbacke M, Clarsund M (2013) Cold-adapted proteases as an emerging class of therapeutics. Infect Dis Ther 2 (1): 15-26

Freeman SA, Peek K, Prescott M, Daniel R (1993) Characterization of a chelator-resistant protease from *Thermus* strain Rt4A2. Biochem J 295 (Pt 2): 463-469

Fujinami S, Fujisawa M (2010) Industrial applications of alkaliphiles and their enzymes-past, present and future. Environ Technol 31 (8-9): 845-856. doi: 10.1080/09593331003762807

Fusek M, Lin XL, Tang J (1990) Enzymic properties of thermopsin. J Biol Chem 265 (3): 1496-1501

George S, Raju V, Krishnan MRV, Subramanian TV, Jayaraman K (1995) Production of protease by *Bacillus amyloliquefaciens* in solid-state fermentation and its application in the unhairing of hides and skins. Process Biochem 30 (5): 457-462

Gessesse A, Hatti-Kaul R, Gashe BA, Mattiasson B (2003) Novel alkaline proteases from alkaliphilic bacteria grown on chicken feather. Enzym Microb Technol 32 (5): 519-524

Giménez MI, Studdert CA, Sánchez JJ, De Castro RE (2000) Extracellular protease of *Natrialba magadii*: purification and biochemical characterization. Extremophiles 4 (3): 181-188

Giri SS, SukumaranV, Sen SS, Oviya M, Banu BN, Jena PK (2011) Purification and partial characterization of a detergent and oxidizing agent stable alkaline protease from a newly isolated *Bacillus subtilis* VSG-4 of tropical soil. J Microbiol 49 (3): 455-461

Glass JD (1981) Enzymes as reagent in the synthesis of peptides. Enzyme Microb Technol 3 (1): 2-8

Gohel SD, Singh SP (2012) Purification strategies, characteristics and thermodynamic analysis of a highly thermostable alkaline protease from a salt-tolerant alkaliphilic actinomycete, *Nocardiopsis alba* OK-5. J Chromatogr B Analyt Technol Biomed Life Sci 889-890: 61-68

Gohel SD, Singh SP (2013) Characteristics and thermodynamics of a thermostable protease from a salt-tolerant alkaliphilic actinomycete. Int J Biol Macromol 56: 20-27

Gomes J, Steiner W (2004) The biocatalytic potential of extremophiles and extremozymes. Food Technol Biotechnol 42 (4): 223-235

Graziano G, Merlino A (2014) Molecular bases of protein halotolerance. Biochim Biophys Acta 1844 (4): 850-858

Grøn H, Bech LM, Branner S, Breddam K (1990) A highly active and oxidation-resistant subtilisin-like enzyme produced by a combination of sidedirected mutagenesis and chemical modification. Eur J Biochem 194 (3): 897-901

Gupta R, Beg QK, Lorenz P (2002) Bacterial alkaline proteases: molecular approaches and indus-trial applica-tion. Appl Microbiol Biotechnol 59 (1): 15-32

Gupta A, Roy I, Patel RK, Singh SP, Khare SK, Gupta MN (2005) One-step purification and characterization of an alkaline protease from haloalkaliphilic *Bacillus* sp. J Chromatogr A 1075 (1-2): 103-108

Halling PJ (1994) Thermodynamic prediction for biocatalysis in nonconventional media: theory, tests, and recommendations for experimental design and analysis. Enzyme Microb Technol 16 (3): 178-206

Hameed A, Keshavarz T, Evans CS (1999) Effect of dissolved oxygen tension and pH on the production of extracellular protease from a new isolate of *Bacillus subtilis* K2, for use in leather processing. J Chem Technol Biotechnol 74 (1): 5-8

Hao JH, Sun M (2015) Purification and characterization of a cold alkaline protease from a psychrophilic *Pseudomonas aeruginosa* HY1215. Appl Biochem Biotechnol 175 (2): 715-722

Hasan AKMQ, Tamiya E (2001) Cold-active protease CP70. Patent US 6200793

Hasbay Ifrij I, Ogel ZB (2002) Production of neutral and alkaline extracellular proteases by the thermophilic fungus, *Scytalidium thermophilum*, grown on microcrystalline cellulose. Biotechnol Lett 24 (13): 1107-1110

Hiraga K, Nishikata Y, Namwong S, Tanasupawat S, Takada K, Oda K (2005) Purification and characterization of serine proteinase from a halophilic bacterium, *Filobacillus* sp. RF2-5. Biosci Biotechnol Biochem 69 (1): 38-44

Horikoshi K (1999) Alkaliphiles: some applications of their products for biotechnology. Microbiol Mol Biol Rev 63 (4): 735-750

Hui Z, Doi H, Kanouchi H, Matsuura Y, Mohri S, Nonomura Y, Oka T (2004) Alkaline serine protease produced by *Streptomyces* sp. degrades PrP (Sc). Biochem Biophys Res Commun 321 (1): 45-50

Huston A, Methe B, Deming J (2004) Purification, characterization, and sequencing of an extracellular cold-active aminopeptidase produced by marine psychrophile *Colwellia psychrerythraea* strain 34H. Appl Environ Microbiol 70 (6): 3321-3328

Ignatova Z, Gousterova A, Spassov G, Nedkov P (1999) Isolation and partial characterization of extracellular keratinase from a wool degrading thermophilic actinomycete strain *Thermoactinomyces candidus*. Can J Microbiol 45 (3): 217-222

Iqbal I, Aftab MN, Afzal M, Ur-Rehman A, Aftab S, Zafat A, Ud-Din Z, Khuharo AR, Iqbal J, Ul-Haq I (2015) Purification and characterization of cloned alkaline protease gene of *Geobacillus stearothermophilus*. J Basic Microbiol 55 (2): 160-171

Izotova LS, Strongin AY, Chekulaeva LN, Sterkin VE, Ostoslavskaya VI, Lyublinskaya LA, Timokhina EA, Stepanov VM (1983) Purification and properties of serine protease from *Halobacterium halobium*. J Bacteriol 155 (2): 826-830

Jang HJ, Kim BC, Pyun YR, Kim YS (2002) A novel subtilisin-like serine protease from *Thermoanaerobacter yonseiensis* KB-1: its cloning, expression, and biochemical properties. Extremophiles 6 (3): 233-243

Jaouadi B, Ellouz-Chaabouni S, Rhimi M, Bejar S (2008) Biochemical and molecular characterization of a detergent stable serine alkaline protease from *Bacillus pumilus* CBS with high cata-lytic efficiency. Biochimie 90 (9): 1291-1305

Jaouadi B, Ellouz-Chaabouni S, Ben Ali M, Ben Messaoud E, Naili B, Dhouib A, Bejar S (2009) Excellent laundry detergent compatibility and high dehairing ability of the *Bacillus pumilus* CBS alkaline proteinase (SAPB). Biotechnol Bioprocess Eng 14: 503-512

Jaouadi B, Abdelmalek B, Fodil D, Ferradji FZ, Rekik H, Zarai N, Bejar S (2010a) Purification and characterization of a thermostable keratinolytic serine alkaline proteinase from *Streptomyces* sp. strain AB1 with high stability in organic solvents. Bioresour Technol 101 (21): 8361-8369

Jaouadi B, Aghajari N, Haser R, Bejar S (2010b) Enhancement of the thermostability and the catalytic efficiency of *Bacillus pumilus* CBS protease by site-directed mutagenesis. Biochimie 92 (4): 360-369

Joo HS, Chang CS (2005) Oxidant and SDS-stable alkaline protease from a halo-tolerant *Bacillus clausii* I-52: enhanced production and simple purification. J Appl Microbiol 98 (2): 491-497

Joo HS, Kumar CG, Park GC, Paik SR, Chang CS (2003) Oxidant and SDS-stable alkaline protease from *Bacillus clausii* I-52: production and some properties. J Appl Microbiol 95 (2): 267-272

Joshi S, Satyanarayana T (2013a) Biotechnology ofcold-active proteases. Biology 2 (2): 755-783

Joshi S, Satyanarayana T (2013b) Characteristics and applications of a recombinant alkaline serine protease from anovel bacterium *Bacillus lehensis*. Bioresour Technol 131: 76-85

Kamekura M, Dyall-Smith ML (1995) Taxonomy of the family Halobacteriaceae and the description of two new genera *Halorubrobacterium* and *Natrialba*. J Gen Appl Microbiol 41 (4): 333-350

Kamekura M, Seno Y (1990) A halophilic extracellular protease from a halophilic archaebacterium strain 172

P1. Biochem Cell Biol 68 (1): 352-359

Kamekura M, Seno Y, Dyall-Smith ML (1996) Halolysin R4, a serine proteinase from the halophilic archaeon *Haloferax mediterranei*; gene cloning, expression and structural studies. Biochim Biophys Acta 1294 (2): 159-167

Karan R, Khare SK (2010) Purification and characterization of a solvent-stable protease from *Geomicrobium* sp. EMB2. Environ Technol 31 (10): 1061-1072

Karan R, Singh SP, Kapoor S, Khare SK (2011) A novel organic solvent tolerant protease from a newly isolated *Geomicrobium* sp. EMB2 (MTCC 10310): production optimization by response surface methodology. New Biotechnol 28 (2): 138-145

Karbalaei-Heidari HR, Shahbazi M, Absalan G (2013) Characterization of a novel organic solvent tolerant protease from a moderately halophilic bacterium and its behavior in ionic liquids. Appl Biochem Biotechnol 170 (3): 573-586

Khan F (2013) New microbial proteases in leather and detergent industries. Innov Res Chem 1 (1): 1-6

Kikani BA, Shukla RJ, Singh SP (2010) Biocatalytic potential of thermophilic bacteria and actinomycetes. In: Mendez-Vilas A (ed) Current research, technology and education topics in applied microbiology and microbial biotechnology. Formatex Research Center, Badajoz, Spain, pp 1000-1007

Kim J, Dordick JS (1997) Unusual salt and solvent dependence of a protease from an extreme halophile. Biotechnol Bioeng 55 (3): 471-479

Kim T, Lei XG (2005) Expression and characterization of a thermostable serine protease (TfpA) from *Thermomonospora fusca* YX in *Pichia pastoris*. Appl Microbiol Biotechnol 68 (3): 355-359

Kim W, Choi K, Kim Y, Park H, Choi J, Lee Y, Oh H, Kwon I, Lee S (1996) Purification and characterization of a fibrinolytic enzyme produced from *Bacillus* sp. strain CK 11-4 screened from Chungkook-Jang. Appl Environ Microbiol 62 (7): 2482-2488

KimuraY, Muraya K, Araki Y, Matsuoka H, Nakanishi K, Matsuno R (1990a) Synthesis peptides consisting of essential amino acids by a reactor system using three proteinases and an organic solvent. Agric Biol Chem 54 (12): 3331-3333

Kimura Y, Nakanishi K, Matsuno R (1990b) Enzymatic synthesis of the precursor of Leu-enkephalin in water-immiscible organic solvent systems. Enzyme Microb Technol 12 (4): 273-280

Kimura Y, Yoshida T, Muraya K, Nakanishi K, Matsuno R (1990c) Continuous synthesis of a tripeptide by successive condensation and transesterification catalyzed by two immobilized proteinases in organic solvent. Agric Biol Chem 54 (6): 1433-1440

Klein UJ, Prykhodzka A, Cerovsky V (2000) The applicability of subtilisin Carlsberg in peptide synthesis. J Pept Sci 6 (11): 541-549

Kleine R (1982) Properties of thermitase, a thermostable serine protease from *Thermoactinomyces vulgaris*. Acta Biol Med Ger 41 (1): 89-102

Klingeberg M, Galunsky B, Sjoholm C, Kasche V, Antranikian G (1995) Purification and properties of a highly thermostable, sodium dodecyl sulfate-resistant and stereospecific proteinase from the extremely thermophilic archaeon *Thermococcus stetteri*. Appl Environ Microbiol 61 (8): 3098-3104

Knight ZA, Garrison JL, Chan K, King DS, Shokat KM (2007) A remodelled protease that cleaves phosphotyrosine substrates. J Am Chem Soc 129 (38): 11672-11673

Kobayashi T, Hakamada Y, Adachi S, Hitomi J, Yoshimatsu T, Koike K, Kawai S, Ito S (1995) Purification and properties of an alkaline protease from alkalophilic *Bacillus* sp. KSM-K16. Appl Microbiol Biotechnol 43 (3): 473-481

Kobayashi T, Lu J, Li Z, Hung VS, Kurata A, Hatada Y, Takai K, Ito S, Horikoshi K (2007) Extremely high alkaline protease from a deep-subsurface bacterium, *Alkaliphilus transvaalensis*. Appl Microbiol Biotechnol 75 (1): 71-80

Kuddus M, Ramteke PW (2009) Cold-active extracellular alkaline protease from an alkaliphilic *Stenotrophomonas maltophilia*: production of enzyme and its industrial applications. Can J Microbiol 55 (11): 1294-1301

Kuddus M, Ramteke PW (2011) Production optimization of an extracellular cold-active alkaline protease from *Stenotrophomonas maltophilia* MTCC 7528 and its application in detergent industry. Afr J Microbiol Res 5 (7): 809-816

Kuddus M, Ramteke PW (2012) Recent developments in production and biotechnological applications of cold-active microbial proteases. Crit Rev Microbiol 38 (4): 330-338

Kudrya VA, Simonenko IA (1994) Alkaline serine proteinase and lectin isolation from the culture fluid of *Bacillus subtilis*. Appl Microbiol Biotechnol 41 (5): 505-509

Kühn D, Dürrschmidt P, Mansfeld J, Ulbrich-Hofmann R (2002) Boilysin and thermolysin in dipeptide synthesis: a comparative study. Biotechnol Appl Biochem 36 (Pt 1): 71-76

Kullmann W (1982) Protease-catalyzed peptide bond formation: application to synthesis of the COOH-terminal octapeptide of cholecystokinin. Proc Natl Acad Sci U S A 79 (9): 2840-2844

Kumar D, Bhalla TC (2005) Microbial proteases in peptide synthesis: approaches and applications. Appl Microbiol Biotechnol 68 (6): 726-736

Kumar D, Savitri, Thakur N, Verma R, Bhalla TC (2008) Microbial proteases and application as laundry detergent additive. Res J Microbiol 3 (12): 661-672

Kumar CG, Tiwari MP, Jany KD (1999) Novel alkaline serine proteases from alkalophilic *Bacillus* spp.: purification and some properties. Process Biochem 34 (5): 441-449

Laane C, Boeren S, Vos K, Veeger C (1987) Rules for optimization of biocatalysis in organic solvents. Biotechnol Bioeng 30 (1): 81-87

Lagzian M, Asoodeh A (2012) An extremely thermotolerant, alkaliphilic subtilisin-like protease from hyperthermophilic *Bacillus* sp. MLA64. Int J Biol Macromol 51 (5): 960-967

Lama L, Romano I, Calandrelli V, Nicolaus B, Gambacorta A (2005) Purification and characterization of a protease produced by an aerobic haloalkaliphilic species belonging to the *Salinivibrio* genus. Res Microbiol 156 (4): 478-484

Langeveld JP, Wang JJ, Van de Wiel DF, Shih GC, Garssen GJ, Bossers A, Shih JC (2003) Enzymatic degradation of prion protein in brain team stem from infected cattle and sheep. J Infect Dis 188 (11): 1782-1789

Li AN, Li DC (2009) Cloning, expression and characterization of the serine protease gene from *Chaetomium thermophilum*. J Appl Microbiol 106 (2): 369-380

Li WF, Zhou XX, Lu P (2005) Structural features of thermozymes. Biotechnol Adv 23 (4): 271-281

Li AN, Ding AY, Chen J, Liu SA, Zhang M, Li DC (2007) Purification and characterization of two thermostable proteases from the thermophilic fungus *Chaetomium thermophilum*. J Microbiol Biotechnol 17 (4): 624-631

Li Q, Yi L, Marek P, Iverson BL (2013) Commercial proteases: present and future. FEBS Lett 587 (8): 1155-1163

Lin X, Tang J (1990) Purification, characterization, and gene cloning of thermopsin, a thermostable acid protease from *Sulfolobus acidocaldarius*. J Biol Chem 265 (3): 1490-1495

Litchfield CD (2011) Potential for industrial products from the halophilic Archaea. J Ind Microbiol Biotechnol 38 (10): 1635-1647

Lü J, Wu X, Jiang Y, Cai X, Huang L, Yang Y, Wang H, Zeng A, Li A (2014) An extremophile *Microbacterium* strain and its protease production under alkaline conditions. J Basic Microbiol 54 (5): 378-385

Majeed T, Tabassum R, Orts WJ, Lee CC (2013) Expression and characterization of *Coprothermobacter proteolyticus* alkaline serine protease. Sci World J. Article ID 396156

Manikandan M, Pašić L, Kannan V (2009) Purification and biological characterization of a halophilic thermostable protease from *Haloferax lucentensis* VKMM 007. World J Microbiol Biotechnol 25 (12): 2247-2256

Marcy RM, Engelhardt TC, Upadhyay JM (1984) Isolation, partial characterization, and some properties of protease I from a thermophilic mold *Thermoascus aurantiacus* var. *levisporus*. Mycopathologia 87 (1-2): 57-65

Margesin R, Dieplinger H, Hofmann J, Sarg B, Lindner H (2005) A cold-active extracellular metalloprotease from *Pedobacter cryoconitis* -production and properties. Res Microbiol 156 (4): 499-505

Martinez R, Jakob F, Tu R, Siegert P, Maurer KH, Schwaneberg U (2013) Increasing activity and thermal resistance of *Bacillus gibsonii* alkaline protease (BgAP) by directed evolution. Biotechnol Bioeng 110 (3): 711-720

Matsuzawa H, Tokugawa K, Hamaoki M, Mizoguchi M, Taguchi H, Terada I, Kwon ST, Ohta T (1988) Purification and characterization of aqualysin I (a thermophilic alkaline serine protease) produced by *Thermus aquaticus* YT-1. Eur J Biochem 171 (3): 441-447

Maurer KH (2004) Detergent proteases. Curr Opin Biotechnol 15 (4): 330-334

Mei HC, Liaw YC, Li YC, Wang DC, Takagi H, Tsai YC (1998) Engineering subtilisin YaB: restriction of substrate specificity by the substitution of Gly124 and Gly151 with Ala. Protein Eng 11 (2): 109-117

Mitsuiki S, Sakai M, Moriyama Y, Goto M, Furukawa K (2002) Purification and some properties of a keratinolytic enzyme from an alkaliphilic *Nocardiopsis* sp. TOA-1. Biosci Biotechnol Biochem 66 (1): 164-167

Miyazaki K, Wintrode PL, Grayling RA, Rubingh DN, Arnold FH (2000) Directed evolution study of temperature adaptation in a psychrophilic enzyme. J Mol Biol 297 (4): 1015-1026

Moreno ML, Garcia MT, Ventosa A, Mellado E (2009) Characterization of *Salicola* sp. IC10, a lipase-and protease-producing extreme halophile. FEMS Microbiol Ecol 68 (1): 59-71

Morihara K, Oka T (1981) Peptide bond synthesis catalyzed by subtilisin, papain, and pepsin. J Biochem 89 (2): 385-395

Morikawa M, Izawa Y, Rashid N, Hoaki T, Imanaka T (1994) Purification and characterization of a thermostable thiol protease from a newly isolated hyperthermophilic *Pyrococcus* sp. Appl Environ Microbiol 60 (12): 4559-4566

MoritaRY (1975) Psychrophilic bacteria. Bacteriol Rev 39 (2): 144-167

Najafi MF, Deobagkar D, Deobagkar D (2005) Potential application of protease isolated from *Pseudomonas aeruginosa* PD 100. Electron J Biotechnol 8 (2): 197-203

Naki D, Paech C, Ganshaw G, Schellenberger V (1998) Selection of a subtilisin-hyperproducing *Bacillus* in a highly structured environment. Appl Microbiol Biotechnol 49 (3): 290-294

Nam GW, Lee DW, Lee HS, Lee NJ, Kim BC, Choe EA, Hwang JK, Suhartono MT, Pyun YR (2002) Native-feather degradation by *Fervidobacterium islandicum* AW-1, a newly isolated keratinase-producing thermophilic anaerobe. Arch Microbiol 178 (6): 538-547

Namwong S, Hiraga K, Takada K, Tsunemi M, Tanasupawat S, Oda K (2006) A halophilic serine proteinase from *Halobacillus* sp. SR5-3 isolated from fish sauce: purification and characterization. Biosci Biotechnol Biochem 70 (6): 1395-1401

Narhi LO, Stabinsky Y, Miller L, Sachdev R, Finley S, Park S, Kolvenbach C, Arakawa T, Zukowski M (1991) Enhanced stability of subtilisin by three point mutations. Biotechnol Appl Biochem 13 (1): 12-24

Narinx E, Baise E, Gerday C (1997) Subtilisin from psychrophilic antarctic bacteria: characterization and site-directed mutagenesis of residues possibly involved in the adaptation to cold. Protein Eng 10 (11): 1271-1279

Neveu J, Regeard C, DuBow MS (2011) Isolation and characterization of two serine proteases from metagenomic libraries of the Gobi and Death Valley deserts. Appl Microbiol Biotechnol 91 (3): 635-644

Nordberg P, von Hofsten B (1969) Proteolytic enzymes from extremely halophilic bacteria. J Gen Microbiol 55 (2): 251-256

Ogino H, Watanabe F, Yamada M, Nakagawa S, Hirose T, Noguchi A, Yasuda M, Ishikawa H (1999a) Purification and characterization of organic solvent-stable protease from organic solvent-tolerant *Pseudomonas aeruginosa* PST-01. J Biosci Bioeng 87 (1): 61-68

Ogino H, Yamada M, Watanabe F, Ichinose H, Yasuda M, Ishikawa H (1999b) Peptide synthesis catalyzed by organic-solvent-stable protease from *Pseudomonas aeruginosa* PST-01 in monophasic aqueous-organic solvent systems. J Biosci Bioeng 88 (5): 513-518

Ong PS, Gaucher GM (1976) Production, purification and characterization of thermomycolase, the extracellular serine protease of the thermophilic fungus *Malbranchea pulchella* var. *sulfurea*. Can J Microbiol 22 (2): 165-176

Pandey S, Rakholiya KD, Raval VH, Singh SP (2012) Catalysis and stability of an alkaline protease from a haloalkaliphilic bacterium under non-aqueous conditions as a function of pH, salt and temperature. J Biosci Bioeng 114 (3): 251-256

Pantoliano MW, Ladner RC, Bryan PN, Rollence ML, Wood JF, Poulos TL (1987) Protein engineering of subtilisin BPN': enhanced stabilization through the introduction of two cysteines to form a disulfide bond. Biochemistry 26 (8): 2077-2082

Pantoliano MW, Whitlow M, Wood JF, Rollence ML, Finzel BC, Gilliland GL, Poulos TL, Bryan PN (1988) The engineering of binding affinity at metal ion binding sites for the stabilization of proteins: subtilisin as a test case. Biochemistry 27 (22): 8311-8317

Paul T, Das A, Mandal A, Jana A, Halder SK, Das Mohapatra PK, Pati BR, Mondal KC (2014) Smart cleaning properties of a multi tolerance keratinolytic protease from an extremophilic *Bacillus tequilensis* hsTKB2: prediction of enzyme modification site. Waste Biomass Valoriz 5 (6): 931-945

Pawar R, Zambare V, Barve S, Paratkar G (2009) Application of protease isolated from *Bacillus* sp. 158 in enzymatic cleansing of contact lenses. Biotechnology 8 (2): 276-280

Pedersen NR, Wimmer R, Matthiesen R, Pedersen LH, Gessesse A (2003) Synthesis of sucrose laurate using a new alkaline protease. Tetrahedron Asymmetry 14 (6): 667-673

Peek K, Daniel RM, Monk C, Parker L, Coolbear T (1992) Purification and characterization of a thermostable proteinase isolated from *Thermus* sp. strain Rt41A. Eur J Biochem 207 (3): 1035-1044

Phadatare SU, Deshpande VV, Srinivasan MC (1993) High activity alkaline protease from *Conidiobolus coronatus* (NCL 86. 8. 20): enzyme production and compatibility with commer-cial detergents. Enzyme Microb Technol 15 (1): 72-76

Phrommao E, Yongsawatdigul J, Rodtong S, Yamabhai M (2011) A novel subtilase with NaCl-activated and oxidant-stable activity from *Virgibacillus* sp. SK37. BMC Biotechnol 11: 65

Puri S (2001) An alkaline protease from a *Bacillus* sp. : production and potential applications in detergent formulation and degumming of silk. MSc thesis, University of Delhi, New Delhi

Purohit MK, Singh SP (2011) Comparative analysis of enzymatic stability and amino acid sequences of thermostable alkaline proteases from two haloalkaliphilic bacteria isolated from Coastal region of Gujarat, India. Int J Biol Macromol 49 (1): 103-112

Pushpam PL, Rajesh T, Gunasekaran P (2011) Identification and characterization of alkaline serine protease from goat skin surface metagenome. AMB Express 1 (1): 3

Qua D, Simidu U, Taga N (1981) Purification and some properties of halophilic protease produced by a moderately halophilic marine *Pseudomonas* sp. Can J Microbiol 27 (5): 505-510

Rai SK, Roy JK, Mukherjee AK (2010) Characterization of a detergent-stable alkaline protease from a novel thermophilic strain *Paenibacillus tezpurensis* sp. nov. AS-S24-II. Appl Microbiol Biotechnol 85 (5): 1437-1450

Rao MB, Tanksale AM, Ghatge MS, Deshpande VV (1998) Molecular and biotechnological aspects of microbial proteases. Microbiol Mol Biol Rev 62 (3): 597-635

Raval VH, Pillai S, Rawal CM, Singh SP (2014) Biochemical and structural characterization of a detergent-stable serine alkaline protease from seawater haloalkaliphilic bacteria. Process Biochem 49 (6): 955-962

Ray MK, Devi KU, Kumar GS, Shivaji S (1992) Extracellular protease from the Antarctic yeast *Candida humicola*. Appl Environ Microbiol 58 (6): 1918-1923

Rheinnecker M, Baker G, Eder J, Fersht AR (1993) Engineering a novel specificity in subtilisin BPN'. Biochemistry 32 (5): 1199-1203

Rheinnecker M, Eder J, Pandey PS, Fersht AR (1994) Variants of subtilisin BPN' with altered specificity profiles. Biochemistry 33 (1): 221-225

Riva S, Nonini M, Ottolina G, Danieli B (1998) Subtilisin-catalyzed esterification of di-and oligosaccharides containing a D-fructose moiety. Carbohydr Res 314 (3-4): 259-266

Ruiz DM, De Castro RE (2007) Effect of organic solvents on the activity and stability of an extra-cellular protease secreted by the haloalkaliphilic archaeon *Natrialba magadii*. J Ind Microbiol Biotechnol 34 (2): 111-115. doi: 10. 1007/s10295-006-0174-4

Ruiz DM, Iannuci NB, Cascone O, De Castro RE (2010) Peptide synthesis catalysed by a haloalkaliphilic serine protease from the archaeon *Natrialba magadii* (Nep). Lett Appl Microbiol 51 (6): 691-696

Saba I, Qazi PH, Rather SA, Dar RA, Qadri QA, Ahmad N, Johri S, Taneja SC, Shawl S (2012) Purification and characterization of a cold active alkaline protease from *Stenotrophomonas* sp. , isolated from Kashmir, India. World J Microbiol Biotechnol 28 (3): 1071-1079

Saeki K, Hitomi J, Okuda M, Hatada Y, Kageyama Y, Takaiwa M, Kubota H, Hagihara H, Kobayashi T, Kawai S, Ito S (2002) A novel species of alkaliphilic *Bacillus* that produces an oxidatively stable alkaline serine protease. Extremophiles 6 (1): 65-72

Saha S, Dhanasekaran D, Shanmugapriya S, Latha S (2013) *Nocardiopsis* sp. SD5: a potent feather degrading rare actinobacterium isolated from feather waste in Tamil Nadu, India. J Basic Microbiol 53 (7): 608-616

Sakaguchi M, Takezawa M, Nakazawa R, Nozawa K, Kusakawa T, Nagasawa T, Sugahara Y, Kawakita M (2008) Role of disulphide bonds in a thermophilic serine protease Aqualysin I from *Thermus aquaticus* YT-1. J Biochem 143 (5): 625-632

Sako Y, Croocker PC, Ishida Y (1997) An extremely heat-stable extracellular proteinase (aeropyrolysin) from the hyperthermophilic archaeon *Aeropyrum pernix* K1. FEBS Lett 415 (3): 329-334

Sana B, Ghosh D, Saha M, Mukherjee J (2006) Purification and characterization of a salt, solvent, detergent and bleach tolerant protease from a new gamma-*Proteobacterium* isolated from marine environment of the *Sundarbans*. Process Biochem 41 (1): 208-215

Sánchez-Porro C, Mellado E, Bertoldo C, Antranikian G, Ventosa A (2003) Screening and characterization of the protease CP1 produced by the moderately halophilic bacterium *Pseudoalteromonas* sp. strain CP76. Extremophiles 7 (3): 221-228

Santos AF, Valle RS, Pacheco CA, Alvarez VM, Seldin L, Santos ALS (2013) Extracellular proteases of *Halobacillus blutaparonensis* strain M9, a new moderately halophilic bacterium. Braz J Microbiol 44 (4): 1299-1304

Sarethy IP, Saxena Y, Kapoor A, Sharma M, Sharma SK, Gupta V, Gupta S (2011) Alkaliphilic bacteria: applications in industrial biotechnology. J Ind Microbiol Biotechnol 38 (7): 769-790

Sayers EW, Barrett T, Benson DA, Bryant SH, Canese K, Chetvernin V, Church DM, DiCuccio M, Edgar R, Federhen S, Feolo M, Geer LY, Helmberg W, Kapustin Y, Landsman D, Lipman DJ, Madden TL, Maglott DR, Miller V, Mizrachi I, Ostell J, Pruitt KD, Schuler GD, Sequeira E, Sherry ST, Shumway M, Sirotkin K, Souvorov A, Starchenko G, Tatusova TA, Wagner L, Yaschenko E, Ye J (2009) Database resources of the National Center for Biotechnology Information. Nucleic Acids Res 37: D5-D15

Schmitt W, Rdest U, Goebel W (1990) Efficient high-performance liquid chromatographic system for the purification of halobacterial serine protease. J Chromatogr A 521 (2): 211-220

Secades P, Alvarez B, Guijarro J (2003) Purification and properties of a new psychrophilic metalloprotease (Fpp2) in the fish pathogen *Flavobacterium psychrophilum*. FEMS Microbiol Lett 226 (2): 273-279

Selim S, Hagagy N, Aziz MA, El-Meleigy ES, Pessione E (2014) Thermostable alkaline halophilicprotease production by *Natronolimnobius innermongolicus* WN18. Nat Prod Res 28 (18): 1476-1479

Sellek GA, Chaudhuri JB (1999) Biocatalysis in organic media using enzymes from extremophiles. Enzyme Microb Technol 25 (6): 471-482

Setati ME (2010) Diversity and industrial potential of hydrolase-producing halophilic/halotolerant eubacteria. Afr J Biotechnol 9 (11): 1555-1560

Setyorini E, Takenaka S, Murakami S, Aoki K (2006) Purification and characterization of two novel halotolerant extracellular protease from *Bacillus subtilis* strain FP-133. Biosci Biotechnol Biochem 70 (2): 433-444

Shao Z, Zhao H, Giver L, Arnold FH (1998) Random-priming in vitro recombination: an effective tool for directed evolution. Nucleic Acids Res 26 (2): 681-683

Shi W, Tang XF, Huang Y, Gan F, Tang B, Shen P (2006) An extracellular halophilic protease SptA from a halophilic archaeon *Natrinema* sp. J7: gene cloning, expression and characterization. Extremophiles 10 (6): 599-606

Shi WL, Zhong CQ, Tang B, Shen P (2007) Purification and characterization of extracellular halophilic protease from haloarchaea *Natrinema* sp. R6-5. Acta Microbiol Sin 47 (1): 161-163

Singh J, Batra N, Sobti RC (2001) Serine alkaline protease from a newly isolated *Bacillus* sp. SSR1. Process Biochem 36 (8-9): 781-785

Singhal P, Nigam VK, Vidyarthi AS (2012) Studies on production, characterization and applications of microbial alkaline proteases. Int J Adv Biotechnol Res 3 (3): 653-669

Sinha R, Khare SK (2012) Isolation of a halophilic *Virgibacillus* sp. EMB13: characterization of its protease for detergent application. Indian J Biotechnol 11: 416-426

Sinha R, Khare SK (2013) Characterization of detergent compatible protease of a halophilic *Bacillus* sp. EMB9: differential role of metal ions in stability and activity. Bioresour Technol 145: 357-361

Sinsuwan S, Rodtong S, Yongsawatdigul J (2010) A NaCl-stable serine proteinase from *Virgibacillus* sp. SK33 isolated from Thai fish sauce. Food Chem 119 (2): 573-579

Son E, Kim JI (2003) Multicatalytic alkaline serine protease from psychrotrophic *Bacillus amyloliquefaciens* S94. J Microbiol 41 (1): 58-62

Stepanov VM, Rudenskaya GN, Revina LP, Gryaznova YB, Lysogorskaya EN, Filippova IY, Ivanova II (1992) A serine proteinase of an archaebacterium, *Halobacterium mediterranei*. A homologue of eubacterial subtilisins. Biochem J 285 (Pt 1): 281-286

Stepanov VM (1996) Proteinases as catalysts in peptide synthesis. Pure Appl Chem 68 (6): 1335-1339

Strausberg SL, Alexander PA, Gallagher DT, Gilliland GL, Barnett BL, Bryan PN (1995) Directed evolution of a subtilisin with calcium-independent stability. Nat Biotechnol 13: 669-673

Studdert CA, Herrera Seitz MK, Plasencia Gil MI, Sanchez JJ, De Castro RE (2001) Purification and biochemical characterization of the haloalkaliphilic archaeon *Natronococcus occultus* extracellular serine protease. J Basic Microbiol 41 (6): 375-383

Suzuki Y, Tsujimoto Y, Matsui H, Watanabe K (2006) Decomposition of extremely hard-to-degrade animal proteins by thermophilic bacteria. J Biosci Bioeng 102 (2): 73-81

Synowiecki J (2010) Some application of thermophiles and their enzymes for protein processing. Afr J Biotechnol 9 (42): 7020-7025

Takagi H, Takahashi T, Momose H, Inouye M, Maeda Y, Matsuzawa H, Ohta T (1990) Enhancement of the thermostability of subtilisin E by introduction of a disulfide bond engineered on the basis of structural comparison with thermophilic serine protease. J Biol Chem 265 (12): 6874-6878

Takai H, Sakato K, Nakamizo N, Isowa Y (1981) Enzymatic synthesis of caerulein peptide. In: Oyama K, Nikimura S (eds) Peptide chemistry. Protein Research Foundation, Osaka, Japan, pp 213-241

Tariq AL, Reyaz AL, Prabakaran J (2011) Purification and characterization of 56kDa cold active protease from *Serratia marcescens*. Afr J Microbiol Res 5 (32): 5841-5847

Thangam EB, Rajkumar GS (2002) Purification and characterization of alkaline protease from *Alcaligenes faecalis*. Biotechnol Appl Biochem 35 (2): 149-154

Thumar J, Singh SP (2007) Two-step purification of a highly thermostable alkaline protease from salt-tolerant alkaliphilic *Streptomyces clavuligerus* strain Mit-1. J Chromatogr B 854 (1-2): 198-203

Thumar JT, Singh SP (2009) Organic solvent tolerance of an alkaline protease from salt-tolerant alkaliphilic *Streptomyces clavuligerus* strain Mit-1. J Ind Microbiol Biotechnol 26 (2): 211-218

Tiberti M, Papaleo E (2011) Dynamic properties of extremophilic subtilisin-like serine-proteases. J Struct Biol 174 (1): 69-83

Toplak A, Wu B, Fusetti F, Quaedflieg PJ, Janssen DB (2013) Proteolysin, a novel highly thermostable and cosolvent-compatible protease from the thermophilic bacterium *Coprothermobacter proteolyticus*. Appl Environ Microbiol 79 (18): 5625-5632

Turkiewicz M, Pazgier M, Kalinowska H, Bielecki S (2003) A cold-adapted extracellular serine proteinase of the yeast *Leucosporidium antarcticum*. Extremophiles 7 (6): 435-442

van den Burg B, Enequist HG, van den Haar ME, Eijsink VG, Stulp BK, Venema G (1991) A highly thermostable neutral protease from *Bacillus caldolyticus*: cloning and expression of the gene in *Bacillus subtilis* and characterization of the gene product. J Bacteriol 173 (13): 4107-4115

Varela H, Ferrari MD, Belobrajdic L, Vazquez A, Loperena ML (1997) Skin unhairing proteases of *Bacillus subtilis*: production and partial characterization. Biotechnol Lett 19 (8): 755-758

Vazquez S, Ruberto L, Mac Cormak W (2005) Properties of extracellular proteases from three psychrotolerant *Stenotrophomonas maltophilia* isolated from Antarctic soil. Polar Biol 28 (4): 319-325

Vidyasagar M, Prakash S, Litchfield C, Sreeramulu K (2006) Purification and characterization of a thermostable, haloalkaliphilic extracellular serine protease from the extreme halophilic archaeon *Halogeometricum borinquense* strain TSS101. Archaea 2: 51-57

Vidyasagar M, Prakash S, Mahajan V, Shouche YS, Sreeramulu K (2009) Purification and characterization of an extreme halothermophilic protease from a halophilic bacterium *Chromohalobacter* sp. TVSP101. Braz J Microbiol 40 (1): 12-19

VijayAnand S, Hemaprita J, Selvin J, Kiran S (2010) Production and optimization of haloalkaliphilic protease by an extremophile-*Halobacterium* sp. Js1, isolated from thalassohaline environment. Glob J Biotechnol Biochem 5 (1): 44-49

Voorhorst WGB, Eggen RIL, Geerling ACM, Platteeuw C, Siezen RJ, de Vos WM (1996) Isolation and characterization of the hyperthermostable serine protease, pyrolysin, and its gene from the hyperthermophilic archaeon *Pyrococcus furiosus*. J Biol Chem 271 (34): 20426-20431

Waghmare SR, Gurav AA, Mali SA, Nadaf NH, Jadhav DB, Sonawane KD (2015) Purification and characterization of novel organic solvent tolerant 98kDa alkaline protease from isolated *Stenotrophomonas maltophilia* strain SK. Protein Expr Purif 107: 1-6

Wang QF, Miao JL, Hou YH, Ding Y, Wang GD, Li GY (2005) Purification and characterization of an extracellular cold-active serine protease from the psychrophilic bacterium *Colwellia* sp. NJ341. Biotechnol Lett 27 (16): 1195-1198

Wang QF, Hou HY, Xu Z, Miao JL, Li GY (2008) Purification and properties of an extracellular cold-active protease from the psychrophilic bacterium *Pseudoalteromonas* sp. NJ276. Biochem Eng J 38 (3): 362-368

Wilson SA, Daniel RM, Peek K (1994) Peptide synthesis with a proteinase from the extremely thermophilic organism *Thermus* Rt41A. Biotechnol Bioeng 44 (3): 337-346

Xiong H, Song L, Xu Y, Tsoi MY, Dobretsov S, Qian PY (2007) Characterization of proteolytic bacteria from the Aleutian deep-sea and their proteases. J Ind Microbiol Biotechnol 34 (1): 63-71

Xu B, Zhong Q, Tang X, Yang Y, Huang Z (2009) Isolation and characterization of a new keratinolytic bacterium that exhibits significant featherdegrading capability. Afr J Biotechnol 8 (18): 4590-4596

YangC, Wang F, Hao J, Zhang K, Yuan N, Sun M (2010) Identification of a proteolytic bacterium, HW08, and characterization of its extracellular cold-active alkaline metalloprotease Ps5. Biosci Biotechnol Biochem 74 (6): 1220-1225

Yang XS, Chen XL, Xu XZ, Zeng RY (2011) Cold-adaptive alkaline protease from the psychrophilic *Planomicrobium* sp. 547: enzyme characterization and gene cloning. Adv Polar Sci 22 (1): 49-54

Yang J, Li J, Mai Z, Tian X, Zhang S (2013) Purification, characterization, and gene cloning of a cold-adapted thermolysin-like protease from *Halobacillus* sp. SCSIO 20089. J Biosci Bioeng 115 (6): 628-632

Yin J, Chen JC, Wu Q, Chen GQ (2014) Halophiles, coming stars for industrial biotechnology. Biotechnol Adv 33 (7): 1433-1442

Yu TX (1991) Proteases of haloalkaliphiles. In: Horikoshi K, Grand WD (eds) Superbugs: microorganisms in extreme environments. Japan Scientific Societies Press/Springer-Verlag, Tokyo/ Berlin, pp 76-83

Yuan Q, Hayashi A, Kitamura Y, Shimada T, Na R, Jin X (2009) Purification and characterization of cold-adapted metalloprotease from deep sea water lactic acid bacteria *Enterococcus faecalis* TN-9. Int J Biol 1 (2): 12-21

Zeng R, Zhang R, Zhao J, Lin N (2003) Cold-active serine alkaline protease from the psychrophilic bacterium *Pseudomonas* strain DY-A: enzyme purification and characterization. Extremophiles 7 (4): 335-337

Zhao H, Arnold FH (1999) Directed evolution converts subtilisin E into a functional equivalent of thermitase. Protein Eng 12 (1): 47-53

Zhao GY, Zhou MY, Zhao HL, Chen XL, Xie BB, Zhang XY, He HL, Zhou BC, Zhang YZ (2012) Tenderization effect of cold-adapted collagenolytic protease MCP-01 on beef meat at low temperature and its mechanism. Food Chem 134 (4): 1738-1744

Zhu HY, Tian Y, Hou YH, Wang TH (2009) Purification and characterization of the cold-active alkaline protease from marine cold-adaptive *Penicillium chrysogenum* FS010. Mol Biol Rep 36 (8): 2169-2174

http: //www. chem. qmul. c. uk/iubmb/enzyme

http: //merops. sanger. ac. uk

http: //www. ncbi. nlm. nih. gov/Taxonomy/CommonTree/wwwcmt. cgi

第十五章

适冷性β-半乳糖苷酶：来源、生化性质及其生物技术应用潜力

Hubert Cieśliński[1]，Marta Wanarska[1]，
Anna Pawlak-Szukalska[1]，Ewelina Krajewska[1]，
Monika Wicka[1]，Józef Kur[1,2]

15.1 β-D-半乳糖苷酶：基本特征

β-D-半乳糖苷酶（EC 3.2.1.23）也称为β-D-半乳糖苷半乳糖水解酶（系统名称），是一种高度特异性的外切糖苷酶，可以催化β-D-半乳糖苷末端非还原性的β-D-半乳糖苷残基水解（Wanarska and Kur 2005）。二糖乳糖是我们最为熟知的β-D-半乳糖苷之一，在牛奶中大量存在并且可以被酶解为D-葡萄糖和D-半乳糖。在高浓度的乳糖存在下，一些β-D-半乳糖苷酶也可以表现出半乳糖基转移酶活性（Cruz et al. 1999）。已经报道了几种β-D-半乳糖苷酶具有其他类型的酶活性。例如，从小红酵母 *Rhodotorula minuta* IFO897 分离的β-D-半乳糖苷酶可水解β-D-葡萄糖苷、β-D-岩藻糖苷和α-L-阿拉伯糖苷（Onishi and Tanaka 1996）。

β-D-半乳糖苷酶广泛存在于细菌、真菌、植物和动物中。不同来源的β-D-半乳糖苷酶在性质方面存在明显的差异，包括分子量、蛋白质结构、对金属离子的需求、底物特异性以及最适酶活温度和pH。细菌来源的β-D-半乳糖苷酶最丰富，一些细菌不只产生一种β-D-半乳糖苷酶。例如青春双歧杆菌 *Bifidobacterium adolescentis* DSM 20083（Van Laere et al. 2000）和 *Bacillus circulans*（环状芽孢杆菌）（Vetere and Paoletti 1998）分别能产生两种和三种不同的β-D-半乳糖苷酶。从大肠杆菌中分离的LacZ β-D-半乳糖苷酶是目前最为熟知的、研究最为明确的β-D-半乳糖苷酶。对该酶广泛的研究有利于了解β-D-半乳糖苷酶的催化机理以及Jacob和Monod操纵子模型中基因表达的调控机制。目前，LacZ能水解X-Gal（5-溴-4-氯-3-吲哚基-β-D-吡喃半乳糖苷）产生蓝色产物，这也是克隆中蓝白斑筛选的分子原理（Juers et al. 2012）。最近，来自其他几种细菌和真菌中的β-D-半乳糖苷酶也被运用于各个科学和工业领域。β-D-半乳糖苷酶及其应用将在下一章详细讨论。

---

[1] Faculty of Chemistry，Department of Molecular Biotechnology and Microbiology，Gdańsk，University of Technology，80-233 Gdańsk，Poland.

[2] e-mail：kur@pg. gda. pl.

15.2 β-D-半乳糖苷酶：工业酶的来源和应用

β-D-半乳糖苷酶已广泛用于多种乳制品和食品中。在乳制品中，β-D-半乳糖苷酶用于水解牛奶和乳清的乳糖。牛奶中乳糖的酶解在工业、经济和医学领域均有着重要的作用。首先，低乳糖牛奶（LLM）是专门用于乳糖不耐受个体的乳制品，乳糖不耐受是指个体在其自身的消化系统中不具有足够高水平的β-D-半乳糖苷酶以实现乳糖的水解。乳糖不耐受症可表现为食用牛奶或一些奶制品后发生腹痛、腹胀、肠鸣和便溏（Juers et al. 2012）。据报道，大部分亚洲人（超过90%）、非洲人（80%～100%）、美洲原住民（超过90%）和南欧洲（超过80%）属于乳糖不耐受（Mlichova and Rosenberg 2006 ）。因此，在这些地区可通过大规模生产LLM作为零售商店销售的商品以部分解决严重乳糖不耐受人群对乳制品的需求。理论上，化学水解法和酶解法均可用于生产LLM。然而，实际应用中对这两种方法的比较发现，酶解法具有几个化学水解法不可比拟的优点，如不产生额外的副产物，不会降解牛奶中其他有用物质，不改变牛奶的颜色、气味和味道。因此，乳糖的化学水解（酸水解）法不用于生产LLM。此外，用β-D-半乳糖苷酶处理的牛奶还能保持原有的营养价值，这是因为乳糖酶解产物，D-葡萄糖和D-半乳糖仍被保留在LLM中（Ladero et al. 2003）。最近，工业规模生产LLM的方法可以基于直接利用酶解法或将酶解法与乳糖分离的物理法结合，如色谱分离或膜分离技术与乳糖酶解结合的方法（Harju et al. 2012）。

由于乳糖的溶解性较差以及对不同食品质量的不利影响，研究人员对乳糖酶解在食品工业中新方法的开发表现出极大的兴趣。例如，用于生产冰淇淋或炼乳的牛奶中含高浓度的乳糖可导致乳糖结晶，造成奶制品呈现沙、粗砂或粉状。乳糖水解可使这些产品的乳脂性和可舀性得到显著改善，同时也能使其更易消化。此外，在发酵培养制作乳制品（包括酸奶和白软干酪）中，乳糖水解产生的单糖比乳糖本身更容易发酵。因此，乳糖的水解可有效缩短达到发酵乳制品所需低pH的时间。此外，LLM中的葡萄糖和半乳糖亦能显著增加这些乳制品的甜度（约50%），这就可以显著地减少乳制品中甜味剂的添加量。

除了牛奶外，乳糖的另一个主要来源是乳酪生产期间产生的乳清。全球每年乳清产量为1.5亿吨，其主要成分为乳糖（44～52g/L）、蛋白质（6～8g/L）和矿物质（4.3～9.5g/L）（Johansen et al. 2002）。如今，干酪乳清在工业中主要用于生产乳清蛋白浓缩物（WPC），它是具有高营养价值的多功能食品成分。这个过程还会产生大量的乳清渗透物，仍含有4%～5%的乳糖（Mlichová and Rosenberg 2006）。结晶纯化的乳糖主要用作食品中的补充剂，以及配制药物的赋形剂。尽管乳清具有许多用途，然而目前全世界每年产生的乳清估计约47%仍未被利用，因此，这是一个重要的废弃物处理问题（Guimaraes et al. 1992；Mlichová and Rosenberg 2006）。乳清中残留的乳糖对乳清利用的不利影响是制约乳清利用的主要问题。乳清生化需氧量（BOD）的90%以上是被乳糖所消耗，因此如何从乳清中回收乳糖对于乳清的有效利用至关重要。在这种背景下，乳糖酶解则会带来几个益处。与乳糖相比，乳糖水解产物半乳糖和葡萄糖可以作为大量微生物（包括重组体）发酵的直接碳源，产生其他有价值的生物产物，如乳酸盐、醋酸盐、乙醇、丁二醇和生物聚合物（Guimaraes et al. 1992；Coté et al. 2004；Mehaia et al. 1993；Nath et al. 2014）。此外，在乳清渗透物中乳糖水解产物也可以用于生产半乳糖-葡萄糖浆（乳糖浆），它在食品工业中用作甜味剂。此外，乳糖的水解也可以防止乳清浓缩过程中的结晶，这将显著减少运输和贮存乳清过程中的成本。

β-D-半乳糖苷酶同时还具有半乳糖基转移酶活性，使之成为食品工业领域中别具吸引力的酶。β-D半乳糖苷酶的半乳糖基转移酶活性可用于乳糖合成两个或多个半乳糖单元的低聚半乳糖（GOSs）。有效合成 GOSs 所需的条件包括反应介质中高浓度的乳糖。此外，在酶促反应过程形成的糖苷键 GOSs 的组成如 β（1，4），β（1，3），β（1，2）和 β（1，6）取决于酶的来源和总体反应条件。目前，GOSs 自然存在于在人体和牛奶中，已经引起食品行业研究人员的高度关注。牛奶中的 GOSs 能刺激双歧杆菌的生长，进而导致人类或其他动物的肠道中的腐败细菌浓度降低。有人认为，GOSs 的益生性质对其消费者具有一定益处，包括：①粪便中腐败产物的数量减少；②降低血液胆固醇含量；③从饮食中吸收更多的 Ca^{2+}；④较低的结肠癌发病率。此外，GOSs 在食品加工过程中的酸性和高温条件下是稳定的，并且能保持优良的口感。为此，GOSs 是具有广泛市场应用前景的碳水化合物功能性食品配料，并且已经进行了大量研究工作以开发高效和经济生产 GOSs 的方法。如今，GOSs 不仅用于益生食品成分，而且还用于化妆品添加成分和低热量甜味剂（Mlichová and Rosenberg 2006）。

除了用于合成 GOSs 之外，β-D-半乳糖苷酶也可以通过将半乳糖基转移到除了乳糖、葡萄糖或半乳糖之外的任何糖类来生产异低聚糖（heterooligosaccharides，HOS）。目前，通常被称为乳果糖的 4-*O*-β-D-吡喃半乳糖基-β-D-呋喃果糖是食品工业中应用最为广泛的 HOS。乳果糖具有益生元特性，因此已经采用与 GOSs 类似的方式用于功能性食品添加成分。另一方面，医药中的乳果糖用于治疗便秘和肝性脑病。此外，HOS 还可以用于结肠病症诊断的氢气呼气试验。当前工业生产乳果糖的方式涉及乳糖的碱性异构化。但是，该生产过程的苛刻条件（pH 为 10.5～11.5 和温度为 70～100℃）可导致乳糖的降解从而形成几种副产物。与此相比，乳果糖的酶促合成可以在非常温和的条件下进行，因此在这个过程中可以使用粗乳糖物质，例如乳清或乳清渗透物。此外，值得注意的是由 β-D-半乳糖苷酶催化的异源半乳糖基转移反应可产生几乎无限多样性的寡糖（Pawlak-Szukalska et al. 2014）。

β-D-半乳糖苷酶的半乳糖基转移酶活性不仅可用于合成 GOSs 和 HOS，还可用于各类 β-糖苷的合成，其中 D-半乳糖残基通过 β-1→4 与各种醇（Stevenson et al. 1993）、抗生素（Scheckermann et al. 1997）、麦角生物碱（Kren et al. 1992）以及黄酮醇苷杨梅毒素（Shimizu et al. 2006）连接。这些结果表明 β-D 半乳糖苷酶可以潜在用于药物筛选和一系列其他生物活性化合物的工业化生产。

15.3　β-D-半乳糖苷酶：工业来源

尽管来自大肠杆菌的 LacZ 酶是研究最为深入的 β-D-半乳糖苷酶，但是这种酶并不适用于食品工业，因为使用大肠杆菌作为酶来源存在较大的风险。因此在工业规模上用于制造 LLM 和其他乳制品的 β-D-半乳糖苷酶主要从 *Kluyveromyces lactis*（乳酸克鲁维酵母）（来自 DSM 的 Maxilact® 产品系列，来自 Novozymes 的纯乳清产品和来自 DuPont 的 GODO-YNL2 乳糖酶）和 *Aspergillus oryzae*（米曲霉）（Tolerase ™L，来自 DSM）中分离出来，这些微生物通常被认为安全的（GRAS 状态）。水解乳糖的 β-D-半乳糖苷酶的选择主要取决于反应条件。例如，使用酵母产生的 β-D-半乳糖苷酶在甜乳清或牛乳中水解乳糖的最佳 pH 范围是 6.5～7.0。然而，使用真菌产生的 β-D-半乳糖苷酶在酸性乳清中水解乳糖的最佳 pH 在 3.0～5.0 内（Boon et al. 2000；Harju 1987）。工业酶的明显缺点是缺乏反应产物的反馈抑制。目前，对 β-D-半乳糖苷酶催化的乳糖水解的机理已被深入研究。最常提出的是 D-半乳糖对来自 *Kluyveromyces* 和 *Aspergillus* 的 β-D-半乳糖苷酶的竞争性抑制的米氏方程动力学模型（Santos et al. 1998 及其参考文献；Jurado et al. 2002）。值得注意的是，这两种工业酶都是由中温微生物产生的，因此它们不能很好地适用于低温或高温条件下酶促反应的新型生物技术的开发。为此，在过去的15年中，大量实验研究具有生物技术应用潜力的热稳定或适冷性β-D-半乳糖苷酶。通常情况下，工业使用热稳定或适冷性β-D-半乳糖苷酶具有许多显著的优点，下一节将重点介绍适冷β-D-半乳糖苷酶在食品加工业中的优势与应用。

15.4 适冷性 β-D-半乳糖苷酶

适冷酶的显著特征，例如低温和中温条件下的高催化效率以及中温或高温条件下的热不稳定性，在新型生物技术的开发中具有独特的优势。这些优点中最重要的包括：减少反应时间；降低能源成本；减少挥发性化合物的损失，包括反应底物和/或产物的损失；进行涉及热敏性化合物的反应的能力；在酶促反应完成后，酶在中温条件下快速灭活。然而，这些酶热稳定性较差也是不利因素，特别是需要中长期储存时。因此，越来越多的研究集中在适冷酶在不同领域的潜在生物技术应用上。在这方面，β-D-半乳糖苷酶已经成为最受重视的“适冷”酶之一，因为它们可以与对应的中温酶相同的方式生产 LLM 或水解乳清和乳清渗透物。然而，与对应的中温酶相反，适冷酶可以在低温（约 10℃）条件下有效水解乳糖。水解乳糖可以在牛奶的运输和贮存期间进行，这将显著减少整个生产过程所需的时间。此外，产物被中温微生物污染的风险也将明显降低。适冷酶的使用还将防止非酶促褐变产物的产生，而使用中温β-D-半乳糖苷酶在较高温度下降解牛奶中的乳酸会形成非酶促褐变产物。

适冷性β-D-半乳糖苷酶的主要来源是嗜冷和耐冷微生物，并且绝大多数从 *Arthrobacter*（节杆菌）属中分离得到。存在这种明显偏好的原因可能是这些菌株大多数是从土壤样品中分离的，而 *Arthrobacter* sp. 是主要的土壤细菌。据我们所知，只有两个来源于真核微生物，耐冷酵母 *Guehomyces pullulans* 的适冷性β-D-半乳糖苷酶被分离鉴定以及进行生化特性研究（Song et al. 2010；Nakagawa et al. 2006b）。

大多数产生适冷性β-D-半乳糖苷酶的细菌是从低温环境中分离的。例如，从荒原土壤中分离的 *Arthrobacter* sp. C2-2（Karasová-Lipovová et al. 2003），南极洲的超盐池塘分离得到的 *Planococcus* sp. SOS Orange（动球菌属）（Sheridan and Brenchley 2000）以及冬天从美国宾夕法尼亚州乳清处理的土壤样品中分离的肉杆菌 *Carnobacterium piscicola* BA（Coombs and Brenchley 1999）。其他产适冷性β-D-半乳糖苷酶的菌株包括分离自南极磷虾消化道的 *Pseudoalteromonas* sp. 22b（假交替单胞菌）（Cieśliński et al. 2005）和从西南格陵兰多伊卡（*Ikka*）石灰岩柱分离的格陵兰菌 *Alkalilactibacillus ikkense*（Schmidt and Stougaard 2010）。然而，这些微生物的生活条件通常不可能在实验室中重建（使用常规方法），并且目前可培养的微生物占比不到自然微生物的1%（White et al. 2000）。为此，Wang 等直接克隆了中国黑龙江省油田收集的环境样品中的 DNA，构建了一系列宏基因组文库。使用这种方法，研究人员成功地识别到编码适冷性β-D-半乳糖苷酶的 *zd410* 基因（Wang et al. 2010）。

除了适冷性β-D-半乳糖苷酶的来源，一个更重要的用于比较它们酶学性质的标准是它们与糖苷水解酶（GH）家族的联系。目前已经鉴定了133个 GH 家族，并且属于同一家族的蛋白质氨基酸序列相似性不低于30%（Poltorak et al. 2007）。目前普遍认为 GH1、GH2、GH35 和 GH42 家族的酶具有β-D-半乳糖苷酶活性。就去除乳制品中乳糖的能力而言，GH2 家族的酶较其他任何家族的酶更有效，因为它们具有更高的乳糖水解活性。然而，值得注意的是，来自 GH42 家族的β-D-半乳糖苷酶具有 GH2 家族不具备的多种其他活性。已经有研究报道 GH42 家族的几种酶具有β-D-岩藻糖苷酶（EC 3.2.1.38）和α-L-阿拉伯糖苷酶（EC 3.2.1.55）活性（Di Lauro et al. 2008；Lee et al. 2011；Kosugi et al. 2002；Saishin et al. 2010；Sheridan and Brenchley 2000）。

大多数已被分离鉴定和进行生化性质研究的适冷性 β-D-半乳糖苷酶都属于 GH2 和 GH42 家族的成员，尽管也发现一些具有 β-D-半乳糖苷酶活性的适冷酶属于 GH35 家族。还有文献报道了一些“适冷”微生物具有多个能够编码具有 β-D-半乳糖苷酶活性的酶的基因。例如，耐冷菌 *Arthrobacter* sp. B7 携带三种具有 β-D 半乳糖苷酶活性的同工酶的基因，分别可以归类于 GH2、GH35 和 GH42 家族（Trimbur et al. 1994；Gutshall et al. 1995，1997）。*Carnobacterium piscicola* BA 可产生两种 β-D-半乳糖苷酶，分别是 GH35 和 GH42 家族的成员（Coombs and Brenchley 1999，2001）。此外，*Arthrobacter* sp. ON14 也含有两种适冷性 β-D-半乳糖苷酶，分别是 GH2 和 GH42 家族的成员（Xu et al. 2011）。Coombs 和 Brenchley（2001）提出，在一种微生物中存在两种或多种 GH 家族酶，可能是由于细菌需要这些酶的协同作用降解具有 α-和 β-半乳糖苷键的多糖，这类多糖在自然环境中最为丰富。该假说是基于 *Carnobacterium piscicola* BA 一个操纵子上同时存在两种适冷性 β-D-半乳糖苷酶和一种 α-半乳糖苷酶的基因。通过对特定糖类的协同降解，这些酶可以为微生物提供良好的碳替代来源（Coombs and Brenchley 2001）。此外，对 *A. ikkense* 编码的适冷性 β-D-半乳糖苷酶的基因研究显示，它仅与 α-半乳糖苷酶基因存在 16bp 的间隔。假定这些基因处于相同启动子的控制之下（Schmidt and Stougaard 2010），这将有助于微生物在“冷”环境中水解多糖复合物。

基于上述数据，本章的剩余部分将集中讨论适冷性 β-D-半乳糖苷酶在不同 GH 家族中的酶学性质。

15.4.1 GH2 家族

GH2 家族的 β-D-半乳糖苷酶对多种 β-D-吡喃半乳糖苷（包括乳糖）显示出高特异性。乳糖水解产物可以抑制一些适冷性 β-D 半乳糖苷酶的活性。D-半乳糖是 *Arthrobacter* sp. 32cB 和 *Arthrobacter* sp. SB β-D-半乳糖苷酶的竞争性抑制剂（Pawlak-Szukalska et al. 2014；Coker et al. 2003）。相比之下，D-葡萄糖抑制 *Pseudoalteromonas* sp. 22b 的 β-D 半乳糖苷酶活性（Makowski et al. 2009），而副球菌 *Paracoccus* sp. 32d 产生的 β-D-半乳糖苷酶则被上述两种单糖抑制（Wierzbicka-Woś et al. 2011）。

GH2 家族适冷性 β-D-半乳糖苷酶亚基的分子质量约 110kDa，尽管这些酶自然状态下以四聚体形式存在，然而也有例外。例如来自 *Flavobacterium* sp. 4214（黄杆菌属）产生的 β-D 半乳糖苷酶（Sørensen et al. 2006），其单体构象是有活性的；来自 *Arthrobacter* sp. 32cB（Pawlak-Szukalska et al. 2014）和 *Paracoccus* sp. 32d（Wierzbicka-Woś et al. 2011）的 β-D-半乳糖苷酶，这两种酶活性状态以二聚体形式存在。GH2 家族的适冷性 β-D-半乳糖苷酶的部分性质见表 15.1。如表 15.1 所示，GH2 家族酶的反应最适温度和 pH 值分别为 10～45℃，pH 6.0～8.5。除了分离自 *Arthrobacter* sp. SB（Coker and Brenchley 2006）和 *Arthrobacter* sp. 32cB（Pawlak-Szukalska et al. 2014）的 β-D-半乳糖苷酶，其他适冷酶在 60℃时均失活。此外，二价金属离子如 Mg^{2+} 和 Mn^{2+} 对酶活性具有促进作用（表 15.1），而重金属离子如 Cu^{2+}、Ni^{2+}、Pb^{2+} 和 Zn^{2+} 则抑制酶的活性。此外，K^+ 和 Na^+ 对 GH2 家族的一些 β-D 半乳糖苷酶酶活性具有促进作用。

表 15.1 GH2 家族的适冷性 β-D-半乳糖苷酶的生化特征

微生物	分子质量/kDa	低聚状态	最适温度/℃	最适 pH	活化剂	抑制剂	热失活条件	参考文献
Alkalilactibacillus ikkense	NR (119.1①)	NR	20～30	8.0	NR	NR	5min,50℃	Schmidt and Stougaard(2010)

续表

微生物	分子质量/kDa	低聚状态	最适温度/℃	最适pH	活化剂	抑制剂	热失活条件	参考文献
Arthrobacter sp. B7(基因 *15*)	NR (111.0①)	NR	40	7.2	Mg^{2+}, Mn^{2+}, 2-ME	EDTA, Cu^{2+}, Ca^{2+}	10min,50℃	Trimbur et al. (1994)
Arthrobacter sp. C2-2 izosyme 1	550 (110.8①)	4	40	7.5	DTT, Mg^{2+}, Mn^{2+}	EDTA, Cu^{2+}, Al^{3+}, Tris	10min,50℃	Karasová-Lipovová et al. (2003)
Arthrobacter sp. 20B	460 (113.7①)	4	25	6.0～8.0	DTT, 2-ME, Na^+, K^+, Mn^{2+}	Pb^{2+}, Zn^{2+}, Cu^{2+}, pCMB	1min,60℃	Białkowska et al. (2009)
Arthrobacter sp. SB	463 (114.0①)	4	18	7.0	Mg^{2+}, Mn^{2+}, K^+	EDTA, D-半乳糖	10min,37℃	Coker and Brenchley (2006), Coker et al. (2003)
Arthrobacter psychrolactophilus F2	548 (111.7①)	4	10	8.0	NR	NR	5min,50℃	Nakagawa et al. (2006a)
Arthrobacter sp. ON14(*gal A*)	NR (111.4①)	NR	15	8.0	Na^+, K^+, Mg^{2+}, Mn^{2+}	Zn^{2+}, Cu^{2+}	20min,50℃	Xu et al. (2011)
Arthrobacter sp. 32 cB	257 (109.6①)	2	28	8.0	Mg^{2+}, DTT	Mn^{2+}, Ni^{2+}, Co^{2+}, Ca^{2+}, EDTA, 谷胱甘肽; 半胱氨酸, TCEP, D-半乳糖	5min,44℃	Pawlak-Szukalska et al. (2014)
Flavobacterium sp. 4214	<66 (114.3①)	1	42	7.5	NR	NR	NR	Sørensen et al. (2006)
Pseudoalteromonas haloplanktis TAE 79	>300 (118.1①)	4	45	8.5	Mg^{2+}, Mn^{2+}, Ca^{2+}, Li^+, 2-ME	EDTA, Zn^{2+}, Cu^{2+}, Ni^{2+}	60min,45℃	Hoyoux et al. (2001)
Paracoccus sp. 32d	161 (81.7①)	2	40	7.5	—	Ca^{2+}, Mn^{2+}, Ni^{2+}, Co^{2+}, DTT, 氧化谷胱甘肽, D-葡萄糖, D-半乳糖	15min,50℃	Wierzbicka-Woś et al. (2011)

续表

微生物	分子质量/kDa	低聚状态	最适温度/℃	最适pH	活化剂	抑制剂	热失活条件	参考文献
Pseudoalteromonas sp. 22b	490 (117.1①)	4	40	6.0～8.0	Na^+, K^+, Mg^{2+}, Mn^{2+}, DTT, 谷胱甘肽, 2-ME	Zn^{2+}, Cu^{2+}, Ni^{2+}, Pb^{2+}, pCMB, EDTA, D-葡萄糖	2min, 50℃	Turkiewicz et al. (2003), Cieśliński et al. (2005), Makowski et al. (2007)

注：NR，未发表；2-ME，2-巯基乙醇；pCMB，4-氯汞苯甲酸；TCEP，三（2-羧乙基）膦；DTT，二硫苏糖醇；EDTA，乙二胺四乙酸；Tris，三（羟甲基）氨基甲烷。

①由氨基酸序列计算得到单体分子。

15.4.2　GH42 家族

目前，已经有多种 GH42 家族的适冷 β-D-半乳糖苷酶被分离并且表征其生化特性。这些适冷 β-D-半乳糖苷酶主要来自耐冷细菌如 *Arthrobacter*（Hildebrandt et al. 2009；Gutshall et al. 1995）、*Planococcus*（Hu et al. 2007；Sheridan and Brenchley 2000）和 *Carnobacterium*（Coombs and Brenchley 1999）以及宏基因组 DNA 中（Wang et al. 2010）。

GH42 家族适冷性 β-D-半乳糖苷酶的部分性质见表 15.2。这些适冷性 β-D-半乳糖苷酶的亚基的分子质量是约 70kDa，并且这些酶通常以二聚体或三聚体形式存在。这些酶的最适温度为 20～50℃和 pH 值为 6.5～7.0。此外，GH42 家族的适冷酶在 40～50℃的温度下酶活丧失。有趣的是，Mg^{2+}、Mn^{2+}、Ca^{2+}、Co^{2+}是 *Arthrobacter* sp. B7（基因 12）β-D-半乳糖苷酶的酶活化剂（Gutshall et al. 1995），却抑制 *Arthrobacter* sp. 32c 产生的 β-D-半乳糖苷酶的活性（Hildebrandt et al. 2009）。此外，Mn^{2+}离子抑制 *Planococcus* sp. SOS Orange（Sheridan and Brenchley 2000）以及来自宏基因组文库的 *zd*410 基因编码的 β-D-半乳糖苷酶的活性。Co^{2+}离子抑制 *Planococcus* sp. L4（Hu et al. 2007）和 *Planococcus* sp. SOS Orange（Sheridan and Brenchley 2000）产生的 β-D-半乳糖苷酶的酶活性。所有 GH42 家族的适冷酶均被 Cu^{2+}离子抑制。

表 15.2　GH42 家族的适冷性 β-D-半乳糖苷酶的生化特征

微生物	分子质量/kDa	低聚状态	最适温度/℃	最适pH	热失活条件	底物特异性	活化剂	抑制剂	参考文献
Arthrobacter sp. 32c	195 (76.1①)	3	50	6.5	NR	乳糖, ONPG, PNPG	乙醇	Cu^{2+}, Zn^{2+}, Mn^{2+}, Ni^{2+}, Fe^{2+}, Co^{2+}, Ca^{2+}, Mg^{2+}, 谷胱甘肽, D-葡萄糖	Hildebrandt et al. (2009)
Arthrobacter sp. B7（基因 12）	NR (71②)	NR	45～50	6.6	低于 15min, 50℃	乳糖, ONPG, PNPG	Mg^{2+}, Mn^{2+}, Ca^{2+}, Co^{2+}, 2-ME	Cu^{2+}, EDTA	Gutshall et al. (1995)
Planococcus sp. L4	156 (77.3①)	2	20	6.8	10min, 45℃	乳糖, ONPG, PNPG	低浓度 Na^+, K^+	Zn^{2+}, Cu^{2+}, Ni^{2+}, Co^{2+}	Hu et al. (2007)

续表

微生物	分子质量/kDa	低聚状态	最适温度/℃	最适pH	热失活条件	底物特异性	活化剂	抑制剂	参考文献
Planococcus sp. SOS Orange	155 (77.5①)	2	42	6.5	10min, 55℃	ONPG, PNPG, PNPF, ONPF	无	Zn^{2+}, Cu^{2+}, Ni^{2+}, Co^{2+}, Mn^{2+}	Sheridan and Brenchley (2000)
宏基因组DNA（基因 *zd410*）	NR (78.6①)	NR	38	7.0	1h, 50℃	乳糖, ONPG, PNPG, pNP-β-D-阿拉伯糖苷, pNP-β-D-葡糖苷酸, pNP-β-D-mannoside	Na^{+}, K^{+}, Ca^{2+}	Cu^{2+}, Fe^{2+}, Zn^{2+}, Mn^{2+}	Wang et al. (2010)
Carnobacterium piscicola BA（基因 *bgaB*）	NR (76.8①)	NR	30	NR	10min, 40℃	ONPG, PNPG, pNP β-D-岩藻糖苷, pNP β-D-半乳糖醛酸	NR	NR	Coombs and Brenchley (1999)

注：NR，未发表；ONPF，邻硝基苯基β-D-吡喃岩藻糖苷；PNPF，对硝基苯基β-D-吡喃岩藻糖苷；ONPG，2-硝基苯基-β-D-吡喃半乳糖苷；PNPG，对硝基苯基β-D-吡喃半乳糖苷；EDTA，二乙胺四乙酸；2-ME，2-巯基乙醇。

① 由氨基酸序列计算得到单体分子。

② 由十二烷基硫酸钠聚丙烯酰胺凝胶电泳估算得到单体分子。

15.4.3 GH43家族

Arthrobacter sp. B7（基因 *14*）（Gutshall et al. 1997）和 *Carnobacterium piscicola* BA（基因 *bgaC*）（Coombs and Brenchley 1999）产生的β-D-半乳糖苷酶是GH35家族中唯一从适冷微生物分离的成员。

Arthrobacter sp. B7（基因 *14*）的β-D-半乳糖苷酶以二聚体形式存在，分子质量为110kDa。该酶在65℃保温10min后失活，在20℃下酶活性能维持2h。该酶的最适pH为6.5，是潜在用于去除牛奶中乳糖的一种β-D-半乳糖苷酶。使用对硝基苯基（pNP）连接底物法测定 *Arthrobacter* sp. B7中GH35酶的底物特异性，结果显示这种酶具有水解pNP-β-吡喃半乳糖苷和pNP-β-吡喃葡萄糖苷的活性。用Galβ1-3GlcNAcb1-3Galβ1-4Glc-7-氨基甲基香豆素作为底物进行测定时发现基因 *14* 的表达产物也能特异切割β-1,3键。

GH35家族的第二种β-D-半乳糖苷酶由 *Carnobacterium piscicola* BA的 *bgaC* 基因编码。与上述其他的GH35家族酶相反，这种酶仅被初步表征，它的最适反应温度为40℃。此外，该酶的半衰期在45℃下为15min。有研究表明，BgaC酶活性可以被 Zn^{2+}、Ni^{2+} 和 Cu^{2+} 抑制（Coombs and Brenchley 2001）。

15.4.4 其他家族

一些适冷性β-D半乳糖苷酶缺乏文献资料，使得无法对这些酶进行更加细致的分类。这些酶的酶学性质如表15.3所示。

比较表15.1、表15.2和表15.3中的数据（列：“分子量”和“低聚状态”），结果显示来源于 *Halomonas* sp. S62（盐单胞菌属）和 *Pseudoalteromonas* sp. TAE 79b的适冷性β-D半乳糖苷酶可分别归类于GH42和GH2家族的成员。此外，*Guehomyces pullulans* 17-1、*Guehomyces pullulans* R1（表15.3）和几种其他适冷β-D-半乳糖苷酶（表15.1，表15.2和

表 15.3）的最佳 pH 存在显著差异表明它们的来源不同。与表中的分离自不同细菌的酶不同，还有两种适冷性 β-D-半乳糖苷酶来自耐冷的类酵母菌。如上所述，最适 pH 偏低也是中温真菌（例如 *Aspergillus oryzae*）的 β-D-半乳糖苷酶的特征。

表 15.3 未分类的适冷性 β-D-半乳糖苷酶的生化特征

微生物	分子质量/kDa	低聚状态	最适温度/℃	最适pH	活化剂	抑制剂	热失活条件	底物特异性	参考文献
Guehomyces pullulans 17-1	335（170①）	2	50	4.0	2.5h，60℃	Li^+	Ca^{2+}、Mg^{2+}、Co^{2+}、Ag^+、Cu^{2+}、Fe^{3+}、Hg^{2+}、Fe^{2+}、碘乙酸、SDS、1，10-菲咯啉、DTT	ONPG，乳糖	Song et al.(2010)
Guehomyces pullulans R1	NR	NR	50	4.0	1h，50℃	NR	NR	乳糖，ONPG	Nakagawa et al.(2006b)
Halomonas sp. S62	130②（63①）	2	45	7.0	5min，60℃	Fe^{2+}，Mn^{2+}，Na^+，Ca^{2+}	Fe^{3+}，Cu^{2+}	ONPG，乳糖	Wang et al.(2013)
Bacillus subtilis KL88	90②	NR	50	6.0	NR	Na^+，K^+，Li^+，2-ME乙醇，DTT	Cu^{2+}、Fe^{2+}、Fe^{3+}、Zn^{2+}、Ca^{2+}、Mg^{2+}、Mn^{2+}、D-葡萄糖、D-半乳糖	乳糖，葡萄糖-β-D-半乳糖苷，甲基 β-D-半乳糖甲酰化物，甲氧基-β-半乳糖苷，ONPG，PNPG	Rahim and Lee（1991），Torres and Lee(1995)
Pseudoaltero monas sp. TAE 79b	513（110①）	4	26	9.0	1h，40℃	Na^+，K^+，Mg^{2+}，Mn^{2+}	Ca^{2+}、Zn^{2+}、Hg^{2+}、Cu^{2+}、H_2O_2、D-半乳糖	ONPG，乳糖	Fernandes et al.(2002)
Rahnella aquatilis 14-1	NR（60①）	NR	35	6.5～7.0	1.5h，45℃	Mn^{2+}	Cu^{2+}、Zn^{2+}、Fe^{3+}、Al^{3+}、Ca^{2+}、Mg^{2+}、Na^+	ONPG，乳糖	Liu et al.(2008)

注：ONPG，邻硝基苯基-β-D-吡喃半乳糖苷；PNPG，对硝基苯基 β-D-吡喃半乳糖苷；DTT，二硫苏糖醇；2-ME，2-巯基乙醇；SDS，十二烷基硫酸钠；NR，未发表。

① 由十二烷基硫酸钠聚丙烯酰胺凝胶电泳估算得到单体分子。

② 由非变性凝胶电泳计算得到单体分子。

15.5 潜在的生物技术应用

15.5.1 牛奶中乳酸的水解

到目前为止已分离鉴定了 26 种适冷性 β-D-半乳糖苷酶。考虑到现有技术状况，科学报告主要集中在使用邻硝基苯基-β-D-吡喃半乳糖苷（ONPG）作为底物来研究这些酶的水解活性的详细特征。此外，所有这些报道都证实这些酶能在低温下有效地催化该底物的水解。然而，我们发现，关于牛奶中乳糖水解的研究仅局限于GH2和GH42家族12种适冷性β-D-半乳糖苷酶（表15.4），缺乏关于GH35家族的适冷性β-D-半乳糖苷酶水解乳糖的文献数据。

表15.4 适冷性β-D-半乳糖苷酶水解牛奶中乳糖的研究结果

酶源	底物特异性/(U/mg)	酶活力(0～20℃)	酶量	牛奶反应条件下乳糖			参考文献
				温度/℃	时间/h	乳糖水解率/%	
Arthrobacter psychrolactophilus F2	33.3	80%～100%(0～10℃)	1U/mL的牛奶	10	24	80	Nakagawa et al.(2006a)
Arthrobacter sp. ON14	25.4	30%～100%(0～15℃)	200μg/mL的牛奶(5.08U/mL的牛奶)	4	8	100	Xu et al.(2011)
Guehomyces pullulans R1	1.25	20%(0℃)	10U/mL的牛奶	10	96	80	Nakagawa et al.(2006b)
Halomonas sp. S62	118.45	50%～70%	0.15U/10μL的牛奶	7	24	60	Wang et al.(2013)
Paracoccus sp. 32d	40.98	0%～30%	1U/mL的牛奶	10	71 24	91 97	Wierzbicka-Woś et al.(2011)
Planococcus sp. L4	数据未知	27%～100%	2.5μg的酶	5	1	36	Hu et al.(2007)
Pseudoalteromonas haloplanktis TAE79	138.2	数据未知	5μg/500μL的牛奶(0.691U/500μL的牛奶)	25 4	0.5 50min	26 33	Hoyoux et al.(2001)
Pseudoalteromonas sp. TAE 79b	256.0	28%(5℃)	1U/mL的牛奶	10	4 8	30 40	Fernandes et al.(2002)
Rahnella aquatilis 14-1	321.7	40%(15℃)	0.5mL/mL的牛奶(6.37U/mL的牛奶)	30	2	80	Liu et al.(2008)
metagenome-derived β-D-galactosidase gene *zd410*	数据未知	11%～54%	20U/g牛奶中的乳糖(1U/mL的牛奶)	4 20 25	1 1 1	4.2 22.8 36.2	Wang et al.(2010)
Arthrobacter sp. SB	25.4	50%～100%	98μg/mL的牛奶	2.5	7.5	80	Coker and Brenchley(2006)
Arthrobacter sp. 32cB	213.3	31%～65%	1U/mL的牛奶 2U/mL的牛奶	10	24	70 90	Pawlak-Szukalska et al.(2014)

此外，牛奶中Ca^{2+}以及D-半乳糖和D-葡萄糖（乳中乳糖水解的产物）对这些酶的水解活性影响的研究仅使用ONPG为底物。其实这类研究也可以使用乳糖作为底物。例如，Ca^{2+}、D-半乳糖和D-葡萄糖是*Paracoccus* sp. 32d适冷性β-D-半乳糖苷酶的抑制剂，但这种酶对于牛奶中的乳糖水解也是非常有效的（Wierzbicka-Woś et al. 2011）。因此，仅通过

Ca^{2+}、D-葡萄糖或 D-半乳糖抑制 ONPG 的研究结果并不能用于评估酶是否能够用于乳糖的水解。

K_m 和 k_{cat} 值是用于评估实现底物（例如，工业过程中牛奶中的乳糖）有效水解所需酶量的关键因素。在评估任何涉及使用酶作为生物催化剂的工业过程的经济可行性时，通常会考虑到这个重要参数。表 15.5 中的 K_m 值的分析结果显示 ONPG 是表中所有酶的优选底物。而且，表中大多数酶以乳糖为作用底物时，酶促反应的 K_m 值较高，表明这些酶对乳糖的亲和力较低。此外，表 15.5 中仅测定一些酶的 k_{cat}/K_m 值（k_{cat}/K_m 是衡量酶有效催化底物转化为特定产物的量度）。

表 15.5 适冷性 β-D-半乳糖苷酶的动力学参数

酶源	T/℃	K_m /(mmol/L)		V_{max} /(U/mg)		k_{cat} /s^{-1}		k_{cat}/K_m /[L/(mol·s)]		参考文献
		ONPG	乳糖	ONPG	乳糖	ONPG	乳糖	ONPG	乳糖	
Arthrobacter sp. B7 基因 *15*	30	0.4	16	1.182	117	NR	NR	NR	NR	Trimbur et al. (1994)
Arthrobacter sp. B7 基因 *12*	30	0.57	4.81	254	3.97	NR	NR	NR	NR	Gutshall et al. (1995)
Arthrobacter sp. C2-2 izosyme 1	10	NR	344.2	NR	NR	NR	324	NR	0.9	Karasová-Lipovová et al. (2003)
Arthrobacter sp. SB	15	0.8	11.5	NR	NR	100	5.3	125	0.5	Coker and Brenchley(2006)
Arthrobacter sp. 32c	10	5.75	77.54	NR	NR	52.4	1.76	9.12	0.023	Hildebrandt et al. (2009)
Arthrobacter sp. 32cB	10	1.52	16.56	16.74	17.44	30.55	31.84	20.16	1.92	Pawlak-Szukalska et al. (2014)
Arthrobacter psychrolactophilus F2	10	2.8	50	NR	NR	93.5	18	33.39	0.36	Nakagawa et al. (2006a)
Carnobacterium piscicola BA 基因 *bgaB*	30	1.7	NR	450	NR	588	NR	NR	NR	Coombs and Brenchley(1999)
Flavobacterium sp. 4214	25	0.65	NR	62	NR	NR	NR	NR	NR	SRrensen et al. (2006)
Guehomyces pullulans 17-1	50	3.3	NR	9.2	NR	NR	NR	NR	NR	Song et al. (2010)
Guehomyces pullulans R1	10	NR	50.5	NR	NR	NR	NR	NR	NR	Nakagawa et al. (2006b)
Halomonas sp. S62	35	2.9	NR	NR	NR	390.3	NR	NR	NR	Wang et al. (2013)
	45	NR	32.06	NR	NR	NR	269.5	NR	NR	
Paracoccus sp. 32d	10	1.17	2.94	NR	NR	71.81	43.23	61.38	15.06	Wierzbicka-Wo. 32dal. (2011)
Planococcus sp. L4	10	3.8	11.2	NR	NR	195	62	51.3	5.5	Hu et al. (2007)

续表

酶源	T/℃	K_m /(mmol/L)		V_{max} /(U/mg)		k_{cat} /s^{-1}		k_{cat}/K_m /[L/(mol·s)]		参考文献
		ONPG	乳糖	ONPG	乳糖	ONPG	乳糖	ONPG	乳糖	
Planoccocus sp. SOS Orange	10	4.5	NR	80	NR	104	NR	23	NR	Sheridan and Brenchley(2000)
Pseudoalteromonas haloplanktis TAE 79	25	NR	2.4	NR	NR	NR	33	NR	13.7	Hoyoux et al. (2001)
Pseudoalteromonas sp. 22b	20	0.28	3.3	NR	NR	312	157	1114	47.5	Turkiewicz et al. (2003)
Pseudoalteromonas sp. TAE 79b	25	0.16	NR	11.7	NR	NR	NR	NR	NR	Fernandes et al. (2002)
来源于宏基因组的β-半乳糖苷酶基因 *zd410*	10	4.7	27.1	77	56	66.7	33	14.2	1.21	Wang et al.(2010)

注：NR 表示未发表。

基于表中所示的数据，可知根据适冷性β-D-半乳糖苷酶的现有研究不足以可靠地比较已知酶在牛奶乳糖水解中的适用性。然而，表 15.4 所示的一些已经报道的适冷酶在低温条件下酶解牛奶中乳糖的效率与 *K. lactis* 的β-D 半乳糖苷酶在中温条件下酶解牛奶中乳糖的效率相当。据我们所知，现有资料中仅有两篇专利使用来自 *Alkalilactibacillus ikkense*（Stougaard and Shmidt，专利号：US 8288143B2）和假交替单胞菌 *Pseudoalteromonas haloplanktis*（Hoyoux et al. 专利号：US 6727084 B1）的适冷性β-D-半乳糖苷酶，用于在低温条件下对乳制品和牛奶中的乳糖进行水解。

适冷性β-D-半乳糖苷酶也用于固定化酶。将 *Pseudoalteromonas* sp. 22b LacZ 酶固定在戊二醛处理的壳聚糖珠上。固定化后酶的最适反应温度明显提高，最适 pH 范围较宽，同时其热稳定性也略有提高。此外，固定化的β-D 半乳糖苷酶不受 D-葡萄糖和 D-半乳糖的抑制。在 15℃的柱反应器中，固定化酶至少能保持 40d 水解乳糖的活性。15℃条件下 18h 内，酶∶乳糖为 30U∶1g 时，乳糖水解度高达到 93%。此外，*Pseudoalteromonas* sp. 22b 的固定化β-D-半乳糖苷酶在 4℃下至少能稳定保存 12 个月（Makowski et al. 2007）。来自 *Pseudoalteromonas* sp. TAE 79b 的β-D 半乳糖苷酶可通过吸附或共价偶联固定在琼脂糖上。固定化酶比可溶性酶更稳定，且它们水解乳糖的能力与可溶性β-D-半乳糖苷酶类似（Fernandes et al. 2002）。总而言之，适冷性β-D-半乳糖苷酶的固定化制备在工业应用方面起重要作用。

15.5.2 β-D-吡喃半乳糖苷的合成

迄今为止关于使用适冷性β-D-半乳糖苷酶合成 GOSs 和其他β-吡喃半乳糖苷的研究很少。1991 年，Rahim 和 Lee 发表利用耐冷 *Bacillus subtilis* KL88 的β-D-半乳糖苷酶合成寡糖的研究结果。在这项研究中，作者指出该适冷酶可以利用乳糖作为底物有效地产生寡糖（Rahim and Lee 1991）。含有 20%乳糖的反应混合物在 10℃下温育 4h 后，反应液中 GOSs 浓度达到最大。在这些条件下，涉及乳糖水解的反应显著减少，而更利于 GOSs 的形成反应。2003 年，Karasová-Lipovová 等分离鉴定到另一种来自南极耐冷细菌 *Arthrobacter* sp. C2-2 的适冷性β-D-半乳糖苷酶，它不仅可以催化乳糖的水解，还可以催化转糖基反应。

在文章中，15℃条件下 0.68mol/L 乳糖（在反应开始时乳糖浓度）约反应 10h 后，混合物中三糖的浓度达到平台期（34mmol/L），但四糖的浓度低得多（6mmol/L），并且在 50h 后才达到平台期。根据这篇报告的作者阐述，这些结果表明来自 *Arthrobacter* sp. C2-2 的 β-D-半乳糖苷酶可以（也可能以固定形式）用于合成 GOSs，也可以直接在低温牛奶、乳清或乳清渗透物中产生 GOSs 作为乳制品中的功能添加剂。作为比较，来自 *K. lactis* 的市售酶在 40℃下产生 40mmol/L 三糖。在另一份报告中，Nakagawa 等（2006a）证明 *Arthrobacter psychrolactophilus* F2 的适冷性 rBglAp β-D-半乳糖苷酶在 10℃下合成 GOSs 三糖的效率类似于 *Arthrobacter* sp. C2-2 相应酶的效率。然而，在这篇文章中缺乏该条件下生产四糖的信息。作为比较，Pawlak-Szukalska 等（2014）报道了在 10℃和 20℃孵育 24h 或在 30℃孵育 6h 后的含 *Arthrobacter* sp. 32cB 适冷性 β-D 半乳糖苷酶的乳糖溶液（292～584mmol/L）中三糖和四糖的有效产量。20℃和 30℃温育反应中还发现少量五糖的产生。此外，与上述提到的酶不同的是，*Arthrobacter* sp. 32cB 的 β-D-半乳糖苷酶也可以催化乳糖或特定单糖合成异低聚糖，例如乳果糖（即半乳糖基-果糖）、半乳糖基-木糖和半乳糖基-阿拉伯糖。最大产量的乳果糖是在 30℃下孵育 8h 获得的。在 D-木糖和 L-阿拉伯糖的半乳糖基化过程中也得到类似的结果。来自 *Arthrobacter* sp. 32cB 的酶还能有效地催化 1-丁醇的半乳糖基化，而其他的糖基化醇，即 2-丙醇、1-己醇和环己醇的产率较低。此外，以乳糖作为半乳糖基供体和 *Arthrobacter* sp. 32cB 的 β-D-半乳糖苷酶作为生物催化剂也能使芳香糖苷水杨苷［2-（羟甲基）苯基-β-D-吡喃葡萄糖苷］半乳糖基化。来自南极 *Pseudoalteromonas* sp. 22b 的适冷性 β-D-半乳糖苷酶也发现具有合成烷基半乳糖苷的能力（Makowski et al. 2009）。该酶能有效催化 C_3-C_6 醇的半乳糖基化衍生物的合成。值得注意的是，在反应混合物含有有机溶剂（体积分数低于 50%）时，*Pseudoalteromonas* sp. 22b 的适冷 β-D 半乳糖苷酶合成烷基半乳糖苷的能力明显增强（在一些情况下产率几乎加倍），该结果与其对应的中温酶结果相似。

总之，基于上述数据，可以得出结论，适冷性 β-D-半乳糖苷酶作为生物催化剂在功能性食品生产、化妆品和药物开发，特别是用于热不稳定化合物的糖基化应用方面，具有巨大的应用潜力。值得注意的是，最近有提交在波兰专利局的关于利用 *Arthrobacter* sp. 32cB 适冷性 β-D-半乳糖苷酶合成 GOSs 的专利申请（Pawlak et al. 的专利申请号：PL 407475）。

15.5.3 D-塔格糖的产生

D-塔格糖是一种稀有的天然己酮糖，是己醛糖 D-半乳糖的异构体。这种单糖在甜度上类似于蔗糖，但热值低得多。在 2001 年，D-塔格糖被美国食品药品监督管理局（FDA）认定为安全（GRAS）产品，这意味着它可以作为低热量和低血糖的甜味剂用于饮料、糖果和其他饮食产品中。有报道指出 D-塔格糖对人体健康十分有益，例如益生元、无龋齿性和抗氧化能力（Levin 2002；Oh 2007）。此外，这种糖还可作为治疗 2 型糖尿病和肥胖的潜在新药。最近 3 期临床试验的结果表明 D-塔格糖可以有效降低 2 型糖尿病患者的糖基化血红蛋白 A1c 的水平。D-塔格糖还能降低空腹血糖水平和总 LDL-胆固醇（低密度-脂蛋白-胆固醇）浓度（Lu et al. 2008；Ensor et al. 2015）。

制备 D-塔格糖的一种经济可行的方法是通过氢氧化钙异构法或 L-阿拉伯糖异构酶使 D-半乳糖异构为 D-塔格糖。D-半乳糖可以通过 β-D-半乳糖苷酶水解纯乳糖或者乳清、乳清渗透物中的乳糖获得。然后将得到的 D-葡萄糖和 D-半乳糖混合物进行分离纯化，纯的 D-半乳糖再进行异构化，得到 D-塔格糖（Beadle et al. 专利号：US 500261226；Oh 2007）。

海洋细菌 *Pseudoalteromonas haloplanktis* 的适冷性 β-D-半乳糖苷酶用于研究它在乳清渗透物中水解乳糖的能力。在 23℃和 pH 7.0 条件下，将乳清渗透物稀释三次（对应于质量分数10%的干物质含量和122g/L乳糖)，即1mL乳清渗透物使用0.1mL细胞粗提液（相当于0.05mg/mL的酶)，24h可获得高于96%的乳糖转化。尽管在D-葡萄糖的存在下效果更好，但*P. haloplanktis*的β-D-半乳糖苷酶活性会被D-葡萄糖和D-半乳糖抑制（Van de Voorde et al. 2014)。D-半乳糖也可以使用来自氯酚节杆菌*Arthrobacter chlorophenolicus*重组菌株*Pichia pastoris*（巴斯德毕赤酵母）的β-D-半乳糖苷酶获得。在乳清渗透物中加入重组酵母菌株会导致乳糖的水解以及D-葡萄糖的利用，而D-半乳糖则保留在培养基中。在含有120g/L乳糖的乳清渗透物中，30℃下孵育168h可实现约90%的乳糖水解效率。在这种情况下，D-葡萄糖的抑制作用可被忽略（Wanarska and Kur 2012；Wanarska et al.专利号：PL 217153)。

使用来源于*A. chlorophenolicus*的β-D-半乳糖苷酶以及表达*Arthrobacter* sp. 22c L-阿拉伯糖异构酶的重组*P. pastoris*可直接从乳清渗透物中生产获得D-塔格糖。在添加了L-阿拉伯糖异构酶（0.2mg/mL）的乳糖含量为110g/L的乳清渗透物中培养重组酵母菌株，可使乳糖转化率达90%，在144h内能完全利用D-半乳糖且30%的D-半乳糖异构成为D-塔格糖（Wanarska and Kur 2012；Wanarska et al.专利号：PL 216683)。

15.6 结论和未来趋势

适冷酶已被用于许多工业和生物技术过程中，并且在许多其他过程也发挥了重要的作用，在许多方面，降低反应温度是十分有益的。这种方法可以节省能量、降低生产成本、改善卫生、保持口感以及其他几种感官特性，同时还降低污染的风险。适冷酶已经用于合成精细化学品，环境生物技术，能源和生物燃料生产，制药、医药、洗涤剂、纺织品和食品工业中（Marx et al. 2007；Cavicchioli et al. 2011)，具体实例包括使用适冷性β-D-半乳糖苷酶。近来，分离表征新型β-D-半乳糖苷酶是一个新的发展趋势。

值得注意的是，有两个巨大的储备库可用于发掘适冷活性物质。目前大量的不可培养微生物可通过改进培养条件或通过使用宏基因组方法来筛选，每一种技术都具有其自身独特的优缺点（Vester et al. 2014，2015)。基于培养的方法可以获得能产生潜在的新型活性酶的微生物，但这种方法重叠率非常高。基于宏基因组学的方法可以大量筛选鉴定可能编码蛋白酶的基因，然而这种方法不能保证基因在异源宿主中的表达活性。功能宏基因组学可以发现具有新功能活性的酶，但这种方法命中率非常低。

目前，重组DNA技术可用于表达和优化不同微生物中已知高活性的β-D-半乳糖苷酶（Oliveira et al. 2011)。本章提及的几乎所有适冷性β-D-半乳糖苷酶都是重组酶，它们在不同的大肠杆菌表达系统中能有效地表达，这大大扩展了β-D-半乳糖苷酶的潜在应用范围及其在工业过程中的经济效益。当代分子生物学技术结合生物工程技术可用于优化蛋白酶生产，构建技术上和经济上有效的酶生产系统。

研究β-D半乳糖苷酶有两种形式：一种涉及新型适冷和嗜热β-D半乳糖苷酶的水解活性，另一种涉及除了水解活性之外的新的蛋白酶活性的表征。尽管β-D-半乳糖苷酶的水解活性通常在乳品工业中用于获得无乳糖产品，但是仍有大量新的开发空间。乳品行业中适冷性β-D-半乳糖苷酶的广泛使用是可以设想的。基于GOSs在功能性食品领域中的关键作用，具有其他活性（例如转糖基作用活性）的β-D-半乳糖苷酶无疑将吸引研究者的关注。考虑到市场上目前GOSs产品生产的不合理设计，仍然需要开发具有高半乳糖基转移活性的β-D-半乳糖苷酶的微生物，并且分析GOSs混合物的具体组分。实际上，这是一个研究热点领域，很可能会继续发展。总之，β-D-半乳糖苷酶将成为非常有前景的合成工具。

尽管适冷性β-D-半乳糖苷酶具有许多优点，使得它们成为在低温下进行的工业过程中有吸引力的生物催化剂，但目前在上述提及的方法中还没有具体使用，无法具体说明出现这种状况的主要原因。文献研究显示，大多数适冷性β-D-半乳糖苷酶可以在不同的大肠杆菌表达系统中以活性形式有效表达。然而，据我们所知，目前少有适冷性β-D-半乳糖苷酶的生产大于实验室规模的报道。因此，难以预测能满足当前工业需求的适冷性β-D-半乳糖苷酶生产的实际成本。

致谢

这项工作得到了波兰国家研究与开发中心（PBS1/A9/7/2012）的资助。

参考文献

Beadle JR，Saunders JP，Wajda TJ Jr. Process for manufacturing tagatose. US Patent 5，002，612，26

Białkowska AM，Cieśliński H，Nowakowska KM，Kur J，Turkiewicz M（2009）A new β-galactosidase with a low temperature optimum isolated from the Antarctic *Arthrobacter* sp. 20B：gene cloning，purification and characterization. Arch Microbiol 191：825-835

Boon MA，Janssen AEM，van't Riet K（2000）Effect of temperature and enzyme origin on the enzymatic synthesis of oligosaccharides. Enzyme Microb Techol 26：271-281

Cavicchioli R，Charlton T，Ertan H，Mohd OS，Siddiqui KS，Williams TJ（2011）Biotechnological uses of enzymes from psychrophiles. Microb Biotechnol 4：449

Cieśliński H，Kur J，Białkowska A，Baran I，Makowski K，Turkiewicz M（2005）Cloning，expression，and purification of a recombinant cold-adapted β-galactosidase from antarctic bacterium *Pseudoalteromonas* sp. 22B. Protein Expr Purif 39：27-34

Coker JA，Brenchley JE（2006）Protein engineering of a cold-active beta-galactosidase from *Arthrobacter* sp. SB to increase lactose hydrolysis reveals new sites affecting low temperature activity. Extremophiles 10：515-524

Coker JA，Sheridan PP，Loveland-Curtze J，Gutshall KR，Auman AJ，Brenchley JE（2003）Biochemical characterization of a β-galactosidase with a low temperature optimum obtained from an Antarctic *Arthrobacter* isolate. J Bacteriol 185：5473-5482

Coombs JM，Brenchley JE（1999）Biochemical and phylogenetic analyses of cold-active β-galactosidase from the lactic acid bacterium *Carnobacterium piscicola* BA. Appl Environ Microbiol 65：5443-5450

Coombs JM，Brenchley JE（2001）Characterization of two new glycosyl hydrolases from the lactic acid bacterium *Carnobacterium piscicola* strain BA. Appl Environ Microbol 67：5094-5099

Coté A，Brown WA，Cameron D，van Walsum GP（2004）Hydrolysis of lactose in whey permeate for subsequent fermentation to ethanol. J Dairy Sci 87：1608-1620

Cruz R，Cruz VD，Belote JG，Khenayfes MD，Dorta C，Oliveira LHD（1999）Properties of a new fungal beta-galactosidase with potential application in the dairy industry. Rev Microbiol 30：265-271

Di Lauro B，Strazzulli A，Perugino G，La Cara F，Bedini E，Corsaro MM，Rossi M，Moracci M（2008）Isolation and characterization of a new family 42 β-galactosidase from the thermoaci-dophilic bacterium *Alicyclobacillus acidocaldarius*：identification of the active site residues. Biochim Biophys Acta 1784：292-301

Ensor M，Banfield AB，Smith RR，Wiliams J，Lodder RA（2015）Safety and efficacy of D-tagatose in glycemic control in subjects with type 2 diabetes. J Endocrinol Diab Obes 3：1065

Fernandes S，Geueke B，Delgado O，Coleman J，Hatti-Kaul R（2002）β-Galactosidase from a cold-adapted bacterium：purification，characterization and application for lactose hydrolysis. Appl Microbiol Biotechnol 58：313-321

Guimaraes WV，Dudey GL，Ingram LO（1992）Fermentation of sweet whey by ethanologenic *Escherichia coli*. Biotechnol Bioeng 40：41-45

Gutshall K，Trimbur D，Kasmir J（1995）Analysis of a novel gene and beta-galactosidase isozyme from a psychrotrophic *Arthrobacter* isolate. J Bacteriol 177：1981-1988

Gutshall K，Wang K，Brenchley JE（1997）A novel *Arthrobacter* β-galactosidase with homology to eucaryotic β-galactosidases. J Bacteriol 179：3064-3067

Harju M（1987）Lactose hydrolysis. Bull Int Dairy Fed 212：50-54

Harju M，Kallioinen H，Tossavainen O（2012）Lactose hydrolysis and other conversions in dairy products：technological aspects. Int Dairy J 22：104-109

Hildebrandt P，Wanarska M，Kur J（2009）A new cold-adapted beta-D-galactosidase from the Antarctic *Arthrobacter* sp. 32c-gene cloning，overexpression，purification and properties. BMC Microbiol 9：151

Hoyoux A，Francois JM，Dubois P，Baise E，Jennes I，Genicot S，Gerday C. Cold-active beta-galactosidase，the process for its preparation and the use thereof. US Patent 6，727，084 B1

Hoyoux A，Jennes I，Dubois P，Genicot S，Dubail F，Franc JM，Baise E，Feller G，Gerday C（2001）Cold-adapted β-galactosidase from the Antarctic psychrophile *Pseudoalteromonas haloplanktis*. Appl Environ Microbiol 67：1529-1535

Hu JM, Li H, Cao LX, Wu PC, Zhang CT, Sang SL, Zhang XY, Chen MJ, Lu JQ, Liu YH (2007) Molecular cloning and characterization of the gene encoding cold-active beta-galactosidase from a psychrotrophic and halotolerant *Planococcus* sp. L4. J Agric Food Chem 55: 2217-2224

Johansen AG, Vegarud GE, Skeie S (2002) Seasonal and regional variation in the composition of whey from Norwegian Cheddar-type and Dutch-type cheeses. Int Dairy J 12: 621-629

Juers DH, Matthews BW, Huber RE (2012) LacZ beta-galactosidase: structure and function of an enzyme of historical and molecular biological importance. Protein Sci 21: 1792-1807

Jurado E, Camacho F, Luzón G, Vicaria JM (2002) A new kinetic model proposed for enzymatic hydrolysis of lactose by a β-galactosidase from *Kluyveromyces fragilis*. Enzyme Microb Technol 31: 300-309

Karasová-Lipovová P, Strnad H, Spiwok V, Malá S, Králová B, Russell NJ (2003) The cloning, purification and characterisation of a cold-active b-galactosidase from the psychrotolerant Antarctic bacterium *Arthrobacter* sp. C2-2. Enzyme Microb Technol 33: 836-844

Kosugi A, Murashima K, Doi RH (2002) Characterization of two noncellulosomal subunits, ArfA and BgaA, from *Clostridium cellulovorans* that cooperate with the cellulosome in plant cell wall degradation. J Bacteriol 184: 6859-6865

Kren V, Sedmera P, Havlicek V, Fiserova A (1992) Enzymatic galactosylation of ergotalkaloids. Tetrahedron Lett 47: 7233-7236

Ladero M, Perez MT, Santos A, Garcia-Ochoa F (2003) Hydrolysis of lactose by free and immo-bilized beta-galactosidase from *Thermus* sp. strain T2. Biotechnol Bioeng 81: 241-252

Lee JH, Kim YS, Yeom SJ, Oh DK (2011) Characterization of a glycoside hydrolase family 42 β-galactosidase from *Deinococcus geothermalis*. Biotechnol Lett 33: 577-583

Levin GV (2002) Tagatose, a new GRAS sweetener and health product. J Med Food 5: 23-36

Liu WY, Shi YW, Wang XQ, Wang Y, Wei CQ, Lou K (2008) Isolation and identification of a strain producing cold-adapted β-galactosidase, and purification and characterisation of the enzyme. Czech J Food Sci 26: 284-290

Lu Y, Levin GV, Donner TW (2008) Tagatose, a new antidiabetic and obesity control drug. Diabetes Obes Metab 10: 109-134

Makowski K, Białkowska M, Szczesna-Antczak M, Kalinowska H, Kur J, Cieslinski H, Turkiewicz M (2007) Immobilized preparation of cold-adaptedand halotolerant antarctic β-galactosidase as a highlystable catalyst in lactose hydrolysis. FEMS Microbiol Ecol 59: 535-542

Makowski K, Białkowska A, Olczak J, Kur J, Turkiewicz M (2009) Antarctic, cold-adapted β-galactosidase of *Pseudoalteromonas* sp. 22b as an effective tool for alkyl galactopyranosides synthesis. Enzyme Microbial Technol 44: 59-64

Marx JC, Collins T, D' Amico S, Feller G, Gerday C (2007) Cold-adapted enzymes from marine Antarctic microorganisms. Marine Biotechnol 9: 293

Mehaia MA, Alverez J, Cheryan M (1993) Hydrolysis of whey permeate lactose in a continuous stirred tank membrane reactor. Int Dairy J 3: 179-192

Mlichová Z, Rosenberg M (2006) Current trends of β-galactosidase application in food technol-ogy. J Food Nutr Res 45: 47-54

Nakagawa T, Fujimoto Y, Ikehata R, Miyaji T, Tomizuka N (2006a) Purification and molecular characterization of cold-active beta-galactosidase from *Arthrobacter psychrolactophilus* strain F2. Appl Microbiol Biotechnol 72: 720-725

Nakagawa T, Ikehata R, Uchino M, Miyaji T, Takano K, Tomizuka N (2006b) Cold-active acid β-galactosidase activity of isolated psychrophilic-basidiomycetous yeast *Guehomyces pullulans*. Microbiol Res 161: 75-79

Nath A, Mondal S, Chakraborty S, Bhattacharjee C, Chowdhury R (2014) Production, purification, characterization, immobilization, and application of β-galactosidase: a review. Asia-Pac J Chem Eng 9: 330-348

Oh DK (2007) Tagatose: properties, applications, and biotechnological processes. Appl Microbiol Biotechnol 76: 1-8

Oliveira C, Guimarães PM, Domingues L (2011) Recombinant microbial systems for improved β-galactosidase production and biotechnological applications. Biotechnol Adv 29: 600

Onishi N, Tanaka T (1996) Purification and properties of a galacto-and gluco-oligosaccharide-producing beta-glycosidase from *Rhodotorula minuta* IFO879. J Ferment Bioeng 82: 439-443

Pawlak A, Wanarska M, Popinigis A, Kur J. β-D-galaktozydaza *Arthrobacter* sp. S3*, β-D-galaktozydaza *Arthrobacter* sp. 32cB, sekwencje nukleotydowe szczepu *Arthrobacter* sp. S3 * i *Arthrobacter* sp. 32cB kodujące β-D-galaktozydazę oraz sposób wytwarzania mleka o obniżonej zawartości laktozy, galaktooligosacharydów, heterooligosacharydów i glikozylowanych związków chemicznych z wykorzystaniem tych enzymów. Patent applica-tion no: PL 407, 475

Pawlak-Szukalska A, Wanarska M, Popinigis AT, Kur J (2014) A novel cold-active β-D-galactosidase with transglycosylation activity from the Antarctic *Arthrobacter* sp. 32cB-gene cloning, purification and characterization. Process Biochem 49: 2122-2133

Poltorak OM, Chukhrai ES, Pilipenko OS, Atyaksheva LF, Beregalov AE (2007) A comparative study of the structure and properties of β-galactosidases. Russ J Phys Chem A 81: 808-812

Rahim KA, Lee BH (1991) Specificity, inhibitory studies, and oligosaccharide formation by beta-galactosidase from psychrotrophic *Bacillus subtilis* KL88. J Dairy Sci 74: 1773-1778

Saishin N, Ueta M, Wada A, Yamamoto I (2010) Properties of β-galactosidase purified from *Bifidobacterium longum* subsp. *longum* JCM 7052 grown on gum arabic. Int J Biol Macromol 10: 23-31

Santos A, Ladero M, García-Ochoa F (1998) Kinetic modeling of lactose hydrolysis by a β-galactosidase from *Kluyveromyces fragilis*. Enzyme Microb Technol 22: 558-567

Scheckermann C, Wagner F, Fischer L (1997) Galactosylation of antibiotics using the beta-galactosidase from *Asper-*

gillus oryzae. Enzyme Microb Technol 20：629-634

Schmidt M，Stougaard P (2010) Identification，cloning an expression of a cold-active β-galactosidase from a novel arctic bacterium，*Alkalilactibacillus ikkense*. Environ Technol 3：1107-1114

Sheridan PP，Brenchley JE (2000) Characterization of a salt-tolerant family 42 β-galactosidase from a psychrophilic antarctic *Planococcus* isolate. Appl Environ Microbiol 66：2438-2444

Shimizu R，Shimabayashi H，Moriwaki M (2006) Enzymatic production of highly soluble myrici-trin glycosides using beta-galactosidase. Biosci Biotechnol Biochem 70：940-948

Song C，Liu G-L，Xu J-L，Chi Z-M (2010) Purification and characterization of extracellular β-galactosidase from the psychrotolerant yeast *Guehomyces pullulans* 17-1 isolated from sea sediment in Antarctica. Process Biochem 45：954-960

Sørensen HP，Porsgaard TK，Kahn RA，Stougaard P，Mortensen KK，Johnsen MG (2006) Secreted β-galactosidase from *Flavobacterium* sp. isolated from low-temperature environment. Appl Microbiol Biotechnol 70：548-557

Stevenson DE，Stanley RA，Furneaux RH (1993) Optimization of alkyl β-D-galactopyranoside synthesis from lactose using commercially available β-galactosidases. Biotechnol Bioeng 42：657-666

Stougaard P，Schmidt M. Cold-active beta-galactosidase，a method of producing same and use of such enzyme. US 8，288，143 B2

Torres MJ，Lee BH (1995) Cloning and expression of β-galactosidase from psychotrophic *Bacillus subtilis* KL88 into Escherichia coli. Biotechnol Lett 17：123-128

Trimbur D，Gutshall K，Prema P，Brenchley JE (1994) Characterization of a psychrotrophic *Arthrobacter* gene and its cold-active-galactosidase. Appl Environ Microbiol 60：4544-4552

Turkiewicz M，Kur J，Białkowska A，Cieśliński H，Kalinowska H，Bielecki S (2003) Antarctic marine bacterium *Pseudoalteromonas* sp. 22b as a source of cold-adapted β-galactosidase. Biomol Eng 20：317-324

Van de Voorde I，Goiris K，Syryn E，Van den Bussche C，Aerts G (2014) Evaluation of the coldactive *Pseudoalteromonas haloplanktis* β-galactosidase enzyme for lactose hydrolysis in whey permeate as primary step of D-tagatose production. Process Biochem 49：2134-2140

Van Laere KMJ，Abee T，Schols HA，Beldman G，Voragen AGJ (2000) Characterization of a novel beta-galactosidase from *Bifidobacterium adolescentis* DSM 20083 active towards transgalac-tooligosaccharides. Appl Environ Microbiol 66：1379-1384

Vester JK，Glaring MA，Stougaard P (2014) Discovery of novel enzymes with industrial potential from a cold and alkaline environment by a combination of functional metagenomics and culturing. Microb Cell Fact 13：72

Vester JK，Glaring MA，Stougaard P (2015) Improved cultivation and metagenomics as new tools for bioprospecting in cold environments. Extremophiles 19：17-29

Vetere A，Paoletti S (1998) Separation and characterization of three beta-galactosidases from *Bacillus circulans*. BBA Gen Subj 1380：223-231

Wanarska M，Hildebrandt PŁ，Kur JW. Sposób otrzymywania D-galaktozy z laktozy. Patent no：PL217，153

Wanarska M，Hildebrandt PŁ，Kur JW. Sposób wytwarzania D-tagatozy z laktozy z wykorzystan-iem rekombinantowego szczepu drożdży Pichia pastoris wytwarzającego β-D-galaktozydazę *Arthrobacter chlorophenolicus* i izomerazy arabinozowej *Arthrobacter* sp. 22c. Patent no：PL216，683

Wanarska M，Kur J (2005) β-D-Galactosidases-sources，properties and applications. BioTechnologia 4：46-62

Wanarska M，Kur J (2012) A method for the production of D-tagatose using a recombinant Pichia pastoris strain secreting β-D-galactosidase from *Arthrobacter chlorophenolicus* and a recombinant L-arabinose isomerase from *Arthrobacter* sp. 22c. Microb Cell Fact 11：113

Wang K，Li G，Qin Y，Znang CT，Liu YH (2010) A novel metagenome-derived β-galactosidase：gene cloning，overexpression，purification and characterization. Appl Microbiol Biotechnol 88：155-165

Wang GX，Gao Y，Hu B，Lu XL，Liu XY，Jiao BH (2013) A novel cold-adapted β-galactosidase isolated from *Halomonas* sp. S62：gene cloning，purification and enzymatic characterization. World J Microbiol Biotechno129：1473-1480

White PL，Wynn-Williams DD，Russell NJ (2000) Diversity of thermal responses of lipid composition in the membranes of the dominant culturable members of an Antarctic fellfield soil bacterial community. Antarct Sci 72：386-393

Wierzbicka-Woś A，Cieśliński H，Wanarska M，Kozłowska-Tylingo K，Hildebrandt P，Kur J (2011) A novel cold-active β-D-galactosidase from the *Paracoccus* sp. 32d-gene cloning，purification and characterization. Microb Cell Fact 10：108

Xu K，Tang X，Gai Y，Mehmood M，Xiao X，Wang F (2011) Molecular characterization of cold-inducible beta-galactosidase from *Arthrobacter* sp. ON14 isolated from Antarctica. J Microbiol Biotechnol 21：236-242

第十六章
极端磷酸三酯酶样内酯酶在生物技术上的应用

Elena Porzio[1,2], Immacolata Del Giudice[1],
Giuseppe Manco[1,3]

16.1 引言

有机磷酸酯化合物（OPs）是杀虫剂（如对氧磷、对硫磷、杀虫磷、二嗪农、乐果和毒死蜱）和许多神经性毒剂［如沙林、塔崩、索曼和 VX（维埃克斯）］的主要组成物质（表 16.1）。

表 16.1　磷酸三酯酶样内酯酶（PLL）作用底物

类型	名称	结构	用途
有机磷农药	对氧磷	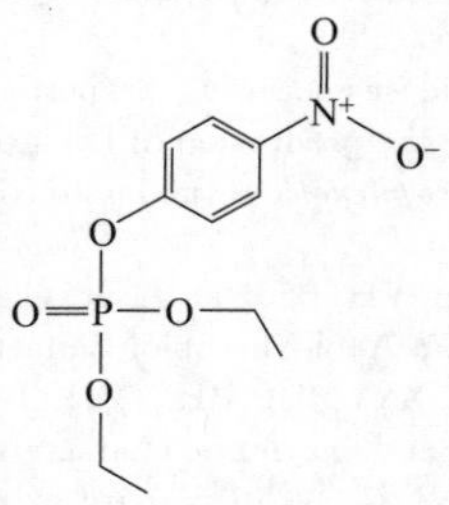	杀虫剂
	对硫磷	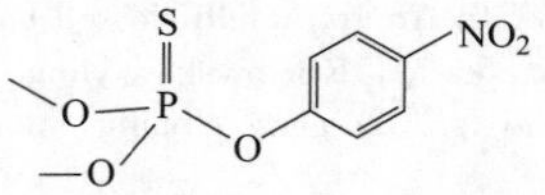	杀虫剂
	香豆磷	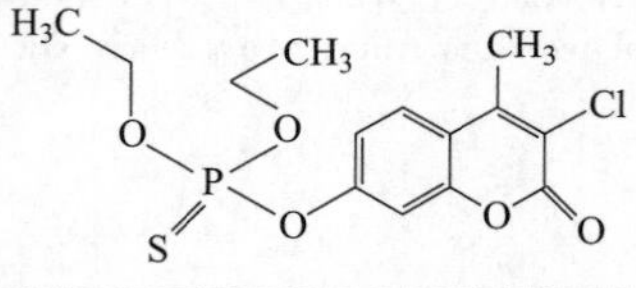	杀螨剂

❶ Institute of Protein Biochemistry，CNR，Via P. Castellino 111，80131 Naples，Italy.
❷ e-mail：e. porzio@ibp. cnr. it.
❸ g. manco@ibp. cnr. it.

续表

类型	名称	结构	用途
有机磷农药	毒死蜱	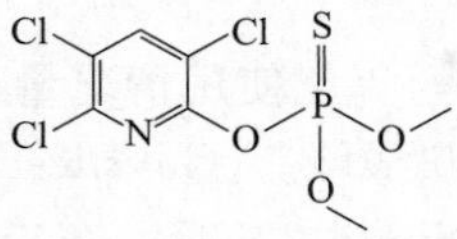	杀虫剂
	二嗪农	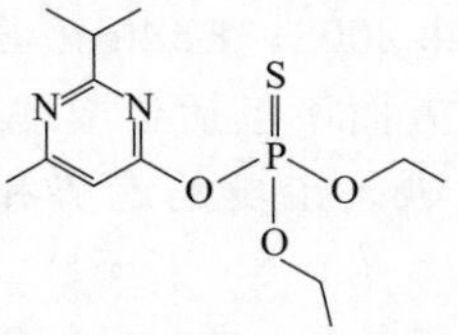	杀虫剂
	乐果	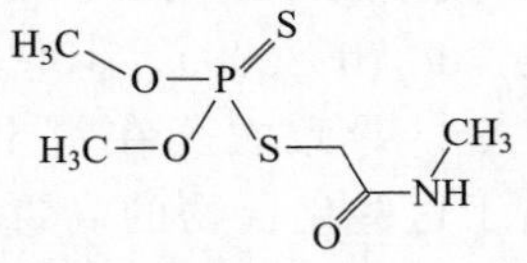	杀虫剂
OP 神经毒气	塔崩	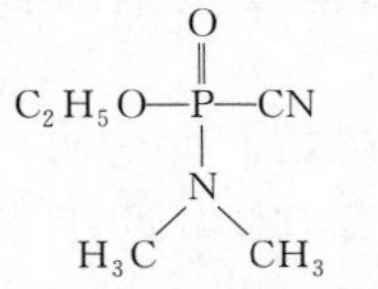	CWA①
	沙林	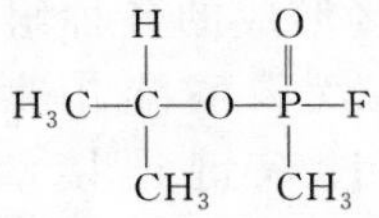	CWA①
	索曼	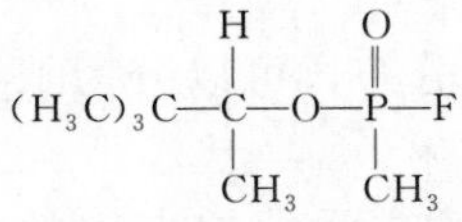	CWA①
	VX	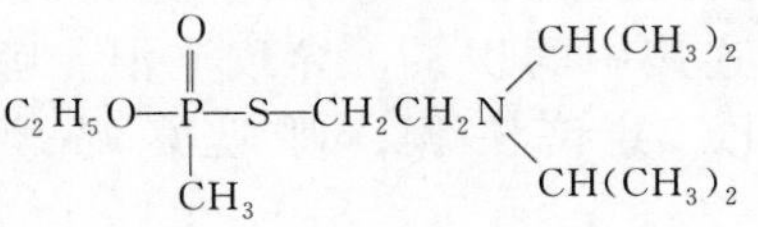	CWA①
内酯	AHL②	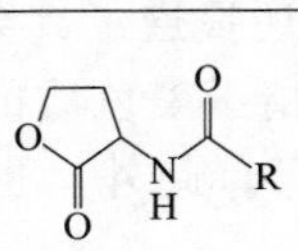	—
	3-oxo-AHL②	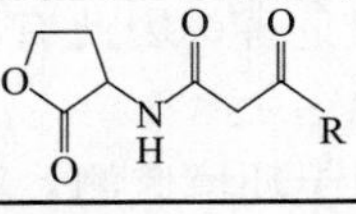	—

注：常用的有机磷化合物和酰基高丝氨酸内酯的一般结构（AHL）。

① 化学战剂。

② R 对应不同大小的酰基链。

农药可控制植物或动物（害虫）数量，用于农业生产可以获得更高的生产力。然而，农药的广泛应用已经对水生系统造成了严重污染，同时对人类健康也构成了威胁。

20 世纪开始使用 OPs，其使用量迅速增加，目前约占全球农药应用总量的 38%（Singh 2009）。

美国环境保护署（EPA）对农药年度使用情况量进行最新估计，年使用量超过 500 万磅（1 磅≈453.59g），其中美国超过 100 万磅（Gavrilescu 2005；Grube et al. 2011）。目前，世界上大约储存了 20 万吨的 OPs（Singh 2009），在农业生产中，经常会引发农民农药中毒（Aardema et al. 2008；Calvert et al. 2008；Eddleston et al. 2005）。

农药对环境的影响取决于几个方面，包括土壤理化性质、地形、天气、农业管理措施以及每种农药的化学性质（水溶性、吸附土壤的趋势和农药持久性）等（Tiryaki and Temur 2010）。

通常情况下，阳光照射、水和微生物水解导致 OPs 快速降解，因此其在土壤中稳定性较低（Caceres et al. 2010；Ragnarsdottir 2000）。但它们的过度使用、储存不当以及许多发展中国家的农药控制方案的不健全，导致农药在水土中的异常积累，并通过食物链富集（OrtizHernàndez et al. 2013）。由于这些化合物的高毒性，联合国粮食及农业组织（FAO）和世界卫生组织（WHO）制定了食品安全国际食品标准和准则（国际食品法典委员会）（FAO，2014）。

OPs 除了作为杀虫剂使用外，在二十世纪初，也被合成为毒性很强的化学武器，并在第二次世界大战中使用（Raushel 2002），近年来，也被用于恐怖袭击（Tokio subway，1995）和投毒（Syria 2013）。

OPs 是磷和氧、碳、硫、氮连接形成的各种组合的酯类物质，共有六个不同的亚类：磷酸酯、膦酸酯、硫代磷酸、二硫代磷酸酯、硫代磷酸酯和氨基磷酸酯（Can 2014）。它们具有类似的结构，其中磷通过双键连接氧原子（oxon-OPs）或硫原子（thion-OPs），并通过烷氧基或氨基的酯键连接到另一个基团（卤素、脂肪族、芳香族或杂环基团）。

基于配体在中心磷原子之间的分布，可以分为 R_P 或 S_P 手性构型，其中 S_P 构型通常是毒性较强的异构体。

OPs 的毒性作用方式主要是通过活性位点内的丝氨酸残基共价磷酸化，不可逆地抑制胆碱传递中的关键酶乙酰胆碱酯酶（AChE）。这种抑制导致神经递质乙酰胆碱在神经元-神经元或神经元-肌肉连接中的积累，导致流泪、唾液及支气管黏液分泌过多、支气管收缩、骨骼肌纤维化、共济失调、呼吸衰竭、低体温，最终死亡（Carey et al. 2013）。

最近，对 OPs 毒理学性质的研究还报道了其会导致基因突变、染色体畸变、DNA 损伤（Ojha and Gupta 2014）、精液质量和精子染色质的改变（Salazar-Arredondo et al. 2008）以及胰岛素抵抗（Lasram et al. 2014）等。此外，也有报道说 OPs 会致癌和导致内分泌紊乱（Gupta 2005）。

OPs 对人类健康和野生动物构成危害，因此有必要制定实施相应风险评估、环境监测、废物处理和污染场地修复的策略。

传统色谱和光谱分析技术通常用于环境监测，其灵敏度高、选择性强，但需要大量专业设备和时间，而且检测费用昂贵。通过生物传感器检测是对传统方法的一个很好的补充或替代，可以简化或消除一些限制步骤，例如样品制备和现场测试更容易，显著降低每个分析的成本（Marrazza 2014）。

许多用于 OPs 去污领域的经典方法，包括化学处理、电化学氧化和还原、挥发、焚烧或光分解（Kiss and Viràg 2009），都花费较高、污染大，并且对低浓度污染物的效果差（LeJeune et al. 1998）。作为用于外源性化合物代谢途径和必要酶的补充，微生物可以在几个有前景的生物技术平台加以应用，包括生物监测和生物修复、神经性毒剂储备以及医学预防中作为清除剂。然而其仍存在一些明显的缺点，包括严格生长温度、氧气和营养供应、恶劣的环境条件下的不稳定性、不同化学品的存在下会抑制生长等问题（Singh and Walker 2006）。

因此，上述应用中，筛选开发水解有机磷酸酯中磷酸酯或磷脂键的酶类是较为理想的方法。水解酶具有稳定的半衰期，底物范围广，特异性高，反应快速，并且对外消旋混合物和单个立体异构体有效（Tsai et al. 2010a，b）。这些酶通常是从不同的微生物中分离和鉴定出来的磷酸三酯酶（PTE；EC 3.1.8.1）（Islam et al. 2010；Singh 2009；Zhang et al. 2005）。

为了改善几种熟知的有机磷酸酶的特性（稳定性、活性、选择性和底物特异性），许多研究者已经结合不同的突变方法（定向进化和合理设计）进行定向改造（Goldsmith et al. 2012；Iyer and Iken 2015；Tsai et al. 2012）。在过去几年中，定向进化，特别是与理论设计方法相结合时（蛋白质结构分析和生物信息学工具）已成为设计关键数据库的最佳方法（Cobb et al. 2013a，b；Denard et al. 2015）。定向进化由随机突变和/或 DNA 重组的迭代循环组成，在靶基因序列中产生多样性，并允许通过高通量筛选鉴定所需的蛋白质变异体（Cobb et al. 2012）。多年来，定向进化已经使得已知特征的蛋白质进行了工程化改进，生成优化的生物催化剂应用于不同的工业。此外，它还可以用于设计或优化在自然界中活性未被发现过的生物催化剂（Denard et al. 2015）。

本文将介绍一些近来在 OP 水解生物催化剂工程中取得的研究成果，其中主要关注磷酸三酯酶样内酯酶（磷酸鸟苷酶样乳酸酶）的混合活性改造，它在 OPs 治理中是具有广阔的应用前景。

值得注意的是，酶的混合功能是非常重要的，因为当环境改变时，可能会出现新活性，并产生新的酶谱。此外，它们为生物技术和药物开发中新功能发掘提供了可能。

16.2 有机磷化合物水解酶

到目前为止，已经鉴定了许多能够水解有机磷化合物的酶，例如有机磷水解酶（OPH）、甲基对硫磷水解酶（MPH）、血清对氧磷酶（PON）和微生物脯氨酸肽酶/有机磷酸酐水解酶类等。

16.2.1 有机磷水解酶

目前已经鉴定了一些不同形式、位置以及宿主来源的有机磷降解相关基因。首先鉴定的飞鱼鞘氨醇菌 *Sphingobium fuliginis* ATCC 27551（以前为黄杆菌属）（Kawahara et al. 2010；Singh and Walker 2006）和短波单胞菌 *Brevundimonas diminuta* GM（以前称为缺陷假单胞菌 GM）（Harper et al. 1988；Segers et al. 1994）*opd* 基因及它们的相关蛋白质产物，这些酶对 OP 化合物，特别是对对氧磷化合物表现出强的水解活性（Dumas et al. 1989a）。

来自 *B. diminuta* 的 OPH（在本章中命名为 *bd*PTE）属于酰胺水解酶家族（亚型Ⅰ）（Seibert and Raushel 2005），包含了在大质粒转座元件中的基因。结构数据显示 *bd*PTE 是具有 $(\beta/\alpha)_8$-桶形折叠结构的同二聚体，包括在 C 末端的催化双核中心（Benning et al. 1994，1995，2001；Vanhooke et al. 1996），含有两个金属离子，锌或钴为最常见的辅因子（Dumas et al. 1989a，b；Benning et al. 2001）。参与催化位点的残基是与两个二价阳离子直接相互作用的四个组氨酸（H55、H57、H201 和 H230）、天冬氨酸（D301）和羧化赖氨酸（K169）（Benning et al. 2000）。

在 *bd*PTE 结构中，以及在酰胺水解酶家族的其他成员的结构中，已经鉴定了两个环结构具有重要功能，一个（环 7）参与底物结合和水解，另一个（环 8）负责底物特异性（Seibert and Raushel 2005）。

*bd*PTE 水解不同的有机磷酸盐神经性毒剂，如近年来用作化学武器的沙林、索曼和 VX（Caldwell and Raushel，1991）。此外该酶是在迄今发现的具有 OP 降解功能基因中效率最高［$k_{cat}/K_m = 4\times10^7$ L/(mol·s)］（Caldwell et al. 1991）。有趣的是，目前检测到 *bd*PTE 存在不同混合活性（羧酸酯酶和内酯酶活性）（Roodveldt and Tawfik 2005），但天然底物活性尚未确定。这表明由于在全世界广泛使用大量的 OP 杀虫剂，其催化活性目前可能已经从最初的酶中得到进化（Raushel and Holden 2000）。

在 Jackson 等最近的综述中，深入探讨了 *bd*PTE 对对氧磷的高催化效率的结构决定因素。*bd*PTE 描述了具有低能结构转变的两个主要构象亚态（“关闭”和“开放”）。“关闭”状态可能促进了底物识别和降低反应活化能，但不促进底物/产物的快速扩散；相反，“开放”状态可能主要促进酶/产物复合物解离而不影响水解反应的速率（Jackson et al. 2008）。这项研究表明了酶催化优化中构象变化的重要性，这一理论也被 Zhang 等（2015）在 G*ka*P 的工作中证明。

从放射性土壤杆菌 P230 中分离出显示有机磷酸酶活性的 *bd*PTE 的同系物，并命名为 OpdA（Horne et al. 2002，2003）。OpdA 能够水解二乙基 OPs（对氧磷、蝇毒磷、对硫磷、二嗪农、香豆磷），并且以比 *bd*PTE 高的速率水解二甲基底物（甲基对硫磷、磷酸盐、倍硫磷、dMUP）（Horne et al. 2002）。

在几种细菌基因组测序结果中也证明了 *opd* 样基因的存在，虽然这些序列比黄杆菌（*Flavobacterium*）、缺陷短波单胞菌（*B. diminuta*）和放射土壤杆菌（*Agrobacterium radiobacter*）中发现的序列同源性更低（在氨基酸水平上只有 30%～40%的一致性）。

另外，已经从大肠杆菌中分离、纯化、鉴定出 OPD（邻苯二胺）样蛋白，并命名为 PHP（磷酸三酯酶同源蛋白）（Buchbinder et al. 1998）。生化分析表明，大肠杆菌（ePHP）的 PHP 是具有两个锌离子的单体，其结构类似于 *bd*PTE 的结构，并且与金属离子配位的残基是保守的（不包括 K169）（Buchbinder et al. 1998）。这个 OPD 同源蛋白的天然功能是未知的。至于 ePHP 突变体，仅报道了弱酯酶和磷酸三酯酶活性（Roodveldt and Tawfik 2005）。

一个根瘤菌（*Mesorhizobium loti*）基因 *mll7664* 翻译的蛋白质具有类似于 PHP 的特征。与 ePHP 类似，该酶具有桥接两种金属离子的谷氨酸（通过相似性），并且主要活性是羧酸酯酶活性，具有混合的磷酸三酯酶和磷酸二酯酶活性。因此，该酶被称为磷酸三酯酶样羧酸酯酶（PLC）。通过将谷氨酸突变为赖氨酸，羧酸酯酶活性完全丧失，而磷酸二酯酶活性成为其主要活性（Mandrich and Manco 2009）。

近年来，许多研究集中在改善已知 OPH 酶的催化效率。例如，已经采用了几种定向进化策略来提高 *bd* PTE 对硫代磷酸酯底物如甲基对硫磷和毒死蜱的水解活性，它们相对于对氧磷的水解速度降低至 $\frac{1}{30}$ 和 $\frac{1}{1000}$（Cho et al. 2002，2004；Dumas et al. 1989a）。通过基于 DNA 混编的突变策略，得到了一株对甲基对硫磷水解活性提高（25 倍）的 *bd*PTE（22A11）突变株（Cho et al. 2002）。

通过使用进化的突变体（22A11）作为模板，将新的诱变方法与毒死蜱降解菌筛选结合，并筛选鉴定出比野生型降解活力高 725 倍的毒死蜱降解菌的突变体（B3561）[k_{cat}/K_m 值为 2.2×10^8 L/(mol·s)]（Cho et al. 2004）。一些替换（A80V、I274N 和 K185R）在两种突变体（B3561 和 22A11）中均有发现，导致构象发生变化进而引起活性位点拓宽和促进在无金属状态下的稳定性（Cho et al. 2004，2006）。

许多研究集中在改变 *bd*PTE 的立体选择性，以使酶对有机磷化合物的 S_P 有毒异构体更具活性。结构研究分析了 *bd*PTE 的三个结合口袋（小、大和解离基团），负责与底物磷中心的取代基相互作用（Chen-Goodspeed et al. 2001a，b）。对 *bd*PTE 立体选择性的决定因素的深入研究已经证明，对手性底物的偏好由小的亚位点的大小决定。实际上，甘氨酸残基被丙氨酸取代减少了小的亚位点体积，从而导致对 S_P 对映异构体偏好的增加（Chen Goodspeed et al. 2001b）。相反，导致小的亚位点扩大的突变（I106A、F132A 和 S308A）降低了对 S_P 异构体的偏好（Chen-Goodspeed et al. 2001b）。

从几个研究中选择收集优先突变体（K185R、H254Q、H254G、H257F、H257W、I274N 和 L303T），用来设计诱变文库并产生有效的 *bd*PTE 突变体。通过对这些文库的筛选，筛选出了 GWT-F5 突变体，显示其对一种 S_P 类似物的神经性毒剂（索曼、塔崩、沙林、环孢菌素、VX）的催化效率范围在 10^2 L/(mol·s) 和 10^6 L/(mol·s) 之间（Benning et al. 1995；Madej et al. 2012；Tsai et al. 2012）。

最近的一项工作也凸显了体外进化的作用，黄杆菌属 OPH 中的 9 个氨基酸定点突变（A80V、I106V、F132D、K185R、D208G、H257W、I274N、S308L 和 R319S）构建的突变体（M9），对 VX 的催化速率相较于野生型增加了 35 倍（Jeong et al. 2014），并通过预测获得 M9 的结构，然后与 VX 底物模拟结合。从这一分析中，确定了两个额外的关键残基（L271 和 Y309），如果用丙氨酸替代，显示出比 M9 高 5 倍的活性（Jeong et al. 2014）。

体外进化的作用也表现在其他有机磷降解酶中。例如，为了增加农杆菌 OpdA 对马拉硫磷的催化活性，研究人员使用了组合活性位点饱和测试（CASTing）（Naqvi et al. 2014）。在该诱变方法中，分析蛋白质三维结构以便在结合口袋中选择几个氨基酸组合，同时随机化产生容易筛选的小诱变文库。用于 OpdA 的马拉硫磷水解活性的 CASTing 策略可以鉴定双突变体（S308L/Y309A），其显示出活性位点的拓宽和对马拉硫磷的催化效率增加了 5000 倍，达到了催化该底物活性报道的最高值（Naqvi et al. 2014）。

16.2.2　甲基对硫磷水解酶

一个独特的有机磷降解途径由甲基对硫磷降解基因（*mpd*）表示，其具有对甲基对氧磷、甲基对硫磷和毒死蜱的水解活性。基于其类似金属-β-内酰胺酶的结构域，*mpd* 基因的蛋白质产物被命名为甲基对硫磷水解酶（MPH）。在无色杆菌属（*Achromobacter*）、苍白杆菌属（*Ochrobactrum*）、嗜麦芽寡养单胞菌（*Sternotrophomonas*）和假单胞菌属（*Pseudomonas*）中均发现了 *mpd* 基因，但是这些基因都未显示与 *opd* 基因或其他有机磷降解基因具有同源性（Cui et al. 2001；Yang et al. 2006；Zhang et al. 2005）。

近些年，由于全基因组测序工程，许多推测的 *mpd* 同系物被发现并克隆，系统发育分析揭示了这些基因是从 *opd* 基因分离进化而来的。

从中国土壤细菌中分离出许多 *mpd* 基因（Liu et al. 2005；Zhang et al. 2005，2006；Zhongli et al. 2001），表明 *mpd* 进化受环境影响。来自苏云金芽孢杆菌（*Bacillus thuringiensis*）的属于 β-内酰胺酶超家族的酰基高丝氨酸乳酸酶（AHL lactonase）显示出一些不同的有机磷酸酶活性，这表明 OPH 和 MPH 可能从不同的内酯酶中产生（Afriat et al. 2006）。

来自假单胞菌 *Pseudomonas* sp. WBC-3 的 MPH 酶是每个亚基含有混合双核锌中心的二聚体，其中一个锌离子通常被镉取代（Dong et al. 2005）。已经证明在活性位点入口处的三个芳香族残基在测定对甲基对硫磷的亲和力中具有关键作用，事实上，在这些位置的任何取代将导致在测试底物上的催化活性的明显缺失（Dong et al. 2005）。

此外，来自假单胞杆菌 *Pseudomonas pseudoalcaligenes* 的新鉴定的 MPH（命名为 OPHC2）显示出意想不到的耐热性和广泛的底物活性。OPHC2 催化内酯、不同的磷酸三酯和酯的水解（Gotthard et al. 2013）。其高 T_m 值（97.8℃）可能是由于存在扩展的二聚化表面和分子内二硫键（热稳定蛋白质典型特征），使得 OPHC2 成为 OP 去污的良好候选物（Gotthard et al. 2013）。

16.2.3 血清对氧磷酶

在 OPs 代谢中显示关键作用的另一个有趣的酶类是哺乳动物来源的内酯酶/芳基酯酶，通常命名为血清对氧磷酶（PONs）（Draganov 2010）。这些酶是高密度脂蛋白的组成成分，其可能涉及脂质氧化的毒性副产物的失活（Blum et al. 2006；Mackness et al. 2000）。PON 蛋白的结构类似于表现出多种有机磷酸酶活性的两种内酯酶：鱿鱼二异丙基氟磷酸酶（DF-Pase）和人衰老标记蛋白-30（Belinskaya et al. 2012）。

在人体中，发现了以下 PON 突变体：PON1、PON2 和 PON3。大量的研究已经证明，仅 PON1 能够水解杀虫剂中的 P—O 键，例如对氧磷和毒死蜱，以及 G-系列神经性毒剂（Draganov 2010；Josse et al. 2001；Rochu et al. 2007）。

鉴于其在药理学应用中作为生物清除剂的重要性，PON1 酶多次被作为进化研究的模板。PON 蛋白在重组表达系统中是不溶的，因此，通过 DNA 混编来诱变人、小鼠和兔的 PON1 基因，以获得高度可溶的重组酶（Aharoni et al. 2004）。一种突变体 rePON1 被鉴定并用作蛋白质工程研究的起点（Aharoni et al. 2004；Harel et al. 2004；Madej et al. 2012）。对 rePON1 的功能分析鉴定了活性位点中的 L69、V346 和 H115 是作为催化中的关键残基（Harel et al. 2004）。用这些残基的替代物作为靶标改变酶的立体选择性。L69V 和 V346A 的取代增强了索曼（soman）和二异丙基氟磷酸（DFP）的 S_P 异构体催化，而 H115W 取代增加了对 P—S 键的特异性（Amitai et al. 2006；Rochu et al. 2007）。

在另一项研究中，使用良好表征的 PON1 催化位点的组合活性位点饱和诱变获得在环孢菌素上具有更高催化效率的突变体［10^7 L/(mol·min)］。该结果与体内适当的预防性保护抵御实验一致。实际上，小鼠中的保护抵御测定表明，在这些突变体存在的情况下，小鼠在暴露于环孢菌素后，其生存能力增加至 6h（Gupta et al. 2011）。还通过蛋白质工程获得了对索曼和环孢菌素具有高催化效率的 PON1 嵌合突变体，但是，其对于氨基磷酸酯残基的活性仍然非常低，并且在 VX 上不存在（Worek et al. 2014）。其他 PON1 性质演变的成功案例，在最近的 Iyer 和 Iken 的一篇综述文章中也得到了很好的描述（Iyer and Iken 2015）。

16.2.4 微生物脯氨酸肽酶/有机磷酸酐水解酶

除了它们的主要功能，许多能够水解 G-系列神经性毒剂中的 P—O 和 P—F 键的酶被鉴定并被命名为有机磷酸酐水解酶（OPAAs）（EC 3.1.8.2）。特别的，从交替单胞菌属中分离出对 G-系列神经性毒剂具有水解活性的脯氨酸肽酶/OPAA 酶（Theriot et al. 2011；Vyas et al. 2010）。

属于脯氨酸肽酶家族的这些酶被发现存在于所有域中（从细菌到人）（Theriot et

al. 2011)，并且作为新型热稳定模板被用于定向进化研究。尽管在 *bd*PTE 和 PON1 的改造中取得了成功，但是通过诱变策略构建高效有机磷降解脯氨酸肽酶仍未得到充分利用。

OPAA 蛋白与脯氨酸肽酶类似，事实上，它们能够水解在羧基末端具有脯氨酸的二肽，同时也在 OPH 和 MPH 酶的许多底物上具有活性。它们与 OPH 或 MPH 的催化和立体选择性具有较高的相似性（Cheng and DeFrank 2000），但没有显著的序列同源性；现在认为 OPAA 可以由脯氨酸肽酶进化而来（Theriot and Grunden 2011；Vyas et al. 2010）。

在超嗜热古菌激烈火球菌 *Pyrococcus furiosus*（*Pf* prol）和掘越氏火球菌 *Pyrococcus horikoshii*（*Ph1* prol）中鉴定出两种脯氨酸肽酶；它们被鉴定并用作定向进化的模板（Theriot et al. 2010a，b，2011；Theriot and Grunden 2011）。

为了在较低温度下获得高活性的嗜热酶，采用了许多诱变方法。对来自 P. furiosus（Pf*prol*）的脯氨酸肽酶的第一个研究集中在 *35℃*、*50℃* 和 *70℃* 下酶催化活性的增加（*Theriot et al. 2010a*）。通过对 P. horikoshii（Ph*1prol*）的随机诱变，获得在更宽的温度范围内具有高热稳定性和活性增加的脯氨酸肽酶和磷酸三酯酶活性的四种突变体（*A*195*T*/*G*306*S*、*Y*301*C*/*K*342*N*、*E*127*G*/*E*252*D* 和 *E*36*V*）（*Theriot et al*. 2011）。负责提升催化作用的突变主要位于酶的环和连接区中，并且可能在远离活性位点的区域中引起结构改变（*Madej et al*. 2012；*Theriot et al*. 2010*b*，2011）。

迄今为止，来自假单胞菌（Pseudomonas）/短波单胞菌（Brevundiminas）的 bd*PTE*、来自农杆菌（Agrobacterium）的 *OpdA* 和来自哺乳动物的 *PON*1 是研究最深入的有机磷酸盐降解酶。这些酶和它们的突变体显示出对 *OP* 和神经性毒剂的显著活性，但是它们在溶液中的低稳定性限制了它们在有机磷生物传感和修复中的应用。

为此，近年来越来越多的研究者关注从极端微生物中发现的有机磷降解酶（磷酸三酯酶样内切酶）。

16.3　磷酸三酯酶样内酯酶

磷酸三酯酶样内酯酶（*PLL*）家族包括一组对内酯和酰基高丝氨酸内酯（*AHL*）具有主要内酯酶活性的酶，另外还有针对有机磷酸酯化合物（*OPs*）的低的混合磷酸三酯酶活性的酶（表 16.1）。

首先在极端嗜热古菌硫矿硫化叶菌 S. solfataricus（Sso*Pox*）和嗜酸热硫化叶菌 S. acidocaldarius（Sac*Pox*）中鉴定了磷酸三酯酶样内酯酶（*PLLs*）（*Afriat et al*. 2006；*Merone et al*. 2005；*Porzio et al*. 2007），该酶同时也在嗜温生物体如结核分枝杆菌 Mycobacterium tuberculosis（*PPH*）、红球菌红细菌 Rhodococcuserythropolis（*AhlA*）、耐辐射异常球菌 Deinococcusradiodurans（Dro*PH*/Dr0930）（*Afriat et al*. 2006；*Hawwa et al*. 2009*b*）和鸟分枝杆菌副结核亚种 *K*-10 Mycobacterium avium *subsp*. Paratuberculosis（*MCP*）（*Chow et al*. 2009）中发现。后来又报道了来自嗜热脂肪地芽孢杆菌 Geobacillus stearothermophilus（*GsP*）（*Hawwa et al*. 2009*a*）和来自地热芽孢杆菌 Geobacilluskaustophilus（*GKL*/*Gka*P）（*Chow et al*. 2010）的新的热稳定性磷酸三酯酶样内酯酶。最近，已经从 Sulflolobus islandicus（Sis*Lac*）（*Hiblot et al*. 2012*b*）和岩化热古菌 Vulcanisaetamoutnovskia（Vmo*Lac*/*VmutPLL*）鉴定出其他类型的磷酸三酯酶样内酯酶（*Kallnik et al*. 2014）（表 16.2）。

表 16.2 PLLs 的分类、主要结构及动力学信息

分类	微生物	酶的名称	结构（PDB 编号）	序列（UniProKB 编号）	化学计量和金属含量	对对氧磷酶的催化效率① $k_{cat} \cdot K_M^{-1}$ /[L/(mol·s)]	主要参考文献②
	Brevundimonas diminuta	*bd*PTE③	1HZY	P0A434	同源二聚体 Zn^{2+}-Zn^{2+}	4×10^7	Benning et al.(2001)
嗜热古菌	*Sulfolobus sol-fataricus*	*Sso*Pox	2VC5	Q97VT7	同源二聚体 Fe^{3+}-Co^{2+}	3.75×10^3	Afriat et al.(2006)，Merone et al.(2005)
嗜热古菌	*Sulfolobus acidocaldarius*	*Sac*Pox	未解析	Q4J6Z8	—	2.66×10^4	Afriat et al.(2006)，Porzio et al.(2007)，(2013)，Bzdrenga et al.(2014)
嗜热古菌	*Sulfolobus islandicus*	*Sis*Lac	4G2D	C4KKZ9	同源二聚体 Fe^{2+}-Co^{2+}	6.98×10^2	Hiblot et al.(2012b)
嗜热古菌	*Vulcanisaeta moutnovskia*	*Vmo*Lac	4RDY	F0QXN6	同源二聚体 Co^{2+}-Co^{2+}	1.86	Kallnik et al.(2014)，Hiblot et al.(2015)
嗜热/极端细菌	*Deinococcus radiodurans*	*Dr*0930/*Dr*OPH	3FDK	Q9RVU2	同源二聚体 Zn^{2+}-Zn^{2+}	1.39	Afriat et al.(2006)，Hawwa et al.(2009b)，Xiang et al.(2009)
嗜热/极端细菌	*Geobacillus stearothermophilus*	GsP	3F4D	D0VX06	单体 Co^{2+}-Co^{2+}	3.28×10^3	Hawwa et al.(2009a)
嗜热/极端细菌	*Geobacillus ka-ustophilus*	GKL/*GkaP*	3ORW	Q5KZU5	同源二聚体 Fe^{3+}-Zn^{2+}	4.5	Chow et al.(2010)，Zeng et al.(2011)
中温细菌	*Mycobacterium tuberculosis*	PPH	4IF2	P9WHN9	单体 Zn^{2+}-Zn^{2+}	8.65	Afriat et al.(2006)，Zhang et al.
中温细菌	*Rhodococcus erythropolis*	AhlA	2R1N	Q93LD7	单体 Fe^{2+}-Co^{2+}	0.50	Afriat et al.(2006)，Jackson et al.(2008)
中温细菌	*Mycobacterium avium subsp. paratuberculosis K-10*	MCP	未解析	GI：41409766④	—	4.1	Chow et al.(2009)

① 对于乙基对氧磷的特异性常数的报道值已在每组酶的最佳温度下进行测定：25℃，*bd*PTE；70℃，来自超嗜热古菌的 PLL；35℃，来自极端细菌的 PLL；25℃，来自嗜温细菌的 PLL。

② 报道了每种酶的主要参考文献，其中可以找到表中总结的关于结构和催化性质的主要信息。

③ 该表中的 *bd*PTE 仅仅是为了比较其催化活性与 PLL 的催化活性。

④ NCBI 代码。

磷酸三酯酶样内酯酶家族从结构和生物化学的角度看与细菌磷酸三酯酶（PTE）相关（Afriat et al. 2006；Elias et al. 2008）。大多数磷酸三酯酶样内酯酶与鉴定出的磷酸三酯酶 *bd*PTE 具有较低的序列同源性（约 30%）。起初它们大多被鉴定为磷酸三酯酶，因为能够降解杀虫剂如对氧磷所以也被称为“对氧磷酶”（Pox）（Merone et al. 2005；Porzio et al. 2007）。

然而，进一步的结构、系统发育和生物化学研究已经证明，这些酶除了弱磷酸三酯酶活性外还具有明显的内酯酶活性（Afriat et al. 2006）。因此，这些新酶被设计为磷酸三酯酶样内酯酶，并且它们与 *bd*PTE 共享保守双核金属中心的 $(\beta/\alpha)_8$ 桶形折叠结构，该结构对于酶催化是必需的（Seibert and Raushel 2005）。PTE 和 PLL 家族都属于酰胺水解酶超家族，但是一些极端磷酸三酯酶样内酯酶在结构上不同于嗜温 PTE，特别是对于环 7，在极端亲和酶中几乎不存在。而对于环 8，从结构的角度来看其中少数序列的差异则显得更为重要。我们通过使用这些 PLL 的已知 3D 结构执行结构比对，并且确认除了来自红球菌红细菌（*R. erythropolis*）的 AhlA，所有磷酸三酯酶样内酯酶和环 7 中的 *bd*PTE 相比具有很大差异（图 16.1）。

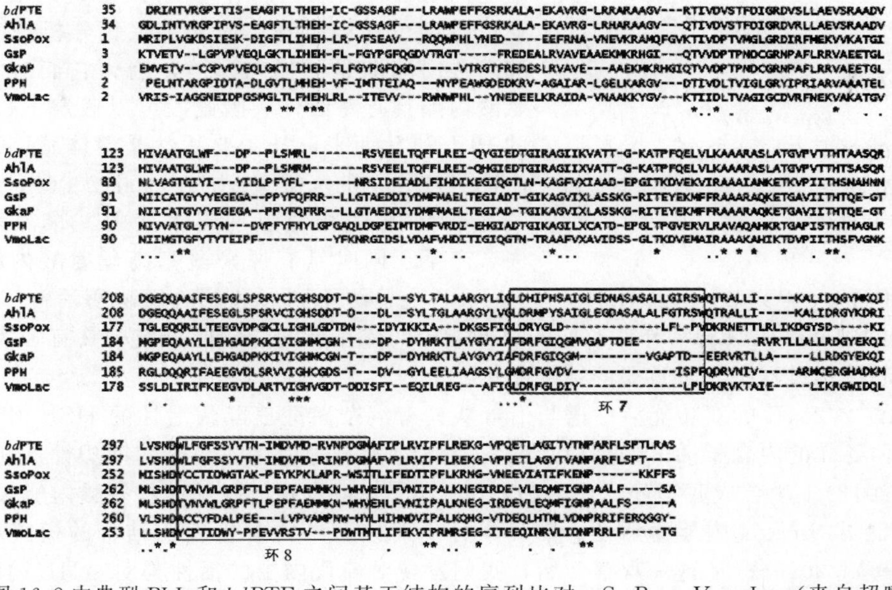

图 16.1　图 16.2 中典型 PLL 和 *bd*PTE 之间基于结构的序列比对。*Sso*Pox，*Vmo* Lac（来自超嗜热古菌）；GsP，*Gka*P（来自极端细菌）；PPH，Ahla（来自嗜温细菌）。结构的比对使用 Swiss-PdbViewer 4.1.0 进行。保守残基用星号标记。环 7 和环 8 由黑色方框表示。构成双核金属中心的保守残基以灰色标记

事实上，得益于 PLL（*Sso*Pox）的 3D 结构的解析，研究发现了非常重要的结构差异（Elias et al. 2008）。古菌磷酸三酯酶样内酯酶的结构显示非典型的活性位点拓扑结构和单一的疏水通道，完全适应内酯的底物，如 AHLs（Elias et al. 2008）。在图 16.2 中，*bd*PTE 的结构与六个 PLL（两个来自超嗜热古菌：*Sso*Pox 和 *Vmo*Lac；两个来自极端细菌：GsP 和 *Gka*P；两个来自嗜温细菌：PPH 和 Ahla）的结构的叠加表明蛋白质的同源性。值得注意的是，所有磷酸三酯酶样内酯酶中的环 7 的长度［除了 AhlA（粉红色）］，均短于 *bd*PTE（灰色）。所有磷酸三酯酶样内酯酶的环 8 具有与 *bd*PTE 几乎相同的长度，但是其拓扑结构对于 *Gka*P（绿色）和 GsP（蓝色）是不同的，其中它比 *bd*PTE 和 *Vmo*PLL 更多地从蛋白质核心分离得到；在后一种情况下，与热稳定 PLL *Sso*Pox（图 16.2）相比，环 8 是刚性的

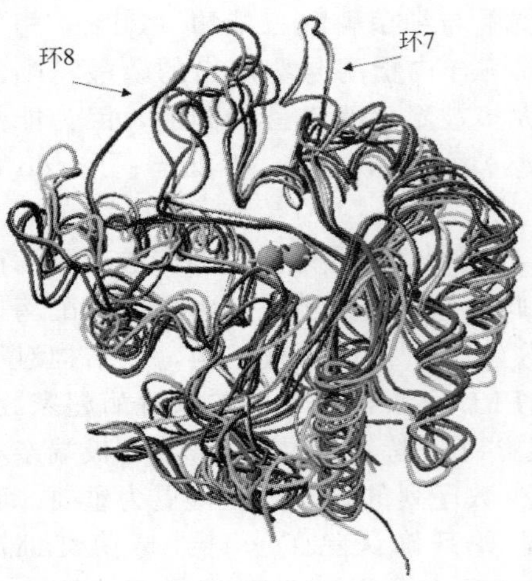

图 16.2 *bd*PTE 与六个 PLLs 的结构叠加（见彩图），结构叠加使用 Swiss-PdbViewer 4.1.0。*B. diminuta*（灰色）来自 *bd*PTE，*Sso*Pox（红色）源自 *S. solfataricus*；GsP（蓝色）为 *G. stearothermophilus*，*G. kaustophilus* 为 *Gka*P（绿色），PPH（青色）为 *M. tuberculosis*，Ahla 为 *R. erythropolis*（粉色）；金属阳离子为灰色球体

并且构造形成 α-螺旋。一些研究已经假定 *bd*PTE 中环 7 的延伸形成短的 α-螺旋，其可以提供使活性位点口缩窄的活性“帽”，从而可以增加对底物对氧磷酶的磷酸三酯酶活性（Afriat et al. 2006）。

所有磷酸三酯酶样内酯酶具有同质或异质双核金属中心，其产生参与 OPs 和内酯的水解和催化氢氧根离子的反应。两种水解活性的主要区别是过渡态几何结构：磷酸三酯的 sp^3 配位和内酯的五配位状态（Elias et al. 2008）。在相同活性位点中两种活性的共存表明二者有相同的催化机制，优化与环境适应相关的底物特异性。

磷酸三酯酶样内酯酶的活性位点包含非常适合于内酯结合的三个亚位点：小亚位点、大亚位点和疏水通道（Elias et al. 2008）。内酯的脂肪链在疏水通道内结合，大亚位点容纳 *N*-酰基链的酰胺基，小亚位点定位在内酯环。此外，生化分析证实磷酸三酯酶样内酯酶是天然内酯酶，因为它们对酰基高丝氨酸内酯具有高活性。因此，可以假定 PTE 从未知的 PLL 进化而来，其主要活性是群体感应高丝氨酸内酯（HSL）的水解（Afriat et al. 2006）。最近报道了与 *bd*PTE 的相邻点一起突变的九个氨基酸的缺失，产生了同时具有原来酰基高丝氨酸内酯酶活性（k_{cat}/K_m 值高）（PTEΔ7-2/254R）（在 *bp*PTE 中检测不到）和低对氧磷酶活性（Afriat et al. 2012）的中间体。具有特异性改变的双功能中间体在新的酶功能分歧中具有强大的作用，这是由于环重塑过程中非常重要的结构变化。

PTEΔ7-2/254R 的双功能活性谱与沿着从具有混杂磷酸三酯酶活性的 HSL 引导至具有弱的、各种各样的内含子活性的高效磷酸三酯酶的轨迹的进化中间体的期望性质相符（Afriat et al. 2012）。用合成底物（硫代内酯）（图 16.3）或真正的酰基高丝氨酸内酯（Afriat et al. 2006）测试 *Sso*Pox 内酯酶活性。然而，分子对接和结构分析表明结合的模式可能不同（Merone et al. 2010）。从这一观察开始，我们发现对硫代内酯的活性与对 AHL 内酯的活性有很大不同，因为特异性常数随着酰基侧链的增加而降低（图 16.3）。

AHL 也称为自体诱导物 1（AI-1），是一类小信号分子，主要由革兰氏阴性细菌产生，并介导称为群体感应（QS）的细胞与细胞之间的交流。当这些分子在细胞外环境中累积直到临界阈值时，细菌的转录谱被改变（Hentzer et al. 2003）。

据报道，一些病原体的毒力和生物膜形成由 QS 调节（Costerton et al. 1999；Dickschat 2010；Jones et al. 2010；Popat et al. 2008），研究表明这种机制的猝灭（群体猝灭）可能是一个针对使用 AHL 的群体感应如铜绿假单胞菌的多重耐药病原菌的很好的策略（Amara et al. 2011；Dong et al. 2000，2001；Ma et al. 2009；Reimmann et al. 2002）。因此，作为内酯酶，磷酸三酯酶样内酯酶可以水解 AHL 并猝灭 QS 机制，如苏云金芽孢杆菌（AiiA），人对氧磷酶（human paraoxonases）（Dong et al. 2001；Ma et al. 2009）和 *Sso*Pox，*Sac*Pox 和 *Dr*0930，能够减少群体多胺氧化酶（PAO）活动力（Mandrich et al. 2011）。此外，*Sso*Pox 最近还显示可在体外降低毒力因子表达和生物膜形成，并降低大鼠肺部感染模型中

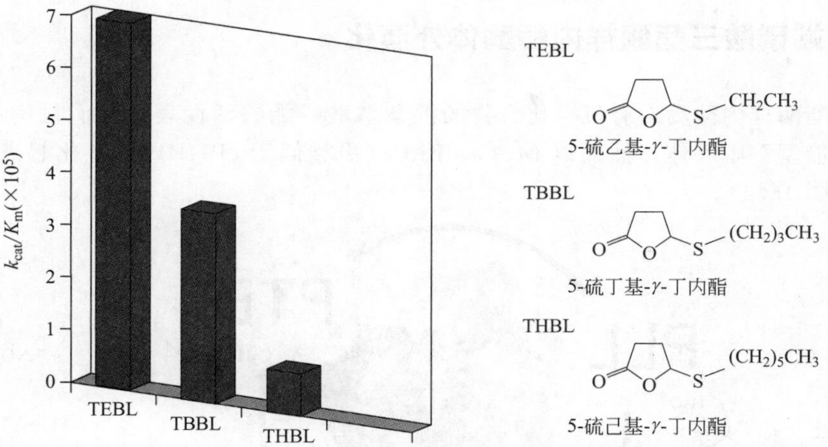

图 16.3　*Sso*Pox 对不同酰基链长度的硫代内酯的内酯酶活性。该测定由 Afriat 等（2006）在 70℃下进行测定。*Sso*Pox 在 TEBL，TBBL 和 THBL 上显示的 k_{cat}/K_m 分别为 7×10^5 L/(mol·s)，3×10^5 L/(mol·s) 和 0.8×10^5 L/(mol·s)

的大鼠死亡率（Hraiech et al. 2014）。

考虑到它们同时具有内酯酶和磷酸三酯酶活性，磷酸三酯酶样内酯酶在生物技术上的应用通常可以作为群体猝灭剂（Dong et al. 2000）或 OPs 生物隔离剂（Singh 2009）的候选物。

然而，目前仍不清楚它们的生物功能是什么，因为在许多情况下不产生 AHL 或由 AHL 调节的微生物中发现磷酸三酯酶样内酯酶，因此，可能它们可以作用于其他微生物的通信机制。此外，磷酸三酯酶样内酯酶家族的一些成员有效地水解 γ 和/或 δ 氧代内酯，但不水解 AHL（Chow et al. 2010；Xiang et al. 2009）。无论如何，对这些酶的应用，特别是在净化领域中的生物技术应用，在于探索磷酸三酯酶活性。在不同的程度上，大多数极端磷酸三酯酶样内酯酶能够水解不同的有机磷农药。作为代表性化合物，报道了所述磷酸三酯酶样内酯酶对乙基对氧磷的催化效率（表 16.2）。在古菌磷酸三酯酶样内酯酶中，*Sso*Pox 对对氧磷的混杂催化活性为 3.75×10^3 L/(mol·s) 是 70℃时的 *Sac*Pox［2.66×10^4 L/(mol·s)］的 $\frac{1}{10}$（Merone et al. 2010；Porzio et al. 2013），高于 70℃下 *Sis*Lac 催化效率［6.98×10^2 L/(mol·s)］（Hiblot et al. 2012b）。

*Sso*Pox 和 *Sac*Pox 均为二聚体形式，这与 *Sis*Lac、*Dr*0930、GsP 和 *Gka*P 的晶体结构一致（表 16.2）。

最近发现来自 *V. mountnovski*（*Vmo*Lac）的超极性 PLL，与 *Dr*0930［1.39L/(mol·s)］（Hawwa et al. 2009b；Mandrich et al. 2013）和 G*ka*P 相比［在 35℃为 4.5L/(mol·s)，在 75℃为 1.1×10^2 L/(mol·s)］对对氧磷催化效率较低，为 1.86L/(mol·s)（Hiblot et al. 2015），即使它们都是高度稳定的酶。在其他极端细菌磷酸三酯酶样内酯酶（PLLs）中，来自 *G. stearothermophilus* 的 GsP 在对氧磷上具有最高的 k_{cat}/K_m 值，在相同的可比温度（35℃）下为 3.28×10^3 L/(mol·s)（Hawwa et al. 2009a）。所有中温磷酸三酯酶样内酯酶，PPH、AhlA 和 MCP，表现出弱的磷酸三酯酶活性。对于通过 AhlA 的对氧磷的水解，评估 k_{cat}/K_m［0.5L/(mol·s)］，MCP 和 PPH 分别表现出更高的值，分别为 4.1L/(mol·s) 和 8.6L/(mol·s)（Afriat et al. 2006；Chow et al. 2009）。

16.4 极端磷酸三酯酶样内酯酶体外演化

磷酸三酯酶样内酯酶对有机磷化合物的混杂磷酸三酯酶活性表明它们在构成 PTE 时可能产生“全能型”中间体。然而以 *M. loti* PLC（更类似于 ePHP）为进化起点的可能性不能被否定（图 16.4）。

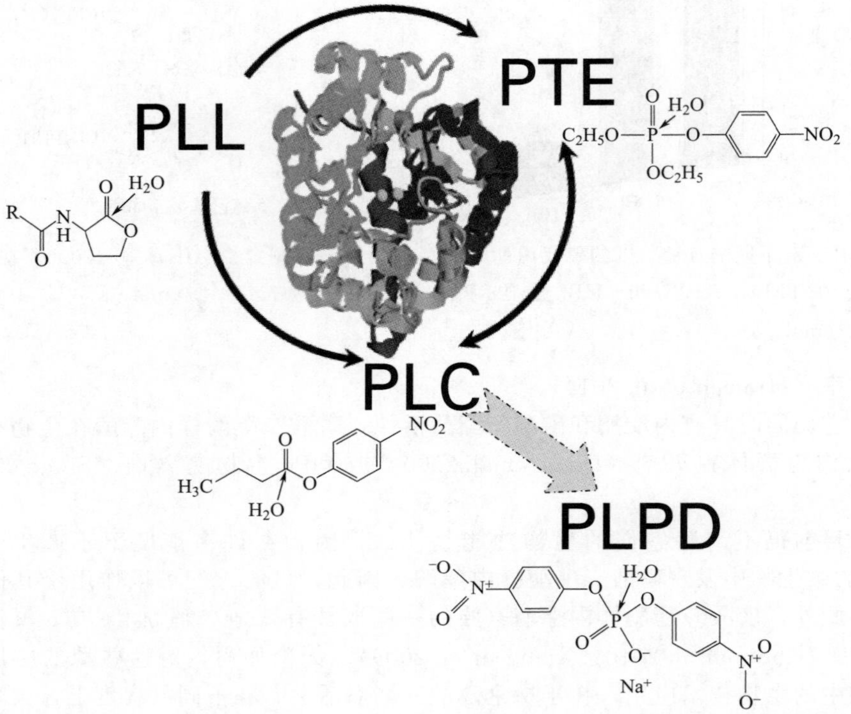

图 16.4 酰胺水解酶超家族成员之间可能的进化路径。PLL：磷酸三酯酶样内酯酶；PTE：磷酸三酯酶；PLC：磷酸三酯酶样羧酸酯酶；PLPD：磷酸三酯酶样磷酸二酯酶，由 PLC 通过单位点突变产生

最近，各种诱变策略，如定点诱变和定向进化，已被用于增强多个磷酸三酯酶样内酯酶的混杂有机磷降解活性。这种方法有效地重塑酶的催化效率和特异性，并为磷酸三酯酶样内酯酶的有机磷去污的演化提供了有力证据。

PLL 和 PTE 酶之间的重要区别在于环 7 和环 8 具有不同的长度和拓扑结构。如上所述，酰胺水解酶超家族中的两个环通常与底物相互作用，并且在底物特异性的测定中起重要作用（Seibert and Raushel 2005）。*Sso*Pox 蛋白和溶剂工程的关联导致动力学参数（k_{cat} 和特异性常数）更接近 PTE，但是它明显更加稳定，因此它代表一个有趣的脱毒装置（Merone et al. 2010）。*Sso*Pox 的活性位点结构显示出对温度变化的低弹性，因为不同突变产生不同的活性/温度曲线（Merone et al. 2010）。

这意味着在活性位点中，在保证硬度需求和耐受高温上达到一定的平衡，同时，灵活性是催化活性的先决条件（Tehei et al. 2005）。详细地说，突变体特征 Y97W、I98F、Y97W/I98F 与野生型非常相似，即使它们的活化作用不同。如在 Elias 等（2008）的报道，单突变体 Y97W 表明位点 97 参与底物的定向催化。突变体 W263F 作为最佳突变体出现，但是所有突变体都能观察到启动效应（表 16.3）。W263F 对乙基对氧磷的降解效率提高了 6 倍，并且也显示出对环磷酰胺试剂的水解速率略有提高（Merone et al. 2010）。

表 16.3 体外进化获得 PLL 突变体的对氧磷酶活性

PLL	重要的鉴定残基及在结构中的相关的位置及作用	与进化相关的变异	对对氧磷催化效率的影响(k_{cat}/K_m)	参考文献
Sso Pox	M79 Y97(与底物相互作用) I98 I261(环 8) W263(环 8)	M79V Y97W I98F W263Q W263F I261F M79V/W263Q I98F/I261F Y97W/I98F Y97W/I261F Y97W/W263Q Y97W/W263F Y97W/	从 1.7 倍的野生酶的活性增加到 6 倍	Merone et al. (2010)
		W263L,M,F,D,I,P,Y,V,A,C,E,G,Q,N,S,H,T,R,K	从 1.7 倍野生酶的活性增加 23 倍	Hiblot et al. (2013)
Sis Lac	E14 Y34	E14K Y34H E14K/Y34Q	从 3 倍野生酶的活性增加 17 倍	Hiblot et al. (2012b)
Dr 0930	F26(小的亚位点) T68(小的亚位点) Y97(金属离子结合) Y98(离去基团) G207 R228(大的亚位点) W269(小的亚位点) W287(大的亚位点) D71(离去基团) A85 E101(离去基团) V35 A43 A44 V64 V65 N75(大的亚位点) F26(小的亚位点) M234(大的亚位点) Y28A(小的亚位点) S48 V235(大的亚位点) R272 A103 L231(大的亚位点)	F26G T68S Y97W Y98F G207D R228A W269A W287G D71G/E101G D71G/A85G/ V165M V35L/A43T/ A44G/V64G/ V65G/D71G/ N75T/E101G F26G/M234R Y28A/Y97W, Y28C/Y97W, Y28G Y28W/S48N/ Y97C D71G/E101G/ M234I D71G/ E101G/M234L D71G/E101G/ V235L D71G/E101G/ R272C/W287R Y97G Y97I A103V L231D L231R/W287G	从 1.5 倍野生酶的活性增加 559 倍	Hawwa et al. (2009b)
		Y28L/D71N/ E101G/E179D/ V235L/L270M	增加到野生酶活性的 100 倍	Meier et al. (2013)

续表

PLL	重要的鉴定残基及在结构中的相关的位置及作用	与进化相关的变异	对对氧磷催化效率的影响(k_{cat}/K_m)	参考文献
Gka P	E15 F28(小的结合口袋) D73 Y99(在离去基团的亚位点) T171 F228(在环 7 的底部) N269(环 8 的内部) V270(环 8 的内部) W271(环 8 的内部) G273(环 8 的内部) F276 I299	Y99L D73Y/Y99L E15V/Y99L/G273D/F276V/I299S Y99L/T171S/V270G/G273D Y99L/W271C F28I/Y99L/T171S/F228L/N269S/V270G/G273D F28I/D73Y/Y99L/T171S/F228L/N269S/V270G/G273D F28I/Y99L/T171S/F228L/N269S/V270G/W271C/G273D F28I/D73Y/Y99L/T171S/F228L/N269S/V270G/W271C/G273D	从 3 倍野生酶的活性增加 230 倍	Zhang et al. (2012)

据报道，W263 残基在 *Sso*Pox 活性位点的底物结合中，在环 8 中二聚体界面处起重要作用，并影响蛋白质的稳定性、酶活性和杂交活性（Hiblot et al. 2013）。对 *Sso*Pox 的最新结构分析显示关于 W263 残基的所有 19 个不同的替代，通过在环 8 定位中引起非常微妙的变化，对非常短的环 7 没有影响，从而提高 *Sso*Pox 对氧磷酶活性（Hiblot et al. 2013）（表 16.3）。

Hiblot 等（2012b）发现，在 *Sis*Lac 和 *Sso*Pox 结构分析的基础上，位点 14 和 34 的变化可能对两种结构之间有很大的影响，可能影响二聚变化（*Sis*Lac 的位点 34）或稳定性（*Sis*Lac 的位点 14、34）。突变体 E14K 和 Y34Q 的 T_m 的分析显示单取代使 *Sis*Lac 不稳定，但它们的组合有助于恢复部分稳定性（Hiblot et al. 2012b）。另外，通过不同底物对设计合理的突变体活性进行测试，观察到突变体（E14K）的活性是野生型 *Sis*Lac 的 17 倍（Hiblot et al. 2012b）（表 16.3）。

为了改善 *Dr*0930 的有机磷酸酶活性，在乙基和甲基对氧磷上，使用了定点诱变、随机诱变和位点诱变的方法。筛选出超过 30000 个有潜力的突变体，并且 26 个突变的酶被纯化并在动力学上表征（Hawwa et al. 2009a）。总的来说，所有的诱变结果都提高了对氧磷的特异性，最高的是野生型的 559 倍（表 16.3）。

这项研究强调，通过使用迭代方法进行诱变，对已经活跃的突变体进行突变时，有可能实现较大的催化速率增加。此外，在动力学参数上的变化影响酶结构和功能信息（Hawwa et al. 2009b）。另外，其他作者描述了合理的蛋白质设计和随机诱变方法，结合有效的体外筛选方法以鉴定具有增强降解能力的高效水解酶突变体（Meier et al. 2013）。最佳的突变体显示比野生型 *Dr*0930 高出两个数量级以上的催化活性（表 16.3）。

关于芽孢杆菌属（*Geobacillus* sp.），通过合理组合和随机诱变策略，经过四轮筛选后从地芽孢杆菌*G. kaustophilus* 中的*Gka*P 成功获得新的突变体。对乙基对氧磷的最佳突变体的催化效率（k_{cat}/K_m）比 D71G/E101G/V235L *Dr*0930 突变体高约 40 倍，与突变体 W263F *Sso*Pox 相当（Zhang et al. 2012）。

*Gka*P 中 8 个氨基酸的变化对乙基对氧磷的催化效率提高了近两个数量级。在八个突变（F28I、Y99L、T171S、F228L、N269S、V270G、W271C 和 G273D）中，其中一些取代位于远离催化活性 β-金属离子结构域的位置，而其他位于催化位点附近。位于双核金属中心入口处并被认为位于远离基团位点的 Y99 是 PLL 系列中保留的残基，可能在稳定内酯环中起关键作用（Elias et al. 2008）。*Gka*P 突变体在热稳定性和催化活性之间呈现强烈的反比关系，即使它们是非常耐热的（Zhang et al. 2012）。

除了体外进化方法外，值得注意的是，活性位点中的金属离子在影响动力学参数方面具有重要作用，如 *bd*PTE（Rochu et al. 2004）。最近我们发现了 *Sac*Pox 的杂交对氧磷酶活性似乎比双核金属中心对金属物质的主要内酯酶活性更敏感（Porzio et al. 2013）。事实上，在 Mn^{2+}（*Sac*Pox-Mn^{2+}）制备的酶相对于用 Cd^{2+}（*Sac*Pox-Cd^{2+}）制备的酶，分别显示对对氧磷或甲基对氧磷效率增加约 30 倍和 19 倍。这一发现表明，一种新的有效的方法可用于增加磷酸三酯酶杂交活性在生物技术的应用。

最后，功能和结构同源性表明 PTE 从一个未知的 PLL 进化，使用其混杂对氧磷酶活性作为基本起点。如 *Sso*Pox 的性质所证明的，PLL 的对氧磷酶活性可以相当高，即使不损害内酯活性，同时也可能获得其他活性如芳基酯酶活性。在这方面，*Sso*Pox 类似于 PTE 可能出现的“全能型”中间体（Aharoni et al. 2005；Afriat et al. 2006）。

如 *bd*PTE 所示，最近的磷酸三酯酶专业化可能通过插入环 7 来完成，并没有完全消除原来的内酯酶活性（Afriat et al. 2006）。在 *bd*PTE 和磷酸三酯酶样内酯酶中观察到的混杂羧酸酯酶和磷酸二酯酶活性，这反过来可以为预测具有混杂酶活性的 PTE 样酶铺平道路。已经发现，来自 *Mesorhizobium loti* 的 ePHP 的同系物是具有混杂磷酸三酯酶活性的有效羧酸酯酶，因此被称为 *Mlo*PLC（磷酸鸟苷酶类羧酸酯酶）（Mandrich and Manco 2009）。它还显示出低混杂的磷酸二酯酶和内酯酶活性，考虑到它维持这两种混杂的活性，我们认为它可能源自祖先的 PLL 样或类似 PTE 的酶（图 16.4）。此外，*Mlo*PLC 的发现支持了从酰胺水解酶超家族中的杂交活性开始的酶进化理论（Afriat et al. 2006；Aharoni et al. 2005；Khersonsky et al. 2006）。然而，通过单一突变将 *Mlo*PLC 转化为优异的磷酸二酯酶，同时完全丧失羧酸酯酶活性，这是通过双核金属中心的稳定化的底物辅助增益功能的一种罕见情况（Mandrich and Manco 2009）。这种情况让我们猜测在 PTE 和 PLL 之间存在一个类似于 PLC（磷脂酶 C）的新中间体，在序列/结构重排进化过程中，可能会出现磷酸二酯酶活性（图 16.4）。所获得的突变体显示羧酸酯酶和内酯酶活性的完全丧失（Mandrich and Manco 2009），表明磷酸二酯酶类（PLPD）可能代表这一进化路线的终点，猜想可能在 PLC 之后存在其他的中间体。

16.5 进化型有机磷降解酶在生物技术上的应用潜力

关于进化型降解有机磷酶数据的快速积累导致了一些生物技术应用新策略的兴起，例如治疗和预防有机磷中毒、有机磷生物监测和有机磷修复等。

16.5.1 治疗和预防有机磷中毒

人体的 PON1 为生物体提供了一些天然保护，不受一些有机磷酸盐（如对氧磷）的低剂量摄入的影响，但在摄入高剂量致命的有机磷酸盐（如化学制剂）的情况下，效能不佳。因此，*bd*PTE、OpdA 和 PON1 的突变体，已经测定并用于 OP 农药和神经药物的体内预防的研究（Masson and Rochu 2009）。尽管这些研究取得了一些成果，但其要应用到治疗上，还需要更深入研究其有效的运输和传递方法，以减少酶活性的降低或完全丧失。此外，关键问题还涉及其表达、药物代谢动力学和伦理问题。

通过这些酶的聚乙二醇化来实现热稳定性的改善和小鼠平均生存时间的延长（从几分钟到几天）（Trovaslet-Leroy et al. 2011；Nachon et al. 2013）。

16.5.2 有机磷生物监测

虽然 ELISA 试剂盒可商业用于检测少量 OPs，但近年来，酶传感器的深入研究已用于快速检测有机磷（Carullo et al. 2015）。

目前使用的有机磷生物传感器是基于抑制的而不是基于催化的。酶抑制需要定期更换酶成分，并通过硫代胆碱、过氧化氢或对氨基苯酚的电流检测来测量（Wanekaya et al. 2008）。这些化合物是通过乙酰胆碱酶（AChE）水解乙酰胆碱（硫代胆碱）或乙酸氨基苯酯（对氨基苯酚）得来的，或者通过胆碱氧化酶水解胆碱氧化（过氧化氢）制备的。

其他有机磷生物传感器的使用是基于乙酸还原引起 pH 变化的电位转换。

基于 AChE 的生物传感器是敏感的，但是具有许多缺点，例如由于 AChE 被几种神经毒素（氨基甲酸酯、重金属）抑制，无法通过活化剂，如吡啶 2-醛肟再生而重复使用，并且不适合实时监控。随着时间的推移，稳定性也是一个问题。近期的研究使用来自酸热脂环酸杆菌（*Alicyclobacillus acidocaldarius*）的热稳定性 EST2 作为 AChE 替代品（Febbraio et al. 2011）。

使用基于催化的系统（通常使用 OPH）对有机磷生物传感非常有吸引力，因为它们具有广泛的底物识别性，并且还可以通过电化学和光学方法定量检测水解产物（对硝基苯酚）（Wanekaya et al. 2008 ）。

使用较高离子强度的缓冲液（其中使酶在所有过程中显示最大活性的措施）代表了使用光学生物传感器而不是电位传感器的主要优势（Wanekaya et al. 2008）。

例如，一个典型的应用是 *bd*PTE 的突变体（H254R/H257L）结合碳纳米管对 V 型神经药物和硫赶式甲基内吸磷（demeton-S）杀虫剂进行灵敏、快速、选择安培电流检测时，显示出对 P—S 键水解的催化速率增加（Joshi et al. 2006）。

16.5.3 有机磷修复

近年来，人们强烈反对使用有污染的化学方法修复 OPs，加快了替代环保设备的研究和开发。因为存在许多问题（生长对温度的依赖性、生长抑制剂如有毒化合物和一些代谢物的存在等），导致微生物的使用具有有限的适用性，因此酶是一种可行而又良好的选择。

已经报道了降解酶在修复中应用的许多实例。

由外膜代表的渗透性屏障的存在阻碍了细胞中 OP/OP 降解酶相互作用（Richins et al. 1997）。为了克服这个问题，使用了不同的表面锚定基序，例如 Lpp-OmpA 嵌合体、冰核蛋白（INP）或自动转运蛋白将这些酶定位到细胞表面（Li et al. 2008；Richins et al. 1997；Shimazu et al. 2001）。

Yang 等（2013）通过使用 INP 锚定基序显示出 *Sphingobium japonicum* UT26 细胞表面的 PTE-GFP 融合功能。含有融合物的菌株显示了在生物修复的时间过程中水解 OPs 并跟踪荧光的能力。

相同的锚定基序成功地用于在大肠杆菌的细胞表面上显示出来自 *B. diminuta* 的改良的

OPH 突变体（Tanga et al. 2014）。该工程生物催化剂在已经发现的 OPH 菌株中显示出最高的对氧磷酶活性，最适温度为55℃，具有高稳定性（室温下 1 个月时为 100%活性）（Tanga et al. 2014）。

Pinjari 等（2013）在假单胞菌属细菌膜上异源表达了 *bd*PTE。Ind01 是从农药生产场所的活性污泥中分离的菌株（Pinjari et al. 2012）。该工程菌株在不同的市售有机磷杀虫剂的修复中非常有效（Pinjari et al. 2013）。

许多方法已被用于固定、包封或捕获有机磷降解酶用于产生和保留有机磷酸酶活性的材料。在最近的一篇论文中，将载玻片上的介孔薄膜用作酶修复剂的宿主基质。这些生物催化剂具有活性，非常敏感（对氧磷浓度低至 15μmol/L），而且易于制备，重复使用数次催化活性无明显损失（Francic et al. 2014）。

Gao（2014）报道了 OpdA 酶在高度多孔的非织造聚酯织物上通过共价固定用于降解 OP。首先用乙二胺活化这些物质，然后用双功能交联剂戊二醛固定酶。在酶固定后，甲基对硫磷的 K_m 增加，pH 曲线变宽，酶稳定性增强。在批量模式下，这种修复系统可以在非缓冲溶液中水解 20μmol/L 甲基对硫磷（Gao et al. 2014）。

虽然有这些应用，但许多数据表明，在高达 35℃的温度下，*bd*PTE 逐渐失活，直到 60℃下完全丧失催化活性（Caldwell and Raushel 1991），所有试图固定酶的方法均与短期稳定性、高成本和可能变化的动力学性质有关（Braatz 1994；Caldwell et al. 1991；Gill and Ballesteros 2000；Grimsley et al. 2001；LeJeune et al. 1997）。

因此，找到替代的高抗性和稳定的酶是至关重要的。如之前描述的，已经发现磷酸三酯酶活性是在不同极端宿主分离的几种水解酶如脯氨酸酯酶、PLL 和 MBL 酶（金属 β-内酰胺酶）中的混杂组分。极端细菌和古菌酶的特殊热稳定性在工业过程中提供了许多生物技术优势，例如耐苛性条件（溶剂或洗涤剂的存在）、最小化的污染潜力、极端的稳定性和酶的寿命增加。此外，在嗜温宿主中表达酶的纯化成本是可承受的，实际上，通过简单的热分馏就可以沉淀源自宿主的污染蛋白质。

由于蛋白质稳定性与进化之间的联系（Bloom et al. 2006），有报道高稳定支架将用于下一代工程技术。

为此，关于使用极端微生物磷酸三酯酶样内酯酶作为修复装置已经付出了许多努力。迄今为止，仅报道了实际应用磷酸三酯酶样内酯酶的两个例子，即使其中一个并不适用于有机磷修复。事实上，第一份报告涉及 *Sso*Pox 对酰基高丝氨酸内酯的内酯酶活性的使用（Ng et al. 2011）。该酶被吸收到纳米氧化铝膜上，为了干扰群体感知信号，用固定化酶处理细菌培养物，结果揭示了生物技术应用的前景，例如，可以使用固定在过滤膜上的内酯来控制水净化系统中有害的微生物活动（Ng et al. 2011）。

在第二次报道中，评价了 *Sso*Pox 的突变体（W263F）作为解毒工具的优点。将酶溶解在不同的缓冲水溶液（30% 乙醇，30% 或 50% 甲醇和 0.1% 十二烷基硫酸钠）中以分析其在应激变性环境下的活性，例如通常从污染土壤中提取毒素（Merone et al. 2010；Hiblot et al. 2012a）。在这种情况下，将结果与来自 *B. diminuta* 的 *bd*PTE 获得的结果进行比较。在大多数测试条件下，W263F 优于 *bd*PTE，在室温下 15min 内，约 99.5%的对氧磷在 30%甲醇和乙醇和 0.1%十二烷基硫酸钠（Merone et al. 2010）中水解。

16.6 结论和未来前景

普遍接受的理论是，热稳定酶是体外进化的最佳候选物，因为它们能够接受不稳定的突变，这可能导致新的或改进的酶功能出现。这里报道的调查概述了磷酸三酯酶样内酯酶的体外进化潜力及其在有机磷解毒中的成功应用。众所周知，对氧磷/对硫磷水解产生的对硝基苯酚是有毒的（是原始有机磷的$\frac{1}{120}$）（Munnecke 1979），但是，在过去几年中，研究了能够使用对硝基苯酚作为碳源的不同土壤细菌，提供了完全消除这些化合物毒性的可能（Spain and Gibson 1991 ）。

我们描述了许多具有活性显著改善的有机磷降解酶的应用实例，其与可以利用 PNP 的细菌结合使用以使有机磷化合物完全降解（Shimazu et al. 2001）。

致谢

这项研究得到意大利大学和研究部（MIUR）的支持（项目 PON01 _ 01585 至 G. M. ）。

参考文献

Aardema H，Meertens JH，Ligtenberg JJ，Peters-Polman OM，Tulleken JE，Zijlstra JG（2008）Organophosphorus pesticide poisoning：cases and developments. Neth J Med 66（4）：149-153

Afriat L，Roodveldt C，Manco G，Tawfik DS（2006）The latent promiscuity of newly identified microbial lactonase is linked to a recently diverged phosphotriesterase. Biochemistry 45：13677-13686

Afriat-Jurnou L，Jackson CJ，Tawfik DS（2012）Reconstructing a missing link in the evolution of a recently diverged phosphotriesterase by active-site loop remodeling. Biochemistry 51：6047-6055

Aharoni A，Gaidukov L，Yagur S，Toker L，Silman I，Tawfik DS（2004）Directed evolution ofmammalian paraoxonases PON1 and PON3 for bacterial expression and catalytic specialization. Proc Natl Acad Sci USA 101（2）：482-487

Aharoni A，Gaidukov L，Khersonsky O，Mc QGS，Roodveldt C，Tawfik DS（2005）The 'evolvability' of promiscuous protein functions. Nat Genet 37：73-76

Amara N，Krom BP，Kaufmann GF，Meijler MM（2011）Macromolecular inhibition of quorum sensing：enzymes，antibodies，and beyond. Chem Rev 111：195-208

Amitai G，Gaidukov L，Adani R，Yishay S，Yacov G，Kushnir M et al（2006）Enhanced stereoselective hydrolysis of toxic organophosphates by directly evolved variants of mammalian serum paraoxonase. FEBS J 273（9）：1906-1919

Belinskaya T，Pattabiraman N，diTargiani R，Choi M，Saxena A（2012）Differences in amino acid residues in the binding pockets dictate substrate specificities of mouse senescence marker protein-30，human paraoxonase1，and squid diisopropylfluorophosphatase. Biochim Biophys Acta 1824（5）：701-710

Benning MM，Kuo JM，Raushel FM，Holden HM（1994）Three-dimensional structure of phosphotriesterase：an enzyme capable of detoxifying organophosphate nerve agents. Biochemistry33：15001-15007

Benning MM，Kuo JM，Raushel FM，Holden HM（1995）Three-dimensional structure of the binu clear metal center of phosphotriesterase. Biochemistry 34：7973-7983

Benning MM，Hong SB，Raushel FM，Holden HM（2000）The binding of substrate analogs to phosphotriesterase. J Biol Chem 275：30556-30560

Benning MM，Shim H，Raushel FM，Holden HM（2001）High resolution X-ray structures of different metal-substituted forms of phosphotriesterase from *Pseudomonas diminuta*. Biochemistry 40：2712-2722

Bloom JD，Labthavikul ST，Otey CR，Arnold FH（2006）Protein stability promotes evolvability. Proc Natl Acad Sci USA 103（15）：5869-5874

Blum MM，Löhr F，Richardt A，Rüterjans H，Chen JC（2006）Binding of a designed substrate analogue to diisopropyl fluorophosphatase：implication for the phosphotriesterase mechanism. J Am Chem Soc 128：12750-12757

Braatz JA（1994）Biocompatible polyurethane-based hydrogels. J Biomater Appl 9：71-96

Buchbinder JL，Stephenson RC，Dresser MJ，Pitera JW，Scanlan TS，Fletterick RJ（1998）Biochemical characterization and crystallographic structure of an *Escherichia coli* protein from the phosphotriesterase gene family. Biochemistry 37：5096-5106

Bzdrenga J，Hiblot J，Gotthard G，Champion C，Elias M，Chabriere E（2014）SacPox from the thermoacidophilic crenarchaeon Sulfolobus acidocaldarius is a proficient lactonase. BMC Res Notes 7：333

Cáceres T，Megharaj M，Venkateswarlu K，Sethunathan N，Naidu R（2010）Fenamiphos and related organophosphorus pesticides：environmental fate and toxicology. Rev Environ Contam Toxicol 205：117-162

Caldwell SR，Raushel FM（1991）Detoxification of organophosphate pesticides using a nylon based immobilized phosphotriesterase from *Pseudomonas diminuta*. Appl Biochem Biotechnol 31：59-63

Caldwell SR，Newcomb JR，Schlecht KA，Raushel FM（1991）Limit of diffusion in the hydrolysis of substrate by the phosphotriesterase from *Pseudomonas diminuta*. Biochemistry 30：7438-7444

Calvert GM，Karnik J，Mehler L，Beckman J，Morrissey B，Sievert J，Barrett R，Lackovic M，Mabee L，Schwartz A，Mitchell Y，Moraga-McHaley S（2008）Acute pesticide poisoning among agricultural workers in the United States，1998-2005. Am J Ind Med 51（12）：883-898

Can A (2014) Quantitative structure-toxicity relationship (QSTR) studies on the organophosphate insecticides. Toxicol Lett 230 (3): 434-443

Carey JL, Dunn C, Gaspari RJ (2013) Central respiratory failure during acute organophosphate poisoning. Respir Physiol Neurobiol 189 (2): 403-410

Carullo P, Cetrangolo GP, Mandrich L, Manco G, Febbraio F (2015) Fluorescence spectroscopy approaches for the development of a real-time organophosphate detection system using an enzymatic sensor. Sensors (Basel) 15 (2): 3932-3951

Cheng TC DeFrank JJ (2000) Hydrolysis of organophosphorus compounds by bacterial prolidases. In: Zwanenburg B, Mikolajczyk M, Kielbasinski P (eds) Enzymes in action: green solutions for chemical problems, vol 33. Kluwer, Dordrecht, pp 243-262

Chen-Goodspeed M, Sogorb MA, Wu F, Hong SB, Raushel FM (2001a) Structural determinants of the substrate and stereochemical specificity of phosphotriesterase. Biochemistry 40 (5): 1325-1331

Chen-Goodspeed M, Sogorb MA, Wu F, Raushel FM (2001b) Enhancement, relaxation, and reversal of the stereoselectivity for phosphotriesterase by rational evolution of active site residues. Biochemistry 40 (5): 1332-1339

Cho CMH, Mulchandani A, Chen W (2002) Bacterial cell surface display of organophosphorus hydrolase for selective screening of improved hydrolysis of organophosphate nerve agents. Appl Environ Microbiol 68 (4): 2026-2030

Cho CMH, Mulchandani A, Chen W (2004) Altering the substrate specificity of organophosphorus hydrolase for enhanced hydrolysis of chlorpyrifos. Appl Environ Microbiol 70 (8): 4681-4685

Cho CMH, Mulchandani A, Chen W (2006) Functional analysis of organophosphorus hydrolase variants with high degradation activity towards organophosphate pesticides. Protein Eng Des Sel 19 (3): 99-105

Chow JY, Wu L, Yew WS (2009) Directed evolution of a quorum-quenching lactonase from *Mycobacterium avium* subsp. *paratuberculosis* K-10 in the amidohydrolase superfamily. Biochemistry 48: 4344-4353

Chow JY, Xue B, Lee KH, Tung A, Wu L, Robinson RC, Yew WS (2010) Directed evolution of a thermostable quorum-quenching lactonase from the amidohydrolase superfamily. J Biol Chem 285 (52): 40911-40920

Cobb RE, Si T, Zhao H (2012) Directed evolution: an evolving and enabling synthetic biology tool. Curr Opin Chem Bio 16: 285-291

Cobb RE, Chao R, Zhao H (2013a) Directed evolution: past, present, and future. AIChE J 59: 1432-1440

Cobb RE, Sun N, Zhao H (2013b) Directed evolution as a powerful synthetic biology tool. Methods 60: 81-90

Costerton JW, Stewart PS, Greenberg EP (1999) Bacterial biofilms: a common cause of persistent infections. Science 284: 1318-1322

Cui ZL, Li SP, Fu GP (2001) Isolation of methyl parathion-degrading strain M6 and cloning of the methyl parathion hydrolase gene. Appl Environ Microbiol 67: 4922-4925

Denard CA, Ren H, Zhao H (2015) Improving and repurposing biocatalysts via directed evolution. Curr Opin Chem Biol 25: 55-64

Dickschat JS (2010) Quorum sensing and bacterial biofilms. Nat Prod Rep 27 (3): 343-69

Dong YH, Xu JL, Li XZ, Zhang LH (2000) AiiA, an enzyme that inactivates the acylhomoserine lactone quorum-sensing signal and attenuates the virulence of *Erwinia carotovora*. Proc Natl Acad Sci USA 97: 3526-3531

Dong YH, Wang LH, Xu JL, Zhang HB, Zhang XF et al (2001) Quenching quorum-sensing dependent bacterial infection by an N-acyl homoserine lactonase. Nature : 813-817

Dong YJ, Bartlam M, Sun L, Zhou YF, Zhang ZP, Zhang CG et al (2005) Crystal structure of methyl parathion hydrolase from *Pseudomonas* sp. WBC-3. J Mol Biol 353: 655-663

Draganov DI (2010) Lactonases with organophosphatase activity: structural and evolutionary perspectives. Chem Biol Interact 187 (1): 370-372

Dumas DP, Caldwell SR, Wild JR, Raushel FM (1989a) Purification and properties of the phos photriesterase from *Pseudomonas diminuta*. J Biol Chem 264 (33): 19659-19665

Dumas DP, Wild JR, Raushel FM (1989b) Diisopropylfluorophosphate hydrolysis by a phosphot riesterase from *Pseudomonas diminuta*. Biotechnol Appl Biochem 11: 235-243

Eddleston M, Eyer P, Worek F, Mohamed F, Senarathna L, von Meyer L, Juszczak E, Hittarage A, Azhar S, Dissanayake W, Sheriff MH, Szinicz L, Dawson AH, Buckley NA (2005) Differences between organophosphorus insecticides in human self-poisoning: a prospective cohort study. Lancet 366 (9495): 1452-1459

Elias M, Dupuy J, Merone L, Mandrich L, Porzio E, Moniot S, Rochu D, Lecomte C, Rossi M, Masson P, Manco G, Chabriere E (2008) Structural basis for natural lactonase and promiscuous phosphotriesterase activities. J Mol Biol 379: 1017-1028

FAO (2014) Tools for the implementation of the international code of conduct on pesticide management Italian Ministry Accessed 12 May 2014

Febbraio F, Merone L, Cetrangolo GP, Rossi M, Nucci R, Manco G (2011) Thermostable esterase 2 from *Alicyclobacillus acidocaldarius* as biosensor for the detection of organophosphate pesticides. Anal Chem 83 (5): 1530-1536. doi: 10. 1021/ac102025z

Francic N, Bellino MG, Soler-Illia GJAA, Lobnik A (2014) Mesoporous titania thin films as efficient enzyme carriers for paraoxon determination/detoxification: effects of enzyme binding and pore hierarchy on the biocatalyst activity and reusability. Analyst 139: 3127-3136

Gao Y, Truong YB, Cacioli P, Butler P, Kyratzis IL (2014) Bioremediation of pesticide contaminated water using an organophosphate degrading enzyme immobilized on nonwoven polyester textiles. Enzyme Microb Technol 54: 38-44

Gavrilescu M (2005) Fate of pesticides in the environment and its bioremediation. Eng Life Sci 6: 497-526

Gill I, Ballesteros A (2000) Bioencapsulation within synthetic polymers (part 2): non-sol-gel protein-polymer composites. Trends Biotechnol 18: 469-479

Goldsmith M, Ashani Y, Simo Y, Ben-David M, Leader H, Silman I et al (2012) Evolved stereose lective hydrolases for broad-spectrum G-type nerve agent detoxification. Chem Biol 19 (4): 456-466

Gotthard G, Hiblot J, Gonzalez D, Elias M, Chabriere E (2013) Structural and enzymatic characterization of the phosphotriesterase OPHC2 from *Pseudomonas pseudoalcaligenes*. PLoS One 8, e77995. doi: 10. 1371/journal. pone. 0077995

Grimsley JK, Singh WP, Wild JR, Giletto A (2001) A novel, enzyme-based method for the wound surface removal and decontamination of organophosphorous nerve agents. Bioact Fibers Polymers 742: 35-49

Grube A, Donaldson D, Kiely T, Wu L (2011) Pesticides industry sales and usage 2006-2007 market estimates. US Environmental Protection Agency, Washington, DC. http: //www. epa. gov/opp00001/pestsales/07pestsales/market _ estimates2007. pdf

Gupta RC (2005) Toxicology of organophosphate and carbamate compounds. Academic, London

Gupta RD, Goldsmith M, Ashani Y, Simo Y, Mullokandov G, Bar H et al (2011) Directed evolution of hydrolases for prevention of G-type nerve agent intoxication. Nat Chem Biol 7 (2): 120-125

Harel M, Aharoni A, Gaidukov L, Brumshtein B, Khersonsky O, Meged R et al (2004) Structure and evolution of the serum paraoxonase family of detoxifying and anti-atherosclerotic enzymes. Nat Struct Mol Biol 11 (5): 412-419

Harper LL, McDaniel CS, Miller CE, Wild JR (1988) Dissimilar plasmids isolated from *Pseudomonas diminuta* MG and a *Flavobacterium* sp. (ATCC 27551) contain identical *opd* genes. Appl Environ Microbiol 54 (10): 2586-2589

Hawwa R, Aikens J, Turner RJ, Santarsiero BD, Mesecar AD (2009a) Structural basis for thermo stability revealed through the identification and characterization of a highly thermostable phos photriesterase-like lactonase from *Geobacillus stearothermophilus*. Arch Biochem Biophys 488: 109-120

Hawwa R, Larsen SD, Ratia K, Mesecar AD (2009b) Structure-based and random mutagenesis approaches increase the organophosphate-degrading activity of a phosphotriesterase homologue from *Deinococcus radiodurans*. J Mol Biol 393: 36-57

Hentzer M, Wu H, Andersen JB, Riedel K, Rasmussen TB et al (2003) Attenuation of *Pseudomonas aeruginosa* virulence by quorum sensing inhibitors. EMBO J 22: 3803-3815

Hiblot J, Gotthard G, Chabriere E, Elias M (2012a) Characterisation of the organophosphate hydrolase catalytic activity *of Sso*Pox. Sci Rep 2: 779

Hiblot J, Gotthard G, Chabriere E, Elias M (2012b) Structural and enzymatic characterization of the lactonase SisLac from *Sulfolobus islandicus*. PLoS ONE 7 (10), e47028

Hiblot J, Gotthard G, Elias M, Chabriere E (2013) Differential active site loop conformations mediate promiscuous activities in the lactonase *Sso*Pox. PLoS ONE 8 (9), e75272

Hiblot J, Bzdrenga J, Champion C, Chabriere E, Elias M (2015) Crystal structure of VmoLac, a tentative quorum quenching lactonase from the extremophilic crenarchaeon *Vulcanisaeta moutnovskia*. J Sci Rep 5: 8372

Horne I, Sutherland TD, Harcourt RL, Russell RJ, Oakeshott JG (2002) Identification of an *opd* (organophosphate degradation) gene in an *Agrobacterium* isolate. Appl Environ Microbiol 68 (7): 3371-3376

Horne I, Qiu X, Russell RJ, Oakeshott JG (2003) The phosphotriesterase gene *opdA* in *Agrobacterium radiobacter* P230 is transposable. FEMS Microbiol Lett 222 (1): 1-8

Hraiech S, Hiblot J, Lafleur J, Lepidi H, Papazian L, Rolain JM, Raoult D, Elias M, Silby MW, Bzdrenga J, Bregeon F, Chabriere E (2014) Inhaled lactonase reduces Pseudomonas aeruginosa quorum sensing and mortality in rat pneumonia. PLoS ONE 9 (10), e107125

Islam SM, Math RK, Cho KM, Lim WJ, Hong SY, Kim JM, Yun MG, Cho JJ, Yun HD (2010) Organophosphorus hydrolase (OpdB) of *Lactobacillus brevis* WCP902 from kimchi is able to degrade organophosphorus pesticides. J Agric Food Chem 58 (9): 5380-5386

Iyer R, Iken B (2015) Protein engineering of representative hydrolytic enzymes for remediation of organophosphates. Biochem Eng J 94: 134-144

Jackson CJ, Foo JL, Kim HK, Carr PD, Liu JW, Salem G, Ollis DL (2008) In crystallo capture of a Michaelis complex and product-binding modes of a bacterial phosphotriesterase. J Mol Biol 375 (5): 1189-1196

Jeong YS, Choi JM, Kyeong HH, Choi JY, Kim EJ, Kim HS (2014) Rational design of organophosphorus hydrolase with high catalytic efficiency for detoxifying a V-type nerveagent. Biochem Biophys Res Commun 449 (3): 263-267

Jin F, Wang J, Shao H, Jin M (2010) Pesticide use and residue control in China. J Pestic Sci 35: 138-142

Jones MB, Peterson SN, Benn R, Braisted JC, Jarrahi B et al (2010) Role of luxS in *Bacillus anthracis* growth and virulence factor expression. Virulence 1: 72-83

Joshi KA, Prouza M, Kum M, Wang J, Tang J, Haddon R, Chen W, Mulchandani A (2006) V-type nerve agent detection using a carbon nanotube-based amperometric enzyme electrode. Anal Chem 78: 331

Josse D, Lockridge O, Xie W, Bartels CF, Schopfer LM, Masson P (2001) The active site of human paraoxonase (PON1). J Appl Toxicol 21 (S1): S7-S11

Kallnik V, Bunescu A, Sayer C, Bräsen C, Wohlgemuth R, Littlechild J, Siebers B (2014) Characterization of a phosphotriesterase-like lactonase from the hyperthermoacidophilic crenarchaeon *Vulcanisaeta moutnovskia*. J Biotechnol 190: 11-17

Kawahara K, Tanaka A, Yoon J, Yokota A (2010) Reclassification of a parathion degrading *Flavobacterium* sp. ATCC 27551 as *Sphingobium fuliginis*. J Gen Appl Microbiol 56 (3): 249-255

Khersonsky O, Roodveldt C, Tawfik DS (2006) Enzyme promiscuity: evolutionary and mechanis tic aspects. Curr Opin Chem Biol 10: 498-508

Kiss A, Virág D (2009) Photostability and photodegradation pathways of distinctive pesticides. J Environ Qual 38 (1): 157-163

Lasram MM, Dhouib IB, Annabi A, El Fazaa S, Gharbi N (2014) A review on the molecular mechanisms involved in insulin resistance induced by organophosphorus pesticides. Toxicology 322: 1-13

LeJeune KE, Mesiano AJ, Bower SB, Grimsley JK, Wild JR, Russell AJ (1997) Dramatically stabilised phosphotriesterase polymers for nerve agent degradation. Biotechnol Bioeng 54: 105-114

LeJeune KE, Wild JR, Russell AJ (1998) Nerve agents degraded by enzymatic foams. Nature 395 (6697): 27-28

Li C, Zhu Y, Benz I, Schmidt MA, Chen W, Mulchandani A, Qiao C (2008) Presentation of functional organophosphorus hydrolase fusions on the surface of *Escherichia coli* by the AIDA-I autotransporter pathway. Biotechnol Bioeng 99: 485-490

Liu H, Zhang JJ, Wang SJ, Zhang XE, Zhou NY (2005) Plasmid-borne catabolism of methyl parathion and p-nitrophenol in *Pseudomonas* sp. strain WBC-3. Biochem Biophys Res Commun 334 (4): 1107-1114

Ma F, Wang Y, Zhang Y, Xiong N, Yang B et al (2009) Heterologous expression of human paraoxonases in *Pseudomonas aeruginosa* inhibits biofilm formation and decreases antibiotic resistance. Appl Microbiol Biotechnol 83: 135-141

Mackness B, Durrington PN, Mackness MI (2000) The paraoxonase gene family and coronary heart disease. Curr Opin Lipidol 13: 357-363

Madej T, Addess KJ, Fong JH, Geer LY, Geer RC, Lanczycki CJ et al (2012) MMDB: 3D structures and macromolecular interactions. Nucleic Acids Res 40 (Database issue): D461-D464

Mandrich L, Manco G (2009) Evolution in the amidohydrolase superfamily: substrate-assisted gain of function in the E183K mutant of a phosphotriesterase-like metal-carboxylesterase. Biochemistry 48 (24): 5602-5612

Mandrich L, Porzio E, Merone L, Febbraio F, Nucci R, Manco G (2011) Exploring thermostable quorum quenching lactonases to counteract bacterial infections in cystic fibrosis. In: Mendez Vilas A (ed) Science and technology against microbial pathogens. Research, development and evaluation. Proceedings of the international conference on antimicrobial research (ICAR2010). World Scientific, pp 150-154

Mandrich L, Di Gennaro S, Palma A, Manco G (2013) A further biochemical characterization of DrPLL the thermophilic lactonase from *Deinococcus radiodurans*. Protein Pept Lett 20 (1): 36-44

Marrazza G (2014) Piezoelectric biosensors for organophosphate and carbamate pesticides: a review. Biosensors 4: 301-317

Masson P, Rochu D (2009) Catalytic bioscavengers against toxic esters, an alternative approach for prophylaxis and treatments of poisonings. Acta Nat 1 (1): 68-79

Meier MM, Rajendran C, Malisi C, Fox NG, Xu C, Schlee S, Barondeau DP, Hocker B, Sterner R, Raushel FM (2013) Molecular engineering of organophosphate hydrolysis activity from a weak promiscuous lactonase template. J Am Chem Soc 135: 11670-11677

Merone L, Mandrich L, Rossi M, Manco G (2005) A thermostable phosphotriesterase from the archaeon *Sulfolobus solfataricus*: cloning, overexpression and properties. Extremophiles 9 (4): 297-305

Merone L, Mandrich L, Porzio E, Rossi M, Müller S, Reiter G, Worek F, Manco G (2010) Improving the promiscuous nerve agent hydrolase activity of a thermostable archaeal lactonase. Bioresour Technol 101 (23): 9204-9212

Munnecke DM (1979) Hydrolysis of organophosphate insecticides by an immobilized-enzyme system. Biotechnol Bioeng 21: 2247-2261

Nachon F, Brazzolotto X, Trovaslet M, Masson P (2013) Progress in the development of enzyme based nerve agent bioscavengers. Chem Biol Interact 206: 536-544

Naqvi T, Warden AC, French N, Sugrue E, Carr PD, Jackson CJ et al (2014) A 5000-fold increase in the specificity of a bacterial phosphotriesterase for malathion through combinatorial active site mutagenesis. PLoS ONE 9 (4), e94177

Ng FSW, Wright DM, Seah SYK (2011) Characterization of a phosphotriesterase-like lactonase from *Sulfolobus solfataricus* and its immobilization for disruption of quorum sensing. Appl Environ Microbiol 77 (4): 1181-1186

Ojha A, Gupta Y (2014) Evaluation of genotoxic potential of commonly used organophosphate pesticides in peripheral blood lymphocytes of rats. Hum Exp Toxicol pii, 0960327114537534. [Epub ahead of print]

Ortiz-Hernández ML, Sánchez-Salinas E, Castrejón Godínez ML, Dantan González E, Popoca Ursino EC (2013) Mechanisms and strategies for pesticide biodegradation: opportunity for waste, soils and water cleaning. Rev Int Contam Ambie 29: 85-104

Pinjari AB, Novikov B, Rezenom YH, Russell DH, Wales ME, Siddavattam D (2012) Mineralization of acephate, a recalcitrant organophosphate insecticide is initiated by a *Pseudomonad* in environmental samples. PLoS One 7, e31963

Pinjari AB, Pandey JP, Kamireddy S, Siddavattam D (2013) Expression and subcellular localization of organophosphate hydrolase in acephate-degrading *Pseudomonas* sp. strain Ind01 and its use as a potential biocatalyst for elimination of organophosphate insecticides A. B. Lett Appl Microbiol 57: 63-68

Popat R, Crusz SA, Diggle SP (2008) The social behaviours of bacterial pathogens. Br Med Bull 87: 63-75

Porzio E, Merone L, Mandrich L, Rossi M, Manco G (2007) A new phosphotriesterase from *Sulfolobus acidocaldarius* and its comparison with the homologue from *Sulfolobus solfataricus*. Biochimie 89 (5): 625-636

Porzio E, Di Gennaro S, Palma A, Manco G (2013) Mn^{2+} modulates the kinetic properties of an archaeal member of the PLL family. Chem Biol Interact 203 (1): 251-256

Ragnarsdottir KV (2000) Environmental fate and toxicology of organophosphate pesticides. J Geol Soc 157: 859-876

Raushel FM (2002) Bacterial detoxification of organophosphate nerve agents. Curr Opin Microbiol 5 (3): 288-295

Raushel FM, Holden HM (2000) Phosphotriesterase: an enzyme in search of its natural substrate. Adv Enzymol Relat Areas Mol Biol 74: 51-93

Reimmann C, Ginet N, Michel L, Keel C, Michaux P et al (2002) Genetically programmed auto inducer destruction reduces virulence gene expression and swarming motility in *Pseudomonas aeruginosa* PAO1. Microbiology 148: 923-932

Richins RD, Kaneva I, Mulchandani A, Chen W (1997) Biodegradation of organophosphorus pesticides by surface-expressed organophosphorus hydrolase. Nat Biotechnol 15: 984-987

Rochu D, Viguie N, Renault F, Crouzier D, Froment MT, Masson P (2004) Contribution of the active site metal cation to catalytic activity and to conformational stability of phosphotriesterase: a thermo- and pH-dependence study. Biochem J 380: 627-633

Rochu D, Chabriere E, Masson P (2007) Human paraoxonase: a promising approach for pretreatment and therapy of organophosphorus poisoning. Toxicology 233 (1): 47-59

Roodveldt C, Tawfik DS (2005) Shared promiscuous activities and evolutionary features in various members of amidohydrolase superfamily. Biochemistry 44: 12728-12736

Salazar-Arredondo E, Solis-Heredia MJ, Rojas-Garcia E, Ochoa IH, Quintanilla Vega B (2008) Sperm chromatin alteration and DNA damage by methyl-parathion, chlorpyrifos and diazinon and their oxon metabolites in human spermatozoa. Reprod Toxicol 25: 455-460

Segers P, Vancanneyt M, Pot B, Torck U, Hoste B, Dewettinck D et al (1994) Classification of *Pseudomonas diminuta* Leifson and Hugh 1954 and Pseudomonas vesicularis Büsing, Döll, and Freytag 1953 in *Brevundimonas* gen. nov. as *Brevundimonas diminuta* comb. nov. And *Brevundimonas vesicularis* comb. nov. , respectively. Int J Syst Bacteriol 44 (3): 499-510

Seibert CM, Raushel FM (2005) Structural and catalytic diversity within the amidohydrolase superfamily. Biochemistry 44 (17): 6383-6391

Shimazu M, Mulchandani A, Chen W (2001) Simultaneous degradation of organophosphorus pesticides and p-nitrophenol by a genetically engineered *Moraxella* sp. with surface-expressed organophosphorus hydrolase. Biotechnol Bioeng 76: 318-324

Singh BK (2009) Organophosphorus-degrading bacteria: ecology and industrial applications. Nat Rev Microbiol 7 (2): 156-164

Singh BK, Walker A (2006) Microbial degradation of organophosphorus compounds. FEMS Microbiol Rev 30 (3): 428-471

Spain JC, Gibson DT (1991) Pathway for biodegradation of p-nitrophenol in a *Moraxella* sp. Appl Environ Microbiol 57: 812-819

Tanga X, Lianga B, Yi T, Manco G, Palchetti I, Liua A (2014) Cell surface display of organophos phorus hydrolase for sensitive spectrophotometric detection of p-nitrophenol substituted organophosphates. Enzyme Microb Technol 55: 107-112

Tehei M, Madern D, Franzetti B, Zaccai G (2005) Neutron scattering reveals the dynamic basis of protein adaptation to extreme temperature. J Biol Chem 280: 40974-40979

Theriot CM, Grunden AM (2011) Hydrolysis of organophosphorus compounds by microbial enzymes. Appl Microbiol Biotechnol 89 (1): 35-43

Theriot CM, Du X, Tove SR, Grunden AM (2010a) Improving the catalytic activity of hyperther mophilic *Pyrococcus* prolidases for detoxification of organophosphorus nerve agents over a broad range of temperatures. Appl Microbiol Biotechnol 87 (5): 1715-1726

Theriot CM, Tove SR, Grunden AM (2010b) Characterization of two proline dipeptidases (prolidases) from the hyperthermophilic archaeon *Pyrococcus horikoshii*. Appl Microbiol Biotechnol 86 (1): 177-188

Theriot CM, Semcer RL, Shah SS, Grunden AM (2011) Improving the catalytic activity of hyperthermophilic Pyrococcus horikoshii prolidase for detoxification of organophosphorus nerveagents over a broad range of temperatures. Archaea 2011: 5651271-5651279

Tiryaki O, Temur C (2010) The fate of pesticide in the environment. J Biol Environ Sci 4 (10): 29-38

Trovaslet-Leroy M, Musilova L, Renault F, Brazzolotto X, Misik J, Novotny L, Froment MT, Gillon E, Loiodice M, Verdier L, Masson P, Rochu D, Jun D, Nachon F (2011) Organophosphate hydrolases as catalytic bioscavengers of organophosphorus nerve agents. Toxicol Lett 206: 14-23

Tsai PC, Bigley A, Li Y, Ghanem E, Cadieux CL, Kasten SA et al (2010a) Stereoselective hydrolysis of organophosphate nerve agents by the bacterial phosphotriesterase. Biochemistry 49 (37): 7978-7987

Tsai PC, Fan Y, Kim J, Yang L, Almo SC, Gao YQ, Raushel FM (2010b) Structural determinants for the stereoselective hydrolysis of chiral substrates by phosphotriesterase. Biochemistry 49 (37): 7988-7997

Tsai PC, Fox N, Bigley AN, Harvey SP, Barondeau DP, Raushel FM (2012) Enzymes for the homeland defense: optimizing phosphotriesterase for the hydrolysis of organophosphate nerve agents. Biochemistry 51 (32): 6463-6475

Vanhooke JL, Benning MM, Raushel FM, Holden HM (1996) Three dimensional structure of the zinc containing phosphotriesterase with the bound substrate analog diethyl 4-methylben zylphosphonate. Biochemistry 35: 6020-6025

Vyas NK, Nickitenko A, Rastogi VK, Shah SS, Quiocho FA (2010) Structural insights into the dual activities of the nerve agent degrading organophosphate anhydrolase/prolidase. Biochemistry 49 (3): 547-559

Wanekaya AK, Chenb W, Mulchandani A (2008) Recent biosensing developments in environmen tal security. J Environ Monit 10：703-712

Worek F，Seeger T，Goldsmith M，Ashani Y，Leader H，Sussman JS，Tawfik D，Thiermann H，Wille T（2014）Efficacy of the rePON1 mutant IIG1 to prevent cyclosarin toxicity in vivo and to detoxify structurally different nerve agents in vitro. Arch Toxicol 88（6）：1257-1266

Xiang DF，Kolb P，Fedorov AA，Meier MM，Fedorov LV et al（2009）Functional annotation and three-dimensional structure of Dr0930 from *Deinococcus radiodurans*，a close relative of phos photriesterase in the amidohydrolase superfamily. Biochemistry 48：2237-2247

Yang C，Liu N，Guo X，Qiao C（2006）Cloning of *mpd* gene from a chlorpyrifos-degrading bacterium and use of this strain in bioremediation of contaminated soil. FEMS Microbiol Lett 265：118-125

Yang C，Liu R，Yuan Y，Liu J，Cao X，Qiao C，Song C（2013）Construction of a green fluores cent protein（GFP）-marked multifunctional pesticide-degrading bacterium for simultaneous degradation of organophosphates and γ-hexachlorocyclohexane. J Agric Food Chem 61（6）：1328-1334

Zhang R，Cui ZL，Jiang J，He J，Gu X，Li S（2005）Diversity of organophosphorus pesticide degrading bacteria in a polluted soil and conservation of their organophosphorus hydrolase genes. Can J Microbiol 51（4）：337-343

Zhang R，Cui ZL，Zhang X，Jiang J，Gu JD，Li SP（2006）Cloning of the organophosphorus pesticide hydrolase gene clusters of seven degradative bacteria isolated from a methyl parathion contaminated site and evidence of their horizontal gene transfer. Biodegradation 17（5）：465-472

Zhang Y，An J，Ye W，Yang G，Qian ZG，Chen HF，Cui L，Feng Y（2012）Enhancing the promiscu ous phosphotriesterase activity of a thermostable lactonase（GkaP）for the efficient degradation of organophosphate pesticides. Appl Environ Microbiol 78（18）：6647-6655

Zhang Y，An J，Yang GY，Bai A，Zheng B，Lou Z，Wu G，Ye W，Chen HF，Feng Y，Manco G（2015）Active site loop conformation regulates promiscuous activity in a lactonase from *Geobacillus kaustophilus* HTA426. PLoS ONE 10（2），e0115130

Zhongli C，Shunpeng L，Guoping F（2001）Isolation of methyl parathion-degrading strain M6 and cloning of the methyl parathion hydrolase gene. Appl Environ Microbiol 67（10）：4922-4925

第十七章

古菌 α-淀粉酶的序列、结构及进化分析

Štefan Janeček[1]

17.1 引言

α-淀粉酶（EC 3.2.1.1）、β-淀粉酶和葡萄糖淀粉酶属于熟知的淀粉水解酶。α-淀粉酶催化淀粉和相关α-葡聚糖中α-1,4-糖苷键的水解（Janecek et al. 2014）。相比之下，外切活性的β-淀粉酶和葡萄糖淀粉酶主要作用于底物的非还原性末端，分别得到β-麦芽糖和β-葡萄糖（Ray and Nanda 1996；Sauer et al. 2000；Kumar and Satyanarayana 2009；Marin-Navarro and Polaina 2011）。α-淀粉酶的酶解产物包含从长的麦芽寡糖到α-糊精等多种形式（Vihinen and Mantsala 1989；MacGregor et al. 2001）。与利用α-糖苷键转化机制的β-淀粉酶和葡萄糖淀粉酶水解不同，α-淀粉酶催化作用的结果产物为α-异头构型（McCarter and Withers 1994；Davies and Henrissat 1995）。三种淀粉水解酶在进化过程中也是不同的，这是因为三种酶在氨基酸序列上存在明显不同，并且裂解α-糖苷键的催化机制也不同（Janecek 1994a；Pujadas et al. 1996；Coutinho and Reilly 1997；Janecek 1997）。

根据酶的来源，一些细菌、真菌甚至哺乳动物的α-淀粉酶除了淀粉水解作用外，还可以在非常小的程度上催化转糖基化反应（Brzozowski and Davies 1997；Aghajari et al. 2002；Ramasubbu et al. 2003；Li et al. 2005）。因此，当将α-淀粉酶特异性归类为淀粉水解酶时，由生物信息学推断得到的一些生物化学特性是需要重点关注的（Janecek et al. 2014；Janecek et al. 1995）。例如，具有转糖基活性的环糊精葡萄糖基转移酶，在较低程度也具有催化水解的辅助功能，与典型的α-淀粉酶相似（Leemhuis et al. 2003a，b）。

关于其分类来源，与α-淀粉酶相比，β-淀粉酶和葡萄糖淀粉酶之间也存在显著差异。β-淀粉酶起源于植物，在细菌和真菌中较少发现（Ray and Nanda 1996；Kim et al. 2000，2001a）。葡萄糖淀粉酶则主要来源于细菌、真菌，古菌中也有所发现（Marin-Navarro and

---

[1] Laboratory of Protein Evolution，Institute of Molecular Biology，Slovak；Academy of Sciences，Dubravska cesta 2，SK-84551 Bratislava，Slovakia.

Faculty of Natural Sciences，Department of Biology，University of SS Cyril and Methodius，Nam. J. Herdu 2，SK-91701 Trnava，Slovakia.

e-mail：Stefan. Janecek@savba. sk；Stefan. Janecek@ucm. sk.

Polaina 2011)。β-淀粉酶和葡萄糖淀粉酶都不是来源于动物的，而 β-淀粉酶不存在于古菌中，葡萄糖淀粉酶在植物中也没有发现。另一方面，α-淀粉酶在生物的三界领域（即细菌、古菌和真核生物）都被鉴定出来，这些酶广泛存在于微生物、植物和动物中（Janecek 1994b；Janecek et al. 1999）。然而，值得注意的是，一些原核生物，尤其是古菌，在其基因组内没有糖苷水解酶基因的存在，也不含有任何 α-淀粉酶基因（Coutinho and Henrissat 1999）。

现代大量的生产和工业需求使得 α-淀粉酶成为研究最为透彻，也是最常用的酶类之一（Leveque et al. 2000b；Pandey et al. 2000；Bertoldo and Antranikian 2002；van der Maarel et al. 2002；Gupta et al. 2003；Sharma and Satyanarayana 2013）。而古菌由于生存在极端环境下，其产生的 α-淀粉酶可能具有更为独特、优越的性能。

17. 1. 1 CAZy 分类系统

α-淀粉酶、β-淀粉酶和葡萄糖淀粉酶具有相似的功能，但在序列结构、酶解机制以及进化方面存在明显差异，这三种酶在碳水化合物活性酶（CAZy）数据库（Lombard et al. 2014）各自定义为独立的酶家族。CAZy 分类系统最初是依据糖苷水解酶（GH）序列信息进行分类的（Henrissat 1991），现在发展成为国际公认的数据库，包含了所有糖苷键形成、降解和修饰相关的酶类（http：//www. cazy. org/）。1991 年，β-淀粉酶和葡糖淀粉酶分别形成其单个特异性家族 GH14 和 GH15，每个家族仅有 5 个序列，而 α-淀粉酶归类于 GH13 家族，其包含了大量不同的多特异性酶类，该家族的 40 多种蛋白酶分别包括了异淀粉酶、支链淀粉酶、Ⅱ型支链淀粉酶、新支链淀粉酶、环糊精葡萄糖基转移酶，α-葡萄糖苷酶、葡聚糖酶和外切淀粉酶（Henrissat 1991）。α-淀粉酶家族酶类的特异性是由实验数据结合生物信息学研究分析得到的（Svensson 1988；MacGregor and Svensson 1989；Jespersen et al. 1991；Takata et al. 1992）。截止到 2015 年 4 月 GH13 α-淀粉酶家族包含了约 24000 个序列，以及超过 30 个不同的酶特异性，GH13 是目前包含了 133 个分类的 GH 蛋白酶中最大的家族（Janecek et al. 2014；Lombard et al. 2014）。

由于 CAZy 分类系统主要是依据氨基酸序列进行归类，因此出现同一种功能的酶类被归类于不同家族并不奇怪。换句话说就是，序列不同的蛋白酶执行相同的酶学活性。这也是数据库中的 α-淀粉酶，除了最初定义的 GH13 家族外，还包含了 GH57 和 GH119 家族的原因（Lombard et al. 2014）。古菌的 α-淀粉酶主要来源于 GH13 家族，少数也存在于 GH57 家族中（Janecek et al. 2014）。

α-淀粉酶最主要属于 GH13 家族蛋白酶（Janecek et al. 2014），同时也包括 GH70 家族环状排列的葡聚糖蔗糖酶（MacGregor et al. 1996；Vujicic-Zagar et al. 2010）、GH77 家族的 4-α-葡聚糖转移酶（Przylas et al. 2000；Godany et al. 2008），其都属于糖苷水解酶 GH-H 系，（MacGregor et al. 2001）。α-淀粉酶包括了 GH13 家族 1，5，6，7，15，24，27，28，36，37 亚家族（Da Lage et al. 2004，2013；van der Kaaij et al. 2007；Hostinova et al. 2010；Lei et al. 2012；Majzlova et al. 2013；Puspasari et al. 2013），以及一些未被定义的蛋白酶（Janecek et al. 2014）。GH13 家族的 α-淀粉酶具有相同的反应机制（MacGregor et al. 2001），包含了 7 个保守序列区（CSRs）（Janecek 2002）以及 $(\beta/\alpha)_8$-桶形结构域（TIM-桶）（Matsuura et al. 1984；Uitdehaag et al. 1999）：天冬氨酸作为催化亲核分子，谷氨酸作为质子供体，天冬氨酸在 β4、β5 和 β7 片层上作为过渡态稳定剂。

GH57 家族的 α-淀粉酶目前包含了 10 种酶类型约 1100 个蛋白酶（Janecek 2005；Lombard et al. 2014；Janecek et al. 2014）。GH57 α-淀粉酶有 5 个保守序列区（Zona et al. 2004）

以及不同的催化机制（Imamura et al. 2003）：谷氨酸作为亲核基团，β4 和 β7 片层的天冬氨酸作为质子供体，采用（β/α）$_7$-桶形结构域（不完整 TIM-桶）构成酶的催化结构。有趣的是，生物信息学分析及结构预测发现（Janecek and Kuchtova 2012），GH119 家族的 α-淀粉酶（只有 10 个成员）与 GH57 家族的酶共享上述特征，但它目前不包含任何古菌序列（Lombard et al. 2014）。

17.2 古菌 GH13 家族α-淀粉酶

20 世纪 90 年代初首次报道了从古菌极端嗜热菌火球菌属（*Pyrococcus*）分离得到的 α-淀粉酶/淀粉水解酶（Koch et al. 1990，1991；Laderman et al. 1993b）。尽管在 1993 年，首个古菌淀粉酶结构在细菌 *Pyrococcus furiosus* 中报道（Laderman et al. 1993a），但它的序列与典型的 α-淀粉酶存在较大的差异，因而归类于 GH13 家族（Janecek 1998）。事实上，它曾经被当作 GH57 家族第二个发现的 α-淀粉酶基本成员（Henrissat and Bairoch 1996）。然而，目前被认为应该属于 4-α-葡聚糖转移酶（Laderman et al. 1993a；Blesak and Janecek 2012）。首个真正确定的氨基酸序列的 GH13 家族 α-淀粉酶分离自 *Pyrococcus* sp. KOD1，其生化特性也同时被研究（Tachibana et al. 1996），然而该菌后来被重新归类为 *Thermococcus kodakaraensis*（Atomi et al. 2004）。

随着越来越多的 α-淀粉酶在古菌中发现，其与植物中的淀粉酶的进化关系出现独立的分枝（Janecek et al. 1999；Jones et al. 1999）。在此之前，植物来源 α-淀粉酶在进化树上存在着接近主体的独立分枝，导致细菌来源的 α-淀粉酶分类较为模糊（Janecek 1994b）。来自古菌、植物和细菌的三种密切相关的 α-淀粉酶（溶解产生或细胞内产生）分别分配到各自的 α-淀粉酶亚家族 GH13_7、GH13_6 和 GH13_5 中（Stam et al. 2006）。目前，还有一种明显区别于 GH13_7 亚家族的 α-淀粉酶，由嗜盐古菌产生，目前还没有对应的 GH13 亚家族分类（Lombard et al. 2014）。遗憾的是，这类淀粉酶只有少数被表达，并且只研究了部分成为潜在的 α-淀粉酶的特性（Kobayashi et al. 1994；Hutcheon et al. 2005）。

除了 α-淀粉酶，也有其他 GH13 家族的古菌淀粉水解酶基因序列被测定，且相关生化性质通过实验描述（Lombardet al. 2014）。例如，环糊精葡萄糖基转移酶（Rashid et al. 2002；Lee et al. 2006；Bautista et al. 2012）、Ⅲ型支链淀粉酶（Niehaus et al. 2000；Ahmad et al. 2014）、支链淀粉酶（Li et al. 2013）、环麦芽糖糊精酶/麦芽糖淀粉酶（Hashimoto et al. 2001；Li et al. 2010；Jung et al. 2012；Park et al. 2013；Sun et al. 2015）、麦芽寡糖基海藻糖水解酶（Kobayashi et al. 1996；Feese et al. 2000）、糖基海藻糖合酶（Maruta et al. 1996；Kobayashi et al. 2003）以及海藻糖合酶（Chen et al. 2006）。其中一些已归类于 GH13 亚家族中，如环糊精葡萄糖基转移酶、环麦芽糖糊精酶/麦芽糖淀粉酶、麦芽寡糖基海藻糖水解酶、糖基海藻糖合酶以及海藻糖合酶——分别为 GH13_2、GH13_20、GH13_10、GH13_26 和 GH13_16 亚家族，但对于其他一些酶，分类信息还未研究确认（Lombard et al. 2014）。

17.2.1 古菌 α-淀粉酶 GH13_7 亚家族

广古细菌门嗜热球菌产生的 α-淀粉酶属于 GH13_7 亚家族（Tachibana et al. 1996；Dong et al. 1997；Jorgensen et al. 1997；Frillingos et al. 2000；Leveque et al. 2000a；Lim et al. 2007；Wang et al. 2008），包括火球菌属（*Pyrococcus*）、巴斯德毕赤酵母菌属（*Thermococcus*）和盐球菌属（*Palaeococcus*）细菌产生的 α-淀粉酶（Lombard et al. 2014）。GH13_

7 亚家族淀粉酶具有 GH13 家族淀粉酶典型的三结构域组织（Janecek et al. 2014），包括 TIM-桶形催化结构域 A，结构域 B 通常具有不规则的二级结构，在第三个 β 折叠片层和第三个 α-螺旋之间的桶形体中伸出一个突出的长环，以及具有典型反向平行的 β 片层结构组成的结构域 C（Linden et al. 2003）。其他的 GH13 家族亚家族及其他类型古菌所产生的淀粉酶则可能还含有其他的附加的结构域（MacGregor et al. 2001；Janecek et al. 2014）。

极端嗜热古菌产生的 α-淀粉酶归类于 GH13 _ 7 亚家族，包含由 Janecek 等人（1999）最初确定的几个特征序列。这些序列只出现在植物来源的 GH13 _ 6 亚家族中（图 17.1）。它们的特征描述如下（以 *Thermococcus hydrothermalis* α-淀粉酶为例）：①CSR-Ⅵ（链 β2）- Ile42（保守甘氨酸右边）以及两个相邻的脯氨酸 Pro48-Pro49 分别位于起始和末端；②CSR-Ⅰ（链 β3）- Ile107 位于保守天冬氨酸后；③CSR-Ⅴ（域 B）- Ile196；④分别位于起始，中间和末端的 CSR-Ⅱ（域 β4）-(Ala194)-Trp195，Tyr199 和 Gly202；⑤CSR-Ⅲ（链 β5）-Ala219 和 Tyr223-分别位于保守色氨酸 Trp224 以及催化质子供体 Glu222 之后；⑥CSR-Ⅳ（链 β7）-Ala286；⑦CSR-Ⅶ（链 β8）-Gln309（保守甘氨酸右边），三肽 Ile312-Phe313-Tyr314 和 Asp316 分别位于起始、中间和末端。

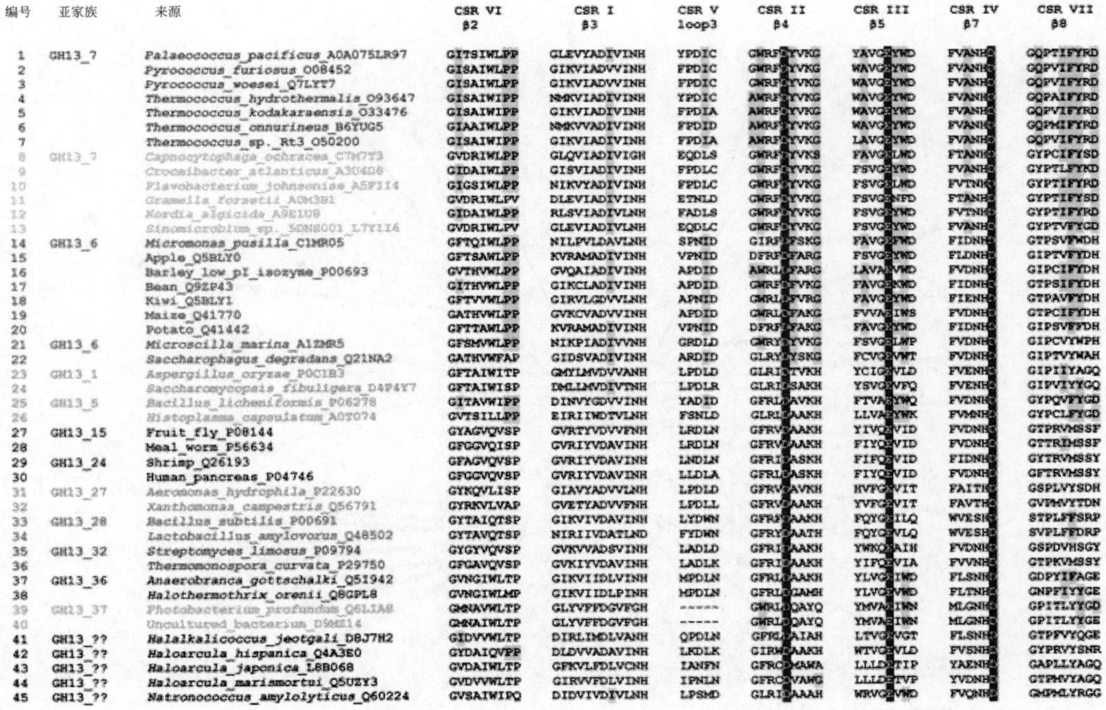

图 17.1 α-淀粉酶 GH13 家族保守区序列（见彩图）。序列包含了多种 GH13 家族不同分类亚家族蛋白酶，尤其是古菌来源的序列。未分类的古菌序列用双问号标注。古菌 GH13 _ 7 α-淀粉酶特征序列用亮黄色标注。催化三联体用黑色标出。保守的序列区域（CSR-Ⅰ至 CSR-Ⅶ）主要在 TIM 桶形结构的单个 β 折叠中（Janecek 2002）。蛋白酶序列由 UniProt 登录号（UniProt Consortium 2015）区分

上述提及的部分氨基酸残基在植物和古菌 α-淀粉酶中都发挥着重要的作用（Kadziola et al. 1998；Linden et al. 2003）。通过阿卡波糖（Kadziola et al. 1998）研究大麦 α-淀粉酶的复杂结构，CSR-Ⅱ末端的甘氨酸残基是 Ca^{2+} 配体，CSR-Ⅲ催化质子供体后第二位的色氨酸残基通过堆叠的阿卡波糖环与活性中心作用。通过 *Pyrococcus woesei* 的 α-淀粉酶的结构证明，甘氨酸和色

氨酸可能在古菌 α-淀粉酶中也发挥相似的作用（Linden et al. 2003）。然而，值得注意的是质子供体后面第二位的色氨酸残基在 GH13 _ 7 古菌 α-淀粉酶中也不是唯一的，因为它经常出现在 GH13 _ 36 亚家族 α-淀粉酶“中间体”（Sivakumar et al. 2006；Majzlova et al. 2013）和/或寡-1，6-葡萄糖苷酶和新支链淀粉酶亚家族成员中，起着类似的结合作用（Oslancova and Janecek 2002；Hondoh et al. 2003，2008；Kobayashi et al. 2015）。

图 17.2 是 GH13 家族亚家族中各 α-淀粉酶进化关系图，着重突出古菌分枝和植物分枝相似的进化距离。考虑到细菌来源的 α-淀粉酶与古菌来源的更为接近，因此，细菌来源的 α-淀粉酶也被归类为 GH13 _ 7 亚家族成员（图 17.2）。有趣的是，它们都是来源于各种黄杆菌属（Lombard et al. 2014）。事实上，黄杆菌属来源的 α-淀粉酶已被体外表达且其生化特性也已被研究。*Sinomicrobium* sp. 5DNS001 产生的 α-淀粉酶尤为重要，因为尽管与极端嗜热古菌的 α-淀粉酶同源，但是黄杆菌属只是弱耐热细菌（Li et al. 2014）。

关于植物 α-淀粉酶，部分细菌来源的酶也具有相同的功能，但都是根据全基因组序列推测的假定酶类（Lombard et al. 2014）。在这种情况下，也应该考虑到具有植物样 α-淀粉酶拷贝的噬糖菌属 *Saccharophagus degradans* 基因组中水平基因转移的可能性（图 17.2）（Da Lage et al. 2004）。

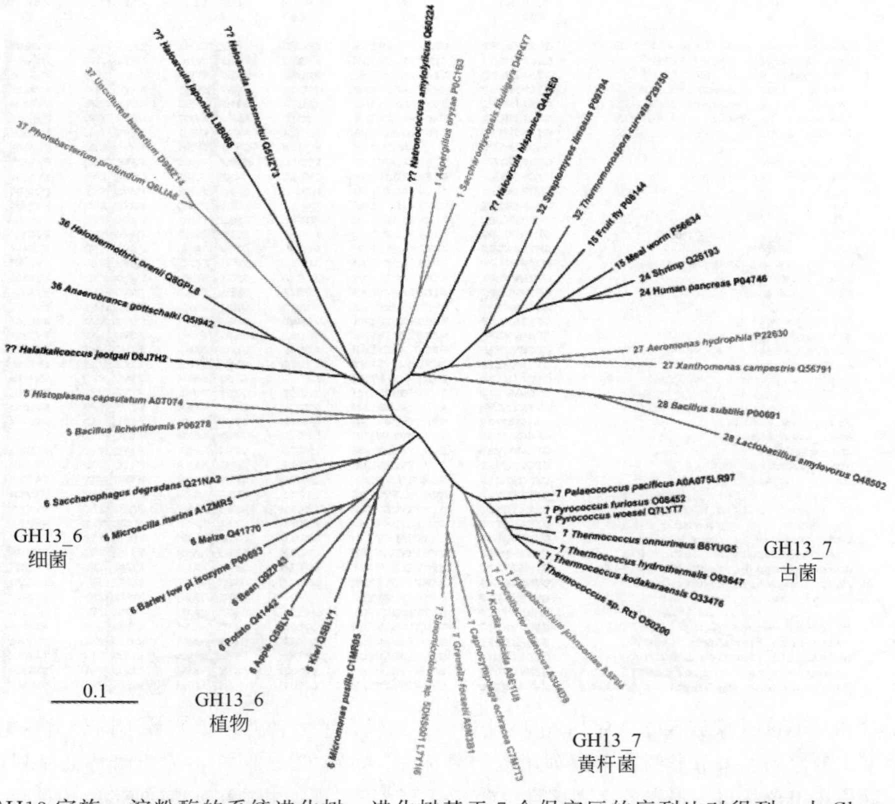

图 17.2　GH13 家族 α-淀粉酶的系统进化树。进化树基于 7 个保守区的序列比对得到。由 Clustal-X（Jeanmougin 等 1998）和 TreeView（Page 1996）软件计算并构建得到

来自古菌和植物的 α-淀粉酶（即亚家族 GH13 _ 7 和 GH13 _ 6）只有在进化树基于其序列的最佳保守区时才在进化树中表现为相邻分支（图 17.2），例如，基于 CSRs 区构建的进化树（图 17.1）。否则的话，古菌和植物 α-淀粉酶之间的相关性将受到 GH13 _ 5 亚家族的影响（Da Lage et al. 2004；van der Kaaij et al. 2007；Godany et al. 2010），包括液化细菌的

α-淀粉酶、来自一些有毒真菌的胞内组分，甚至是来源于一些潜在的古菌如产甲烷八叠球菌的未知特性物质（Janecek et al. 2014；Lombard et al. 2014）。

值得一提的是，古菌、黄杆菌属以及植物来源的 α-淀粉酶具有相同序列特征以区分其他的亚家族（图 17.1），但在三级结构上没有明显的差异（图 17.3），除了来自 GH13 _ 5 亚

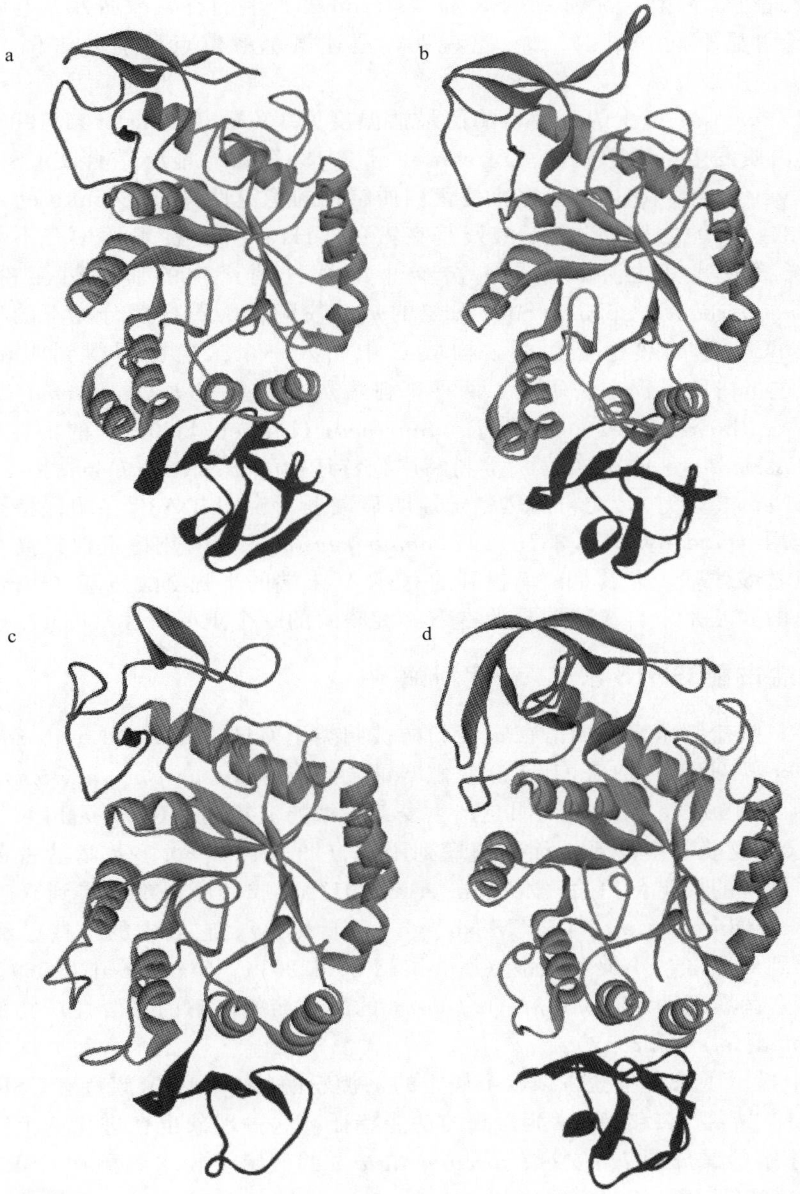

图 17.3　古菌 α-淀粉酶及其同源酶体的蛋白质三级结构（见彩图）。(a) GH13 _ 7-*Pyrococcus woesei*（PDB code：1MWO；Linden 等 2003）；(b) GH13 _ 7-*Sinomicrobium* sp. 5DNS001（UniProt accession No.：L7Y1I6；Li 等 2014）；(c) GH13 _ 6-*Hordeum vulgare* - barley isozyme AMY-1（PDB code：1P6W；Robert 等 2003）；(d) GH13 _ 5-*Bacillus licheniformis*（PDB 码：1BLI；Machius 等 1998）。蛋白酶 $(\beta/\alpha)_8$-桶形催化结构域：绿色；结构域 B：红色；结构域 C：蓝色。黄杆菌 α-淀粉酶以 *P. woesei* α-淀粉酶（PDB 码：1MWO；Deshpande 等 2005）为模型，利用 Phyre-2 同源模拟得到（Kelley 和 Sternberg 2009），并通过 WebLabViewerLite 软件（Molecular Simulations，Inc.）展示

家族的细菌 α-淀粉酶中存在较长的结构域 B（图 17.3d），而古菌的 α-淀粉酶则拥有独特的序列特征。跟踪分析及区分古菌 α-淀粉酶中来自黄杆菌和植物的序列特征，有利于揭示古菌 α-淀粉酶具有极高的热稳定性（最适酶活温度＞80℃）的可能原因（Leveque et al. 2000b；Linden and Wilmanns 2004）。CSR-Ⅳ（strand β7）上的丙氨酸残基位点与植物来源的淀粉酶明显不同，此外，古菌 α-淀粉酶 CSR-Ⅶ（strand β8）上的谷氨酰胺残基与黄杆菌属保守的酪氨酸残基明显不同（图 17.1）。这说明，上述氨基酸位点可能与古菌淀粉酶嗜热性有关。

例如，*P. furiosus* 产生的 α-淀粉酶的最适温度为 100℃（Dong et al. 1997），然而温度超过 106℃蛋白质发生不可逆变性（Brown et al. 2013），Zn^{2+} 能够结合在 CSR-Ⅴ 最后一个氨基酸残基 Cys165 上（图 17.1）来维持蛋白质结构的稳定性（Savchenko et al. 2002；Linden et al. 2003）。有趣的是，半胱氨酸残基在古菌 GH13_7 α-淀粉酶中并不是绝对保守氨基酸，但半胱氨酸残基也存在于同一分类下的黄杆菌产生的非嗜热淀粉酶中（Li et al. 2014）。*Sinomicrobium* sp. 5DNS001 分泌的 α-淀粉酶作为黄杆菌分类下的代表酶，最适酶活温度为 50℃，它的热稳定性在结构域 C 中引入一个二硫键后得到明显改善（Li et al. 2014）。通过蛋白质工程方法研究了两种来自嗜热古球菌 *T. hydrothermalis*（Leveque et al. 2000a；Horvathova et al. 2006）和 *T. onnurineus*（Lim et al. 2007）的 α-淀粉酶。模仿热袍菌属细菌 *Thermotoga maritime* α-淀粉酶序列特征构建 *T. hydrothermalis* α-淀粉酶 Y39I 突变体（Liebl et al. 1997），淀粉酶热稳定性明显减弱，说明 Tyr39 在酶保持热稳定性中发挥着重要的作用（Godany et al. 2010）；*T. onnurineus* α-淀粉酶热稳定性提高可以通过恢复锌离子结合氨基酸残基，尤其是前面提到的 CSR-V 末端的半胱氨酸残基（Lim et al. 2007）。因此，半胱氨酸残基是来自火球菌属的古菌 α-淀粉酶的一个重要特性（图 17.1）。

17.2.2 嗜盐古菌 GH13 家族 α-淀粉酶

最近几年，嗜盐细菌来源的潜在 α-淀粉酶被归类于 GH13 家族（Lombard et al. 2014）。起初，GH13 家族只有一种来自淀粉分解球菌 *Natronococcus amylolyticus* 的 α-淀粉酶（Kobayashi et al. 1994），后来被证明是产生麦芽三糖的淀粉酶（Kobayashi et al. 1992）。与 GH13_7 亚家族极端嗜热古菌 α-淀粉酶序列不同（图 17.1），大多数嗜盐古菌淀粉酶是从基因组测序中发现的假定蛋白酶（Zorgani et al. 2014），或其序列难以得到（Perez-Pomares et al. 2003；Fukushima et al. 2005；Moshfegh et al. 2013）。它们中的一些已经作为 α-葡萄糖苷酶被归类到 GH13_31 亚家族中（Lombard et al. 2014；Stam et al. 2006）。这也可以通过来自嗜盐小盒菌属细菌 *Haloarcula japonica* 的酶对淀粉水解的可溶性产物的分析来加以验证说明（Onodera et al. 2013）。

图 17.1 中显示了来自嗜盐古菌 5 个假定的 α-淀粉酶的 GH13 家族特有 CSRs。这些酶明显不含有 GH13_7 亚家族古菌 α-淀粉酶的功能特征，这一现象也在进化树上得以表现（图 17.2）。来自于嗜盐碱球菌 *Halalkalicoccus jeogtali* 的假定淀粉水解酶在 CSR-Ⅴ 区出现寡-1,6-葡萄糖苷酶特征片段（QPDLN）（Oslancova and Janecek 2002）。它在进化树中的位置位于 GH13_36 亚家族分枝（图 17.2），与寡-1,6-葡萄糖苷酶支链淀粉酶亚族紧密相关（Majzlova et al. 2013）。然而日本盐盒菌 *Haloarcula japonica*（yielding mainly maltose；Onodera et al. 2013）和死海盐盒菌 *Haloarcula marismortui* 来源的淀粉水解酶属于 GH13_37 亚家族（Lei et al. 2012），来源于海洋细菌的 GH13_37 亚家族淀粉酶一般缺失结构域 B（Janecek et al. 2014），西班牙盐盒菌 *Haloarcula hispanica* 的淀粉酶在进化树上呈现单独的分枝，且与细菌和动物来源的 GH13_15，24，27，28 和 32 亚家族相毗邻（图

17.2)。亚硝化球菌属 *Natronococcus amylolyticus* 的淀粉酶拥有麦芽三糖酶的特性（Kobayashi et al. 1992，1994），与真菌来源的 GH13 _ 1 α-淀粉酶属同一分枝（图 17.2），如曲霉菌 *Aspergillus oryzae* 产的 Taka 淀粉酶 A（Matsuura et al. 1984）。有趣的是，来源于嗜盐古菌 5 个假定的 α-淀粉酶利用 Phyre-2 进行同源模建（Kelley and Sternberg 2009），预测淀粉酶三级结构使用的模板来源于 *Haloarcula japonica*（GH13 _ 5）和 *Natronococcus amylolyticus*（GH13 _ 24，GH13 _ 5 和 GH13 _ 15）产生的酶，而这些寡-1,6-葡萄糖苷酶（GH13 _ 31）和新支链淀粉酶（GH13 _ 20）亚家族分别被确定为 *Halalkalicoccus jeogtali*、*Haloarcula japonica* 和 *Haloarcula marismortui* 产生的酶。因此，为了阐明来自嗜盐古菌淀粉酶的特性和催化机制，需要对其生物化学特性进行更深入研究。

17.3 古菌 α -淀粉酶 GH57 家族

GH57 家族的创建是基于两个从极端嗜热原核生物分离得到的“α-淀粉酶”序列，且这两个序列没有 GH13 家族 α-淀粉酶的主要特征（Henrissat and Bairoch 1996；Janecek 1998）。*Dictyoglomus thermophilum*（Fukusumi et al. 1988）和 *Pyrococcus furiosus*（Laderman et al. 1993a）产生的淀粉酶最初被认为属于 GH13 家族 α-淀粉酶，但基于生物信息学分析则都表现为 4-α-葡聚糖转移酶（Laderman et al. 1993b；Nakajima et al. 2004；Blesak and Janecek 2012)。然而，应该考虑到，尽管最近的生物信息学分析揭示了 GH57 家族中的 50～60 个真正潜在的 α-淀粉酶序列（Janecek and Blesak 2011；Blesak and Janecek 2012)，但是关于 α-淀粉酶特异性存在的唯一证据是来自产甲烷古菌詹氏甲烷球菌 *Methanococcus jannaschii* 的 α-淀粉酶的生物化学表征，显示能够降解可溶性淀粉（Kim et al. 2001b)。该淀粉酶是通过细菌基因组测序首次发现（Bult et al. 1996），关于其属于专一性 α-淀粉酶的讨论是基于其降解支链淀粉的速率比降解淀粉快 80%（Kim et al. 2001b)。除了淀粉水解的特性（Janecek et al. 2014），目前 GH57 家族有两个密切相关的淀粉酶（Blesak and Janecek 2013)，一个是麦芽糖化淀粉酶（或麦芽糖外切淀粉酶）（Jeon et al. 2014；Jung et al. 2014；Park et al. 2014），另一个为非特异性淀粉酶（Wang et al. 2011）。

詹氏甲烷球菌 α-淀粉酶为代表的 GH57 家族 α-淀粉酶大约有 500 个氨基酸残基，包含 2 个该家族必不可少的结构区域（Janecek and Blesak 2011)：一束 3～4 个 α 螺旋区，以及不完整的 TIM 桶形催化结构域。尽管催化机制是保守的，活性中心在桶形结构的 β4 片层（CSR-3）和 β7 片层（CSR-4）之间（图 17.4），但螺旋结构被认为属于 GH57 家族的催化区域，因为其包含了 CSR-5 功能必需的氨基酸残基（Palomo et al. 2011)。其他 GH57 特征酶可能还包含其他的结构域（Blesak and Janecek 2012)。由于 CSR-1 十分接近 N 端结构（图 17.4），因此 α-淀粉酶可能不存在信号肽。

上述提及的大多数结构序列特征同样适用于另一组 GH57 家族成员，该组酶与 α-淀粉酶密切相关，称为 α-淀粉酶样类似物（Janecek et al. 2014)。最重要的区别是 α-淀粉酶样类似物缺乏一种或两种催化氨基酸残基，也就是说它们的催化机制不完全或完全丧失（Janecek and Blesak 2011)。因此，在詹氏甲烷球菌所产的 α-淀粉酶中催化的亲核残基 Glu145 和质子供体 Asp237 通常被丝氨酸和谷氨酸所取代（图 17.4)。

尽管 α-淀粉酶样类似物中酶活性最终丧失，但是它们显然与其酶促活性对应物（即家族 GH57α-淀粉酶）具有密切的进化相关性（Blesak and Janecek 2013)。这种进化上的相似性也在保守 CSRs 序列上显现出来（图 7.4），表明可以定义一个指纹序列来指示同一家族中的特异性（Blesak and Janecek 2012)。α-淀粉酶以及 α-淀粉酶样类似物，位点 1（CSR-1）

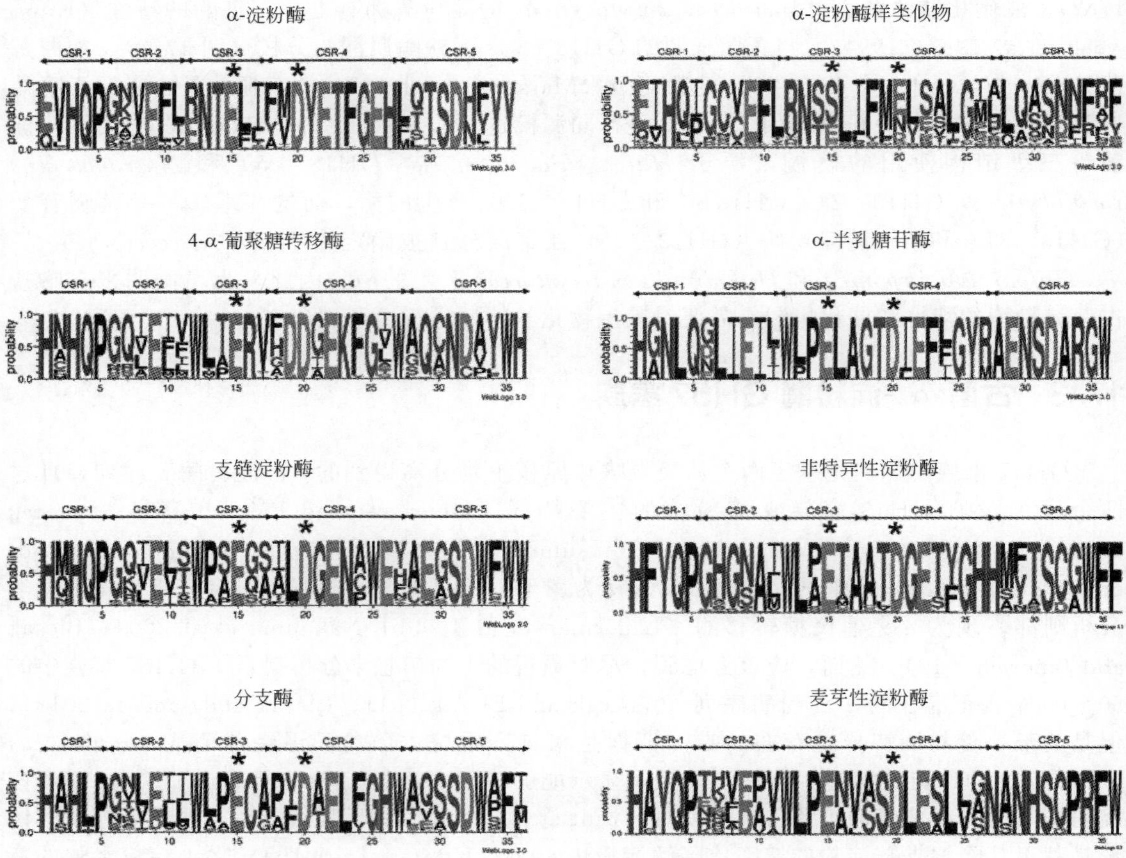

图 17.4　GH57 家族特征性酶序列特征分析。α-淀粉酶，56 个序列；α-淀粉酶样类似物，48 个序列；4-α-葡聚糖转移酶，46 个序列；支链淀粉酶，99 个序列；分支酶，158 个序列；α-半乳糖苷酶，8 个序列；非特异性淀粉酶，127 个序列；麦芽性淀粉酶，12 个序列。CSR-1，氨基酸序列 1-5；CSR-2，氨基酸序列 6-11；CSR-3，氨基酸序列 12-17；CSR-4，氨基酸序列 18-27；CSR-5，氨基酸序列 28-36。催化亲核试剂（No. 15，谷氨酸）和质子供体（No. 20，天冬氨酸）用星号表示（Janecek 和 Blesak 2011；Blesak 和 Janecek 2012，2013）

主要包含谷氨酸，位点 12（CSR-2）主要为精氨酸（或谷氨酸），其他的特性菌种在对应位点上分别为组氨酸和色氨酸（图 17.4）。位点 13（CSR-2）基本上由天冬酰胺残基占据。另一方面，位点 21（CSR-4）和 35-36（CSR-5）稳定存在的酪氨酸残基应该为主要功能位点，可用来区分 α-淀粉酶以及 α-淀粉酶样类似物（Janecek and Blesak 2011）。

值得注意的是，在分类谱中也可能有两种蛋白质发生：α-淀粉酶主要来源于古菌而 α-淀粉酶样类似物主要存在于细菌分类中（Janecek and Blesak 2011）。因为类似的现象在主要的 GH13 家族 α-淀粉酶中也被观察到（Janecek et al. 1997；Fort et al. 2007；Gabrisko and Janecek 2009），但不能被排除 GH57 家族 α-淀粉酶样类似物可能发挥不同于淀粉水解（酶）活性的作用。

17.4　结论

淀粉分解酶种类繁多，α-淀粉酶、β-淀粉酶和葡糖淀粉酶具有不同的演化历史（Janecek 1994a）。尽管 α-淀粉酶必须遵从 α-1,4-糖苷键的酶解机制，然而 α-淀粉酶仍是一个巨大的酶家族。淀粉酶可能具有更多的序列结构模板，目前已知的是 GH13 和 GH57（GH119）家族。古菌 α-淀粉酶主要存在于 GH13 和 GH57 家族中，具有其独特的功能和特性（Janecek et al. 2014）。α-淀粉酶、糖苷水解酶和葡聚糖转移酶的研究已相当深入，且在实际应用方面取得了一定成效（Bissaro et al. 2015）。

古菌 α-淀粉酶单独应用或与其他淀粉酶联合使用，例如，支链淀粉酶、Ⅱ型支链淀粉酶在多个工业生产应用中具有极大吸引力（Leveque et al. 2000b；Hii et al. 2012；Nisha and Satyanarayana 2013）。这些耐热淀粉酶可以替换掉来自细菌和真菌的热不稳定性淀粉酶。更重要的是，Ⅱ型支链淀粉酶与古菌 α-淀粉酶类似，在 GH13 和 GH57 家族中均有发现（Zona et al. 2004；Blesak and Janecek 2012；Jiao et al. 2013）。上面已经提及，GH57 家族包含其他与 α-淀粉酶相似的由古菌生产的淀粉酶，命名为麦芽糖化淀粉酶（或麦芽糖外切淀粉酶）（Jeon et al. 2014；Jung et al. 2014；Park et al. 2014），以及一些非特异性淀粉酶（Wang et al. 2011）。Elleuch 等人（2014，2015）关于极端微生物产生的酶类在工业上的应用的综述表明海洋极端微生物的研究值得科学家特别关注（Dalmaso et al. 2015）。

致谢

感谢斯洛伐克科研机构 VEGA（grant No. 2/0150/14）和斯洛伐克研究开发机构 APVV（contract No. LPP-0417-09）提供资金支持。

参考文献

Aghajari N，Roth M，Haser R（2002）Crystallographic evidence of a transglycosylation reaction：ternary complexes of a psychrophilic α-amylase. Biochemistry 41（13）：4273-4280

Ahmad N，Rashid N，Haider MS，Akram M，Akhtar M（2014）Novel maltotriose-hydrolyzing thermoacidophilic type III pullulan hydrolase from *Thermococcus kodakarensis*. Appl Environ Microbiol 80（3）：1108-1115

Atomi H，Fukui T，Kanai T，Morikawa M，Imanaka T（2004）Description of *Thermococcus kodakaraensis* sp. nov.，a well studied hyperthermophilic archaeon previously reported as *Pyrococcus* sp. KOD1. Archaea 1（4）：263-267

Bautista V，Esclapez J，Perez-Pomares F，Martinez-Espinosa RM，Camacho M，Bonete MJ（2012）Cyclodextrin glycosyltransferase：a key enzyme in the assimilation of starch by the halophilic archaeon *Haloferax mediterranei*. Extremophiles 16（1）：147-159

Bertoldo C，Antranikian G（2002）Starch-hydrolyzing enzymes from thermophilic archaea and bacteria. Curr Opin Chem Biol 6（2）：151-160

Bissaro B，Monsan P，Fauré R，O'Donohue MJ（2015）Glycosynthesis in a waterworld：new insight into the molecular basis of transglycosylation in retaining glycoside hydrolases. Biochem J 467（1）：17-35

Blesak K，Janecek S（2012）Sequence fingerprints of enzyme specificities from the glycoside hydrolase family GH57. Extremophiles 16（3）：497-506

Blesak K，Janecek S（2013）Two potentially novel amylolytic enzyme specificities in the prokaryotic glycoside hydrolase α-amylase family GH57. Microbiology 159（12）：2584-2593

Brown I，Dafforn TR，Fryer PJ，Cox PW（2013）Kinetic study of the thermal denaturation of a hyperthermostable extracellular α-amylase from *Pyrococcus furiosus*. Biochim Biophys Acta 1834（12）：2600-2605

Brzozowski AM，Davies GJ（1997）Structure of the *Aspergillus oryzae* α-amylase complexed with the inhibitor acarbose at 2.0 Å resolution. Biochemistry 36（36）：10837-10845

Bult CJ，White O，Olsen GJ，Zhou L，Fleischmann RD，Sutton GG，Blake JA，FitzGerald LM，Clayton RA，Gocayne JD，Kerlavage AR，Dougherty BA，Tomb JF，Adams MD，Reich CI，Overbeek R，Kirkness EF，Weinstock KG，Merrick JM，Glodek A，Scott JL，Geoghagen NS，Venter JC（1996）Complete genome sequence of the methanogenic archaeon，*Methanococcus jannaschii*. Science 273（5278）：1058-1073

Chen YS，Lee GC，Shaw JF（2006）Gene cloning，expression，and biochemical characterization of a recombinant trehalose synthase from *Picrophilus torridus* in *Escherichia coli*. J Agric Food Chem 54（19）：7098-7104

Coutiho PM，Henrissat B（1999）Life with no sugars J Mol Microbiol Biotechnol 1（2）：307-308

Coutinho PM，Reilly PJ（1997）Glucoamylase structural，functional，and evolutionary relationships. Proteins 29（3）：334-347

Da Lage JL，Feller G，Janecek S（2004）Horizontal gene transfer from Eukarya to bacteria and domain shuffling：the α-amylase model. Cell Mol Life Sci 61（1）：97-109

Da Lage JL，Binder M，Hua-Van A，Janecek S，Casane D（2013）Gene make-up：rapid and massive intron gains after horizontal transfer of a bacterial α-amylase gene to Basidiomycetes. BMC Evol Biol 13：40

Dalmaso GZ, Ferreira D, Vermelho AB (2015) Marine extremophiles: a source of hydrolases for biotechnological applications. Mar Drugs 13 (4): 1925-1965

Davies G, Henrissat B (1995) Structures and mechanisms of glycosyl hydrolases. Structure 3 (9): 853-859

Deshpande N, Addess KJ, Bluhm WF, Merino-Ott JC, Townsend-Merino W, Zhang Q, Knezevich C, Xie L, Chen L, Feng Z, Green RK, Flippen-Anderson JL, Westbrook J, Berman HM, Bourne PE (2005) The RCSB protein data bank: a redesigned query system and relational database based on the mmCIF schema. Nucleic Acids Res 3 (1): D233-D237

Dong G, Vieille C, Savchenko A, Zeikus JG (1997) Cloning, sequencing, and expression of the gene encoding extracellular α-amylase from *Pyrococcus furiosus* and biochemical characterization of the recombinant enzyme. Appl Environ Microbiol 63 (9): 3569-3576

Elleuche S, Schröder C, Sahm K, Antranikian G (2014) Extremozymes-biocatalysts with unique properties from extremophilic microorganisms. Curr Opin Biotechnol 29: 116-123

Elleuche S, Schäfers C, Blank S, Schröder C, Antranikian G (2015) Exploration of extremophiles for high temperature biotechnological processes. Curr Opin Microbiol 25: 113-119

Feese MD, Kato Y, Tamada T, Kato M, Komeda T, Miura Y, Hirose M, Hondo K, Kobayashi K, Kuroki R (2000) Crystal structure of glycosyltrehalose trehalohydrolase from the hyperthermophilic archaeum *Sulfolobus solfataricus*. J Mol Biol 301 (2): 451-464

Fort J, de la Ballina LR, Burghardt HE, Ferrer-Costa C, Turnay J, Ferrer-Orta C, Uson I, Zorzano A, Fernandez-Recio J, Orozco M, Lizarbe MA, Fita I, Palacin M (2007) The structure of human 4F2hc ectodomain provides a model for homodimerization and electrostatic interaction with plasma membrane. J Biol Chem 282 (43): 31444-31452

Frillingos S, Linden A, Niehaus F, Vargas C, Nieto JJ, Ventosa A, Antranikian G, Drainas C (2000) Cloning and expression of α-amylase from the hyperthermophilic archaeon *Pyrococcus woesei* in the moderately halophilic bacterium *Halomonas elongata*. J Appl Microbiol 88 (3): 495-503

Fukushima T, Mizuki T, Echigo A, Inoue A, Usami R (2005) Organic solvent tolerance of halophilic alpha-amylase from a Haloarchaeon, *Haloarcula* sp. strain S-1. Extremophiles 9 (1): 85-89

Fukusumi S, Kamizono A, Horinouchi S, Beppu T (1988) Cloning and nucleotide sequence of a heat-stable amylase gene from an anaerobic thermophile, *Dictyoglomus thermophilum*. Eur J Biochem 174 (1): 15-21

Gabrisko M, Janecek S (2009) Looking for the ancestry of the heavy-chain subunits of heteromeric amino acid transporters rBAT and 4F2hc within the GH13 α-amylase family. FEBS J 276 (24): 7265-7278

Godany A, Vidova B, Janecek S (2008) The unique glycoside hydrolase family 77 amylomaltase from *Borrelia burgdorferi* with only catalytic triad conserved. FEMS Microbiol Lett 284 (1): 84-91

Godany A, Majzlova K, Horvathova V, Vidova B, Janecek S (2010) Tyrosine 39 of GH13 α-amylase from *Thermococcus hydrothermalis* contributes to its thermostability. Biologia 65 (3): 408-415

Gupta R, Gigras P, Mohapatra H, Kumar Goswami V, Chauhan B (2003) Microbial α-amylases: a biotechnological perspective. Process Biochem 38 (11): 1599-1616

Hashimoto Y, Yamamoto T, Fujiwara S, Takagi M, Imanaka T (2001) Extracellular synthesis, specific recognition, and intracellular degradation of cyclomaltodextrins by the hyperthermophilic archaeon *Thermococcus* sp. strain B1001. J Bacteriol 183 (17): 5050-5057

Henrissat B (1991) A classification of glycosyl hydrolases based on amino acid sequence similarities. Biochem J 208 (2): 309-316

Henrissat B, Bairoch A (1996) Updating the sequence-based classification of glycosyl hydrolases. Biochem J 316 (2): 695-696

Hii SL, Tan JS, Ling TC, Ariff AB (2012) Pullulanase: role in starch hydrolysis and potential industrial applications. Enzyme Res 2012: 921362

Hondoh H, Kuriki T, Matsuura Y (2003) Three-dimensional structure and substrate binding of *Bacillus stearothermophilus* neopullulanase. J Mol Biol 326 (1): 177-188

Hondoh H, Saburi W, Mori H, Okuyama M, Nakada T, Matsuura Y, Kimura A (2008) Substrate recognition mechanism of α-1, 6-glucosidic linkage hydrolyzing enzyme, dextran glucosidase from *Streptococcus mutans*. J Mol Biol 378 (4): 913-922

Horvathova V, Godany A, Sturdik E, Janecek S (2006) α-Amylase from *Thermococcus hydrother- malis*: re-cloning aimed at the improved expression and hydrolysis of corn starch. Enzyme Microb Technol 39 (6): 1300-1305

Hostinova E, Janecek S, Gasperik J (2010) Gene sequence, bioinformatics and enzymatic characterization of α-amylase from *Saccharomycopsis fibuligera* KZ. Protein J 29 (5): 355-364

Hutcheon GW, Vasisht N, Bolhuis A (2005) Characterisation of a highly stable α-amylase from the halophilic archaeon *Haloarcula hispanica*. Extremophiles 9 (6): 487-495

Imamura H, Fushinobu S, Yamamoto M, Kumasaka T, Jeon BS, Wakagi T, Matsuzawa H (2003) Crystal structures of 4-α-glucanotransferase from *Thermococcus litoralis* and its complex with an inhibitor. J Biol Chem 278 (21): 19378-19386

Janecek S (1994a) Parallel β/α-barrels of α-amylase, cyclodextrin glycosyltransferase and oligo-1, 6-glucosidase versus the barrel of β-amylase: evolutionary distance is a reflection of unrelated sequences. FEBS Lett 353 (2): 119-123

Janecek S (1994b) Sequence similarities and evolutionary relationships of microbial, plant and animal α-amylases. Eur J Biochem 224 (2): 519-524

Janecek S (1997) α-Amylase family: molecular biology and evolution. Prog Biophys Mol Biol 67 (1): 67-97

Janecek S (1998) Sequence of archaeal *Methanococcus jannaschii* α-amylase contains features of families 13 and 57 of glycosyl hydrolases: a trace of their common ancestor? Folia Microbiol 43 (2): 123-128

Janecek S (2002) How many conserved sequence regions are there in the α-amylase family Biologia 57 (11): 29-41

Janecek S (2005) Amylolytic families of glycoside hydrolases: focus on the family GH-57. Biologia 60 (16): 177-184

Janecek S, Blesak K (2011) Sequence-structural features and evolutionary relationships of family GH57 α-amylases and their putative α-amylase-like homologues. Protein J 30 (6): 429-435

Janecek S, Kuchtova A (2012) In silico identification of catalytic residues and domain fold of the family GH119 sharing the catalytic machinery with the α-amylase family GH57. FEBS Lett 586 (19): 3360-3366

Janecek S, Svensson B, MacGregor EA (1995) Characteristic differences in the primary structure allow discrimination of cyclodextrin glucanotransferases from α-amylases. Biochem J 305 (2): 685-686

Janecek S, Svensson B, Henrissat B (1997) Domain evolution in the α-amylase family. J Mol Evol 45 (3): 322-331

Janecek S, Leveque E, Belarbi A, Haye B (1999) Close evolutionary relatedness of α-amylases from Archaea and plants. J Mol Evol 48 (4): 421-426

Janecek S, Svensson B, MacGregor EA (2014) α-Amylase-an enzyme specificity found in various families of glycoside hydrolases. Cell Mol Life Sci 71 (7): 1149-1170

Jeanmougin F, Thompson JD, Gouy M, Higgins DG, Gibson TJ (1998) Multiple sequence alignment with Clustal X. Trends Biochem Sci 23 (10): 403-405

Jeon EJ, Jung JH, Seo DH, Jung DH, Holden JF, Park CS (2014) Bioinformatic and biochemical analysis of a novel maltose-forming α-amylase of the GH57 family in the hyperthermophilic archaeon *Thermococcus* sp. CL1. Enzyme Microb Technol 60: 9-15

Jespersen HM, MacGregor EA, Sierks MR, Svensson B (1991) Comparison of the domain-level organization of starch hydrolases and related enzymes. Biochem J 280 (1): 51-55

Jiao YL, Wang SJ, Lv MS, Fang YW, Liu S (2013) An evolutionary analysis of the GH57 amylopullulanases based on the DOMON _ glucodextranase _ like domains. J Basic Microbiol 53 (3): 231-239

Jones RA, Jermiin LS, Easteal S, Patel BK, Beacham IR (1999) Amylase and 16S rRNA genes from a hyperthermophilic archaebacterium. J Appl Microbiol 86 (1): 93-107

Jorgensen S, Vorgias CE, Antranikian G (1997) Cloning, sequencing, characterization, and expression of an extracellular α-amylase from the hyperthermophilic archaeon *Pyrococcus furiosus* in *Escherichia coli* and *Bacillus subtilis*. J Biol Chem 272 (26): 16335-16342

Jung TY, Li D, Park JT, Yoon SM, Tran PL, Oh BH, Janecek S, Park SG, Woo EJ, Park KH (2012) Association of novel domain in active site of archaic hyperthermophilic maltogenic amylase from *Staphylothermus marinus*. J Biol Chem 287 (11): 7979-7989

Jung JH, Seo DH, Holden JF, Park CS (2014) Maltose-forming α-amylase from the hyperthermophilic archaeon *Pyrococcus* sp. ST04. Appl Microbiol Biotechnol 98 (5): 2121-2131

Kadziola A, Søgaard M, Svensson B, Haser R (1998) Molecular structure of a barley α-amylase-inhibitor complex: implications for starch binding and catalysis. J Mol Biol 278 (1): 205-217

Kelley LA, Sternberg MJ (2009) Protein structure prediction on the Web: a case study using the Phyre server. Nat Protoc 4 (3): 363-371

Kim HO, Park JN, Sohn HJ, Shin DJ, Choi C, Im SY, Lee HB, Chun SB, Bai S (2000) Cloning and expression in *Saccharomyces cerevisiae* of a β-amylase gene from the oomycete *Saprolegnia ferax*. Biotechnol Lett 22 (18): 1493-1498

Kim HO, Park JN, Shin DJ, Lee HB, Chun SB, Bai S (2001a) A gene encoding *Achlya bisexualis* β-amylase and its expression in *Saccharomyces cerevisiae*. Biotechnol Lett 23 (14): 1101-1107

Kim JW, Flowers LO, Whiteley M, Peeples TL (2001b) Biochemical confirmation and characterization of the family-57-like α-amylase of *Methanococcus jannaschii*. Folia Microbiol 46 (6): 467-473

Kobayashi T, Kanai H, Hayashi T, Akiba T, Akaboshi R, Horikoshi K (1992) Haloalkaliphilic maltotriose-forming α-amylase from the archaebacterium *Natronococcus* sp. strain Ah-36. J Bacteriol 174 (11): 3439-3444

Kobayashi T, Kanai H, Aono R, Horikoshi K, Kudo T (1994) Cloning, expression, and nucleotide sequence of the α-amylase gene from the haloalkaliphilic archaeon *Natronococcus* sp. strain Ah-36. J Bacteriol 176 (16): 5131-5134

Kobayashi K, Kato M, Miura Y, Kettoku M, Komeda T, Iwamatsu A (1996) Gene cloning and expression of new trehalose-producing enzymes from the hyperthermophilic archaeum *Sulfolobus solfataricus* KM1. Biosci Biotechnol Biochem 60 (11): 1882-1885

Kobayashi M, Kubota M, Matsuura Y (2003) Refined structure and functional implications of trehalose synthase from *Sulfolobus acidocaldarius*. J Appl Glycosci 50 (1): 1-8

Kobayashi M, Saburi W, Nakatsuka D, Hondoh H, Kato K, Okuyama M, Mori H, Kimura A, Yao M (2015) Structural insights into the catalytic reaction that is involved in the reorientation of Trp238 at the substrate-binding site in GH13 dextran glucosidase. FEBS Lett 589 (4): 484-489

Koch R, Zablowski P, Spreinat A, Antranikian G (1990) Extremely thermostable amylolytic enzyme from the archaebacterium *Pyrococcus furiosus*. FEMS Microbiol Lett 71 (1-2): 21-26

Koch R, Spreinat A, Lemke K, Antranikian G (1991) Purification and properties of a hyperthermoactive α-amylase from the archaeobacterium *Pyrococcus woesei*. Arch Microbiol 155 (6): 572-578

Kumar P, Satyanarayana T (2009) Microbial glucoamylases: characteristics and applications. Crit Rev Biotechnol 29 (3): 225-255

Laderman KA, Asada K, Uemori T, Mukai H, Taguchi Y, Kato I, Anfinsen CB (1993a) α-Amylase from the hyperthermophilic archaebacterium *Pyrococcus furiosus*. Cloning and sequencing of the gene and expression in *Escherichia co-*

li. J Biol Chem 268 (32): 24402-24407

Laderman KA, Davis BR, Krutzsch HC, Lewis MS, Griko YV, Privalov PL, Anfinsen CB (1993b) The purification and characterization of an extremely thermostable α-amylase from the hyperthermophilic archaebacterium *Pyrococcus furiosus*. J Biol Chem 268 (32): 24394-24401

Lee HS, Shockley KR, Schut GJ, Conners SB, Montero CI, Johnson MR, Chou CJ, Bridger SL, Wigner N, Brehm SD, Jenney FE Jr, Comfort DA, Kelly RM, Adams MW (2006) Transcriptional and biochemical analysis of starch metabolism in the hyperthermophilic archaeon *Pyrococcus furiosus*. J Bacteriol 188 (6): 2115-2125

Leemhuis H, Rozeboom HJ, Dijkstra BW, Dijkhuizen L (2003a) The fully conserved Asp residue in conserved sequence region I of the α-amylase family is crucial for the catalytic site architecture and activity. FEBS Lett 541 (1-3): 47-51

Leemhuis H, Rozeboom HJ, Wilbrink M, Euverink GJ, Dijkstra BW, Dijkhuizen L (2003b) Conversion of cyclodextrin glycosyltransferase into a starch hydrolase by directed evolution: the role of alanine 230 in acceptor subsite +1. Biochemistry 42 (24): 7518-7526

Lei Y, Peng H, Wang Y, Liu Y, Han F, Xiao Y, Gao Y (2012) Preferential and rapid degradation of raw rice starch by an α-amylase of glycoside hydrolase subfamily GH13 _ 37. Appl Microbiol Biotechnol 94 (6): 1577-1584

Leveque E, Haye B, Belarbi A (2000a) Cloning and expression of an α-amylase encoding gene from the hyperthermophilic archaebacterium *Thermococcus hydrothermalis* and biochemical characterisation of the recombinant enzyme. FEMS Microbiol Lett 186 (1): 67-71

Leveque E, Janecek S, Belarbi A, Haye B (2000b) Thermophilic archaeal amylolytic enzymes. Enzyme Microb Technol 26 (1): 2-13

Li C, Begum A, Numao S, Park KH, Withers SG, Brayer GD (2005) Acarbose rearrangement mechanism implied by the kinetic and structural analysis of human pancreatic α-amylase in complex with analogues and their elongated counterparts. Biochemistry 44 (9): 3347-3357

Li D, Park JT, Li X, Kim S, Lee S, Shim JH, Park SH, Cha J, Lee BH, Kim JW, Park KH (2010) Overexpression and characterization of an extremely thermostable maltogenic amylase, with an optimal temperature of 100 degrees C, from the hyperthermophilic archaeon *Staphylothermus marinus*. N Biotechnol 27 (4): 300-307

Li X, Li D, Park KH (2013) An extremely thermostable amylopullulanase from *Staphylothermus marinus* displays both pullulan- and cyclodextrin-degrading activities. Appl Microbiol Biotechnol 97 (12): 5359-5369

Li C, Du M, Cheng B, Wang L, Liu X, Ma C, Yang C, Xu P (2014) Close relationship of a novel Flavobacteriaceae α-amylase with archaeal α-amylases and good potentials for industrial applications. Biotechnol Biofuels 7 (1): 18

Liebl W, Stemplinger I, Ruile P (1997) Properties and gene structure of the *Thermotoga maritima* α-amylase AmyA, a putative lipoprotein of a hyperthermophilic bacterium. J Bacteriol 179 (3): 941-948

Lim JK, Lee HS, Kim YJ, Bae SS, Jeon JH, Kang SG, Lee JH (2007) Critical factors to high thermostability of an α-amylase from hyperthermophilic archaeon *Thermococcus onnurineus* NA1. J Microbiol Biotechnol 17 (8): 1242-1248

Linden A, Wilmanns M (2004) Adaptation of class-13 α-amylases to diverse living conditions. Chembiochem 5 (2): 231-239. doi: 10. 1002/cbic. 200300734

Linden A, Mayans O, Meyer-Klaucke W, Antranikian G, Wilmanns M (2003) Differential regula- tion of a hyperthermophilic α-amylase with a novel (Ca, Zn) two-metal center by zinc. J Biol Chem 278 (11): 9875-9884

Lombard V, Golaconda Ramulu H, Drula E, Coutinho PM, Henrissat B (2014) The carbohydrateactive enzymes database (CAZy) in 2013. Nucleic Acids Res 42 (Database Issue): D490-D495 MacGregor EA, Svensson B (1989) A super-secondary structure predicted to be common to several α-1, 4-D-glucan-cleaving enzymes. Biochem J 259 (1): 145-152

MacGregor EA, Jespersen HM, Svensson B (1996) A circularly permuted α-amylase-type α/β-- barrel structure in glucan-synthesizing glucosyltransferases. FEBS Lett 378 (3): 263-266

MacGregor EA, Janecek S, Svensson B (2001) Relationship of sequence and structure to specificity in the α-amylase family of enzymes. Biochim Biophys Acta 1546 (1): 1-20

Machius M, Declerck N, Huber R, Wiegand G (1998) Activation of *Bacillus licheniformis* α-amylase through a disorder→order transition of the substrate-binding site mediated by a calcium-sodium-calcium metal triad. Structure 6 (3): 281-292

Majzlova K, Pukajova Z, Janecek S (2013) Tracing the evolution of the α-amylase subfamily GH13 _ 36 covering the amylolytic enzymes intermediate between oligo-1, 6-glucosidases and neopullulanases. Carbohydr Res 367: 48-57

Marin-Navarro J, Polaina J (2011) Glucoamylases: structural and biotechnological aspects. Appl Microbiol Biotechnol 89 (5): 1267-1273

Maruta K, Mitsuzumi H, Nakada T, Kubota M, Chaen H, Fukuda S, Sugimoto T, Kurimoto M (1996) Cloning and sequencing of a cluster of genes encoding novel enzymes of trehalose biosynthesis from thermophilic archaebacterium *Sulfolobus acidocaldarius*. Biochim Biophys Acta 1291 (3): 177-181

Matsuura Y, Kusunoki M, Harada W, Kakudo M (1984) Structure and possible catalytic residues of Taka-amylase A. J Biochem 95 (3): 697-702

McCarter JD, Withers SG (1994) Mechanisms of enzymatic glycoside hydrolysis. Curr Opin Struct Biol 4 (6): 885-892

Moshfegh M, Shahverdi AR, Zarrini G, Faramarzi MA (2013) Biochemical characterization of an extracellular polyextremophilic α-amylase from the halophilic archaeon *Halorubrum xinjiangense*. Extremophiles 17 (4): 677-687

Nakajima M, Imamura H, Shoun H, Horinouchi S, Wakagi T (2004) Transglycosylation activity of *Dictyoglomus thermophilum* amylase A. Biosci Biotechnol Biochem 68 (11): 2369-2373

Niehaus F, Peters A, Groudieva T, Antranikian G (2000) Cloning, expression and biochemical characterisation of a unique thermostable pullulan-hydrolysing enzyme from the hyperthermophilic archaeon *Thermococcus aggregans*. FEMS Microbiol Lett 190 (2): 223-229

Nisha M, Satyanarayana T (2013) Recombinant bacterial amylopullulanases: developments and perspectives. Bioengineered 4 (6): 388-400

Onodera M, Yatsunami R, Tsukimura W, Fukui T, Nakasone K, Takashina T, Nakamura S (2013) Gene analysis, expression, and characterization of an intracellular α-amylase from the extremely halophilic archaeon *Haloarcula japonica*. Biosci Biotechnol Biochem 77 (2): 281-288

Oslancova A, Janecek S (2002) Oligo-1, 6-glucosidase and neopullulanase enzyme subfamilies from the α-amylase family defined by the fifth conserved sequence region. Cell Mol Life Sci 59 (11): 1945-1959

Page RD (1996) TreeView: an application to display phylogenetic trees on personal computers. Comput Appl Biosci 12 (4): 357-358

Palomo M, Pijning T, Booiman T, Dobruchowska JM, van der Vlist J, Kralj S, Planas A, Loos K, Kamerling JP, Dijkstra BW, van der Maarel MJ, Dijkhuizen L, Leemhuis H (2011) *Thermus thermophilus* glycoside hydrolase family 57 branching enzyme: crystal structure, mechanism of action, and products formed. J Biol Chem 286 (5): 3520-3530

Pandey A, Nigam P, Soccol CR, Soccol VT, Singh D, Mohan R (2000) Advances in microbial amylases. Biotechnol Appl Biochem 31 (2): 135-152

Park JT, Song HN, Jung TY, Lee MH, Park SG, Woo EJ, Park KH (2013) A novel domain arrangement in a monomeric cyclodextrin-hydrolyzing enzyme from the hyperthermophile *Pyrococcus furiosus*. Biochim Biophys Acta 1834 (1): 380-386

Park KH, Jung JH, Park SG, Lee ME, Holden JF, Park CS, Woo EJ (2014) Structural features underlying the selective cleavage of a novel exo-type maltose-forming amylase from *Pyrococcus* sp. ST04. Acta Crystallogr D Biol Crystallogr 70 (6): 1659-1668. doi: 10.1107/S1399004714006567

Perez-Pomares F, Bautista V, Ferrer J, Pire C, Marhuenda-Egea FC, Bonete MJ (2003) α-Amylase activity from the halophilic archaeon *Haloferax mediterranei*. Extremophiles 7 (4): 299-306

Przylas I, Tomoo K, Terada Y, Takaha T, Fujii K, Saenger W, Straeter N (2000) Crystal structure of amylomaltase from *Thermus aquaticus*, a glycosyltransferase catalysing the production of large cyclic glucans. J Mol Biol 296 (3): 873-886

Pujadas G, Ramirez FM, Valero R, Palau J (1996) Evolution of β-amylase: patterns of variation and conservation in subfamily sequences in relation to parsimony mechanisms. Proteins 25 (4): 456-472

Puspasari F, Radjasa O, Noer A, Nurachman Z, Syah Y, van der Maarel M, Dijkhuizen L, Janecek S, Natalia D (2013) Raw starch degrading α-amylase from *Bacillus aquimaris* MKSC 6.2: isolation and expression of the gene, bioinformatics and biochemical characterization of the recombinant enzyme. J Appl Microbiol 114 (1): 108-120

Ramasubbu N, Ragunath C, Mishra PJ (2003) Probing the role of a mobile loop in substrate binding and enzyme activity of human salivary amylase. J Mol Biol 325 (5): 1061-1076

Rashid N, Cornista J, Ezaki S, Fukui T, Atomi H, Imanaka T (2002) Characterization of an archaeal cyclodextrin glucanotransferase with a novel C-terminal domain. J Bacteriol 184 (3): 777-784

Ray RR, Nanda G (1996) Microbial β-amylases: biosynthesis, characteristics, and industrial applications. Crit Rev Microbiol 22 (3): 181-199

Robert X, Haser R, Gottschalk TE, Ratajczak F, Driguez H, Svensson B, Aghajari N (2003) The structure of barley α-amylase isozyme 1 reveals a novel role of domain C in substrate recognition and binding: a pair of sugar tongs. Structure 11 (8): 973-984

Sauer J, Sigurskjold BW, Christensen U, Frandsen TP, Mirgorodskaya E, Harrison M, Roepstorff P, Svensson B (2000) Glucoamylase: structure/function relationships, and protein engineering. Biochim Biophys Acta 1543 (2): 275-293

Savchenko A, Vieille C, Kang S, Zeikus JG (2002) *Pyrococcus furiosus* α-amylase is stabilized by calcium and zinc. Biochemistry 41 (19): 6193-6201

Sharma A, Satyanarayana T (2013) Microbial acid-stable α-amylases: characteristics, genetic engineering and applications. Process Biochem 48 (2): 201-211

Sivakumar N, Li N, Tang JW, Patel BK, Swaminathan K (2006) Crystal structure of AmyA lacks acidic surface and provide insights into protein stability at poly-extreme condition. FEBS Lett 580 (11): 2646-2652

Stam MR, Danchin EG, Rancurel C, Coutinho PM, Henrissat B (2006) Dividing the large glycoside hydrolase family 13 into subfamilies: towards improved functional annotations of α-amylase-related proteins. Protein Eng Des Sel 19 (12): 555-562

SunY, Lv X, Li Z, Wang J, Jia B, Liu J (2015) Recombinant cyclodextrinase from *Thermococcus kodakarensis* KOD1: expression, purification, and enzymatic characterization. Archaea 2015: 397924

Svensson B (1988) Regional distant sequence homology between amylases, α-glucosidases and transglucanosylases. FEBS Lett 230 (1-2): 72-76

Tachibana Y, Mendez Leclere M, Fujiwara S, Takagi M, Imanaka T (1996) Cloning and expression of the α-amylase gene from the hyperthermophilic archaeon Pyrococcus sp. KOD1, and characterization of the enzyme. J Ferment Bioeng 82 (3): 224-232

Takata H, Kuriki T, Okada S, Takesada Y, Iizuka M, Minamiura N, Imanaka T (1992) Action of neopullulanase. Neopullulanase catalyzes both hydrolysis and transglycosylation at α- (1, 4) - and α- (1, 6) -glucosidic linkages. J Biol Chem 267 (26): 18447-18452

Uitdehaag JC, Mosi R, Kalk KH, van der Veen BA, Dijkhuizen L, Withers SG, Dijkstra BW (1999) X-ray structures along the reaction pathway of cyclodextrin glycosyltransferase elucidate catalysis in the α-amylase family. Nat Struct Biol 6 (5): 432-436

UniProt Consortium (2015) UniProt: a hub for protein information. Nucleic Acids Res 43 (Database Issue): D204-D212

van der Kaaij RM, Janecek S, van der Maarel MJEC, Dijkhuizen L (2007) Phylogenetic and biochemical characterization of a novel cluster of intracellular fungal α-amylase enzymes. Microbiology 153 (12): 4003-4015

van der Maarel MJEC, van der Veen B, Uitdehaag JC, Leemhuis H, Dijkhuizen L (2002) Properties and applications of starch-converting enzymes of the α-amylase family. J Biotechnol 94 (2): 137-155

Vihinen M, Mantsala P (1989) Microbial amylolytic enzymes. Crit Rev Biochem Mol Biol 24 (4): 329-418

Vujicic-Zagar A, Pijning T, Kralj S, Lopez CA, Eeuwema W, Dijkhuizen L, Dijkstra BW (2010) Crystal structure of a 117kDa glucansucrase fragment provides insight into evolution and product specificity of GH70 enzymes. Proc Natl Acad Sci U S A 107 (50): 21406-21411

Wang S, Lu Z, Lu M, Qin S, Liu H, Deng X, Lin Q, Chen J (2008) Identification of archaeonproducing hyperthermophilic α-amylase and characterization of the α-amylase. Appl Microbiol Biotechnol 80 (4): 605-614

Wang H, Gong Y, Xie W, Xiao W, Wang J, Zheng Y, Hu J, Liu Z (2011) Identification and charac- terization of a novel thermostable *gh*-57 gene from metagenomic fosmid library of the Juan de Fuca Ridge hydrothemal vent. Appl Biochem Biotechnol 164 (8): 1323-1338

Zona R, Chang-Pi-Hin F, O'Donohue MJ, Janecek S (2004) Bioinformatics of the family 57 glyco- side hydrolases and identification of catalytic residues in amylopullulanase from *Thermococcus hydrothermalis*. Eur J Biochem 271 (14): 2863-2872

Zorgani MA, Patron K, Desvaux M (2014) New insight in the structural features of haloadaptation in α-amylases from halophilic *Archaea* following homology modeling strategy: folded and stable conformation maintained through low hydrophobicity and highly negative charged sur- face. J Comput Aided Mol Des 28 (7): 721-734

第十八章

极端嗜热菌DNA复制所需的蛋白质：PCR与其他技术

Dennis W. Grogan[1]

18.1 概述

极端嗜热菌和超嗜热菌的固有热稳定蛋白质已被用于开发一系列广泛应用的生物技术，相关技术也用于实验分子生物学、环境科学、法医学及医学相关的诊断。本章以 DNA 复制相关蛋白质作为参照粗略概述这些生物技术的基本原理。首先总结了细胞中 DNA 复制的基本特征，然后介绍传统的 PCR 方法。这一比较有利于认识 PCR 技术的优化以及技术的发展过程，所有这些方法均利用了嗜热古菌和细菌的固有热稳定蛋白质（表 18.1）。尽管单亚基 DNA 聚合酶仍然是本章所描述的大多数 PCR 技术的核心，但技术的创新通常意味着用其他嗜热菌（包括一些从未培养过的）的蛋白质来补充或修饰相应工具酶。

表 18.1　DNA 合成中的固有热稳定蛋白质

类型		技术用途	相关性质
DNA 聚合酶类	细菌，A 家族	常规 PCR 方法	稳健伸长
	细菌，噬菌体	全基因组扩增	链置换聚合
	古菌，B 家族	高保真 PCR	校对（3′外切核酸酶）；结合模板 dU
	古菌，Y 家族	受损 DNA 的 PCR	绕过病变，非连续性
	DNA 结合蛋白	融合到 DNA 聚合酶	持续的增加
	DNA 连接酶	单核苷酸基因分型	高度特异性用于正确的碱基配对
准确性/修复蛋白质	MutS 蛋白	SNP 检测	结合双链 DNA 的错配
	脱氧尿苷三磷酸酶（dUTPase）	高保真 PCR	防止 dUTP 掺入
	尿嘧啶 DNA 糖基化酶	高保真 PCR	从 DNA 中提取 dU

[1] Department of Biological Sciences，University of Cincinnati，614 Rieveschl Hall，Cincinnati，OH 45221-0006，USA. e-mail：grogandw@ucmail. uc. edu.

18.2 生物的 DNA 复制

细胞通常会对 DNA 合成的时间进行严格控制。这种受控启动的生化策略在生物体中不同，但都涉及蛋白质的调控，这些蛋白质能在一个或多个特定位点（复制起点）解开双链 DNA，以响应某种形式的细胞信号。尽管古菌、细菌和真核生物使用酶不同，但其后续步骤在所有细胞中是非常相似的。O'Donnell 等（2013）总结了 DNA 复制过程分子特征的详细信息以及它们在三域生物之间的不同点。

一旦启动复制，具有种属特异性的蛋白质解开双链 DNA 形成巨大的复制泡以允许解旋酶进一步装配。这种解旋酶是一种蛋白质复合物，围绕一条 DNA 链并由 ATP 水解驱动沿着单链 DNA 移动，从而使 DNA 双链解开。每条 DNA 单链上各结合一个解旋酶复合物，解旋酶复合物从原点出发，逐渐扩大复制泡的大小。每个解旋酶成为相应复制叉的中心。接下来着重描述了由解旋酶产生的含两个臂的复制叉。

聚合反应在复制叉的两个臂上同时进行，但方向相反，这取决于反向平行双链 DNA 的方向。在 1 条 DNA 链上，“前导链”以与复制叉前进相同的方向聚合，并且该过程原则上可以无限继续。相比之下，“后随链”延伸方向与复制叉的前进方向相反，这一过程由于解旋酶不断产生新的 ssDNA（单链 DNA），需要频繁重复下面的合成反应：①DNA 引物酶使用 ssDNA 作为模板合成短链 RNA；②DNA 聚合酶识别 RNA 引物并从其 3′端复制延伸，环状蛋白质复合物（“滑卡”）环绕 DNA 以维持模板上的聚合酶；③DNA 聚合酶到达 RNA 引物结合位置并分离；④RNA 引物脱离并由 DNA 替换，仅留下片段间切口；⑤最后由 DNA 连接酶连接封闭缺口（O'Donnell et al. 2013）。

两条单链复制完成后，还需进行最后一个过程，即合成片段的 DNA 修复（DNA 错配修复）过程。错配修复蛋白质（大肠杆菌 MutS 和 MutL 蛋白的同源物）扫描双链 DNA。当发现错配时，这些蛋白质破坏 DNA 链，产生约 1 kb 长的间隙，再通过 DNA 聚合酶重新合成间隙片段，从而校正初始合成产生的误差（Li 2008）。

18.3 聚合酶链式反应

尽管聚合酶链式反应（PCR）与 DNA 生物合成存在相似性，但与 DNA 生物合成相比，常规 PCR 更为简单，仅涉及三个步骤和一个蛋白质。与 DNA 生物合成相似，双链 DNA 不是 DNA 合成的底物，因此在 PCR 中，当组分混合（模板 DNA、引物、dNTP 和热稳定性 DNA 聚合酶）时反应并不能进行，混合体系须加热至 95℃使所有双链体 DNA 变性后才能开始 PCR 反应。作为单分子反应，变性在几秒钟内完成，然后冷却至退火温度使引物（通常两个合成的寡核苷酸）特异性地与 ssDNA 片段互补。由于引物浓度通常相对较高，所以第二步反应也是快速的，遵循“准一级”动力学。

引物经退火后与模板链通过碱基配对结合，然后被 DNA 聚合酶单体识别并开始延伸。引物退火的配置对于成功扩增模板 DNA 的区间至关重要。寡核苷酸必须结合到模板链中几个相距较近（通常几千碱基对）的位点，进而允许从一个位点成功延伸到另一个位点，并且两个 3′末端必须定向到该区间的内部。在第三步反应中，体系温度升至 72℃，有利于聚合酶的延伸反应，该过程必须持续到扩增片段 3′末端，即延伸到反向引物的结合位点再结束。然后重复这三步循环（变性、引物退火、引物延伸），每次循环会使体系中目标 DNA 片段的量加倍，其长度为两个引物结合位点之间的距离。

PCR 与基因组复制尽管具有相应的功能结果，但其实现过程存在明显不同。在 PCR 中，高温代替体内的解旋酶实现了模板链的解链，并且外源引物将复制限制在固定区域内。该区域通常仅对应于模板 DNA 序列的一小部分，但其复制速率非常高。PCR 的性能可以通过来自极端嗜热菌和超嗜热菌的某些蛋白质来加以提升（表 18.1），而其他热稳定蛋白质提供不同于 PCR 的其他 DNA 分析技术，具有它们自己的特殊用途。下文中我们将讨论这些蛋白质及其技术用途，同时还对它们影响和模拟天然 DNA 复制方面进行讨论。

18.4 时间控制和“热启动”技术

虽然起始的精确时间对大多数 PCR 应用来说并不存在问题，但在某些情况下，低温下（即在设置期间）残留的聚合酶活性可能会影响引物的特异性。例如，如果退火时引物 3′末端与 DNA 模板非靶位点或另一引物中的几个碱基对配对，并且延伸，则可能会产生新的引物，导致异常产物的产生。这样的问题已经通过各种“热启动”PCR 技术解决。这些方法中的大多数通过可逆抑制形式将 DNA 聚合酶添加到反应中，其仅在 PCR 变性加热阶段才恢复催化活性。在 PCR 中使用固有热稳定性 DNA 聚合酶可以使用高温进行可逆抑制，但是抑制方法可以有多种形式，包括抗体结合、试剂螯合和化学修饰等（Paul et al. 2010；Barnes and Rowlyk 2002；Kermekchiev et al. 2003；Kaboev et al. 2000；Louwrier and van de Valk 2005；Liu and Liu 2011）。

18.5 序列特异性及其对全基因组扩增与基因分型的影响

传统 PCR 的主要优点是其能特异性扩增较短片段的 DNA。然而，如果目标是不加选择地复制 DNA，即进行全基因组扩增，则必须克服该区域特异性问题。体外全基因组合成的最常用的方法之一是在恒温下无差别合成双链 DNA 的“多重置换扩增”（MDA）技术（Dean et al. 2002）。MDA 使用具有链置换（SD）活性的嗜热 DNA 聚合酶，这个技术以不同于 DNA 生物复制（ATP 依赖性解旋酶）和 PCR（热变性）的策略解决了模板链分离问题。SD 聚合酶自身就能起到解旋酶的作用。当引物延伸反应发生时，SD 聚合酶能置换它遇到的任何 DNA 链，将其转化成单链 DNA。被置换的 DNA 链成为新的模板，促使恒温 DNA 合成持续进行。该反应通常在恒定但适中的高温下进行，相应的聚合酶已从嗜热细菌、极端嗜热细菌及热环境里的噬菌体中分离和克隆得到（Hafner et al. 2001；Chander et al. 2014）。

要实现持续恒温合成，需及时提供和补充 DNA 合成引物。缺口的双链 DNA 可以作为新一轮合成的底物，并且可以通过添加核酸内切酶或类似活性酶产生新的片段缺口来触发新一轮的合成。在这些条件下，SD 聚合酶将以原始的两条链为模板以恒定速率催化单链 DNA（具有不均匀的长度分布）的连续产生。除了切口活性之外，如果加入低特异性（“随机”）引物，该过程产生的双链体 DNA 将以指数倍增长，这是因为通过合成置换的单链 DNA 为新一轮合成提供了引发位点（Dean et al. 2002）。然而，与 DNA 的生物体复制不同，新合成技术的连续性、随机起始位点等导致具有复杂结构、高度分支化的 DNA 的产生，这使得基因组产物的一些用途更加复杂化（对于 SD 和相关的等温 DNA 合成方法的完整概述参见 Yan et al. 2014）。

在某些遗传分析中，如对单核苷酸多态性（SNPs）进行评分，需要提供比普通 PCR 更精准的位点（即单个碱基对）和更高的严谨度，这是 PCR 特异性的另一个极端。原则上

Taq 酶不会延伸错配的 3′末端，并且这已经用于设计各种形式的等位基因特异性 PCR。然而实际上，在特定位置上一个碱基和另一个碱基之间的“有或无”的区分往往难以捉摸。DNA 连接酶则可在碱基之间提供更为可靠的鉴别手段，因为它们对于错配的 3′末端不具有连接酶活性。因此，模板上两个相邻的寡核苷酸探针仅仅在与上游探针的 3′末端相对的模板 DNA 中出现特定碱基时才能彼此成功连接。运行多个连接反应，每一个反应的上游探针的 3′末端具有不同的碱基，因此可以用于鉴别该位点哪个碱基发生了变化（或者在杂合二倍体的情况下是哪两个碱基发生了变化）。超嗜热古菌的连接酶在这些技术的应用中具有独特的优越性（Wiedmann et al. 1994）。这些酶的热稳定性允许其发生多轮热变性和连接，使连接产物进一步积累，更易于检测。此外，古菌连接酶通过消耗 ATP 形成磷酸二酯键，这相对于使用昂贵且不稳定的 NAD 作为供能物质的细菌热稳定连接酶，技术上更具优势。反应体系含有模板 DNA、杂交探针、ATP 和连接酶。类似于 PCR，连接反应通过热循环进行多轮退火和连接。如果只使用两个彼此相邻的探针进行循环反应，则每个后续循环的连接产物呈线性增加，该技术被称为连接检测反应（LDR）。此外，如果另外加入查询位点互补链的相应探针对，则连接产物将如同 PCR 中那样呈指数积累，因为在一轮中产生的连接产物可用作下一轮的模板，这种指数模式称为连接酶链反应（LCR）（Wiedmann et al. 1994）。

因为连接的产物短，所以通常利用毛细管或聚丙烯酰胺凝胶电泳技术进行检测。或者，在循环次数足够使产物（如果有的话）积累的情况下，可以使用溶解曲线检测产物是否存在。这种方法最重要的是连接产物具有与未连接探针和模板差异较大的 T_m 值。

另一种检测方法是使用不同的探针拓扑，其中检测位点的两侧分别与一条相对较长的单链 DNA 的两端互补，这种长链 DNA 称为“锁式探针”。滚环扩增（RCA）技术（Lizardi et al. 1998 ；Nilsson et al. 2006）如图 18.1 所示。如果探针末端对应模板链上的多态性位点，则该探针两端与模板 DNA 的碱基配对并通过连接酶产生小的单链环。尽管热稳定连接酶和热循环提供了增加循环产物的方法，但是需要使用单链 DNA 环作为模板来实现检测所需的扩增。在这种情况下，锁式探针的中心区域或“主链”的常规引物通过热稳定的 SD DNA 聚合酶启动滚环状 DNA 合成（图 18.1d，e）。探针通过连接而环化后，这种等温过程会产

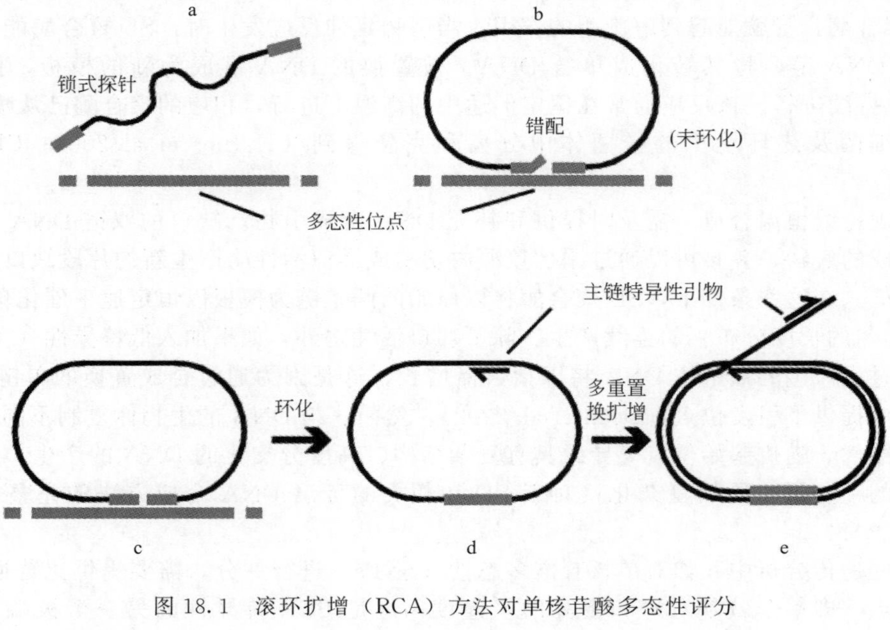

图 18.1　滚环扩增（RCA）方法对单核苷酸多态性评分

生探针序列的长多联体。根据这一原理的环化/滚环扩增变化的检测方法已经被设计出来并应用于高通量基因分型研究（Alsmadi et al. 2003；Yan et al. 2014；Chang et al. 2015)。

18.6　复制精度及其技术重要性

Taq 仍然是常规 PCR 最有效的 DNA 聚合酶之一，部分原因在于其缺乏 3′核酸外切酶活性，这使得链延伸快速且稳定，但这将导致扩增出现相对较高的错误率，尤其是在扩增产物包含待克隆的功能基因时，该问题显得尤为重要。例如，Taq 酶的复制错配率约为 3×10^{-4}（Tindall and Kunkel 1988)，这意味着 1 kbp 的靶基因扩增产生的产物链中将有高达 30%的片段会出现碱基错配。因为 PCR 中每个循环的产物会被用作下一个循环的模板，25 次循环的 PCR 反应将产生总计约 $(0.7)^{25}\approx10^{-4}$ 的预期正确的 DNA 分子。因此，要获得更为准确的基因扩增序列，则需要比 Taq 酶聚合精度更高的聚合酶。

早期解决这个问题的方案是使用具有校对功能的热稳定性 DNA 聚合酶替换 Taq 酶。一般是来自极端或超嗜热古菌的 B-家族 DNA 聚合酶。与 Taq 酶相反，这些 DNA 聚合酶具有强的 3′核酸外切酶（校对）活性。这将导致扩增过程中较低的错配率，虽然它也可能存在造成 PCR 过程中引物缩短的风险。由于超嗜热古菌（HA）比栖热菌属（*Thermus*）生长温度要高，因此它们的聚合酶通常热稳定性更强，在 PCR 程序的后期循环中活性损失较少。

古菌 DNA 聚合酶的这些优点使 PCR 及其相关技术复杂化，然而对于其生物学意义仍然是未知的。古菌 DNA 聚合酶能紧密结合模板链中的脱氧尿苷（dU)，从而阻碍新生链合成（Lasken et al. 1996)。研究者已经以各种方式克服了该技术上的障碍，提高了 PCR 的性能，并阐明了 dU 引起的终止合成的生物学意义。通过在 HA 的 DNA 聚合酶（例如 *Pfu* 高保真酶）中加入热稳定脱氧尿苷三磷酸酶（dUTPase）能明显改善 PCR 的性能（Hogrefe et al. 2002)。该技术成功表明，在 PCR 过程中遇到的大部分模板 dU 是由胞嘧啶脱氨而引起，而不是在引入之后才有的。这与体外研究一致，表明 HA 的 DNA 聚合酶对 dUTP 的识别能力较差（Gruz et al. 2003)。另一种有效的方法是使用热稳定的尿嘧啶 DNA 糖基化酶（UDG）作为补充剂加入古菌聚合酶中（Liu and Liu 2011)。由于 UDG 的作用是在循环反应中破坏含有 dU 的模板链，这也意味着由 dU 结合引起的问题之一就是其他活性 DNA 聚合酶分子的活性丧失。这也表明非产品化的聚合酶，dU 复合物是非常稳定的，并且提出这种复合物在体内将具有什么影响的问题。解决上述问题的第三种方法是通过鉴定 HA 的 DNA 聚合酶中的介导反应延滞的特异性 dU 结合位点（Fogg et al. 2002)。构建 HA 聚合酶突变体去除该袋状结构域将使得聚合酶具备极强的热稳定和准确性，同时该突变体对其模板中 dU 不敏感（Gaidamaviciute et al. 2010)。

活细胞中 DNA 复制的极高准确度是由于复制后错配修复（MMR）的作用。与具有校正功能的聚合酶相比，体内复制可以将错误率降至 $\frac{1}{1000}\sim\frac{1}{100}$。MMR 的关键步骤由大肠杆菌 MutS 和 MutL 蛋白的同源物介导，尽管整个过程仍然难以在体外模拟（Li 2008)。复制后错配修复的第一步是 MutS 同源物与错配碱基部位结合，该结合方式可以用于体外复制错配的检测。通过荧光标记 *Thermus* 属的热稳定 MutS 蛋白，其结合位点通常是碱基错配位置，虽然不同的错配类型检测效率不同（Cho et al. 2007)。然而，超嗜热古菌不具有 MutS 和 MutL 蛋白，并且这种缺失的生物学意义仍不清楚（White and Grogan 2008)。在其他特定条件下，例如定向实验进化中，需要在基因中引入多种变化，常规 PCR 使用 Taq 酶的精度足以满足这类实验的需求。然而，反应中加入 Mn^{2+} 且单种 dNTP 的浓度不均一将使得

Taq 酶介导的 PCR 反应存在较高的错配率。突变的程度可以通过适当调整反应条件来控制，并且据报道这种方法的突变率要高于纯化学或纯生物诱变（Rasila et al. 2009）。

原则上，生物体内产生的极端和超嗜热聚合酶的突变体可作为诱变 PCR 的替代途径。目前对易错热稳定性 DNA 聚合酶研究较多，例如典型的超嗜热古菌的 Y-家族 DNA 聚合酶，目前已经获得其与多种 DNA 底物结合的结晶体并广泛用于生物化学的研究。大的催化位点和校正活性的缺失允许这些酶插入不同 DNA 损伤的核苷酸，这也暗示着其生物学功能可能就是介导转录损伤 DNA 合成（TLS）。这些 TLS 聚合酶的催化特征还限制了复制完整模板的准确性，使得其中一些（例如嗜酸热硫化叶菌 *Sulfolobus acidocaldarius* 产生的 Dbh）成为已知的最易错的 DNA 聚合酶（Yang and Woodgate 2007；Pata 2010）。然而，与限制损伤旁路遗传成本的生物学必要性一致，所有 TLS 聚合酶（包括极其热稳定的）均具有低内在持续合成能力，因此每次只合成少数碱基对（Pata 2010；Waters et al. 2009）。原则上，通过蛋白质修饰提高酶的合成效率或简单地提高反应体系中的酶浓度可以为易错 PCR 提供一种替代形式。然而，观察到 *Sulfolobus* 酶的一些特性（Potapova et al. 2002；Wilson and Pata 2008），认为它们可能产生高比例的移码突变，这对于产生功能基因多样性是不可接受的。这种限制是否适用于其他固有热稳定的 TLS 聚合酶（例如来自极端嗜热细菌的 TLS 聚合酶）还需进一步确定。

18.7 损伤旁路和损伤 DNA 的扩增

尽管超嗜热古菌的 Y-家族热稳定聚合酶尚未被用作诱变剂，但它们在某些领域显示着巨大的应用潜力，如较弱的特异性有利于片段的延伸并且使得分析难扩增的样品成为可能，这与 TLS 聚合酶支持 DNA 损伤复制的生物学功能一致。添加这类聚合酶到标准 PCR 反应中可以改善损伤 DNA 的扩增（McDonald et al. 2006）。虽然目前只报道了其在紫外线损伤中的应用，但这种策略原则上也适用于脱嘌呤、氧化和相关的降解过程损坏的模板 DNA。通过工程改造常规（复制）聚合酶（Ghadessy et al. 2004；d'Abbadie et al. 2007；Jozwiakowski et al. 2014），使用区域化自我复制等方法，目前已经解决了扩增损伤 DNA 的问题。

18.8 非天然碱基的插入

与损伤 DNA 扩增相关的挑战在于正常模板的非天然核苷酸插入。这对于诸如引物或探针的核酸荧光标记、DNA 测序、合成后共价吸附等过程来说是必需的。这些应用要求聚合酶能够插入化学修饰的 dNTP，这些修饰的 dNTP 通常比天然 dNTP 要大并且通常无法结合到聚合酶活性位点。这一领域的许多进展来自 Taq 聚合酶的工程改造，以降低其对单体底物（dNTP）的特异性。利用分子进化方法（见下文）已经产生了多种 Taq 酶的变体，大大增加了末端错配引物的延伸或荧光标记的 dNTPs 的掺入（Laos et al. 2013；Gardner and Jack 2002）。这些变体没有表现出错配率的增加，可见这些变体仍保留了天然 dNTP 的精确选择。

"松弛"形式的热稳定 DNA 聚合酶展现出一系列性质，例如改造后的 *Thermococcus* 聚合酶可以从 DNA 模板合成 RNA（Cozens et al. 2012）。类似的，这类聚合酶还可以实现多种功能的增强或新的技术产生。例如改进用于高通量测序技术的可逆染料终止子的掺入，原始 DNA 的扩增（见下文）以及使用人工碱基来产生新的碱基对等，其可以与天然的 A：T

和 G：C 碱基对一起被复制（Laos et al. 2014）。

18.9 PCR 抑制剂耐受

生物 DNA 复制发生在活细胞内稳定的、良好的生物化学环境中，而 PCR 则需要具有干扰体外扩增过程中存在的一系列物质。PCR 的用途显然随着其稳健性的增加而增加，也就是说取决于其延伸反应的持续合成能力。DNA 聚合酶本身的持续性已经通过形成某些嵌合体而得到增强（Yamagami et al. 2014），而其他研究小组已经探索了通过包括滑卡在体外再生的生物学策略的方法（Motz et al. 2002；Ribble et al. 2015）。类似的并且在商业上成功使用的方法有硫化叶菌（*Sulfolobus*）的双链 DNA 结合蛋白在与 DNA 聚合酶融合时模拟滑卡（Wang et al. 2004）的应用。这又反过来大大增强了 PCR 的整体性能，使得其可以用于未纯化的样品，例如全血样品的基因扩增（Trombley-Hall et al. 2013）。

将广义蛋白质进化技术应用于聚合酶中，PCR 在含有各种抑制剂的基质中的性能也得到了显著改善（Kermekchiev et al. 2009）。区域化自我复制（CSR）是一种特别有效的技术，与热稳定聚合酶独特相关，因为它使这些酶能够扩增它们自己的基因。如图 18.2 所示，CSR 技术包括在合适的表达载体中克隆诱变的 DNA 聚合酶基因，并用所得质粒文库转化大肠杆菌（Ghadessy et al. 2001）。将各自表达文库不同变体的单个细胞用 dNTP、引物和缓冲液包被在油包水乳液的液滴中。乳液通过热循环裂解液滴区域内的细胞，使得每种聚合酶根据其催化性质扩增其自身的基因。将所得产物汇集并重新克隆以产生适合第二轮选择的第二代文库（Ghadessy et al. 2001），该过程可以重复几个循环。

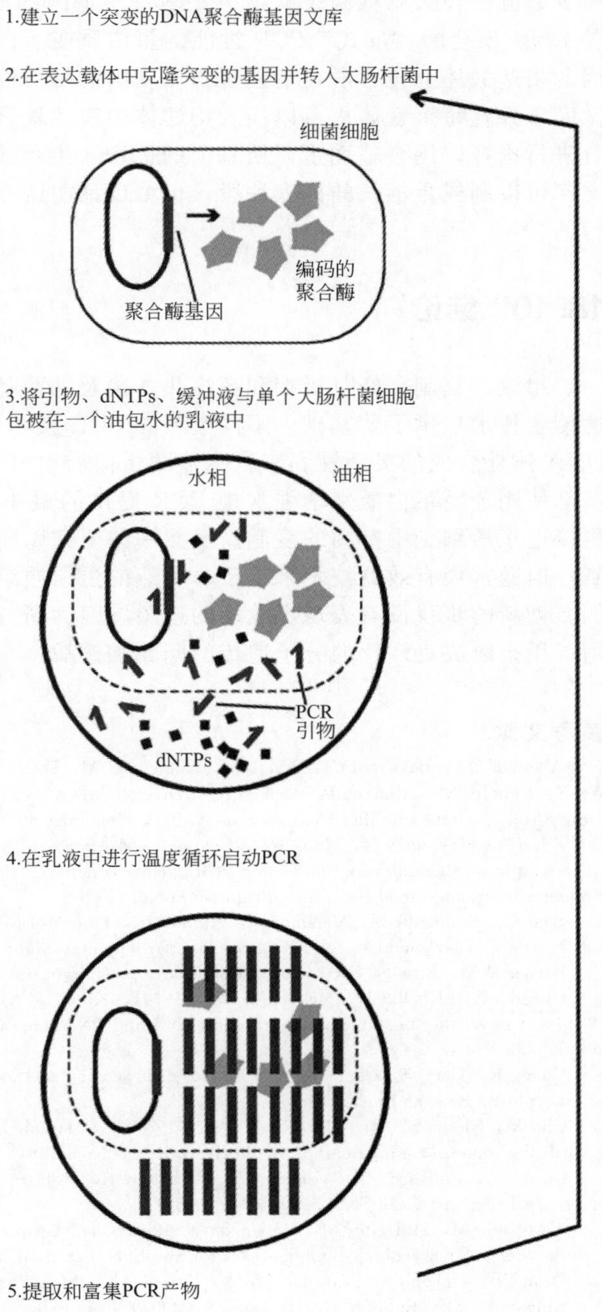
1.建立一个突变的DNA聚合酶基因文库
2.在表达载体中克隆突变的基因并转入大肠杆菌中
细菌细胞
编码的
聚合酶
聚合酶基因
3.将引物、dNTPs、缓冲液与单个大肠杆菌细胞
包被在一个油包水的乳液中
水相
油相
PCR
引物
dNTPs
4.在乳液中进行温度循环启动PCR
5.提取和富集PCR产物

图 18.2 DNA 聚合酶的区域化自我复制

该方法不仅可以根据所编码酶的性能选择改良的聚合酶基因，而且还可以在存在抑制剂等特定条件下使用。通过

这种方式已经筛选出多个氨基酸取代修饰的 Taq 聚合酶，使得酶在不利的条件下的性能得到大大提升。酶氨基酸序列的内在可塑性以及 CSR 恢复氨基酸序列变化的能力均能提高 PCR 条件下的选择性能，这表明可根据特定应用来定制特定性质的聚合酶变体。例如可用于扩增原始 DNA 序列（d'Abbadie et al. 2007 ；Baar et al. 2011）、耐受荧光标记的 dNTP（Ramsay et al. 2010）和缩短的 PCR 循环等（Arezi et al. 2014）。

CSR 作为蛋白质工程方法似乎仅局限于热稳定的 DNA 聚合酶或 SD 聚合酶，因为这些是少数能够在大肠杆菌裂解物中复制自身基因的酶类。然而，原则上 CSR 能够选择任何改善 DNA 聚合酶"反式"PCR 性能的蛋白质或蛋白质变体。这反映了在具有 DNA 聚合酶基因的表达载体上包含潜在增强蛋白质基因的可行性。因此，可以通过辅助构建蛋白质突变体文库，并且每个变体也可以插入构建体中表达聚合酶。然后通过合适的引物对构建表达体集合进行选择以用于辅助蛋白质基因的扩增，类似于常规的 CSR 程序。或者说，如果突变体文库可以制成足够长的插入片段，也可以证明诱变对聚合酶和辅助蛋白质基因进行选择是可行的。

18.10 结论

组装、复制和修复极端和超嗜热古菌及细菌基因组的热稳定蛋白质为一系列复杂的分子遗传学技术提供了酶基础。不同类型的 PCR 技术仍继续主导着这些技术领域，其中强调了 DNA 体外合成的灵活性和来自嗜热菌蛋白质对其相关酶促过程的稳定性。这些方法中的许多都使用辅助蛋白质来增强常规 PCR 提供的最小 DNA 复制模式，并且在某些方面的增强更接近于活细胞中基因的复制。虽然一些研究以明确的仿生方式追求完整、准确的基因组复制，但通常最有效的技术来自于非自然的蛋白质组合，并且该过程没有任何明确的生物学先例。创新的步伐没有表现出放缓的迹象，因为新方法、新酶和新的蛋白质工程技术被结合应用，用来解决更广泛的分子遗传的问题和挑战。

参考文献

Alsmadi OA，Bornarth CJ，Song W，Wisniewski M，Du J，Brockman JP，Faruqi AF，Hosono S，Sun Z，Du Y，Wu X，Egholm M，Abarzua P，Lasken RS，Driscoll MD（2003）High accuracy genotyping directly from genomic DNA using a rolling circle amplification based assay. BMC Genomics 4：21

Arezi B，McKinney N，Hansen C，Cayouette M，Fox J，Chen K，Lapira J，Hamilton S，Hogrefe H（2014）Compartmentalized self-replication under fast PCR cycling conditions yields Taq DNA polymerase mutants with increased DNA-binding affinity and blood resistance. Front Microbiol 5：408

Baar C，d'Abbadie M，Vaisman A，Arana ME，Hofreiter M，Woodgate R，Kunkel TA，Holliger P（2011）Molecular breeding of polymerases for resistance to environmental inhibitors. Nucleic Acids Res 39，e51

Barnes WM，Rowlyk KR（2002）Magnesium precipitate hot start method for PCR. Mol Cell Probes 16：167-171

Chander Y，Koelbl J，Puckett J，Moser MJ，Klingele AJ，Liles MR，Carrias A，Mead DA，Schoenfeld TW（2014）A novel thermostable polymerase for RNA and DNA loop-mediated isothermal amplification（LAMP）. Front Microbiol 5：395

Chang K，Deng S，Chen M（2015）Novel biosensing methodologies for improving the detection of single nucleotide polymorphism. Biosens Bioelectron 66：297-307

Cho M，Chung S，Heo SD，Ku J，Ban C（2007）A simple fluorescent method for detecting mismatched DNAs using a MutS-fluorophore conjugate. Biosens Bioelectron 22：1376-1381

Cozens C，Pinheiro VB，Vaisman A，Woodgate R，Holliger P（2012）A short adaptive path from DNA to RNA polymerases. Proc Natl Acad Sci USA 109：8067-8072

d'Abbadie M，Hofreiter M，Vaisman A，Loakes D，Gasparutto D，Cadet J，Woodgate R，Paabo S，Holliger P（2007）Molecular breeding of polymerases for amplification of ancient DNA. Nat Biotechnol 25：939-943

Dean FB，Hosono S，Fang L，Wu X，Faruqi AF，Bray-Ward P，Sun Z，Zong Q，Du Y，Du J，Driscoll M，Song W，Kingsmore SF，Egholm M，Lasken RS（2002）Comprehensive human genome amplification using multiple displacement amplification. Proc Natl Acad Sci USA 99：5261-5266

Fogg MJ，Pearl LH，Connolly BA（2002）Structural basis for uracil recognition by archaeal family- B DNA polymerases. Nat Struct Biol 9：922-927

Gaidamaviciute E，Tauraite D，Gagilas J，Lagunavicius A（2010）Site-directed chemical modification of archaeal Thermococcus litoralis Sh1B DNA polymerase：acquired ability to read through template-strand uracils. Biochim Biophys Acta 1804：1385-1393

Gardner AF，Jack WE（2002）Acyclic and dideoxy terminator preferences denote divergent sugar recognition by archaeon and Taq DNA polymerases. Nucleic Acids Res 30：605-613

Ghadessy FJ，Ong JL，Holliger P（2001）Directed evolution of polymerase function by compartmentalized self-replication. Proc Natl Acad Sci USA 98：4552-4557

Ghadessy FJ，Ramsay N，Boudsocq F，Loakes D，Brown A，Iwai S，Vaisman A，Woodgate R，Holliger P（2004）Generic expansion of the substrate spectrum of a DNA polymerase by directed evolution. Nat Biotechnol 22：755-759

Gruz P，Shimizu M，Pisani FM，De Felice M，Kanke Y，Nohmi T（2003）Processing of DNA lesions by archaeal DNA polymerases from *Sulfolobus solfataricus*. Nucleic Acids Res 31：4024-4030

Hafner GJ，Yang IC，Wolter LC，Stafford MR，Giffard PM（2001）Isothermal amplification and multimerization of DNA by Bst DNA polymerase. Biotechniques 30：852-856

Hogrefe HH，Hansen CJ，Scott BR，Nielson KB（2002）Archaeal dUTPase enhances PCR amplifications with archaeal DNA polymerases by preventing dUTP incorporation. Proc Natl Acad Sci USA 99：596-601

Jozwiakowski SK，Keith BJ，Gilroy L，Doherty AJ，Connolly BA（2014）An archaeal family-B DNA polymerase variant able to replicate past DNA damage：occurrence of replicative and translesion synthesis polymerases within the B family. Nucleic Acids Res 42：9949-9963

Kaboev OK，Luchkina LA，Tretiakov AN，Bahrmand AR（2000）PCR hot start using primers with the structure of molecular beacons（hairpin-like structure）. Nucleic Acids Res 28，e94

Kermekchiev MB，Tzekov A，Barnes WM（2003）Cold-sensitive mutants of Taq DNA polymerase provide a hot start for PCR. Nucleic Acids Res 31：6139-6147

Kermekchiev MB，Kirilova LI，Vail EE，Barnes WM（2009）Mutants of Taq DNA polymerase resistant to PCR inhibitors allow DNA amplification from whole blood and crude soil samples. Nucleic Acids Res 37，e40

Laos R，Shaw R，Leal NA，Gaucher E，Benner S（2013）Directed evolution of polymerases to accept nucleotides with nonstandard hydrogen bond patterns. Biochemistry 52：5288-5294

Laos R，Thomson JM，Benner SA（2014）DNA polymerases engineered by directed evolution to incorporate nonstandard nucleotides. Front Microbiol 5：565

Lasken RS，Schuster DM，Rashtchian A（1996）Archaebacterial DNA polymerases tightly bind uracil-containing DNA. J Biol Chem 271：17692-17696

Li GM（2008）Mechanisms and functions of DNA mismatch repair. Cell Res 18：85-98

Liu XP，Liu JH（2011）Characterization of family IV UDG from Aeropyrum pernix and its application in hot-start PCR by family B DNA polymerase. PLoS ONE 6，e27248

Lizardi PM，Huang X，Zhu Z，Bray-Ward P，Thomas DC，Ward DC（1998）Mutation detection and single-molecule counting using isothermal rolling-circle amplification. Nat Genet 19：225-232

Louwrier A，van der Valk A（2005）Thermally reversible inactivation of Taq polymerase in an organic solvent for application in hot start PCR. Enzyme Microb Technol 36：947-952 McDonald JP，Hall A，Gasparutto D，Cadet J，Ballantyne J，Woodgate R（2006）Novel thermostable Y-family polymerases：applications for the PCR amplification of damaged or ancient DNAs. Nucleic Acids Res 34：1102-1111

Motz M，Kober I，Girardot C，Loeser E，Bauer U，Albers M，Moeckel G，Minch E，Voss H，Kilger C，Koegl M（2002）Elucidation of an archaeal replication protein network to generate enhanced PCR enzymes. J Biol Chem 277：16179-16188

Nilsson M，Dahl F，Larsson C，Gullberg M，Stenberg J（2006）Analyzing genes using closing and replicating circles. Trends Biotechnol 24：83-88

O'Donnell M，Langston L，Stillman B（2013）Principles and concepts of DNA replication in bac- teria，Erchaea，and Eukarya. Cold Spring Harb Perspect Biol 5：pii：a010108. doi：10. 1101/cshperspect. a010108

Pata JD（2010）Structural diversity of the Y-family DNA polymerases. Biochim Biophys Acta 1804：1124-1135

Paul N，Shum J，Le T（2010）Hot start PCR. Methods Mol Biol 630：301-318

Potapova O，Grindley ND，Joyce CM（2002）The mutational specificity of the Dbh lesion bypass polymerase and its implications. J Biol Chem 277：28157-28166

Ramsay N，Jemth AS，Brown A，Crampton N，Dear P，Holliger P（2010）CyDNA：synthesis and replication of highly Cy-dye substituted DNA by an evolved polymerase. J Am Chem Soc 132：5096-5104

Rasila TS，Pajunen MI，Savilahti H（2009）Critical evaluation of random mutagenesis by error- prone polymerase chain reaction protocols，*Escherichia coli* mutator strain，and hydroxylamine treatment. Anal Biochem 388：71-80

Ribble W，Kane SD，Bullard JM（2015）Long-range PCR amplification of DNA by DNA polymerase III holoenzyme from Thermus thermophilus. Enzyme Res 2015：837-842

Tindall KR，Kunkel TA（1988）Fidelity of DNA synthesis by the *Thermus aquaticus* DNA polymerase. Biochemistry 27：6008-6013

Trombley-Hall A，McKay Zovanyi A，Christensen DR，Koehler JW，Devins Minogue T（2013）Evaluation of inhibitor-resistant real-time PCR methods for diagnostics in clinical and environ- mental samples. PLoS One 8，e73845

Wang Y，Prosen DE，Mei L，Sullivan JC，Finney M，Vander Horn PB（2004）A novel strategy to engineer DNA polymerases for enhanced processivity and improved performance in vitro. Nucleic Acids Res 32：1197-1207

Waters LS，Minesinger BK，Wiltrout ME，D'Souza S，Woodruff RV，Walker GC（2009）Eukaryotic translesion polymerases and their roles and regulation in DNA damage tolerance，Microbiol Mol Biol Rev 73：134-154

White MF，Grogan DW（2008）DNA stability and repair. In：Robb FT，Antranikian G，Grogan DW，Driessen AJ（eds）Thermophiles：biology and technology at high temperatures. CRC Press，Boca Raton，pp 179-188

Wiedmann M，Wilson WJ，Czajka J，Luo J，Barany F，Batt CA（1994）Ligase chain reaction（LCR）- overview and applications. PCR Methods Appl 3：S51-S64

Wilson RC，Pata JD（2008）Structural insights into the generation of single-base deletions by the Y-family DNA polymerase Dbh. Mol Cell 29：767-779

Yamagami T，Ishino S，Kawarabayasi Y，Ishino Y（2014）Mutant Taq DNA polymerases with improved elongation ability as a useful reagent for genetic engineering. Front Microbiol 5：461 Yan L，Zhou J，Zheng Y，Gamson AS，Roembke BT，Nakayama S，Sintim HO（2014）Isothermal amplified detection of DNA and RNA. Mol Biosyst 10：970-1003

Yang W，Woodgate R（2007）What a difference a decade makes：insights into translesion DNA synthesis. Proc Natl Acad Sci USA 104：15591-15598

第十九章

嗜热微生物降解蔬菜废弃物生物质技术进展

Annarita Poli[1]，Ilaria Finore[1]，
Annabella Tramice[1]，Paola Di Donato[2]，
Barbara Nicolaus[1][3] Licia Lama[1]

19.1 世界蔬菜废弃物生物量：木质纤维素类生产概述

生物质是用于可持续产生能量和有用分子的主要可再生原料。生物质用于生产生物能源和化工产品可以大大减少温室气体排放，因此它代表了目前基于化石燃料经济的一种很有价值的替代品。在这一方面，生物质的开发可以代表一种生物基础经济的可持续环境实践，其中没有资源（耕地、作物）从食物链转移到生物燃料生产。在这一框架中，根据综合生物炼制方法，绿色化学是目前将生物质转化为生物能源（生物燃料和生物能量）和生物化工产品最有希望的方法。实际上，通过化学、物理和生物化学技术的结合，不同种类的生物质用作生产化工产品、生物聚合物、生物技术酶/微生物和能量的原始材料。通常基于不可再生化石资源的能源和化工产品的产生转向可持续且环境影响较低的系统。

蔬菜废弃物和残渣是植物和农作物在种植和工业加工的过程中产生的典型生物质。这些残渣可以定义为：主要废弃物，即采伐剩余物，如农作物残渣（秸秆、海藻柄、甘蔗渣）和农林剩渣；二次废弃物，如农产品工业产生的二次废弃物（果壳、废纸浆、果皮和种子）或林业产品工业产生的二次废弃物（木屑）。

这些废弃物主要为木质纤维素残余物，其目前用于几个方面，例如生物燃料生产或增值化合物（天然抗氧化剂、脂质、寡糖、纤维和蛋白质）的提取。

特别的，木质纤维素残渣是由纤维素、半纤维素和木质素构成，其在天然来源中为互连

---

[1] Institute of Biomolecular Chemistry ICB，CNR，via Campi Flegrei 34，80078 Pozzuoli，NA，Italy.

[2] Institute of Biomolecular Chemistry ICB，CNR，via Campi Flegrei 34，80078 Pozzuoli，NA，Italy；Department of Sciences and Technologies，Centro Direzionale，University of Naples ‘Parthenope’，Naples，Italy.

[3] e-mail：barbara. nicolaus@icb. cnr. it.

大分子结构，即原纤维。由于纤维素、半纤维素和木质素在不同来源的植物中的比例不同，它们的异质结构可能是开发尽可能多的通用多步骤程序的主要障碍之一，其包括机械和化学预处理以及基于酶的生物转化的方法。其重点是发酵从生物质原料获得的糖，除产生其他有用的分子之外还可产生乙醇（FitzPatrick et al. 2010 ）。

纤维素作为生物能源的重要可再生来源，经处理转化为可发酵葡萄糖（Bhalla et al. 2013）。不同的是，半纤维素的内在结构，也可以提供在营养领域非常有用的低聚糖益生元（Azevedo Carvalho et al. 2013）。

然而，这些残渣仍然是未充分利用的原料，其通过生物工艺可用于生产多种生物产品，例如生物技术上有用的极端微生物和酶。实际上，由于其复杂的化学组成，这些残渣可以作为唯一的碳源来促进和维持微生物生长。因此，废弃生物质可以提供廉价和环境友好的方式来培养有用的微生物（例如极端微生物）及其相关的酶。自从它们在生产第二代生物燃料和增值化合物方面有多种应用，例如可用于生产其他有用化工产品的生物活性寡糖，后者成为一个越来越热门的课题。

每年有数百万吨的这种废弃物产生，因此损失了大量的纤维素和半纤维素。如表19.1所示，这种廉价的可再生原材料有巨大的空间可以利用。废弃食物如水果、蔬菜和谷物残渣等来自于罐头和包装企业的农产品工业。苹果、黑加仑、樱桃、野樱莓和梨等通常被用于加工果汁和果酱，水果处理过程中产生的废渣的主要组成成分是纤维素和半纤维素。苹果渣是众多废渣中的一种，是纤维素和半纤维素的丰富来源，两种成分分别占苹果渣干重43.6%和24.4%。表中其他水果产生的果渣大约50万吨。黑加仑、樱桃、野樱莓和梨的混合果渣中纤维素和半纤维素的平均组成大概为34.6%和33.5%（Nawirska and Kwasniewska 2005）。废弃物柑橘类水果产生大量废弃物也是纤维素和半纤维素的廉价来源，在柑橘中纤维素和半纤维素大概分别占22.6%和6.0%，其中，在橘子渣中，纤维素和半纤维素分别占干重的37.1 %和11.0 %（Sanchez-Vazquez et al. 2013）。胡萝卜和麸皮中也具有较高含量的纤维素。废弃物胡萝卜选材和加工生产胡萝卜汁过程产生的废弃物中纤维素和半纤维素的质量分数分别为51.6%和12.3%；玉米糠由干磨玉米粒生成，大约有780万吨，其中由质量分数为70.0%的纤维素和质量分数为28.0%的半纤维素组成。另外，加工番茄罐头的废弃物也可以是半纤维素和纤维素的来源，两者分别占这些废弃物质量的7.5%～11%和9.1%，主要包括果皮和种子。

表19.1 主要废弃生物质的可再生能源纤维素和半纤维素的年产量

类别	废弃物类型	年产量	参考文献
食物废弃物	苹果渣	3.42百万吨	Nawirska and Kwasniewska(2005)
	黑醋栗，樱桃，苦莓，梨渣	0.5百万吨	Nawirska and Kwasniewska(2005)
	橙皮	7.4百万吨	Sanchez-Vazquez et al. (2013)
	橘皮	3.1百万吨	—
	胡萝卜残留物	约11百万吨	Nawirska and Kwasniewska(2005)，Sanchez-Vazquez et al. (2013)
	玉米糠	7.8百万吨	Rose et al. (2010)
	番茄	11百万吨	Tommonaro et al. (2008)，Das and Sing(2004)

续表

类别	废弃物类型	年产量	参考文献
农作物残留物	木薯渣	约 200 百万吨	Das and Sing(2004)
	甘蔗渣	300 百万吨	Cardona et al. (2010)
	玉米秸秆	696 百万吨	Sanchez-Vazquez et al. (2013)
	大麦秸秆	约 5.6 百万～9.8 百万吨	Das and Sing(2004)
	燕麦秸秆	约 10.5 百万～15.7 百万吨	Das and Sing(2004)
	稻草秸秆	731 百万吨	Das and Sing(2004), Binod et al. (2010)
	高粱秸秆	约 28 百万～43 百万吨	Das and Sing(2004)
	小麦秸秆	550 百万吨	Kuan and Liong(2008), Das and Sing(2004)
林业/木材残留物	木材残留物	210 万立方米	FAO 数据
	木屑	22 百万吨	
	碎片和颗粒	250 万立方米	

农作物残渣包括收获后留在地上的植物材料，即秸秆、叶子、稻草（平均占农作物的 50 % 以上）或甘蔗渣。这些废弃物含有大量的纤维素和半纤维素，因此它们在世界范围内是主要的可再生木质纤维素（Sanchez-Vazquez et al. 2013）。木质纤维素的主要废弃物来源：含有质量分数约 33.0%纤维素和质量分数 26.0 %半纤维素的玉米秸秆（包括叶、壳和茎）（Sanchez-Vazquez et al. 2013）；含有质量分数约 27.8%纤维素和质量分数 32.0%半纤维素的稻草（Das and Sing 2004；Binod et al. 2010）；纤维素和半纤维素分别占干重 39.0 %和 36.0 %的小麦秸秆（Kuan and Liong 2008；Das and Sing 2004）。大麦、燕麦和高粱秸秆的产量较低，但纤维素和半纤维素的平均含量分别相当于废弃物干重的 38.6 %和 24.3 %（Das and Sing 2004）。

林业残渣几乎全是木质纤维素材料，一般来说，它们包括收获后森林中残留的生物质和木材加工所产生的所有废弃物（锯屑、碎屑和颗粒）。根据粮农组织统计数据库显示，2013 年全世界森林木材残留总量约 2.1 亿立方米，其中生物质包括采伐残渣、过剩的小树、粗糙或腐烂的死木。通常这些残留物都留在森林里或被烧毁。在木材加工废弃物方面，2013 年生产的木屑约 2200 万吨：锯屑、木材碾磨的残余物或木材、家具和建筑物制造的废料生产，通常是作为颗粒燃料开发的。最后，廉价木质纤维素储备的其他主要木质残余物变为切片和颗粒，被还原成小块并适用于制浆、纤维板的生产。这些废弃物由于它们的高纤维素含量，也可以用作微生物和酶生产的廉价发酵培养基。

19.1.1　木质纤维素类废弃生物质预处理的主要技术

食物废弃物、农作物和林业/木材残留物都是富含多糖的材料，如前所述，可以用于多种生物技术应用。事实上，由于它们丰富的化学成分，可以用作生长培养基来生产有用的微生物（例如极端微生物）及其相关的酶。在这方面，文献中已有几个这些废弃物有效用于发酵培养基的实例证明（Das and Sing 2004；Di Donato et al. 2011）。然而，木质纤维素类废弃物是难利用的生物质，因为植物细胞壁的纤维素深深地插入由半纤维素和木质素组成的网络（网状物）中，这阻碍了微生物或酶进入多糖基质。因此，需要一种预处理才能打开木质素鞘，以便进一步处理纤维素。文献中描述了不同的预处理技术，它们包括物理、化学、生物或组合处理的方法（表 19.2）。由于处理方法的不同，可能会发生半纤维素的部分水解以及木质素基质的降解。合适的预处理的方法由木质纤维素生物质和所需产品的性质决定，因此确定最佳技术是不可能的。在表19.2中列出了目前实验室或工业规模下用于不同生物质预处理的一些最具代表性的技术。

表 19.2 废木质纤维素生物预处理的主要技术

预处理型	方法	生物质	参考文献
物理	粉碎法	农作物残留物(小麦秸秆,稻草,甘蔗渣);林业残留物	Zheng et al. (2014)
	挤压法	作物残渣(大麦秸秆,玉米)	
	高温液态水法	作物残渣(小麦和稻草;甘蔗渣)	
化学	酸水解法	作物残渣(秸秆;甘蔗渣)	Sun and Cheng(2002)
	碱水解法	食物废弃物(果渣);作物残渣(小麦和稻草,玉米秸秆,甘蔗渣);林业残留物(叶子,木材残留物)	Chang and Holtzapple (2000), Zhang et al. (2007), Kim et al. (2003)
	有机溶剂法	农作物残留物(小麦和稻草,甘蔗渣);林业残留物(木材残留物)	Sun and Cheng(2002)
物理化学相结合	氨纤维爆破法	作物残渣(小麦和稻草,玉米秸秆);林业残留物(木屑)	Sun and Cheng(2002)
	催化蒸汽爆破法	作物残渣(小麦秸秆,玉米秸秆);林业残留物(木屑)	
生物	真菌降解法	作物残渣(小麦秸秆);林业残留物(木屑)	Sun and Cheng(2002)

粉碎是通过铣削或研磨机械实现的物理预处理，用于降低生物质颗粒大小。虽然这种技术会影响纤维素的超微结构，但能有效增加聚合物的可及表面积；此外，它可以减少结晶度和纤维素的聚合，从而改善酶降解能力。挤压机内部结构是由驱动螺杆沿着筒体移动，通过该技术，可以对生物质施加不同的机械处理，即摩擦、混合和剪切力。挤压使纤维素更易于利用，并且在一些情况下，除了半纤维素和木质素基质的降解之外，它还可以引起纤维素的部分解聚。在高温液态水（LHW）技术中，生物质处于高压下，高温液态水渗透植物细胞壁，从而水化纤维素。利用这种方式，可以溶解半纤维素，同时木质素被部分去除。因此，通过增加纤维素的可接触表面积，LHW 可以通过微生物和酶提高随后的纤维素降解(Zheng et al. 2014)。

酸水解是一种化学方法，可通过使用稀酸或浓酸来实现。稀酸的预处理中，用硫酸、马来酸或富马酸喷洒生物质，然后将其加热至160～220℃处理几分钟。这种处理通常可以去除半纤维素以及增加孔隙率，以这种方式，酶水解提供寡糖或单体糖用于进一步的应用（例如发酵成乙醇)。用浓酸的预处理中，硫酸或盐酸是使用最广泛的酸。尽管这种技术提高了纤维素解聚产率（不需要进一步的酶水解)，但是由于设备的腐蚀和试剂回收的需要，成本更高（Sun and Cheng 2002)。

碱水解可以通过在低温或更高温度下使用氢氧化钙、氢氧化钠或氨水来进行（Kim et al. 2003)。这些处理需要较长的反应时间，主要是木质素的溶解（70%～80 %）和有限的糖降解（Chang and Holtzapple 2000)。最后，有机溶剂法则通过用纯有机溶剂（乙醇、甲醇、丙酮、乙二醇）或其混合物处理生物质来去除木质素。根据生物质类型在低温或高温下进行处理，主要的优点是通过木质素和半纤维素解聚的增溶作用，从而提高纤维素的纯度和酶或微生物的可及性（Sun and Cheng 2002）。

最有名的综合物理化学处理是氨纤维爆破（AFEX）和催化蒸汽爆破。AFEX 技术基于在高温和高压下进行的液氨处理约 30min（Bals et al. 2010）。尽管氨的成本极大地限制了它的使用，但这种方法提供了较小的半纤维素解聚，能有效地去除木质素。另一方面，催化蒸汽爆破在经济和环境方面更为方便。事实上，这种方法只需要少量的化工产品和较低的能量输入，因此是一种对环境影响较小的技术。通常，用高压和1% H_2SO_4作为催化剂的饱和蒸汽处理生物质。该方法在 160～260℃和 0.69～4.83 MPa 下持续几分钟，然后恢复至大气压。通过该过程，半纤维素降解和木质素溶解相当完整，纤维素酶水解的潜力显著增加（Sun and Cheng 2002）。

生物预处理是通过利用白色、棕色或软腐烂真菌降解半纤维素和木质素基质。这些方法处理条件温和，能量输入低，然而由于木质素聚合物和半纤维素的生物降解速率低，半纤维素的糖被真菌同化，其应用仍然受限于实验室规模（Sun and Cheng 2002）。目前，催化蒸汽爆破法是除了热处理和酸水解预处理技术之外，唯一具有成本效益的方法。

19.1.2 植物生物质木聚糖提取流程

有效利用生物质如谷物秸秆和草本植物的条件之一是通过相对温和的过程分离主要成分，如多糖和木质素，能够保证其最小的物理和化学修饰以及有效的提取过程。很多完全提取半纤维素来获得木聚糖的过程用于酶解应用已经被详细描述（Aachary and Prapulla 2009）。

然而，在开始预处理之前，原料用有机溶剂（乙醇或乙酸乙酯）洗涤来分离杂质和其他分子（例如蜡和果胶），可以使接下来的程序（如木糖-寡糖步骤）更简单（Brienzo et al. 2009）。在一些情况下，醇和酮可以应用于可溶性木聚糖的收集中，或者在酶或化学反应后浓缩释放的 XOS（低聚木糖）（Akpinar et al. 2009）。

稀酸处理、稀碱提取、热水预水解、酶水解是目前从农业残留物中分离半纤维素聚合物的通用预处理方法（图 19.1）（Azevedo Carvalho et al. 2013；Chapla et al. 2012；Aachary and Prapulla 2009；Akpinar et al. 2010 ）。在某些情况下，有机溶剂用作提取助溶剂，这些方法虽然效率高，但并不完全是环保和经济的。每一钟方法都有其优点和缺点。

图 19.1 木聚糖提取的预处理和 XOS 生产流程

热水预水解是通过半纤维素的热水解产生乙酸来进行木聚糖的脱乙酰化。该方法使用专门的设备达到高温和高压（Kabel et al. 2002）。

目前，水热法应联合使用微波加热或蒸汽爆破。根据工作的温度和时间，可以调节所得到的木聚糖/低聚木糖的去聚合度和分子量间隔（Aguedo et al. 2014）。半纤维素材料通常以液相回收，而沉淀固体富含纤维素，可用于其他目的（Azevedo Carvalho et al. 2013）。不过，通过该方法从木质纤维素材料获得的产物含几种不需要的组分如木质素、糠醛等，从而需要进一步的纯化（Zhu et al. 2006）。

酸预处理是将半纤维素暴露于酸溶液中。无论植物初始生物量如何，这些提取物在酸性条件下温和进行（通常为 0.01mol/L 至 0.5mol/L 的 H_2SO_4），以限制木糖的释放量并避免或减少所产生的腐蚀。然而，酸性方法总是与木质纤维素材料的蒸煮相结合（>

120℃）来去除木质素，在这种情况下，处理时间对于调节木聚糖降解到木糖和避免其完全解聚，避免形成有毒糠醛和羟甲基糠醛（HMF）是很关键的（Akpinar et al. 2009）。事实上，单糖和这些有毒成分的产生是酸处理的主要障碍，因此这些提取方法也需要更大程度的纯化。

碱处理已被广泛应用，因为它们破坏细胞壁结构并切割氢键，与乙酰基单元和羟基肉桂酸的酯键和与细胞壁基质中的木质素的共价键（主要是α-苄基醚键），在水性介质中释放半纤维素聚合物。限制性就是碱性预水解不保留木聚糖链上的乙酰基（Lama et al. 2014；Bian et al. 2012）。钾、钠、钡、钙和氢氧化锂的水溶液已广泛用于不同温度，特别是4℃至40℃下回收半纤维素（Bian et al. 2012）。然而，最受欢迎的碱是氢氧化钾，因为在碱性提取物的中和过程中产生的乙酸钾比其他乙酸盐更易溶于用于沉淀的醇（Lawther et al. 1996）。根据木质纤维素材料的性质，使用的碱的含量可能不同。此外，有时还可以使用共溶剂（例如乙醇）（Li et al. 2015）。木聚糖回收较好的方法是高浓度碱法和热水解法，如玉米芯生物质（Samanta et al. 2012）。

在低温下（20～60℃），用过氧化氢（2%～4 %）进行的含皂碱预处理对于甘蔗渣半纤维素提取、脱木素和半纤维素从秸秆和草本植物中溶解，并保持其化学结构具有良好的效果（Brienzo et al. 2009）。

与碱溶液偶联的微波辐射可能是木质纤维素材料在较低温度下的另一种预处理方法，以克服酶解过程中高温下（热水预水解）产生的非期望的或有害的化合物（Zhu et al. 2006）。

用选定的酶进行预处理可能更有吸引力，因为它不会产生非期望的副产物或大量的单糖，并且不需要用高温工作的特殊仪器。不过，它们仅用于易感材料，通过直接酶处理含木聚糖材料生产XOS，如柑橘皮（Aachary and Prapulla 2011）。

目前，微波辅助酶水解被认为通过减少反应时间并避免非期望的产物来生产低聚木糖的可能方法，通过使用商业木聚糖酶从麦麸中生产益生元（Wang and Lu 2013）。

木质纤维素材料预处理提取的木聚糖可以通过强酸性反应转化为单糖，若想制备低聚糖可通过酶水解反应（Azevedo Carvalho et al. 2013），而实际上酶水解被认为食品工业上用来生产低聚木糖（XOS）的最佳选择。

XOS的制备由内切木聚糖酶进行，内切木聚糖酶水解木聚糖主链中的β-1,4键。如果酶复合物用于该目的，则它们必须具有低的木聚糖酶活性以避免或减少木糖的产生，抑制XOS的形成（图19.1）（Vázquez et al. 2005）。

19.2 废弃生物质作为极端微生物生长基质的替代碳源

废弃生物质改良作为微生物生长的替代营养源仍未完全探索。这些微生物可以被认为是一个不可估量的分子工厂，在多个领域和不同的工业过程中有许多用途。因此，微生物发酵所需化学成分的减少导致成本的降低，在工业规模上更是如此，具有很好的发展前景。另一方面，这种变化可以恢复废弃物的商业价值，从而延长起始材料的使用寿命周期。此外，这也是一种节约废弃物成本的方式（Di Donato et al. 2011）。当考虑到极端微生物时，上述概念被放大，这是因为极端微生物偏好于以一个或多个应力因子为特征的环境生态位，例如高温或低温、压力、pH、缺氧、辐射、高盐和金属浓度。因此，它们能够在压力条件下很好的生长，其使用效果比嗜温微生物更有利。

了解这些生物体使我们能够更好地了解地球上的生命起源和发展，同时也使我们了解如何改进某些过程或从已知过程获得更好的结果（Mastascusa et al. 2014）。事实上，这些微生物在这种不利条件下生存和发育，已经形成了特殊的性质，如细胞膜特有的化学结构和成分、特殊的代谢过程及能量转化机制、细胞内环境调节（Antranikian and Egorova 2007）。嗜热菌目前用于生产醇类和其他生物活性化合物（类胡萝卜素、氨基酸、抗生素），从废弃固体或水中去除金属离子和有机化合物。嗜热酶在食品工业中应用高含量糖（淀粉酶、木糖异构酶、支链淀粉酶）的糖浆生产，并提高感官特性（果胶酶）或某些食物的消化率（β-半乳糖苷酶）；几种嗜热活性蛋白酶和脂肪酶应用于洗涤剂行业和烘烤过程中；来自嗜热芽孢杆菌（*Bacillus thermoproteolyticus*）的嗜热蛋白酶用于二肽阿斯巴甜这种低热量甜味剂的合成（Ogino et al. 1995）。嗜热酶也用于医学和生物学药物前体，或者在 PCR（聚合酶链式反应）过程中作为 DNA 聚合酶和 DNA 连接酶，用于诊断医学、分子生物学和分类学中。

极端微生物的适应策略使它们即使是存在化学物理胁迫因素的情况下也能够进行所有代谢过程。这些策略在制药、食品、遗传、饲料、环境等方面的生物技术研究方面派上用场。事实上，各种工业生产过程的基础是将其极端潜力及生物大分子（酶、聚合物、脂质等）一起开发（Elleuche et al. 2014）。对于所有这些目的，显而易见的是寻找更经济的方法来增殖极端微生物。这个问题的研究受到了极大的关注，并已有文献报道了一些相关研究成果。

有一个有趣的大分子是聚羟基链烷酸酯（PHAs），它们基本上是由在高盐浓度存在下生长的嗜盐细菌生产的可生物降解的聚合物。PHAs 以微粒的形式作为储备能量积累在微生物细胞中，并作为对比不平衡生长条件的策略。PHAs 的生物相容性使其在几个领域（包装、医药、农业、药物传递等）中的利用率达到工业水平（Shrivastav et al. 2013 ）。

Danis 等（2015）研究了利用较便宜的废弃物如玉米淀粉、蔗糖、乳清、甜瓜、苹果和番茄废弃物作为唯一碳源和能量来源的嗜盐古菌钠线菌属（*Natrinema*）1KYS1 的生产聚羟基链烷酸酯。在所研究的废弃物中，含有玉米淀粉的培养基最适合于 PHA 细胞的积累，其产率为干细胞质量的 53.14 %（Danis et al. 2015）。研究还测试了嗜盐古菌（*Haloterigena hispanica*）利用番茄和胡萝卜废弃物生长和生产多羟基丁酸（PHB）的能力。结果表明，最好的底物是胡萝卜，其 PHB 产率相当于用标准复合肉汤［（1.25±0.05）mg/g（以干细胞计）和（1.35±0.06 ）mg/g（以干细胞计）］获得的产量（Di Donato et al. 2014; Huang et al. 2006）。另一个极端嗜盐的古菌（*Haloferax mediterranei*）通过使用预处理的米糠和淀粉，重复分批发酵相结合，增加 PHA 积累到细胞中的机会。最高的 PHA 产率达到干细胞质量的 55.6 %。

在聚合物领域，许多尝试主要是针对微生物胞外多糖（EPS）进行的，其中生物技术应用已经被公认，并且它们从食品延伸到药物、药理学、保健品、化妆品、除草剂和杀虫剂（Nwodo et al. 2012 ）。由替代碳源生产的 EPS 有几种，果聚糖在极端微生物领域可以被引用，它是果糖的均聚物，其可用于食品、医药、制药和农业（Rhee et al. 2005）。从细胞外果聚糖生产过程得到的底物发酵的一个例子就是嗜盐微生物（*Halomonas smyrniensis*）AAD6 的研究。经过多种预处理，通过使用淀粉糖蜜和甜菜糖蜜，作为蔗糖的替代来源，来使微生物增殖，使得甜菜糖蜜从微生物中更好地用于 EPS 释放（12.4g/L）（Küçükaşik et al. 2011）。

在微生物可能来源的分子中，酶可能是工业生产过程中利用最充分的。特别是嗜热酶，因为它们的热稳定性，对有机溶剂和金属的耐受性总是能在一些领域发挥用途。深入研究的是淀粉酶、纤维素酶和木聚糖酶活性，它们可以分别水解淀粉、纤维素和木聚糖聚合物。多糖在自然界中是最常见的，可产生高经济潜力的副产物（单糖和寡糖）。淀粉酶商业应用包括制药、造纸、食品、纺织、燃料、洗涤剂和淀粉工业（de Souza and Magalhães 2010 ）。因此，嗜热淀粉芽孢杆菌（*Anoxybacillus amylolyticus*）在不同碳源的液态和固态农业废弃物发酵中生长，以研究利用它们生长和产生 α-淀粉酶的能力。测试并添加到基本培养基（质量浓度 10g/L）的废弃物是来自芦竹（*Arundo donax L.*）的根茎，来自朝鲜蓟（*Cynara carduncculus*）的茎叶以及马铃薯皮废弃物。将生长率和酶产率与从常规化工产品如酵母提取物和可溶性淀粉获得的产率进行比较。深层发酵条件下，芦竹根茎块被认为 *A. amylolyticus* 诱导淀粉酶的最佳底物，其活性是利用酵母提取物培养的 2.2 倍（Finore et al. 2014）。用另一种嗜热微生物热熔梭菌属亚种杆菌（*Geobacillus thermoleovorans* subsp. *stromboliensis*）测试其利用废弃物的能力，它产生的胞外 α-淀粉酶也能在高乙醇浓度存在下水解生淀粉，它使得该酶成为异源宿主（例如酿酒酵母 *Saccharomyces cerevisiae*）中克隆程序的合适候选物，同时进行糖化和发酵，以便从未处理的淀粉中直接获得乙醇（Finore et al. 2011；Kasav et al. 2012 ）。因此 *G. thermolovorans* 可在柠檬、番茄、茴香、胡萝卜和芦竹根茎等废弃物存在的条件下进行分批和透析发酵。所研究的淀粉酶活性在来自芦竹的根茎存在时达到最高值，其活性比标准复合培养基高出 1.1 倍；而与复合培养基相比，柠檬废弃物仅具有 39%的比活性（Di Donato et al. 2014 ）。木聚糖酶的工业过程很多，涉及动物饲料、造纸工业、纤维素漂白等（Polizeli et al. 2005）。木热纤维芽孢杆菌（*Thermobacillus xylanilyticus*）是具有完全稳定的半纤维素酶的嗜热细菌。在麦麸和秸秆培养基上进行测试，并与葡萄糖和木聚糖进行比较，实验证明在任何情况下微生物可利用所有的测试碳源，最旺盛的活动是合成木聚糖酶，而培养基的组成影响脱支酶（淀粉-1,6-葡萄糖苷酶）的相对含量。更详细地说，当麦杆存在时，产生更多酯酶，当细菌在稻草上生长时，阿拉伯呋喃糖苷酶活性更丰富（Rakotoarivonina et al. 2012）。热葡糖苷酶地芽孢杆菌（*Geobacillus thermantarcticus*）也是产生胞外嗜热木聚糖酶和 β-木糖苷酶的嗜热微生物（Lama et al. 2004 ）。对于这种细菌，含有朝鲜蓟茎叶的廉价培养基木聚糖酶释放量是标准生长条件下测量的 1.6 倍（Di Donato et al. 2014）。

将茴香、胡萝卜和番茄废弃物作为嗜碱盐芽孢杆菌生长的可能底物，并研究了细胞内 α-葡糖苷酶的产生。在透析和分批发酵中，在基本培养基中加入质量浓度 10g/L 的废弃生物质。实验发现所有蔬菜废弃物都支持微生物的生长，并且在茴香废弃物存在下，发现酶的产量与在标准生长条件下测定的活性相当（Di Donato et al. 2014）。

19.3 用于生物质转化获得生物燃料的极端酶系

通过酸或酶处理，纤维素和半纤维素被转化成己糖和戊糖。（半）纤维素酶的酶水解是一种更好的方法，可以达到更高的转化率，同时相比酸水解的毒性较小，更环保。酵母或细菌可以将所有游离糖发酵成乙醇。如今，在乙醇生产过程中，使用商业酶对半纤维素和纤维素进行转化，是生物乙醇生产中最昂贵的环节（Elleuche et al. 2014）。事实上，可利用的酶已经可以由嗜温生物体产生，并且酶促反应在 ≤50℃ 条件下进行。这导致酶水解缓慢且不完全，木素纤维素生物质的生物转化率低；此外，高含量的酶是必需的，同时也具有高的污染风险（Bhalla et al. 2013）。

为了解决这些问题，在改进水解步骤上已经作出了许多努力，增加纤维素酶活性，优化反应参数，如调整合适的酶和底物组合、级联反应、酶再利用等方面的研究。

在最近为优化木质纤维素生物质水解的研究中，应用统计学方法加上因子设计，将各种来源的酶合并，适量混合（Zhou et al. 2009）。为了从葡聚糖和木聚糖中分别获得葡萄糖和木糖，大量减少蛋白质的量（减少了$\frac{1}{2}$），数据经统计学方法验证，二者分别达到了 99%和

88％ 的转化率（Berlin et al. 2005 ）。目前，很多研究正在努力降低成本并最大限度地提高酶的产量。

可以通过增加比活性策略（通过定向进化和定点诱变）提高酶催化能力，从而提高发酵期间的纤维素酶催化效率来减少酶的用量或降低酶的生产成本（通过使用廉价的工艺工程方法，使用廉价的底物，如生物质，通过生物质精炼生产酶或在植物中表达酶）。生物精炼中的最佳酶系应该在原位产生并可用于连续培养，可以提高活性，在高温和抑制化合物存在（醛）等工艺条件下活性稳定，并具有相同的半衰期。

19.3.1　嗜热细菌和热稳定酶

通过利用极端微生物及其热稳定酶可以解决生物质转化生物燃料的相关问题。

极端微生物在生物能源的框架中一直在应用。最近，这些特殊形式的生物引起新的关注。实际上，极端微生物能够在独特的环境条件下如低或高的 pH、温度、高盐或压力下存活（Bhalla et al. 2013）。此外，由于代谢和生理的环境适应，这些微生物在类似的生态环境中快速发展。极端酶的动力学表明，其能够在压力条件下进行整合。

在了解了有关极端微生物在生物燃料工艺生产过程中应用的知识时，大多数微生物来源基本上是嗜热的。这并不令人震惊，因为嗜热菌能够耐受 pH、温度和环境变化，这些特性使得这些微生物应用到具有较高商业价值的燃料生产过程中（Bhalla et al. 2013）。

使用极端酶，特别是耐热酶，具有生物转化的巨大潜力，是商业化过程的关键步骤。据文献广泛报道，生物燃料生产中的极端酶主要分布在各种原料的聚合物水解中。

生物燃料工业需要高度底物特异性的酶，即使在工艺参数变化过程中，也不会受到最终产品的积累和稳定的影响。嗜酸性和嗜热微生物产生的酶由于其特性，受到广泛关注。

事实上，利用热稳定酶在木质纤维素生物质水解过程中具有许多优点。高温保证了反应物和产物的主要溶解性，因此更有效，特别是在反应速度方面，对酶的量有所要求（Zhang et al. 2011）；污染风险较低，从而提高生产力；释放挥发性化合物（例如乙醇），更容易回收；与热处理后的冷却相关的能量成本降低；更高的稳定性，有更长的水解时间和更灵活的生产过程。

众所周知，木素纤维素解聚中，生产热稳定纤维素酶和木聚糖酶的细菌是最有效的微生物（Liang et al. 2011）。

目前的研究对具有同时耐受酸和热的酶具有极大的兴趣，这两种特性可以改善木质纤维素的加工。这些酶活性可以通过嗜热微生物/极端微生物（Bhalla et al. 2013）得到。在生物燃料工业中，这些微生物与其酶的适用潜力可能会增加。

19.3.2　纤维素解构

纤维素是植物中最丰富的聚合物和全球含量最丰富的有机分子。纤维素是由 β-1,4-糖苷键和 β-D-吡喃葡萄糖单元连接在一起的葡聚糖同聚多糖。纤维素酶水解结晶和非结晶型纤维素的糖苷键（Mischnick and Momcilovic 2010），并且根据纤维素酶不同，与纤维素的不同部分和链的结合是靶向的。

纤维素转化为葡萄糖单体需要内切纤维素酶（EC 3.2.1.4），外切纤维素酶（纤维二糖水解酶，CBH，EC 3.2.1.91；葡聚糖水解酶，EC 3.2.1.74）和 β-葡糖苷酶（EC 3.2.1.21）。内切纤维素酶随机地作用于内部糖苷键，它导致聚合物延伸的快速降低和还原糖的逐渐增强。外切纤维素酶活性通过从还原末端和非还原末端开始从基本上去除纤维二糖而切割纤维素。以这种方式，其可以快速增加还原糖的浓度，而聚合物长度的变化却很小。

内切纤维素酶和外切纤维素酶的水解作用是协同作用的，并导致低聚糖和纤维二糖释放，然后通过β-葡糖苷酶将其转化为葡萄糖（Kumar et al. 2008）。

木质纤维素生物质的转化过程是至关重要的，有必要寻找新的和更有效的酶。初始酶对结晶纤维素的有序和不溶性结构的降解是纤维素生物燃料生产中最困难和最缓慢的步骤。

不同水解活性的组合对于各种来源的半纤维素的完全降解是必需的。即使纤维素酶水解一种类型的结合物，由于结晶基质的结合范围很广，因此需要一组酶或称为多纤维素酶体的多组分体系的参与（Kumar et al. 2008）。它们由携带至少一个糖类结合模块（CBM）和各种内聚模块组成。简而言之，CBM是非催化模块，有助于酶对不溶性聚合物的靶向以及通过内聚-停泊聚蛋白质相互作用介导催化模块的结合，主要在于提高降解原木质纤维素生物质的效率。在速度和成本方面，纤维素更有效的降解将产生环境和经济优势，而开发酶混合物以及工程细胞株目前仍然是研究的热点。

表19.3是许多来自嗜热和超嗜热细菌的热稳定内切葡聚糖酶的主要性质。据报道，大量的细菌和真菌是热稳定纤维素酶的来源。从属于根霉属、木霉属、曲霉属和小菌核，*Thermoascus thermophile*，嗜热链霉菌，嗜热丝孢菌，木薯属（Barnard et al. 2010）的嗜热和嗜温真菌属中提取纤维素酶，但纤维素的水解常不完全。例如，木霉属的主要不利方面之一是其含有较少的β-葡糖苷酶（Rahman et al. 2009）。据报道，在许多种嗜热真菌（Dimarogona et al. 2012）中发现的纤维素酶、纤维二糖脱氢酶和糖苷水解酶61（GH61）蛋白质家族的结合促进木质纤维素水解的增加（Horn et al. 2012）。

表19.3 来自各种嗜热和超嗜热细菌的热稳定内切葡聚糖酶的特性

生物	酶	最适pH	最适温度/℃	稳定性	参考文献
Thermotoga sp.	内切葡聚糖酶	6.8～7.8	100～105	在108℃半衰期为7min	—
T. neapolitana	内切葡聚糖酶	6.0	95	在106℃半衰期为2.16 h	Bok et al.(1998)
T. neapolitana	内切葡聚糖酶	6.0～6.6	100	在110℃半衰期为0.43h	Bok et al.(1998)
Rhodothermus marinus	内切葡聚糖酶	7.0	95	在100℃,3.5 h后保留50%的活性,在90℃下16h后保留80%的活性	Hreggvidsson et al.(1996)
Alicyclobacillus acidocaldarius ATCC 27009	内切葡聚糖酶	4.0	80	在80℃温育1h后保留60%的活性	Eckert and Schneider(2003)
Caldibacillus cellulovorans	内切葡聚糖酶	6.5～7.0	80	在80℃半衰期为0.53h,在85℃为0.03h,在70℃下3h后保持83%活性	Huang and Monk (2004)
Geobacillus thermoleovorans T4	内切葡聚糖酶	7.0	70	在90℃和100℃下1h后保留10%以上的活性	Tai et al.(2004)
Geobacillus sp. DUSEL R7 NA	内切葡聚糖酶	5.0	75	在60℃持续300h保持26%的活性	Rastogi et al.(2009)
Alicycloba cillus sp. A4	内切葡聚糖酶	2.6	65	在60℃下1h后保留90%的活性	Bai et al.(2010a)

续表

生物	酶	最适pH	最适温度/℃	稳定性	参考文献
Clostridium thermocellum	内切葡聚糖酶	7.0	70	在60℃下48h后，保持50%的活性	Romaniec et al. (1992)
Clostridium stercorarium	内切葡聚糖酶	6.0～6.5	90	稳定很多天	Bronnenmeier and Staudenbauer (1990)
Rhodothermus marinus	内切葡聚糖酶	7.0	95	在100℃下3.5h后保持50%活性，在90℃下16h后保持80%活性	Hreggvidsson et al. (1996)
Acidothermus cellulolyticus	内切葡聚糖酶	5.0	83	在110℃下失活	Himmer et al. (1994)
Thermosipho sp. strain 3	内切葡聚糖酶	5.6	70	在存在 Ca^{2+} 的条件下，在70℃下90 h后保持50%活性	Dipasquale et al. (2014)
E. coli expressing gene from *Thermonanaerobacter tengcongensis* MB4	内切葡聚糖酶	6.0～6.5	75～80	在82℃下半衰期为0.5 h	Liang et al. (2011)
E. coli expressing gene from *Clostridium thermocellum*	内切葡聚糖酶	6.4	80	在90℃下0.5 h后的活性是在80℃下的15%	Zverlov et al. (2005)

热球菌属 *Pyrococcus*（Kim and Ishikawa 2010），硫化叶菌属 *Sulfolobus*（Girfoglio et al. 2012），热袍菌属 *Thermotoga*（Hong et al. 2007），土芽孢杆菌属 *Geobacillus*（Rastogi et al. 2011）和栖热菌属 *Thermus*（Antranikian and Egorova 2007）代表了产耐热内切葡聚糖酶的古菌和细菌。此外，在 *Pyrococcus* 和 *Thermus* spp.（Chang et al. 2001；Xiangyuan et al. 2001）中已经描述了外切纤维素酶和葡糖苷酶。通过分子生物学技术实现了酶特异性和活性的改善（表19.3）。由于木质纤维素材料的降解需要大量的酶（Lynd et al. 2005），所用的纤维素酶来自需氧真菌的重组菌株，如里氏木霉 *Trichoderma reesei*（红褐内座菌 syn. *Hypocrea jecorina*）和特异腐质霉（*Humicola insolens*）（Karlsson et al. 2002）。

使用多结构超嗜热纤维素酶其在高于90℃的温度下的良好的木质素降解能力，GBM结构域的缺失可能导致几种超嗜热微生物无法在高于75℃的温度下有效地分解结晶纤维素（Graham et al. 2011）。

热稳定的内切葡聚糖酶可以在不同的最佳pH值下操作。解纤维热酸菌（*Acidothermus cellulolyticus*）表达内切葡聚糖酶，其最适pH为5.0，温度为80℃（Lindenmuth and McDonald 2011）。热嗜酸性脂环芽孢杆菌（*Alicyclobacillus*）A4表达极其适酸性的β-1,4葡聚糖酶（最适pH 2.6和最适温度65℃）（Bai et al. 2010a）。芽孢杆菌KSM-S237表达耐热碱性内切葡聚糖酶（最适pH 8.6～9.0），并且在100℃的温度下，在pH 9.0下暴露0.17h后，呈现超过原始30 %的活性（Hakamada et al. 1997）。目前，使用这些热嗜酸性和嗜碱性的酶是非常普遍的。例如，木质纤维素生物质必须通过酸或碱处理，然后在降解酶作用之前进行中和步骤（Zambare et al. 2011）。另一方面，当热嗜酸性和热嗜碱性的酶用于木质纤维素生物质的水解时，在中和步骤可以被去除（Zambare et al. 2011）。

19.3.3 半纤维素解构

半纤维素是一种复合聚合物，其可以是均聚物和杂聚物，含有构成主链的木糖-甘露糖、葡萄糖和半乳糖-吡喃糖的单元，含有25%～35%的木质纤维素生物质（Jain et al. 2014）。除了支链中的其他取代基之外，戊糖诸如D-木糖、D-阿拉伯糖，己糖如D-甘露糖、D-葡萄糖、D-半乳糖和糖酸使得该杂聚物对于每种植物资源来说都是非常复杂和独特的。木聚糖是半纤维素的主要成分，其丰富度是地球上可获得的可再生次级高分子生物质的三分之一（Dhiman et al. 2008）。从结构的角度来看，木聚糖是一种杂聚物，其中主链由重复单元β-1,4-连接的木糖构成，并以4-*O*-甲基葡萄糖醛酸吡喃糖基、α-L阿拉伯呋喃糖基、α-D-葡萄糖醛酸残基乙酰基、阿魏酰基和对香豆酰基单元进行修饰（Sun et al. 2005）。

硬木材是木聚糖作为*O*-乙酰基-4-*O*-甲基葡萄糖醛酸的的天然来源，而软木材是阿拉伯糖-4-*O*-甲基葡萄糖醛酸的来源。它们与4-*O*-甲基葡萄糖醛酸相邻、用α-(1,3)-L-阿拉伯呋喃糖残基替代。在具有高淀粉含量和谷物的胚乳外部中，可以用α-L-阿拉伯呋喃糖醛酸代替C2或C3位置的β-(1,4)-D-吡喃糖主链来回收阿拉伯木聚糖，其可以通过与酚酸和/或4-*O*-甲基-D-葡糖醛酸的酯化连接。在草和谷物的木质部分中分离出几种聚合物，例如通过乙酰化和阿魏酸酯化而连接的（葡萄糖醛酸）阿拉伯木聚糖。不同的是，在谷物茎、种子和橡胶中，这些杂聚物比单糖或寡糖的许多取代更为复杂（Sedlmeyer 2011）。分别存在木聚糖和葡甘露聚糖的硬木材和软木材也是木葡聚糖、葡甘露聚糖、半乳甘露聚糖和阿拉伯半乳聚糖的来源。

为了将半纤维素的复合结构完全水解成发酵糖如戊糖（D-木糖和D-阿拉伯糖），己糖（D-葡萄糖、D-半乳糖和D-甘露糖）和糖酸（Subramaniyan and Prema 2000，2002）一些酶是必需的。糖苷水解酶、多糖裂解酶、糖类酯酶是半纤维素水解消化酶的一些实例，这些酶共同作用以破坏糖苷键和酯键，并去除链中的取代基（Sweeney and Xu 2012）。涉及的酶主要有内切-β-1,4木聚糖酶（EC 3.2.1.8），木聚糖1,4-β-木糖苷酶（EC 3.2.1.37），α-L-阿拉伯呋喃糖苷酶（EC 3.2.1.55），α-葡糖苷酸酶（EC 3.2.1.139），乙酰木聚糖酯酶（EC 3.1.1.72），阿魏酸酯酶（EC 3.1.1.73），甘露聚糖内切-1,4-β-甘露聚糖酶（3.2.1.78），β-1,4-甘露糖苷酶（EC 3.2.1.25）和阿拉伯聚糖内切-1,5-α-L-阿拉伯糖苷酶（EC 3.2.1.99）（Collins et al. 2005）。

在半纤维素的水解中，内切木聚糖酶和外切木聚糖酶可用于开始断裂交联的聚合物，然后β-木糖苷酶将低聚木糖转化成木糖与各种长度的低聚物。另一方面，α-阿拉伯呋喃糖苷酶以呋喃糖和吡喃糖形式破坏阿拉伯糖单位，甲基葡萄糖醛酸替代物被α-葡萄糖醛酸酶水解，而乙酰木聚糖酯酶和阿魏酸酯酶水解乙酰基取代基、阿拉伯糖和阿魏酸。要注意的是，其中的几种酶水解其他化合物，当消耗木质纤维素材料时，很难知道它们的酶活性。现在有许多具有良好活性和高抗抑制性的蛋白质组合的实例，其含有来自真菌或细菌的水解酶。

在表19.4中列出了从嗜热和超嗜热微生物分离的木聚糖酶。热稳定的木聚糖酶主要由细菌和真菌产生（Collins et al. 2005）。生产热稳定性木聚糖酶的真菌有嗜热子囊菌 *Thermoascus aurantiacus*（Zhang et al. 2011），米色根霉 *Rhizomucor miehei*（Fawzi 2011），嗜热棉毛菌 *Thermomyces lanuginosus*（Singh et al. 2003），非曲霉 *Nonomuraea flexuosa*（Zhang et al. 2011），硫磺菌 *Laetiporus sulphureus*（Lee et al. 2009），嗜热篮状菌属 *Talaromyces thermophiles*（Maalej et al. 2009）。由细菌产生的木聚糖酶具有较高的最适温度，并且用于木质纤维素材料降解时相对于来自真菌的木聚糖酶有较高的热稳定性（Bhalla et al. 2013)。已经发现的具有热稳定性木聚糖酶的细菌有脂环酸芽孢杆菌属 *Alicyclobacillus*、厌氧杆菌属 *Anoxybacillus*、拟芽孢杆菌属 *Paenibacillus*、热厌氧杆菌属 *Thermoanaerobacterium*、放线菌属 *Actinomadura*、涅斯捷连科氏菌属 *Nesterenkonia*、肠杆菌属 *Enterobacter*、嗜酸性杆菌属 *Acidothermus*、纤维单胞菌属 *Cellulomonas*、芽孢杆菌属 *Bacillus*、地衣芽孢杆菌属 *Geobacillus* 和热酵母属 *Thermotoga*。

表 19.4 来自各种嗜热和超嗜热细菌的热稳定木聚糖酶的特性

生物	酶	最适pH	最适温度/℃	稳定性	参考文献
Thermotoga sp. strain FjSS3-B. 1	β-1,4-木聚糖酶	5.5	80	在 95℃,100℃和 105℃的半衰期分别为 1.5h,0.13h 和小于 0.03 h	Simpson et al. (1991)
Bacillus stearothermophilus T-6	β-1,4-木聚糖酶	6.5	75	在 70℃和 75℃的半衰期分别为 14.5h 和 0.33 h	Khasin et al. (1993)
B. flavothermus strain LB3A	β-1,4-木聚糖酶	7.0	70	在 80℃的半衰期为 0.16h	Sunna et al. (1997)
B. thermoleovorans strain K-3d	β-1,4-木聚糖酶	7.0	70～80	在 80℃ 的半衰期为 0.3h	Sunna et al. (1997)
Bacillus sp. strain SPS-0	β-1,4-木聚糖酶	6.0	75	木聚糖存在的情况下 70℃,4h 保持 80 % 活性,无木聚糖存在的情况下保持 20%活性	Bataillon et al. (2000)
B. licheniformis	β-1,4-木聚糖酶Ⅰ	7.0	70	X-Ⅰ 和 X-Ⅱ 分别在 50℃ 和 60℃,1h 保持 90%活性	Damiano et al. (2006)
	β-1,4-木聚糖酶Ⅱ	8.0～10.0	75		
B. subtilis	β-1,4-木聚糖酶	8.0	60	在 60℃,70℃和 80℃的半衰期分别为 16.2h,9.6h 和 2.8 h	Saleem et al. (2011)
Clostridium sp. TCW1	β-1,4-木聚糖酶	6.0	75	无数据	Lo et al. (2011)
Bacillus sp.	β-1,4-木聚糖酶	6.5,8.5 和 10.5	50	在 50℃,23h 可以保持全部活性	Sapre et al. (2005)
Enterobacter sp. MTCC	β-1,4-木聚糖酶	9.0	100	在 60℃和 70℃,18h 分别保持 85%和 64% 的活性	Khandeparkar and Bhosle(2006)
Paenibacillus macerans IIPSP3	β-1,4-木聚糖酶	4.5	60	在 60℃ 的半衰期为 6 h,在 90℃的半衰期为 2 h	Dheeran et al. (2012)
E. coli expressing *Alicyclobacillus* sp. A4	β-1,4-木聚糖酶	7.0	55	在 60℃,65℃和 70℃的半衰期分别为 6.5h,0.28h 和 0.05 h	Bai et al. (2010b)
Geobacillus thermantarcticus	β-1,4-木聚糖酶	5.6	80	在 70℃ 的半衰期为 24h,在 80℃的半衰期为 50min	Lama et al. (2004)
	β-木聚糖酶	6.0	70	在 60℃保持全部活性 1h	

从南极洲收集的嗜热菌微生物——嗜热土芽孢杆菌 *Geobacillus thermantarcticus*，产生

细胞外木聚糖酶和β-木糖苷酶。这些酶的部分活性显示了其应用于生物技术的潜力，例如最佳pH和温度、热稳定性、缺少纤维素降解活性和高含量的低分子量的木糖寡聚物。当*G. thermantarcticus*使用木聚糖作为唯一的有机来源时，通过调节生长条件和物理化学参数，有利于回收水解产物的两个步骤中特征性的木聚糖消化系统（Lama et al. 2004）。

热稳定性木聚糖酶的其他来源是超嗜热菌、热嗜酸菌和嗜热菌。厌氧嗜热杆菌*Thermoanaerobacterium saccharolyticum* NTOU1是一种海洋嗜盐细菌，在2mol/L NaCl存在下孵育24h，其木聚糖酶保留71 %的活性（Hung et al. 2011）。脂环酸芽孢杆菌*Alicyclobacillus* A4在pH 3.8至9.4的范围内具有很好的木聚糖酶稳定性，这种酶已经克隆到大肠杆菌中，在60℃下孵育1h后可保留90 %酶活性（Bai et al. 2010b）。肠杆菌属*Enterobacter* sp. MTCC 5112中也发现其他热稳定性木聚糖酶（Khandeparkar and Bhosle 2006）。

19.3.4 木质素解构

木质素结构是由三种酚类前体（即松柏醇、芥子酰和对香豆素）的氧化偶联产生的复杂网状结构。前体分别形成了鸟苷酸、丁香基和羟基苯基苯丙素亚基。由于木质素阻碍与纤维素和半纤维素聚合物的有效接触，不利于发酵，因此必须除去木质素以有效地进行生物质处理。已经研究了几种能够通过酶或化学氧化机制解聚木质素的微生物。真菌（例如担子菌）可以通过不同种类的过氧化物酶（EC 1.11.1.13），木质素酶（EC 1.11.1.14），多功能过氧化物酶（EC 1.11.1.16）和酚氧化酶（漆酶）（EC 1.10.3.2）来水解木质素。还需要其他酶来进行木质素的氧化降解，如纤维二糖脱氢酶（EC 1.1.99.18）、乙二醛氧化酶（EC 1.2.3.5）、芳基醇氧化酶（EC 1.1.3.7）和纤维二糖/醌氧化还原酶（EC 1.1.5.1）。这些酶能够以非特异性的方式产生破坏木质素聚合物的H_2O_2（Wong 2009）。

19.3.5 热稳定酶：木质纤维素降解的过表达

大多数嗜热细菌甚至在最佳生长条件下也不会产生大量的酶。例如*Geobacillus* spp.在60℃具有0.0113U/mL的内切葡聚糖酶活性（Tai et al. 2004），最高为0.058U/mL（Rastogi et al. 2009）。类似的，在55℃时土芽孢杆菌产生的内切葡聚糖酶活性为0.064U/mL（Abdel-Fattah et al. 2007）。因此，为了在工业规模中使用这种热稳定性酶，有必要通过在合适的宿主（通常是大肠杆菌）中过表达，有些物种则需要芽孢杆菌属（例如枯草芽孢杆菌和巨大芽孢杆菌）来过量产生。此外，巴斯德毕赤酵母*Pichia pastoris*已经被描述为重组纤维素酶和木聚糖酶的良好生产者（Sriyapai et al. 2011；Lindenmuth and McDonald 2011）。在65℃下显示最佳活性的嗜热芽孢杆菌菌种70PC53产生的内切葡聚糖酶在大肠杆菌中有效表达（Ng et al. 2009）。还利用大肠杆菌从新型嗜热菌株*Geobacillus* sp. MT-1中表达木聚糖酶基因，除了在20℃至90℃的温度范围内具有类似活性外，重组和野生型木聚糖酶都在70℃显示出最佳活性（Wu et al. 2006）。有趣的是其他内切葡聚糖酶如枯草芽孢杆菌菌株I15（Yang et al. 2010），*Bacillus* sp.（Afzal et al. 2010），枯草芽孢杆菌（Li et al. 2000，2009），热厌氧杆菌MB4（Liang et al. 2011）和结核杆菌*Fervidobacterium nodosum* Rt17-B1（Wang et al. 2010）均在大肠杆菌重组表达。重组纤维素酶和木聚糖酶的生产也通过其他宿主如巨大芽孢杆菌，枯草芽孢杆菌和巴斯德毕赤酵母（Lindenmuth and McDonald 2011）进行实施。对来自*Actinomadura* sp. S14的热稳定木聚糖酶，通过在巴斯德毕赤酵母中表达获得的重组酶（在80℃下孵育2h后保留50 %活性）比在大肠杆菌中表达获得更强的热稳定性（在80℃下孵育2h后保留30 %活性）（Sriyapai et al. 2011）。这样的结果表明，嗜温系统中的表达可能导致不同的翻译后修饰，如可能的蛋白质折叠或糖基化（Gao et al. 2012）。如今，关于木质纤维素解聚酶在嗜热宿主表达的数据仍然很少。

19.3.6 水解淀粉的极端微生物

淀粉是许多谷物如玉米、小麦、稻米、燕麦、马铃薯、木薯等的主要碳水化合物能量储备，是可持续能源（作为气体或液体生物燃料）和化工产品的主要潜在原料（Kumar et al. 2007）。

淀粉降解酶属于 α-淀粉酶超家族。这个家族包括许多在初级序列中显示高度相似性的酶，并通过保留催化机制起作用，从而释放 α 构型中的糖（Sinnott 1990）。α-淀粉酶超家族归为糖苷水解酶 GH-H 家族，包括 GH13 家族的三个序列相关家族。这种酶的特异性变异能力是由于具体的共有序列和可变数目的结构域，其又导致不同的水解或转移酶活性以及不同的底物特异性。

淀粉转化为葡萄糖的工业过程是两步法，即需要首先进行能量密集的液化步骤，然后进行糖化（Sivaramakrishnan et al. 2006）。液化在高温（高于 100℃）下进行，因此需要使用高度耐热的酶。*Methanocaldococcus*、*Pyrococcus*、*Sulfolobus* 和 *Thermococcus* 属的超嗜热古菌产生的嗜热 α-淀粉酶已被广泛研究（Kim and Peeples 2006；Van et al. 2007；Yang et al. 2004）。除了热稳定性之外，这些酶通常在约为 90℃ 时显示最佳活性，而在 120℃下处理 4h 之后，活性也保持不变。最引人注目的例子是由热革火球菌（T_{opt}100℃；$T_{\frac{1}{2}}=98$℃，13h），由 *Methanocaldococcus jannaschii*（T_{opt} 120℃；$T_{\frac{1}{2}}=100$℃，50h）产生的淀粉酶活性；由 *Thermococcus kadakaraensis*（T_{opt} 90℃；$T_{\frac{1}{2}}=70$℃，24h）产生的淀粉酶活性（Antranikian and Egorova 2007）。地衣芽孢杆菌产生的中度热稳定淀粉酶也被淀粉工业广泛利用（Bravo Rodriguez et al. 2006）。此外，还研究了来自嗜盐物种的糖苷水解酶，因为它们对高盐和溶剂浓度具有耐受性（Antranikian and Egorova 2007）。特别的，描述了一些嗜盐古菌淀粉酶，因为它们在几种溶剂（例如甲苯、苯或氯仿）中表现出显著的活性，并且在高盐浓度（高达 4.5mol/L NaCl）或高 pH 值（高达 10）表现出显著的活性（Antranikian and Egorova 2007）。还需要其他水解酶来完成淀粉的降解，例如葡糖淀粉酶（EC 3.2.1.3）和 α-葡糖苷酶（EC 3.2.1.20）。葡糖淀粉酶从聚合物的非还原末端释放 α-1,4 连接的 D-葡萄糖单元，耐热和嗜酸性葡糖淀粉酶是由属于硫化叶菌属 *Sulfolobus* 的一些古菌属（Kim et al. 2004），嗜酸菌属 *Picrophilus*（T_{opt}90℃；$T_{\frac{1}{2}}=90$℃，24h；$pH_{opt}=2$）和热原体属 *Thermoplasma*（T_{opt} 75～90℃；$T_{\frac{1}{2}}=60$～90℃，24～40h；$pH_{opt}=2$～5）产生的（Serour and Antranikian，2002）。α-葡糖苷酶对 D-葡萄糖的二聚体、三聚体和四聚体中的 α-1,4 键起作用，这些酶的典型例子是一些产生高耐热性酶的硫磺属（*Sulfolobus*）、铁原体属（*Ferroplasma*）、热原球菌（*Pyrococcus*）和热球菌属（*Thermococcus*）的物种（Chang et al. 2001；Piller et al. 1996；Schiraldi et al. 2000）。

19.4 蔬菜生物质降解的进展

在第二代生物燃料（也称为高级生物燃料）的范围内，从植物生物质原料中回收的脱淀粉多糖和寡糖物质的化学特征对于在增值化合物生产中至关重要。特别是木质纤维素材料（作为农业残留物和林业生物质）是重要的低成本和丰富的可再生生物质资源，具有相当大的生物转化为特殊生物产品的潜力（Kamm and Kamm 2004）。

在本节中，将介绍从木质纤维素材料获得特征化学产物程序的概述，将重点介绍使用极端酶制备的寡糖和糖缀合物。

19.4.1 半纤维素提取物的单糖组成

为了了解从植物生物质提取的半纤维素的性质，必须分析其单体组成。阿拉伯糖/木糖或葡糖醛酸/木糖比例被认为是半纤维素分支的线性程度（Verbruggen et al. 1995）。

从植物生物质提取的半纤维素的单体组成分析对于了解其性质至关重要。

通常，在不同的实验条件下，它们被硫酸彻底水解。例如，麦麸中提取半纤维素，需要在 1mol/L H_2SO_4 中 100℃下水解 3h；对于桉树细胞壁和大麦秸秆的半纤维素，需在 105℃下水解 2.5h。

中和后，通过 HPAEC-PAD（具有电流检测的高效阴离子交换色谱）系统在适当的洗脱条件下分析总单糖（木糖、鼠李糖、阿拉伯糖、甘露糖、葡萄糖、半乳糖、葡糖醛酸和半乳糖醛酸）（Aguedo et al. 2014；Lina et al. 2006）。

在甘蔗渣半纤维素的情况下，中性单糖在 120℃，2h，2mol/L 三氟乙酸（TFA）条件下水解，在丙二醇中经过乙酰化转化并进行气相色谱（GC）分析（Sun et al. 2004）。

通过在 105℃下用 TFA 2 N[1-(N-TFA)2-甲基哌啶] 水解 4h，从麦麸中分解阿拉伯木聚糖中的尿酸（Aguedo et al. 2014）。按照经典的检测糖链一级结构的方法，通过 GC-MS 测定单糖组成为对乙酰化甲基苷。芦竹根茎的半纤维素，需在 1.25mol/L HCl/MeOH，80℃和过夜进行甲醇分解（Lama et al. 2014）。

值得注意的是，有时完全水解的木聚糖提取物的单糖组成可以通过在适当的洗脱条件（*n*-BuOH/AcOH/H_2O 体积比为 6∶2∶2 或 EtOAc/H_2O/AcOH/2-丙醇/HCOOH，体积比为 25∶15∶10∶5∶1）的薄层色谱（TLC）分析进行定性和快速检测（Lama et al. 2014）。

19.4.2 由半纤维素农业废弃物生产寡糖的色谱特性

高效阴离子交换色谱（HPAEC）是通过木聚糖提取物的酶促或化学水解获得的低聚木糖的色谱分析的典型程序。通过使用适当的标准来鉴定寡糖及恢复其浓度值。

使用 Shearzyme（GH10 内切-1,4-*β*-D-木聚糖酶）和两种 *α*-L-阿拉伯呋喃糖苷酶（AXH-m 和 AXHd3）从小麦阿拉伯糖基木聚糖（AX）产生的阿拉伯木聚糖寡糖（AXOS）进行 HPAEC-PAD 分析。

HPAEC-PAD 分析提供了这些反应的色谱图，其中不仅预先从柱中洗脱出最短的低聚木糖，而且在单和双 α-L-阿拉伯呋喃糖基（Ara*f*）取代阿拉伯木聚糖寡糖（AXOS）的收集物内遵循相应的洗脱顺序，此外，*β*-D-吡喃木糖基（Xylp）单元上的 α-L-阿拉伯呋喃糖基连接位置也影响洗脱顺序（Pastell et al. 2008）。

最近，来自嗜碱芽孢杆菌 TSEV1 的细胞外内切木聚糖酶，其在热和碱性 pH 条件下稳定，在大肠杆菌中克隆并表达，然后开发应用于几种农业残留物的低聚木糖生产中。通过配备有差示折射率检测器的高压液相色谱（HPLC）进行糖化的监测（Kumar and Satyanarayana 2014 ）。

然而，有时通过 TLC 分析的定性监测可以产生足以鉴定由农业残留物产生的低聚木糖，通过使用来自嗜热细菌木霉菌 TSAA1 的热稳定的内切木聚糖酶来制备低聚糖生产的小麦糠半纤维素［TLC 系统溶剂为正丁醇∶乙醇∶水（5∶3∶2）］的 TLC 的降解研究是有用的（Anand et al. 2013）。此外，使用芦竹半纤维素提取物和不同的嗜热性酶制剂（*T. neapolitana*

和 *Thermoanaerobacterium thermostercoris* 的无细胞提取液，嗜热芽孢杆菌 *Geobacillus thermantarcticus* 的细胞外悬浮液，商业木聚糖酶的嗜热棉毛菌 *Thermomyces lanuginosus* 和海栖热袍菌 *Thermotoga maritima*）将未知寡糖的个体化进入混合物反应，然后进行光谱检测（Lama et al. 2014）。

19.4.3　酶消化前后半纤维素馏分的光谱研究：NMR、MS、FT-IR 分析

在低聚木糖或其增值衍生物的设计方法出现之前，木质纤维素来源的木聚糖多糖的 1D 和 2D 核磁共振（NMR）研究是必不可少的。

NMR 分析提供关于鉴定几种半纤维素提取物的结构所必需的单糖的性质、构型和相对含量的结构信息。在大多数情况下，多糖提取物的均匀性对应着具有良好分辨信号的相对简单的光谱。

通常，在阿拉伯糖（葡糖醛酸）木聚糖及其寡糖衍生物的质子和碳 NMR 谱中，根据糖链的支化来检测一些鉴定信号，其强度是可变的（Jin et al. 2009）。

在表 19.5 中，我们总结了属于半纤维素结构的化学位移的特殊值。

从芦竹的根茎分离的半纤维素，在 40℃下，半纤维素底物在 *d*-DMSO 中的 ^{13}C NMR 光谱显示出如表 19.5 所示的归因于木聚糖主链的强烈的特征信号，这些信号通常表示（1→4)-连接的 β-木聚糖的主要信号（Bendahou et al. 2007）。此外，在芦竹半纤维素提取物的 ^{13}C NMR 中，存在对应于阿拉伯糖残基的信号。这些值证实了多糖提取物的阿拉伯木聚糖结构，其强度表明有部分脱分支骨架。

对来自蔗渣不同温度下的半纤维素提取物（表 19.5）进行 ^{13}C 和 ^{1}H NMR 谱（在 D_2O）分析，揭示了 β-木糖链和阿拉伯糖残基的信号和其他属于 4-*O*-甲基-α-D-葡萄糖醛酸单元的更小强度的信号（Bian et al. 2012）。

表 19.5　木质纤维素来源的木聚糖在 ^{1}H 和 ^{13}C 光谱中的诊断化学位移

化合物	结构	核磁共振(NMR)	
		$^{1}H/10^{-6}$	$^{13}C/10^{-6}$
（乙酰基）阿拉伯糖醛木聚糖（Hoffmann et al. 1992a,b）	在 C-2 和 C-3(二取代的),C-3(单取代的)或未取代的取代的 β-D-木糖残基异头位点	4.5∶4.8	
	Ara*f* 的异头位点属于短侧链	～5.4	
	位点 2,5 属于木糖和阿拉伯糖残基	3.1∶4.3	
	甲基葡糖醛酸的部位	5.3	
	甲基葡糖醛酸的甲基信号	1.8	
	木聚糖链的乙酰基	2.6	
从 *A. donax* 的根茎中分离出来的阿拉伯木聚糖	木聚糖主链的异头位点		101.7
	1,4-β-连接的位点		75.3
	1,4-β-连接的木糖中的木糖 2,3,5 位点		73.9,72.5,63.2
	阿拉伯呋喃糖残基的异头位点		107.1
	α-L-Araf 的位点 2,5 与木糖残基的 3 位连接		87.3,86.0,80.2,77.8,61.8

续表

化合物	结构	核磁共振(NMR)	
		$^{1}H/10^{-6}$	$^{13}C/10^{-6}$
甘蔗渣中的半纤维素(Bian et al. 2012)	位点 2,5 的主链的 1,4-β-连接的木糖残基	4.34(H-1), 3.96(H-5 eq), 3.66(H-4), 3.40(H-3),	102.32,75.89, 74.91,73.28, 63.27
	α-L-Ara*f* 的位点 2,5		109.50,86.46, 80.23,78.37, 61.72
	位点 1,6 和 OCH_3-4 取代 4-*O*-甲基-α-D-葡糖醛酸残基		177.01(C-6),97.46(C-1), 75.3(C-2)72.12(C-3), 73.86(C-5),79.26(C-4), 59.53(OCH_3-4)
桉树的半纤维素(Li et al. 2015)	位点 1,2,5,6 和 OCH_3-4 取代 4-*O*-甲基-α-D-葡糖醛酸残基	5.14(H-1), 4.15(H-5), 3.49(H-2), 3.35(-OCH3)	

类似的，在最近研究的桉树细胞壁的半纤维素提取物的质子光谱中，尽管 β-木聚糖主链的信号占主导地位，但是检测到了小的信号并归因于 4-*O*-甲基-α-葡糖醛酸残基（Li et al. 2015）。

2D-NMR 实验，如 COSY（相关光谱）、TOCSY（全相关光谱）、HSQC（异核单量子相关）、HSQC-EDITED（多重编码 HSQC）、HMBC（异核多键相关谱）、NOESY（核过氧化物光谱）对于每个单糖残基内每个位置的质子和碳值信号的分配是必要的，它将多糖链的一级结构和/或单糖单元的序列建立成寡糖骨架。

COSY、TOCSY 和 HSQC 实验对于识别自旋系统至关重要，通过评估 HMBC 实验中的长程相关性 C—H 或 NOE（奥弗豪塞尔核效应）实验中的偶极耦合 H—H，可以沿着木糖骨架定位酸性残基或阿拉伯糖。

实际上，通过详细的 2D-NMR 光谱研究，必须表征通过酶解消化芦竹根茎的半纤维素提取物获得的五糖（图 19.2，化合物 1）和四糖（图 19.2，化合物 2）的结构，通过使用商业嗜热丝孢菌（*Thermomyces lanuginosus*）木聚糖酶和来自 *Thermoanaerbacterium thermoercoris* 的无细胞提取物的木聚糖酶（Lama et al. 2014）。

通过质谱法平行分析从木聚糖释放的寡糖，并支持 NMR 研究。通常，矩阵辅助激光解吸/电离飞行时间质谱（MALDI-TOF-MS）和电喷雾质谱（ESI-MS）是最常见的分析方法（Aachary and Prapulla 2011；Pastell et al. 2008）。

在某些情况下，这些分析与分离程序如尺寸排阻色谱（SEC）、高效阴离子交换色谱（HPAEC）和反相高效液相色谱（RPHPLC）相结合进行。

最近，有提议使用 ESI 的阳性串联质谱法用于表征未衍生的或乙酰化的中性和酸性 XOS（Reis et al. 2005）。

此外，在通过酶消化产生的低聚木糖的混合物的情况下，MS 研究提供了低聚木糖分子量，其结合关于初始木聚糖的主要结构的信息，提出可能的结构。另一方面，确定的结构只能通过串联质谱法（ESI-MS/MS）获得（Reis et al. 2005）。

1

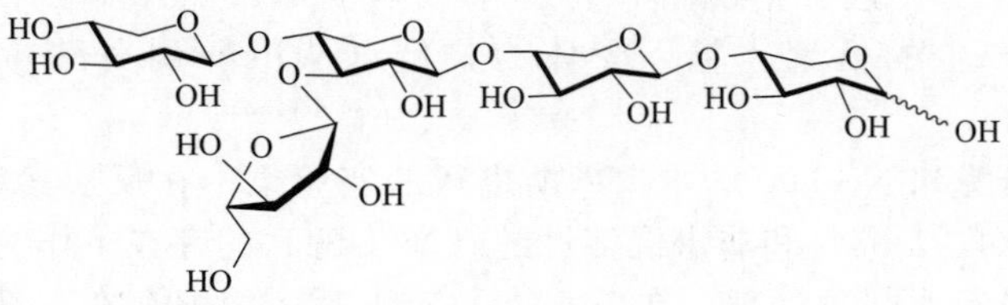

2

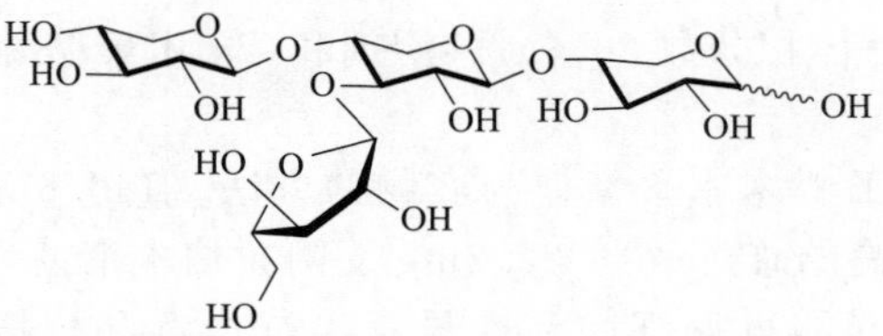

图 19.2 五糖 β-D-Xylp-(1-4)-[α-L-Araf-(1-3)]-β-D-Xylp-(1-4)-β-D-Xylp-(1-4)-β-D-Xylp (1) 和四糖和 β-D-Xylp-(1-4)-[α-L-Araf-(1-3)]-β-D-Xylp-(1-4)-D-Xylp (2)

红外光谱已广泛用于研究多糖的官能团（Li et al. 2015）。

在半纤维素的 IR 光谱中，在约 $1745cm^{-1}$ 处记录属于乙酰基，糖醛酸和阿魏酸酯基团的 C ═O 的拉伸信号以及在酯基中对应于—C—O— 的 $1249cm^{-1}$ 处的弯曲信号 。

此外，在羰基伸展区域，$1463cm^{-1}$、$1426cm^{-1}$、$1382cm^{-1}$ 和 $1318cm^{-1}$ 处的信号分别归因于—CH_2 对称弯曲，CH 和 OH 弯曲，OH 平面内弯曲和—CH 摇摆（Sun et al. 1996）。然而，在 $903cm^{-1}$ 处的条带的存在表明半纤维素提取物中糖单元之间的主要 *β*-糖苷键（Robert et al. 2005）。

在 1120～$1000cm^{-1}$ 区域，具有多个峰的信号是高度取代的阿拉伯木聚糖的独特特征，木糖残基的 C-3 位点的阿拉伯糖取代在 1173 cm^{-1} 处提供条带（Subba and Muralikrishna 2004）。

一般来说，$990cm^{-1}$ 和 $900cm^{-1}$ 之间的谱带强度强烈依赖于木聚糖骨架上 Ara*f* 单位的数量（Robert et al. 2005）。

19.4.4 转糖基作用过程与半纤维素提取物

利用木质纤维素生物质进行生物燃料回收的替代方案可以是高附加值分子的合成。在这种情况下，将木聚糖天然水解成低聚氧化物和木糖单体的木聚糖酶以及将寡聚核苷酸水解成木糖单糖的木糖苷酶在生产木糖共轭体的转糖基作用过程中非常重要。

如我们所知，烷基取代的糖代表非离子表面活性剂，其目前添加到液体和粉末洗涤剂、药物制剂和个人护理用品中。作为重要基团的戊糖基表面活性剂的实例，烷基氧化物是具有刺激性的表面活性剂（Xu et al. 2012），并且这些分子可以由木质纤维素生物质生产，通过酶合成在环境友好中降低生产成本。

在本节中，来自热嗜酸芽孢杆菌 TSAA1 的重组 *β*-木糖苷酶最近用于生产烷基木糖苷，起始于具有短链（C1～C5）作为受体的脂族醇和 XO 的混合物，其由脱粒麦麸提取物作为供体。

在适当的反应条件下，在甲醇和麦麸水解产物反应 16h 后，在 TLC 板上检测到良好的甲基木糖苷的相应谱带，此外，在 HPLC 色谱图中，除了甲基木糖苷以外，记录了对应于甲基低聚糖的峰的存在（Jain et al. 2014）。

在木聚糖酶（Tx-xylanase）和商业木聚糖酶（Novozymes NS-50030）的存在下进行更

有效的转糖基反应。在这种情况下，将桦木或燕麦木聚糖用作供体，或者由水热预处理和去饱和的麦麸产生的低聚木糖，受体为长链（从甲醇到癸醇）的脂肪族醇（Ochs et al. 2011）。

用 Tx-木聚糖酶和 NS-50030 木聚糖酶进行的实验集中在反应参数的分析上，目的是在部分水混溶性醇（戊-1-醇）和非水混溶性醇（辛-1-醇）的存在下作为受体。

破坏的麦麸进行水热预处理，在 135℃反应 1h 后，溶解存在于麸皮中的阿拉伯木聚糖。将过滤和冻干后的寡糖溶液用作浓度为阿拉伯木聚糖当量 20g/L 的供体。该供体在与辛-1-醇（体积分数 20%），叔丁醇（体积分数 20%）共溶剂和 Tx-木聚糖酶（20IU/mL）的反应 1h 中使用。

令人惊奇的是，使用 20g/L 桦木木聚糖作为底物的方法（146.6mg），使用从预处理的麦麸获得的上清液中记录到最高的总产量（222.2mg/g 阿拉伯木聚糖当量）。

之前报道，9-芴基甲基苷对卵巢癌细胞活性显示出抗病毒和抗增殖作用（Tramice et al. 2008，2009）。

最近，将来自芦竹茎叶的半纤维素提取物用作转糖基化过程中的供体，目的是通过使用来自 *T. neapolitana* 的马尾松木聚糖酶以及来自 *T. thermercoris* 和 *G. thermantarcticus* 的木聚糖酶/β-木糖苷酶活性来生产 9-芴甲醇木糖苷（Lama et al. 2014）。用各种酶系统进行反应的甲醇提取物通过 MS 光谱分析，揭示 9-芴基甲醇的单和二木糖化衍生物的产生。记录 m/z 351 $[M+Na]^+$ 和 483 $[M+Na]^+$ 的信号。此外，在与 *G. thermantarcticus* 粗酶液的反应中，通过在 m/z 651 $[M+Na]^+$ 的反应 MS 谱中的信号确定存在于三羟基化的 9-芴基甲醇的反应介质中。

19.5 极端微生物在废弃生物质开发利用中的生物技术应用：生物燃料和生物氢生产

极端微生物的主要应用之一是用于转化来自农业和林业的废弃物以获得生物燃料。特别是代表第二代生物燃料原料的木质纤维素农林废弃物是不与食品竞争的原材料，与食品价格上涨无关。鉴于此，寻找具有热稳定酶的嗜热菌是对这些废弃生物质的最佳利用。为了获得纤维素和半纤维素，木质纤维素起始材料需要几种预处理，通常是通过热机械预处理来打开木质素鞘，然后进行如表 19.2 所述的化学处理。之后，需要酶解过程以释放可发酵的糖，通常是纤维素酶、半纤维素酶和木聚糖酶。这些酶解过程通常在小于 50℃下进行，在该条件下，水解速度较慢，酶浓度必须高，并且有与微生物污染的风险，一般来说，可发酵糖的产率似乎较低。在这种情况下，嗜热微生物及其热稳定酶的使用对于克服木质纤维素生物质转化的限制和改善整个过程是非常有用的（Bhalla et al. 2013）。事实上，因为它们通常需要 50℃至 80℃的工作温度，所以使用这些热稳定性酶意味着较短的水解时间，较高的试剂和产品溶解度，较高的水解速率，较低的微生物污染风险和促进挥发性产品的恢复（Zhang et al. 2011；Liang et al. 2011）。

由于嗜热酶具有很好的适应不同 pH 值的能力，这些酶成为需要酸性或碱性化学预处理的情况下极端微生物进行木质纤维素转化的理想参与者。木质纤维素生物质转化可以通过使用能够进行有效的木质纤维素解构的嗜热酶来实现。使用热稳定酶与产乙醇嗜热菌的组合，增加水解产率，并且在该条件下可以在更高的温度下仅使用一种发酵罐进行发酵（嗜热同步糖化发酵，SSF）（Podkaminer et al. 2011；Shaw et al. 2008）。例如 Shaw 等人（2008）使用了该生物过程，其使用嗜热酵母杆菌菌株 ALK2，与在 37℃下工作的酿酒酵母进行的 SSF

相比，在 50℃下实现 SSF 方法过程的等效水解所需的酶量降至$\frac{2}{5}$。在文献中有几个发酵嗜热细菌（发酵温度范围为 45℃至 70℃）的实例，其能够使用木糖、葡萄糖、纤维二糖、半乳糖、甘露糖以及衍生自木质纤维素生物质如玉米秸秆的生物分解的混合糖，以获得乙醇。在这种情况下，即使可以实现其他产物如乙酸盐、乳酸盐、丙酮酸盐和琥珀酸盐的产生，1g 糖类或纤维素能产生 0.16g 至 0.47g 乙醇（Cai et al. 2011；Cripps et al. 2009；Georgieva et al. 2008）。然而，使用预处理的木质纤维素生物质的蔗糖分解发酵嗜热菌的综合生物处理（CBP）可以获得最好的结果（Olson et al. 2012）。与 CBP 相关的最具挑战性的任务是选择合适的微生物或微生物群落（consortium），可以分解产生乙醇的木质纤维素材料。为此目的，原材料不要求任何预处理，如化学、物理或酶作用，只要减少粒径就足够了（Paulová et al. 2014）。在 CBP 中，水解酶的生产、木质纤维素废弃物的降解和通过发酵释放的单糖的使用在单个反应器中有效地发生，与 SSCF（同步糖化和共发酵）相比，成本降至$\frac{1}{4}$（Lynd et al. 2005）。CBP 中有两类微生物：Ⅰ类纤维素酶生产者和Ⅱ类乙醇生产者。属于第一类的例如热梭菌（*Clostridium thermocellum*）、热葡萄糖苷酶地芽孢杆菌（*Geobacillus thermoglucosidans*）、烷嗜热厌氧杆菌（*Thermoanaerobacter mathranii*）和纤维素分解真菌（丝状真菌瑞氏木霉 *Tricoderma reesei*、宛氏拟青霉 *Paecilomyces variotii*）（Paulová et al. 2014）。第二类 CBP 生产者包括方便工程化的乙醇生产者微生物，如酿酒酵母、马克斯克鲁维酵母和运动发酵单胞菌（Paulová et al. 2014）。这些蔗糖分解嗜热菌已经通过基因工程得到改进，提供含有一组有趣的水解酶的构建微生物。这是热纤维梭菌（*Clostridium thermocellum*）能够通过 CBP 将植物生物质转化为乙醇的实例。为了提高乙醇耐受性并在乙醇生产中获得高产量，在 Brown 等人（2011）报道的醇脱氢酶水平上进行诱变。还报告了结合使用两类 CBP 生产者（Ⅰ和Ⅱ）的微生物群落的可能性。还研究了使用由酿酒酵母、里氏木霉和舍弗勒氏菌属组成的微生物群落来研究乙醇生产。观察到使用酸性预处理小麦秸秆的最大乙醇浓度为 9.8g/L，理论乙醇收率为 69%（以乙醇理论收率 0.51g/g 为 100%，代表酿酒酵母在葡萄糖的乙醇发酵过程中产率）。减少与木质纤维素转化为生物燃料相关成本的其他方法是使用固定化的生物催化剂，如纤维素分解酶和微生物。显然，与纤维素酶固定相比，微生物细胞固定得到更好的结果，因为在酶固定化后酶活性降低（Paulová et al. 2014）。

此外，已经开发了几个有效的氢电池工厂，其使用的为产生热稳定的纤维素分解和木聚糖分解酶的厌氧嗜热菌。碳纤维素酶解糖酵解菌代表了使用木质纤维素生物质如甘蔗渣和甜高粱的合适的氢生产候选物（VanFossen et al. 2009）。当在单个反应器中使用 *Caldicellulosiruptor bescii* DSM 6725 时，木质纤维素生物质生物转化为生物燃料的整个过程的有效性达到最大表达。实际上，这种微生物的最佳生长温度为 80℃，显示出将未处理的生物质如百慕达草和柳枝稷转化为氢（Yang et al. 2009）。在这种情况下，*C. bescii* 的高生长温度有助于从生物反应器中回收挥发性物质（乙醇）（Chang and Yao 2011）。

19.6 结论

能够转化木质纤维素材料的热稳定酶的新嗜热细菌的研究可以改善该方法的生产率和能量消耗。除经典方法之外，根据对降解木质纤维素底物的嗜热微生物基因组草图分析，可揭示编码纤维素或木质素降解酶的基因，进而帮助探索生物燃料生产的过程。事实上，即使使用关于废弃物组成和微生物途径的所有信息，也并不是所有的微生物生长利用废弃物的尝试都能成功。全基因组知识可以对用于废弃生物质利用的合适的极端微生物进行初步筛选（Studholme 2015）。

在世界能源方案中，绝对的确定性包括化石燃料快速消耗，加上人口增长，在各个层面上都加强了一种高效和有价值的选择。目前正在检验不同的可再生能源。在本章中，蔬菜废弃生物质被认为增值产品的来源。普遍认为，它们的量很大，需要在广泛的做法中适当正确处置，制定旨在重用和增值有用废料的程序。因此，所有工业和农业残留物的创新理念是倾向于零废弃物排放。如图 19.3 所示，利用嗜热微生物获得聚合物组分（淀粉、木质素、纤维素、木聚糖）及其用于生物燃料生产的单糖和寡糖的废弃物生物量（芦竹的根茎）的降解。此外，植物生物质可以被用作嗜热菌生长及其酶超量生产的有机和能量支撑材料（Finore et al. 2014；Lama et al. 2014）。在这个集成系统中，每种处理的残留物代表进一步转化的起始材料，其最终目的是实现零排放过程。

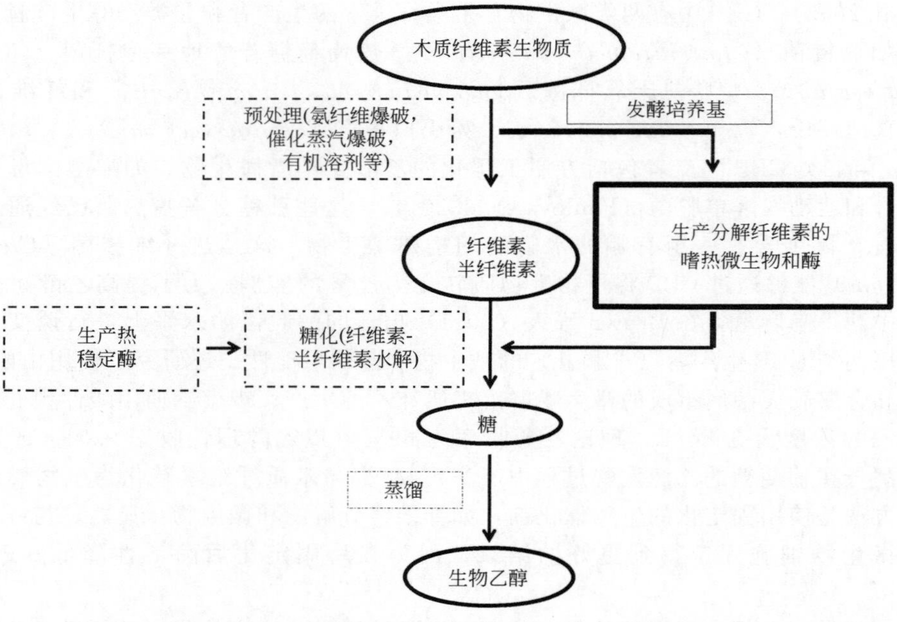

图 19.3　芦竹根茎木质纤维素生物质综合处理方案

当今竞争中新兴的合成生物学提供了克服木质纤维素转化相关挑战的巨大潜力。合成生物学被认为具有重构生物系统的有趣学科（Keasling 2008）。而在可再生能源领域，它代表了获得生物燃料和/或化学物质的经济方式（Nieves et al. 2015）。这个学科基于从酶、代谢途径、遗传回路到整个细胞的新生物成分的设计和构建，以组装可以解决具体问题的综合系统。例如，在木质纤维素转化的情况下，微生物催化剂的主要目的是木糖的良好利用和对呋喃醛的抗性（Sandoval et al. 2012；Wang et al. 2013）。下一步将是建立能够提供木质纤维素转化的细胞，这要归功于利用木质纤维素降解嗜热菌的基因组序列的新开发的遗传技术。

无利益冲突声明

Annarita Poli，Ilaria Finore，Annabella Tramice，Paola Di Donato，Barbara Nicolaus

和 Licia Lama 声明他们没有利益冲突。

致谢

这项工作部分得到 PON03PE _ 00107 _ 1 BioPoliS 项目的支持，该项目开发绿色技术，用于生产生物化学药品，用于合成和工业应用 POLImerici 材料，化学材料来自从坎帕尼亚地区的可持续耕种系统获得的农业资源。

参考文献

Aachary AA，Prapulla SG（2009）Value addition to corncob：production and characterization of xylo-oligosaccharides from alkali pretreated lignin-saccharide complex using *Aspergillus oryzae* MTCC 5154. Bioresour Technol 100：991-995

Aachary AA，Prapulla SG（2011）Xylo-oligosaccharides（XOS）as an emerging prebiotic：microbial synthesis，utilization，structural characterisation，bioactive properties，and applications. Compr Rev Food Sci Food Saf 10（1）：2-16

Abdel-Fattah YR，El-Helow ER，Ghanem KM，Lotfy WA（2007）Application of factorial designs for optimization of avicelase production by a thermophilic *Geobacillus* isolate. Res J Microbiol 2：13-23

Afzal S，Saleem M，Yasmin M，Naz M，Imran M（2010）Pre and post cloning characterization of a β-1，4-endoglucanase from *Bacillus* sp. Mol Biol Rep 37：1717-1723

Aguedo M，Fougnies C，Dermiencec M，Richel A（2014）Extraction by three processes of arabinoxylans from wheat bran and characterization of the fractions obtained. Carbohydr Polym 105：317-324

Akpinar O，Erdogan K，Bostanci S（2009）Production of xylo-oligosaccharides by controlled acid hydrolysis of lignocellulosic materials. Carbohydr Polym 344：660-666

Akpinar O，Erdogan K，Bakir U，Yilmaz L（2010）Comparison of acid and enzymatic hydrolysis of tobacco stalk xylan for preparation of xylo-oligosaccharides. LWT-Food Sci Technol Int 43（1）：119-125

Anand A，Kumar V，Satyanarayana T（2013）Characteristics of thermostable endoxylanase and β-xylosidase of the extremely thermophilic bacterium *Geobacillus thermodenitrificans* TSAA1 and its applicability in generating xylooligosaccharides and xylose from agro-residues. Extremophiles 17：357-366

Antranikian G，Egorova K（2007）Extremophiles，a unique resource of biocatalysts for industrial biotechnology. In：Gerday C，Glansdorff N（eds）Physiology and biochemistry of extremophiles. ASM Press，Washington，DC，pp 361-406

Azevedo Carvalho AF，de Oliva Neto P，da Silva Fernandes D，Pastore GM（2013）Xylooligosaccharides from lignocellulosic materials：chemical structure，health benefits and production by chemical and enzymatic hydrolysis. Food Res Int 51：75-85

Bai Y，Wang J，Zhang Z，Pengjun Shi P，Luo H，Huang H，Luo C，Yao B（2010a）Expression of an extremely acidic beta-1，4-glucanase from thermoacidophilic Alicyclobacillus sp. A4 in Pichia pastoris is improved by truncating the gene sequence. Microb Cell Fact 9：33

Bai Y，Wang J，Zhang Z，Yang P，Shi P，Luo H，Meng K，Huang H，Yao B（2010b）A new xylanase from thermoacidophilic *Alicyclobacillus* sp. A4 with broad-range pH activity and pH stability. J Ind Microbiol Biotechnol 37：187-194

Bals B，Rogers C，Jin M，Balan V，Dale B（2010）Evaluation of ammonia fibre expansion（AFEX）pretreatment for enzymatic hydrolysis of switchgrass harvested in different seasons and locations. Biotechnol Biofuels 3：1-11

Barnard D，Casanueva A，Tuffin M，Cowan D（2010）Extremophiles in biofuel synthesis. Environ Technol 31：871-888

Bataillon M，Cardinali APN，Castillon N，Duchiron F（2000）Purification and characterization of a moderately thermostable xylanase from *Bacillus* sp. strain SPS-0. Enzyme Microb Technol 26：187-192

Bendahou A，Dufresne A，Kaddami H，Habibi Y（2007）Isolation and structural characterization of hemicelluloses from palm of Phoenix dactylifera L. Carbohydr Polym 68（3）：601-608

Berlin A，Gilkes N，Kurabi A，Bura R，Tu MB，Kilburn D，Saddler J（2005）Weak lignin binding enzymes. A novel approach to improve the activity of cellulases for hydrolysis of lignocellulosics. Appl Biochem Biotechnol 121：163-170

Bhalla A，Bansal N，Kumar S，Bischoff KM，Sani RK（2013）Improved lignocellulose conversion to biofuels with thermophilic bacteria and thermostable enzymes. Bioresour Technol 128：751-759

Bian J，Peng F，Peng XP，Xu F，Sun RC，Kennedy JF（2012）Isolation of hemicelluloses from sugarcane bagasse at different temperatures：structure and properties. Carbohydr Polym 88：638-645

Binod P，Sindhu R，Singhania RR，Vikram S，Devi L，Nagalakshmi S，Kurien N，Sukumaran RK，Pandey A（2010）Bioethanol production from rice straw：an overview. Bioresour Technol 101：4767-4774

Bok J，Dienesh A，Yernool D，Eveleigh D（1998）Purification，characterization and molecular analysis of thermostable cellulases CelA and CelB from*Thermotoga neapolitana*. Appl Environ Microbiol 64：4774-4781

Bravo Rodriguez V，Jurado Alameda E，Martinez Gallegos JF，Reyes Requena A，Garcia Lopez AI（2006）Enzymatic hydrolysis of soluble starch with an alpha-amylase from*Bacillus licheniformis*. Biotechnol Prog 22（3）：718-722

Brienzo M，Siqueira AF，Milagres AMF（2009）Search for optimum conditions of sugarcane bagasse hemicellulose extraction. Biochem Eng J 46：199-204

Bronnenmeier K，Staudenbauer W（1990）Cellulose hydrolysis by a highly thermostable endo-1，4-glucanase（Avicelase I）from Clostridium stercorarium. Enzyme Microb Technol 12：431-436

Brown SD, Guss AM, Karpinets TV, Parks JM, Smolin N, Yang S, Land ML, Klingeman DM, Bhandiwad A, Rodriguez M Jr, Raman B, Shao X, Mielenz JR, Smith JC, Keller M, Lynd LR (2011) Mutant alcohol dehydrogenase leads to improved ethanol tolerance in *Clostridium thermocellum*. Proc Natl Acad Sci U S A 108: 13752-13757

Cai Y, Lai C, Li S, Liang Z, Zhu M, Liang S, Wang J (2011) Disruption of lactate dehydrogenase through homologous recombination to improve bioethanol production in *Thermoanaerobacterium aotearoense*. Enzyme Microb Technol 48: 155-161

Cardona CA, Quintero JA, Paz IC (2010) Production of bioethanol from sugarcane bagasse: status and perspectives. Bioresour Technol 101: 4754-4766

Chang VS, Holtzapple MT (2000) Fundamental factors affecting biomass enzymatic reactivity. Appl Biochem Biotechnol 84-86: 5-37

Chang T, Yao S (2011) Thermophilic, lignocellulolytic bacteria for ethanol production: current state and perspectives. Appl Microbiol Biotechnol 92: 13-27

Chang ST, Parker KN, Bauer MW, Kelly RM (2001) Alpha-glucosidase from *Pyrococcus furiosus*. Methods Enzymol 330: 260-269

Chapla D, Pandit P, Shah A (2012) Production of xylooligosaccharides from corncob xylan by fungal xylanase and their utilization by probiotics. Bioresour Technol 115: 215-221

Collins T, Gerday C, Feller G (2005) Xylanases, xylanase families and extremophilic xylanases. FEMS Microbiol Rev 29: 3-23

Cripps RE, Eley K, Leak DJ, Rudd B, Taylor M, Todd M, Boakes S, Martin S, Atkinson T (2009) Metabolic engineering of *Geobacillus thermoglucosidasius* for high yield ethanol production. Metab Eng 11: 398-408

Damiano VB, Ward R, Gomes E, Alves-Prado HF, Da Silva R (2006) Purification and character ization of two xylanases from alkalophilic and thermophilic *Bacillus licheniformis* 77-2. Appl Biochem Biotechnol 129-132: 289-302

Danis O, Ogan A, Tatlican P, Attar A, Cakmakci E, Mertoglu B, Birbir M (2015) Preparation of poly (3-hydroxybutyrate-co-hydroxyvalerate) films from halophilic archaea and their potential use in drug delivery. Extremophiles. doi: 10. 1007/s00792-015-0735-4

Das H, Sing SK (2004) Useful byproducts from cellulosic wastes of agriculture and food indus try-a critical appraisal. CRC Cr Rev Food Sci 44: 77-89

de Souza PM, Magalhães PDO (2010) Application of microbial α-amylase in industry-a review. Braz J Microbiol 41 (4): 850-861

Dheeran P, Nandhagopal N, Kuma S, Jaiswal YK, Adhikari DK (2012) A novel thermostable xyla nase of *Paenibacillus macerans* IIPSP3 isolated from the termite gut. J Ind Microbiol Biotechnol 39 (6): 851-860

Dhiman SS, Sharma J, Battan B (2008) Industrial aspects and future prospects of microbial xylanases: a review. BioResources 3: 1377-1402

Di Donato P, Fiorentino G, Anzelmo G, Tommonaro G, Nicolaus B, Poli A (2011) Re-use of vegetable wastes as cheap substrates for extremophile biomass production. Waste Biomass Valoriz 2: 103-111

Di Donato P, Finore I, Anzelmo G, Lama L, Nicolaus B, Poli A (2014) Biomass and biopolymer production using vegetable wastes as cheap substrates for extremophiles. Chem Eng Trans 38: 163-168

Dimarogona M, Topakas E, Olsson L, Christakopoulos P (2012) Lignin boosts the cellulase performance of a GH-61 enzyme from *Sporotrichum thermophile*. Bioresour Technol 110: 480-487

Dipasquale L, Romano I, Picariello G, Calandrelli V, Lama L (2014) Characterization of a native cellulase activity from an anaerobic thermophilic hydrogen-producing bacterium *Thermosipho* sp. strain 3. Ann Microbiol 64: 1493-1503

Eckert K, Schneider E (2003) A thermoacidophilic endoglucanase (CelB) from *Alicyclobacillus acidocaldarius* displays high sequence similarity to arabinofuranosidases belonging to family 51 of glycoside hydrolases. Eur J Biochem 270: 3593-3602

Elleuche S, Schröder C, Sahm K, Antranikian G (2014) Extremozymes-biocatalysts with unique properties from extremophilic microorganisms. Curr Opin Biotechnol 29: 116-123 Fawzi EM (2011) Highly thermostable xylanase purified from *Rhizomucor miehei* NRL 3169. Acta Biol Hung 62: 85-94

Finore I, Kasavi C, Poli A, Romano I, Toksoy Oner E, Kirdar B, Dipasquale L, Nicolaus B, Lama L (2011) Purification, biochemical characterization and gene sequencing of a thermostable raw starch digesting alpha-amylase from *Geobacillus thermoleovorans* subsp. Stromboliensis subsp. nov. World J Microbiol Biotechnol 27: 2425-2433

Finore I, Di Donato P, Poli A, Kirdar B, Kasavi C, Toksoy EO, Nicolaus B, Lama L (2014) Use of agro waste biomass for α-amylase production by *Anoxybacillus amylolyticus*: purification and properties. J Microb Biochem Technol 6: 320-326

FitzPatrick M, Champagne P, Cunningham MF, Whitney RA (2010) A biorefinery processing perspective: treatment of lignocellulosic materials for the production of value-added products. Bioresour Technol 101 (23): 8915-8922

Gao L, Gao F, Wang L, Geng C, Chi L, Zhao J, Qu Y (2012) N-glycoform diversity of cellobiohydrolase I from *Penicillium decumbens* and synergism of nonhydrolytic glycoform in cellulose degradation. J Biol Chem 287: 15906-15915

Georgieva TI, Mikkelsen MJ, Ahring BK (2008) Ethanol production from wetexploded wheat straw hydrolysate by thermophilic anaerobic bacterium *Thermoanaerobacter* BG1L1 in a continuous immobilized reactor. Appl Biochem Biotechnol 145: 99-110

Girfoglio M, Rossi M, Cannio R (2012) Cellulose degradation by *Sulfolobus solfataricus* requires a cell-anchored endo-β-1-4-Glucanase. J Bacteriol 194 (18): 5091-5100

Graham JE, Clark ME, Nadler DC, Huffer S, Chokhawala HA, Rowland SE, Blanch HW, Clark DS, Robb FT

(2011) Identification and characterization of a multidomain hyperthermophilic cellulase from an archaeal enrichment. Nat Commun 2：375

Hakamada Y，Koike K，Yoshimatsu T，Mori H，Kobayashi T，Ito S（1997）Thermostable alkaline cellulase from an alkaliphilic isolate，*Bacillus* sp. KSMS237. Extremophiles 1：151-156

Himmel M，Adney W，Tucker M，Grohmann K（1994）Thermostable purified endoglucanase from *Acidothermus cellulolyticus* ATCC 43068. US Patent 5，275，944

Hoffmann RA，Geijtenbeek T，Kamerling JP，Vliegenthart JF（1992a）1H-N. M. R. study of enzymically generated wheat-endosperm arabinoxylan oligosaccharides：structures of hepta- to tetradeca-saccharides containing two or three branched xylose residues. Carbohydr Res 223：19-44

Hoffmann RA，Kamerling JP，Vliegenthart JFG（1992b）Structural features of a water-soluble arabinoxylan from the endosperm of wheat. Carbohydr Res 226（2）：303-311

Hong SY，Lee JS，Cho KM，Math RK，Kim YH，Hong SJ，Cho YU，Cho SJ，Kim H，Yun HD（2007）Construction of the bifunctional enzyme cellulase-beta-glucosidase from the hyperthermophilic bacterium *Thermotoga maritima*. Biotechnol Lett 29（6）：931-936

Horn SJ，Vaaje-Kolstad G，Westereng B，Eijsink VG（2012）Novel enzymes for the degradation of cellulose. Biotechnol Biofuels 5：45

Hreggvidsson GO，Kaiste E，Holst O，Eggertsson G，Palsdottir A，Kristjansson JK（1996）An extremely thermostable cellulase from the thermophilic eubacterium *Rhodothermus marinus*. Appl Environ Microbiol 62：3047-3049

Huang XP，Monk C（2004）Purification and characterization of a cellulose（CMCase）from a newly isolated thermophilic aerobic bacterium *Caldibacillus cellulovorans* gen. nov.，sp. nov. World J Microbiol Biotechnol 20：85-92

Huang TY，Duan KJ，Huang SY，Chen CW（2006）Production of polyhydroxyalkanoates from inexpensive extruded rice bran and starch by *Haloferax mediterranei*. J Ind Microbiol Biotechnol 33：701-706

Hung KS，Liu SM，Tzou WS，Lin FP，Pan CL，Fang TY，Sun KH，Tang SJ（2011）Characterization of a novel GH10 thermostable，halophilic xylanase from the marine bacterium *Thermoanaerobacterium saccharolyticum* NT-OU1. Process Biochem 46：1257-1263

Jain I，Kumar V，Satyanarayana T（2014）Applicability of recombinant beta-xylosidase from the extremely thermophilic bacterium *Geobacillus thermodenitrificans* in synthesizing alkylxylo sides. Bioresour Technol 170：462-469

Jin AX，Ren JL，Peng F，Xu F，Zhou GY，Sun RC，Kennedy JF（2009）Comparative characteriza tion of degraded and non-degradative hemicelluloses from barley straw and maize stems：com position，structure，and thermal properties. Carbohydr Polym 78：609-619

Kabel MA，Carvalheiro F，Garrote G，Avgerinos E，Koukios E，Parajò JC，Girio FM，Schols HA，Voragen AGJ（2002）Hydrothermally treated xylan rich by-products yield different classes of xylo-oligosaccharides. Carbohydr Polym 50（1）：47-56

Kamm B，Kamm M（2004）Biorefinery-systems. Chem Biochem Eng Q 18（1）：1-6

Karlsson J，Momcilovic D，Wittgren B，Schulein M，Tjerneld F，Brinkmalm G（2002）Enzymatic degradation of carboxymethyl cellulose hydrolyzed by the endoglucanases Cel5A，Cel7B，and Cel45A from *Humicola insolens* and Cel7B，Cel12A and Cel45Acore from Trichoderma reesei. Biopolymers 63（1）：32-40

Kasavi C，Finore I，Lama L，Nicolaus B，Oliver SG，Toksoy EO，Kirdar B（2012）Evaluation of industrial *Saccharomyces cerevisiae* strains for ethanol production from biomass. Biomass Bioenergy 45：230-238

Keasling JD（2008）Synthetic biology for synthetic chemistry. ACS Chem Biol 3：64-76

Khandeparkar R，Bhosle NB（2006）Purification and characterization of thermoalkalophilic xyla nase isolated from the *Enterobacter* sp. MTCC 5112. Res Microbiol 157：315-325

Khasin A，Alchanati I，Shoham Y（1993）Purification and characterization of a thermostable xyla nase from *Bacillus stearothermophilus* T-6. Appl Environ Microbiol 59：1725-1730

Kim HW，Ishikawa K（2010）Complete saccharification of cellulose at high temperature using endo cellulase and beta-glucosidase from *Pyrococcus* sp. J Microbiol Biotechnol 20（5）：889-892

Kim JW，Peeples TL（2006）Screening extremophiles for bioconversion potentials. Biotechnol Prog 22（6）：1720-1724

Kim TH，Kim JS，Sunwoo C，Lee YY（2003）Pretreatment of corn stover by aqueous ammonia. Bioresour Technol 90（1）：39-47

Kim MS，Park JT，Kim YW，Lee HS，Nyawira R，Shin HS，Park CS，Yoo SH，Kim YR，Moon TW，Park KH（2004）Properties of a novel thermostable glucoamylase from the hyperthermophilic archaeon Sulfolobus solfataricus in relation to starch processing. Appl Environ Microbiol 70（7）：3933-3940

Kuan YH，Liong MT（2008）Chemical and physicochemical characterization of agrowaste fibrous materials and residues. J Agric Food Chem 56：9252-9257

Küçükaşik F，Kazak H，Güney D，Finore I，Poli A，Yenigün O，Nicolaus B，Toksoy EO（2011）Molasses as fermentation substrate for levan production by *Halomonas* sp. Appl Microbiol Biotechnol 89：1729-1740

Kumar V，Satyanarayana T（2014）Secretion of recombinant thermo-alkali-stable endoxylanase of polyextremophilic *Bacillus halodurans* TSEV1 and its utility in generating xylooligosaccha rides from renewable agro-residues. Process Biochem 49（11）：1875-1883

Kumar S，Kumar P，Satyanarayana T（2007）Production of raw starch-saccharifying thermostable and neutral glucoamylase by the thermophilic mold thermomucor indicae-seudaticae in sub merged fermentation. Appl Biochem Biotechnol 142：221-230

Kumar R, Singh S, Singh OV (2008) Bioconversion of lignocellulosic biomass: biochemical and molecular perspective. J Ind Microbiol Biotechnol 35: 377-391

Lama L, Calandrelli V, Gambacorta A, Nicolaus B (2004) Purification and characterization of thermostable xylanase and β-xylosidase by the thermophilic bacterium *Bacillus thermantarcticus*. Res Microbiol 155 (4): 283-289

Lama L, Tramice A, Finore I, Anzelmo G, Calandrelli V, Pagnotta E, Tommonaro G, Poli A, Di Donato P, Nicolaus B, Fagnano M, Mori M, Impagliazzo A, Trincone A (2014) Degradative actions of microbial xylanolytic activities on hemicelluloses from rhizome of *Arundo donax*. AMB Expr 4: 55-64

Lawther JM, Sun RC, Banks WB (1996) Effects of extraction conditions and alkali type on yield and composition of wheat straw hemicelluloses. J Appl Polym Sci 60 (11): 1827-1837

Lee JW, Park JY, Kwon M, Choi IG (2009) Purification and characterization of a thermostable xylanase from the brown-rot fungus *Laetiporus sulphureus*. J Biosci Bioeng 107 (1): 33-37

Li W, Zhang WW, Yang MM, Chen YL (2000) Cloning of the thermostable cellulase gene from newly isolated *Bacillus subtilis* and its expression in *Escherichia coli*. Mol Biotechnol 40: 195-201

Li W, Huan X, Zhou Y, Ma Q, Chen Y (2009) Simultaneous cloning and expression of two cellulase genes from *Bacillus subtilis* newly isolated from Golden Takin (Budorcas taxicolor Bedfordi). Biochem Biophys Res Commun 383: 397-400

Li HY, Sun SN, Zhou X, Peng F, Sun RC (2015) Structural characterization of hemicelluloses and topochemical changes in Eucalyptus cell wall during alkali ethanol treatment. Carbohydr Polym 123: 17-26

Liang C, Xue Y, Fioroni M, Rodriguez-Ropero F, Zhou C, Schwaneberg U, Ma Y (2011) Cloning and characterization of a thermostable and halo-tolerant endoglucanase from Thermoanaerobacter tengcongensis MB4. Appl Microbiol Biotechnol 89: 315-326

Lina L, Ping Z, Yaqi C, Mou S (2006) High-performance anion exchange chromatography with pulsed amperometric detection for simultaneous determination of monosaccharides and uronic acids. Chin J Anal Chem 34 (10): 1371-1374

Lindenmuth BE, McDonald KA (2011) Production and characterization of Acidothermus cellulolyticus endoglucanase in *Pichia pastoris*. Protein Expr Purif 77: 153-158

Lo YC, Huang CY, Cheng CL, Lin CY, Chang JS (2011) Characterization of cellulolytic enzymes and bioH_2 production from anaerobic thermophilic Clostridium sp. TCW1. Bioresour Technol 102: 8384-8392

Lynd LR, van Zyl WH, McBride JE, Laser M (2005) Consolidated bioprocessing of cellulosic biomass: an update. Curr Opin Biotechnol 16: 577-583

Maalej I, Belhaj I, Masmoudi NF, Belghith H (2009) Highly thermostable xylanase of the thermophilic fungus *Talaromyces thermophilus*: purification and characterization. Appl Biochem Biotechnol 158: 200-212

Mastascusa V, Romano I, Di Donato P, Poli A, Della Corte V, Rotundi A, Bussoletti E, Quarto M, Pugliese MG, Nicolaus B (2014) Extremophiles survival to simulated space conditions: an astrobiology model study. Orig Life Evol Biosph 44 (3): 231-237

Mischnick P, Momcilovic D (2010) Chemical structure analysis of starch and cellulose derivatives. Adv Carbohydr Chem Biochem 64: 117-210

Nawirska A, Kwasniewska M (2005) Dietary fibre fractions from fruit and vegetable processing waste. Food Chem 91: 221-225

Ng IS, Li CW, Yeh YF, Chen PT, Chir JL, Ma CH, Yu SM, Ho TH, Tong CG (2009) A novel endoglucanase from the thermophilic bacterium *Geobacillus* sp. 70PC53 with high activity and stability over a broad range of temperatures. Extremophiles 13: 425-435

Nieves LM, Panyon LA, Wang X (2015) Engineering sugar utilization and microbial tolerance toward lignocellulose conversion. Front Bioeng Biotechnol 3: 17

Nwodo UU, Green E, Okoh AI (2012) Review bacterial exopolysaccharides: functionality and prospects. Int J Mol Sci 13 (11): 14002-14015

Ochs M, Muzard M, Plantier-Royon R, Boris E, Rémond C (2011) Enzymatic synthesis of alkyl β-d-xylosides and oligoxylosides from xylans and from hydrothermally pretreated wheat bran. Green Chem 13 (9): 2380-2388

Ogino H, Yasui K, Shiotani T, Ishihara T, Ishikawa H (1995) Organic solvent-tolerant bacterium which secretes an organic solvent-stableproteolytic enzyme. Appl Environ Microbiol 61: 4258-4262

Olson DG, McBride JE, Shaw AJ, Lynd LR (2012) Recent progress in consolidated bioprocessing. Curr Opin Biotechnol 23 (3): 396-405

Pastell H, Tuomainen P, Virkki L, Tenkanen M (2008) Step-wise enzymatic preparation and structural characterization of singly and doubly substituted arabinoxylo-oligosaccharides with non- reducing end terminal branches. Carbohydr Res 343 (18): 3049-3057

Paulová L, Patáková P, Branská B, Rychtera M, Melzoch K (2014) Lignocellulosic ethanol: tech nology design and its impact on process efficiency. Biotechnol Adv. doi: 10. 1016/j. biotechadv. 2014. 12. 002

Piller K, Daniel RM, Petach HH (1996) Properties and stabilization of an extracellular alpha glucosidase from the extremely thermophilic archaebacteria *Thermococcus* strain AN1: enzyme activity at 130 degrees C. Biochim Biophys Acta 1292 (1): 197-205

Podkaminer KK, Shao X, Hogsett DA, Lynd LR (2011) Enzyme inactivation by ethanol and devel opment of a kinetic model for thermophilic simultaneous saccharification and fermentation at 50 C with Thermoanaerobacterium saccharolyticum ALK2. Biotechnol Bioeng 108: 1268-1278

Polizeli ML, Rizzatti R, Monti HF, Terenzi JA, Jorge JA, Amorim DSJ (2005) Xylanases from fungi: properties and industrial applications. Appl Microbiol Biotechnol 67：577-591

Rahman Z，Shida Y，Furukawa T，Suzuki Y，Okada H，Ogasawara W，Morikawa Y（2009）Application of Trichoderma reesei cellulase and xylanase promoters through homologous recombination for enhanced production of extracellular beta-glucosidase I. Biosci Biotechnol Biochem 73：1083-1089

Rakotoarivonina H，Hermant B，Monthe N，Rémond C（2012）The hemicellulolytic enzyme arse nal of *Thermobacillus xylanilyticus* depends on the composition of biomass used for growth. Microb Cell Fact 11：159

Rastogi G，Muppidi GL，Gurram RN，Adhikari A，Bischoff KM，Hughes SR，Apel WA，Bang SS，Dixon DJ，Sani RK（2009）Isolation and characterization of cellulose-degrading bacteria from the deep subsurface of the Homestake gold mine，Lead，South Dakota，USA. J Ind Microbiol Biotechnol 36：585-598

Rastogi G，Bhalla A，Adhikari A，Bischoff KM，Hughes SR，Christopher LP，Sani RK（2011）Characterization of thermostable cellulases produced by *Bacillus* and *Geobacillus* strains. Bioresour Technol 101（22）：8798-8806

Reis A，Pinto P，Evtuguin DV，Neto CP，Domingues P，Ferrer-Correia AJ，Domingues MRM（2005）Electrospray tandem mass spectrometry of underivatised acetylated xylo-oligosaccharides. Rapid Commun Mass Spectrom 19（23）：3589-3599

Rhee SK，Song KB，Kim CH，Park BS，Jang EK，Jang KH（2005）Levan. Biopolymers 5. doi：10. 1002/3527600035. bpol5014

Robert P，Marquis M，Barron C，Guillon F，Saulnier L（2005）FT-IR investigation of cell wall polysaccharides from cereal grains. Arabinoxylan infrared assignment. J Agric Food Chem 53（18）：7014-7018

Romaniec M，Fauth U，Kobayashi T，Huskisson N，Barker P，Demain A（1992）Purification and characterization of a new endoglucanase from Clostridium thermocellum. Biochem J 283：69-73

Rose DJ，Inglett GE，Liu SX（2010）Utilisation of corn（*Zea mays*）bran and corn fiber in the production of food components. J Sci Food Agric 90：915-924

Saleem M，Aslam F，Akhtar MS，Tariq M，Rajoka MI（2011）Characterization of a thermostable and alkaline xylanase from *Bacillus* sp. and its bleaching impact on wheat straw pulp. World J Microbiol Biotechnol 28（2）：513-522

Samanta AK，Jayapal N，Kolte AP，Senani S，Suresh KP，Sampath KT（2012）Enzymatic production of xylooligosaccharides from alkali solubilized xylan of natural grass（*Sehima nervosum*）. Bioresour Technol 112：199-205

Sanchez-Vazquez SA，Hailes HC，Evans JRG（2013）Hydrophobic polymers from food waste：resources and synthesis. Polym Rev 53：627-694

Sandoval NR，Kim JY，Glebes TY，Reeder PJ，Aucoin HR，Warner JR et al（2012）Strategy for directing combinatorial genome engineering in *Escherichia coli*. Proc Natl Acad Sci U S A 109：10540-10545

Sapre MP，Jha H，Patil MB（2005）Purification and characterization of a thermoalkalophilic xylanase from *Bacillus* sp. World J Microbiol Biotechnol 21：649-654

Schiraldi C，Martino A，Acone M，Di Lernia I，Di Lazzaro A，Marulli F，Generoso M，Carteni M，De Rosa M（2000）Effective production of a thermostable alpha-glucosidase from *Sulfolobus solfataricus* in *Escherichia coli* exploiting a microfiltration bioreactor. Biotechnol Bioeng 70（6）：670-676

Sedlmeyer FB（2011）Xylan as by-product of biorefineries：characteristics and potential use for food applications. Food Hydrocoll 25（8）：1891-1898

Serour E，Antranikian G（2002）Novel thermoactive glucoamylases from the thermoacidophilic Archaea *Thermoplasma acidophilum*，*Picrophilus torridus* and *Picrophilus oshimae*. Anton van Leeuw 81（1-4）：73-83

Shaw AJ，Podkaminer KK，Desai SG，Bardsley JS，Rogers SR，Thorne PG，Hogsett DA，Lynd LR（2008）Metabolic engineering of a thermophilic bacterium to produce ethanol at high yield. Proc Natl Acad Sci U S A 105：13769-13774

Shrivastav A，Kim HY，Kim YR（2013）Advances in the applications of polyhydroxyalkanoate nanoparticles for novel drug delivery system. BioMed Res Int 581684. doi：10. 1155/2013/581684

Simpson HD，Haufler UR，Daniel RM（1991）An extremely thermostable xylanase from the thermophilic eubacterium *Thermotoga*. Biochem J 277：413-417

Singh S，Madlala AM，Prior BA（2003）*Thermomyces lanuginosus*：properties of strains and their hemicellulases. FEMS Microbiol Rev 27：3-16

Sinnott ML（1990）Catalytic mechanisms of enzymic glycosyl transfer. Chem Rev 90：1171-1202

Sivaramakrishnan S，Gangadharan D，Nampoothiri KM，Soccol CR，Pandey A（2006）α-amylases from microbial sources-an overview on recent developments. Food Technol Biotechnol 44：173-184

Sriyapai T，Somyoonsap P，Matsui K，Kawai F，Chansiri K（2011）Cloning of a thermostable xylanase from *Actinomadura* sp. S14 and its expression in *Escherichia coli* and *Pichia pastoris*. J Biosci Bioeng 111：528-536

Studholme DJ（2015）Some（bacilli）like it hot：genomics of *Geobacillus* species. Microb Biotechnol 8（1）：40-48

Subba MVSST，Muralikrishna G（2004）Structural analysis of arabinoxylans isolated from native and malted finger millet（*Eleusine coracana*，ragi）. Carbohydr Res 339（14）：2457-2463

Subramaniyan S，Prema P（2000）Cellulase-free xylanases from *Bacillus* and other microorganisms. FEMS Microbiol Lett 183（1）：1-7

Sun Y，Cheng J（2002）Hydrolysis of lignocellulosic materials for ethanol production：a review. Bioresour Technol 83（1）：1-11

Sun RC，Lawther JM，Banks WB（1996）Fractional and structural characterization of wheat straw hemicelluloses. Carbohydr Polym 29（4）：325-331

Sun JX，Sun XF，Sun RC，Su YQ（2004）Fractional extraction and structural characterization of sugarcane bagasse hemicelluloses. Carbohydr Polym 56（2）：195-204

Sun XF, Sun RC, Fowler P, Baird MS (2005) Extraction and characterization of original lignin and hemicelluloses from wheat straw. J Agric Food Chem 53 (4): 860-870

Sunna A, Moracci M, Rossi M, Antranikian G (1997) Glycosyl hydrolases from hyperthermophiles. Extremophiles 1: 2-13

Sweeney MD, Xu F (2012) Biomass converting enzymes as industrial biocatalysts for fuels and chemicals: recent developments. Catalysts 2: 244-263

Tai SK, Lin HP, Kuo J, Liu JK (2004) Isolation and characterization of a cellulolytic *Geobacillus thermoleovorans* T4 strain from sugar refinery wastewater. Extremophiles 8: 345-349

Tommonaro G, Poli A, De Rosa S, Nicolaus B (2008) Tomato derived polysaccharides for biotechnological applications: chemical and biological approaches. Molecules 13 (6): 1384-1398

Tramice A, Arena A, De Gregorio A, Ottanà R, Maccari R, Pavone B, Arena N, Innello D, Vigorita MG, Trincone A (2008) Facile biocatalytic access to 9-fluorenyl methyl polyglycosides: evaluation of antiviral activity on immunocompetent cells. ChemMedChem 3 (9): 1419-1426

Tramice A, Melck D, Virno A, Randazzo A, Motta A, Trincone A (2009) Enzymatic synthesis and 3-D structure of anti-proliferative acidic (MeGlcA) xylotetrasaccharide. J Mol Catal B Enzym 61 (3-4): 129-135

Van TT, Ryu SI, Lee KJ, Kim EJ, Lee SB (2007) Cloning and characterization of glycogendebranching enzyme from hyperthermophilic archaeon *Sulfolobus shibatae*. J Microbiol Biotechnol 17 (5): 792-799

VanFossen AL, Verhaart MR, Kengen SM, Kelly RM (2009) Carbohydrate utilization patterns for the extremely thermophilic bacterium *Caldicellulosiruptor saccharolyticus* reveal broad growth substrate preferences. Appl Environ Microbiol 75: 7718-7724

Vázquez MJ, Garrote G, Alonso JL, Domiguez H, Parajò JC (2005) Refining of autohydrolysis liquors for manufacturing xylo-oligosaccharides: evaluation of operational strategies. Bioresour Technol 96 (8): 889-896

Verbruggen MA, Beldman G, Voragen AGJ (1995) The selective extraction of glucuronoarabinoxylans from Sorghum endosperm cell walls using barium and potassium hydroxide solutions. J Cereal Sci 21 (3): 271-282

Wang TH, Lu S (2013) Production of xylooligosaccharide from wheat bran by microwave assisted enzymatic hydrolysis. Food Chem 138 (2-3): 1531-1535

Wang J, Bai Y, Yang P, Shi P, Luo H, Meng K, Huang H, Yin J, Yao B (2010) A new xylanase from thermoalkaline *Anoxybacillus* sp. E2 with high activity and stability over a broad pH range. World J Microbiol Biotechnol 26: 917-924

Wang X, Yomano LP, Lee JY, York SW, Zheng H, Mullinnix MT, Shanmgam KT, Ingram LO (2013) Engineering furfural tolerance in *Escherichia coli* improves the fermentation of lignocellulosic sugars into renewable chemicals. Proc Natl Acad Sci U S A 110: 4021-4026

Wong DW (2009) Structure and action mechanism of ligninolytic enzymes. Appl Biochem Biotechnol 157 (2): 174-209

Wu S, Liu B, Zhang X (2006) Characterization of a recombinant thermostable xylanase from deepsea thermophilic *Geobacillus* sp. MT-1 in East Pacific. Appl Microbiol Biotechnol 72: 1210-1216

Xiangyuan H, Shuzheng Z, Shoujun Y (2001) Cloning and expression of thermostable betaglycosidase gene from *Thermus nonproteolyticus* HG102 and characterization of recombinant enzyme. Appl Biochem Biotechnol 94: 243-255

Xu W, Osei-Prempeh G, Lema C, Devis Oldham E, Aguilera RJ, Parkin S, Rankin SE, Knutson BL, Lehmler HJ (2012) Synthesis, thermal properties and cytotoxicity evaluation of hydrocarbon and fluorocarbon alkyl-β-D xylopyranoside surfactants. Carbohydr Res 349: 12-23

Yang SJ, Lee HS, Park CS, Kim YR, Moon TW, Park KH (2004) Enzymatic analysis of an amylolytic enzyme from the hyperthermophilic archaeon Pyrococcus furiosus reveals its novel catalytic properties as both an alpha-amylase and a cyclodextrin-hydrolyzing enzyme. Appl Environ Microbiol 70 (10): 5988-5995

Yang SJ, Kataeva I, Hamilton-Brehm SD, Engle NL, Tschaplinski TJ, Doeppke C, Davis M, Westpheling J, Adams MW (2009) Efficient degradation of lignocellulosic plant biomass, without pretreatment, by the thermophilic anaerobe *Anaerocellum thermophilum* DSM 6725. Appl Environ Microbiol 75: 4762-4769

Yang D, Weng H, Wang M, Xu W, Li Y, Yang H (2010) Cloning and expression of a novel thermostable cellulase from newly isolated *Bacillus subtilis* strain I15. Mol Biol Rep 37: 1923-1929

Zambare VP, Bhalla A, Muthukumarappan K, Sani RK, Christopher LP (2011) Bioprocessing of agricultural residues to ethanol utilizing a cellulolytic extremophile. Extremophiles 15 (6): 11-618

Zhang Y-HP, Ding SY, Mielenz JR, Cui J-B, Elander RT, Laser M, Himmel ME, McMillan JR (2007) Fractionating recalcitrant lignocellulose at modest reaction conditions. Biotechnol Bioeng 97 (2): 214-223

Zhang J, Siika-Aho M, Puranen T, Tang M, Tenkanen M, Viikari L (2011) Thermostable recombinant xylanases from *Nonomuraea fl exuosa* and *Thermoascus aurantiacus* show distinct properties in the hydrolysis of xylans and pretreated wheat straw. Biotechnol Biofuels 4: 12

Zheng Y, Zhao J, Xu F, Li Y (2014) Pretreatment of lignocellulosic biomass for enhanced biogas production. Prog Energy Combust Sci 42: 35-53

Zhou J, Wang Y-H, Chua J, Luoa L-Z, Zhuanga Y-P, Zhanga S-L (2009) Optimization of cellulase mixture for efficient hydrolysis of steam-exploded corn stover by statistically designed experiments. Bioresour Technol 100: 819-825

Zhu S, Wu Y, Yu Z, Zhang X, Li H, Gao M (2006) The effect of microwave irradiation on enzymatic hydrolysis of rice straw. Bioresour Technol 97 (15): 1964-1968

Zverlov VV, Schantz N, Schmitt-Kopplin P, Schwarz WH (2005) Two new major subunits in the cellulosome of Clostridium thermocellum: xyloglucanase Xgh74A and endoxylanase Xyn10D. Microbiology 151: 3395-3401

第二十章 通过重建古生物的祖先序列设计热稳定酶的策略

Satoshi Akanuma[1], Akihiko Yamagishi[2]

20.1 引言

"我们最古老的祖先是如何生活的"是一个长期存在争议的问题，因为该问题的答案不仅有助于了解生物共同祖先的特征和生活环境，而且更重要的是还能帮助了解它们是如何在变化的地球上不断进化生存下来的。最近通过细菌和古菌基因组数据的扩展能够进行物种间系统发育分析，揭示地球上生命的演变。此外，还可以推断祖先基因序列，并根据该序列重组祖先蛋白质（Thornton 2004）。由于蛋白质的特性能够反映出微生物的特征和生活环境，重建祖先蛋白质序列并分析它们的性质是预测祖先的特征和生活环境的有力手段（Gaucher et al. 2010；Boussau and Gouy，2012）。

本章内容主要介绍最近通过实验解决的一个问题：现存生命的最后一个共同祖先的生长温度是多少（Akanuma et al. 2013a）？用于回答这个问题的方法主要包括：构建系统发育树来推断古菌和细菌祖先的核苷二磷酸激酶（NDK）的氨基酸序列、表达编码推断序列的基因并表征基因产物的解折叠温度。该研究最终为"所有现存生命的共同祖先是在非常高的温度下生活的（极端）嗜热生物"的观点提供了第一个实验支持。此外，祖先序列重建的步骤可作为生产极端嗜热蛋白质的可靠手段（Gaucher et al. 2003，2008；Akanum et al. 2011，2013a；Butzin et al. 2013）。

20.2 最后一个共同祖先"Commonote"

地球上所有现存的生命可以分为三个域：古菌、细菌和真核生物（Woese 1987；Woese

[1] Department of Applied Life Sciences，Tokyo University of Pharmacy and Life Sciences，1432-1，Horinouchi，Hachioji，Tokyo 192-0392，Japan；Faculty of Human Sciences，Waseda University，Shinjuku，Japan.

[2] Department of Applied Life Sciences，Tokyo University of Pharmacy and Life Sciences，1432-1，Horinouchi，Hachioji，Tokyo 192-0392，Japan. e-mail：yamagish@toyaku. ac. jp.

et al. 1990)。比较分析小亚基核糖体 RNA (rRNA) 序列构建的系统发育树，结果表明所有的现存生命都是一个共同祖先的后代，即存在最后一个共同祖先（图 20.1)。共同的祖先又被称为 LUCA、LCA 或 senancestor（前辈），在这里，称为“Commonote”（Yamagishi et al. 1998)。Commonote 是怎样的有机体？这个问题是全人类所关注的，因为它的答案将帮助了解共同的祖先以及祖先居住的早期地球的环境。

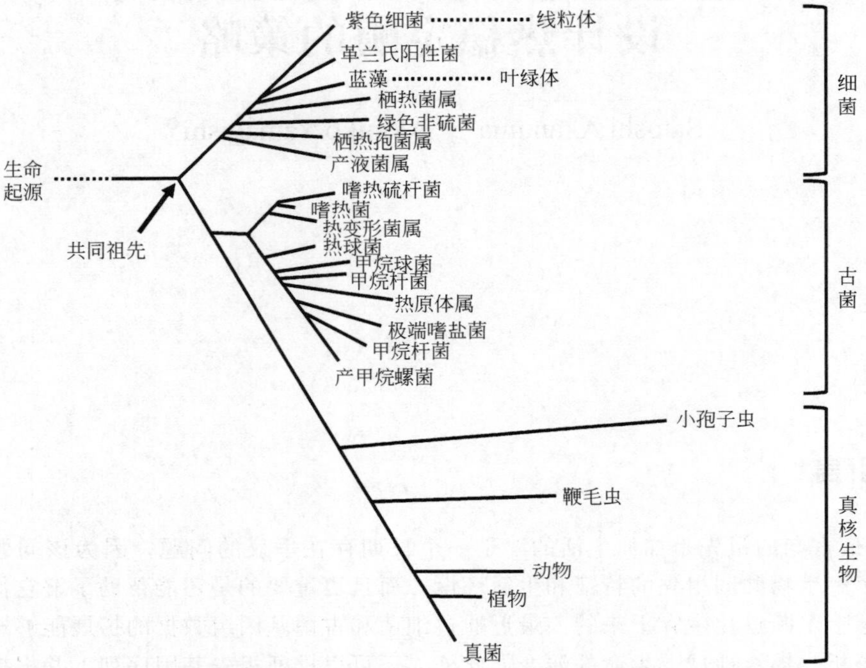

图 20.1 用 SSU-rRNA 序列构建的系统发育树表明 Commonote 的进化位置（Woese 等 1990；Yamagishi 等 1998)

20.3 Commonote 生存的环境温度的早期研究进展

古生物生存的环境温度一直是令人感兴趣的话题。已经有许多研究人员阐述了 Commonote 的环境温度。有研究者提出古菌和细菌的共同祖先是极端嗜热生物（Pace 1991；Stetter 1996)，因为在最常用的 rRNA 序列构建的系统进化树中，极端嗜热古菌和细菌位于进化树的基底附近（Woese 1987；Achenbach-Richter et al. 1988)。因此，Commonote 也被简单地预测为极端嗜热生物。然而，Greaves 和 Warwicker 认为最古老的祖先可能生活在较冷的环境中，后来才逐渐适应了较高的温度（Greaves and Warwicker 2007)。此外，Brochier 和 Philippe (2002) 在分析细菌系统发育时提出进化树上的第一个门并不是极端嗜热生物。他们还提出，细菌域内的极端嗜热细菌的出现是第二次适应高温的结果。

逆旋转酶的进化历程也表明原始生物不可能是极端嗜热生物。逆旋转酶是依赖 ATP 的 Ⅰ型 DNA 拓扑异构酶，仅在嗜热生物体中发现。因为所有已知的极端嗜热生物体都含有逆旋转酶基因（Heine and Chandra 2009)，所以这种蛋白质被认为在适应高温环境方面起着重要作用（Forterre 2002；Atomi et al. 2004)。逆旋转酶由不相关的两个结构域组成，拓扑异构酶结构域和解旋酶结构域（Declais et al. 2000)。如果这些结构域是先独立起源、进化，之后才融合形成逆旋转酶的话，那么第一生物体则不可能是极端嗜热菌（Forterre 1996)。

但这一论点并不能排除“逆旋转酶出现在“Commonote”之前”这一可能性。

20.4 计算分析祖先序列

祖先的核苷酸序列反映了非嗜热祖先的生活。这种观点的基本概念是现有原核核糖体RNA主干区的（G+C）含量与其宿主的环境温度相关。Galtier等估计核糖体RNA序列中祖先（G+C）含量，这与高温生物体不相容（Galtier et al. 1999）。然而，使用不同的计算算法导致了相互矛盾的结论（Di Giulio et al. 2000，2003a，b）。

祖先蛋白质的氨基酸组成也被用来估算古生物生活的环境温度。Brooks等人（2004）使用最大期望法来计算共同祖先一系列蛋白质的氨基酸组成。相较现存的中温菌，祖先蛋白质的这些氨基酸组成与现存嗜热生物的氨基酸组成更为相似，这表明共同祖先是嗜热的。Groussin和Gouy（2011）对古菌系统发育进行了综合分析。他们估算出祖先rRNA的（G+C）含量以及蛋白质的氨基酸组成，并推断与古菌系统发育的每个内部节点相对应的古菌祖先的最佳生长温度。最终，他们推测最古老的祖先是极端嗜热的，现存的中温生物已经适应了较冷的环境。另一项研究通过计算祖先rRNA序列中的（G+C）含量和祖先蛋白质序列中的氨基酸组成表明，古菌和细菌的最后一个共同祖先是嗜热生物，但普适祖先是中温生物（Boussau et al. 2008）。Groussin等人（2013）不仅计算了蛋白质的原始氨基酸组成，还估算了早期生物体的最佳环境温度。虽然他估算的普适祖先的环境温度（33～68℃）高于Boussau等人（2008）计算的（1～37℃），但仍可得知普适祖先比古菌和细菌的共同祖先生活的环境温度要低。

大量的计算工作都集中在古代生物居住的环境温度上，然而并未从这些研究中获得表征这些温度的确凿证据。此外，与古代生物体的环境温度直接相关的祖先（G+C）含量的假设可能太过简单。还不能确定现代生物体的最佳环境温度与其核糖体RNA的（G+C）含量之间的相关性是否适用于估计初始地球的古生物体的环境温度。众所周知蛋白质的热稳定性通常对几个氨基酸的替换非常敏感。因此没有实验验证，从祖先蛋白质序列中的氨基酸组成分析得出的结论也只是推论。下文中，将介绍最近的祖先序列重建研究，该研究为（极端）嗜热普适共同祖先的存在提供了实验证据（Akanuma et al. 2013a）。

20.5 核苷二磷酸激酶的祖先序列重建

重建了古菌和细菌的最后一个共同祖先可能拥有的NDK序列。普遍认为NDK家族的祖先是在进化早期出现的，因为大多数现存的生物，从细菌到人类，都含有编码这个蛋白质家族成员的基因。现存的NDK序列相对保守，因此预测祖先的NDK中氨基酸残基的准确率较高。选择NDK的一个更重要的原因是，NDK的解折叠温度与其宿主生物体的最佳环境温度高度相关（Akanuma et al. 2013a）（如图20.2）。换句话说，每个NDK的热稳定性都可作为其主体自然环境温度的分子温度计。

图20.3描述了重建祖先氨基酸序列的过程。首先，从公共数据库检索现有的同源氨基酸序列。然后，生成一个多重序列比对。多重序列比对中的插入和空位通常会影响推断祖先序列。幸运的是，在NDK多重序列比对中只发现少量的插入/缺失。尽管如此，仍手动调整纠正了插入缺失的位置。使用所得到的序列比对构建系统发育树。最大似然法（ML）（Yang et al. 1995）和贝叶斯法（Yang and Rannala 1997）用于计算树形拓扑和祖先序列。文中只使用ML方法，因为ML系统发育算法可能会更准确地重建祖先序列（Hanson-

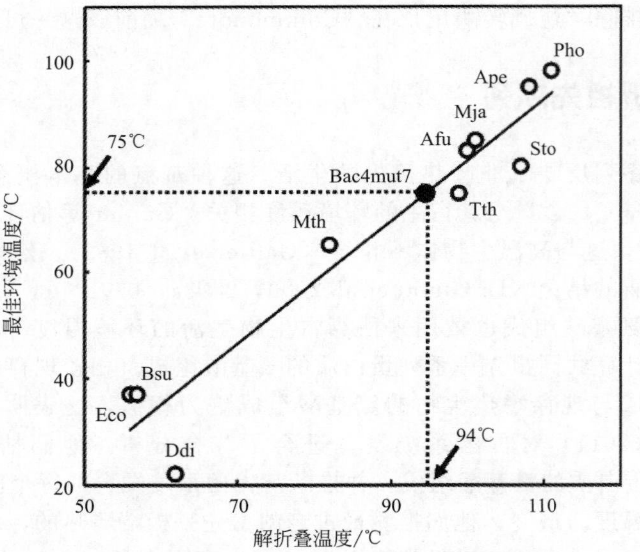

图 20.2 微生物 NDK 序列的解折叠温度与其宿主的最佳环境温度之间的关系。运用校准曲线和 Bac4mut7 解折叠温度（94℃）揭示 Commonote 的最低环境温度估计值（75℃）。Ddi，盘基网柄菌；Eco，大肠杆菌；Bsu，枯草芽孢杆菌；Mth，嗜热自养甲烷杆菌；Tth，嗜热菌；Afu，闪烁古球菌；Mja，詹氏甲烷球菌；Sto，极端嗜热古菌；Ape，敏捷气热菌；Pho，极端嗜热火球菌

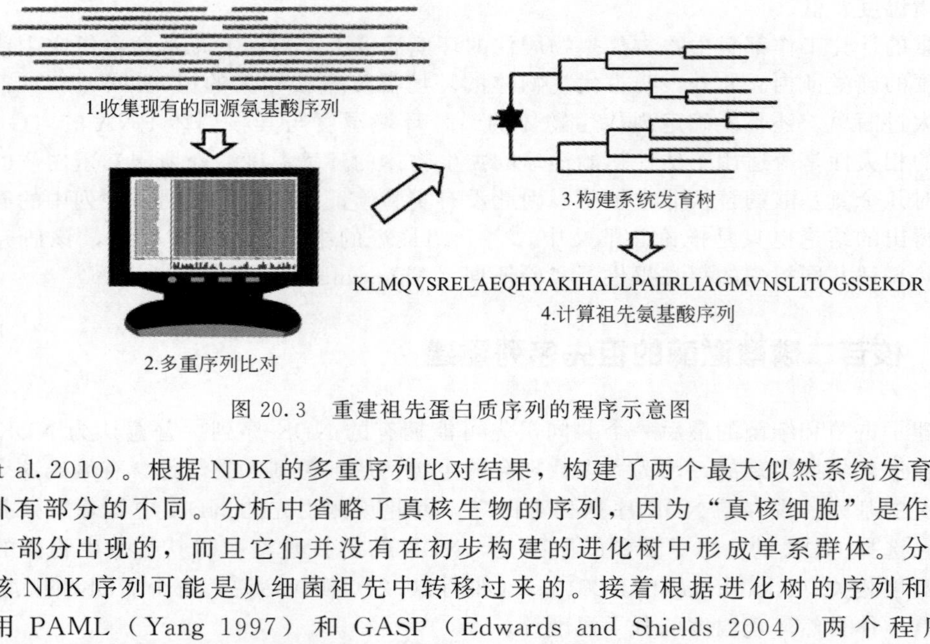

图 20.3 重建祖先蛋白质序列的程序示意图

Smith et al. 2010）。根据 NDK 的多重序列比对结果，构建了两个最大似然系统发育树，其树形拓扑有部分的不同。分析中省略了真核生物的序列，因为“真核细胞”是作为“细菌”的一部分出现的，而且它们并没有在初步构建的进化树中形成单系群体。分析中使用的真核 NDK 序列可能是从细菌祖先中转移过来的。接着根据进化树的序列和拓扑结构，使用 PAML（Yang 1997）和 GASP（Edwards and Shields 2004）两个程序中的 CODEML 来推断对应于古菌和细菌的最后共同祖先的 NDK 序列。前一种模式用于给出祖先序列，后者模型用于给出空位的位置。古菌祖先被命名为 Arc3 和 Arc4，细菌祖先被命名为 Bac3 和 Bac4。

Groussin 等人（2015）提出，计算进化中基因重复、水平转移和遗传丢失可以提高祖先序列推测的准确性。可见重建蛋白质的物理性质可能受到用于推断祖先序列的树形拓扑结构的影响。两种 NDK 进化树的拓扑结构与 rRNA 序列构建的进化树有所不同（Woese

1987)。因此，构建了用于估算 Arc3/4 和 Bac3/4 的 NDK 序列的物种的小亚基 rRNA 序列的进化树。根据由 rRNA 序列构建的系统发育树的拓扑结构，并利用其相应物种的 NDK 序列代替每个 rRNA 序列，推断出与古菌和细菌的最后共同祖先相对应的节点序列，分别命名为 Arc5 和 Bac5。

20.6 通过祖先的核苷二磷酸激酶序列估算古生物生活的环境温度

为了估算 Commonote 的环境温度，通过 PCR 介导的全基因合成技术合成了编码祖先氨基酸序列的基因。在大肠杆菌中表达扩增得到的祖先基因序列，然后分离纯化祖先的 NDK。纯化的祖先 NDK 在 100℃甚至更高的温度下仍稳定并具有催化活性（如图 20.4）。该结果用于分析重建祖先序列的树形拓扑结构是可信的。根据重建的 NDK 的解折叠温度和校准曲线（图 20.2），可得出细菌和古菌最后的共同祖先生活的最佳环境温度分别为 80～93℃和 81～97℃（Akanuma et al. 2013a）。

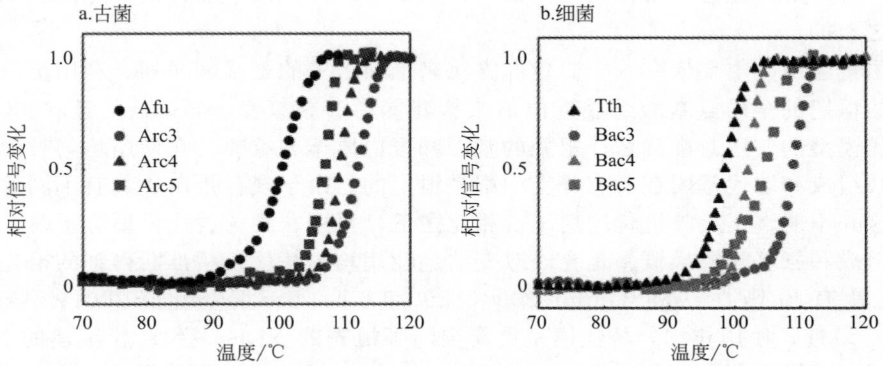

图 20.4 通过 222nm 处椭圆率变化监测现存的嗜热菌和祖先 NDK 的解折叠温度。Afu，闪烁古球菌的 NDK；Tth，嗜热菌的 NDK

20.7 估算 Commonote 生存的环境温度

因为推断祖先序列的进化树是“无根”的，所以我们的研究不能证明 Commonote 的 NDK 序列的精确性。为了确定 Commonote 在系统发育树上的位置，我们应该建立一个含有两个或更多旁系同源蛋白质的复合树，这些旁系同源蛋白质在 Commonote 之前就已经与它们最原始的共同祖先蛋白质分开（Akanuma et al. 2013b）。期望 Commonote 的位置能在古菌的根源和细菌的根源之间。此外，在祖先 NDK 序列重建的 139 个氨基酸残基中，有 115 个氨基酸在 6 个重建 NDK（Arc3/4/5 和 Bac3/4/5）中的排序完全相同。因此，这 115 个氨基酸残基也可能存在于 Commonte 的 NDK 序列上相同的位置。此外，“Commonte”的序列只是简单地被期望还存在古菌和细菌最后一个共同祖先的序列上。可见，既然存在 24 个非保守位点，那么 Commonte 的 NDK 序列至少残留在推断的古菌或细菌祖先的其中一个 NDK 序列中。因此，我们试图找出所有可能的 Commonote 的 NDK 序列中的稳定性最弱的序列。

在所有重建的祖先 NDK 中，Bac4 的解折叠温度最低，因此单独将 Bac4 的这 24 个氨基酸残基用其他祖先 NDK 中相同位置上的残基替换。还测试了位于 5Å（1Å$=10^{-10}$m）之间的氨基酸替代组合对 Bac4 稳定性的影响，结果发现替换 7 个氨基酸能降低 Bac4 的热稳定性。接着将所有不稳定氨基酸取代引入到 Bac4，得到了 Bac4mut7。Bac4mut7 的解折叠温度是 94℃，是估算的 Commonote 的 NDK 的解折叠温度的最低值。同时，根据图 20.2 所示的校准曲线计算，估计 Commonote 的环境温度可能为 75℃或者要高于 75℃。总之，Commonote 可能是一种（极端）嗜热生物，生活在 75℃或更高的温度中（Akanuma et al. 2013a）。

如上所述，必须先建立可能在 Commonote 出现之前就已经分散的旁系同源蛋白质的复合系统发育树，从而确定树的根部。延伸因子 Tu/1（Iwabe et al. 1989；Baldauf et al. 1996）和 H^+ ATP 酶（Gogarten et al. 1989）构建的两个复合进化树表明 Commonote 位于连接古菌的共同祖先和细菌的共同祖先的分支上。3-异丙基麦芽糖脱氢酶、异柠檬酸脱氢酶（Miyazaki et al. 2001）和脂肪族氨酰基 tRNA 合成酶构建的复合树（Brown and Doolitle 1995；Fournier et al. 2011）也支持根的位置。还有研究表明 Commonote 是细菌（Cavalier-Smith 2002，2006a，b，2010；Lake et al. 2008，2009）。而我们假设正确的 Commonote 节点位于连接古菌和细菌进化枝的分支上。由于 Commonote 序列的准确性在很大程度上依赖于树根的位置，所以注意，结论是建立在“Commonote 位于连接细菌和古菌根部的分支上”这个假说之上的。

蛋白质重建研究主要依靠一个简单的方法来推断祖先的氨基酸序列。利用蛋白质进化和跨谱系中的恒定的全局氨基酸组成近似值来构建同源替代模型。Boussau 等（2008）指出，非均匀替代模型的使用对准确估计祖先的基因和蛋白质至关重要。在使用均匀替代模型分析时，氨基酸组成和取代基团在所有分支中都是恒定的。由于现存蛋白质具有不同的氨基酸组成，所以这并不能真正反映进化的情况。相比之下，使用非均匀替代模型分析时，进化过程中氨基酸组成和氨基酸置换概率是允许改变的。使用均匀替代模型推断得到的祖先序列显示普适祖先具有嗜热性（Di Giulio 2000，2003a，b；Brooks et al. 2004；Akanuma et al. 2013a）。最近，使用非均匀替代模型重新推断了祖先的 NDK 序列。新重建的 NDK 的热稳定性表明，除了古菌和细菌的祖先之外，Commonote 的 NDK 同样具有极高的热稳定性。因此，即使使用不同谱系进化树中的所有不同氨基酸组成来放宽该替代模型的约束，也同样可以得出结论（该成果将发表在其他地方）。

20.8 其他合成祖先蛋白质的研究

有研究者通过重新计算祖先氨基酸序列并合成祖先蛋白质，以实证经验为基础的生物技术为估算古菌生存的环境温度提供了实验依据。Gaucher 等（2003）转录了祖先延伸因子 Tu，并报道了细菌祖先是嗜热的，而不是极端嗜热或中温菌。他们还对细菌系统发育进行了综合分析，并估计了与整个细菌域内部节点相对应的环境温度。他们还观察到随着古代海洋温度的降低，嗜热祖先开始逐渐适应较冷的环境（Gaucher et al. 2008）。此外，重建的延伸因子 Tu 在现存的嗜热生物体中能正常运作（Zhou et al. 2012）。同样，Hart 等人（2014）分别研究了现代中温和嗜热核糖核酸酶 H1（RNH）是如何适应低温和高温的。祖先 RNH 的热稳定性在中温和嗜热 RNH 之间。此外，在嗜热谱系的进化分析中发现中间祖先的解折叠温度逐渐升高，而在中温谱系进化中，中间祖先的解折叠温度先是急剧下降，随后再缓慢下降。另一个祖先序列重建实验证明热袍菌目 Thermotogales 最近的共同祖先生活在比后代更高的温度中（Butzin et al. 2013）。

祖先序列重建也用于研究通用遗传密码表的进化。Fournier 和 Alm（2015）重构了酪氨酰和色氨酰-tRNA 合成酶共同祖先的氨基酸序列。在最后一个普适共同祖先之前的祖先合成酶氨基酸序列中可能含有酪氨酸但不含色氨酸，这表明色氨酸可能是后来添加到氨基酸序列中的。

20.9 重建祖先序列设计热稳定蛋白质

设计高热稳定性的蛋白质仍具有挑战。传统的方法主要依赖高分辨率的三维结构来设计突变位点，从而提高蛋白质热稳定性（Ulmer 1983）。目前已经有几种改善蛋白质热稳定性的合理方案。这些方法包括提高疏水相互作用和蛋白质内部核心部分的包装（Clark et al. 2004；Dong et al. 2008），增加蛋白质表面的离子对和离子对网络（Christodoulou et al. 2003；Tanaka et al. 2006），引入二硫键（Ivens et al. 2002），并增加离子对的数量或改善亚基之间的疏水作用（Kirino et al. 1994；Cheung et al. 2005）。然而，通常单个修饰的效果很小，并且高度依赖于其结构背景。基于计算的蛋白质设计法被用于提高蛋白质的热稳定性（Korkegian et al. Science 2005）。本文表述的重建祖先序列法为表达具有较好的热稳定性的突变蛋白质提供了新途径。如果祖先生物是（极端）嗜热的，那么与非祖先氨基酸残基相比，祖先氨基酸残基将更大程度负责蛋白质的热稳定性。表达和表征了系统发育预测的“祖先”序列中的一个或少数氨基酸残基取代了原始氨基酸残基形成的突变体，用以验证上述方法的可靠性。极端嗜热菌的两种同源酶被用于该研究中。亮氨酸生物合成酶、3-异丙基苹果酸脱氢酶（IPMDH）和 TCA 循环中的异柠檬酸脱氢酶（ICDH）被认为在 Commonote 之前的酶的共同祖先中复制而来的（Hurley and Dean 1994；Zhang and Koshland 1995；Suzuki et al. 1997）。构建了两个旁系同源酶的复合树，推断了祖先 IPMDH 和 ICDH 的氨基酸序列可能由 Commonote 控制。然后，将推测得到的祖先氨基酸序列引入极端嗜热古菌（*Sulfolobus tokodaii*）和嗜热菌（*Thermus thermophilus*）热稳定的 IPMDH 中，另外还引入来源于极端嗜热古菌（*Caldococcus noboribetus*）的 ICDH 中。在这些研究表明（表 20.1）45%～67%的突变体比相应的野生型酶的热稳定性更好（Miyazaki et al. 2001；Iwabata et al. 2005；Watanabe et al. 2006）。

在研究参与嗜热链球菌翻译的酶时也观察到了类似的趋势。我们预测了 α_2 型甘氨酰-tRNA 合成酶（GlyRS）的祖先氨基酸序列，然后将推测的祖先氨基酸序列的一种或多种氨基酸引入嗜热链球菌的 GlyRS 中。热稳定性测量结果显示（表 20.1）与野生型 GlyRS 相比，测试的 8 个突变体中有 6 个热稳定性增强（Shimizu et al. 2007）。这些突变体的研究结果清楚地表明了，引入重建的祖先氨基酸序列中的氨基酸残基能进一步提高来源于（极端）嗜热菌的热稳定酶的热稳定性。最近有研究证明，仅使用含有真核生物序列的数据集推断祖先氨基酸的引入也可以提高真菌酶的稳定性（Semba et al. 2015）。此外，祖先蛋白质突变体的催化效果通常与野生型酶的效果相同，或者更高。

表 20.1 点突变或全基因合成序列与重建的祖先序列相结合表达的蛋白质的解折叠温度

目的蛋白质	野生型/突变体/祖先	T_m/℃	参考文献
S. tokodaii IPMDH	野生型	96	Miyazaki et al. (2001)
	Met91Leu + Ile95Leu	99①	
	Lys152Arg + Gly154Ala	97①	
	Lys152Arg	96	
	Gly154Ala	98①	
	Ala259Ser+Phe261Pro	97①	
	Tyr282Leu	95	

续表

目的蛋白质	野生型/突变体/祖先	T_m/℃	参考文献
T. thermophiles IPMDH	野生型	87	Watanabe et al. (2006)
	Phe53Leu	86	
	Pro56Glu	89[①]	
	Arg58Leu	84	
	Val61Ile	87	
	Leu134Asn	91[①]	
	His179Lys	87	
	Val181Thr	89[①]	
	Asp184His	88[①]	
	Ser261Asn	86	
	Pro324Thr	87	
	Ala35Glu	88[①]	
C. noboribetus ICDH	野生型	88	Iwabata et al. (2005)
	Tyr309Ile + Ile310Leu	88	
	Ile321Leu	89[①]	
	Ala325Pro+Gly326Ser	91[①]	
	Ala336Phe	74	
T. thermophiles GlyRS	野生型	80	Shimizu et al. (2007)
	Gln29Ala	82[①]	
	Met167Leu + Val173Ile	84[①]	
	Asp202Glu + Ser205Arg	78	
	Val232Thr	82[①]	
	Tyr242Phe	81[①]	
	Phe314Tyr + Gly317Ser	77	
	Ala455Cys	81[①]	
	Val479Ile	83[①]	
延伸因子 Tu	*T. thermophilus* 延伸因子 Tu	77	Gaucher et al. (2008)
	细菌的共同祖先	65	
	细菌的共同祖先	73	
DNA 促旋酶 ATP 酶域	*T. thermophilus* ATP 酶结构域	90	Akanuma et al. (2011)
	祖先 ATP 酶结构域	90	
NDK	*Pyrococcus horikoshii* NDK	111	Akanuma et al. (2013a)
	古菌的共同祖先 Arc1	114	
	古菌的共同祖先 Arc2	109	
	古菌的共同祖先 Arc3	112	
	古菌的共同祖先 Arc4	109	
	古菌的共同祖先 Arc5	108	
	细菌的共同祖先 Bac1	99	
	细菌的共同祖先 Bac2	98	
	细菌的共同祖先 Bac3	109	
	细菌的共同祖先 Bac4	102	
	细菌的共同祖先 Bac5	107	

续表

目的蛋白质	野生型/突变体/祖先	T_m/℃	参考文献
肌醇-3-磷酸合酶(MIPS)	*Thermococcus sibiricus* MM 739 MIPS *Thermotoga* sp. str. RQ2 MIPS *T. maritima* MSB8 MIPS Ancestor of *Thermotoga* ATM _ T1 Ancestor of *Thermotoga* ATM _ T1 Ancestor of *Thermotoga* ATM _ T1 Ancestor of *Thermotoga* ATM_T1 *Thermococcus* ACM_C1 的祖先 *Thermococcus* ACM_C1 的祖先 *Thermotoga* 和 *Thermococcus* AAM_ A1 的共同祖先 *Thermotoga* 和 *Thermococcus* AAM_ A2 的祖先	81 85 81 89 89 89 89 >99 >99 >99 >99	Butzin et al. (2013)
RNH	*T. thermophiles* RNH *E. coli* RNH *T. thermophilus* 和 *E. coli* RNHs Anc1 的共同祖先 嗜热世系 Anc2 的祖先 嗜热世系 Anc3 的祖先 嗜温世系 AncA 的祖先 嗜温世系 AncB 的祖先 嗜温世系 AncC 的祖先 嗜温世系 AncD 的祖先	89 68 77 77 83 70 68 67 68	Hart et al. (2014)

① 与原酶相比，突变体的 T_m 有所提高。

使用祖先序列重建技术设计由 16 个细菌 DNA 促旋酶的 ATP 酶结构域构建的系统发育树的最深节点位置的序列（Akanuma et al. 2011），然后合成和表征设计的祖先序列。祖先的 ATP 酶结构域的热稳定性与嗜热链球菌 DNA 促旋酶的相应结构域的热稳定性相当（表 20.1）。此外，当蛋白质浓度大于 15μmol/L 时，祖先的 ATP 酶结构域的催化活性比嗜热链球菌的 ATP 酶结构域高。因此，即便只有小部分同源序列可用，重建方法也能创建完整的蛋白质序列，产生具有高热稳定性、高催化活性的蛋白酶。

利用祖先序列重建技术设计热稳定蛋白质，不需要知道任何关于目的蛋白酶结构的知识，它仅依赖于同源酶的氨基酸序列，而这些序列在大多数情况下可以直接在不断扩大的公共数据库中获得。最近扩大的细菌和古菌的基因组计划将为系统发育树的构建提供必要的资源信息，这些信息对计算祖先序列至关重要。由于极端嗜热蛋白质序列包含在用于推断祖先序列的系统发育树中，因此通过祖先序列重建技术通常可以产生比现存的极端嗜热蛋白质更稳定的蛋白质（Akanuma et al. 2013a；Butzin et al. 2013）。尤其在同源嗜热模板不可用时，该方法更具优越性。因此，祖先序列重建技术为生产热稳定蛋白质提供了一种通用的、有效的途径。

定向进化是改变蛋白质热稳定性或其他性质的另一种方法（Arnold et al. 2001）。定向进化方法功能强大，因为它不需要任何有关靶蛋白质的结构和功能的信息，仅依赖于建立一个拥有数百万独立变量的大型数据库，然而，这可能需要大量成本和时间。当然，基于系统发育方法的蛋白质重建技术可以利用已经被自然进化过程限制的序列空间（Gaucher 2007）。利用蛋白质家族的系统发育信息重建进化自适应路径（REAP）方法，可以找到可能改变蛋白质功能的突变位点。联合 REAP 和分子定向进化技术可以减少所需的序列空间，从而识别能够接受非标准核苷的 DNA 聚合酶突变体（Chen et al. 2010；Cole and Gaucher 2011）。

20.10 结论

研究早期生命对全面了解生命与地球环境的共同演化至关重要。为了解决这个问题，重新构建了古生物可能拥有的 NDK 序列，并优化它们的环境，表征其特性。重建的可能存在3500 万～3800 万年前的 NDK 的热稳定性与现存的（极端）嗜热蛋白质的热稳定性相当。因此，共同祖先可能在很高的温度下茁壮成长。鉴于古生物的（极端）嗜热性，祖先序列重建方法仅依赖同源氨基酸序列就能为产生热稳定酶提供可靠的途径。

无利益冲突声明

Satoshi Akanuma 和 Akihiko Yamagishi 声明他们没有利益冲突。

参考文献

Achenbach-Richter L，Gupta R，Zillig W，Woese CR（1988）Rooting the archaebacterial tree：the pivotal role of *Thermococcus celer* in archaebacterial evolution. Syst Appl Microbiol 10：231-240

Akanuma S，Iwami S，Yokoi T，Nakamura N，Watanabe H，Yokobori S，Yamagishi A（2011）Phylogeny-based design of a B-subunit of DNA gyrase and its ATPase domain using a small set of homologous amino acid sequences. J Mol Biol 412：212-225

Akanuma S，NakajimaY，Yokobori S，Kimura M，Nemoto N，Mase T，Miyazono K，Tanokura M，Yamagishi A（2013a）Experimental evidence for the thermophilicity of ancestral life. Proc Natl Acad Sci U S A 110：11067-11072

Akanuma S，Yokobori S，Yamagishi A（2013b）Thermophilic microbes in environmental and industrial biotechnology. In：Satyanarayana T，Litterchild J，Kawarabayasi Y（eds）Comparative genomics of thermophilic bacteria and archaea. Springer，Berlin，pp 331-349

Arnold FH，Wintrode PL，Miyazaki K，Gershenson A（2001）How enzymes adapt：lessons from directed evolution. Trends Biochem Sci 26：100-106

Atomi H，Matsumi R，Imanaka T（2004）Reverse gyrase is not a prerequisite for hyperthermophilic life. J Bacteriol 186：4829-4833

Baldauf SL，Palmer JD，Doolittle WF（1996）The root of the universal tree and the origin of eukaryotes based on elongation factor phylogeny. Proc Natl Acad Sci U S A 93：7749-7754 Boussau B，Gouy M（2012）What genomes have to say about the evolution of the Earth. Gondwana Res 21：483-494

Boussau B，Blanquart S，Necsulea A，Lartillot N，Gouy M（2008）Parallel adaptations to high temperatures in the Archaean eon. Nature 456：942-945

Brochier C，Philippe H（2002）Phylogeny：a non-hyperthermophilic ancestor for bacteria. Nature 417：244

Brooks DJ，Fresco JR，Singh M（2004）A novel method for estimating ancestral amino acid com- position and its application to proteins of the Last Universal Ancestor. Bioinformatics 20：2251-2257

Brown JR，Doolittle WF（1995）Root of the universal tree of life based on ancient aminoacyl- tRNA synthetase gene duplications. Proc Natl Acad Sci U S A 92：2441-2445

Butzin NC，Lapierre P，Green AG，Swithers KS，Gogarten JP，Noll KM（2013）Reconstructed ancestral Myo-inositol-3-phosphate synthases indicate that ancestors of the Thermococcales and Thermotoga species were more thermophilic than their descendants. PLoS One 8：e84300

Cavalier-SmithT（2002）The neomuran origin of archaebacteria，the negibacterial root of the universal tree and bacterial megaclassification. Int J Syst Evol Microbiol 52：7-76

Cavalier-Smith T（2006a）Cell evolution and Earth history：stasis and revolution. Philos Trans R Soc Lond B Biol Sci 361：969-1006

Cavalier-Smith T（2006b）Rooting the tree of life by transition analyses. Biol Direct 1：19 Cavalier-Smith T（2010）Deep phylogeny，ancestral groups and the four ages of life. Philos Trans R Soc Lond B Biol Sci 365：111-132

Chen F，Gaucher EA，Leal NA，Hutter D，Havemann SA，Govindarajan S，Ortlund EA，Benner SA（2010）Reconstructed evolutionary adaptive paths give polymerases accepting reversible terminators for sequencing and SNP detection. Proc Natl Acad Sci U S A 107：1948-1953

Cheung YY，Lam SY，Chu WK，Allen MD，Bycroft M，Wong KB（2005）Crystal structure of a hyperthermophilic archaeal acylphosphatase from Pyrococcus horikoshii-structural insights into enzymatic catalysis，thermostability，and dimerization. Biochemistry 44：4601-4611

Christodoulou E，Rypniewski WR，Vorgias CR（2003）High-resolution X-ray structure of the DNA-binding protein HU from the hyper-thermophilic Thermotoga maritima and the determi- nants of its thermostability. Extremophiles 7：111-122

Clark AT，McCrary BS，Edmondson SP，Shriver JW（2004）Thermodynamics of core hydrophobicity and packing in the hyperthermophile proteins Sac7d and Sso7d. Biochemistry 43：2840-2853

Cole MF，Gaucher EA（2011）Exploiting models of molecular evolution to efficiently direct protein engineering. J Mol Evol 72：193-203

Declais AC，Marsault J，Confalonieri F，de La Tour CB，Duguet M（2000）Reverse gyrase，the two domains intimately cooperate to promote positive supercoiling. J Biol Chem 275：19498-19504

Di Giulio M（2000）The universal ancestor lived in a thermophilic or hyperthermophilic environment. J Theor Biol 203：203-213

Di Giulio M（2003a）The universal ancestor and the ancestor of bacteria were hyperthermophiles. J MolEvol 57：721-730

Di Giulio M（2003b）The universal ancestor was a thermophile or a hyperthermophile：tests and further evidence. J Theor Biol 221：425-436

Dong H，Mukaiyama A，Tadokoro T，Koga Y，Takano K，Kanaya S（2008）Hydrophobic effect on the stability and folding of a hyperthermophilic protein. J Mol Biol 378：264-272

Edwards RJ，Shields DC（2004）GASP：Gapped Ancestral Sequence Prediction for proteins. BMC Bioinf 5：123

Forterre P（1996）A hot topic：the origin of hyperthermophiles. Cell 85：789-792

Forterre P（2002）A hot story from comparative genomics：reverse gyrase is the only hyperthermophile- specific protein. Trends Genet 18：236-237

Fournier GP，Alm EJ（2015）Ancestral reconstruction of a pre-LUCA aminoacyl-tRNA synthetase ancestor supports the late addition of Trp to the genetic code. J Mol Evol 80：171-185

Fournier GP，Andam CP，Alm EJ，Gogarten JP（2011）Molecular evolution of aminoacyl tRNA synthetase proteins in the early history of life. Orig Life Evol Biosph 41：621-632

Galtier N，Tourasse N，Gouy M（1999）A nonhyperthermophilic common ancestor to extant life forms. Science 283：220-221

Gaucher EA（2007）Ancestral sequence reconstruction as a tool to understand natural history and guide synthetic biology：realizing（and extending）the vision of Zukerkandl and Pauling. Oxford University Press，Oxford，pp 20-33

Gaucher EA，Thomson JM，Burgan MF，Benner SA（2003）Inferring the palaeoenvironment of ancient bacteria on the basis of resurrected proteins. Nature 425：285-288

Gaucher EA，Govindarajan S，Ganesh OK（2008）Palaeotemperature trend for Precambrian life inferred from resurrected proteins. Nature 451：704-707

Gaucher EA，Kratzer JT，Randall RN（2010）Deep phylogeny-how a tree can help characterize early life on Earth. Cold Spring Harb Perspect Biol 2：a002238

Gogarten JP，Kibak H，Ditrrich P，Taiz L，Bowman EJ，Bowman BJ，Manolsono MF，Poole RJ，Date T，Oshima T，Konishi J，Dendai K，Yoshida M（1989）Evolution of the vacuolar H^+- ATPase：implications for the origin of eukaryotes. Proc Natl Acad Sci U S A 86：6661-6665

Greaves RB，Warwicker J（2007）Mechanisms for stabilisation and the maintenance of solubility in proteins from thermophiles. BMC Struct Biol 7：18

Groussin M，Gouy M（2011）Adaptation to environmental temperature is a major determinant of molecular evolutionary rates in archaea. Mol Biol Evol 28：2661-2674

Groussin M，Boussau B，Charles S，Blanquart S，Gouy M（2013）The molecular signal for the adaptation to cold temperature during early life on Earth. Biol Lett 9：20130608

Groussin M，Hobbs JK，Szollosi GJ，Gribaldo S，Arcus VL，Gouy M（2015）Toward more accurate ancestral protein genotype-phenotype reconstructions with the use of species tree-aware gene trees. Mol Biol Evol 32：13-22

Hanson-Smith V，Kolaczkowski B，Thornton JW（2010）Robustness of ancestral sequence reconstruction to phylogenetic uncertainty. Mol Biol Evol 27：1988-1999

Hart KM，Harms MJ，Schmidt BH，Elya C，Thornton JW，Marqusee S（2014）Thermodynamic system drift in protein evolution. PLoS Biol 12：e1001994

Heine M，Chandra SB（2009）The linkage between reverse gyrase and hyperthermophiles：a review of their invariable association. J Microbiol 47：229-234

Hurley JH，Dean AM（1994）Structure of 3-isopropylmalate dehydrogenase in complex with NAD+：ligand-induced loop closing and mechanism for cofactor specificity. Structure 2：1007-1016

Ivens A，Mayans O，Szadkowski H，Jurgens C，Wilmanns M，Kirschner K（2002）Stabilization of a（betaalpha）8-barrel protein by an engineered disulfide bridge. Eur J Biochem 269：1145-1153

Iwabata H，Watanabe K，Ohkuri T，Yokobori S，Yamagishi A（2005）Thermostability of ancestral mutants of *Cal-*

dococcus noboribetus isocitrate dehydrogenase. FEMS Microbiol Lett 243：393-398

Iwabe N，Kuma K，Hasegawa M，Osawa S，Miyata T（1989）Evolutionary relationship of archaebacteria，eubacteria，and eukaryotes inferred from phylogenetic trees of duplicated genes. Proc Natl Acad Sci U S A 86：9355-9359

Kirino H，Aoki M，Aoshima M，Hayashi Y，Ohba M，Yamagishi A，Wakagi T，Oshima T（1994）Hydrophobic interaction at the subunit interface contributes to the thermostability of 3-isopropylmalate dehydrogenase from an extreme thermophile *Thermus thermophilus*. Eur J Biochem 220：275-281

Korkegian A，Black ME，Baker D，Stoddard BL（2005）Computational thermostabilization of an enzyme. Science 308：857-860

Lake JA，Servin JA，Herbold CW，Skophammer RG（2008）Evidence for a new root of the tree of life. Syst Biol 57：835-843

Lake JA，Skophammer RG，Herbold CW，Servin JA（2009）Genome beginnings：rooting the tree of life. Philos Trans R Soc Lond B Biol Sci 364：2177-2185

Miyazaki J，Nakaya S，Suzuki T，Tamakoshi M，Oshima T，Yamagishi A（2001）Ancestral residues stabilizing 3-isopropylmalate dehydrogenase of an extreme thermophile：experimental evidence supporting the thermophilic common ancestor hypothesis. J Biochem 129：777-782

Pace NR（1991）Origin of life-facing up to the physical setting. Cell 65：531-533

Semba Y，Ishida M，Yokobori SI，Yamagishi A（2015）Ancestral amino acid substitution improves the thermal stability of recombinant lignin-peroxidase from white-rot fungi，Phanerochaete chrysosporium strain UAMH 3641. Protein Eng Des Sel. doi：10. 1093/protein/gzv023

Shimizu H，Yokobori S，Ohkuri T，Yokogawa T，Nishikawa K，Yamagishi A（2007）Extremely thermophilic translation system in the common ancestor commonote：ancestral mutants of Glycyl-tRNA synthetase from the extreme thermophile *Thermus thermophilus*. J Mol Biol 369：1060-1069

Stetter KO（1996）Hyperthermophilic procaryotes. FEMS Microbiol Rev 18：149-158

Suzuki T，Inoki Y，Yamagishi A，Iwasaki T，Wakagi T，Oshima T（1997）Molecular and phylogenetic characterization of isopropylmalate dehydrogenase of a thermoacidophilic archaeon，Sulfolobus sp. strain 7. J Bacteriol 179：1174-1179

Tanaka T，Sawano M，Ogasahara K，Sakaguchi Y，Bagautdinov B，Katoh E，Kuroishi C，Shinkai A，Yokoyama S，Yutani K（2006）Hyper-thermostability of CutA1 protein，with a denaturation temperature of nearly 150℃. FEBS Lett 580：4224-4230

Thornton JW（2004）Resurrecting ancient genes：experimental analysis of extinct molecules. Nat Rev Genet 5：366-375

Ulmer KM（1983）Protein engineering. Science 219：666-671

Watanabe K，Ohkuri T，Yokobori S，Yamagishi A（2006）Designing thermostable proteins：ancestral mutants of 3-isopropylmalate dehydrogenase designed by using a phylogenetic tree. J Mol Biol 355：664-674

Woese CR（1987）Bacterial evolution. Microbiol Rev 51：221-271

Woese CR，Winker S，Gutell RR（1990）Architecture of ribosomal RNA：constraints on the sequence of "tetraloops". Proc Natl Acad Sci U S A 87：8467-8471

Yamagishi A，Kon T，Takahashi G，Oshima T（1998）From the common ancestor of living organisms to protoeukaryotic cell. In：Wiegel J，Adams M（eds）Thermophiles：the keys to molecu- lar evolution and the origin of life? Taylor & Francis，London，pp 287-295

Yang Z（1997）PAML：a program package for phylogenetic analysis by maximum likelihood. Comput Appl Biosci 13：555-556

Yang Z，Rannala B（1997）Bayesian phylogenetic inference using DNA sequences：a Markov Chain Monte Carlo method. Mol Biol Evol 14：717-724

Yang Z，Kumar S，Nei M（1995）A new method of inference of ancestral nucleotide and amino acid sequences. Genetics 141：1641-1650

Zhang T，Koshland DE Jr（1995）Modeling substrate binding in *Thermus thermophilus* isopropyl- malate dehydrogenase. Protein Sci 4：84-92

Zhou Y，Asahara H，Gaucher EA，Chong S（2012）Reconstitution of translation from *Thermus thermophilus* reveals a minimal set of components sufficient for protein synthesis at high temperatures and functional conservation of modern and ancient translation components. Nucleic Acids Res 40：7932-7945

第二十一章

关于细菌应对温度变化的系统生物学观点

Marco Fondi[1]，Emanuele Bosi[1]，
Angelina Lo Giudice[2]，Renato Fani[1,3]

21.1 引言

“极端环境（extreme environments）”被定义为化学或物理条件严苛的生态位。其对大多数生物体而言是难以生存的，因此极端环境下物种多样性较低。尽管极端微生物在三域系统中均存在，其中包括病毒，但最普遍存在的还是微生物，主要原因在于微生物能适应急剧变化的环境、寡营养环境以及具有代谢多样性的特征。

栖息在极端环境下的生物体已经形成了独特的机制来应对极端条件，可以说它们标志了地球生命的化学物理界限。

一般而言，适应（adaptation）和驯化（acclimation）是两个不同的定义。适应，是指有机体为了应对特定生态位而逐代积累的遗传变化；驯化，是指生物为了应对环境的短暂变化而表现的短期的身体调节（Morgan-Kiss et al. 2006）。

从应用生物学和基础生物学的角度来看，研究极端微生物克服生态位中自然选择压力的机制很有意义。例如，适冷微生物可以用于低温蛋白质的表达，促进温度不稳定性蛋白质的过表达（Papa et al. 2007；Miyake et al. 2007）。此外，微生物中的适冷酶已被应用于生物技术（如食品生产中的酶制剂）和分子生物学（如碱性磷酸酶被用于克隆前 DNA 载体的去磷酸）多种领域（Cavicchioli et al. 2001）。

21.1.1 冷适应机制

低温是最常见的极端条件。事实上，海洋作为地球生物圈的重要组成部分，大部分海洋（约 90%）温度不高于 5℃（Russell 1990）。此外，极地地区占地球表面的 15%。极端寒冷

[1] Department of Biology，University of Florence，Via Madonna del Piano 6，50019 Sesto Fiorentino，Firenze，Italy.

[2] Institute for Coastal Marine Environment，National Research Council (IAMC-CNR)，Spianata San Raineri 86，I-98124 Messina，Italy；Department of Biological and Environmental Sciences，University of Messina，Viale F. Stagnod' Alcontrès 31，I-98166 Messina，Italy.

[3] e-mail：renato. fani@unifi . it.

的环境在某种程度上限制了微生物的多样性，这导致适应该环境的微生物缺乏生存竞争(Pearce 2012)。尽管如此，这些有机体的代谢仍受到一些因素限制，比如较低的催化活性、较低的底物亲和力（Gerday et al. 1997)、较低的热能和反应速率（Collins et al. 2008)。与中温环境中的微生物一致，上述因素以及水黏度增加和膜流动性下降（Graumann and Marahiel 1996）会共同导致细胞生长减缓。此外，当温度过低而结冰时，还会导致微生物额外的应力效应，如机械破损、氧化损伤和渗透压失衡（Tanghe et al. 2003)。

为了应对上述情况，适冷有机体进化出多种不同的策略。总的来说，在分子水平上的适应主要包括以下方面：分泌低温酶，合成抗冻剂，合成特异性脂质以改变膜流动性，以及增加响应相关基因的表达。下一部分将举例阐述微生物如何运用这些机制来应对温度变化。

21.1.1.1 冷活性酶

酶的进化使其在低温下具有很高的转换率和催化效率，虽然在较高温度下其稳定性明显降低。这些蛋白酶的特征是活性位点和整体水平上的结构柔性增加，进而使酶促反应的活化能降低。但柔性增加的缺点是导致酶在中温或高温下不稳定（Feller and Gerday 2003)。

上述特征由多种因素造成：与中温同源酶相比较，冷活性酶的脯氨酸（Pro）和精氨酸(Arg）残基减少，甘氨酸（Gly）残基增加。例如，类 TAB5 碱性磷酸酶的诱变实验已证明这些氨基酸对于酶的嗜冷性起重要作用（Mavromatis et al. 2002)。因此，当温度降低时，指导这些氨基酸合成的生物合成途径会更加活跃（例子见第 2 章第 2.3 节)。

21.1.1.2 抗冻剂分子

在低温环境中，水结冰及冰晶体的形成会降低自由水的含量，细胞因渗透压变化而破裂。事实上，尽管胞内的冰晶可充当溶质吸引水分子跨膜进入细胞，但胞外形成的冰会改变渗透压导致细胞膜破裂（Mavromatis et al. 2002)。

为了消除上述影响，冷适应生物体合成不同的抗冻剂分子（Kawahara 2008)。抗冻剂分子有以下类别：①小分子（如糖、二醇类和胺类)，此类抗冻剂能降低水的凝结温度(Bouvet and Ben 2003)；②抗冻蛋白质，此类抗冻剂能结合到初期冰晶表面以阻止它们扩散；③冷冻保护剂蛋白质类，此类抗冻剂能通过与不稳定蛋白质表面结合来防止蛋白质冷变性（Kawahara 2008)，或能协助其他关键蛋白质的折叠。

21.1.1.3 脂类的生成

如前所述，冷休克其中一个主要影响是膜弹性降低，使膜更易破裂以及机械溶解。为了避免该情况并维持膜的液晶态，细胞通过调节特定酶（如脂肪酸还原酶和脱饱和酶）的表达来改变膜的脂肪酸构成。细胞对脂肪酸的调节包括：通过脱饱和酶引入双键，缩短脂肪酸链长度和引入甲基分支。不同微生物具有不同数量和类型的脱饱和酶，通常有一种属于管家酶，还有其他冷诱导酶（Klein et al. 1999)。

21.1.2 热适应机制

高温可导致细胞的弱键断裂进而使大分子（如蛋白质和核酸）变性。为了在高温环境下生存，一些嗜热生物进化出独特性质的酶，举例来说，一些酶热稳定性和最适温度大于70℃，在一些情况下甚至为 110℃。而且，为了抵消热休克的影响，细胞可激活热休克反应(heat-shock response)。这种应激反应最主要的效应器是一组保守蛋白质，也就是热休克蛋白（HSPs)，热休克蛋白可作为分子伴侣帮助其他蛋白质折叠，并保持正确的结构，也可作为蛋白酶，降解未折叠的蛋白质。细菌中热休克反应的调节已在大肠杆菌模型中进行了很好的表征，已发现一种可替代 σ 因子的 $\sigma32$ 因子，它作为 HSPs 合成的正调节物增加了 RNA

聚合酶与HSP启动子的亲和力（Morimoto et al. 1997）。

热休克反应由基于σ32因子活性调节的复杂分子机制来完成，其中σ32因子活性在翻译水平和翻译后水平受调节。在翻译水平上的调节由σ32因子的mRNA序列（编码基因*rpoH*）决定。中温环境下，该mRNA被折叠成二级结构，其部分折叠遮盖了翻译起始位点，导致翻译率低，而在较高温度下，*rpoH* mRNA的三维结构发生改变，翻译起始位点不再被遮挡，因此允许*rpoH*的高效表达（Morita et al. 1999）。在翻译后水平上的调节则存在一个反馈回路，即组成型表达的HSPs（DnaK，KnaJ和GrpE）将σ32因子结合到金属蛋白酶FtsH上。随着温度升高，变性蛋白质的数量增加，阻隔了HSPs的作用，由此σ32因子水平上升（Arsene et al. 2000）。极端嗜热的生物能在近100℃下生存，因此，它们的大分子要比中温生物的热稳定性更强（Sterner and Liebl 2001）。由于前述特点，嗜热生物的酶可作为研究模型被运用到多个学科中，如生物、化学和物理，以探索酶进化的一个关键问题，即功能酶所能承受的温度上限是多少（Vieille and Zeikus 2001）。

如前所述，一些旨在解密生物应对短期温度变化（即驯化）的分子策略的研究所用的模式生物都是中温生物（如*E. coli*）。但是，需要特别指出的是，冷适应和热适应微生物的驯化机制可能不同（或根本不存在）。有趣的是，至少在几个例子中发现，随着温度下降，中温微生物和极端微生物的代谢调整在一般趋势（表型）上有相似之处（详见第4章第4.3节）。此外，对冷适应机制的研究显示，不同的适冷微生物可表达不同的冷诱导蛋白质，这也说明了不同适冷微生物使用不同策略来应对低温环境。即便如此，冷适应机制的共性（至少在嗜冷微生物中）也能被归纳总结，包括RNA分子伴侣CspA和（不同构象的）肽基脯氨酰异构酶（PPIase）的诱导和RNA聚合酶的调节（Kawamoto et al. 2007）。

21.2 研究应对温度变化反应的组学方法

新技术的出现使获得大量生物学数据成为了可能，比如高通量测序技术（next generation sequencing，NGS）。生物学的关注点也已经从单个特定对象（单个基因/操纵子/蛋白质等的特征性描述）转移到生物分子集合的表征和量化上，这也被称之为“组学”。从更广泛的意义来说，生物学的研究方法已从曾经广泛流行的还原论法（reductionist approach）变为整体论法（holistic approach），前者成功地辨析出复杂生物结构中的单个元件，后者的重点则在于研究不同生物组成成分之间的关系。以下所述的是组学方法在温度变化适应方面的最新研究进展，并根据所采用的组学技术进行分类归纳，分别有基因组学、转录组学、蛋白质组学和表型组学。

21.2.1 基因组学

高通量测序技术的出现使快速的全基因组测序成为可能，且测序费用不断降低，因此测序项目对于大多数实验室都具可行性。随着时间的推移，生物数据库已经被大量生物的基因组序列充斥并达到饱和。这些数据库中包含成千上万个已注释的基因序列，可利用这些序列去为新发现的未知基因做注释。例如，热休克蛋白信息资源数据库（heat shock protein information resource，HSPIR）（Ratheesh et al. 2012）包含来自227个原核和真核生物物种全基因组的9902个蛋白质记录，冷休克结构域数据库（cold shock domain database，CSDBase）（Weber et al. 2002）是含有冷休克结构域的蛋白质和细菌冷休克反应蛋白质信息的资源库。在分析基因组序列时，充分利用已注释序列及数据库，可鉴定出与特定应激反应相关的推定基因，再将其与文献进行比较。随后，再通过不同的计算机技术或体外实验进一步研究该未知基因，加强证据的可信性。基因组学技术已被广泛用作微生物温度变化适应研究的基因组学基础，其中具代表性的是对来自南极海洋的 *Pseudoalteromonas haloplanktis* TAC125（Medigue et al. 2005）和 *Colwellia psychrerythraea* 34H（Methe et al. 2005）的基因序列的分析。在上述两个例子中，冷适应生活方式的基因标签的特性都得到确定，包括：缺失生产活性氧的整个通路，以应对低温下氧气溶解度的增加；扩充与细胞膜合成或影响耐寒性物质的摄入和合成相关的基因家族。

通常，一个基因组序列的可用性还需要结合表型的实验数据来看，因为表型数据可以：①确定基因的功能；②识别功能模块；③深入认识基因与基因表达产物之间的网络关系。综上，确定一个微生物的基因组序列通常是研究工作的第一步，其中运用了不同组学技术来进行分析，也有可能需要整合多种技术（详见第 3 章第 3.2 节）。

21.2.2 转录组学

通过分析一个生物体在不同条件产生的整套转录组（transcriptome），可以探究基因集差异表达与不同实验条件间的相关性。对于应激反应，基因的差异表达模式能够了解细胞应激事件中基因的作用（如温度变化）。在分析转录物组的变化时，需要在作出正确结论之前考虑一些问题。第一，研究已表明，即使在持续刺激下，基因表达反应也是短暂的，这意味着新的 mRNA 水平达到稳态时也与相对正常状态下的相似（Lopez-Maury et al. 2008）。对于应激过程中生理性变化的基因表达，其代表性特征是某些特定的应激元件（基因簇）的激活和生长相关基因的下调。第二，某个特定基因在转录水平上的改变不一定与基因表达水平上的显著改变直接相关（Griffin et al. 2002）。因为调控仍可发生在翻译水平及翻译后水平（比如 mRNA 和/或蛋白质的降解）。换句话说，在最终的分析中，除了转录水平的量化外，还可以使用蛋白质水平的定量分析来测得真实有效的基因表达水平。

转录组学已被广泛用于温度急剧变化（升或降）对多种（非极端）微生物影响的研究上，比如说大肠杆菌（*E. coli*）（Jozefczuk et al. 2010）、枯草芽孢杆菌（*Bacillus subtilis*）（Kaan et al. 2002）和希瓦氏菌（*Shewanella oneidensis*）（Gao et al. 2004，2006），这些研究通常都结合了其他组学技术，如蛋白质组学和代谢组学。这些研究工作的结果得出了一些共性，如与产能及生物合成（生长基因）相关的代谢基因下调，编码应激反应效应子的特定应激基因上调，以及调节基因（如 *hylU* 和 *narQ*）的差异性表达。

21.2.3 蛋白质组学

实验已证实，在酿酒酵母（*Saccharomyces cerevisiae*）中，mRNA 水平的变化并不与蛋白质表达水平的变化呈线性关系（Griffin et al. 2002）。从数学的角度来看，即便是最简单的基因网络，想要详细理解这些网络的调控关系，也需要 mRNA 和蛋白质水平两方面的数据作为支撑（Hatzimanikatis and Lee 1999）。可见，除了对所表达蛋白质组进行定量分析外，还需要整合其他组学的数据。这种方法已被用作深入研究冷/热适应和冷/热驯化。比如说，Ting 对阿拉斯加鞘氨醇单胞菌（*Sphingopyxis alaskensis*）蛋白质组进行定量分析时发现一些低温下共同的生理学特征（Ting et al. 2010），包括参与合成氨基酸（如色氨酸、组氨酸和脯氨酸等）的酶大量增加，这些氨基酸在多种生物的冷休克反应中具重要作用，以及脂质脱饱和酶及特定的冷休克反应蛋白的过表达（Liu et al. 2002；Kawamura et al. 1994）。值得注意的是，这些分析中一些尚未被表征的蛋白质的过表达将促使人们对其进行进一步的分析。

最近有两个研究报道了 *P. haloplanktis* TAC125 细胞在 4℃和 18℃下的生长动力学及蛋白质组学（Piette et al. 2010，2011）。这些细菌在 18℃下的倍增时间是 100min，而在 4℃下这个值增加到了 4 h。同时，这些研究发现两个温度下导致了 123 个基因差异性表达（其中有 83 个下调，40 个上调）。总的来说，这些蛋白质组学分析揭示了细菌适应低温的特征是新陈代谢水平下调，即与化合物分解、生物合成及能量生成相关的蛋白质低表达，冷休克反应蛋白的激活。细菌应对冷休克的另一个效应是铁代谢的下调和冷特异性氧化应激反应蛋白质（cold-specific oxidative stress response proteins）的上调。上述这些效应都可以得到解释，因为这种下调很大程度上关闭了三羧酸循环和呼吸链，所以对三价铁的需求降低，而在新陈代谢水平较低时，氧化应激效应因子之一的谷胱甘肽因低水平的 NADPH 而含量较低（Piette et al. 2010，2011）。

21. 2. 4 表型组学

表型组学（高通量表型分析）的发展使得自动且大规模地探索特定生物体的表型空间（phenotypic space）及探究其表型能力（如代谢特定碳源的能力）成为可能。典型的研究方法是表型微阵列技术（phenotype microarrays，PMs）（Bochner et al. 2001），该技术利用细胞呼吸作用（如 NADH 的减少），并最终以紫色作为报告标识显示总体的代谢活动（Fondi and Lio 2015）。迄今为止尚无关于冷适应细菌（或其他极端微生物）在不同温度下生长的表型组学研究。然而，从不同环境生态位（水层和南极海绵）中分离得到的两株南极 *P. haloplanktis* TAC125 和 TB41 的初步研究数据显示，在不同生长温度下（4℃和 15℃）下，两菌株的表型差异较大。事实上，即便这两株菌在各自的自然生长温度下有一些相似的表型特征，但它们在不同的底物中仍具不同的生长速率，这表明与温度相关的因素能影响菌体代谢物质的范围。同时，这些分析显示了尽管是同属一个物种的不同菌株也存在一定程度上的代谢可塑性。

21. 3 研究温度改变时响应的系统生物学

尽管不同组学技术各有其重要性，但没有任何一种组学技术仅凭一己之力就能破解微生物学的复杂性（Zhang et al. 2010）。因此，多组学技术被广泛运用于多个不同研究领域，包括生物源燃料（Zhu et al. 2013）、生物制药（Schaub et al. 2012）、医学研究（Wiench et al. 2013）和宿主-病原体的相互作用研究（Ansong et al. 2013）。由于数据的格式多样性、高维度性以及对数据标准化的需要（Fondi and Lio 2015），这种多数据类型的整合与诠释也许可被视为当今生物信息学（bioinformatics）的关键挑战之一。

现在，这些数据类型与基因组衍生信息的整合被证明是一种最有效的诠释高通量数据的方法之一。特别是考虑到表型特征与微生物基本代谢之间的联系，基因组规模的代谢重建在这点上可被高效利用起来。

下一部分，将讲述代谢网络重建与模型建立的概要、代谢模型与多组学数据的整合方法，以及近期一个运用上述方法研究低温驯化的案例。

21. 3. 1 全基因组范围代谢重建与代谢建模

严格来说，某个特定微生物的代谢模型（metabolic model）（又称代谢网络，metabolic network）是所有可能发生的生化反应的集合。通过序列的相似性，微生物的代谢反应集合（the metabolic repertoire）可以由注释服务器来自动计算，比如 RAST（Overbeek et al. 2014a)、KBASE 和 Metacyc(Caspi et al. 2014)。如今,一旦得到准确可靠的基因组注释,便可以相对简易快捷地完成这些工作(图 21. 1a)。

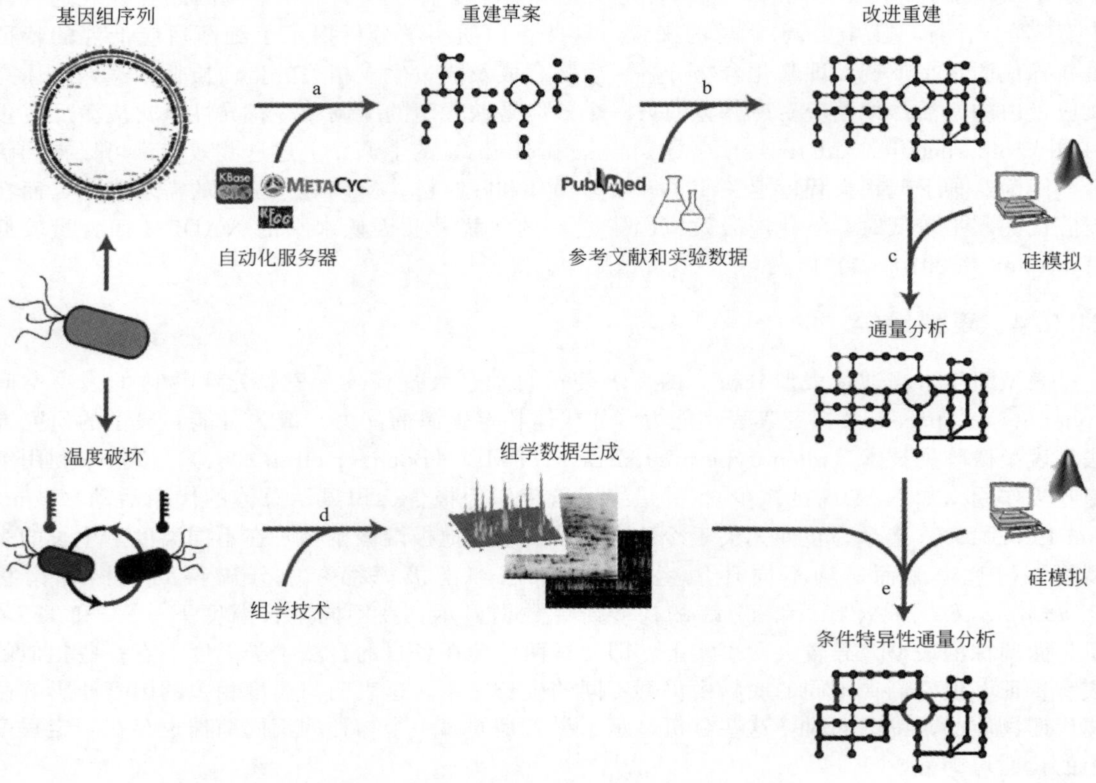

图 21. 1　运用代谢建模与组学整合技术研究温度刺激实验的假想途径

这些从头重建的方法很少能成功表示某个微生物所有可能的功能和生理状态,主要是因为:①关键的代谢步骤通常不包含在重建中,因此无法维持细胞分裂速率的预测(代谢间隙常与错误或不完整的基因组功能注释相关);②嵌入反应的关键问题,如化学计量,方向性和命令的丢失或出错;③由于开始时模型主要是依据与其他微生物的同源性进行重建,因此模型会忽略一些有机体特有的代谢途径(而它们通常控制关键的表型特征)。基于以上原因,要重建一个能代表有机体真正代谢空间(metabolic space)的模型,通常少不了大量的人工优化工作。

目前用于重建代谢网络计算模型的综合方案由 Thiele 和 Palsson(Thiele and Palsson 2010)提出,方案中描述的步骤(至少)包含了 96 步,其中大部分是对现有文献和实验数据的广泛修订(图 21. 1b)。在这个阶段,组学方法得出的数据十分有用,对空缺数据的填充和扩展模型方法都有指导作用。例如,基因组学、蛋白质组学和转录组学,它们既有助于优化代谢模型的草图,也能使模型更接近有机体的真正表型(详见 Fondi 和 Lio 2015 的综述)。

一旦空缺数据完成填充,就可以开始运用重建的代谢网络进行模拟(见图 21. 1c)。大致来说,模拟细胞代谢流的分配有两个可能的方法:化学计量法和动力学模型法。动力学模型包含了能被常微分方程(ODEs)或偏微分方程(PDEs)描述的一系列反应(Tomar and De 2013)。但考虑到用此方法模拟的生物体模型或反应非常有限,而且需要大量的参数,因此动力学模型法只适用于小型的代谢系统中。相反,化学计量法(基于约束的)模型可适用于更大型的系统（如基因组水平的），因为它只需要代谢反应的化学计量学信息（Oberhardt et al. 2009）。流平衡分析（flux balance analysis，FBA）是目前运用最广泛的基于约束的建模技术，它被用于计算细胞中所有可预测发生的化学反应的平衡结果。换而言之，一旦确定了边界条件（即可用营养来源和它们的吸收流量），就可利用 FBA 获取一系列可能的稳定态流量以优化一个确定的细胞对象，通常是指在代谢网络中的最大生物产量。总的来说，FBA 预测结果可被用于推测关键生理特征，比如生长率（一定时间内产生的生物量）、可代谢化合物种类以及随环境变化（如温度变化或生长培养基的组成变化）而在系统水平上发生的流量分配的变化。一般来说，应该将模型所得结果与实验数据进行全面比对后才能得出可靠结论。在预测生长表型时，要评估代谢模型的可靠性，常将其与大规模生长试验（如 Biolog Phenotype Microarray）或实验计算的生长速率进行对比。

虽然代谢重建的目标确定了现阶段它对准确性的要求不苛刻，但仍需要在表型的实验数据与模拟的数据间达成广泛的一致性。

21.3.2　数据整合

随着生物信息学的发展，组学数据的使用也越来越广泛，从协助提高代谢模型的预测准确性，到在系统水平上了解细胞行为。在温度休克研究的假定中，可以在细胞接受热/冷休克后收集组学数据（图 21.1d）。过去十年，这种方法被广泛运用于冷休克反应（Gao et al. 2006；Kaan et al. 2002；Garnier et al. 2010）和热休克/热适应（Garcia-Descalzo et al. 2014；Li et al. 2011；Wang et al. 2014）的研究。虽然这些研究为定义和认识温度刺激下的适应/驯化在基因组水平上的趋势提供了有用的见解，但值得注意的是，没有任何一种单一的组学分析方法可以完全解释基础微生物学的复杂性（Zhang et al. 2010）。就此而言，基因组水平的代谢模型（连同建模方法，如 FBA）提供了一个强大而有意义的平台，来预测温度刺激后可能的代谢变化。事实上，多种多样的实验数据都可以与代谢模型相融合（图 21.1e），以更真实和全面地认识细胞代谢通路和其在环境变化后的变化，要实现以上目标，可以通过整合基因产物、代谢产物的定量测试，以及它们之间的相互关系来实现（Zhang et al. 2010；Fondi and Lio 2015）。

21.3.3　整合表达数据与代谢建模以研究温度改变反应

在至少四个研究实例中，利用整合表达数据与代谢建模的方法研究温度刺激后微生物的代谢变化（Tong et al. 2013；Fondi et al. 2014；Topfer et al. 2012；Navid and Almass 2012）。Topfer 等人（2012）通过整合时间分辨的转录组学数据与基于通量的方法，揭示了 *Escherichia coli* 在冷和热休克诱导下的代谢驯化。在该研究中，研究人员能同时定性细菌在温度变化下的整体以及某一个特定反应的过渡性行为。特别是研究人员发现在不同的时间点下，有两类反应存在显著的活动形式差异（冷/热休克之后均有此现象）。第一类反应在细菌的所有生长阶段都会发生，而且估计其与生长过程必不可少的代谢过程密切相关（如氨基酸的生物合成，核苷酸、核苷和碱基的相互转换，以及 TAC 循环）。第二类反应在细菌接受环境压力时使用量不稳定。从功能上来说，研究表明这些反应与（某些）氨基酸、有机酸和葡萄糖的分解，丙酮酸向乙酰 CoA 的转化，以及天冬氨酸家族氨基酸的生物合成等过程相关。此外，研究显示细菌在应对冷刺激和热刺激时，其反应模式之间存在部分相同之处。但是，研究人员也指出，虽然应对冷和热温度胁迫两者涉及相同的生物过程，但激活（或失活）这一短暂过程的应用上可能有些许差别，这些差别随后进一步放大了应对冷/热刺激的特定基因的效应（Topfer et al. 2012）。

Tong 等人（2013）结合代谢网络建模与蛋白质组数据，用于研究培养温度下降对嗜热厌氧细菌 *Thermoanaerobacter tengcongensis* 代谢产生的影响。研究者将培养温度由 75℃降至 55℃，收集差异蛋白质组学数据，并将之与代谢网络模型以及稳健性分析（对代谢网络中反应通量的微扰导致的细胞生长速率变动趋势的预测）相结合。此研究展示了稳健性分析在大部分情况下（73%）能正确预测微扰相关的蛋白质丰度变化对细胞整体生长及生理的影响。在一些研究中，FBA 预测和蛋白质表达数据之间存在歧义，研究者推测原因在于反应活性和酶含量之间无关联。这些反应大多与氨基酸的生物合成有关（即缬氨酸、异亮氨酸、亮氨酸和苏氨酸），同时研究者推测存在着多种水平的调节（包括温度、缓冲液、蛋白质修饰和/或变构结构）。

研究人员利用 *Yersinia pestis*（鼠疫耶尔森氏菌）的基因表达水平研究温度变化下的代谢改变（Navid and Almaas 2012）。对此菌株而言，当温度从 26℃升高至 37℃时，菌株从非毒性表型转变为毒性表型（Chromy et al. 2005；Konkel and Tilly 2000）。研究人员收集两株菌（*Y. pestis* 201 和 KIM5）的基因表达数据，并用于分析不同表型行为的代谢基础。在 *Y. pestis* 中，细胞主要的代谢反馈包括降低非关键通路通量来节约能量，比如嘌呤的生成，因为它可以通过较慢的核苷酸降解途径代偿。伴随着活性氧（ROS）产生的减少，细胞通过增加氧化还原反应通量（如过氧化氢酶-过氧化物酶）来保护细胞大分子。而利用氨基酸分解代谢作为碳原子的来源是预测细胞活跃度的另一种策略。即使如此，计算机模拟、基因表达的信息与实验观察到的结果间拥有很好的一致性。其中，实验观察结果是指菌株应对不同温度变动下的生理学特征。

在近期的研究中（Fondi et al. 2014），研究人员将多种组学数据（包括表型组学、蛋白质组学和基因组学）与代谢模型进行了整合，从而全面地探究 *P. haloplanktis* TAC125 在低温下生长的代谢调整。下一章节将详细讲述此研究方法。

如本节所示，到目前为止，通过整合组学数据和代谢模型的方法研究了两种极端微生物对温度变化的响应。鉴于此方法较好的预测准确性，未来可以用来阐明温度变化下生物适应和驯化的系统水平的代谢情况。

21.4 *P. haloplanktis* TAC125 研究案例

从南极洲的海水样本中分离得到的 *P. haloplanktis* TAC125（*Ph* TAC125）是冷适应细菌的模型生物之一。由于它在低温下快速繁殖的能力，研究人员提出将其作为可溶性异源蛋白质过表达的载体（Duilio et al. 2004；Wilmes et al. 2010；Rippa et al. 2012；Corchero et al. 2013）。另外，这种细菌能在较宽的温度范围（从 4℃到接近 30℃）内正常生长。由于以上原因，深入了解其代谢概况能有效开发利用其生物技术潜能，并了解其系统层面的细胞生理功能。因此，重建 *Ph* TAC125 基因组水平的代谢模型则显得十分必要，这也使得定义其表型空间在系统水平的特征成为可能，包括在环境变动（如生长培养基营养成分的改变和温度骤变）下的反应。

21.4.1 模型概述

人工优化工作的第一步是利用 RAST 得到 *Ph* TAC125 代谢重建模型的草图，并利用 ModelSeed 进行后期处理（Overbeek et al. 2014b），随后优化的主要步骤如 Thiele 和 Palsson（2010）所述。此外，代谢重建模型是通过由 *Ph* TAC125 原基因组注释（Medigue et al. 2005）和多个不同的功能性数据库信息的整合来共同完成的，这些数据库包括 KEGG

(Kanehisa 2002)、BRENDA(Scheer et al. 2011)和 MetCyc(Caspi et al. 2006)。表 21.1 列出了 *Ph*TAC125 代谢重建模型的主要特征。*Ph*TAC125 基因组水平的代谢模型［根据标准命名方案命名为 iMF721(Reed et al. 2003)］包括 721 个基因［ORFs(DNA 序列开放阅读框);*Ph*TAC125 20.7%的蛋白质编码基因］,1133 种代谢产物和 1322 个生化反应。

表 21.1 *P. haloplanktis* TAC125 代谢重建模型的特征

基因组	基因组大小(染色体 1+染色体 2)(bp)	3850272
	编码蛋白质数量	3484
模型	基因数量(蛋白质编码基因所占比例,%)	721(20.7)
	反应数量	1322
	与基因相关	1146
	与基因无关(交换反应)	176(85)
	代谢产物	1133

值得注意的是,这个模型包含了在 16℃ 复合培养基中所表达的 97% 的代谢基因(Wilmes et al. 2011),因此,原则上它几乎代表了在这种生活条件下的微生物的完整代谢潜能。

21.4.2 代谢模型的测试

建立好的模型需要利用多种大量的参考数据进行检验,尤其是需要将模型预测得到的生长速率与实验数据进行对比(Wilmes et al. 2010;Giuliani et al. 2011)。

图 21.2 是模型预测的生长速率与实验数据的对比,可以看出两者在整体上具有一致性。另外,运用 Biolog Phenotype Microarray(PM)数据来评估和修正 iMF721 的生长预测数据,具体是利用生物合成反应中的模拟通量和表型微阵列实验测得的细胞呼吸活动的比对来完成的。要模拟微生物在 64 种碳源上的生长,方法是分别利用每一个碳源作为唯一碳源,这要求将其吸收率设置在 1mmol/(g·h^{-1}) 任意值(有氧条件)。在多数情况下,模型预测结果和 PM 数据之间的差异可以填补模型的空缺,比如增加一个代谢反应到网络中,增加或缺失某些转运反应。用每一个碳源重复上述步骤,在 84%(54/64)的情况下,电脑模拟结果和活体实验结果相一致。

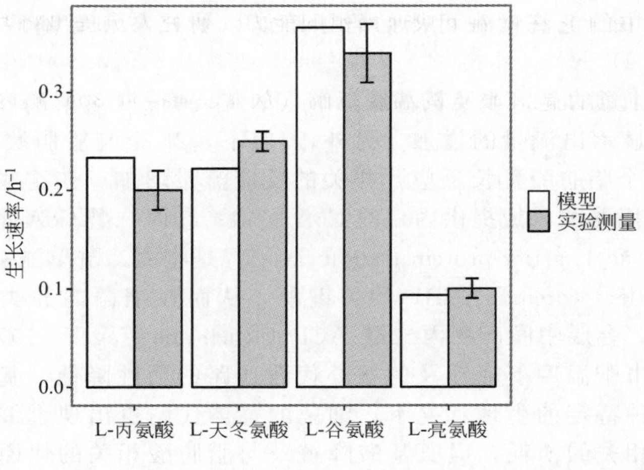

图 21.2 *Ph*TAC125 模拟(白色条形)和实验所得(灰色条形)的生长速率

模型的定性和定量测试结果大多数都在代谢重建模型的可行范围内（例子详见 Schatschneider et al. 2013；Bartell et al. 2014；Fang et al. 2011；Durot et al. 2008），这说明 iMF721 是模拟 *Ph*TAC125 中心代谢的可靠模型。

21.4.3 冷休克反应的研究

由于实验数据和模型模拟结果在整体上存在一致性，iMF721 模型原则上可以用于研究 *Ph*TAC125 在环境变化（如温度变化）下系统水平的代谢反应。整合了蛋白质丰度数据与基于约束的 iMF721 模型，获取了 *Ph*TAC125 在 4℃和 18℃下的蛋白质组学和生长速率的数据（Piette et al. 2010，2011），发现了在温度下调后出现差异表达的 123 种基因（83 种下调和 40 种上调）。最重要的是，iMF721 模型包含了大部分（代谢）基因在两种条件下的基因-蛋白质规律（Gene-protein rules，GPR）（占 60.90％中的 54％）。

如前所述（见本章 21.3.2），通过调节与相应酶表达水平有关反应的可接受通量，可将蛋白质丰度数据（以及基因表达数据）和代谢模型结合起来。在这个研究案例中，将上调和下调的比率与具有不同表达的代谢调节（metabolic adjustment by different expression，MADE）的 iMF721 代谢模型结合（Jensen and Papin 2011）。利用基因/蛋白质表达值可构建以二进制表达的状态（“1”和“0”分别表示反应的“发生”和“未发生”），它与所有测量所得的表达变化情况相匹配，可以并入到代谢重建模型中。若有了初始的基因表达值，MADE 便生成了模拟细胞真实代谢功能状态的模型。沿用上述方法，可以得到两个截然不同的代谢功能状态，即原始的 iMF721 模型和“关闭”下调基因作用从而模拟低温生长情况的衍生模型。在 4℃的衍生模型中，33 个基因（与模型中 54 个反应相关）被关闭（因它们在 4℃下被抑制），同时，12 个基因（与 35 个反应相关）被开启。

运用这种方法，可以研究基因表达水平上温度改变的代谢效应，还能深入分析 *Ph*TAC125 整体流量分配的变化。经过分析流量分配的变化，科学家发现蛋白质丰度变化最显著的缺点是 *Ph*TAC125 的整体代谢调整抑制了中心代谢途径的活性。总体上，209 个反应在两种温度下的流量分配并不一致，其中 141 个在 4℃下缩小了流量，另外 68 个则显示出流量增加。这个结果与两种温度下诱导基因与抑制基因的相对数量以及 *Ph*TAC125 在 4℃下低生长速率的观测结果一致（Piette et al. 2010，2011）。

与生物合成和产能相关的途径（图 21.3）大多是指在低温下受抑制的反应。这些途径囊括了大多数与嘌呤和嘧啶前体的生物合成、糖酵解反应和磷酸戊糖途径相关的反应。在这些途径中的低流量很可能妨碍了产能以及核酸和氨基酸组装的重要中间产物的合成。

引起研究人员注意的是，涉及氨基酸降解（如胱硫醚、β-裂解酶和苏氨酸脱水酶）的反应在温度降低时显示出流量的增加。另外，4℃下，23 个与脂肪酸合成/代谢（7 个脂肪酸合成反应，16 个脂肪酸延长反应）相关的反应流量增加。在这些反应中发现：①乙酰 CoA 连接酶，为脂肪酸合成提供丙二酰 CoA 底物；②丙二酰 CoA-酰基载体蛋白酰基转移酶（malonyl CoA-acyl carrier protein transacylase），是将丙二酰单酰基转移到全酰基载体蛋白（holo-acyl carrier protein，ACP）的关键酶，从而在细菌的Ⅱ类脂肪酸合成（FAS Ⅱ）的起始步骤中，合成中间产物丙二酰-ACP（Ruch and Vagelos 1973）。根据模型计算结果显示，似乎是由于温度下降涉及的核心代谢通路的活性降低，基因表达出现细菌的代谢重编程。在这种特定的生理状况下，细菌的整体代谢也出现重编程以减少与氨基酸和核苷酸生物合成相关的消耗，但氨基酸降解及与脂肪酸相关的代谢过程的重要性则有所提升。

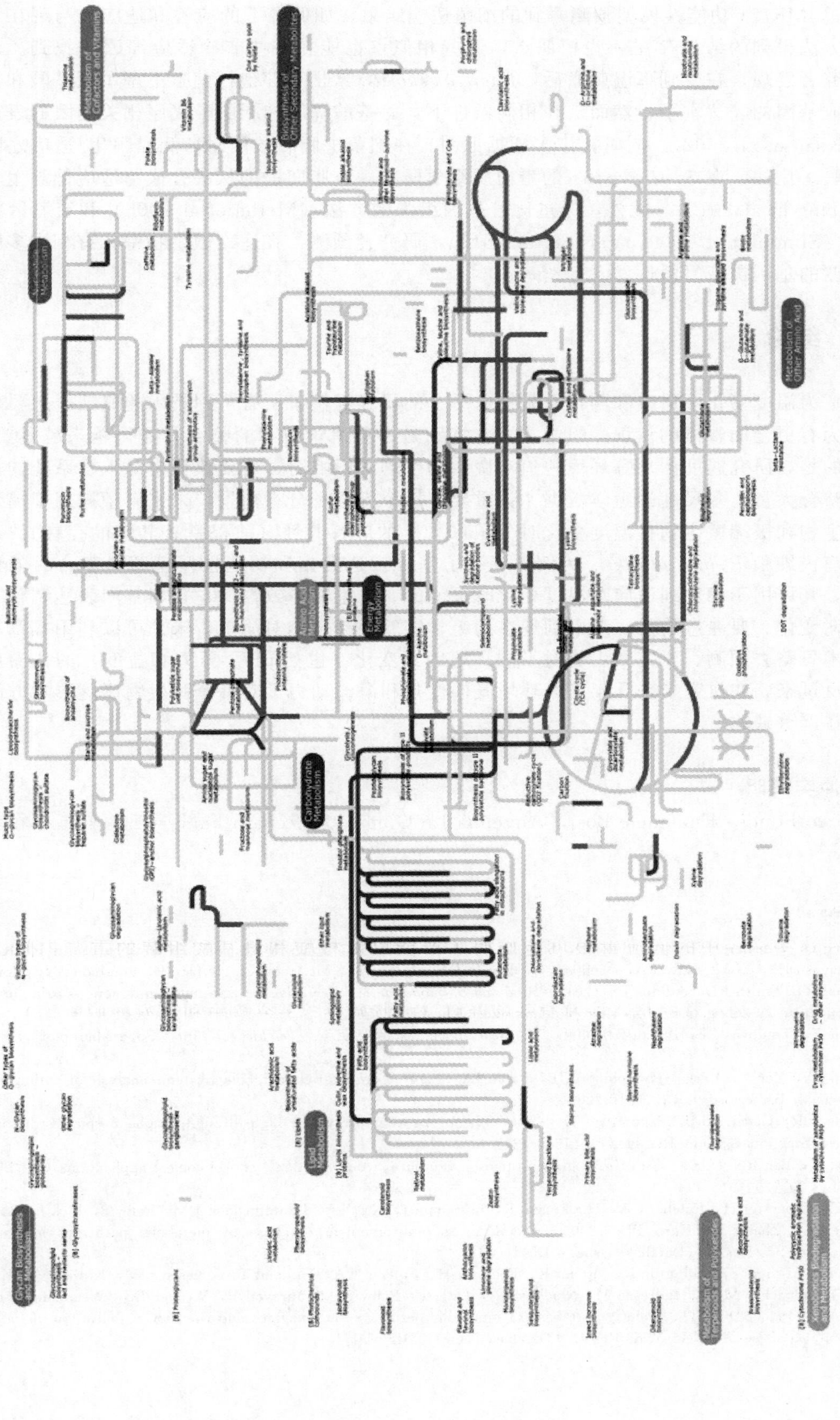

图 21.3　代谢示意图(见彩图):温度下降后 *Ph*TAC125 代谢流量的分配出现变化。红色和蓝色线分别表示反应流量的下降和增加(起码有因素 2),灰色线表示反应的流量值无明显改变

通过 iMF721 功能性模型预测得到的细菌适应结果，如低温下的应答和适应，与现在通过组学方法得到的结果存在一些相似点，这些相似点也能在一些非冷适应细菌中找到。例如，研究者发现，在受到降温刺激后，*Bacillus subtilis*（枯草芽孢杆菌）下调了氨基酸和核苷酸合成基因的表达水平，然而，在相同条件下，氨基酸降解和脂肪酸代谢相关基因的表达增加（Kaan et al. 2002）。同时，许多证据证明，在细菌中脂肪酸代谢与低温驯化/适应之间存在关联。比如，在 *S. alaskensis* 的蛋白质组学研究中，脂肪酸的从头合成与适应低温生长有关（Ting et al. 2010）。此外，通过比较基因组学的方法（Methe et al. 2005）和宏基因组学方法（Simon et al. 2009；Varin et al. 2012），研究者预测，在适冷微生物中，有相当多脂肪酸代谢的基因被恢复活性。

21.5 结论

本章为温度变化下适应和驯化在系统水平上的研究提供了所有可用技术的概括。想要对细胞行为有更全面整体的认识，组学技术（特别是多种组学技术的整合）显得越来越重要。要强调的是，研究温度休克等环境胁迫对微生物生理的影响时，若要获得更深入、系统的认识，则整合大量、多尺度的组学数据尤为重要。这个方法十分有前景，未来有望被用于揭示中温微生物和极端微生物在温度变化时分子水平上的应答共性（前提是有共性的存在）。最后，以南极细菌 *P. haloplanktis* 为例，展示了基因表达数据和功能性模型两者整合后的预测潜能，其可用于确定细菌应对温度变化时可能的、具生物一致性的，且与基因表达调整有关的代谢变化。要注意的是，考虑到现在的组学实验做起来相对简单，未来可以利用本章所述的技术思路去预测、研究微生物对不同环境因素变化（包括 pH、压力和盐度）的代谢反应。但反过来，在研究未来几年在大环境气候变化和海洋动力学对整个微生物群落影响方面可能存在严重缺陷。

无利益冲突声明

Marco Fondi，Emanuele Bosi，Angelina Lo Giudice 和 Renato Fani 声明他们不存在利益冲突。

参考文献

Ansong C，Deatherage BL，Hyduke D，Schmidt B，McDermott JE，Jones MB，Chauhan S，Charusanti P，Kim YM，Nakayasu ES，Li J，Kidwai A，Niemann G，Brown RN，Metz TO，McAteer K，Heffron F，Peterson SN，Motin V，Palsson BO，Smith RD，Adkins JN（2013）Studying *Salmonellae* and *Yersiniae* host-pathogen interactions using integrated 'omics and modeling. Curr Top Microbiol Immunol 363：21-41

Arsene F，Tomoyasu T，Bukau B（2000）The heat shock response of *Escherichia coli*. Int J Food Microbiol 55（1-3）：3-9

Bartell JA，Yen P，Varga JJ，Goldberg JB，Papin JA（2014）Comparative metabolic systems analysis of pathogenic *Burkholderia* . J Bacteriol 196（2）：210-226

Bochner BR，Gadzinski P，Panomitros E（2001）Phenotype microarrays for high-throughput phenotypic testing and assay of gene function. Genome Res 11（7）：1246-1255

Bouvet V，Ben RN（2003）Antifreeze glycoproteins：structure，conformation，and biological applications. Cell Biochem Biophys 39（2）：133-144

Caspi R，Foerster H，Fulcher CA，Hopkinson R，Ingraham J，Kaipa P，Krummenacker M，Paley S，Pick J，Rhee SY，Tissier C，Zhang P，Karp PD（2006）MetaCyc：a multiorganism database of metabolic pathways and enzymes. Nucleic Acids Res 34（Database issue）：D511-D516

Caspi R，Altman T，Billington R，Dreher K，Foerster H，Fulcher CA，Holland TA，Keseler IM，Kothari A，Kubo A，Krummenacker M，Latendresse M，Mueller LA，Ong Q，Paley S，Subhraveti P，Weaver DS，Weerasinghe D，Zhang P，Karp PD（2014）The MetaCyc database of metabolic pathways and enzymes and the BioCyc collection of Pathway/Genome Databases. Nucleic Acids Res 42（Database issue）：D459-D471

Cavicchioli R，Charlton T，Ertan H，Mohd Omar S，Siddiqui KS，Williams TJ（2011）Biotechnological uses of enzymes from psychrophiles. Microbial Biotechnol 4（4）：449-460

Chromy BA，Choi MW，Murphy GA，Gonzales AD，Corzett CH，Chang BC，Fitch JP，McCutchen-Maloney SL（2005）Proteomic characterization of *Yersinia pestis* virulence. J Bacteriol 187（23）：8172-8180

Collins T，Roulling F，Piette F，Marx J-C，Feller G，Gerday C，D'Amico S（2008）Fundamentals of cold-adapted enzymes. In：Marx J-C，Gerday C（eds）Psychrophiles：from biodiversity to biotechnology. Springer，Berlin，pp 211-227

Corchero JL，Gasser B，Resina D，Smith W，Parrilli E，Vazquez F，Abasolo I，Giuliani M，Jantti J，Ferrer P，Saloheimo M，Mattanovich D，Schwartz S Jr，Tutino ML，Villaverde A（2013）Unconventional microbial systems for the cost-effi cient production of high-quality protein therapeutics. Biotechnol Adv 31（2）：140-153

Duilio A，Tutino ML，Marino G（2004）Recombinant protein production in Antarctic Gramnegative bacteria. Methods Mol Biol 267：225-237

Durot M，Le Fevre F，de Berardinis V，Kreimeyer A，Vallenet D，Combe C，Smidtas S，Salanoubat M，Weissenbach J，Schachter V（2008）Iterative reconstruction of a global metabolic model of *Acinetobacter baylyi* ADP1 using high-throughput growth phenotype and gene essentiality data. BMC Syst Biol 2：85

Fang K，Zhao H，Sun C，Lam CM，Chang S，Zhang K，Panda G，Godinho M，Martins dos Santos VA，Wang J（2011）Exploring the metabolic network of the epidemic pathogen *Burkholderia cenocepacia* J2315 via genome-scale reconstruction. BMC Syst Biol 5：83

Feller G，Gerday C（2003）Psychrophilic enzymes：hot topics in cold adaptation. Nat Rev Microbiol 1（3）：200-208

Fondi M，Lio P（2015）Multi -omics and metabolic modelling pipelines：challenges and tools for systems microbiology. Microbiol Res 171C：52-64

Fondi M，Maida I，Perrin E，Mellera A，Mocali S，Parrilli E，Tutino ML，Lio P，Fani R（2014）Genome-scale metabolic reconstruction and constraint-based modelling of the Antarctic bacterium *Pseudoalteromonas haloplanktis* TAC125. Environ Microbiol. doi：10. 1111/1462-2920. 12513

Gao H，Wang Y，Liu X，Yan T，Wu L，Alm E，Arkin A，Thompson DK，Zhou J（2004）Global transcriptome analysis of the heat shock response of *Shewanella oneidensis*. J Bacteriol 186（22）：7796-7803

Gao H，Yang ZK，Wu L，Thompson DK，Zhou J（2006）Global transcriptome analysis of the cold shock response of *Shewanella oneidensis* MR-1 and mutational analysis of its classical cold shock proteins. J Bacteriol 188（12）：4560-4569

Garcia-Descalzo L，Garcia-Lopez E，Alcazar A，Baquero F，Cid C（2014）Proteomic analysis of the adaptation to warming in the Antarctic bacteria *Shewanella frigidimarina*. Biochim Biophys Acta 1844（12）：2229-2240

Garnier M，Matamoros S，Chevret D，Pilet MF，Leroi F，Tresse O（2010）Adaptation to cold and proteomic responses of the psychrotrophic biopreservative *Lactococcus piscium* strain CNCM I-4031. Appl Environ Microbiol 76（24）：8011-8018

Gerday C，Aittaleb M，Arpigny JL，Baise E，Chessa JP，Garsoux G，Petrescu I，Feller G（1997）Psychrophilic enzymes：a thermodynamic challenge. Biochim Biophys Acta 1342（2）：119-131

Giuliani M，Parrilli E，Ferrer P，Baumann K，Marino G，Tutino ML（2011）Process optimization for recombinant protein production in the psychrophilic bacterium *Pseudoalteromonas haloplanktis*. Process Biochem J 46：953-959

Graumann P，Marahiel MA（1996）Some like it cold：response of microorganisms to cold shock. Arch Microbiol 166（5）：293-300

Griffi n TJ，Gygi SP，Ideker T，Rist B，Eng J，Hood L，Aebersold R（2002）Complementary profiling of gene expression at the transcriptome and proteome levels in *Saccharomyces cerevisiae*. Mol Cell Proteomics 1（4）：323-333

Hatzimanikatis V，Lee KH（1999）Dynamical analysis of gene networks requires both mRNA and protein expression information. Metab Eng 1（4）：275-281

Jensen PA，Papin JA（2011）Functional integration of a metabolic network model and expression data without arbitrary thresholding. Bioinformatics 27（4）：541-547

Jozefczuk S，Klie S，Catchpole G，Szymanski J，Cuadros-Inostroza A，Steinhauser D，Selbig J，Willmitzer L（2010）Metabolomic and transcriptomic stress response of Escherichia coli . Mol Syst Biol 6：364

Kaan T，Homuth G，Mader U，Bandow J，Schweder T（2002）Genome-wide transcriptional profiling of the *Bacillus subtilis* cold-shock response. Microbiology 148（Pt 11）：3441-3455

Kanehisa M（2002）The KEGG database. Novartis Found Symp 247：91-101；discussion 101-103，119-128，244-152

Kawahara H（2008）Cryoprotectants and ice-binding proteins. In：Margesin R，Schinner F，Marx J-C，Gerday C（eds）Psychrophiles：from biodiversity to biotechnology. Springer，Heidelberg，pp 229-246

Kawamoto J，Kurihara T，Kitagawa M，Kato I，Esaki N（2007）Proteomic studies of an Antarctic cold-adapted bacterium，*Shewanella livingstonensis* Ac10，for global identifi cation of cold- inducible proteins. Extremophiles 11（6）：819-826

Kawamura D，Yamashita I，Nimi O，Toh-e A（1994）Cloning and nucleotide sequence of a gene conferring ability to grow at a low temperature on *Saccharomyces cerevisiae* tryptophan auxotrophs. J Biosci Bioeng 77（1）：1-9

Klein W，Weber MH，Marahiel MA（1999）Cold shock response of Bacillus subtilis：isoleucine- dependent switch in the fatty acid branching pattern for membrane adaptation to low tempera- tures. J Bacteriol 181（17）：5341-5349

Konkel ME，Tilly K（2000）Temperature-regulated expression of bacterial virulence genes. Microbes Infect 2（2）：157-166

Li JS，Bi YT，Dong C，Yang JF，Liang WD（2011）Transcriptome analysis of adaptive heat shock response of *Streptococcus thermophilus*. PLoS One 6（10）：e25777

Liu S，Graham JE，Bigelow L，Morse PD，Wilkinson BJ（2002）Identifi cation of Listeria monocy- togenes genes expressed in response to growth at low temperature. Appl Environ Microbiol 68 (4): 1697-1705

Lopez-Maury L, Marguerat S, Bahler J (2008) Tuning gene expression to changing environments: from rapid responses to evolutionary adaptation. Nat Rev Genet 9 (8): 583-593

Mavromatis K, Tsigos I, Tzanodaskalaki M, Kokkinidis M, Bouriotis V (2002) Exploring the role of a glycine cluster in cold adaptation of an alkaline phosphatase. Eur J Biochem 269 (9): 2330-2335

Medigue C, Krin E, Pascal G, Barbe V, Bernsel A, Bertin PN, Cheung F, Cruveiller S, D'Amico S, Duilio A, Fang G, Feller G, Ho C, Mangenot S, Marino G, Nilsson J, Parrilli E, Rocha EP, Rouy Z, Sekowska A, Tutino ML, Vallenet D, von Heijne G, Danchin A (2005) Coping with cold: the genome of the versatile marine Antarctica bacterium *Pseudoalteromonas haloplanktis* TAC125. Genome Res 15 (10): 1325-1335

Methe BA, Nelson KE, Deming JW, Momen B, Melamud E, Zhang X, Moult J, Madupu R, Nelson WC, Dodson RJ, Brinkac LM, Daugherty SC, Durkin AS, DeBoy RT, Kolonay JF, Sullivan SA, Zhou L, Davidsen TM, Wu M, Huston AL, Lewis M, Weaver B, Weidman JF, Khouri H, Utterback TR, Feldblyum TV, Fraser CM (2005) The psychrophilic lifestyle as revealed by the genome sequence of *Colwellia psychrerythraea* 34H through genomic and proteomic analyses. Proc Natl Acad Sci U S A 102 (31): 10913-10918

Miyake R, Kawamoto J, Wei YL, Kitagawa M, Kato I, Kurihara T, Esaki N (2007) Construction of a low-temperature protein expression system using a cold-adapted bacterium, *Shewanella* sp. strain Ac10, as the host. Appl Environ Microbiol 73 (15): 4849-4856

Morgan-Kiss RM, Priscu JC, Pocock T, Gudynaite-Savitch L, Huner NP (2006) Adaptation and acclimation of photosynthetic microorganisms to permanently cold environments. Microbiol Mol Biol Rev 70 (1): 222-252

Morimoto RI, Kline MP, Bimston DN, Cotto JJ (1997) The heat-shock response: regulation and function of heat-shock proteins and molecular chaperones. Essays Biochem 32: 17-29

Morita MT, Tanaka Y, Kodama TS, Kyogoku Y, Yanagi H, Yura T (1999) Translational induction of heat shock transcription factor sigma32: evidence for a built-in RNA thermosensor. Genes Dev 13 (6): 655-665

Navid A, Almaas E (2012) Genome-level transcription data of *Yersinia pestis* analyzed with a new metabolic constraint-based approach. BMC Syst Biol 6: 150

Oberhardt MA, Chavali AK, Papin JA (2009) Flux balance analysis: interrogating genome-scale metabolic networks. Methods Mol Biol 500: 61-80

Overbeek R, Olson R, Pusch GD, Olsen GJ, Davis JJ, Disz T, Edwards RA, Gerdes S, Parrello B, Shukla M, Vonstein V, Wattam AR, Xia F, Stevens R (2014a) The SEED and the rapid annotation of microbial genomes using Subsystems Technology (RAST). Nucleic Acids Res 42 (Database issue): D206-D214

Overbeek R, Olson R, Pusch GD, Olsen GJ, Davis JJ, Disz T, Edwards RA, Gerdes S, Parrello B, Shukla M, Vonstein V, Wattam AR, Xia F, Stevens R (2014b) The SEED and the rapid annotation of microbial genomes using subsystems technology (RAST). Nucleic Acids Res 42 (Database issue): D206-D214

Papa R, Rippa V, Sannia G, Marino G, Duilio A (2007) An effective cold inducible expression system developed in Pseudoalteromonas haloplanktis TAC125. J Biotechnol 127 (2): 199-210

Pearce DA (2012) Extremophiles in Antarctica: life at low temperatures. Springer, Dordrecht

Piette F, D'Amico S, Struvay C, Mazzucchelli G, Renaut J, Tutino ML, Danchin A, Leprince P, Feller G (2010) Proteomics of life at low temperatures: trigger factor is the primary chaperone in the Antarctic bacterium Pseudoalteromonas haloplanktis TAC125. Mol Microbiol 76 (1): 120-132

Piette F, D'Amico S, Mazzucchelli G, Danchin A, Leprince P, Feller G (2011) Life in the cold: a proteomic study of cold-repressed proteins in the antarctic bacterium *Pseudoalteromonas haloplanktis* TAC125. Appl Environ Microbiol 77 (11): 3881-3883

Ratheesh RK, Nagarajan SN, Arunraj PA, Sinha D, Veedin Rajan VB, Esthaki VK, D'Silva P (2012) HSPIR: a manually annotated heat shock protein information resource. Bioinformatics 28 (21): 2853-2855

Reed JL, Vo TD, Schilling CH, Palsson BO (2003) An expanded genome-scale model of *Escherichia coli* K-12 (iJR904 GSM/GPR). Genome Biol 4 (9): R54

Rippa V, Papa R, Giuliani M, Pezzella C, Parrilli E, Tutino ML, Marino G, Duilio A (2012) Regulated recombinant protein production in the Antarctic bacterium *Pseudoalteromonas haloplanktis* TAC125. Methods Mol Biol 824: 203-218

Ruch FE, Vagelos PR (1973) The isolation and general properties of Escherichia coli malonyl coenzyme A-acyl carrier protein transacylase. J Biol Chem 248 (23): 8086-8094

Russell NJ (1990) Cold adaptation of microorganisms. Philos Trans R Soc Lond B Biol Sci 326 (1237): 595-608

Schatschneider S, Persicke M, Watt SA, Hublik G, Puhler A, Niehaus K, Vorholter FJ (2013) Establishment, in silico analysis, and experimental verifi cation of a large-scale metabolic network of the xanthan producing *Xanthomonas campestris* pv. campestris strain B100. J Biotechnol 167 (2): 123-134

Schaub J, Clemens C, Kaufmann H, Schulz TW (2012) Advancing biopharmaceutical process development by system-level data analysis and integration of omics data. Adv Biochem Eng Biotechnol 127: 133-163

Scheer M, Grote A, Chang A, Schomburg I, Munaretto C, Rother M, Sohngen C, Stelzer M, Thiele J, Schomburg D (2011) BRENDA, the enzyme information system in 2011. Nucleic Acids Res 39 (Database issue): D670-D676

Simon C, Wiezer A, Strittmatter AW, Daniel R (2009) Phylogenetic diversity and metabolic potential revealed in a glacier ice metagenome. Appl Environ Microbiol 75 (23): 7519-7526

Sterner R, Liebl W (2001) Thermophilic adaptation of proteins. Crit Rev Biochem Mol Biol 36 (1): 39-106

Tanghe A, Van Dijck P, Thevelein JM (2003) Determinants of freeze tolerance in microorganisms, physiological importance, and biotechnological applications. Adv Appl Microbiol 53: 129-176

Thiele I, Palsson BO (2010) A protocol for generating a high-quality genome-scale metabolic reconstruction. Nat Protoc 5 (1): 93-121

Ting L, Williams TJ, Cowley MJ, Lauro FM, Guilhaus M, Raftery MJ, Cavicchioli R (2010) Cold adaptation in the marine bacterium, *Sphingopyxis alaskensis*, assessed using quantitative proteomics. Environ Microbiol 12 (10): 2658-2676

Tomar N, De RK (2013) Comparing methods for metabolic network analysis and an application to metabolic engineering. Gene 521 (1): 1-14

Tong W, Chen Z, Cao Z, Wang Q, Zhang J, Bai X, Wang R, Liu S (2013) Robustness analysis of a constraint-based metabolic model links cell growth and proteomics of *Thermoanaerobacter tengcongensis* under temperature perturbation. Mol Biosyst 9 (4): 713-722

Topfer N, Jozefczuk S, Nikoloski Z (2012) Integration of time-resolved transcriptomics data with flux-based methods reveals stress-induced metabolic adaptation in *Escherichia coli*. BMC Syst Biol 6: 148

Varin T, Lovejoy C, Jungblut AD, Vincent WF, Corbeil J (2012) Metagenomic analysis of stress genes in microbial mat communities from Antarctica and the High Arctic. Appl Environ Microbiol 78 (2): 549-559

Vieille C, Zeikus GJ (2001) Hyperthermophilic enzymes: sources, uses, and molecular mechanisms for thermostability. Microbiol Mol Biol Rev 65 (1): 1-43

Wang ZX, Zhou XZ, Meng HM, Liu YJ, Zhou Q, Huang B (2014) Comparative transcriptomic analysis of the heat stress response in the fi lamentous fungus *Metarhizium anisopliae* using RNA-Seq. Appl Microbiol Biotechnol 98 (12): 5589-5597

Weber MH, Fricke I, Doll N, Marahiel MA (2002) CSDBase: an interactive database for cold shock domain-containing proteins and the bacterial cold shock response. Nucleic Acids Res 30 (1): 375-378

Wiench B, Chen YR, Paulsen M, Hamm R, Schroder S, Yang NS, Efferth T (2013) Integration of different "-omics" technologies identifies inhibition of the IGF1R-Akt-mTOR signaling cascade involved in the cytotoxic effect of shikonin against leukemia cells. Evid Based Complement Alternat Med 2013: 818709

Wilmes B, Hartung A, Lalk M, Liebeke M, Schweder T, Neubauer P (2010) Fed-batch process for the psychrotolerant marine bacterium *Pseudoalteromonas haloplanktis*. Microb Cell Fact 9: 72

Wilmes B, Kock H, Glagla S, Albrecht D, Voigt B, Markert S, Gardebrecht A, Bode R, Danchin A, Feller G, Hecker M, Schweder T (2011) Cytoplasmic and periplasmic proteomic signatures of exponentially growing cells of the psychrophilic bacterium Pseudoalteromonas haloplanktis TAC125. Appl Environ Microbiol 77 (4): 1276-1283

Zhang W, Li F, Nie L (2010) Integrating multiple 'omics' analysis for microbial biology: application and methodologies. Microbiology 156 (Pt 2): 287-301

Zhu H, Ren X, Wang J, Song Z, Shi M, Qiao J, Tian X, Liu J, Chen L, Zhang W (2013) Integrated OMICS guided engineering of biofuel butanol-tolerance in photosynthetic *Synechocystis* sp. PCC 6803. Biotechnol Biofuels 6 (1): 106

第二十二章
极端微生物的实验微生物进化

Paul Blum[1,2], Deepak Rudrappa[1], Raghuveer Singh[1], Samuel McCarthy[1], Benjamin Pavlik[3]

22.1 实验微生物进化理论及在模式生物中的应用

实验微生物进化（EME）是研究经人为因素干扰多代后微生物种群的进化关系（Kussell 2013）。实验室适应进化（ALE）是选择在实验室施加条件下存活的细菌（Bennett and Hughes 2009；Dragosits and Mattanovich 2013）。实验进化研究侧重于遗传漂变和自然突变率在进化中所作的贡献。因此，在非选择性条件下进行实验进化研究，以避免由选择所导致的变化对实验的干扰（Hindré et al. 2012）。

为了了解实验进化方法在极端微生物中的应用，必须考虑近十年来使用的非极端模式微生物的增长情况。实验微生物进化领域的发展反映了实验微生物进化在验证进化假说方面起的重要作用，尤其是在分析实验操作导致的不同代之间的遗传变化。自然界中很多重要的过程都由微生物推动，因此了解和探索微生物群落是如何通过自然选择产生的是当前重要的课题之一。然而，目前很多微生物生态的研究主要是通过单物种的研究来揭示微生物生态的结构多样性及物种丰度（Prosser et al. 2007）。因此，了解物种这一概念是很有必要的。目前对物种概念普遍认可的遗传学定义是指能够杂种繁殖产生可育后代的个体的组合，不同物种之间存在生殖隔离（Prosser et al. 2007）。另一种生态物种的定义是具有所有相同的相关生态特征的个体的集合（Cohan 2001），由于微生物种群进化在物种丰度以及进化深度上的复杂性，这一生态物种的定义显得尤为重要。Cohan 推测“在没有其他生态位物种干扰下，细菌生态位的分离以及周期选择将降低同一生态位中的遗传多样性。”细菌、古菌包括极端微生物在内的基因组包含了核心基因组及辅助基因组两个部分，因此研究微生物基因的交流机制十分重要（Cohan 2001）。核心基因组包含了生物体功能所必需的全部基因，而辅助基因组则由能通过功能获得或丧失适应环境变化的相关基因组成。同一物种的菌株可以在辅助基

---

[1] School of Biological Science, University of Nebraska, Lincoln, NE, USA.

[2] e-mail: pblum1@unl.edu.

[3] Department of Chemical and Biomolecular Engineering, University of Nebraska, Lincoln, NE, USA.

因的组成上有所不同，因此它们能够适应不断变化的生态系统（Cohan 2001；Tettelin et al. 2005；Gill et al. 2005）。生态多样性还存在于质粒、转座子和致病性岛中，这些辅助遗传物质可以在有利环境下进行传递，但在其他的生态环境下发现的相同物种则不存在这一现象（Wertz et al. 2003）。这种现象是研究 ALE 和群落微生物生态的主要挑战，因此需要建立一个将微生物群落的流体性质与其生态学联系起来的合适理论（Wertz et al. 2003；Coleman et al. 2006）。了解种群基因频率的性质与作用是种群与进化遗传学的研究重点（Prosser et al. 2007），也是理解实验进化的关键。Tatum 和 Lederberg（1947）在大肠杆菌中发现横向转移的基因。而对 *Helicobacter pylori*（幽门螺杆菌）的两株不同细菌进行全基因组测序发现超过 6%的基因是不相同的（Alm et al. 1999）。当重组罕见并且只局限于少数基因时，大多数其他基因将会通过垂直遗传传递。突变在此期间内积累很慢，但将导致不可逆的子代分离。当重组事件较为普及的时候，基因会从基因组中释放，与自适应等位基因相关的基因的多样性将通过选择作用被清除（Polz et al. 2006）。这种混交群体结构在 *Neisseria gonorrhoeae*（奈瑟氏菌属细菌）和 *Rhizobium meliloti* lineages（根瘤菌细菌）中均有报道（Smith et al. 1993）。

实验微生物进化是通过与原始菌株比较进而量化进化细菌的基因型和表型改变。例如，特定基因或全基因组 DNA 测序技术就被用于确定进化菌株的遗传变异（Kussell 2013）。ALE 是 EME 研究中普遍使用的方法，可用于深入研究分子进化的基本机理以及长时间干预条件下微生物种群积累的适应性变化（Bennett et al. 1990）。在过去的二三十年间，大量的实验用于研究微生物 ALE 的适应性变化，尤其是对 *E. coli* 和酿酒酵母（*Saccharomyces cerevisiae*）的研究（Bennett et al. 1990；Paquin and Adams 1983）。微生物对新环境的适应主要通过两种不同的机制：没有遗传物质改变的基因选择性调控或者发生能产生新性状的稳定突变。实验进化允许表型随实验施加的生长条件的变化而变化（Hardison 2003）。EME 研究为进化生物学提供了重要见解和实验依据（Bennett et al. 1990；Paquin and Adams 1983）。Lenski 课题组在大肠杆菌 EME 研究方面长期保持领先地位，单组平行大肠杆菌适应实验已经超过了 50000 代（Sniegowski et al. 1997；Lenski et al. 1998；Cooper and Lenski 2000）。这些研究及其他类似的实验在提高微生物遗传基础（Barrick et al. 2009）、进化过程中的意义（Woods et al. 2011）、对种群规模的理解、可进化性（Bloom et al. 2007；Elena et al. 2007；Draghi et al. 2010）以及克隆干扰（Kao and Sherlock 2008）方面提供了有用的见解。

实验进化允许微生物通过繁殖进化，依靠微生物的遗传能力进而引入突变（Sauer 2001）。繁殖快、易维持种群规模、易储存等优点使得微生物十分适用于实验室 ALE 研究（Elena and Lenski 2003）。然而，在有限的自然突变率下，选择预期的表型需要较长的培养周期。此外，由于中性突变的积累，识别导致特定表型变化所必需的突变存在一定困难，因此利用“组学”方法来研究表型与突变的关系是必要的（Bro and Nielsen 2004）。此外，与基于反向遗传学的方法相比，实验进化更具优势，反向遗传学方法是有针对性地激活或失活靶基因，而 ALE 可以导致意料外的组合突变，使得有机体功能的增加（Conrad et al. 2011）。随着人们对一些模式生物的生理生化特性的日益了解，以及在二十世纪八九十年代开发的基因和绘图技术的使用，为实验进化的研究提供一种新途径（Helling et al. 1987；Rosenzweig et al. 1994；Treves et al. 1998；Kinnersley et al. 2014；Ferenci 2007；Adams et al. 1992）。这些相同的方法被用于研究实验进化及自然产生的微生物，包括其基因组内基因或序列的扩增、缺失、插入和重排等（Cooper et al. 2003；Philippe et al. 2007；Kadam et al. 2008；Bachmann et al. 2012；Gresham et al. 2008；Wenger et al. 2011）。

22.2 实验微生物进化的相关技术

对于极端微生物和非极端微生物而言，培养方法是最基础的技术条件之一，因为培养条件决定生理状态进而导致突变形成。为此，细菌培养是实验进化研究最重要的环节。在选择环境下延长微生物的生长时间也是必要的。在这种条件下，微生物作为进化模型是有利的，因为微生物具有种群数量较大、繁殖快、基因组相对较小和突变的供应量普遍较高的优点（Gresham and Dunham 2014）。极端微生物可能表现出较长的繁殖周期但仍然保持正常的突变率（Grogan et al. 2001），这就需要花更长的培养时间来获得相同数量的突变。对连续培养和连续批量培养的认识和应用进一步扩大了微生物系统的使用，进而解决进化问题。尽管实验进化至关重要，但目前这些培养技术在极端微生物 ALE 研究中使用并不广泛，因此，发展极端微生物培养技术的研究是很有必要的。

利用连续长时间培养技术获取突变菌群和应变演化的研究在二十世纪中期很流行。恒化器是首次由 Monod（1942）提出的连续培养装置。Novick 和 Szilard 发明了现代恒化器，并且用于微生物遗传变异的研究（Novick and Szilard 1950）。这种设备的主要原理是持续移除培养基并用新鲜培养基替代，微生物种群增长率与培养基的替换频率成正比。该设备的现代改良版已用于实现更多数量的实验重复以及用于单细胞的分析（Dénervaud et al. 2013）。恒化器的使用是依据微生物对环境胁迫的生理响应。这些研究包括：种间和种内变异、突变率、突变接受、种群变化、群体感应、基因组重排、多样性分析、代谢和能量以及膜运输等（Ferenci 2008）。恒化器为极端微生物实验进化的研究提供方法手段，目前主要应用于生物进程的研究（Lorantfy et al. 2014）。

与恒化器类似的是恒浊器，由 Bryson 和 Szybalski（1952）推广应用于微生物进化研究。主要的区别是营养素不受限制，新鲜培养基的引入与最大培养浊度成正比，使得微生物生长保持最大生长速率（Gresham and Dunham 2014）。恒化器与恒浊器维持一个相对稳定的环境，但分别代表最小和最佳资源利用条件。这些条件的不同导致基于养分胁迫或生长增强和整体适应改进的不同突变类型。恒浊器目前已经应用于微生物对抗菌物质的进化应答方面（Avrahami-Moyal et al. 2012；Toprak et al. 2012）。与恒化器相同，恒浊器在极端微生物进化的应用中同样受到严重的限制。

用于微生物进化的长期批量培养技术需要在一个已知稀释梯度和持续时间下连续稀释培养（Atwood et al. 1951）。这种技术为微生物的生长提供了动态环境，条件一致性的要求也比连续培养要低。在这种培养条件下，细菌在其生长的各个阶段都可以被选择。而且，在连续培养装置的情况下，这里引入了一个非连续选择的复杂环境。这种方法最近被应用于极端微生物 *Sulfolobus solfataricus*（硫化叶菌）的实验进化研究中。在不知道连续培养过程中营养变化对生理反应影响的情况下，这是一种慎重的方法（McCarthy 2015）。

22.2.1 极端微生物基因型和表型进化的高通量方法

随着“组学”技术的发展，微生物基因组进化分析的通量及分辨率均得到极大的提升。无论微生物的种类如何，都必须考虑该技术手段。高通量技术如全基因组测序、转录组学、基因芯片和染色质免疫沉淀（ChIP-seq）可用于研究菌落特征（性状）与其进化分子机制之间的关系。高通量技术已经成为 EME 研究不可或缺的技术。DNA 高通量测序技术的发展极大地发掘极端微生物在实验微生物进化领域的潜在研究价值（Araya et al. 2010；Barrick et al. 2009；Hong and Gresham 2014；Kvitek and Sherlock 2013；Lang et al. 2013；Herring et al. 2006)。从第一个古菌 *Methanocaldococcus jannaschii*（詹氏甲烷球菌）全基因组测序（Bult et al. 1996）到现在，许多极端微生物全基因组信息被发掘（Allers and Mevarech 2005）。基因表达数据也可以通过 RNA 测序（RNA-Seq）或微阵列的高通量形式得到（DeRisi et al. 1997；Lashkari et al. 1997；Dunham et al. 2002；Gresham et al. 2006；Wurtzel et al. 2010；Rigaut et al. 1999）。此外，蛋白质-蛋白质相互作用（Schwikowski et al. 2000）和蛋白质-DNA 相互作用也用于 EME 模式生物及极端微生物的研究中（Gresham et al. 2006；Dunham et al. 2002；Wilbanks et al. 2012）。

22.3 极端微生物的实验进化

22.3.1 极端微生物的特殊挑战

实验进化要在极端微生物中应用则必须克服与培养条件或极端环境下产生的相关分子的生物物理方面的困难。实验室极端微生物培养需要特殊的技术和设备，以维持细菌所需的极端环境和复杂的营养需求。与其他微生物模式相比，许多极端微生物生长速率较慢。这意味着要得到相同的世代数，极端微生物 ALE 研究需要更长的时间。例如，在能量受限影响复制率的情况下，培养的自养极端微生物尤其如此（McCarthy et al. 2014；Maezato et al. 2012）。

在大分子存在的条件下，EME 影响了它们的鉴定，蛋白质构成特定的壁垒，而脂质和各种生物小分子也会影响实验结果。来源于嗜热微生物的蛋白质需要在高温条件下进行分析，因此需要使用专业仪器设备。来源于嗜盐极端微生物的蛋白质难以纯化和鉴定，这是因为在低盐浓度下这类蛋白质不稳定，而高盐浓度的电泳、离子交换法又与它们不相容，导致其分离受到阻碍（Anfinsen et al. 1995）。在非极端微生物宿主如大肠杆菌体内重组表达极端微生物蛋白质面临着相似的问题：脱离其天然的极端条件，嗜盐蛋白质在外部环境容易发生错误折叠和聚集（Allers 2010）。由于这些原因，研究极端生物大分子在自然极端宿主中的生物性能是目前研究的关键，同时，这也取决于遗传系统的有效性（Maezato et al. 2012）。

22.3.2 极端微生物的培养和菌种保存

分离自意大利卡诺岛（Vulcano）地热海洋沉积物的极端嗜热厌氧细菌海栖热袍菌（*Thermotoga maritima*）（Huber and Stetter 1998）是典型的极端生活方式的例子。尽管大多数 Thermotogales 细菌来自热泉和深海热液喷口，但由于这些生态系统中部分氧化的热沉积物和流体，很少有成员能够适应部分氧化条件（Rusch et al. 2005）。*T. maritima* 虽归类为严格厌氧菌，但其仍然能在体积分数 0.5%氧气含量下生长（Le Fourn et al. 2008）。在海洋热液喷口的缺碳环境下，可能导致该物种广泛代谢多种糖类，没有明显的选择性。例如，在 55～90℃，pH 5.5～9.5 条件下均能使用简单和复杂糖类。该物种比大多数其他 *T. maritima* 及其同源细菌 *T. neapolitana*，*Thermotoga* sp. strain RQ2，*T. naphthophila* 和 *T. petrophila* 拥有明显更高数量的 ABC 型的底物转运体（Nelson et al. 1999；Ren et al. 2007；Chhabra et al. 2003；Conners et al. 2005）。*T. maritima* 可采用添加不同碳源的厌氧管或血清瓶批量培养。厌氧管可以用丁基橡胶塞密封（Bellco Biotechnology），金属圈卷曲，顶部空间可用氮气交换。培养基接种可以用 1mL 无菌注射器连接 20½G 针头置于 80℃下厌氧培养过夜。

T. maritima 存在明显的基因水平转移现象以及基因组持续缩减的趋势（Nelson et al. 1999；Singh et al. 2015）。此外，由于其产氢的速率超过其他微生物，有可能作为工程菌

用于工业生产中（Schröder et al. 1994）。正在进行的实验进化研究通过使用染色体重组的基因瞬时失活结合净化选择来扩大这种能力，从而建立了新的具有更强产氢能力的细胞系（Singh et al. 2015）。基因瞬时失活是指连续单交叉事件导致的染色体位点的暂时性基因中断。净化选择是一种经典的遗传方法，包括在分离克隆种群的同时保持选择性生长条件。

另一个极端生物的例子是分离自意大利和美国火山温泉中的极端嗜热嗜酸菌 *Sulfolobus solfataricus*（Brock et al. 1972）。该细菌生长的温度范围为 65～90℃，pH 值范围为 2.5～5.0，最佳生长条件为温度 80℃，pH 3.0（Brock et al. 1972；Grogan 1989）。该生物在实验室条件下使用改良的基础盐类培养基培养（Brock et al. 1972）：复合培养基中补充 0.2%蛋白胨，基本培养基中则多添加 0.2%葡萄糖。培养物在 80℃条件下带螺旋盖的玻璃瓶中温育并在轨道水浴箱中曝气培养（Rolfsmeier and Blum 1995；Bini et al. 2002；Worthington et al. 2003），吸收波长 540 nm 处测量细菌生长浓度。*S. solfataricus* 作为古菌模式生物，具有已建立的遗传体系（Maezato et al. 2011）。这使得它成为理想的 ALE 研究菌株，在实验中发现该菌的遗传变化可以在野生型细胞系中重建以验证其效应。*S. solfataricus* 的热稳定和酸稳定酶已经被应用在多种工业过程中（如海藻糖生产及纤维素降解）（Antranikian et al. 2005）。实验进化研究已经获得了该菌酸稳定性增强菌株（McCarthy 2015）。

极端微生物在冰冻状态下可以长时间保存，这对在 ALE 研究期间保持进化变化的中间阶段的不同微生物种群是至关重要的。为了得到长时间储存的细菌，细菌培养至指数增长期时通过离心收集，用含体积分数为 7%的二甲基亚砜（DMSO）新鲜培养基洗涤，然后乙醇-干冰冰浴条件下将样品混合并快速冷冻，－80℃长期存储（Maezato et al. 2012）。

22.4 极端微生物实验微生物进化实例

基因组变化是非极端微生物进化的基础。基因重复、缺失、易位、倒位、横向基因转移和转座是导致基因组进化的原因（Kunin and Ouzounis 2003；Suyama and Bork 2001；Garcia-Vallvé et al. 2000；Nelson et al. 1999；Kidwell and Lisch 2000；Fraser-Liggett 2005）。极端微生物基因组的进化可以以类似的方式发生。因为不同的遗传因素，同向重复序列（DR），反向重复序列（IR）和转座子均存在于极端微生物中。由于 DRs 和 IRs 具有潜在操纵基因组动力学的能力（Ussery et al. 2004），极端微生物能够通过这些序列导致微生物基因的进化。非嗜热菌的 DR 的起源是来自基因水平转移，碱基错配和基因组跳跃（Achaz et al. 2002；Romero and Palacios 1997）。目前已经对 Thermotogales 纲细菌水平基因转移事件进行了系统的研究，因此这些极端微生物可以用来研究基因组的进化。通过 RecA 独立重组可消除具有较小间隔区的紧密重复序列或串联重复序列（Bi and Liu 1994；Lovett et al. 1993），而依赖于选择压力的具有较长间隔的同向重复序列可以稳定地保持在基因组中，不太可能被非常规重组删除（Lovett et al. 1994；Chédin et al. 1994）。重复序列的生物学意义是提供两种表型变异的方式。位于基因调控区的重复序列可以调控基因的表达（van Ham et al. 1993），而当基因重排导致编码区内的重复丢失时可能导致蛋白质的截断。重复序列的保留也取决于选择性压力，这方面的基因组不稳定性可以通过实验测试。

转座因子通过突变、复制和基因组重排提供的遗传多样性通常被认为是有害的，只有少数是对宿主有利（Kidwell and Lisch 2000；Schneider and Lenski 2004）。由于转座因子广泛分布在极端微生物中，它们的活动又与微生物的进化变化密切相关。Lenski 等的大肠杆菌进化实验中，发现在进化的细菌中几个关键的变化与转座有关，但转座率与细菌的适应率无关：适应实验显示细菌适应率在早期快速增加，然后急剧减少，然而转座率保持相对稳定

(Barrick et al. 2009；Schneider and Lenski 2004)。在另一个大肠杆菌热适应 EME 实验中，适热菌株表型只有约一半可以由相似基因组区域的重复事件及重复基因的上调进行解释。然而，另一半的适热菌株则没有发生重复或转座事件，同时这些基因的表达也没有变化，表明这些适应性状是通过不同的且未知的途径产生 (Riehle et al. 2001)。这些研究可用于预测极端微生物类似的适应性进化机制。

22.4.1　极端嗜热厌氧菌 *Thermotoga maritima* 基因缺失的形成

T. maritima 是 Thermotogae 门中研究最为透彻的细菌之一。Thermotogae 门包含了大量种类的嗜热和厌氧极端微生物 (Huber and Stetter 1998)。有研究发现 *T. maritima* 基因组中存在古菌水平转移产生的基因 (Nelson et al. 1999)，这使得这株细菌成为研究实验基因组进化的模式细菌。同时存在 8 kb 的片段缺失支持着这株细菌的基因组进化 (Zhaxybayeva et al. 2009)。寻找同向重复序列 (DRs) 的生物信息学分析证实该菌中各种大的或小的 DRs 可能促进生物基因组的进化 (图 22.1)。为了确定可能有助于基因组进化的 DRs 的位置，使用 REPuter 扫描 *T. maritima* 基因组，发现了 50～1500bp 的 DRs 序列信息 (Kurtz et al. 2001)。Nelson 报道的 *T. maritima* 基因组被用于描述基因坐标和基因座标签 (Nelson et al. 1999)。在 *T. maritima* 中，一个可变间隔区被一组重复序列包围，各种 DRs 聚类产生较大的 DR 区域。TM1299 和 TM1332 之间的通过聚类形成的一个更大的 DR 被确定为一个潜在的基因进化点。*T. maritima* 的多种 DRs 的排列方式见图 22.2。最大的 DRs (921 bp) 存在于 TM1322 中 (1340942～1341862)，同时，TM1332 (1350970～1351890) 周围存在一个 10 kb 的间隔区。考虑到 DRs (921 bp) 发生在基因组中的位置，它们可以成为中间区域的缺失基因位点。这种情况在细菌基因组进化中广泛存在。一种用于产生基因缺失的实验进化方法已被证明可以在细菌基因组中缺失 8 kb 的片段 (Singh et al. 2015)。通过染色体重组的基因瞬时失活结合净化选择的方法建立一种新的具有独特性能的细胞系 (Singh)。基因组重测序发现 TM1322 和 TM1332 之间存在一个 10 kb 的基因缺失，该缺失是由侧翼 DRs (921 bp) 删除插入区域 (134186～1351890) 而产生的。这是 *T. maritima* 的实验微生物基因组进化的首次报道。*T. maritima* 的 10 kb 缺失菌株被命名为 Tma200，由于其产氢量的改变，可能为细菌极端特征的进一步进化提供了策略 (Singh et al. 2015)。

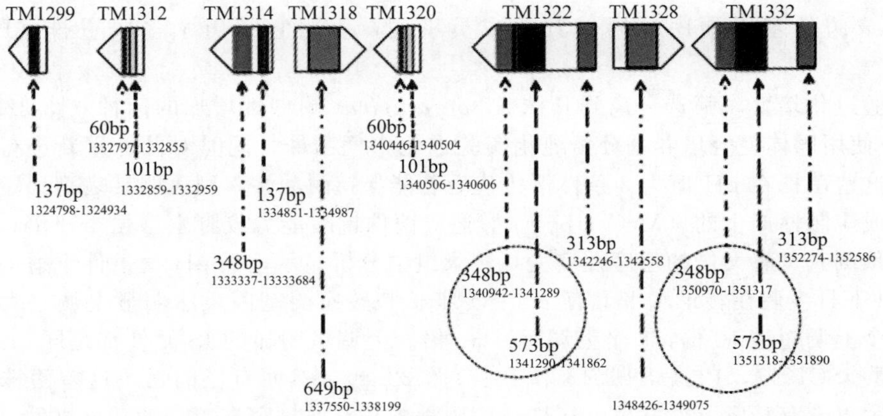

图 22.1　*T. maritima* 潜在的基因组进化位点 (见彩图)。特定的颜色块代表保守的同向重复序列长度。基因组坐标和基因座标记根据 *T. maritima* 基因组的报道 (Nelson 等 1999)

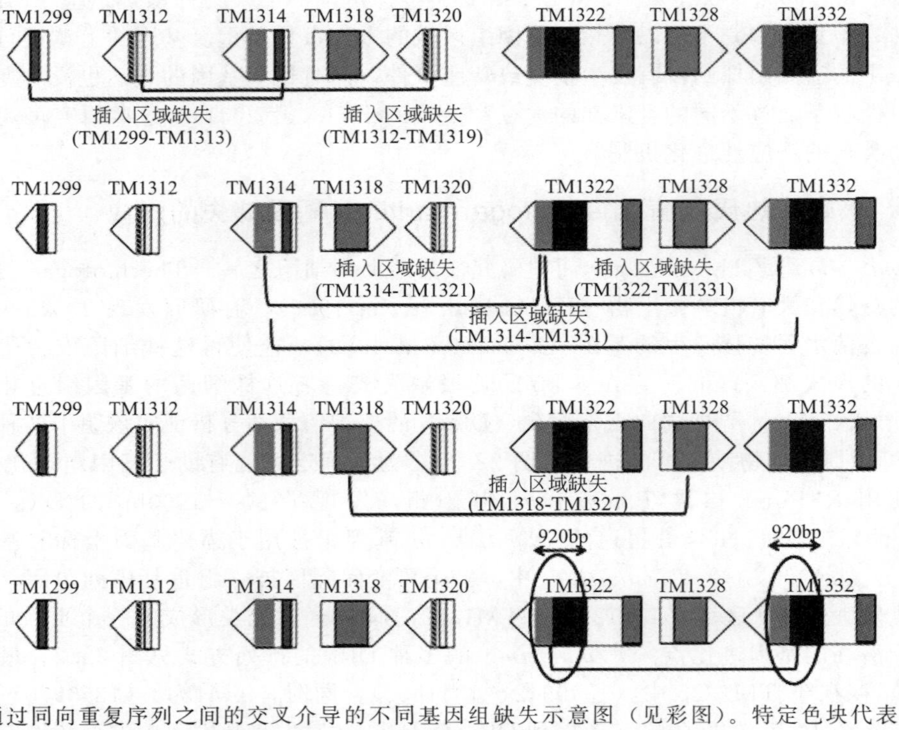

图 22.2 通过同向重复序列之间的交叉介导的不同基因组缺失示意图（见彩图）。特定色块代表保守同向重复序列的长度。相同的同向重复序列可以进行同源重组，图中用方括号表示

22.4.2 古菌 *Sulfolobus solfataricus* 硫化叶菌及其插入序列元件的作用

Sulfolobus solfataricus 是一种北极 Crenarchaeotal 门极端嗜热嗜酸古菌，目前正用于嗜热嗜酸进化研究且已经获得了一系列新的细胞系类型（McCarthy）。这里介绍了与 ALE 有关的谱系的各个方面。与该属其他物种不同，*S. solfataricus* 基因组富含插入序列片段（IS），超过 200 个的 IS 元件（占基因组 10%）以及由这些转座因子导致的重排使得其基因组不断发生变化（Redder and Garrett 2006；Brügger et al. 2004）。该菌经历了高频率的转座，同时该菌的基因顺序与同属其他成员不同，表明它经历了多次重排（Brügger et al. 2004）。

经实验进化培育，筛选分离到几株 *S. solfataricus* 耐酸性增加的菌种（图 22.3）（McCarthy）。使用固体复合培养基分离纯化实验进化中期菌株，同时对菌株重新进行耐酸性筛选验证。这些菌株在 pH 值为 3 条件下传代后若能保持耐酸性，则表明其表型稳定遗传而非瞬态响应或应激性产生的（McCarthy）。酸适应菌株抗酸能力较野生型强 150 倍，根据转录组学研究预测其具有大量的适应性突变。转录组学分析显示，低 pH 值条件下适应株与最适 pH 值条件下野生菌在转录水平上有 27 个转座子和转座酶基因表达明显上调（大于 5 倍）。其中有 8 个上调超过 10 倍，8 个上调超过 30 倍。上调最明显的 IS 元件都是同一家族成员，包括转座酶 ISC1217，1234/ST1916 和 IS1（图 22.4）。然而奇怪的是，这些菌株的基因组重测序显示很少有转座事件发生。三株适应菌种中只有两株存在单一的转座事件，且第三株则没有表现出转座。尽管许多转座酶基因的表达在低 pH 值下增加，然而在适应株中转座频率减少。在适应株中转座频率的降低表明转座不是引起适应性的原因。这也是在适应株基因组中存在着高频 IS 元件会令人惊讶的原因。

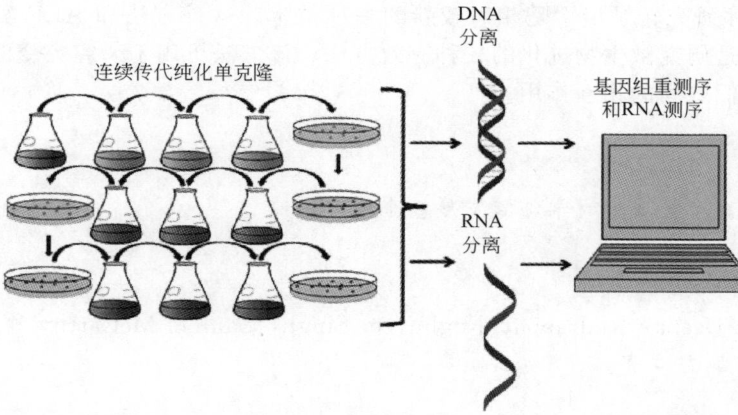

图 22.3 实验微生物进化的连续传代方法示意图。细胞传代培养在一个更极端的条件下（由介质变化的颜色表示），传代多次直到它们适应这个条件，然后进行条件调整和重复。单独适应性培养以及筛选具有独特表型的单克隆。对适应菌基因组和转录组进行检测以确定其适应原因

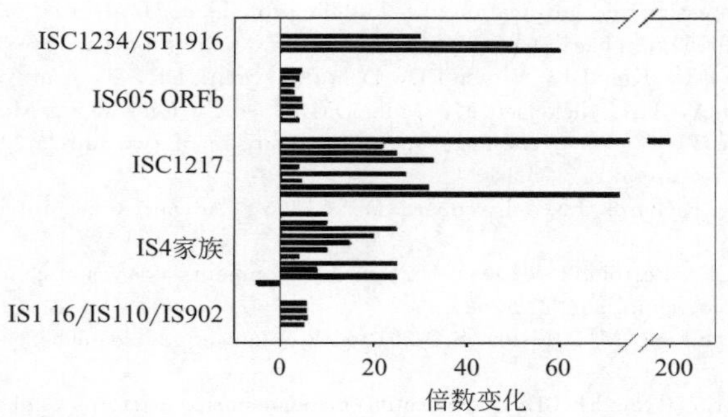

图 22.4 *S. solfataricus* 强酸适应菌株插入元件转录组分析。数据表示 pH 3.00 条件下强酸适应菌株与野生菌株转座酶基因表达差异分析。大多数插入元件在适应株中被上调，上调最明显的是 ISC1217 和 ISC1234/ST1916 家族转座元件

酸适应菌株的整体转录组分析给其耐酸机理提供了额外的线索。观察发现假定转运蛋白的表达、预计参与信号转导的基因以及整体代谢均发生较大的变化（图 22.5）。在原核生物

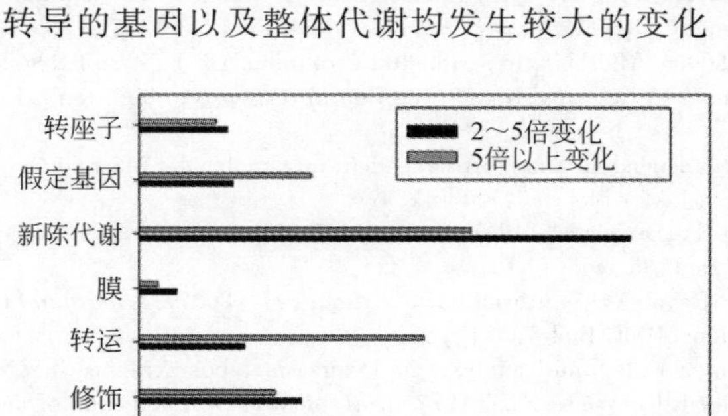

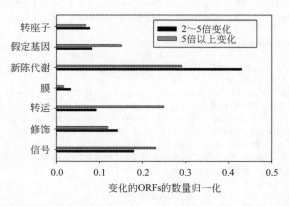

图 22.5 *S. solfataricus* 适应株基因表达变化的 ORFs 分类。在 pH 1 生长条件下，不同功能下有表达差异小的（2～5 倍）以及差异大的（>5 倍）的 ORFs 基因总数统计。倍数变化是相对于 pH 3 的野生型 *S. solfataricus*。所有的 RNA 测序样品生物复制的皮尔逊相关系数大于 0.96

中使用 RNA 测序研究转录组学是相对较新的一项技术。这种方法和 ALF 实验基因组学研究相结合，将会是研究微生物进化的一种有效方法，该方法可用于研究突变的作用、基因表达的变化以及它们之间的相互作用。

致谢

本章得到了内布拉斯加大学细胞发展基金的资助。

无利益冲突声明

Paul Blum，Deepak Rudrappa，Raghuveer Singh，Samuel McCarthy 和 Benjamin Pavlik 声明他们没有利益冲突。

参考文献

Achaz G，Rocha EP，Netter P，Coissac E（2002）Origin and fate of repeats in bacteria. Nucleic Acids Res 30：2987-2994

Adams J，Puskas-Rozsa S，Simlar J，Wilke CM（1992）Adaptation and major chromosomal changes in populations of *Saccharomyces cerevisiae*. Curr Genet 22：13-19

Allers T（2010）Overexpression and purification of halophilic proteins in *Haloferax volcanii*. Bioeng Bugs 1：288-290

AllersT，Mevarech M（2005）Archaeal genetics-the third way. Nat Rev Genet 6：58-73

Alm RA，Ling LS，Moir DT，King BL，Brown ED，Doig PC，Smith DR，Noonan B，Guild BC，deJonge BL，Carmel G，Tummino PJ，Caruso A，Uria-Nickelsen M，Mills DM，Ives C，Gibson R，Merberg D，Mills SD，Jiang Q，Taylor DE，Vovis GF，Trust TJ（1999）Genomic-sequence comparison of two unrelated isolates of the human gastric pathogen *Helicobacter pylori*. Nature 397：176-180

Anfinsen CB，Edsall JT，Richards FM，Eisenberg DS（1995）Advances in protein chemistry. Academic Press，San Diego

Antranikian G，Vorgias CE，Bertoldo C（2005）Extreme environments as a resource for microorganisms and novel biocatalysts. Adv Biochem Eng Biotechnol 96：219-262

Araya CL，Payen C，Dunham MJ，Fields S（2010）Whole-genome sequencing of a laboratoryevolved yeast strain. BMC Genomics 11：88

Atwood KC，Schneider LK，Ryan FJ（1951）Selective mechanisms in bacteria. Cold Spring Harb Symp Quant Biol 16：345-355

Avrahami-Moyal L，Engelberg D，Wenger JW，Sherlock G，Braun S（2012）Turbidostat culture of *Saccharomyces cerevisiae* W303-1A under selective pressure elicited by ethanol selects for mutations in SSD1 and UTH1. FEMS Yeast Res 12：521-533

Bachmann H，Starrenburg MJ，Molenaar D，Kleerebezem M，van Hylckama Vlieg JE（2012）Microbial domestication signatures of *Lactococcus lactis* can be reproduced by experimental evolution. Genome Res 22：115-124

Barrick JE，Yu DS，Yoon SH，Jeong H，Oh TK，Schneider D，Lenski RE，Kim JF（2009）Genome evolution and adaptation in a long-term experiment with *Escherichia coli*. Nature 461：1243-1247

Bennett AF，Hughes BS（2009）Microbial experimental evolution. Am J Physiol Regul Integr Comp Physiol 297：29

Bennett AF，Dao KM，Lenski RE（1990）Rapid evolution in response to high-temperature selection. Nature 346：79-81

Bi X，Liu LF（1994）recA-independent and recA-dependent intramolecular plasmid recombination. Differential homology requirement and distance effect. J Mol Biol 235：414-423

Bini E，Dikshit V，Dirksen K，Drozda M，Blum P（2002）Stability of mRNA in the hyperthermophilic archaeon *Sulfolobus solfataricus*. RNA 8：1129-1136

Bloom JD，Lu Z，Chen D，Raval A，Venturelli OS，Arnold FH（2007）Evolution favors protein mutational robustness in sufficiently large populations. BMC Biol 5：29

Bro C，Nielsen J（2004）Impact of 'ome' analyses on inverse metabolic engineering. Metab Eng 6：204-211

Brock TD，Brock KM，Belly RT，Weiss RL（1972）*Sulfolobus*：a new genus of sulfur-oxidizing bacteria living at low pH and high temperature. Arch Mikrobiol 84：54-68

Brügger K，Torarinsson E，Redder P，Chen L，Garrett RA（2004）Shuffling of *Sulfolobus* genomes by autonomous and non-autonomous mobile elements. Biochem Soc Trans 32：179-183

BrysonV，Szybalski W（1952）Microbial selection. Science 116：45-46

Bult CJ，White O，Olsen GJ，Zhou L，Fleischmann RD，Sutton GG，Blake JA，FitzGerald LM，Clayton RA，Gocayne JD，Kerlavage AR，Dougherty BA，Tomb JF，Adams MD，Reich CI，Overbeek R，Kirkness EF，Weinstock KG，Merrick JM，Glodek A，Scott JL，Geoghagen NS，Venter JC（1996）Complete genome sequence of the methanogenic archaeon，*Methanococcus jannaschii*. Science 273：1058-1073

Chédin F，Dervyn R，Ehrlich SD，Noirot P（1994）Frequency of deletion formation decreases exponentially with distance between short direct repeats. Mol Microbiol 12：561-569

Chhabra SR，Shockley KR，Conners SB，Scott KL，Wolfinger RD，Kelly RM（2003）Carbohydrateinduced differential gene expression patterns in the hyperthermophilic bacterium *Thermotoga maritima*. J Biol Chem 278：7540-7552

Cohan FM（2001）Bacterial species and speciation. Syst Biol 50：513-524

Coleman ML，Sullivan MB，Martiny AC，Steglich C，Barry K，Delong EF，Chisholm SW（2006）Genomic islands and the ecology and evolution of *Prochlorococcus*. Science 311：1768-1770

Conners SB，Montero CI，Comfort DA，Shockley KR，Johnson MR，Chhabra SR，Kelly RM（2005）An expression-driven approach to the prediction of carbohydrate transport and utilization regulons in the hyperthermophilic bacterium *Thermotoga maritima*. J Bacteriol 187：7267-7282

Conrad TM，Lewis NE，Palsson BØ（2011）Microbial laboratory evolution in the era of genomescale science. Mol Syst Biol 7：42

Cooper VS，Lenski RE（2000）The population genetics of ecological specialization in evolving *Escherichia coli* populations. Nature 407：736-739

Cooper TF，Rozen DE，Lenski RE（2003）Parallel changes in gene expression after 20，000 generations of evolution in *Escherichia coli*. Proc Natl Acad Sci USA 100：1072-1077

Dénervaud N，Becker J，Delgado-Gonzalo R，Damay P，Rajkumar AS，Unser M，Shore D，Naef F，Maerkl SJ（2013）A chemostat array enables the spatio-temporal analysis of the yeast proteome. Proc Natl Acad Sci USA 110：15842-15847

DeRisi JL，Iyer VR，Brown PO（1997）Exploring the metabolic and genetic control of gene expression on a genomic scale. Science 278：680-686

Draghi JA，Parsons TL，Wagner GP，Plotkin JB（2010）Mutational robustness can facilitate adaptation. Nature 463：353-355

Dragosits M，Mattanovich D（2013）Adaptive laboratory evolution-principles and applications for biotechnology. Microb Cell Fact 12：1475-2859

Dunham MJ，Badrane H，Ferea T，Adams J，Brown PO，Rosenzweig F，Botstein D（2002）Characteristic genome rearrangements in experimental evolution of *Saccharomyces cerevisiae*. Proc Natl Acad Sci USA 99：16144-16149

Elena SF，Lenski RE（2003）Evolution experiments with microorganisms：the dynamics and genetic bases of adaptation. Nat Rev Genet 4：457-469

Elena SF，Wilke CO，Ofria C，Lenski RE（2007）Effects of population size and mutation rate on the evolution of mutational robustness. Evolution 61：666-674

Ferenci T（2007）Bacterial physiology，regulation and mutational adaptation in a chemostat environment. In：Poole RK（ed）Advances in microbial physiology，vol 52. Elsevier Academic Press，San Diego

Ferenci T（2008）Bacterial physiology，regulation and mutational adaptation in a chemostatronment. In：Poole RK（ed）Advances in microbial physiology，vol 53. Elsevier Academic Press，San Diego

Fraser-Liggett CM（2005）Insights on biology and evolution from microbial genome sequencing. Genome Res 15：1603-1610

Garcia-Vallvé S，Palau J，Romeu A（2000）Horizontal gene transfer in bacterial and archaeal complete genomes. Genome Res 10：1719-1725

Gill SR，Fouts DE，Archer GL，Mongodin EF，Deboy RT，Ravel J，Paulsen IT，Kolonay JF，Brinkac L，Beanan M，Dodson RJ，Daugherty SC，Madupu R，Angiuoli SV，Durkin AS，Haft DH，Vamathevan J，Khouri H，Utterback T，Lee C，Dimitrov G，Jiang L，Qin H，Weidman J，Tran K，Kang K，Hance IR，Nelson KE，Fraser CM（2005）Insights on evolution of virulence and resistance from the complete genome analysis of an early methicillin-resistant *Staphylococcus aureus* strain and a biofilm-producing methicillin-resistant *Staphylococcus epidermidis* strain. J Bacteriol 187：2426-2438

Gresham D，Dunham MJ（2014）The enduring utility of continuous culturing in experimental evolution. Genomics 104：399-405

Gresham D，Ruderfer DM，Pratt SC，Schacherer J，Dunham MJ，Botstein D，Kruglyak L（2006）Genome-wide detection of polymorphisms at nucleotide resolution with a single DNA microarray. Science 311：1932-1936

Gresham D，Desai MM，Tucker CM，Jenq HT，Pai DA，Ward A，DeSevo CG，Botstein D，Dunham MJ（2008）The repertoire and dynamics of evolutionary adaptations to controlled nutrientlimited environments in yeast. PLoS Genet 4：e1000303

Grogan DW（1989）Phenotypic characterization of the archaebacterial genus *Sulfolobus*：comparison of five wild-type strains. J Bacteriol 171：6710-6719

Grogan DW，Carver GT，Drake JW（2001）Genetic fidelity under harsh conditions：analysis of spontaneous mutation in the thermoacidophilic archaeon *Sulfolobus acidocaldarius*. Proc Natl Acad Sci USA 98：7928-7933

Hardison RC（2003）Comparative genomics. PLoS Biol 1：e58

Helling RB，Vargas CN，Adams J（1987）Evolution of *Escherichia coli* during growth in a constant environment. Genetics 116：349-358

Herring CD，Raghunathan A，Honisch C，Patel T，Applebee MK，Joyce AR，Albert TJ，Blattner FR，van den Boom D，Cantor CR，Palsson BØ（2006）Comparative genome sequencing of *Escherichia coli* allows observation of bacterial evolution on a laboratory timescale. Nat Genet 38：1406-1412

Hindré T，Knibbe C，Beslon G，Schneider D（2012）New insights into bacterial adaptation through in vivo and in silico experimental evolution. Nat Rev Microbiol 10：352-365

Hong J, Gresham D (2014) Molecular specificity, convergence and constraint shape adaptive evolution in nutrient-poor environments. PLoS Genet 10: e1004041

Huber H, Stetter KO (1998) Hyperthermophiles and their possible potential in biotechnology. J Biotechnol 64: 39-52

Kadam SV, Wegener-Feldbrügge S, Søgaard-Andersen L, Velicer GJ (2008) Novel transcriptome patterns accompany evolutionary restoration of defective social development in the bacterium *Myxococcus xanthus*. Mol Biol Evol 25: 1274-1281

Kao KC, Sherlock G (2008) Molecular characterization of clonal interference during adaptive evolution in asexual populations of *Saccharomyces cerevisiae*. Nat Genet 40: 1499-1504

Kidwell MG, Lisch DR (2000) Transposable elements and host genome evolution. Trends Ecol Evol 15: 95-99

Kinnersley M, Wenger J, Kroll E, Adams J, Sherlock G, Rosenzweig F (2014) Clonal reinforcement drives evolution of a simple microbial community. PLoS Genet 10: e1004430

Kunin V, Ouzounis CA (2003) The balance of driving forces during genome evolution in prokaryotes. Genome Res 13: 1589-1594

Kurtz S, Choudhuri JV, Ohlebusch E, Schleiermacher C, Stoye J, Giegerich R (2001) REPuter: the manifold applications of repeat analysis on a genomic scale. Nucleic Acids Res 29: 4633-4642 Kussell E (2013) Evolution in microbes. Annu Rev Biophys 42: 493-514

Kvitek DJ, Sherlock G (2013) Whole genome, whole population sequencing reveals that loss of signaling networks is the major adaptive strategy in a constant environment. PLoS Genet 9: e1003972

Lang GI, Rice DP, Hickman MJ, Sodergren E, Weinstock GM, Botstein D, Desai MM (2013) Pervasive genetic hitchhiking and clonal interference in forty evolving yeast populations. Nature 500: 571-574

Lashkari DA, DeRisi JL, McCusker JH, Namath AF, Gentile C, Hwang SY, Brown PO, Davis RW (1997) Yeast microarrays for genome wide parallel genetic and gene expression analysis. Proc Natl Acad Sci USA 94: 13057-13062

Le Fourn C, Fardeau ML, Ollivier B, Lojou E, Dolla A (2008) The hyperthermophilic anaerobe *Thermotoga maritima* is able to cope with limited amount of oxygen: insights into its defence strategies. Environ Microbiol 10: 1877-1887

Lenski RE, Mongold JA, Sniegowski PD, Travisano M, Vasi F, Gerrish PJ, Schmidt TM (1998) Evolution of competitive fitness in experimental populations of *E. coli*: what makes one genotype a better competitor than another? Antonie van Leeuwenhoek 73: 35-47

Lorantfy B, Seyer B, Herwig C (2014) Stoichiometric and kinetic analysis of extreme halophilic Archaea on various substrates in a corrosion resistant bioreactor. New Biotechnol 31: 80-89

Lovett ST, Drapkin PT, Sutera VA Jr, Gluckman-Peskind TJ (1993) A sister-strand exchangemechanism for recA-independent deletion of repeated DNA sequences in *Escherichia coli*. Genetics 135: 631-642

Lovett ST, Gluckman TJ, Simon PJ, Sutera VA Jr, Drapkin PT (1994) Recombination between repeats in *Escherichia coli* by a recA-independent, proximity-sensitive mechanism. Mol Gen Genet 245: 294-300

Maezato Y, Dana K, Blum P (2011) Engineering thermoacidophilic archaea using linear DNA recombination. Methods Mol Biol 765: 435-445

Maezato Y, Johnson T, McCarthy S, Dana K, Blum P (2012) Metal resistance and lithoautotrophy in the extreme thermoacidophile *Metallosphaera sedula*. J Bacteriol 194: 6856-6863

McCarthy S, Ai C, Wheaton G, Tevatia R, Eckrich V, Kelly R, Blum P (2014) Role of an archaeal PitA transporter in the copper and arsenic resistance of *Metallosphaera sedula*, an extreme thermoacidophile. J Bacteriol 196: 3562-3570

McCarthy S, Johnson T, Pavlik B, Payne S, Schackwitz W, Martin J, Lipzen A, Keffler E, Blum P (2015) Expanding the limits of thermoacidophily in the archaeon Sulfolobus solfataricus by adaptive evolution. Appl Environ Microbiol. PII: AEM. 03225-15 [Epub ahead of print], PMID: 26590281

Monod J (1942) Recherche sur la croissance des cultures bactériennes. Hermann and Cie, Paris

Nelson KE, Clayton RA, Gill SR, Gwinn ML, Dodson RJ, Haft DH, Hickey EK, Peterson JD, Nelson WC, Ketchum KA, McDonald L, Utterback TR, Malek JA, Linher KD, Garrett MM, Stewart AM, Cotton MD, Pratt MS, Phillips CA, Richardson D, Heidelberg J, Sutton GG, Fleischmann RD, Eisen JA, White O, Salzberg SL, Smith HO, Venter JC, Fraser CM (1999) Evidence for lateral gene transfer between Archaea and bacteria from genome sequence of *Thermotoga maritima*. Nature 399: 323-329

Novick A, Szilard L (1950) Description of the chemostat. Science 112: 715-716

Paquin C, Adams J (1983) Frequency of fixation of adaptive mutations is higher in evolving diploid than haploid yeast populations. Nature 302: 495-500

Philippe N, Crozat E, Lenski RE, Schneider D (2007) Evolution of global regulatory networks during a long-term experiment with *Escherichia coli*. Bioessays 29: 846-860

Polz MF, Hunt DE, Preheim SP, Weinreich DM (2006) Patterns and mechanisms of genetic and phenotypic differentiation in marine microbes. Philos Trans R Soc Lond B Biol Sci 361: 2009-2021

Prosser JI, Bohannan BJM, Curtis TP, Ellis RJ, Firestone MK, Freckleton RP, Green JL, Green LE, Killham K, Lennon JJ, Osborn AM, Solan M, van der Gast CJ, Young JPW (2007) The role of ecological theory in microbial ecology. Nat Rev Microbiol 5: 384-392

Redder P, Garrett RA (2006) Mutations and rearrangements in the genome of *Sulfolobus solfataricus* P2. J Bacteriol 188: 4198-4206

Ren Q, Chen K, Paulsen IT (2007) TransportDB: a comprehensive database resource for cytoplasmic membrane transport systems and outer membrane channels. Nucleic Acids Res 35: D274-D279

Riehle MM, Bennett AF, Long AD (2001) Genetic architecture of thermal adaptation in *Escherichia coli*. Proc Natl Acad Sci USA 98: 525-530

Rigaut G, Shevchenko A, Rutz B, Wilm M, Mann M, Seraphin B (1999) A generic protein purification method for protein complex characterization and proteome exploration. Nat Biotechnol 17: 1030-1032

Rolfsmeier M, Blum P (1995) Purification and characterization of a maltase from the extremely thermophilic crenarchaeote *Sulfolobus solfataricus*. J Bacteriol 177: 482-485

Romero D, Palacios R (1997) Gene amplification and genomic plasticity in prokaryotes. Annu Rev Genet 31: 91-111

Rosenzweig RF, Sharp RR, Treves DS, Adams J (1994) Microbial evolution in a simple unstructured environment: genetic differentiation in *Escherichia coli*. Genetics 137: 903-917

Rusch A, Walpersdorf E, de Beer D, Gurrieri S, Amend JP (2005) Microbial communities near the oxic/anoxic interface in the hydrothermal system of Vulcano Island, Italy. Chem Geol 224: 169-182

Sauer U (2001) Evolutionary engineering of industrially important microbial phenotypes. In: Nielsen J, Eggeling L, Dynesen J, Gárdonyi M, Gill RT, Graaf AA, Hahn-Hägerdal B, Jönsson LJ, Khosla C, Licari R, McDaniel R, McIntyre M, Miiller C, Nielsen J, Cordero Otero RR, Sahm H, Sauer U, Stafford DE, Stephanopoulos G, Wahlbom CE, Yanagimachi KS, Zyl WH (eds) Metabolic engineering. Springer, Berlin/Heidelberg

Schneider D, Lenski RE (2004) Dynamics of insertion sequence elements during experimental evolution of bacteria. Res Microbiol 155: 319-327

Schröder C, Selig M, Schönheit P (1994) Glucose fermentation to acetate, CO_2 and H_2 in the anaerobic hyperthermophilic eubacterium *Thermotoga maritima*: involvement of the EmbdenMeyerhof pathway. Arch Microbiol 161: 460-470

Schwikowski B, Uetz P, Fields S (2000) A network of protein-protein interactions in yeast. Nat Biotechnol 18: 1257-1261

Singh R, Gradnigo J, White D, Lipzen A, Martin J, Schackwitz W, Moriyama E, Blum P (2015) Complete genome sequence of an evolved *Thermotoga maritima* isolate. Genome Announc 3: e00557-15

Singh R, White D, Kelly R, Noll K, Blum P. Uncoupling fermentative synthesis of molecular hydrogen from biomass formation in *Thermotoga maritima*. (in preparation)

Smith JM, Smith NH, O'Rourke M, Spratt BG (1993) How clonal are bacteria? Proc Natl Acad Sci USA 90: 4384-4388

Sniegowski PD, Gerrish PJ, Lenski RE (1997) Evolution of high mutation rates in experimental populations of *E. coli*. Nature 387: 703-705

Suyama M, Bork P (2001) Evolution of prokaryotic gene order: genome rearrangements in closely related species. Trends Genet 17: 10-13

Tatum EL, Lederberg J (1947) Gene recombination in the bacterium *Escherichia coli*. J Bacteriol 53: 673-684

Tettelin H, Masignani V, Cieslewicz MJ, Donati C, Medini D, Ward NL, Angiuoli SV, Crabtree J, Jones AL, Durkin AS, Deboy RT, Davidsen TM, Mora M, Scarselli M, Margarit Rosy I, Peterson JD, Hauser CR, Sundaram JP, Nelson WC, Madupu R, Brinkac LM, Dodson RJ, Rosovitz MJ, Sullivan SA, Daugherty SC, Haft DH, Selengut J, Gwinn ML, Zhou L, Zafar N, Khouri H, Radune D, Dimitrov G, Watkins K, O'Connor KJ, Smith S, Utterback TR, White O, Rubens CE, Grandi G, Madoff LC, Kasper DL, Telford JL, Wessels MR, Rappuoli R, Fraser CM (2005) Genome analysis of multiple pathogenic isolates of Streptococcus agalactiae: implications for the microbial "pan-genome". Proc Natl Acad Sci USA 102: 13950-13955

Toprak E, Veres A, Michel JB, Chait R, Hartl DL, Kishony R (2012) Evolutionary paths to antibiotic resistance under dynamically sustained drug selection. Nat Genet 44: 101-105

Treves DS, Manning S, Adams J (1998) Repeated evolution of an acetate-crossfeeding polymorphism in long-term populations of *Escherichia coli*. Mol Biol Evol 15: 789-797

Ussery DW, Binnewies TT, Gouveia-Oliveira R, Jarmer H, Hallin PF (2004) Genome update: DNA repeats in bacterial genomes. Microbiology 150: 3519-3521

vanHam SM, van Alphen L, Mooi FR, van Putten JP (1993) Phase variation of *H. influenzae* fimbriae: transcriptional control of two divergent genes through a variable combined promoter region. Cell 73: 1187-1196

Wenger JW, Piotrowski J, Nagarajan S, Chiotti K, Sherlock G, Rosenzweig F (2011) Hunger artists: yeast adapted to carbon limitation show trade-offs under carbon sufficiency. PLoS Genet 7: e1002202

Wertz JE, Goldstone C, Gordon DM, Riley MA (2003) A molecular phylogeny of enteric bacteria and implications for a bacterial species concept. J Evol Biol 16: 1236-1248

Wilbanks EG, Larsen DJ, Neches RY, Yao AI, Wu CY, Kjolby RA, Facciotti MT (2012) A workflow for genome-wide mapping of archaeal transcription factors with ChIP-seq. Nucleic Acids Res 40 (10): e74

Woods RJ, Barrick JE, Cooper TF, Shrestha U, Kauth MR, Lenski RE (2011) Second-order selection for evolvability in a large *Escherichia coli* population. Science 331: 1433-1436

Worthington P, Blum P, Perez-Pomares F, Elthon T (2003) Large-scale cultivation of acidophilic hyperthermophiles for recovery of secreted proteins. Appl Environ Microbiol 69: 252-257

Wurtzel O, Sapra R, Chen F, Zhu Y, Simmons BA, Sorek R (2010) A single-base resolution map of an archaeal transcriptome. Genome Res 20: 133-141

Zhaxybayeva O, Swithers KS, Lapierre P, Fournier GP, Bickhart DM, DeBoy RT, Nelson KE, Nesbø CL, Doolittle WF, Gogarten JP, Noll KM (2009) On the chimeric nature, thermophilic origin, and phylogenetic placement of the Thermotogales. Proc Natl Acad Sci USA 106: 5865-5870

第二十三章

固相结合多肽：在生物技术中极端微生物催化剂的固定化策略

Andrew Care[1]，Peter L. Bergquist[2]，Anwar Sunna[3]

23.1 引言

生物催化是一种广泛使用的与纯粹化学方法互补的化学合成工具。酶是一种具备高效和多功能特性的生物催化剂。在不同操作条件下，酶显示出高区域、对映、化学和立体选择性。与化学催化剂相比，生物催化不需要对官能团进行保护或脱保护，并且能通过减少副反应产生废弃产物，缩短获得最终产品的途径。因此，随着对“绿色化学”的重视，酶催化工艺在一些大规模工业领域已经变得越来越重要。

在过去，酶通常在溶液状态或者固定在固相支持物上进行研究，在一系列固相支持物上已经取得了重大的进展（Brady and Jordaan 2009；Hanefeld et al. 2009）。游离酶类在工业中的应用由于一系列因素而受到限制，如高温加工时不稳定、非水相体系中性能较差、回收与再利用困难，这些都导致了催化成本的提高。另一方面，相比游离酶，固定化酶具有几个优势，在单次和连续生物过程中重复使用、可重复使用多个周期、超过可溶性游离形式酶的稳定性、快速终止反应、容易把酶移除以及消除酶对产品的污染（Raghuvanshi and Gupta 2010）。然而，由于其异构性和活性三级结构在变性（和无活性）无规卷曲结构上的边际稳定性，蛋白质的固定化仍然是极具挑战性的任务。最近也有归纳了这一领域的进展和状况（Camarero 2008；Hanefeld et al. 2009）。

❶ ARC Centre of Excellence for Nanoscale BioPhotonics (CNBP)，Macquarie University，Sydney，NSW，Australia.

❷ Department of Molecular Medicine and Pathology，Medical School，University of Auckland，Auckland，New Zealand；Department of Chemistry and Biomolecular Sciences，Macquarie University，Sydney，NSW，Australia；Biomolecular Frontiers Research Centre，Macquarie University，Sydney，NSW，Australia.

❸ Department of Chemistry and Biomolecular Sciences，Macquarie University，Sydney，NSW，Australia；Biomolecular Frontiers Research Centre，Macquarie University，Sydney，NSW，Australia. e-mail：anwar. sunna@mq. edu. au.

23.1.1 蛋白质固定化

蛋白质固定化（通常不仅仅是特定的酶）的方法可以分为：吸附，共价结合，包埋/微囊化，交联。Cowan 和 Fernandez-Lafuente（2011）发表了一篇关于固定化嗜热酶的综合报道，主要强调了在固相支持物上的化学固定化作用。选择固定化的嗜热酶的各种方法汇总在表 23.1 中。

表 23.1 用于嗜热酶类固定化的方法

酶	生物体	固定化方法	参考文献
乙醇脱氢酶	*Thermoanaerobacter brockii*（嗜热厌氧杆菌）	吸附	Miroliaei(2007)
乙醇脱氢酶	*Thermus thermophilus* HB27（嗜热栖热菌）	共价结合吸附	Rocha-Martin et al. (2009)
乙醇氧化还原酶	*Sulfolobus solfataricus*（硫化叶菌）	共价结合	Raia et al. (1995)
氨基酰化酶	*Thermococcus litoralis*（嗜热球菌）	包埋/微囊化吸附	Ngamsom et al. (2010)
L-阿拉伯糖异构酶	*Thermoanaerobacter mathranii*（嗜热厌氧菌）	交联酶集合体	Jørgensen et al. (2004)
β-葡萄糖苷酶	*Caldicellulosiruptor saccharolyticus*（解糖热解纤维素菌）	共价结合	Tran et al. (2013)
Caldolysin（一种蛋白酶）	*Thermus aquaticus* T351（栖热水生菌）	共价结合 吸附	Cowan et al. (1987)
过氧化氢酶	*Thermus thermophilus*（嗜热栖热菌）	共价结合 交联酶集合体	Hidalgo et al. (2003)
外切葡聚糖酶	*Caldicellulosiruptor saccharolyticus*	共价结合	Tran et al. (2013)
脱卤素酶	*Sulfolobus tokodaii*（硫化叶菌）	吸附	Bachas-Daunert et al. (2009)
鸟苷环化酶	*Thermotoga maritime*（海栖热袍菌）	包埋/微囊化	Pasunooti et al. (2010)
α-半乳糖苷酶	*Thermus sp. strain* T2	共价结合吸附	Filho et al. (2008)
β-半乳糖苷酶	*Sulfolobus solfataricus*	共价结合	Jørgensen et al. (2004)
	Thermus sp. strain T2	共价结合吸附	Pessela et al. (2004)
α-葡萄糖苷酶	*Thermococcus* AN1	化学修饰	Piller et al. (1996)
β-葡萄糖苷酶	*Pyrococcus furiosus*（火球菌）	共价结合	Fischer et al. (1996)
谷氨酸脱氢酶	*Thermus thermophilus*	共价结合	Bolivar et al. (2009)
氢化酶	*Pyrococcus furiosus*	吸附	Greiner et al. (2003)
(+)-λ-内酰胺酶	*Sulfolobus solfataricus* MT4	交联酶集合体	Hickey et al. (2009)
L-乳酸脱氢酶	*Clostridium thermohydrosulfuricum*（嗜热硫化氢梭菌）	包埋/微囊化	Lehn and Schmidt(1997)
	Thermus caldophilus GK24	化学修饰	Taguchi et al. (1984)

续表

酶	生物体	固定化方法	参考文献
脂肪酶	*Thermus thermophilus*	吸附	Palomo et al.(2004)
	Thermus aquaticus	吸附	Palomo et al.(2004)
蛋白酶	*Thermus* Rt41A	共价结合	Wilson et al.(1994)
焦磷酸酶	*Pyrococcus furiosus*	共价结合	Dong et al.(2014)
木聚糖酶	*Thermotoga* FjSS3-B.1	共价结合	Simpson et al.(1991)
D-木糖异构酶	*Thermus aquaticus*	共价结合 交联酶集合体	Fernandez-Lafuente(2009)
	Thermotoga maritima	共价结合	Bandlish et al.(2002)
	Thermotoga neapolitana 5068	共价结合	Bandlish et al.(2002)

23.1.1.1 吸附

大部分传统的将蛋白质固定化在固相支持物上的方法都依赖于非特异性物理吸附。不同类型的作用力介导蛋白质吸附在非水溶性载体的表面，包括氢键、范德瓦尔斯力、离子键和疏水作用（Hanefeld et al. 2009）。该技术主要优势是操作简单，能在温和的条件中进行以及酶无需进行预处理和化学修饰。然而，该方法的主要缺点是在水相介质中吸附的酶可能会由于固定化过程中的结合作用较弱而从载体上渗出。

许多热稳定脂肪酶已经通过物理吸附作用成功固定在疏水支持物上（Palomo et al. 2004；Wilson et al. 2006）。例如，将来自 *Thermus thermophilus*（嗜热栖热菌）和 *Thermus aquaticus*（栖热水生菌）的脂肪酶固定化在十八烷基树脂上，显示在中温下催化活性提高了 10 倍并且功能稳定性也得到提升（Palomo et al. 2004）。相反，通过离子作用吸附和共价结合固定的脂肪酶酶活则下降了 20%～30%。

23.1.1.2 共价结合

共价结合固定化技术是通过对酶进行化学修饰进而促进酶与载体之间共价键的形成。共价结合主要由酶上的功能基团介导（例如氨基、羟基、硫醇基、羰基、硫基）。作为这种技术的先决条件，修饰必须在不影响酶催化功能的前提下进行。尤其酶的活性位点不应该受化学修饰或试剂的影响。共价结合会防止酶从支持物上脱离，但这种方法的缺点是酶分子可能会在化学加工时部分或完全变性，导致固定化酶比溶液中的蛋白质效力更低。很难通过以共价键为基础的固定化技术来控制固相支持物上酶的组织与定位。缺乏控制是指如果位于活性位点附近的必需氨基酸残基参与酶的固定化时，酶活性可能会降低，而且，缺乏定位控制可能会限制底物靠近酶活性位点。因此，与游离酶相比，固定化酶通常可能会显示更低的活性和更高的米氏常数（Kress et al. 2002；Cao 2006）。

一种来自古菌激烈火球菌（*Pyrococcus furiosus*）的嗜热无机焦磷酸酶以戊二醛作为偶联剂共价连接在氨基官能化的硅胶珠上（Dong et al. 2014）。与游离的酶相反，该固定化酶具有更高的活性以及 pH 与热稳定性，并且该固定化酶在 95℃连续进行 12 次无机焦磷酸水解反应后仍保持超过 50%的初始活力。

共价偶联反应的另一种方法是等离子体浸没离子注入（PⅢ），这是一种物理过程，在材料表面生成活性自由基，与生物分子形成共价键，而无需化学连接器。该技术最近已经实现将来自嗜热细菌 *Caldicellulosiruptor saccharolyticus*（解糖热解纤维素菌）的纤维素酶 CelB 和 β-葡萄糖苷酶 BglA 固定化在聚苯乙烯板上（Tran et al. 2013）。该固定化酶保持了它们的功能，并且两者共同将羧甲基纤维素水解为葡萄糖。然而，需要注意的是，用 PⅢ处理很耗时并且大规模使用不太符合实际。

23.1.1.3　包埋/微囊化

酶的包埋涉及诸如凝胶（聚丙烯酰胺、海藻酸钙和明胶）、硅溶胶-凝胶、硅笼或中空纤维酯类的多聚物网络，以及在酶存在下多聚体基质的合成。多聚体网络必须通过反应物和产物的扩散来发生反应，同时避免酶渗入到介质中。通常情况下，往往因为物理条件太弱而不能阻止酶的渗出，可能需要使用共价连接。上述方法的主要优势是减少低分子量底物渗出，提高其稳定性和自由扩散性。但是，底物、产物的传质阻力以及抑制剂会限制包埋作用，而限制大分子底物的接触则是另外一个缺点。该固定化技术比其他技术的处理要更为细致（Betancor and Luckarift 2008；Hudson et al. 2008；Hartmann and Kostrov 2013；Magner 2013）。一种从 *Geobacillus* sp.（芽孢杆菌属）中分离的嗜热嗜碱性酯酶固定化在海藻酸钙珠上，然后用一层硅酸盐涂覆（Gülay and Şanlı-Mohamed 2012）。该包被层阻止了酶的外漏从而提高了酶的载量、固定化率以及重复利用性。而且，截留的酯酶比游离酶有着更好的操作性、热稳定性和储存稳定性。

23.1.1.4　交联

双官能团试剂（如戊二醛）进行交联的方法已经被用于制备交联酶结晶（CLECs），但近年来应用更多的是交联酶集合体（CLEAs），如无载体酶集合体等。目前，已经被开发用于简化程序以及酶纯化的效果（Sheldon 2011；Sheldon and van Pelt 2013）。CLEAs 是添加常用蛋白质沉淀剂（如硫酸铵、聚乙二醇、乙醇）使蛋白质聚集而制备出来的。这些物理聚集体通过非共价结合聚集在一起，并且当聚集蛋白质分散到水系介质中能够被重新溶解。随后的交联步骤通过形成共价键确保了聚集物的不可溶和结构完整性，因此保存了酶集合体的催化活性。该技术的主要优点是生产具有高度活性浓缩酶、低产品成本以及相对稳定的强大系统。然而，不同酶形成 CLEAs 的理想条件有所不同，并且必须进行测试以避免催化活性的降低。

来自 *Thermoanaerobacter mathranii*（产甲烷嗜热厌氧杆菌）的 L-阿拉伯糖异构酶以及 *Sulfolobus solfataricus*（硫化叶菌）的 β-葡萄糖苷酶通过与戊二醛和聚乙烯亚胺交联进行固定化后获得热稳定的 CLEAs（Jørgensen et al. 2004），这些 CLEAs 把在 65℃中进行的 D-半乳糖经过两步转化为 D-塔格糖的反应简化为一个转化率为 38%的单一反应。

23.2　利用固相结合多肽进行固定化

固相结合多肽（SBPs）是短氨基酸序列，它能选择性地识别并牢固结合于多种固体材料表面，包括金属（Naik et al. 2002b；Hnilova et al. 2008；Heinz et al. 2009；Li et al. 2009；Chiu et al. 2010；Forbes et al. 2010），碳材料（Wang et al. 2003；Kulp et al. 2005；Pender et al. 2006；Cui et al. 2010），半导体（Whaley et al. 2000；Lee et al. 2002；Estephan et al. 2011），聚合物和矿物（Li et al. 2002；Serizawa et al. 2007；Gungormus et al. 2008；Roy et al. 2008）。这些独特的多肽能够作为分子接头指导生物分子（如蛋白质和酶）定向固定在固体基质上而不损害其生物活性。而且，它们具有调节不同复杂性功能的纳米结构的合成和组装的能力。通过最近的一些关于 SBPs 的综述可以全面了解该领域的现状（Sarikaya et al. 2003；Baneyx and Schwartz 2007；Tamerler et al. 2010；Care et al. 2015）。最近已经有人研究利用这些多肽在温和环境下合成、组装和形成无机材料的过程（Ball

2001；Seeman and Belcher 2002；Sarikaya et al. 2003）。因此，SBPs已经用作许多纳米生物技术应用中的分子工具，例如纳米材料的功能化（Lu et al. 2014）、无机纳米复合物（Inoue et al. 2014）和复合生物材料（Chen et al. 2014）的合成以及纳米生物电子学（Dang et al. 2011；Sano et al. 2013）。

SBPs对各自的固相具有高亲和力和选择性，结合亲和力（K_A）达到nmol/L到亚μmol/L范围内。它们还显示出较大的负结合能（Tamerler et al. 2006；Seker et al. 2007；Hnilova et al. 2008；Kacar et al. 2009a；So et al. 2009；Tang et al. 2013）。这种强结合是多重非共价相互作用的综合结果，包括静电力、疏水作用、极性和氢键。然而，由于肽-材料界面的动态性和复杂性，许多SBPs识别、选择性和亲和力的机制尚不清楚。迄今为止，多肽的多样性（如组成、序列、构象、结构、化学和电荷）、结合固体材料的表面（如氧化态、电荷、晶体定位、缺损）以及溶液类型（如水或有机溶剂）都被证明有助于结合的相互作用（Naik et al. 2002b；Goede et al. 2004；Nel et al. 2009；Sapsford et al. 2013；Tang et al. 2013；Slocik and Naik 2014；Care et al. 2015）。进一步的复杂性来自组装肽-材料表面界面的动态性质。肽在这个界面不断地运动，倾向于扩散，调整并使自身适应最低的能量结构。

大部分SBPs已经利用组合展示技术得以分离（Brown 1997；Whaley et al. 2000；Naik et al. 2002a；Sarikaya et al. 2003；Thai et al. 2004），特别是噬菌体展示技术（Smith 1985；Hoess 2001）和细胞表面展示技术（Charbit et al. 1986；Wittrup 2001，图23.1）。这些方法相对简单、稳定，并且允许对庞大的多肽文库进行筛选获得活性结合材料。一旦选定，整个显示多肽的噬菌体颗粒（Whaley et al. 2000；Huang et al. 2005）或者细菌胞体（Park et al. 2009）就能够用作固相结合剂。这些多肽能够使用固相合成的方式合成（Nochomovitz et al. 2010），或者该肽作为融合蛋白的一部分用于促进生物分子固定在固体表面（Ko et al. 2009）。最近开发出利用计算模拟的方法来选择新型SBPs（如基于知识的设计、分子模建以及生物信息学）。然而，该方法需要生物信息学工具和算法结合大量的实验数据，从而来生产具有改进性能的肽（Oren et al. 2007，2010；Evans et al. 2008；Notman et al. 2010；Masica et al. 2010；Walsh 2014）。但是直到现在，SBPs的实际应用还没有扩展到大型工业生物技术中。

研究表明，SBPs能够控制酶定位吸附在固相表面而不会破坏其天然构象和催化功能（Yang et al. 2011，表23.2）。例如，Kacar等（2009b）研究了通过金结合肽（GBP1）将碱性磷酸酶（AP）固定化在金表面。与非修饰的AP相比，GBP1-AP对金具有特异的结合亲和力，当进行固定化时能组装成密集均匀的单分子膜，与没有GBP1的酶相比，GBP1-AP每单位面积显示出更好的定位以及更高的酶活性。GBP1也显示有助于酶固定化而不影响它们的天然构象（Yang et al. 2011）。具有结合硅胶（Naik et al. 2002a）或氧化铁（Brown 1992）亲和力的SBPs已经被用于促进生物修复相关酶类（卤代烷烃脱卤酶）固定在硅胶、涂覆或未涂覆氧化铁纳米磁性颗粒上（Johnson et al. 2008）。在这两个例子中，酶都保持了它的活性，酶黏附在磁性纳米颗粒上使得它们在多个反应中可以直接回收和再利用（Puri et al. 2013）。

23.2.1 固相结合多肽在生物催化上的用途

许多不可溶的材料可以用作酶固定化的载体。然而，酶与载体之间的作用模式对固定化酶的稳定性和催化效率有重要影响。在大部分条件下SBPs结合它们相应的固相载体，不需要任何化学修饰或者物理处理，这不仅保证了酶的完整性还证明了它们在生物催化过程中对

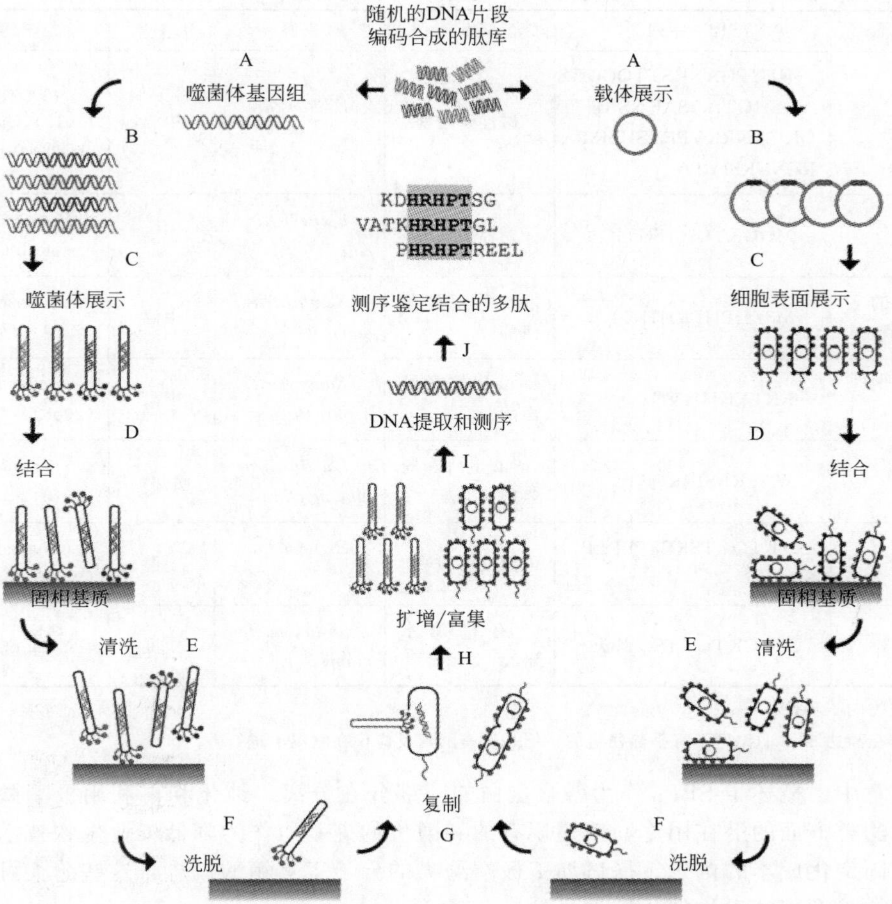

图 23.1　噬菌体和细胞表面展示技术用于分离结合到固定相底物上的多肽。为了筛选固相结合肽，随机合成的 DNA 片段编码的庞大和多样的肽库（10^8～10^9）被用于融合到在噬菌体颗粒和细菌胞体表面表达的蛋白质中（A～C）。因此，每个噬菌体或细胞都展示出具有唯一肽段的克隆，这些唯一的肽段与其包装在内部的编码基因紧密联系。对固相底物具有高亲和力的肽段能通过将噬菌体库或细胞展示多肽与固相底物孵育这一选择性过程被分离出来（D），没有结合的噬菌体或细菌会通过清洗被移除（E），那些紧密结合到底物上的则被洗脱下来（F），洗脱得到的噬菌体或细菌随后被扩增（G～H）并且重复结合和扩增（C～H）来富集结合多肽。经过 3～4 轮的筛选后，单克隆或空斑的 DNA 通过测序鉴定能够紧密结合固相底物的多肽氨基酸序列（I～J）（改编自 Sarikaya 等. 2003）

环境友好。工业规模的生物催化通常需要在高温、高压或极端 pH 值中稳定的廉价材料。无机材料因其具有良好的结构、操作稳定以及不容易被微生物所降解等特点，使其成为极为优秀的载体。然而，其需要繁琐的试误法来鉴定特定酶/载体组合的理想固定化策略（Garcia-Galan et al. 2011；Sassolas et al. 2012）。另一方面，通过组合展示技术，SBPs 能够被选择用于几乎任何一种材料，使酶附着到任何预选基质上。尽管有着这些理想的特点，但 SBPs 的实际应用目前仍只限于在化学/生物传感器中酶的固定化和环境条件下起作用的生物鉴定（Yang et al. 2011；Cetinel et al. 2013，表 23.2）。这些固定化的报道通常使用进口的和昂贵的实验室基质，并且不可能实现经济的大规模的工艺。关于廉价基质的报道很少，并且对于 SBPs 在嗜热酶固定化上的应用以及在工业规模生物催化中的可行性了解较少。

表 23.2 利用 SBPs 固定化酶

结合基质	SBPs 序列	酶	来源	嗜热性①	参考文献
类金刚石	HFYPGANRSTTQGGGS ANLHQTAASAKNSAP-QK-SENRKVPFYSHSRT RENNRSIYTA	碱性磷酸酶	*Escherichia coli*	中温	Gabryelczyk et al. (2013), Gabryelczyk et al. (2014)
金	MHGKTQATSGTIQS	碱性磷酸酶	*Escherichia coli*	中温	Brown (1997), Kacar et al. (2009b)
硅包被的氧化铁纳米颗粒	MSPHPHPRHHHT	卤代烷烃脱卤酶	*Xanthobacter autotrophicus*	中温	Johnson et al. (2008)
氧化铁纳米颗粒	RRTVKHHVN	卤代烷烃脱卤酶	*Xanthobacter autotrophicus*	中温	Johnson et al. (2008)
金	WALRRSIRRQSY	L-乳酸脱氢酶	*Bacillus stearothermophilus*	嗜热②	Cetinel et al. (2013)
合成沸石	VKTQATSREEPPRLP SKHRP	碱性磷酸酶	*Escherichia coli*	中温	Nygaard et al. (2002)
纳米金	MHGKTQATSGTIQS	有机磷水解酶	*Flavobacterium* sp	中温	Yang et al. (2011)

① 源生物的嗜热性。

② 该酶在高达 60℃的温度下都是热稳定的，但是所有酶促反应仅在室温下进行。

在本章中，概述了 SBPs 作为融合蛋白的一部分在分离、纯化和再利用极端微生物、嗜热微生物酶类方面的潜在用途，使用现有的和廉价的硅基质作为降低工业生物技术工程中生物催化剂固定化成本的例子。探讨加工极端酶类的分子工具箱的概念，这些酶类可用于许多应用生物技术领域的生物催化模块设计。

23.2.2 硅结合固相结合多肽

硅固相基质，比如硅胶、沸石和介孔硅胶，被认为是最适合工业中酶固定化的无机基质(Blanco et al. 2004；Ho et al. 2004；Pierre 2004)。这些无机载体具有独特的性能，包括高机械强度和稳定性，化学惰性以及可在宽范围的操作压力和条件下使用。它们还具有较高表面积，能够进行化学修饰和官能团配体连接。

硅胶吸附是最简单和最经济的酶固定化途径之一，但其弱吸附作用可能导致酶在水相介质中外漏。此外，与其他非特异性蛋白质黏附的情况一样，蛋白质可能不会以一个正确的定位方式固定。最近，一种基于硅胶特异性 SBPs 的选择性和更具体的技术被开发用于将蛋白质和酶固定化在无机硅基质上（Brown 1992；Taniguchi et al. 2007；Sunna et al. 2013a；Abdelhamid et al. 2014；Coyle and Baneyx 2014）。例如，Sunna 等（2013a）报道的 SBPs 对天然沸石和商用级别合成沸石、硅胶、含硅材料显示出高结合亲和力。他们将该 SBPs 序列［(VKTQATSREEPPRLPSKHRPG)$_4$ VKTQTAS（Nygaard et al. 2002）］与绿色荧光蛋白（GFP，表 23.5）C 端进行融合，融合蛋白显示出选择性结合天然沸石的能力（图 23.2）。类似的，他们将 SBPs 与抗体结合蛋白融合，结果显示在几分钟内融合蛋白可介导抗体定向固定在含硅材料中，而无需复杂的表面化学修饰（Sunna et al. 2013b；Care et al. 2014a；Lu et al. 2014）。该系统对被硅胶包被且掺和了稀土的高集成度抗体纳米晶体进行官能化，促进水性原生动物

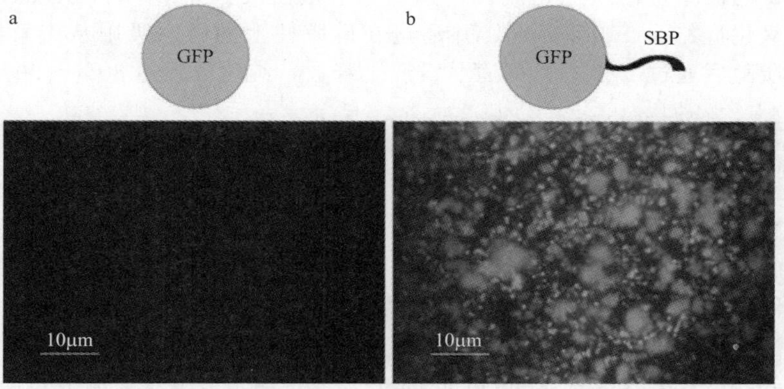

图 23.2　绿色荧光蛋白 GFP 和 GFP-SBP（表 23.5）与天然沸石结合及荧光显微镜图（见彩图）。(a) 不含有 SBP 的 GFP 不能结合到天然沸石上且沸石没有显示荧光；(b) 只有当 GFP 融合了 SBP 后的重组表达蛋白质才能与天然沸石结合并显示荧光

病原体 *Giardia lambia* 的结合及标签发光（Lu et al. 2014）。

硅结合 SBPs 已经被用于硅基蛋白质的亲和纯化（图 23.3）。Abdelhamid 等（2014）将来自 *Bacillus cereus*（蜡样芽孢杆菌）的孢子外壳蛋白 CotB1 C 端 14 个氨基酸的多肽 CotB1p（SGRARAQRQSSRGR）与 mCherry 融合，作为一个融合蛋白模型来证明这种亲和纯化方法的效率。使用 CotB1p 硅胶结合 SBPs 后的纯化和产率能与传统的亲和标签报告的值相当（Lichty et al. 2005）。与此相似的是，Coyle 和 Baneyx（2014）报道了一种新的十二肽纯化标签（DSARGFKKPGKR），使未修饰的二氧化硅对其融合蛋白具有亲和力。他们

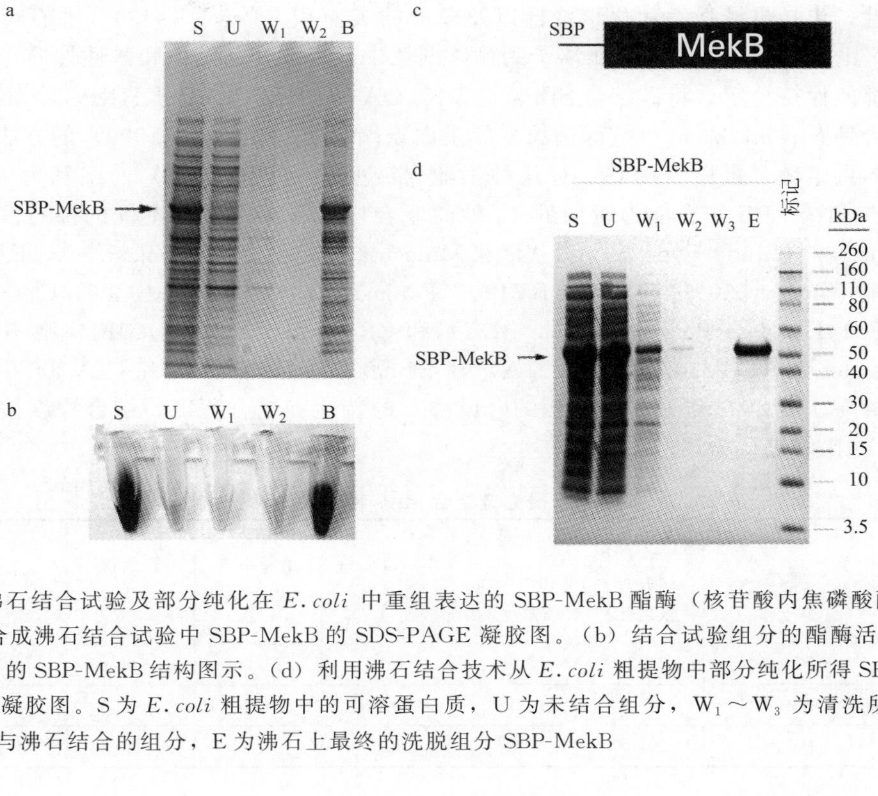

图 23.3　沸石结合试验及部分纯化在 *E. coli* 中重组表达的 SBP-MekB 酯酶（核苷酸内焦磷酸酶磷酸二酯酶）。(a) 合成沸石结合试验中 SBP-MekB 的 SDS-PAGE 凝胶图。(b) 结合试验组分的酯酶活力。(c) N 端含有 SBP 的 SBP-MekB 结构图示。(d) 利用沸石结合技术从 *E. coli* 粗提物中部分纯化所得 SBP-MekB 的 SDS-PAGE 凝胶图。S 为 *E. coli* 粗提物中的可溶蛋白质，U 为未结合组分，W_1～W_3 为清洗所得组分 1、2、3，B 为与沸石结合的组分，E 为沸石上最终的洗脱组分 SBP-MekB

开发了一种廉价的硅基蛋白质纯化流程，其中重组蛋白质（如 GFP、mCherry 和麦芽糖结合蛋白）用二氧化硅结合 SBPs 标记，并通过在硅胶柱上的特定吸附从裂解物中选择性回收。标签蛋白质随后通过 L-赖氨酸孵育从硅胶上释放，纯化效果达 80%～90%。

23.2.3 用于硅结合固相结合多肽的基质

在蛋白质纯化中非修饰硅胶的使用相对少（Ghose et al. 2004），将生物识别分子固定在硅胶或硅基传感基底上的方法则使用较为频繁。参与表面共价固定化方法通常有三个步骤：首先，对修饰的表面添加特定的官能团；其次，通过其中一个反应基团与交联剂共价连接；最后，识别元件与交联剂剩余活性位点共价连接。一些方法已经在生物传感器结合方法中有提到，这些方法允许生物识别并结合到感应面上且保留其全部功能，例如，抗体定向固定到硅晶片（Taniguchi et al. 2007；Ikeda et al. 2009）或者纳米线上作为灵敏和定向检测生物分析物的电子设备（Patolsky et al. 2006）。

大部分早期的酶固定化研究涉及微孔硅胶和沸石。最近的研究聚焦在介孔基质上，该基质由于孔的尺寸太小而不允许蛋白质的进入，因此酶结合在外表面（Hudson et al. 2008；Popat et al. 2011；Zhou and Hartmann 2012；Magner 2013）。这些复合物能够轻易有序合成，可以规定孔结构、限制孔径分布，并且可以对官能团进行化学修饰。缺点为很容易让酶外漏，但它们能够以共价固定化来提高稳定性（Gaffney et al. 2012；Puddu and Perry 2012；Hartmann and Kostrov 2013）。Puddu 和 Perry（2012）强调，纳米硅颗粒上的多肽吸附涉及疏水作用并且受肽本身支配。他们的工作显示可以通过改变结合环境或改变硅胶的表面特性来改变硅胶上的多肽结合。

沸石是具有高度有序结构的天然的无机材料。它们也能被合成为具有孔径和管道系统的分子尺寸在 0.3～3.0 nm 范围内的纳米晶体。它们有着作为蛋白质固定化的无机载体材料的理想特性，包括机械强度和化学惰性以及较大的表面积（Breck 1973）。沸石可以通过改变其 Si/Al 比例或者引入不同金属离子到晶体框架中，来改变其碱性和酸性性质，并且在干湿条件下都能保持稳定。将硅结合 SBPs [$(VKTQATSREEPPRLPSKHRPG)_4$VKTQTAS] 的特性融合到 *Streptococcus*（链球菌属）G 蛋白顶端（Goward et al. 1990）的方法，广泛测试了得到的重组融合蛋白（LPG）对几种工业含硅基质（包括沸石）的亲和力（Sunna et al. 2013a）。测试 LPG 结合能力所用的 13 种商业合成沸石属于 8 个不同的沸石家族。LPG 对于 Valfor 100（Linde Type A，LTA）和 Molecular Sieve 13X（Faujasite-X，FAU-X）的亲和力较弱（40%～60%），对于 CBV400（Faujasite-Y，FAU-Y）、CP914C（Ferrierite，FER）、CP814E（Beta Polymorph A，BEA）和 CBV2314（ZSM-5，MFI）则不具有亲和力，对 Mordenite（CBV10A 和 CBV21A）和 Faujasite-Y（CBV100 和 CBV300）类型的沸石亲和力最高。表 23.3 汇总了被检测的合成沸石的特性和结合程度，结合程度按结合与未结合 LPG 的比例进行描述。

表 23.3 部分用于研究硅结合 SBPs 特异性的沸石的性质①

类型	沸石	SiO_2/Al_2O_3（物质的量之比）	阳离子	Na_2O 质量分数/%	单元小室尺寸/Å	表面积/(m^2/g)	结合情况② U B
八面沸石（*FAU-Y*）	CBV 100	5.1	钠	13.0	24.65	900	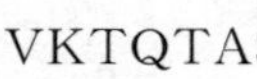
	CBV 300	5.1	铵	2.8	24.68	925	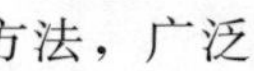
	CBV 400	5.1	氢	2.8	24.50	730	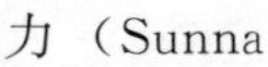

续表

类型	沸石	SiO_2/Al_2O_3（物质的量之比）	阳离子	Na_2O 质量分数/%	单元小室尺寸/Å	表面积/(m^2/g)	结合情况[②] U B
八面沸石（*FAU-X*）	Molecular sieve 13X	5.1	钠	14.6	24.94	700	
发光沸石（MOR）	CBV 10A	13.0	钠	65.0	数据未提供	425	
	CBV 21A	20.0	铵	0.08	数据未提供	500	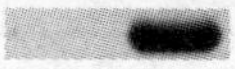
镁碱沸石（FER）	CP 914c	20.0	铵	0.05	数据未提供	400	
β晶型（BEA）	CP 814E	25.0	铵	0.05	数据未提供	680	
ZSM-5（MFI）	CBV 2314	23.0	铵	0.05	数据未提供	425	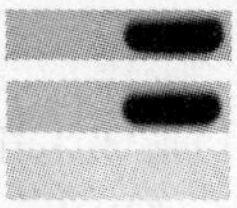
Linden type A(LTA)	Valfor 100	1.1	钠	15.0	数据未提供	71[③]	

① 除非特别说明，所有数据来自制造商（LPG 与材料结合的 SDS-PAGE 除外）。
② LPG 与材料结合的 SDS-PAGE：U，未结合部分；B，结合部分；洗脱部分未显示。
③ Hui 和 Chao（2008）。

测试了 12 种不同的市售硅胶材料固定化 LPG 的能力。LPG 对 10 个样品具有高亲和力（90%～100%），但不能结合 Silica LC 60A（50%结合力）和 Silica Gel 60（30%结合力）。表 23.4 汇总了被检测硅胶样品的主要特性。

表 23.4 部分用于研究硅结合 SBP 特异性的硅石的性质[①]

硅石	SiO_2/%	颗粒尺寸/μm	孔尺寸/Å	表面积/(m^2/g)	结合[②] U B
Silica(precipitated)	99.0[③]	数据未提供	Non-porous[④]	62	
Silica gel grade 646	99.4	250～500	107～189	275～375	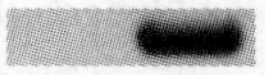
Silica Davisil LC 60A	99.4	40～63	60	500～600	
Silica gel 60 Scharlau	99.4	60～200	60	500	
Silica Rhodoline HP 34 M	94.0	2.6	数据未提供	157	
Silica Tixosil 38	94.0	15	数据未提供	190	
Silica Tixosil 38A	94.0	60	数据未提供	230	
Silica Tixosil 68	94.0	250	数据未提供	160	
Silica Perkasil SM 660	98.0	17	数据未提供	200	

续表

硅石	SiO_2/%	颗粒尺寸/μm	孔尺寸/Å	表面积/(m^2/g)	结合[②] U　B
Silica Perkasil KS 300-PD	98.0	14.5	数据未提供	125	
Silica Elfadent SM 514	98.0	12	数据未提供	125	
Silica Durafill 200[⑤]	82.0	6.0	数据未提供	80	

① 除非特别说明，所有数据来自制造商（LPG 与材料结合的 SDS-PAGE 除外）。

② LPG 与材料结合的 SDS-PAGE：U，未结合部分；B，结合部分；洗脱部分未显示。

③ Çaykara 和 Güven（1998）。

④ Furlong（1982）。

⑤ 硅酸镁铝钠盐（8.5% Al_2O_3，2%MgO）。

使用适当的沸石和硅胶粉末作为酶固定化的基质有经济上的优势。沸石在需要批量供应的工业加工中被频繁使用，如水净化的气体分离、核废料再处理、洗涤剂业、建筑工业、土壤处理以及其他农业加工。化学合成沸石被用于石油的催化裂化和水化学裂化以及分子筛。对于沸石用途更加全面的综述可以在 Naber 等（1994）和 Rhodes（2010）找到。同样，工业上常用硅胶粉末，如用于钢铁精炼、铝铸件、化学和半导体以及电子工业、作为食品中的抗结块剂、涂料增稠剂、建筑工业的几种混凝土组成成分。低纯度硅胶的报价在 10～500 美元/吨。因此，用于工业酶固定化的基质的成本不是主要的因素。感兴趣的读者可以在 Kogel 等（2006）的文章以及在 www.monographs.iarc.fr/ENG/Monographs/vol100C-14.pdf 中找到更多详细的用途。

23.2.4　硅结合固相结合多肽作为嗜热酶融合蛋白的一部分

目前几乎没有关于利用 SBP 将热稳定酶固定化在固体材料上的报道，大部分涉及硅结合 SBP 的研究已经在之前有所描述，有部分已经由 Care 等（2014b）报道过，另外的在本节中以未发表的结果的形式总结。

23.2.4.1　固定化

到目前为止，Sunna 等（2013a）报道的硅结合 SBP 已经被成功融合到超过 20 种蛋白质或酶中（Care et al. 2015）。为测试该 SBP 的效果，将其引入到含有不同嗜热酶基因的质粒当中，这些酶的最佳活性温度范围在 50℃到 100℃之间。一系列融合蛋白的构建都含有一个选择在 N 端或 C 端上通过遗传学手段融合到工业相关的热稳定酶类中的 SBP（表 23.5）。所有重组 SBP 融合蛋白都在 *E. coli* 中以可溶蛋白质形式表达并且保持了它们的催化功能（图 23.4）。在所有的案例中，硅结合 SBP 都能介导融合蛋白特异性结合到沸石基质当中。沸石结合试验使用相应的不含融合 SBP 的酶，结果显示酶无特异性结合或仅仅非特异性吸附到沸石中（图 23.4b）。尽管蛋白质分子质量与吸附效率没有直接联系，但它们在沸石上的吸附已经被证明是 pH 值依赖性的，当 pH 刚好等于或者低于蛋白质等电点时吸附量最大（Klint and Eriksson 1997；Tavolaro et al. 2006）。当 pH 低于其等电点时，蛋白质带正电并通过静电吸引吸附到沸石（总体带负电荷）的外表面。在 pH 值高于等电点时，蛋白质带负电，吸附在沸石上的量最少，这是由于负电荷蛋白质与沸石之间存在排斥力。在这些条件下的任何吸附都可能是其他吸引力的结果（如疏水作用），SBP 已被报道在较宽的 pH 范围内（5～9）可介导蛋白质特异性结合到沸石上，这表明了这是一个真正的非 pH 依赖的结合亲和性（Sunna et al. 2013a）。

表 23.5 *E. coli* 中重组表达的 SBP 融合蛋白

结构	功能	来源	SBP 位置	是否维持功能	是否结合沸石	参考文献
TPL—SBP	脂肪酶	宏基因组富集	C①	是	是	Bell et al. (2002)
SBP—XynAd2	木聚糖酶	*Caldibacillus cellulovorans*	N②	是	是	Sunna et al. (2000)
SBP—XynA	木聚糖酶	环境 DNA	N	是	是	Sunna and Bergquist(2003)
SBP—XynB	木聚糖酶	*Dictyoglomus thermophilum*	N	是	是	Morris et al. (1998), Sunna et al. (2013a)
SBP—GH26	甘露聚糖酶	*Caldicellulosiruptor saccharolyticus*	N	是	是	Sunna(2010)
ManA—SBP	甘露聚糖酶	*Dictyoglomus thermophilum*	C	是	是	Gibbs et al. (1999), Care et al. (2014b)
SBP—BglA	β-葡萄糖苷酶	*Caldicellulosiruptor saccharolyticus*	N	是	是	Love et al. (1988)
SBP—Cel5A	内葡聚糖酶	*Thermotoga maritima*	N	是	是	Chhabra et al. (2002)
SBP—EndH	内葡聚糖酶	*Dictyoglomus thermophilum*	N	是	是	Coil et al. (2014)

注：显示了 SBP 掺入后酶活性功能是否维持及融合蛋白与沸石结合的亲和力。

① C，SBP 结合在融合蛋白的 C 末端。

② N，SBP 结合在融合蛋白的 N 末端。

SBPs 序列通常带正电，因为它们含有大量的碱性氨基酸残基。SBPs 也显示出对于能够促进不稳定的有序结构的氨基酸残基的偏好性（Sarikaya et al. 2003；Ikeda and Kuroda 2011；Sunna et al. 2013a）。SBPs 这种固有的紊乱状态维持其柔性和可塑性，促进了 SBP 的正电荷残基与含硅材料表面的负电荷之间的离子相互作用（Collino and Evans 2008）。因此可以预计 SBP 持续保持非折叠状态，这便于它们在结合界面适应底物（如沸石和硅胶）的吸附和构象（Ikeda and Kuroda 2011）。最近有提及聚酯、聚酰胺或棉花纺织物可作为生物分子固定化的新型替代基质（Opwis et al. 2014）。这些织物代表了相对廉价且具有柔性、开放结构、机械强度和大表面积的支撑材料。它们的开放结构允许大分子底物转化。多种用于将酶固定在织物上的化学（交联）和非化学（光激活交联）策略已经显示出该技术的潜力，使得织物具有高酶载量、高活性和酶再利用的特点（Opwis et al. 2005，2014）。

另一种在织物载体上固定化酶的策略是直接将载体热沉积在织物上。该方法的潜力最近

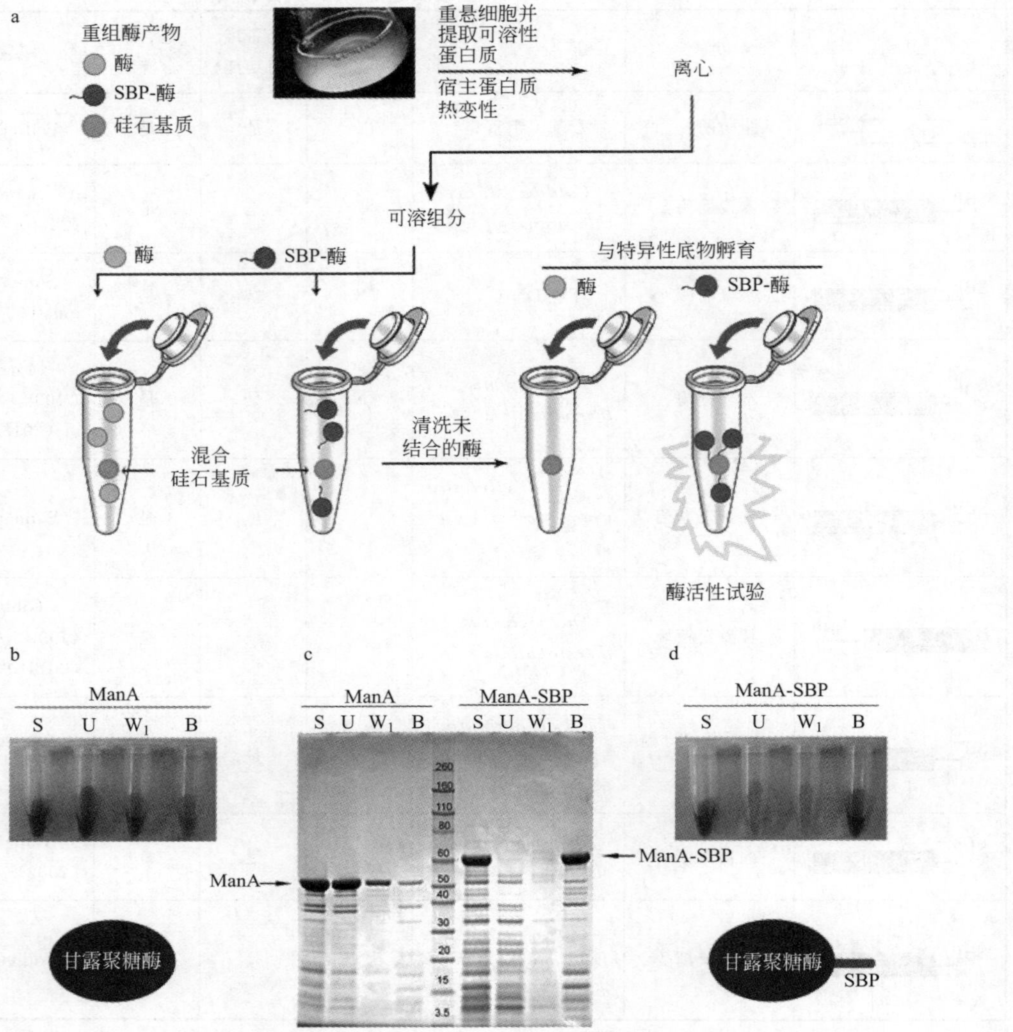

图 23.4　硅石机制结合及活性试验（见彩图）。(a) 从 *E. coli* 中提取重组可溶蛋白质及后续硅石基质结合试验的标准流程图。(b) 来自 *Dictyoglomus thermophilum* 的热稳定甘露聚糖酶 ManA（表 23.5）结合试验中各组分活性。(c) 部分纯化的 ManA 和 ManA-SBP 与沸石结合试验的 SDS-PAGE 凝胶图。(d) ManA-SBP 结合试验各组分的甘露聚糖酶活性。甘露聚糖酶活性试验是以刺槐豆胶（LBG）为底物在 80℃下反应 10min。从 LBG 中释放出来的还原性糖用 DNS 法（二甲基水杨酸法）进行定量。S 代表 *E. coli* 可溶蛋白质粗提物，U 为非结合组分，W_1 位第一次冲洗组分（清洗组分 2、3 没有进样），B 为沸石结合蛋白组分

已经通过直接固定含有硅结合 SBP 的绿色荧光蛋白（GFP）融合蛋白得以体现（图 23.5）。该融合蛋白通过 SBP 对硅胶的强亲和力被成功固定在硅载织物上（Sunna 未发表结果）。

23.2.4.2 再利用/再循环

尽管在生物技术和蛋白质工程上有所进展，酶在特定工业中广泛应用，但由于它们缺乏长期操作的稳定性、繁琐的回收和较差的再利用特点而受到限制（Sheldon and van Pelt 2013）。将酶固定在不可溶基质或支持物上提供了以固体形式代替液体中游离形式处理酶的优势，包括酶制剂更方便的处理、易于与反应产物分离、降低了下游工艺的成本以及简化了酶的应用。此外，固定化常常显示出能提高酶的稳定性和对 pH、温度变化的抵抗性（Cher-

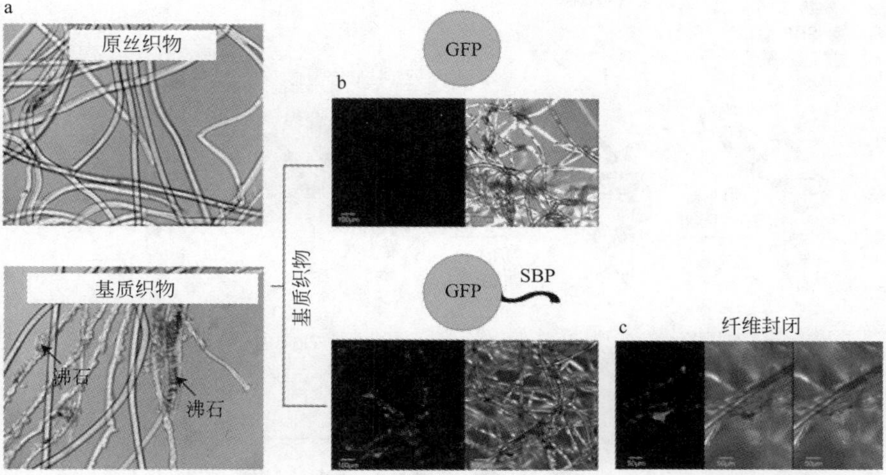

图 23.5　GFP 及 GFP-SBP 结合原丝及沸石织物（见彩图）。(a) 原丝和沸石织物纤维的相差显微镜图像。(b) 原丝和沸石织物纤维在与 GFP 和 GFP-SBP 共孵育后的共聚焦显微镜图像。左边面板显示 GFP 荧光，右边面板为相应的 DIC（数字图像相关法）图。(c) 用 GFP-SBP 处理后的沸石织物在共聚焦显微镜下的特写。左边为 GFP 荧光，中间为相应的 DIC（数字图像相关法）图，右边为 GFP 荧光与 DIC 图的重叠

ry and Fidantsef 2003)。与游离的酶相比，固定化酶能够轻易回收和重新投入到反应中，从而实现生物催化剂的低成本的多次再利用。通过固定化回收利用具有成本上的优势，尤其是在需要高成本的生物催化剂的操作方面，并且它还支持固定化酶的持续加工。酶的成本占工业生物过程的总体费用的最大一部分。从制造成本的角度，固定化酶活性的产率是由固定化方式和使用的可溶性酶的数量所决定的。改进生物催化剂操作性能和回收利用固定化酶可获得更多的催化产物（kg 产物/kg 酶），这对每千克产品所需的酶的成本有直接的正面作用(Sheldon and van Pelt 2013)。

理论上讲，酶固定化的大规模工业应用应该是一个简单的过程，不需要高纯度酶制剂或者昂贵的催化剂支撑基质。因此，使用低成本支持物如硅胶和沸石的 SBPs 介导的固定化在固定工业应用的酶上显示出显著的潜力。通过与 SBPs 融合固定在含硅的材料上的嗜热性酶已经显示出能在高操作温度下保持功能，并且可全部回收利用几个循环（图 23.6，Care et al. 2015）。

23.2.4.3　挑战

通常 SBP 的结合亲和力是通过其在实验室条件下严格控制各项因素的缓冲液中进行鉴定的。因此，许多 SBPs 在苛刻的工业条件下并不稳定，导致其从相应的固体中解离或完全降解。*Filamentous* M13 噬菌体已经显示能在 pH 值 3～11 和温度低于 80℃ 时保持活性(Branston et al. 2013)。因此，已市售 M13 噬菌体（如来自 New England Biolabs）代表了一种分离 SBPs 的潜在方法，该方法兼容工业流程。为了促进其工业化应用，提出了利用组合显示库来选择 SBPs，SBPs 会在工业应用之前或是应用过程中所接触到的条件下（如极端 pH 和高温）结合固相材料。这一行为会确保分离得到的 SBPs 在工业条件下能够保持其稳定性和结合功能，使其可以与嗜热酶进行融合，并且提高它们在工业生物催化剂中的适用性。

通常而言，SBPs 通过基因手段与感兴趣的蛋白质进行融合，并在大肠杆菌（*E. coli*）中产生融合蛋白。尽管 *E. coli* 是重组蛋白质生产的常用宿主，但它并不能天然分泌大量蛋

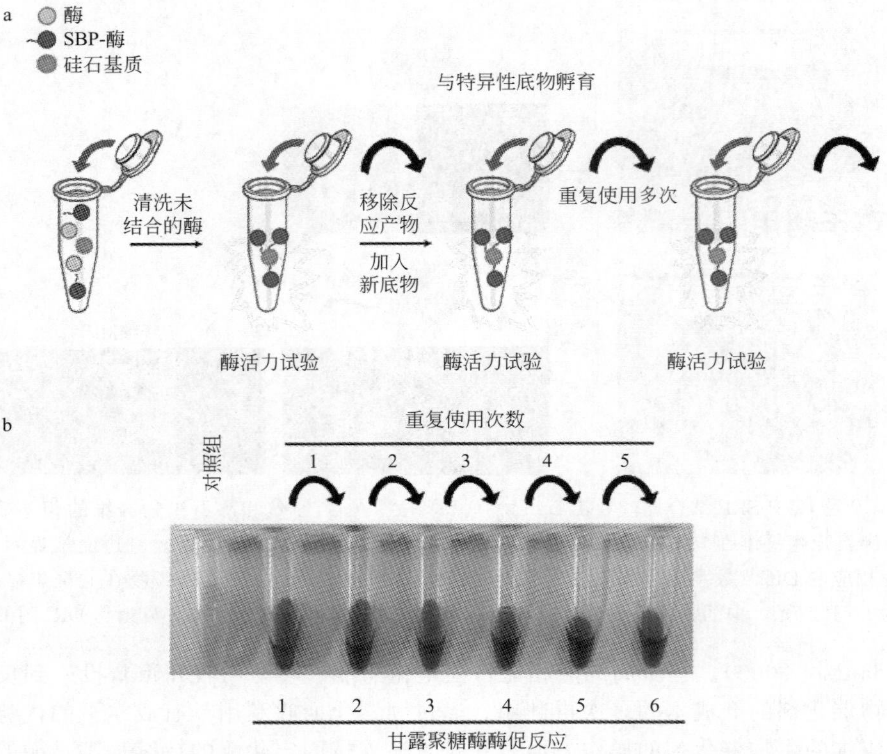

图 23.6 硅石基质结合与酶的重复使用。(a) 从 *E.coli* 中部分纯化重组可溶蛋白质及随后硅石基质结合试验和酶重复利用的标准流程图。(b) *D.thermophilum* 热稳定甘露聚糖酶 ManA-SBP（表 23.5）被固定化在合成沸石上，并且在 80℃中进行了 6 次酶促反应，超过 60min（10min/次）。甘露聚糖酶活性试验以刺槐豆胶（LBG）为底物，以 DNS 法测定释放出的还原性糖

白质。因此需要部分或完全破碎细胞来回收所有 *E.coli* 产生的蛋白质，这是其作为工业生产宿主应用的主要障碍。丝状真菌例如曲霉属（*Aspergillus*）和木霉属（*Trichoderma*）是工业生产蛋白质的常用工具，因为它们分泌到培养基中的酶产量高、活性高。最近，测试了丝状真菌里氏木霉（*Trichoderma reesei*）作为代替宿主来获得更高产量和胞外热稳定的 β-甘露聚糖酶-SBP 融合蛋白、ManA-SBP（表 23.5 和图 23.4，Care et al. 2014b）。据显示大部分由 *T.reesei* 分泌的胞外蛋白质都不能结合沸石，而且 *E.coli* 产的 SBP 会被 *T.reesei* 上清液中的胞外宿主蛋白酶降解。有人认为 *T.reesei* 真核表达系统可能产生一种糖基化形式的 SBP 来抵抗蛋白酶水解。然而，将 ManA-SBP 融合蛋白的优化密码子导入到 *T.reesei* 之后，在生长培养基中只检测到没有融合 SBP 的热稳定 ManA 酶。随后，研究者运用一个合理的办法去生产多接头且期望能抵抗 *T.reesei* 蛋白酶的衍生物，它们在 *E.coli* 中生产并且检测了其沸石结合的特异性和蛋白酶稳定性。但所有衍生物显示对沸石没有特异性亲和力而且只有一个对 *T.reesei* 蛋白酶具有抗性。

因此，为了减少 SBP 序列被 *T.reesei* 蛋白酶降解，建议在筛选步骤中利用组合显示技术来分离在 *T.reesei* 蛋白酶存在的情况下（如含有 *T.reese* 胞外上清液时）可以结合沸石的多肽。该方式应该可以让选定的 SBPs 在 *T.reesei* 生长培养基中分泌后其抵抗蛋白酶和结合沸石的能力仍保留下来。然而，并不能保证这些 SBPs 在 *T.reesei* 中可以有效表达，显然需要通过实验来进行评估。

值得注意的是，组合显示技术也有它们的缺点。所有肽库都具有一定的组成、位置和表达偏倚，这可能会导致具有良好特性和功能的肽的低表达（Umlauf et al. 2014）。此外，据显示筛选程序对通过静电作用结合的正电荷多肽存在固有的偏好，而对于通过非静电作用结合的多肽则不敏感，因此会有许多具有良好结合能力的 SBPs 可能没有筛选到（Puddu and Perry 2012）。

23.3　对于利用固相结合多肽固定化热稳定酶的展望

23.3.1　合成生物学

随着合成生物学的发展，大部分项目已经聚焦于如何将代谢工程与系统生物学设计相结合。这种体内（自上而下）的合成生物学方法极其复杂、困难和费力（Kwok 2010；Hodgman and Jewett 2012）。另一方面，体外（自下而上）无细胞合成生物学提供了更多的工程灵活性，而且消除了体内方法中遇到的细胞存活限制和物理障碍（细胞壁和细胞膜）。

热稳定酶的生产成本和产率仍旧是它们用于大规模生物制造的根本障碍（Turner et al. 2007）。解决这些因素的一个策略是使用重组技术在中温宿主中表达嗜热酶类，这使得宿主蛋白在高温下容易变性，这是一种获得相对较纯的蛋白质的简单方法，而且不需要昂贵和繁琐的过程。该策略包括自动优化使用的密码子，这是目前商业基因合成供应商标准选项，来改善异源重组蛋白质的表达和产率。此外，蛋白质工程的发展，尤其是计算蛋白质设计，已经形成新的设计方法来产生想要的具有特异性、活性和稳定性的酶（参见 Khare and Fleishman 2013）。最近，Ye 等人（2012）证明了热稳定酶在无细胞合成生物学中的适用性。它们通过合成代谢工程构建了一个简单的无细胞途径，包括四个主要步骤：①选择热稳定酶；②在*E. coli* 中表达；③高温灭活宿主内源蛋白酶并且部分纯化热稳定酶；④合理的组合酶以实现化学计量数的底物转化。根据这些原则，他们用其中来自 *Thermus thermophilus*（嗜热栖热菌）、*Pyrococcus horikoshii*（掘越氏热球菌）和小宝岛嗜热球菌 *Thermococcus kodakarensis* 的七种糖酵解酶类构建了一个合成嵌合 EM 途径，在 *E. coli* 进行重组表达。该合成途径利用葡萄糖生成了化学计量数的乳酸，整体 ATP 转换数达 31。

尽管取得了这些进展和对无细胞合成生物学中嗜热酶的开发，但是这些酶在生物技术进程中的广泛使用经常受到产量、操作稳定性和回收与再利用的问题的阻碍。Zhang 等人（2011）以及 You 和 Zhang（2013）最近已经假设了使用固定在固相基质上的热稳定酶作为无细胞合成生物学的替代方法，用于低成本生物制造的生物催化模块。结合固相基质上固定化的酶和酶的再利用，有利于热稳定酶的大规模操作，并且具有巨大的潜力来降低最终产物的成本和提高每个工艺的产量。SBPs 已经通过遗传手段融合到热稳定酶中（表 23.5），并且形成的融合蛋白（SBP-酶）已经被成功固定化在包含硅胶和沸石的固相无机基质上（Care et al. 2014b；Care et al. 2015）。固定在固相基质上的 SBP-酶显示能够保留其固有的热稳定性和催化功能。

和无机物一样，有机物也能够作为纳米级别的模块，能够被组合形成集合功能的纳米材料。通过合理设计稳定的具有预定功能和催化特性的基本模块（单一固定化酶），SBP 能够被引入到分子工具箱中。分子工具箱（图 23.7）有三个主要的可互换的模块，低成本无机基质（如硅胶、沸石）、基质特异性的 SBP 以及热稳定酶（如糖苷水解酶）。通过合理组合 SBP、固相基质和具有功能的热稳定酶，研究者就能灵活地通过自然和非自然途径来选择和组合多种模块，并且/或者可以设计生物催化模块用于有效降解生物多聚体。

这样一个分子工具箱会依赖于生物信息学软件的持续开发、大规模数据（数据库）的共

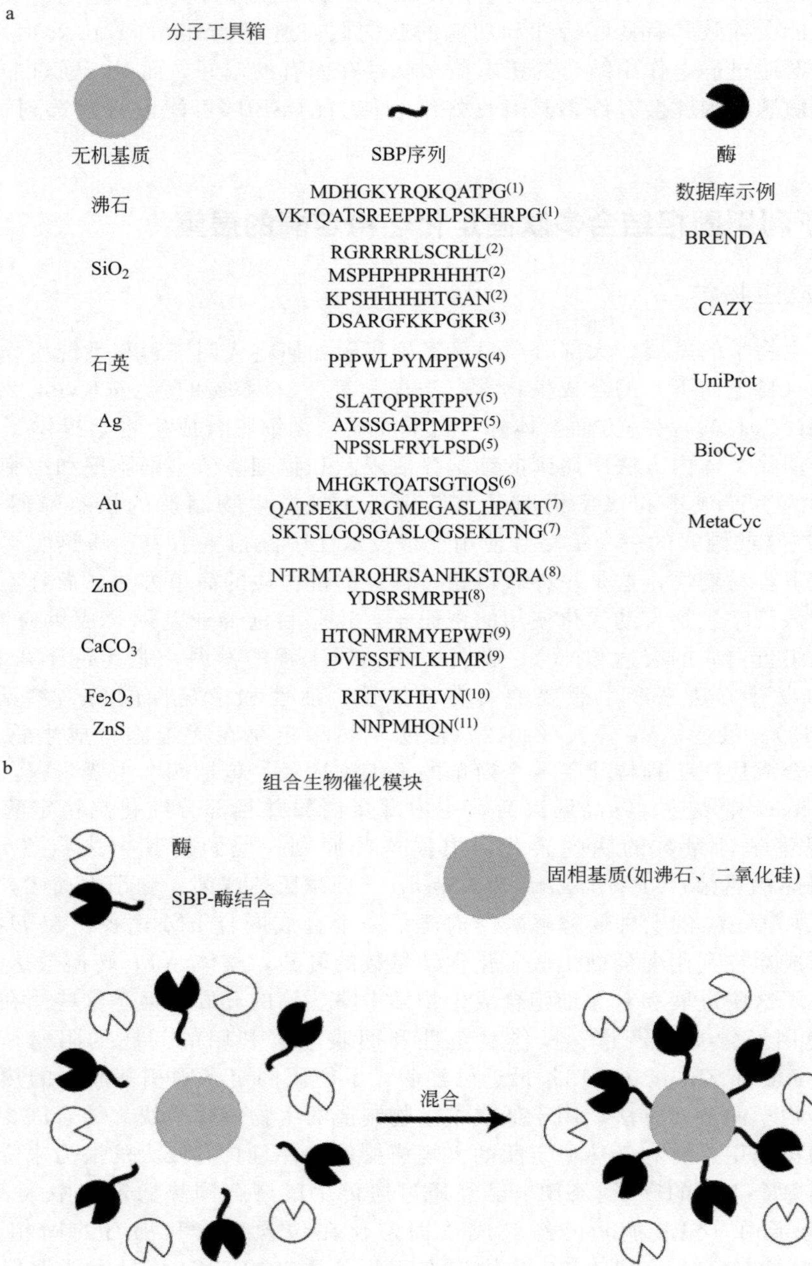

图 23.7 分子工具箱。(a) 包含的三种主要的可互换分子工具；低成本无机基质，基质特异性 SBP，热稳定性酶。基质特异性 SBP 相关参考文献 (1) Nygaard et al. 2002；(2) Naik et al. 2002a；(3) Coyle and Baneyx 2014；(4) Oren et al. 2007；(5) Naik et al. 2002b；(6) Brown 1997；(7) Brown et al. 2000；(8) Kjaergaard et al. 2000；(9) Gaskin et al. 2000；(10) Brown 1992；(11) Lee et al. 2002。(b) 组合生物催化模块与预定义的功能和催化性能

享以及无需天然模板的情况下短期内将数字序列转化为遗传物质（从头合成）的能力 (Mueller et al. 2009)。该策略同时也受到持续下降的基因合成成本的推动。目前基因合成的成本类似于柱合成寡核苷酸所需的成本。然而，从基因合成过渡到基于阵列的寡核苷酸预计

成本会下降到大约每 10^3～10^5 bp 1 美元（Kosuri and Church 2014）。

一些与该方法相关的生物学数据库（生物化学、基因组、蛋白质或酶、途径）已经可以在公共区域获得。BRENDA（braunschweig enzyme database，不伦瑞克数据库）具有超过 6500 个酶条目的功能和属性数据。另一个重要的蛋白质数据库是 UniProt（universal protein resource），这是一个蛋白质序列和注释数据的集合。CAZy（Carbohydrate-ActiveenZYmes，糖类活性酶）是一个可以提供关于参与合成、代谢和糖类水解的酶的结构和生化信息的数据库。目前 CAZy 具有超过 3895 个条目。途径数据库描述了特定代谢过程的生化过程，而且包含了化学途径、反应和成分的主要数据。途径数据库像 BioCyc 就含有 5500 个途径/基因组信息的数据库，以及通路图的图形显示软件工具。代谢途径数据库 MetaCyc 从 2600 个不同有机体中收集了 2260 个通路，它储存了每一个实验阐明的途径的代表性例子。

23.3.2 生产生物燃料

对于化石燃料的依赖以及化石燃料使用所排放的温室气体已成为一个全球问题，对全球环境造成严重的后果。油价的波动造成商业的不稳定，这两个因素促使了人们从可再生生物资源中寻找燃料和能源的替代物。早期能源替代来源是谷物乙醇，但谷物乙醇生产过程中能量输出的检测已经成为一个争议话题。这种第一代生物燃料是基于传统酵母发酵的成熟工业。第二代生物燃料使用生物质能，第二代生物燃料的生产不是来源于粮食，而是来自大农场和林业废料，所有这些都归类为“纤维素乙醇”。相关的技术利用酶和全细胞催化体系来生产生物降解产物，从而产生比化石资源更少的废物和更低的碳排放，并且可长期持续。最近在代谢工程、测序、系统生物学以及合成生物学上的进展已经为参与乙醇发酵的微生物的遗传与生理开辟了新的机遇。为实际应用而设计的微生物细胞工厂已被组装用于天然和从头设计的途径，以将碳流向燃料和其他所需的生物来源的化合物（Dellomonaco et al. 2010；Hollinshead et al. 2014）。

生物燃料生产的主要问题是降解生物质聚合物的酶的成本。第二代生物燃料，尤其是木质纤维素乙醇、五碳糖（戊糖）和六碳糖（己糖）以及木质素是主要的生物质成分，酶成本已经成为阻碍生物燃料开发的一个主要的经济因素。为此研究者们提出的解决方案包括联合生物加工（CBP），该方案利用厌氧嗜热细菌同时对生物质进行同步糖化发酵（SSF）（Lynd et al. 2002），以及使用基因工程嗜热产乙醇细菌，如 *Bacillus thermoglucosidasius*（现在为 *Geobacillus thermoglucosidans*），这种菌的乙酸和乳酸合成途径已经通过基因手段敲除（Cripps et al. 2009）。

还有研究报道，可将柳枝稷木质纤维素生物质进行解构，这些生物质将可溶的纤维素、半纤维素和木质素在 78℃ 中用超嗜热厌氧菌 *Caldicellulosiruptor bescii* 进行预处理，除了主要产品有机酸以外还产生少量的乙醇（Kataeva et al. 2013；Basen et al. 2014）。Blumer-Schuette 等人（2014）已经对该领域进行了综合性概述，他以 *Caldicellulosiruptor* 家族酶目录作 CBP 测试。这些水解纤维素的酶存在一些值得注意的缺陷，所以应该尽可能使用单一的嗜热菌株作为“基因库”来补充具有这些缺陷的菌株。由于大部分嗜热菌株的遗传系统都是不完全的，这有利于为缺乏酶的 CBP 菌株进行额外补充而不是对产乙醇菌株进行基因改造。这些基因能够在遗传驯化的宿主如 *E. coli* 中克隆表达，与本章所提的硅结合 SBP 连接，使得酶可以结合在廉价的天然沸石上从而实现回收和循环利用。

Puri 等人（2013）简要回顾了酶在生物燃料生产中的应用前景，但他建议的用于固定化的基质对于生产而言太过昂贵。最近 Cho 等人（2012）提到在掺和了金的纳米磁性硅胶上共固定了三种纤维素酶用来降解纤维素。来自 *Trichoderma* 的一种内纤维素酶和一种外纤维素酶以及来自 *Thermotoga maritima*（海栖热袍菌）的 β-半乳糖苷酶被带上半胱氨酸标签，而且吸附在纳米金颗粒和掺和了金的纳米磁性颗粒上，并检测了它们水解可溶性和不可溶性底物的能力。掺和了金的纳米磁性颗粒看起来更加适合于固定化，但其活性随着时间推移而下降，据推测这是由酶外漏和内切纤维素酶与外切纤维素酶在实验温度 80℃中不稳定而失活造成。

通常而言，使用通过硅结合 SBP 固定化在天然硅胶和沸石的酶将有助于固定化酶在作用完后的简单移除和回收而且能够再次使用。这样的系统允许单个基因的简单转移和作用，使酶能从适当的生物质中产生糖，如阿拉伯糖和甘露糖，而且可以根据生物质的来源，混合和匹配酶比例来调整水解条件。例如，与硬木相比，来自松属的纤维素乙醇生产可能需要增加甘露聚糖酶和甘露糖苷酶的份量。同样的，在同一沸石基质上构筑水解途径是可行的，或者制造一个含有理想特性固定化酶的人工混合基质，这取决于底物选择和组合的方式。

酶解纤维素的经典方案含有三个主要的水解酶：内切葡聚糖酶、纤维素生物水解酶（外切纤维素酶）和 β-葡萄糖苷酶。所有这些酶都是作用于多糖中 β-1,4-糖苷键的水解酶（Wood and Garcia-Campayo 1990；Teeri 1997）。高温中的酶解可用于将多糖转化为可发酵糖类，用于生物技术应用，如生物燃料，但酶成本在工业生物转化和加工中是一个关键的因素。因此，通过 SBP 进行酶固定化代表了一种能提供具有功能的和稳定的生物催化剂的可行技术。图 23.8 显示参与经典的纤维素降解为葡萄糖过程的主要酶类，以及在上一节中所描述的生物催化模块如何引入多糖的一个简单的降解途径的简化例子。加工过程中的产物抑制（纤维二糖）可能会对参与酶的作用有不利影响。功能催化模块中，结合 SBP 的能力、廉价的固相基质（如沸石、硅胶）以及热稳定纤维素降解酶可使得加工更加简单并且有更好的操作稳定性与灵活性，尤其是对于连续水解和酶解回收。该方法不仅能更好地控制产物抑制效应，还可以提高酶的经济价值。

23.3.3 气相催化和 CO_2 捕获

酶作为催化性蛋白质已经在工业中广泛应用。然而，一个主要的限制是许多酶热不稳定，导致蛋白质变性，酶活消失。前期研究已经表明，当酶干燥时（含水量少于 3%），它们明显具有热稳定性。例如，干燥的胰蛋白酶在超过 100℃时表现出长期稳定性，而同样的蛋白酶充分水合后（含水量超过 20%）在 60℃中会快速变性（Mullaney 1966）。人们普遍认为蛋白酶水化是酶催化发生的必要条件。然而，当酶的水化足以让它们活化时，它们同样会变得不稳定。而当酶足够干燥而变得稳定时，它们又失去了活性。使蛋白酶具有活性所必需的水化程度有多种估算，公认的阈值大概为每克蛋白酶 0.2gH_2O，相当于 40%～50%的水化程度（Dunn and Daniel 2004）。

气相反应物进行生物催化是一种相对较新的开发的技术，许多研究已经聚焦在卤代化合物的水解上。气相催化有以下这些优点：避免添加溶剂，比水相系统更能限制副产物的产生，简化了纯化和加工。气相中的传质比溶液中的传质更有效，反应不可能被扩散所限制，而抑制性、挥发性和有毒性的物质可以被持续移除。而且，在脱水形式下的酶更能够抵抗热失活（Trivedi et al. 2006），因此与在水相系统相比可以用于更高的温度中。此外，非常规的反应也成为了可能，比如用水解酶进行转移反应或合成反应（Lamare and Legoy 1993）。

除了 Dunn 和 Daniel（2004）在酶活力所需的水解程度中报道的之外，几乎没有关于气相基质的基本催化机制的信息。Lind 等（2004）已经研究了以气相丁酸乙酯为底物的水合脂肪酶作为酯化酶的活性。这些实验可以直接比较不同水合物的酶活性且没有扩散限制，这

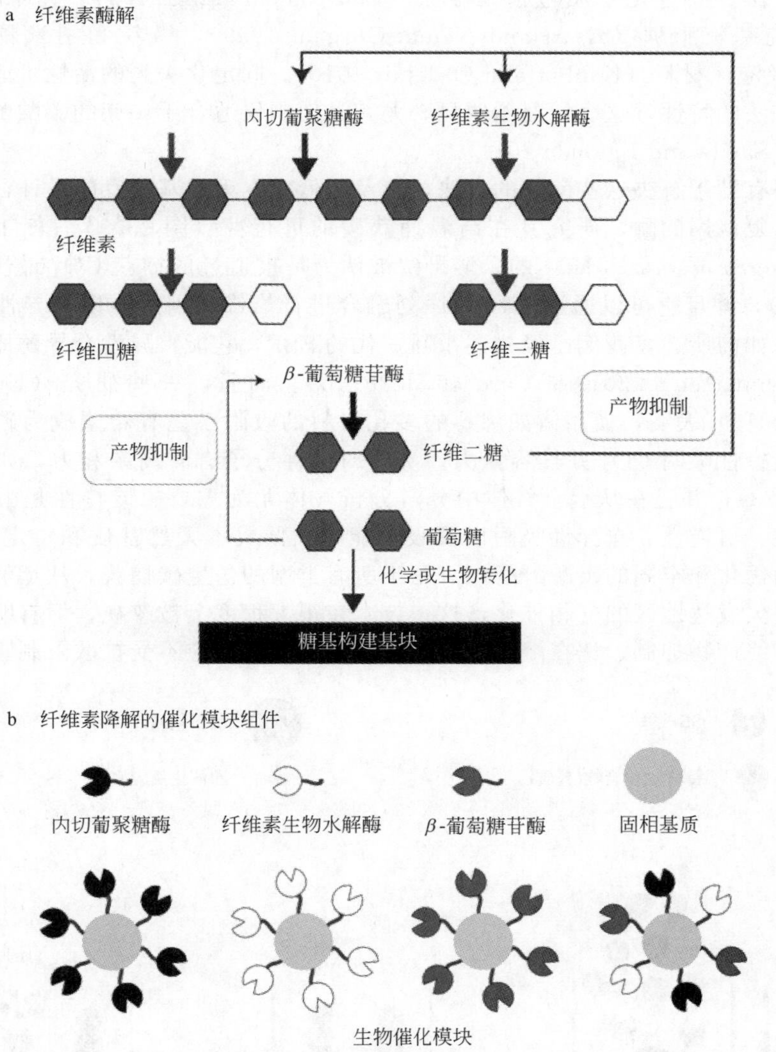

图 23.8　纤维素转化为葡萄糖。(a) 纤维素酶解的经典流程显示主要参与的三个酶：内切葡聚糖酶，纤维素生物水解酶和 β-葡萄糖苷酶。(b) 纤维素降解的催化模块组件的功能

些酶活性被证明比之前报道的酶活性低得多。

有大量证据表明化石燃料的燃烧是全球气候变化的主要原因，煤占世界能源供应量的 25%左右和 40%的碳排放，显著降低世界范围碳排放的煤碳捕获和封存的规模大得惊人。碳酸酐酶已经被鉴定具有加速从主要排放者中捕获 CO_2 的潜力，如仿生工艺中的燃煤发电机（Fisher et al. 2012）。最近对酶辅助碳捕获的研究进行了综述（Pierre 2012；Boone et al. 2013）。然而，目前要增强酶技术平台而不是取代现有的燃烧后溶剂加工来捕获碳。

碳酸酐酶可催化 CO_2 水化形成重碳酸和一个质子，转换数约为 10^6/s（Ho et al. 2008）。不幸的是，大部分天然分离的碳酸酐酶在燃烧后的苛刻条件下（温度在 50℃ 到超过 125℃，高浓度有机胺、重金属、氮硫氧化物）稳定性和活性都较差，限制了这些酶的应用，目前酶辅助捕获 CO_2 和储存工艺更进一步限制不良传质和抑制剂的快速积累。一种改良的仿生捕获 CO_2 系统需要：①更好或者理想的非限制界面传质；②不会导致构象活性丢失的固定化酶；③原位去除抑制剂。碳酸酐酶的固定化是最常见的用于稳定酶和限制暴露在碳捕获时的

变性条件的方法。固定化基质包括聚酰胺（Belzil and Parent 2005）、壳聚糖（Sharma et al. 2011）和烷基琼脂糖（Azari and Nemat-Gorgani 1999），微粒和纳米颗粒（Yadav et al. 2011）以及泡沫材料（Kanbar and Ozdemir 2010）。固定化丢失的活性可能较大，但由于使用的试验方法的特性，这个情况不明显，尤其是使用仅适合于α-族的碳酸酐酶的对硝基苯酚酯酶活性（Savile and Lalonde 2011）。

另一种将有助于解决这些问题的方法，涉及用低成本天然基质专门设计在连续气相过程中稳定和可重复使用的酶，研究其在高温捕获碳的可行性（图 23.9）。来自超嗜热微生物 *Sulphurhydrogenibium azorense* 的碳酸酐酶被认为是报道过的捕获 CO_2 最快的酶（Capasso et al. 2012），并且它可以通过与 SBP 序列融合进行修饰，因此可以特异性固定化在固态无机材料中。如前所述，我们已经描述了新一代的 SBP，它能特异性介导酶固定在合成和天然沸石上（Sunna et al. 2013a；Care et al. 2014b）。而且，一些研究（Kim et al. 2012；Hasan et al. 2013）表明，使用诸如沸石的多孔材料的吸附工艺有希望成为碳捕获的替代方案，因为它选择性吸附 CO_2 并且容量大。沸石对极性分子有着高亲和力，如对水和二氧化碳（Breck 1973），并且在从混合气体中分辨分析气体尤为重要。该特性是由于它独特的分子孔径、极性、可逆性、选择性吸附以及吸附能力。低成本天然基质和特定 SBP 组合会形成一种新型的优化和牢固的碳酸酐酶，可用于所有类型的仿生碳捕获，从烟气到发酵液。另外，它将促进欠发达地区的气相催化，这一技术在很大程度上被忽视，但有明显的优势，比如速度、没有终产物抑制、比在溶液中更有效传质，以及反应不受扩散限制。

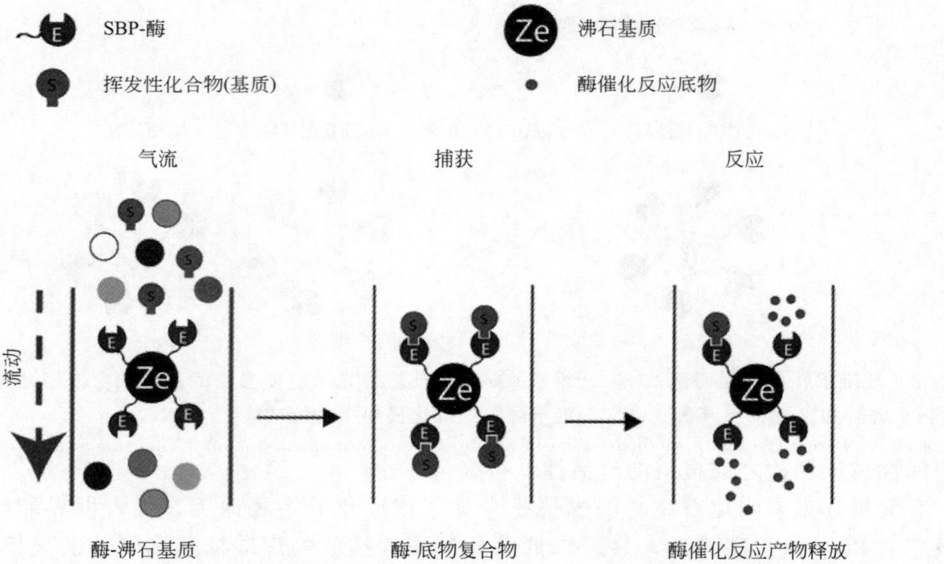

图 23.9 SBP-酶固定化技术用于以沸石和/或二氧化硅固体基质进行常规气相催化反应的应用概念

23.4 结论

SBP 可选择性识别和结合大部分固体材料。近年来，人们对与硅基固体基质（如二氧化硅、沸石和介孔二氧化硅）具有结合亲和力的 SBP 产生了相当大的兴趣，这些基质被认为是工业加工中最适合用于酶固定化的无机基质。在文献中有大量的关于 SBP 与硅基合成材料结合使用的信息，大部分用于纳米生物技术领域。然而，大部分涉及 SBP 介导蛋白质/

酶固定化的研究都已聚焦在生物传感器、构建纳米材料和官能化上。硅基支持基质如沸石和硅胶常用于许多工业用途，而且批量供应的准备也已经建立完善。低成本开采和获得的天然沸石代表一种重要的批量供应基质，能被多种应用工艺使用。然而，除了一些发表的关于使用天然沸石作为固定化基质的研究，几乎没有关于它们在附加应用中的信息。

在热稳定酶固定化方面已经提及硅结合 SBP 在高温和广泛 pH 范围内可保持稳定。将 SBP 序列与热稳定酶序列融合的基因构建已经证明了可以成功保留酶的催化能力以及重组融合蛋白中（SBP-酶）的 SBP 的固相结合能力，这可以被用于蛋白质回收和生物催化剂的再次使用。

SBP 的进一步应用被设想为分子工具箱的一部分，用来设计有预设功能和催化特性的稳定的基本构件（单一固定化酶），该构件可用于构建生物催化模块。这些生物催化模块能够被引入构建简单的水解途径或者组合在一起以适应多种底物成分，例如复合酶固定化技术生产糖类用于酒精发酵的应用。通过 SBPs 将超嗜热碳酸酐酶固定在硅基基质上，这种方法可用于碳捕获。这可以导致温室气体的减少，至少开发出一个工艺把 CO_2 当作“商业气体”的副产物来进行纯化，并且与地下封存和矿化相比，这貌似是一个更好的解决方案。然而，在碳捕获过程中气相催化剂的适用性还有待进一步研究。虽然 SBP 的特殊性和其生物医学相关方面已经占据了研究领域的主导地位，但仍会有许多环境友好的“绿色化学”机会，这可能会更适用于适当集中的研究项目中去。

参考文献

Abdelhamid MA，Motomura K，Ikeda T et al（2014）Affinity purification of recombinant proteins using a novel silica-binding peptide as a fusion tag. Appl Microbiol Biotechnol 98（12）：5677-5684

Azari F，Nemat-Gorgani M（1999）Reversible denaturation of carbonic anhydrase provides a method for its adsorptive immobilization. Biotechnol Bioeng 62（2）：193-199

Bachas-Daunert P，Sellers Z，Wei Y（2009）Detection of halogenated organic compounds using immobilized thermophilic dehalogenase. Anal Bioanal Chem 395（4）：1173-1178

Ball P（2001）Life's lessons in design. Nature 409（6818）：413-416

Bandlish RK，Michael Hess J，Epting KL et al（2002）Glucose-to-fructose conversion at high temperatures with xylose（glucose）isomerases from *Streptomyces murinus* and two hyperthermophilic *Thermotoga* species. Biotechnol Bioeng 80（2）：185-194

Baneyx F，Schwartz DT（2007）Selection and analysis of solid-binding peptides. Curr Opin Biotechnol 18（4）：312-317

Basen M，Rhaesa AM，Kataeva I et al（2014）Degradation of high loads of crystalline cellulose and of unpretreated plant biomass by the thermophilic bacterium *Caldicellulosiruptor bescii*. Bioresour Technol 152：384-392

Bell PJL，Sunna A，Gibbs MD et al（2002）Prospecting for novel lipase genes using PCR. Microbiology 148（8）：2283-2291

Belzil A，Parent C（2005）Methods of chemical immobilization of an enzyme on a solid support. Biochem Cell Biol 83（1）：70-77

Betancor L，Luckarift HR（2008）Bioinspired enzyme encapsulation for biocatalysis. Trends Biotechnol 26（10）：566-572

Blanco RM，Terreros P，Fernández-Pérez M et al（2004）Functionalization of mesoporous silica for lipase immobilization：characterization of the support and the catalysts. J Mol Catal B：Enzym 30（2）：83-93

Blumer-Schuette SE，Brown SD，Sander KB et al（2014）Thermophilic lignocellulose deconstruction. FEMS Microbiol Rev 38（3）：393-448

Bolivar JM，Rocha-Martin J，Mateo C et al（2009）Coating of soluble and immobilized enzymes with ionic polymers：full stabilization of the quaternary structure of multimeric enzymes. Biomacromolecules 10（4）：742-747

Boone CD，Gill S，Habibzadegan A et al（2013）Carbonic anhydrase：an efficient enzyme with possible global implications. Int J Chem Eng 2013：6

Brady D，Jordaan J（2009）Advances in enzyme immobilisation. Biotechnol Lett 31（11）：1639-1650

Branston S，Stanley E，Ward J et al（2013）Determination of the survival of bacteriophage M13 from chemical and physical challenges to assist in its sustainable bioprocessing. Biotechnol Bioprocess Eng 18（3）：560-566

Breck DW（1973）Zeolite molecular sieves：structure，chemistry，and use. Wiley，New York

Brown S（1992）Engineered iron oxide-adhesion mutants of the *Escherichia coli* phage lambda receptor. Proc Natl Acad Sci U S A 89（18）：8651-8655

Brown S (1997) Metal-recognition by repeating polypeptides. Nat Biotechnol 15 (3): 269-272

Brown S, Sarikaya M, Johnson E (2000) A genetic analysis of crystal growth. J Mol Biol 299 (3): 725-735

Camarero JA (2008) Recent developments in the site-specific immobilization of proteins onto solid supports. Pept Sci 90 (3): 450-458

Cao L (2006) Carrier-bound immobilized enzymes: principles, application and design. Wiley-VCH, Weinheim

Capasso C, De Luca V, Carginale V et al (2012) Biochemical properties of a novel and highly thermostable bacterial alpha-carbonic anhydrase from *Sulfurihydrogenibium yellowstonense* YO3AOP1. J Enzyme Inhib Med Chem 27 (6): 892-897

Care A, Chi F, Bergquist P et al (2014a) Biofunctionalization of silica-coated magnetic particles mediated by a peptide. J Nanopart Res 16 (8): 1-9

Care A, Nevalainen H, Bergquist P et al (2014b) Effect of *Trichoderma reesei* proteinases on the affinity of an inorganic-binding peptide. Appl Biochem Biotechnol 173 (8): 2225-2240

Care A, Bergquist PL, Sunna A (2015) Solid-binding peptides: smart tools for nanobiotechnology. Trends Biotechnol 33 (5): 259-268

Çaykara T, Güven O (1998) Effect of preparation methods on thermal properties of poly (acrylic acid) /silica composites. J Appl Polym Sci 70 (5): 891-895

Cetinel S, Caliskan HB, Yucesoy DT et al (2013) Addressable self-immobilization of lactate dehydrogenase across multiple length scales. Biotechnol J 8 (2): 262-272

Charbit A, Boulain JC, Ryter A et al (1986) Probing the topology of a bacterial membrane protein by genetic insertion of a foreign epitope; expression at the cell surface. EMBO J 5 (11): 3029-3037

Chen AY, Deng Z, Billings AN et al (2014) Synthesis and patterning of tunable multiscale materials with engineered cells. Nat Mater 13 (5): 515-523

Cherry JR, Fidantsef AL (2003) Directed evolution of industrial enzymes: an update. Curr Opin Biotechnol 14 (4): 438-443

Chhabra SR, Shockley KR, Ward DE et al (2002) Regulation of endo-acting glycosyl hydrolases in the hyperthermophilic bacterium *Thermotoga maritima* grown on glucan- and mannan- based polysaccharides. Appl Environ Microbiol 68 (2): 545-554

Chiu CY, Li Y, Huang Y (2010) Size-controlled synthesis of Pd nanocrystals using a specific mul- tifunctional peptide. Nanoscale 2 (6): 927-930

Cho EJ, Jung S, Kim HJ et al (2012) Co-immobilization of three cellulases on Au-doped magnetic silica nanoparticles for the degradation of cellulose. Chem Commun 48 (6): 886-888

Coil DA, Badger JH, Forberger HC et al (2014) Complete genome sequence of the extreme thermophile *Dictyoglomus thermophilum* H-6-12. Genome Announc 2 (1): e00109-e00114

Collino S, Evans JS (2008) Molecular specifications of a mineral modulation sequence derived from the aragonite-promoting protein n16. Biomacromolecules 9 (7): 1909-1918

Cowan DA, Fernandez-Lafuente R (2011) Enhancing the functional properties of thermophilic enzymes by chemical modification and immobilization. Enzyme Microb Technol 49 (4): 326-346

Cowan DA, Daniel RM, Morgan HW (1987) Some observations on the inhibition and activation of a thermophilic protease. Int J Biochem 19 (5): 483-486

Coyle BL, Baneyx F (2014) A cleavable silica-binding affinity tag for rapid and inexpensive protein purification. Biotechnol Bioeng 111 (10): 2019-2026

Cripps RE, Eley K, Leak DJ et al (2009) Metabolic engineering of *Geobacillus thermoglucosidasius* for high yield ethanol production. Metab Eng 11 (6): 398-408

Cui Y, Kim SN, Jones SE et al (2010) Chemical functionalization of graphene enabled by phage displayed peptides. Nano Lett 10 (11): 4559-4565

Dang X, Yi H, Ham M-H et al (2011) Virus-templated self-assembled single-walled carbon nanotubes for highly efficient electron collection in photovoltaic devices. Nat Nanotechnol 6 (6): 377-384

Dellomonaco C, Fava F, Gonzalez R (2010) The path to next generation biofuels: successes and challenges in the era of synthetic biology. Microb Cell Fact 9 (1): 3

Dong Q, Yan X, Zheng M et al (2014) Immobilization of a thermostable inorganic pyrophosphatase from the archaeon *Pyrococcus furiosus* onto amino-functionalized silica beads. J Appl Polym Sci 131 (17): 8669-8675

Dunn RV, Daniel RM (2004) The use of gas-phase substrates to study enzyme catalysis at low hydration. Phil Trans R Soc Lond 359 (1448): 1309-1320

Estephan E, Saab MB, Martin M et al (2011) Phages recognizing the indium nitride semiconductor surface via their peptides. J Pept Sci 17 (2): 143147

Evans JS, Samudrala R, Walsh TR et al (2008) Molecular design of inorganic binding polypeptides. MRS Bull 33: 514-518

Fernandez-Lafuente R (2009) Stabilization of multimeric enzymes: strategies to prevent subunit dissociation. Enzyme Microb Technol 45 (6-7): 405-418

Filho M, Pessela BC, Mateo C et al (2008) Immobilization-stabilization of an α-galactosidase from *Thermus* sp. strain T2 by covalent immobilization on highly activated supports: selection of the optimal immobilizationstrategy. Enzyme Microb Technol 42 (3): 265-271

Fischer L, Bromann R, Kengen SW et al (1996) Catalytical potency of beta-glucosidase from the extremophile *Pyro-*

coccus furiosus in glucoconjugate synthesis. Biotechnology (NY) 14 (1): 88-91

Fisher Z, Boone CD, Biswas SM et al (2012) Kinetic and structural characterization of thermostabilized mutants of human carbonic anhydrase Ⅱ. Protein Eng Des Sel 25 (7): 347-355

Forbes LM, Goodwin AP, Cha JN (2010) Tunable size and shape control of platinum nanocrystals from a single peptide sequence. Chem Mater 22 (24): 6524-6528

Furlong D (1982) Adsorption of Tris (2, 2′-bipyridine) ruthenium (Ⅱ) cations at silica/aqueous solution interfaces. Aust J Chem 35 (5): 911-917

Gabryelczyk B, Szilvay GR, Salomaki M et al (2013) Selection and characterization of peptides binding to diamond-like carbon. Colloids Surf B Biointerfaces 110: 66-73

Gabryelczyk B, Szilvay GR, Linder MB (2014) The structural basis for function in diamond-like carbon binding peptides. Langmuir 30 (29): 8798-8802

Gaffney D, Cooney J, Magner E (2012) Modification of mesoporous silicates for immobilization of enzymes. Top Catal 55 (16-18): 1101-1106

Garcia-Galan C, Berenguer-Murcia Á, Fernandez-Lafuente R et al (2011) Potential of different enzyme immobilization strategies to improve enzyme performance. Adv Synth Catal 353 (16): 2885-2904

Gaskin DJH, Starck K, Vulfson EN (2000) Identification of inorganic crystal-specific sequences using phage display combinatorial library of short peptides: a feasibility study. Biotechnol Lett 22 (15): 1211-1216

Ghose S, McNerney TM, Hubbard B (2004) Preparative protein purification on underivatized silica. Biotechnol Bioeng 87 (3): 413-423

Gibbs MD, Reeves RA, Sunna A, Bergquist PL (1999) Sequencing and expression of a β-mannanase gene from the extreme thermophile *Dictyoglomus thermophilum* Rt46B. 1, and characteristics of the recombinant enzyme. Curr Microbiol 39 (6): 351-357

Goede K, Busch P, Grundmann M (2004) Binding specificity of a peptide on semiconductor surfaces. Nano Lett 4 (11): 2115-2120

Goward CR, Murphy JP, Atkinson T et al (1990) Expression and purification of a truncated recombinant streptococcal protein G. Biochem J 267 (1): 171-177

Greiner L, Schroder I, Muller DH et al (2003) Utilization of adsorption effects for the continuous reduction of NADP+ with molecular hydrogen by *Pyrococcus furiosus* hydrogenase. Green Chem 5 (6): 697-700

Gülay S, Şanlı-Mohamed G (2012) Immobilization of thermoalkalophilic recombinant esterase enzyme by entrapment in silicate coated Ca-alginate beads and its hydrolytic properties. Int J Biol Macromol 50 (3): 545-551

Gungormus M, Fong H, Kim IW et al (2008) Regulation of in vitro calcium phosphate mineralization by combinatorially selected hydroxyapatite-binding peptides. Biomacromolecules 9 (3): 966-973

Hanefeld U, Gardossi L, Magner E (2009) Understanding enzyme immobilisation. Chem Soc Rev 38 (2): 453-468

Hartmann M, Kostrov X (2013) Immobilization of enzymes on porous silicas: benefits and challenges. Chem Soc Rev 42 (15): 6277-6289

Hasan MMF, First EL, Floudas CA (2013) Cost-effective CO_2 capture based on in silico screening of zeolites and process optimization. Phys Chem Chem Phys 15 (40): 17601-17618

Heinz H, Farmer BL, Pandey RB et al (2009) Nature of molecular interactions of peptides with gold, palladium and pd-au bimetal surfaces in aqueous solution. J Am Chem Soc 131 (28): 9704-9714

Hickey AM, Ngamsom B, Wiles C et al (2009) A microreactor for the study of biotransformations by a cross-linked γ-lactamase enzyme. Biotechnol J 4 (4): 510-516

Hidalgo A, Betancor L, Lopez-Gallego F et al (2003) Design of an immobilized preparation of catalase from *Thermus thermophilus* to be used in a wide range of conditions: structural stabilization of a multimeric enzyme. Enzyme Microb Technol 33 (2-3): 278-285

Hnilova M, Oren EE, Seker UOS et al (2008) Effect of molecular conformations on the adsorption behavior of gold-binding peptides. Langmuir 24 (21): 12440-12445

Ho L-F, Li S-Y, Lin S-C et al (2004) Integrated enzyme purification and immobilization processes with immobilized metal affinity adsorbents. Process Biochem 39 (11): 1573-1581

Ho MT, Allinson GW, Wiley DE (2008) Reducing the cost of CO_2 capture from flue gases using membrane technology. Ind Eng Chem Res 47 (5): 1562-1568

Hodgman CE, Jewett MC (2012) Cell-free synthetic biology: thinking outside the cell. Metab Eng 14 (3): 261-269

Hoess RH (2001) Protein design and phage display. Chem Rev 101 (10): 3205-3218

Hollinshead W, He L, Tang YJ (2014) Biofuel production: an odyssey from metabolic engineering to fermentation scale-up. Front Microbiol 5: 344

Huang Y, Chiang C-Y, Lee SK et al (2005) Programmable assembly of nanoarchitectures using genetically engineered viruses. Nano Lett 5 (7): 1429-1434

Hudson S, Cooney J, Magner E (2008) Proteins in mesoporous silicates. Angew Chem Int Ed 47 (45): 8582-8594

Hui KS, Chao CYH (2008) Methane emissions abatement by multi-ion-exchanged zeolite A prepared from both commercial-grade zeolite and coal fly ash. Environ Sci Technol 42 (19): 7392-7397

Ikeda T, Kuroda A (2011) Why does the silica-binding protein "Si-tag" bind strongly to silica surfaces? Implications of conformational adaptation of the intrinsically disordered polypeptide to solid surfaces. Colloids Surf B Biointerfaces 86 (2): 359-363

Ikeda T, Hata Y, Ninomiya K-i et al (2009) Oriented immobilization of antibodies on a silicon wafer using Si-tagged protein A. Anal Biochem 385 (1): 132-137

Inoue I, Watanabe K, Yamauchi H et al (2014) Biological construction of single-walled carbon nanotube electron transfer pathways in dye-sensitized solar cells. ChemSusChem 7 (10): 2805-2810

Johnson A, Zawadzka A, Deobald L et al (2008) Novel method for immobilization of enzymes to magnetic nanoparticles. J Nanopart Res 10 (6): 1009-1025

Jørgensen F, Hansen OC, Stougaard P (2004) Enzymatic conversion of d-galactose to d-tagatose: heterologous expression and characterisation of a thermostable l-arabinose isomerase from *Thermoanaerobacter mathranii*. Appl Microbiol Biotechnol 64 (6): 816-822

Kacar T, Ray J, Gungormus M et al (2009a) Quartz binding peptides as molecular linkers towards fabricating multifunctional micropatterned substrates. Adv Mater 21 (3): 295-299

Kacar T, Zin MT, So C et al (2009b) Directed self-immobilization of alkaline phosphatase on micro-patterned substrates via genetically fused metal-binding peptide. Biotechnol Bioeng 103 (4): 696-705

Kanbar B, Ozdemir E (2010) Thermal stability of carbonic anhydrase immobilized within polyurethane foam. Biotechnol Prog 26 (5): 1474-1480

Kataeva I, Foston MB, Yang S-J et al (2013) Carbohydrate and lignin are simultaneously solubilized from unpretreated switchgrass by microbial action at high temperature. Energy Environ Sci 6 (7): 2186-2195

Khare SD, Fleishman SJ (2013) Emerging themes in the computational design of novel enzymes and protein-protein interfaces. FEBS Lett 587 (8): 1147-1154

Kim J, Lin L-C, Swisher JA et al (2012) Predicting large CO_2 adsorption in aluminosilicate zeolites for postcombustion carbon dioxide capture. J Am Chem Soc 134 (46): 18940-18943

Kjaergaard K, Sørensen JK, Schembri MA et al (2000) Sequestration of zinc oxide by fimbrial designer chelators. Appl Environ Microbiol 66 (1): 10-14

Klint D, Eriksson H (1997) Conditions for the adsorption of proteins on ultrastable zeolite Y and its use in protein purification. Protein Expr Purif 10 (2): 247-255

Ko S, Park TJ, Kim H-S et al (2009) Directed self-assembly of gold binding polypeptide-protein A fusion proteins for development of gold nanoparticle-based SPR immunosensors. Biosens Bioelectron 24 (8): 2592-2597

Kogel JE, Trivedi NC, Barker JM et al (2006) Industrial minerals & rocks: commodities, markets, and uses, 7th edn. Society for Mining, Metallurgy, and Exploration, Littleton

Kosuri S, Church GM (2014) Large-scale de novo DNA synthesis: technologies and applications. Nat Methods 11 (5): 499-507

Kress J, Zanaletti R, Amour A et al (2002) Enzyme accessibility and solid supports: which molecular weight enzymes can be used on solid supports? An investigation using confocal Raman microscopy. Chem Eur J 8 (16): 3769-3772

Kulp JL, Shiba K, Evans JS (2005) Probing the conformational features of a phage display polypeptide sequence directed against single-walled carbon nanohorn surfaces. Langmuir 21 (25): 11907-11914

Kwok R (2010) Five hard truths for synthetic biology. Nature 463 (7279): 288-290

Lamare S, Legoy M-D (1993) Biocatalysis in the gas phase. Trends Biotechnol 11 (10): 413-418

Lee SW, Mao C, Flynn CE et al (2002) Ordering of quantum dots using genetically engineered viruses. Science 296 (5569): 892-895

Lehn C, Schmidt H-L (1997) Stability and stabilization of enzymes from mesophilic and thermophilic organisms in respect to their use in FIA-systems for the determination of L-lactate and acetate. J Chem Technol Biotechnol 69 (2): 161-166

Li CM, Botsaris GD, Kaplan DL (2002) Selective in vitro effect of peptides on calcium carbonate crystallization. Cryst Growth Des 2 (5): 387-393

Li YJ, Whyburn GB, Huang Y (2009) Specific peptide regulated synthesis of ultrasmall platinum nanocrystals. J Am Chem Soc 131 (44): 15998-15999

Lichty JJ, Malecki JL, Agnew HD et al (2005) Comparison of affinity tags for protein purification. Protein Expr Purif 41 (1): 98-105

Lind PA, Daniel RM, Monk C et al (2004) Esterase catalysis of substrate vapour: enzyme activity occurs at very low hydration. Biochim Biophys Acta 1702 (1): 103-110

Love DR, Fisher R, Bergquist PL (1988) Sequence structure and expression of a cloned β-glucosidase gene from an extreme thermophile. Mol Gen Genet 213 (1): 84-92

Lu Y, Zhao J, Zhang R et al (2014) Tunable lifetime multiplexing using luminescent nanocrystals. Nat Photon 8 (1): 32-36

Lynd LR, Weimer PJ, van Zyl WH et al (2002) Microbial cellulose utilization: fundamentals and biotechnology. Microbiol Mol Biol Rev 66 (3): 506-577

Magner E (2013) Immobilisation of enzymes on mesoporous silicate materials. Chem Soc Rev 42 (15): 6213-6222

Masica DL, Schrier SB, Specht EA et al (2010) De novo design of peptide-calcite biomineralization systems. J Am Chem Soc 132: 12252-12262

Miroliaei M (2007) Studies on the activity and stability of immobilized thermophilic alcohol dehydrogenase. Sci Iranica 14 (2): 112-117

Morris DD, Gibbs MD, Chin CW et al (1998) Cloning of the xynB gene from *Dictyoglomus thermophilum* Rt46B. 1 and action of the gene product on kraft pulp. Appl Environ Microbiol 64 (5): 1759-1765

Mueller S, Coleman JR, Wimmer E (2009) Putting synthesis into biology: a viral view of genetic engineering through de novo gene and genome synthesis. Chem Biol 16 (3): 337-347

Mullaney PF (1966) Dry thermal inactivation of trypsin and ribonuclease. Nature 210 (5039): 953

Naber JE, de Jong KP, Stork WHJ et al (1994) Industrial applications of zeolite catalysis. In: Weitkamp J, Karge HG, Pfeifer H, Hölderich W (eds) Studies in surface science and catalysis, vol 84. Elsevier, Amsterdam, pp 2197-2219

Naik RR, Brott LL, Clarson SJ et al (2002a) Silica-precipitating peptides isolated from a com- binatorial phage display peptide library. J Nanosci Nanotechnol 2 (1): 95-100

Naik RR, Stringer SJ, Agarwal G et al (2002b) Biomimetic synthesis and patterning of silver nanoparticles. Nat Mater 1 (3): 169-172

Nel AE, Madler L, Velegol D et al (2009) Understanding biophysicochemical interactions at the nano-bio interface. Nat Mater 8 (7): 543-557

Ngamsom B, Hickey AM, Greenway GM et al (2010) Development of a high throughput screening tool for biotransformations utilising a thermophilic l-aminoacylase enzyme. J Mol Catal B: Enzym 63 (1-2): 81-86

Nochomovitz R, Amit M, Matmor M et al (2010) Bioassisted multi-nanoparticle patterning using single-layer peptide templates. Nanotechnology 21 (14): 145305

Notman R, Oren EE, Tamerler C et al (2010) Solution study of engineered quartz binding peptides using replica exchange molecular dynamics. Biomacromolecules 11 (12): 3266-3274

Nygaard S, Wendelbo R, Brown S (2002) Surface-specific zeolite-binding proteins. Adv Mater 14 (24): 1853-1856

Opwis K, Knittel D, Bahners T et al (2005) Photochemical enzyme immobilization on textile carrier materials. Eng Life Sci 5 (1): 63-67

Opwis K, Straube T, Kiehl K et al (2014) Various strategies for the immobilization of biocatalysts on textile carrier materials. Chem Eng Trans 38: 223-238

Oren EE, Tamerler C, Sahin D et al (2007) A novel knowledge-based approach to design inorganic binding peptides. Bioinformatics 23 (21): 2816-2822

Oren EE, Notman R, Kim IW et al (2010) Probing the molecular mechanisms of quartz-binding peptides. Langmuir 26: 11003-11009

Palomo JM, Segura RL, Mateo C et al (2004) Improving the activity of lipases from thermophilic organisms at mesophilic temperatures for biotechnology applications. Biomacromolecules 5 (1): 249-254

Park TJ, Zheng S, Kang YJ et al (2009) Development of a whole-cell biosensor by cell surface display of a gold-binding polypeptide on the gold surface. FEMS Microbiol Lett 293 (1): 141-147

Pasunooti S, Surya W, Tan SN et al (2010) Sol-gel immobilization of a thermophilic diguanylate cyclase for enzymatic production of cyclic-di-GMP. J Mol Catal B: Enzym 67 (1-2): 98-103

Patolsky F, Zheng G, Lieber CM (2006) Fabrication of silicon nanowire devices for ultrasensitive, label-free, real-time detection of biological and chemical species. Nat Protoc 1 (4): 1711-1724

Pender MJ, Sowards LA, Hartgerink JD et al (2006) Peptide-mediated formation of single-wall carbon nanotube composites. Nano Lett 6 (1): 40-44

Pessela BCC, Mateo C, Fuentes M et al (2004) Stabilization of a multimeric β-galactosidase from *Thermus* sp. strain T2 by immobilization on novel heterofunctional epoxy supports plus aldehyde-dextran cross-linking. Biotechnol Prog 20 (1): 388-392

Pierre AC (2004) The sol-gel encapsulation of enzymes. Biocatal Biotransfor 22 (3): 145-170

Pierre AC (2012) Enzymatic carbon dioxide capture. ISRN Chem Eng 2012: 22

Piller K, Daniel RM, Petach HH (1996) Properties and stabilization of an extracellular alpha- glucosidase from the extremely thermophilic archaebacteria *Thermococcus* strain AN1: enzyme activity at 130 degrees C. Biochim Biophys Acta 1292 (1): 197-205

Popat A, Hartono SB, Stahr F et al (2011) Mesoporous silica nanoparticles for bioadsorption, enzyme immobilisation, and delivery carriers. Nanoscale 3 (7): 2801-2818

Puddu V, Perry CC (2012) Peptide adsorption on silica nanoparticles: evidence of hydrophobic interactions. ACS Nano 6 (7): 6356-6363

Puri M, Barrow CJ, Verma ML (2013) Enzyme immobilization on nanomaterials for biofuel production. Trends Biotechnol 31 (4): 215-216

Raghuvanshi S, Gupta R (2010) Advantages of the immobilization of lipase on porous supports over free enzyme. Protein Pept Lett 17 (11): 1412-1416

Raia CA, D'Auria S, Guagliardi A et al (1995) Characterization of redox proteins from extreme thermophilic archaebacteria: studies on alcohol dehydrogenase and thioredoxins. Biosens Bioelectron 10 (1-2): 135-140

Rhodes CJ (2010) Properties and applications of zeolites. Sci Prog 93 (Pt 3): 223-284

Rocha-Martin J, Vega DE, Cabrera Z et al (2009) Purification, immobilization and stabilization of a highly enantioselective alcohol dehydrogenase from *Thermus thermophilus* HB27 cloned in *E. coli*. Process Biochem 44 (9): 1004-1012

Roy MD, Stanley SK, Amis EJ et al (2008) Identification of a highly specific hydroxyapatitebinding peptide using phage display. Adv Mater 20 (10): 1830-1836

Sano K-I, Miura A, Yoshii S et al (2013) Nonvolatile flash memory based on biologically integrated hierarchical nanostructures. Langmuir 29 (40): 12483-12489

Sapsford KE, Algar WR, Berti L et al (2013) Functionalizing nanoparticles with biological molecules: developing chemistries that facilitate nanotechnology. Chem Rev 113 (3): 1904-2074

Sarikaya M, Tamerler C, Jen AKY et al (2003) Molecular biomimetics: nanotechnology through biology. Nat Mater 2 (9): 577-585

Sassolas A, Blum LJ, Leca-Bouvier BD (2012) Immobilization strategies to develop enzymatic biosensors. Biotechnol Adv 30 (3): 489-511

Savile CK, Lalonde JJ (2011) Biotechnology for the acceleration of carbon dioxide capture and sequestration. Curr Opin Biotechnol 22 (6): 818-823

Seeman NC, Belcher AM (2002) Emulating biology: building nanostructures from the bottom up. Proc Natl Acad Sci U S A 99 (2): 6451-6455

Seker UOS, Wilson B, Dincer S et al (2007) Adsorption behavior of linear and cyclic genetically engineered platinum binding peptides. Langmuir 23 (15): 7895-7900

Serizawa T, Techawanitchai P, Matsuno H (2007) Isolation of peptides that can recognize syndiotactic polystyrene. ChemBioChem 8 (9): 989-993

Sharma A, Bhattacharya A, Shrivastava A (2011) Biomimetic CO_2 sequestration using purified carbonic anhydrase from indigenous bacterial strains immobilized on biopolymeric materials. Enzyme Microb Technol 48 (4-5): 416-426

Sheldon RA (2011) Characteristic features and biotechnological applications of cross-linked enzyme aggregates (CLEAs). Appl Microbiol Biotechnol 92 (3): 467-477

Sheldon RA, van Pelt S (2013) Enzyme immobilisation in biocatalysis: why, what and how. Chem Soc Rev 42 (15): 6223-6235

Simpson HD, Haufler UR, Daniel RM (1991) An extremely thermostable xylanase from the thermophilic eubacterium *Thermotoga*. Biochem J 277 (Pt 2): 413-417

Slocik JM, Naik RR (2014) Peptide-nanoparticle strategies, interactions and challenges. In: Walsh TR, Knecht MR (eds) Bio-inspired nanotechnology. Springer, New York, pp 1-16

Smith GP (1985) Filamentous fusion phage: novel expression vectors that display cloned antigens on the virion surface. Science 228 (4705): 1315-1317

So CR, Kulp JL, Oren EE et al (2009) Molecular recognition and supramolecular self-assembly of a genetically engineered gold binding peptide on Au {111}. ACS Nano 3 (6): 1525-1531

Sunna A (2010) Modular organisation and functional analysis of dissected modular β-mannanase *Cs*Man26 from *Caldicellulosiruptor* Rt8B. 4. Appl Microbiol Biotechnol 86 (1): 189-200

Sunna A, Bergquist PL (2003) A gene encoding a novel extremely thermostable 1, 4-β-xylanase isolated directly from anenvironmental DNA sample. Extremophiles 7 (1): 63-70

Sunna A, Gibbs MD, Bergquist PL (2000) A novel thermostable multidomain 1, 4-β-xylanase from '*Caldibacillus cellulovorans*' and effect of its xylan-binding domain on enzyme activity. Microbiology 146 (11): 2947-2955

Sunna A, Chi F, Bergquist PL (2013a) A linker peptide with high affinity towards silica-containing materials. N Biotechnol 30 (5): 485-492

Sunna A, Chi F, Bergquist PL (2013b) Efficient capture of pathogens with a zeolite matrix. Parasitol Res 112 (7): 2441-2452

Taguchi H, Matsuzawa H, Ohta T (1984) l-Lactate dehydrogenase from *Thermus caldophilus* GK24, anextremely thermophilic bacterium. Eur J Biochem 145 (2): 283-290

Tamerler C, Oren EE, Duman M et al (2006) Adsorption kinetics of an engineered gold binding peptide by surface plasmon resonance spectroscopy and a quartz crystal microbalance. Langmuir 22 (18): 7712-7718

Tamerler C, Khatayevich D, Gungormus M et al (2010) Molecular biomimetics: GEPI-based biological routes to technology. Biopolymers 94 (1): 78-94

Tang Z, Palafox-Hernandez JP, Law WC et al (2013) Biomolecular recognition principles for bionanocombinatorics: an integrated approach to elucidate enthalpic and entropic factors. ACS Nano 7 (11): 9632-9646

Taniguchi K, Nomura K, Hata Y et al (2007) The Si-tag for immobilizing proteins on a silica surface. Biotechnol Bioeng 96 (6): 1023-1029

Tavolaro A, Tavolaro P, Drioli E (2006) Influence of synthesis parameters on vanadium-silicalite-1 crystal growth prepared with fluoride-containing media. J Cryst Growth 289 (2): 609-616

Teeri TT (1997) Crystalline cellulose degradation: new insight into the function of cellobiohydrolases. Trends Biotechnol 15 (5): 160-167

Thai CK, Dai H, Sastry MSR et al (2004) Identification and characterization of Cu_2O- and ZnO- binding polypeptides by *Escherichia coli* cell surface display. J Biotech Bioeng 87 (2): 129-137

Tran CTH, Nosworthy NJ, Kondyurin A et al (2013) CelB and β-glucosidase immobilization for carboxymethyl cellulose hydrolysis. RSC Adv 3 (45): 23604-23611

Trivedi AH, Spiess AC, Daussmann T et al (2006) Study on mesophilic and thermophilic alcohol dehydrogenases in gas-phase reaction. Biotechnol Prog 22 (2): 454-458

Turner P, Mamo G, Karlsson E (2007) Potential and utilization of thermophiles and thermostable enzymes in biorefining. Microb Cell Fact 6 (1): 9

Umlauf BJ, McGuire MJ, Brown KC (2014) Introduction of plasmid encoding for rare tRNAs reduces amplification bias in phage display biopanning. Biotechniques 58 (2): 81-84

Walsh TR (2014) Fundamentals of peptide-materials interfaces. In: Walsh TR, Knecht MR (eds) Bio-inspired nanotechnology. Springer, New York, pp 17-36

Wang SQ, Humphreys ES, Chung SY et al (2003) Peptides with selective affinity for carbon nanotubes. Nat Mater 2 (3): 196-200

Whaley SR, English DS, Hu EL et al (2000) Selection of peptides with semiconductor binding specificity for directed nanocrystal assembly. Nature 405 (6787): 665-668

Wilson SA, Peek K, Daniel RM (1994) Immobilization of a proteinase from the extremely thermophilic organism *Thermus* Rt41A. Biotechnol Bioeng 43 (3): 225-231

Wilson L, Palomo JM, Fernández-Lorente G et al (2006) Improvement of the functional properties of a thermostable lipase from *Alcaligenes* sp. via strong adsorption on hydrophobic supports. Enzyme Microb Technol 38 (7): 975-980

Wittrup KD (2001) Protein engineering by cell-surface display. Curr Opin Biotechnol 12 (4): 395-399

Wood T, Garcia-Campayo V (1990) Enzymology of cellulose degradation. Biodegradation 1 (2-3): 147-161

Yadav R, Satyanarayanan T, Kotwal S et al (2011) Enhanced carbonation reaction using chitosanbased carbonic anhydrase nanoparticles. Curr Sci 100 (4): 520-524

Yang M, Choi BG, Park TJ et al (2011) Site-specific immobilization of gold binding polypeptide on gold nanoparticle-coated graphene sheet for biosensor application. Nanoscale 3 (7): 2950-2956

Ye X, Honda K, Sakai T et al (2012) Synthetic metabolic engineering-a novel, simple technology for designing a chimeric metabolic pathway. Microb Cell Fact 11 (1): 120

You C, Zhang YHP (2013) Cell-free biosystems for biomanufacturing. In: Zhong J-J (ed) Future trends in biotechnology, vol 131. Springer, Berlin, pp 89-119

Zhang YHP, Myung S, You C et al (2011) Toward low-cost biomanufacturing through in vitro synthetic biology: bottom-up design. J Mater Chem 21 (47): 18877-18886

Zhou Z, Hartmann M (2012) Recent progress in biocatalysis with enzymes immobilized on mesoporous hosts. Top Catal 55 (16-18): 1081-1100

第二十四章
嗜冷酶结构与功能关系的分子动力学研究

Elena Papaleo[1][2], Matteo Tiberti[3],
Gaetano Invernizzi[1]

24.1 嗜冷酶的稳定性和结构柔性

阐明嗜冷酶热稳定性、活性与动力学结构机理之间的关系在嗜冷酶的基础研究和工业应用方面是极为重要的（Gerday et al. 2000）。事实上，适冷微生物产生的酶类通常在低温条件下才能保持极高活性，同时其热不稳定性和特异性，使得嗜冷酶适用于大范围的工业应用，因此，酶的冷适应研究也有着更广泛、更普遍的意义。此外，也必须考虑到适冷微生物能够在非常具有挑战性和限制性的生境中生长和繁殖。值得注意的是，适冷微生物的适冷进化特征的代谢通量与中温微生物在最佳生长温度下展现出的代谢通量相当（van den Burg 2003；Struvay and Feller 2012）。不同适冷微生物的不同适应机制已经有一些研究报道。已有研究发现，生物在适应极端低温环境中会产生大量嗜冷酶类。事实上，如果将温度作为一个基本的环境因子，当环境温度从37℃降低到10℃时反应速率会显著降低（Feller and Gerday 2003；Siddiqui and Cavicchioli 2006）。然而嗜冷微生物在温度低至－20℃时仍能记录到其具有生物活性（Cary et al. 2010），因此嗜冷微生物需要进化和优化出嗜冷酶类以适应极端寒冷环境。

嗜冷酶相对于中温酶的一般特性是其具有较强的热不稳定性以及较低温度下具有较高的催化效率（k_{cat}/K_m）（Siddiqui and Cavicchioli 2006；Struvay and Feller 2012）。嗜冷酶的最适温度较中温酶明显下移，表明嗜冷酶的热稳定性差，在中温情况下容易解构并失活。同时，嗜冷酶的冷适应通常是“不完整的”，因为大多数的嗜冷酶在0℃时的酶活性虽然很高，但通常还是比中温酶在37℃时的酶活性低（Siddiqui and Cavicchioli 2006）。

k_{cat}/K_m 的优化通常是通过增大 k_{cat} 和减小 K_m 或在不同的低温酶中同时优化两个动力

[1] Structural Biology and NMR Laboratory, Department of Biology, University of Copenhagen, Copenhagen, Denmark.

[2] e-mail：elena. papaleo@bio. ku. dk.

[3] Department of Biotechnology and Biosciences, University of Milano-Bicocca, Piazza della Scienza 2, 20126 Milan, Italy.

参数（Siddiqui and Cavicchioli 2006；Struvay and Feller 2012）。k_{cat} 的增加与一个催化反应的活化自由能的降低有关，特别是在减少活化焓的方面。据推测，在结构上减少反应过渡态的焓驱动的交互反应数目就能达到减少活化焓的目的。这些方面都可视为嗜冷酶具有低稳定性以及高结构韧性的依据，至少是嗜冷酶在催化位点附近区域所具有的特性。这是首次提出嗜冷酶的活性和热稳定性之间的关系，同时是从具有更高结构韧性的蛋白质三维结构方面反映出来的（Fields 2001；Somero 2004）。

尽管如此，适冷酶活性、柔性以及热稳定性之间的关系仍然存在争议。适冷酶的热不稳定性以及其在低温时酶活性的保持表明这两者间存在一定的联系。值得注意的是，在低温情况下酶活性的保持需要更弱的分子间相互作用，这就使得蛋白质的稳定性降低。然而，这种关系并不清晰和直观，因为热稳定性的差异也可能与遗传性漂移相关。这可能只是低温环境下稳定的酶缺乏进化压力的结果（Wintrode and Arnold 2000；Leiros et al. 2007；Fedøy et al. 2007）。与此同时，具有热稳定性和高催化活性的低温酶（Leiros et al. 2007；Fedøy et al. 2007）证明在体外具有分解稳定性与酶活性的研究也有报道（Wintrode and Arnold 2000；Jónsdóttir et al. 2014）。所有这些结果都使酶活性-稳定性-结构柔性之间关系的定义更具挑战性。

关于适冷酶的结构重排、蛋白质柔性、催化活性以及热稳定性之间联系的假设促使科学界逐步开展与适冷酶原子细节以及关键氨基酸残基替换（如基因突变）有关的机制研究。

24.2 局部结构柔性与酶的适冷性

适冷酶柔性的增加已被证明不一定需要蔓延至整个三维结构，柔性结构可以只存在特定的区域，从而影响催化位点周围结构的柔性变化（Olufsen et al. 2005；Papaleo et al. 2006，2008，2011a，b；Chiuri et al. 2009；Pasi et al. 2009；Mereghetti et al. 2010；Tiberti and Papaleo 2011；Martinez et al. 2011；Isaksen et al. 2014）。

嗜冷酶较差的热稳定性支持了其柔性假说，但柔性的定义和测定的困难，导致验证十分困难。事实上，可以依据嗜冷酶动态运动以及在时间尺度上的变化或者在特定温度下相关结构的变形程度来定义其结构的柔性。

在原子水平研究溶液中蛋白质的柔性是一项极具挑战性的任务，只有少数实验技术可以用于该研究，如核磁共振（NMR）、电子顺磁共振（EPR）或中子散射以及最近 X 射线晶体学中的电子密度分析（Fraser and Jackson 2011）。在本章中不会详细介绍这些技术，而是重点讲述分子动力学模拟的使用，例如计算手段用于适冷酶研究。尽管如此，上述实验技术（Tehei et al. 2005；Heidarsson et al. 2009；van den Bedem and Fraser 2015），尤其是 NMR 在结构模拟与整合方面更有前途，但不能完全依赖这些方法，至于原因将在接下来的内容中进行介绍。

24.3 蛋白质结构的集合描述和动力学的重要性

在过去十年里，在冷适应领域中研究者开始使用实验和计算手段解决酶冷适应的难题，越来越多的证据支持蛋白质结构与功能以及酶动力学与活性之间的关系（Henzler-Wildman et al. 2007；Henzler-Wildman and Kern 2007；Nashine et al. 2010）。此外，蛋白质是动态而不是静态实体的概念开始被蛋白质研究界广泛接受，它能够更好地描述蛋白质在溶液中的构象（Hilser et al. 2006；Acuner Ozbabacan et al. 2010；Woldeyes et al. 2014）。

事实上，蛋白质不是静态的分子，它们的构象在不同的时间尺度上存在着不同的状态。其中有些运动也被称为蛋白质结构的“呼吸运动”。不同状态之间的转换通常取决于折叠构象残基的协同运动，这种铰链和摇摆运动的时间跨度一般从 10^{-8}s 到 10^{3} s。这表明蛋白质侧链和主链可存在多种构象，包括旋转、拉伸或扭转运动（10^{-12} s）从而可以构成大的结构改变（Eisenmesser et al. 2005)。

因此，在适冷酶的结构和计算研究中，考虑使用时间尺度的研究技术至关重要。传统的 MD（分子动力学方法）模拟可以达到微秒级（10^{-6}s），在特殊情况下可以达到毫秒级（10^{-3} s)，然而当涉及同种蛋白质在溶液中时间尺度上与动力学模拟的比较时，这种方法仍然存在局限性。事实上，MD 模拟经常会使蛋白质在局部有限空间中储存时间过长。为了克服传统 MD 模拟的局限性，增强采样方法耦合到原子的 MD 物理模型被提出（Spiwok et al. 2014；Barducci et al. 2015)，例如温度或哈密顿变换、变化动力学方法或使用 NMR 化学位移的限制模拟方法（Camilloni et al. 2012，2013；Camilloni and Vendruscolo 2014)。后面两种方法在毫秒级别上描述蛋白质动力学更具优势（Camilloni et al. 2012；Sutto and Gervasio 2013；Palazzesi et al. 2013；Papaleo et al. 2014b)。

如上所述，酶催化涉及的蛋白质特定部位结构的变化，使得酶可以容纳底物以及达到酶促反应的过渡态。这种构象变化的发生可能是催化效率的决定因素。许多蛋白质即使是在自由状态下（未结合底物)，都可以观察到类似于功能状态的构象，包括底物结合、配体结合或过渡态的构象（Boehr et al. 2009；Ma and Nussinov 2010；Kosugi and Hayashi 2012)。这些普遍的状态只占溶液中游离酶所有构象的一小部分（少于 10%)。然而这些构象的具体细节难以被探知（Baldwin and Kay 2009；Mittermaier and Kay 2009)。

在蛋白质动力学领域的发现也会影响研究适冷结构的决定因素的方法。蛋白质结构的整体研究方案已经广泛应用于研究适冷酶并作为这些酶的生物化学表征的补充技术。嗜冷酶的分子动力学模拟在适冷研究中处于领先地位。

24.4 蛋白质分子动力学模拟：概述、局限性和优势

蛋白质结构的整体性描述以及不同时间尺度上酶学动力学的研究对蛋白质功能研究至关重要，这也有助于从原子水平上对适冷酶进行理解。分子模拟就是开展这些研究比较合适的计算方法。在该领域的不同技术中，显式溶剂分子动力学是其中最有前途的技术之一。事实上，这种技术可以对蛋白质溶液的原子层面描述以及在不同的时间尺度上（从皮秒到毫秒级）对样品进行动态采样。这种方法多年前就已经应用于蛋白质研究（McCammon et al. 1977)，最近，三位杰出的科学家在蛋白质模拟领域获得了诺贝尔奖（Smith and Roux 2013；Nussinov 2014)。一些综述和书籍章节里已经详细地介绍过这种技术（Rapaport 1998；Dror et al. 2012；Leach 2001；van Gunsteren and Berendsen 1990)。尽管如此，还是需要回顾下这些重要的实践技术，这对于将这些方法用于适冷酶研究的研究者来说是至关重要的。分子模拟的简要方案在图 24.1 中展示并讨论。

分子模拟是依据现有研究系统的模型开展的，即尽可能精准地展示其系统行为。原则上，可以利用计算机借助物理学第一原理模拟原子运动模型。然而，对于上千原子组成的蛋白质以及不同时间尺度上蛋白质存在的不同状态来说，使用量子力学（QM）理论研究分子模拟是不大可能实现的。因此，急需开发其他的替代方法。事实上，被广泛接受的玻恩-奥本海默近似法允许在实验条件下研究解耦系统的原子核和电子的行为。这是因为重核的瞬时配置中的电子云能快速达到平衡。这就可以避免考虑精确的电子分布，并且原子核的运动可

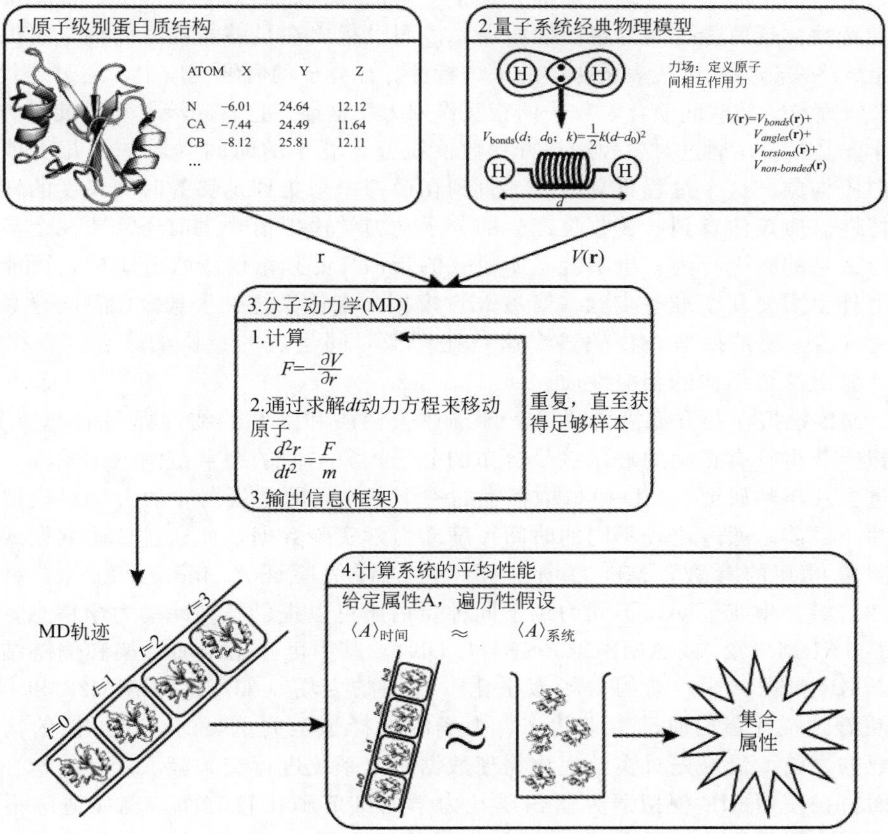

图 24.1 MD 模拟流程示意图。每一个模拟都开始于两条必需的信息。第一条是模板蛋白质的具体结构信息（1），模板蛋白质所有原子或重原子的三维坐标必须明确。一旦模板蛋白质选定，需要对模拟蛋白质添加溶剂分子（通常为水）、离子（通常为 Na^+ 或 Cl^-）使得模拟条件更接近真实情况。第二步是选取适应的原子间相互作用模型（2）。在分子力学框架内，原子间相互作用通常依赖于传统力学，因此，静电作用不需要被明确考虑。例如，势能定义为符合胡克定律的共价键。由许多潜在的势能值共同建立起来的总势能函数 V，这完全取决于一组参数（统称为所谓的“力场”）和原子的位置。一旦满足这些要求，可以进行几个准备步骤以满足系统所需的热力学条件（在这里不显示）。最后，运行分子动力学模拟（3），以及一些连续的构象被输出（MD 轨迹）（4）。运行计算后，系统的平均性能可以根据轨迹数据计算出，好比根据玻尔兹曼分布采样收集获得一个独立的结构（5）。这可能是因为遍历性假设，该假设指出在足够长的时间，模拟的蛋白质的计算性能最终是在热力学中代表蛋白质的平均性能

以用一个核势能面（PES）表示。得到 PES 数据，就可以使用经典力学来理解原子核的动力学，进而研究感兴趣的蛋白质或生物分子的动力学。但是，由于已经舍弃第一原理，所以就需要以某种方式来定义核是如何相互作用的。因此需要一个势能函数 V（$\boldsymbol{r}$，p），利用一些参数 p 和原子的三维坐标 $\boldsymbol{r}$ 确定系统的能量。这种势函数被称为力场。力场的设计能充分体现关键的系统物理研究，这通常是通过建立力场来达到的，以便它逼近从头计算奥本海默表面的相关区域，或者正确地再现实验数据。生物分子动力学模拟的典型力场的函数形式是许多有用点的总和，每个有用点代表单个分子的行为。原子间的键长和键角围绕平衡值振动，符合胡克定律。这种周期性的变化能够引起或产生扭距。库伦力产生的势能需要考虑孤对电子的静电相互作用，而原子间的范德瓦尔斯相互作用则是通常通过伦纳德-琼斯势模型

表现出来。分子模拟的一个目标是能够在给予一定的选择的热力学条件下，计算蛋白质原子运动的平均特性。这可能是因为遍历性假设，该假设指出在足够长的时间内，模拟的蛋白质的计算性能最终在热力学中代表蛋白质的平均特性。在分子动力学模拟中，通过迭代求解牛顿运动方程来研究粒子的时间演化，粒子的相互作用力场描述。这首先涉及从势能函数作用于原子的力场计算。然后，通过对牛顿运动方程数值积分计算小的时间（通常在几飞秒）跳跃从而计算蛋白质构象。这个过程重复多次，直到在模拟中采集到足够的时间尺度的样本来研究感兴趣的特性。应该注意到，在收集产生的分子动力学模拟信号之前还需要几个步骤，以保证系统进行适当的物理行为。事实上，所研究的蛋白质必须溶解于水分子中，可能还需要添加离子，另外也需要几个准备步骤（简短 MD 模拟）来使系统处于选定的热力学条件下。

MD 模拟的主要结果是 MD 轨迹，这是在模拟时间生成的结构的集合。根据遍历性假设，可以计算出这个结构的相应性质。

总之，MD 模拟在两方面需要明确，即描述蛋白质和溶剂的物理模型的选择（即力场）以及在模拟中获得具有正确的玻尔兹曼分布的分子构象空间的覆盖范围（即采样）。

根据过去几年的研究，MD 模拟中存在的一个主要问题是蛋白质动力学对模拟所选的力场的敏感性。目前，通过比较不同的时间尺度动力学实验数据，如通过核磁共振或其他生物物理光谱测量得到的参数，MD 力场已被广泛的评估验证（Lindorff-Larsen et al. 2012；Best et al. 2012）。事实证明，不同力场下同种蛋白质可以提供不同的动力学信息，一些新的力场（如 CHARMM22* 或 AMBER99SB*-ILDN），其中可修正添加骨架和侧链结构，参比蛋白质的 NMR 数据，这些新的力场似乎比一些旧的力场（如 CHARMM22 和 OLPS versions）能更好地描述蛋白质动力学状态。力场的选择是至关重要的，在可能的情况下，在得出模拟数据之前，最好先对实验生物物理数据研究系统进行交叉验证。

一方面，在使用 MD 模拟研究蛋白质动力学的细节和比较方面，例如适应不同条件的不同类型的酶或同系物（如极端酶），一些力场已经达到的高精确度是非常令人欣慰。事实上，最精确的力场已被证明能够将微小快速折叠结构域展开并重新折叠为其天然结构（Lindorff Larsen et al. 2011；Piana et al. 2013），对折叠蛋白质的天然空间构象采样（Lindorff-Larsen et al. 2012；Martín-García et al. 2015）并定量估计酶的最小状态数以及这些状态突变引起的影响（Papaleo et al. 2014b）。另一方面，原子的经典力场不完善，仍存在许多局限性，例如高估溶液中盐桥的影响（Debiec et al. 2014；Jónsdóttir et al. 2014），尤其是不涉及静电网络或溶剂暴露的带电残基（Jónsdóttir et al. 2014），甚至处理与金属蛋白酶或酶结合的金属离子（Calimet and Simonson 2006；Zhu et al. 2013；Li et al. 2013）以及导致极化效应（Halgren and Damm 2001）。因此，需要在这些方向上进行改进，特别是在适冷酶的研究中，其中许多研究目标是金属结合酶或盐桥作用是否可以提高对温度的适应性。

24.5 适冷酶的折叠漏斗模型

D'Amico 等（2003）提出在嗜冷酶折叠漏斗模型中，整合嗜冷酶生物化学和生物物理数据，该折叠漏斗模型描述了适冷酶的折叠-展开反应。他们提出适应不同的温度酶可以呈现出不同形状的漏斗（图 24.2）。漏斗的高度表示蛋白质折叠所需的自由能，这与构象稳定性有关。漏斗顶部表示蛋白质未折叠状态。在这个模型中，适冷蛋白漏斗的边缘比适温蛋白更大，相当于未折叠状态蛋白质在适冷蛋白质中占更大的比例。蛋白质折叠进程中，自由能降低，同时整体构象更接近于蛋白质结构，这与传统蛋白质折叠漏斗理论相一致（Onuchic and Wolynes 2004）。漏斗外壁在嗜冷酶和嗜热酶中呈现不同程度的粗糙度（图 24.2）。嗜热蛋白的漏斗模型整体皱褶明显。由于适冷酶结构一般不存在中间体，分子内相互作用较少，因此漏斗外壁陡峭且平滑。适温蛋白折叠隧道的底部具有单个全局（能量）最小值（区），这个最小区与其他众多的能量次小区之间有大的能障，并且折叠隧道底部整体上的构象自由度很小。适冷蛋白的折叠漏斗模型底部更宽，并且粗糙不平，因为这代表了一组由低能障隔开的构象，使得蛋白质更脆弱更有柔性。根据一个构象的选择方案（Ma and Nussinov 2010），底物将与酶发生相互作用并结合到部分酶分子上，因此可以观察到这部分酶转化成为易与底物结合的构象，最终导致整体“活性”结构。在适冷酶的折叠漏斗底部崎岖不平的情况下，上述不同构象状态的相互转换要求适度的自由能，从而解释在增加柔性中的作用，例如促进底物的结合。

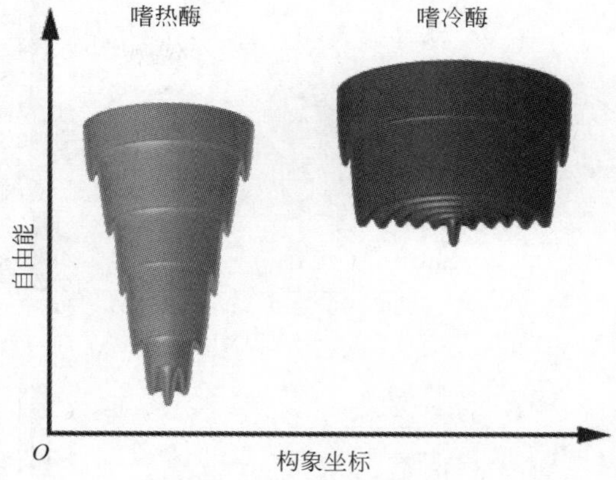

图 24.2　嗜热、嗜冷酶的折叠漏斗模型。蛋白质折叠的自由能由函数纵坐标表示。漏斗的顶部表示蛋白质未折叠构象，而底部表示天然构象。不同高度的漏斗和底部的粗糙程度表明不同的构象状态之间相互转换的能障，以及两极端微生物之间的结构性波动程度不同

适冷漏斗模型的计算研究是一项具有挑战性的任务，因为它要求在蛋白质折叠和未折叠状态下适当的采样以及描述蛋白质在折叠和展开状态下的表面自由能。此外，蛋白质的自由能形貌图（free energy landscape，FEL）是一个复杂的多维图形，因此在模拟中需要将其减少到两或三个坐标（反应坐标）。描述 FEL 的反应坐标必须是集体变量，即它们需要捕捉整体构象的主要特征。FEL 的描述是严格依赖于选择的反应坐标。应用坐标的合适描述可以是特殊分类下残基的均方根偏差（Rmsd）（即在一个循环中预计有不同的适冷/适温偏好构象的同系物）、回转半径、主成分分析（PCA）的第一主成分或联系网络。在一个平衡的热力学系统中，自由能可以用一个或多个反应坐标的概率密度函数来估计（Poland 2001）。

在冷适应酶中，已经介绍了丝氨酸蛋白酶和尿嘧啶-DNA 糖基化酶的多复制全原子 MD 模拟的 FEL 的定性表示（Papaleo et al. 2009；Mereghetti et al. 2010）。这项研究仅限于描述适冷和适温蛋白的底部折叠漏斗模型，也就是天然状态下与构象波动的相关区域。希望这项研究可以促进这类研究的进一步开展，旨在描述适应不同温度酶的未折叠状态。在研究中，使用不同的属性，例如分子模拟 PCA 的主成分（Amadei et al. 1993；Garcia et al. 1992），催化部位周围回路的均方根偏差以及蛋白质回转半径作为反应坐标（图 24.3）。这些计算研究允许测试上述适冷酶折叠漏斗模型的假设。事实上，适冷酶和适温酶分子动力学模拟 FEL 图表明了适冷蛋白变体有固定的趋势，以发现更多的结构构象和更崎岖的底部构造，这有利于几个亚稳态构象之间的相互转换。

最近，许多增强蛋白质构象的空间定量采样方法被提出（Spiwok et al. 2014；Barducci et al. 2015），这些方法可以与原子 MD 力场结合使用。因此，未来一个有趣的方向将是从 MD 实验研究扩展到整个 FEL 研究，包括适冷、适温同源蛋白质折叠-展开的研究，前提是能在温度高于当前参数化的溶剂模型的情况下优化 MD 模拟的溶剂模型。直到多温度范围的溶剂模型被开发，才能在不同的温度条件下依靠 MD 模拟研究极端酶。同时另一个良好的研究策略是将这种计算方法与实验结合起来（Invernizzi et al. 2009；Ganjalikhany et al. 2012）。

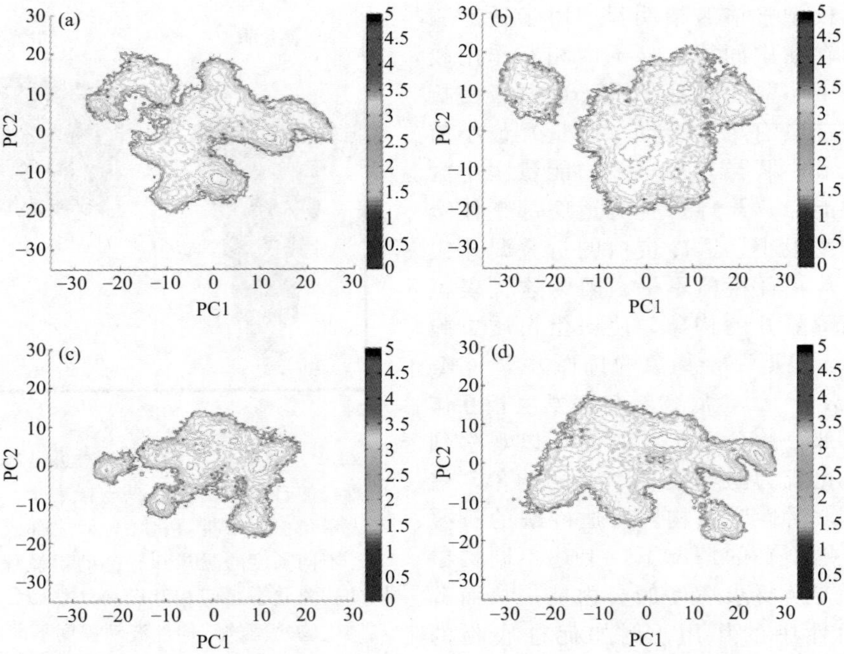

图 24.3 适冷酶和适温酶的二维 FEL 图（见彩图）。以 MD 轨迹主成分分析的两个主分量作为反应坐标，中温（a，b）和嗜冷（c，d）丝氨酸蛋白酶分别在 283K（a，c）和 310 K（b，d）下（Mereghetti 等. 2010）

24.6 结构需求

由于冷适应酶的结构数据的可用性相对较低，适冷蛋白用于比较目的的有限结构数据是一个巨大的约束和限制。必须存在几种适应不同温度和高分辨原子结构的同源酶，以便进行适当的比较。MD 目前多数用于中温酶和嗜冷酶的比较研究，当然其对高温酶的三维结构分析也是可用的（Tiberti and Papaleo 2011；Sigtryggsdóttir et al. 2014）。

为了克服嗜冷酶及其同系物的三维结构的缺乏问题，同源模拟及其他模拟技术（Eswar et al. 2008）原则上在扩大用于比较可用结构的集合上是有用的。然而，结构模拟需要注意许多事情。事实上，同源模型得到的结构较实际结构并不非常准确，同时其精确性随着目标和模板之间的序列同一性和相似性的降低而显著降低。如果认为适温酶和适冷酶之间的差异是由细微结构差别导致的，那么同源模型本身和缺少实验验证的情况下并不是最好的策略。这个问题在 MD 应用方面更显重要，因为开始于模型的模拟受到许多限制，经常遇到样本空间的构象与有限的功能和结构相关的影响（Fan and Mark 2004；Raval et al. 2012）。因此，当采用模型来描述适冷酶的结构和动力学时，结合实验是必要的（Parravicini et al. 2013；Papaleo et al. 2014a）。

也应该记住，模型只是一种用于克服缺少实验结构的一种方法，但实验方法解决嗜冷微生物的未知蛋白质结构也是必需的。

24.7 适冷酶的分子动力学模拟回顾

表 24.1 报告了 1999～2015 年已发表的经典 MD 模拟研究。就模拟长度、力场选择、重

复数、温度和分析方法等方面对目前为止的适冷酶或适冷酶与适温酶的比较进行了简要总结。

表 24.1 1999～2015 年发表的低温酶模拟研究综述

蛋白酶	长度和重复数	力场	温度/K	分析方法	参考文献
胰蛋白酶（嗜冷和中温酶）	0.5～1.2ns	CHARMM22	300	rmsf、2D 结构	Br and sdal et al.（1999）
胰蛋白酶（嗜冷和中温酶）	6ns（4 重复）	GROMOS87	300	rmsf、2D 结构、SAS（统计分法软件）、氢键	Papaleo et al.（2005）
尿嘧啶-DNA-糖基酶（嗜冷、中温酶和突变型）	1～2ns	AMBER95	300	rmsf	Olufsen et al.（2005）
嗜热菌素（嗜冷、中温和嗜热酶）	3ns	AMBER94	273 和 300	rmsf、盐桥、阳离子-π、氢键、腔	Adekoya et al.（2006）
弹性蛋白酶（嗜冷和中温酶）	12ns（4 重复）	GROMOS87	283 和 310	聚类、2D 结构、氢键、rmsf、SAS、盐桥、S^2 序参量	Papaleo et al.（2006，2007）
尿嘧啶-DNA-糖基酶（嗜冷和中温酶）	6～9ns	Parm99	375，400 和 425	SAS、2D 结构、rmsf、氢键	Olufsen et al.（2007）
α-淀粉酶、柠檬酸合成酶、苹果酸脱氢酶、碱性蛋白酶、木聚糖酶（嗜冷、中温和嗜热酶）	1～1.4ns	GROMOS87	300 280～330	rmsf、PCA（主分量分析）	Spiwok et al.（2007）
胰蛋白酶（嗜冷和中温酶）	10ns（4 重复）	GROMOS96	300	rmsf、PCA、2D 结构、S^2 序参量	Papaleo et al.（2008）
嗜热菌素（嗜冷、中温和嗜热酶）	30 ns	CHARMM22	280 和 310	rmsf、氢键、盐桥	Xie et al.（2009）
弹性蛋白酶（嗜冷酶）	10ns	GROMOS96	277，293 和 318	rmsf、SAS、2D 结构、氢键	D'Auria et al. 2009 Aurilia et al.（2009）
α-淀粉酶（嗜冷、中温和嗜热酶）	6～11ns（6 重复）	GROMOS87	298	氢键、2D 结构、rmsf、各向异性温度因子、自相关运动组织映射、PCA	Pasi et al.（2009）
弹性蛋白酶和尿嘧啶-DNA-糖基酶（嗜冷和中温酶）	12ns（10～12 重复）	GROMOS96	283 和 310	PCA、FEL、聚类、位形熵	Mereghetti et al.（2010）
枯草杆菌蛋白酶（嗜冷、中温和嗜热酶）	20ns（4～8 重复）	GROMOS96 43a1	283，300 和 343	PCA2D 结构、氢键、SAS、rmsf、盐桥网络	Tiberti（2009）

续表

蛋白酶	长度和重复数	力场	温度/K	分析方法	参考文献
枯草杆菌蛋白酶（嗜冷、耐热酶和突变型）	50ns	未标明	283 和 363	PCA、rmsf、2D 结构、聚类	Martinez et al. (2011)
α-淀粉酶（嗜冷、中温酶和突变型）	6～10ns（6 重复）	GROMOS96	300	2D 结构、疏水指数、PCA、氢键、盐桥、rmsf、各向异性温度因子、自组织图	Papaleo et al. (2011)，Fraccalvieri et al. (2012)
GH5 纤维素酶（嗜冷、中温、耐热和嗜热酶）	10ns	AMBERff03	300,400 和 500	2D 结构、rmsf、SAS	Badieyan et al. (2012)
几丁质酶（嗜冷酶）	10ns	GROMOS96	273,288 和 300	聚类、rmsf、SAS、盐桥、氢键	Ramli et al. (2012)
酯酶（嗜冷酶）	30ns（3 重复）	AMBERff99SB	278,308 和 323	rmsf、水的径向分布函数、NMA、PCA、距离波动图	Ganjalikhany et al. (2012)
碱性磷酸酶（嗜冷酶）	40～60ns（4 重复）	GROMOS96 43a1	303	2D 结构、PCA、rmsf、盐桥、PSN、相关运动	Papaleo et al. (2013)
酰基氨基肽酶（嗜冷酶）	60ns	GROMOS96 43a1	290	2D 结构、盐桥	Parravicini et al. (2013)
酰基氨基肽酶（嗜冷酶）	90ns（7 重复）	CHARMM22-CMAP	290	2D 结构、分子内网络、PSN	Papaleo et al. (2014)
枯草杆菌蛋白酶（嗜冷和嗜热酶）	100ns（2 重复）	AMBER ff99SB-ILDN	300	rmsf、SAS	Sigtryggsdóttir et al. (2014)
胰蛋白酶（嗜冷和中温酶）	100ns	OPLS2005	300	rmsf	Isaksen et al. (2014)

注：在此总结了用经典力场对适冷酶进行原子模拟的方法，其他类型的计算技术（如量子力学方法或粗粒度方法）不在此清单中，同时也不包括对没有酶活性的冷适应蛋白酶的研究。

最早适冷酶的 MD 研究可追溯到 1999 年，在 1～2ns 时间尺度上使用单一 MD 运行适冷酶和适温酶。

最近许多 MD 研究中，多复制 MD 模拟的使用保证了更广泛的构象采样和识别适冷/温酶柔性差异可能。结果的可重复性是 MD 模拟的一个关键点，使用多复制 MD 模拟不仅可以弥补采样不足的问题，也允许验证两个不同的系统的差异是否具有统计学意义还是仅仅只是一个模型的独特性。

到目前为止发表的适冷酶 MD 模拟分析使用了许多不同的工具，不仅研究了其柔性特征，还包括其长距离通信，分子内和分子间的相互作用网络变化，结构变化以及不同状态的构象。

在分析 MD 数据以及应用于适冷酶研究的技术中，已经提到了 PCA（也称基本动力学，ED），其目的是通过降低轨迹整体的复杂性和隔离较大振幅的运动来提取多维空间中有用的运动方向。

极端酶 MD 模拟中常用的柔性评价指标是每个氨基酸残基的波动均方根（rmsf）（图 24.4）或者是 MD 轨迹的各向异性温度因子，该因子可作为模拟的平均结构或 S2 反计算广义序参量。使用 rmsf 或 β 因子评价 MD 模拟蛋白质的柔性需要谨慎。它们十分依赖于平均结构作为参考，而获取柔性数据更优的方法应该是计算这些指标在较短的时间内的完整模拟长度，看两个不同的蛋白质的灵活性是否有统计学差异。事实上，由于适冷和适温同源蛋白

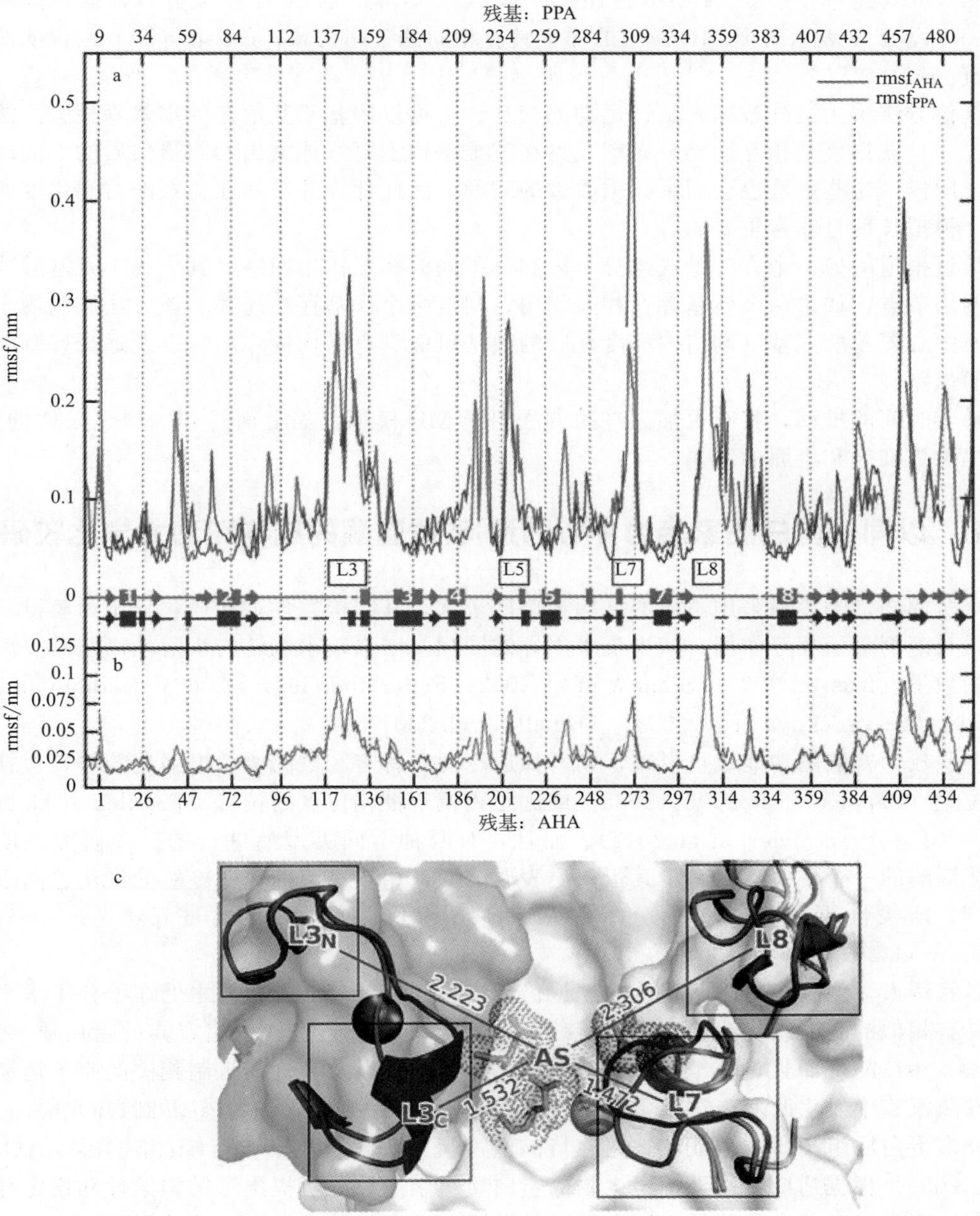

图 24.4　适冷蛋白（AHA）和适温蛋白（PPA）rmsf 以及 Cα 原子 MD 的柔性分析（见彩图）。根据两种蛋白质结构比对对两种蛋白质 rmsf 进行比较分析，对应的对齐间隙断裂处用粗线表示。(a) 在这里展示整个级联轨迹 rmsf 分析（即合并所有重复区域）。(b) rsmf 超过 100 ps 间隔显示。(c) 柔性环在 3D 结构中突出显示。距离是整体平均数，计算基团到中心的距离，距离单位 nm（Pasi 等 2009）

质具有相似的三级结构，以及少数氨基酸的替换，rmsf 或 β 因子的差异更可能是波动强度的差异，而不是某些峰的存在或不存在的差异。因此，如果 rmsf 强度的差异不是由分析的噪音引起而是真实存在的，那么该变化位点是至关重要的。

其他分析方法旨在定义适冷和适温蛋白质变体构象的相似性，即定义两个蛋白质构象之间的重复区域。这种分析方法使用了很长时间，比方说在 PCA 分析基础上计算均方根内积（rmsip）或其他重叠参数。rmsip 往往是计算 PCA 矩阵的前 20 个主成分，数据范围从 0 到 1。rmisp 等于 1 表示两个 MD 模拟的样本具有相同的空间构象；rmisp 等于 0 则说明两个构象垂直。

其他方法采用矩阵数据来估计运动的交互性，可以用基于互信息的皮尔森相关方法进行研究。一旦获得残基相互运动分布图，可在三维结构上更好地突出不同蛋白质的不同动力学模式。同样，这些方法也会面临数据收敛的问题，因此计算并平均在更短的时间尺度所涵盖的整个模拟长度是非常重要的。

最近报道的另一个方法是描述分子内和分子间的相互作用网络，其中每个氨基酸残基是一个网络节点，如果两个残基存在相互作用，则这两个残基存在连线。该方法可以用于识别蛋白质中心氨基酸残基（网络内高度互联的残基可能具有结构性作用）以及远端残基间的远距离通讯路径。

随着时间的推移，将有更加先进的方法用于 MD 模拟，如上面提到的增强采样的方法，可以提供更准确和全面的结果。

24.8 以同一蛋白质家族为中心的适冷/适温酶的结构和动力学比较研究

众所周知，嗜冷酶采用不同的结构策略用于增强其结构柔性，如弱化分子内氢键、优化蛋白质与溶剂分子相互作用、减少疏水中心的堆积、增强疏水侧链的水溶性以及减少离子对网络的数量（Russell 2000；Gianese et al. 2002；Feller and Gerday 2003；Siddiqui and Cavicchioli 2006；Adekoya et al. 2006；Tronelli et al. 2007）。

一些较为普遍的功能也可以被识别，如适冷酶在特定区域的柔性增强的趋势，包括催化中心位置（Papaleo et al. 2006，2008）或远距离调节催化位点的位置（Papaleo et al. 2011a，2012，2013；Fraccalvieri et al. 2012）。而且，对两种不同类型酶进行的三项研究表明，适冷和适温酶的一个主要不同在于其蛋白质表面结构，相较适温酶，适冷酶的进化趋向于更加“柔软”，即柔性更强和更少的分子内相互作用（Pasi et al. 2009；Papaleo et al. 2011a；Isaksen et al. 2014）。

尽管诱人，但一个适冷酶普适的理论仍无法建立。策略上的相似之处在于分子水平上蛋白质具有相似的空间结构或相似的功能残基，如某些属于同家族或超家族的蛋白质（Papaleo et al. 2011b）。而且，分子间和分子内相互作用网络以及与这些机制相关的原子运动可以在不同酶家族中发生改变，这与蛋白质空间折叠密切相关（Papaleo et al. 2011a，b）。

许多蛋白质组学研究也证实缺乏独特的适冷机制理论。事实上，Gu 和 Hilser（Gu and Hilser 2009）根据适应不同环境微生物的蛋白质组学组分分析提出了结构柔性和稳定性的同源调制，而且温度适应的分子机制在不同种类蛋白质中仍然可能有显著差异。

到目前为止所有的研究显示，每种酶采用不同的结构策略来适应低温，这种策略往往难以被准确识别。

在这种情况下，应该牢记比较研究，就像用 MD 模拟实现的那样，嗜冷蛋白质与中温蛋白质序列的相似性应该相对较高才能使得比较研究有意义。如果蛋白质结构存在明显差异，精细的结构效应难以被评估。寻找足以使蛋白质适应寒冷的微妙修饰是很重要的。为了解决这些问题，序列相似（60%～70%）且适应不同温度的同源酶被用于比较MD分析，并且已在某些情况下应用（Papaleo et al. 2008；Sigtryggsdóttir et al. 2014）。鼓励在原子水平上进行适冷研究，使之成为更常规且普遍的方法。

例如，适冷丝氨酸蛋白酶的MD研究，阐明了该超家族的不同成员是如何解决低温对蛋白质的活性和稳定性的不利影响的（Papaleo et al. 2006，2007，2008；Isaksen et al. 2014）。嗜冷和中温胰蛋白酶（或弹性蛋白酶）的分离是在胰蛋白酶与弹性蛋白酶分离之后进行的，这表明适冷胰蛋白酶和弹性蛋白酶独立进化出相同的策略，即在较低的温度下优化结构的柔性，这是分子趋同进化的一个引人注目的例子（Papaleo et al. 2008）。

本章提到的许多结构和计算研究表明，尽管无法提出一个适冷酶适冷机制的普适理论，但具有相似功能和空间构象的酶采用相同的方案改变它们的结构柔性和稳定性，使得它们能够在极低温环境下行使功能，这表明同源适冷酶在结构和动力学上具有的进化趋同性。这些证据表明即使共同的结构策略在适冷酶中无法确立，但以家族为中心的观点在适冷、适温酶的比较研究中是非常有用的。

24.9 未来前景

在过去二十年中，MD模拟展现了其在适冷酶结构和动力学性质研究方面的潜能。现在到了该领域取得更大进展的时候了。使用新的更准确的力场以及新的方法增强模拟研究中的采样准确性或解释来自NMR数据的实验限制，例如不同时间尺度上的动力学探针计算原子的化学位移。无论是超出微秒尺度的采样亦或是结果的可重复性评估，在传统MD的应用中是必须改进的。新的方法也可以分析MD数据，用于识别如蛋白质在溶液状态下不同构象转变的差异以及构象集合中主要和次要状态的差异。MD模拟在嗜冷酶的应用研究方面也有很大的潜力。MD模拟提供了一个依赖于计算的技术，其可用于筛选和设计增加局部柔性或增强蛋白质稳定性的突变体。

无利益冲突声明

Elena Papaleo，Matteo Tiberti 和 Gaetano Invernizzi 声明他们没有利益冲突。

参考文献

Acuner Ozbabacan SE，Gursoy A，Keskin O，Nussinov R（2010）Conformational ensembles，signal transduction and residue hot spots：application to drug discovery. Curr Opin Drug Discov Devel 13：527-537

Adekoya OA，Helland R，Willassen N-P，Sylte I（2006）Comparative sequence and structure analysis reveal features of cold adaptation of an enzyme in the thermolysin family. Proteins 62：435-449

Amadei A，Linssen AB，Berendsen HJ（1993）Essential dynamics of proteins. Proteins 17：412-425

Aurilia V，Rioux-Dube JF，Marabotti A，Pezolet M，D'Auria S（2009）Structure and dynamics of cold-adapted enzymes as investigated by FT-IR spectroscopy and MD. The case on an esterase from Pseudoalteromonas haloplanktis. J Phys Chem B 113：7753-7761

Badieyan S，Bevan DR，Zhang C（2012）Study and design of stability in GH5 cellulases. Biotechnol Bioeng 109：31-44

Baldwin AJ，Kay LE（2009）NMR spectroscopy brings invisible protein states into focus. Nat Chem Biol 5：808-814

Barducci A，Pfaendtner J，Bonomi M（2015）Tackling sampling challenges in biomolecular simulations. Methods Mol Biol 1215：151-171

Best RB，Zhu X，Shim J et al（2012）Optimization of the additive CHARMM all-atom protein force field targeting improved sampling of the backbone φ，ψ and side-chain χ（1）and χ（2）dihedral angles. J Chem Theory Comput 8：3257-3273

Boehr DD，Nussinov R，Wright PE（2009）The role of dynamic conformational ensembles in biomolecular recognition. Nat Chem Biol 5：789-796

Brandsdal BO，Heimstad ES，Sylte I，Smalås AO（1999）Comparative molecular dynamics of mesophilic and psychrophilic protein homologues studied by 1.2 ns simulations. J Biomol Struct Dyn 17：493-506

Calimet N，Simonson T（2006）Cys（x）His（y）-Zn2＋ interactions：possibilities and limitations of a simple pair-wise force field. J Mol Graph Model 24：404-411

Camilloni C，Vendruscolo M（2014）Statistical mechanics of the denatured state of a protein using replica-averaged metadynamics. J Am Chem Soc 136：8982-8991

Camilloni C，Robustelli P，De Simone A et al（2012）Characterization of the conformational equilibrium between the two major substates of RNase A using NMR chemical shifts. J Am Chem Soc 134：3968-3971

Camilloni C，Cavalli A，Vendruscolo M（2013）Replica-averaged metadynamics. J Chem Theory Comput 9：5610-5617

Cary SC，McDonald IR，Barrett JE，Cowan DA（2010）On the rocks：the microbiology of Antarctic Dry Valley soils. Nat Rev Microbiol 8：129-138

Chiuri R，Maiorano G，Rizzello A et al（2009）Exploring local flexibility/rigidity in psychrophilic and mesophilic carbonic anhydrases. Biophys J 96：1586-1596

D'Amico S，Marx J-C，Gerday C，Feller G（2003）Activity-stability relationships in extremophilic enzymes. J Biol Chem 278：7891-7896

D'Auria S，Aurilia V，Marabotti A，Gonnelli M，Strambini G（2009）Structure and dynamics of cold-adapted enzymes as investigated by phosphorescence spectroscopy. J Phys Chem B 113：13171-13178

Debiec KT，Gronenborn AM，Chong LT（2014）Evaluating the strength of salt bridges：a comparison of current biomolecular force fields. J Phys Chem B. doi：10.1021/jp500958r

Dror RO，Dirks RM，Grossman JP et al（2012）Biomolecular simulation：a computational microscope for molecular biology. Annu Rev Biophys 41：429-452

Eisenmesser EZ，Millet O，Labeikovsky W et al（2005）Intrinsic dynamics of an enzyme underlies catalysis. Nature 438：117-121

Eswar N，Eramian D，Webb B et al（2008）Protein structure modeling with MODELLER. Methods Mol Biol 426：145-159

Fan H，Mark AE（2004）Refinement of homology-based protein structures by molecular dynamics simulation techniques. Protein Sci 13：211-220

Fedøy A-E，Yang N，Martinez A et al（2007）Structural and functional properties of isocitrate dehydrogenase from the psychrophilic bacterium Desulfotalea psychrophila reveal a cold-active enzyme with an unusual high thermal stability. J Mol Biol 372：130-149

Feller G，Gerday C（2003）Psychrophilic enzymes：hot topics in cold adaptation. Nat Rev Microbiol 1：200-208

Fields PA（2001）Review：protein function at thermal extremes：balancing stability and flexibility. Comp Biochem Physiol A Mol Integr Physiol 129：417-431

Fraccalvieri D，Tiberti M，Pandini A（2012）Functional annotation of the mesophilic-like character of mutants in a cold-adapted enzyme by self-organising map analysis of their molecular dynamics. Mol Biosyst 8：2680-2691

Fraser JS，Jackson CJ（2011）Mining electron density for functionally relevant protein polysterism in crystal structures. Cell Mol Life Sci 68：1829-1841

Ganjalikhany MR，Ranjbar B，Taghavi AH，Tohidi Moghadam T（2012）Functional motions of Candida antarctica lipase B：a survey through open-close conformations. PLoS One 7：e40327

Garcia AE（1992）Large-amplitude nonlinear motions in proteins. Phys Rev Lett 68：2696-2699

Gerday C，Aittaleb M，Bentahir M et al（2000）Cold-adapted enzymes：from fundamentals to biotechnology. Trends Biotechnol 18：103-107

Gianese G，Bossa F，Pascarella S（2002）Comparative structural analysis of psychrophilic and meso- and thermophilic enzymes. Proteins 47：236-249

Gu J，Hilser VJ（2009）Sequence-based analysis of proteinenergy landscapes reveals nonuniform thermal adaptation within the proteome. Mol Biol Evol 26：2217-2227

HalgrenTA，Damm W（2001）Polarizable force fields. Curr Opin Struct Biol 11：236-242

Heidarsson PO，Sigurdsson ST，Asgeirsson B（2009）Structural features and dynamics of a cold-adapted alkaline phosphatase studied by EPR spectroscopy. FEBS J 276：2725-2735

Henzler-Wildman K，Kern D（2007）Dynamic personalities of proteins. Nature 450：964-972

Henzler-Wildman KA，Lei M，Thai V et al（2007）A hierarchy of timescales in protein dynamics is linked to enzyme catalysis. Nature 450：913-916

Hilser VJ，García-Moreno EB，Oas TG et al（2006）A statistical thermodynamic model of the protein ensemble. Chem Rev 106：1545-1558

Invernizzi G，Papaleo E，Grandori R et al（2009）Relevance of metal ions for lipase stability：structural rearrangements induced in the Burkholderia glumae lipase by calcium depletion. J Struct Biol 168：562-570

Isaksen GV，Åqvist J，Brandsdal BO（2014）Protein surface softness is the origin of enzyme cold- adaptation of trypsin. PLoS Comput Biol 10：e1003813

Jónsdóttir LB，Ellertsson BÖ，Invernizzi G et al（2014）The role of salt bridges on the temperature adaptation of aqualysin I，a thermostable subtilisin-like proteinase. Biochim Biophys Acta. doi：10.1016/j. bbapap. 2014.08.011

Kosugi T，Hayashi S（2012）Crucial role of protein flexibility in formation of a stable reaction transition state in an α-amylase catalysis. J Am Chem Soc 134：7045-7055

Leach AR（2001）Molecular modelling. Principles and applications，2nd edn. Prentice Hall，Harlow/New York

Leiros H-KS, Pey AL, Innselset M et al (2007) Structure of phenylalanine hydroxylase from Colwellia psychrerythraea 34H, a monomeric cold active enzyme with local flexibility around the active site and high overall stability. J Biol Chem 282: 21973-21986

Li P, Roberts BP, Chakravorty DK, Merz KM (2013) Rational design of particle mesh Ewald compatible Lennard-Jones parameters for +2 metal cations in explicit solvent. J Chem Theory Comput 9: 2733-2748

Lindorff-Larsen K, Piana S, Dror RO, Shaw DE (2011) How fast-folding proteins fold. Science 334: 517-520

Lindorff-Larsen K, Maragakis P, Piana S et al (2012) Systematic validation of protein force fields against experimental data. PLoS One 7: e32131

Ma B, Nussinov R (2010) Enzyme dynamics point to stepwise conformational selection in catalysis. Curr Opin Chem Biol 14: 652-659

Martinez R, Schwaneberg U, Roccatano D (2011) Temperature effects on structure and dynamics of the psychrophilic protease subtilisin S41 and its thermostable mutants in solution. Protein Eng Des Sel 24: 533-544

Martín-García F, Papaleo E, Gomez-Puertas P et al (2015) Comparing molecular dynamics force fields in the essential subspace. PLoS One 10: e0121114

McCammon J, Gelin B, Karplus M (1977) Dynamics of folded proteins. Nature 267: 585-590

Mereghetti P, Riccardi L, Brandsdal BO et al (2010) Near native-state conformational landscape of psychrophilic and mesophilic enzymes: probing the folding funnel model. J Phys Chem B 114: 7609-7619

Mittermaier AK, Kay LE (2009) Observing biological dynamics at atomic resolution using NMR. Trends Biochem Sci 34: 601-611

Nashine VC, Hammes-Schiffer S, Benkovic SJ (2010) Coupled motions in enzyme catalysis. Curr Opin Chem Biol 14: 644-651

Nussinov R (2014) The significance of the 2013 Nobel Prize in Chemistry and the challenges ahead. PLoS Comput Biol 10: e1003423

Olufsen M, Smalås AO, Moe E, Brandsdal BO (2005) Increased flexibility as a strategy for cold adaptation: a comparative molecular dynamics study of cold- and warm-active uracil DNA glycosylase. J Biol Chem 280: 18042-18048

Olufsen M, Brandsdal BO, Smalås AO (2007) Comparative unfolding studies of psychrophilic and mesophilic uracil DNA glycosylase: MD simulations show reduced thermal stability of the cold-adapted enzyme. J Mol Graph Model 26: 124-134

Onuchic JN, Wolynes PG (2004) Theory of protein folding. Curr Opin Struct Biol 14: 70-75

Palazzesi F, Barducci A, Tollinger M, Parrinello M (2013) The allosteric communication pathways in KIX domain of CBP. Proc Natl Acad Sci USA 110: 14237-14242

Papaleo E, Fantucci P, De Gioia L (2005) Effects of calcium binding on structure and autolysis regulation in trypsins. A molecular dynamics investigation. J Chem Theory Comput 1: 1286-1297

Papaleo E, Riccardi L, Villa C et al (2006) Flexibility and enzymatic cold-adaptation: a comparative molecular dynamics investigation of the elastase family. Biochim Biophys Acta 1764: 1397-1406

Papaleo E, Olufsen M, De Gioia L, Brandsdal BO (2007) Optimization of electrostatics as a strategy for cold-adaptation: a case study of cold- and warm-active elastases. J Mol Graph Model 26: 93-103

Papaleo E, Pasi M, Riccardi L et al (2008) Protein flexibility in psychrophilic and mesophilic trypsins. Evidence of evolutionary conservation of protein dynamics in trypsin-like serineproteases. FEBS Lett 582: 1008-1018

Papaleo E, Mereghetti P, Fantucci P, Grandori R, De Gioia L (2009) Free-energy landscape, principal component analysis, and structural clustering to identify representative conformations from molecular dynamics simulations: the myoglobin case. J Mol Graph Model 27: 889-899

Papaleo E, Pasi M, Tiberti M, De Gioia L (2011a) Molecular dynamics of mesophilic-like mutants of a cold-adapted enzyme: insights into distal effects induced by the mutations. PLoS One 6: e24214

Papaleo E, Tiberti M, Invernizzi G et al (2011b) Molecular determinants of enzyme cold adaptation: comparative structural and computational studies of cold- and warm-adapted enzymes. Curr Protein Pept Sci 12: 657-683

Papaleo E, Renzetti G, Tiberti M (2012) Mechanisms of intramolecular communication in a hyperthermophilic acylaminoacyl peptidase: a molecular dynamics investigation. PLoS One 7: e35686

Papaleo E, Renzetti G, Invernizzi G, Asgeirsson B (2013) Dynamics fingerprint and inherent asymmetric flexibility of a cold-adapted homodimeric enzyme. A case study of the Vibrio alkaline phosphatase. Biochim Biophys Acta 1830: 2970-2980

Papaleo E, Parravicini F, Grandori R et al (2014a) Structural investigation of the cold-adapted acylaminoacyl peptidase from Sporosarcina psychrophila by atomistic simulations and biophysical methods. Biochim Biophys Acta 1844: 2203-2213

Papaleo E, Sutto L, Gervasio FL, Lindorff-Larsen K (2014b) Conformational changes and free energies in a proline isomerase. J Chem Theor Comput 10: 4169-4174. doi: 10.1021/ct500536r, 140820163733003

Parravicini F, Natalello A, Papaleo E et al (2013) Reciprocal influence of protein domains in the cold-adapted acyl aminoacyl peptidase from Sporosarcina psychrophila. PLoS One 8: e56254

Pasi M, Riccardi L, Fantucci P et al (2009) Dynamic properties of a psychrophilic alpha-amylase in comparison with a mesophilic homologue. J Phys Chem B 113: 13585-13595

Piana S, Lindorff-Larsen K, Shaw DE (2013) Atomic-level description of ubiquitin folding. Proc Natl Acad Sci USA 110: 5915-5920

Poland D (2001) Freeenergy distributions in proteins. Proteins 45: 325-336

Ramli AN, Mahadi NM, Shamsir MS, Rabu A, Joyce-Tan KH, Murad AM, Illias RM (2012) Structural prediction of a novel chitinase from the psychrophilic glaciozyma antarctica Pl12 and an analysis of its structural properties and function. J Comput Aided Mol Des 26: 947-961

Rapaport D (1998) The art of molecular dynamics simulation. Cambridge University Press, Cambridge, UK

Raval A, Piana S, Eastwood MP et al (2012) Refinement of protein structure homology models via long, all-atom molecular dynamics simulations. Proteins 80: 2071-2079

Russell NJ (2000) Toward a molecular understanding of cold activity of enzymes from psychrophiles. Extremophiles 4: 83-90

Siddiqui KS, Cavicchioli R (2006) Cold-adapted enzymes. Annu Rev Biochem 75: 403-433

Sigtryggsdóttir AR, Papaleo E, Thorbjarnardóttir SH, Kristjánsson MM (2014) Flexibility of cold-and heat-adapted subtilisin-like serine proteinases evaluated with fluorescence quenching and molecular dynamics. Biochim Biophys Acta 1844: 705-712

Smith JC, Roux B (2013) Eppur si muove! The 2013 Nobel Prize in Chemistry. Structure 21: 2102-2105

Somero GN (2004) Adaptation of enzymes to temperature: searching for basic "strategies". Comp Biochem Physiol B Biochem Mol Biol 139: 321-333. doi: 10. 1016/j. cbpc. 2004. 05. 003

Spiwok V, Lipovová P, Skálová T, Dusková J, Dohnálek J, Hasek J, Russell NJ, Králová B (2007) Cold-active enzymes studied by comparative molecular dynamics simulation. J Mol Model 13: 485-497

Spiwok V, Su ćur Z, Hošek P (2014) Enhanced sampling techniques in biomolecular simulations. Biotechnol Adv. doi: 10. 1016/j. biotechadv. 2014. 11. 011

Struvay C, Feller G (2012) Optimization to low temperature activity in psychrophilic enzymes. Int J Mol Sci 13: 11643-11665

Sutto L, Gervasio FL (2013) Effects of oncogenic mutations on the conformational free-energy landscape of EGFR kinase. Proc Natl Acad Sci 110: 10616-10621. doi: 10. 1073/pnas. 1221953110

Tehei M, Madern D, Franzetti B, Zaccai G (2005) Neutron scattering reveals the dynamic basis of protein adaptation to extreme temperature. J Biol Chem 280: 40974-40979

Tiberti M, Papaleo E (2011) Dynamic properties of extremophilic subtilisin-like serine-proteases. J Struct Biol 174: 69-83

Tronelli D, Maugini E, Bossa F, Pascarella S (2007) Structural adaptation to low temperatures— analysis of the subunit interface of oligomeric psychrophilic enzymes. FEBS J 274: 4595-4608 Van den Bedem H, Fraser JS (2015) Integrative, dynamic structural biology at atomic resolution—it's about time. Nat Methods 12: 307-318

Van den Burg B (2003) Extremophiles as a source for novel enzymes. Curr Opin Microbiol 6: 213-218

Van Gunsteren WF, Berendsen HJC (1990) Computer simulation of molecular dynamics: method- ology, applications and perspective. Angew Chem Int Eng Ed 29: 992-1023

Wintrode PL, Arnold FH (2000) Temperature adaptation of enzymes: lessons from laboratory evolution. Adv Protein Chem 55: 161-225

Woldeyes RA, Sivak DA, Fraser JS (2014) E pluribus unum, no more: from one crystal, many conformations. Curr Opin Struct Biol 28C: 56-62

Xie BB, BianF, Chen XL, He HL, Guo J, Gao X, Zeng YX, Chen B, Zhou BC, Zhang YZ (2009) Cold adaptation of zinc metalloproteases in the thermolysin family from deep sea and arctic sea ice bacteria revealed by catalytic and structure properties and molecular dynamics: new insights into relationship between conformational flexibility and hydrogen bonding. J Biol Chem 284: 9257-9269

Zhu T, Xiao X, Ji C, Zhang JZH (2013) A new quantum calibrated force field for zinc—protein complex. J Chem Theory Comput 9: 1788-1798

第二十五章
嗜盐杆菌表达系统表达恶性疟原虫环子孢子蛋白

Wolf T. Pecher[1], Jong-Myoung Kim[2],
Priya DasSarma[3], Ram Karan[3],
Photini Sinnis[4], Shiladitya DasSarma[3][5]

25.1 引言

极端嗜盐杆菌 *Halobacterium* sp. NRC-1 在抗原和疫苗的生产方面有许多优势，可被用于构建抗原生产和疫苗开发的创新平台（DasSarma 2007；DasSarma et al. 2013）。研究表明该菌无毒，不含脂多糖（LPS），且作为盐和咸味调味品的天然成分，该菌还具有生物相容性。目前已对该菌进行了全基因组测序，同时还运用遗传学、转录组学和蛋白质组学方法对其生物学性质进行了详细研究（Berquist et al. 2006；DasSarma 2004；Ng et al. 1998，2000）。高水平启动子也已用于天然及外源蛋白质表达载体的开发（Karan et al. 2013，2014）。此外，*Halobacterium* 菌体在低渗条件下会裂解，可稳定释放表达的蛋白质。

值得注意的是，*Halobacterium* sp. NRC-1 表达系统目前已用于表达长约 300 nm 的抗原蛋白质纳米颗粒，又称为气体囊泡（GVNPs）。该柠檬状细胞器是生物工程化和高度佐剂化的，可以用于细胞浮选（DasSarma and Arora 1997；DasSarma et al. 2013；Stuart et al. 2001，2004）。GVNPs 仅由蛋白质组成，排列在 GVNPs 纳米颗粒外表面上的一种蛋白

[1] Department of Microbiology and Immunology，University of Maryland School of Medicine，Baltimore，MD，USA；Institute of Marine and Environmental Technology，University System of Maryland，Baltimore，MD，USA；College of Arts and Sciences，University of Baltimore，Baltimore，MD，USA.

[2] Department of Microbiology and Immunology，University of Maryland School of Medicine，Baltimore，MD，USA；Institute of Marine and Environmental Technology，University System of Maryland，Baltimore，MD，USA；Department of Marine Bio－Materials and Aquaculture，PuKyong National University，Pusan，Republic of Korea.

[3] Department of Microbiology and Immunology，University of Maryland School of Medicine，Baltimore，MD，USA；Institute of Marine and Environmental Technology，University System of Maryland，Baltimore，MD，USA.

[4] Department of Molecular Microbiology and Immunology，Bloomberg School of Public Health，Johns Hopkins University，Baltimore，MD，USA.

[5] e-mail：sdassarma@som. umaryland. edu.

质（GvpC）可与外源抗原融合。GVNPs已经成功用于显示SIV、衣原体（*Chlamydia*）和沙门氏菌（*Salmonella*）抗原，并且能被巨噬细胞缓慢处理（Childs and Webley 2012；DasSarma and DasSarma 2015；DasSarma et al. 2014；Sremac and Stuart 2008，2010；Stuart et al. 2001，2004）。结合了 *Salmonella* SopB抗原的GVNPs能有效减小小鼠免疫器官中的细菌负荷（DasSarma et al. 2014）。

最新的研究运用 *Halobacterium* sp. NRC-1系统表达疟原虫子孢子的表面蛋白质，这种蛋白质是蚊子传播给人类的恶性疟原虫（*Plasmodium falciparum*）的红细胞前体。疟疾是迫切需求疫苗的疾病之一，疟疾主要由 *P. falciparum* 感染引发，每年受感染人数高达2亿，死亡人数超过50万。环子孢子蛋白（CSP）具有较强的免疫原性，由疟原虫特有的单拷贝基因编码（Nussenzweig and Nussenzweig 1989）。CSP由N末端含信号肽，四肽重复中心和C末端含糖基磷脂酰肌醇（GPI）锚定区域组成（图25.1）（Coppi et al. 2011；Dame et al. 1984；del Portillo et al. 1987）。根据来源不同，四肽重复的数目在25至49个拷贝之间（Bowman et al. 2013）。毕赤酵母（*Pichia pastoris*）中表达的疟疾疫苗RTS中亚单位S就是以 *P. falciparum* 环子孢子蛋白抗原为基础研发的。该疫苗是将CSP的19个四肽重复中心和C末端功能域结合到乙肝病毒表面抗原上。在不同的免疫刺激化合物的作用下，该疫苗具有一定保护效果。但是该疫苗对婴儿的免疫保护效果较差，同时产生保护力的时间十分有限（Campo et al. 2014；Mo and Augustine 2014）。红细胞前体疫苗的最终目标是通过静脉注射辐照处理的子孢子使人获得无菌免疫（Seder et al. 2013）。

旨在利用 *Halobacterium* sp. NRC-1系统表达完整的 *P. falciparum* CSP。为了使 *Halobacterium* sp. 中CSP的表达量最大化，设计了密码子优化的基因序列并将其克隆到表达载体中用于单独生产CSP蛋白或作为与GVNP的融合体，该表达蛋白质包含了疟原虫成熟蛋白质中不含有的N末端和C末端区域，这些区域可能有助于增强免疫原性或免疫力（Kastenmüller et al. 2013；Sedegah et al. 2013）。本论文将为生物合成完整的CSP提供理论依据。

25.2 材料与方法

25.2.1 *P. falciparum* 3D7菌株CSP基因的设计及克隆

首先从NCBI中获得 *P. falciparum* 3D7株CSP基因序列（NCBI编号XM_001351086.1）。用 *Halobacterium* sp. NRC-1中预测基因的密码子选用表代替稀有密码子（Kennedy et al. 2001）。对目的基因进行密码子修饰（AAC> GGC）以添加独特的 *KasI* 酶切位点，该修饰导致201位天冬酰胺突变为甘氨酸（图25.1）。将前后都添加了 *AfeI* 酶切位点的优化后的CSP基因克隆到 *Halobacterium* sp. NRC-1表达载体中，进行合成（Life Technologies，Carlsbad，CA，USA）。

25.2.2 表达菌株的构建及培养

将优化后的添加了 *AfeI* 酶切位点的完整CSP基因克隆到 *Halobacterium* 表达载体pDRK和pSD中，这些载体都有独特的 *AfeI* 限制酶的切割位点（Karan et al. 2014）。将野生型 *Halobacterium* sp. NRC-1菌株中pNRC100和pNRC200中的活性 *gvp* 基因簇A敲除，将构建好的pDRKcsp6质粒（图25.2a）转化到 *Halobacterium* sp. SD109菌株中，构建 *Halobacterium* sp. SD109（pDRKcsp6）菌株（DasSarma et al. 1988，1995）。pSDcsp20质

```
AfeI        .         .    Signal peptide .         .         .
AGCGCTATGATGCGCAAGCTCGCCATCCTCAGCGTCAGCAGCTTCCTCTTCGTCGAGGCC
  1         M  M  R  K  L  A  I  L  S  V  S  S  F  L  F  V  E  A     18
            .         .         .         .         .         .
CTCTTCCAGGAGTACCAGTGCTACGGCAGCAGCAGCAACACCCGCGTCCTCAACGAGCTC
 19   L  F  Q  E  Y  Q  C  Y  G  S  S  S  N  T  R  V  L  N  E  L     38
            .         .         .         .         .         .
AACTACGACAACGCCGGCACCAACCTCTACAACGAGCTCGAGATGAACTACTACGGCAAG
 39   N  Y  D  N  A  G  T  N  L  Y  N  E  L  E  M  N  Y  Y  G  K     58
            .         .         .         .         .         .
CAGGAGAACTGGTACAGCCTCAAGAAGAACAGCCGCAGCCTCGGCGAGAACGACGACGGC
 59   Q  E  N  W  Y  S  L  K  K  N  S  R  S  L  G  E  N  D  D  G     78
            .         .         .         .         .         .
AACAACGAGGACAACGAGAAGCTCCGCAAGCCCAAGCACAAGAAGCTCAAGCAGCCCGCC
 79   N  N  E  D  N  E  K  L  R  K  P  K  H  K  K  L  K  Q  P  A     98
            .         .    Tetrapeptide repeats .         .
GACGGCAACCCCGACCCCAACGCCAACCCCAACGTCGACCCCAACGCCAACCCCAACGTC
 99   D  G  N  P  D  P  N  A  N  P  N  V  D  P  N  A  N  P  N  V    118
            .         .         .         .         .         .
GACCCCAACGCCAACCCCAACGTCGACCCCAACGCCAACCCCAACGCCAACCCCAACGCC
119   D  P  N  A  N  P  N  V  D  P  N  A  N  P  N  A  N  P  N  A    138
            .         .         .         .         .         .
AACCCCAACGCCAACCCCAACGCCAACCCCAACGCCAACCCCAACGCCAACCCCAACGCC
139   N  P  N  A  N  P  N  A  N  P  N  A  N  P  N  A  N  P  N  A    158
            .         .         .         .         .         .
AACCCCAACGCCAACCCCAACGCCAACCCCAACGCCAACCCCAACGCCAACCCCAACGCC
159   N  P  N  A  N  P  N  A  N  P  N  A  N  P  N  A  N  P  N  A    178
            .         .         .         .         .         .
AACCCCAACGCCAACCCCAACGCCAACCCCAACGCCAACCCCAACGCCAACCCCAACGTC
179   N  P  N  A  N  P  N  A  N  P  N  A  N  P  N  A  N  P  N  V    198
      KasI            .         .         .         .         .
GACCCCGGCGCCAACCCCAACGCCAACCCCAACGCCAACCCCAACGCCAACCCCAACGCC
199   D  P  G  A  N  P  N  A  N  P  N  A  N  P  N  A  N  P  N  A    218
            .         .         .         .         .         .
AACCCCAACGCCAACCCCAACGCCAACCCCAACGCCAACCCCAACGCCAACCCCAACGCC
219   N  P  N  A  N  P  N  A  N  P  N  A  N  P  N  A  N  P  N  A    238
            .         .         .         .         .         .
AACCCCAACGCCAACCCCAACGCCAACCCCAACGCCAACCCCAACGCCAACCCCAACGCC
239   N  P  N  A  N  P  N  A  N  P  N  A  N  P  N  A  N  P  N  A    258
            .         .         .         .         .         .
AACCCCAACGCCAACCCCAACGCCAACCCCAACGCCAACCCCAACAAGAACAACCAGGGC
259   N  P  N  A  N  P  N  A  N  P  N  A  N  P  N  K  N  N  Q  G    278
            .         .         .         .         .         .
AACGGCCAGGGCCACAACATGCCCAACGACCCCAACCGCAACGTCGACGAGAACGCCAAC
279   N  G  Q  G  H  N  M  P  N  D  P  N  R  N  V  D  E  N  A  N    298
            .         .         .         .         .         .
GCCAACAGCGCCGTCAAGAACAACAACAACGAGGAGCCCAGCGACAAGCACATCAAGGAG
299   A  N  S  A  V  K  N  N  N  N  E  E  P  S  D  K  H  I  K  E    318
            .         .         .         .         .         .
TACCTCAACAAGATCCAGAACAGCCTCAGCACCGAGTGGAGCCCCTGCAGCGTCACCTGC
319   Y  L  N  K  I  Q  N  S  L  S  T  E  W  S  P  C  S  V  T  C    338
            .         .         .         .         .         .
GGCAACGGCATCCAGGTCCGCATCAAGCCCGGCAGCGCCAACAAGCCCAAGGACGAGCTC
339   G  N  G  I  Q  V  R  I  K  P  G  S  A  N  K  P  K  D  E  L    358
            .         .         .         .         .         .
GACTACGCCAACGACATCGAGAAGAAGATCTGCAAGATGGAGAAGTGCAGCAGCGTCTTC
359   D  Y  A  N  D  I  E  K  K  I  C  K  M  E  K  C  S  S  V  F    378
            .         .     GPI anchor  .         .         .
AACGTCGTCAACAGCAGCATCGGCCTCATCATGGTCCTCAGCTTCCTCTTCCTCAACTAA
379   N  V  V  N  S  S  I  G  L  I  M  V  L  S  F  L  F  L  N  *    398
AfeI
AGCGCT
```

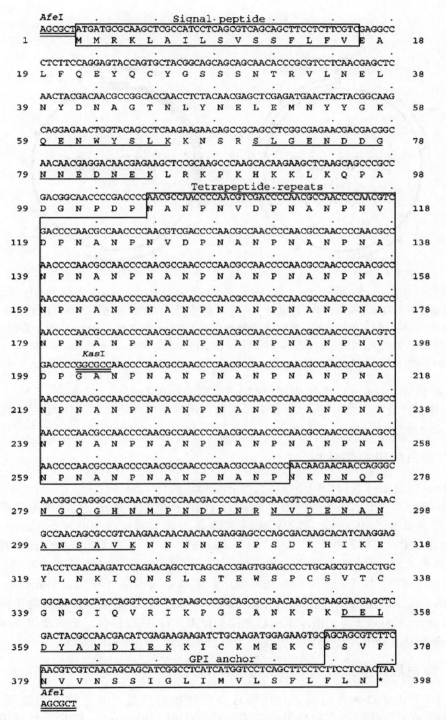

图 25.1　优化后的 *P. falciparum* CSP 基因序列以及在 *Halobacterium* sp. 中的表达产物。双下划线标记的是 *AfeI* 和 *KasI* 酶切位点。下划线标记的是 LC-MS/MS 分析鉴定的肽片段。方框标记的是信号肽、四肽重复序列和 GPI 锚定序列

粒（图 25.2b）也转化到 SD109 中，构建 *Halobacterium* sp. SD109（pSDcsp20）。使用引物步移测序法在两条链中对每个质粒的 CSP 基因区进行完全测序。将含表达质粒的 *Halobac-*

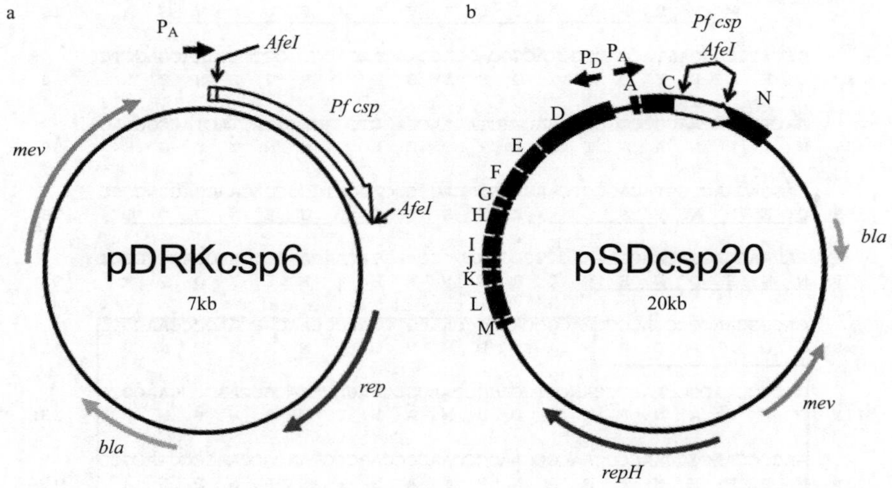

图 25.2　*P. falciparum* CSP 基因表达质粒 pDRKcsp6（a）和 pSDcsp20（b）。*AfeI* 酶切位点间的宽白箭头表示的是编码 *P. falciparum* CSP（标记为 *Pf csp*）的基因。窄黑箭头指示的是启动子。窄灰箭头指示的是洛伐他汀（*mev*）和 β-内酰胺酶（*bla*）的选择标记基因和复制酶基因（*rep*，*repH*）的位置。pDRKcsp6 质粒上的 CSP 基因 5′-端的小白框表示的是 His 标签的位置。pSDcsp20 质粒上的黑框表示的是 *gvpACN* 和 *gvpDEFGHIJKLM* 基因

terium 接种到含 20μg/mL 洛伐他汀（Merck，Sharp 和 Dohme，Rahway，NJ，USA）的 CM^+ 培养基中，42℃，220r/min 培养至稳定期。细胞裂解和 GVNP 制备如先前所述（DasSarma et al. 1995，2013）。

25.2.3　蛋白质印迹法分析

Western 印迹法（蛋白质印迹法）按照先前所述进行（Shukla and DasSarma 2004）。在 9％或 12％的 SDS（十二烷基硫酸钠）-PAGE（聚丙烯酰胺）凝胶上电泳细胞裂解物，再将蛋白质转移至孔径为 0.45μm 的 PVDF（聚偏氟乙烯）膜上（Millipore Corp.，Boston，MA，USA）。用 PTM 缓冲液（PBS，0.1％Tween 20，5％脱脂奶粉）封闭适当时间，然后将膜放在含有 CSP 特异性单克隆抗体 2A10（1∶1000 稀释）的稀释液中，4℃孵育过夜（Wirtz et al. 1987）。用 PTM 缓冲液洗涤膜后，用碱性磷酸酶（Sigma Aldrich，St. Louis，MO，USA）标记的兔抗鼠二抗（1∶2500 稀释）孵育。然后用化学发光底物 1-Step NBT/BCIP（Thermo Fisher Scientific，Rockford，IL，USA）进行显色。最后使用 Image J 对蛋白质条带进行定量分析。

25.2.4　蛋白质组学分析

先用 DTT 还原目的蛋白质，并用碘乙酰胺使蛋白质烷基化，随后用胰蛋白酶水解（Shevchenko et al. 1996）。水解后得到的多肽样本先用 stage-tip C_{18} 脱盐，然后用含 0.1％三氟乙酸的 60％乙腈溶液洗脱，接着重悬在 0.1％甲酸中，用于 LC-MS/MS 分析。运用与 Proxionnano-LC 系统相连的 Q-Exactive 仪器（Thermo Scientific，Rockford，IL，USA）进行多肽鉴定。利用 Thermo Scientific MS2 处理器和 Xtract 软件，根据卷积和去卷积算法从 MS 和 MS/MS 光谱的同位素解析数据中推算多肽质量从而鉴定多肽序列。使用与 Sequest 搜索引擎接口的 Proteome Discoverer 1.4 软件对获得的 MS/MS 数据进行

分析鉴定。

25.3 结果

25.3.1 含优化 CSP 基因的盐杆菌表达菌株的构建

由于 *P. falciparum* 和 *Halobacterium* sp. 之间 GC 含量差异较大（GC 含量分别为 19% 和 66%），根据野生型 NRC-1 菌株的密码子使用数据设计合成了能在 *Halobacterium* 中表达的 CSP 基因（图 25.1）。这个合成的 CSP 基因 GC 含量达 62.3%，它包含来自 *P. falciparum* 3D7 菌株 CSP 基因的完整编码区，包括 N 端信号序列和 C 端锚定序列，此外，该合成基因前后都添加了 *AfeI* 酶切位点。

将合成的 CSP 基因插入质粒 pDRK 的 *gvp*A 启动子下游的 *AfeI* 酶切位点处，构建表达质粒 pDRKcsp6（图 25.2a）。由 pDRKcsp6 编码的 CSP 的 N 端多了一个含有 His 标签的 15 个氨基酸残基片段（MHHHHHHLKRLPRSA），随后是有 397 个氨基酸残基组成的完整的 CSP 蛋白（图 25.1）。另外将合成的 CSP 基因克隆到表达质粒 pSD 的 *gvp*C 启动子下游的 *AfeI* 酶切位点处，再转入 *Halobacterium* 中表达 GVNP（图 25.2b）。所得的 pSDcsp20 能编码由 293 个氨基酸组成的融合蛋白 GvpC 以及完整的 CSP 蛋白。

25.3.2 *P. falciparum* CSP 基因在盐杆菌中的表达

将表达质粒 pDRKcsp6 和 pSDcsp20 转化入气体囊泡基因缺失 *Halobacterium* sp. SD109 菌株中。菌株培养至稳定期，离心弃上清，使用低渗条件裂解细胞。重组体的生长速率与亲本菌株的生长速率相当（数据未显示）。将细胞裂解物进行 SDS-PAGE 电泳，转膜，并使用 CSP 单克隆抗体 2A10 检测。发现菌株 SD109（pSDcsp20）出现了 120kDa 的蛋白质条带，菌株 SD109（pDRKcsp6）也检测到 60kDa 的蛋白质条带，实验结果与预期相符（图 25.3a）。使用胰蛋白酶消化 *Halobacterium* sp. SD109（pDRKcsp6）的蛋白质，然后进行 LC-MS/MS 分析。结果显示在表达的蛋白质中能检测到 *P. falciparum* CSP 蛋白的五种独特的肽段（图 25.1）。鉴定的肽涵盖 64/397 个氨基酸，占 CSP 氨基酸总数的 16%。值得注意的是，整个四肽重复区缺少胰蛋白酶酶切位点，所以不能用该方法检测。

运用 Western 印迹法，*P. falciparum* 子孢子裂解物作为内参，确定 CSP 在 *Halobacterium* sp. 中的表达量。子孢子是从成年雌性按蚊的唾液腺中获得的。将 *P. falciparum* 子孢子和 *Halobacterium* sp. SD109（pDRKcsp6）裂解并富集，再将裂解物进行 SDS-PAGE 电泳，转膜，并使用 CSP 单克隆抗体 2A10 检测（图 25.3b）。子孢子体中的 CSP 条带在 55kDa 左右（泳道 1～4），*Halobacterium* 则存在 60kDa 的 CSP 蛋白条带（泳道 5～8），这种差异可能是嗜盐杆菌中表达的 CSP 蛋白存在 N 端信号肽，C 端的 GPI 锚定序列以及 His 标签。根据 Western 印迹中显示的这些主要条带的强度，确定 10^3 个 *P. falciparum* 子孢子中的 CSP 含量和 8.3×10^5 个 *Halobacterium* sp. SD109（pDRKcsp6）中表达的量相差无几。可见每升 *Halobacterium* sp. SD109（pDRKcsp6）培养物可得到 3.3 mg 的游离 CSP。另外，在 pSDcsp20 中表达的显示在 GVNPs 上的 CSP 融合蛋白（大小在 120kDa 左右）的量只有 SD109 中表达量的 15%，大概每升培养物仅可获得 0.5 mg CSP（图 25.3a，数据未显示）。

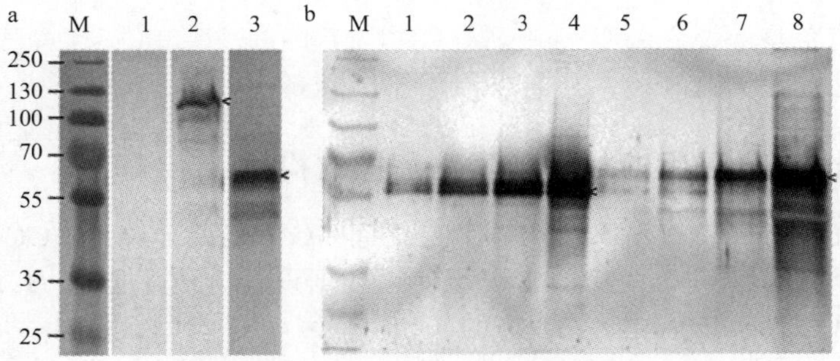

图 25.3 Western 印迹分析 CSP 表达株，结果采用 CSP 单克隆抗体 2A10 检测。a：1～3 泳道分别表示 *Halobacterium* sp. SD109，SD109（pSDcsp20），和 SD109（pDRKcsp6）裂解物。b：1～4 泳道分别表示 1.9×10^2 个、7.5×10^2 个、3×10^3 个和 1.2×10^4 个 *P. falciparum* 子孢子的裂解物。b：5～8 泳道分别表示 8×10^4 个、3×10^5、1.3×10^6 个和 5×10^6 个 *Halobacterium* sp. SD109（pDRKcsp6）的裂解物。M：分子质量标准物，大小用 kDa 表示。箭头指示的是 CSP 蛋白条带

25.4 讨论

利用 *Halobacterium* sp. NRC-1 表达系统成功表达了 *P. falciparum* CSP 基因。这种极端表达系统不仅能够表达足够的 CSP 用于免疫学研究，同时在疫苗开发方面也有巨大的发展潜力。*Halobacterium* sp. NRC-1 表达系统同时还具有很好的生物相容性、无毒性、不含 LPS 等优点。此外该菌最显著的优点是细胞在低渗条件下容易裂解，释放内容物，不需要机械或酶处理。*Halobacterium* sp. 细胞内含有大量的 GVNPs，GVNPs 具有高度佐剂活性，能引起强烈的、长期的免疫应答（Childs and Webley 2012；DasSarma and DasSarma 2015；DasSarma et al. 2013，2014；Sremac and Stuart 2008，2010；Stuart et al. 2001，2004）。该表达系统能够表达游离的或有序的锚定在 GVNPs 上的 CSP 蛋白。在 *Halobacterium* 中表达的 CSP 蛋白可被 CSP 单克隆抗体 2A10 识别，可将其抗原性保留在宿主细胞中。这些结果均表明，极端 *Halobacterium* sp. NRC-1 表达系统在开发免疫学检测和疫苗方面有很好的发展前景。

从不同的疟原虫菌株中获得的 CSP 基因序列，一部分已经在大肠杆菌、酵母和其他生物体中成功表达（Dame et al. 1984；Kastenmüller et al. 2013；Kolodny et al. 2001；Plassmeyer et al. 2009；Young et al. 1985）。然而，在这些成功表达的例子中，获得的 CSP 都是经过剪接的，要么缺少 N 端信号肽或 C 端的 GPI 锚定区，要么就是两者都缺。1985 年 Young 等人尝试在大肠杆菌中表达完整 CSP，但重组蛋白质不仅表达水平低，而且还不稳定（Young et al. 1985）。在实验中，无论是游离还是融合到 GvpCs 上的 CSP，都能在 *Halobacterium* sp. 中完整、稳定表达。大量的表达完整的 CSP 分子的例子证明该表达系统在替代其他表达系统方面有很大的潜力。

最近研究表明，与缺少整个 N 端的 CSP 疫苗 RTS/S 相比，缺少信号序列和 GPI 锚定序列的 CSP 能在小鼠和灵长类动物中引发更强的抗原反应（Kastenmüller et al. 2013）。在小鼠体内，佐剂对抗体、Th1 和 CD4＋T 细胞免疫保护应答显示出不同的介导作用。*Salmonella* 的 SopB 抗原-GVNPs 与减毒活细菌一起使用时，能成功减少细菌负担，可见嗜盐杆菌表达的 GVNPs 能显著提高抗原免疫反应（DasSarma et al. 2014）。免疫小鼠体内 IFN-γ、IL-2 和 IL-9 的水平显著提升，更加有力地证实了 SopB-GVNPs 的免疫刺激效应。

与成熟的子孢子 CSP 相比，*Halobacterium* sp. 表达的完整的 CSP 序列含有附加序列。这些序列可能具有免疫原性或是免疫刺激剂，只是目前在截短的蛋白质中没有被检测到。如 Sedegah 等人发现 CSP 的信号肽含有能被来自疟疾流行地区的个体的 CD8＋T 细胞识别的抗原表位（Sedegah et al. 2013）。此外，细胞介导组合物（即 CD4＋Th1，CD8＋T 细胞）和体液应答对提供保护性免疫至关重要（Ménard et al. 2013；Schofield et al. 1987）。因此有必要进行进一步的研究以确定 *Halobacterium* 中表达的游离或融合到 GvpCs 上的 CSP 的免疫原性效应。

Halobacterium 细胞和 GVNPs 的生物相容性是该系统的另一个显著特征（DasSarma et al. 2010）。有报道显示，在 *Halobacterium* 中表达的 GNVPs 与抗原融合后能长时间在没有冷藏条件下稳定储存，这表明对发展中国家来说，是潜在的疾病疫苗替代品（DasSarma et al. 2010，2014）。虽然可通过浮选初步纯化 GVNP，但直接使用含有 CSP 的细胞或该细胞裂解物也可以进行初步纯化，而且还具有免疫原性的优点。在某些情况下，His 标签标记 N 末端有助于宿主中表达蛋白质的分离纯化，当需要纯化蛋白质时，也可以利用该方法（Kolodny et al. 2001；Plassmeyer et al. 2009）。最后，如果将来需要的话，*Halobacterium* sp. CSP 表达系统也可以用于修饰表达的变异蛋白质，包括较短或截短形式的蛋白质。

无利益冲突声明

Wolf T. Pecher，Jong-Myoung Kim，Priya DasSarma，Ram Karan，Photini Sinnis 和 Shiladitya DasSarma 声明他们没有利益冲突。

致谢　这项工作得到了 Bill 和 Melinda Gates 基金会的资助 OPP1061509，美国国家卫生研究院对 SD 的资助 R03 AI107634，美国国家卫生研究院对 PS 的资助 R01 AI056840。JK 得到了 PKNU 国外研究基金 CD-2013-0914 的支持。感谢 Susan Barnes 和 Folasade Ekulona 的技术支持，感谢 Stefanie Trop 和 Peter Dumoulin 的宝贵讨论，以及 F. Zavala 教授对手稿的指导。

参考文献

Berquist BR，Müller JA，DasSarma S（2006）Chapter 27. Genetic systems for halophilic archaea. In：Oren A，Rainey F（eds）Methods in microbiology，vol 35. Elsevier/Academic Press，Amsterdam，pp 649-680

Bowman NM，Congdon S，Mvalo T，Patel JC，Escamilla V，Emch M，Martinson F，Hoffman I，Meshnick SR，Juliano JJ（2013）Comparative population structure of *Plasmodium falciparum* circumsporozoite protein NANP repeat lengths in Lilongwe，Malawi. Sci Rep 3：1990

Campo JJ，Sacarlal J，Aponte JJ，Aide P，Nhabomba AJ，Dobaño C，Alonso PL（2014）Duration of vaccine efficacy against malaria：5th year of follow-up in children vaccinated with RTS，S/AS02 in Mozambique. Vaccine 32（19）：2209-2216. doi：10. 1016/j. vaccine. 2014. 02. 042

Childs TS，Webley WC（2012）In vitro assessment of halobacterial gas vesicles as a *Chlamydia* vaccine display and delivery system. Vaccine 30（41）：5942-5948

Coppi A，Natarajan R，Pradel G，Bennett BL，James ER，Roggero MA，Corradin G，Persson C，Tewari R，Sinnis P（2011）The malaria circumsporozoite protein has two functional domains，each with distinct roles as sporozoites journey from mosquito to mammalian host. J Exp Med 208（2）：341-356

Dame JB，Williams JL，McCutchan TF，Weber JL，Wirtz RA，Hockmeyer WT，Maloy WL，Haynes JD，Schneider I，Roberts D，Sanders GS，Reddy EP，Diggs CL，Miller LH（1984）Structure of the gene encoding the immunodominant surface antigen on the sporozoite of the human malaria parasite *Plasmodium falciparum*. Science 225（4662）：593-599

DasSarma S（2004）Genome sequence of an extremely halophilic archaeon. In：Fraser CM，Read T，Nelson KE（eds）Microbial genomes. Humana Press，Totowa，pp 383-399

DasSarma S（2007）Extreme microbes：the salty side of life. Am Sci 95（3）：224-231

DasSarma S，Arora P（1997）Genetic analysis of the gas vesicle gene cluster in haloarchaea. FEMS Microbiol Lett 153（1）：1-10

DasSarma S，DasSarma P（2015）Gas vesicle nanoparticles for antigen display. Vaccines 3（3）：686-702

DasSarma S, Halladay JT, Jones JG, Donovan JW, Giannasca PJ, Tandeau de Marsac N (1988) High-frequency mutations in a plasmid-encoded gas vesicle gene in *Halobacterium halobium*. Proc Natl Acad Sci U S A 85 (18): 6861-6865

DasSarma S, Robb FT, Place AR, Sowers KR, Schreier HJ, Fleischmann EM (eds) (1995) Archaea: a laboratory manual-halophiles. Cold Spring Harbor Laboratory Press, Plainview

DasSarma P, Coker JA, Huse V, DasSarma S (2010) Halophiles, industrial applications. In: Flickinger MC (ed) Encyclopedia of industrial biotechnology: bioprocess, bioseparation, and cell technology. Wiley, Hoboken, pp 1-43

DasSarma S, Karan R, DasSarma P, Barnes S, Ekulona F, Smith B (2013) An improved genetic system for bioengineering buoyant gas vesicle nanoparticles from haloarchaea. BMC Biotechnol 13: 112. doi: 10. 1186/1472-6750-13-112

DasSarma P, Negi VD, Balakrishnan A, Karan R, Barnes S, Ekulona F, Chakravortty D, DasSarma S (2014) Haloarchaeal gas vesicle nanoparticles displaying *Salmonella* SopB antigen reduce bacterial burden when administered with live attenuated bacteria. Vaccine 32 (35): 4543-4549

Del Portillo HA, Nussenzweig RS, Enea V (1987) Circumsporozoite gene of a *Plasmodium falciparum* strain from Thailand. Mol Biochem Parasitol 24 (3): 289-294

Karan R, Capes MD, DasSarma P, DasSarma S (2013) Cloning, overexpression, purification, and characterization of a polyextremophilic β-galactosidase from the Antarctic haloarchaeon Halorubrum lacusprofundi. BMC Biotechnol 13: 3. doi: 10. 1186/1472-6750-13-3

Karan R, DasSarma P, Balcer-Kubiczek E, Weng RR, Liao C-C, Goodlett DR, Ng WV, DasSarma S (2014) Bioengineering radioresistance by overproduction of RPA, a mammalian-type singlestranded DNA-binding protein, in a halophilic archaeon. Appl Microbiol Biotechnol 98 (4): 1737-1747

Kastenmüller K, Espinosa DA, Trager L, Stoyanov C, Salazar AM, Pokalwar S, Singh S, Dutta S, Ockenhouse CF, Zavala F, Seder RA (2013) Full-length *Plasmodium falciparum* circumsporozoite protein administered with long-chain poly (I · C) or the Toll-like receptor 4 agonist glucopyranosyl lipid adjuvant-stable emulsion elicits potent antibody and CD4+ T cell immunity and protection in mice. Infect Immun 81 (3): 789-800

Kennedy SP, Ng WV, Salzberg SL, Hood L, DasSarma S (2001) Understanding the adaptation of *Halobacterium* species NRC-1 to its extreme environment through computational analysis of its genome sequence. Genome Res 11: 1641-1650

Kolodny N, Kitov S, Vassell MA, Miller VL, Ware LA, Fegeding K, De La Vega P, Sacci JB Jr, Lanar DE (2001) Two-step chromatographic purification of recombinant *Plasmodium falciparum* circumsporozoite protein from Escherichia coli. J Chromatogr B Biomed Sci Appl 762 (1): 77-86

Ménard R, Tavares J, Cockburn I, Markus M, Zavala F, Amino R (2013) Looking under the skin: the first steps in malarial infection and immunity. Nat Rev Microbiol 11 (10): 701-712

Mo AX, Augustine AD (2014) NIAID meeting report: improving malaria vaccine strategies through the application of immunological principles. Vaccine 32 (10): 1132-1138

Ng WV, Ciufo SA, Smith TM, Bumgarner RE, Baskin D, Faust J, Hall B, Loretza C, Seto J, Slagel J, Hood L, DasSarma S (1998) Snapshot of a large dynamic replicon in a halophilic archaeon: megaplasmid or minichromosome? Genome Res 8: 1131-1141

Ng WV, Kennedy SP, Mahairas GG, Berquist B, Pan M, Shukla HD, Lasky SR, Baliga NS, Thorsson V, Sbrogna J, Swartzell S, Weir D, Hall J, Dahl TA, Welti R, Goo YA, Leithauser B, Keller K, Cruz R, Danson MJ, Hough DW, Maddocks DG, Jablonski PE, Krebs MP, Angevine CM, Dale H, Isenbarger TA, Peck RF, Pohlschroder M, Spudich JL, Jung K-H, Alam M, Freitas T, Hou S, Daniels CJ, Dennis PP, Omer AD, Ebhardt H, Lowe TM, Liang P, Riley M, Hood L, DasSarma S (2000) Genome sequence of *Halobacterium* species NRC-1. Proc Natl Acad Sci U S A 97 (22): 12176-12181

Nussenzweig V, Nussenzweig RS (1989) Rationale for the development of an engineered sporozo- ite malaria vaccine. Adv Immunol 45: 283-334

Plassmeyer ML, Reiter K, Shimp RL Jr, Kotova S, Smith PD, Hurt DE, House B, Zou X, Zhang Y, Hickman M, Uchime O, Herrera R, Nguyen V, Glen J, Lebowitz J, Jin AJ, Miller LH, MacDonald NJ, Wu Y, Narum DL (2009) Structure of the *Plasmodium falciparum* circumsporozoite protein, a leading malaria vaccine candidate. J Biol Chem 284: 26951-26963

Schofield L, Villaquiran J, Ferreira A, Schellekens H, Nussenzweig R, Nussenzweig V (1987) γ interferon, CD8+T cells and antibodies required for immunity to malaria sporozoites. Nature 330 (6149): 664-666

Sedegah M, Kim Y, Ganeshan H, Huang J, Belmonte M, Abot E, Banania JG, Farooq F, McGrath S, Peters B, Sette A, Soisson L, Diggs C, Doolan D, Tamminga C, Villasante E, Hollingdale MR, Richie T (2013) Identification of minimal human MHC-restricted CD8+ T-cell epitopes within the *Plasmodium falciparum* circumsporozoite protein (CSP). Malar J 12: 185

Seder RA, Chang LJ, Enama ME, Zephir KL, Sarwar UN, Gordon IJ, Holman LA, James ER, Billingsley PF, Gunasekera A, Richman A, Chakravarty S, Manoj A, Velmurugan S, Li ML, Ruben AJ, Li T, Eappen AG, Stafford RE, Plummer SH, Hendel CS, Novik L, Costner PJM, Mendoza FH, Saunders JG, Nason MC, Richardson JH, Murphy J, Davidson SA, Richie T, Sedegah M, Sutamihardja A, Fahle GA, Lyke KE, Laurens MB, Roeder M, Tewari K, Epstein JE, Sim BKL, Ledgerwood JE, Graham BS, Hoffman SL, the VRC 312 Study Team (2013) Protection against malaria by intravenous immunization with a nonreplicating sporozoite vaccine. Science 341 (6152): 1359-1365

Shevchenko A, Wilm M, Vorm O, Mann M (1996) Mass spectrometric sequencing of proteins silver-stained polyacrylamide gels. Anal Chem 68 (5): 850-858

Shukla HD, DasSarma S (2004) Complexity of gas vesicle biogenesis in *Halobacterium* sp. strain NRC-1: identification of five new proteins. J Bacteriol 186（10）：3182-3186

Sremac M，Stuart ES（2008）Recombinant gas vesicles from *Halobacterium* sp. displaying SIV peptides demonstrate biotechnology potential as a pathogen peptide delivery vehicle. BMC Biotechnol 8：9

Sremac M，Stuart ES（2010）SIVsm Tat，Rev，and Nef1：functional characteristics of r-GV internalization on isotypes，cytokines，and intracellular degradation. BMC Biotechnol 10：54

Stuart ES，Sremac M，Morshed F，DasSarma S（2001）Antigen presentation using novel particulate organelles from halophilic archaea. J Biotechnol 88（2）：119-128

Stuart ES，Morshed F，Sremac M，DasSarma S（2004）Cassette-based presentation of SIV epitopes with recombinant gas vesicles from halophilic archaea. J Biotechnol 114（3）：225-237

Wirtz RA，Zavala F，Charoenvit Y，Campbell GH，Burkot TR，Schneider I，Esser KM，Beaudoin RL，Andre RG（1987）Comparative testing of monoclonal antibodies against *Plasmodium falciparum* sporozoites for ELISA development. Bull World Health Organ 65（1）：39-45

Young JF，Hockmeyer WT，Gross M，Ballou WR，Wirtz RA，Trosper JH，Beaudoin RL，Hollingdale MR，Miller LH，Diggs CL，Rosenberg M（1985）Expression of *Plasmodium falciparum* circumsporozoite proteins in Escherichia coli for potential use in a human malaria vaccine. Science 228（4702）：958-962

索　引

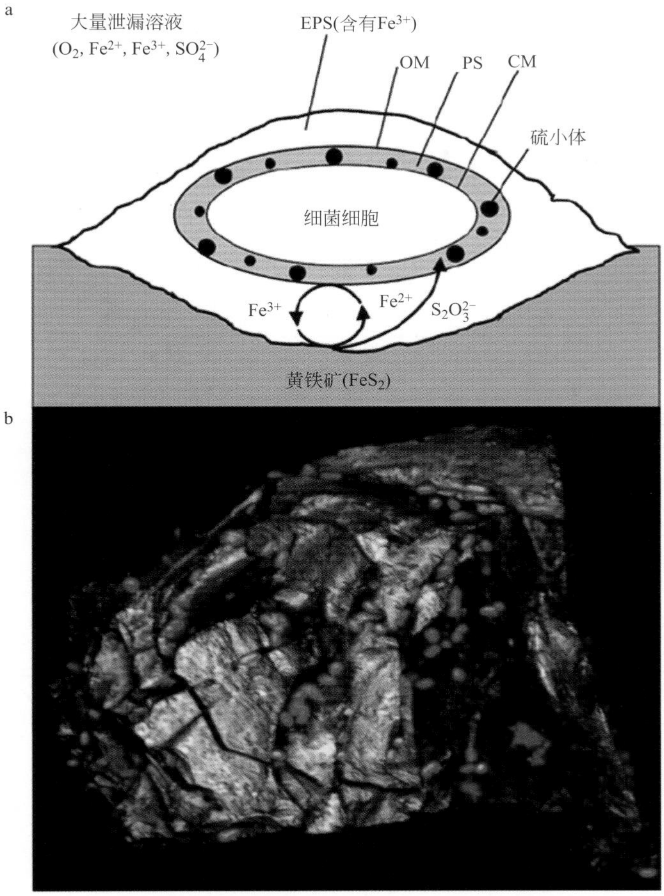

图 6.1 细菌细胞催化的接触浸出模型（来自 Vera 等，2013b）。(a) 表示嵌入在附着于黄铁矿上的胞外聚合物（EPS）层中的生物膜单元。图中显示在金属硫化物（MS）溶解过程中存在的化合物，如铁（二价）/（三价）、硫代硫酸盐。CM 表示细胞质膜；PS 表示周质空间；OM 表示外膜。(b) CLSM 图像显示的是孵育 1 周后氧化亚铁硫杆菌细胞定植的黄铁矿颗粒（50～100μm）的 3D 投影。采用 Syto 9（绿色）对细胞中的核酸和伴刀豆球蛋白 A（Con A）进行双重染色。绿色为 Syto 9，红色为 Con A-四甲基罗丹明异硫氰酸酯（TRITC），灰色为反射。图中显示了三个通道的合并图像。细菌定植模式与表面缺陷密切相关

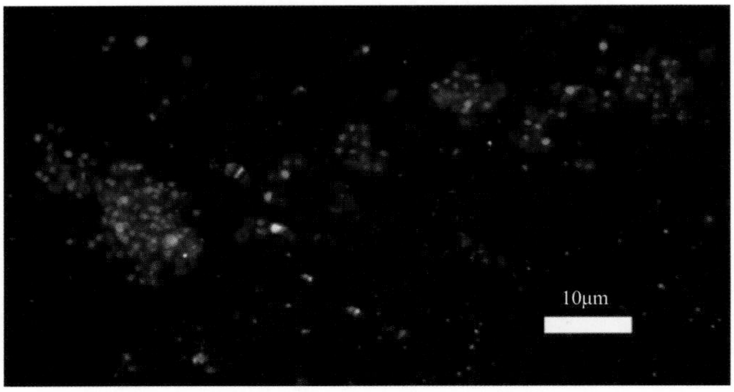

图 6.2 附着硫元素的嗜酸热古菌 DSM 29099 生物膜的最大强度投影。采用 Con A 和 SybrGreen 对细胞和生物膜基质进行双重染色。颜色分配：绿色为 SybrGreen，红色为 Con A-四甲基罗丹明异硫氰酸酯（TRITC），灰色为反射。微生物形式的生物膜细胞嵌入含有甘露糖和葡萄糖的 EPS 基质中

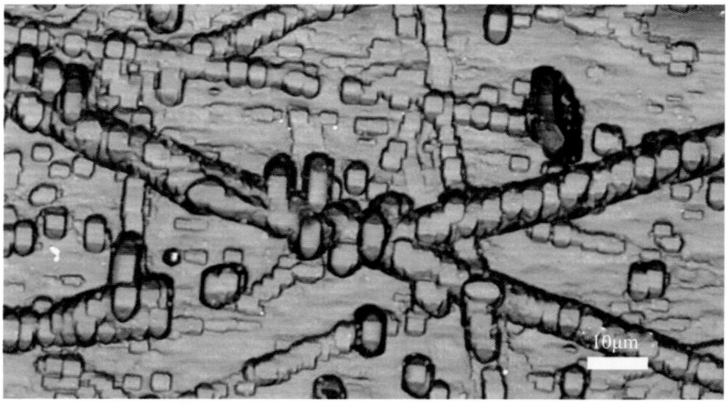

图 6.4　附着于黄铁矿上的 *Acidianus* sp. DSM 29099 生物膜的 CLSM 图像。采用凝集素 AAL 对细胞进行染色。颜色分配：绿色为 AAL-异硫氰酸荧光素（FITC）；黄铁矿表面以反射模式（灰色）显示。细胞优先附着到黄铁矿表面有缺损的位点

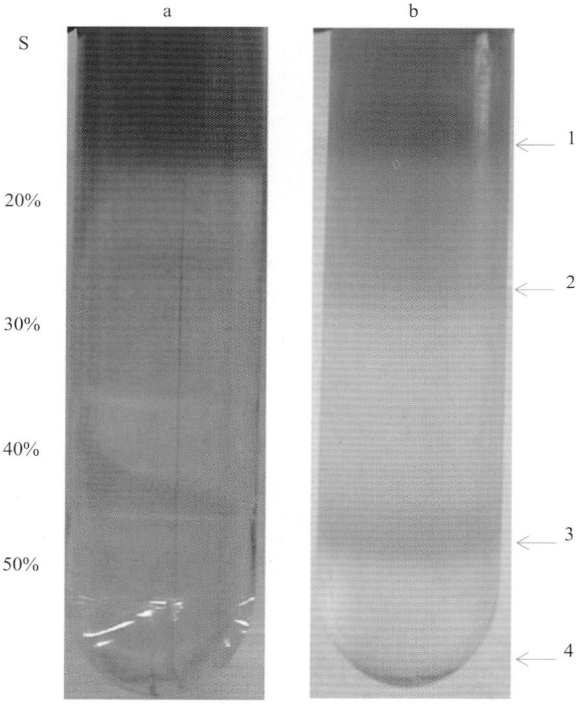

图 10.7　蔗糖密度梯度（20%～50%）离心。细胞裂解物加载在液体顶部（a）。超速离心后，膜蛋白集中在第 3 和第 4 部分（b）

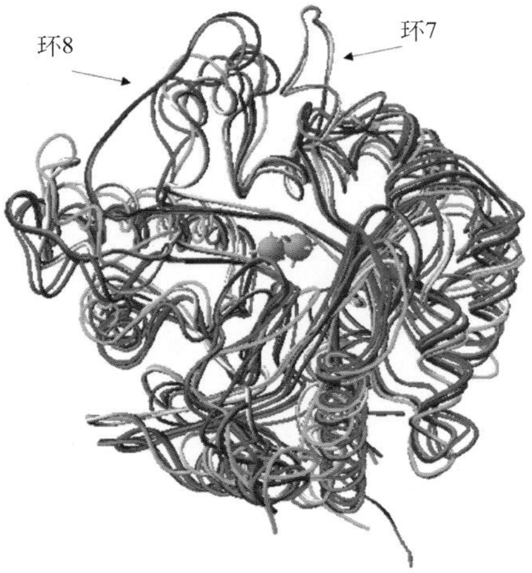

图 16.2 *bd*PTE 与六个 PLLs 的结构叠加，结构叠加使用 Swiss-PdbViewer 4.1.0。*B. diminuta*（灰色）来自 *bd*PTE，*Sso*Pox（红色）源自 *S. solfataricus*；GsP（蓝色）为 *G. stearothermophilus*，*G. kaustophilus* 为 *Gka*P（绿色），PPH（青色）为 *M. tuberculosis*，Ahla 为 *R. erythropolis*（粉色）；金属阳离子为灰色球体

编号	亚家族	来源	CSR VI β2	CSR I β3	CSR V loop3	CSR II β4	CSR III β5	CSR IV β7	CSR VII β8
1	GH13_7	Palaeococcus_pacificus_A0A075LR97	GITSIWLPP	GLEVYADIVINH	YPDIC	GWRFDYVKG	YAVGEYWD	FVANHD	GQPTIFYRD
2		Pyrococcus_furiosus_O08452	GISAIWLPP	GIKVIADVVINH	FPDIC	GWRFDYVKG	WAVGEYWD	FVANHD	GQPVIFYRD
3		Pyrococcus_woesei_Q7LYT7	GISAIWLPP	GIKVIADVVINH	FPDIC	GWRFDYVKG	WAVGEYWD	FVANHD	GQPVIFYRD
4		Thermococcus_hydrothermalis_O93647	GISAIWIPP	NMKVIADIVINH	YPDIC	AWRFDYVKG	WAVGEYWD	FVANHD	GQPAIFYRD
5		Thermococcus_kodakaraensis_O33476	GISAIWIPP	GIKVIADIVINH	FPDIA	AWRFDYVKG	WAVGEYWD	FVANHD	GQPVIFYRD
6		Thermococcus_onnurineus_B6YUG5	GIAAIWLPP	NMKVVADIVINH	FPDID	AWRFDYVKG	WAVGEYWD	FVANHD	GQPMIFYRD
7		Thermococcus_sp._Rt3_O50200	GISAIWIPP	GIKVIADIVINH	FPDIA	AWRFDYVKG	LAVGEYWD	FVANHD	GQPVIFYRD
8	GH13_7	Capnocytophaga_ochracea_C7M7T3	GVDRIWLPP	GLQVIADIVIGH	EQDLS	GWRFDYVKS	FAVGELWD	FTANHD	GYPCIFYSD
9		Croceibacter_atlanticus_A3U4D8	GIDAIWLPP	GISVIADIVINH	FPDLC	GWRFDYVKG	FSVGEYWD	FVANHD	GYPTLFYRD
10		Flavobacterium_johnsoniae_A5FII4	GIGSIWLPP	NIKVYADIVINH	FPDLC	GWRFDYVKG	FSVGELWD	FVTNHD	GYPTIFYRD
11		Gramella_forsetii_A0M3B1	GVDRIWLPV	DLEVIADIVINH	ETNLD	GWRFDYVKG	FSVGENFD	FTANHD	GYPTIFYSD
12		Kordia_algicida_A9E1U8	GIDAIWLPP	RLSVIADIVLNH	FADLS	GWRFDYVKG	FSVGEYWD	FVTNHD	GYPTIFYRD
13		Sinomicrobium_sp._5DNS001_L7Y1I6	GVDRIWLPV	GLEVIADIVLNH	EQDLC	GWRFDYVKG	FSVGEYWD	FVANHD	GYPTVFYGD
14	GH13_6	Micromonas_pusilla_C1MR05	GFTQIWLPP	NILPVLDAVLNH	SPNID	GIRFDFSKG	FAVGEFWD	FIDNHD	GTPSVFWDH
15		Apple_Q5BLY0	GFTSAWLPP	KVRAMADIVINH	VPNID	DFRFDFARG	FSVGEYWD	FLDNHD	GIPTVFYDH
16		Barley_low_pI_isozyme_P00693	GVTHVWLPP	GVQAIADIVINH	APDID	AWRLDFARG	LAVAEVWD	FVDNHD	GIPCIFYDH
17		Bean_Q9ZP43	GITHVWLPP	GIKCLADIVINH	APDID	GWRFDFVKG	FAVGEKWD	FIDNHD	GTPSIFYDH
18		Kiwi_Q5BLY1	GFTVVWLPP	GIRVLGDVVLNH	APNID	GWRLDFVRG	FAVGEYWD	FIENHD	GTPAVFYDH
19		Maize_Q41770	GATHVWLPP	GVKCVADVVINH	APDID	GWRLDFAKG	FVVAEIWS	FVDNHD	GTPCIFYDH
20		Potato_Q41442	GFTTAWLPP	KVRAMADIVINH	VPNID	DFRFDFAKG	FAVGEYWD	FIDNHD	GIPSVFFDH
21	GH13_6	Microscilla_marina_A1ZMR5	GFSMVWLPP	NIKPIADIVVNH	GRDLD	GWRYDYVKG	FSVGELWP	FVDNHD	GIPCVYWPH
22		Saccharophagus_degradans_Q21NA2	GATHVWFAP	GIDSVADIVINH	ARDID	GLRYDYSKG	FCVGEVWT	FVDNHD	GIPTVYWAH
23	GH13_1	Aspergillus_oryzae_P0C1B3	GFTAIWITP	GMYLMVDVVANH	LPDLD	GLRIDTVKH	YCIGEVLD	FVENHD	GIPIIYAGQ
24		Saccharomycopsis_fibuligera_D4P4Y7	GFTAIWISP	DMLLMVDIVTNH	LPDLR	GLRIDSAKH	YSVGEVFQ	FVENHD	GIPVIYYGQ
25	GH13_5	Bacillus_licheniformis_P06278	GITAVWIPP	DINVYGDVVINH	YADID	GFRLDAVKH	FTVAEYWQ	FVDNHD	GYPQVFYGD
26		Histoplasma_capsulatum_A0T074	GVTSILLPP	EIRIIWDTVLNH	FSNLD	GLRIDAAKH	LLVAEYWK	FVMNHD	GYPCLFYGD
27	GH13_15	Fruit_fly_P08144	GYAGVQVSP	GVRTYVDVVFNH	LRDLN	GFRVDAAKH	YIVQEVID	FVDNHD	GTPRVMSSF
28		Meal_worm_P56634	GFGGVQISP	GVRIYVDAVINH	LRDLN	GFRVDAAKH	FIYQEVID	FVDNHD	GTTRIMSSF
29	GH13_24	Shrimp_Q26193	GFAGVQVSP	GVRIYVDAVINH	LNDLN	GFRIDASKH	FIFQEVID	FIDNHD	GYTRVMSSY
30		Human_pancreas_P04746	GFGGVQVSP	GVRIYVDAVINH	LLDLA	GFRLDASKH	FIYQEVID	FVDNHD	GFTRVMSSY
31	GH13_27	Aeromonas_hydrophila_P22630	GYKQVLISP	GIAVYADVVLNH	LPDLD	GFRVDAVKH	HVFGEVIT	FAITHD	GSPLVYSDH
32		Xanthomonas_campestris_Q56791	GYRKVLVAP	GVETYADVVFNH	LPDLL	GFRVDAAKH	YVFGEVIT	FAVTHD	GVPMVYTDN
33	GH13_28	Bacillus_subtilis_P00691	GYTAIQTSP	GIKVIVDAVINH	LYDWN	GFRFDAAKH	FQYGEILQ	WVESHD	STPLFFSRP
34		Lactobacillus_amylovorus_Q48502	GYTAVQTSP	NIRIIVDATLND	FYDWN	GFRYDAATH	FQYGEVLQ	WVESHD	SVPLFFDRP
35	GH13_32	Streptomyces_limosus_P09794	GYGYVQVSP	GVKVVADSVINH	LADLD	GFRIDAAKH	YWKQEAIH	FVDNHD	GSPDVHSGY
36		Thermomonospora_curvata_P29750	GFGAVQVSP	GVKIYVDAVINH	LADLK	GFRIDAAKH	YIFQEVIA	FVVNHD	GTPKVMSSY
37	GH13_36	Anaerobranca_gottschalki_Q5I942	GVNGIWLTP	GIKVIIDLVINH	MPDLN	GFRLDAAKH	YLVGEIWD	FLSNHD	GDPYIFAGE
38		Halothermothrix_orenii_Q8GPL8	GVNGIWLMP	GIKVIIDLPINH	MPDLN	GFRLDGAMH	YLVGEVWD	FLTNHD	GNPFIYYGE
39	GH13_37	Photobacterium_profundum_Q6LIA8	GMNAVWLTP	GLYVFFDGVFGH	-----	GWRLDQAYQ	YMVAEIWN	MLGNHD	GPITLYYGD
40		Uncultured_bacterium_D9MZ14	GMNAIWLTP	GLYVFFDGVFGH	-----	GWRLDQAYQ	YMVAEIWN	MLGNHD	GPITLYYGE
41	GH13_??	Halalkalicoccus_jeotgali_D8J7H2	GIDVVWLTP	DIRLIMDLVANH	QPDLN	GFRLDAIAH	LTVGEVGT	FLSNHD	GTPFVYQGE
42	GH13_??	Haloarcula_hispanica_Q4A3E0	GYDAIQVPP	DLDVVADAVINH	LKDLK	GIRMDAAKH	WTVGEVLD	FVSNHD	GYPRVYSNR
43	GH13_??	Haloarcula_japonica_L8B068	GVDAIWLTP	GFKVLFDLVCNH	IANFN	GFRCDMAWA	LLLDETIP	YAENHD	GAPLLYAGQ
44	GH13_??	Haloarcula_marismortui_Q5UZY3	GVDVVWLTP	GIRVVFDLVINH	IPNLN	GFRCDVAMG	LLLDETVP	YVDNHD	GTPMVYAGQ
45	GH13_??	Natronococcus_amylolyticus_Q60224	GVSAIWIPQ	DIDVIVDIVLNH	LPSMD	GLRIDAAAH	WRVGEVWD	FVQNHD	GMPMLYRGG

图 17.1 α-淀粉酶 GH13 家族保守区序列。序列包含了多种 GH13 家族不同分类亚家族蛋白酶，尤其是古菌来源的序列。未分类的古菌序列用双问号标注。古菌 GH13 _ 7 α-淀粉酶特征序列用亮黄色标注。催化三联体用黑色标出。保守的序列区域（CSR-Ⅰ至 CSR-Ⅶ）主要在 TIM 桶形结构的单个 β 折叠中（Janecek 2002)。蛋白酶序列由 UniProt 登录号（UniProt Consortium 2015）区分

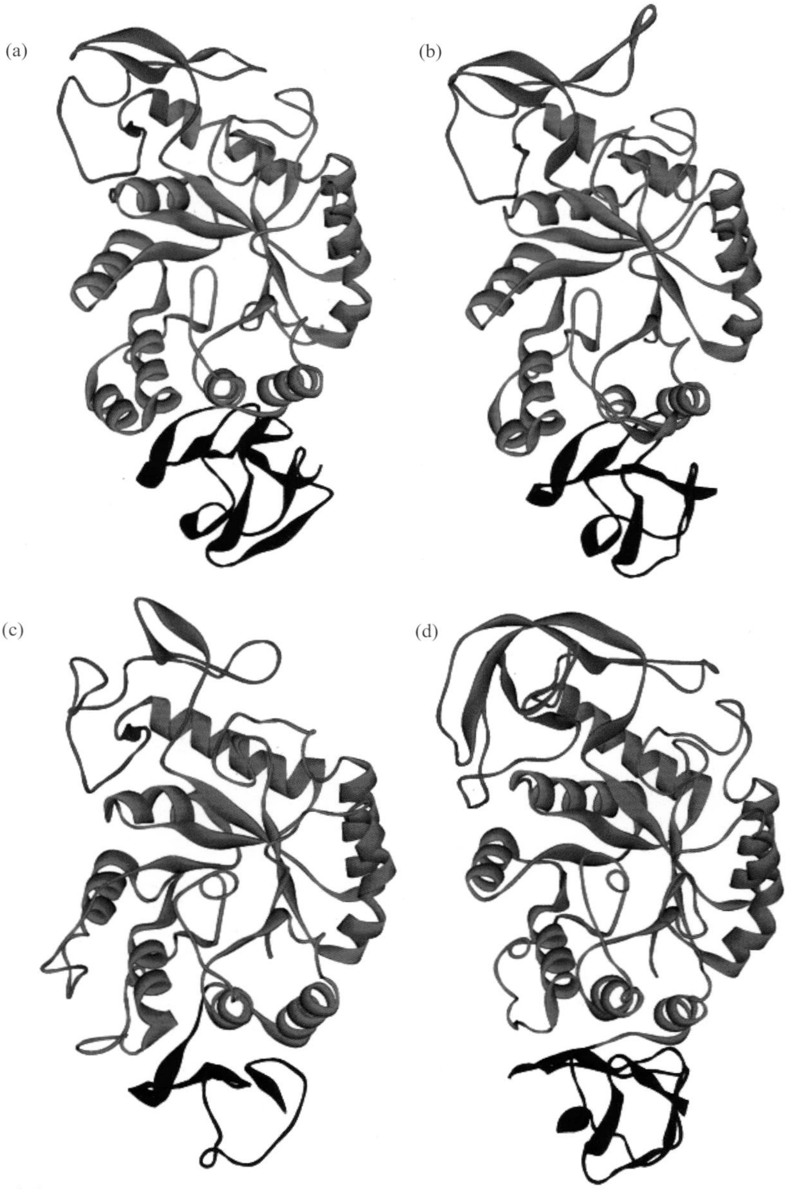

图 17.3　古菌 α-淀粉酶及其同源酶体的蛋白质三级结构。（a）GH13 _ 7-*Pyrococcus woesei*（PDB code：1MWO；Linden 等 2003）；（b）GH13 _ 7-*Sinomicrobium* sp. 5DNS001（UniProt accession No.：L7Y1I6；Li 等 2014）；（c）GH13 _ 6-*Hordeum vulgare* - barley isozyme AMY-1（PDB code：1P6W；Robert 等 2003）；（d）GH13 _ 5-*Bacillus licheniformis*（PDB 码：1BLI；Machius 等 1998）。蛋白酶（β/α）$_8$-桶形催化结构域：绿色；结构域 B：红色；结构域 C：蓝色。黄杆菌 α-淀粉酶以 *P. woesei* α-淀粉酶（PDB 码：1MWO；Deshpande 等 2005）为模型，利用 Phyre-2 同源模拟得到（Kelley 和 Sternberg 2009），并通过 WebLabViewerLite 软件（Molecular Simulations，Inc.）展示

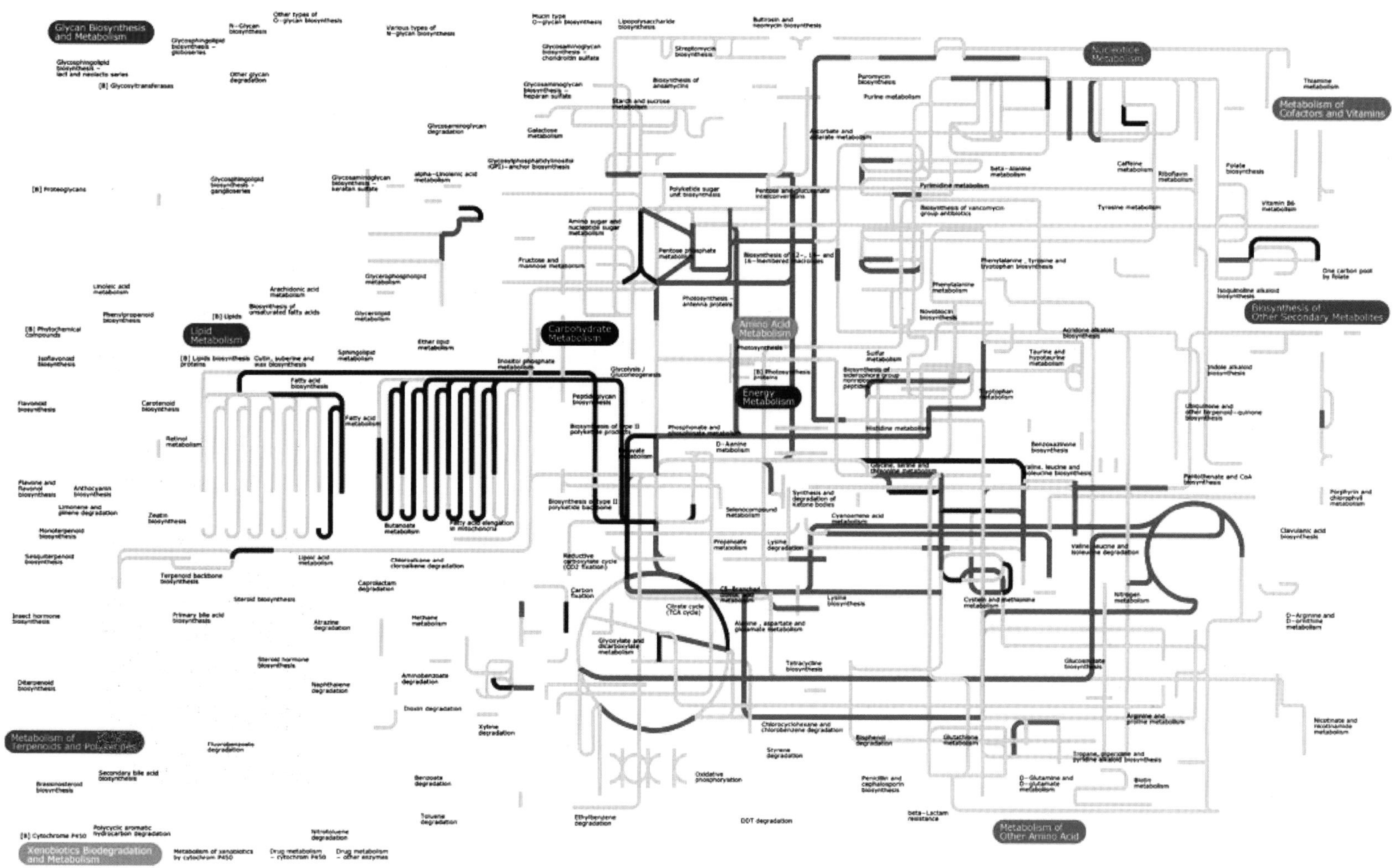

图 21.3　代谢示意图:温度下降后 *Ph*TAC125 代谢流量的分配出现变化。红色和蓝色线分别表示反应流量的下降和增加(起码有因素 2),灰色线表示反应的流量值无明显改变

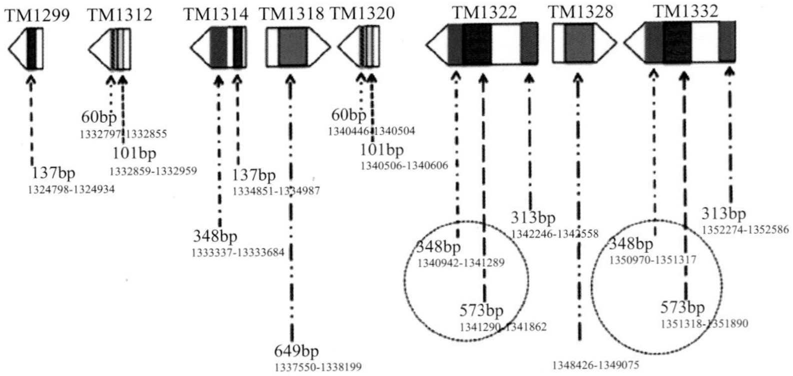

图 22.1　*T. maritima* 潜在的基因组进化位点。特定的颜色块代表保守的同向重复序列长度。基因组坐标和基因座标记根据 *T. maritima* 基因组的报道（Nelson 等 1999）

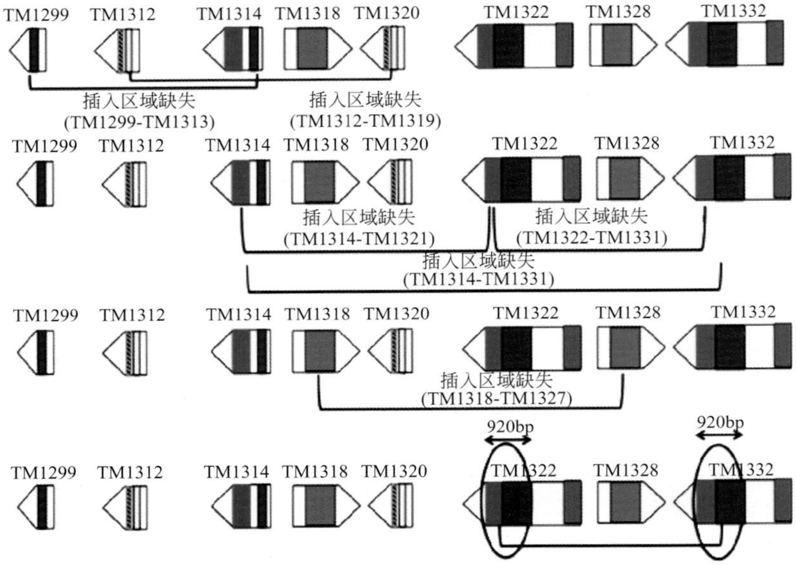

图 22.2　通过同向重复序列之间的交叉介导的不同基因组缺失示意图。特定色块代表保守同向重复序列的长度。相同的同向重复序列可以进行同源重组，图中用方括号表示

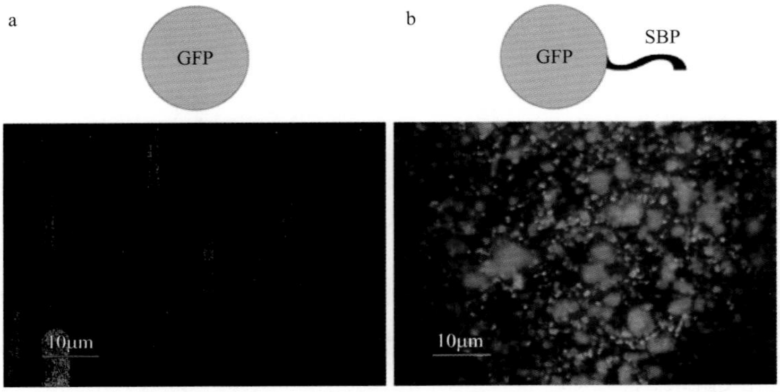

图 23.2　绿色荧光蛋白 GFP 和 GFP-SBP（表 23.5）与天然沸石结合及荧光显微镜图。(a）不含有 SBP 的 GFP 不能结合到天然沸石上且沸石没有显示荧光；(b）只有当 GFP 融合了 SBP 后的重组表达蛋白质才能与天然沸石结合并显示荧光

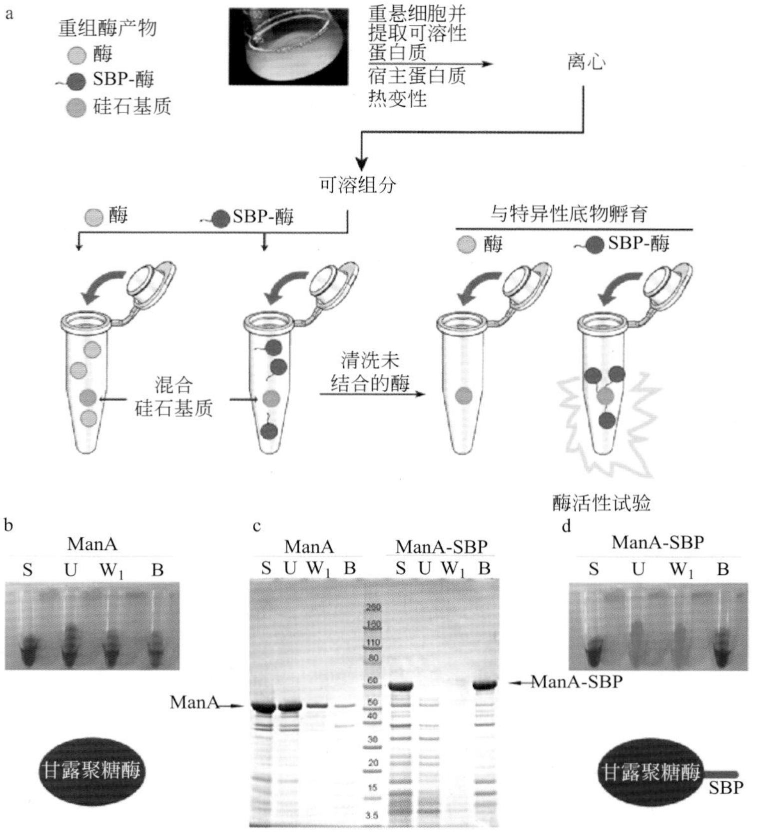

图 23.4 硅石机制结合及活性试验。(a) 从 *E. coli* 中提取重组可溶蛋白质及后续硅石基质结合试验的标准流程图。(b) 来自 *Dictyoglomus thermophilum* 的热稳定甘露聚糖酶 ManA (表 23.5) 结合试验中各组分活性。(c) 部分纯化的 ManA 和 ManA-SBP 与沸石结合试验的 SDS-PAGE 凝胶图。(d) ManA-SBP 结合试验各组分的甘露聚糖酶活性。甘露聚糖酶活性试验是以刺槐豆胶 (LBG) 为底物在 80℃下反应 10min。从 LBG 中释放出来的还原性糖用 DNS 法 (二甲基水杨酸法) 进行定量。S 代表 *E. coli* 可溶蛋白质粗提物，U 为非结合组分，W_1 位第一次冲洗组分 (清洗组分 2、3 没有进样)，B 为沸石结合蛋白组分

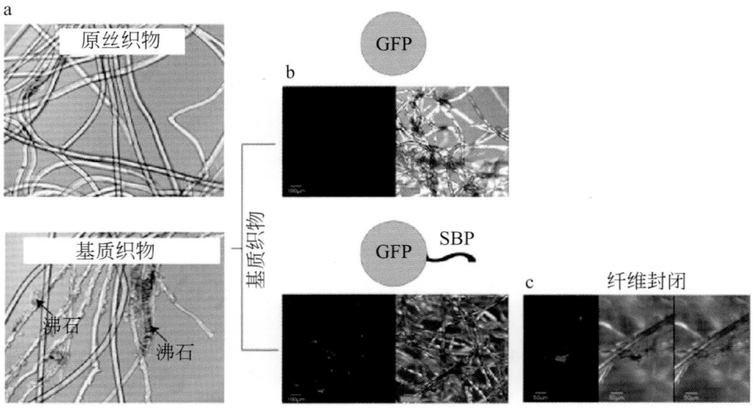

图 23.5 GFP 及 GFP-SBP 结合原丝及沸石织物。(a) 原丝和沸石织物纤维的相差显微镜图像。(b) 原丝和沸石织物纤维在与 GFP 和 GFP-SBP 共孵育后的共聚焦显微镜图像。左边面板显示 GFP 荧光，右边面板为相应的 DIC (数字图像相关法) 图。(c) 用 GFP-SBP 处理后的沸石织物在共聚焦显微镜下的特写。左边为 GFP 荧光，中间为相应的 DIC (数字图像相关法) 图，右边为 GFP 荧光与 DIC 图的重叠

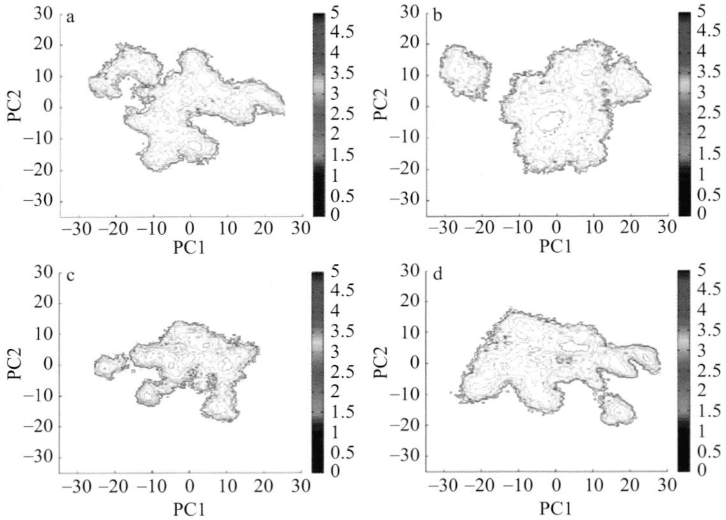

图 24.3　适冷酶和适温酶的二维 FEL 图。以 MD 轨迹主成分分析的两个主分量作为反应坐标，中温（a，b）和嗜冷（c，d）丝氨酸蛋白酶分别在 283K（a，c）和 310 K（b，d）下（Mereghetti 等 2010）

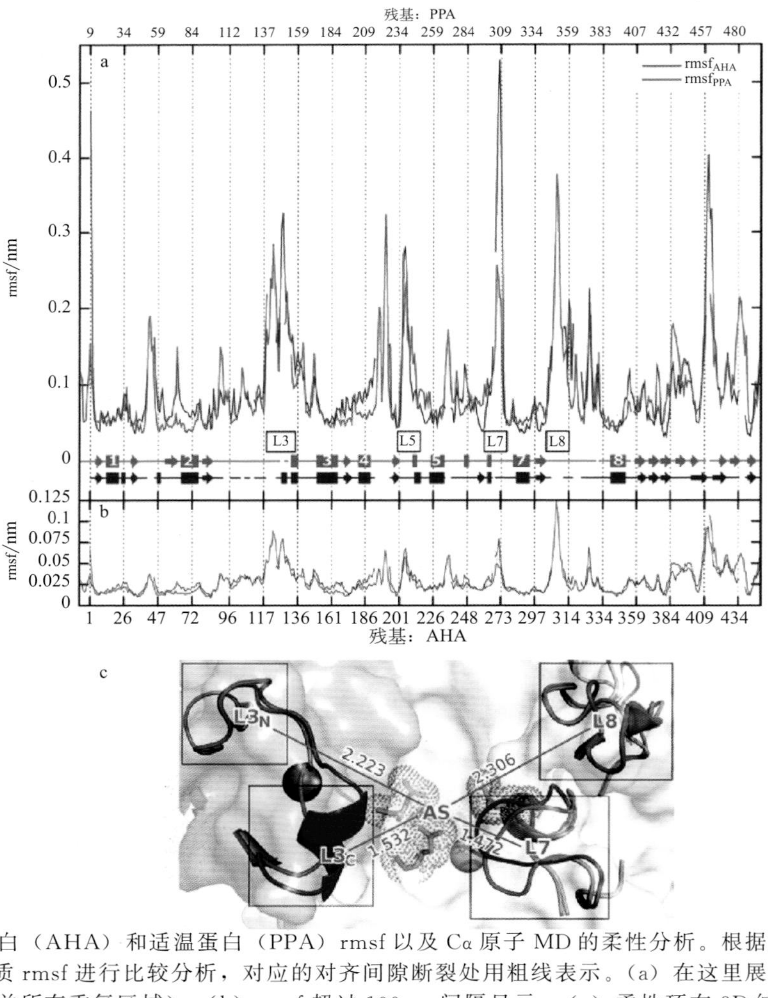

图 24.4　适冷蛋白（AHA）和适温蛋白（PPA）rmsf 以及 Cα 原子 MD 的柔性分析。根据两种蛋白质结构比对对两种蛋白质 rmsf 进行比较分析，对应的对齐间隙断裂处用粗线表示。(a) 在这里展示整个级联轨迹 rmsf 分析（即合并所有重复区域）。(b) rsmf 超过 100 ps 间隔显示。(c) 柔性环在 3D 结构中突出显示。距离是整体平均数，计算基团到中心的距离，距离单位 nm（Pasi 等 2009）